Thrips as Crop Pests

Thrips . . .
Tiny, shiny
Black snips
Of life
Whose main claim
To fame
In the world
Rests
Because they are
Injurious pests
Infesting cereal
Crops,
Onions, peas, fruit
And ornamental flowers.
The thunderflies
That fill
Sultry summer skies
With innumerable hordes
Occasionally
Causing strife
By alarming plant
Growers,
Irritating skin
And getting in
Ears, nose and eyes . . .
Whilst most
Contrary of all,
Squeezing themselves
Behind picture frame
And glass
There to die . . .
A bizarre
Habit
Which can annoy
Nearly
Beyond words.

Hugh D. Loxdale, 1993

THRIPS AS CROP PESTS

Edited by
Trevor Lewis

Institute of Arable Crops Research–Rothamsted,
Harpenden, Herts, UK

CAB INTERNATIONAL

CABI is a trading name of CAB International

CABI Head Office
Nosworthy Way
Wallingford
Oxfordshire OX10 8DE
UK

Tel: +44 (0)1491 832111
Fax: +44 (0)1491 833508
Email: cabi@cabi.org
Web site: www.cabi.org

CABI North American Office
875 Massachusetts Avenue
7th Floor
Cambridge, MA 02139
USA

Tel: +1 617 395 4056
Fax: +1 617 354 6875
Email: cabi-nao@cabi.org

A catalogue record for this book is available from the British Library, London, UK

Library of Congress Cataloging-in-Publication data
Thrips as crop pests / edited by Trevor Lewis.
p. cm.
Includes index.
ISBN 0-85199-178-5 (alk. paper)
1. Thrips. 2. Thrips-Integrated control. 1. Lewis, Trevor. 1933-
SB945.T6T47 1997 97-24320
632´.758-dc21 CIP

ISBN-13: 978-0-85199-178-8
ISBN-10: 0-85199-178-5

First published 1997
Transferred to print on demand 2006

Printed and bound in the UK by CPI Antony Rowe, Eastbourne.

Contents

Contributors vii

Preface ix

Acknowledgements xi

1 **Pest Thrips in Perspective** 1
T. Lewis

2 **Structure, Growth and Development** 15
G. Moritz

3 **Host Selection, Communication and Reproductive Behaviour** 65
L.I. Terry

4 **Feeding** 119
W.D.J. Kirk

5 **Flight and Dispersal** 175
T. Lewis

6 **Biological Diversity** 197
L.A. Mound

7 **Distribution, Abundance and Population Dynamics** 217
W.D.J. Kirk

8 **Predation by Insects and Mites** 259
M.W. Sabelis and P.C.J. Van Rijn

9 **Interactions with Hymenopterous Parasitoids and Parasitic Nematodes** 355
A.J.M. Loomans, T. Murai and I.D. Greene

10 **Fungal Pathogens of Thrips** 399
T.M. Butt and M. Brownbridge

11 **Field and Laboratory Techniques** 435
T. Lewis

12 **Culturing Thrips and Parasitoids** 477
A.J.M. Loomans and T. Murai

13 **Feeding and Oviposition Injuries to Plants** 505
C.C. Childers

14 **Thrips as Vectors of Plant Pathogens** 539
D.E. Ullman, J.L. Sherwood and T.L. German

15 **Chemical Control** 567
T. Lewis

16 **Integrated Pest Management (IPM) in Field Crops** 595
M.P. Parrella and T. Lewis

17 **Integrated Pest Management (IPM) in Tree Crops** 615
B.L. Parker and M. Skinner

18 **Integrated Pest Management (IPM) in Glasshouses** 639
R.J. Jacobson

Appendix I Thrips Species Cited, with Authorities and Common Names 667

Appendix II Major Crops Infested by Thrips with Main Symptoms and Predominant Injurious Species 675

Appendix III Tree Crops, Associated Thrips and Components of Control 711

Index 719

The colour plates appear between pages 196 and 197

Contributors

Michael Brownbridge, *University of Vermont, Entomology Research Laboratory, PO Box 53400, Burlington, Vermont 05405–3400, USA.*

Tariq M. Butt, *Institute of Arable Crops Research–Rothamsted, Harpenden, Herts AL5 2JQ, UK.*

Carl C. Childers, *University of Florida, Citrus Research and Education Center, 700 Experiment Station Road, Lake Alfred, Florida 33850, USA.*

Thomas L. German, *University of Wisconsin-Madison, Department of Plant Pathology, Russell Laboratory, 1630 Linden Drive, Madison, Wisconsin 53706, USA.*

Ian D. Greene, 6735 Mesquite Way, Salinas, California 93907, USA.

Robert J. Jacobson, *Horticulture Research International, Stockbridge House, Cawood, Selby, North Yorkshire YO8 0TZ, UK.*

William D.J. Kirk, *Department of Biological Sciences, Keele University, Keele, Staffs ST5 5BG, UK.*

Trevor Lewis, *Institute of Arable Crops Research – Rothamsted, Harpenden, Herts AL5 2JQ, UK.*

Antoon J.M. Loomans, *Section Entomology, Plant Protection Service, Geertjesweg 15, PO Box 9102, 6700 HC Wageningen, The Netherlands.*

Gerald Moritz, *Martin-Luther-Universitat, Halle-Wittenberg, Institut für Zoologie, Domplatz 4, Halle (Saale), 06108, Germany.*

Laurence A. Mound, *Division of Entomology, CSIRO, GPO Box 1700, Canberra, ACT 2601, Australia.*

Tamotsu Murai, *Research Institute for Bioresources, Okayama University, 2-20-1 Chuo, Kurashiki 710, Japan.*

Bruce L. Parker, *University of Vermont, Entomology Research Laboratory, PO Box 53400, Burlington, Vermont 05405-3400, USA.*

Michael P. Parrella, *Department of Entomology, University of California, Davis, California 95616-8584, USA.*

Maurice W. Sabelis, *University of Amsterdam, Department of Pure and*

Applied Ecology, Section Population Biology, Kruislaan 320, 1098 SM Amsterdam, The Netherlands.

John L. Sherwood, University of Georgia, Department of Plant Pathology, Athens, Georgia 3060 2–7274, USA.

Margaret Skinner, *University of Vermont, Entomology Research Laboratory, P.O. Box 53400, Burlington, Vermont 05405-3400, USA.*

L. Irene Terry, *University of Utah, Department of Biology, 201 South Biology Building, Salt Lake City, Utah 84112, USA.*

Diane E. Ullman, *University of California, Department of Entomology, Davis, California 95616, USA.*

Paul C.J. Van Rijn, *University of Amsterdam, Department of Pure and Applied Ecology, Section Population Biology, Kruislaan 320, 1098 SM Amsterdam, The Netherlands.*

Preface

I began studying thrips over 40 years ago when there were barely 20 serious thysanopterists worldwide, most of whom were taxonomists and systematists. A background in practical agriculture led me to concentrate on the applied biology and ecology of these insects in which I became progressively more interested over some 15 years of active research, culminating in the publication by Academic Press in 1973 of *Thrips, Their Biology, Ecology and Economic Importance*. Thereafter, although diverted by other entomological activities and latterly by scientific administration, I retained and fostered my early fascination in thrips as more people became aware of their importance to crop production in many countries. In the last decade this awareness has grown rapidly and there are now more thrips-related problems in agriculture, and more entomologists and growers concerned with their impact on a wide range of crops, than ever before.

On formal retirement I resolved to condense and present the vast amount of new information on pest thrips into a second book, bringing together the wealth of fundamental and applied knowledge that has accrued over the past quarter of a century. It was soon apparent that to cover the many facets of the subject authoritatively, I should need to enlist the help of specialists in several disciplines. I was delighted and grateful when a number of eminent thysanopterists responded to my invitation to contribute towards a treatise on the status of thrips as pests worldwide. The outcome of our joint endeavours is this book which I believe will provide background and detailed knowledge about thrips and how to control them to growers and advisers, and act as a point of reference and stimulus for further research on these challenging and intriguing insects, especially to young workers.

The approach has been to assemble reviews of 18 topics, each combining some historical information with recent research findings. Much hitherto unpublished information is also included so that overall the book is the most comprehensive compendium available on pest thrips. The text is arranged in three parts. First (Chapters 2–6) there is a grounding in basic biology, structure and behaviour of thrips, emphasizing aspects particularly relevant to pest species. A second part (Chapters 7–12) concentrates on the principles of population biology and ecology, considering in detail the many factors that contribute to thrips abundance and distribution on crops, and methods devised to study them in the field and laboratory. A third part (Chapters 13–18) builds on this information to consider direct and indirect crop injury, and control by chemicals alone or combined with natural agents and cultural practices within integrated pest management programmes in field, forestry and glasshouse crops. Several appendices have been compiled for convenient reference including a catalogue of significant pests worldwide and the symptoms they cause on plants. Where appropriate, the text is generously illustrated with diagrams and photographs.

Trevor Lewis
Institute of Arable Crops Research – Rothamsted
April 1997

Acknowledgements

It is a pleasure to acknowledge advice, help and material received from many sources during preparation of this book. In addition to those of contributors, coloured plates were provided by: P. Bielza, Nüñez Morgado 6, Madrid, Spain; J. Cho, University of Hawaii; G. Del Bene, Instituto Sperimentale per la Zoologica Agraria, Firenze, Italy; P. Del Estel, Tenerife; Margery Daughtrey, University of Hawaii; A. Dewar and A. Ferguson, Institute of Arable Crops Research, UK; D. Grasselly, Centre Technique Interprofessionel des Fruits et Légumes, Bellegard, France; A. Lacasa, Department Proteccion Vegetal, CIDA, La Alberca, Spain; Hanna Legutowska, Agricultural University of Warsaw, Poland; K. Lynch, New Brunswick Department of Agriculture, Canada; Rita Marullo, Universita degli Studi della Basilicata, Potenza, Italy; Nancy Matteson, University of Idaho, USA; J.G. Morse, University of California, USA; G.V. Pollard, University of West Indies, Trinidad; D.A.J. Teulon, Crop and Food Research, Christchurch, New Zealand; R. Tremblay, New Brunswick Department of Agriculture, Canada; Susanne Vestergaard, Chr. Hansens BioSystems, Denmark; M. Wysoki, Volcani Institute, Israel. The photography and digital composite for Plate I.1 was by Jack K. Clark, Davis, California, from thrips colonies of Diane E. Ullman, University of California, Davis. Additional black and white photographs were provided by I.F. Chisholm, Institute of Arable Crops Research – Rothamsted, UK, C. Ellington, University of Cambridge, UK, and R. Samson, Centraal Bureau voor Schimmelcultures, Baarn, The Netherlands. The cover illustrations were provided by Jennifer M. Palmer, D.A.J. Teulon and T. Lewis.

The authors of Chapter 8 thank A. Janssen, R. van den Meiracker, A. Pallini and M. Venzon for comments and literature suggestions. The authors of Chapter 10 received advice from Suzanne Vestergaard Royal Veterinary and Agricultural University, Copenhagen, R. Segers, University of Louvain, Belgium, and G. Schreiter, University of Halle, Germany. They also received partial funding from the American Floral Endowment, Bedding Plants

Foundation Inc., Horticultural Research Institute, Ohio Floriculture Foundation, The Northeast Region SARE Program, USA and the North American Maple Syrup Council (University of Vermont). The author of Chapter 13 was grateful for material and/or discussion with J.V. Edelson, Oklahoma State University, USA; C.M. Felland, Pennyslvania State University, USA; J.V. French, Texas A. and M. University, Kingsville, USA; A. Muckenfuss, Clemson University, USA; R.D. Oetting, University of Georgia, USA; P.A. Stanley, University of Florida, USA; J.W. Todd, University of Georgia, USA; Susan E. Webb, University of Florida, USA. B. Heming, University of Edmonton, Canada, offered advice to several authors on aspects of thrips morphology, anatomy and behaviour, and S. Nakahara, USDA Beltsville, USA, contributed to Appendix II.

Thanks are due to the Councils, Editors and Publishers of scientific journals and publications from which figures have been redrawn or photographs reproduced, including: Academic Press, London; *Annals of Applied Biology*; *Applied Entomology and Zoology*; *Biologische Rundschau*; *Bulletin of Entomological Research*; Commonwealth Agricultural Bureau; *Canadian Journal of Forestry Research*; *Canadian Journal of Zoology*; Elsevier, New York; *Entomologia Experimentalis et Applicata*; *Entomological Society of America Bulletin*; *Environmental Entomology*; *Florida Entomologist*; *Hilgardia*; *IOBC/WPRS Bulletin*; *Journal of Animal Ecology*; *Journal of Economic Entomology*; *Journal of Experimental Biology*; *Journal of Insect Behaviour*; *Journal of Insect Physiology*; *Pacific Insects*; *Plant Protection Bulletin*; Plenum Press, New York; *Redia*; Zenkoku Noson Kyoika Publishing Co.; *Zoomorphology*; *Zoologischer Jahrbücher, Anatomie*.

The editor was greatly assisted by colleagues at IACR – Rothamsted, particularly J.N. Perry (statistics), I. Denholm (insecticide resistance), Lynda M. Castle, Nicola Seymour and P.C. Swatton (graphics and photography), Jackie Fountain, Sally Lawson, Linda Perryment and Helen Richardson (secretarial) and Angela Arnold (information retrieval).

Financial assistance towards the cost of including colour plates in the book was provided by the Lawes Agricultural Trust, Rothamsted, and BASF (America). The editor thanks the Lawes Agricultural Trust and IACR – Rothamsted for support and facilities. IACR – Rothamsted receives grant-aided support from the Biotechnology and Biological Sciences Research Council.

1

Pest Thrips in Perspective

Trevor Lewis
Institute of Arable Crops Research – Rothamsted, Harpenden, Herts AL5 2JQ, UK

Introduction

Thrips are minute, slender insects, usually only a few millimetres long. Most species have band-like wings delicately fringed with long cilia from which the name of the order, Thysanoptera, is derived. They occur worldwide with a preponderance of tropical species, many temperate ones and a few inhabitating cool, even arctic, regions. Most feed, using their characteristic piercing and sucking mouthparts, either on flowers and leaves of green plants, or on fungi; some are predatory, and a few feed on mosses and detritus.

These insects were first described by De Geer in 1744 under the name of *Physapus*, but in 1758 Linnaeus ignored this name and placed the four species known to him in a genus which he called Thrips. The group was elevated to the rank of an order by the English entomologist, Haliday, in 1836, but Linnaeus' name is still applied to any insect in the order, incidentally retaining the 's' in both the singular and plural forms.

Relatively few early entomologists specialized in studying thrips, perhaps because many species can only be found by diligent searching and specimens need careful and individual preparation and mounting for microscopic identification (Palmer *et al.*, 1989), and doubtless a century or more ago there were many larger and more obvious insects attracting the attention of applied entomologists. Interest in them increased rapidly after about 1900 with a peak period of descriptive activity and naming in the 1920s and 1930s. Of an estimated 8000 extant species, just over 5000 are now recognized and placed into two suborders and eight families (Gaston and Mound, 1993) (see Chapter 6) exhibiting a wide range of fascinating biology, form and behaviour. Among these, only relatively few species, mainly in the family Thripidae, are serious crop pests. Just a few of these have aroused a recent strong resurgence of interest in the order, raised its economic profile and attracted many more workers and resources than hitherto. Notwithstanding the immense biological and behavioural diversity exhibited by the Thysanoptera as a whole (Lewis, 1991) (see Chapter 6), this book is confined to exploring the biology, ecology and control of pest species. It contains little taxonomic or systematic information, which is well covered in other publications (see below), though the fundamentals of these disciplines clearly underpin much of what is included. Also, studies on pest thrips are collectively the greatest source of biological information on the order to date, so even though the subject matter relates directly to only a small proportion of known species, aspects are relevant to many more.

A citation index of more than half the described species is provided by Jacot-Guillarmod (1970–1979) and Jacot-Guillarmod and Brothers (1986). Introductions or checklists to the adult thrips of particular parts of the world include: North America (Stannard, 1957, 1968); Panama (Mound and Palmer, 1989); Central and South America (Mound and Marullo, 1996); Jamaica (Sakimura, 1986); Britain (Mound *et al.*, 1976); Europe (Priesner, 1964; Schliephake and Klimt, 1979); Africa (Pitkin and Mound, 1973; Palmer, 1990); India (Ananthakrishnan and Sen, 1980); Japan (Miyazaki and Kudô, 1988); Australia (Mound and Houston, 1987); New Zealand (Mound and Walker, 1982, 1986). Introductions and keys to limited groups of larvae have been produced by Heming (1991), Kirk (1987), Miyazaki and Kudô (1986), Priesner (1964) and Vance (1974). Palmer *et al.* (1989) and Moritz (1994) provide excellent keys to significant pest species.

The life history of typical terebrantian (Thripidae) pest species consists of an egg, two active feeding larval instars, and two relatively inactive non-feeding pupal instars designated propupa and pupa (Plate I.1). The tubuliferan family, Phlaeothripidae, to which a few pests belong, has an additional pupal instar. Thrips eggs are relatively large in relation to the size of the female. Those of Terebrantia are smooth shelled and are inserted into plant tissue with a saw-like ovipositor (see Chapter 2). Phlaeothripids, lacking an ovipositor, stick their eggs, which often have sculptured shells, onto the surface of vegetation. Individual females lay between 30 and 300 eggs depending on the species and quality of food available (see Chapter 7).

Males are derived from unfertilized eggs but, in many unrelated species over the whole or part of their geographical range, they are rare. In such species reproduction is often parthenogenic. The feeding larvae remain on their host plants but pupation may occur on a plant or in the soil beneath, depending on species. The complete life cycle usually lasts between 10 and 30 days, depending very much on temperature (Table 1.1), with adults surviving for a similar period. In warm regions and in greenhouses where breeding may be continuous, up to 12 or 15 generations may be completed each year. In cooler regions there may be only one or two generations with winter being spent as a larva in a soil cell, or as an adult among dead plant litter, tree bark or crop debris (see Chapter 7) (Lewis, 1973).

Changing Emphases on Thrips Pests

The features of thrips biology that predispose them to become pests are: their ability to cause direct feeding damage or, for a few species, spread virus diseases to food, fibre and ornamental crops; their tendency to spread and colonize widely, mainly by natural flight but also secreted within vegetation aboard planes, ships and vehicles; and their ability to multiply rapidly in favourable environments. During this century the geographical and agricultural foci of interest in pest thrips have changed continually as different regions of the world have been opened up and exploited for commercial food production, and trade patterns have altered. As a result, endemic species have transferred to new hosts and alien species have been introduced to new areas (Mound, 1983).

Another reason for the changing emphasis relates to where the relatively few thysanopterists, in the past far fewer than one per cent of the entomological profession, have happened to live and work, and to the perceived harmful species of the time. This is reflected in the literature: *Thrips tabaci* has probably received more attention over the years than any other single species, until the mid 1980s when much emphasis was switched to other pests, especially *Frankliniella occidentalis* (Mound and Teulon, 1995). As cereal production expanded in the former USSR in the 1930s, 1940s and 1950s, much effort was expended on cereal thrips (*Haplothrips* spp.) (Rubtzov, 1935; Pavlov, 1937;

Table 1.1. Average developmental periods (days) for each stage in the life history of *Frankliniella occidentalis* at three different constant temperatures* (from Lublinkhof and Foster, 1977) (see Plate I.1).

	Temperature (°C)		
Stage	15	20	30
Egg incubation	11.2	6.4	4.3
Larva I	4.9	2.3	1.1
Larva II	9.1	5.2	4.3
Propupa	2.9	2.2	1.4
Pupa	5.6	2.9	1.6
Adult pre-oviposition	10.4	2.4	2.4
Total of above	44.1	21.4	15.1
Longevity	70.8	56.8	27.5
No. eggs laid	24.0	95.5	43.8
Egg to egg	44.1	22.4	15.0

* The duration of all stages would increase at fluctuating temperatures typically occurring on a daily cycle in the field. Further examples of the duration of life cycles for other species at different temperatures are cited by Lewis (1973). Modern methods of relating growth and temperature are described by van Rijn *et al.* (1996) (see also Chapter 7).

Grivanov, 1939). At the same time grass dwellers affecting seed production (*Chirothrips* spp.; *Anaphothrips obscurus*) increased in prominence in Scandinavia (Hukkinen, 1936; Johansson, 1946) and North America (Bailey, 1948; Roney, 1949; Riherd, 1954; Kamm, 1971). Also in North America, species of *Thrips*, *Frankliniella*, *Scirtothrips* and *Hercinothrips* on fruit and vegetables received much attention as agrobusinesses were expanded during the middle of the century (Bailey, 1938; Quayle, 1938), and this impetus has been maintained as conventional chemical control methods have gradually failed. Likewise, in southern Africa in the 1930s, *Scirtothrips* on citrus attracted attention as this region raised production for export (Hall, 1930) and in the 1950s *Selenothrips* on cocoa and cashew in the Caribbean were studied in depth (Fennah, 1955, 1963, 1965). The discovery in 1927 that thrips could transmit a 'spotted wilt' to plants (Pittman, 1927) and that it had a viral aetiology (Samuel *et al.*, 1930) widened the coverage to include thrips as plant virus disease vectors, a topic which is still evolving and expanding. In the last decade the focus of interest has again moved with the enormous increase worldwide in airborne transport of fresh flowers, fruit and vegetables; the establishment of mass plant propagation enterprises, especially for greenhouse ornamentals in western Europe and North America; and the increase in intensive vegetable production in East and South Asia (Talekar, 1991). These developments have particularly facilitated the recent rapid spread of many species, especially *Frankliniella occidentalis* and *Thrips palmi*, with serious consequences to some of the most productive and lucrative agricultural and horticultural areas in the world (Mantel and Vierbergen, 1996).

There has also been intermittent interest in thrips attacking forest trees, for example *Thrips laricivorus* on larch in central Europe in the middle of the century (Kratochvil and Farsky, 1942; Ewald and Burst, 1959; Maksymov, 1965; Zenther Møller, 1965), and in the late 1980s and 1990s *Taeniothrips inconsequens* on sugar maple in north-eastern USA (Parker, 1991; Teulon *et al.*, 1993).

The current degree of interest in pest thrips is reflected in the consistent number of publications appearing in the scientific literature. Between 1985 and 1994 over 1000 are cited in the CAB Abstracts Database, excluding conference proceedings, and almost as many in the previous decade. The various chapters in this book cite some 3000 references spanning nearly a century, but concentrating on the last 20 years.

Alongside changing geographical and crop interests there has been an evolution of control measures. Early in the century they consisted mainly of simple cultivations, stubble burning, regard for planting and harvesting times, and the use of crude chemicals and natural plant extracts sprayed at high volumes. These were gradually superseded by sophisticated synthetic insecticides applied at low volume by increasingly refined equipment. As many pest species responded to these pesticides by developing resistance (see Chapter 15), there has been intense interest in finding ways of encouraging beneficial organisms, including fungal and nematode pathogens and a wide range of mites and insects, not least predatory thrips, either alone or more commonly as components of integrated pest management (IPM) systems (see Chapters 8, 9,

10, 16, 17, 18). The aim has been to recombine some of the older cultural approaches with modern chemicals compatible with biological control agents and tolerant cultivars to produce more effective and environmentally benign control.

As well as being driven by economic pressures, the recent pace of progress has also been encouraged by the many new techniques available (see Chapter 11). For example, improved stereoscan microscopy has revealed hitherto unappreciated morphological and anatomical features associated with feeding and sensory perception (see Chapter 2). Advanced serological and immuno-assay technology has revolutionized understanding of host plant/vector/virus relationships (see Chapter 14). Improved photography and video recording have been used to study flight, mating and fighting (see Chapters 3, 4, 5). Microanalytical equipment has been used to detect and identify semiochemicals with a role in inter- and intraspecific behaviour. DNA-based methods are beginning to be used to study taxonomic and systematic affinities, genetic diversity and the degree of mixing and rate of spread of pest and beneficial populations (Lewis *et al.*, 1991; Gillings *et al.*, 1996).

All these topics relating to pest thrips and the many options for controlling them are covered in depth in various chapters in this book. At this juncture, suffice to state that thrips are a serious and increasing threat to the production of high quality food and ornamental produce worldwide. There are now at least 1200 entomologists studying these insects and the crop damage they cause to whom this book should appeal (Parker and Skinner, 1994), either as a compendium of knowledge, a stimulus and aid to research, or as a practical guide to plant damage and pest control.

The Scale of Thrips Damage

The status of thrips as pests differs greatly between crops and geographical areas. A few crops are attacked by the same species in widely separated parts of the world. *Thrips tabaci* is a cosmopolitan pest of onions grown between sea level and 2000 m and *Hercinothrips femoralis* may occur on bananas almost wherever they are grown. By contrast, some crops grown in different regions are infested by different species in each place. In southern Africa *Scirtothrips aurantii* is the most harmful thrips on citrus, in California *S. citri* fills this niche, and in Florida *Frankliniella bispinosa* and *Heliothrips haemorrhoidalis*, among others, attack citrus flowers and fruit. In East Africa *Caliothrips impurus* and *C. sudanensis* attack cotton whereas *Frankliniella fusca* and *F. tritici* cause comparable damage in North America (see Appendix II).

Direct feeding damage to leaves, flowers and fruit caused by the ingestion of sap (see Chapters 4 and 13) is most harmful in dry climates and seasons when heavily attacked plants lose moisture rapidly. In these conditions, infestation can seriously deplete yields and sometimes render crops uneconomic. Indeed, there is evidence that some crops under water stress not only develop symptoms more rapidly than adequately watered ones, but provide thrips with more nutritious food and therefore encourage heavier infestations

(Fennah, 1963; Silva, 1964; see also Chapter 4, Distribution on Plants). On the other hand, many crops can outgrow damage to seedlings and young plants, especially in cooler climates, and eventual yield losses are small or undetectable. The overall impact of infestation of a crop thus depends on many factors: the size of the thrips population, the plant growth stage, its vulnerability to direct feeding and oviposition damage or to virus infection, the duration of the infestation, and the suitability of weather for population growth. Generally, the more important the visual appearance of a crop, the more likely is the grower to suffer substantial economic losses. Many orchard- and field-grown fruits become scarred and distorted, lowering their value or making them unmarketable. High-value greenhouse crops, such as vegetables, fruit and especially ornamentals, are particularly vulnerable because of the high capital investment and production costs involved and the requirements for totally unblemished produce (see Chapter 18). The scale of damage can be appreciated from three viewpoints: the area of crops affected, decline in yields, and loss of revenue.

In field and plantation crops, initial infestations usually by airborne immigrants (see Chapter 5) can spread rapidly to cover extensive areas. There are records of 20,000 ha of sugar cane in Taiwan being seriously damaged annually by *Fulmekiola serrata* (Wang, 1952) and the author has seen vistas of cassava in the Andean foothills of Colombia almost totally defoliated by *Corynothrips stenopterus*. Partial defoliation of sugar maple trees caused by *Taeniothrips inconsequens* extended over 40,000 ha in Pennsylvania and 200,000 ha in Vermont in 1988 (Parker, 1991). More limited infestations, but nevertheless extensive and often first affecting the edges of plantings, can spread through large arable fields (e.g. cabbages, cereals, cotton, onions, soybean, and tobacco) and plantations (e.g. citrus, coffee, tea, and stonefruit) to reach densities of tens, hundreds or even thousands of millions of individuals per hectare (see Chapter 5, Exodus flight and dispersal).

Measured or estimated crop losses expressed as percentages of attainable yield vary enormously and because of this are not a reliable quantitative indicator of the overall importance of any one species, but collectively they illustrate the damaging potential of thrips. Before modern control methods were available, in some years up to 80 per cent of the citrus crop in California (Quayle, 1938) and 50 per cent in Southern Rhodesia (now Zimbabwe) (Hall, 1930) were spoilt by *Scirtothrips* spp. Even now in California, yield increases in navel oranges ranging from 8 to 25 per cent can be obtained by controlling *Scirtothrips citri* (Elmer *et al.*, 1975; Elmer and Brawner, 1988). Considerable yield losses caused by *Thrips tabaci* to onions still occur widely despite the numerous studies made for decades on this problem; in 1988 and 1989 losses of 34 per cent and 43 per cent, respectively, were recorded in Canada (Fournier *et al.*, 1996). Table 1.2 lists examples of other studies illustrating the range of losses incurred in arable and plantation crops.

Typically, grain losses in wheat, barley and rye have been assessed as 2 to 10 per cent in Europe (Rubtzov, 1935; Vappula, 1965; Mantel, 1985) and slightly more in North America (Post, 1957), but the economic relevance of such records depends on whether they are determined using grain numbers or weight, because damaged heads usually bear heavier individual grains. Recent evidence from wheat infested with *Haplothrips tritici* in Spain indicates that the more larvae per ear, the lighter in weight is a given volume of grain (Bielza and Torres-Vila, 1995). Reported losses in grass seed crops are usually greater than in cereals, often reaching 30 per cent (Doull, 1956; Wetzel, 1964; Kamm, 1971). In New Zealand, Morrison (1961) found that each 1 per cent increase in infestation of *Chirothrips pallidicornis* in cocksfoot seed crops corresponded to a 0.76 per cent yield loss.

The ultimate measure of crop loss is revenue to the grower, which is determined by production costs and market forces at time of sale. As marketing becomes more sophisticated and international, high quality produce is at a premium. Monetary values commensurate with the scale of percentage yield losses listed in Table 1.2, or the cost of preventing them, will vary with factors such as crop, season, availability, demand and distribution costs, but clearly losses of these magnitudes would be serious for individual growers and the wider industry.

As illustrations, damage to sweet peppers in Florida caused by *Frankliniella occidentalis* and *Thrips palmi* in 1993 exceeded $US10 million (Nuessly and Nagata, 1995); scarring of citrus fruit in California by *Scirtothrips citri* contributes to a loss of approximately $US2 per carton of fruit (Morse, 1995); *Frankliniella occidentalis* can cause up to 90 per cent loss of summer-replanted cucumbers in UK glasshouses worth up to £50,000 ($US75,000) ha^{-1} each year

Table 1.2. Examples of percentage loss of yield in field crops due to direct damage by thrips.

Crop	Species	Loss (%)	Location	Reference
Alfalfa	*Thrips tabaci*	33	Iowa	Wolfenbarger and Hibbs (1958)
Carrot seed	*Frankliniella* spp.	50	California	Bailey (1938)
Cassava	*Corynothrips stenopterus*	8–15	Colombia	Schoonhoven and Peña (1976)
Cowpea	*Megalurothrips sjostedti*	30–90	Senegal	Bal (1991)
Onion	*Thrips tabaci*	34–43	Canada	Fournier *et al.* (1995)
Panicum maximum	*Chirothrips crassus*	50	Cuba	Paz *et al.* (1990)
Peas	*Kakothrips pisivorus*	20	Holland	Franssen (1960)
Red clover	*Haplothrips niger*	10	Romania	Perju and Muset (1982)
Rice	*Haplothrips gangebaueri*	50–63	W. Bengal	Mukhopadhyay (1989)
Rye	*Limothrips* spp. *Haplothrips aculeatus*	3–13	Sweden	Larsson (1988)
Tea	*Scirtothrips* spp. *Heliothrips haemorrhoidalis*	100	Kenya	Benjamin (1968)
Tobacco	*Scirtothrips aurantii*	75	S. Africa	Strofberg (1948)

(R. Jacobson, Horticulture Research International, UK, 1996, personal communication); resowing UK sugar beet fields after early spring attack by *Thrips angusticeps*, as occurred widely in 1996, costs about £145 ($US218) ha^{-1} (A. Dewar, Institute of Arable Crops Research, UK, 1996, personal communication).

Infection of plants with tomato spotted wilt virus (TSWV) transmitted by thrips may likewise be costly to growers. In a survey of greenhouse ornamental plants in Pennsylvania during 1989 to 1990, plants infested with the *Impatiens* strain of TSWV with a retail value of £450,000 ($US675,000) were destroyed under the direction of the States Department of Agriculture to prevent further spread of the virus and vectors. In addition, growers routinely destroyed plants in their own disease management programmes (Hausbeck *et al.*, 1992). Similarly, in Denmark, plant health inspectors visit exporting greenhouse nurseries unannounced approximately every 3 weeks. If *Frankliniella occidentalis* is found on plants, or on sticky traps (see Chapter 11), the grower's export licence is temporarily suspended until the thrips are controlled (H.F. Brødsgaard, Denmark, 1993, personal communication; Shipp, 1995). In Finland, the government spent £260,000 ($US390,000) between 1987 and 1990 to eradicate *Frankliniella occidentalis* from glasshouses growing vegetables and ornamentals. The consequences to the industry of failing to do this were calculated as costing £4 million ($US7.5 million) over 4 years, so the eradication programme appears to have been worthwhile (Rautapää, 1992).

In summary, considering the range of crops attacked, the type and degree of direct damage inflicted and the virus diseases spread, the hundred or so significant pests in the thysanopteran family Thripidae and the smaller number of pest phlaeothripids are now arguably collectively comparable in importance to world crop production with the whiteflies (Aleurodidae), scale insects and mealy bugs (Coccoidea), or possibly the aphids (Aphididae). For individual growers of many crops they can cause serious problems requiring expensive control measures which may sometimes fail.

References

Ananthakrishnan, T.N. and Sen, S. (1980) Taxonomy of Indian Thysanoptera. *Zoological Survey of India, Handbook Series* 1, 1–234.

Bailey, S.F. (1938) Thrips of economic importance in California. *Circular of the California Agriculture Experiment Station* 346.

Bailey, S.F. (1948) Grain and grass infesting thrips. *Journal of Economic Entomology* 41, 701–705.

Bal, A.B. (1991) Action threshold for flower thrips on cowpea (*Vigna unguiculata* (L.) Welp.) in Senegal. *Tropical Pest Management* 37, 363–367.

Benjamin, D.M. (1968) Insects and mites on tea in Africa and adjacent islands. *East Africa Agriculture and Forestry Journal* 34, 1–16.

Bielza, P. and Torres-Vila, L.M. (1995) Efecto del nivel poblacional de *Haplothrips tritici* Kurd. (Thysanoptera; Phlaeothripidae) en la producion de trigo en Castille – La Mancha. *XVI Jornalus de la Asociacion Española de Entomología*, Cuenca, Spain.

Doull, K.M. (1956) Thrips infesting cocksfoot in New Zealand. II The biology and economic importance of the cocksfoot thrips *Chirothrips manicatus* Haliday. *New Zealand Journal of Science and Technology* (A) 38, 56–65.

Elmer, H.S. and Brawner, O.L. (1988) The effect of citrus thrips and citrus red mite on navel orange fruit production in California's San Joaquin Valley. In: Goren, R. and Mendel, K. (eds) *Proceedings of the 6th International Citrus Congress, Tel Aviv, Israel*. Balabon Publishers, Philadelphia, Pennsylvania, pp. 1089–1095.

Elmer, H.S., Ewart, W.H. and Brawner, O.L. (1975) Effect of citrus populations on navel orange fruit yield and tree growth in central valley. *Citrograph* 61, 9–10, 20–22.

Ewald, G. and Burst, R. (1959) Untersuchungen über Larchenblasenfuss und Larchenminiermotte. *Angewande Forst-u Jagdztq* 130, 173–181. (*Review of Applied Entomology* (A) 49, 576–577.)

Fennah, R.G. (1955) The epidemiology of cacao-thrips on cacao in Trinidad. *Report of Cacao Research* 24, 7–26.

Fennah, R.G. (1963) Nutritional factors associated with seasonal population increases of cacao thrips, *Selenothrips rubrocinctus* (Giard) (Thysanoptera), on cashew *Anacardium occidentale*. *Bulletin of Entomological Research* 53, 681–713.

Fennah, R.G. (1965) The influence of environmental stress on the cacao tree in predetermining the feeding sites of cacao thrips, *Selenothrips rubrocinctus* (Giard), on leaves and pods. *Bulletin of Entomological Research* 56, 333–349.

Fournier, F., Boivin, G. and Stewart, R.K. (1995) Effect of *Thrips tabaci* (Thysanoptera: Thripidae) on yellow onion yields and thresholds for its management. *Journal of Economic Entomology* 88, 1401–1407.

Fournier, F., Boivin, G. and Stewart, R.K. (1996) Impact and economic threshold of *Thrips tabaci* on onions. In: Parker, B.L., Skinner, M. and Lewis, T. (eds) *Thrips Biology and Management*. Plenum Press, New York, pp. 71–76.

Franssen, C.J.H. (1960) Biology and control of the pea thrips. *Verslag Landbouwk undig Onderzoek Ryklandbproefstn* 66.

Gaston, K.J. and Mound, L.A. (1993) Taxonomy, hypothesis testing and the biodiversity crisis. *Proceedings of the Royal Society, London* (B) 251, 139–142.

Gillings, M.R., Rae, D., Herron, G.A. and Beattie, G.A.C. (1996) Tracking thrips populations using DNA-based methods. *Proceedings of the 1995 Australia and New Zealand Thrips Workshop*. NSW Agriculture, Gosford, pp. 97–103.

Grivanov, K.P. (1939) Deep ploughing instead of burning the stubble for the control of wheat thrips (*Haplothrips tritici* Kurd.). (In Russian.) *Review of Applied Entomology* (A) 28, 69–70.

Hall, W.J. (1930) The South African citrus thrips in Southern Rhodesia. *Report of Mazoe Citrus Experiment Station* 1, 5–55.

Hausbeck, M.K., Welliver, R.A., Derr, M.A. and Gildow, M.A. (1992) Tomato spotted wilt virus survey among greenhouse ornamentals in Pennsylvania. *Plant Disease* 76, 795–800.

Heming, B.S. (1991) Order Thysanoptera. In: Stehr, F.W. (ed.) *Immature Insects*, Vol. 2. Kendall/Hunt Publishing Company, Dubeque, Iowa, pp. 1–21.

Hukkinen, Y. (1936) Tutimuksia Nurmipuntarpään (*Alopecurus pratensis* L.) Siementuholaisista. 1. *Chirothrips hamatus* Tryb. *Valtion Maatalouskoetoiminnan Julkaisuja* 81.

Jacot-Guillarmod, C.F. (1970–1979) Catalogue of the Thysanoptera of the world (parts 1–4, Terebrantia; parts 5–6, Tubulifera). *Annals of the Cape Provincial Museums (Natural History)* 7, 1–1724.

Jacot-Guillarmod, C.F. and Brothers, D.J. (1986) Catalogue of the Thysanoptera of the

World (Part 7). *Annals of the Cape Provincial Museums (Natural History)* 17, 1–93.

Johansson, E. (1946) Studies och försök rörande de pa gras och sadesslag levande tripsarnis biologi och skadegörelse. II Tripsarnis frekvers och spridning i jämforelse med andra sugande insekters samt deras fruskadegörande betydelse. *Meddelingen St VaxtskAmst* 46, 1–59. (*Review of Applied Entomology* A 37, 454.)

Kamm, J.A. (1971) Silvertop of bluegrass and bentgrass produced by *Anaphothrips obscurus*. *Journal of Economic Entomology* 64, 1385–1387.

Kirk, W.J.D. (1987) A key to the larvae of some common Australian flower thrips (Insecta: Thysanoptera), with a host plant survey. *Australian Journal of Zoology* 35, 173–185.

Kratochvil, J. and Farsky, O. (1942) Das Absterben der diesjährigen terminalen Larchentriebe. *Zeitschrift für angewandte Entomologie* 29, 177–218.

Larsson, H. (1988) Economic damage caused by cereal thrips in winter rye in Sweden. *Acta Phytopathologica et Entomologica Hungarica* 23, 291–293.

Lewis, T. (1973) *Thrips, their Biology, Ecology and Economic Importance*. Academic Press, London.

Lewis, T. (1991) An introduction to the Thysanoptera: a survey of the group. In: Parker, B.L., Skinner, M. and Lewis, T. (eds) *Towards Understanding Thysanoptera*. General Technical Report, NE-147, USDA, Forest Service, Radnor, Pennsylvania, pp. 3–19.

Lewis, T., Loxdale, H.D. and Brookes, C.P. (1991) Prospects for studying persistence and dispersal of pear thrips using genetic markers (allozymes). In: Parker, B.C. and Doane, E.N. (eds) *The 1990 Conference on Thysanoptera: Abstracts*. Vermont Agriculture Experiment Station Bulletin 698, pp. 1–7.

Lublinkhof, J. and Foster, D.E. (1977) Development and reproductive capacity of *Frankliniella occidentalis* (Thysanoptera: Thripidae) reared at three temperatures. *Journal of the Kansas Entomological Society* 50, 313–316.

Maksymov, J.K. (1965) Die Überwinterung des Larchenblasenfusses *Taeniothrips laricivorus* Kratochvil und Farsky. *Mitteilungen der Schweizerische Anstalt für das forstliche Versuchswesen* 41, 1–17.

Mantel, W.P. (1985) Thrips in winter wheat in The Netherlands. In: Holman, J., Pelikan, J., Dixon, A.F.G. and Weissmann, L. (eds) *Population Structure, Genetics and Taxonomy of Aphids and Thysanoptera*. SPB Academic Publishing, The Hague, pp. 448–453.

Mantel, W.P. and Vierbergen, G. (1996) Additional species to the Dutch list of Thysanoptera and new intercepted Thysanoptera on imported plant material. *Folia Entomologica Hungarica* 57 (suppl.), 91–96.

Miyazaki, M. and Kudô, I. (1986) Descriptions of thrips larvae which are noteworthy on cultivated plants (Thysanoptera: Thripidae). Species occurring on solanaceous and cucurbitaceous crops. *Akitu* N.S. No. 79.

Miyazaki, M. and Kudô, I. (1988) Bibliography and host plant catalogue of Thysanoptera of Japan. *Miscellaneous Publication of the National Institute of Agro-Environmental Sciences* 3, 1–246.

Moritz, G. (1994) Pictorial key to the economically important species of Thysanoptera in central Europe. *Bulletin de l'Organisation Europiénne et Mediterraniénne pour la Protection des Plants* 24, 181–208.

Morrison, L. (1961) Thrips infestation of cocksfoot seed crops. *New Zealand Journal of Agricultural Research* 4, 246–252.

Morse, J.G. (1995) Prospects for IPM of citrus thrips in California. In: Parker, B.L., Skinner, M. and Lewis, T. (eds) *Thrips Biology and Management*. Plenum Press, New York, pp. 371–379.

Mound, L.A. (1983) Natural and disrupted patterns of geographical distribution in Thysanoptera (Insecta). *Journal of Biogeography* 10, 119–133.

Mound, L.A. and Houston, K.J. (1987) An annotated check-list of Thysanoptera from Australia. *Occasional Papers on Systematic Entomology* 4, 1–28.

Mound, L.A. and Marullo, R. (1996) The thrips of Central and South America: an introduction. *Memoires on Entomology, International* 6, 1–488.

Mound, L.A. and Palmer, J.M. (1989) *The Thysanoptera of Panama: a Biological Catalogue and Bibliography.* G.B. Fairchild Commemorative Volume, University of Panama Museum.

Mound, L.A., Morison, G.D., Pitkin, B.R. and Palmer, J.M. (1976) Thysanoptera. *Handbooks for the Identification of British Insects* 1 (2), 1–79.

Mound, L.A. and Teulon, D.J. (1995) Thysanoptera as phytophagous opportunists. In: Parker, B.L., Skinner, M. and Lewis, T. (eds) *Thrips Biology and Management.* Plenum Press, New York, pp. 3–19.

Mound, L.A. and Walker, A.K. (1982) Terebrantia (Insecta: Thysanoptera). *Fauna of New Zealand* 1, 1–113.

Mound, L.A. and Walker, A.K. (1986) Tubulifera (Insecta: Thysanoptera). *Fauna of New Zealand* 10, 1–140.

Mukhopadhyay, A.K. (1989) Incidence, extent of damage and new food plants of rice inflorescence thrips in West Bengal. *Environment and Ecology* 7, 466–468.

Nuessly, G.S. and Nagata, R.T. (1995) Pepper varietal response to thrips feeding. In: Parker, B.L., Skinner, M. and Lewis, T. (eds) *Thrips Biology and Management.* Plenum Press, New York, pp. 115–118.

Palmer, J.M. (1990) Identification of the common thrips of Tropical Africa (Thysanoptera: Insecta). *Tropical Pest Management* 36, 27–49.

Palmer, J.M., Mound, L.A. and du Heaume, G.J. (1989) *CIE Guides to Insects of Importance to Man. 2. Thysanoptera.* CAB International, Wallingford, UK.

Parker, B.L. (1991) The pear thrips problem. In: Parker, B.L., Skinner, M. and Lewis, T. (eds) *Towards Understanding Thysanoptera.* General Technical Report NE-147, USDA Forest Service, Radnor, Pennsylvania, pp. 179–192.

Parker, B.L. and Skinner, M. (1994) *The Thrips Network. A worldwide directory of specialists and individuals concerned with the Thysanoptera.* Asian Vegetable Research and Development Center, Taipei, AVRDC Publication No. 94–428.

Pavlov, I.F. (1937) Noxiousness of *Haplothrips tritici* Kurd. (In Russian.) *Review of Applied Entomology* (A) 25, 143–144.

Paz, G. de la, Miret, R. and Delgado, A. (1990) Severe attack of florets and date on which thrips affect inflorescences of *Panicum maximum* Jacq. (In Spanish.) *Pastos y Forrajes* 13, 279–283.

Perju, T. and Muset, D. (1982) Red clover thrips (*Haplothrips niger* Osb. Phlaeothripidae – Thysanoptera): bioecology and control. *Probleme de Protectia Plantelor* 10, 113–122.

Pitkin, B.R. and Mound, L.A. (1973) A catalogue of West African Thysanoptera. *Bulletin de l'Institut Fondamental de l'Afrique Noire* 35, 407–449.

Pittman, H.A. (1927) Spotted wilt of tomatoes. Preliminary note concerning the transmission of the 'spotted wilt' of tomatoes by an insect vector (*Thrips tabaci* Lind.). *Journal of the Council of Scientific and Industrial Research (Australia)* 1, 74–77.

Post, R.L. (1957) The barley thrips in North Dakota. *The North Dakota Seed Journal* 26 (March), 4.

Priesner, H. (1964) Ordnung Thysanoptera (Fransenflügler, Thripse). In: Franz, H. (ed.) *Bestimmungsbucher zur Bodenfauna Europas* 2, Akademie Verlag, Berlin, pp. 1–242.

Quayle, H.J. (1938) *Insects of Citrus and other Tropical Fruit.* Comstock, Ithaca, New York.

Rautapää, J. (1992) Eradication of *Frankliniella occidentalis* and tomato spotted wilt virus in Finland: a case study in costs and benefits. *Bulletin de l'Organisation Europiénne et Mediterraniénne pour la Protection des Plantes* 22, 545–550.

Riherd, P.T. (1954) Thrips as a limiting factor in grass seed production. *Journal of Economic Entomology* 47, 709–719.

Roney, J.N. (1949) Bermuda seed grass insects in Arizona. *Journal of Economic Entomology* 42, 555.

Rubtzov, I.A. (1935) *Haplothrips tritici* Kurd. and the coefficient of its injury. (In Russian.) *Review of Applied Entomology* (A) 23, 565–566.

Sakimura, K. (1986) Thrips in and around the coconut plantations in Jamaica, with a few taxonomical notes (Thysanoptera). *Florida Entomologist* 69, 348–363.

Samuel, G., Bald, J.G. and Pittman, H.A. (1930) Investigations on 'spotted wilt' of tomatoes in Australia. *Commonwealth Council of Scientific and Industrial Research Bulletin* No. 44.

Schliephake, G. and Klimt, K. (1979) Thysanoptera, Fransenflügler. *Die Tierwelt Deutschlands* 66, 1–477.

Schoonhoven, A. van and Peña, J. (1976) Estimation of yield losses in cassava following attack from thrips. *Journal of Economic Entomology* 69, 514–516.

Shipp, J.L. (1995) Monitoring of western flower thrips on glasshouse vegetable crops. In: Parker, B.L., Skinner, M. and Lewis, T. (eds) *Thrips Biology and Management.* Plenum Press, New York, pp. 547–555.

Silva, P. (1964) Tripes do cacaueiro – causador do 'queima' da fôlha e da 'ferrugem' du fruto. *Cacau Atual* 1, 9–10.

Stannard, L.J. (1957) The phylogeny and classification of the North American genera of the suborder Tubulifera (Thysanoptera). *Illinois Biological Monographs* 25, 1–200.

Stannard, L.J. (1968) The thrips, or Thysanoptera, of Illinois. *Bulletin of the Illinois State Natural History Survey* 29, 215–552.

Strofberg, F.J. (1948) New host of *Scirtothrips aurantii* (Faure). *Journal of the Entomological Society of South Africa* 10, 196–197.

Talekar, N.S. (ed.) (1991) Thrips in Southeast Asia. *Proceedings of a regional consultation workshop, Bangkok, Thailand, 13 March 1991.* Asian Vegetable Research and Development Center, Taipei, AVRDC Publication No. 91–342.

Teulon, D.A.J., Kolb, T.E., Cameron, E.A., McCormick, L.H. and Hoover, G.A. (1993) Pear thrips, *Taeniothrips inconsequens* Uzel. (Thysanoptera: Thripidae), on sugar maple, *Acer saccharum* Marsh: a review. *Zoology (Journal of Pure and Applied Zoology)* 4, 355–380.

Vance, T.C. (1974) Larvae of the Sericothripini (Thysanoptera: Thripidae), with reference to other larvae of the Terebrantia of Illinois. *Illinois Natural History Survey Bulletin* 31, 145–208.

van Rijn, P.C.J., Mollema, C. and Steinhuis-Broers, G.M. (1996) Comparative life history studies of *Frankliniella occidentalis* and *Thrips tabaci* (Thysanoptera: Thripidae) on cucumber. *Bulletin of Entomological Research* 85, 285–297.

Vappula, N.A. (1965) Pests of cultivated plants in Finland. *Annales Agricultura Fennici* 1, (suppl. 1), 1–239.

Wang, C.S. (1952) Study of cane thrips. *Taiwan Sugar Industry Seasonal Journal for 1952*, 90–107.

Wetzel, T. (1964) Untersuchungen zum Auftreten, zur Schadwirkung und zur Bekämpfung von Thysanoptera in Grassamenbeständen. *Beitrage Entomologie* 14, 427–500.

Wolfenbarger, D. and Hibbs, E.T. (1958) Onion thrips (*Thrips tabaci* Lind.) infesting cabbage. *Journal of Economic Entomology* 51, 394–396.
Zenther-Møller, O. (1965) *Taeniothrips lavicivorus* (Krat.) (Thripidae, Thysanoptera) in Danish stands of *Larix decidua* (Mill.) *Oikos* 16, 58–69.

Structure, Growth and Development

2

Gerald Moritz
Martin-Luther-Universitat, Halle-Wittenberg, Institut für Zoologie, Domplatz 4, Halle (Saale), 06108 Germany

Introduction

An appreciation of the structure, growth and development of thrips helps in the understanding of their ability to disperse and locate crops, how they feed and reproduce on them, and in some instances how they transmit plant viruses. It also provides a background for devising control methods based on conventional or novel behavioural approaches. This chapter provides a detailed account of the structure and functioning of thrips, concentrating on the suborder Terebrantia in which most pests occur. The terminology used assumes a familiarity with insect morphology and anatomy, but the emphasis of the presentation is pictorial comprising a series of stereoscan photomicrographs and careful drawings amplified by the text.

External Morphology and Sexual Dimorphism of Adults

The small size of thrips is an important factor contributing to the success of the order and their status as crop pests. Most species are only about 1 to 2 mm in length; some Tubulifera, mainly tropical or subtropical, may attain 15 mm. The colour of the dorsoventrally flattened, shiny body varies from white to brown, dark brown or black. To the naked eye thrips resemble black, brown or yellow dashes but under magnification can clearly be differentiated into a separate head with forward-directed antennae, a prothorax, a fused pterothorax (long-winged species with two pairs of fringed wings) and an 11-segmented abdomen, which bears in terebrantian females a well-developed ovipositor. Figure 2.1 gives an impression of the more obvious features.

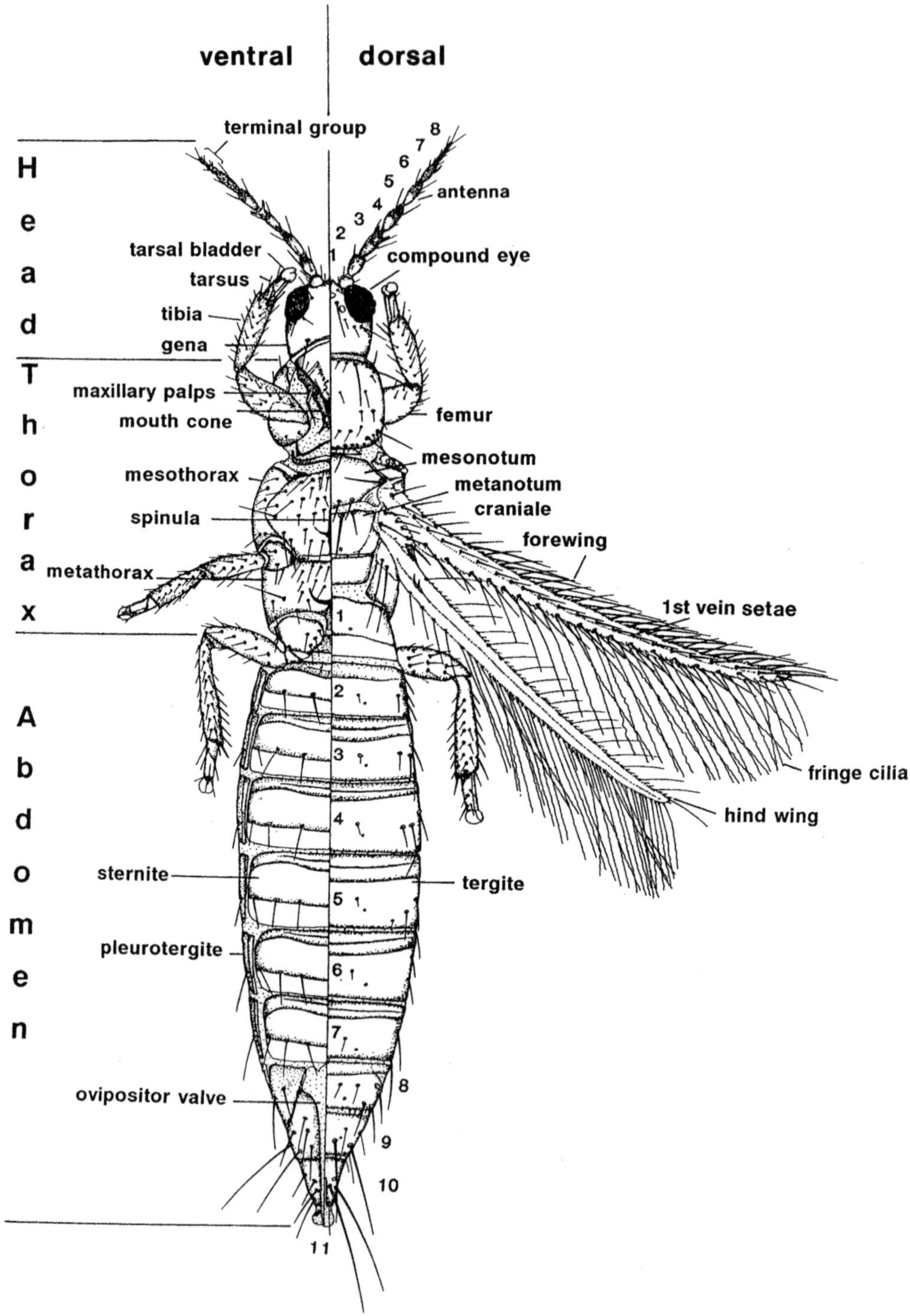

Fig. 2.1. Composite drawing of *Frankliniella tenuicornis* illustrating the principal features on the dorsal and ventral sides of a typical adult thripid.

Head

Antennae

The antennae of Thysanoptera comprise four to nine, but usually seven to eight, segments (Fig. 2.2: 1–4). The degree of fusion of the terminal antennal segments is variable but aberrations occur quite frequently (Fig. 2.3: 1a, 1b) (Mound and Walker, 1982). The cube-shaped scape articulates proximally

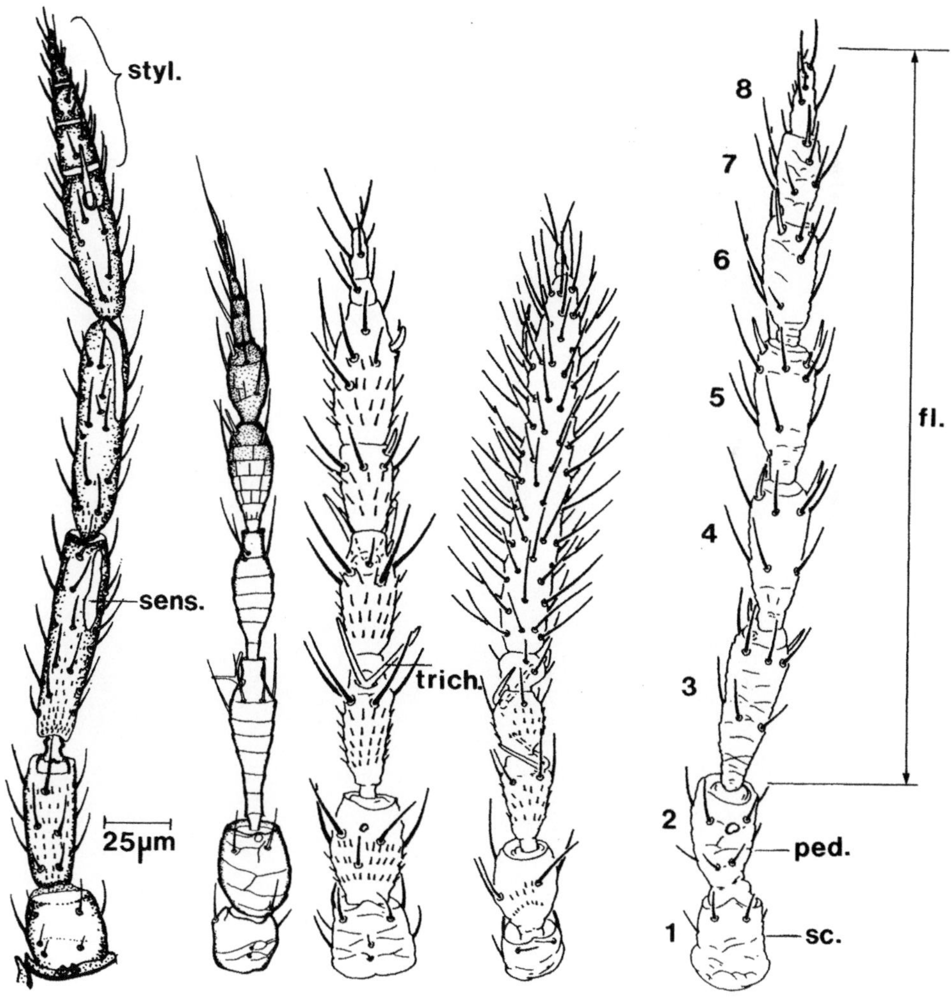

Fig. 2.2. Examples of thrips antennae: 1. *Aeolothrips intermedius*, ♀ ventral; 2. *Hercinothrips femoralis*, ♀ dorsal; 3a. *Mycterothrips consociatus*, ♀ dorsal; 3b. *Mycterothrips consociatus*, ♂ lateral; 4. *Cryptothrips nigripes*, ♀ dorsal (after Mound *et al.*, 1976). fl. = flagellum, ped. = pedicel, sc. = scape, sens. = sensoria, styl. = stylus, trich. = trichome.

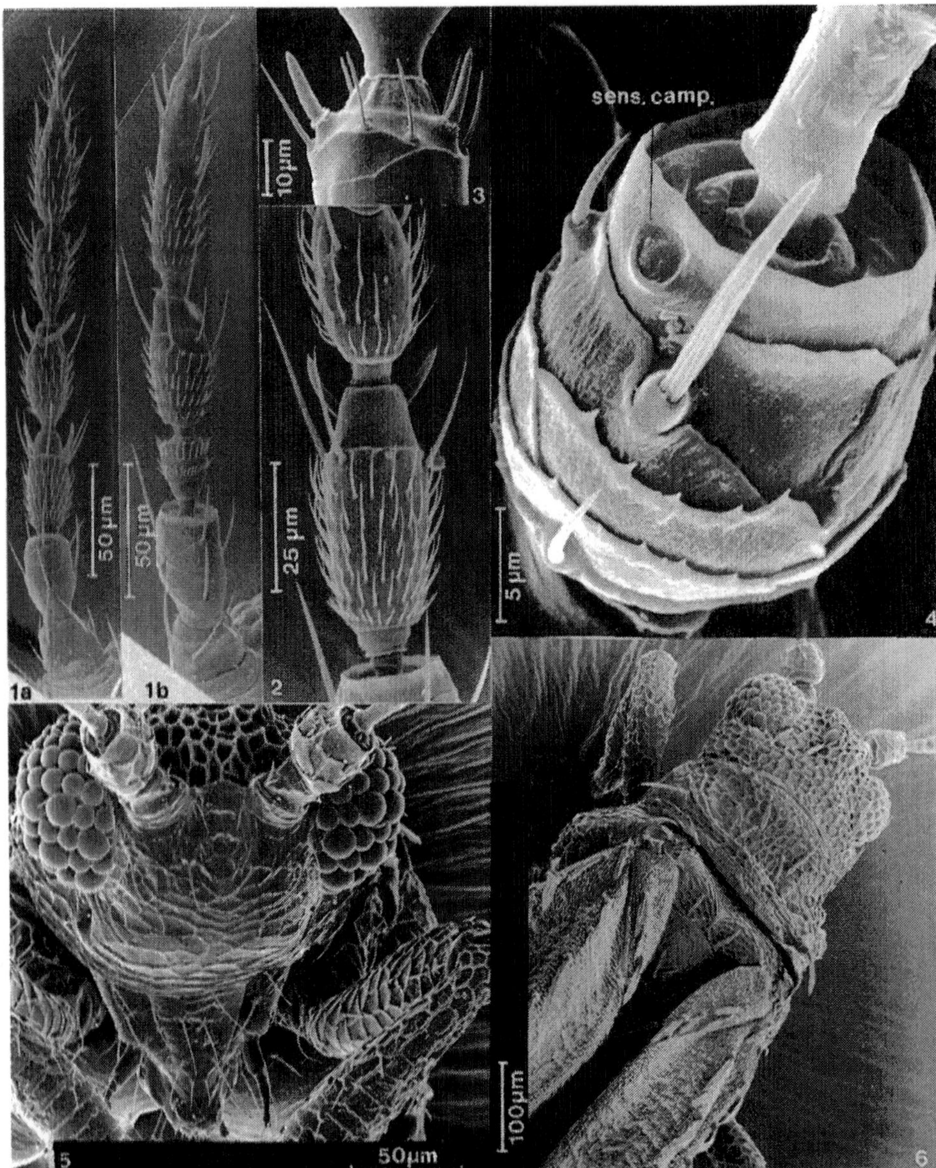

Fig. 2.3. 1a. *Frankliniella occidentalis*, ♀ right antenna, ventral; 1b. *Frankliniella occidentalis*, ♀ right antenna, dorsal, aberration of flagellum segments; 2. *Frankliniella occidentalis*, ♀ right antenna, ventral, connection between pedicellus and 3rd segment; 3. *Gynaikothrips ficorum*, ♀ 4th segment with sense cones; 4. *Hercinothrips femoralis*, ♀ pedicel with sensillum campaniforme and connection to flagellum; 5. *Parthenothrips dracaenae*, ♀ frontal view of head; 6. *Parthenothrips dracaenae*, ♀ dorsal view of head and thorax.

with the head capsule and distally with a dorsal and ventral condyle at the base of the pedicel (Fig. 2.2: 2, 4; Fig. 2.4: 2c) (Risler, 1957; Mickoleit, 1963). In 7- to 9-segmented antennae, the distal segments often form a uniform terminal group, the stylus (Fig. 2.2: 1). The flagellar segments and the pedicel are connected by small inner cuticular rings (= anellus and postanellus, Moritz, 1982b), which increase the mobility of the flagellum. The antennae bear sense organs of different size, shape and position (Heming, 1975; Moritz. 1982b, 1989b). Forked or simple thin-walled chemoreceptors (trichomes) typically occur on the dorsal side of antennal segment 3 and on the ventral side of antennal segment 4 in some terebrantian and tubuliferan genera, whereas the Aeolothripidae, Merothripidae and Heterothripidae have different shaped sensoria (Fig. 2.2: 1–4; Fig. 2.3: 1–3) (Slifer and Sekhon, 1974). The function of these external proprioreceptors is unknown, although several chemicals appear to stimulate them. Other antennal sense organs include small thin-walled sense cones, coeloconic pegs, microtrichia, and tactile hairs (for positions see Table 2.1). The pedicel bears dorsally the typical external structure of a sensillum campaniforme (Fig. 2.2: 3, 4; Fig. 2.3: 4). Terebrantian males are usually smaller than females and have shorter antennae. Distinct sexual dimorphism may occur, for example in the flagellum of *Mycterothrips consociatus* whose males have a greatly enlarged 6th antennal segment (Fig. 2.2: 3a, 3b) (Mound *et al.*, 1976).

Head capsule, compound eyes, ocelli and tentorium

The antennae are inserted into a socket on the head capsule. In many species the median rim of the socket forms a process between the first antennal segments (prominent in *Ankothrips*) or the head is more or less prolonged in front of the eyes (*Chirothrips*, *Limothrips*). The sclerites of the head capsule are completely fused and many sutures indistinct (Peterson, 1915; Risler, 1957; Mickoleit, 1963). In the family Aeolothripidae the dorsal vertex is separated from the occiput by an occipital suture, whereas a postoccipital suture with a small postocciput occurs in nearly all other species (Fig. 2.3: 6; Fig. 2.4: 1a, 2a, 2c). Ventrally the clypeofrons is bounded by the clypeal suture (Fig. 2.4: 1b, 2b, 2c; Fig. 2.5: 3). In the Terebrantia the paired genae and subgenae are separated, on the left side only, by the clypeogenal suture and the subgenal suture. The triangular anteclypeal membrane lies between the clypeofrontal suture and the labral border. It allows typical movements of the ventrally directed mouthcone during the pro- and retraction of the mandible (Fig. 2.4: 2c; Fig. 2.5: 3).

The paired compound eyes lie more or less dorsolaterally or dorsoventrally between vertex, clypeofrons and genae (Fig. 2.3: 5, 6; Fig. 2.4: 1, 2; Fig. 2.5: 2, 3; Fig. 2.6: 1). The number of facets differ from almost holoptic to as few as three in some apterae (Mound and Heming, 1991). In adults, the lenses of a few ventral facets are orange or dark brown and have a different optical density compared with unpigmented ommatidia (Moritz, 1995). Nakahara (1988) described about 300 terebrantian species with such two to seven pigmented facets arranged characteristically in different species. In aeolothripids and phlaeothripids the eyes of some species are prolonged ventrally and often a few of the ventral ommatidia are remarkably large (Fig. 2.4: 2b) (Mickoleit, 1963).

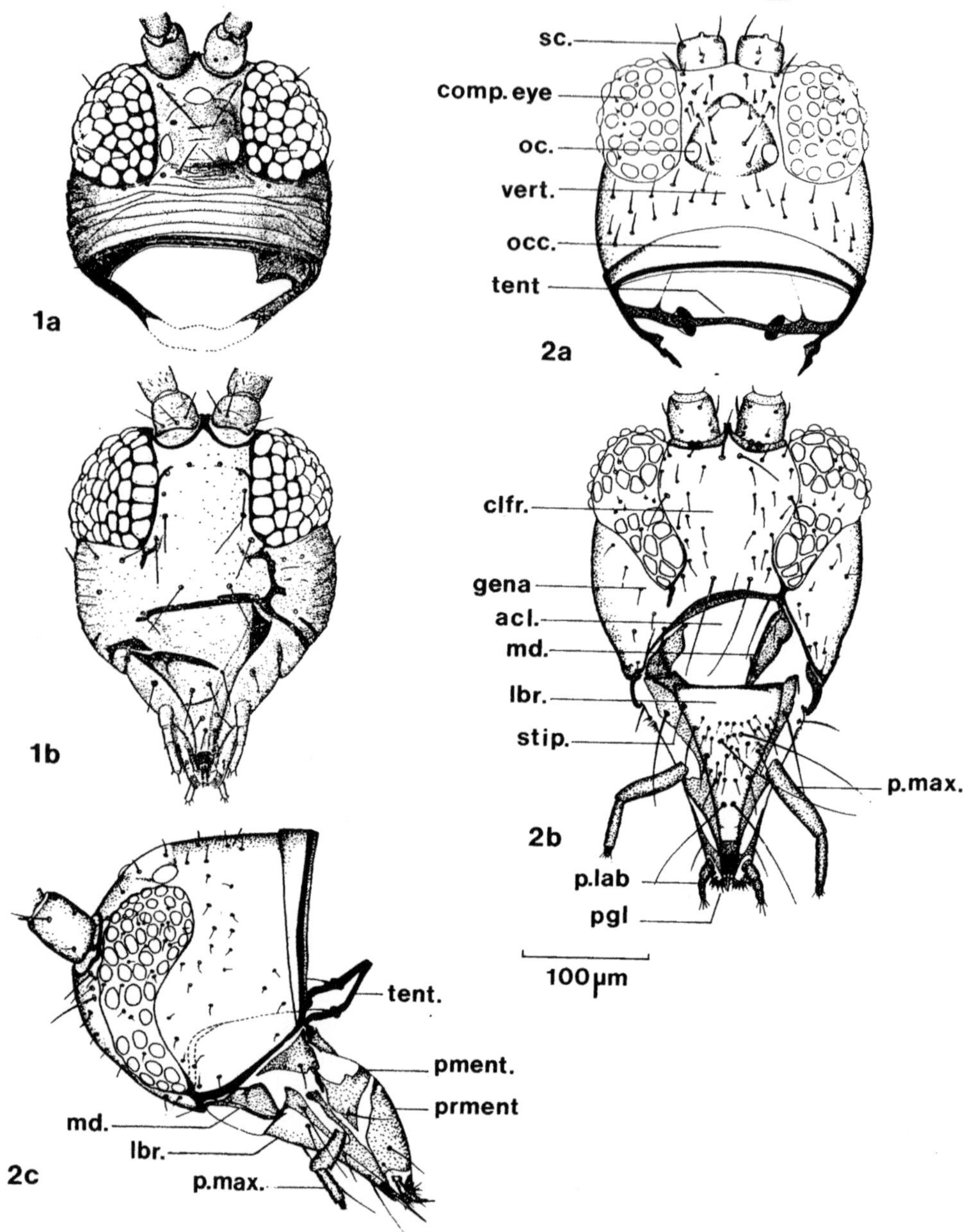

Fig. 2.4. Features of the head. 1a. *Thrips physapus*, ♀ dorsal; 1b. ♀ ventral (modified after Risler, 1957); 2a. *Aeolothrips intermedius*, ♀ dorsal; 2b. ♀ ventral; 2c. ♀ lateral (Moritz, 1982b). acl. = anteclypeal membrane, clfr. = clypeofrons, comp. eye = compound eye, lbr. = labrum, md. = mandible, oc. = ocellus, occ. = occiput, pgl. = paraglossa, p.lab. = labial palp, p.max. = maxillary palp, pment. = postmentum, prment. = prementum, sc. = scape, stip. = stipes, tent. = tentorium, vert. = vertex.

Table 2.1. Sense organs of the antennae of the adult females of *Frankliniella* (*F.*), *Hercinothrips* (*He*), *Bagnalliella* (*B.*) and *Haplothrips* (*Ha*) (after Slifer and Sekhon, 1974; Heming, 1975; Moritz, 1989b).

Segment	Species	Sensillum chaeticum (tactile hairs)	Sensilum campaniforme	Sensillum microtrichoideum (thin-walled receptors)	Sensillum trichodeum (emergent & forked trichomes)	Johnston's organ	Sensilum coelocapitularis (coeloconic pegs)
1	*F. fusca*	11	0	0	0	0	0
	F. tenuicornis	11	0	0	0	0	0
	He. femoralis	9	0	0	0	0	0
	B. yuccae	10	0	0	0	0	0
	Ha. verbasci	11	0	0	0	0	0
2	*F. fusca*	7	1	0	0	1	0
	F. tenuicornis	7	1	0	0	1	0
	He. femoralis	7	1	1	0	1	0
	B. yuccae	11	1	0	0	?	0
	Ha. verbasci	12	1	0	0	1	0
3	*F. fusca*	5	0	0	1	0	1
	F. tenuicornis	5	0	0	1	0	1
	He. femoralis	5	0	0	1	0	1
	B. yuccae	10	0	2	0	0	1
	Ha. verbasci	13	0	2	0	0	1
4	*F. fusca*	5	0	1	1	0	0
	F. tenuicornis	5	0	1	1	0	0
	He. femoralis	4	0	1	1	0	0
	B. yuccae	9	0	4	0	0	0
	Ha. verbasci	11	0	4	0	0	0
5	*F. fusca*	6	0	2	0	0	1
	F. tenuicornis	5–6	0	2	0	0	1
	He. femoralis	4	0	2	0	0	1
	B. yuccae	13–14	0	2–3	0	0	1
	Ha. verbasci	13	0	3	0	0	1
6	*F. fusca*	8	0	3	0	0	1
	F. tenuicornis	7	0	3	0	0	1
	He. femoralis	3	0	2	0	0	1
	B. yuccae	10	0	2	0	0	0
	Ha. verbasci	10–12	0	3	0	0	0
7	*F. fusca*	3	0	1	0	0	1
	F. tenuicornis	3	0	1	0	0	1
	He. femoralis	3	0	1	0	0	1
	B. yuccae	12	0	1	0	0	1
	Ha. verbasci	12	0	1	0	0	1
8	*F. fusca*	8(7)	0	0	0	0	0
	F. tenuicornis	7	0	0	0	0	0
	He. femoralis	7	0	0	0	0	1
	B. yuccae	11–12	0	1	0	0	1
	Ha. verbasci	12–13	0	1	0	0	1

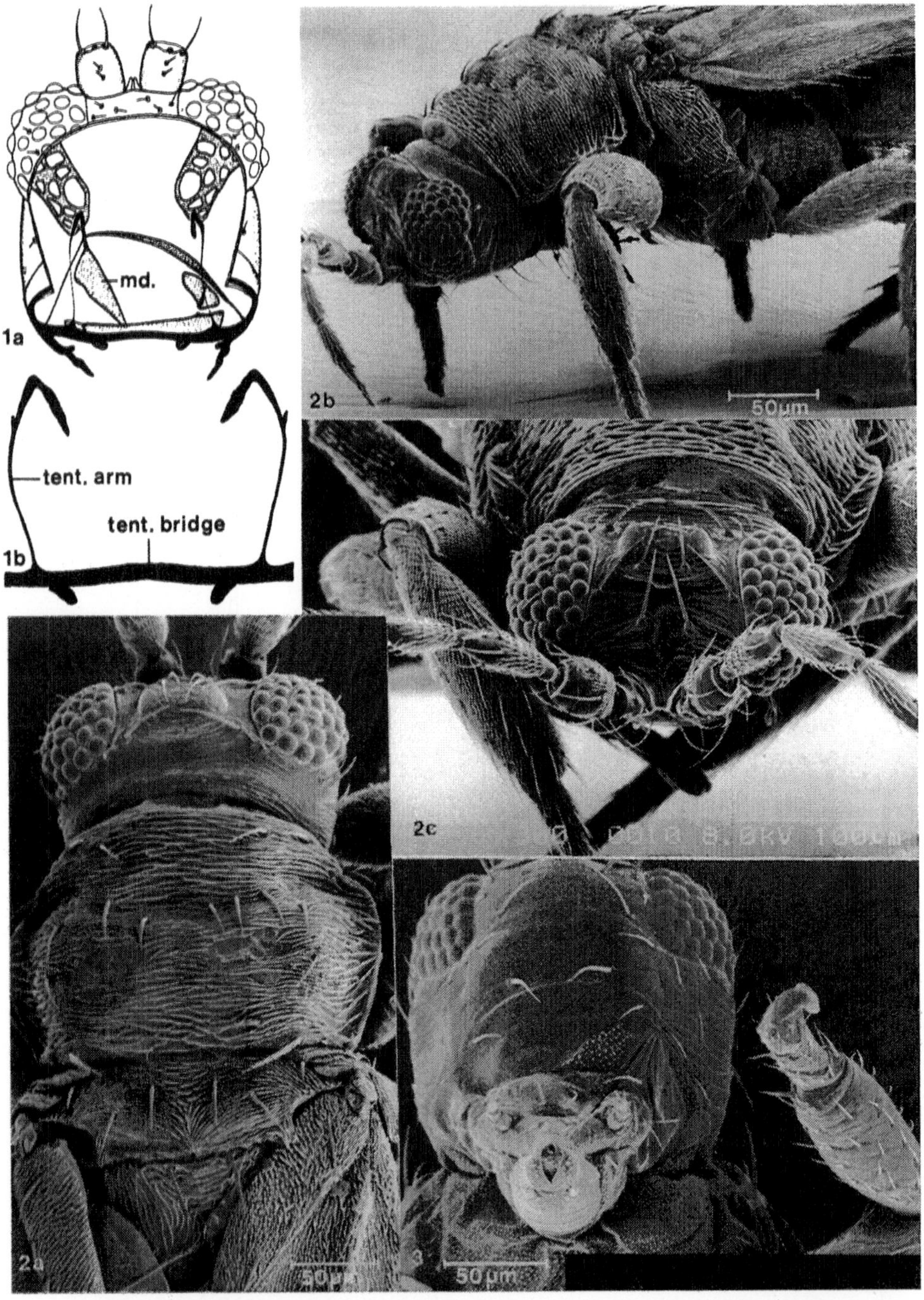

Fig. 2.5. 1a. *Aeolothrips intermedius*, ♀ dorsal opened head capsule with endoskeletal structures; 1b. *Aeolothrips intermedius*, ♀ tentorium (Moritz, 1982a); 2a. *Sericothrips* sp., ♀ dorsal, head and thorax; 2b. *Sericothrips* sp., ♀ lateral, head and thorax; 2c. *Sericothrips* sp., ♀ frontal, head; 3. *Frankliniella occidentalis*, ♀ ventral, head. md. = mandible, tent. = tentorium.

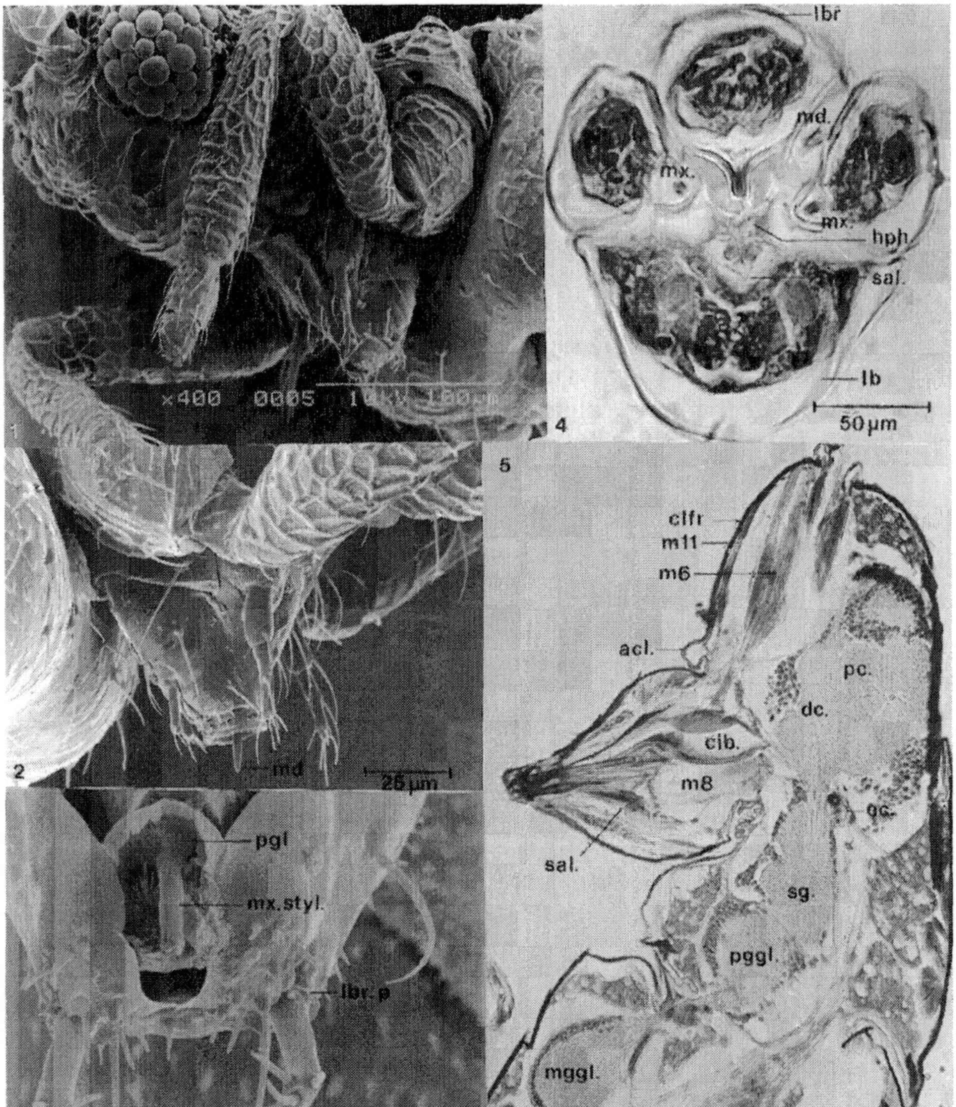

Fig. 2.6. 1. *Parthenothrips dracaenae*, ♀ lateral, lower head and mouthcone; 2. *Parthenothrips dracaenae*, ♀ lateral, mouthcone; 3. *Frankliniella occidentalis*, ♀ frontal, mouthcone, paraglossae and stylets; 4. *Aeolothrips intermedius*, ♀ t.s. mouthcone; 5. *Aeolothrips intermedius*, ♀ l.s. head and prothorax. acl. = anteclypeal membrane, cc. = corpora cardiacae, cib. = cibarium, clfr. = clypeofrons, dc. = deutocerebrum, hph. = hypopharynx, lb. = labium, lbr. = labrum, lbr.p. = labral palp, md. = mandible, mggl = mesothoracic ganglion, mx. = maxilla, mx.styl. = maxillary stylets, pc. = protocerebrum, pggl. = prothoracic ganglion, pgl. = paraglossae, sal. = salivary duct, sg. = suboesophageal ganglion, m6 = clypeo-epipharyngealis secundus muscle, m8 = salivary muscle, m11 = labral retractor muscle.

In winged species, three ocelli are situated dorsally on the head capsule between the compound eyes to form a triangle (Fig. 2.3: 6; Fig. 2.4: 1a, 2a; Fig. 2.5: 2a, 2b, 2c). The length, number and position of the setae on the vertex and the ocellar region are important taxonomic characters. In males of some merothripids the vertex is perforated by the openings of canals arising from a glandular area lying dorsally in the head (= vertex gland: Moritz, 1984a).

The complete head endoskeleton (tentorium) consists of several invaginated ridges and dorsal, ventral and hind arms as well as the tentorial bridge which are visible as dark structures in microscopic preparations of uzelothripids, aeolothripids and adiheterothripids (Fig. 2.4: 2a, 2c; Fig. 2.5: 1a, 1b) (Mound *et al.*, 1980). Dorsal tentorial arms are present in various species of *Melanthrips* and *Ankothrips* (Moritz, 1982a). A similar but less stoutly developed tentorium is visible in merothripids, but most species of this group have the tentorial bridge interrupted medially. In all other families the tentorium is reduced to rudimentary anterior arms which form important endoskeletal apodemes for antennal muscles. These structures are hollow.

Mouthparts

The asymmetric 'punch and suck' mouthparts whereby thrips cause so much damage to plants (see Chapters 4 and 13) together form a ventrally directed mouth cone with only one complete left mandible, two well-developed maxillae, the labrum in front and the labium behind (prementum and postmentum) (Risler, 1957; Mickoleit, 1963; Moritz, 1982b, 1989b; Heming, 1985, 1993a; Hunter and Ullman, 1992) (Fig. 2.3: 5; Fig. 2.4: 1b, 2b, 2c; Fig. 2.5: 3; Fig. 2.6: 1–3).

The left mandible is almost straight and attached proximally to the junction between the clypeal (epistomal), subgenal and clypeogenal sutures (Chisholm and Doncaster, 1982; Chisholm and Lewis, 1984). The right mandible is reduced to a small basal plate and degenerates almost completely during embryogenesis and metamorphosis. In *Frankliniella bispinosa* the mandible appears solid except for a small innervated channel (Childers and Achor, 1991) but in *Frankliniella occidentalis* (Hunter and Ullman, 1992), *Limothrips cerealium* (Chisholm and Lewis, 1984), *Aeolothrips intermedius* (Moritz, 1982b), and *Scirtothrips citri* (Wiesenborn and Morse, 1988) it is hollow. The main channel within the mandibular stylet usually contains three sensory dendrites, with an additional dendrite in each of the side channels present in some species (Heming, 1978; Childers and Achor, 1991; Hunter and Ullman, 1992). Four sensory dendrites, probably mechanosensors, occur within the maxillary stylets (Fig. 2.7: 3). The base of each maxillary stylet articulates with a lever which is joined to the cardo. Both maxillary stylets have ridge and groove coadaptions enabling them to combine to form a sucking tube with a subterminal opening (Mound, 1971; Wiesenborn and Morse, 1988). The position of the opening and the rapid probing of the paired stylets allows a steady food intake. In contrast to the mandible, the styliform maxillae are movable independently and without the thrusting of the head capsule. The two- or three-segmented maxillary palpi are inserted into the stipites and the smaller labial palpi are at the tip of the easily movable prementum. The (post)mentum

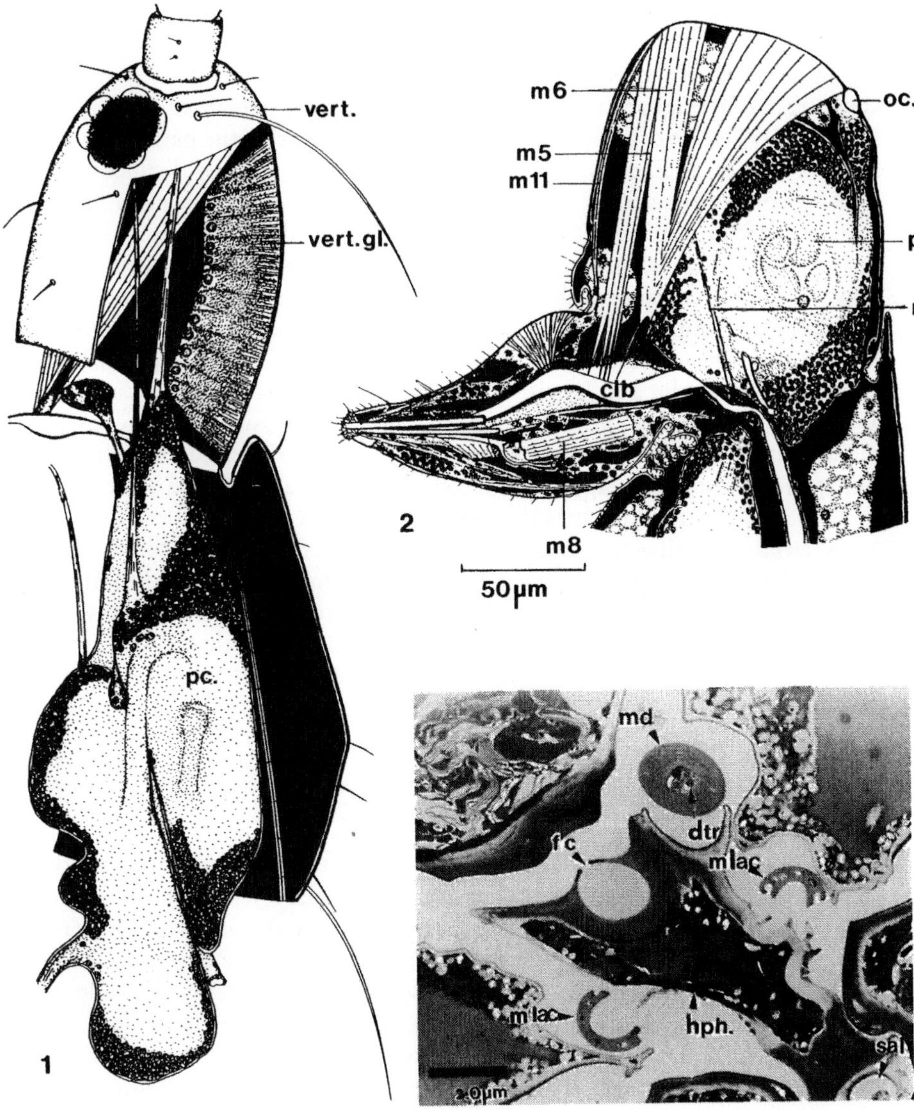

Fig. 2.7. 1. *Merothrips brunneus*, ♂ l.s. head with vertex gland and translocated brain in prothorax (Moritz, 1984); 2. *Aeolothrips intermedius*, ♀ l.s. head; 3. *Frankliniella occidentalis*, ♀ t.s. mouthcone (after Childers and Achor, 1991). cib. = cibarium, dtr. = dendrites within mandible, fc. = food canal, hph. = hypopharynx, md. = mandible, mlac. = maxillary laciniae, oc. = ocellus, pc. = protocerebrum, sal. = salivarium, vert. = vertex, vert.gl. = vertex gland, m5 = clypeo-epipharyngealis primus muscle, m6 = clypeo-epipharyngealis secundus muscle, m7 = pharyngeal muscle, m8 = salivary muscle, m11 = labral retractor muscle.

and prementum of the labium are fused medially (Fig. 2.4: 2c; Fig. 2.5: 3; Fig. 2.6: 2).

In addition to the paired mouthparts the cibarial and salivary pumps plus an elongated hypopharynx form the principal feeding structures (Fig. 2.6: 4,5; Fig. 2.7: 2). The tip of the mouthcone is covered at rest by the paraglossae (Fig. 2.5: 3; Fig. 2.6: 3). Each paraglossa has three morphologically distinct types of nine to ten sensillae (two kinds of s. basiconica and s. trichodea) (Moritz, 1982b, 1989b; Chisholm and Lewis; 1984, Childers and Achor, 1991), possibly for smelling and/or tasting. The number and arrangement of these sensory structures seems to be similar in all species investigated: *Frankliniella occidentalis* (Hunter and Ullman, 1989), *Hercinothrips femoralis*, *Aeolothrips intermedius* (Moritz, 1982b, 1989b), *Frankliniella bispinosa*, *Limothrips cerealium* (Chisholm and Lewis, 1984). When opened, the paraglossae form a V-shaped area and a horseshoe-shaped labral pad supports the piercing mandibular and maxillar stylets. Strengthening sclerites radiating from the centre of the labral pad support the stylets while they are inserted in a plant and serve to hold the tip of the mouthcone tightly against the substrate (Fig. 2.6: 3). More slender papillae situated at the lower part of the pad perhaps serve to equalize pressure during sucking. The strong left mandible is used to make an initial puncture in the plant tissue by characteristic up- and downward movements of the head. The paired maxillary stylets are then thrust into the substrate, the grooved interiors of both forming the food channel.

Thorax

Prothorax

Because the head and prothorax often function as a unit, there is a special articulation between the pterothorax and the prothorax. The dorsal region of the prothorax is stout and consists of a solid pronotum with a strong ligament to the (post)occiput, whereas the ventral structures are mainly membranous (Fig. 2.8: 1a–c; Fig. 2.9: 1, 4). The forelegs attached to the cranial part of the pterothorax and the ventral membranous regions of the prothorax allow the typical up and down movements of the head–prothorax complex during feeding (Fig. 2.9: 1). The ventral sclerites are more caudally arranged (Mickoleit, 1961; Moritz, 1989c) or much reduced (Bhatti, 1988). Also the lateral region of the first thoracic segment has fewer, smaller sclerites and is fused with the pronotum (Fig. 2.8: 1b, 1c). The crista pleuralis articulates ventrally with the lateral coxal suture. A well-developed trochantinus is present in Aeolothripidae (Fig. 2.8: 1b, 1c). The caudal membrane between the pronotum and the cranial pterothoracic sclerites forms the functional articulation between these thoracic tagma (Fig. 2.8: 1a–c).

Pterothorax

In winged species the meso- and metathorax are enlarged relative to the prothorax and form a pterothorax, bearing the wings and associated flight musculature (Fig. 2.8: 1a–c). Because of this the pterothorax has movable dorsal sclerites and membranes but strong fused plates in the ventral and

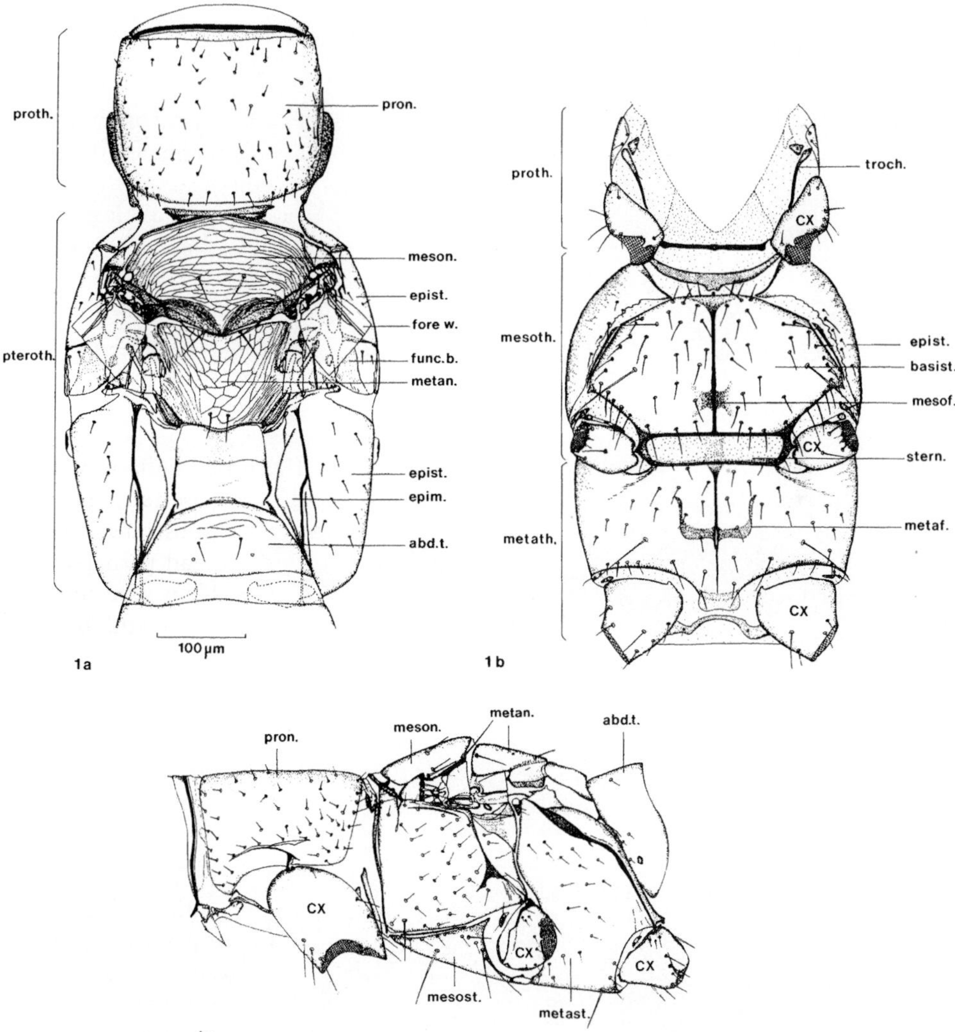

Fig. 2.8. Thorax of *Aeolothrips intermedius*, 1a. ♀ dorsal, 1b. ♀ ventral, 1c. ♀ lateral (Moritz, 1982c). abd.t. = abdominal tergite, basist. = basisternum, cx. = coxa, epim. = epimeron, epist. = episternum, fore w. = fore wing, func.b. = functional basalare, mesof. = mesofurca, metaf. = metafurca, meson. = mesonotum, mesost. = mesosternum, mesoth. = mesothorax, metan. = metanotum, metast. = metasternum, metath. = metathorax, pron. = pronotum, proth. = prothorax, pteroth. = pterothorax, stern. = sternellum, troch. = trochantinus.

pleural regions (meso- and metasternum, episterna: Fig. 2.8: 1a–c; Fig. 2.9: 2–5). The bases of the wings are displaced cranially to accommodate the dorsal invagination of the first abdominal segment (Fig. 2.8: 1c; Fig. 2.10: 1). In contrast with this the sternal plates are extended caudally between the hind legs with a consequent reduction of the first abdominal sternite (Fig. 2.8: 1b; Fig. 2.9: 5). The meso- and metanotal exoskeleton is thus well adapted for flight

Fig. 2.9. 1. *Parthenothrips dracaenae*, ♀ lateral, head and prothorax; 2. *Frankliniella occidentalis*, ♀ dorsal, pterothorax (Moritz, 1982c); 3. *Parthenothrips dracaenae*, ♀ dorsal, base of left fore and hind wing; 4. *Frankliniella occidentalis*, ♀ dorsal, head and prothorax; 5. *Parthenothrips dracaenae*, ♀ ventral, thorax showing indentations of the endoskeletal meso- and metathoracic furcae.

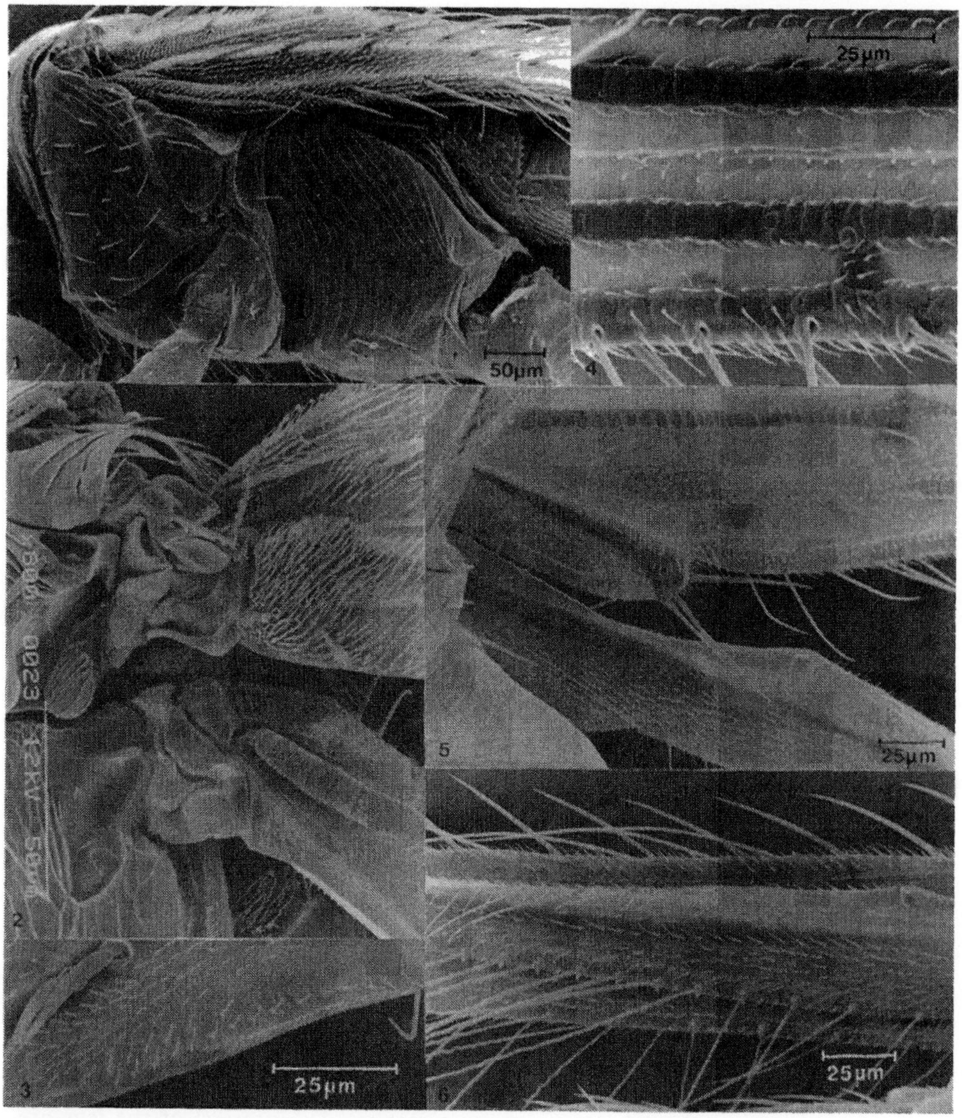

Fig. 2.10. Wing structure in *Frankliniella occidentalis*, 1. ♀ lateral, pterothorax with functional basalare; 2. ♀ dorsal, base of fore and hind wing; 3. ♀, special bristles and sclerites at the proximal end of the hind wing; 4. ♀, sensilla on the wing surface and bristles with special sockets that retain them in different positions at rest and in flight; 5. ♀, wing bases with adaptive structures between alulae and fore and hind wings; 6. ♀, fore and hind wings held together at rest.

(Fig. 2.8: 1a; Fig. 2.9: 2). The terebrantian metanotum is divided into a metanotum craniale and caudale. In some taxa small sclerites are present between the main plates.

The basisterna of the meso- and metathorax form robust plates to which the powerful flight muscles are attached. They derive extra stability through the fusion of lateral episternal sclerites. The crista pleuralis divides the lateral sclerites into a dorsal epimeron and a ventral episternum (Fig. 2.8: 1c; Fig. 2.10: 1). The shape of apodemal structures or furcal constructions differs within the genera and often the medial process of the furcae forms a median spine (Fig. 2.8: 1b). The position of the meso- and metacoxae, the latter sometimes folded under the body, enables the hind legs to be thrust forwards allowing many thrips to jump. The epimeral sclerites are bordered dorsally by the membranous region of the wing joints. A basalare is inserted between the mesothoracic epimeron and the metathoracic episternum (Fig. 2.10: 1).

The conformation of the fore and hind wings of phlaeothripids is quite different from terebrantian species (Mound and O'Neill, 1974; Mound *et al.*, 1980). Phlaeothripid wings have no veins, veinal setae, or microtrichia on the wing membrane and the fringes do not arise from sockets (Bhatti, 1991). Wingless species have fewer and fused thoracic sclerites. In winged thrips the wing structure is greatly modified, because for such small insects the air is akin to a viscous medium with Reynold's numbers between 1 and 20 (Moritz, 1989c). Physical criteria like small body size, low flight speed and high wing beat frequency have led to the evolution of wings resembling small membranous bands. At rest the hind wing fits into the fore wing (Fig. 2.10: 6). Functionally the wing surface is increased by the fringes, which can change their position within the asymmetrically constricted sockets during flight and rest in terebrantian species (Fig. 2.10: 4,6) (see Chapter 5). The base of the fore and hind wings has a well-developed alula (Fig. 2.9: 3; Fig. 2.10: 2) with special adaptions for coupling the fore and hind wing during rest and in flight so that they beat synchronously. This is achieved by the linkage of a smooth proximal area and special flattened bristles at the base of the alula and the hind wing (Fig. 2.10: 3,5). Fossil Thysanoptera with a phylogenetic age of over 130 million years show remarkably comparable structures (Zur Strassen, 1973). In addition, the microtrichia of the wing membranes hook into each other to join the wings in the manner of a 'Velcro' fastener. The caudal margin of the fore wing and the cranial margin of the hind wing are also linked by special bristles and sclerites. On the underside of the wings several sensillae and a row of sclerotized ridges are visible on the veins (Fig. 2.10: 4).

Each leg of an adult thrips consists of a basal coxa, trochanter, femur, tibia, one- or two-segmented tarsus, and a pretarsus. The pretarsi of the legs are characterized by a large, protrusible vesicle, the arolium (Physapoda: Heming, 1971). During walking, this alone is in contact with the substrate (Fig. 2.11: 1,2), allowing thrips to grip smooth surfaces and avoid being blown off vegetation by wind (Heming, 1972). In some species males differ from females in having swollen fore femora (*Merothrips*) or legs with tubercles, teeth or claws on the arolium (*Taeniothrips*), the tarsus (*Oxythrips*), the tibia (*Odontothrips*) or the femora (Mound *et al.*, 1976; Crespi, 1986), which are probably used for fighting or in some species to assist emergence from protective cells formed within soil for pupation or overwintering.

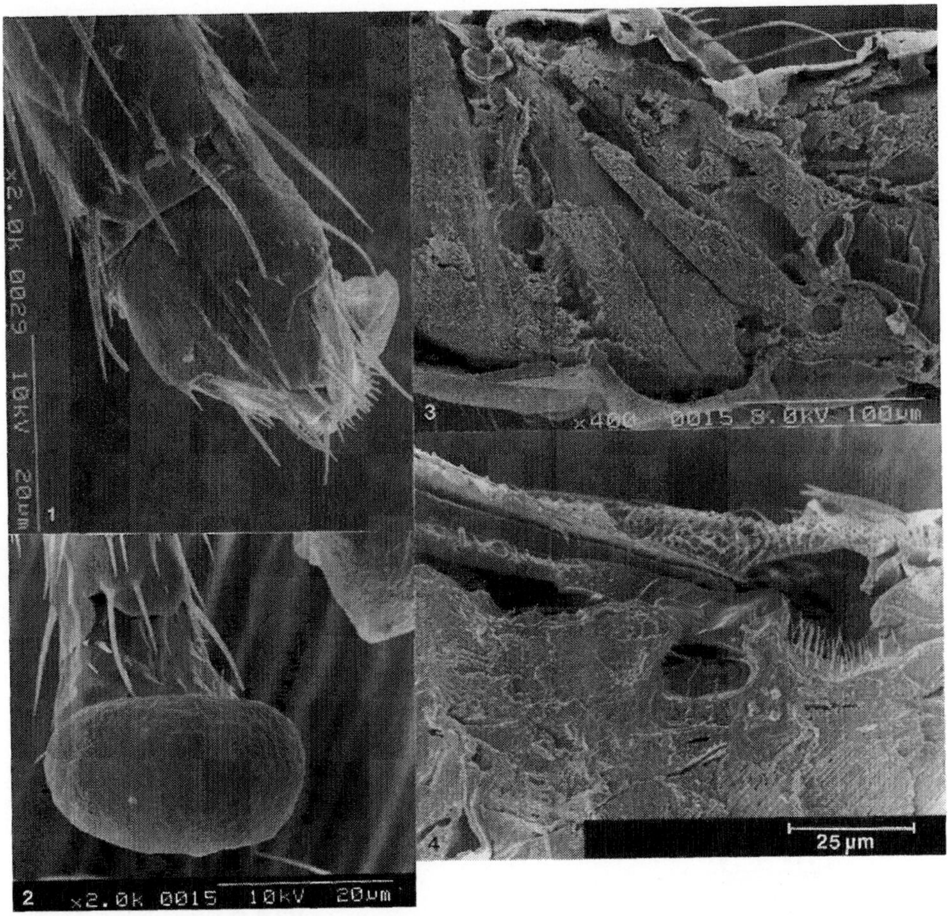

Fig. 2.11. *Frankliniella occidentalis*, ♀ 1. tip of tarsus with arolium (foot bladder) retracted; 2. tip of tarsus with arolium extruded; 3. l.s. through flight muscle system; 4. l.s. through bases of fore and hind wings. (Moritz, 1995.)

Abdomen

The abdomen consists of 11 segments (Fig. 2.12: 1; Fig. 2.13: 1, 2). The pregenital segments are uniform except the first, which has a well-developed tergite but greatly reduced sternite. Small lateral pleurosternites and/or pleurotergites connect the tergites and sternites of the more or less dorsoventrally flattened abdomen. In many species males have glandular areas on some abdominal sternites (Stannard, 1968; Pitkin, 1978); depending on the species the pore plates differ in their position, size and shape. Abdominal segments II and VIII of adults bear stigmata. The lateral parts of the tergites carry special structures, the ctenidia, with which the fringed wings are combed (Fig. 2.13: 3). The caudal margin of tergite VIII often has a complete comb (Fig. 2.13: 1). In aeolothripids, abdominal tergite VIII of males has a pair of clasper-like processes and in some genera the genital segments also bear thorn-like setae.

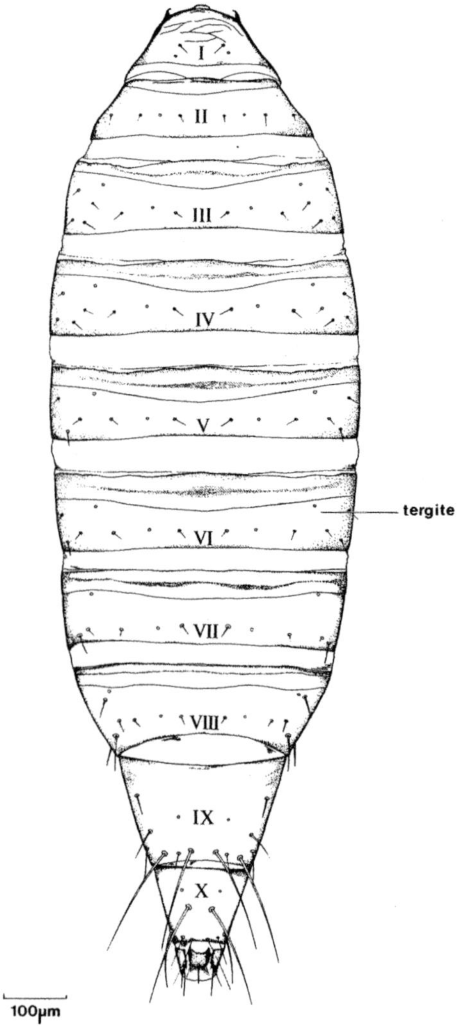
I
II
III
IV
V
VI
tergite
VII
VIII
IX
X
100µm

1a

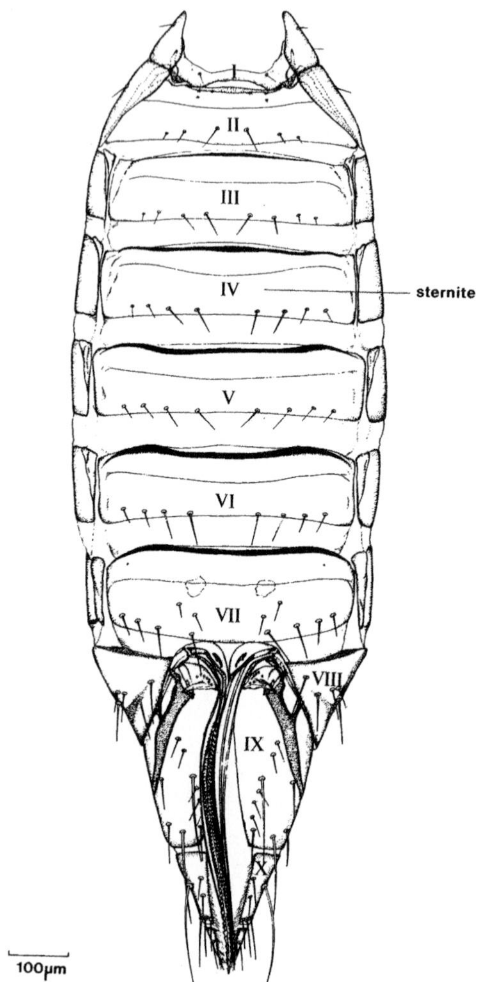
I
II
III
IV
sternite
V
VI
VII
VIII
IX
X
100µm

1b

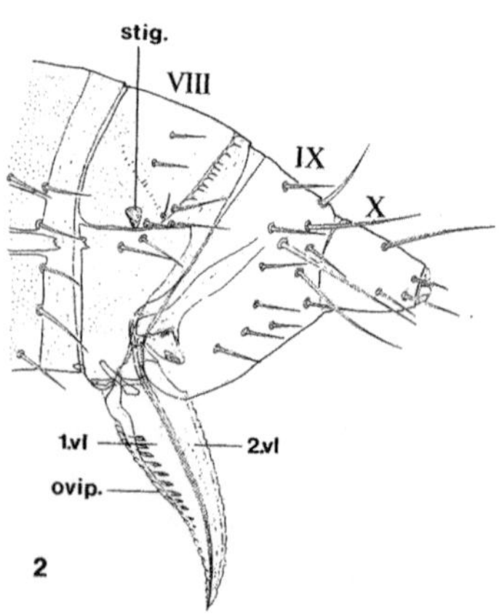
stig.
VIII
IX
X
1.vl
2.vl
ovip.

2

The abdominal tergites IX to XI are extended ventrally to form a cavity protecting the ovipositor (Fig. 2.13: 2). The terebrantian ovipositor lacks the third valvulae, but resembles that of other female pterygote insects (Fig. 2.13: 2, 4a, 4b; Fig. 2.14: 6) (Heming, 1970a, Bode, 1975, Moritz, 1982d, 1989d). A large and well-developed ovipositor is needed for laying eggs within the tissue of green plants. Smooth egg shells (Fig. 2.21: 2) and the saw-like valvulae support the ovipositor while the eggs are being laid (Fig. 2.13: 4a). The 2nd valvifer has three sensilla trichodea dorsally and three sensilla campaniformia ventrally. The 1st and 2nd valvulae also have external sensillae at regular intervals, which sense the movement and the depth to which the ovipositor penetrates during oviposition (Fig. 2.13: 4a, 4b). Basal sclerites (1st and 2nd valvifer, apodeme IX) articulate with the valvulae to deposit the egg within the plant tissue.

The genital segments of the males consist of a phallobase, a median aedeagus and laterally situated parameres (Heming, 1970b). The phallobase extends posteriorly to form a membranous phallotheca. A protrusible vesicle, the endotheca, covers the male genitals and when the phallus is at rest the endotheca remains invaginated (Fig. 2.14: 1).

Anatomy of Adults

Nervous system

The cerebral, suboesophageal and prothoracic ganglia are fused. The origins and routes of some major nerves point to the subdivision of the cerebral ganglion into the proto-, deuto- and tritocerebrum (Fig. 2.6: 5; Fig. 2.7: 1,2; Fig. 2.14: 5). The protocerebrum, as the largest part of this complex, innervates the ommatidia and contains the optic lobes and the chiasma, which connects the lamina ganglionaris with the medullae externa et interna (Fig. 2.15: 3). Both hemispheres are connected through the optical and protocerebral commissure, as well as the protocerebral bridge. The deutocerebrum is characterized by the antennal nerves arising cranially and is attached ventrocaudally to the tritocerebrum lying beside the pharynx. The paired nerve cords on the labrum lie in this region (Staub, 1979).

In merothripid males the head capsule contains a large vertex gland, displacing the brain which remains in the larval position during the adult stage (Fig. 2.7: 1; Fig. 2.14: 5) (Moritz, 1984b). The suboesophageal ganglion consists of three pairs of ganglia with one unpaired mandibular nerve on the left side and paired maxillary and labial nerves.

Fig. 2.12. *Opposite* Abdominal morphology, 1a. *Aeolothrips intermedius*, ♀ dorsal; 1b. *Aeolothrips intermedius*, ♀ ventral (Moritz, 1982d); 2. *Thrips validus*, ♀ lateral, abdominal segments VIII–X with extruded ovipositor (after Bode, 1975). ovip. = ovipositor, st. = sternite, stig. = stigmata, tg. = tergite, vl. = valves of ovipositor.

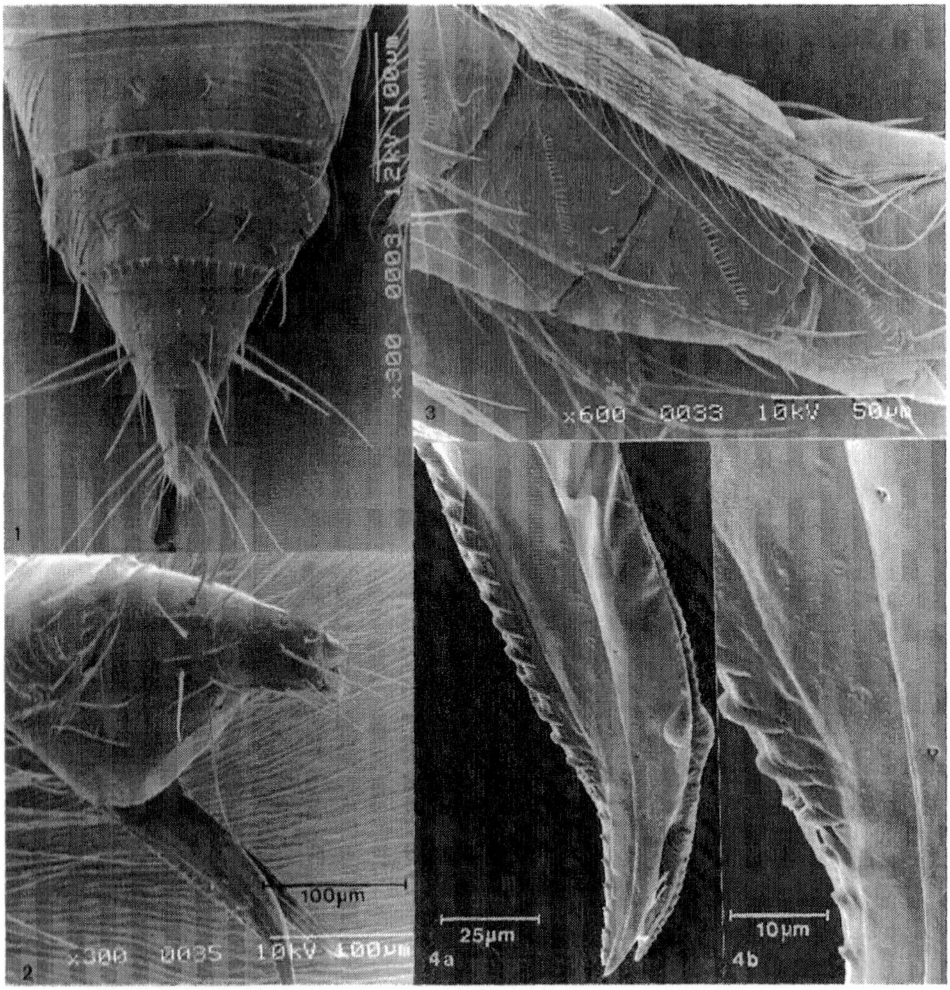

Fig. 2.13. Abdomen and ovipositor of *Frankliniella occidentalis* ♀, 1. abdominal segment VIII with typical structures and ctenidia; 2. genital segments of the abdomen and ovipositor; 3. lateral, showing wing retaining setae and ctenidia; 4a. valvulae of the ovipositor; 4b. rows of sensillae to detect the depth of penetration of ovipositor into plant tissue.

The prothoracic ganglion is broadly fused with the suboesophageal ganglion and the pterothoracic ganglia are embedded between the furcal constructions of the sterna. The abdominal ganglia fuse within abdominal segments I to III, in some genera these ganglia are fused with the metathoracic ganglion (*Aeolothrips*, *Taeniothrips*, *Thrips*, *Haplothrips*, *Gynaikothrips*) and in other genera joined by long connectives (*Limothrips*, *Anaphothrips*, *Frankliniella*, *Parthenothrips*, *Chaetanaphothrips*, *Merothrips*) (Moritz, 1984b). Abdominal segments V to XI are innervated by a combined nerve cord extending to the tip of the abdomen, whereas segments I to IV are each innervated by separate nerves (Fig. 2.15: 1).

Important sense organs

Little is known about thrips sense organs or their physiological function. For example, the sense cones of antennal segments 3 and 4 are known as typical chemoreceptors (Slifer and Sekhon, 1974), but there is no understanding as to how they function. Some thrips exhibit intense flight activity before thunderstorms (see Chapter 5) so, possibly, the bifurcated sense cones are proprioceptors to detect electrical fields. The chordotonal organ (Johnston's organ) by which antennal movements are perceived is present in larvae and adults within the pedicel of the second antennal segment (Bode, 1986).

A non-ciliary proprioceptor consisting of two bipolar, multiterminal sense cells, lies between the tentorium and the hind mandibular margin (Honomichl, 1982). Hunter and Ullman (1992) describe in *Frankliniella occidentalis* 4 precibarial and 20 cibarial chemosensilla, of which 18 are on the epipharyngeal sclerite and 2 on the hypopharyngeal sclerite.

Several structures used as taxonomic features are probably sense organs, such as many antennal setae (Fig. 2.2: 1–4; Fig. 2.3: 1–3), the haplopores or sensilla campaniformia of the metanotum craniale (Fig. 2.8: 1a; Fig. 2.9: 2) or a trichobothrium-like seta on the hind wings (Fig. 2.10: 3). Sensilla campaniformia occur on the underside of the wings (Fig. 2.10: 4), on the abdominal tergites (Fig. 2.12: 1a; Fig. 2.13: 1), and on the 1st and 2nd valvulae (Fig. 2.13: 4a, 4b).

Winged adults have three dorsal ocelli arranged in a triangle lying between the compound eyes (Fig. 2.4: 1a,2a; Fig. 2.5: 2c). They may register cyclical changes in light intensity to control biological rhythms, but are present only in winged species. The number of ommatidia in the compound eyes varies from 3 to more than 100 (most species have 30 to 50). In adults, some facets are pigmented orange brown and have a different optical density from the unpigmented facets (Nakahara, 1988; Moritz, 1995).

Musculature

Thrips have more than 170 different, and in most cases paired, muscles within the head (Risler, 1957; Davies, 1958; Mickoleit, 1963; Heming, 1975, 1978; Moritz, 1982b, 1989b), thorax (Mickoleit, 1961; Davies, 1969; Moritz, 1982c, 1989c) and abdomen (Davies, 1961, 1969; Moritz, 1982d). The various muscles (M1, M2, etc.) are named after their sites of origin and insertion or by reference to their function (Table 2.2). In the figures depicting muscles, only those lying in the plane of the section are labelled.

Head muscles

The antennae are moved by extrinsic muscles arising from tentorial structures and inserted on the base of the scape (M1, M2), and by intrinsic muscles connecting the scape with the pedicel (M3, M4).

Powerful cibarial and pharyngeal muscles (M5, M6, M7) run from the clypeofrons to the dorsal wall of the cibarial pump and the pharynx. These muscle groups act alternately, thereby drawing food through the extremely narrow stylet tube. Between the labial sclerites and the hypopharynx lies the salivarium, into which the salivary glands deposit saliva, and a special salivary muscle controls its

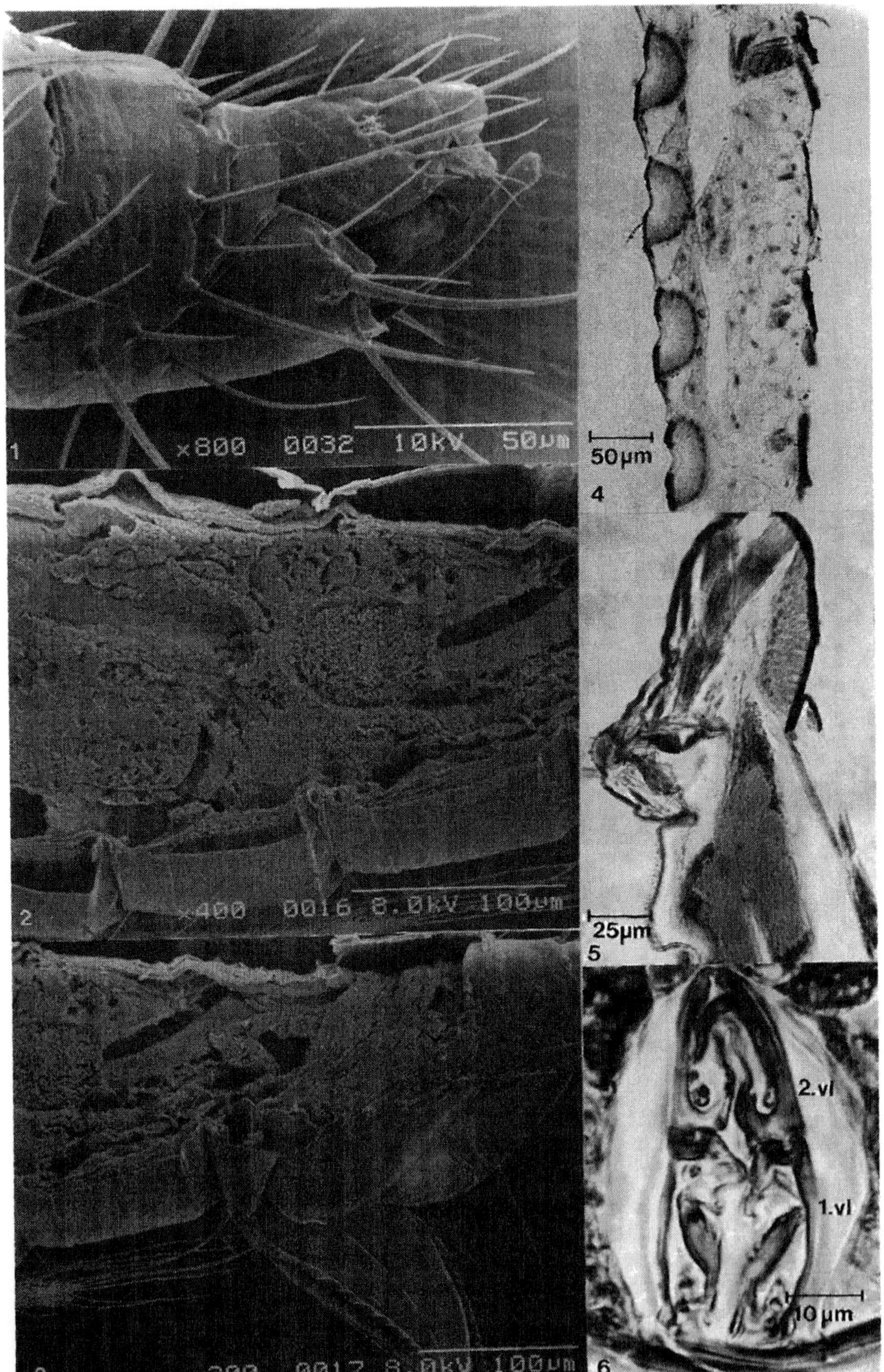
1
×800 0032 10kV 50μm
2
×400 0016 8.0kV 100μm
3
×300 0017 8.0kV 100μm
50μm
4
25μm
5
2.vl
1.vl
10 μm
6

flow (M8). The re- (M9) and protractors (M10) of the hypopharynx originate on the sclerites of the salivarium (Fig. 2.6: 5; Fig. 2.7: 2).

Retractor muscles within the mouthcone serve to protrude the mandible indirectly and a labral retractor (M11) likewise assists protrusion of the maxillary stylets. Two muscles arising from the gena and the clypeofrons are inserted on the mandibular base and act as direct pro- (M12) and retractors (M13) of the mandible (Fig. 2.6: 5).

The maxillary stylets have three pairs of protractors (M14) which originate on the stipes wall, as well as retractors (M15) rising from the gena and inserted on the articulation lever. The maxillary palps are moved by small flexors (M16). The labial muscles are divided into a premental (M17) and glossal and paraglossal retractor (M18).

Thoracic muscles

Functionally the thoracic muscle system is divided into a prothoracic and pterothoracic region. The former has dorsal (M19, M20, M21, M22) and ventral longitudinal muscles (M23) as levators and depressors of the head and the head–prothorax complex. Ventral oblique muscles (M24), arising from the sternal apophysis, are attached to the spinasternum to support the head depressors. Leg muscles, comprising extrinsic and intrinsic groups, arise from the tergum and are inserted on the trochantin or the coxa (Mickoleit, 1961; Davies, 1969; Moritz, 1982c, 1989c) (Fig. 2.16: 1).

In winged adults the main parts of the pterothoracic muscles are specially developed for flight with a short sarcomer length of 1.5 μm compared with 5 μm in normal muscles. This ultrastructure is typical for asynchronous flight muscles in insects with a high wing beat frequency. In such small insects, nervous impulses and relaxation between action potentials would not allow the wings to beat fast enough for successful flight, so in thrips the muscles contract as a direct response to being stretched. The flight muscle system is triggered into action by the metafurcal depressor of the trochanter (M33), which is activated by the jumping behaviour before take off (Ellington, 1980) (see Fig. 5.4). The flight muscle fibrils contain more wing depressors than raisers allowing flying thrips, despite their small size, to exert some control in choosing where to alight (Moritz, 1988c, 1989e). The dorsal longitudinal muscles of the mesothorax act mainly as wing depressors (M25), while in the metathorax they are used to raise the abdomen (M26, M27).

The mesothoracic ventral muscles raise the head–prothorax complex (M28) and the meso- and metathoracic ventral muscles lower the abdomen

---

Fig. 2.14. *Opposite* 1. *Frankliniella occidentalis*, ♂ lateral, genital segments; 2. *Frankliniella occidentalis*, ♀ l.s., pregenital segments with developing eggs and chorion; 3. *Frankliniella occidentalis*, ♀ l.s., genital segments with muscles to ovipositor; 4. *Taeniothrips inconsequens*, ♂ l.s., abdominal glands opening on to sternites; 5. *Merothrips brunneus*, ♂ l.s., head showing vertex gland; 6. *Aeolothrips intermedius*, ♀ t.s., ovipositor. (Moritz, 1984a.) 1.vl = first valve, 2.vl = second valve.

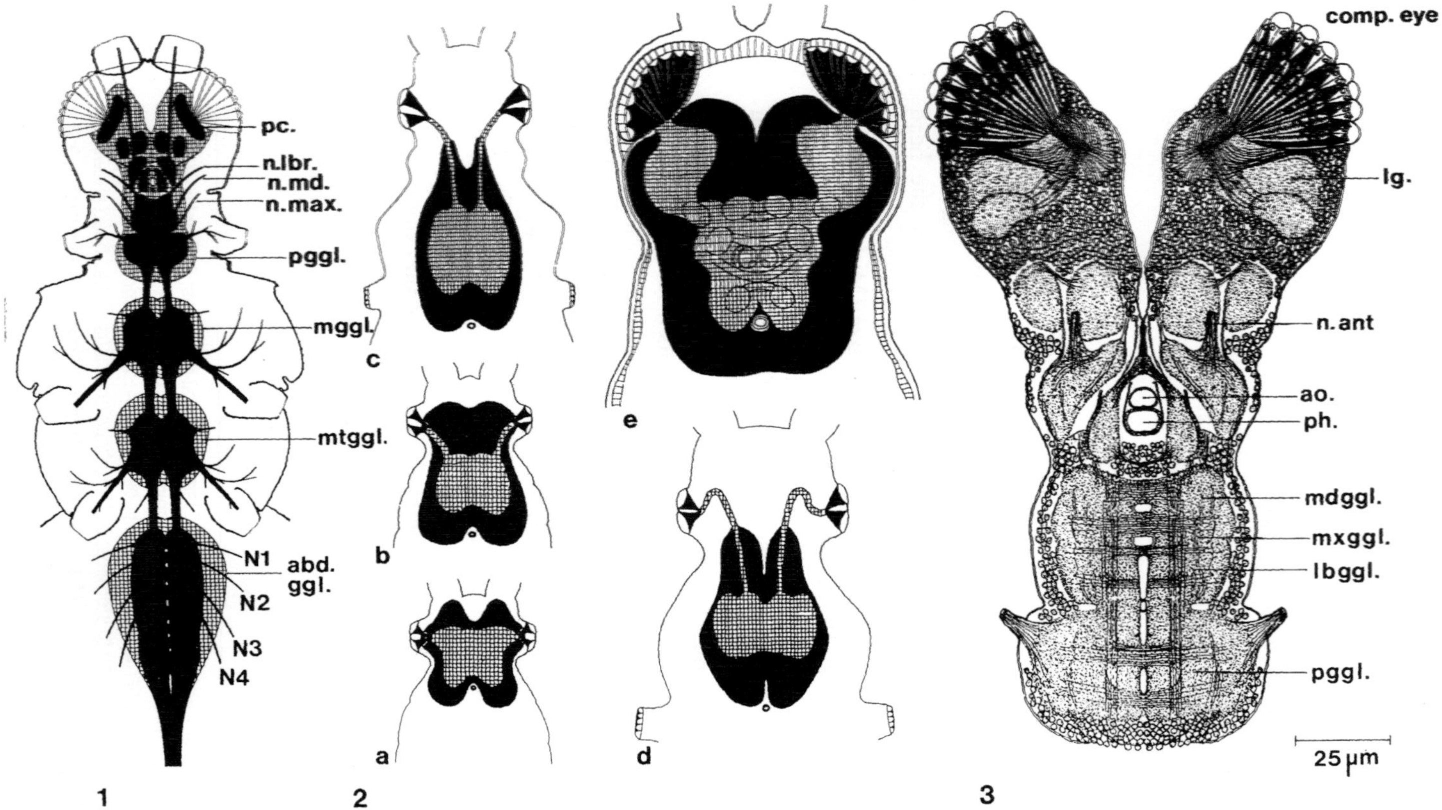

Fig. 2.15. Nervous system of *Hercinothrips femoralis*, 1. ♀ frontal section, adult; 2a–e. ♀ frontal section, development of supraoesophageal ganglion through larval and pupal stages to adult (modified after Moritz, 1988b, 1989a); 3. ♀ frontal section, supra-, sub-, and prothoracic ganglia of adult (modified after Moritz, 1989b). abd.ggl. = abdominal ganglion, ao. = aorta, comp.eye = compound eye, lbggl. = labial ganglion, lg. = lamina ganglion, mdggl. = mandibular ganglion, mggl. = mesothoracic ganglion, mtggl. = metathoracic ganglion, mxggl. = maxillar ganglion, N1–N4 = abdominal nerve, n.ant. = antennal nerve, n.lbr. = labral nerve, n.md. = mandibular nerve, n.max. = maxillary nerve, pc. = protocerebrum, pggl. = prothoracic ganglion, ph. = pharynx.

Table 2.2. Some important thrips muscles (those with asterisks are labelled in the accompanying figures).

No.	Name	No.	Name
M1	M. depressor scapi	M25/26*	M. dorsalis rectus
M2	M. levator scapi	M27*	M. metanoto-abdominotergalis
M3	M. pedicelli medialis	M28*	M. mesofurco-profurcalis
M4	M. pedicelli lateralis	M29*	M. mesofurco-abdominosternalis
M5	M. clypeo-epipharyngealis primus	M30*	M. metofurco-abdominosternalis
M6*	M. clypeo-epipharyngealis secundus	M31*	M. sterno-notalis
M7*	M. radialis pharyngis	M32*	M. praecoxo-basilaris
M8*	M. salivarii	M33*	M. depressor trochanteris basilaris
M9	M. retractor hypopharyngis	M34/35*	M. remotor coxae tergalis
M10	M. protractor hypopharyngis	M36/37*	M. promotor coxae furcalis
M11*	M. clypeo-labralis medialis	M38/39*	M. remotor coxae furcalis
M12	M. clypeo-labralis lateralis	M40	M. mesofurco-metacoxalis
M13	M. retractor mandibulae	M41/42	M. levator trochanteris coxalis
M14	M. protractor laciniae	M43/44*	M. depressor trochanteris furcalis
M15	M. retractor laciniae	M45	M. protractor valvulae primae
M16	M. flexor palpi maxillaris	M46*	M. protractor valvulae secundae
M17	M. retractor praementi	M47*	M. retractor valvulae primae
M18	M. retractor glossae et paraglossae	M48*	M. retractor valvulae secundae
M19*	M. levator capitis prophragmaticus	M49*	M. depressor terebrae
M20*	M. levator capitis tergalis	M50*	M. longitudinalis abdominis dorsalis interior
M21	M. depressor capitis tergalis	M51*	M. longitudinalis abdominis dorsalis exterior
M22*	M. dorsalis obliquus	M52*	M. longitudinalis abdominis ventralis
M23	M. depressor capitis furcalis	M53*	M. tergo-sternalis
M24	M. profurco-prospinalis		

(M29, M30). The dorsoventral flight muscles of the pterothorax are divided into inner and outer groups, the former acting as indirect wing levators (M31) and the latter as direct depressors of the fore and hind wings (M32). From the origin of the wing starter muscle on the metathoracic trochantin (M44) there arises an oscillatory muscle (M33) which activates the basalare (Fig. 2.11: 3, 4; Fig. 2.16:1, 2).

The legs are moved by the contraction of extrinsic dorsoventral muscles (M34, M35), arising from the tergum and inserted on the coxa or trochantin (Heming, 1971, 1972, 1973). Ventral muscles serve as pro- (M36, M37) and

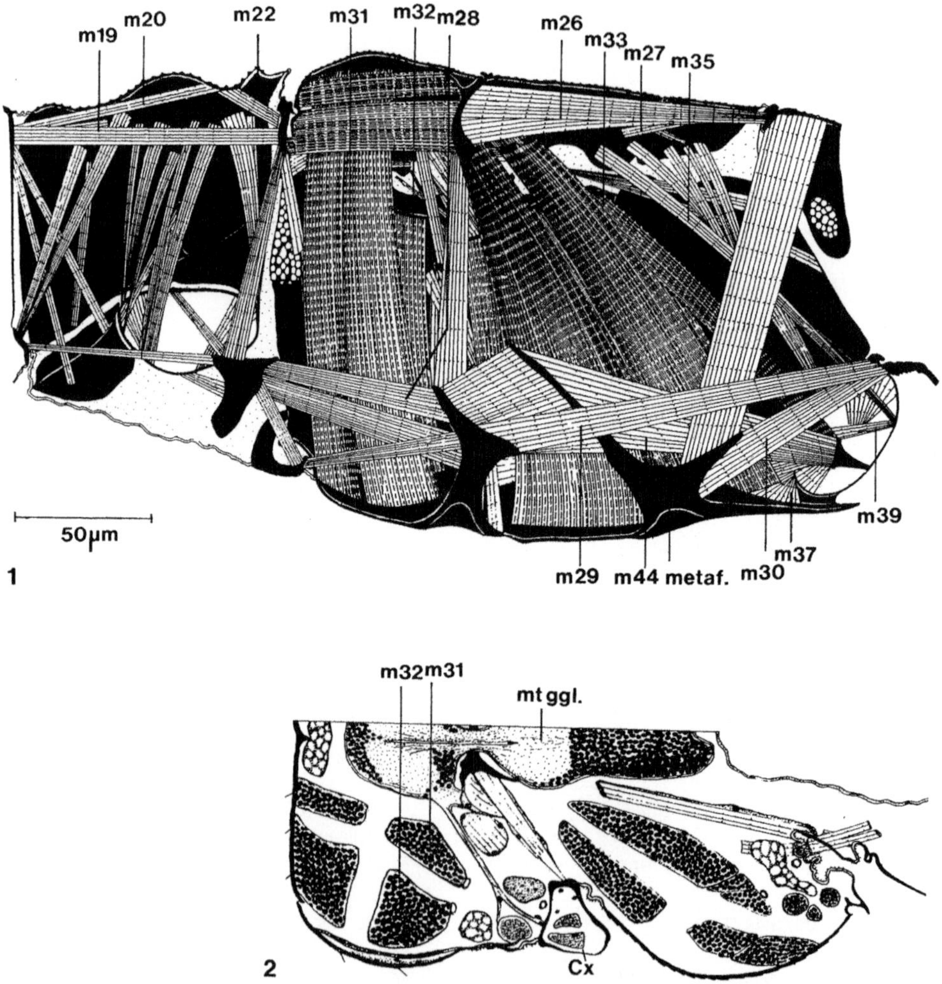

Fig. 2.16. 1. *Hercinothrips femoralis*, ♀, lateral, muscle system of thorax with marked oscillatory wing muscles (modified after Moritz, 1989c); 2. *Aeolothrips intermedius*, ♀ t.s., dorsoventral flight muscles. cx. = coxa, metaf. = metafurca, mtggl. = metathoracic ganglion (the names of the numbered muscles are listed in Table 2.2).

remotors (M38, M39, M40) of the meso- and metacoxa as well as raisers (M41, M42) and depressors (M43, M44) of the meso- and metatrochantin.

Abdominal muscles

The musculature of the abdomen is much simpler, but there are special muscles associated with the ovipositor and male genitalia. Sternal, tergal and pleural muscles of the VIIIth and IXth abdominal segments serve as pro- (M45, M46) and retractors (M47, M48, M49) of the ovipositor valves as well as for special movements of the oviduct, the vagina and accessory gland, and the ovipositor sheath (Fig. 2.14: 3; Fig. 2.17: 1–3). Interior and exterior dorsal longitudinal muscles (M50, M51), originate and are inserted on to the antecosta or intersegmental folds. Together with the ventral longitudinal muscles (M52) these groups raise, lower and flex the abdomen. Within each segment there are tergosternal muscles (M53) which compress the abdomen to assist respiration.

Respiratory system

As in almost all insects, gaseous exchange occurs through internal air-filled tracheae (Fig. 2.18: 1a, 1b). Only in the semi-aquatic *Organothrips bianchi* is there a more or less closed tracheal system with reduced and interrupted tracheae just behind the spiracles (Moritz, 1985). In all other thrips species the respiratory system is open and ventilated with paired spiracles situated on the pleura of the meso- and metathorax and on tergites I and VIII of the abdomen (Priesner, 1968). The spiracles in all instars have typical sclerite structures (peritreme; Fig. 2.18: 1c) allowing plastron respiration during rain (Moritz, 1985). A pleural muscle (M. occlusor spiraculi) regulates the oxygen requirements into each spiracle. The taenidia, which allow the branches to be flexible but resist compression, are visible as spiral ridges on the small tracheal opening directly below the peritreme. In the pre-adult stages the metathoracic spiracle is absent. The dorsolateral trunks of the tracheal system are connected within the thorax by commissures. The tracheoles end in the tissues with many tracheal capillaries in the flight musculature, the salivary glands and the alimentary canal as well as numerous branches within the rectal papillae.

Circulatory system

The dorsal vessel is the conduction organ of the circulatory system and is divided into the heart and the aorta (Fig. 2.18: 2). The flattened heart is situated dorsally within the abdominal segments VII and VIII. Its diameter is about 50 μm and it has three paired aliform muscles, which originate on tergites VI to VIII. During diastole, haemolymph is drawn into the heart through two paired ostia. Systole forces the blood forward in the aorta to discharge over the corpora allata and cardiaca as well as the supraoesophageal ganglion. The pulsation rate is influenced by endocrine, neurocrine and external factors. However, in adult thrips (*Hercinothrips femoralis*) the measured heart rate is about 100 min^{-1} at 24°C. In first instar larvae the rate under comparable conditions is 60 beats min^{-1} rising to about 110 in second instars.

Alimentary tract and associated organs

The alimentary canal is divided into three regions: the fore intestine as an anterior, and the hind intestine as a similar posterior, ectodermal ingrowth, and the mid-intestine, formed as an endodermal tube (mesenteron) (Fig. 2.19: 1–5). This organization arises early in embryogenesis starting with the stomo- and proctodaeum and persists through the larval and pupal stages (Fig. 2.22: 1, 2).

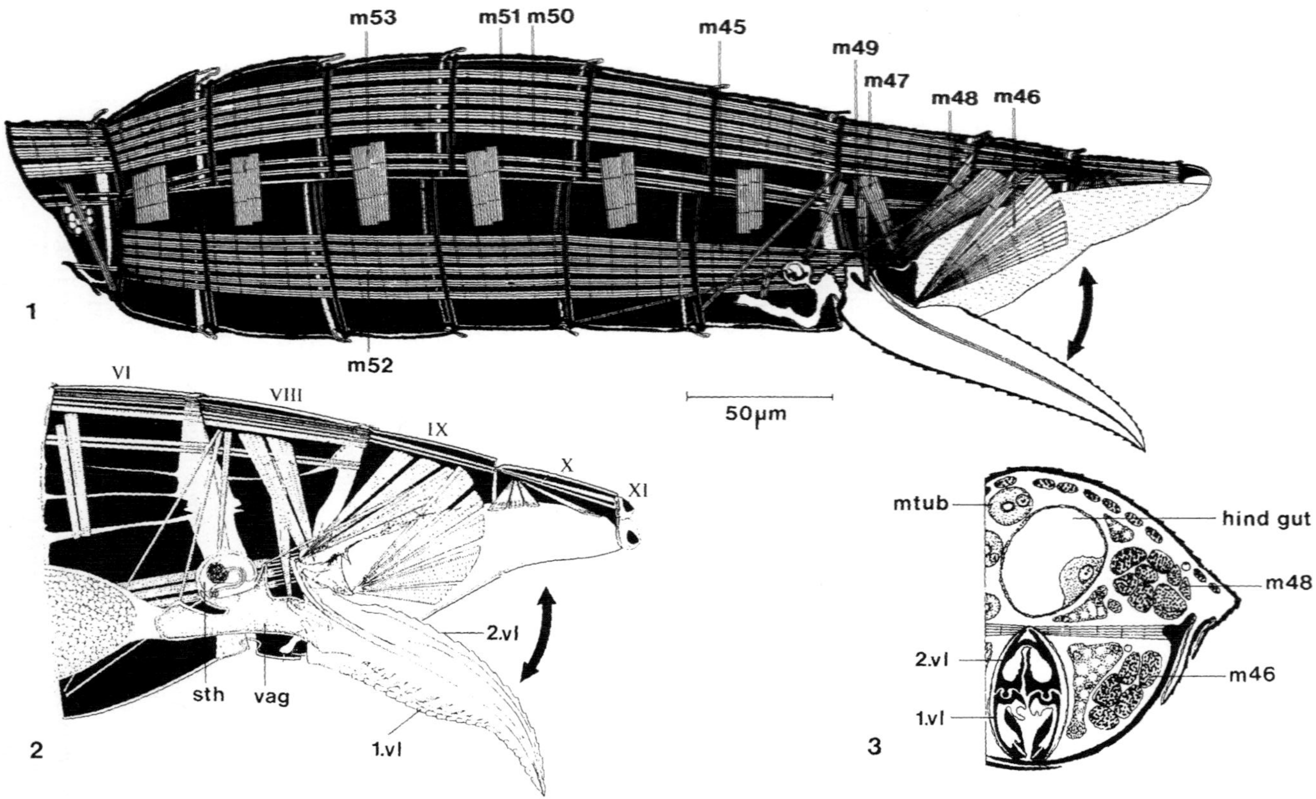

Fig. 2. 17. Abdominal musculature, 1. *Hercinothrips femoralis,* ♀ l.s., muscle system of the abdomen (modified after Moritz, 1989d); 2. *Thrips validus,* ♀ l.s., muscle system of the genital segments (after Bode, 1975); 3. *Aeolothrips intermedius,* ♀, transverse section of segments VIII and IX with resting ovipositor. 1.vl. = 1. (ventral) valve, 2.vl. = 2. (dorsal) valve, mtub. = Malphigian tubule, sth. = spermatheca, vag. = vagina (the names of the numbered muscles are listed in Table 2.2).

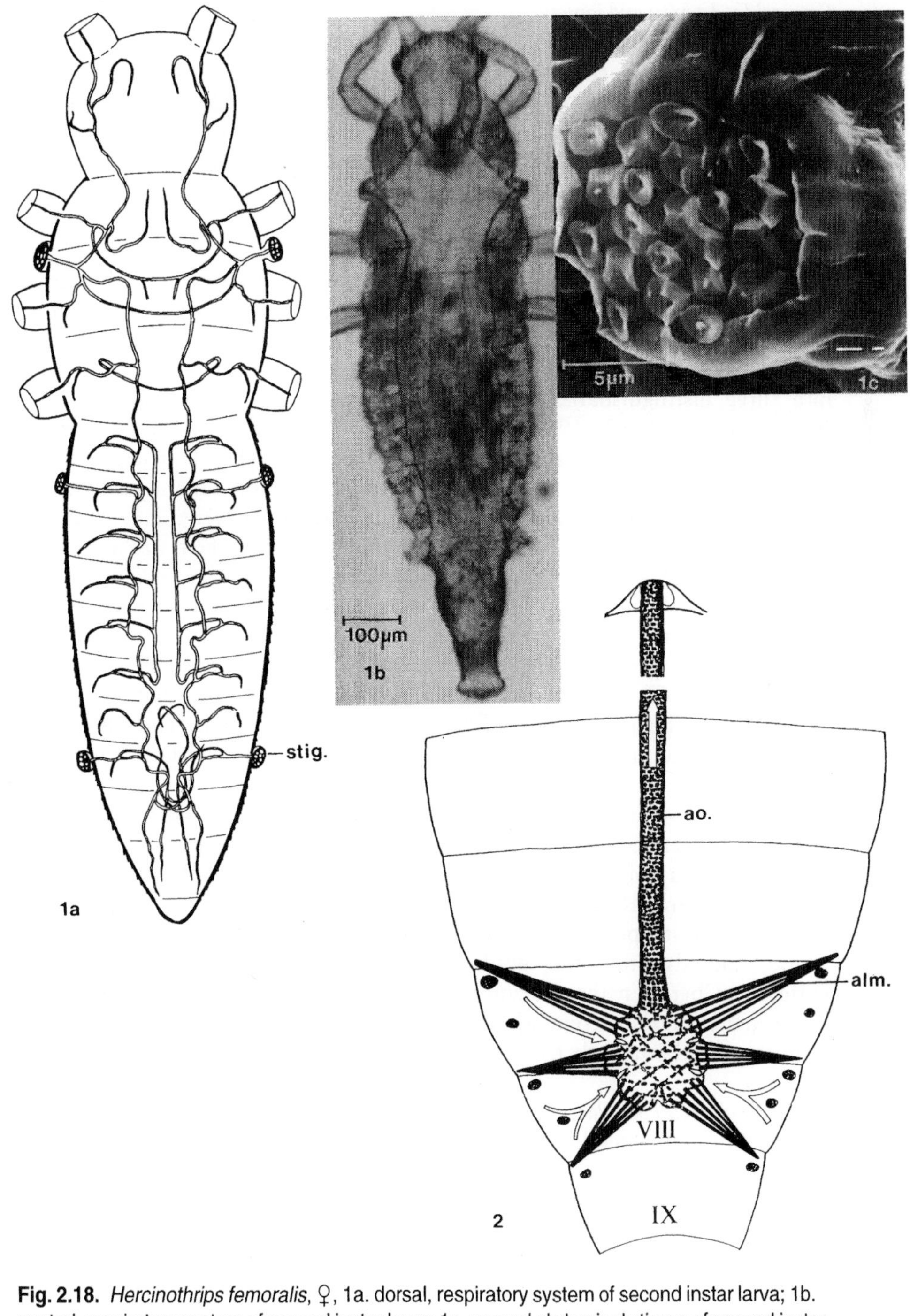

Fig. 2.18. *Hercinothrips femoralis*, ♀, 1a. dorsal, respiratory system of second instar larva; 1b. ventral, respiratory system of second instar larva; 1c. second abdominal stigma of second instar larva; 2. dorsal, circulatory system, heart and aorta. alm = aliform muscles, ao = aorta, stig. = stigmata.

Pre-oral food cavity and foregut

The labrum forms the roof of the pre-oral cavity and the hypopharynx divides this into a dorsal cibarium and a ventral salivarium (Fig. 2.6: 4, 5; Fig. 2.7: 2). The foregut is flexible and can expand and contract through the action of a network of striated muscle fibres (Fig. 2.19: 1). The epithelial cell layer of the oesophagus secretes the cuticular lining as far as the cardiac valve (valvula cardiaca) (Fig. 2.19: 2).

Midgut

The midgut is not lined with cuticle and becomes progressively shorter as moults progress through the instars. In the pupa this part of the alimentary canal forms a straight tube consisting of a flattened cell layer. The adult midgut is a long tube of variable diameter and consists of two loops, which divide it into three distinguishable histological regions (Fig. 2.19: 6, mgut1, 2, 3) (Ullman *et al.*, 1989, 1992). All regions are formed by a single layer of columnar epithelial cells each with a basally located polyploid nucleus (Moritz, 1989d). The cardiac valve forms the junction between foregut and midgut, and the pyloric valve the junction between midgut and hindgut (Fig. 2.19: 2, 5, 6).

Numerous microvilli extend into the midgut lumen, their size and shape differing in each region (Kitajima, 1975). Small regenerative cells are visible among the epithelial cells. The cytoplasm is rich in endoplasmic reticulum, mitochondria and dense lamellated bodies or sphaerocrystals, structures associated with excretion and osmoregulation (Ullman *et al.*, 1989; Del Bene *et al.*, 1991). The occurrence of large apical secretory vacuoles differs depending on the age and physiological status of individuals. The gut cells lie on a thin basal membrane which is encircled by an alternating series of longitudinal and circular muscle fibres (Fig. 2.19: 5, 6).

The peritrophic membrane in the midgut contains no detectable chitin and consists of layers of membranous laminae about 80 Å thick (Bode, 1979); it does not have a uniform structure and resembles the plexiform surface coat of some Hemiptera (Reger, 1971; Richards and Richards, 1977; Ryerse *et al.*, 1992). The peritrophic membrane extends beyond the cardiac valve as the larva grows, and also thickens, gradually becoming less permeable to viruses. Differences between these membranes in larval and adult stages, as well as between adults of different species, partly account for their variable ability to acquire and subsequently transmit virus (see Chapter 14).

Hindgut and rectal papillae

The pyloric sphincter lies between the mid- and hindgut, and close to the junction of these two regions the four Malpighian tubules are inserted. Like the midgut, the hindgut is lined with numerous microvilli and is divisible into an anterior tube (ileum), a colon, and a wider rectum, which opens exteriorly at the anus. The cellular layer of the rectum is greatly thickened to form four rectal papillae (five in genus *Melanthrips*), which maintain the proper salt and water balance in the insect (Bode, 1977; Dallai *et al.*, 1990).

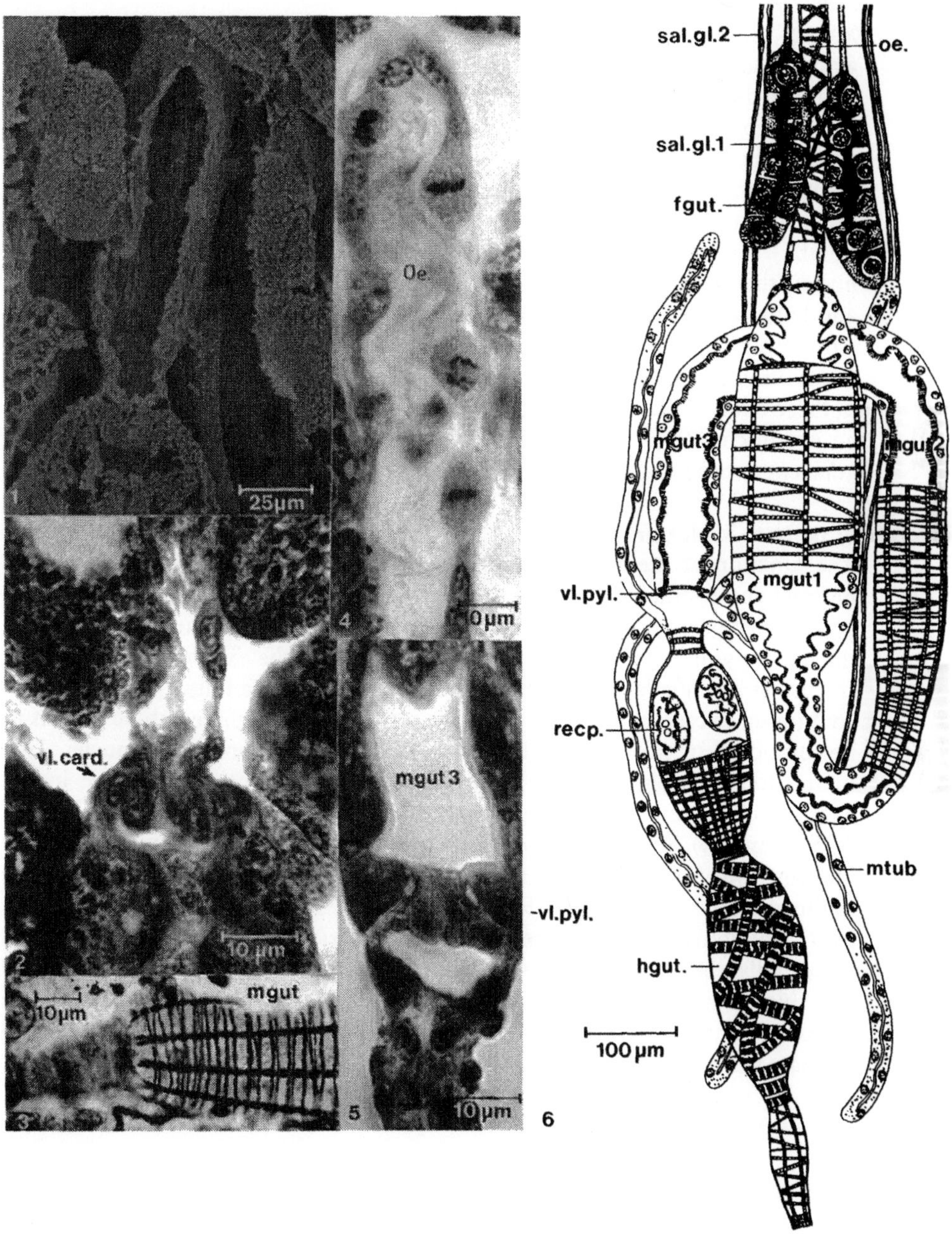

Fig. 2.19. 1. *Frankliniella occidentalis*, t.s., oesophagus and valvula cardiaca of second instar; 2. *Aeolothrips intermedius*, ♀ l.s., valvula cardiaca; 3. *Thrips tabaci*, ♀ l.s., muscle system of midgut; 4. *Hercinothrips femoralis*, prepupa, oesophageal cells in mitosis; 5. *Hercinothrips femoralis*, ♀ t.s., valvula pylorica and midgut–hindgut region; 6. *Hercinothrips femoralis*, ♀ t.s., alimentary system. fgut. = foregut, hgut. = hindgut, oe. = oesophagus, mgut. = midgut, mtub. = Malpighian tubule, rec.p. = rectal papilla, sal.gl. = salivary gland, vl.card. = valvula cardiaca, vl.pyl. = valvula pylorica.

Salivary glands

Several glands are associated with the intake and digestion of food. These glands usually originate as paired invaginations of the ectoderm associated with the developing labial structures. In addition to the unpaired mandibular gland, two paired labial glands discharge their secretion into the pre-oral food cavity, where it mixes with the food as it is ingested. Their main ducts fuse to form a common outlet discharging into the salivarium. One pair of large lobed salivary glands lies within the pro- and mesothoracic segment (Fig. 2.19: 6; sal.gl.1). The lobes consist of large, loosely-aggregated cells with very large nuclei; their cytoplasm is rich in lipid droplets, Golgi bodies, coated vesicles and lamellar infoldings (Del Bene *et al.*, 1991; Ullman *et al.*, 1992). A second paired tubular salivary gland leads from the first loop of the midgut to a small reservoir within the head–prothorax complex (Fig. 2.19: 6; sal.gl.2). The lumen of these ducts is 1 to 3 μm in diameter and the cells surrounding it are lined with an epicuticle 40 to 60 nm thick. Caudally, these tubular salivary glands lie adjacent to the muscle fibres of the midgut.

Excretory system: Malpighian tubules

The function of the excretory system is to maintain a more or less constant internal environment in the insect body. This involves the removal of the end products of protein breakdown and the regulation of the ionic composition of the haemolymph. The Malpighian tubules, part of the excretory system, each consist of a single layer of epithelial cells surrounding a lumen of 0.5 to 1 μm containing excretory fluid. In cross-section each Malpighian tubule is seen to be composed of two or three cells with microvilli and prominent nuclei. Two of the four Malpighian tubules are directed anteriorly and are free in the haemocoel beyond their insertion shortly behind the valvula pylorica (Fig. 2.19: 6; mtub.). The other two, directed posteriorly, are also mainly free of the intestine, but for a short distance both these tubules sometimes tightly adhere to the hindgut wall (Sharga, 1933; Ullman *et al.*, 1989; Dallai *et al.*, 1991). The four Malpighian tubules open into the pyloric gut region (Fig. 2.19: 6). Proximally, the cells of the tubules do not have microvilli but are lined by a 20 nm thick cuticle indicating a transitional area near to the hindgut. The proximity of the hind tubules to the rectal papillae allows both these organs to regulate ion and water transport using the haemolymph.

Reproductive system

Females

The paired ovaries of the female each consist of four egg tubes, or ovarioles, which discharge the matured eggs into a pair of tubular lateral oviducts (Fig. 2.14: 2, 3; Fig. 2.20: 2). In ovipositing adults these lateral sections are short and surrounded by outer circular and inner longitudinal muscle fibres, whereas the common oviduct and vagina have no muscle fibres (Heming, 1970a). The ovariole consists of a terminal filament, a germarium and a vitellarium (Fig. 2.20: 2). The germarium contains the primordial germ cells, the oogonia, which later become differentiated into oocytes and follicle cells. The oocytes pass down the vitellarium and grow as yolk is deposited in them, forming a series of developing eggs in each ovariole. Each egg is enclosed in a layer of follicular epithelial cells which secrete the chorion on to the vitelline membrane (Tsutsumi, 1996).

Thrips ovarioles are typical panoistic, because they have no nurse cells (Fig. 2.20: 2). The development of clusters of cells in thrips germaria is common and interpreted by Pritsch and Büning (1989) as indicating a secondary type of panoistic ovary, the cells arising by mitosis from a stem oogonium, as occurs in insects having meroistic ovarioles. By contrast, Tsutsumi *et al.* (1993) described the homogeneous cell clusters as a plesiomorphic character found in nearly all insects. Heming (1995) believes that clonal proliferation of the oogonia in neopanoistic ovaries is the result of adaptation of ancestors with polytrophic ovaries to small adult body size. Mature eggs are large in relation to the insect abdomen. The number of follicles within the four paired ovarioles usually ranges between 4 and 10, but in some cases about 20 egg chambers occur (Moritz, 1982d, 1989d). In each germarium the cytoplasm of the oocytes increases in volume as yolk is incorporated. During this process the elongated chromosomes of early oocyte stages become unclear, because the large quantities of yolk hinder adequate fixation (Fig. 2.21: 3a–c). In fully developed eggs of tubuliferan species there appears to be an anterior and posterior mass of 'symbiotic bacteria' within the yolk (Bournier, 1961; Haga and Matsuzaki, 1980; Haga, 1985; Heming, 1995), but later ultrastructural and histochemical examinations of this 'mycetome' indicate that it is merely an aggregation of lysosomes (Tsutsumi *et al.*, 1994).

The spermatheca is roughly spherical opening into the common oviduct through a spermathecal duct. The spermatozoa are stored in the spermatheca after copulation as a small twisted ball (= spermatophore after Heming, 1970a). When the valve of the spermathecal duct opens, the spermatozoa fertilize the eggs. In some species small muscle fibres in the vagina and spermatheca control the fertilization procedure. The accessory gland secretes mucoprotein for lubrication during oviposition and/or for cementing eggs within or on to the host plant (Bode, 1975).

Males

The male reproductive system comprises a pair of testes, two lateral ducts, or vasa deferentia, seminal vesicles and paired accessory glands. The posterior end of each vas deferens enters the seminal vesicle and is fused to the ejaculatory duct. The testes are orange to brown and lie usually within abdominal segments VII to IX. Most thrips have a haploid male germ line, male meiosis is aberrant. This means meiosis I is mitotic and meiosis II is reduced to the ejection of a chromatin mass from each secondary spermatocyte (Heming, 1995). The flagellum of a terebrantian spermatozoon consists of 27 microtubular elements (9 microtubular doublets with dynein arms, 9 microtubular doublets without dynein arms and 9 microtubular singlets with dynein arms; Bode, 1983, 1990; Dallai *et al.*, 1991; Heming, 1995). During the premating period terebrantian spermatozoa complete their differentiation within the testes, whereas tubuliferan spermatozoa complete development later.

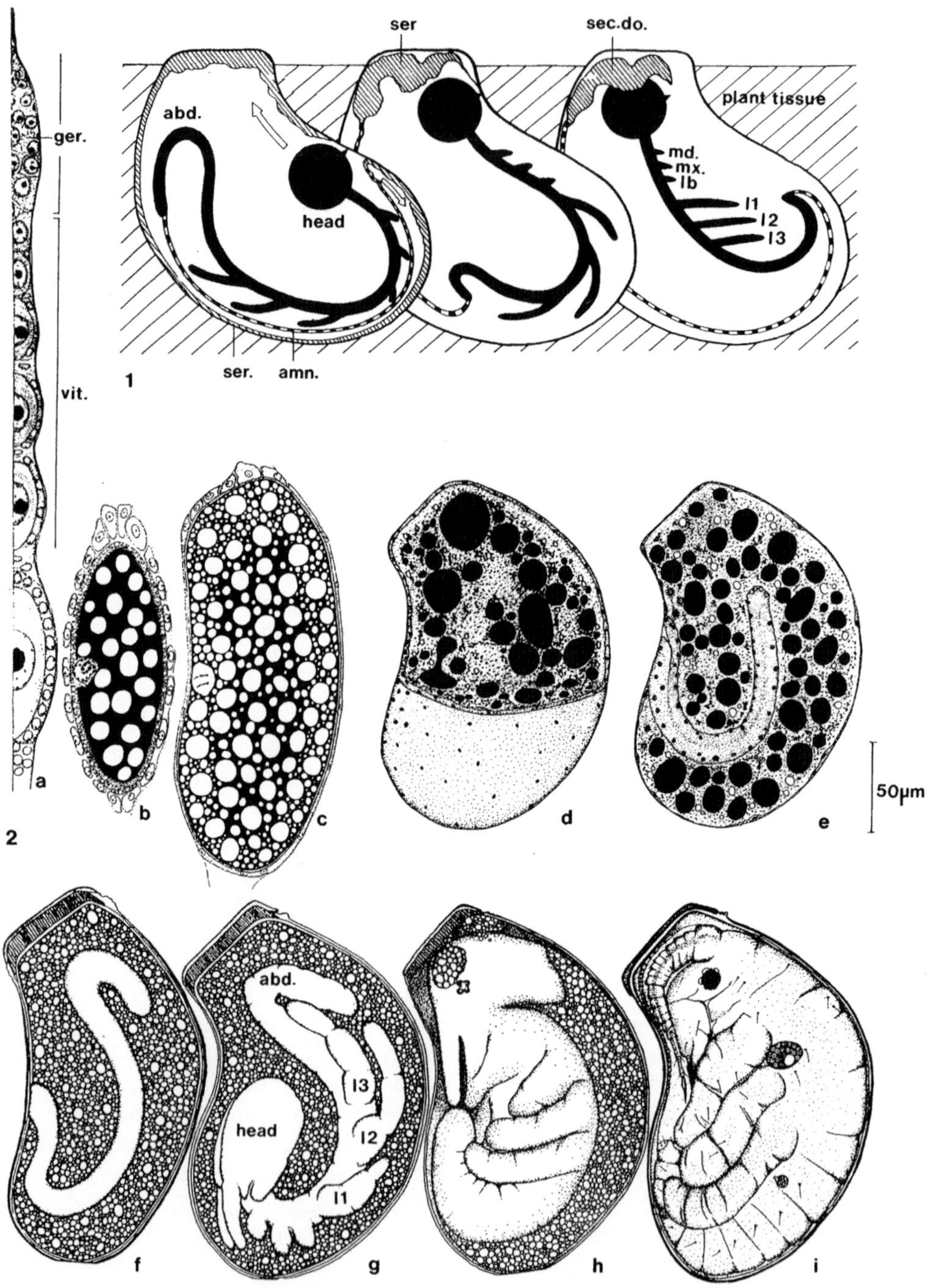
ser
sec.do.
plant tissue
abd.
head
md.
mx.
lb
l1
l2
l3
ser.
amn.
1
ger.
vit.
a
b
c
d
e
50µm
2
abd.
l3
head
l2
l1
f
g
h
i

The wall of the paired accessory glands is two-layered with a thin, non-pigmented, peritoneal sheath and an inner secretory epithelium (Heming, 1975). In contrast to the Tubulifera, the investigated terebrantian species lack an acrosome, but the sperm structure in both suborders differs totally from that of other hemipteroids (Bode, 1988). Males of the Thripinae have sternal abdominal glands, which are commonly considered to secrete pheromones (Fig. 2.14: 5). These glands have cells with mitochondria, microbodies and agranular, tubular endoplasmic reticulum. The principal components of their secretion are lipids, but acid mucopolysaccharides are also present (Bode, 1978). A comparable glandular area is common in males of merothripids in the dorsal region of the head (Fig. 2.14: 6; Moritz, 1984a).

Types of reproduction, sex ratios and essential genetics

In most Thysanoptera, reproduction requires copulation between adults of the opposite sex. As in Hymenoptera, females lay two kinds of eggs. Those which are fertilized have a full diploid number of chromosomes and produce only females, whereas unfertilized eggs are haploid and give only males (arrhenotoky). The usual method of sex determination by sex chromosomes is replaced by this apparently simple mechanism, but there are some indications that information relating to sex is enclosed within the autosomes. Later, different cell groups undergo endopolyploid development leading to larger chromosome sets, particularly in the salivary glands and the midgut.

In a few thrips species, obligate parthenogenesis occurs resulting in only female progeny (thelytoky) or very rarely males (deuterotoky). The gonads of these males produce typical sperm (Bode, 1988, 1990) and they copulate normally with females. In *Thrips tabaci*, sexual and asexual populations are common (Zawirska, 1976) and virgin females of *Apterothrips apteris* produce both sexes (Mound, 1992). Possibly, the regulation of chromosome sets occurs through fusion of the egg nucleus with the second polar body to produce females. The occurrence of males is reminiscent of an ancestral status, but the presence of bacteria of the genus *Wolbachia* in a haplodiploid thrips species, *Taeniothrips inconsequens*, suggests that the regulation of sex may be associated with a microbe, as in *Trichogramma* (Stouthamer *et al.*, 1993; Stouthamer and Werren, 1993). Furthermore, the frequency of males in thelytokous populations often depends on the temperature at which their

Fig. 2.20. *Opposite* 1. Lateral aspect of the egg of a panchaetothripine (*Hercinothrips femoralis*) summarizing the steps anatrepsis and katatrepsis of the embryo (modified after Moritz ,1988a), abd. = abdomen, amn. = amnion, l1.–l3. = thoracic legs, lb. = labium, md. = mandible, mx. = maxilla, sec.do. = secondary dorsal organ, ser. = serosa; 2. successive stages in the embryogenesis of a terebrantian thrips: a. ovariole structure and oogenesis, ger. = germarium, vit. = vitellarium, b. ovocyte during chorion formation, c. ovocyte with cleavage nucleus, d. formation of blastoderm after cleavage, e. anatrepsis and development of amniotic folds, f. maximal extension of the germ band, g. embryo during organogenesis with appendages, abd. = abdomen, l1–l3. = thoracic legs, h. embryo after katatrepsis during primary dorsal closure, i. ready-to-hatch embryo with chaetotaxis of the first instar larva.

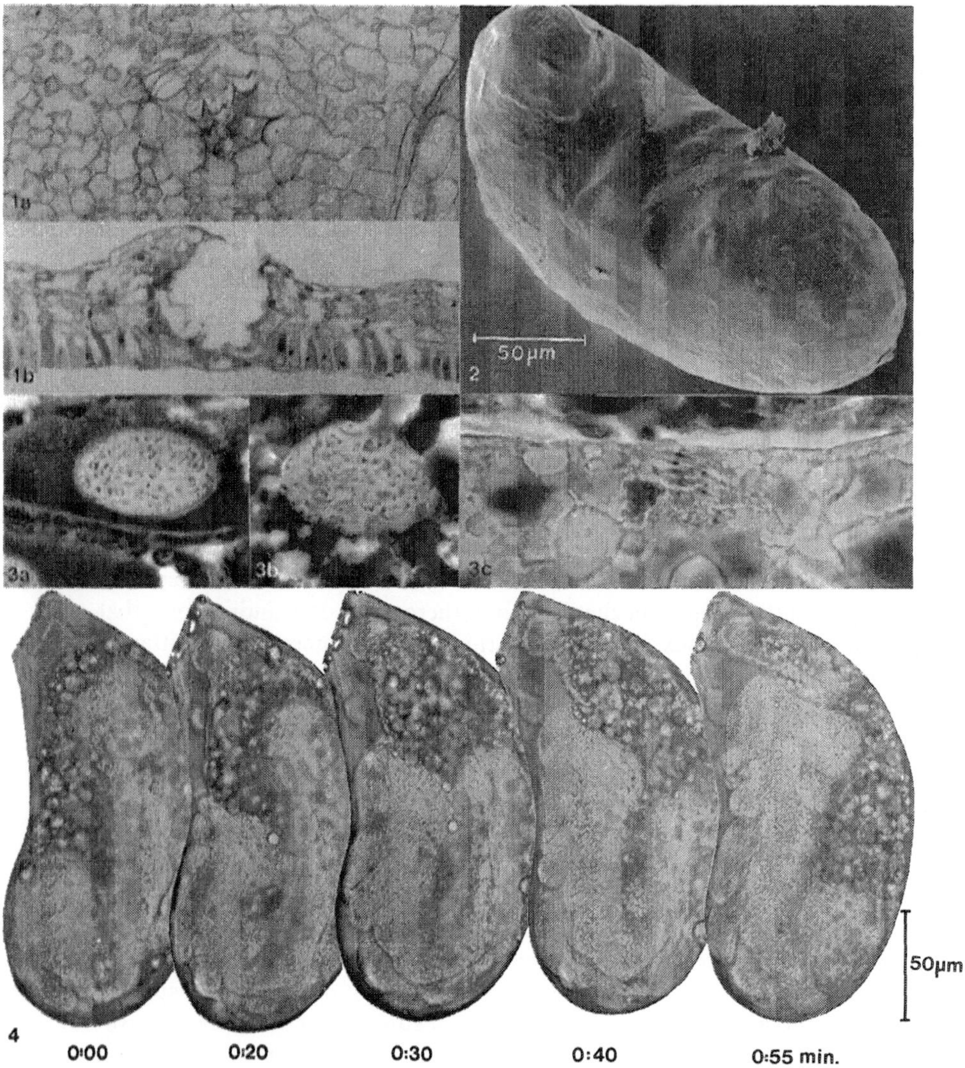

Fig. 2.21. Leaf of *Saintpaulia*: 1a. leaf surface with an egg chamber of *Frankliniella occidentalis*; 1b. section of a leaf with an empty egg shell after hatching; 2. ventral surface of an egg of *Frankliniella occidentalis*; 3a. ovocyte during vitellogenesis with condensed chromocentres, 3b. karyolysis, 3c. first cleavage after karyolysis; 4. successive steps of katatrepsis in *Aeolothrips astutus*.

mothers developed, higher temperatures producing more males. This reversion to arrhenotoky may be caused by temperature susceptibility of the microbes.

In some arrhenotokous and thelytokous species (e.g. *Gynaikothrips ficorum*, *Hercinothrips femoralis*) embryonic development of unfertilized, male-producing eggs starts within the terminal region of the vitellarium. In these circumstances the 'decision' on sex determination must be made by the female before the egg passes the spermatheca and must be the result of interactions between the follicle, the oocyte and the haemocoel. As yet, there is no detailed knowledge of how females control fertilization of their eggs. Indeed, knowledge of thrips reproductive strategies is sparse and the cytological mechanisms involved in sex determination are unknown.

Description of Eggs, Immature Stages and Development

Oviposition and egg structure

The manner by which female-producing eggs are fertilized is unknown; only in a phlaeothripid species, *Haplothrips verbasci*, has a micropyle on the ventral side of the egg been described (Heming, 1979). Thrips eggs are pale white, yellow or darkly coloured, cylindrical or kidney-shaped and covered by an outer shell, or chorion. The outer sculpture is related to the oviposition strategy. In terebrantian species, the egg shell is smooth like a plastic bag because they lay their eggs singly in an incision within the plant tissue made by the ovipositor (Fig. 2.20:1; Fig. 2.21: 1a, 1b, 2). In phlaeothripids, which lay their eggs on the substrate, the shell is often penta- or hexagonal sculptured and fixed to the host plant with a sticky patch (Lewis, 1973; Haga, 1985). In terebrantians, the eggs have an anterior operculum which is removed by a saw-shaped oviruptor at hatching. For several hours after laying the terebrantian egg needs the support of the surrounding plant tissue. Thereafter, metabolic changes within the egg and the hardening of the shell give it a more stable structure, but terebrantian eggs placed experimentally on agar or moist paper still develop like eggs in normal positions in their host plant tissues.

Embryogenesis

The length, and even more the width, of the eggs increases during embryogenesis as a result of yolk forming over embryonic structures (Moritz, 1988a, 1995). After maturation, syngamy and first cleavage take place in the vitellaria and vagina of haplodiploid species (Fig. 2.21: 3a–c). Immediately after oviposition, the cleavage energids enter the periplasma to create the uniform plasmodial preblastoderm (Fig. 2.20: 2a–d). After blastoderm cell differentiation, asynchronous mitotic activity divides the blastoderm into presumptive embryonic and serosal layers. The posterior embryonic cells of the germ anlage invaginate into the yolk. This structure then bends towards the anterior pole of the egg and starts to lengthen (Fig. 2.20: 2e, 2f). Later the embryo separates totally from the serosa and forms an S-shaped germ band, two cell layers thick. Segmentation begins in the thoracic region and proceeds cephalad and caudad. The buds of the thoracic legs and antennae arise approximately simultaneously and are followed by the gnathal appendages (Fig. 2.20: 2g, 2h). After gastrulation, the germ band shows typical organization, consisting of an ectoblastem (outer layer) and a hypoblastem (inner layer), which is subdivided into the ento- and mesoblastem. Between the flexed abdominal segments a few extra-embryonic germ cells are enclosed. Next, and as a result of the growing embryo, the orientation of the embryo–yolk system changes completely because the extra-embryonic membranes fuse and rupture adjacent to the head of the embryo, which still lies at the posterior pole of the egg. The serosal cells at the anterior end of the egg contract and form the secondary dorsal organ (Fig. 2.20: 1). The fate of this glandular organ is unclear, because it becomes an integral part of the early prelarva (Fig. 2.21: 4). The amnion rolls back and replaces the serosa as a primary dorsal closure, covering the rest of the yolk mass within the midgut. The pleuropodia appear as paired glandular cell structures shortly before katatrepsis occurs. They may function in synthesis, transport and secretion of factors inducing katatrepsis as well as inducing changes in the embryonic and serosal cuticle (Heming, 1980, 1993a, b). Katatrepsis takes 1 to 2 hours in the investigated species (*Hercinothrips femoralis*, *Frankliniella occidentalis*, *Aeolothrips intermedius*; Moritz, 1988a). After katatrepsis, the cells of nearly all organ systems are visible (nervous system, digestive tract, muscle system, circulatory system, sensory system) and begin to function physiologically (Fig. 2.22: 1, 2). The death of cells destined to form the right mandible is induced by the invagination of the mouthparts during the embryonic evolution (Moritz, 1988a, 1995).

First and second instar larvae

As in many other insects, the body size of the thrips larvae increases rapidly after hatching (Fig. 2.22: 3, 4). The first and second instars resemble a miniature version of the adult, apart from the presence of wings and genital appendages, with the organs adapted to a smaller body size (Fig. 2.23: 1–5). The cuticle of both instars is often yellow or pale and membranous. The stability of the chitin plates results more from the inner blood pressure of the larvae than their inherent strength. Unusually, both stages grow in length and girth as they feed, so late first instar and newly moulted second instar larvae appear very similar. The structure of the chitin surface allied to the shape of the bristles forms a hydrophobic area for plastron respiration during rain, differing from adults in this feature (Fig. 2.23: 5).

Both stages have 'punch and suck' mouthparts consisting of two laciniae and the left mandible enclosed within a hypognathous mouthcone (Fig. 2.23: 4; Fig. 2.25: 2). The antennae resemble the final adult version; the pedicellus in both larval stages has a campaniform sensillum and the flagellum shows similar chaetotaxy and sense cones (Fig. 2.23: 1, 2, Fig. 2.25: 2). The visual system usually consists of four stemmata, but ocelli are absent in all larval stages (Fig. 2.23: 2, 4). The small head containing the active cibarial muscles results in the brain being displaced far into the thorax of the first instar larva (Fig. 2.25: 1). In the late second instar larva this repositioning causes looping of the optic nerves within the brain (Staub, 1979; Moritz, 1984b). The arolium of the pretarsus differs slightly between instars and adults (Fig. 2.23: 3) (Heming, 1971, 1972, 1973). The wings and genital appendages occur only in adults, but their internal organization starts with the proliferation of the fore and hind wing discs in the late second instar larva. The fore- and hindgut are fused with the midgut through the cardiacic and pyloric valves (Fig. 2.25: 1) and fore- and midgut cells often have two nuclei. As in adults, the larval hindgut contains rectal papillae (Fig. 2.25: 4). Larvae also have two labial glands, four Malpighian tubules and a pygidial gland (Fig. 2.24: 4). The paired

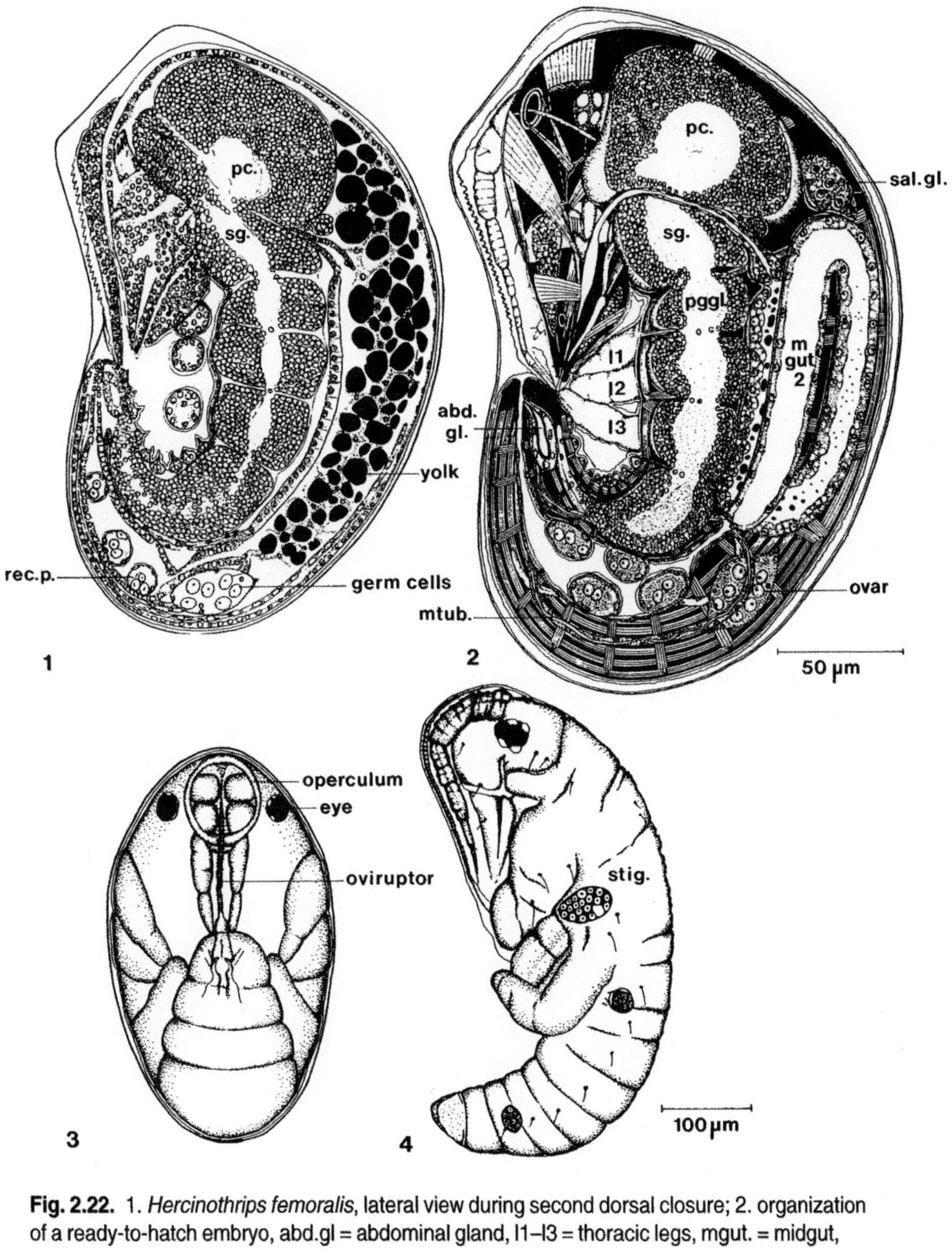

Fig. 2.22. 1. *Hercinothrips femoralis*, lateral view during second dorsal closure; 2. organization of a ready-to-hatch embryo, abd.gl = abdominal gland, l1–l3 = thoracic legs, mgut. = midgut, ovar = ovary, pggl. = prothoracic ganglion, pc. = protocerebrum, rec.p. = rectal papilla, sal.gl. = salivary gland, sg. = suboesophageal ganglion; 3. ventral view of a prelarval stage with oviruptor and operculum; 4. first instar larva immediately after hatching and rupture of the embryonic cuticle, stig. = stigma.

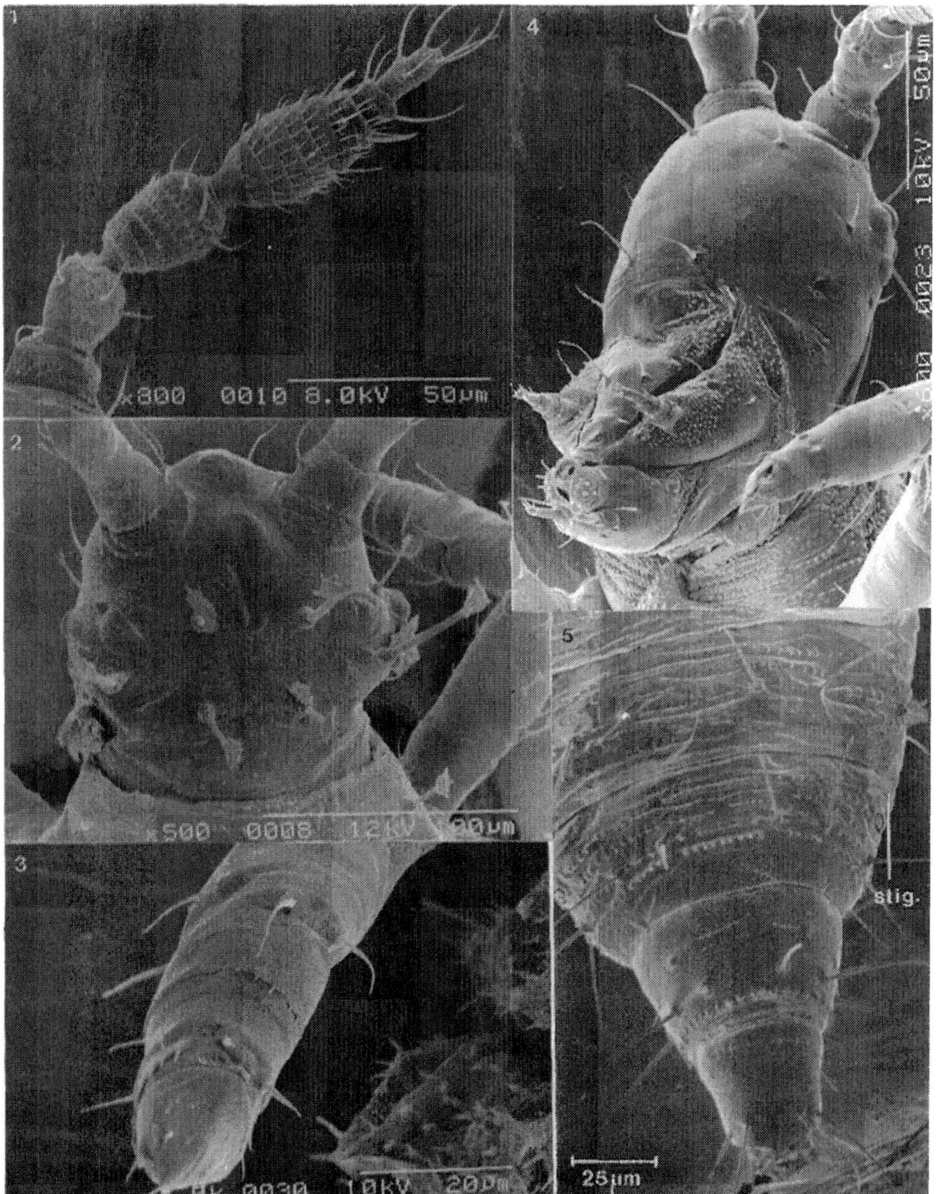

Fig. 2.23. Larval stages: 1. *Thrips tabaci*, second instar larva, right antenna; 2. *Parthenothrips dracaenae*, second instar larva, dorsal, head; 3. *Frankliniella occidentalis*, second instar larva, foreleg; 4. *Frankliniella occidentalis*, second instar larva, head, ventrolateral; 5. *Frankliniella occidentalis*, second instar larva, VIIth–Xth segment of the abdomen, SEM 1400×.

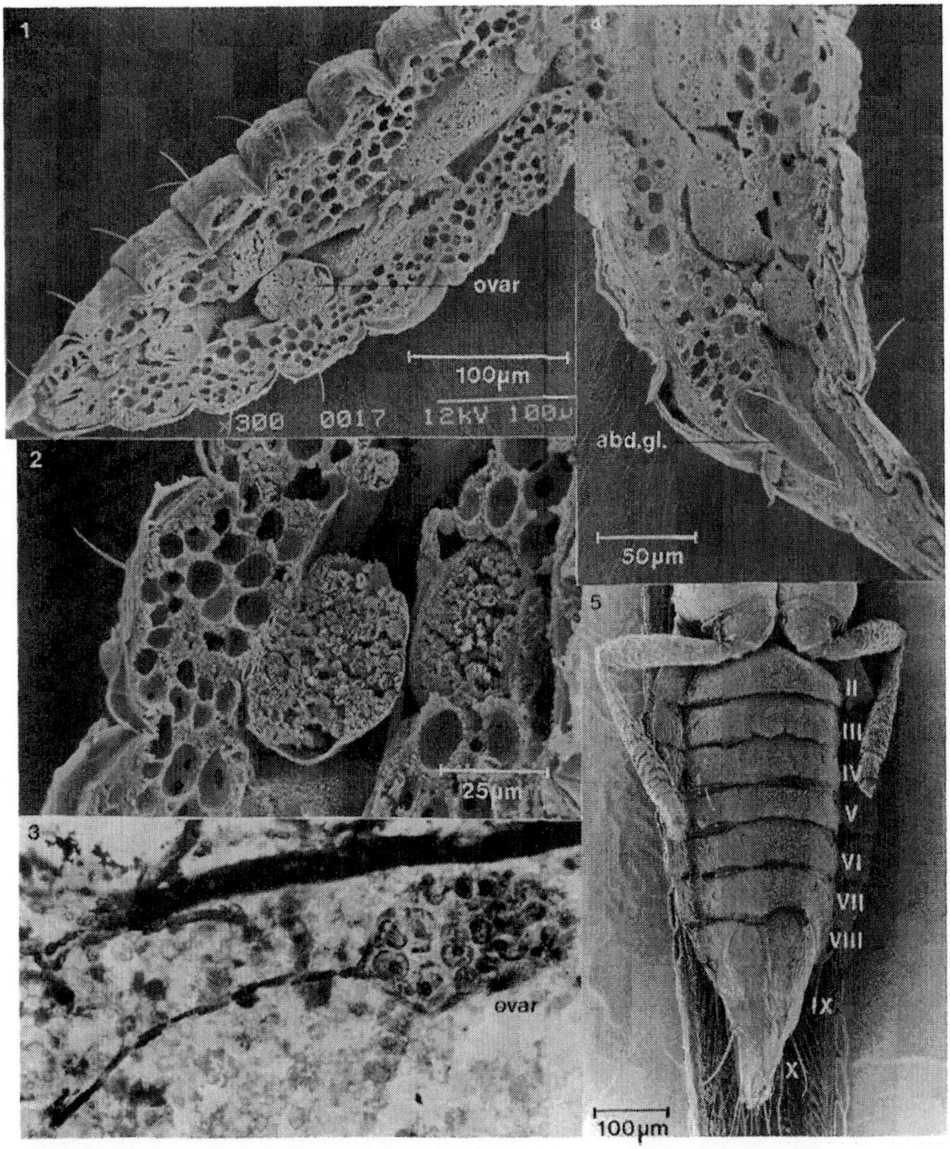

Fig. 2.24. 1. *Frankliniella occidentalis*, lateral, second instar larva, abdomen, development of the paired ovary; 2. *Frankliniella occidentalis*, second instar larva, VIth abdominal segment with ovary; 3. *Hercinothrips femoralis*, lateral, second instar larva, ovary with developing lateral oviduct; 4. *Frankliniella occidentalis*, lateral, second instar larva, hindgut with rectal glands and pygidial (= abdominal) gland, abd.gl. = abdominal gland; 5. *Frankliniella occidentalis*, ventral, pupa, abdomen with covered ovipositor.

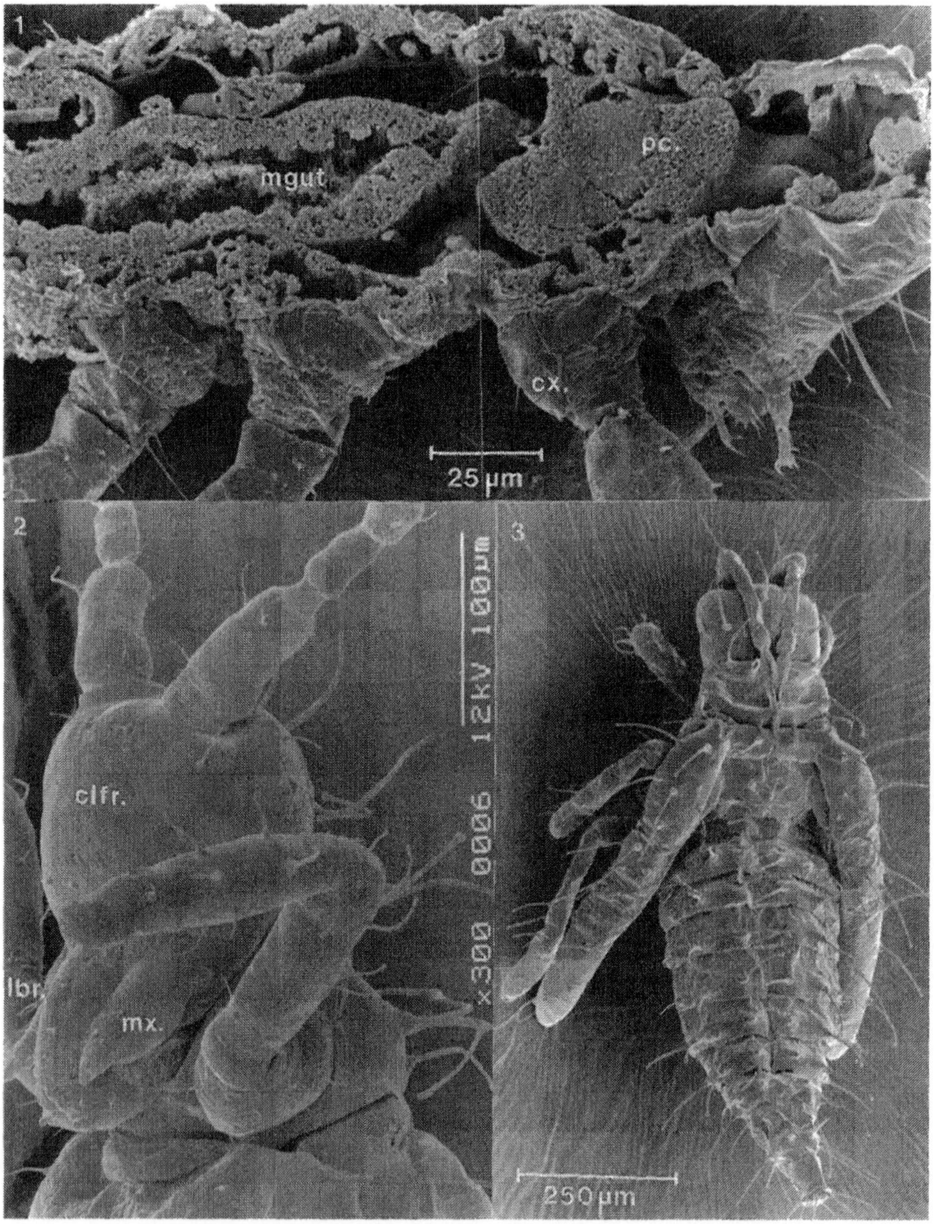

Fig. 2.25. *Frankliniella occidentalis*, 1. second instar larva, view of the head and thoracic segments with brain and midgut; 2. second instar larva, head with developing mouthcone, clfl. = clypeofrons, lbr. = labrum, mx. = maxilla; 3. pupa, dorsal view with wing pads.

elliptical ovaries lie in abdominal segments V and VI and are connected with the genital discs through the lateral oviducts (Fig. 2.24: 1, 2, 3). During these metaphase stages, large centromeres appear and small chromomeres are concentrated in the nuclear membrane (Fig. 2.24: 3).

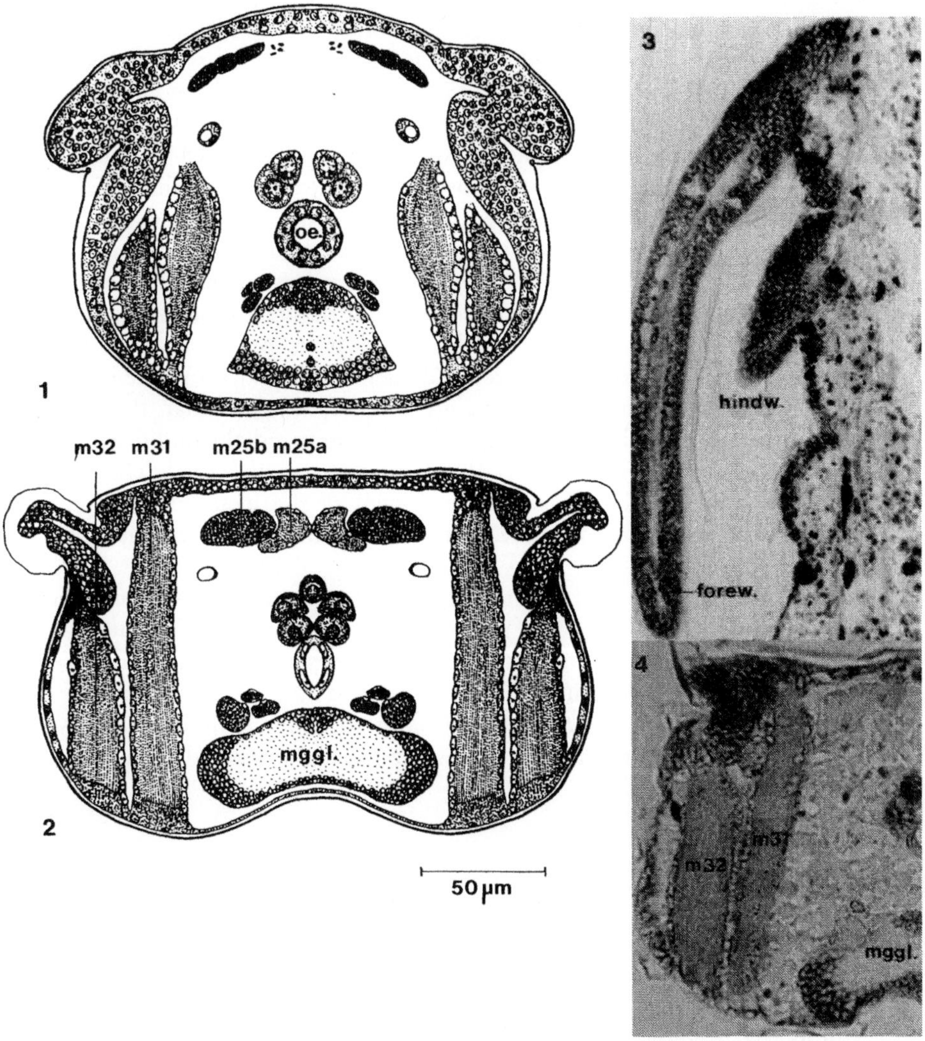

Fig. 2.26. *Hercinothrips femoralis*, 1. prepupa, t.s., mesothorax with developing transversal oscillatory muscles; 2. pupa, t.s., mesothorax, with transversal oscillatory muscles attached to the nota, mggl. = mesothoracic ganglion; 3. pupa, f.s., fore and hind wing pads; 4. pupa, t.s., mesothorax with direct and indirect flight muscles (for numbered muscles see Table 2.2.).

Propupa and pupa

The two or three quiescent, non-feeding and inactive instars, propupa and pupa (pupa I and II in the Tubulifera), are covered by a smooth, membranous, colourless and unsclerotized cuticle often with long setae inserted on raised bases (Fig. 2.25: 3). The segmentation of the antennae and legs is reduced and lobes of the developing ovipositor are present on abdominal sternites VIII and IX. The developing compound eyes, ocelli, gnathal stylets, fore and hind wings and valves of the ovipositor of the adult are visible through the pupal cuticle in pharate adults (Fig. 2.24: 5; Fig. 2.25: 3). The larval meso- and metathorax fuse more firmly to the adult pterothorax and the hind coxae shift from their pleural position to a more sternal region to allow the insect to jump. The changes in the muscle system are enormous. For example, within the head, myofibrils change their fibril type, and within the thoracic and genital region many muscles are added, whilst some disappear. The development of the oscillatory wing musculature of the pterothorax starts immediately after the propupal moult (Fig. 2.26: 1–4) as does the repositioning of the supraoesophageal ganglion, the development of the visual parts of the brain, and the differentiation of the compound eyes and ocelli. The changes in the digestive system during the quiescent instars are drastic; in particular the midgut shortens and the epithelium disappears. The larval pygidial gland degenerates in the pharate pupa (Fig. 2.24: 4). The first abdominal spiracle changes its position from tergite II to I, and development of the metathoracic spiracle begins. The paired rudiments of the ovary begin to divide from the anterior end into four ovarioles.

Ecologically, larval and adult stages occupy similar niches, but winged adults are more likely to invade new habitats. During the quiescent instars, extensive structural change occurs comparable with that found in endopterygote insects. From this point of view the terms ‘pupa’ and ‘propupa’ are preferable to ‘nymph’, a term more relevant to hemimetabolous insects, in which changes during the resting stages are far less profound than in thrips (Nüesch, 1987).

References

Bhatti, J.S. (1988) The orders Terebrantia and Tubulifera of the superorder Thysanoptera (Insecta). A critical appraisal. *Journal of Pure and Applied Zoology* 1, 167–240.

Bhatti, J.S. (1991) Surface patterns of wings in the order Tubulifera (Insecta). *Journal of Pure and Applied Zoology* 3, 1–95.

Bode, W. (1975) Der Ovipositor und die weiblichen Geschlechtswege der Thripiden (Thys., Terebrantia). *Zeitschrift für Morphologie und Ökologie der Tiere* 81, 1–53.

Bode, W. (1977) Die Ultrastruktur der Rektalpapillen von *Thrips* (Thysanoptera, Terebrantia). *Zoomorphology* 86, 251–270.

Bode, W. (1978) Ultrastructure of the sternal glands in *Thrips validus* Uzel (Thys., Terebrantia). *Zoomorphologie* 90, 53–65.

Bode, W. (1979) Peritrophische Membranen bei *Thrips validus* Uzel (Thysanoptera).

Zoologische Jahrbücher, Anatomie 102, 491–509.

Bode, W. (1983) Spermienstruktur und Spermatohistogenese bei *Thrips validus* Uzel (Insecta, Thysanoptera). *Zoologische Jahrbücher, Anatomie* 109, 301–318.

Bode, W. (1986) Fine structure of the scolopophorous organs in the pedicel of three species of Thysanoptera (Insecta). *International Journal of Insect Morphology and Embryology* 15, 139–154.

Bode, W. (1988) The spermatozoa of Thysanoptera and their relevance for systematics. *Acta Phytopathologica et Entomologica Hungarica* 23, 267–273.

Bode, W. (1990) Spermienultrastruktur bei Thysanopteren und ihre Bedeutung für die Systematik. *Mitteilungen für Deutschen Gesellschaft allgemeine und angewandte Entomologie* 7, 535–540.

Bournier, A. (1961) Sur l'existence et l'évolution d'un mycétome au cours de l'embryogènese de *Caudothrips buffai* Karny. *Proceedings of the 11th International Congress of Entomology* 1, 352–354.

Childers, C.C. and Achor, D.S. (1991) Structure of the mouthparts of *Frankliniella bispinosa* (Morgan) (Thysanoptera: Thripidae). In: Parker, B.L., Skinner, M. and Lewis, T. (eds) *Towards Understanding Thysanoptera.* Vermont Agricultural Experimental Station. General Technical Report NE 147, 71–94.

Chisholm, I.F. and Doncaster, C.C. (1982) Studying and recording the feeding behaviour of thrips. *Entomologia experimentalis et Applicata* 31, 324–327.

Chisholm, I.F. and Lewis, T. (1984) A new look at thrips (Thysanoptera) mouthparts, their action and effects of feeding on plant tissue. *Bulletin of Entomological Research* 74, 663– 675.

Crespi, B.J. (1986) Territoriality and fighting in a colonial thrips, *Hoplothrips pedicularius*, and sexual dimorphism in Thysanoptera. *Ecological Entomology* 11, 119–130.

Dallai, R., Del Bene, G. and Marchini, D. (1990) Fine structure of the hindgut and rectal pads of *Frankliniella occidentalis* (Pergande) (Thysanoptera: Thripidae). *Redia* 74, 29–44.

Dallai, R., Del Bene, G. and Marchini, D. (1991) The ultrastructure of malpighian tubules and hindgut of *Frankliniella occidentalis* (Pergande) (Thysanoptera: Thripidae). *International Journal of Insect Morphology and Embryology* 20, 223–233.

Davies, R.G. (1958) Observations on the morphology of the head and the mouthparts in the Thysanoptera. *Proceedings of the Royal Entomological Society of London* 33, 97–106.

Davies, R.G. (1961) The postembryonic development of the female reproductive system in *Limothrips cerealium* Haliday (Thys., Thripidae). *Proceedings of the Zoological Society of London* 136, 411–437.

Davies, R.G. (1969) The skeletal musculature and its metamorphosis in *Limothrips cerealium* Haliday (Thys., Thripidae). *Transactions of the Royal Entomological Society of London* 121, 167–233.

Del Bene, G., Dallai, R. and Marchini, D. (1991) Ultrastructure of the midgut and the adhering tubular salivary glands of *Frankliniella occidentalis* (Pergande) (Thysanoptera: Thripidae). *International Journal of Insect Morphology and Embryology* 20, 15–24.

Ellington, C.P. (1980) Wing mechanics and take-off preparation of thrips (Thysanoptera). *Journal of Experimental Biology* 85, 129–136.

Haga, K. (1985) Oogenesis and embryogenesis of the idolothripine thrips, *Bactrothrips brevitubus* (Thysanoptera, Phlaeothripidae). In: Ando, H. and Miya, K. (eds) *Recent Advances in Insect Embryology in Japan*. ISEBU Co. Ltd, Tsukuba, pp. 45–106.

Haga, K. and Matsuzaki, M. (1980) Ovarian structure and oogenesis in the idolothripine thrips, *Bactrothrips brevitubus* (Thysanoptera, Phlaeothripidae). *Abstracts of the International Congress of Entomology* 2p-2, 53.

Heming, B.S. (1970a) Postembryonic development of the female reproductive system in *Frankliniella fusca* (Thripidae) and *Haplothrips verbasci* (Phlaeothripidae) (Thysanoptera). *Miscellaneous Publications of the Entomological Society of America* 7, 197–234.

Heming, B.S. (1970b) Postembryonic development of the male reproductive system in *Frankliniella fusca* (Thripidae) and *Haplothrips verbasci* (Phlaeothripidae) (Thysanoptera). *Miscellaneous Publications of the Entomological Society of America* 7, 235–272.

Heming, B.S. (1971) Functional morphology of the thysanopteran pretarsus. *Canadian Journal of Zoology* 49, 91–108.

Heming, B.S. (1972) Functional morphology of the pretarsus in larval Thysanoptera. *Canadian Journal of Zoology* 50, 751–766.

Heming, B.S. (1973) Metamorphosis of the pretarsus in *Frankliniella fusca* (Hinds) (Thripidae) and *Haplothrips verbasci* (Osborn)(Phlaeothripidae)(Thysanoptera). *Canadian Journal of Zoology* 51, 1211–1234.

Heming, B.S. (1975) Antennal structure and metamorphosis in *Frankliniella fusca* (Hinds)(Thripidae) and *Haplothrips verbasci* (Osborn)(Phlaeothripidae)(Thysanoptera). *Quaestiones Entomologicae* 11, 25–68.

Heming, B.S. (1978) Structure and function of the mouthparts in larvae of *Haplothrips verbasci* (Osborn)(Thysanoptera, Tubulifera, Phlaeothripidae). *Journal of Morphology* 156, 1–38.

Heming, B.S. (1979) Origin and fate of germ cells in male and female embryos of *Haplothrips verbasci* (Osborn)(Insecta, Thysanoptera, Phlaeothripidae). *Journal of Morphology* 160, 323–344.

Heming, B.S. (1980) Development of the mouthparts in embryos of *Haplothrips verbasci* (Osborn). *Journal of Morphology* 164, 235–263.

Heming, B.S. (1985) *Thrips* (Thysanoptera) in Alberta. *Agriculture and Forestry Bulletin* 8, 19–24.

Heming, B.S. (1993a) *Structure, Function, Ontogeny, and Evolution of Feeding in Thrips (Thysanoptera).* Thomas Say Publications in Entomology: *Proceedings, Entomological Society of America,* Lanham, Maryland, pp. 3–41.

Heming, B.S. (1993b) Origin and fate of pleuropodia in embryos of *Neoheegeria verbasci* (Osborn) (Thysanoptera: Phlaeothripidae). *Journal of Pure and Applied Zoology* 4, 205–223.

Heming, B.S. (1995) History of germ line in male and female thrips. In: Parker, B.L, Skinner, M. and Lewis, T. (eds) *Thrips Biology and Management.* Plenum, New York, pp. 505–535.

Honomichl, K. (1982) Fine structure of a mandibular stretch receptor in *Thrips physapus* L. (Thysanoptera, Insecta). *Zoomorphology* 101, 177–186.

Hunter, W.B. and Ullman, D.E. (1989) Analysis of mouthpart movements during feeding of *Frankliniella occidentalis* (Pergande) and *F. schultzei* Trybom (Thysanoptera: Thripidae). *International Journal of Insect Morphology and Embryology* 18, 161–171.

Hunter, W.B. and Ullman, D.E. (1992) Anatomy and ultrastructure of the piercing-sucking mouthparts and paraglossal sensilla of *Frankliniella occidentalis* (Pergande) (Thysanoptera: Thripidae). *International Journal of Insect Morphology and Embryology* 21, 17–35.

Kitajima, E.W. (1975) A peculiar type of glycocalyx on the microvilli of the midgut epithelial cells of the thrips *Frankliniella* sp. (Thys.,Thripidae). *Cytobiologie* 11, 299–303.

Lewis, T. (1973) *Thrips, their Biology, Ecology and Economic Importance.* Academic Press, London.

Mickoleit, G. (1961) Zur Thoraxmorphologie der Thysanoptera. *Zoologische Jahrbücher, Anatomie* 79, 1–92.

Mickoleit, E. (1963) Untersuchungen zur Kopfmorphologie der Thysanopteren. *Zoologische Jahrbücher, Anatomie* 81, 101–150.

Moritz, G. (1982a) Zur Morphologie des Kopfinnenskeletts (Tentorium) bei den Thysanoptera. *Deutsche entomologische Zeitschrift* 29, 17–26.

Moritz, G. (1982b) Zur Morphologie und Anatomie des Fransenflüglers *Aeolothrips intermedius* Bagnall. 1. Mitteilung: Der Kopf. *Zoologische Jahrbücher, Anatomie* 107, 557–608.

Moritz, G. (1982c) Zur Morphologie und Anatomie des Fransenflüglers *Aeolothrips intermedius* Bagnall. 2. Mitteilung: Thorax. *Zoologische Jahrbücher, Anatomie* 108, 55–106.

Moritz, G. (1982d) Zur Morphologie und Anatomie des Fransenflüglers *Aeolothrips intermedius* Bagnall. 3. Mitteilung: Das Abdomen. *Zoologische Jahrbücher, Anatomie* 108, 293–340.

Moritz, G. (1984a) Zum Vorkommen einer exokrinen Vertexdrüse bei den Männchen der Gattung *Merothrips* Hood. 1914. (Merothripidae, Thysanoptera). *Zoologische Jahrbücher, Anatomie* 111, 1–13.

Moritz, G. (1984b) Differenzierungsformen der imaginalen Bauchganglienkette bei Thysanoptera (Insecta). *Entomologische Nachrichten Berlin* 28, 27–29.

Moritz, G. (1985) Mikromorphologische Chitinstrukturen als Plastronbildner bei terrestrisch lebenden Thysanopteren (Insecta). *Biologische Rundschau* 23, 161–170.

Moritz, G. (1988a) Die Ontogenese der Thysanoptera (Insecta) unter besonderer Berücksichtigung des Fransenflüglers *Hercinothrips femoralis* (O.M. Reuter, 1891) (Thys., Thripidae, Panchaetothripinae) 1. Mitt. Embryonalentwicklung. *Zoologische Jahrbücher, Anatomie* 117, 1–64.

Moritz, G. (1988b) Die Ontogenese der Thysanoptera (Insecta) unter besonderer Berücksichtigung des Fransenflüglers *Hercinothrips femoralis* (O.M. Reuter, 1891) (Thys., Thripidae, Panchaetothripinae) 2. Mitt. Erst- und Zweitlarve. *Zoologische Jahrbücher, Anatomie* 117, 299–351.

Moritz, G. (1988c) On the genesis of the wings and the wing musculature of *Hercinothrips femoralis* (O.M. Reuter, 1891)(Thysanoptera, Insecta). *Acta Phytopathologica et Entomologica Hungarica* 23, 313–319.

Moritz, G. (1989a) Die Ontogenese der Thysanoptera (Insecta) unter besonderer Berücksichtigung des Fransenflüglers *Hercinothrips femoralis* (O.M. Reuter, 1891)(Thys., Thripidae, Panchaetothripinae) 3. Mitt. Praepupa und Pupa. *Zoologische Jahrbücher, Anatomie* 118, 15–54.

Moritz, G. (1989b) Die Ontogenese der Thysanoptera (Insecta) unter besonderer Berücksichtigung des Fransenflüglers *Hercinothrips femoralis* (O.M. Reuter, 1891) (Thys., Thripidae, Panchaetothripinae) 4. Mitt. Imago Kopf. *Zoologische Jahrbücher, Anatomie* 118, 273–307.

Moritz, G. (1989c) Die Ontogenese der Thysanoptera (Insecta) unter besonderer Berücksichtigung des Fransenflüglers *Hercinothrips femoralis* (O.M. Reuter, 1891) (Thys., Thripidae, Panchaetothripinae) 5. Mitt. Imago Thorax. *Zoologische Jahrbücher, Anatomie* 118, 391–427.

Moritz, G. (1989d) Die Ontogenese der Thysanoptera (Insecta) unter besonderer

Berücksichtigung des Fransenflüglers *Hercinothrips femoralis* (O.M. Reuter, 1891) (Thys., Thripidae, Panchaetothripinae) 6. Mitt. Imago Abdomen. *Zoologische Jahrbücher, Anatomie* 119, 157–217.

Moritz, G. (1989e) Zur Morphogenese des Flugapparates von *Hercinothrips femoralis* (O. M. Reuter, 1891) (Thysanoptera, Insecta). *Verhandlungen IX. SIEEC Gotha 1986 (Dresden 1989)* 1, 374–378.

Moritz, G. (1995) Morphogenetic development of some species of the order Thysanoptera (Insecta). In: Parker, B.L., Skinner, M. and Lewis, T. (eds) *Thrips Biology and Management*, NATO ASI, Series A 276, Plenum, New York, pp. 489–504.

Mound, L. A. (1971) The feeding apparatus of thrips. *Bulletin of Entomological Research* 60, 547–548.

Mound, L.A. (1992) Patterns of sexuality in Thysanoptera. In: Cameron, E.A., Teulon, D.A., McCormick, L.H. and Kolb, T.E. (eds) *The 1991 Conference on Thrips (Thysanoptera): Insect and Disease Considerations in Sugar Maple Management.* GTR, NE 161, US Dept. Agriculture, Radnor, pp. 2–14.

Mound, L.A. and Heming, B.S. (1991) Thysanoptera (Thrips). In: *The Insects of Australia.* Melbourne University Press, pp. 510–515.

Mound, L.A. and O'Neill, K. (1974) Taxonomy of the Merothripidae, with ecological and phylogenetic considerations (Thysanoptera). *Journal of Natural History* 8, 481–509.

Mound, L.A. and Walker, A.K. (1982) Antennal aberration in Thysanoptera. *Entomologists Monthly Magazine* 118, 175–176.

Mound, L.A., Morison, G.D., Pitkin, B.R. and Palmer, J.M. (1976) *Thysanoptera Handbooks for the Identification of British Insects. Royal Entomological Society of London* Vol. 1, Part 11, pp. 1–79.

Mound, L.A., Heming, B.S. and Palmer, J.M. (1980) Phylogenetic relationships between the families of recent Thysanoptera (Insecta). *Zoological Journal of the Linnean Society* 69, 111–141.

Nakahara, S. (1988) Preliminary study of facetal pigmentation in the compound eyes of *Terebrantia* (Thysanoptera). *Acta Phytopathologica et Entomologica Hungarica* 23, 321–329.

Nüesch, H. (1987) Metamorphose bei Insekten. Direkte und indirekte Entwicklung bei Apterygoten und Exopterygoten. *Zoologische Jahrbücher, Anatomie* 115, 453–487.

Peterson, A. (1915) Morphological studies on the head and mouth-parts of the Thysanoptera. *Annals of the Entomological Society of America* 8, 20–59.

Pitkin, B. R. (1978) A revision of the Australian species of *Anaphothrips* Uzel (Thysanoptera: Thripidae). *Australian Journal of Zoology* 26, 349–371.

Priesner, H. (1968) Thysanoptera. In: Kükenthal: *Handbuch der Zoologie* 4(2)2/19, 1–32.

Pritsch, M. and Büning, J. (1989) Germ cell cluster in the panoistic ovary of Thysanoptera (Insecta). *Zoomorphology* 108, 309–313.

Reger, J.F. (1971) Fine structure of the surface coat of midgut epithelial cells in the Homopteran, *Phylloscelis atra* (Fulgorid.). *Journal of Submicroscopic Cytology* 3, 353–358.

Richards, A.G. and Richards, P.A. (1977) The peritrophic membranes of insects. *Annual Review of Entomology* 22, 219–240.

Risler, H. (1957) Der Kopf von *Thrips physapus* L. (Thysanoptera, Terebrantia). *Zoologische Jahrbücher, Anatomie* 76, 251–302.

Ryerse, J.S., Purcell, J.P., Sammons, R.D. and Lavrik, P.B. (1992) Peritrophic membrane structure and formation in the larva of a moth, *Heliothis. Tissue and Cell* 24, 751–771.

Sharga, U.S. (1933) On the internal anatomy of some Thysanoptera. *Transactions of the Royal Entomological Society of London* 81, 185–240.

Slifer, E.H. and Sekhon, S.S. (1974) Sense organs on the antennae of two species of *Thrips* (Thysanoptera, Insecta). *Journal of Morphology* 143, 445–455.

Stannard, L.J. (1968) The thrips, or Thysanoptera, of Illinois. *Bulletin of the Illinois Natural History Survey* 29, 215–552.

Staub, R.F. (1979) Die postembryonale Entwicklung des Nervensystems von *Parthenothrips dracaenae* Heeger (Thysanoptera, Terebrantia). *Revue suisse de Zoologie* 86, 367–394.

Stouthamer, R. and Werren, J.H. (1993) Microbes associated with parthenogenesis in wasps of the genus *Trichogramma*. *Journal of Invertebrate Pathology* 61, 6–9.

Stouthamer, R., Breeuwer, J.A.J., Luck, R.F. and Werren, J.H. (1993) Molecular identification of microorganisms associated with parthenogenesis. *Nature, London* 361, 66–67

Tsutsumi, T. (1996) Formation of the egg membranes of an idolothripine thrips, *Bactrothrips brevitubus* (Insecta: Thysanoptera). *Proceedings of the Arthropod Embryological Society of Japan* 31, 9–13.

Tsutsumi, T., Machida, R. and Haga, K. (1993) How can the panoism in the Thysanoptera be understood? *Proceedings of the Arthropod Embryological Society of Japan* 28, 9–12.

Tsutsumi, T., Matsuzaki, M. and Haga, K. (1994) New aspect of the 'Mycetome' of a thrips, *Bactrothrips brevitubus* Takahashi (Insecta: Thysanoptera). *Morphology* 221, 235–242.

Ullman, D.E., Westcot, D.M., Hunter, W.B. and Mau, R.F.L. (1989) Internal anatomy and morphology of *Frankliniella occidentalis* (Pergande) (Thysanoptera, Thripidae) with special reference to interactions between thrips and tomato spotted wilt virus. *International Journal of Insect Morphology and Embryology* 18, 289–310.

Ullman, D.E., Cho, J.J., Mau, R.F.L., Hunter, W.B., Westcot, D.M. and Custer, D.M. (1992) Thrips–tomato spotted wilt virus interactions: morphological, behavioral and cellular components influencing thrips transmission. *Advances in Disease Vector Research* 9, 195–240.

Wiesenborn, W.D. and Morse, J.G. (1988) The mandible and maxillary stylets of *Scirtothrips citri* (Moulton) (Thysanoptera: Thripidae). *Pan-Pacific Entomologist* 64, 39–42.

Zawirska, I. (1976) Untersuchungen über zwei biologische Typen von *Thrips tabaci* Lind. (Thysanoptera, Thripidae) in der VR Polen. *Archiv Phytopathologie Pflanzensch* 12, 411–422.

Zur Strassen, R. (1973) Fossile Fransenflügler aus mesozoischen Bernstein des Libanon (Insecta: Thysanoptera). *Stuttgarter Beitrage zur Naturkkunde, Series A*, 256, 1–51.

3

Host Selection, Communication and Reproductive Behaviour

L. Irene Terry
University of Utah, Department of Biology, 201 South Biology Building, Salt Lake City, Utah 84112, USA

Introduction

Thrips exploit a diversity of hosts with many complex host utilization patterns. Some species are strictly sporophagous, mycophagous, or herbivorous, and others omnivorous or primarily predatory. Some, that live in long-term habitats with overlapping generations, have evolved subsocial and eusocial behaviour, whereas others only aggregate briefly for mating. Host detection and selection processes, mating behaviour, and choice of oviposition site are closely tied to their hosts, i.e. plants and foliage for herbivores, tree bark and fungi for mycophagous species and prey or prey hosts for predatory species. Such a diversity of hosts has generated a broad range of cues used for searching and detecting them. This chapter concentrates on the mechanisms whereby phytophagous species of Terebrantia find and choose their hosts for feeding, mating and oviposition; information on a few tubuliferan pests is also included. The subsequent growth and dynamics of populations on their chosen hosts, especially crops, is covered in Chapter 7.

Host Plant Detection and Selection

Host selection generally follows several sequential or simultaneous decision processes. These comprise initially long- and then short-range cues that lead to detection and landing on the host, which is then tested for the appropriate mating, feeding and/or oviposition cues to see whether it is acceptable. Larval Thysanoptera cannot move far to locate other hosts so host selection is primarily made by their mothers. The cues used by thrips can be directly assessed by examining various potential cues and determining landing rates due to these (e.g. volatiles, colour), and also by examining landing rates on hosts themselves. Other cues may be indirectly assessed by determining the number of adults present on various host species or cultivars, which is a combined measure of those landing and remaining on the host. If numbers of immatures are assessed, this provides a measure of the acceptability of the plant as a reproductive host.

Like other insects, thrips locate hosts using colour, shape, size and volatiles associated with them. Cues for detection of hosts may be general for polyphagous species or very specific for more monophagous species. Mating, feeding and oviposition may occur on the same host so cues used for detection of feeding sites may also serve for detection of hosts for reproduction. Once a potential host is contacted then odour, tactile and gustatory cues may predominate as the animal tests its acceptability for feeding or oviposition. Polyphagous and some anthophilous species use floral and vegetative parts of the host for both feeding and reproduction. Others use only the floral parts of some hosts (e.g. *Heterothrips azaleae* on azalea) or may feed and aggregate on a particular plant but not breed there. Within a host plant species or cultivar, some traits may make the plant undesirable or resistant to thrips. Many host plant resistance studies have identified those traits that promote or inhibit the herbivore from finding, colonizing, and accepting the host (antixenosis), or traits that either promote or inhibit herbivore development or reproduction on the host (antibiosis).

The frequency with which thrips disperse to new hosts varies between species and their type of host. Some thrips live on the same host for generations, others move from host to host during the day. Some thrips eclose within or near the host, especially the more host-specialized species or brachypterous or apterous species. These latter categories walk to new hosts or host clones so short-range cues are used. The ability of anthophilous species, whose floral hosts are ephemeral, to detect subtle changes in host quality (e.g. floral age, pollen quality) and then to disperse and find new hosts or host plant parts may be the key to their fitness (Murai and Ishii, 1982; Kirk, 1985a; Trichilo and Leigh, 1988; Teulon and Penman, 1991). Individuals may move to different hosts several times during their larval development or throughout their adult life. For some species this dispersal may occur daily or throughout the day both within plants, such as *Thrips tabaci* on onions (Sites *et al.*, 1992) and *Kakothrips pisivorus* on *Vicia faba* (Kirk, 1985a), and among hosts, such as *Frankliniella occidentalis* in cotton (Matteson and Terry, 1992). Kirk (1985a) suggested the term 'index of nomadism' to describe the ratio between the duration of a thrips stage to the lifespan of the individual plant or plant part on which it was feeding. This indicates the approximate number of host plants the insect must find during its development or adult lifetime and is useful for comparing life history strategies. For example, essential flower parts die before *Kakothrips pisivorus* larvae complete their larval development necessitating movement between senescing and newly opening flower buds (Kirk, 1985a). An index of three for *K. pisivorus* larvae on field beans (i.e. 24 days for larval development and 8 days for flower longevity) was calculated. By contrast, gall-producing and spore-feeding species of the Tubulifera may live colonially with overlapping generations, and only require to select new hosts after several generations.

Visual ecological cues

Visual cues include colour, shape, architecture, size, contrasts with background components (water, soil, rocks, etc.) and angle of orientation of the host relative to the perceiving organism (Prokopy and Owen, 1983). Characteristics associated with colour appear to be of primary importance for many thrips species.

Colour

Colour and colour contrast are used by insects to distinguish between a host and the surrounding environment. Colour alone can attract some thrips species, although some improved trapping of thrips is obtained by adding plant volatiles to coloured surfaces (see Chapter 11). The components of colour that are critical for discriminating between hosts and non-hosts are hue (the dominant wavelength remitted by the surface, λ_{max}), saturation (purity of the hue) and brightness (total energy, percentage reflectance at peak wavelengths). To evaluate these characteristics, and to determine which qualities are most attractive to thrips, many researchers have tested painted surfaces of traps (water traps, sticky cards or cylinders) or filtered surfaces and evaluated the number trapped on different colours. Such data only indicate those characters that affect thrips alighting on a host and do not give information on the residence time or subsequent leaving rates. Other studies have compared the number of adults found on various flower colours within a host species.

In a precedent-setting study, Kirk (1984a) established an ecological correlate for thrips and their response to hues in terms of the types of host they utilize. Grass feeders showed little preference for any hue (UV white, non-UV white, black, blue-green, yellow, red, or black). By contrast, all anthophilous thrips were trapped by colours that match flowers, i.e. low-UV white, blue and yellow, and few were trapped by green, red and black. All previous and subsequent studies examining colour confirm this generalization (Table 3.1). In addition, the difference between hues that attract the maximum and minimum numbers of anthophilous species are dramatically higher (4 to 615 times) than those of grass-feeding ones (around 2 to 3 times). Grass feeders may utilize other cues for detecting and landing on hosts. Of the species examined, namely *Frankliniella tenuicornis*, *Aeolothrips intermedius*, *Thrips major* and *Thrips pillichi* (Kirk, 1984a), *Frankliniella occidentalis* (Gillespie and Vernon, 1990; Vernon and Gillespie, 1990a,b; Matteson and Terry, 1992), *Ceratothrips frici* and *Thrips obscuratus* (Teulon and Penman, 1992), males and females had similar responses to colour.

Among attractive hues, brighter ones captured more flower thrips than darker ones (Brødsgaard, 1989; Vernon and Gillespie, 1990a,b; Gomez and Hernandez, 1992; Matteson and Terry, 1992; Chiu and Wu, 1993). Matteson and Terry (1992) found that the degree of a colour's attractiveness to *Frankliniella occidentalis* corresponded to the brightness in the blue wavelengths. Similar relationships were found by Vernon and Gillespie (1990b). The attractiveness of blues, violets, yellows and whites to *F. occidentalis* was reduced either by decreasing the brightness at these wavelengths with the addition of black paint or by desaturating the colour with the addition of non-

Table 3.1. Thrips response to hue.

Species	Test site	Relative attractiveness rank of hue[A]							Max/min	Reference
		White + UV	White – UV	Blue	Green	Yellow	Red	Black		
<u>Grass feeders</u>										
Chirothrips manicatus	Wheat	NT	1a	NT	1b	NT	NT	1c	1	Lewis (1959)[C]
Frankliniella tenuicornis	Grass	1b	2a	1a	1a	2a	2b	2c	<2	Kirk (1984a)
Limothrips cerealium	Grass	1a	1b	1e	1g	1d	1c	1f	<2	Kirk (1984a)
"	Field	NT	1c	1d[B]	1a	1d	1e	1b	<2	Teulon and Penman (1992)[B]
"	Wheat	NT	1c	NT	1a	NT	NT	1b	<2	Lewis (1959)[C]
Limothrips denticornis	Grass	1a	1b	2a	1c	1d	2c	2b	<3	Kirk (1984a)
"	Crops	NT	1	1	1	1	2	NT	2	Czencz (1987)[C]
Stenothrips graminum	Crops	NT	1	1	1	1	2	NT	<3	Czencz (1987)[C]
<u>Predators</u>										
Aeolothrips intermedius	Grass	2	1b	1c	3	1a	3	3	32	Kirk (1984a)
"	Crops	NT	1b	1c	2	1a	2	NT	30	Czencz (1987)[C]
<u>Polyphagous foliage, flower and fruit feeders</u>										
Frankliniella fusca	Tomato	1	NT	1[B]	NT	1	NT	NT	<2	Cho *et al.* (1995)[B]
Scirtothrips citri	Orange	2c	NT	2b[B]	2a	1	2d	2e	72	Moreno *et al.* (1984)[B]
"	Orange	NT	1	3b	3a	2	3c	4	256	Beavers *et al.* (1971)[C]
Scirtothrips dorsalis	Orange	NT[E]	2b	2–3	1–2	1	2–3	3	14	Tsuchiya *et al.* (1995c)[E]
Neohydatothrips variabilis	Rye	2b	2b	2c[B]	2a	1	2e	2d	20	Walker (1974)[B]
Thrips atratus	Crops	NT	2	1	3	3	3	NT	10	Czencz (1987)[C]
Taeniothrips inconsequens	Sugar maple	NT	1b	2[B]	2	1a	1c	NT	10	Coli *et al.* (1992)[B]

Thrips simplex	Rye	3e	3d	3c[B]	3a	2–3	2a[D]	3b	10	Walker (1974)[BD]
Thrips angusticeps	Crops	NT	1b	1a	2	1c	3	NT	25	Czencz (1987)[C]
Thrips tabaci	Grass	3	1	2a	4	2b	5	5	32	Kirk (1984a)
	Field	NT	2	3a[B]	3b	1	4b	4a	152	Teulon and Penman (1992)[B]
	Crops	NT	1a	1c	2	1b	3	NT	52	Czencz (1987)[C]
	Onion	NT	2	1	3	3	4	5	17	Lu (1990)
<u>Anthophilous species</u>										
Ceratothrips frici	Field	NT	2	2[B]	2	1	2	2	7	Teulon and Penman (1992)[B]
Frankliniella bispinosa	Citrus	NT	2	1	NT	3	4	NT	53	Childers and Brecht (1996)
Frankliniella intonsa	Crops	NT	2	3	5	1	4	NT	100	Czencz (1987)[C]
Frankliniella occidentalis	Glasshouse	4	2	1a	3	1b	NT	5	200	Gillespie and Vernon (1990)
"	Cotton	5	2	1a	4	3–4	5	NT	14	Matteson and Terry (1992)
"	Lettuce	NT	1b	1a	2a	1c	2b	NT	5	Yudin *et al.* (1987)[E]
"	Orange	NT[E]	2[E]	1	3d	3a	3b	3c	7	Tsuchiya *et al.* (1995b)[E]
"	Glasshouse	NT[E]	2[E]	1a	3	1b	NT	NT	4	Robb (1989)[E]
"	Tomato	2	NT	1[B]	NT	1	NT	NT	5	Cho *et al.* (1995)[E]
"	Glasshouse	NT	2b	1	NT	2a	NT	NT	12	Brødsgaard (1989)
"	Orange	NT	1	NT	NT	2	NT	NT	6	Moreno *et al.* (1984)
"	Pear	NT	1	NT	NT	2	NT	NT	11	Moffitt (1964)[C]
Frankliniella tritici	Rye	3b	1	2b[B]	3a	2a	3c	3d	14	Walker (1974)
"	Tomato	2–3	NT	1–2[B]	NT	1a	NT	NT	6	Cho *et al.* (1995)[B]
Megalurothrips usitatus	Adzuki bean	?[E]	1–2[E]	1a	2–3	2	NT	NT	9	Chang (1990)[E]
Thrips atratus	Grass	4	2	1	3	3	3	4	>5	Kirk (1984a)
Thrips hawaiiensis	Citrus	NT	2b	1	2–3	2a	3	NT	615	Chiu and Wu (1993)[E]
Thrips imaginis	Garden	2d	1a	1b	2a	1c	2c	2b	7	Kirk (1984b)

Table 3.1. Continued.

		Relative attractiveness rank of hue[A]								
Species	Test site	White + UV	White – UV	Blue	Green	Yellow	Red	Black	Max/min	Reference
Thrips linarius	Crops	NT	1	2	4	3	5	NT	14	Czencz (1987)[C]
Thrips major	Grass	3c	1a	1b	3a	2	3a	3b	20	Kirk (1984a)
"	Crops	NT	1	2	4	2	3	NT	7	Czencz (1987)[C]
Thrips obscuratus	Field	NT	1	2[B]	2	2	2	2	29	Teulon and Penman (1992)[B]
Thrips physapus	Crops	NT	2	1	5	3	4	NT	17	Czencz (1987)[C]
Thrips pillichi	Grass	5	2	3	4c	1	4a	4b	14	Kirk (1984a)
Thrips vulgatissimus	Grass	5	3	1	4	2	5	6	21	Kirk (1984a)

[A] Ranking within a row, hues with the same number are not significantly different; 1 = colours with highest thrips numbers; lower case letters within a number rank indicate non-significant differences of mean ranks; a = highest mean; some tests include additional hues (e.g. orange, purple, violet) not presented here. [B] Blue reflectance at λ_{max} between 450 and 480 nm ≤ 30 per cent. [C] No reflectance data given. [D] An orange hue trapped most thrips in this test. [E] UV reflectance not reported; the tests reported by Robb (1989) and Cho *et al.* (1995) indicate that there is UV reflectance of the white card used. NT, hue not tested.

UV white. No effect was observed by darkening green, but all green traps tested had less brightness in green-yellow wavelengths (less than 60 per cent at 530 to 570 nm) than the attractive white or yellow standards (more than 80 per cent). However, desaturation of green, which increased the reflectance in both the blue (more than 40 per cent at 440 to 470 nm) and yellow (more than 80 per cent at 550 nm) range, increased thrips capture compared with non-UV white. Highly reflective blues attracted more than six times as many thrips as the most attractive white. Vernon and Gillespie (1990b) concluded that blue and yellow attraction appeared to be based on separate responses or sensitivities.

Some of the low response to blue and violet traps by anthophilous and polyphagous thrips reported in Table 3.1 could be due to the low brightness in the blue and violet range (for example see Walker, 1974; Moreno *et al.*, 1984; Yudin *et al.*, 1987; Coli *et al.*, 1992; Teulon and Penman, 1992). Also, some violet and purple colours tested additionally reflected in the red range, which could reduce the response to these hues.

For hues that are attractive to anthophilous and polyphagous thrips, the relative reflectance of UV wavelengths (350 to 390 nm) is also important in determining whether or not these thrips alight on a host. If UV reflectance is very high, then anthophilous thrips are repelled from the surface of attractive colours, whereas grass-feeding thrips are not affected by the UV reflectance. Low reflectance in UV (less than 35 per cent) did not inhibit trap catch of *Frankliniella occidentalis* in blue, yellow or white painted traps (Matteson and Terry, 1992). However, white traps that reflected 50 to 70 per cent in the UV wavelengths trapped almost no thrips. Further, Vernon and Gillespie (1990b) demonstrated that brighter attractive visible hues could reduce somewhat the repellent effects of UV reflectance. Walker (1974) used cut-off filters and painted surfaces to determine the effect of the brightness of specific UV wavelengths on the response of *F. tritici* to attractive hues. Filters that eliminated wavelengths shorter than 370 nm or blue-green transmitting filters placed over a highly reflective UV lead-white trap, increased the capture of thrips dramatically (up to 38 times in the UV lead-white trap). These results indicate that a highly reflective UV surface is repellent, and more specifically, those UV wavelengths between 350 and 370 nm.

Due to the repellent effect of UV on polyphagous and flower thrips, some success has been achieved in reducing the thrips infestation, disease or damage in crops mulched with UV-reflective sheets (Table 3.2). The methods were most effective when the ground was well covered with the UV-reflective surface relative to leaf area, as in low density plantings or early in the season. Also, the type of transmission of glasshouse cover affected the movement, distribution and build up of *Thrips palmi* (Kawai *et al.*, 1985; Kawai, 1986a). There was more emigration from glasshouse plots covered with UVA-absorbing plastic compared with UV-transmitting covers. Similarly, fewer *Frankliniella occidentalis* were trapped in plants grown in walk-in tunnels covered with UV-absorbing vinyl than those covered with UV-transmitting vinyl (Antignus *et al.*, 1996), and aluminium surfaces near entrances and vents reduced thrips entry into greenhouses (McIntyre *et al.*, 1996).

Colour contrasts between background and trap colour and border colour

Table 3.2. Impact of UV-reflective mulching and UV-absorbing vinyl on thrips in agricultural settings.

Thrips species	UV source	Crop/site	Decrease or increase[A]	Reference
Frankliniella tritici	Aluminium foil	Roses, outdoor	↓ thrips, ↓ damage	Ota and Smith (1968)
Thrips simplex	Aluminium foil mulch	Gladioli flowers, field	↓ damage to saleable flowers	Smith *et al.* (1972)
Thrips spp.	Aluminium coloured plastic	Tomato, pepper, field	↓ thrips	Greenough *et al.* (1990)
Thrips palmi	UVA absorbing vinyl film	Cucumber, eggplant, sweet pepper, glasshouse cover	↑ emigration from UVA area, ↓ thrips UVA	Kawai (1986a)
Frankliniella fusca, Frankliniella tritici, Neohydatothrips variabilis	Aluminium painted plastic	Tomato, field	↓ thrips early in season	Scott *et al.* (1989)
Thrips tabaci	Silver mulch	Onion, field	↓ thrips in low density planting	Lu (1990)
Thrips palmi	Silver-coloured polyethylene, 54–61% ground covered	Chrysanthemum, greenhouses	↓ thrips first month after transplant	Ohno *et al.* (1990)
Frankliniella occidentalis, Frankliniella tritici, Frankliniella fusca	Aluminium painted plastic	Tomato, field	↓ thrips early, no difference in damage	Brown and Brown (1992)
Thrips (no species given)	Aluminium painted plastic mulch	Pepper, tomato, field	↓ thrips trapped, ↓ thrips on leaves	Kring and Schuster (1992)
Scirtothrips dorsalis	Reflective sheet (> 84% at 290–800 nm)	Mandarin, trees	↓ thrips, ↓ damage	Tsuchiya *et al.* (1995a)
Thrips palmi	UV absorbing film	Vegetables, glasshouse cover	↓ thrips	Nonaka and Nagai (1983)
"	Silver polyethylene film	Cucumber, field	↓ thrips	Makino (1984)
"	Silver polyethylene film	Cucumber, field	↓ thrips	Suzuki and Miyara (1984)
"	Silver film	Cucumber, field	↓ thrips	Nagai (1981)
Thrips parvispinus	Silver plastic	Hot peppers, field	↓ thrips and disease delay	Vos *et al.* (1995)
Frankliniella occidentalis	UV absorbing polyethylene	Cucumber, walk-in tunnels	↓ thrips	Antignus *et al.* (1996)

[A] Effects compared with either no mulch, black mulch or with non-UV absorbing glasshouse cover.

affect thrips' response to hues. With *Frankliniella tritici*, highly reflective UV borders reduced the catch of a low-UV white centre (Walker, 1974). Czencz (1987) showed that background colour of a crop, due to growth stage and floral colour, affected the ability of a particular hue to trap thrips. Generally, background colours of the same hue as the trap reduced the trap catch of many species. A yellow background enhanced trap catches of most hues, and green backgrounds weakened the effect of white and blue traps. Similarly, Vernon and Gillespie (1995) found that yellow and violet traps placed against backgrounds of the same colour caught fewer *Frankliniella occidentalis* than did yellow or violet traps on contrasting backgrounds. Both violet and blue backgrounds more than doubled the catch on yellow traps relative to green and yellow backgrounds, and yellow backgrounds slightly enhanced the catch of violet traps compared with green, blue or violet backgrounds.

Based on the behavioural response of *Frankliniella occidentalis* to UV, blue and yellow traps, Vernon and Gillespie (1990b) suggested that flower thrips have a photosystem, similar to bees, with three photoreceptors (i.e. photopigments) tuned to 350 to 360 nm in the UV, 440 to 450 nm in the blue, and 540 to 570 nm in the yellow wavelengths. However, slightly different information was obtained from electroretinograms (ERG) of male and female *F. occidentalis*, which established spectral efficiency curves (Fig. 11.6, Matteson *et al.*, 1992). Both sexes have a similar pattern of efficiency, with one peak of efficiency in the UV tested at 365 nm and another in the green-yellow region around 540 nm (Fig. 3.1). This evidence with the behavioural data suggest that there are only two types of photopigments residing in two types of photoreceptors, one sensitive to UV and one to green-yellow wavelengths. There is no physiological evidence for a third photopigment sensitive to blue wavelengths. Matteson *et al.* (1992) suggested that thrips use peak sensitivity to green-yellow in long-range orientation to plants and then use contrasts within plants to find flowers. Discrimination of blue spectral hues between UV and green-yellow physiological peaks is possible when both receptors are simultaneously stimulated. The physiological and behavioural data also suggest that *Frankliniella occidentalis* is red colour blind like honey bees (Kuhn, 1927).

No physiological data exist for cereal thrips or any other species. The visual pigment chromophore, retinal, has been isolated from the eyes of *Thrips* spp. (collected from both roses and magnolia), although the authors did not indicate the species nor whether thrips were collected from blossoms or foliage (Smith and Goldsmith, 1990). Numbers of cereal thrips, *Limothrips cerealium* (which show no preference for hue), trapped by colour traps were less than 5 per cent of those caught by suction traps (4815 individuals), whereas five times more *Taeniothrips* spp., a polyphagous genus, were caught on white traps than in a suction trap (1045 individuals) (Lewis, 1959). Such a difference in numbers responding to colour relative to the aerial sample suggests either that grass thrips use other host cues to find hosts or that other cues are stronger than colour.

These physiological and behavioural findings are consistent with the sites occupied by polyphagous and anthophilous thrips. The hues of many insect-pollinated flowers match those of the attractive trap colours, and flower

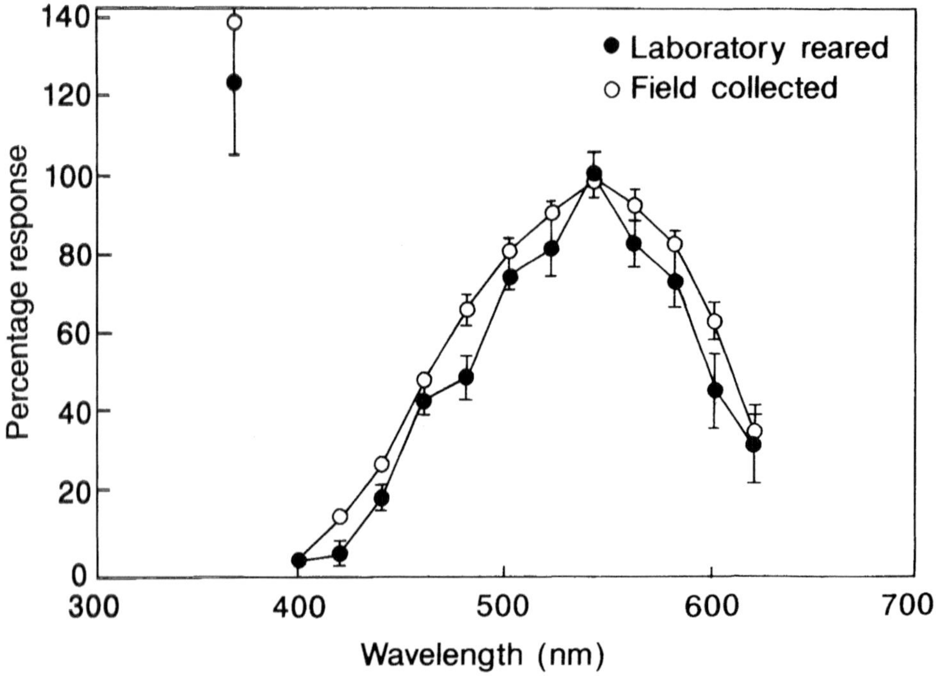

Fig. 3.1. Spectral efficiency curves (± 95 per cent CL) of laboratory-reared and wild-caught female *Frankliniella occidentalis* (after Matteson *et al.*, 1992).

colours contrast with background colours to attract their pollinators. Contrasts of flowers with sand, rock, limestone, water and leaves which may have high UV reflectance are best achieved by flowers that are white without UV, blue and yellow (Menzel, 1990). Many day-blooming white flowers and flowers of temperate and desert regions in New and Old World have these reflectance patterns with low UV emission (Menzel, 1990). Even within flowers, there are contrasting hues. Some allogamous flowers have moderate UV-reflecting parts (Mulligan and Kevan, 1973), which may contrast with the non-UV-reflecting floral areas to guide pollinators to floral reproductive parts.

Many researchers have noted that flower thrips are commonly found on blue, violet, white and yellow flowers. More *Frankliniella occidentalis* were found on bright blue *Saintpaulia ionatha* than on darker blue ones (Brødsgaard, 1989). Of the 16 cultivars each of rose and carnation examined in a greenhouse by Robb (1989), there was a high correlation between the number of *Frankliniella occidentalis* adults found per flower and the brightness of floral reflectance at 475 nm (blue-green). Correlations were high between the adult and larval counts indicating that attractive rose and carnation cultivars were also good reproductive hosts. In Hawaii, Hata *et al.* (1991) found that *Dendrobium* cultivars with pigmented flowers usually had more *Thrips palmi* present inside the flowers than non-pigmented ones. Adults and larvae of

Heterothrips azaleae, a specialist thrips on flowers of North American deciduous azaleas, and *Frankliniella tritici*, were found on all species of azalea except *Rhododendron prunifolium*, which is deep orange-red (Braman and Beshear, 1994). All other perennial azaleas have yellow, pink, white or yellow-orange flowers. Kirk (1984b) found that adult *Thrips imaginis* were found in low-UV white, blue and yellow flowers of a number of species including jacaranda, *Medicago sativa* and *Echium plantagineum*. Flowers with the least thrips were red. Several plants with pink-blue flowers including lucerne and jacaranda had numerous adults but very few larvae, suggesting a difference between floral attractiveness and its appropriateness for thrips reproduction. In a survey of 25 rose cultivars, all the red, orange and copper flowered roses and some pink ones had fewer adults and larvae than yellow or blue to purple ones (Gaum *et al.*, 1994). Based on floral and plant trait comparisons that affect *Kakothrips pisivorus* on *Vicia faba*, Stoddard (1986) suggested that flower colour is the primary pre-alighting factor and that other traits affect post-alighting host acceptance. White flowered and the more lush lines of faba bean had greater infestations of thrips than those with dark diffuse-pigmented flowers or more restricted growth patterns; but long-podded lines had greater larval and egg populations, which may reflect more accessible flowers due to the greater number of ovules (Stoddard, 1986). On gladiolus flowers, most *Frankliniella occidentalis* were found on lilac and violet coloured blooms with fewest on red and orange ones (Chyzik *et al.*, 1995). Both *Thrips hawaiiensis* and *Haplothrips tenuipennis* in *Lantana camara* are invariably found in yellow flowers within floral clusters that include magenta, scarlet, orange and low reflectance pink flowers (Mathur and Mohan Ram, 1978, 1986).

Exceptionally, *Frankliniella schultzei* was found on a red flowered ornamental species, *Malvaviscus arboreus*, in Australia, and adults and larvae developed well on diets of pollen or petals of this plant (Milne *et al.*, 1996). Adult and larval *F. schultzei* were present in flowers but not on leaves, though they were evenly distributed among different floral parts. In the laboratory, when female thrips were given a choice of pollen, leaf disc or petal disc, there was no significant difference in the plant part first encountered, suggesting no specific orientation to a particular organ. However, over the entire observation period, more thrips encountered petals than leaves or pollen.

The importance of flowers in host selection by anthophilous thrips has been well established. Many population censuses of polyphagous and anthophilous thrips have demonstrated that adults or larvae are found almost exclusively on blooms of particular hosts, for example, *Heterothrips azaleae* on wild azalea species (Braman and Beshear, 1994), *Frankliniella occidentalis* on apple (DeGrandi-Hoffman *et al.*, 1988; Terry and DeGrandi-Hoffman, 1988) and on pistachios (Frank and Huber, 1987); or that populations are much higher and peak during the flowering period, for example, *Thrips hawaiiensis* on bananas (Tsai *et al.*, 1992), *Frankliniella bispinosa* and *F. kelliae* on citrus in Florida (Childers *et al.*, 1990; Childers and Beshear, 1992), *T. hawaiiensis*, *Megalurothrips usitatus* and *F. intonsa* on citrus in Taiwan (Chiu *et al.*, 1991), *Frankliniella occidentalis*, *F. tritici* and *F. fusca* on tomato (Salguero Navas *et al.*, 1991; Kumar *et al.*, 1995a).

Comparing phenological stages of hosts, Yudin *et al.* (1988) clearly demonstrated the importance of flowers in attracting thrips. All hosts of *Frankliniella occidentalis* tested had at least twice as many thrips per plant on flowering than on their non-flowering counterparts (both preflowering hosts and ones of the same age with flowers removed). The least preferred preflowering host, *Verbesina enceliodes*, was the most densely populated flowering host, supporting 60 times more thrips after flowering than before. This species has large yellow flowers and their removal drastically reduced the number of thrips. Similar results with *F. occidentalis* were found between flowering and non-flowering weed species (Bautista and Mau, 1994) and chrysanthemum (de Jager *et al.*, 1993).

Kumar *et al.* (1995a) showed that landing rates of *Frankliniella occidentalis* on foliage of all but one of 13 *Lycopersicon* accessions tested were much lower than landing rates on flowers. Landing rates on leaves were similarly low on all cultivars including resistant ones, but landing rates on flowers did differ between cultivars. By contrast, there were differences in *Thrips tabaci* landing rates on leaves of different accessions. *Thrips tabaci* discriminated between host vegetation more than *F. occidentalis*.

In addition to attracting thrips to potential food, flowers have also been shown to be good meeting places for the opposite sex of some flower thrips (see Chapter 3, Mating behaviour, territoriality and fighting).

Host shape and size

In addition to floral colour, other morphological features appear to encourage adults to land or lay eggs. For example, several growth characteristics of *Vicia faba* cultivars affect *Kakothrips pisivorus* (Stoddard, 1986). Those with lush growth had larger infestations than short cultivars or those with terminal inflorescences that had brief blooming cycles. Irrigation in the spring increased infestation probably because of the greater lushness of irrigated plants, but irrigation during winter had no effect. Long-podded cultivars had larger infestations, possibly because the flowers were more accessible and had more ovules.

Differences in flower morphology are associated with the number of *Thrips palmi* found in *Dendrobium* orchids (Hata *et al.*, 1991). More larvae were present in flowers of *Phalaenanthe* morphotype cultivars than in the semi-phalaenopsis, *Ceratobium*, or wild bamboo orchids, even those of similar colour. Although most of the *Phalaenanthe* types tested had purple to lavender-white flowers, the deep purple *Ceratobium* morphotype contained up to ten times fewer larvae than deep purple *Phalaenanthe* types. Unfortunately, neither the relative surface areas of each morphotype nor any other characteristic of the flowers were reported.

Among chrysanthemum cultivars, spider flower types had fewer thrips per flower than single, decorative/pompom, or anemone types (Broadbent and Allen, 1995; de Jager *et al.* 1995b). Spider flower types may be less preferred because they have only ray florets which may be more difficult to feed on.

Thrips responses to colour, and to a lesser extent size and shape, have been exploited in trap designs for sampling field populations (see Chapter 11).

Host odours

Vision and olfaction are the primary cues used by insects to orient to their plant and floral hosts and sometimes the two cues work in concert (Dobson, 1994). Orientation to flowers can be driven by odour alone or olfactory cues may lower the threshold for visual orientation (Pellmyr and Thien, 1986). After alighting on a host flower, thrips may be further guided to a particular floral region or plant part by close range visual or olfactory cues.

Floral volatiles

Over 700 compounds have been identified from floral volatiles (Knudsen *et al.*, 1993). Floral fragrances are generally complex blends of isoprenoids, fatty acid derivatives, benzenoids, or aminoid compounds each of which have many different functional groups. Common isoprenoids found in floral fragrances of many species are monoterpenes such as sesquiterpenes and benzenoids. Examples include geraniol, linalool, germacrene, benzaldehyde and eugenol. Other compounds such as fatty acids, nitrogenous compounds, or sulphur-containing compounds are less universal in floral volatiles, but they may be important components in certain plant taxa. Many minor components may also be important in attracting specific insects (Dobson, 1994). Composition and concentration of floral volatiles are affected by many temporal and spatial factors; concentrations are important because optima may be critical for attraction.

There is both circumstantial and direct evidence that thrips use floral fragrances and other plant odours for host location, even in the absence of colour. Many plants with floral scents appear to attract more flower thrips than odourless ones (Annand, 1926; Appanah and Chan, 1981). Aldehydes were the first floral volatiles shown to attract thrips; traps baited with cinnamaldehyde, benzaldehyde, salicaldehyde and anisaldehyde caught more thrips than controls (Howlett, 1914; Morgan and Crumb, 1928). However, Evans (1932) found no increase in *Thrips imaginis* catches in traps containing anisaldehyde, salicaldehyde, benzaldehyde, citral or geraniol.

More recently, numerous studies have examined several floral volatiles that commonly occur in plants (Table 3.3). Kirk (1985d) found that of the floral scents tested, geraniol, myrcene, eugenol and anisaldehyde, only anisaldehyde attracted three to eight times more flower thrips (*Thrips tabaci*, *T. vulgatissimus*, *T. major*, *T. pillichi* and *Frankliniella intonsa*) than white water-pan traps alone. None of the scents affected catches of cereal thrips (*Limothrips cerealium*, *L. denticornis* or *Frankliniella tenuicornis*) or the predatory *Aeolothrips intermedius*.

Numerous researchers have tried to improve trapping of *F. occidentalis* using odours (Table 3.3). Frey *et al.* (1994) found in the laboratory that blue-coloured trap captures of *F. occidentalis* could be increased about 1.9 times by adding geraniol, and other odours alone were as attractive as blue sticky traps alone. The addition of the odour and colour together resulted in an additive catch effect. However, under greenhouse conditions, the addition of geraniol did not improve trap catches reliably, possibly due to competition from plant flowers, lack of adequate odour diffusion or more complicated odour diffusion patterns. Similar results were found by Brødsgaard (1990) in a greenhouse using anisaldehyde to

Table 3.3. Chemical volatiles or plant parts recorded as attracting or repelling thrips (× = trap catch compared with unbaited control).

Chemical/plant part	Thrips attracted	Thrips not attracted or repelled[a]	Reference
Ethyl nicotinate	*Thrips obscuratus* (> 100×) *Thrips tabaci* (2–5×) *Frankliniella occidentalis* (5×)	*Limothrips cerealium* *Thrips fuscipennis* *Anaphothrips obscurus* *Frankliniella intonsa*	Penman *et al.* (1982)[b]
	Thrips obscuratus (> 100×)		Teulon *et al.* (1993)[b]
	Frankliniella occidentalis (2–3×)		Frey *et al.* (1994)
Anisaldehyde	*Thrips imaginis* (1.8–3.3×)	Grass thrips[d], *Aeolothrips intermedius*[e]	Kirk (1987)
	Flower thrips[c] (3–8×)	*Limothrips cerealium*	Kirk (1985d)
	Thrips obscuratus (6–30×) *Thrips tabaci* (> 100×) *Frankliniella intonsa* (5–10×) *Thrips fuscipennis* (5–10×)	*Anaphothrips obscurus*	Teulon *et al.* (1993)
	Frankliniella occidentalis (> 5×)		Brødsgaard (1990) Frey *et al.* (1994)
	Frankliniella occidentalis (1.2–12×)		Hollister *et al.* (1995)
Eugenol, geraniol	*Frankliniella occidentalis* (< 3×)	Flower thrips[e]	Frey *et al.* (1994) Kirk (1985d)
Myrcene		Flower thrips[(R)]	Kirk (1985d)
Benzaldehyde	*Thrips tabaci* (4–10×) (New Zealand) *Thrips obscuratus* (10–35×) (summer) *Frankliniella intonsa* (10–30×)	*Limothrips cerealium* *Thrips obscuratus* (spring) *Anaphothrips obscurus* *Thrips fuscipennis*	Teulon *et al.* (1993)
Single rose bud (susceptible)		*Frankliniella occidentalis*	Gaum *et al.* (1994)
6 rose buds (mixed cultivars)		*Frankliniella occidentalis* (R)	

Cowpea leaf bud	*Megalurothrips sjostedti* (2×)		Kyamanywa *et al.* (1993)
Maize leaves		*Megalurothrips sjostedti* (R)	Kyamanywa *et al.* (1993)
Mixed cowpea/maize leaves		*Megalurothrips sjostedti* (R)	Kyamanywa *et al.* (1993)

[a] (R) indicates repulsion by this compound or plant component.

[b] Peaches/peach juice caught the same number of thrips as unbaited controls.

[c] Includes flower or polyphagous thrips, *Thrips vulgatissimus, T. major, T. flavus, T. tabaci, T. pillichi, Frankliniella intonsa; T. flavus* is repelled by myrcene.

[d] Grass thrips, *Limothrips cerealium, L. denticornis, Frankliniella tenuicornis.*

[e] Kirk 1985d, numbers of each species caught were very low, but not significantly different from controls.

attract *F. occidentalis*. In greenhouse peppers, anisaldehyde-baited yellow sticky traps increased female *F. occidentalis* captures by 1.8 to 5.7 times, but had less effect on trap capture of males (Teulon *et al.*, 1993).

Certain floral volatiles play no role in attraction and may hinder host plant location (Table 3.3). Although morphological and chemical factors appear to be involved in the resistance of roses to thrips, either by antibiosis or antixenosis, floral volatiles of all cultivars of roses tested, even those that support large numbers of thrips, did not attract *Frankliniella occidentalis* (Gaum *et al.*, 1994). Volatiles of some rose buds repelled adult *F. occidentalis* compared with blank controls in Y-tube olfactometer tests, and rose volatiles at high concentrations (i.e. six buds of mixed susceptible cultivars in an olfactometer) repelled thrips.

Non-floral odours

Because some flower thrips are also associated with ripe fruit or vegetative hosts, non-floral olfactory cues may be used to find these (Table 3.3). Penman *et al.* (1982) and Teulon *et al.* (1993) demonstrated that ethyl nicotinate, a chemical found in carambola fruit but not in floral volatiles, increased trap catches of *Thrips obscuratus*, a species found in flowers and on ripe peaches and apricots, by more than 100 times compared with unbaited traps. However, ethyl nicotinate had little to no effect on *Thrips tabaci*, *Limothrips cerealium* or *Anaphothrips obscuratus*, and attracted fewer *Frankliniella occidentalis*, *F. intonsa* or *Thrips fuscipennis* than *p*-anisaldehyde or benzaldehyde (Teulon *et al.*, 1993).

Olfactometer and field studies suggest that thrips use odours of leaf buds and foliage to orient to the host, and that non-host odours interfere with close range settling on host plants. *Megalurothrips sjostedti* was attracted to cowpea leaf buds in olfactometer tests, but the addition of maize leaves to leaf buds interfered with the attraction (Kyamanywa *et al.*, 1993). In field choice tests, more thrips infested cowpeas without maize interplanted than cowpeas mixed with maize. Similarly, Rösingh (1980) found less damage by this species to cowpeas when they were intercropped with maize, though this may have been due to the effects of shelter rather than odour.

Post-alighting host effects

Most host selection studies have examined cues that affect landing, whereas relatively few compare plants or hosts for rates of leaving after alighting. Some studies regard the number of adults on the host as the combined effect of landing and leaving rates, and the presence of eggs or larvae as an indication that the host is acceptable for reproduction. Post-alighting host cues that determine whether thrips remain to feed and oviposit on the host include floral traits, pollen or plant odours, texture, morphology, presence or absence of feeding stimulants, gustatory repellents, and proper nutrition for suitable development.

Close range plant odours and structures

Although close range odours produced by the host could affect its acceptability, there has been little work on the effect of host odours on the residence time on a host, although specific pollen (size, volume, species) has been shown to influence the duration of meals (Kirk, 1984c, 1985c).

Pollen and especially pollenkitt (the oily coating of pollen grains) produce species-specific aromas (Dobson, 1987) which bees can distinguish. Flower thrips also show some discrimination between host species pollen (Kirk, 1984c, 1985c). *Kakothrips pisivorus* (a specialist on *Vicia*, *Pisum* and *Lathyrus*) and *Ceratothrips ericae* (a specialist on *Calluna* and *Erica*) recognized pollen of these flowers before probing them, and avoided pollen of non-hosts, suggesting that there are host-specific pollen odours. Specialists such as *K. pisivorus* fed longer and faster and laid more eggs when fed on their own host pollen than they did when fed on non-hosts, whereas a generalist, *Thrips fuscipennis*, laid a similar number of eggs when feeding on each species of pollen. Pollen grains with exine spines had no effect on feeding, but adhesive pollen grains interfered with it.

Plant chemicals and morphological factors

Tactile, mechanical and chemical receptors for detecting olfactory, gustatory and morphological differences between plants are found on antennae and mouthparts (see Chapter 2) and they also help thrips discriminate between hosts and non-hosts. Plant morphology, such as trichome length and width, exudates associated with them, and the thickness of leaf cuticle and waxes, can have a profound effect on the acceptability of a plant (Ananthakrishnan and Gopichandran, 1993). Plants also vary in their primary and secondary metabolites, which may affect the suitability of the host for either feeding or oviposition. Plant defensive compounds such as phenolics, tannins and alkaloids have been shown to increase resistance. They can be either constitutive components or induced by environmental factors such as UV light, herbivory, or pathogen infection, and may act by deterring insect infestation. One plant hormone, ethylene, induced by at least three different thrips on different hosts, causes abscission and senescence of particular plant tissues (Wien and Rösingh, 1980; Kendall and Bjostadt, 1990; Rieske and Raffa, 1995), (see Chapter 13, Phytohormone induction). Flavonoids and carotenoids impart particular floral hues to plants (Ananthakrishnan and Gopichandran, 1993) and these may also affect the attractiveness as well as the acceptability and suitability of the host.

Plant cues that induce polyphagous thrips to feed may be very general as several species will feed on nutrient or sucrose solution even when presented through plastic (stretched Parafilm M, Clingfilm, Plan-B-Loon) or fish scale membranes which have no surface feeding stimulants or resemblance to plant morphology. Such membrane feeding has been reported in *Scirtothrips citri* (Wiesenborn and Morse, 1985, 1986); *Thrips imaginis* (Day and Irzykiewicz, 1954; Teulon and Penman, 1986, 1991); *Frankliniella intonsa* and *T. hawaiiensis* (Murai and Ishii, 1982; Kirk, 1985c); *Limothrips cerealium* (Chisholm and Doncaster, 1982; Chisholm and Lewis, 1984; Lewis, 1987); *F. occidentalis* (Hunter and Ullman, 1989; Teulon, 1992); *F. schultzei* (Hunter and Ullman, 1989); *Hercinothrips femoralis* (Laughlin, 1971) and *Thrips tabaci* (Sakimura and Carter, 1934). *Limothrips cerealium* and *Scirtothrips citri* have sensilla on the paraglossae that are exposed even when thrips are not feeding (Chisholm and Lewis, 1984; Wiesenborn and Morse, 1988). These sensilla appear to have a chemosensory function, but the morphological analysis of *F. occidentalis* and *F. schultzei* (Hunter and Ullman, 1989, 1992) suggests that the paraglossal sensilla have both mechanical and chemical receptors that allow thrips to detect both plant surface topography (e.g. toughness, protrusions) and chemistry (stimulants or repellents). Because these species will feed through membranes, feeding stimulants or arrestants appear not to be required for thrips to begin feeding, although surface stimulants may induce faster initial feeding or more protracted feeding bouts. Thrips will also oviposit through artificial membranes (see Chapter 3, Oviposition sites and process, and Chapter 12).

Thrips palmi, a generalist, did not feed on tomato plants even though it landed and walked on the leaves (Hirano *et al.*, 1993), which suggested that the plant derived protection from a toxic or antifeedant component in the sap. No eggs are laid on tomato leaves (Kawai, 1986b), and female longevity is reduced. When adult female *Thrips palmi* were reared on filter paper discs impregnated with methanol extracts of tomato leaf plus sucrose, they died in about 5 days, whereas most survived longer when reared on extracts of either aubergine or cucumber leaves plus sucrose. Further fractionation of tomato leaf extract indicated a toxic component in the *n*-butanol fraction (Yasumi *et al.*, 1991). A steroidal glycoalkoid, α-tomatine, isolated from the *n*-butanol fraction deterred feeding and dramatically reduced longevity of *Thrips palmi* (Hirano *et al.*, 1994) but the effect only lasted during exposure to the chemical (Yasumi *et al.*, 1991), suggesting an antifeedant.

Several tomato cultivars and wild *Lycopersicon* species were compared for mechanical and thrips transmission of tomato spotted wilt virus (TSWV-MT2). Transmission by *Frankliniella occidentalis* on one tomato species (*L. hirsutum glabratum*) via the leaves was less effective than mechanical inoculation (Kumar *et al.*, 1993). This was presumably a result of either less feeding or toxicity; the glandular trichomes on the leaves of this species are known to confer some resistance to arthropods (Kennedy, 1984). Because adult thrips landing rates on leaves were similar on all tomato cultivars tested, differences in thrips transmission of virus are probably due to differential feeding after landing (Kumar *et al.*, 1995a). All tomato cultivars avoided by larval thrips had leaves with dense glandular trichomes (Kumar *et al.*, 1995b). If larvae moved onto these leaves they were either hindered or trapped. Because most larvae avoided leaves before touching them, some non-contact cues may also be involved. Generally, those tomato species that were most antixenotic to larvae had least adult feeding damage.

Frankliniella occidentalis does not thrive on leaves of some species, although it may be abundant on their flowers. This applies to leaves of chrysanthemum (de Jager *et al.*, 1993, 1995a), peanut (Lowry *et al.*, 1992), and some cotton cultivars (Trichilo and Leigh, 1988). *Frankliniella occidentalis* larvae confined to chrysanthemum leaves do not survive well to adulthood (de Jager *et al.*, 1993). However, Robb (1989) found that survival of larvae and pupae on Polaris, a cultivar sensitive to TSWV, was greater than 87 per cent when they were reared on leaves plus pollen, 72 per cent on leaves without pollen, but only 14 per cent when reared on virus-infected leaves. Thus either morphological or chemical leaf constituents of some cultivars appear to affect feeding and development adversely, and virus-infected plants may add to that adversity.

Several studies have examined a number of plant and physical characters associated with chrysanthemum resistance to *Frankliniella occidentalis*. Of 42 cultivars studied, de Jager *et al.* (1995b) found that oviposition varied 30-fold among preflowering plants. The most resistant plants preflowering were the most resistant after flowering. The number of eggs and leaf feeding damage, both silvering and distortion, were correlated. No correlation was found between leaf toughness, hairiness, or surface area and the number of eggs laid; however, there was a positive correlation between the number of eggs and the plant height or the number of leaves, presumably arising from the differences in the surface area available for thrips to lay in. However, it has been shown in some plant systems that the better defended plants spend more energy on chemical defence and less on growth (Coley, 1988). A chemical defence is associated with the sap of resistant chrysanthemum's leaves and flowers (de Jager *et al.*, 1995c) and may account for the reduced oviposition and feeding damage. In contrast, van Dijken *et al.* (1994) found that the smallest of 13 chrysanthemum lines had the most silvering damage, but the smallest was a different species not examined in other studies. In non-choice experiments, de Jager *et al.* (1995b) found that flower colour (white, yellow, orange, purple, green) was not associated with the number of thrips; but the flower production per plant and flower weight were positively correlated with the number of thrips per flower. Spider types, with only ray flowers, had fewer thrips per flower and less feeding damage than single, decorative/pompom, or anemone types (Broadbent and Allen, 1995; de Jager *et al.*, 1995b).

Zeier and Wright (1995) found that *Gladiolus* spp. varied in their resistance to thrips, primarily *Thrips simplex* and *Frankliniella occidentalis*. With some exceptions, damage ratings on leaves were slightly negatively correlated with the density of epidermal folds and with the amount of epicuticular wax.

Five different wild peanut species (*Arachis* spp.) sustained less damage from *Frankliniella fusca* than did cultivated peanut (*Arachis hypogea*) in which resistance differed between cultivars (Yang *et al.*, 1993; Snook *et al.*, 1994). Some differences in cuticular lipids of blooms and foliage were found, particularly the phenolic acids and acetates which are general digestibility inhibitors (Yang *et al.*, 1993). Resistance to thrips was not associated with increased levels of L-chicoric acid, the only major polyphenol found in peanut terminals (Snook *et al.*, 1994).

Frankliniella occidentalis larvae generally caused less damage to leaves of cucumber, tomato and peppers containing low concentrations of aromatic amino acids than those with higher concentrations of these compounds (Mollema and Cole, 1996). Exceptional among the crops tested was lettuce; cultivars with both low and high concentrations of aromatic amino acids were similarly damaged.

Numerous studies of cotton cultivars (either *Gossypium hirsutum* or *G. barbadense*) have shown variation in susceptibility to thrips. Many traits likely to affect infestation levels and damage have been examined, especially chemicals such as gossypol, and morphological characters including epidermal thickness and pubescence. Several species of thrips damage cotton, especially early in the season, including *Frankliniella occidentalis*, *F. fusca*, *F. tritici* and *Thrips tabaci* in the USA. Pilous and/or highly pubescent lines reduce infestation and damage (Wardle and Simpson, 1927; Watts, 1936; Ballard, 1951; Quisenberry and Rummel, 1979; Rummel and Quisenberry, 1979; Walker *et al*., 1979). Ballard (1951) also noted that the most resistant cultivars were densely pubescent and the most damaged were glabrous. In addition, cultivars with intermediate hairiness varied in their resistance and susceptibility, suggesting that other morphological or chemical factors can also affect resistance. Similar results have been reported for *Thrips palmi* damage to cotton in the Philippines; cultivars with low pilosity and low density of hairs sustain much more damage than those with many hairs (Bournier, 1983), which presumably interfere with thrips movement and acceptance of the host for feeding and reproduction.

Thrips tabaci feeds on the lower surface of cotton (both *G. hirsutum* and *G. barbadense*) in Egypt (Gawaad and Soliman, 1972; Gawaad *et al*., 1973). The numbers on different cotton lines were negatively correlated with the thickness of the lower epidermis. Generally, the thinner the layer, the more thrips per seedling. For most cultivars tested, resistant lines had epidermal thickness of around 10 to 13 μm compared with 6.6 μm in susceptible ones. No differences were found in the thickness of the upper epidermis. Gawaad *et al*. (1973) suggested that the thick epidermis limited the reach of the mandibles and maxillae during feeding, but several studies have shown that maxillary stylets of several species can extend beyond this distance (Heming, 1978; Wiesenborn and Morse, 1988; Hunter and Ullman, 1989; Childers and Achor, 1991) (see also Chapter 4, Feeding behaviour).

Gopichandran *et al*. (1992) found that of five cotton cultivars examined, two were susceptible and three were highly resistant to *Retithrips syriacus*. No thrips were found on cultivars with high trichome density, high levels of secondary plant metabolites (gossypol, tannins and flavonoids) and low primary metabolites (carbohydrates, proteins, amino acids and nitrogen sources). Most striking of the chemical differences was the phenolic, gallic acid, which was absent from susceptible cultivars.

Some rice cultivars resistant to planthoppers are also resistant to *Stenchaetothrips biformis* (Velusamy and Saxena, 1991; Velusamy *et al*., 1994). Damage by adult thrips decreased and oviposition rates were lower on cultivars with planthopper resistance genes. Susceptible cultivars dipped in steam distillates of a resistant line were toxic to larvae, which suggests a physiological antibiosis. In onions, several studies have shown that some genotypes of bulb onions are resistant to *Thrips tabaci* (Verma, 1966; Singh *et al*., 1977; Coudriet *et al*., 1979; Brar *et al*., 1993) depending on leaf structure, growth form (tightness of leaves, uprightness of growth) and bulb colour, features which might affect thrips landing as well as subsequent feeding (see Chapters 7 and 16).

Cabbage, a late season crop in the north-eastern USA, is susceptible to *Thrips tabaci* infestations and damage. Even though numerous thrips species were found in aerial traps either adjacent to or in cabbage fields, most of them were not present on cabbage, suggesting that many species discriminate between this crop and their preferred hosts (North and Shelton, 1986a,b; Shelton and North, 1986). However, North and Shelton (1986a) did find some thrips on the outer leaves (*Anaphothrips obscurus* and *Frankliniella tritici*), but these did not penetrate the inner head. Thus cabbage seems to present the proper cues to attract these species, but not for them to remain. Some cabbage cultivars are resistant to *Thrips tabaci*; Shelton *et al.* (1983) found differences in the damage to the outer ten leaves among 16 cultivars tested. Neither date of maturity nor amount of dry matter was associated with differences in damage. Shelton *et al.* (1988) suggested that antibiosis, antixenosis and tolerance were all involved, but Stoner and Shelton (1988a,b,c) found that *Thrips tabaci* resistance was primarily due to non-preference after alighting on the host. Some resistant cultivars had more adult thrips on the outer frame leaves than susceptible cultivars, but they had fewer thrips in the head. The effect of attractiveness of the outer frame leaves was independent of either the acceptability or suitability of the head. The amount of tightly enclosed leaf area in the inner head of cabbage could prevent thrips entry. Among cultivars tested in Ohio, Hoy and Kretchman (1991) found that only some were consistently resistant or susceptible from year to year and their vulnerability was altered by the growing conditions.

Some virus-infected hosts support more thrips adults and larvae than non-infected plants. This applies to *Frankliniella occidentalis* on TSWV-infected lettuce (Yudin *et al.*, 1987), *Thrips tabaci* on *Emilia sonchifolia* (Carter, 1939), *Thrips palmi* on cucumber (Culliney, 1990) and *F. occidentalis* on weed hosts (Bautista *et al.*, 1995). Preference for diseased hosts may be due to both adult attraction and reproduction; most infected plants have lighter coloured leaves which are more attractive to adults (Yudin *et al.*, 1987; Culliney, 1990). In contrast, larval and pupal mortality of *Frankliniella occidentalis* was higher on virus-infected chrysanthemum (Robb, 1989) than on healthy leaves.

Some cucumber cultivars studied by Soria and Mollema (1995) were partially resistant to *Frankliniella occidentalis* because females were less fecund by 36–50 per cent on susceptible cultivars. In this instance the cause may be physiologial rather than due to interferences with oviposition by leaf morphology. Using an electrical recording technique (electrical penetration graph, Harrewijn *et al.*, 1996a) that distinguishes between different electrical waveforms associated with stylet penetration, salivation and ingestion, Harrewijn *et al.* (1996b) compared *F. occidentalis* feeding patterns on susceptible and resistant cucumber leaves. No difference in feeding was observed during the first 24 h after placement on leaves. However, after 24 h, thrips probing, penetration and ingestion on resistant leaves was greatly reduced, also suggesting a physiological component to resistance in cucumber.

Bautista and Mau (1994) showed that, although adult *Frankliniella occidentalis* prefers flowering romaine lettuce over other hosts, adults and larvae become trapped by the sticky sap during feeding or oviposition. This suggests that host landing cues are not disrupted by any cue associated with sap or surface characteristics.

In addition to the direct effects on thrips, leaf structures can also indirectly affect thrips through effects on their predators. The attack rate on *Frankliniella occidentalis* by *Orius insidiosus* is lower on tomatoes than on beans or pepper plants (Coll and Ridgway, 1995) because the glandular hairs of tomato inhibit search efficiency of *Orius* and so reduce its predation rate on thrips.

Pheromones and Allelochemicals

As for many other groups of insects, the widespread nature and importance of communication by semiochemicals in the Thysanoptera is well established. Defensive secretions and allelochemicals have been identified in terms of both their specific chemistry (Table 3.4) and the behaviour that they mediate. The roles of the defensive chemicals discovered so far in Thysanoptera fall into several categories, primarily alarm pheromones, arrestants, attractants, and irritant or repellant allomones. On the other hand, little is known about production of sexual pheromones beyond circumstantial evidence related to the behaviour and to the morphology of structures that resemble pheromone-producing glands.

Sexual pheromone production

No chemical has yet been identified that is known to mediate mating behaviour of Thysanoptera, either for mate finding or sexual stimulation after potential partners have met. Neither is there any circumstantial evidence for the production of a long-distance mate-attracting pheromone, although there is some evidence for a short-range attractant (Milne, 1997). The opposite sexes probably locate each other using other cues. Many species of Tubulifera live in aggregations or colonies, or in galls, and some have overlapping generations (Lewis, 1973; Kiester and Strates, 1984; Ananthakrishnan, 1990; Crespi, 1993). Because reproductive adults are found within these colonies or aggregations, only short-range behavioural or chemical cues are needed to advertise sexual readiness. Within anthophilous or grass-feeding terebrantian species, aggregations of mixed sexes also occur on or near the host. Therefore, instead of producing a long-distance pheromone for locating reproductive individuals of the opposite sex, appropriate cues are used to find hosts and then short-range chemical and visual cues possibly used to find receptive mates. Mating aggregations of *Frankliniella occidentalis* occur on some white, blue, violet and yellow artificial surfaces similar to flower colours (Terry and Gardner, 1990; Matteson and Terry, 1992) in the absence of any other volatile, visual or tactile cues, thus proving the sufficiency of colour alone to attract males and females.

On the other hand, short-range sex pheromones have been implicated as influencing the behaviour of some female Terebrantia during mating. Evidence reviewed by Lewis (1973) as well as more recent morphological evidence of pheromone production is primarily circumstantial. No chemical has been isolated that mediates any specific behaviour. In summary, males and some females within the families Phlaeothripdae and Thripidae have glands on the sternites of some abdominal segments. On males they generally occur on abdominal segments III to VII or VIII; they vary in shape (elliptical, oval, round, dumbbell, reniform), type of gland (individual cells or aggregations of cells), pore size and number (see Chapter 2). This diverse morphology is useful for differentiating between genera and species (Stannard, 1968). Klocke (1926) proposed that a secretion from these glands was an adhesive used to hold the male onto the female during copulation, based on the location of the glands and the position of males during mating. Pelikán (1951), while observing *Pezothrips dianthi* mating behaviour through a microscope, noted that male sternites did not touch the female's abdomen during copulation and therefore the secretion could not be an adhesive. Instead, he believed it to be a lipoid volatile pheromone that calmed the female during mating. It also attracted other females to the mating pair.

During copulation, *Frankliniella occidentalis* couples have a calm period (Terry and Schneider, 1993) detected on replays of video-taped matings (see p. 91, Mating behaviour, territoriality and fighting). Similar calm periods during copulation have been observed in *F. schultzei* (Milne, 1997). However, by the time the calm period began, the male had already climbed onto the female's abdomen, tapped her antennae with his, and begun copulation. If this calm period is representative of many Terebrantia, a pheromone could be associated with calming the female during copulation until the male completes the transfer of the spermatophore, rather than making the female receptive initially. Because a female must be receptive before she will allow a male to mount her, the pheromone alone probably does not calm her as the male approaches. A female will continually buck a male if she is not receptive (Terry and Schneider, 1993). This occurred with males that successfully mated with another female immediately after being rejected, indicating the capability of such males to calm and mate other females. More recently, Milne (1997) demonstrated that male *Frankliniella schultzei* attract females over at least a short distance both in the laboratory and field.

As well as these behaviours that suggest pheromonal calming and short-distance attractant, the morphology of the sternal glands lends support to a pheromone-producing and secreting function (see Chapter 2). Male sternal glands in *Thrips validus* are modified epidermal cells that have the necessary cellular organelles and structures in sufficient numbers to produce pheromones (Bode, 1978). The cells of these glands, which are of only one type, meet the requirements for a secretory function: abundance of mitochondria for its production, microtubules, and agranular endoplasmic reticulum for modification of the product. Histochemically, the primary substance in the cytoplasm of these cells is a lipoid with some minor mucopolysaccharide components. There are no vacuoles in these cells where the lipids collect, but lipids are excreted to the outside of the body through short epicuticular ducts and pores resembling those of other insect exocrine glands. In Merothripidae, within the Terebrantia, Moritz (1984) found a vertex gland in the head capsule of males in the genus *Merothrips* (see Fig. 2.7:1). These glands closely resemble the sternal glands of male Thripidae in morphology. The secretions of this gland presumably are associated with a pheromone function.

Table 3.4. Chemical constituents of whole-body extracts or anal exudates of thrips.

Class of chemical	Constituent	Biological activity	*Eurylaphrothrips crassus*	*Elaphrothrips tuberculatus*	*Gynaikothrips ficorum*	*Gynaikothrips uzeli*	*Bagnalliella yuccae*	*Liothrips kuwanai*	*Liothrips piperinus*	*Haplothrips leucanthemi*	*Hoplothrips japonicus*	*Holothrips japonicus*	*Holothrips hagai*	*Varshneyia pasania*	*Thlibothrips isunoki*	*Bactrothrips brevitubus*	*Ponticulothrips diospyrosi*	*Frankliniella occidentalis*
Aliphatic hydrocarbon	dodecane		0	0	0	0	0	1	1	0	0	0	0	1	0	0	0	0
	tridecane		0	0	1	1	0	3	3	0	0	1	2	3	0	0	0	0
	tetradecane		0	0	1	1	0	1	1	0	0	1	1	1	0	0	0	0
	(*Z*)-7-pentadecane		0	0	0	0	0	2	0–1	0	0	1	0	0	0	0	0	0
	pentadecane	RI	0	0	3	3	0	3	3	0	0	3	3	2	0	0	0	0
	α-heptadecadiene		0	0	0	0	0	0	0	0	0	3	3	0	0	0	0	0
	(*Z*)-8-heptadecene		0	0	0	2	0	1	0	0	0	3	2	1	0	0	0	0
	heptadecane		0	0	1	1	0	0	0	0	0	0	0	0	0	0	0	0
Lactone	γ-decalactone	RI	0	0	0	0	3	0	0	0	0	0	0	0	0	0	0	0
	4-octadec-9-enolide		2?	0	0	0	0	0	0	0	0	0	0	0	0	0	0	0
Pyranone	3-butanol-4-hydroxy 6-methyl-2*H*-pyran-2-one		0	0	0	0	0	0	0	0	0	3	2	0	0	0	0	0
Fatty acid	decanoic acid		2?	0	0	0	0	0	0	0	0	0	0	0	0	3	0	0
	2-methylbutyric acid	R	0	0	0	0	0	0	0	0	0	0	0	2	0	0	0	0
	(*E*)-4-decanoic acid		0	0	0	0	0	0	0	0	0	0	0	0	0	3	0	0
	dodecanoic acid		2?	2?	0	0	0	0	0	0	0	0	0	0	0	1	0	0
	(*Z*)-5-dodecanoic acid	A	0	0	0	0	0	0	0	0	1	0	0	0	0	3	0	0
	(*E*)-3-dodecanoic acid	AAg	0	0	0	0	0	0	0	0	3	0	0	0	0	2	0	0
	(*Z*)-5-tetradecanoic acid		0	2?	0	0	0	0	0	0	0	0	0	0	0	0	3	0
	α-tetradecadianoic acid		0	0	0	0	0	0	0	0	0	0	0	0	0	0	1	0

Napthoquinone	juglone	R	0	3	0	0	0	0	0	0	0	0	0	0	0	2	0	0
	2-methyljuglone	R	0	0	0	0	0	0	0	0	0	0	0	0	0	0	3	0
	7-methyljuglone		0	0	0	0	0	0	0	0	0	0	0	0	0	0	2	0
Terpene	perillene	AR	0	0	0	0	0	2	2	0	0	0	0	3	0	0	0	0
	myrcene	AR	0	0	0	0	0	0	0	0	0	0	0	0	3	0	0	0
	β-acaridial		0	0	0	3	0	3	0	0	0	0	0	3	0	0	0	0
	meillon	RI	0	0	0	0	0	0	0	3	0	0	0	0	0	0	0	0
Aliphatic ester	decyl acetate	AAr	0	0	0	0	0	0	0	0	0	0	0	0	0	0	0	3
	dodecyl acetate	AAr	0	0	0	0	0	0	0	0	0	0	0	0	0	0	0	3
	tetradecyl acetate		0	0	1	1	0	2	0	0	0	0	0	3	0	0	0	0
	hexadecyl acetate	RI	0	0	3	3	0	1–2	0	0	0	0	0	1–2	0	0	0	0
	octadecyl acetate		0	0	0	1	0	0	0	0	0	0	0	0	0	0	0	0
Reference			Blum (1991)	Crespi (1990) and Blum (1991)	Howard *et al.* (1987)	Suzuki *et al.* (1989)	Howard *et al.* (1983)	Suzuki *et al.* (1988, 1989)	Suzuki *et al.* (1988)	Blum *et al.* (1992)	Haga *et al.* (1989)	Suzuki *et al.* (1993)	Suzuki *et al.* (1993)	Suzuki *et al.* (1986, 1988, 1989)	Haga *et al.* (1990)	Suzuki *et al.* (1990)	Suzuki *et al.* (1995)	Teerling *et al.* (1993a,b)

R = repellent, I = irritant, FD = feeding deterrent, A = alarm pheromone, Ar = arrestant, Ag = aggregation pheromone.
Level of constituent: 0 = not detected or reported, 1= trace or minor, 2 = low (5–15%), 3 = main constituent (> 15%), ? = specific levels not given.

Alarm pheromones and defensive allomones

Observations of thrips behaviour have suggested the production of defensive anal secretions (Buffa, 1911; Mukaigawa, 1912; Hodson, 1935; Lewis, 1973) long before any specific chemical was isolated and identified from an exudate. *Heliothrips haemorrhoidalis* secrete a drop of fluid with excrement at the abdominal tip and spread it when confronted by a predator (Buffa, 1911). Anal excretions of *Varshneyia pasanii* in Japan produce an acid response on litmus paper (Mukaigawa, 1912). *Liothrips vaneecki* raise their abdomens and release a pungent odour when disturbed (Hodson, 1935); individuals then race around, spray predators and move to more cryptic areas. *Thrips tabaci* wag their abdomens as well as emit tiny anal droplets as an effective defence against predatory phytoseiid mites (e.g. *Amblyseius cucumeris*) (Bakker and Sabelis, 1987, 1989). If contacted by the droplet, mites withdraw from the prey and begin grooming. Mites successfully attack second instar thrips by capturing larvae by the thorax, which avoids the wagging and anal fluid (see Chapter 8). *Bagnalliella yuccae*, the first species whose exudate chemicals were identified (Howard *et al.*, 1983), expose a special applicator tip comprising the tubular distal segments of the abdomen, which discharges the fluid. In many species, setae on the tube hold the material, which may be resorbed (Howard *et al.*, 1987).

The wealth of diversity of natural products in thysanopteran defensive secretions is mostly attributable to studies of the Tubulifera (Table 3.4). This may reflect either the difference in capability of Tubulifera to produce defensive compounds compared with Terebrantia or the difficulty of studying the smaller Terebrantia. If Tubulifera are more prolific producers of defensive compounds, it could be that their colonial and gall-dwelling lifestyles select for and promote this type of defence. To maintain safety and integrity within the colony, a strong repellent or alarm against predators would be effective. Most terebrantians could simply fly to new hosts if predators are near.

Extracts of either whole bodies of thrips adult males, females and/or larvae or anal exudates of many species have been analysed by gas chromatography and mass spectral analysis. Although there are some similarities of components across species, there is great diversity of compounds produced despite their similarity in function (see Table 3.4). There is no clear similarity of chemistry among related taxa. Compounds range from aliphatic hydrocarbons, lactones and fatty acids to many straight-chain esters. Some species produce a single active component in the volatile exudate, for example *Bagnalliella yuccae* (Howard *et al.*, 1983) and *Haplothrips leucanthemi* (Blum *et al.*, 1992), while most produce a suite of different compounds, which may include combinations of many aliphatic hydrocarbons and straight-chain esters. Many biological-responses have been ascribed to these constituents. Most of the compounds that have been bioassayed (mainly against predatory ants) show repellent or contact irritant properties, causing predators to move away or to drag and groom themselves after contact with the chemical. A few compounds serve as an alarm pheromone and/or ovipositional deterrent. Pentadecane is one of the most common constituents found, in terms of the number of species that produce it as well as its concentration in the exudate. This aliphatic hydrocarbon repels and irritates ants that come in contact with it. The biological activity elicited by many of the constituents, especially the minor ones, has not been demonstrated. Howard *et al.* (1987) found that the major components (pentadecane and hexadecyl acetate) in exudates of *Gynaikothrips ficorum* did not account for the full effect of the exudate and suggested that minor constituents are necessary to complete the activity.

The anal exudates of only one terebrantian, *Frankliniella occidentalis*, have been identified to date. Teerling *et al.* (1993a,b), identified a two-component extract, dodecyl acetate and decyl acetate, present in whole-body extracts of adults and first instars and in anal droplets of second instars. These compounds, either alone or in combination, caused alarm responses, including inducing second instars to fall from plants, and reduced oviposition. Further, the droplets acted as arrestants/attractants for predatory ants, *Orius*, and a predatory mite, *Amblyseius cucumeris* (Teerling, 1995). However, this response was weak and thrips only moved a short distance relative to the control response. This type of pheromonal activity is unique among the Thysanoptera yet studied, and the compounds produced by this species have not been found in any other species. In this species, an alarm pheromone/kairomone could be useful for thrips within aggregations. First, it could serve as an alarm to cue adults to fly and larvae to drop immediately from a plant or move to refugia when a predator is near. Female thrips would not spend time in an area laying an egg when predators were near. Second, a short-range arrestant or attractant to predators would cause predators to stay in an area recently vacated by thrips.

No specific glands or glandular cells that definitely produce and/or secrete anal exudates have been located. There has been no specialized exocrine structure found in the wall of either the hindgut, the alimentary canal, or the abdomen of any of the species examined (Suzuki *et al.*, 1986; Howard *et al.*, 1987; Blum *et al.*, 1992), but exudates are concentrated in the hindgut (Howard *et al.*, 1987; Blum *et al.*, 1992). Perhaps either the epithelial cells of the hindgut or papillate cells of the Malpighian tubules that extend into the gut are involved in production/secretion of these compounds (see Howard *et al.*, 1987). Perillene collects in the rectum of *Varshneyia pasania*, which have many rectal glands, but it is not known whether these glands actually produce the compound (Suzuki *et al.*, 1986).

Understanding the role and chemistry of any defensive secretion could help to predict the success of a predator release programme and/or selection of races of predators capable of overcoming or resisting a kairomone used for defence. For example, Bakker and Sabelis (1987, 1989) demonstrated that first instar larvae of *Thrips tabaci* are the primary food source of the predatory mites because second instars defend themselves more effectively. The age distribution of thrips is a key factor in the maintenance and success of the mite predators of this species.

Mating Behaviour, Territoriality and Fighting

Details on the mating behaviour of sexual species of Terebrantia are limited mainly to descriptions of copulation once sexes have found each other, and very little is known about their mating systems (i.e. mate finding, choice, advertisement, guarding, or alternative mating tactics) (Franssen and Mantel, 1964; Lewis, 1973; Ananthakrishnan, 1990; Terry and Dyreson, 1996). This lack of information on Terebrantia may be due in part to characters that make behavioural observations difficult: their small size, the ephemerality of their floral hosts, and their claustral nature. By contrast, the mating behaviour of several tubuliferan species that have more long-lived hosts or stable habitats, and are generally much larger, is better understood (Kiester and Strates, 1984; Crespi, 1986a,b; 1988a,b,c,d; 1989, 1990, 1993). Some background on these species of Tubulifera may provide insight with which to compare and contrast the mating behaviour of Terebrantia. The habitat stability relative to generation time of Tubulifera (in some species more then one generation per host) may promote a more colonial or subsocial lifestyle (Evans, 1977), which may include cooperation in caring for young, overlap of at least two generations to contribute to colony labour, and reproductive division of labour (Michener, 1974). Some Tubulifera studied by Kiester and Strates (1984) and Crespi (1990, 1992a,b) exhibit some of these traits. Among the Tubulifera as a whole, there is a great diversity of feeding preferences and habitats, morphology and degree of wing development, and mating behaviour. Their mating systems are complex, comprising alternative male mating tactics, male territoriality, maternal care, and oviparity or viviparity depending upon the species, environment and sometimes the sex produced (see Crespi, 1986a,b; 1988a,b,c,d; 1993; Ananthakrishnan, 1990).

Sexes of the colonial tubuliferans live together in the same colony, so that a mate may be found there, although in some species males disperse to new and developing colonies to search for receptive females. In species where males are wingless, females are winged and hosts are patchily distributed, females may mate with siblings before dispersing from the brood site. Such species generally have a more female-biased sex ratio as expected under local mate competition (Hamilton, 1967; Crespi, 1991, 1993). Some fungus-feeding species have communal oviposition sites (e.g. *Hoplothrips karnyi*, *H. pedicularius* and *Sporothrips amplus*, Crespi, 1986a, 1988c,d, 1993) where many females of the colony lay their eggs. Males of these species have enlarged fore legs which they use to fight other males in defence of territories consisting of the oviposition sites. They mate with females as they arrive just before they oviposit. In *Hoplothrips*, males vary in wingedness and fore leg size. The extremes of size are sometimes referred to as 'gynecoid', in males with small fore legs and 'oedymerous' in males with enlarged legs (Ananthakrishnan, 1990), although this terminology is misleading and does not reflect the continuous variation of traits and the complexity of the sexual differences. Larger males are superior fighters and fights may escalate to the death in *Hoplothrips karnyi* (Crespi, 1988d). There is a trade-off in wingedness and fore leg size and an association of morphology with alternative mating tactics (Crespi, 1988c). Apterous males have large legs with which they defend oviposition sites; small legged, alate males disperse to new sites searching for other colonies either already established or those being founded. Small males avoid fights and mate by sneak mating. Large dominant males perform most of the matings (about 80 per cent)

(Crespi, 1986a, 1988d). Other species, such as *Anactinothrips gustaviae* (Kiester and Strates, 1984) and *Merothrips* and *Cryptothrips* (Mound and O'Neill, 1974), that are colonial or parasocial have communal oviposition sites and males with enlarged fore legs. These could have similar systems, but particular behaviours have not been reported. The females of some species, such as *Elaphrothrips tuberculatus* (Crespi, 1986b, 1988b, 1990), have individual oviposition sites and guard their clutch of eggs until hatching. Males of this species guard and fight over the individual egg-guarding females. Large, non-guarding males challenge the guarding males. Small non-guarding patrolling males copulate with females by sneak mating.

These tubuliferan males fight using their enlarged forelegs (Crespi, 1986a,b, 1988b,d; Ananthakrishanan, 1990) and abdomens. Forelegs may be armed with spines and teeth. In *Elaphrothrips tuberculatus*, males line up in parallel and flick the tips of the abdomens at each other (Crespi, 1986b). Fights can escalate to one male grabbing and stabbing his opponent. In *Hoplothrips karnyi*, the escalation can proceed to the death of a stabbed opponent (Crespi, 1988d). Many of the Australian gall-forming thrips also engage in lethal fighting that involves female–female and male–male fights during gall development and colony founding or during an invasion by other thrips and species (Crespi, 1992a,b).

Because terebrantian lifestyles vary considerably, their mating systems may be quite diverse as they are shaped by specific selection pressures imposed by their environment. Some species are completely parthenogenetic; sexual species are typically haplodiploid (see Chapter 2). Both sexes of most flower thrips species are winged and can disperse, so mating probably occurs after dispersal (Lewis, 1973; Crespi, 1993) and the primary sex ratio is close to 1:1 as expected for most panmictic species (Hamilton, 1967; Crespi, 1993). Wingless species have a more female-biased sex allocation. The focus herein will be to review some of the generalities of the mating systems of Terebrantia, with details of the better understood species.

Aggregations are common in Thysanoptera (Ananthakrishnan, 1984), including Terebrantia, but many of these are presumed to be associated with finding common resources such as floral parts for food and/or mating. Notably, several species of flower thrips aggregate on flowers and mating takes place there. The primary purpose of these aggregations appears to be location of mates as most action in these aggregations on flowers appears to be mating, with minimal feeding and ovipositional activity (Morison, 1947–1949; Kirk, 1985b; Terry and Gardner, 1990; Terry and Dyreson, 1996; Milne, 1997).

Three British Thripidae, *Thrips fuscipennis* and *T. major* and a few *T. atratus*, aggregate and mate on flowers of *Calystegia sepium*, and dispersal to and from these sites occurs throughout the day (Kirk, 1985b). Thrips orientated towards and landed on the freshly opened flowers during the morning and the numbers diminished later in the day as the flowers closed and faded in the evening. Males accumulated on the surface of the corolla while they patrolled over the surface. If two males encountered each other, they immediately parted without incident. No aggressiveness among males or females was observed. Females that landed among the males were immediately noticed by males a few millimetres away; males approached them and mated. Copulation lasted just over 2 min for the two most common species. Counts of *T. fuscipennis* were female-biased, but *T. major* was male-biased within aggregations. Most females observed did not mate a second time and left within a few minutes. Flowers on the surface of the vegetation had more thrips than those hidden within the plants. Females already on the flower did not mate, but remained deep within the corolla and fed (Kirk, 1985b). Because the thrips were primarily engaged in mating and there was no aggression or oviposition at these specific sites, this may represent a lek-type mating behaviour (Crespi, 1993), although some vertebrate biologists, by definition, exclude those species whose males aggregate at food resources or oviposition sites, or where females do not exhibit a choice among displaying males (see Bradbury, 1985). The male-biased sex ratios within aggregations contrast with the more usual female-bias in adult sex ratios of flower thrips (Crespi, 1993). The local male-biased aggregations are a result of the different behaviour of males and females.

Similar aggregating behaviour has been observed in two other species of Thripidae (Terry and Gardner, 1990; Terry and Dyreson, 1996; Milne, 1997), and such behaviour may be common in other anthophilous thrips. *Frankliniella schultzei* males aggregate on several species of flowers (*Hibiscus rosasinensis*, *Gossypium hirsutum*, *Malvaviscus arboreus* and *Ipomoea indica*) throughout the day (0900 to 1400 h) in summers in Queensland, Australia (Milne, 1997). The highest density within aggregations occurs generally between 1100 and 1200 h. Males patrol the petals of these flowers and attempt to mate with females as they land in the aggregations. Females always land on petals and on no other floral parts. A small proportion of the females mate, ranging from 18 to 30 per cent depending upon the plant host. Those that do not mate after landing either hide in the flower or fly away. All aggregations with more than 30 individuals were highly male biased, about 80 per cent males.

Frankliniella occidentalis male aggregations have been observed on coloured or white surfaces similar to flowers and also on many flowers, including cotton, magnolia and gloxinia (Terry and Gardner, 1990; Matteson and Terry, 1992; Terry and Dyreson, 1996). Shortly after dawn and occasionally throughout the day, males aggregate at these sites and patrol the blossoms. Almost all females that enter the aggregations mate and then leave immediately. This contrasts with the lower percentage mating of *F. schultzei* females entering aggregations.

This male aggregating behaviour has been indicated before when highly fluctuating *Frankliniella occidentalis* sex ratios have been observed on numerous hosts (Rosenheim *et al.*, 1990; Chamberlin *et al.*, 1992; Higgins and Myers, 1992; Terry and Kelly, 1993; Chyzik *et al.*, 1995). Although in cucumber all samples were female biased, less female bias was observed in flowers than in other plant parts (Rosenheim *et al.*, 1990). The proportion of females on floral parts averaged 38 per cent whereas the proportion of males averaged 11 per cent on leaves and up to 24 per cent on fruit. Sex ratios of *Thrips palmi* in cucumber were less variable and averaged 19 per cent males. Highly fluctuating adult sex ratios within apple blossom clusters did not appear to be associated with differential sex allocation (Terry and Kelly, 1993), as sex ratios of larvae were similar across cluster stages and were close to the 1:1 expected sex ratio for a panmictic species. Thus most of this variation in adult sex ratio appeared to be due to the aggregating behaviour of males associated with mating activities.

Matteson and Terry (1992) observed differences in male compared with female dispersal among *Frankliniella occidentalis* throughout the day associated with mating aggregations. During the favoured period of the day for aggregations, up to three times as many males as females were landing. However, when aggregations were not observed, sex ratios of landing adults were female biased. There was a sixfold difference in male landing rates between periods. Aggregations seemed to be encouraged by particular weather conditions (high dewpoints, low vapour pressure deficits pertaining to thunderstorms, and during monsoon conditions in the desert south-west of the USA).

In contrast to the *Thrips* spp. (Kirk, 1985b) and *Frankliniella schultzei* (Milne, 1997) aggregations in which no male–male aggressiveness was observed, aggressiveness among *F. occidentalis* males on flowers and artificial surfaces was common (Terry and Dyreson, 1996). The more aggressive males, once they encountered another male, lined up in parallel with heads pointing in the same direction (Fig. 3.2A 1–3), and then flicked their abdomens at each other. This abdominal flicking has been observed in several species of both Terebrantia and Tubulifera, with aggressive interactions (see Dyreson and Terry, 1996) among larvae and adults. A more aggressive interaction between a pair of male *F. occidentalis* included rounds of abdominal sparring, followed by one male climbing on the other's thorax and abdomen and grabbing his opponent (Fig. 3.2B 1–4). The opponent would attempt to flick the grabbing male with his abdomen. Several bouts of abdominal sparring and grabbing/flipping behaviour commonly ensued before one of the following consequences occurred: one or both males left the fighting area; one male stayed and continued to chase other males from the area; one or both males were flicked off the floral or artificial surface (Terry and Dyreson, 1996). The 'winner' of an escalated fight (i.e. the male who stayed in the fighting area and chased away other males) seemed to be the individual who spent more time grabbing his opponent (Terry, 1995). The most aggressive interactions (grabbing and flipping) among males were relatively rare events, as most males would immediately part or exchange just a few abdominal flicks. Behaviour changed drastically as densities within the swarms increased and the most aggressive interactions were never observed when male densities were very high (Terry, 1995). Changes in male behaviour associated with density were also observed by Kirk (1985b) within aggregations of *Thrips* spp. in Britain.

The fights among male *Frankliniella occidentalis* ostensibly occurred to clear a small area on the floral corolla lobes where most females landed. Females generally mated with the first male they encountered within the aggregation and then refused most other males (Terry and Dyreson, 1996). Within gloxinia blossoms, most males whether fighting or in mating pairs were found on floral lobes, whereas females were more evenly distributed on floral parts. No fights or mating occurred around the stamens. This caused local differences in sex ratios: highly male-biased ratios on floral lobes and female-

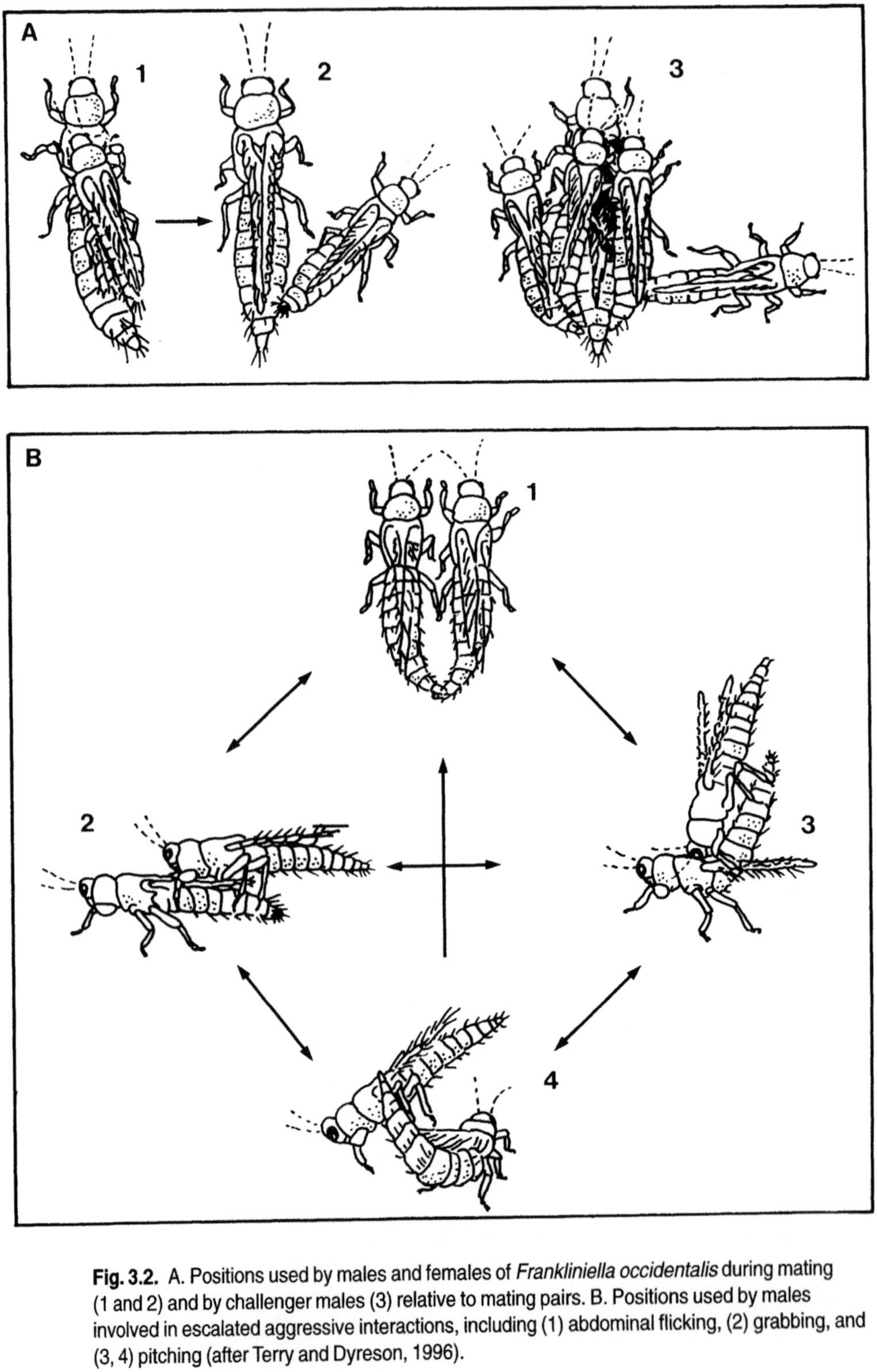

Fig. 3.2. A. Positions used by males and females of *Frankliniella occidentalis* during mating (1 and 2) and by challenger males (3) relative to mating pairs. B. Positions used by males involved in escalated aggressive interactions, including (1) abdominal flicking, (2) grabbing, and (3, 4) pitching (after Terry and Dyreson, 1996).

biased ratios near the stamens. There is strong selection on *F. occidentalis* males to be the first to mate because females generally rejected males for many days after mating (Terry and Schneider, 1993). Based on these observations, Terry and Dyreson (1996) suggested that this mating system is similar to a resource-based defence polygyny rather than a lek or landmark based system (Thornhill and Alcock, 1983; Andersson, 1994).

Terry and Dyreson (1996) compared the morphology of fighting males to those observed not fighting. Based on results of a discriminant function analysis, fighting males had larger abdomens and legs and were heavier than non-fighters. Adjusted for size, fighters had even larger than expected legs and abdomen widths than non-fighters, who had relatively longer but narrower abdomens for their size. This greater abdominal width of fighters would increase torque when they engaged in abdomen flicking. Larger legs would improve their grabbing ability during offensive actions. The larger leg, abdomen width and weight would give a greater moment of inertia for defence when being flicked. Mating males fit the morphology of both the fighters and non-fighters, which suggests that both fighters and non-fighters obtained mates. It is not clear whether these are alternative tactics (dominant males fighting to clear an area, and a subordinate patrolling tactic) and the success of each has not been determined.

Another thrips species may have a similar behaviour. Appanah and Chan (1981) noted that *Megalurothrips*, a flower thrips that pollinates dipterocarps in Malaysia, was territorial in flowers it inhabitated and defended areas of petals against other thrips.

In wingless Terebrantia, or in species where only the females are winged and hosts are patchily distributed, mating is expected to occur at brood sites among siblings before dispersal and this will lead to female-biased sex allocation due to local mate competition (Hamilton, 1967). This generally holds for those species examined that fit these criteria (Crespi, 1993; Strauss and Karban, 1994a,b). Only enough males should be produced within a brood to ensure adequate insemination of females. Unusual mating behaviour has been observed in *Limothrips cerealium* and *L. denticornis* species with wingless males (Pussard-Radulesco, 1931; Lewis, 1973; Bates and Weiss, 1991b). The males eclose in cereals before the winged females, and some males try to mate with female pupae on the leaf sheath, though the majority of matings are with newly emerged adult females.

It is not known whether sexes find and recognize each other within aggregations by visual or chemical means. Kirk (1985b) observed that *Thrips* spp. males appear to detect females within 2 mm. Lewis (1973) suggested that sexes find each other by odours using the sense cones on the antennae. Milne (1997) demonstrated that *Frankliniella schultzei* males attract females by pheromones. Translucent traps containing males placed within the host near aggregations attracted females (8.8 to 10.9 females per trap) and not males, while control traps with no males caught almost no females (0.1 to 0.7 per trap). In the laboratory, Milne (1997) also reported changes in virgin male and female behaviour when near the opposite sex. When an open vial containing a virgin male was placed next to a virgin female, females walked faster, turned more and vibrated their wings more frequently. Almost all females walked into the vial with the male. Once a female entered the male's vial, the male also increased walking, vibrating wings and rubbing his leg over his abdomen. These activities suggest that a short-range sex pheromone may help females find males. It is not clear whether males detect females through visual or chemical cues.

Terry and Schneider (1993), using playbacks (at one-fifth to one-tenth speed) of video recordings of *Frankliniella occidentalis* matings through a dissecting microscope, in combination with an event recorder, documented the details of copulatory behaviour. Males most often (68 per cent) initially contacted the female head-to-head with a brief antennal contact between them. Otherwise, the male approached the female's abdomen and touched her with his antennae, then attempted to mount by climbing on her dorsum then curling his abdomen beneath hers. He then inserted his aedeagus into her genitalia. If the female rejected the male, she continually bucked him until he retreated. He usually attempted to mate again. If she accepted the male, she lowered her abdomen and thorax slightly as he climbed on her. After copulation began, the male palpated the base of the female antennae with the tips of his. At the same time, the male stroked the female thorax and abdomen with his mesothoracic leg. This continued for approximately 18 per cent of the initial copulation period. This period was followed by a relatively calm period with little or no movement lasting about 35 per cent of the copulation time. Just before the sexes parted, a female (especially if disturbed by other males) would walk around, sometimes dragging the male behind her. Within aggregations (Fig. 3.2A 3), other males climbed on the female or attempted to disrupt the mating pair (Terry and Dyreson, 1996). After separating, some males and most females groomed. Males generally tried to mate again with the same female, but females almost always rejected the same or other males for more than 5 days (Terry and Schneider, 1993). The behaviour of males rejected by virgin females was compared with those not rejected. Generally, rejections occurred immediately before or after the male climbed on the female. The most common cause of failure by males after mounting was omitting to tap the female's antennae early in copulation. These females subsequently mated with other males and produced female offspring.

This description resembles those for other Terebrantia including *Frankliniella schultzei* (Milne, 1997), *Rhipiphorothrips cruentatus* and *Retithrips syriacus* (Ananthakrishanan, 1990), *Aeolothrips intermedius* and other Terebrantia (Buffa, 1907; Franssen and Mantel, 1964; Lewis, 1973) and the tubuliferan *Tiarothrips subramanii* (Ananthakrishnan and James, 1983; Ananthakrishnan, 1990) which have been studied in terms of approaches and positions used during copulation. There are many similarities in positions and approaches between sexes of both suborders, but differences in mating frequencies and which sex controls the opportunities for mating.

Frequency of mating and copulation duration

Both tubuliferan and terebrantian males are promiscuous (Lewis, 1973; Gilstrap and Oatman, 1976; Varadarasan and Ananthakrishnan, 1982; Strauss and Karban 1994a,b), but few studies have documented the mating frequencies and success of individuals. Adults of Thysanoptera generally mate within 2 to 3 days of eclosion from the pupal stage (Lewis, 1973); some males mate within 6 h of eclosion. Tubuliferan males, *Hoplothrips pedicularius*, *Elaphrothrips tuberculatus*, and *H. karnyi*, all mate several times with ovipositing females, some preferentially with females that have not yet oviposited (Crespi, 1986a, 1988b, 1988d). Male *Frankliniella occidentalis* can mate in the first 24 h after eclosion and throughout their life (Terry and Schneider, 1993) with only a slight decline in activity with age. In laboratory studies, males attempted to mate with up to 15 females within a 3 h period and inseminated up to ten females during this period. In *Thrips palmi*, the male precopulatory period lasts until 6 h after eclosion (Chu and Wang, 1990; Wang and Chu, 1990). Males produced from virgin females were slightly more active and copulated more frequently during the first 10 days after eclosion than those produced from mated mothers. Mating ability did not significantly decline during the first 10 days after eclosion and ranged from one to two matings per day. Males of the wingless species, *Apterothrips apteris*, on seaside daisy, mate with several females per day (Strauss and Karban, 1994b). Male *Scolothrips sexmaculatus* mate with up to three females per 1.5 h observation period (Gilstrap and Oatman, 1976).

Females vary greatly in their mating frequency. Several sporophagous/mycophagous tubuliferan females, *Hoplothrips pedicularius*, *Elaphrothrips tuberculatus*, and *H. karnyi* mate up to four times with the same guarding male prior to ovipositing (Crespi, 1986a, 1988b,d). Females were never observed rejecting matings (Crespi, 1988d). By contrast, females of some Indian tubuliferan gall-forming thrips remain quiet for at least 1 or 2 days after eclosion, then aggregate within the gall, copulate and refuse further matings (Varadarasan and Ananthakrishanan, 1982). Studies of terebrantians suggest restriction on mating by females (Terry and Schneider, 1993; Milne, 1997). Female *Frankliniella occidentalis* can successfully mate within 12 h of eclosion, but females refuse males for more than 5 days after the initial mating. Because males mate frequently and females are receptive infrequently, there is strong selection for males to find and mate virgin females (Terry and Schneider, 1993). Similar behaviours were observed in another anthophilous thripid, *F. schultzei* (Milne, 1997). Virgin females would accept males for the initial mating, but then would not mate again within the 1 h observation period, although males attempted to mate. Wang and Chu (1990) found that *Thrips palmi* females mated immediately after eclosion, but only produced male eggs on the first 2 days, and then produced mixed sexes thereafter. Female mating frequencies were not reported.

The duration of copulation varies both within and among species of Thysanoptera. Among Tubulifera, some very brief copulation times have been observed. *Hoplothrips pedicularius* matings last from 7 to 70 s, with an average of 26.3 s, and those of *H. karnyi* range from 8 to 42 s with a 19.2 s average (Crespi, 1986a, 1988d). By contrast, some idolothripines copulate for 3 to 15 min (Ananthakrishnan, 1984). Among Terebrantia, average copulation times generally range from 3 to 15 min (Lewis, 1973; Kirk, 1985b; Ananthakrishnan,

1990). Some Thripidae remain copulating for a long time; more than 60 min has been observed in *Apterothrips apteris* on the seaside daisy, with an average of 35 min (Strauss and Karban, 1995). Many aeolothripids have been observed mating for 20 to 30 min (Buffa, 1907). In the laboratory, when *Frankliniella occidentalis* pairs were tested in the absence of other males or females, successful copulations lasted from 114 to 595 s, with an average of 240 s (Terry and Schneider, 1993). Females that mated for less than 69 s produced only male offspring, suggesting no insemination in these shorter coupling times. In separate field studies with aggregations, copulation duration averaged 229 s (Terry and Gardner, 1990) and 176 s (Terry and Dyreson, 1996). In the field, couples were often disturbed by other males attempting to mate with the female. This may cause the couple to depart sooner. Similar variability was observed with *F. schultzei* (Milne, 1997). Within aggregations in the field, matings lasted an average of 87 to 92 s depending upon the flower species. In the laboratory, matings lasted an average of 200 s and some matings lasted much longer than others, depending upon the rearing host. Copulation duration did not vary between diapausing and non-diapausing males and females of *F. intonsa* (Murai, 1987), but the diapause state did affect the pre-mating times (i.e. from first contact to initiation of copulation). Non-diapausing pairs began mating within 26 s of the first contact, whereas diapausing pairs took over 8 min.

Oviposition Sites and Process

Morphological characters associated with oviposition clearly distinguish the two suborders of Thysanoptera. Terebrantian females have a large well-developed ovipositor, which is curved upwards in Aeolothripidae and downwards in most other Terebrantia (see Chapter 2). The ovipositor of Terebrantia is used to pierce plant tissue and deposit the egg, whereas tubuliferan ovipositors are reduced to a U-shaped chute (Heming, 1995) and eggs are laid on the surface of a substrate (Lewis, 1973). Terebrantia lay eggs singly into plant tissue but Tubulifera may lay clusters of eggs at oviposition sites that in some species are guarded by males or females. Eggs of Terebrantia are oval to kidney shaped and the chorion is smooth to reticulate and pale to brown in colour (Lewis, 1973; Ananthakrishnan, 1990; Heming, 1995, see Chapter 2). Melanthripid and aeolothripid eggs are brown and are cut off obliquely at their anterior ends (Ananthakrishnan, 1990). Kirk (1985e) describes thripid eggs as having an anterior operculum that is removed when hatching. Most mature eggs are large relative to the abdomen size (Heming, 1995) and measure 200 × 100 μm. Tubuliferan eggs are longer, approximately 350 to 550 μm, and more cylindrical in shape.

Terebrantia will oviposit into most plant tissues, although some species have preferred sites. For example, *Thrips calcaratus* push their ovipositor into the sides of the main veins on the underneath side of the leaf of American basswood (Raffa *et al.*, 1992). Others scatter their eggs on various floral and vegetative tissues, depending upon plant species and phenological stage.

Tubulifera lay eggs in a variety of places depending upon the species. Gall dwellers generally lay inside the gall wall, others glue the eggs to the walls or between phyllodes that are glued together (Varadarasan and Ananthakrishan, 1981; Sota, 1988; Ananthakrishnan, 1990; Crespi, 1992a). Some inquiline thrips predators of gall producers lay eggs near their host eggs (Varadarasan and Ananthakrishnan, 1981, 1982; Crespi, 1992a). Tubulifera that feed on fungi lay eggs near their fungal hosts, under bark, in bark crevices, or on old leaves.

Terebrantian eggs are embedded into tissue with their long axis at a shallow angle to the surface or a steeper angle of 30 to 45° (Lewis, 1973). In many cases oviposition causes direct damage to the plant, usually a cosmetic blemish but nevertheless an economic loss to producers of many crops (Childers and Achor, 1995) (see Fig. 13.6).

Oviposition

To lay, the ovipositing terebrantian female raises the tip of the abdomen which contracts and pushes the egg through the oviduct. While the abdomen is arched upward and the tip touches the surface, the female feels the substrate with the terminal abdominal setae for an oviposition site. With the ovipositor slightly protruding beyond the abdomen, the female stabs its tip vertically into the tissue. When the abdomen is pushed backwards, the ovipositor pivots on its basal articulation and then further protrudes from the sheath and becomes perpendicular to the abdomen and plant tissue. The abdomen works the ovipositor into the plant and then saw-like blades cut further into the tissue, then separate to allow the egg to be expelled into the incision (see Chapter 2 and Fig. 2.13: 2, 4). Some species seal off the egg with a drop of fluid as the ovipositor retracts from the tissue; others have their eggs more deeply embedded into the plant which closes over the egg as the ovipositor is removed from the plant (see Fig. 2.21: 1).

Eggs of Terebrantia not completely buried or covered could be more subject to desiccation and predation as found in *Megalurothrips sjostedti* (Tamò *et al.*, 1993a). As an exception in Terebrantia, *Chirothrips* eggs are only partly embedded into tissue and one species, *Chirothrips crassus*, lays its eggs on the leaf surface as do the Tubulifera (Pesson, 1951). The time required to oviposit completely usually takes a few minutes and females either feed or rest after laying each egg (Lewis, 1973).

Eggs tend to be laid in older, non-expanding leaf tissue to avoid eggs being crushed as cells enlarge and proliferate. In citrus leaves, cork cells are produced in reaction to eggs of *Heliothrips haemorrhoidalis*. This corking can protect the egg from excessive water but can crush or squeeze it out of its cavity if the leaf is young and cells are rapidly expanding (Rivnay, 1935; Lewis, 1973). In India, *Retithrips syriacus* and *Rhipiphorothrips cruentatus* oviposit on slightly mature or mature leaves of their respective hosts and rarely on very young or senescent leaves (Ananthakrishnan, 1990). Techniques used to visualize eggs inside plant tissue are described in Chapter 11.

Several thrips will oviposit as well as feed on artificial surfaces lacking any specific plant surface stimulants, such as stretched parafilm membranes (see

Chapter 12). These include many generalists such as *Thrips tabaci* (Sakimura and Carter, 1934), *T. imaginis* (Teulon and Penman, 1986, 1991), *Hercinothrips femoralis* (Laughlin, 1971), *Frankliniella intonsa* and *T. hawaiiensis* (Murai and Ishii, 1982), *F. occidentalis* (Teulon, 1992), *F. schultzei* (Milne *et al.*, 1996), *T. fuscipennis* (Kirk, 1985c), and two host specialists, *Kakothrips pisivorus* and *Ceratothrips ericae* (Kirk, 1985c).

Distribution of eggs on plants

Eggs are laid generally in leaf or floral tissue but some species have distinct preferences. Overwintering *Taeniothrips inconsequens* adult females lay in buds of sugar maple. Approximately ten times as many eggs were found in flowering as in vegetative buds, but no preference was observed between male flower buds and mixed male and female flower buds (Leskey *et al.*, 1995). Within leaf buds, most eggs are laid on the main veins and petioles; within flower buds, petioles are preferred, bud scales and leaf stems less so. No particular aspect of the tree (quadrat position, upper or lower part of the crown) seems to be particularly favoured (Teulon and Cameron, 1995).

On citrus, *Frankliniella bispinosa* lays eggs in the pistil, calyx, petals and filaments of both flowers and swollen (unopened) floral buds (Childers and Achor, 1991). Almost equal proportions hatch from open as unopened buds. Within flowers, most larvae emerge from the pistil–calyx region of open flowers, followed by the petals, filaments and anthers.

Eggs of *Frankliniella occidentalis* are laid in apple blossom of all ages: unopened pink buds, open blossoms, and petalless clusters (Terry, 1991). Although adults are most common in open blossoms, more eggs are found on the petalless clusters, where unhatched eggs accumulate as the cluster matures from pink to petalless stages. Most eggs are laid on the king bud, i.e. the central bud surrounded by the five or six lateral buds, and fewer on sepals and stems where they cause no blemish.

Within-plant distribution of *Thrips tabaci* on leek varies with insect stage, suggesting differences in their preferences for plant regions. Adults appear to lay where they feed, moving continually toward the centre of the plant on to younger tissue (Theunissen and Legutowska, 1991).

A number of studies have examined the distribution of eggs and larvae of *Kakothrips pisivorus* on beans and peas (Buhl, 1937; Kirk, 1985a,c; Stoddard, 1986). Adult thrips enter the closed flowers at night just before anthesis to lay on the staminal sheath along its line of contact with the keel petal (Buhl, 1937). Stoddard (1986) studied oviposition sites on many different lines of faba beans. Most eggs were found in the neck of the pistil near its line of emergence from the staminal sheath; a few were laid on other parts of the ovarian wall. Only a small minority of flowers had the stylar canal blocked by an egg which would have prevented fertilization of the flower.

The polyphagous *Megalurothrips sjostedti* damages floral parts and induces shedding on cowpea. Eggs are laid in specific organs and ages of tissue (Rösingh, 1980; Salifu, 1986; Tamò *et al.*, 1993b). Potted plants exposed to field thrips infestations for a short time had more eggs on leaf petioles than on stipules, young inflorescences or leaves (Rösingh, 1980). However, within-plant distribution of eggs in field plants was analysed over time (Tamò *et al.*, 1993b) and changes in preference occurred over phenological stages. Early in the season, very young inflorescences are unattractive oviposition sites because of their small size, and leaf petioles are preferred. After peduncle elongation, inflorescences are preferred, especially older ones (Salifu, 1986; Tamò *et al.*, 1993b). On adzuki bean, *Megalurothrips usitatus* lays on foliage, petals and sepals (Chang, 1992). Larval aggregations result from the concentration of eggs in the flowers.

Limothrips denticornis lays in the interveinal spaces of the inner surface of cereal leaf sheaths (Bournier, 1956). Most eggs on spring barley are laid during the period between the boot stage and completion of inflorescence emergence (Bates and Weiss, 1991a). The middle third of the sheath is preferred; the upper third is more exposed to heat and drying and the lower section collects water.

Within flowers of *Malvaviscus arboreus*, *Frankliniella schultzei* lays most eggs on the mid-section and base of petals with fewer on petal apices and staminal columns (Milne *et al.*, 1996). *Frankliniella occidentalis*, *F. tritici* and *F. fusca*, the most common thrips in tomato flowers (Salguero Navas *et al.*, 1994), rarely occurred on other plant parts (Salguero Navas *et al.*, 1991). Eggs of *F. occidentalis* laid into young tomato fruit cause halo cosmetic damage. Details of the damage produced directly by oviposition are given in Chapter 13.

Oviposition by *Haplothrips tritici*, a phlaeothripid that infests wheat in central Europe, appears to be affected by plant morphological traits (Mikhailova, 1981; Czencz, 1992). Eggs are found on the grain base or within the glumes. Species of wheat with dense hairs at the base of the grains (e.g. *Triticum spelta*) were ten times more heavily infested with eggs than sparsely haired grains of *T. monococcum* (Mikhailova, 1981). Cultivars of *Triticum aestivum* with hairy glumes also appear to be attractive oviposition sites for *Haplothrips tritici* (Czencz, 1992).

Summary

This chapter reviews mechanisms by which Thysanoptera find and select their hosts and mates, and behaviours involved in these processes. No general pattern of host selection can be assigned to this taxon. How species select their hosts can only be broadly grouped into associations with their food preferences. Anthophilous species clearly use specific colour and odour cues associated with floral qualities for long- and short-range host finding and then finalize their selection after landing on and tasting potential food sources. Attractive colours or odours will induce these thrips to land in the absence of other cues. Other host qualities may keep them there or drive them away after landing, including such traits as degree of pubescence and epidermal thickness, as well as close-range odours and feeding allelochemicals. In contrast, cereal and grass feeders do not appear to use particular colours or floral volatiles for host finding. Other odours may be involved but have not been adequately investigated. Very specific associations also apply. *Thrips obscuratus* that feeds on flowers and fruit of stonefruit is highly attracted to a non-floral volatile, ethyl nicotinate, but several generalists show little or no reaction to this compound. Some very successful generalists may do well on leaves and reproductive parts of many species, but fail to thrive on leaves of certain plants that a more specialist thrips may do well on.

The reproductive behaviour of Thysanoptera has had limited study from which to draw generalities. Specific details of the copulatory behaviour and oviposition process bear many similarities among thrips, but how most species attract and select mates remains largely speculative or unknown. Aggregations are common throughout the order, and some of these are strictly for mating. For some anthophilous thripids, floral parts are used as aggregation sites for males, who intercept landing females for mating. There is evidence among these thripids of a male sex pheromone that attracts females. Within Thysanoptera there is a diverse range of reproductive biology. Some species reproduce parthenogenetically and others are haplodiploid (see Chapter 2), some have extremely female-biased sex ratios and siblings mate within their home colony, some have complex mating systems with alternative mating tactics associated with male fighting and sneak mating, and many are panmictic with approximately 1:1 sex allocation. This taxon offers a wealth of different behaviours among species for comparison and, so far, the mating system of only a few tubuliferans has been thoroughly studied.

References

Ananthakrishnan, T.N. (1984) *Bioecology of Thrips*. Indira Publishing House, Oak Park, Michigan.

Ananthakrishnan, T.N. (1990) *Reproductive Biology of Thrips*. Indira Publishing House, Oak Park, Michigan.

Ananthakrishnan, T.N. and Gopichandran, R. (1993) *Chemical Ecology in Thrips–Host Plant Interactions*. International Science Publisher, New York.

Ananthakrishnan, T.N. and James, P.W. (1983) Feeding preferences in some sporophagous Idolothripinae (Tubulifera: Thysanoptera). In: *Proceedings of the Indian National Science Academy*, B. 49, 86–92.

Andersson, M. (1994) *Sexual Selection*. Princeton University Press, Princeton, New Jersey.

Annand, P.N. (1926) Thysanoptera and the pollination of flowers. *American Naturalist* 60, 177–182.

Antignus, Y., Mor, N., Ben Joseph, R., Lapidot, M. and Cohen, S. (1996) Ultraviolet-absorbing plastic sheets protect crops from insect pests and from virus diseases vectored by insects. *Environmental Entomology* 25, 919–924.

Appanah, S. and Chan, H.T. (1981) Thrips: pollinators of some dipterocarps. *Malaysian Forester* 44, 234–252.

Bakker, F.M. and Sabelis, M.W. (1987) Attack success of *Amblyseius mckenzei* and the stage related defensive capacity of thrips larvae. *Mededelingen van de Faculteit Landbouwetenschappen Rijksuniversiteit Gent* 51/3a, 1041–1044.

Bakker, F.M. and Sabelis, M.W. (1989) How larvae of *Thrips tabaci* reduce the attack success of phytoseid predators. *Entomologia Experimentalis et Applicata* 50, 47–51.

Ballard, W.W. (1951) Varietal differences in susceptibility to thrips injury in upland cotton. *Agronomy Journal* 43, 37–44.

Bates, B.A. and Weiss, M.J. (1991a) The spatial distribution of *Limothrips denticornis* Haliday (Thysanoptera: Thripidae) eggs on spring barley. *Canadian Entomologist* 123, 205–210.

Bates, B.A. and Weiss, M.J. (1991b) Seasonal abundance of *Limothrips denticornis* (Thysanoptera: Thripidae) on spring barley. *Environmental Entomology* 20, 577–581.

Bautista, R.C. and Mau, R.F.L. (1994) Preferences and development of western flower thrips (Thysanoptera: Thripidae) on plant hosts of tomato spotted wilt tospovirus in Hawaii. *Environmental Entomology* 23, 1501–1507.

Bautista, R.C., Mau, R.F.L., Cho, J.J. and Custer, D.M. (1995) Potential of tomato spotted wilt tospovirus plant hosts in Hawaii as virus reservoirs for transmission by *Frankliniella occidentalis* (Thysanoptera: Thripidae). *Phytopathology* 85, 953–958.

Beavers, J.B., Shaw, J.G. and Hampton, R.B. (1971) Color and height preference of the citrus thrips in a navel orange grove. *Journal of Economic Entomology* 64, 1112–1113.

Blum, M.S. (1991) Chemical ecology of the Thysanoptera. In: Parker, B.L., Skinner, M. and Lewis, T. (eds) *Towards Understanding Thysanoptera*. USDA Forest Service General Technical Report NE-147, pp. 95–112.

Blum, M.S., Foottit, R. and Fales, H.M. (1992) Defensive chemistry and function of the anal exudate of the thrips (*Haplothrips leucanthemi*). *Comparative Biochemistry and Physiology, Comparative Pharmacology and Toxicology* 102, 209–211.

Bode, W. (1978) Ultrastructure of the sternal glands in *Thrips validus* Uzel (Thysanoptera, Terebrantia). *Zoomorphologie* 90, 53–65.

Bournier, A. (1956) Contribution à l'étude de la parthenogenese des Thysanoptères et de sa cytologie. *Archives de Zoologie Experimentale et Generale Notes et Revue* 93, 280–292.

Bournier, J.P. (1983) Un insecte polyphage: *Thrips palmi* (Karny) important ravageur du cotonnier aux Philippines. *Coton et Fibres Tropicale* 38, 286–289.

Bradbury, J.W. (1985) Contrasts between insects and vertebrates in the evolution of male display, female choice, and lek mating. *Fortschritte der Zoologie*, Bd. 31 Holldobler/Lindauer (Hrsg.): *Experimental Behavioral Ecology*. G. Fischer Verlag, Stuttgart, pp. 273–289.

Braman, S.K. and Beshear, R.J. (1994) Seasonality of predaceous plant bugs (Heteroptera: Miridae) and phytophagous thrips (Thysanoptera: Thripidae) as influenced by host plant phenology of native azaleas (Ericales: Ericaceae). *Environmental Entomology* 23, 712–718.

Brar, K.S., Sidhu, A.S. and Chadha, M.L. (1993) Screening onion varieties for resistance to *Thrips tabaci* Lind. and *Helicoverpa armigera* (Hubner). *Journal of Insect Science* 6, 123–124.

Broadbent, B.A. and Allen, W.R. (1995) Interactions within the western flower thrips/tomato spotted wilt virus/host plant complex epidemiology. In: Parker, B.L., Skinner, M. and Lewis, T. (eds) *Thrips Biology and Management*. Plenum, New York, pp. 433–438.

Brødsgaard, H.F. (1989) Coloured sticky traps for *Frankliniella occidentalis* (Pergande) (Thysanoptera: Thripidae) in glasshouses. *Journal of Applied Entomology* 107, 136–140.

Brødsgaard, H.F. (1990) The effect of anisealdehyde as a scent attractant for *Frankliniella occidentalis* (Thysanoptera: Thripidae) and the response mechanism involved. *International Organization of Biological Control/West Palaeartic Regional Section Bulletin* XIII/5, 36–38.

Brown, S.L. and Brown, J.E. (1992) Effect of plastic mulch color and insecticides on thrips populations and damage to tomato. *HortTechnology* 2, 208–211.

Buffa, P. (1907) Trentuna specie di Tisanotteri Italiani. *Atti della Societa Toscana di Sciencze Naturali Residente in Pisa: Memoire* 23, 227–303.

Buffa, P. (1911) Studi intorno al ciclo partenogenetico dell'*Heliothrips haemorrhoidalis* (Bouché). *Redia* 7, 71–109.

Buhl, C. (1937) Beitrage zur Kenntnis der Biologie, wirtschaftlichen Bedeutung und Bekampfung von *Kakothrips robustus* Uz. *Zeitschrift für angewandte Entomologie* 23, 65–113.

Carter, W. (1939) Populations of *Thrips tabaci*, with special reference to virus transmission. *Journal of Animal Ecology* 8, 261–271.

Chamberlin, J.R., Todd, J.W., Beshear, R.J., Culbreath, A.K. and Demski, J.W. (1992) Overwintering hosts and wingform of thrips, *Frankliniella* spp., in Georgia (Thysanoptera: Thripidae): implications for management of spotted wilt disease. *Environmental Entomology* 21, 121–128.

Chang, N.T. (1990) Color preference of thrips in adzuki bean field. *Plant Protection Bulletin* 32, 307–316.

Chang, N.T. (1992) Dispersion patterns of bean flower thrips, *Megalurothrips usitatus* (Bagnall), (Thysanoptera: Thripidae) on flowers of adzuki bean. *Plant Protection Bulletin* 34, 41–53.

Childers, C.C. and Achor, D.S. (1991) Feeding and oviposition injury to flowers and developing floral buds of 'navel' orange by *Frankliniella bispinosa* (Thysanoptera: Thripidae) in Florida. *Annals of the Entomological Society of America* 84, 272–282.

Childers, C.C. and Achor, D.S. (1995) Thrips feeding and oviposition injuries to economic plants, subsequent damage and host responses to infestation. In: Parker, B.L., Skinner, M. and Lewis, T. (eds) *Thrips Biology and Management*. Plenum, New York, pp. 31–52.

Childers, C.C. and Beshear, R.J. (1992) Thrips (Thysanoptera) species associated with developing citrus flowers in Florida and a key to adult terebrantian females. *Journal of Entomological Science* 27, 392–412.

Childers, C.C. and Brecht, J.K. (1996) Colored sticky traps for monitoring *Frankliniella bispinosa* (Morgan) (Thysanoptera: Thripidae) during flowering cycles in citrus. *Journal of Economic Entomology* 89, 1240–1249.

Childers, C.C., Beshear, R.J., Brushwein, J.R. and Denmark, H.A. (1990) Thrips (Thysanoptera) species, their occurrence and seasonal abundance on developing buds and flowers of Florida (USA) citrus. *Journal of Entomological Science* 25, 601–614.

Chisholm, I.F. and Doncaster, C.C. (1982) Studying and recording the feeding behaviour of thrips. *Entomologia Experimentalis et Applicata* 31, 324–327.

Chisholm, I.F. and Lewis, T. (1984) A new look at thrips (Thysanoptera) mouthparts, their action and effects of feeding on plant tissue. *Bulletin of Entomological Research* 74, 663–675.

Chiu, H.T. and Wu, M.Y. (1993) Attractiveness of color trap to *Thrips hawaiiensis* (Morgan) (Thysanoptera: Thripidae) in the field. *Chinese Journal of Entomology* 13, 229–234.

Chiu, H.T., Shen, S.M. and Wu, M.Y. (1991) Occurrence and damage of thrips in citrus orchards in southern Taiwan. *Chinese Journal of Entomology* 11, 310–316.

Cho, K.J., Eckel, C.S., Walgenbach, J.F. and Kennedy, G.G. (1995) Comparison of colored sticky traps for monitoring thrips populations (Thysanoptera, Thripidae) in staked tomato fields. *Journal of Entomological Science* 30, 176–190.

Chu, Y.I. and Wang, C.L. (1990) The reproductive mechanism of *Thrips palmi* Karny. II. Comparison of male reproduced parthenogenetically and bisexually. *Chinese Journal of Entomology/Zhonghua Kunchong* 10, 79–87.

Chyzik, R., Klein, M., Yair B. and Cohen, A. (1995) Differential population density of western flower thrips in various flower colors of gladiolus. In: Parker, B.L., Skinner, M. and Lewis, T. (eds) *Thrips Biology and Management.* Plenum, New York, pp. 449–452.

Coley, P.D. (1988) Effects of plant growth rate and leaf lifetime on the amount of antiherbivore defense. *Oecologia* 74, 531–536.

Coli, W.M., Hollingsworth, C.S. and Maier, C.T. (1992) Traps for monitoring pear thrips (Thysanoptera: Thripidae) in maple stands and apple orchards. *Journal of Economic Entomology* 85, 2258–2262.

Coll, M. and Ridgway, R.L. (1995) Functional and numerical responses of *Orius insidiosus* (Heteroptera: Anthocoridae) to its prey in different vegetable crops. *Annals of the Entomological Society of America* 88, 732–738.

Coudriet, D.L., Kishaba, A.N., McCreight, J.D. and Bohn, G.W. (1979) Varietal resistance in onion to thrips. *Journal of Economic Entomology* 72, 614–615.

Crespi, B.J. (1986a) Territoriality and fighting in a colonial thrips, *Hoplothrips pedicularius*, and sexual dimorphism in Thysanoptera. *Ecological Entomology* 11, 119–130.

Crespi, B.J. (1986b) Size assessment and alternative fighting tactics in *Elaphrothrips tuberculatus* (Insecta: Thysanoptera). *Animal Behavior* 34, 1324–1335.

Crespi, B.J. (1988a) Sex-ratio selection in a bivoltine thrips. I. Conditional sex ratio manipulation and fitness variation. *Evolution* 42, 1199–1211.

Crespi, B.J. (1988b) Alternative male mating tactics in a thrips: effects of sex ratio variation and body size. *American Midland Naturalist* 119, 83–92.

Crespi, B.J. (1988c) Adaptation, compromise and contraint: the development, morphometrics, and behavioural basis of a fighter–flier polymorphism in male *Hoplothrips karnyi* (Insecta: Thysanoptera). *Behavioral Ecology and Sociobiology* 23, 93–104.

Crespi, B.J. (1988d) Risks and benefits of lethal male fighting in the colonial, polygynous thrips, *Hoplothrips karnyi* (Insecta: Thysanoptera). *Behavioral Ecology and Sociobiology* 22, 293–301.

Crespi, B.J. (1989) Sexual selection and assortive mating in subdivided populations of the thrips *Elaphrothrips tuberculatus* (Insecta: Thysanoptera). *Ethology* 83, 265–278.

Crespi, B.J. (1990) Subsociality and female reproductive success in a mycophagous thrips: an observational and experimental analysis. *Journal of Insect Behavior* 3, 61–74.

Crespi, B.J. (1991) Heterozygosity in the haplodiploid Thysanoptera. *Evolution* 45, 458–464.

Crespi, B.J. (1992a) Behavioural ecology of Australian gall thrips (Insecta, Thysanoptera). *Journal of Natural History* 26, 769–809.

Crespi, B.J. (1992b) Eusociality in Australian gall thrips (Insecta, Thysanoptera). *Nature (London)* 359, 724–726.

Crespi, B.J. (1993) Sex allocation ratio selection in Thysanoptera. In: Wrensch, D.L. and Ebbert, M.A. (eds) *Evolution and Diversity of Sex Ratio in Insects and Mites.* Chapman & Hall, New York, pp. 214–234.

Culliney, T.W. (1990) Population performance of *Thrips palmi* (Thysanoptera: Thripidae) on cucumber infected with a mosaic virus. *Proceedings of the Hawaiian Entomological Society* 30, 85–90.

Czencz, K. (1987) The role of coloured traps in collecting thrips fauna. In: Holman, J.,

Pelikan, J., Dixon, A.F.G. and Weisman, L. (eds) *Population Structure, Genetics and Taxonomy of Aphids and Thysanoptera. Proceedings of International Symposium, Smolenice, Czechoslavakia, September 1985*. SPB Academic Publishing, The Hague, pp. 426–435.

Czencz, K. (1992) A comparative investigation of the infestation of some winter wheat types by *Haplothrips tritici* Kurd (Thysanoptera, Phlaeothripidae). *Journal of Applied Entomology* 113, 209–213.

Day, M.F. and Irzykiewicz, H. (1954) Physiological studies in thrips in relation to transmission of tomato spotted wilt virus. *Australian Journal of Biological Science* 7, 274–281.

DeGrandi-Hoffman, G., Terry, I. and Huber, R.T. (1988) Incorporating fruit set estimates with thrips management to create a decision support system for apples. *Hortscience* 23, 571–574.

de Jager, C.M., Butot, R.P.T., Dejong, T.J., Klinkhamer, P.G.L. and Van Der Meijden, E. (1993) Population growth and survival of western flower thrips *Frankliniella occidentalis* Pergande (Thysanoptera, Thripidae) on different chrysanthemum cultivars: two methods for measuring resistance. *Journal of Applied Entomology* 115, 519–525.

de Jager, C.M., Butot, R.P.T. and Guldemond, A. (1995a) Genetic variation in chrysanthemum for resistance to western flower thrips and *Thrips tabaci*. In: Parker, B.L., Skinner, M. and Lewis, T. (eds) *Thrips Biology and Management*. Plenum, New York, pp. 403–406.

de Jager, C.M., Butot, R.P.T., Klinkhamer, P.G.L., de Jong, T.J., Wolff, K. and van der Meijden, E. (1995b) Genetic variation in chrysanthemum for resistance to *Frankliniella occidentalis*. *Entomologia Experimentalis et Applicata* 77, 277–287.

de Jager, C.M., Butot, R.P.T., Klinkhamer, P.G.L. and van der Meijden, E. (1995c) Chemical characteristics of chrysanthemum cause resistance to *Frankliniella occidentalis* (Thysanoptera: Thripidae). *Journal of Economic Entomology* 88, 1746–1753.

Dobson, H.E.M. (1987) Role of flower and pollen aromas in host-plant recognition by solitary bees. *Oecologia* 72, 618–623.

Dobson, H.E. (1994) Floral volatiles in insect biology. In: Bernays, E.A. (ed.) *Insect–Plant Interactions*, Vol. 5. CRC Press, Florida, pp. 47–81.

Evans, H.E. (1977) Extrinsic versus intrinsic factors in the evolution of insect sociality. *Bioscience* 27, 613–616.

Evans, J.W. (1932) *The Bionomics and Economic Importance of* Thrips imaginis. Council of Scientific and Industrial Research, Australia.

Frank, W.A. and Huber, R.T. (1987) Seasonal population trends and damage associated with western flower thrips, *Frankliniella occidentalis* Pergande on Arizona pistachios. *Southwestern Entomologist* 12, 299–303.

Franssen, C.H.J. and Mantel, W.P. (1964) Notes on copulation of thrips. *Tijdshrift voor Entomologie* 107, 341–344.

Frey, J.E., Cortada, R.V. and Helbling, H. (1994) The potential of flower odours for use in population monitoring of western flower thrips *Frankliniella occidentalis* Perg. (Thysanoptera: Thripidae). *Biocontrol Science and Technology* 4, 177–186.

Gaum, W.G., Giliomee, J.H. and Pringle, K.L. (1994) Resistance of some rose cultivars to the western flower thrips, *Frankliniella occidentalis* (Thysanoptera: Thripidae). *Bulletin of Entomological Research* 84, 487–492.

Gawaad, A.A.A. and Soliman, A.S. (1972) Studies on *Thrips tabaci* Lindeman. IX. Resistance of nineteen varieties of cotton to *Thrips tabaci* L. and *Aphis gossypii* G. *Zeitschrift für angewandte Entomologie* 70, 93–98.

Gawaad, A.A.A., El-Gayer, F.H. and Soliman, A.S. (1973) Studies on *Thrips tabaci* Lindeman. X. Mechanism of resistance to *Thrips tabaci* L. in cotton varieties. *Zeitshrift für angewandte Entomologie* 73, 251–255.

Gillespie, D.R. and Vernon, R.S. (1990) Trap catch of western flower thrips (Thysanoptera: Thripidae) as affected by color and height of sticky traps in mature greenhouse cucumber crops. *Journal of Economic Entomology* 83, 971–975.

Gilstrap, F.E. and Oatman, E.R. (1976) The bionomics of *Scolothrips sexmaculatus* (Pergande) (Thysanoptera: Thripidae), an insect predator of spider mites. *Hilgardia* 44, 27–59.

Gomez, P.R. and Hernandez, A.C. (1992) The effect of the colour and the height of sticky traps of *Frankliniella occidentalis* (Perg.) (Thysanoptera: Thripidae) on pumpkins in greenhouses. *Investigacion Agraria Produccion y Proteccion Vegetales* 7, 443–449.

Gopichandran, R., Gurusubramanian, G. and Ananthakrishnan, T.N. (1992) Influence of chemical and physical factors on the varietal preference of *Retithrips syriacus* (Mayet) (Panchaetothripinae: Thysanoptera) on different cultivars of cotton. *Phytophaga* 4, 1–9.

Greenough, D.R., Black, L.L. and Bond, W.P. (1990) Aluminium-surfaced mulch: an approach to the control of tomato spotted wilt virus in solanaceous crops. *Plant Disease* 74, 805–808.

Haga, K., Suzuki, T., Kodama, S. and Kuwahara, Y. (1989) Secretion of thrips III. Secretions of acid-emitting thrips, *Hoplothrips japonicus* (Thysanoptera: Phlaeothripidae). *Japanese Journal of Applied Entomology and Zoology* 24, 242–244.

Haga, K., Suzuki, T., Kodama, S. and Kuwahara, Y. (1990) Secretion of thrips VI. Identification of β-myrcene from *Thlibothrips isunoki* (Thysanoptera: Phlaeothripidae). *Japanese Journal of Applied Entomology and Zoology* 25, 138–139.

Hamilton, W.D. (1967) Extraordinary sex ratios. *Science* 156, 477–488.

Harrewijn, P., Piron, P.G.M. and Mollema, C. (1996a) Electrical recording of plant penetration by western flower thrips. *Entomologia Experimentalis et Applicata* 79, 345–353.

Harrewijn, P., Piron, P.G.M. and Mollema, C. (1996b) Electrically recorded probing behaviour of thrips species on optimal and suboptimal hosts. *Entomologia Experimentalis et Applicata* 80, 43–45.

Hata, T.Y., Hara, A.H. and Hansen, J.D. (1991) Feeding preference of melon thrips on orchids in Hawaii. *Hortscience* 26, 1294–1295.

Heming, B.S. (1978) Structure and function of the mouthparts in larvae of *Haplothrips verbasci* (Osborn) (Thysanoptera, Tubulifera, Phlaeothripidae). *Journal of Morphology* 156, 1–38.

Heming, B.S. (1995) History of the germ line in male and female thrips. In: Parker, B.L., Skinner, M. and Lewis, T. (eds) *Thrips Biology and Management*. Plenum, New York, pp. 505–535.

Higgins, C.J. and Myers, J.H. (1992) Sex ratio patterns and population dynamics of western flower thrips (Thysanoptera: Thripidae). *Environmental Entomology* 21, 322–330.

Hirano, C., Yasumi, K., Itoh, E., Kim, C.S. and Horiike, M. (1993) Short-distance walking responses of *Thrips palmi* Karny (Thysanoptera: Thripidae) to tomato leaves. *Japanese Journal of Applied Entomology and Zoology* 38, 233–234.

Hirano, C., Yasumi, K., Itoh, E., Kim, C.S. and Horiike, M. (1994) A feeding deterrent for *Thrips palmi* Karny (Thysanoptera: Thripidae) found in tomato leaves: isolation and identification. *Japanese Journal of Applied Entomology and Zoology* 38, 109–120.

Hodson, W.E.H. (1935) The lily thrips (*Liothrips vaneeckei* Priesner). *Bulletin of Entomological Research* 26, 469–474.

Hollister, B., Cameron, E.A. and Teulon, D.A. (1995) Effect of *p*-anisaldehyde and a yellow color on behavior and capture of western flower thrips. In: Parker, B.L., Skinner, M. and Lewis, T. (eds) *Thrips Biology and Management.* Plenum, New York, pp. 571–574.

Howard, D.F., Blum, M.S. and Fales, H.M. (1983) Defense in thrips: forbidding fruitiness of a lactone. *Science* 220, 335–336.

Howard, D.F., Blum, M.S., Jones, T.H., Fales, H.M. and Tomalski, M.D. (1987) Defensive function and chemistry of the anal exudate of the Cuban laurel thrips, *Gynaikothrips ficorum* (Marchal). *Phytophaga* 1, 163–170.

Howlett, F.M. (1914) A trap for thrips. *Journal of Economic Biology* 9, 21–23.

Hoy, C.W. and Kretchman, D.W. (1991) Thrips (Thysanoptera: Thripidae) injury to cabbage cultivars in Ohio. *Journal of Economic Biology* 84, 971–977.

Hunter, W.B. and Ullman, D.E. (1989) Analysis of mouthpart movements during feeding of *Frankliniella occidentalis* (Pergande) and *F. schultzei* Trybom (Thysanoptera: Thripidae). *International Journal of Insect Morphology and Embryology* 18, 161–171.

Hunter, W.B. and Ullman, D.E. (1992) Anatomy and ultrastructure of the piercing-sucking mouthparts and paraglossal sensilla of *Frankliniella occidentalis* (Pergande) (Thysanoptera: Thripidae). *International Journal of Insect Morphology and Embryology* 21,17–35.

Kawai, A. (1986a) Studies on the population ecology of *Thrips palmi* Karny. 13. Comparison of the adult movement in the plastic greenhouse depending on the covering materials. *Proceedings of the Association for Plant Protection of Kyushu* 32, 163–165.

Kawai, A. (1986b) Studies on the population ecology of *Thrips palmi* Karny. X. Differences in population growth on various crops. *Japanese Journal of Applied Entomology and Zoology* 30, 7–11.

Kawai, A., Kitamura, C. and Yoshihara, T. (1985) Studies on the population ecology of *Thrips palmi* Karny. VIII. Population trends examined by the direct count and the sticky trap method. *Bulletin of the Vegetable and Ornamental Crops Research Station Japan*, Ser. C, 8, 81–86.

Kendall, D.M. and Bjostadt, L.B. (1990) Phytohormone ecology: herbivory by *Thrips tabaci* induces greater ethylene production in intact onions than mechanical damage alone. *Journal of Chemical Ecology* 16, 981–991.

Kennedy, G.W. (1984) 2-Tridecanone, tomatoes and *Heliothis zea*: potential incompatibility of plant antibiosis with insecticidal control. *Entomologia Experimentalis et Applicata* 35, 305–311.

Kiester, A.R. and Strates, E. (1984) Social behavior in a thrips from Panama. *Journal of Natural History* 18, 303–314.

Kirk, W.D.J. (1984a) Ecologically selective coloured traps. *Ecological Entomology* 9, 35–41.

Kirk, W.D.J. (1984b) Ecological studies on *Thrips imaginis* Bagnall (Thysanoptera) in flowers of *Echium plantagineum* L. in Australia. *Australian Journal of Ecology* 9, 9–18.

Kirk, W.D.J. (1984c) Pollen feeding in thrips (Insecta: Thysanoptera). *Journal of Zoology*, 204, 107–117.

Kirk, W.D.J. (1985a) Floral display in *Vicia faba*, and the distribution of a flower thrips, *Kakothrips pisivorus. Entomologia Experimentalis et Applicata* 38, 233–238.

Kirk, W.D.J. (1985b) Aggregation and mating of thrips in flowers of *Calystega sepium*. *Ecological Entomology* 10, 433–440.

Kirk, W.D.J. (1985c) Pollen-feeding and the host specificity and fecundity of flower thrips (Thysanoptera). *Ecological Entomology* 10, 281–289.

Kirk, W.D.J. (1985d) The effects of some floral scents on host finding by thrips (Insecta: Thysanoptera). *Journal of Chemical Ecology* 11, 35–43.

Kirk, W.D.J. (1985e) Egg-hatching in thrips (Insecta: Thysanoptera). *Journal of Zoology* (A), 207, 181–190.

Kirk, W.D.J. (1987) Effects of trap size and scent on the catches of *Thrips imaginis* Bagnall (Thysanoptera: Thripidae). *Journal of the Australian Entomological Society* 26, 299–302.

Klocke, F. (1926) Bietrage zur Anatomie und Histologie der Thysanopteren. *Zeitschrift für Wissenschaftl Zoologie* 128, 1–36.

Knudsen, J.T., Tollsten, L. and Bergstrom, L.G. (1993) Floral scents – a checklist of volatile compounds isolated by headspace techniques. *Phytochemistry* 33, 253–280.

Kring, J.B. and Schuster, D.J. (1992) Management of insects on pepper and tomato with UV-reflective mulches. *Florida Entomologist* 75, 119–129.

Kuhn, A. (1927) Der Farbensinn der Bienen. *Zeitschrift für VerGleichende Physiologie* 5, 762–800.

Kumar, N.K.K., Ullman, D.E. and Cho, J.J. (1993) Evaluation of *Lycopersicon* germ plasm for tomato spotted wilt tospovirus resistance by mechanical and thrips transmission. *Plant Disease* 77, 938–941.

Kumar, N.K.K., Ullman, D.E. and Cho, J.J. (1995a) *Frankliniella occidentalis* (Thysanoptera: Thripidae) landing and resistance to tomato spotted wilt tospovirus among *Lycopersicon* accessions with additional comments on *Thrips tabaci* (Thysanoptera, Thripidae) and *Trialeurodes vaporariorum* (Homoptera, Aleyrodidae). *Environmental Entomology* 24, 513–520.

Kumar, N.K.K., Ullman, D.E. and Cho, J.J. (1995b). Resistance among *Lycopersicon* species to *Frankliniella occidentalis* (Thysanoptera: Thripidae). *Journal of Economic Entomology* 88, 1057–1065.

Kyamanywa, S., Baliddawa, C.W. and Ampofo, K.J.O. (1993) Effect of maize plants on colonization of cowpea plants by bean flower thrips, *Megalurothrips sjostedti*. *Entomologia Experimentalis et Applicata* 69, 61–68.

Laughlin, R. (1971) A culture method for *Hercinothrips fermoralis* (Reuter) (Thysanoptera). *Journal of the Australian Entomological Society* 10, 301–303.

Leskey, T.C., Teulon, D.A.J. and Cameron, E.A. (1995) Position and abundance of pear thrips eggs in sugar maple flowering and vegetative buds. In: Parker, B.L., Skinner, M. and Lewis, T. (eds) *Thrips Biology and Management*. Plenum, New York, pp. 93–96.

Lewis, T. (1959) A comparison of water traps, cylindrical sticky traps and suction traps for sampling thysanopteran populations at different levels. *Entomologia Experimentalis et Applicata* 2, 204–215.

Lewis, T. (1973) *Thrips, Their Biology, Ecology and Economic Importance*. Academic Press, London.

Lewis, T. (1987) Feeding of *Limothrips cerealium* (Hal.): the mouthparts, their action, and effects on plant tissue. In: Holman, J., Pelikan, J., Dixon, A.F.G. and Weisman, L. (eds) *Population Structure, Genetics and Taxonomy of Aphids and Thysanoptera, Proceedings of International Symposium, Smolenice, Czechoslavakia, September 1985*, SPB Academic Publishing, The Hague, pp. 439–447.

Lowry, V.K., Smith, J.W. Jr. and Mitchell, F.L. (1992) Life-fertility tables for *Frankliniella fusca* (Hinds) and *F. occidentalis* (Pergande) (Thysanoptera: Thripidae) on

peanut. *Annals of the Entomological Society of America* 85, 744–754.
Lu, F.M. (1990) Color preference and using silver mulches to control the onion thrips, *Thrips tabaci* Lindeman. *Chinese Journal of Entomology/Zhonghua Kunchong* 10, 337–342.
Makino, S. (1984) Control of *Thrips palmi* by the mulching with polyethylene film. (In Japanese.) *Kyushu Agricultural Research* 46, 126.
Mathur, G. and Mohan Ram, Y. (1978) Significance of petal colour in thrips-pollinated *Lantana camara* L. *Annals of Botany* 42, 1473–1476.
Mathur, G. and Mohan Ram, Y. (1986) Floral biology and pollination of *Lantana camara*. *Phytomorphology* 36, 79–100.
Matteson, N. and Terry, L.I. (1992) Response to color by male and female *Frankliniella occidentalis* during swarming and non-swarming behavior. *Entomologia Experimentalis et Applicata* 63, 187–201.
Matteson, N., Terry, I., Ascoli, C.A. and Gilbert, C. (1992) Spectral efficiency of the western flower thrips, *Frankliniella occidentalis*. *Journal of Insect Physiology* 38, 453–459.
McIntyre, J.A., Hopper, D.A. and Cranshaw, W.S. (1996) Aluminized fabric deters thrips from entering greenhouses. *Southwestern Entomologist* 21, 135–141.
Menzel, R. (1990) *Color Vision in Flower Visiting Insects*. Institut für Neurobiologie de Freien, Universitat Berlin.
Michener, C.D. (1974) *The Social Behavior of Bees*. Belknap Press of Harvard University, Cambridge, MA.
Mikhailova, N.A. (1981) Influence of peculiarities of different species of wheat on the attack by *Haplothrips tritici* and *Trigonotylus coelestialium*. *Acta Phytopathologica Academiae Scientiarum Hungaricae* 1–2, 199–202.
Milne, J.R., Walter, G.H., Kaonga, D. and Sabio, G.C. (1996) The importance of non-pollen plant parts as food sources for the common blossom thrips, *Frankliniella schultzei*. *Entomologia Experimentalis et Applicata* 78, 271–281.
Milne, M. (1997) Host plant relationships and significance of prey in the diet of two phytophagous thrips species, *Frankliniella schultzei* and *Thrips tabaci* (Thysanoptera: Thripidae). PhD thesis, University of Queensland, Australia.
Moffit, H.R. (1964) A color preference of the western flower thrips, *Frankliniella occidentalis*. *Journal of Economic Entomology* 57, 604–605.
Mollema, C.A. and Cole, R.A. (1996) Low aromatic amino acid concentrations in leaf proteins determine resistance to *Frankliniella occidentalis* in four vegetable crops. *Entomologia Experimentalis et Applicata* 78, 325–333.
Moreno, D.S., Gregory, W.A. and Tanigoshi, L.K. (1984) Flight response of *Aphytus melinus* (Hymenoptera: Aphelinidae) and *Scirtothrips citri* (Thysanoptera: Thripidae) to trap color, size and shape. *Environmental Entomology* 13, 935–940.
Morgan, A.C. and Crumb, S.E. (1928) Notes on the chemotrophic responses of certain insects. *Journal of Economic Entomology* 21, 913–920.
Morison, G.D. (1947–1949) Thysanoptera of the London Area. *London Naturalist* 59 (Suppl.) 1–131.
Moritz, G. (1984) Exocrine gland in males of the genus *Merothrips* Hood, 1914, (Merothripidae, Thysanoptera, Insecta). *Zoologische Jahrbücher Anatomie* 111, 1–13.
Mound, L. and O'Neill, K. (1974) Taxonomy of the Merothripidae, with ecological and phylogenetic considerations (Thysanoptera). *Journal of Natural History* 8, 481–509.
Mukaigawa, Y. (1912) Shii-onaga-azamiuma (*Cryptothrips pasanii*). (In Japanese.) *Insect World* 16, 481–484.

Mulligan, G.A. and Kevan, P.G. (1973) Color, brightness and other floral characteristics attracting insects to the blossoms of some Canadian weeds. *Canadian Journal of Botany* 51, 1939–1952.

Murai, T. (1987) Reproductive diapause of flower thrips, *Frankliniella intonsa*. In: Holman, J., Pelikan, J., Dixon, A.F.G. and Weisman, L. (eds) *Population Structure, Genetics and Taxonomy of Aphids and Thysanoptera. Proceedings of International Symposium, Smolenice, Czechoslavakia*, September 1985. SPB Academic Publishing, The Hague, pp. 467–479.

Murai, T. and Ishii, T. (1982) Simple rearing method for flower thrips on pollens. *Japanese Journal of Applied Entomology and Zoology* 26, 149–154.

Nagai, K. (1981) Control of *Thrips palmi* Karny. In: *Proceedings of a Symposium on the Control of Pest and Disease of Vegetables*. Japan Plant Protection Association, Tokyo, pp. 35–43.

Nonaka, K. and Nagai, T. (1983) Ecology and control of thrips infesting fruit vegetables. 7. Control of *Thrips palmi* using cut UV film. (In Japanese.) *Kyushu Agricultural Research* 45, 119–129.

North, R.C. and Shelton, A.M. (1986a) Colonization and intraplant distribution of *Thrips tabaci* (Thysanoptera: Thripidae) on cabbage. *Journal of Economic Entomology* 79, 219–223.

North, R.C. and Shelton, A.M. (1986b) Ecology of Thysanoptera within cabbage fields. *Environmental Entomology* 15, 520–526.

Ohno, T., Ichikawa, K., Asayama, T. and Hikota, K. (1990) Control measure of *Thrips palmi* on chrysanthemum. *Research Bulletin of the Aichi-ken Agricultural Research Center* 22, 211–218.

Ota, A.K. and Smith, F.F. (1968) Aluminium foil – thrips repellent. *American Rose Annual* 53, 135–139.

Pelikán, J. (1951) On carnation thrips *Taeniothrips dianthi* Pr. *Entomologicke Listy* 14, 5–38.

Pellmyr, O. and Thien, L.B. (1986) Insect reproduction and floral fragrances: keys to the evolution of the angiosperms. *Taxon* 35, 76–85.

Penman, D.R., Osborne, G.O., Worner, S.P., Chapman, R.B. and McLaren, G.F. (1982) Ethyl nicotinate: a chemical attractant for *Thrips obscuratus* (Thysanoptera: Thripidae) in stonefruit in New Zealand. *Journal of Chemical Ecology* 8, 1299–1303.

Pesson, P. (1951) Ordre des Thysanoptera. In: Grasse, P.P. (ed.) Masson, Paris. *Traite de Zoologie* 10, 1805–1869.

Prokopy, R.J. and Owens, E.D. (1983) Visual detection of plants by herbivorous insects. *Annual Review of Entomology* 28, 337–364.

Pussard-Radulesco, E. (1931) Biological and cytological studies on Thysanoptera. *Annals Epiphytica* 16, 103–188.

Quisenberry, J.E. and Rummel, D.R. (1979) Natural resistance to thrips injury in cotton as measured by differential leaf area reduction. *Crop Science* 19, 879–881.

Raffa, K.F., Hall, D.J., Kearby, W. and Katovich, S. (1992) Seasonal life history of introduced basswood thrips (Thysanoptera: Thripidae) in Wisconsin, with observations on associated thrips species. *Environmental Entomology* 21, 771–779.

Rieske, L.K. and Raffa, K.F. (1995) Ethylene emission by a deciduous tree, *Tilia americana*, in response to feeding by introduced basswood thrips, *Thrips calcaratus*. *Journal of Chemical Ecology* 21, 187–197.

Rivnay, E. (1935) Ecological studies of the greenhouse thrips, *Heliothrips haemorrhoidalis*, in Palestine. *Bulletin of Entomological Research* 26, 267–278.

Robb, K. (1989) Analysis of *Frankliniella occidentalis* (Pergande) as a pest of floriculture crops in California greenhouses. PhD thesis, University of California, Riverside.

Rosenheim, J.A., Welter, S.C., Johnson, M.W., Mau, R.F.L. and Gusukuma-Minuto, L.R. (1990) Direct feeding damage on cucumber by mixed species infestations of *Thrips palmi* and *Frankliniella occidentalis* (Thysanoptera: Thripidae). *Journal of Economic Entomology* 83, 1519–1525.

Rösingh, C. (1980) Untersuchungen uber die Resistenz der Kuherbse, *Vigna unuiculata* (L.) Walp., gegen *Megalurothrips sjostedti* (Trybom) (Thysanoptera: Thripidae). PhD thesis, University of Hohenheim, Germany.

Rummel, D.R. and Quisenberry, J.E. (1979) Influence of thrips injury on leaf development and yield of various cotton genotypes. *Journal of Economic Entomology* 72, 706–709.

Sakimura, K. and Carter, W. (1934) The artificial feeding of Thysanoptera. *Annals of the Entomological Society of America* 27, 341–342.

Salguero Navas, V.E., Funderburk, J.E., Beshear, R.J., Olson, S.M. and Mack, T.P. (1991) Seasonal patterns of *Frankliniella occidentalis* (Thysanoptera: Thripidae) in tomato flowers. *Journal of Economic Entomology* 84, 1818–1822.

Salguero Navas, V.E., Funderburk, J.E., Beshear, R.J., Olson, S.M. and Mack, T.P. (1994) Aggregation indices and sample size curves for binomial sampling of flower-inhabiting *Frankliniella occidentalis* (Thysanoptera: Thripidae) on tomato. *Journal of Economic Entomology* 87, 1622–1626.

Salifu, A.B. (1986) Studies on the biology of the flower thrips, *Megalurothrips sjostedti* (Trybom) (Thysanoptera: Thripidae) with particular reference to resistance in host cowpea, *Vigna unguiculata* (L.). PhD thesis, Wye College, University of London.

Scott, S.J., McLeod, P.J., Montgomery, F.W. and Hander, C.A. (1989) Influence of reflective mulch on incidence of thrips (Thysanoptera: Thripidae) in staked tomatoes. *Journal of Entomological Science* 24, 422–427.

Shelton, A.M. and North, R.C. (1986) Species composition and phenology of Thysanoptera within field crops adjacent to cabbage fields. *Environmental Entomologist* 15, 513–519.

Shelton, A.M., Becker, R.F. and Andalaro, J.T. (1983) Varietal resistance to onion thrips (Thysanoptera: Thripidae) in processing cabbage. *Journal of Economic Entomology* 76, 85–86.

Shelton, A.M., Hoy, C.W., North, R.C., Dickson, M.H. and Barnard, J. (1988) Analysis of resistance in cabbage varieties to damage by Lepidoptera and Thysanoptera. *Journal of Economic Entomology* 81, 634–640.

Singh, D., Ramzan, M. and Gill, S.P.S. (1977) Field screening of onion cultivars at Ludhiana. *Punjab Horticultural Journal* 23, 242–243.

Sites, R.W., Chambers, W.S. and Nichols, B.J. (1992) Diel periodicity of thrips (Thysanoptera: Thripidae) dispersion and the occurrence of *Frankliniella williamsi* on onions. *Journal of Economic Entomology* 85, 100–105.

Smith, F., Boswell, A.L. and Webb, R.E. (1972) Repellent mulches for control of the gladiolus thrips. *Environmental Entomology* 1, 672–673.

Smith, W.C. and Goldsmith, T.H. (1990) Phyletic aspects of the distribution of 3-hydroxyretinal in the class Insecta. *Journal of Molecular Ecology* 30, 72–84.

Snook, M.E., Lynch, R.E., Culbreath, A.K. and Costello, C.E. (1994) 2,3-Di-(*E*)-caffeoyl-(*2R,3R*)-(+)-tartaric acid in terminals of peanut (*Arachis hypogaea* L.) varieties with different resistances to late leaf spot disease (*Cercosporidium personatum*) (Berk. and M.A. Curtis Deighton) and the insects tobacco thrips (*Frankliniella fusca* (Hinds)) and potato leafhopper (*Empoasca fabae* (Harris)). *Journal of Agricultural and Food Chemistry* 42, 1572–1574.

Soria, C. and Mollema, C. (1995) Life history parameters of western flower thrips on susceptible and resistant cucumber genotypes. *Entomologia Experimentalis et Applicata* 74, 177–184.

Sota, T. (1988) Ecology of a gall-forming thrips, *Ponticulothrips diospyrosi*: colony development and gall-associated arthropod community (Thysanoptera: Phlaeothripidae). *Japanese Journal of Applied Entomology and Zoology* 23, 345–352.

Stannard, L.J. (1968) The thrips or Thysanoptera of Illinois. *Bulletin of Illinois State Natural History Survey* 29, 215–552.

Stoddard, F.L. (1986) The distribution of immature thrips among flowers of faba beans in commercial crops and experimental plots. *Annals of Applied Biology* 109, 61–69.

Stoner, K.A. and Shelton, A.M. (1988a) Role of nonpreference in the resistance of cabbage varieties to the onion thrips (Thysanoptera: Thripidae). *Journal of Economic Entomology* 81, 1062–1067.

Stoner, K.A. and Shelton, A.M. (1988b) Effect of planting date and timing of growth stages on damage to cabbage by onion thrips (Thysanoptera: Thripidae). *Journal of Economic Entomology* 81, 1186–1189.

Stoner, K.A. and Shelton, A.M. (1988c) Influence of variety on abundance and within-plant distribution of onion thrips (Thysanoptera: Thripidae) on cabbage. *Journal of Economic Entomology* 81, 1190–1195.

Strauss, S.Y. and Karban, R. (1994a) The significance of outcrossing in an intimate plant/herbivore relationship. I. Does outcrossing provide an escape for plants from herbivores adapted to the parental phenotype? *Evolution* 48, 454–464.

Strauss, S.Y. and Karban, R. (1994b) The significance of outcrossing in an intimate plant/herbivore relationship. II. Does outcrossing pose a problem for thrips adapted to the host plant clone? *Evolution* 48, 465–476.

Strauss, S. and Karban, R. (1995) Selection, dispersal and mode of reproduction: attributes contributing to local adaptation by *Apterothrips apteris* to *Erigeron glaucus*. In: Parker, B.L., Skinner, M. and Lewis, T. (eds) *Thrips Biology and Management*. Plenum, New York, pp. 537–544.

Suzuki, H. and Miyara, A. (1984) Studies on ecology and control of the southern yellow thrips, *Thrips palmi* Karny, (Thysanoptera: Thripidae). (1) Physical control using agricultural covering materials. (In Japanese with an English summary.) *Bulletin of Okinawa Agricultural Experiment Station* 9, 85–93.

Suzuki, T., Haga, K. and Kuwahara, Y. (1986) Anal secretion of thrips. I. Identification of perillene from *Leeuwenia pasanii* (Thysanoptera: Phlaeothripidae). *Japanese Journal of Applied Entomology and Zoology* 21, 461–466.

Suzuki, T., Haga, K., Kodama, S., Watanabe, K. and Kuwahara, Y. (1988) Secretion of thrips II. Secretions of three gall-inhabiting thrips (Thysanoptera: Phlaeothripidae). *Japanese Journal of Applied Entomology and Zoology* 23, 291–297.

Suzuki, T., Haga, K., Leal, W.S., Kodama, S. and Kuwahara, Y. (1989) Secretion of thrips IV. Identification of β-acaridial from three gall-forming thrips (Thysanoptera: Phlaeothripidae). *Japanese Journal of Applied Entomology and Zoology* 24, 222–228.

Suzuki, T., Haga, K., Kodama, S., Nemoto, T. and Kuwahara, Y. (1990) Secretions of thrips V. Secretion of idolothripine thrips, *Bactothrips brevitubus* (Thysanoptera: Phlaeothripidae). *Japanese Journal of Applied Entomology and Zoology* 25, 1–7.

Suzuki, T., Haga, K., Izuno, M.Y., Matsuyama, S. and Kuwahara, Y. (1993) Secretion of thrips VII. Identification of 3-butanoyl-4-hydroxy-6-methyl-2*H*-pyran-3-one from *Hoplothrips japonicus* and *H. hagai* (Thysanoptera: Phlaeothripidae). *Japanese Journal of Applied Entomology and Zoology* 28, 108–112.

Suzuki, T., Haga, K., Katoaka, M., Tsutsumi, T., Nakano, Y., Matsuyama, S. and Kuwahara, Y. (1995) Secretion of thrips VIII. Secretions of the two *Ponticulothrips* species (Thysanoptera: Phlaeothripidae). *Japanese Journal of Applied Entomology and Zoology* 30, 509–519.

Tamò, M., Baumgärtner, J., Delucchi, V. and Herren, H.R. (1993a) Assessment of key factors responsible for the pest status of the bean flower thrips *Megalurothrips sjostedti* (Thysanoptera: Thripidae) in West Africa. *Bulletin of Entomological Research* 83, 251–258.

Tamò, M., Baumgärtner, J. and Arodokoun, D.Y. (1993b) The spatio-temporal distribution of *Megalurothrips sjostedti* (Trybom) (Thysanoptera, Thripidae) life stages on cowpea and development of sampling plans. *Mitteilungen der schweizerischen entomologischen Gesellschaft* 66, 15–34.

Teerling, C.R. (1995) Chemical ecology of western flower thrips. In: Parker, B.L., Skinner, M. and Lewis, T. (eds) *Thrips Biology and Management*. Plenum, New York, pp. 439–447.

Teerling, C.R., Gillespie, D.R. and Borden, J.H. (1993a) Utilization of western flower thrips alarm pheromone as a prey-finding kairomone by predators. *Canadian Entomologist* 125, 431–437.

Teerling, C.R., Pierce, Jr., H.D., Borden, J.H. and Gillespie, D.R. (1993b) Identification and bioactivity of alarm pheromone in the western flower thrips, *Frankliniella occidentalis*. *Journal of Chemical Ecology* 19, 681–697.

Terry, L.I. (1991) *Frankliniella occidentalis* (Thysanoptera: Thripidae) oviposition in apple buds: role of bloom state, blossom phenology, and population density. *Environmental Entomology* 20, 1568–1576.

Terry, L.I. (1995) Competition in western flower thrips males: effects of density on behaviour. In: Parker, B.L., Skinner, M. and Lewis, T. (eds) *Thrips Biology and Management*. Plenum, New York, pp. 433–438.

Terry, L.I. and DeGrandi-Hoffman, G. (1988) Monitoring western flower thrips (Thysanoptera: Thripidae) in "Granny Smith" apple blossom clusters. *Canadian Entomologist* 120, 1003–1016.

Terry, L.I. and Dyreson, E. (1996) Behavior of *Frankliniella occidentalis* (Thysanoptera: Thripidae) within aggregations, and morphometric correlates of fighting. *Annals of the Entomological Society of America* 89, 589–602.

Terry, I. and Gardner, G. (1990) Male mating swarms in *Frankliniella occidentalis* (Pergande) (Thysanoptera: Thripidae). *Journal of Insect Behavior* 3, 133–141.

Terry, L.I. and Kelly, C.K. (1993) Patterns of change in secondary and tertiary sex ratios of the Terebrantian thrips, *Frankliniella occidentalis*. *Entomologia Experimentalis et Applicata* 66, 213–225.

Terry, I. and Schneider, M. (1993) Copulatory behavior and mating frequency of the western flower thrips, *Frankliniella occidentalis* (Insecta: Thysanoptera). (Advances in Thysanopterology). *Journal of Pure and Applied Zoology* 4, 339–354.

Teulon, D.A.J. (1992) Laboratory technique for rearing western flower thrips (Thysanoptera: Thripidae). *Journal of Economic Entomology* 85, 895–899.

Teulon, D.A.J. and Cameron, E.A. (1995) Within-tree distribution of pear thrips (Thysanoptera, Thripidae) in sugar maple. *Environmental Entomology* 24, 233–238.

Teulon, D.A.J. and Penman, D.R. (1986) A rearing method for the New Zealand flower thrips. *Proceedings of the New Zealand Weed and Pest Control Conference* 39, 244–245.

Teulon, D.A.J. and Penman, D.R. (1991) Effects of temperature and diet on oviposition rate and development time of the New Zealand flower thrips, *Thrips obscuratus*.

Entomologia Experimentalis et Applicata 60, 143–155.
Teulon, D.A.J. and Penman, D.R. (1992) Colour preferences of New Zealand thrips (Terebrantia: Thysanoptera). *New Zealand Entomologist* 15, 8–13.
Teulon, D.A.J. and Penman, D.R. (1996) Thrips (Thysanoptera) seasonal flight activity and infestation of ripe stonefruit in Canterbury, New Zealand. *Journal of Economic Entomology* 89, 722–734.
Teulon, D.A.J., Penman, D.R. and Ramakers, P.M.J. (1993) Volatile chemicals for thrips (Thysanoptera: Thripidae) host-finding and applications for thrips pest management. *Journal of Economic Entomology* 86, 1405–1415.
Theunissen, J. and Legutowska, H. (1991) *Thrips tabaci* Lindeman (Thysanoptera, Thripidae) in leek: within-plant distribution. *Journal of Applied Entomology* 112, 309–316.
Thornhill, R. and Alcock, J. (1983) *The Evolution of Insect Mating Systems*. Harvard University Press, Cambridge, MA.
Trichilo, P.J, and Leigh, T.F. (1988) Influence of resource quality on the reproductive fitness of flower thrips (Thysanoptera: Thripidae). *Annals of the Entomological Society of America* 81, 64–70.
Tsai, Y.P., Hwang, M.T. and Chen, H.P. (1992) Occurrence and damage of *Thrips hawaiiensis* in banana orchards. *Chinese Journal of Entomology* 12, 231–237.
Tsuchiya, M., Furuhashi, K. and Masui, S. (1995a) Control of yellow tea thrips *Scirtothrips dorsalis* Hood by reflective sheet in satsuma mandarin (*Citrus unshia* Marc.) orchard. *Japanese Journal of Applied Entomology and Zoology* 39, 219–225.
Tsuchiya, M., Masui, S. and Kuboyama, N. (1995b) Color attraction of western flower thrips (*Frankliniella occidentalis* Pergande). *Japanese Journal of Applied Entomology and Zoology* 39, 313–319.
Tsuchiya, M., Togawa, M., Furuhashi, K. and Maisu, S. (1995c) Color attraction of yellow tea thrips (*Scirtothrips dorsalis* Hood). *Japanese Journal of Applied Entomology and Zoology* 39, 299–303.
van Dijken, F.R., Dik, M.T.A., Gebala, B., De Jong, J. and Mollema, C. (1994) Western flower thrips (Thysanoptera: Thripidae) effects on chrysanthemum cultivars: plant growth and leaf scarring in nonflowering plants. *Journal of Economic Entomology* 87, 1312–1317.
Varadarasan, S. and Ananthakrishnan, T.N. (1981) Population dynamics and prey/predator/parasite relationships of gall-forming thrips. *Proceedings of the Indian National Academy of Sciences* B47, 321–340.
Varadarasan, S. and Ananthakrishnan, T.N. (1982) Biological studies on some gall thrips. *Proceedings of the Indian National Academy of Sciences* B48, 35–43.
Velusamy, R. and Saxena, R.C. (1991) Genetic evaluation for resistance to rice thrips (Thysanoptera: Thripidae) in leafhopper and planthopper resistant rice varieties. *Journal of Economic Entomology* 84, 664–668.
Velusamy, R., Kumar, M.G., Johnson, T., Edward, Y.S. and Sundara Babu, P.C. (1994) Resistance to thrips in traditional rice varieties. *International Rice Research Newsletter* 19, 13–14.
Verma, S.K. (1966) Studies on the host preference of the onion thrips, *Thrips tabaci* Lindeman, to the varieties of onion. *Indian Journal of Entomology* 28, 396–398.
Vernon, R.S. and Gillespie, D.R. (1990a) Response of *Frankliniella occidentalis* (Thysanoptera: Thripidae) and *Trialeurodes vaporariorum* (Homoptera: Aleyrodidae) to fluorescent traps in a cucumber greenhouse. *Journal of the Entomological Society of British Columbia* 87, 38–41.
Vernon, R.S. and Gillespie, D.R. (1990b) Spectral responsiveness of *Frankliniella*

occidentalis (Thysanoptera: Thripidae) determined by trap catches in greenhouses. *Environmental Entomology* 19, 1229–1241.

Vernon, R.S. and Gillespie, D.R. (1995) Influence of trap shape, size and background color on captures of *Frankliniella occidentalis* (Thysanoptera: Thripidae) in a cucumber greenhouse. *Journal of Economic Entomology* 88, 288–293.

Vos, J.G.M., Uhan, T.S. and Sutarya, R. (1995) Integrated crop management of hot pepper (*Capsicum* spp.) under tropical lowland conditions: effects of rice straw and plastic mulches on crop health. *Crop Protection* 14, 445–452.

Walker, J.K. Jr, Hart, E.R., Gannaway, J.R. and Niles, G.A. (1979) Boll weevil and thrips resistance in pilose cotton. *Southwestern Entomologist* 4, 132–140.

Walker, W.F. (1974) Responses of selected Thysanoptera to colored surfaces. *Environmental Entomology* 3, 295–304.

Wang, C.L. and Chu, Y.I. (1990) The reproductive mechanisms of *Thrips palmi* Karny. III. Development and sex ratio of individuals reproduced parthenogenetically and bisexually. *Chinese Journal of Entomology* 10, 125–132.

Wardle, R.A. and Simpson, R. (1927) The biology of Thysanoptera with special reference to the cotton plant. 3. The relation between feeding habits and plant lesions. *Annals of Applied Biology* 14, 513–528.

Watts, J.G. (1936) A study of the biology of the flower thrips, *Frankliniella tritici* (Fitch) with special reference to cotton. *South Carolina Agricultural Experiment Station Bulletin* 306.

Wien, H.C. and Rösingh, C. (1980) Ethylene evolution by thrips infested cowpea provides a basis for thrips resistance screening with ethophon sprays. *Nature* 283, 192–194.

Wiesenborn, W.D. and Morse, J.G. (1985) Feeding rate of Thysanoptera estimated using ^{14}C inulin. *Journal of Economic Entomology* 78, 1515–1518.

Wiesenborn, W.D. and Morse, J.G. (1986) Feeding rate of *Scirtothrips citri* (Moulton) (Thysanoptera: Thripidae) as influenced by life stages and temperature. *Environmental Entomology* 15, 763–766.

Wiesenborn, W.D. and Morse, J.G. (1988) The mandible and maxillary stylets of *Scirtothrips citri* (Moulton) (Thysanoptera: Thripidae). *Pan-Pacific Entomologist* 64, 39–42.

Yang, G., Espelie, K.E., Todd, J.W., Culbreath, A.K., Pittman, R.N. and Demski, J.W. (1993) Cuticular lipids from wild and cultivated peanuts and the relative resistance of these peanut species to fall armyworm and thrips. *Journal of Agricultural and Food Chemistry* 41, 814–818.

Yasumi, K., Shinohara, T., Horiike, M. and Hirano, C. (1991) Effect of tomato leaf constituents on survival of *Thrips palmi* Karny (Thysanoptera: Thripidae). *Japanese Journal of Applied Entomology and Zoology* 35, 311–316.

Yudin, L.S., Mitchell, W.C. and Cho, J.J. (1987) Color preference of thrips (Thysanoptera: Thripidae) with reference to aphids (Homoptera: Aphididae) and leafminers in Hawaiian lettuce farms. *Journal of Economic Entomology* 80, 51–55.

Yudin, L.S., Tabashnik, B.E., Cho, J.J. and Mitchell, W.C. (1988) Colonization of weeds and lettuce by thrips (Thysanoptera: Thripidae). *Environmental Entomology* 17, 522–526.

Zeier, P. and Wright, M.G. (1995) Thrips resistance in *Gladiolus* spp.: potential for IPM and breeding. In: Parker, B.L., Skinner, M. and Lewis, T. (eds) *Thrips Biology and Management*, NATO ASI, Series A, 276. Plenum Press, New York, pp. 411–416.

Feeding 4

William D.J. Kirk
Department of Biological Sciences, Keele University, Keele, Staffs ST5 5BG, UK

Introduction

Knowledge of how thrips feed is fundamental to understanding how they affect crops. Virtually all thrips damage to crops results from feeding, either directly (see Chapter 13) or indirectly, as with virus transmission (see Chapter 14). The presence of thrips on crops may also sometimes be beneficial. They can be useful natural biological-control agents, feeding on pest insects and mites, and thrips in flowers can be useful as pollinators, increasing crop yield. The feeding behaviour of flower thrips brings them into frequent contact with pollen and thus enhances their effectiveness as pollinators. Some thrips produce a combination of these effects. For example, *Frankliniella occidentalis* causes direct damage to crops, transmits virus, controls mites, and pollinates (Carlson, 1964; Allen and Broadbent, 1986; Trichilo and Leigh, 1986), although not all in the same crop at the same time. An understanding of all these consequences of feeding and the balance between them requires detailed knowledge of thrips' feeding behaviour.

The impact of thrips on crops depends not just on the nature of their harmful and beneficial effects, but on their abundance. The availability of food and its nutritional value is an important element in determining the rate of growth of thrips populations.

This chapter reviews feeding, its effects on thrips and its beneficial effects on crops through biological control and pollination.

The Mouthparts

The mouthparts of thrips are broadly similar across all families and between larvae and adults (see Chapter 2), and are used in a similar way, although there are many differences in detail (Heming, 1993). The mouthparts form a mouthcone on the underside of the head (see Chapter 2). In order to feed, an individual protracts a single mandible and a pair of maxillary stylets through the tip of the mouthcone. The mandible resembles a microscopic needle (Fig. 4.1) and is used to pierce a hole in solid material, such as cell walls. The maxillary stylets (laciniae) are roughly C-shaped in cross-section and join together to form a tube (Fig. 4.1), which can enter tissues through or next to the hole made by the mandible and can be used to suck up liquids. This process of piercing with the mandible and sucking through the maxillary stylets is used for a wide variety of food types, so the feeding process in thrips is best described

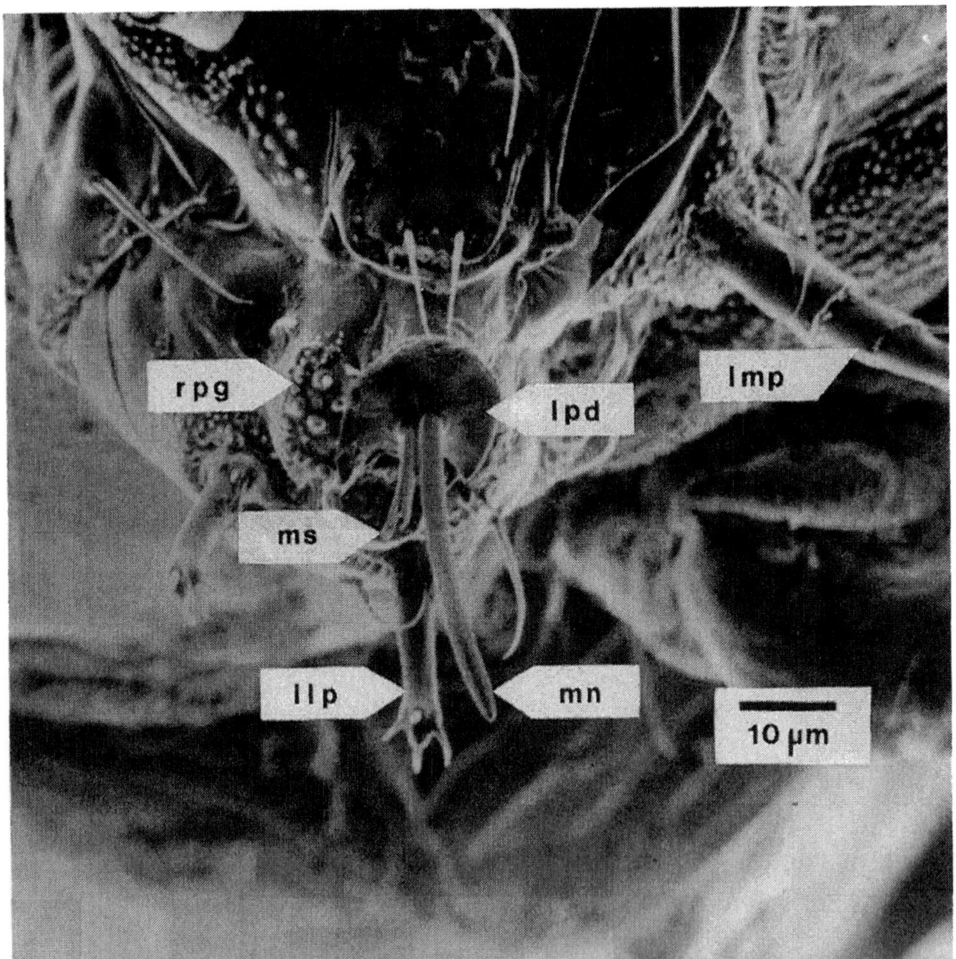

Fig. 4.1. Scanning electron micrograph of an oblique frontal view of the mouthcone and exserted mouthparts of an adult female *Limothrips cerealium*. llp = left labial palp, lmp = left maxillary palp, lpd = labral pad, mn = mandible, ms = maxillary stylets, rpg = right paraglossal lobe (photo: I.F. Chisholm).

as 'piercing-sucking' (Hunter and Ullman, 1992) or 'punch and suck' (Moritz, 1982; Heming, 1993). Some early descriptions of feeding claimed instead that thrips damage the leaf surface by 'rasping' or 'gashing' and then suck up the juices at the surface (Borden, 1915; Foster and Jones, 1915; Wardle and Simpson, 1927), but this has not been supported by subsequent studies. Unfortunately, this misleading description has been perpetuated in many reviews and textbooks (e.g. Miles, 1972; Pfadt, 1978; Gullan and Cranston, 1994).

The mouthparts can be used to feed on liquids that are either exposed or enclosed by a thin wall, so thrips are able to feed on many types of food. There is no obvious reason why the single mandibular stylet should be a particular adaptation to feeding on pollen grains or spores, rather than on other foods, as claimed by Grinfel'd (1959). Thrips have, in principle, a general-purpose feeding tool for obtaining access to the liquid contents of plants and animals. They feed regularly on the cell contents of leaves, petals, fruit and seeds, on pollen grains, and on the eggs and active stages of small arthropods, all of which can be considered as liquids enclosed by a thin wall.

The structure and function of the mouthparts have been studied in detail for a range of mainly pest species (Table 4.1). The similarity of the descriptions of feeding from these studies and from less detailed studies of other species suggests that the main findings are likely to hold generally, at least for the relatively unspecialized feeders infesting crops.

The structure of the mouthparts of the Thripidae is outlined below (for details see Chapter 2 and Figs 2.6 and 2.7). Most of the description probably also applies to the other families of the Terebrantia (Heming, 1993), but some similarities and differences in the Phlaeothripidae are pointed out.

The mouthcone

Viewed externally, the mouthcone consists of the labrum in front and a maxillary stipes on each side, with the labium behind. Above the labrum is a triangular clypeo-labral membrane which is flexible to allow shortening of the mouthcone. The maxillary palps each have five or six curved blunt-tipped sensilla at the tip that are probably contact chemoreceptors, while the labial palps each have four sensilla at the tip (Heming, 1978; Moritz, 1982; Childers and Achor, 1991a; Hunter and Ullman, 1992). Each paraglossal lobe has nine or ten sensilla (Chisholm and Lewis, 1984; Moritz, 1988, 1989; Wiesenborn and Morse, 1988; Hunter and Ullman, 1989, 1992; Childers and Achor, 1991a), consisting of three distinct types which are probably chemoreceptors and mechanoreceptors involved in food selection. These sensilla appear to be absent in phlaeothripids (Heming, 1993). An additional longer sensillum trichoideum is present on the anterior part of each of the paraglossae and is probably also involved in feeding (Moritz, 1988; G. Moritz, Halle, 1994, personal communication).

When a thrips is about to feed, the paraglossal lobes separate and expose the labral pad, which is then unfolded rapidly. The pad is a horseshoe-shaped structure, about 11 μm across, with a posterior gap, that is applied to the substrate during feeding (Fig. 4.1) (Chisholm and Lewis, 1984; Childers and

Table 4.1. A list of thrips species for which the structure or function of the mouthparts has been studied.

Thrips	Reference
Aeolothripidae	
Aeolothrips intermedius	Mickoleit (1963)
Aeolothrips intermedius	Moritz (1982)
Thripidae	
Chirothrips hamatus	Jones (1954)
Frankliniella bispinosa	Childers and Achor (1991a,b)
Frankliniella occidentalis	Hunter and Ullman (1989, 1992)
Frankliniella schultzei	Hunter and Ullman (1989)
Heliothrips haemorrhoidalis	Reyne (1927)
Hercinothrips femoralis	Moritz (1988, 1989)
Limothrips cerealium	Chisholm and Lewis (1984)
Scirtothrips aurantii	Milne and Manicom (1978)
Scirtothrips citri	Wiesenborn and Morse (1988)
Selenothrips rubrocinctus	Reyne (1927)
Thrips angusticeps	Bonnemaison and Bournier (1964)
Thrips palmi	Yasumi *et al.* (1994)
Thrips physapus	Risler (1957)
Phlaeothripidae	
Bagnalliella yuccae	Peterson (1915)
Haplothrips statices	Mickoleit (1963)
Haplothrips verbasci	Heming (1978)
Phlaeothrips coriaceus	Mickoleit (1963)

Achor, 1991a). In *Scirtothrips aurantii* and *Limothrips cerealium* adults, the pad appears to have radial cuticular striations, with a number of small papillae immediately posteriorly (Milne and Manicom, 1978; Chisholm and Lewis, 1984). In *Frankliniella occidentalis* and *Frankliniella bispinosa* adults, the pad is fringed with many small papillae and has thinner, longer papillae posteriorly, but these do not seem to be innervated (Hunter and Ullman, 1989; Childers and Achor, 1991a). The pad may stabilize the mouthcone on the substrate and prevent loss of plant fluids during feeding (Chisholm and Lewis, 1984; Hunter and Ullman, 1989). It is absent or greatly reduced in phlaeothripids (Heming, 1993). Subtending the labral pad proximally is the labral ring, which is the heavily sclerotized apex of the labrum. This rigid structure forces the maxillary stylets together as they are protracted and then supports the exserted mandible and maxillary stylets (Heming, 1993).

The mandible

Although the base of the mandible is on the left side of the head, this asymmetry is not apparent externally when the mandible is protracted. The protracted part of the mandible is virtually straight, round in cross-section, about 1.3 to 2.0 μm in diameter, and pointed at the tip (Wiesenborn and Morse, 1988; Childers and Achor, 1991b). It has no opening and so cannot be used to suck up food (Reyne, 1927; Chisholm and Lewis, 1984). The maximum recorded lengths of protraction are 41 μm for adult *Frankliniella bispinosa,* 20 μm for adult *Limothrips cerealium*, 11 μm for adult *Thrips tabaci*, 40 and 30 μm for larvae II and adults of *Selenothrips rubrocinctus*, and 10, 21 and 15 μm respectively for larvae I, larvae II and adults of *Scirtothrips citri* (Reyne, 1927; Wardle and Simpson, 1927; Chisholm and Lewis, 1984; Wiesenborn and Morse, 1988; Childers and Achor, 1991a). Many of these lengths were measured on dead specimens, so they may not represent the normal length of protraction in living specimens. Protraction of up to 25 μm has been recorded in first instar larvae of a phlaeothripid (Heming, 1978).

The mandible cannot be protracted by direct muscular action because it is fused proximally to the exoskeleton and has no protractor muscle. Instead, muscles within the head shorten the mouthcone dorsoventrally so that the mandible emerges. This indirect method of protraction limits the distance that the mandible can be exposed to the length by which the mouthcone can be compressed, which is a small fraction of actual mandibular length. The mandible is retracted by direct muscular action on its base. In the Phlaeothripidae, it is protracted by the left clypeo-labral muscle, which pulls the mandibular guide of the sheath over the arched base of the mandible and so forces it distally, without compression of the head (Heming, 1978). The protracted mandible is supported at the tip of the mouthcone by the labral ring. No chemosensory organs have been found on the mandible.

The maxillary stylets

The left and right maxillary stylets fit together by means of a 'tongue-and-groove' system with grooves on the left stylet and tongues (ridges) on the right stylet (Fig. 4.2) (Mound, 1971; Heming, 1978; Chisholm and Lewis, 1984; Hunter and Ullman, 1992). They are separate within the mouthcone, but combine during protraction and separate during retraction as they pass over the tip of the hypopharynx about 12 μm inside the tip of the mouthcone (Hunter and Ullman, 1992). They emerge from the tip of the mouthcone, through the labral pad, immediately posterior to the mandible, although the mandible is offset slightly to the left as a result of its asymmetrical origin (Figs 4.1 and 4.2; plates in Childers and Achor, 1991a; Heming, 1993). The stylets are protracted and retracted by the actions of muscles attached to the maxillary levers (Risler, 1957; Moritz, 1982, 1989). The combined stylets are 1 to 2 μm in diameter in both larvae and adults (Mound, 1971; Chisholm and Lewis, 1984; Wiesenborn and Morse, 1988), and the maximum recorded lengths of protraction are 50 μm in adult *Frankliniella bispinosa*, 30 μm in adult *Frankliniella occidentalis*, 60 μm in adult *Limothrips cerealium*, 27 μm in adult *Thrips tabaci*, 40 and 50 μm in larvae II and adults of *Selenothrips rubrocinctus*, and 33, 35 and 23 μm respectively in larvae I, larvae II and adults of *Scirtothrips citri* (Reyne, 1927; Wardle and Simpson, 1927; Chisholm and Lewis, 1984; Wiesenborn and Morse, 1988; Hunter and Ullman, 1989; Childers and Achor, 1991a), though as with the mandibular

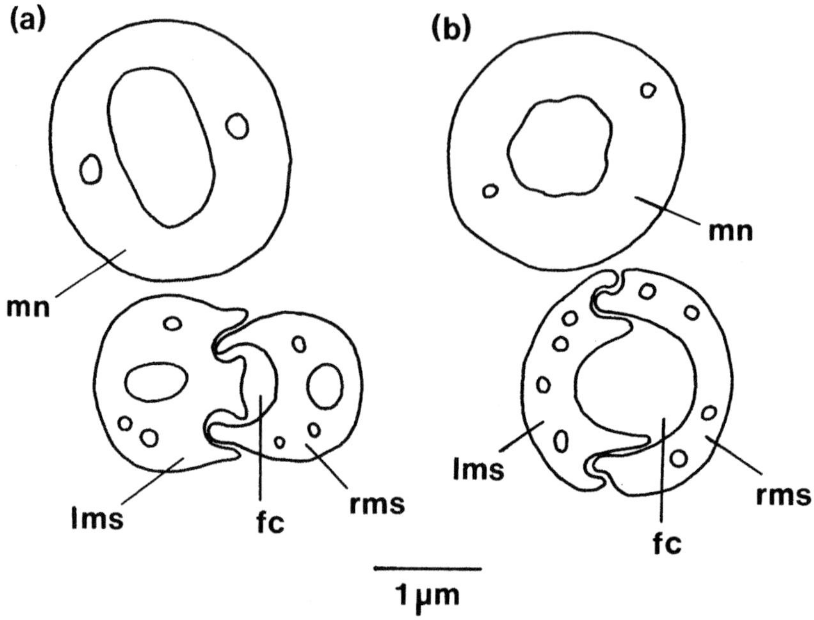

Fig. 4.2. Cross-section of the simultaneously protracted mandible and maxillary stylets, just beyond the tip of the mouthcone. (a) Larva II of *Haplothrips verbasci* (after Heming, 1978), (b) adult of *Frankliniella occidentalis* (after Hunter and Ullman, 1992). The dendrites within the lumina have been omitted. fc = food canal, lms = left maxillary stylet, mn = mandible, rms = right maxillary stylet.

protractions cited, many of these lengths were measured on dead specimens, so they may not represent the normal exserted lengths in living specimens. The retracted stylets are confined to the mouthcone, but in phlaeothripids can be retracted further back into the head capsule and can be much longer. Some phlaeothripids protract their maxillary stylets up to 107 µm (Heming, 1978), possibly much further in specialist feeders on *Casuarina* trees (Mound, 1970). Spore-feeding idolothripine phlaeothripids also have wider stylets, about 5 to 10 µm across, presumably as an adaptation to feeding on intact spores (Mound, 1971).

Each stylet can be operated independently so that left and right sides can slide against each other. The tips are pointed and asymmetrical, differing in size and shape between left and right sides, and when both stylets are equally extended there is a sub-apical aperture (Risler, 1957; Chisholm and Lewis, 1984; Wiesenborn and Morse, 1988; Childers and Achor, 1991a). This is reminiscent of the design of the tip of a syringe needle and probably serves the same purpose of allowing it to pierce solid material without the aperture becoming blocked. Adult *Limothrips cerealium* can feed with the stylets unequally extended and with liquid entering the canal well away from the tip of the longer stylet (Chisholm, 1983). Very little is known about the stylet tips and their function in pest species. Processes on the tips could be involved in piercing internal cell walls and chloroplasts, filtering debris, separating and re-engaging the tongue and groove system during differential protraction of the stylets, orienting food particles such as chloroplasts before ingestion, or steering the stylets. Complex stylet tips have been observed in some non-pest aeolothripids and spore-feeding idolothripines (Reyne, 1927; Mound, 1971), but the stylet tips of thripids are smaller so harder to see. In larvae of the phlaeothripid *Haplothrips verbasci*, the right stylet tip fits into a pre-apical socket on the left stylet (Heming, 1978), but a socket has not been found in thripids (Chisholm and Lewis, 1984; Childers and Achor, 1991a). The socket would allow differential tensions in the stylets to bend the combined stylets to the left or right (Heming, 1978). Observations of thrips feeding have shown that the stylets can bend rapidly in species with an apical socket (Heming, 1978), but they can also bend in species without this feature (Wiesenborn and Morse, 1988). Wiesenborn and Morse (1988) have proposed that thrips may use the same method of indirect stylet control, without an apical socket, as is used by some phytophagous Hemiptera (Pollard, 1969). This requires stylets with inwardly directed apices and differential flexibility along their length. Protraction of one stylet beyond the other bends both to the side of the shorter stylet.

The food canal

A simplified diagram of the food canal within the head is shown in Fig. 4.3 (for details see Figs 2.6 and 2.7). Fluid that passes up the maxillary stylets enters the food canal at a point about 12 μm inside the mouthcone where the stylets separate (Hunter and Ullman, 1994). Here, the food canal is formed posteriorly from a longitudinal groove in the hypopharyngeal sclerite and anteriorly from the epipharyngeal sclerite, with the two sclerites interlocking (Heming, 1978; Hunter and Ullman, 1994). The food canal leads from the precibarium to the cibarium, which forms a pump. In *Frankliniella occidentalis*, there are four precibarial chemosensilla and 20 cibarial chemosensilla (Hunter and Ullman, 1994). Liquid is drawn up the food canal by the pumping action of the cibarium, which has a valve at both ends. Hunter and Ullman (1994) have interpreted the role of the distal cibarial valve as regulation of food intake and separation of precibarial and cibarial sensilla for independent discrimination of plant fluids. It might also serve to ensure that contraction of the cibarium does not force fluids back into the precibarium. Fluids pass from the cibarium into the pharynx and then into the oesophagus.

The salivary glands

Ducts from the salivary (labial) glands in the thorax and abdomen open into the salivarium, which pumps saliva into the preoral cavity at the tip of the mouthcone, but does not deliver it directly between the maxillary stylets as in Hemiptera (Bonnemaison and Bournier, 1964; Heming, 1978; Moritz, 1982, 1988). Thrips differ from Hemiptera in having a single tube for food and saliva, rather than two separate tubes, which means that thrips cannot use the maxillary stylets to pump saliva into their food at the same time as sucking fluids out. There is strong indirect evidence that saliva passes down the stylet

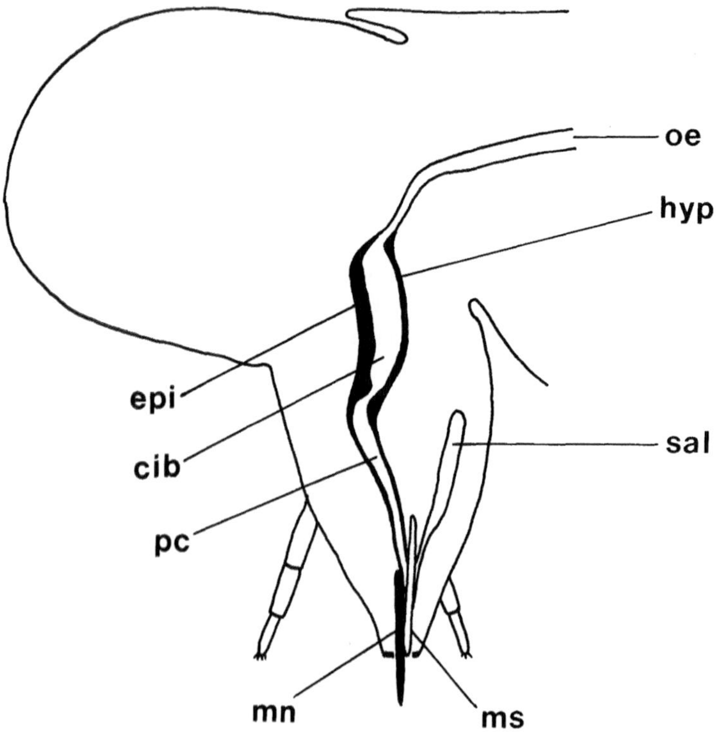

Fig. 4.3. A diagram of a longitudinal section of the head of a terebrantian showing the food canal and some associated structures, (after Risler (1957), Moritz (1988), and Hunter and Ullman (1994). Only the distal parts of the mandible and maxillary stylets are visible in the section. cib = cibarium, epi = epipharyngeal sclerite, hyp = hypopharyngeal sclerite, mn = mandible, ms = maxillary stylets, oe = oesophagus, pc = precibarium, sal = salivarium.

tube and enters food. Kloft and Ehrhardt (1959) fed adult *Thrips tabaci* on a radiolabelled plant and then left the thrips to feed on an unlabelled sheet of epidermal cells. Radiolabelled material was detected within the epidermal cells at the feeding sites, which suggests that saliva was injected down the stylet tube, but this result could instead have been produced by saliva introduced into the cells on the surface of the mandible and maxillary stylets. The saliva within the cells had a predominantly localized distribution, which suggests that a component of the saliva does not mix with cell contents and is possibly rather viscous or is gelling. However, some of the saliva was diffusely distributed through the cells, suggesting that a component of the saliva is watery. Kumar *et al.* (Figs 2 and 3, 1995) have photographed a substance, presumably saliva, on the leaf surface and sheathing the puncture within the cell after feeding by *Frankliniella occidentalis*, while Mitchell *et al.* (1995) have found large amounts of a substance that stains with safranine, and so is also probably saliva, throughout the wound site in peanut leaflets fed on by *Frankliniella fusca*. Adults produced far more of the substance than larvae. Chisholm (1983) filmed adult *Limothrips cerealium* feeding on 0.35 μm-diameter polystyrene latex particles and found that the particles built up in an agglomerate around the stylet opening. Ingestion was able to continue, so the agglomerate was not forming as a result of mechanical blockage; perhaps the saliva at the aperture acted as a filter. He found no visible evidence of any digestion by the saliva in the area around the stylets, although Childers and Achor (1991b) suspected that saliva was causing some of the damage they observed, and proteases, amylases and lipases have been found in the salivary glands of a leaf-feeding gall thrips (Raman *et al.*, 1991). Onions fed on by *Thrips tabaci* produce higher levels of ethylene than onions with artificial mechanical damage, and Kendall and Bjostad (1990) suggested that the thrips were introducing ethylene-inducing cues in the saliva.

The position within the mouthcone of the opening from the salivarium suggests that the saliva is not just pumped down the maxillary stylets, but coats the outer surface of the mandible and maxillary stylets and emerges at the tip of the mouthcone. This saliva could serve several purposes, including lubricating the mandible and stylets by flooding the area around the point of insertion, preventing leakage of plant-cell contents, and providing adhesion of the mouthcone to the leaf surface and adhesion between the protracted stylets (Heming, 1978). 'Saliva plugs' at the point of insertion have been photographed by Childers and Achor (1991b) and Kumar *et al.* (Fig. 3, 1995) (see Chapter 13). The discovery of these plugs preserved on the mouthparts of thrips provides further evidence that at least some of the saliva is highly viscous or gels during use. The plugs appear to be directly analogous to the salivary flanges, formed from gelling saliva, that are found in Homoptera. They probably perform a similar function, which is thought primarily to be sealing the stylets into a cell under turgal pressure (Miles, 1972).

There are likely to be many similarities between phytophagous Thysanoptera and Hemiptera in the composition and function of the saliva, because of their ecological and taxonomic affinity. Homoptera produce at least two kinds of saliva: one that solidifies rapidly on ejection and one that remains watery (Miles, 1972, 1987). There is some evidence for both types in thrips (see above). Hemipteran saliva is thought to have many functions, which include breakdown of food with hydrolysing (digestive) enzymes, overcoming plant chemical defences and lubricating the stylets as they move against plant tissue (Miles, 1972, 1987; Smith, 1985; Peng and Miles, 1988).

Movement of the salivary and cibarial pumps can be observed clearly in larvae that have transparent cuticle. In *Haplothrips verbasci*, salivation occurs during probing and then stops when the cibarial pump starts. However, when the contents of plant cells are becoming exhausted, salivary and cibarial pumps can be active at the same time (Heming, 1978). Presumably when both pumps are active, the saliva is emerging at the tip of the mouthcone or being ingested rather than being injected into the cell. Electronic monitoring of *Frankliniella occidentalis* and *Thrips tabaci* shows that salivation occurs during or immediately after probing, but before cibarial pumping (Hunter *et al.*, 1993; Harrewijn *et al.*, 1996a,b).

The mandibular gland, which occurs in larvae and adults of Terebrantia, but not Tubulifera, may have a role similar to that of the salivary glands. It opens into the mandibular sheath, which encloses the base of the mandible, and may lubricate the movement of the mandible against the posterior wall of the sheath during protraction and retraction (Heming, 1978; B. Heming, Edmonton, 1995, personal communication).

Feeding Behaviour

The damage that thrips cause to crops is usually the result of feeding on leaves, fruit or petals. This form of feeding has received most attention because of its relevance to understanding the nature of crop damage. However, thrips on crops may also be feeding on other foods (Kirk, 1995). Little is known about the breadth of diet of individual thrips in the field, the behaviour involved in switching between food types or the relative amounts of each type that are consumed. This section describes the main feeding behaviours found in the Thripidae, but where there is little or no information about thripids, reference is also made to studies of the Aeolothripidae and Phlaeothripidae.

Feeding on leaves

Leaf-feeding behaviour has been studied by direct observation of thrips on leaves (Heming, 1978; Hunter and Ullman, 1989), observation of thrips feeding through artificial membranes (Chisholm and Doncaster, 1982; Chisholm, 1983), and electronic monitoring (Hunter *et al.*, 1993). The behaviour on leaves appears to be similar to that on fruit or petals (Childers and Achor, 1991b; Achor and Childers, 1995), although little is known about either of these other types of feeding.

An individual that is about to feed walks over a leaf surface waving its antennae (Hunter and Ullman, 1994), sometimes scraping the surface with each foreleg in turn, scraping the antennae with each foreleg, and dragging the antennae in an S-like pattern over the surface (Hunter and Ullman, 1989; Ullman *et al.*, 1992; Hunter *et al.*, 1993). These behaviours are probably involved in examining the surface for appropriate chemical and physical cues. Gustatory hairs at the tip of the antennae may be used when the antennae are dragged across the substrate (Walbank, 1996). The thrips then stands still in a characteristic posture with its forelegs apart and its antennae still and applies the tip of its mouthcone to the surface (Fig. 4.4). The paraglossal lobes move apart and the labral pad is unfolded and pressed against the surface. Sensilla on the paraglossae and on the tips of the maxillary and labial palps contact the surface and are probably used to test its chemical and physical properties. In *Limothrips cerealium* adults, a pair of setae on the clypeofrons also touch the surface (Chisholm and Lewis, 1984). The thrips may nod its head up and down a few times within a second or two, with the mandible not quite, or only just, breaking through the surface (Heming, 1978; Chisholm and Lewis, 1984), possibly to test the strength of the leaf or allow a small amount of sap to exude and be tasted. Nothing is known about what signals are required to initiate feeding. The thrips sometimes moves on without probing and repeats this 'labial dabbing', presumably because the chemical

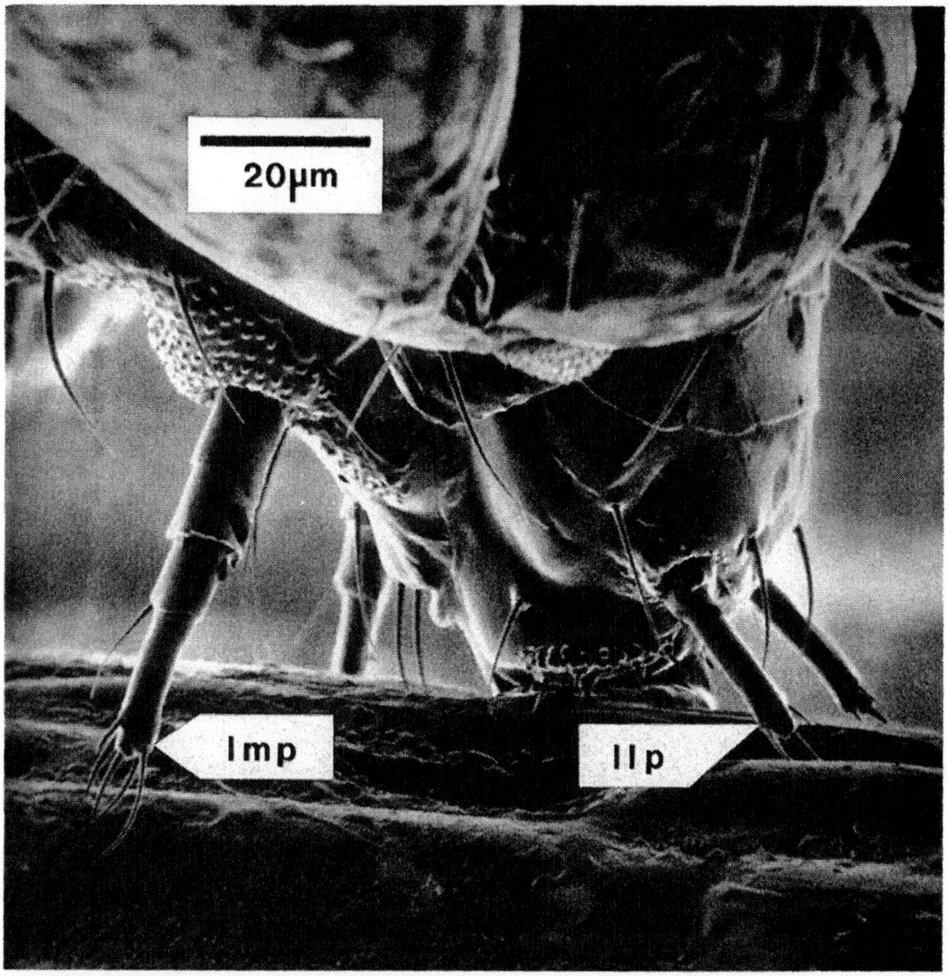

Fig. 4.4. Scanning electron micrograph of the left side of the mouthcone of an adult female *Limothrips cerealium* feeding on a wheat leaf. llp = left labial palp, lmp = left maxillary palp (photo: I.F. Chisholm).

and physical stimuli are not suitable (Hunter and Ullman, 1994).

When the stimuli are right, probing is initiated. The thrips nods its whole head vigorously downwards and backwards (Fig. 4.5) (Kloft and Ehrhardt, 1959; Chisholm and Lewis, 1984; Hunter and Ullman, 1989). Some of the force for the movement may be applied through the forelegs. The tip of the mouthcone remains still on the leaf surface and acts as a pivot for the rest of the mouthcone, which shortens by closing up the flexible clypeo-labral membrane. This shortening of the mouthcone causes the rapid protraction of the mandible, which punctures a distinct hole in the leaf surface. Early observers of these nodding movements mistakenly believed that the thrips were rasping or rooting (Borden, 1915; Foster and Jones, 1915). There can be several

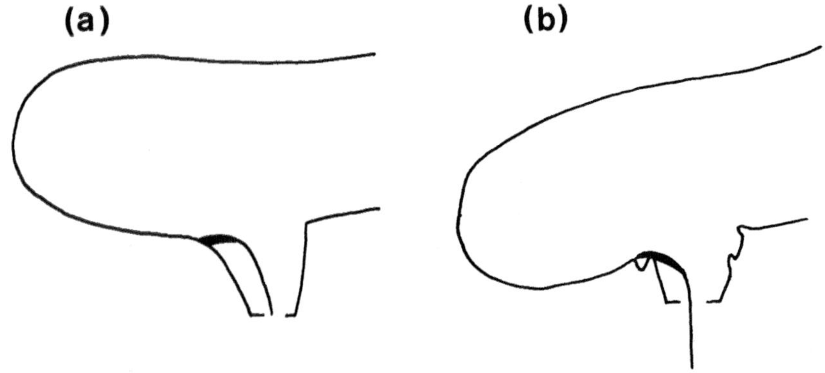

Fig. 4.5. Diagram of a longitudinal section of the head to show the movement of the mandible and folding of membranes in a thripid: (a) mandible retracted, (b) mandible protracted.

minor thrusts of the mandible before the main penetrative stab (Chisholm and Lewis, 1984). The vigour and speed of mandibular probing suggest that thrips are punching a hole in the wall of the epidermal cell, rather than piercing it gently. The mandible extends about 20 μm through the cuticle into an epidermal cell and is then withdrawn within about a second. It is not clear whether the mandible is withdrawn completely or just as far as the cell wall.

After retraction of the mandible, the maxillary stylets are inserted together in a series of rapid thrusts and withdrawals. In *Haplothrips verbasci* (Phlaeothripidae), the maxillary stylets are protracted while the mandible is still moving, so the mandible and stylets must both be inserted in the leaf at the same time, at least momentarily (Heming, 1978). The maxillary stylets emerge at the tip of the mouthcone immediately posterior to the mandible (Figs 4.1 and 4.3), so it would be difficult for the mandible and stylets to enter the leaf through exactly the same hole. Feeding marks on the surface of cereal leaves have a hole shaped like a figure eight of about 2.5 to 5 μm across (Fig. 4.6), indicating that the mandible and maxillary stylets enter the leaf through adjacent but connected holes. Childers and Achor (1991b) also observed feeding marks about 2.5 to 5 μm across, which is greater than the width of the mandible or the maxillary stylets. However, Kumar *et al.* (Fig. 2, 1995) and D.E. Ullman (California, 1995, personal communication) found a mixture of round, elongate oval, 8-shaped and T-shaped holes. The hole shape may vary according to whether the leaf was just tested, probed shallowly or probed deeply. The function of the mandible seems to be more complicated than simply punching a single hole through which the maxillary stylets can follow. Although the mandible no doubt helps by piercing the first hole, the maxillary stylets may sometimes have to widen the hole, at least at the surface, and possibly all the way though the cell wall, thus leaving an 8-shaped hole. If the mandible were not completely retracted after probing, the maxillary stylets might be able to widen the mandibular hole by sliding against the mandible's posterior surface (Fig. 4.2), thereby preventing the stylets from buckling by using the mandible as a guide and support, as suggested by Reyne (1927). An

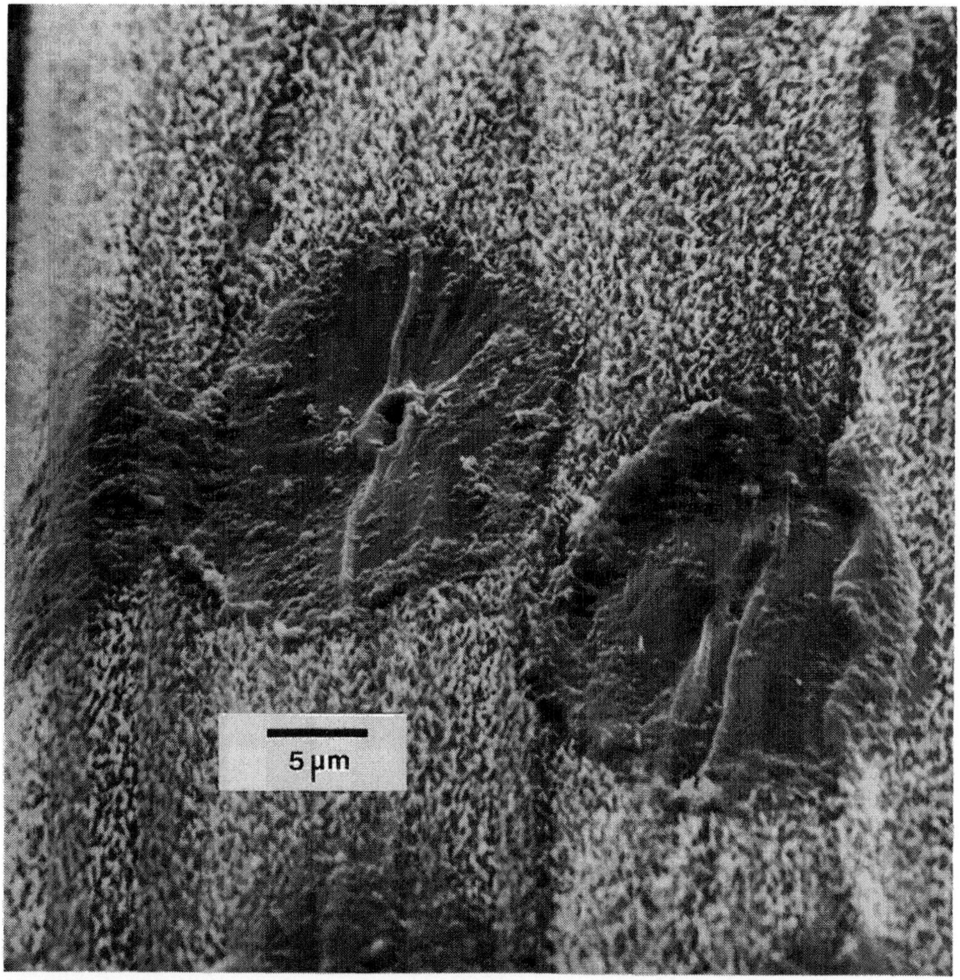

Fig. 4.6. Scanning electron micrograph of marks left on a wheat leaf after feeding by an adult female *Limothrips cerealium.* Each mark consists of a large disc-shaped area with a small figure-of-eight shaped hole in the centre (photo: I.F. Chisholm).

alternative explanation for the 8-shaped holes is that a single hole is used at first, but that it is extended by the mandible to help release the maxillary stylets at the end of feeding (see below). The mandible might pierce the leaf several times in slightly different positions to form an 8-shaped or T-shaped hole.

The maxillary stylets can be protracted up to about 60 μm. In *Haplothrips verbasci*, the maxillary stylets are inserted in a series of rapid thrusts, lasting 30 to 40 s, that slash and puncture epidermal, palisade and spongy mesophyll cells, with each thrust moving the stylets in a slightly different direction (Heming, 1978). This 'jabbing-whipping motion' has been observed in *Scirtothrips citri* feeding through a membrane (Wiesenborn and Morse, 1988), but not in

Limothrips cerealium (Chisholm and Lewis, 1984). The mandible pierces the toughest part, which is the epidermal cuticle and cell wall, but the maxillary stylets apparently have to pierce the walls of internal cells without the aid of the mandible.

The initial probing phase is accompanied by salivation (Hunter *et al.*, 1993; Harrewijn *et al.*, 1996a,b). The salivary pump can be seen vibrating rapidly within the head of larvae of *Haplothrips verbasci* (Heming, 1978). When salivary pumping stops, the cibarial pump takes over. Observations of pulsation of the head capsule, distances moved by particles about to be ingested and electronic monitoring have indicated cibarial pumping rates in the range 2 to 8 Hz (Heming, 1978; Chisholm and Lewis, 1984).

Light microscopy and transmission electron microscopy have shown that thrips feed on the contents of epidermal, palisade and spongy mesophyll cells, leaving collapsed cell walls or destroyed cells with scattered cell contents (see Chapter 11; Wardle and Simpson, 1927; Chisholm, 1983; Chisholm and Lewis, 1984; Childers and Achor, 1991b; Achor and Childers, 1995; Kumar *et al.*, 1995; Mitchell *et al.*, 1995). Variation in depth and extent of feeding is reflected in the time spent at each feeding hole, which ranges from a few seconds to more than an hour (Chisholm and Lewis, 1984; Hunter *et al.*, 1993; Harrewijn *et al.*, 1996a,b). Phlaeothripids have longer maxillary stylets that can reach further, so they are likely to be able to feed for longer at one hole. Feeds lasting more than an hour have been recorded in *Haplothrips verbasci* (Heming, 1978). Thrips can imbibe continuously for longer on fresher tissue, apparently because they are able to feed on more sap that is released into mesophyll air spaces or onto the tissue surface (Harrewijn *et al.*, 1996a,b). Sakimura (1962) believed thrips have two types of feeding. In shallow feeding of short duration, damage is limited to epidermal cells and a few adjacent mesophyll cells, whereas in deep feeding the damage extends to deeper mesophyll cells and possibly the vascular tissues (Sakimura, 1963). This appears to be confirmed by electronic monitoring of *Frankliniella occidentalis*, which has revealed two modes of feeding: frequent probes that are only a few seconds long, and less frequent probes of longer duration (Hunter *et al.*, 1993; Harrewijn *et al.*, 1996a,b). Sakimura (1962) also stated that some species of thrips feed deeply, whereas others feed shallowly, particularly under field conditions. Zimmerman (1901) found that *Heliothrips haemorrhoidalis* fed only on epidermal cells of coffee leaves and left the cells underneath intact and Klee (1958) observed *Thrips laricivorus* making many short shallow feeds. However, such feeds might just be test probes, possibly involving discrimination by the precibarial and cibarial sensilla, which are then followed by deeper feeding if appropriate signals are received (Hunter and Ullman, 1994). Adult *Frankliniella occidentalis* feed on both epidermal and mesophyll cells of acceptable tomato species, but feed only on epidermal cells of highly antixenic species (Kumar *et al.*, 1995). Adult female *Thrips palmi* die after about 5 days if their food contains 0.9 mg ml^{-1} of α-tomatine, a component of tomato leaves (Hirano *et al.*, 1994). This substance is thought to act as a feeding deterrent and prevent tomatoes from being attacked in glasshouses. Resistance to *Frankliniella occidentalis* has been attributed to a chemical that occurred in the leaves but not in the pollen of a cotton variety

(Trichilo and Leigh, 1988) and chemicals in the leaf and flower sap of chrysanthemums (De Jager *et al.*, 1995). Feeding studies have tended to concentrate on second instar larvae and adults, which are of similar size, but first instar larvae are much smaller and may be unable to probe as deeply. There is still no firm evidence as to whether thrips ever feed from vascular bundles. Information on depth of feeding is particularly relevant to understanding virus transmission (see Chapter 14).

Detailed observations of the action of the maxillary stylets have been made for adult *Limothrips cerealium* feeding through a PVC membrane on 0.35 μm-diameter polystyrene latex particles suspended in sucrose solution (Chisholm and Doncaster, 1982; Chisholm, 1983; Chisholm and Lewis, 1984). The depth of stylet penetration varied throughout a feed from a few micrometres to 60 μm, sometimes appearing to change as a response to blockage of the stylet tube. Ingestion was apparent from the jerky movement of the particles towards the stylets, caused by action of the cibarial pump. Thrips sometimes fed with the stylets unequally extended and particles accumulated around the tip of the shorter stylet, which only reached about half-way along the length of the more extended stylet. Thrips were also observed feeding through an artificial membrane on isolated wheat protoplasts (about 32 μm across) or chloroplasts (about 6 μm across) in sucrose solution. The maxillary stylets easily pierced the plasmalemma enclosing the protoplasts before sucking up the contents. Chloroplasts are wider than the internal diameter of the stylets, which is only about 1 μm. However, chloroplasts up to 6 μm across were ingested intact after apparently being distorted into sausage shapes. Chloroplasts moved towards the stylet opening from distances of up to 125 μm. Despite being much larger than the width of the stylet tube, the chloroplasts did not accumulate at the stylet opening in the way that the 0.35 μm-diameter particles had done.*

Towards the end of feeding, thrips sometimes rotate around the point of insertion of the mouthparts, which suggests that they have difficulty in extracting the mouthparts from the plant. More usually, they are withdrawn rapidly, within about 0.2 s, or more gradually by sliding one stylet against the other. Stylet withdrawal is often accompanied by a similar nodding movement to that used for mandible insertion. Perhaps the mandible is being used to widen the feeding hole to release the stylets (Chisholm and Lewis, 1984). Similar mandibular actions have been observed in *Haplothrips verbasci*. When this species feeds on a leaf that has been cut to remove the cuticle and expose the cell contents, the stylets often buckle. The mandible is then inserted alongside the stylets to widen the feeding hole and release them (Heming, 1978). Salivation at the end of feeding may provide lubrication for mandibular movement and maxillary retraction.

Feeding marks are sometimes visible on the surface of leaves at low magnification (Reyne, 1927). The marks left by *Limothrips cerealium* on wheat leaves are particularly clear in SEM photographs (Fig. 4.6). Surface wax is

*A film of these experiments entitled *Thunderfly (thrips): how the pest feeds* is available from the Higher Education Film and Video Library, Concord Video and Film Council Ltd, 201 Felixstowe Road, Ipswich IP3 9BJ, UK.

removed, probably by pressure from the labral pad at the tip of the mouthcone, to leave a round mark about 20 μm across (Chisholm and Lewis, 1984). Similar round marks are left by *Frankliniella occidentalis* on tomato (Kumar *et al.*, 1995). The action of saliva from salivary plugs might also be involved in producing feeding marks. McKenzie *et al.* (1993) found that the feeding marks left by *Thrips tabaci* on onion leaves were about 15 μm across (range 12 to 20 μm) and proposed that thrips were using the whole mouthcone tip, not just the labral pad, to penetrate the thick wax layer on onions before piercing underlying cells.

After feeding at one hole, a thrips may change to another behaviour, such as resting, or may repeat the whole procedure and carry on feeding at a new hole. Mollema *et al.* (1995) found that feeding holes were grouped together on susceptible varieties of cucumber, but when on resistant varieties, thrips appeared restless, spending a larger proportion of their time walking, and the feeding holes were scattered individually. *Frankliniella fusca* leaves trails of feeding sites on peanut leaflets (Mitchell *et al.*, 1995).

Gall thrips feed on leaves, presumably in a broadly similar way to non-galling leaf thrips, and can be pests of horticultural and aesthetic landscape plants (Mound and Walker, 1986 p. 82; Paine *et al.*, 1991; Pelikan, 1991). They have particularly large salivary glands (Heming, 1993), containing proteases, amylases and lipases (Raman *et al.* 1991), and injected saliva is likely to be involved in stimulating the galling response of the plant, although this remains to be demonstrated (Mound, 1994).

Feeding on pollen

The behaviour when feeding on individual pollen grains, which are typically only 25 to 40 μm across, is particularly difficult to observe because it usually takes only a few seconds and the pollen grains are usually obscured by the head, the mouthcone, the forelegs and the substrate. The method of feeding is similar to that for epidermal leaf cells in that the outer wall is pierced and the liquid contents are removed, leaving an empty or collapsed outer wall.

The following account of pollen feeding behaviour is based mainly on the work of Grinfel'd (1959) and Kirk (1984a,b,c, 1985a, 1987a, 1988a). Grinfel'd observed thrips in flowers illuminated by sunlight, while Kirk used a clear gelatine capsule as a transparent substrate through which to observe the underside of the insect's head (Kirk, 1987a). Other transparent substrates have also been used (Annadurai and Morrison, 1987; Kirk, 1988a; Raizada and Nangia, 1988).

Thrips in flowers feed on pollen in anthers and on pollen scattered over floral surfaces. If they feed at anthers, they do not have to search for each new pollen grain so can stay relatively still and feed much more rapidly than when feeding on scattered pollen. The close contact with many pollen grains in an anther means that thrips often become smothered with pollen, which is likely to enhance pollination. Since only a few thrips can fit on an anther at a time, access to pollen may be reduced in crowded flowers, which could lead to a density-dependent reduction in growth and fecundity and affect population dynamics (Kirk, 1994). The feeding behaviour at anthers may have been overlooked, as it sometimes occurs particularly at night (Kirk, 1984a).

A thrips turns to approach pollen grains from distances of up to about a millimetre. Antennal waving may be involved in detection (Raizada and Nangia, 1988). When an oligophagous thrips is offered pollen from non-host plants, it may feed on a few grains at first but then it will pause with its mouthcone poised above pollen grains and move on without touching them. After a few minutes of rejecting grains it will stop and groom. Presumably, olfactory cues are being used to distinguish acceptable pollens.

A thrips that is offered host pollen will approach a grain, stand with its mouthcone above it and immediately place the tip of its mouthcone on the upper surface (Fig. 4.7). The labral pad is unfolded and the mandible immediately pierces the wall of the grain, which is about 1 to 5 μm thick, and is then withdrawn. It is not clear whether the mandible is withdrawn completely or only as far as the pollen wall since this might help to keep the feeding hole open. The nodding movement that accompanies mandibular protraction when terebrantians feed on leaf cells was not observed by Kirk (1984b), but the movement may not be so obvious if the mandible is not protracted as far. A thrips will sometimes insert its mandible and then withdraw it at once and move on without feeding, but usually a thrips probes with its mandible, inserts its maxillary stylets and then starts to feed. The stylets have not been observed within pollen grains and their use is assumed. Pulsation within the head of the thrips, which is often matched by pulsation of the grain wall, shows that the cibarial pump is sucking up the liquid contents of the grain. In some species of pollen, the wall crumples during feeding, while in others, usually those with larger grains, the shape is retained. At the end of feeding on smaller grains, the thrips nods its head and appears to wipe off the remains of the grain against the substrate, but larger grains are abandoned without this movement. The whole process from first touching the grain to leaving it behind normally lasts only a few seconds. Mean feeding times are in the range of 3 to 120 s, and vary with thrips species, instar, grain volume (Fig. 4.8), and temperature (Grinfel'd, 1959; Kirk, 1984b, 1987a; Annadurai and Morrison, 1987). The time to pierce the grain initially and to remove it from the mouthcone at the end of feeding is negligible. Virtually the whole time is spent ingesting. Once a thrips has abandoned a grain, it usually moves on at once to feed on another one nearby.

Thrips might be able to pierce the wall of a pollen grain more easily by using the apertures (pores and furrows) in the surface, but the apertures cannot be seen easily with a dissecting microscope, so this hypothesis cannot be investigated directly. However, thrips virtually always apply the mouthcone to the uppermost part of the grain's surface and it is improbable that an aperture occurs so frequently in exactly this position. The speed with which a thrips applies its mouthcone to the pollen grain would also seem to preclude any searching for an aperture on the grain surface. It is therefore unlikely that thrips need to feed in this way.

The forelegs are sometimes used to handle pollen grains. Although the thrips species observed by Kirk (1984b) rarely used their forelegs, they did use them for certain species of pollen. Pollen grains that tended to stick together in lumps were swept towards the mouthparts and individual pollen tetrads of

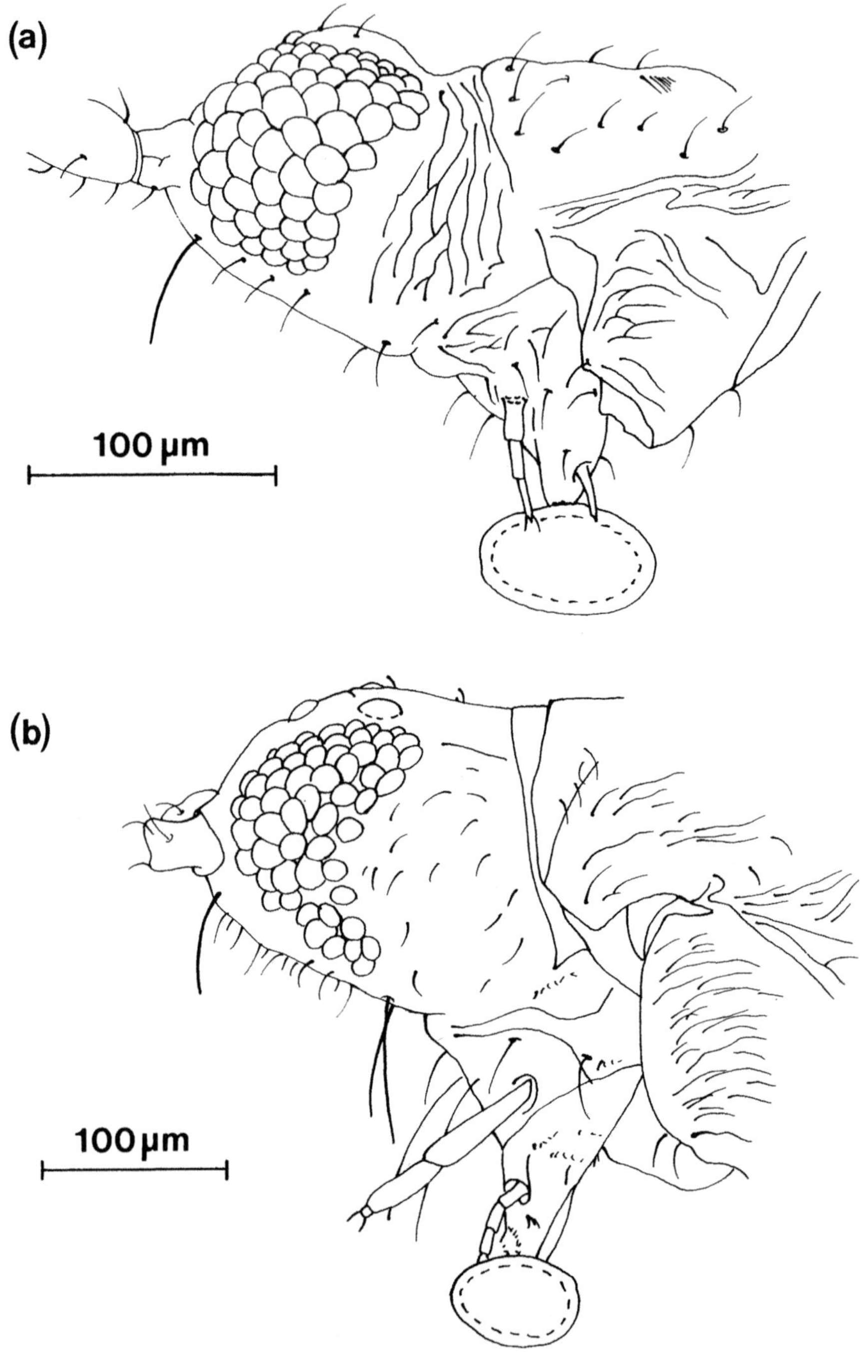

Fig. 4.7. Profile of head and mouthparts of thrips feeding on pollen grains (after Kirk, 1984b). (a) A thripid (*Thrips vulgatissimus*), (b) an aeolothripid (*Aeolothrips intermedius*). The forelegs normally obscure the view and so have been omitted. The broken line on the pollen grain shows the approximate thickness of the pollen grain wall.

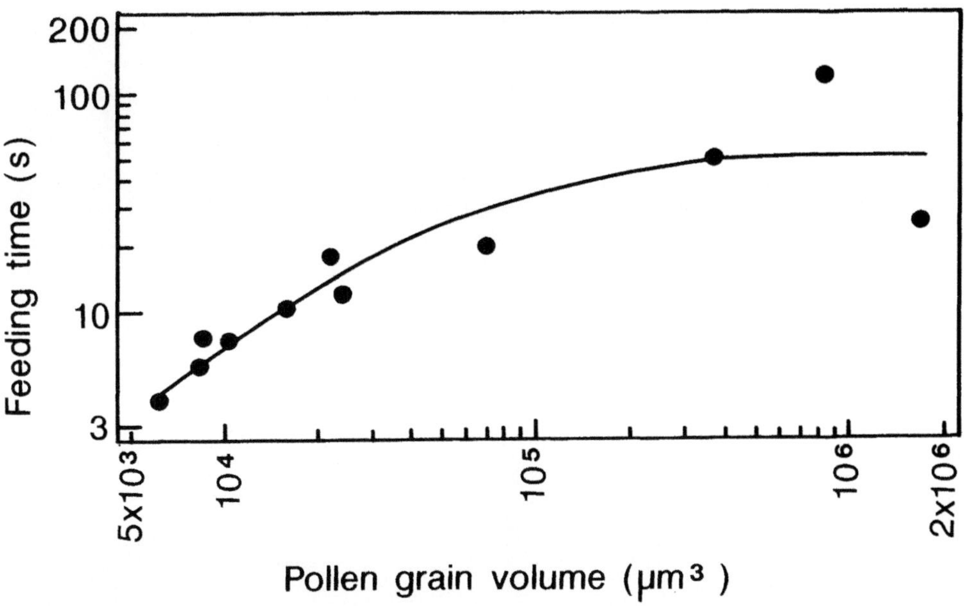

Fig. 4.8. A graph of mean feeding time per grain against pollen grain volume for adult female *Thrips fuscipennis* (after Kirk, 1984b).

Calluna vulgaris were usually turned around before thrips fed on different parts of the tetrad. Thrips have been observed manipulating individual grains and holding them while feeding (Grinfel'd, 1959; Raizada and Nangia, 1988). Occasionally, the forelegs are used to remove grains from the tip of the mouthcone after feeding.

A thrips often feeds with the grain remaining on the substrate, but frequently it raises its head with the grain still at the tip of the mouthcone and even walks around with the grain in this position. Pulsation within the head and buckling of the grain show that the contents are still being ingested. Grinfel'd (1959) claimed that the grain is held by the maxillary and labial palps, but the palps do not seem long enough to be able to do this (Fig. 4.7), although they do touch the upper surface of the grain. The nodding movement of the head and sometimes the use of the forelegs that are required to remove the grain after feeding suggest that the grain does not need to be held. As the grain contracts, the maxillary stylets are likely to be gripped increasingly firmly by the sides of the feeding hole, making it difficult for the thrips to extract them. The feeding hole may even close up completely after feeding, which may explain the lack of success in finding feeding holes in pollen grains emptied by thrips (Fig. 4.9) (Kirk, 1984c). Walking with the grain in mid-air may allow a thrips to locate the next grain while still feeding and have a better view of approaching predators.

Polyphagous thrips feed on pollen of many sizes from a wide taxonomic range. However, some pollen species present particular problems. Very large grains and tetrads are pierced, but are often abandoned early, presumably

Fig. 4.9. Scanning electron micrograph of a pollen grain of *Malva sylvestris* buckled after being fed on by an adult female *Thrips fuscipennis* (photo: W.D.J. Kirk).

because of the difficulty of extracting all the contents. Thrips sometimes break from feeding on very large grains and move the mouthcone to a new position, perhaps because the maxillary stylets cannot reach the full length of the grain. Pollen grains vary enormously in size between species, and thrips have been observed feeding on grains differing in volume by a factor of 800, but despite this there is no evidence that oligophagous flower thrips have adapted in body size to the size of the pollen grains on which they feed, even though body size does correlate with the flower size of the host (Kirk, 1991). The pollen grains of the hollyhock (*Alcea rosea*), in the family *Malvaceae*, are particularly sticky and thrips spend most of their time trying to remove grains from their body rather than feeding on them. Thrips are unable to feed on single hollyhock grains, but can sometimes feed on them if they are in a lump and are therefore anchored. Newly hatched larvae of *Haplothrips kurdjumovi* appear to be desiccated and killed when smothered with pollen (Putman, 1965). Plant species that have highly sticky or very large grains may reduce or prevent pollen damage by thrips, but spines on pollen grains, even up to 9 µm long, do not seem to deter feeding. Most pollen grains that have been fed on have lost their entire liquid contents and are no longer viable.

Thrips feed on pollen in bouts lasting several minutes. Kirk (1984b) observed one particularly long bout in which a thrips fed on 106 grains in 15 min. At the end of a bout, a thrips will stop walking around looking for new grains and stand still, often grooming. After a few minutes, it will quickly deposit a rectal droplet from the end of its abdomen and almost immediately walk away and start feeding again. Rectal droplets are also produced at intervals by thrips feeding on leaves and petals and since they frequently evaporate to leave a residue that appears as a small dark spot, they provide characteristic symptoms of thrips damage (Bournier, 1983; Kirk, 1996). When thrips feed on leaves, the spot is dark green, presumably consisting of concentrated chlorophyll. When spots occur on petals, they are the same colour as, but darker than, the petal pigment. The 'egestion time', or time between the first ingestion of test food and the production of the first droplet, for second instar larvae of *Scirtothrips citri* feeding through a membrane on an artificial diet is about 30 min (Wiesenborn and Morse, 1985). The non-feeding period before producing a droplet may be necessary to allow the gut to process and dispose of excess liquid and so make room for further liquid to be ingested. An increase in the production of urine (diuresis) following feeding, which disposes of excess liquid, is well known in both blood-feeding and plant-feeding insects. The interval that is needed before a droplet is produced and the retention of pigments from leaves or petals suggest that rectal droplets contain both urine (liquid excreted by the Malpighian tubules) and faeces (material passing undigested through the gut).

Feeding on prey

Thrips that are specialist predators are often fast runners, so they appear to possess at least some behavioural adaptations to predation. However, there is no apparent difference in the mouthparts between predatory and phytophagous thrips. Many specialist predators also feed on plant tissue and pollen (Putman, 1965; Bournier *et al.*, 1979; Parrella *et al.*, 1982) and some species that are usually phytophagous, such as *Frankliniella occidentalis* and *Thrips tabaci*, can also be predatory (Trichilo and Leigh, 1986; Wilson *et al.*, 1996).

Specialist predators are common in both the Terebrantia and the Tubulifera (see Chapter 8 and Appendix 3c in Lewis, 1973), and mites of all stages appear to be their commonest prey. Mite eggs are located by chance and recognized by contact (Putman, 1942). Predatory thrips apply the tip of the mouthcone to individual eggs, which are about 130 µm across, and suck out the contents, leaving the crumpled remains behind. MacPhee (1953) observed *Haplothrips kurdjumovi* pierce eggs with the mandible, aided by a slight rotation of the head, and then insert the maxillary stylets and move them with a jabbing-whipping motion. The feeding hole is seldom visible afterwards (Putman, 1965). Adult female *Scolothrips sexmaculatus* take an average of 1.2 min to feed on a recently deposited egg, but 2.1 min to feed on an egg near eclosion (Gilstrap and Oatman, 1976). Second instar larvae take two to three times as long (Fig. 4.10).

Active stages of mites are harder to feed on than eggs, because they have to be caught before they can be consumed. *Scolothrips sexmaculatus* larvae and adults hunt by moving slowly through webbing of the carmine spider mite (*Tetranychus cinnabarinus*), sometimes passing within 0.5 to 1.0 mm of a mite without attacking it. In contrast, *Frankliniella occidentalis* larvae, which are less aggressive than *Scolothrips sexmaculatus*, are deterred by mite webbing (Trichilo and Leigh, 1986). Physical contact with prey seems to be necessary before thrips will attack active stages (Gerlach, 1985), but even after contact, a thrips will only pursue a mite for 2 to 5 mm (Gilstrap and Oatman, 1976).

Scolothrips sexmaculatus captures, holds and manipulates its prey with its forelegs and feeds on it by piercing the middle of the dorsal surface (Gilstrap and Oatman, 1976). Mites sometimes drag their larval captors through the

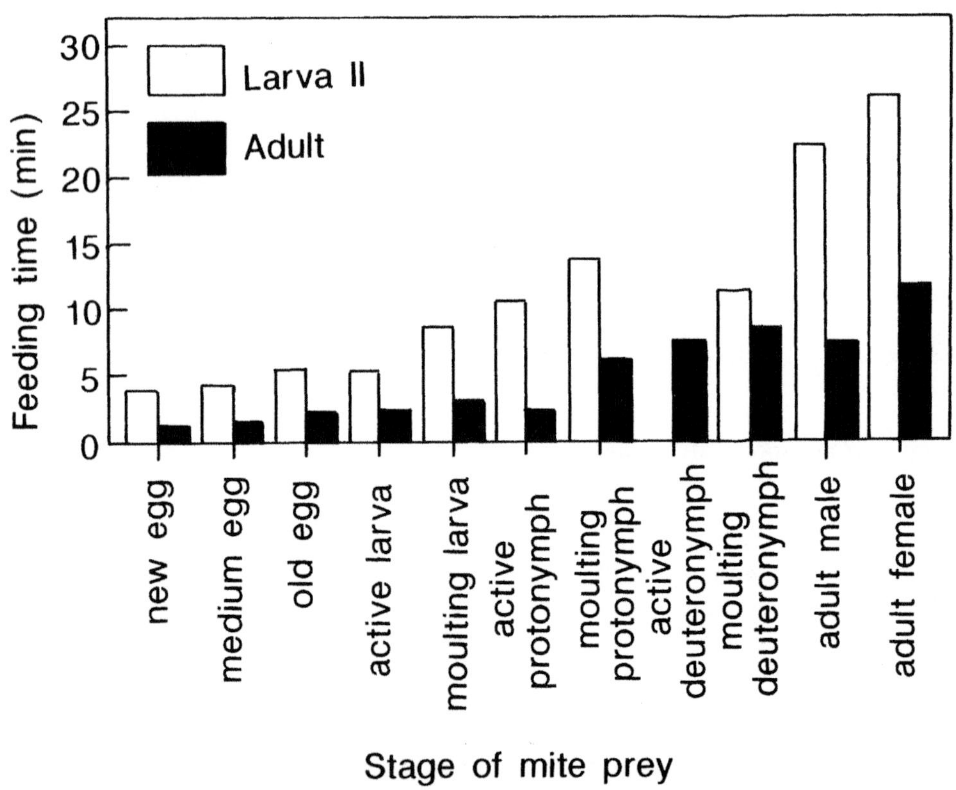

Fig. 4.10. A graph of mean feeding time per individual mite against stage of the mite prey (*Tetranychus cinnabarinus*) for larva II and adult females of *Scolothrips sexmaculatus* (from the data of Gilstrap and Oatman, 1976).

webbing after feeding has started, but the mites stop moving their legs after about 10 to 20 s. The prey is fed on until the body collapses, which can take up to about 26 min, depending on the stage of the thrips and the size of the prey (Fig. 4.10). The relationship between feeding time and prey size parallels that between feeding time and pollen grain size (Fig. 4.8). *Scolothrips longicornis* manipulates its prey with its forelegs and pierces it on the dorsal surface, but it is then turned over and fed on from the ventral surface (Gerlach, 1985). Immature mites die after 30 to 60 s and adults after 60 to 90 s. Thrips are wasteful and often inflict a higher mortality than is necessary for feeding. Although some mites are completely sucked out, others are only fed on until the mite dies. If a thrips is distracted by a passing mite, it will abandon the prey it is eating and attack the passing prey, without always returning later to the original prey. *Leptothrips mali* often kills mites without feeding on them (Parrella and Horsburgh, 1983).

Aeolothrips intermedius has a rather unusual method of feeding on mites by piercing them on the legs and around the mouthparts and then sucking from the dorsal surface (Bournier *et al.*, 1979). *Trichinothrips breviceps*, which feeds only on psocids, runs after prey, seizes it and then mounts its back to feed on it (Seshadri, 1953). After sucking most of the body contents, the psocid is rolled around with the forelegs to allow feeding from other angles. It can easily feed on a psocid twice its size, although it may take as much as 90 min to complete the meal.

A pattern of feeding in bouts with periods of rest in between is found in several species of predatory thrips (Bailey, 1939; Gerlach, 1985). *Scolothrips sexmaculatus* usually spends at least half its time resting, while the percentage of the time spent feeding tends to increase with the size and density of the prey (Gilstrap and Oatman, 1976; Gerlach, 1985). The short time spent searching and the need to contact prey before attacking it means that effective search areas are only a few square centimetres per hour (Gilstrap and Oatman, 1976). The functional response, or relationship between feeding rate and prey density, can be type II (feeding rate rises with prey density, but gradually decelerates to a plateau) or type III (has a sigmoid relationship with prey density) (Parrella and Horsburgh, 1983; Gerlach, 1985; Bailey and Caon, 1986; Trichilo and Leigh, 1986; Wilson *et al.*, 1996).

The phlaeothripid *Aleurodothrips fasciapennis* is an important biological-control agent of scale insects on citrus (Beattie, 1985). When feeding on coconut scales, it feeds continually for long periods. Most scales are fed on for only about 30 s, not long enough to empty the contents, but a few are completely emptied, which takes several minutes. Sometimes they seize a host with their forelegs, pull it out from under the scale cover and hold it up while feeding on it 'as a mantis holds a moth' (Taylor, 1935).

There are few records of predatory thrips attacking aphids (see Appendix 3c in Lewis, 1973). In the laboratory, *Aeolothrips intermedius* will not attack aphids, despite killing a wide range of other small arthropods (Bournier *et al.*, 1979). Aphids may sometimes be too large for thrips to handle. Some thrips show distinct prey preferences. *Haplothrips victoriensis* larvae have been observed eating mite eggs, but not the mobile stages, in the field. They also prefer eggs when offered a choice, but will eat mobile stages if there is no alternative food (Bailey and Caon, 1986). However, they will not eat moth eggs, even if deprived of other food. In contrast, *Haplothrips kurdjumovi* feeds on moth eggs as well as mite eggs (Putman, 1942).

Thrips are sometimes predators of other species of thrips. For example, *Aeolothrips intermedius* kills thrips larvae by piercing them on the thorax and then feeds on them through the dorsal surface of the abdomen (Bournier *et al.*, 1979).

Both predatory and phytophagous thrips can become cannibalistic at high densities and when alternative food is not available (Bailey, 1939; MacPhee, 1953; Loan and Holdaway, 1955; Gilstrap and Oatman, 1976; Bournier *et al.*, 1979). Individuals attack smaller or weaker individuals, such as larvae that are hatching or moulting, sometimes having to fight to overcome their victim (Bailey, 1939; MacPhee, 1953; Gilstrap and Oatman, 1976). Larvae and pupae of *Scolothrips sexmaculatus* ward off cannibals by flicking their abdomens from side to side (Bailey, 1939). Some phlaeothripids cannibalize eggs (Putman, 1942; MacPhee, 1953), although this probably occurs less often among terebrantians, which have their eggs buried within plant tissue rather than exposed on the plant surface.

Feeding on liquids

Thrips can feed on exposed liquids such as water droplets, sugar solution or nectar (Grinfel'd, 1959; Heming, 1978). *Limothrips cerealium* adults and larvae can feed on exposed liquids, either with the stylets completely retracted or with the maxillary stylets protracted up to 30 μm (Chisholm, 1983). *Haplothrips verbasci* larvae apply the tip of the mouthcone to droplets of water and suck up the liquid, probably through the opening at the tip of the maxillary stylets within the mouthcone (Heming, 1978). When this species feeds at cut leaf surfaces where the cell walls are missing, it extends its maxillary stylets to suck up cell contents without protracting its mandible, even from 10 to 60 μm above the leaf surface.

Diet and Nutrition

The available foods

Primitive thrips are presumed to have lived in leaf litter, where they fed on fungal hyphae. From there they moved to flowers and then to leaves (Mound *et al.*, 1980; Heming, 1993). A few have become specialist predators. Only just over half of all species of thrips feed on green leaves (Mound, 1994), so the leaf-feeding habit does not predominate within the Thysanoptera as a whole, even though it predominates among crop pests. The rest of the Thysanoptera consists mainly of fungus-feeding species in the Phlaeothripidae, many of them tropical. Table 4.2 gives a breakdown of the major sources of food in the three main families of thrips.

Fungus feeding is largely confined to the family Phlaeothripidae. About half of the subfamily Phlaeothripinae live in leaf litter or on dead wood and feed

Table 4.2. The major sources of food within the three largest families of thrips (Mound *et al.*, 1980; Heming, 1993). The information is broken down by subfamily, with selected genera of economic significance as examples. The table is based on incomplete knowledge and is only intended as a rough guide.

	Number of species	Leaves	Pollen	Fungal hyphae or products	Fungal spores	Prey
Aeolothripidae	220		x			x
Thripidae						
Panchaetothripinae	110	x				
Thripidae						
Thripinae	1500	x	x			x
Frankliniella		x	x			
Scolothrips						x
Thrips		x	x			
Phlaeothripidae						
Phlaeothripinae	2350	x	x	x		x
Haplothrips		x	x			x
Phlaeothripidae						
Idolothripinae	670				x	

on fungal hyphae or possibly the external digestion products of fungal decay (Mound and Palmer, 1983; Mound, 1994). This form of feeding does not appear to have been studied, but it is probably unimportant for crop pests. However, Yarwood (1943) thought that *Thrips tabaci* was feeding on fungal hyphae of mildewed grape leaves and the possibility of hyphal feeding should not be overlooked. The subfamily Idolothripinae feed on intact fungal spores (Ananthakrishnan and James, 1983; Mound and Palmer, 1983; Ananthakrishnan, 1984), and their wide maxillary stylets appear to be an adaptation to do this (see above). However, the narrower maxillary stylets of thripids may still allow ingestion of small spores. Various spores have been recorded from the guts of pest thripids (Bailey, 1935; Marullo, 1995), but it is not known whether fungal spores ever form a significant part of the diet or whether thripids can forage specifically for them.

Phytophagous thrips can feed on most accessible non-woody parts of plants. These include leaf cells (Heming, 1978; Chisholm and Lewis, 1984), petal, stamen and style cells (Bournier *et al.*, 1979; Kirk, 1984a), pollen grains (Grinfel'd, 1959; Kirk, 1984b), cells of developing seeds (Doull, 1956), and cells of fruit (Lewis, 1973; Bournier, 1983). They can also feed on exposed nectar (Grinfel'd, 1959), water (Heming, 1978) or insect secretions (Downey, 1965). The species that are crop pests are particularly well represented in the large phytophagous genera *Thrips*, *Taeniothrips* and *Frankliniella* in the family Thripidae.

Predatory thrips feed on the eggs or slow-moving stages of a wide range of small arthropods, including mites, moths, psocids, bugs, flies, beetles, sawflies and thrips (Appendix 3c in Lewis, 1973; Palmer and Mound, 1990). Many of these predators are in the genera *Aeolothrips*, *Aleurodothrips*, *Apterygothrips*, *Franklinothrips*, *Haplothrips*, *Karnyothrips*, *Leptothrips*, *Podothrips*, *Scolothrips* and *Xylaplothrips* (zur Strassen, 1995). Cannibalism occurs in both phytophagous and predatory species: *Aeolothrips intermedius* (Bournier *et al.*, 1979); *Frankliniella occidentalis* (personal observation); *Scolothrips longicornis* (Gerlach, 1985); *Scolothrips sexmaculatus* (Gilstrap and Oatman, 1976); *Haplothrips niger* (Loan and Holdaway, 1955); *Haplothrips kurdjumovi* (Putman, 1942); and *Haplothrips victoriensis* (Andrewartha, 1936).

The composition of the diet

Studies of feeding behaviour have often been prompted by damage to a particular crop, and ensuing research has concentrated on the type of feeding responsible for the damage. However, this may give an incomplete picture of the overall feeding behaviour and nutritional requirements of the insect. For example, the diets of leaf-feeding thrips may be supplemented by feeding on pollen grains, small arthropods or fungi.

The range of foods consumed by a particular species is constrained by where it lives. For example, some leaf thrips have only leaf cells available, but others, such as flower thrips, have a potentially wider choice of food. The predatory thrips *Aeolothrips intermedius* can consume a wide range of foods in the laboratory: leaf cells, petal cells, pollen, larvae of its own species, larvae of three other species of thrips, adults of five species of mite, all stages of an aleurodid, and eggs and nymphs of two species of psyllid (Bournier *et al.*, 1979).

The main clues to what a species feeds on are the site occupied on the host plant and the visible signs of damage. However, these can be misleading. Leaf damage is often obvious, but empty pollen grains, empty mite eggs and shrivelled prey are hard to see so are likely to be overlooked when leaf damage is present. Petal damage may also be misleading. *Thrips imaginis* is abundant in flowers of *Medicago sativa* (Kirk, 1984a). Although visible petal damage demonstrates considerable petal feeding, *Thrips imaginis* is also known to feed readily on the pollen of *Medicago sativa* (Kirk, 1987a), and pollen is an important component of its diet (Andrewartha, 1935).

Many thrips have mixed diets. Some can be predatory and phytophagous: *Aeolothrips intermedius* (Bournier *et al.*, 1979), *Frankliniella occidentalis* (Trichilo and Leigh, 1988), *Frankliniella schultzei* (Wilson *et al.*, 1996), *Thrips imaginis* (Wilson *et al.*, 1996), *Thrips tabaci* (Lewis, 1973; Wilson *et al.*, 1996), *Haplothrips kurdjumovi* (Putman, 1965), *Leptothrips mali* (Parrella *et al.*, 1982), *Haplothrips niger* (Loan and Holdaway, 1955), and *Haplothrips victoriensis* (Bailey and Caon, 1986). It is not surprising that predatory thrips should also be able to feed on plant tissue, because it is a useful stand-by when prey is short. It is, perhaps, more surprising that several species usually considered exclusively phytophagous should have been recorded as predators.

Among predominantly phytophagous thrips, there is no clear demarcation between leaf feeders and pollen feeders. The terms 'leaf thrips' and 'flower thrips' are widely used for convenience but they do not imply exclusive diets of leaf cells or pollen grains, respectively. Some species that breed in flowers and feed on pollen also breed on leaves or leaf buds, apparently without any pollen. These include *Frankliniella occidentalis* (Trichilo and Leigh, 1988; Ullman *et al.*, 1992), *Thrips tabaci* and *Thrips fuscipennis* (Mound *et al.*, 1976; Kirk, 1984b).

In species known to have a mixed diet, the mix can be unequal, with one principal component accounting for most of the volume ingested, though minor components may still have a large effect on growth or egg-laying and cannot be ignored. Many predominantly phytophagous species benefit from the addition of pollen to the diet (see below); likewise additional pollen may also benefit mainly predatory species (Putman, 1965; Parrella *et al.*, 1982).

Many thrips may have wider diets than has been realized and unappreciated components may play an important role in the ecology and population dynamics of a species. In particular, extra components in the diet of leaf thrips are most likely to have passed unnoticed because leaf-cell damage is usually obvious, whilst the remains of pollen or prey are hard to find. Extra components may only be available for part of the year or in some years and not others. It is not known whether the breadth of diet found in the laboratory occurs in the field, and whether the breadth is shown by individuals or is a result of variation between populations each with their own narrow diet.

Nutrition

Many studies have compared the effects of diet on rates of growth, mortality and oviposition.

The addition of pollen to a diet of plant tissue is beneficial for many species (Andrewartha, 1935; Putman, 1965; Bournier *et al.*, 1979; Murai and Ishii, 1982; Parrella *et al.*, 1982; Kirk, 1985a; Trichilo and Leigh, 1988; Teulon and Penman, 1991), as is the addition of insects or mites as prey (Putman, 1965; Parrella *et al.*, 1982; Trichilo and Leigh, 1988). Supplementary pollen or prey may increase growth and oviposition rates and decrease mortality rate. Changes in oviposition rate in response to changes in diet are detectable within a day or two (Murai and Ishii, 1982; Kirk, 1985a; Teulon and Penman, 1991). Egg production must be very nutritionally demanding because of the high rate of egg production and the large size of eggs in relation to body weight, so a rapid response to changes in diet is not surprising. However, pollen or prey alone may not provide an adequate diet; diets of pollen and water without plant tissue can cause high mortality (Murai and Ishii, 1982; Teulon and Penman, 1991). Differences in the effect on thrips between species of pollen (Murai and Ishii, 1982; Teulon and Penman, 1991) may be nutritional, but can also arise from different rates of feeding on each pollen species (Kirk, 1985a).

For some flower thrips, pollen appears to be essential for larval growth, sexual maturation or egg laying (Andrewartha and Kilpatrick, 1951; Bournier *et al.*, 1979; Murai and Ishii, 1982), but these findings were based on a small number of experimental diets. Further experiments might well discover adequate non-pollen diets. Teulon and Penman (1986) claimed that pollen was essential for larval development in *Thrips obscuratus*, but later found that while pollen ensured the fastest development and lowest mortality, development could be completed successfully on other diets (Teulon and Penman, 1991).

A nutritious diet is also important for oviposition in leaf-feeding thrips. The oviposition rate of adult female *Hercinothrips femoralis* dropped from about one egg per 12 h to almost nothing over a period of 2 to 3 days after the thrips were removed from leaf discs of sugar beet and fed on only sucrose solution (Laughlin, 1971). *Thrips tabaci* lays more eggs and develops faster on fresh onion leaves than on the bulbs (Gawaad and Shazli, 1970).

The quantity of food available affects the oviposition rate of predatory thrips. When adult female *Scolothrips longicornis* are provided with more mite eggs per day, they consume more and their oviposition rate increases (Gerlach, 1985). The effect is apparent within a day or two.

Plant tissue may be just nutritionally adequate, because insects have a higher content of protein nitrogen than the plants on which they feed (McNeill and Southwood, 1978). As nitrogen content increases, assimilation and growth efficiency often improve and these in turn raise the rates of survival, growth and reproduction (Strong *et al.*, 1984). Insects are affected by the form and availability of nitrogen, but the total nitrogen content is still a good general indicator of nutritional value.

General surveys of the nutritional value of insect foods have concentrated on nitrogen and water content. Figure 4.11 shows the total nitrogen content by wet weight and the water content of various types of food eaten by thrips. The data are transformed from results for nitrogen content by dry weight and water content summarized by Slansky and Scriber (1985). Insect prey, pollen and some seeds have high nitrogen, whereas leaves and fleshy fruits have low nitrogen contents. Leaf buds have higher nitrogen contents than mature leaves.

The nitrogen content by wet weight provides an interpretation of thrips' feeding patterns. Some thrips feed particularly on the high-nitrogen foods of insect prey and pollen, while thrips on low-nitrogen leaves can increase their nitrogen intake by feeding on insects and pollen when these are available. McNeill and Southwood (1978) list a range of predominantly leaf-feeding insects that appear to obtain additional nitrogen by occasional predation or cannibalism. This behaviour is also found in other insects with low-nitrogen diets (Mattson, 1980). The occasional predation often occurs at times when insects have high nitrogen requirements, such as when growing fast or maturing eggs (McNeill and Southwood, 1978). Seeds can have a particularly high nitrogen content, but many seeds will be too hard for thrips to feed on or will be inaccessible within fruits or shells.

The higher nitrogen content of leaf buds may help to explain why so many leaf thrips breed there rather than on mature leaves. The herbivore syndrome of 'early-season feeding', with the life cycle adapted to growth flushes of the host is well known, not just in insects (Mattson, 1980).

Figure 4.11 shows data for whole tissues, whereas thrips feed on the contents and not the surrounding walls. Since the walls, at least for plants, are relatively low in nitrogen, the nitrogen content of the parts that thrips feed on are probably higher.

The aromatic amino acids (AAAs), phenylalanine and tyrosine, play a major role in the physiological processes of insects. Mollema and Cole (1996) have studied the effects of variation in the leaf concentration of AAAs between

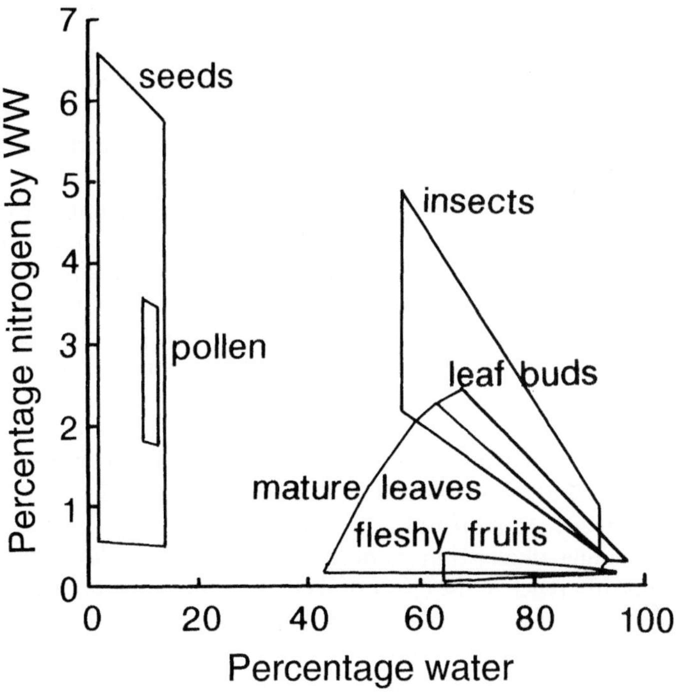

Fig. 4.11. The total nitrogen content by wet weight and the water content of various foods of thrips (modified from Slansky and Scriber, 1985).

genebank accessions of lettuce, tomato, pepper and cucumber. They found that there was a strong correlation between the AAA concentration as a proportion of total protein and the amount of damage from larvae of *Frankliniella occidentalis*. A 66 per cent reduction in AAA concentration gave a 50 per cent reduction in larval damage. It was concluded that higher concentrations of AAAs in plant proteins are important for successful thrips development.

Symbionts can also be a source of amino acids (McNeill and Southwood, 1978). Rod-shaped bacteria, which may be symbionts, occur in the hindgut lumen of *Frankliniella occidentalis* (Ullman *et al.*, 1989, 1992) and mycetomes, organs that harbour symbiotic microorganisms, possibly occur in a fungus thrips (Bournier, 1961; Bournier and Louis, 1971), although further evidence suggests that the observed structures may instead be lysosomal aggregations (Tsutsumi *et al.*, 1994). Mycetomes have not been found in any phytophagous species. Nothing is known about whether symbionts have a role in thrips nutrition. Several artificial diets for rearing thrips are described in Chapter 12.

Sugar appears to stimulate feeding in some thrips. It was added to tartar emetic, which was used to control *Scirtothrips citri* on citrus in the USA in the 1940s, because it improved the effectiveness of the poison by encouraging thrips to feed on it (Boyce and Persing, 1939), and it is still used for this species in combination with botanical baits (Morse, 1995). Some growers have had marked success in controlling *Frankliniella occidentalis* by adding sugar, as a feeding stimulant, to insecticides (Parrella, 1995). Adults and larvae can also feed from dry crystals of sucrose (granulated sugar), leaving a pitted surface (Munger, 1942). The process has not been studied, but presumably the thrips use saliva to dissolve the sugar before ingesting it.

Intraspecific variation between host plants can affect thrips feeding behaviour, nutrition and fecundity. Chemical analysis of seven varieties of rice susceptible to *Stenchaetothrips biformis* showed that they had higher levels of reducing sugars and free amino acids and lower levels of phenolics than six resistant varieties (Thayumanavan *et al.*, 1990). In plant breeding trials, blue lupins (*Lupinus angustifolius*) segregating for alkaloid content into high (bitter) and low (sweet) forms were readily distinguished by two polyphagous species of *Frankliniella*. The thrips fed on the pollen of both forms, but after consuming all the pollen, moved to the foliage where they fed on the leaves, young stems and green pods of the sweet forms, but not the bitter forms. The effect was so distinct that the feeding damage to the leaves could be used to screen the plants for alkaloid content (Forbes and Beck, 1954). In contrast, a monophagous species, *Haplothrips senecionis*, appears to be unaffected by the level of pyrrolizidine alkaloids in its host plant, *Senecio jacobaea* (Vrieling *et al.*, 1991). This difference in the effect of secondary plant compounds between monophagous (specialist) feeders and polyphagous (generalist) feeders is found in other insects (Bernays and Chapman, 1994).

Changes in the nutritional status of plants may affect colonization and population increase. It is common to find more thrips on stressed or weakened plants (see Lewis, 1973), but it is not always clear whether the plant stress has caused the increase, perhaps through nutritional changes in the plant, or whether the thrips themselves have caused the stress. In controlled experiments that ruled out an influence of the thrips, New Zealand spinach plants given more nitrogenous fertilizer received less damage from *Heliothrips haemorrhoidalis* (Wittwer and Haseman, 1946). Thrips are sometimes more numerous on healthier plants than stressed ones. Carey *et al.* (1992) found larger populations of *Taeniothrips inconsequens* on trees with more starch reserves and the numbers of thrips on meadow grass increased with more applied nitrogen, although applied potassium had little effect (Czencz and Gerber, 1986). Field beans that were irrigated or had a varietal characteristic of lush growth were more infested by *Kakothrips pisivorus* (Stoddard, 1986). In such experiments, the treatment will affect the chemical and physical properties of the plants in many ways, so it is difficult to establish the particular factors that have had a direct effect.

Evolutionary patterns

Many phytophagous species of thrips are highly host specific, breeding only on host plants within one genus or family. However, comparison with phytophagous species in other insect orders shows that similar degrees of monophagy or oligophagy are widespread (Chapman, 1982). Figure 4.12 gives a breakdown of the size of the host range for British phytophagous species of insect in the Thysanoptera, Hemiptera, Coleoptera and Diptera. The Thysanoptera have slightly wider host ranges on average than the Hemiptera or Diptera, but a slightly narrower host range than the Coleoptera. The percentages of the 88 British species of phytophagous Thripidae with host plants in one, two, three, four or five plant families are very close to the percentages from a survey of almost 7000 British species of phytophagous insects and mites in 183 families (Ward and Spalding, 1993). For example, 76.1 per cent of British phytophagous thripids have hosts restricted to one family, whereas the equivalent figure across 6933 species of insect and mite is virtually the same at 75.8 per cent. Thysanoptera appear to be typical in the narrowness of their host range.

The degree of polyphagy affects the number of pest species within an order. Ward and Spalding (1993) found that there was a significantly higher proportion of pests among polyphagous species than among non-polyphagous

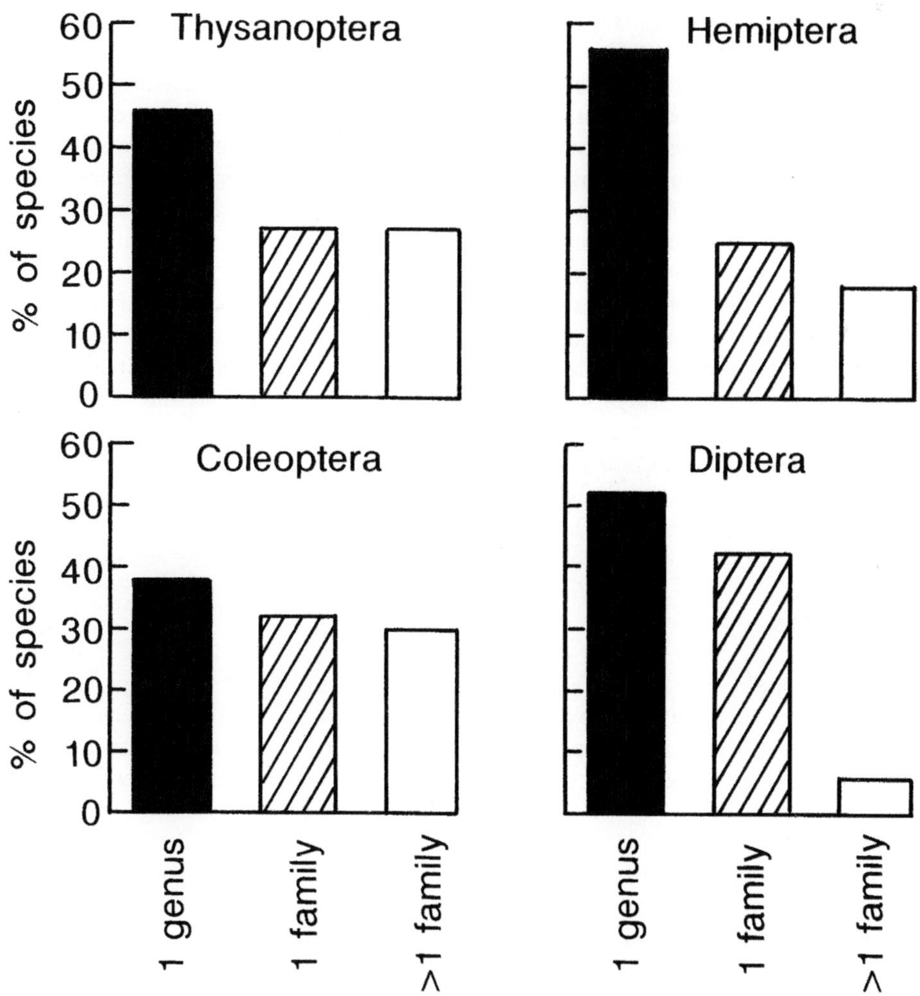

Fig. 4.12. The proportions of four orders of British phytophagous insect species with hosts restricted to one genus, one family or more than one family (after Chapman, 1982).

species. The pattern was present in five orders of insects, including the Thysanoptera. Some of the best-known cosmopolitan thrips pests are highly polyphagous, for example *Frankliniella occidentalis* and *Thrips tabaci* (Bryan and Smith, 1956; Zawirska, 1978).

If thrips that live on leaves can enhance their diet by feeding on pollen grains or prey, the availability of such food could have influenced evolutionary host shifts. Plant species that tend to retain pollen grains or support prey on their leaves are more likely to have been favoured. Wind-pollinated plants release large amounts of pollen into the air, much of which lands on leaves and stems, while insect-pollinated plants produce smaller amounts and do not release it into the air. Kirk (1995) found that among British phytophagous thrips, 21 per cent of flower dwellers had wind-pollinated hosts, compared with 67 per cent of leaf dwellers. About 28 per cent of British flowering plant species are wind pollinated, suggesting that leaf-dwelling thrips favour wind-pollinated hosts.

An alternative interpretation of an interaction between feeding site and host pollination mechanism is that flower thrips have favoured insect-pollinated hosts because the floral colours and scents make the flowers easier to locate for airborne individuals. Many species of flower thrips respond in flight to floral colours and scents (see Chapter 3), so this is quite likely. However, there may be an opposing selective force. Insect-pollinated plants usually have sticky pollen (Faegri and van der Pijl, 1979), which can present special difficulties for thrips. Sticky grains require extra handling and may even prevent feeding. Retention of pollen in the anther restricts access for feeding so could reduce pollen availability. Unfortunately, the two alternative interpretations of an association between main feeding site and the pollination mechanism of the host cannot be disentangled statistically. Other factors, such as species abundance and plant habitat, are likely to have had a large influence in host shifts, and there is evidence that the dimensions of flowers may play an important role (Kirk, 1991).

Feeding Rates

Feeding rates can be related either to the food intake of the thrips (rate of ingestion) or to the damage done (rate of attack). Many factors, such as species, instar, sex, size, age, temperature and type of food, would be expected to influence feeding rates, but the literature contains few detailed studies.

Rates of ingestion

Ingestion rates have been measured over a range of intervals from seconds up to days (Table 4.3). Measurements over a few seconds have involved timing the ingestion of a unitary food item of known volume, such as a protoplast (Chisholm, 1983), pollen grain (Kirk, 1987a, 1993) or mite egg (Gilstrap and Oatman, 1976). The calculated rates are therefore the ingestion rates during feeding. These rates cannot be extrapolated to longer periods because thrips do not spend all their time feeding. However, rates that are measured over longer periods that include feeding and non-feeding periods can perhaps be extrapolated with caution.

The ingestion rates of unitary food items are calculated by dividing the whole volume of the item by the feeding time. The volume of the outer shell is usually ignored. Probing and withdrawal times before and after ingestion are also short (Kirk, 1984b). The calculated feeding rate would be marginally slower if the shell volume was taken into account, but marginally faster if probing and withdrawing times were included; ignoring both small effects probably makes little difference. Comparison of feeding times on pollen grains suggests that ingestion rates increase with body size between species (Kirk, 1984b, 1993). In *Thrips imaginis*, second instar larvae ingest pollen grains faster than adult females and ingestion rates increase with temperature over most of the range 15 to 30°C, but possibly plateau or decrease at the higher temperatures (Kirk, 1987a). Ingestion rates may also depend on the type of food. *Kakothrips pisivorus* breeds particularly on legumes, such as the field

Table 4.3. Mean ingestion rates of individual thrips expressed in nanolitres per hour.

Thrips	Stage	Food	Duration	Temperature (°C)	Rate (nl h^{-1})	Reference
Aeolothrips intermedius	A	Pollen	seconds*	20–25	7.2	1
Limothrips cerealium	A	Wheat protoplasts	seconds*	20–25	5.1	2
	A	10% Sucrose	1–3 h	20–25	18.0	2
Scirtothrips citri	A	3% Sucrose	8 h	36	0.9	4
	LII	3% Sucrose	6 h	28	1.2	3
	LII	3% Sucrose	8 h	28	1.6	4
	LI	3% Sucrose	8 h	28	0.4	4
Scolothrips sexmaculatus	A	Mite eggs	minutes*	27	49.0	5
	A	Mite eggs	days	29	2.5	5
	LII	Mite eggs	minutes*	27	16.0	5
Thrips fuscipennis	A	Pollen	seconds*	20–25	3.6	1
Thrips imaginis	A	Pollen	seconds*	20–21	3.9	6
	A	Pollen	24 h	20	0.1	6
Thrips tabaci	A	5% Sucrose	6–18 min	–	120.0	7
	A	Leaf	30 min	–	6.0	7
	Larva	5% Sucrose	5–18 min	–	72.0	7
	Larva	Leaf	30 min	–	3.6	7

Conversions between masses and volumes have been made by assuming that the food has the same density as water (1 g ml^{-1}). Feeding rates for mite eggs have been calculated from literature data, assuming an egg diameter of 130 μm. An asterisk indicates that the rate is based on a period of continuous feeding, rather than an extended period during which the thrips are likely to have spent some time not feeding. Stage: A = adult female; LII = second instar larva; LI = first instar larva. References: 1 = Kirk (1993); 2 = Chisholm and Lewis (1984) and T. Lewis, Rothamsted, 1995, personal communication; 3 = Wiesenborn and Morse (1985); 4 = Wiesenborn and Morse (1986); 5 = Gilstrap and Oatman (1976); 6 = Kirk (1987a); 7 = Day and Irzykiewicz (1954).

bean, and its ingestion rate is more than twice as rapid on pollen from this species as on non-legume pollens (Kirk, 1985a). When feeding on mite eggs, *Scolothrips sexmaculatus* second instar larvae have consistently longer feeding times and therefore slower ingestion rates than adult females (Gilstrap and Oatman, 1976). Ingestion rates are slower for older eggs, possibly because their contents become more viscous.

Radiolabelled sucrose solutions have been used to measure ingestion rates over periods of minutes to hours. Day and Irzykiewicz (1954) starved thrips and then allowed them to feed for 30 min. The thrips probably fed for most of the experiment and may have ingested faster as a result of prior starvation. Ingestion rates were calculated from the level of radioactivity in the thrips at the end of the experiment, but no allowance was made for any sucrose egested in excretory droplets. Wiesenborn and Morse (1985, 1986) measured ingestion rates in non-starved *Scirtothrips citri* over 6 to 8 h and allowed for egested material. The long duration of the experiment would have included periods of feeding and non-feeding, which probably explains why the rates are slower than those obtained by most other authors. Ingestion rates increased linearly with temperature for adult females over the range 20 to 36°C, but in first and second instar larvae increased up to about 30°C and then decreased. Second instar larvae ingested faster than adult females, which in turn ingested faster than first instar larvae.

The ingestion rates can be converted into proportions of body weight per hour: 17 per cent for adult female *Thrips tabaci* (Day and Irzykiewicz, 1954); 13 per cent for adult female *Limothrips cerealium* (Chisholm, 1983); 18 per cent, 18 per cent and 6 per cent respectively for larvae I, larvae II and adult females of *Scirtothrips citri* (Wiesenborn and Morse, 1985). Excretory droplets are clearly necessary to remove excess water. Egestion plus respiratory water loss and saliva loss must, together, virtually equal the rate of ingestion.

Rates of attack

Damage to cells can be measured on a gross scale by recording the area of damage. For example, Reyne (1927) estimated the area of damage from *Selenothrips rubrocinctus* at about 30 mm^2 day^{-1} for adults and 5 mm^2 day^{-1} for larvae. At the microscopic scale, it is possible to count the number of feeding holes, either directly or from an impression (Shumway *et al.*, 1991). Sugar-maple leaf discs exposed to *Frankliniella occidentalis* were damaged at a rate equivalent to 15 ± 8 feeding sites per thrips per day (Shumway *et al.*, 1991), but estimates for other thrips have been much higher, of the order of thousands of epidermal cells per day (Reyne, 1927; Klee, 1958).

Thrips have been observed to feed on tens of pollen grains in a few minutes (Grinfel'd, 1959; Syed 1978; Kirk, 1984b). However, it is difficult to extrapolate from this to feeding rates over a day or more because the diel patterns of feeding are not known. Pollen grain volumes vary enormously between plant species, so feeding rates in terms of numbers of pollen grains are likely to vary considerably between plant species. Kirk (1987a) measured the number of grains consumed by thrips confined in small cells for 24 h. A length of stamen filament was provided as an alternative food. Adult female *Thrips imaginis* ate a mean of 843 grains day^{-1} of *Echium plantagineum*, which has a grain volume of 2600 μm^3, but only 29 grains day^{-1} of *Jacaranda acutifolia*, which has a grain volume of 39,600 μm^3. Most of the variation in grain number rate was explained by grain volume. Daily feeding rates were equivalent to 0.2 to 0.7 per cent of the average total pollen production of a flower per thrips per day.

The most detailed information about feeding rates is provided by studies of predatory thrips feeding on mite eggs. Feeding rate increases with temperature. Newly-emerged adult female *Scolothrips sexmaculatus* feed daily on about 20 eggs at 18°C, 60 eggs at 30°C, and 140 eggs at 41°C (Gilstrap and Oatman, 1976). Adult females eat more eggs during their reproductive phase, and their feeding rate declines throughout adult life as their egg-laying rate declines (Gilstrap and Oatman, 1976; Gerlach and Şengonca, 1986), presumably reflecting nutritional needs. Adult females feed, on average, more rapidly than males, with adult males feeding at about the same rate as old females after oviposition has ceased. Females also feed faster than males during the second, but not the first, larval instar (Gilstrap and Oatman, 1976; Sşngonca and Weigand, 1988). Adult females feed faster than second instar larvae, which in turn feed more rapidly than first instar larvae (Gilstrap and Oatman, 1976; Gerlach, 1985; Bailey and Caon, 1986). Facultative predators of mites, such as *Frankliniella schultzei* and *Thrips imaginis*, consume many fewer eggs, but this may still be enough to reduce populations of pest mites (Wilson *et al.*, 1996). Bournier *et al.* (1979) detail the feeding rates of larvae of *Aeolothrips intermedius* for a range of prey, including other thrips larvae (3 to 5 day^{-1}) and adult mites (5 to 10 day^{-1}).

The diel periodicity of feeding in predatory thrips does not appear to have been studied. *Scolothrips longicornis* adult females eat fewer mites in the dark than in the light (Gerlach, 1985), but it is not clear whether this results from a decrease in activity or the difficulty of locating and handling prey in the dark.

Distribution on Plants

Many factors influence the distribution of thrips on a plant (see Chapters 3 and 7), but food availability is of great importance.

Phytophagous thrips can inhabit all the non-woody aerial parts of plants where there is access to food, especially flowers and leaves, though a given species often favours particular organs. Predatory thrips occur on the same parts because that is where they find their phytophagous prey. Most species live and breed in characteristic places and knowledge of these can be used to search for a particular pest. For example, species in the genera *Frankliniella*, *Haplothrips*, *Megalurothrips* and *Odontothrips* are frequently flower dwellers, while species of *Aptinothrips*, *Heliothrips* and *Liothrips* are usually leaf feeders. Some thripids, such as *Scirtothrips* species, feed particularly on young leaves, whereas many panchaetothripines, including *Selenothrips*, favour old leaves. Both leaf and flower thrips can sometimes be pests on fruit.

The feeding sites of thrips are often poorly known, either because thrips have been collected indiscriminately or because collectors have only labelled specimens with the name of the source plant. Many identification keys provide little or no host information; details of the feeding sites of many pest species are provided in Appendix II.

As crops grow and eventually produce flowers and fruits, the best feeding site for a species may change. Adults can locate new food resources rapidly within a plant. For example, *Frankliniella occidentalis* and *Kakothrips pisivorus* adults move from leaves to flowers as soon as they are available (Buhl, 1937; Pickett *et al.*, 1988), presumably to feed on pollen. When a field-bean flower fades, *Kakothrips pisivorus* larvae migrate along a raceme to a fresh flower, but when none is available they move to developing pods, sometimes retiring within the calyx (Buhl, 1937; Kirk, 1985b). When lupin flowers are exhausted of pollen, thrips move from the flowers to leaves, stems and pods (Forbes and Beck, 1954).

The feeding sites of *Selenothrips rubrocinctus* on cashew and cacao can indicate subtle changes in the nutritional status of the leaves (Fennah, 1963, 1965). The thrips feed on leaf tissues that are mature or stressed, but not young or senescent. Leaves are colonized rapidly after stress from fire below or if the branch they occupy is cut from the tree. Characteristic patterns of thrips can be seen on individual leaves according to leaf age and within-leaf variation in the physiological condition of the tissue (Fig. 4.13a–c). Thrips also gather around damaged tissue (Fig. 4.13d). These all appear to be responses to the balance of available nutrients and the level of soluble nitrogen, which increases in mature, stressed or damaged leaves. Thrips distributions may also reflect the distribution of facultative prey. Wilson *et al.* (1996) found that larvae of *Thrips tabaci* showed a preference for cotton leaves infested with mites.

Adults and larvae sometimes inhabit different parts of a plant. This may reflect the lower mobility of larvae, which are unable to locate new resources

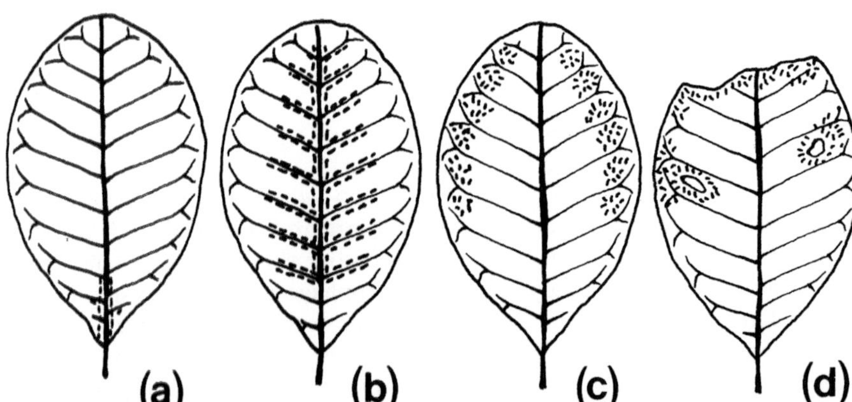

Fig. 4.13. Diagram of four distribution patterns of *Selenothrips rubrocinctus* on cashew leaves. Each individual is represented by a short heavy line. (a) Basal primary subvascular; (b) distal primary subvascular; (c) submarginal intervenal; (d) peritraumatic (after Fennah, 1963).

as quickly as the adults, but may also reflect different nutritional requirements. Adults are often found higher up the plant than larvae, sometimes moving into flowers while the larvae remain on leaves (Bailey, 1933; Kirk, 1985b; Trichilo and Leigh, 1988; Kawai, 1990; Higgins, 1992).

Distributions on plants may change as thrips move between resting sites and feeding or oviposition sites. There are few studies of activity patterns over 24 h, but there are indications that at least some thrips retreat to concealed places by day and then become more active at night when it is more humid (Kirk, 1984a, 1985b; Felland *et al.*, 1994). When this happens, the distribution of thrips on a crop by day may differ from the distribution of damage.

Thrips as Predators

Thrips have long been known as useful predators of mites and Lepidoptera on crops (Table 4.4), but this benefit has not been widely recognized because it has been overshadowed by their reputation as pests. The thrips concerned are nearly always specialist predators, but one notable exception is *Frankliniella occidentalis*, which is best known for its direct feeding damage and virus transmission to many crops, but also acts as a beneficial predator by eating mites on cotton in the USA (Gonzalez *et al.*, 1982; Pickett *et al.*, 1988).

There are several reasons why predatory thrips might be expected to be poor biological-control agents for mites. They often lay few eggs, have a long development time, and are sometimes cannibalistic, so populations cannot increase fast enough to keep up with the mite population (Bailey, 1939; Parrella and Horsburgh, 1983). They may also have low prey requirements, so many thrips would be needed to have much impact. Some thrips can hinder biological control by feeding non-selectively on beneficial predatory mites as well as harmful ones (Rice and Jones, 1972), and susceptibility to pesticides may further limit their potential to be part of an integrated pest management strategy.

However, specialist predatory thrips are well adapted to hunting for tetranychids amongst webbing, and larvae can be very efficient predators at low mite densities (Gilstrap and Oatman, 1976). They can also fly to locate new prey, whereas predatory mites cannot, and can recolonize fields within days of spraying (Bailey and Caon, 1986). Gilstrap (1995) considers that *Scolothrips sexmaculatus* approximates an ideal predator for biological control of mites and details its numerous advantages over other predators. Some thrips would perhaps have more potential if used in conjunction with other predators; when *Leptothrips mali* is combined with the ladybird *Stethorus punctum*, the effect is greater than for each species alone, but there is no benefit from combining it with the bug *Orius insidiosus*, because the bug eats the thrips (Parrella *et al.*, 1980). The many aspects of thrips biology that determine the effectiveness of predatory species as biological-control agents, such as reproductive rates and relative susceptibility to pesticides, also affect the ability of potential invertebrate predators to control pest thrips (see Chapter 8).

The economic benefits of predatory thrips in field crops are hard to

Table 4.4. Some field crops in which thrips are thought to be useful predators of pests.

Crop plant	Pest	Thrips	Region	Reference
Apple	+*Cydia pomonella*	*Haplothrips kurdjumovi*	Canada	MacPhee (1953)
	+*Spilonota ocellana*	*Haplothrips kurdjumovi*	Canada	MacPhee (1953)
	**Calepitrimerus baileyi*	*Leptothrips mali*	USA	Bailey (1940)
	**Panonychus ulmi*	*Haplothrips kurdjumovi*	Canada	MacPhee (1953)
	**Panonychus ulmi*	*Leptothrips mali*	USA	Parrella *et al.* (1982)
Cacao	^*Selenothrips rubrocinctus*	*Franklinothrips tenuicornis*	Trinidad	Callan (1943)
Citrus	**Eotetranychus sexmaculatus*	*Scolothrips sexmaculatus*	USA	Muma (1955)
	‡*Aonidiella aurantii*	*Aleurodothrips fasciapennis*	China	Beattie (1985)
Coconut	‡*Aspidiotus destructor*	*Aleurodothrips fasciapennis*	Java	Taylor (1935)
Cotton	**Tetranychus urticae*	*Frankliniella schultzei*	Australia	Wilson *et al.* (1996)
	**Tetranychus urticae*	*Thrips imaginis*	Australia	Wilson *et al.* (1996)
	**Tetranychus urticae*	*Thrips tabaci*	Australia	Wilson *et al.* (1996)
	**Tetranychus* spp.	*Frankliniella occidentalis*	USA	Gonzalez *et al.* (1982)
	**Tetranychus* spp.	*Neohydatothrips variabilis*	USA	Lincoln *et al.* (1953)
	*Tetranychidae	*Scolothrips acariphagus*	Central Asia	Dyadechko (1977)
	*Tetranychidae	*Scolothrips sexmaculatus*	Ukraine	Dyadechko (1977)
Lucerne	**Tetranychus urticae*	*Haplothrips victoriensis*	Australia	Bailey and Caon (1986)
Peach	**Panonychus ulmi*	*Haplothrips kurdjumovi*	Canada	Putman (1965)
	**Tetranychus pacificus*	*Scolothrips sexmaculatus*	USA	Rice and Jones (1972)
Rhubarb	**Tetranychus urticae*	*Scolothrips sexmaculatus*	USA	Oatman (1970)
Sorghum	**Oligonychus indicus*	*Scolothrips sexmaculatus*	India	Manjunatha and Puttaswamy (1992)
Strawberry	**Tetranychus urticae*	*Scolothrips sexmaculatus*	USA	Oatman and McMurtry (1966)

* denotes eggs or motile stages of mites; +, eggs of Lepidoptera; ‡, scale insects; and ^, thrips. Chazeau (1985) provides a similar list, but includes all records of thrips as predators of tetranychid mites on crops, whether or not the thrips are thought to have any noticeable effect.

quantify, but a knowledge of their role may sometimes allow a management strategy with reduced spraying. Bailey and Caon (1986) suggest that spraying to control mites on lucerne in South Australia is unnecessary if the ratio of *Haplothrips victoriensis* to *Tetranychus urticae* is at least 1:10 when mite numbers begin to increase.

Thrips do not appear, as yet, to have been used commercially as biological-control agents in glasshouses. However, the potential of *Scolothrips longicornis* for control of tetranychid mites in glasshouses or in the field has been much studied (Gerlach, 1985; Gerlach and Şengonca, 1985, 1986; Şengonca and Gerlach, 1984; Selhorst *et al.*, 1991). Experimental data on the lifetime rates of survival and fecundity for *Scolothrips longicornis* and *Tetranychus cinnabarinus* have been fitted to Weibull distributions and incorporated into Leslie matrices to obtain a model of the population dynamics of the predator–prey system. In simulations, the thrips were sometimes able to hold the mite population below the damage threshold, depending on the timing of introduction.

Beattie (1985) studied the predators and parasites of citrus red scale (*Aonidiella aurantii*) on citrus in southern China and found that *Aleurodothrips fasciapennis* (Phlaeothripidae) was the most effective predator and probably the most important natural enemy. He recommended investigating the possibility of introducing it into Australia to improve biological control. This thrips was brought from Java to Fiji for control of coconut scale (*Aspidiotus destructor*) on coconut, but then not released because a similar species was found to be present already (Taylor, 1935).

Thrips as Pollinators

Although flower thrips are often considered to be ineffective pollinators, several aspects of their biology contradict this view. They have probably been systematically overlooked because they do not fit the preconception of an effective pollinator. Unfortunately, there are few rigorous experimental studies with adequate controls that quantify the direct effect of an insect species as a pollinator of a particular plant (Spears, 1983; Corbet *et al.*, 1991), and none appears to have been published for thrips.

Plants that have adapted to pollination by the same class of pollinator tend to share similar floral characteristics as a result of convergent evolution. The floral characteristics typically associated with particular classes of pollinator are called pollination syndromes (Faegri and van der Pijl, 1979; Wyatt, 1983; Endress, 1994), and they are frequently used as a shortcut method of identifying the main pollinators of a plant species. When a plant's main pollinators are not known, they can often be indicated by finding the pollination syndrome that most closely matches the plant's floral characteristics. The flowers can also be observed to check that the indicated pollinator is a frequent visitor. The approach works well for certain classes of pollinator, but it is unlikely to indicate thrips, because they do not have a distinctive syndrome and they are so small that frequent flower visits go unnoticed.

A thrips pollination syndrome (thripophily) has been proposed by Kirk

(1988b) and Endress (1994), and added to by Williams and Adam (1994). Thrips flowers are likely to be medium sized, white to yellow, sweetly scented, with or without nectar, with compact floral structures or globose or urceolate blossoms providing shelter, and with small- to medium-sized pollen grains, possibly with nocturnal pollen presentation. This syndrome is not distinctive and it overlaps several others, including the bee pollination syndrome (Wyatt, 1983). Endress (1994) considers that the thrips pollination syndrome is closest to that of beetles. Since bees and beetles are well-known pollinators, a flower that fits the thrips pollination syndrome is probably more likely to be assigned to the bee or beetle pollination syndromes. Some tropical trees and shrubs fit the thrips pollination syndrome well (Kirk, 1988b; Williams and Adam, 1994), but few crops. The use of syndromes can give the misleading impression that every species of plant is pollinated by only one class of pollinator. However, many plants are pollinated by two or more classes (Grant and Grant, 1965), not just the one indicated by a syndrome, so thrips can still be important supplementary pollinators of plants that do not fit the thrips syndrome.

Flower thrips have been suggested as likely pollinators of a wide range of crops (Table 4.5). In most of these, thrips are thought to be minor pollinators giving a small amount of self-pollination that is useful when major pollinators are absent. In a few crops, the thrips, which are mostly in the genera *Thrips* and *Frankliniella*, are thought to be major pollinators. Several are species that can also be leaf pests, which complicates the assessment of whether they are of net benefit to the crop. Most records are for crops growing in or near the tropics, but this may simply reflect the wider range of entomophilous crops from these areas; plants in cooler regions can also be pollinated by thrips (Hagerup and Hagerup, 1953). Kirk (1988b, 1993) and Endress (1994) review thrips pollination in non-crop plants.

Widespread assumptions about the unsuitability of thrips as pollinators abound. Flowers are often considered as having a 'lock and key' fit to their pollinators (Grant and Grant, 1965), so that thrips, which are usually much smaller than the flowers they occupy, are liable to be dismissed as too small to pollinate (Hagerup, 1950). Thrips are often not recognized as pollen foragers and so appear to be less intimately involved with flowers when compared with other pollinating insects, such as bees or flies (Knox, 1979). Thrips are often described as rarely leaving flowers (McGregor, 1976), so they are likely to be dismissed as possible cross-pollinators. Such inaccurate or incomplete information about the biology of flower thrips has hindered recognition of thrips as potential pollinators.

The importance of pollen in the diet of flower thrips has not been widely recognized so has often not been considered as a factor that makes thrips pollination more likely. For example, even the classic study of pollination of *Calluna vulgaris* by *Ceratothrips ericae* (Hagerup, 1950), which is frequently quoted as the main example of thrips pollination (Faegri and van der Pijl, 1979), only mentions nectar and petals as food. Pollen is now known to be important for egg laying in this species (Kirk, 1985a). Hagerup (1950) particularly attributed the active movement of the females in the flowers to their seeking out males, but the explanation that they were searching for pollen now seems more likely. Flower thrips larvae and adults frequently feed on pollen and thus repeatedly come in contact with it, so they are likely to carry many grains on their bodies. The number and distribution of grains carried by thrips when they are on plants or in flight has often been recorded (Shaw, 1914; Annand, 1926; Veer, 1978; Syed, 1979; Appanah and Chan, 1981; Ananthakrishnan, 1982; Velayudhan *et al.,* 1985; Velayudhan and Annadurai, 1986; Annadurai and Velayudhan, 1986). The average number of grains per thrips is often less than ten, but can reach hundreds. Individuals can be smothered with vast numbers of grains while they feed, although some of these grains are likely to be groomed off at the end of feeding. As thrips move up and down stamens and pistils searching for pollen, they are likely to cause self-pollination. This activity may be obvious by day, but can sometimes be restricted to the night (Kirk, 1984a) and be easily overlooked. Even a few larvae and adults could cause much self-pollination.

Cross-pollination requires movement between plants. Little is known about the frequency of interplant flights in flower thrips. Close observation has sometimes revealed thrips frequently coming and going (Henry and Tu, 1928; Appanah, 1979; Syed 1979). The number of thrips in flowers provides a minimum number of arrivals since the flower opened, but this may seriously underestimate the arrival rate if there is a high turnover. Traps can be used to measure arrival rates. Water traps with a floral colour and scent and a diameter of 165 mm, at least 50 m from any flowers, caught over 100 *Thrips imaginis* trap^{-1} h^{-1} in Australia (Kirk, 1987b). Gopinathan *et al.* (1981) found that the more thrips there were in a flower, the shorter the time each stayed. The rate of interfloral movement is likely to be higher the denser the population. Thrips usually fly by day, but adult thrips arrive at flowers throughout the night in Malaysia (Appanah, 1979) and in the evening in India (Raizada and Nangia, 1989). *Thrips australis* has been caught in suction traps at all hours of the night in warm weather in Australia (Laughlin, 1977). Nocturnal flight may be restricted to areas with hot, humid nights. Arrival and departure at flowers are easily overlooked, but may well be frequent.

Experiments with traps have shown clearly that flower thrips can orient in flight to floral colours and scents (see Chapters 3 and 5), so a thrips leaving one plant is able to find flowers of another plant of the same species. Cameron and Treherne (1918) observed *Taeniothrips inconsequens* flying actively for 23 m or more and Kutter and Winterhalter (1933) observed direct flights of 4 to 6 m by *Kakothrips pisivorus.* Rust (1980) recorded *Heterothrips arisaemae* flying up to 1 m between flowers and Jones and Horton (1911) saw *Scirtothrips citri* flying 6 m from one orange tree to another. In crop monocultures, flights between flowers on different plants up to a few metres apart seem likely. Thrips are often abundant in flowers and frequent flights by many individuals may compensate for the small numbers of pollen grains each carries.

The characteristic grooming behaviour of thrips can contribute to self- and cross-pollination. Thripids arrange the fringe hairs on their wings before and after flight (see Chapter 5), which tends to shed pollen grains from the body. In many flowers, the stigma is prominent and is frequently used for landing and take-off by thrips, so the grooming behaviour before and after flight sheds pollen

Table 4.5. Some crops in which self- or cross-pollination by thrips occurs or is considered likely.

Crop plant	Thrips	Region	Reference
Aubergine (egg plant)	*Frankliniella schultzei*	India	*Velayudhan and Annadurai (1986)
	Taeniothrips major	India	*Velayudhan and Annadurai (1986)
Beet	*Thrips tabaci*	USA	*Shaw (1914)
Bird chilli	Not specified	USA	Odland and Porter (1941)
	Frankliniella schultzei	India	*Velayudhan and Annadurai (1986)
	Taeniothrips major	India	*Velayudhan and Annadurai (1986)
Black gram	*Megalurothrips distalis* etc.	India	*Annadurai and Velayudhan (1986)
Cacao	*Frankliniella parvula*	Trinidad	+Billes (1941)
	Not specified	S. America	Soria Vasco (1970)
Common bean	*Frankliniella occidentalis*	USA	+Mackie and Smith (1935)
	Megalurothrips distalis etc.	India	*Annadurai and Velayudhan (1986)
	Not specified	Not specified	Bliss (1980)
Cowpea	*Megalurothrips distalis* etc.	India	*Annadurai and Velayudhan (1986)
Flax	Not specified	USA	+Henry and Tu (1928)
Hyacinth bean	*Megalurothrips distalis*	India	Velayudhan *et al.* (1985)
	Frankliniella schultzei	India	Velayudhan *et al.* (1985)
Lentil	Not specified	Not specified	Muehlbauer *et al.* (1980)
Lima bean	*Frankliniella* spp.	USA	+Allard (1954)
Lucerne (alfalfa)	Not specified	Not specified	Faegri and van der Pijl (1979)
Mahogany	*Frankliniella* spp.	USA	+Howard *et al.* (1995)
Mango	Thripidae	USA	Wolfenbarger (1977)
Oil palm	*Thrips hawaiiensis*	Malaysia	+Syed (1978, 1979)
Oilseed rape	*Thrips flavus*	India	Veer (1978)
Onion	*Frankliniella occidentalis*	USA	*Carlson (1964)
Papaya (papaw)	Not specified	S. Africa	Hofmeyr (1938)

Peanut	*Frankliniella fusca*	USA	Hammons and Leuck (1966)
	Not specified	Congo	*Pelerents (1957)
	Megalurothrips distalis etc.	India	*Annadurai and Velayudhan (1986)
Pigeon pea	*Megalurothrips distalis*	India	+Sen and Sur (1964)
	Frankliniella schulzei	India	+*Yadav *et al.* (1974)
	Taeniothrips nigricornis	India	+*Yadav *et al.* (1974)
	Not specified	India	Williams (1977)
	Megalurothrips distalis etc.	India	*Annadurai and Velayudhan (1986)
Pyrethrum	*Microcephalothrips* spp.	Kenya	Bullock (1962)
	Taeniothrips spp.	Kenya	Bullock (1962)
	Thrips tabaci	Kenya	Bullock (1962)
Radish	*Thrips hawaiiensis*	India	*Raizada and Saxena (1989)
Rubber	*Craspedothrips minor*	Malaysia	Rao (1961)
	Thrips hawaiiensis	Malaysia	Rao (1961)
	Thrips florum	Malaysia	Rao (1961)
Yam	*Larothrips dentipes*	Nigeria	Pitkin (1973)
	Larothrips dentipes	Nigeria	Segnou *et al.* (1992)
	Larothrips dentipes	Ivory Coast	Bournier (1994)

+ indicates studies in which thrips were considered likely to be major pollinators; * indicates references that include some form of quantitative data for the effects of thrips pollination on yield.

grains directly on to it (Hagerup, 1950; Appanah and Chan, 1981). Thrips can transfer pollen precisely from anther to stigma, without conforming to the 'lock and key' model of pollination. However, in flowers of West Indies mahogany (*Swietenia mahagoni*), thrips enter and leave the staminal tube through a marginal cleft that forces them into contact with an anther and the edge of the stigma, so a close morphological adaptation of the flower to thrips may be present (Howard *et al.*, 1995).

The best evidence for thrips pollination comes from direct observation. Charles Darwin (1876, p. 417) wrote: 'I have seen more than once a minute Thrips; with pollen adhering to its body, fly from one flower to another of the same kind; and one was observed by me crawling about within a convolvulus with four grains of pollen adhering to its head, which were deposited on the stigma'. Hagerup (1950) observed *Ceratothrips ericae* pollinating *Calluna vulgaris*. The abundance of pollen-covered thrips in flowers and the scarcity of other possible pollinators has also been taken as evidence for thrips pollination (Syed, 1979; Rust, 1980), but there are few quantitative measurements. Cruden *et al.* (1990) compared open-pollinated plots, bagged inflorescences with thrips, and caged plots sprayed with malathion insecticide. More seed was set in the bagged inflorescences with thrips than in the caged plots without thrips, indicating self-pollination by thrips. The possible effect of the difference between bagging and caging or the effect of malathion on the seed set of the plant was not investigated.

Thrips frequently come to the notice of pollination biologists because of their ability to get inside cages and net bags that are being used to exclude pollinators. Darwin (1877) frequently mentioned the problems caused to his pollination experiments by thrips. The difficulty of excluding them under field conditions, even with cages, bags and repeated insecticide sprays, is a major problem for quantitative studies.

Although thrips can pollinate in the absence of other pollinators, their importance in open-pollinated crops depends on whether other insects normally pollinate first. Thrips have an advantage in being able to enter unopened buds and so pollinate before bees visit (Mackie and Smith, 1935), but the peak numbers of thrips may occur after the peak in bee visits (Kirk, 1984a). A further complication is that thrips, like bees, feed on pollen as well as pollinating. Heavily infested flowers can be depleted of pollen (Forbes and Beck, 1954; Appanah and Chan, 1981). Little is known about either the damage from pollen grain consumption or the benefit from pollination so it is difficult to predict whether net effects will be beneficial, neutral or harmful. It would be useful to know how many of the grains carried by thrips and other pollinators have been emptied by feeding thrips and are therefore inviable.

References

Achor, D.S. and Childers, C.C. (1995) Fixation techniques for observing thrips morphology and injury with electron microscopy. In: Parker, B.L., Skinner, M. and Lewis, T. (eds) *Thrips Biology and Management*. Plenum Press, New York, pp. 595–600.

Allard, R.W. (1954) Natural hybridization in Lima beans in California. *Proceedings of the American Society for Horticultural Science* 64, 410–416.

Allen, W.R. and Broadbent, A.B. (1986) Transmission of tomato spotted wilt virus in Ontario greenhouses by *Frankliniella occidentalis*. *Canadian Journal of Plant Pathology* 8, 33–38.

Ananthakrishnan, T.N. (1982) Thrips and pollination biology. *Current Science* 51, 168–172.

Ananthakrishnan, T.N. (1984) *Bioecology of Thrips*. Indira Publishing House, Oak Park, Michigan.

Ananthakrishnan, T.N. and James, P.W. (1983) Feeding preferences in some sporophagous Idolothripinae (Tubulifera: Thysanoptera). *Proceedings of the Indian National Science Academy* B49, 86–92.

Andrewartha, H.G. (1935) Thrips investigation 7. On the effect of temperature and food upon egg production and the length of adult life of *Thrips imaginis* Bagnall. *Journal of the Council for Scientific and Industrial Research, Australia* 8, 281–288.

Andrewartha, H.G. and Kilpatrick, D.T. (1951) The apple thrips. *Journal of Agriculture, South Australia* 54, 586–592.

Andrewartha, H.V. (1936) Thrips investigation 8. The influence of temperature on the rate of development of the immature stages of *Thrips imaginis* Bagnall and *Haplothrips victoriensis* Bagnall. *Journal of the Council for Scientific and Industrial Research, Australia* 9, 57–64.

Annadurai, R.S. and Morrison, M.N. (1987) Impact of pollen food on the fecundity and feeding time in two species of thrips (Insecta: Thysanoptera) infesting flowers of Solanaceae. *Current Science* 56, 1032–1033.

Annadurai, R.S. and Velayudhan, R. (1986) Pollination potential of thrips (Insecta, Thysanoptera) in some fabaceous plants. *Proceedings of the Indian Academy of Sciences (Animal Sciences)* 95, 745–750.

Annand, P.N. (1926) Thysanoptera and the pollination of flowers. *American Naturalist* 60, 177–182.

Appanah, S. (1979) The ecology of insect pollination of some tropical rain forest trees. PhD thesis, University of Malaya, Kuala Lumpur.

Appanah, S. and Chan, H.T. (1981) Thrips: pollinators of some dipterocarps. *Malaysian Forester* 44, 234–252.

Bailey, P. and Caon, G. (1986) Predation of two-spotted mite, *Tetranychus urticae* Koch (Acarina: Tetranychidae) by *Haplothrips victoriensis* Bagnall (Thysanoptera: Phlaeothripidae) and *Stethorus nigripes* Kapur (Coleoptera: Coccinellidae) on seed lucerne crops in South Australia. *Australian Journal of Zoology* 34, 515–525.

Bailey, S.F. (1933) The biology of the bean thrips. *Hilgardia* 7, 467–522.

Bailey, S.F. (1935) Thrips as vectors of plant diseases. *Journal of Economic Entomology* 28, 856–863.

Bailey, S.F. (1939) The six-spotted thrips, *Scolothrips sexmaculatus* Perg. *Journal of Economic Entomology* 32, 43–47.

Bailey, S.F. (1940) The black hunter, *Leptothrips mali* (Fitch.). *Journal of Economic Entomology* 33, 539–544.

Beattie, G.A.C. (1985) Ecology of citrus red scale, *Aonidiella aurantii* (Maskell); Report on a study tour to China, 6th June, 1983 to 22nd May, 1984. Unpublished Report for the New South Wales Department of Agriculture, Rydalmere, New South Wales, Australia.

Bernays, E.A. and Chapman, R.F. (1994) *Host-Plant Selection by Phytophagous Insects*. Chapman & Hall, New York.

Billes, D.J. (1941) Pollination of *Theobroma cacao* L. in Trinidad, B.W.I. *Tropical Agriculture (Trinidad)* 18, 151–156.

Bliss, F.A. (1980) Common bean. In: Fehr, W.R. and Hadley, H.H. (eds) *Hybridization of Crop Plants*. American Society of Agronomy, Madison, pp. 273–284.

Bonnemaison, L. and Bournier, A. (1964) Les thrips du lin: *Thrips angusticeps* Uzel et *Thrips linarius* Uzel (Thysanoptères). *Annales des Épiphyties* 15, 97–169.

Borden, A.D. (1915) The mouthparts of the Thysanoptera and the relation of thrips to the non-setting of certain fruits and seeds. *Journal of Economic Entomology* 8, 354–360.

Bournier, A. (1961) Sur l'existence et l'evolution d'un mycetome au cours de l'embryogenèse de *Caudothrips buffai* Karny. *XI. Internationaler Kongress für Entomologie, Wien 1960* 1, 352.

Bournier, A. (1983) *Les Thrips, Biologie, Importance Agronomique.* Institut National de la Recherche Agronomique, Paris.

Bournier, A. and Louis, C. (1971) Confirmation ultrastructurale de la présence d'un mycétome dans les cellules germinales femelles de *Caudothrips buffai* (Thys. Tubulifera). *Annales de la Société Entomologique de France (N.S.)* 7, 721–727.

Bournier, A., Lacasa, A. and Pivot, Y. (1979) Régime alimentaire d'un thrips prédateur *Aeolothrips intermedius* (Thysanoptera, Aeolothripidae). *Entomophaga* 24, 353–361.

Bournier, J. (1994) *Larothrips dentipes* Pitkin (Thysanoptera, Thripidae): description du mâle et de son polymorphisme. *Journal of African Zoology* 108, 493–501.

Boyce, A.M. and Persing, C.O. (1939) Tartar emetic in the control of citrus thrips on lemons. *Journal of Economic Entomology* 32, 153.

Bryan, D.E. and Smith, R.F. (1956) The *Frankliniella occidentalis* (Pergande) complex in California (Thysanoptera: Thripidae). *University of California Publications in Entomology* 10, 359–410.

Buhl, C. (1937) Beiträge zur Kenntnis der Biologie, wirtschaftlichen Bedeutung und Bekämpfung von *Kakothrips robustus* Uz. *Zeitschrift für angewandte Entomologie* 23, 65–113.

Bullock, J.A. (1962) Fertilisation of the pyrethrum flower. *Pyrethrum Post* 6, 39.

Callan, E.McC. (1943) Natural enemies of the cacao thrips. *Bulletin of Entomological Research* 34, 313–321.

Cameron, A.E. and Treherne, R.C. (1918) The pear thrips (*Taeniothrips inconsequens* Uzel) and its control in British Columbia. *Bulletin of the Department of Agriculture Canada, Entomology Branch* 15.

Carey, E., Van Driesche, R.G., Elkinton, J.S., Bellows, T.S. and Burnham, C. (1992) Influence of sugar maple health on fecundity of pear thrips in Massachusetts. *Canadian Entomologist* 124, 1129–1138.

Carlson, E.C. (1964) Effect of flower thrips on onion seed plants and a study of their control. *Journal of Economic Entomology* 57, 735–741.

Chapman, R.F. (1982) Chemoreception: the significance of receptor numbers. *Advances in Insect Physiology* 16, 247–356.

Chazeau, J. (1985) Predaceous insects. In: Helle, W. and Sabelis, M.W. (eds) *Spider Mites. Their Biology, Natural Enemies and Control*, Vol. 1B. Elsevier, Amsterdam, pp. 211–246.

Childers, C.C. and Achor, D.S. (1991a) Structure of the mouthparts of *Frankliniella bispinosa* (Morgan) (Thysanoptera: Thripidae). In: Parker, B.L., Skinner, M. and Lewis, T. (eds) *Towards Understanding Thysanoptera. Proceedings of the International Conference on Thrips, Burlington, Vermont USA, 1989*. USDA, Radnor PA, pp. 71–94.

Childers, C.C. and Achor, D.S. (1991b) Feeding and oviposition injury to 'Navel' orange flowers and developing buds by *Frankliniella bispinosa* (Thysanoptera: Thripidae) in Florida. *Annals of the Entomological Society of America* 84, 272–282.

Chisholm, I.F. (1983) Feeding mechanism, behaviour and effects of *Limothrips* sp. on cereals. PhD thesis, University of London.

Chisholm, I.F. and Doncaster, C.C. (1982) Studying and recording the feeding behaviour of thrips. *Entomologia Experimentalis et Applicata* 31, 324–327.

Chisholm, I.F. and Lewis, T. (1984) A new look at thrips (Thysanoptera) mouthparts, their action and effects of feeding on plant tissue. *Bulletin of Entomological Research* 74, 663–675.

Corbet, S.A., Williams, I.H. and Osborne, J.L. (1991) Bees and the pollination of crops and wild flowers in the European Community. *Bee World* 72, 47–59.

Cruden, R.W., Baker, K.K., Cullinan, T.E., Disbrow, K.A., Douglas, K.L., Erb, J.D., Kirsten, K.J., Malik, M.L., Turner, E.A., Weier, J.A. and Wilmot, S.R. (1990) The mating systems and pollination biology of three species of *Verbena* (Verbenaceae). *Journal of the Iowa Academy of Science* 97, 178–183.

Czencz, K. and Gerber, K. (1986) Die Wirkung der Mineraldüngung auf die Abundanz von Gramineenthripsen. *Acta Phytopathologica et Entomologica Hungarica* 21, 341–346.

Darwin, C. (1876) *The Effects of Cross and Self Fertilization in the Vegetable Kingdom*. Murray, London.

Darwin, C. (1877) *The Different Forms of Flower on Plants of the Same Species*. Murray, London.

Day, M.F. and Irzykiewicz, H. (1954) Physiological studies on thrips in relation to transmission of tomato spotted wilt virus. *Australian Journal of Biological Sciences* 7, 274–281.

De Jager, C.M., Butôt, R.P.T., Klinkhamer, P.G.L. and van der Meijden, E. (1995) Chemical characteristics of chrysanthemum cause resistance to *Frankliniella occidentalis* (Thysanoptera: Thripidae). *Journal of Economic Entomology* 88, 1746–1753.

Doull, K.M. (1956) Thrips infesting cocksfoot in New Zealand. II. The biology and economic importance of the cocksfoot thrips *Chirothrips manicatus* Haliday. *New Zealand Journal of Science and Technology* 38, 56–65.

Downey, J.C. (1965) Thrips utilize exudations of Lycaenidae. *Entomological News* 76, 25–27.

Dyadechko, N.P. (1977) *Thrips or Fringe–Winged Insects (Thysanoptera) of the European Part of the USSR*. (Translated from the Russian by R.S. Chakravarthy.) Amerind Publishing Company, New Delhi.

Endress, P.K. (1994) *Diversity and Evolutionary Biology of Tropical Flowers*. Cambridge University Press, Cambridge.

Faegri, K. and van der Pijl, L. (1979) *The Principles of Pollination Ecology*, 3rd edn. Pergamon Press, Oxford.

Felland, C.M., Goulart, B. and Hellerick, B. (1994) Strawberry thrips plague of 1994. *Proceedings of the 70th Cumberland-Shenandoah Fruit Workers Conference, Winchester, Virginia, USA, 1994*. Rutgers University, Cream Ridge, New Jersey.

Fennah, R.G. (1963) Nutritional factors associated with seasonal population increase of cacao thrips, *Selenothrips rubrocinctus* (Giard) (Thysanoptera), on cashew, *Anacardium occidentale*. *Bulletin of Entomological Research* 53, 681–713.

Fennah, R.G. (1965) The influence of environmental stress on the cacao tree in predetermining the feeding sites of cacao thrips, *Selenothrips rubrocinctus* (Giard) on leaves and pods. *Bulletin of Entomological Research* 56, 333–349.

Forbes, I. and Beck, E.W. (1954) A rapid biological technique for screening blue lupine populations for low-alkaloid plants. *Agronomy Journal* 46, 528–529.

Foster, S.W. and Jones, P.R. (1915) The life history and habits of the pear thrips in California. *United States Department of Agriculture Bureau of Entomology Bulletin* 173.

Gawaad, A.A.A. and Shazli, A.Y. (1970) Effect of food on the life cycle of *Thrips tabaci* (Thysanoptera). *Bulletin de la Société Entomologique d'Égypte* 53, 449–452.

Gerlach, S. (1985) Biologie, Ökologie und Wirksamkeit der räuberischen Thrips, *Scolothrips longicornis* Priesner (Thysanoptera: Thripidae). Dr. agr. Dissertation, Rheinische Friedrich-Wilhelms-Universität, Bonn.

Gerlach, S. and Şengonca, Ç. (1985) Comparative studies on the effectiveness of the predatory mite, *Phytoseiulus persimilis* Athias-Henriot, and the predatory thrips, *Scolothrips longicornis* Priesner. *Zeitschrift für Pflanzenkrankheiten und Pflanzenschutz* 92, 138–146.

Gerlach, S. and Şengonca, Ç. (1986) Fraßaktivität und Wirksamkeit des räuberischen Thrips, *Scolothrips longicornis* Priesner (Thysanoptera, Thripidae). *Zeitschrift für angewandte Entomologie* 101, 444–452.

Gilstrap, F.E. (1995) Six-spotted thrips: a gift from nature that controls spider mites. In: Parker, B.L., Skinner, M. and Lewis, T. (eds) *Thrips Biology and Management.* Plenum Press, New York, pp. 305–316.

Gilstrap, F.E. and Oatman, E.R. (1976) The bionomics of *Scolothrips sexmaculatus* (Pergande) (Thysanoptera: Thripidae), an insect predator of spider mites. *Hilgardia* 44, 27–59.

Gonzalez, D., Patterson, B.R., Leigh, T.F. and Wilson, L.T. (1982) Mites: a primary food source for two predators in San Joaquin Valley cotton. *California Agriculture* 36 (2–3), 8–20.

Gopinathan, K., Varatharajan, R. and Ananthakrishnan, T.N. (1981) Incidence of *Microcephalothrips abdominalis* (Crawford) (Thysanoptera: Insecta) in relation to the pollination biology of the weed *Ageratum conyzoides* L. (Compositae). *Proceedings of the Indian National Science Academy* B47, 505–509.

Grant, V. and Grant, K.A. (1965) *Flower Pollination in the Phlox Family.* Columbia University Press, New York.

Grinfel'd, E.K. (1959) The feeding of thrips (Thysanoptera) on pollen of flowers and the origin of asymmetry in their mouthparts. (In Russian.) *Entomologicheskoe Obozrenie* 38, 798–804.

Gullan, P.J. and Cranston, P.S. (1994) *The Insects: an Outline of Entomology.* Chapman & Hall, London.

Hagerup, E. and Hagerup O. (1953) Thrips pollination of *Erica tetralix. New Phytologist* 52, 1–7.

Hagerup, O. (1950) Thrips pollination in *Calluna. Biologiske Meddelelser, Kongelige Danske Videnskabernes Selskab* 18 (4), 1–16.

Hammons, R.O. and Leuck, D.B. (1966) Natural cross pollination of the peanut, *Arachis hypogaea* L., in the presence of bees and thrips. *Agronomy Journal* 58, 396.

Harrewijn, P., Tjallangi, W.F. and Mollema, C. (1996a) Electrical recording of plant penetration by western flower thrips. *Entomologia Experimentalis et Applicata* 79, 345–353.

Harrewijn, P., Piron, P.G.M. and Mollema, C. (1996b) Electrically recorded probing behaviour of thrips species on optimal and suboptimal hosts. *Entomologia Experimentalis et Applicata* 80, 43–45.

Heming, B.S. (1978) Structure and function of the mouthparts in larvae of *Haplothrips verbasci* (Osborn) (Thysanoptera, Tubulifera, Phlaeothripidae). *Journal of Morphology* 156(1), 1–37.

Heming, B.S. (1993) Structure, function, ontogeny and evolution of feeding in thrips (Thysanoptera). In: Schaefer, C.W. and Leschen, R.A.B. (eds) *Functional Morphology of Insect Feeding*. Thomas Say Publications in Entomology: Proceedings. Entomological Society of America, Maryland, pp. 3–41.

Henry, A.W. and Tu, C. (1928) Natural crossing in Flax. *Journal of the American Society of Agronomy* 20, 1183–1192.

Higgins, C.J. (1992) Western flower thrips (Thysanoptera: Thripidae) in greenhouses: population dynamics, distribution on plants, and associations with predators. *Journal of Economic Entomology* 85, 1891–1903.

Hirano, C., Yasumi, K., Itoh, E., Kim, C.S. and Horiike, M. (1994) A feeding deterrent for *Thrips palmi* Karny (Thysanoptera, Thripidae) found in tomato leaves – isolation and identification. (In Japanese.) *Japanese Journal of Applied Entomology and Zoology* 38, 109–120.

Howard, F.W., Nakahara, S. and Williams, D.S. (1995) Thysanoptera as apparent pollinators of West Indies mahogany, *Swietenia mahagoni* (Meliaceae). *Annales des Sciences Forestières* 52, 283–286.

Hofmeyr, J.D.J. (1938) Genetical studies of *Carica papaya*. *Union of South Africa Department of Agricultural Science Bulletin* 187.

Hunter, W.B. and Ullman, D.E. (1989) Analysis of mouthpart movements during feeding of *Frankliniella occidentalis* (Pergande) and *F. schultzei* Trybom (Thysanoptera: Thripidae). *International Journal of Morphology and Embryology* 18, 161–171.

Hunter, W.B. and Ullman, D.E. (1992) Anatomy and ultrastructure of the piercing-sucking mouthparts and paraglossal sensilla of *Frankliniella occidentalis* (Pergande) (Thysanoptera: Thripidae). *International Journal of Morphology and Embryology* 21, 17–35.

Hunter, W.B. and Ullman, D.E. (1994) Precibarial and cibarial chemosensilla in the western flower thrips, *Frankliniella occidentalis* (Pergande) (Thysanoptera: Thripidae). *International Journal of Morphology and Embryology* 23, 69–83.

Hunter, W.B., Ullman, D.E. and Moore, A. (1993) Electronic monitoring: characterizing the feeding behavior of the western flower thrips (Thysanoptera: Thripidae). In: Ellsbury, M.M., Backus, E.A. and Ullman, D.E. (eds) *History, Development, and Application of AC Electronic Insect Feeding Monitors*. Thomas Say Publications in Entomology: Proceedings. Entomological Society of America, Maryland, pp. 73–85.

Jones, P.R. and Horton, J.R. (1911) The orange thrips. *United States Department of Agriculture Bureau of Entomology Bulletin* 99 (1).

Jones, T. (1954) The external morphology of *Chirothrips hamatus* (Trybom) (Thysanoptera). *Transactions of the Royal Entomological Society of London* 105, 163–187.

Kawai, A. (1990) Life cycle and population dynamics of *Thrips palmi* Karny. *Japanese Agricultural Research Quarterly* 23, 282–288.

Kendall, D.M. and Bjostad, L.B. (1990) Phytohormone ecology. Herbivory by *Thrips tabaci* induces greater ethylene production in intact onions than mechanical damage alone. *Journal of Chemical Ecology* 16, 981–991.

Kirk, W.D.J. (1984a) Ecological studies on *Thrips imaginis* Bagnall (Thysanoptera) in flowers of *Echium plantagineum* L. in Australia. *Australian Journal of Ecology* 9, 9–18.

Kirk, W.D.J. (1984b) Pollen-feeding in thrips (Insecta: Thysanoptera). *Journal of Zoology* 204, 107–117.

Kirk, W.D.J. (1984c) The ecology of thrips in flowers. PhD thesis, University of Cambridge.
Kirk, W.D.J. (1985a) Pollen-feeding and the host specificity and fecundity of flower thrips (Thysanoptera). *Ecological Entomology* 10, 281–289.
Kirk, W.D.J. (1985b) Floral display in *Vicia faba*, and the distribution of a flower thrips, *Kakothrips pisivorus*. *Entomologia Experimentalis et Applicata* 38, 233–238.
Kirk, W.D.J. (1987a) How much pollen can thrips destroy? *Ecological Entomology* 12, 31–40.
Kirk, W.D.J. (1987b) Effects of trap size and scent on catches of *Thrips imaginis* Bagnall (Thysanoptera: Thripidae). *Journal of the Australian Entomological Society* 26, 299–302.
Kirk. W.D.J. (1988a) Assessing the effects of flower thrips. *Acta Phytopathologica et Entomologica Hungarica* 23, 295–300.
Kirk, W.D.J. (1988b) Thrips and pollination biology. In: Ananthakrishnan, T.N. and Raman, A. (eds) *Dynamics of Insect–Plant Interaction. Recent Advances and Future Trends*. Oxford and IBH, New Delhi, pp. 129–135.
Kirk, W.D.J. (1991) The size relationship between insects and their hosts. *Ecological Entomology* 16, 351–359.
Kirk, W.D.J. (1993) Thrips and pollen. *Zoology (Journal of Pure and Applied Zoology)* 4, 251–258.
Kirk, W.D.J. (1994) The effects of density on the oviposition rate of flower thrips. *Courier Forschungsinstitut Senckenberg* 178, 69–73.
Kirk, W.D.J. (1995) Feeding behavior and nutritional requirements. In: Parker, B.L., Skinner, M. and Lewis, T. (eds) *Thrips Biology and Management*. Plenum Press, New York, pp. 21–29.
Kirk, W.D.J. (1996) *Thrips*. Naturalists' Handbook 25. Richmond Publishing Co., Slough.
Klee, O. (1958) Über die Biologie und Saugtätigkeit des Thysanopteren *Taeniothrips laricivorus* Krat. und das Lärchenwipfelsterben. *Waldhygiene* 2, 166–181.
Kloft, W. and Ehrhardt, P. (1959) Zur Frage der Speichelinjektion beim Saugakt von *Thrips tabaci* Lind. (Thysanoptera, Terebrantia). *Naturwissenschaften* 46, 586–587.
Knox, R.B. (1979) *Pollen and Allergy*. Institute of Biology's Studies in Biology No. 107. Arnold, London.
Kumar, N.K.K., Ullman, D.E. and Cho, J.J. (1995) Resistance among *Lycopersicon* species to *Frankliniella occidentalis* (Thysanoptera: Thripidae). *Journal of Economic Entomology* 88, 1057–1065.
Kutter, H. and Winterhalter, W. (1933) Untersuchungen über die Erbsenschädlinge im St. Gallischen Rheintale. *Landwirtschaftliches Jahrbuch der Schweiz* 47, 273–338.
Laughlin, R. (1971) A culture method for *Hercinothrips femoralis* (Reuter) (Thysanoptera). *Journal of the Australian Entomological Society* 10, 301–303.
Laughlin, R. (1977) The gum tree thrips *Isoneurothrips australis* Bagnall: survival at different temperatures and humidities and its relation to capacity for dispersal. *Australian Journal of Ecology* 2, 391–398.
Lewis, T. (1973) *Thrips. Their Biology, Ecology and Economic Importance*. Academic Press, London.
Lincoln, C., Williams, F.J. and Barnes, G. (1953) Importance of a thrips in red spider control. *Journal of Economic Entomology* 46, 899–900.
Loan, C. and Holdaway, F.G. (1955) Biology of the red clover thrips, *Haplothrips niger* (Osborn) (Thysanoptera: Phlaeothripidae). *Canadian Entomologist* 87, 210–219.
Mackie, W.W. and Smith, F.L. (1935) Evidence of field hybridization in beans. *Journal*

of the American Society of Agronomists 27, 903–909.
McGregor, S.E. (1976) *Insect Pollination of Cultivated Crop Plants*. Agricultural Handbook No. 496. United States Department of Agriculture, Washington.
McKenzie, C.L., Cartwright, B., Miller, M.E. and Edelson, J.V. (1993) Injury to onions by *Thrips tabaci* (Thysanoptera: Thripidae) and its role in the development of purple blotch. *Environmental Entomology* 22, 1266–1277.
McNeill, S. and Southwood, T.R.E. (1978) The role of nitrogen in the development of insect/plant relationships. In: Harborne, J.B. (ed.) *Biochemical Aspects of Plant and Animal Coevolution*. Academic Press, London, pp. 77–98.
MacPhee, A.W. (1953) The influence of spray programs on the fauna of apple orchards in Nova Scotia. V. The predacious thrips *Haplothrips faurei* Hood. *Canadian Entomologist* 85, 33–40.
Manjunatha, M. and Puttaswamy (1992) Seasonal activity of *Oligonychus indicus* (Acari: Tetranychidae) and its predators on sorghum. *Journal of Biological Control* 6, 18–21.
Marullo, R. (1995) Possible dissemination of pest fungi by thrips. In: Parker, B.L., Skinner, M. and Lewis, T. (eds) *Thrips Biology and Management*. Plenum Press, New York, pp. 201–202.
Mattson, W.J. (1980) Herbivory in relation to plant nitrogen content. *Annual Review of Ecology and Systematics* 11, 119–161.
Mickoleit, E. (1963) Untersuchungen zur Kopfmorphologie der Thysanoptera. *Zoologisches Jahrbuch der Anatomie* 81, 101–150.
Miles, P.W. (1972) The saliva of Hemiptera. *Advances in Insect Physiology* 9, 183–255.
Miles, P.W. (1987) Feeding process of Aphidoidea in relation to effects on their food plants. In: Minks, A.K. and Harrewijn, P. (eds) *Aphids, their Biology, Natural Enemies and Control*, Vol. A. Elsevier, Amsterdam, pp. 321–339.
Milne, D.L. and Manicom, B.Q. (1978) Feeding apparatus of the South African citrus thrips, *Scirtothrips aurantii* Faure. *Citrus Subtropical Fruit Journal* 535, 6–11.
Mitchell, F.L., Lowry, V.K., Kresta, K.K. and Smith, J.W. (1995) Histological study of tobacco thrips feeding on peanut foliage. In: Parker, B.L., Skinner, M. and Lewis, T. (eds) *Thrips Biology and Management*. Plenum Press, New York, pp. 175–178.
Mollema, C. and Cole, R.A. (1996) Low aromatic amino acid concentrations in leaf proteins determine resistance to *Frankliniella occidentalis* in four vegetable crops. *Entomologia Experimentalis et Applicata* 78, 325–333.
Mollema, C., Steenhuis, G. and Inggamer, H. (1995) Genotypic effects of cucumber responses to infestation by western flower thrips. In: Parker, B.L., Skinner, M. and Lewis, T. (eds) *Thrips Biology and Management*. Plenum Press, New York, pp. 397–401.
Moritz, G. (1982) Zur Morphologie und Anatomie des Fransenflüglers *Aeolothrips intermedius* Bagnall, 1934 (Aeolothripidae, Thysanoptera, Insecta) 1. Mitteilung: Der Kopf. *Zoologisches Jahrbuch der Anatomie* 107, 557–608.
Moritz, G. (1988) Die Ontogenese der Thysanoptera (Insecta) unter besonderer Berücksichtigung des Fransenflüglers *Hercinothrips femoralis* (O.M. Reuter, 1891) (Thysanoptera, Thripidae, Panchaetothripinae) II. Mitteilung: Erst- und Zweitlarve. *Zoologisches Jahrbuch der Anatomie* 117, 299–351.
Moritz, G. (1989) Die Ontogenese der Thysanoptera (Insecta) unter besonderer Berücksichtigung des Fransenflüglers *Hercinothrips femoralis* (O.M. Reuter, 1891) (Thysanoptera, Thripidae, Panchaetothripinae) IV. Mitteilung: Imago – Kopf. *Zoologisches Jahrbuch der Anatomie* 118, 273–307.
Morse, J.G. (1995) Prospects for IPM of citrus thrips in California. In: Parker, B.L., Skinner, M. and Lewis, T. (eds) *Thrips Biology and Management*. Plenum Press, New York, pp. 371–379.

Mound, L.A. (1970) Convoluted maxillary stylets and the systematics of some phlaeothripine Thysanoptera from *Casuarina* trees in Australia. *Australian Journal of Zoology* 18, 439–463.

Mound, L.A. (1971) The feeding apparatus of thrips. *Bulletin of Entomological Research* 60, 547–548.

Mound, L.A. (1994) Thrips and gall induction: a search for patterns. In: Williams, M.A.J. (ed.) *Plant Galls*. Systematics Association Special Volume No. 49. Clarendon Press, Oxford, pp. 131–149.

Mound, L.A. and Palmer, J.M. (1983) The generic and tribal classification of spore-feeding Thysanoptera (Phlaeothripidae: Idolothripinae). *Bulletin of the British Museum (Natural History), Entomology* 46, 1–174.

Mound, L.A. and Walker, A.K. (1986) Tubulifera (Insecta: Thysanoptera). *Fauna of New Zealand* 10.

Mound, L.A., Morison, G.D., Pitkin, B.R. and Palmer, J.M. (1976) *Handbooks for the Identification of British Insects*, Vol. 1, Part 11, Thysanoptera. Royal Entomological Society of London, London.

Mound, L.A., Heming, B.S. and Palmer, J.M. (1980) Phylogenetic relationships between the families of recent Thysanoptera (Insecta). *Zoological Journal of the Linnean Society* 69, 111–142.

Muehlbauer, F.J., Slinkard, A.E. and Wilson, V.E. (1980) Lentil. In: Fehr, W.R. and Hadley, H.H. (eds) *Hybridization of Crop Plants*. American Society of Agronomy, Madison, pp. 417–426.

Muma, M.H. (1955) Factors contributing to the natural control of citrus insects and mites in Florida. *Journal of Economic Entomology* 48, 432–438.

Munger, F. (1942) Reactions of the citrus thrips to sugar in poisoned baits. *Journal of Economic Entomology* 35, 51–53.

Murai, T. and Ishii, T. (1982) Simple rearing method for flower thrips (Thysanoptera; Thripidae) on pollen. (In Japanese.) *Japanese Journal of Applied Entomology and Zoology* 26, 149–154.

Oatman. E.R. (1970) Integration of *Phytoseiulus persimilis* with native predators for control of the two-spotted mite on rhubarb. *Journal of Economic Entomology* 63, 1177–1180.

Oatman, E.R. and McMurtry, J.A. (1966) Biological control of the two-spotted mite on strawberry in southern California. *Journal of Economic Entomology* 59, 433–439.

Odland, M.L. and Porter, A.W. (1941) A study of natural crossing in peppers (*Capsicum frutescens*). *Proceedings of the American Society of Horticultural Science* 38, 585–588.

Paine, T.D., Malinoski, M.K. and Robb, K.L. (1991) Reducing aesthetic injury or controlling insect populations: dilemma of insecticide use against Cuban laurel thrips (Thysanoptera: Phlaeothripidae) in landscape-grown *Ficus*. *Journal of Economic Entomology* 84, 1790–1795.

Palmer, J.M. and Mound, L.A. (1990) Thysanoptera. In: Rosen, D. (ed.) *The Armored Scale Insects, Their Biology, Natural Enemies and Control*, Vol. B. Elsevier Science Publishers, Amsterdam, pp. 67–76.

Parrella, M.P. (1995) IPM – approaches and prospects. In: Parker, B.L., Skinner, M. and Lewis, T. (eds) *Thrips Biology and Management*. Plenum Press, New York, pp. 357–363.

Parrella, M.P. and Horsburgh, R.L. (1983) Functional responses of the black hunter thrips, *Leptothrips mali* (Thysanoptera: Phlaeothripidae), to densities of *Panonychus ulmi* (Acari: Tetranychidae). *Environmental Entomology* 12, 429–433.

Parrella, M.P., McCaffrey, J.P. and Horsburgh, R.L. (1980) Compatibility of *Lepto-*

thrips mali with *Stethorus punctum* and *Orius insidiosus* – predators of *Panonychus ulmi*. *Environmental Entomology* 9, 694–696.

Parrella, M.P., Rowe, D.J. and Horsburgh, R.L. (1982) Biology of *Leptothrips mali*: a common predator in Virginia apple orchards. *Annals of the Entomological Society of America* 75, 130–135.

Pelerents, C. (1957) L'arachide à Yangambi. *Bulletin d'Information de l'Institut National pour l'Étude Agronomique de Congo Belge (INEAC)* 6, 243–255.

Pelikan, J. (1991) The Cuban laurel thrips (*Gynaikothrips ficorum* Marchal, 1908) in greenhouses in Czechoslovakia. (In Czech.) *Ochrana Rostlin* 27, 287–291.

Peng, Z. and Miles, P.W. (1988) Studies on the salivary physiology of plant bugs – function of the catechol oxidase of the rose aphid. *Journal of Insect Physiology* 34, 1027–1033.

Peterson, A. (1915) Morphological studies on the head and mouth-parts of the Thysanoptera. *Annals of the Entomological Society of America* 8, 20–67.

Pfadt, R.E. (1978) *Fundamentals of Applied Entomology*, 3rd edn. Macmillan, New York.

Pickett, C.H., Wilson, L.J. and Gonzalez, D. (1988) Population dynamics and within-plant distribution of the western flower thrips (Thysanoptera, Thripidae), an early season predator of spider mites infesting cotton. *Environmental Entomology* 17, 551–559.

Pitkin, B.R. (1973) *Larothrips dentipes* (Thysanoptera, Thripidae), a new genus and species of thrips from yam flowers in Nigeria. *Bulletin of Entomological Research* 62, 415–8.

Pollard, D.G. (1969) Directional control of the stylets in phytophagous Hemiptera. *Proceedings of the Royal Entomological Society of London* (*A*) 44, 173–185.

Putman, W.L. (1942) Notes on the predaceous thrips *Haplothrips subtilissimus* Hal. and *Aeolothrips meleleucus* Hal. *Canadian Entomologist* 74, 37–43.

Putman, W.L. (1965) The predacious thrips *Haplothrips faurei* Hood (Thysanoptera: Phlaeothripidae) in Ontario peach orchards. *Canadian Entomologist* 97, 1208–1221.

Raizada, U. and Nangia, A. (1988) Observations on pollen feeding by *Scirtothrips dorsalis* Hood (Insecta: Thysanoptera) on *Capsicum annuum* L. *Zoology (Journal of Pure and Applied Zoology)* 1, 155–157.

Raizada, U. and Nangia, A. (1989) Cyclic opening and abscission of *Nyctanthes* flowers with a note on visiting thrips species. *Current Science* 58, 93–96.

Raizada, U. and Saxena, P. (1989) Experimental studies in cross pollination of radish (*Raphanus sativus* L.) by *Thrips flavus* Schrank and *T. hawaiiensis* (Morgan) (Thysanoptera: Insecta). *Ecology (Journal of Pure and Applied Zoology)* 2 (2), 111–119.

Raman, A., Rajadurai, S., Mani, T. and Balakrishna, T. (1991) On the salivary enzymes and soluble proteins of the *Mimusops* gall thrips, *Arrhenothrips ramakrishnae* Hood (Tubulifera: Thysanoptera: Insecta). *Zeitschrift für angewandte Zoologie* 78, 131–136.

Rao, B.S. (1961) Pollination of *Hevea* in Malaya. *Journal of the Rubber Research Institute Malaya* 17, 14–18.

Reyne, A. (1927) Untersuchungen über die Mundteile der Thysanoptera. *Zoologisches Jahrbuch der Anatomie* 49, 391–500.

Rice, R.E. and Jones, R.A. (1972) Experimental miticides on European red and Pacific mites. *California Agriculture* 26 (3), 12–13.

Risler, H. (1957) Der Kopf von *Thrips physapus* L. (Thysanoptera, Terebrantia). *Zoologisches Jahrbuch der Anatomie* 76, 251–302.

Rust, R.W. (1980) Pollen movement and reproduction in *Arisaema triphyllum*. *Bulletin of the Torrey Botanical Club* 107, 539–542.

Sakimura, K. (1962) The present status of thrips-borne viruses. In: Maramorosch, K. (ed.) *Biological Transmission of Disease Agents*. Academic Press, New York, pp. 33–40.

Sakimura, K. (1963) *Frankliniella fusca*, an additional vector for the tomato spotted wilt virus, with notes of *Thrips tabaci*, another vector. *Phytopathology* 53, 412–415.

Segnou, Fatokun, C.A., Akoroda, M.O. and Hahn, S.K. (1992) Studies on the reproductive biology of white yam (*Dioscorea rotundata* Poir.). *Euphytica* 64, 197–203.

Selhorst, T., Söndgerath, D. and Weigand, S. (1991) A model describing the predator–prey interaction between *Scolothrips longicornis* and *Tetranychus cinnabarinus* based upon the Leslie theory. *Ecological Modelling* 59, 123–138.

Sen, S.K. and Sur, S.C. (1964) A study on vicinism in pigeonpea (*Cajanus cajan*) (L.) Millsp.). *Agricultura (Louvain)* 12, 421–426.

Şengonca, Ç. and Gerlach, S. (1984) Einfluss der Blattoberfläche auf die Wirksamkeit des räuberischen Thrips, *Scolothrips longicornis* (Thysan.: Thripidae). *Entomophaga* 29, 55–61.

Şengonca, Ç. and Weigand, S. (1988) Biology of the predatory thrips, *Scolothrips longicornis* Priesner (Thysanoptera, Thripidae). *Acta Phytopathologica et Entomologica Hungarica* 23, 343–349.

Seshadri, A.R. (1953) Observations on *Trichinothrips breviceps* (Bagnall), a little known predatory thrips from South India. *Indian Journal of Agricultural Science* 23, 27–39.

Shaw, H.B. (1914) Thrips as pollinators of beet flowers. *Department Bulletin of the United States Department of Agriculture* No. 104, 1–12.

Shumway, D.L., Teulon, D.A.J. and Kolb, T.E. (1991) Identification of thrips (Thysanoptera: Thripidae) feeding sites on sugar maple leaves. *Journal of Economic Entomology* 84, 1771–1773.

Slansky, F. and Scriber, J.M. (1985) Food consumption and utilization. In: Kerkut, G.A. and Gilbert, L.I. (eds) *Comprehensive Insect Physiology, Biochemistry and Pharmacology*, Vol. 4. Pergamon Press, Oxford, pp. 87–163.

Smith, J.J.B. (1985) Feeding mechanisms. In: Kerkut, G.A. and Gilbert, L.I. (eds) *Comprehensive Insect Physiology, Biochemistry and Pharmacology*, Vol. 4. Pergamon Press, Oxford, pp. 33–85.

Soria Vasco (1970) Studies on *Forcipomyia* spp. midges (Diptera, Ceratopogonidae) related to the pollination of *Theobroma cacao* L. *Dissertation Abstracts* 31, 2744B.

Spears, E.E. (1983) A direct measure of pollinator effectiveness. *Oecologia* 57, 196–199.

Stoddard, F.L. (1986) The distribution of immature thrips among flowers of faba beans in commercial crops and experimental plots. *Annals of Applied Biology* 109, 61–69.

Strong, D.R., Lawton, J.H. and Southwood, T.R.E. (1984) *Insects on Plants: Community Patterns and Mechanisms*. Blackwell Scientific Publications, Oxford.

Syed, R.A. (1978) Thrips pollination of oil palm in West Malaysia. Unpublished Commonwealth Institute of Biological Control Report, Slough.

Syed, R.A. (1979) Studies on oil palm pollination by insects. *Bulletin of Entomological Research* 69, 213–224.

Taylor, T.H.C. (1935) The campaign against *Aspidiotus destructor*, Sign., in Fiji. *Bulletin of Entomological Research* 26, 1–102.

Teulon, D.A.J. and Penman, D.R. (1986) A rearing method for the New Zealand flower thrips. *Proceedings of the 39th N.Z. Weed and Pest Control Conference* 244–245.

Teulon, D.A.J. and Penman, D.R. (1991) Effects of temperature and diet on oviposition rate and development time of the New Zealand flower thrips, *Thrips obscuratus*. *Entomologia Experimentalis et Applicata* 60, 143–155.

Thayumanavan, B., Velusamy, R. and Sadasivam, S. (1990) Phenolic compounds, reducing sugars, and free amino acids in rice leaves of varieties resistant to rice thrips. *International Rice Research Newsletter* 11, 14–15.

Trichilo, P.J. and Leigh, T.F. (1986) Predation on spider mite eggs by the western flower thrips, *Frankliniella occidentalis* (Thysanoptera: Thripidae), an opportunist in a cotton agroecosystem. *Environmental Entomology* 15, 821–825.

Trichilo, P.J. and Leigh, T.F. (1988) Influence of resource quality on the reproductive fitness of flower thrips (Thysanoptera: Thripidae). *Annals of the Entomological Society of America* 81, 64–70.

Tsutsumi, T., Matsuzaki, M. and Haga, K. (1994) New aspect of the 'mycetome' of a thrips, *Bactrothrips brevitubus* Takahashi (Insecta: Thysanoptera). *Journal of Morphology* 221, 235–242.

Ullman, D.E., Westcot, D.M., Hunter, W.B. and Mau, R.L.F. (1989) Internal anatomy and morphology of *Frankliniella occidentalis* (Pergande) (Thysanoptera: Thripidae) with special reference to interactions between thrips and tomato spotted wilt virus. *International Journal of Insect Morphology and Embryology* 18, 289–310.

Ullman, D.E., Cho, J.J., Mau, R.F.L., Hunter, W.B., Westcot, D.M. and Custer, D.M. (1992) Thrips–tomato spotted wilt virus interactions: morphological, behavioral and cellular components influencing thrips transmission. *Advances in Disease Vector Research* 9, 195–240.

Veer, V. (1978) *Thrips flavus* (Schrank) as a pollinating agent. *Indian Journal of Forestry* 1 (4), 278.

Velayudhan, R. and Annadurai, R.S. (1986) Pollination potential of thrips (Insecta: Thysanoptera) in some solanaceous plants. *Proceedings of the Indian Academy of Sciences (Animal Science)* 95, 109–116.

Velayudhan, R., Gopinathan, K. and Bakthavatsalam, N. (1985) Pollination potential, population dynamics and dispersal of thrips species (Thysanoptera: Insecta) infesting flowers of *Dolichos lablab* L. (Fabaceae). *Proceedings of the Indian National Science Academy* B51, 574–580.

Vrieling, K., Soldaat, L.L. and Smit, W. (1991) The influence of pyrrolizidine alkaloids of *Senecio jacobaea* on *Tyria jacobaeae*, *Brachycaudus cardii* and *Haplothrips senecionis*. *Netherlands Journal of Zoology* 41, 228–239.

Walbank, M.H. (1996) The sensory biology of the western flower thrips, *Frankliniella occidentalis* (Pergande) (Thysanoptera: Thripidae) in relation to host location. PhD thesis, University of Birmingham.

Ward, L.K. and Spalding, D.F. (1993) Phytophagous British insects and mites and their food-plant families: total numbers and polyphagy. *Biological Journal of the Linnean Society* 49, 257–276.

Wardle, R.A. and Simpson, R. (1927) The biology of Thysanoptera with reference to the cotton plant. III. The relation between feeding habits and plant lesions. *Annals of Applied Biology* 14, 513–528.

Wiesenborn, W.D. and Morse, J.G. (1985) Feeding rate of Thysanoptera estimated using ^{14}C inulin. *Journal of Economic Entomology* 78, 1515–1518.

Wiesenborn, W.D. and Morse, J.G. (1986) Feeding rate of *Scirtothrips citri* (Moulton) (Thysanoptera: Thripidae) as influenced by life stage and temperature. *Environmental Entomology* 15, 763–766.

Wiesenborn, W.D. and Morse, J.G. (1988) The mandible and maxillary stylets of *Scirtothrips citri* (Moulton) (Thysanoptera: Thripidae). *Pan-Pacific Entomologist* 64, 39–42.

Williams, G. and Adam, P. (1994) A review of rainforest pollination and plant–pollinator interactions with particular reference to Australian subtropical rainforests. *Australian Zoologist* 29, 177–212.

Williams, I.H. (1977) Behaviour of insects foraging on pigeon pea (*Cajanus cajan* (L.) Millsp.) in India. *Tropical Agriculture (Trinidad)* 54, 353–363.

Wilson, L.J., Bauer, L.R. and Walter, G.H. (1996) 'Phytophagous' thrips are predators of twospotted spider mites (Acari: Tetranychidae) on cotton in Australia. *Bulletin of Entomological Research* 86, 297–305.

Wittwer, S.H. and Haseman, L. (1946) Soil nitrogen and thrips injury on spinach. *Journal of Economic Entomology* 38, 615–617.

Wolfenbarger, D.O. (1977) Comments on mango pollination. *Proceedings of the Florida State Horticultural Society* 90, 240–241.

Wyatt, R. (1983) Pollinator–plant interactions and the evolution of breeding systems. In: Real, L. (ed.) *Pollination Biology.* Academic Press, London, pp. 51–95.

Yadav, H.S., Grandrade, G.A. and Jakhmola, S.S. (1974) Note on the relationship between thrips and pod setting in the flowers of pigeon-pea. *Indian Journal of Agricultural Science* 44, 555–556.

Yarwood, C.E. (1943) Association of thrips with powdery mildews. *Mycologia* 35, 189–191.

Yasumi, K., Ohba, K. and Hirano, C. (1994) Scanning electron microscopic observations of the mouthparts and feeding marks of *Thrips palmi* Karny (Thysanoptera: Thripidae). (In Japanese.) *Japanese Journal of Applied Entomology and Zoology* 38, 23–28.

Zawirska, I. (1978) Studies of *Thrips tabaci* Lindeman (Thysanoptera, Thripidae). (In Polish.) *Prace Naukowe Instytutu Ochrony Roslin* 20, 15–138.

Zimmerman, A. (1901) Über einige durch Thiere verursachte Blattflecken. *Annales du Jardin Botanique Buitenzorg* 17, 102–124.

zur Strassen, R. (1995) Binomial data of some predacious thrips. In: Parker, B.L., Skinner, M. and Lewis, T. (eds) *Thrips Biology and Management.* Plenum Press, New York, pp. 325–328.

5 Flight and Dispersal

Trevor Lewis
Institute of Arable Crops Research – Rothamsted, Harpenden, Herts AL5 2JQ, UK

Introduction

The ability of thrips to disperse widely, either naturally by flight or within fresh vegetation transported in commercial agriculture and horticulture, contributes greatly to their pest status. Although weak flyers, their fringed wings enable them to remain airborne long enough to travel easily between neighbouring fields and frequently over far greater distances. In addition, intensive global marketing of food and ornamental crops, especially when produce is flown within and between countries and continents, now provides many opportunities for widespread artificial dispersal. This chapter concentrates on the biological and physical aspects of thrips flight leading to crop infestation, then reviews other means of artificially assisted dispersal which latterly have become so economically important.

Flight Mechanism and Capability

In terebrantian thrips, the slender membranous wings, usually with only a few longitudinal setae-bearing veins, are covered in minute microtrichia and characteristically fringed with long, straight or wavy cilia emanating from basal sockets sited on, or slightly beneath, the wing edges (Fig. 5.1). In most normal-winged (macropterous) forms the wings lie nearly parallel at rest, with fringes folded, forewings covering hind wings and almost the full length of the abdomen (Fig. 5.2a) (see also Chapter 2, Fig. 2.10). Some species may have short-winged (brachypterus) forms which are flightless; a few species are entirely without wings in males or both sexes.

By contrast, the wing structure and deportment in tubuliferans is quite different; their wing fringe cilia are straight and embedded deeply and immovably between the upper and lower surfaces of the wing membrane; the

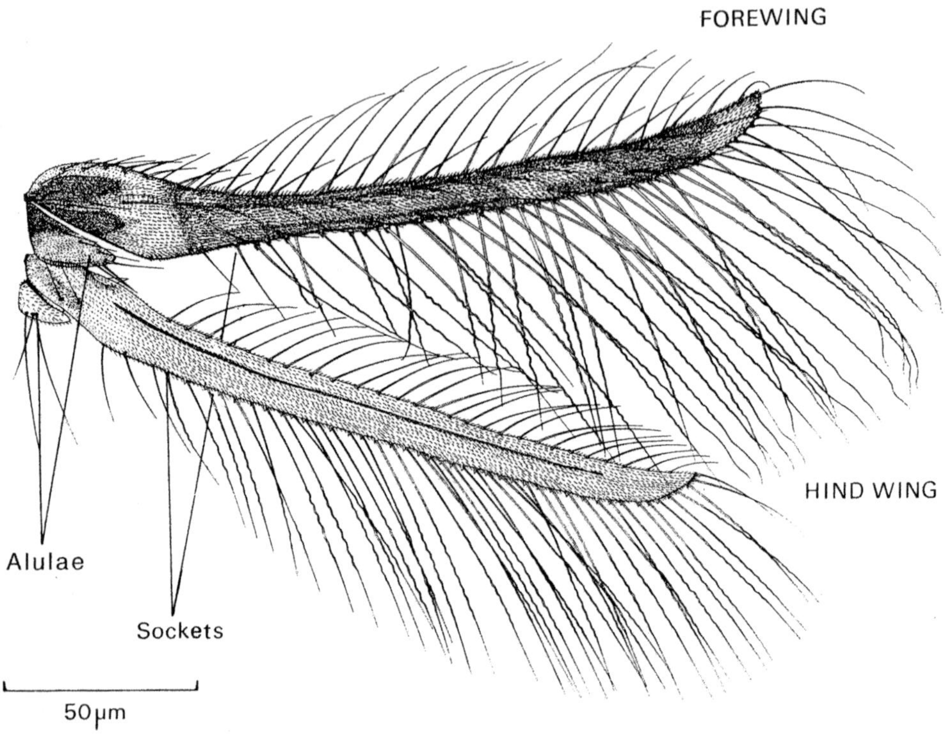

Fig. 5.1. Right fore and hind wing of *Hercinothrips femoralis* female. Note 'wavy' cilia on trailing edge of wings and smooth cilia on leading edges (after Moritz, 1989).

wings overlap at rest along the abdomen (Kuethe, 1975) (Fig. 5.2b). This group is not considered further.

Immediately before and during flight the conformation of the wings is changed by a rearrangement of the fringe cilia to increase the wing area. A generalized description of how this is achieved is based on studies of *Thrips physapus* (Ellington, 1980). The lower leading edge of the forewing bears a row of straight cilia each originating from partially opened 'figure of eight' sockets. These cilia can be protruded forward at an angle of about 70° to the wing and remain locked in that position, held by the elasticity of the socket cuticle. On the hind edge of the forewing, two rows of longer, wavy cilia inserted into more elongated sockets can likewise be opened to form an angle of about 55° or 80° to the wing membrane. The leading edge of the hind wing bears a row of immovable cilia, whereas those on the trailing edge of the hind wing also arise from figure-of-eight sockets by which they can be locked open at almost 75° to the membrane. The whole effect is to convert the resting parallel wings into extended aerodynamically effective surfaces some three to four times their 'resting' size, thus preparing the thrips for flight in what, for such small insects, is essentially a viscous medium. Setae on the scale-like alulae at the base of the wings ensure that fore- and hind wings remain interlocked and synchronized

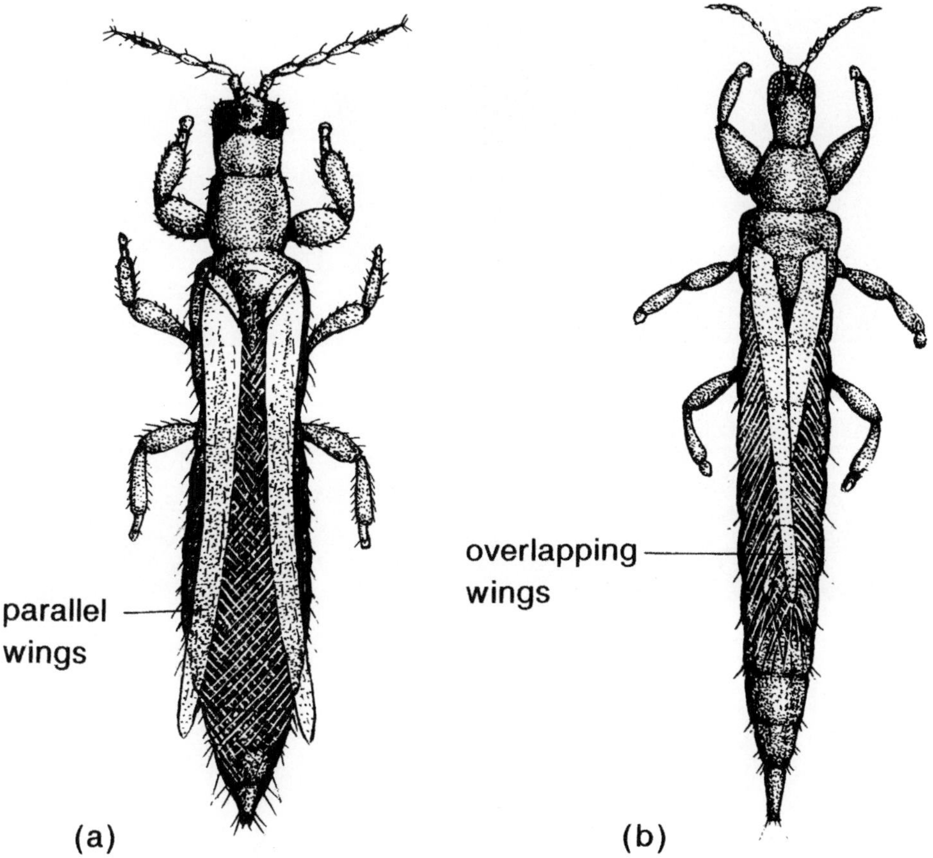

Fig. 5.2. Wing resting position of (a) terebrantian and (b) tubuliferan thrips.

during flight with the hind wings lagging behind the forewings in the beat cycle.

This apparently complex wing preparation can be achieved in less that 0.5 s, or it may last several minutes. It is brought about by the thrips flexing its abdomen to draw the wings through combs and setae on the sides of the abdominal tergites V to VIII; this action opens the fringing cilia (Fig. 5.3). A remarkable series of photographs taken at four frames s^{-1} shows this typical combing sequence in *Thrips physapus* before a final clap of the wings generates lift for take-off aided by a thrusting jump with the hind legs that merges imperceptibly into the beginning of flight when the forewings fling open for the first downstroke (Fig. 5.4) (Ellington, 1980). The body moves about 0.15 mm during the 1.2 ms of the jump suggesting a final jump velocity of 0.25 m s^{-1} which is comparable with the typical flight speed of this species. During flight the wings beat through a stroke angle of about 135° in a plane approximately perpendicular to the body axis. The wings clap together at the dorsal end of the cycle, then fling open to begin the subsequent downstroke (C.P. Ellington, G.G.

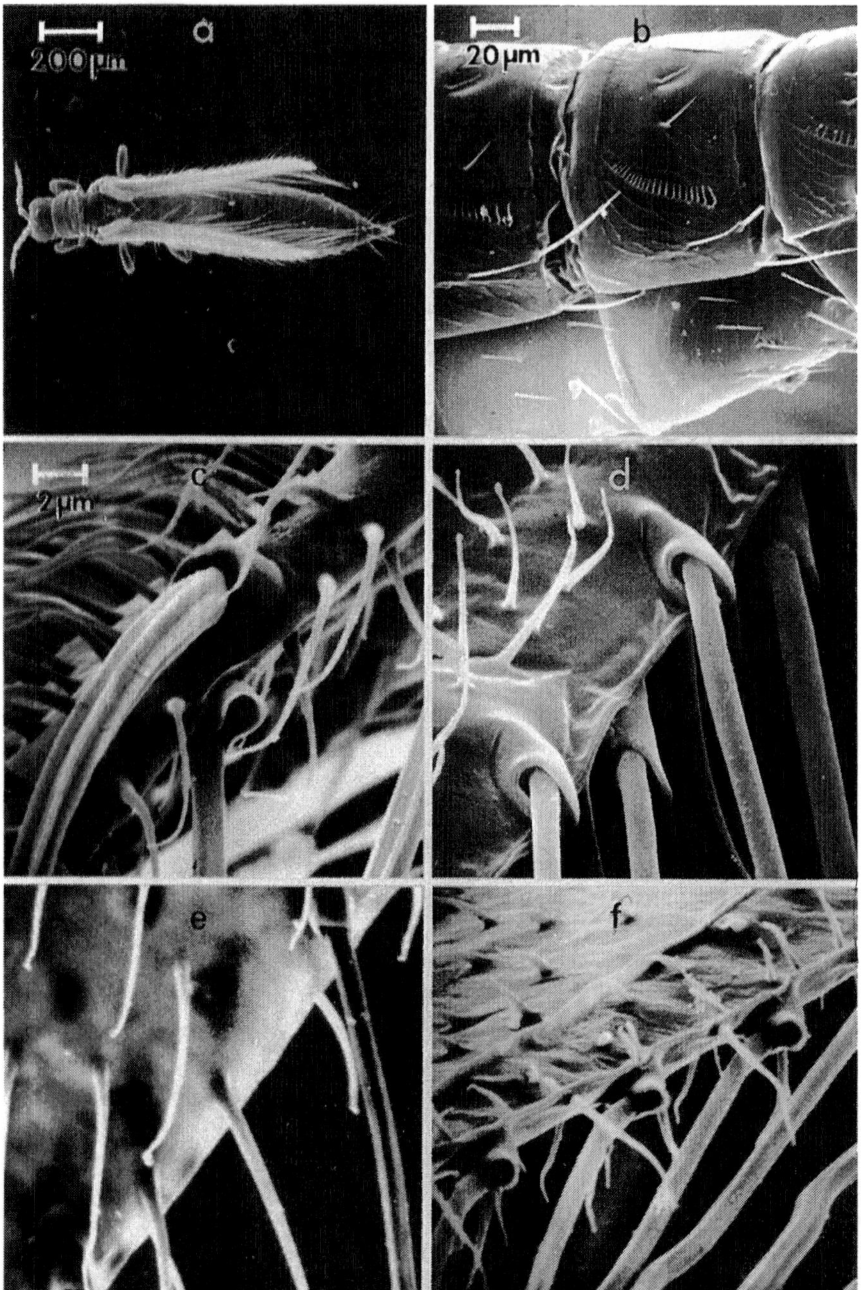
200 µm
a
20 µm
b
2 µm
c
d
e
f

Runnalls and K.E. Machin, Cambridge, 1980, personal communication). Wing movement in flight is maintained by the complex interaction of an array of oscillatory muscles (see Chapter 2, Figs 2.11 and 2.16) which set up high frequency vibrations in the pterothoracic resonance system; this is intensified by a bistable clicking mechanism of the wings (Moritz, 1989). Some 150 to 200 wing beats per second produce flight speeds in still air of about 0.1 to 0.5 m s^{-1} depending on species; individuals must be capable of sustaining this for several hours to travel the long distances recorded (see Chapter 5, Exodus flight and dispersal).

Adults need a short period of teneral development after emergence from the pupae before the wing muscles can function. For *Limothrips cerealium* this lasts about 5 hours at 20°C (Lewis, 1973) and for *Frankliniella intonsa* up to 12 h at similar temperatures (Murai, 1988); for tropical species it may be considerably less. Overwintered adults of some temperate species appear to need a second period of flight maturation in spring to reactivate their flight muscles after several months of inactivity (Lewis, 1963).

Flying thrips cannot feed, so their ability to resist desiccation is crucial in determining how long they can remain airborne and thus how far they may spread. The survival of starved and watered *Thrips australis* has been related to different saturation deficits arising from typical temperatures and relative humidities in southern Australia (Laughlin, 1977). In cool weather this thrips could probably survive over 24 h in the air with little risk of death from desiccation. In hot dry summers survival time was about 6 h, but could fall to as little as 3 h on hot days. However, about 5 per cent of thrips kept in very dry air (14°C, 0 per cent relative humidity) but provided with water survived for 5 days, suggesting that at least a few, perhaps young, individuals in a population could survive long enough to recolonize distant hosts as long as they alighted occasionally to drink either by probing non-host plants or perhaps by imbibing dew (Preisner, 1964). Overwintering *Limothrips cerealium* are even more tolerant of dry conditions (Lewis, 1962), and they probably have the ability along with many other species to survive and travel for longer periods in suitable weather.

Flight capability may also differ between sexes, and in females it is related to the sexual cycle and ovarian development. Clearly, in species with wingless males only the females can fly, but even in some fully winged species (e.g. *Stenothrips graminum*) females predominate in aerial populations. In bisexual species the flying females are usually mated but have immature ovaries;

Fig. 5.3. *Opposite* Scanning electron micrographs of *Thrips physapus*. (a) Dorsal view. (b) Lateral view of VIIth abdominal segment showing combs and setae. (c) Edge-on view of forewing leading edge. A large costal seta is above the centre. The socket of a leading edge cilium is visible on the lower wing edge; cilium is in the closed position. (d)–(f) Scale is the same as (c). (d) Sockets along the forewing trailing edge viewed from the lower surface, showing the double row of cilia. (e) View of the lower surface of hind wing leading edge. Two cilia are visible near borders; they insert rigidly into sockets. (f) Hind wing trailing-edge cilia and sockets viewed from slightly above the wing plane. Cilia are in the closed position (after Ellington, 1980).

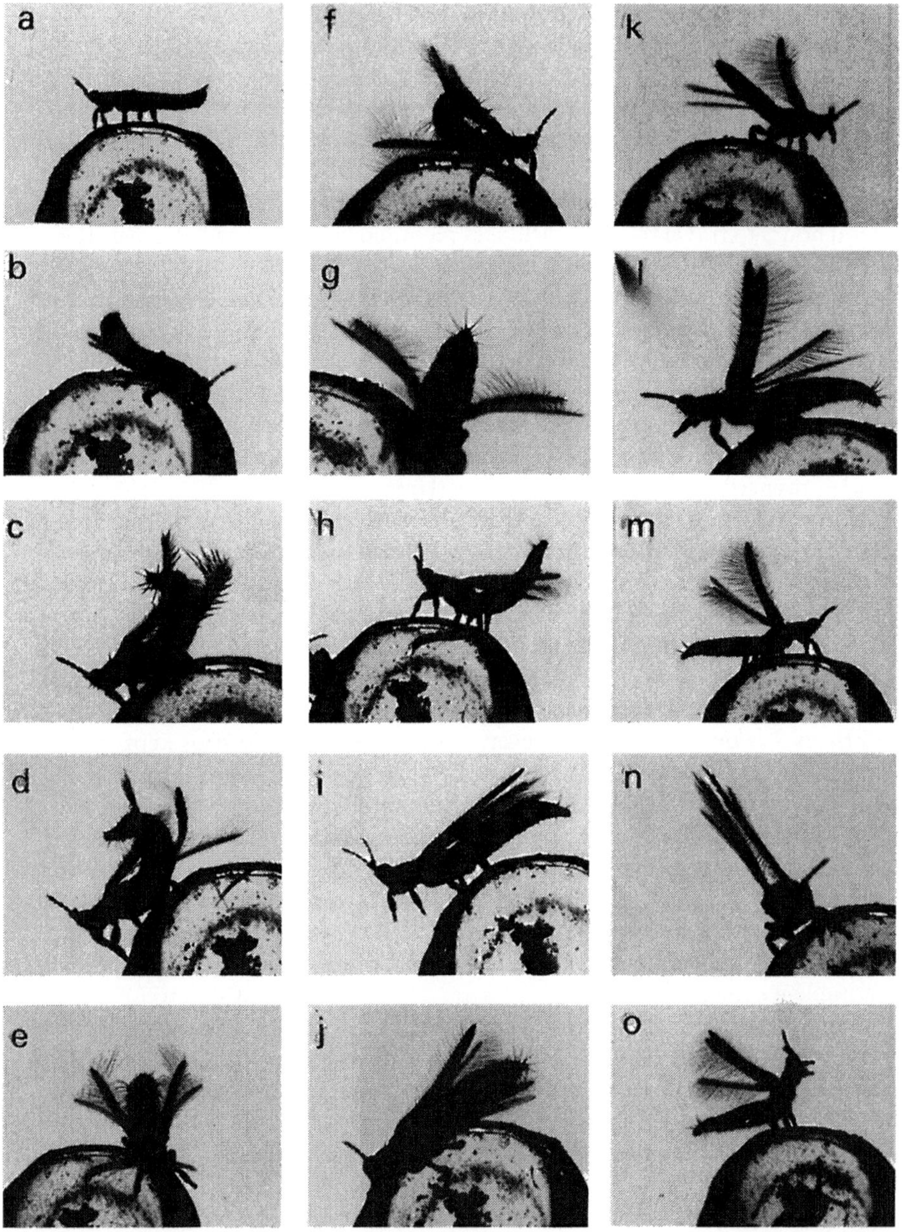

Fig. 5.4. A series of photographs showing abdominal combing during flight preparation, which opens the wing fringes prior to take-off posture and the final jump into the air (o) (after Ellington, 1980).

evidence from *Limothrips cerealium, Limothrips denticornis, Chirothrips manicatus, Anaphothrips obscurus* and *Frankliniella intonsa* indicates that immature individuals predominate in airborne populations distant from host crops during spring and autumn migrations, whereas gravid females fly more locally among host plants in summer, presumably as they seek oviposition sites. Gravid females of some *Thrips* and *Taeniothrips* species seem less inhibited flyers (Lewis, 1965).

The Dispersal Process

Despite the importance of flight and airborne dispersal for the spread of pest species, the topic has received limited detailed study except for a great many descriptive surveys centred on monitoring thrips activity in various crops, usually using water or sticky traps (see Lewis, 1973 for examples). In reality, flight dominates the adult lives of most pest species and the aerial flux of adult thrips that takes place, largely unseen, often for days at a time, involves countless millions of insects being mixed and scattered over vast areas (Johnson, 1969). Individual aspects of dispersal, namely take-off, the exodus flight, swarming and settling, are here considered separately with particular emphasis on the effects of weather on the different stages.

Take-off

After completion of the teneral period and in various species, mating, partial ovarian development, or post-overwintering quiescence, most winged adults are ready to take-off to begin dispersal. Deteriorating host plants also provide a stimulus, but whatever the cause, take-off and the subsequent duration and distance of flight are strongly influenced by weather. Temperature, light and wind are very important; high relative humidity seems to stimulate some species; atmospheric pressure and electricity have a doubtful effect. Most temperate species can take-off when temperatures reach 17 to 21°C. Laboratory and field methods for measuring take-off thresholds are described by Lewis (1963). In field situations where temperature is rising rapidly, adults that emerge from pupae the previous day or overnight are usually ready to take-off as soon as the threshold is reached on the next day. Other, more recently emerged individuals will take-off when ready at temperatures above the critical threshold, thus producing an extended period of exodus from the host crop, sometimes with peaks of activity in morning and afternoon (Lewis, 1973; Murai, 1988).

There is no evidence of thrips taking-off at night, except for some *Thrips* spp. in Malaysian rainforest (Appanah, 1979), though some already in the air may be trapped as stragglers during the hours of darkness and in continental regions, where low-level jet streams develop above nocturnal inversions, a substantial proportion of the flying population may remain airborne into the night (Glick, 1939). As far as is known, the minimum light intensity for take-off in temperate species is about 1000 lux, so low light intensity is unlikely to restrict take-off in early morning when temperature is more likely to be the constraining factor. By contrast, in the tropics low light levels are more likely to be inhibitory than temperature.

For medium-size species such as *Limothrips*, mean hourly wind speeds at vegetation level of up to 12 to 15 km h^{-1} do not prevent take-off; slightly smaller *Frankliniella* spp. may be deterred by mean winds exceeding about 9 km h^{-1}. However, actual launching probably occurs in temporary lulls, so mean wind speeds measured over hours can only be a general guide.

In the field it is virtually impossible to observe and monitor with the naked eye the rapid sequence of events immediately before, at the moment of, and after take-off. In an attempt to detect flight of small airborne insects over a crop canopy, Bent (1982) devised a high powered, infra-red, remote sensing system which was used with some success by L. Purchase (Cranfield, 1988, personal communication) to study take-off and flight by *Limothrips cerealium*. The equipment emitted intense pulsed light, with characteristic wavelengths different from sunlight, from a xenon source for brief periods of 30 μs at intervals of 20 ms. When linked to suitable electronic and computing equipment the characteristic wavelengths, which were reflected from insects flying within the beam, were detected and stored on videotape for frame-by-frame analysis. The flight tracks of insects with a wing area of greater than 0.75 mm^2 were traceable for up to 15 m above a cereal crop and, by inserting a macrolens into the camera, thrips were detectable. When linked with high resolution anemometers, the system permitted study of the relationship between wind and take-off on a fine temporal and spatial scale.

In both field and laboratory, take-off was usually associated with some air movement, and short lulls in the wind of 10 s or so encouraged typical flexing, wing combing and take-off. Under controlled air flows in the laboratory the mean wind speed conducive to preflight flexing and combing was 1.2 m s^{-1} (4.3 km h^{-1}), but for actual take-off 0.6 m s^{-1} (2.1 km h^{-1}) was preferred. In the field, most individuals taking-off from wheat ears initially flew upwards for a few centimetres towards the light and into the oncoming wind, presumably thereby gaining extra lift for less expenditure of energy; thereafter, they probably reorientate to face downwind and appear to be able to tolerate much stronger winds without harm. Individuals of most species can launch themselves from flat surfaces such as petals or leaf blades, but often choose a protruding narrow edge from which to jump.

Direct rain and heavy dew prevent take-off because the wing setae become clogged and small species cannot overcome the surface tension of water droplets. Relative humidity appears to have little effect on take-off (Lewis, 1963) though it may restrict the duration of flight and viability in dry climates. It is considered as a component of stormy weather below.

Exodus flight and dispersal

As most take-off occurs during the warmest part of the day, emigrants rapidly encounter convective turbulence and winds exceeding their own flight speed, so are carried upwards and blown away from the point of departure. Quantification of the abundance and distribution of airborne populations thereafter is possible currently only from traps held aloft by masts, kites, balloons or aircraft; radar, as used to detect other flying insects (Riley, 1989; Riley and Reynolds, 1993) is too insensitive to detect individual thrips. Opportunities for these types of sampling are limited and expensive and, with a few exceptions, most information on thrips has been obtained from general surveys of the airborne fauna.

The literature on thrips pests abounds with claims that particular species 'prefer' to fly at a characteristic height (e.g. Körting, 1931; Murai, 1988). This seems improbable, except possibly in almost still air, given their slow airspeed compared with the winds normally encountered; the impression is more likely an artefact produced by using impaction traps or nets (see Chapter 11, Sticky traps). In reality, density is usually greatest immediately above the source vegetation and for all species declines with height. Small weak-flying species are relatively less common high in the air than stronger flyers, which can presumably climb higher by their own efforts and thus enter stronger ascending currents (Lewis, 1973). Weak flyers and even wingless species and larvae can nevertheless be blown from vegetation, but careful observations of the wingless *Apterothrips apteris* colonizing *Erigeron glaucus* in windy coastal areas of California showed that individuals moved only short distances, and then mostly by walking (Strauss and Karaban, 1995). Mound (1983) believed that species habitually exposed to the wind, as a result of crawling up grass stems or twigs, were more widespread than similar species that always live near the ground, whether winged or wingless. However, the marked differences in vertical density profiles recorded between species (Lewis, 1973) (Fig 5.5) indicate that dispersal of the winged majority of pest species is an active process determined for at least part of its duration by the flight capability of the adults.

Most of the published survey data on airborne thrips populations are summarized by Lewis (1973). The recorded heights attained by pest species include *Kakothrips pisivorus* at 600 m (Hardy and Milne, 1938), *Limothrips cerealium* and *Limothrips denticornis* at 300 m (Lewis, 1965), *Frankliniella tritici* at 3100 m and *Caliothrips phaseoli* at 1200 m (Glick, 1939) facilitating lateral spread over great distances in seasonal or prevailing winds. Catches made on ships at sea indisputably illustrate this. For example, between 1957 and 1967, 94 individual thrips, mainly Thripidae, were trapped in the Atlantic, Indian, Pacific and Southern Oceans up to hundreds of kilometres from land (Gressitt *et al.*, 1960; Holzaptel and Harrel, 1968; Yoshimoto *et al.*, 1962). Thrips have also been recovered from plankton nets towed on the sea surface in the eastern Pacific off California and Oregon, and in the English Channel (Cheng and Birch, 1977). There is strong circumstantial evidence from comparative faunal studies in Australia and New Zealand that the 1500-km wide Tasman Sea is frequently traversed not only by winged species such as *Thrips imaginis*, the 'plague thrips' so abundant in Australia, but also by small wingless species including *Anaphothrips woodi*, along with plant seeds and spores (Close *et al.,* 1978; Mound, 1983), although many of these immigrants fail to establish in New Zealand presumably because the climate is unsuitable (Mound and Walker, 1982). The thrips fauna of Central America centred on studies in Costa Rica includes species originating far to the north and south of that region, but probably carried regularly in winds associated with the annual

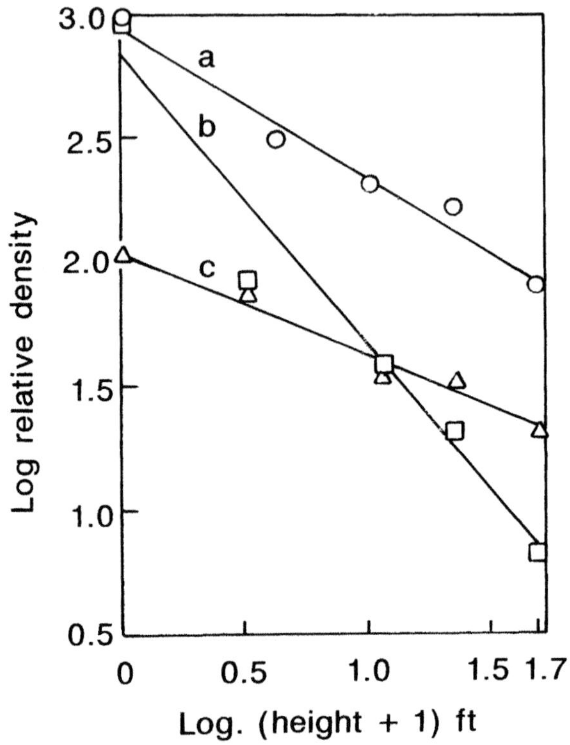

Fig. 5.5. Linear profiles of relative density on height (log scales) for three species caught over exposed grassland in England, (a) *Limothrips cerealium*, (b) *Chirothrips manicutus*, (c) *Limothrips denticornis*. The density of all populations decreased with increasing height, but the rate of decrease differed between species (after Lewis, 1958). Conversion factor : 1ft = 30.48 cm.

movements of the intertropical convergence zone (Mound and Marullo, 1996). The rapid spread of *Thrips palmi* to many parts of the subtropics since the mid-1980s (Mound *et al.*, 1992) may well have been assisted by airborne dispersal, particularly where violent storms occur as in the Caribbean. This species is now widespread along the island chain from Trinidad to Puerto Rico, but not as yet in Costa Rica, which is off the main hurricane track. Firm evidence of long-distance dispersal entirely over land is more difficult to ascertain, but the annual appearance of *Frankliniella tritici* in Illinois where it does not overwinter (Stannard, 1968) indicates a northward movement on frontal winds each spring. Heavy infestations of this species and damage to strawberries in the north-eastern USA have been attributed to airborne invasions carried on spring frontal systems (Felland *et al.*, 1994).

The individual thrips that travel such long distances may only represent a small fraction of those leaving the original breeding site, but such vast numbers are produced on some host crops that there is a realistic chance of enough migrants arriving far downwind to colonize fresh natural habitats or infest annual crops. As an illustration, an estimated 100 to 150 million *Limothrips cerealium*, weighing 4 to 6 kg, emigrate daily from a 10 ha English wheat field with a plant density of 600 ears m^{-2}, over a typical 7 to 10 day period as the crop ripens. The overall exodus of mixed species from intensively cropped arable and horticultural regions in many parts of the world must be prodigous, and there is no evidence that their ability to feed eventually on suitable hosts, or their fecundity, is impaired by the journey, except possibly in dry air (see above, Laughlin, 1977).

As flying thrips are unobservable beyond a few metres their behaviour in the upper air is open to speculation. It is conceivable that the wings are used principally to provide sufficient lift at launching to propel individuals into ascending air (see above). Most will then remain in air warm enough to sustain flight, but with a typical daytime temperature lapse rate of about 5°C for each 500 m rise in altitude (Pedgley, 1982) some may be carried into air too cool for their wing muscles to function. In these circumstances they presumably respond as inert particles, and only after passive descent or circulation into warmer air are the wings able to beat again to generate lift. Thus, some long-distance dispersal might consist of alternate periods of wing beating in warm air and inert tossing in cool air, with final controlled orientation to vegetation and alighting. Only high resolution radar as yet unavailable, or other refined remote sensing equipment is likely to clarify the process. Whatever the detailed mechanism, the net result is that many species are intensely migratory and capable of reaching new habitats and crops far from their source of origin.

An interesting, though in a pest context probably unimportant, contribution to long-distance dispersal is phoresy on bird's plumage. Morison (1973) first recorded a specimen of *Apterothrips secticornis* in Britain on the feathers of a migratory lesser whitethroat (*Sylvia curruca*). Other close associations recorded between thrips and birds which may occasionally lead to casual phoresy include *Aeolothrips* sp. in nests of the highly migratory house martin (*Delichon urbica*) (Lewis, 1973) and several grass-dwelling species, *Anaphothrips obscurus*, *Aptinothrips stylifer* and *Baliothrips dispar*, in the nests of ring-billed gulls (*Larus delawarensis*) in Lake Superior (Freitag and Ryder, 1973).

The intense flight activity, or 'mass flights' that sometimes occur near crop level have given rise to the popular terms, thunder flies, storm flies or thunder bugs, widely used for thrips. There are also frequent references in the literature to 'swarming' behaviour. These two phenomena are essentially different; 'mass flights' refer to the exodus of thrips, often largely or exclusively females, from dense breeding populations, whereas 'swarming' relates to predominantly male mating aggregations that appear transiently on flowers or other attractive surfaces. Both types of behaviour appear to be affected by weather.

Mass flights, defined as at least a doubling of the average hourly aerial density during a species' flight period, are most noticeable when populations on heavily infested crops reach flight maturity over a short time. For example, the hourly density of cereal thrips immediately over wheat in England can rise to 2 m^{-3} for a few days in July or August (Lewis, 1973). Perhaps the most spectacular mass flights recorded relate to spring migrations of *Taeniothrips inconsequens* in California in 1910 (Foster and Jones, 1915). In late March, flying thrips were so abundant that as one of the authors drove his horsedrawn vehicle he had 'great difficulty in keeping both hands on the reins on account of the great number of flying thrips which ... filled his eyes and covered his clothes'. Two days later he encountered a 'swarm' estimated to be 12 km long and 6 km wide, extending 1 to 5 m above the ground, and at its densest part it 'appeared literally like a black, glistening, seething mass moving up and down like heat waves'.

The most detailed attempts to relate such activity to weather, made by Hurst (1964) and Lewis (1964), established that mass flights of six species in southern England were encouraged by dry settled conditions, a maximum temperature of at least 20°C, daytime dew point between 5 and 15°C, and a stable atmosphere above about 1600 m with unstable conditions below caused by sunny weather. Surprisingly, in view of traditional wisdom, thundery weather was generally discouraging. In Hungary, mass flights of *Frankliniella intonsa* were stimulated by similar factors (Jenser, 1973). However, Kittel (1958) working mainly with *Limothrips denticornis*, *Chirothrips manicatus* and *Thrips pini* in the Harz Mountains, central Germany, insisted that mass flights were associated with thunder because he believed that only rapidly changing factors such as electrical potential, temperature or light, could stimulate the thrips to behave collectively in this way. Unfortunately, no convincing experimental or meteorological evidence was produced to support this claim. There are two aspects of atmospheric electricity that might affect insects; one is the potential gradient or electric field, and the other the density of unipolar ions in the air (Maw, 1962, 1963; Johnson, 1969). In fine weather the Earth is usually negatively charged and the atmosphere positively charged. The potential thus produced increases with height, and the rate of change (known as the potential gradient) may be more than 10^6 W m^{-1} in a thunderstorm, resulting in a current of positive charges towards the Earth's surface caused by the unipolar ions in the atmosphere. There is no confirmation that the thrips respond in any way to changes in either the electric field or the density of air ions. Oppressive, thundery weather often coincides with the ripening of summer crops, and there is no doubt that turbulence associated with thunderstorms will assist dispersal (see above), but thunder does not appear *per se* to be an essential meteorological trigger to mass flights.

By contrast, in Arizona, mating 'swarms' of *Frankliniella occidentalis*, which occur largely as a result of males flying to attractive flowers or artificial surfaces, do seem to be stimulated by high humidity (Terry and Gardner, 1990). Females enter swarms continuously but leave after mating (see Chapter 3, Mating behaviour, territoriality and fighting). Dense aggregations of up to 180 individuals observed on white plastic cups in cotton and alfalfa fields peaked soon after sunrise, especially when clouds obscured the sun and later in the day if storms were imminent. Swarms rarely appeared in the dry season when the dew point fell below 12.5°C.

An unforeseen 'pest' problem arising from the dense airborne populations common during summer migrations, especially in cereal growing districts, is the tendency of thrips to trigger false alarms in fire detection systems. Some models of smoke detector based on an infra-red light source and photo cell can be activated accidentally by even a single thrips entering the device and interrupting the light beam; other types of ionization detectors are also vulnerable if thrips enter and cause a short circuit in the electronics within the ionization chamber. Migrating thrips, especially *Limothrips cerealium*, habitually seek out narrow crevices and smoke sensors in buildings near concentrated sources of insects; commonly up to ten thrips can be found inside a single sensor, and as many as 300 individuals have been recorded in one removed from a building in south-east England. The frequent outcome of these invasions is a false alarm which detracts from the credibility of systems, especially in hospitals and public places. They also prove costly where alarm systems are directly connected to fire stations resulting in unnecessary dispatch of fire tenders (Cuthbertson, 1988).

Settling and infestation patterns in crops

Thrips are so small (0.5 to 1 mm long) and light (20 to 50 μg) it is widely assumed that, once borne aloft by winds exceeding their flight speed, individuals have minimal control over their flight path and destination. Similarly, the combination of slight dimensions and weight would be expected to endow a very low terminal velocity (free-falling rate), which would make landing on a preselected site exceedingly difficult except in almost imperceptible breezes. Despite this rationale, there is abundant circumstantial evidence that, at least near vegetation level, settling involves a degree of control sufficient to allow thrips to choose to alight on host crops and even on individual plants. Unlike take-off, settling has not been photographed, but on the rare occasions when it is possible to register with the naked eye the moment of alighting, individuals apparently land feet first on vegetation, instantly fold their wings, indicating that they have been in use, and quiver the antennae as if sensing the substrate. Only in strong winds do they give the impression of having been blown haphazardly against the substrate and thereafter of having to struggle to attain a foothold; on most occasions, therefore, at some distance above the settling site they must adopt a landing posture. The distance over which this is feasible, i.e. the depth of the 'boundary layer' within which sensory mechanisms and behaviour permit active orientation to vegetation (Johnson, 1969; Taylor, 1958), will fluctuate with changing wind speed and at times may be no more than a few millimetres above the plant canopy.

Responses to colour, UV reflectance and scent as well as wind all seem to be involved. Without an ability to make a choice, the very different responses of similar-sized species to various coloured and UV reflecting traps so widely reported (e.g. Kirk, 1984; Czencz, 1987; Matteson and Terry, 1992; Teulon and Penman, 1992) would not occur (see Chapter 3). Up to 20-fold differences in catches have been reported between flower and grass-dwelling thripids in different colour traps only 5 m apart (Kirk, 1984). Likewise, the addition of scent (anisaldehyde) to otherwise uniform water traps increased catches of flower dwellers compared with grass dwellers by three- to eight-fold (Kirk, 1985) either as thrips reacted to an odour-induced visual response or to pure anemotaxis (Brødsgaard, 1990); ethyl nicotinate increased catches of *Thrips obscuratus* by up to two orders of magnitude over unscented traps (Teulon,

1988). Wind-directed sticky traps exposed in soybean fields caught a preponderance (73 per cent) of *Frankliniella tritici* on their leeward surface, but most *Neohydatothrips variabilis* on their windward surface (94 per cent). This suggests that *Frankliniella tritici* attempts to orientate into the wind and/or flies in lulls (less than 2 km h^{-1}) when it can direct its flight; whereas *Neohydatothrips variabilis* is blown more readily and/or flies in conditions when it is less able to control its direction and landing response (Irwin and Yeargan, 1980). These examples cannot be explained without there being different substrate-perception and alighting behaviour between different species of immigrants. An indication that individuals can exert a degree of choice over their landing site is provided by the different internal structure of muscles responsible for raising and lowering the wings. In *Terebrantia*, the wing-lowering muscles contain about twice as many myofibrils as the wing raisers (Moritz, 1989). This suggests that thrips require more power for downward than for upward wing movement. In a behavioural context the implication is that individuals use their wings to orientate towards perceived landing sites rather than being simply blown on to them. The distance over which a significant influence can be exerted will, nevertheless, vary with windspeed, but totally random deposition and subsequent accumulation on hosts caused by differential re-take-off from favoured and unsuitable plants would not seem to be the main mechanisms whereby immigrants build-up on crops, though there may be some local, host-searching, secondary flights among a crop after the initial landing.

The ability of thrips to exert a degree of choice on where to alight, notwithstanding the dependence of this on windspeed, may be reflected in their patterns of distribution in crops. By using sticky-coated green tile traps with a reflectance spectrum similar to soybean leaves, Irwin and Yeargan (1980) measured landing activity within soybean fields and on surrounding grass verges, over an entire growing season. The two soybean-colonizing species, *Neohydatothrips variabilis* and *Frankliniella tritici*, were more often caught within the field, whereas *Frankliniella fusca*, which does not infest soybeans, was more abundant in the grass verges. After immigration, each species redistributed differently on individual soybean plants, but the initial distribution across the field was fairly uniform (Fig. 5.6). By contrast, where field margins contain host plants, the crop itself may receive immigrants from these, creating an initial edge infestation (Lewis, 1973; Morishita and Azuma, 1989) which, for species capable of transmitting virus, may be reflected in the distribution of disease in the crop (Kobatake, 1984).

In regions where fields are surrounded by significant shelter from trees, hedges or even tall herbaceous plants, the wind shadow so formed encourages the accumulation of airborne thrips and also provides more frequent opportunities for them to alight than in the open field, thus producing heavier infestation around sheltered field margins (Fig. 5.7) (Lewis, 1970, 1973; Yudin *et al.*, 1991). Consequently, early immigrants or sparse populations are more likely to be detected on plants in sheltered areas, and gappy, uneven crops tend to be more heavily infested than uniform stands. Similarly, trees on the edges of plantations and orchards are often prone to heavier infestation than those within a stand, if the source of thrips is external (Zenther Møller, 1965). When thrips emerge from

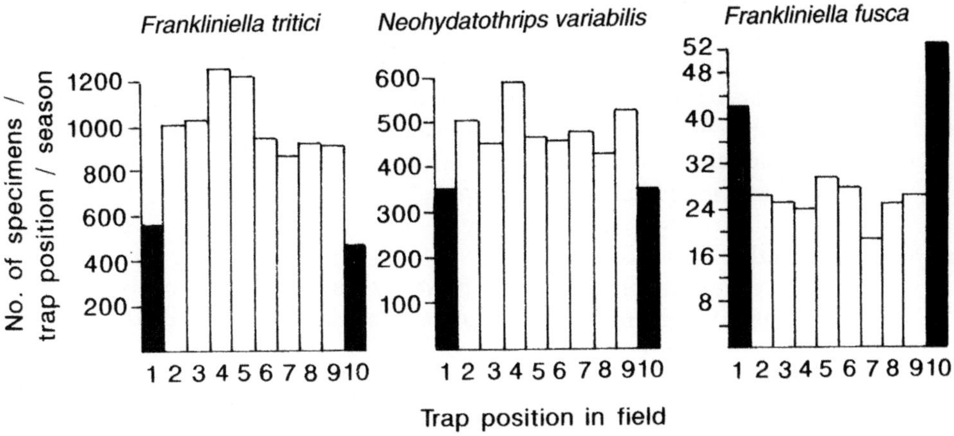

Fig. 5.6. Distribution of three species of thrips in a transect of a soybean field in Illinois. Columns represent total catches of each species on horizontal, green, sticky traps during the growing season. Solid columns represent traps placed outside the soybean crop in grass borders, open columns represent traps in the field (after Irwin and Yeargan, 1980).

soil within a plantation and fly upwards to colonize trees, as occurs with *Taeniothrips inconsequens* infesting sugar maple in spring in New England, edge effects are not apparent (Skinner and Parker, 1991). As a general rule, isolated trees and trees in hedgerows tend to be more heavily infested than trees within woodland, presumably because they are exposed to flying thrips from all

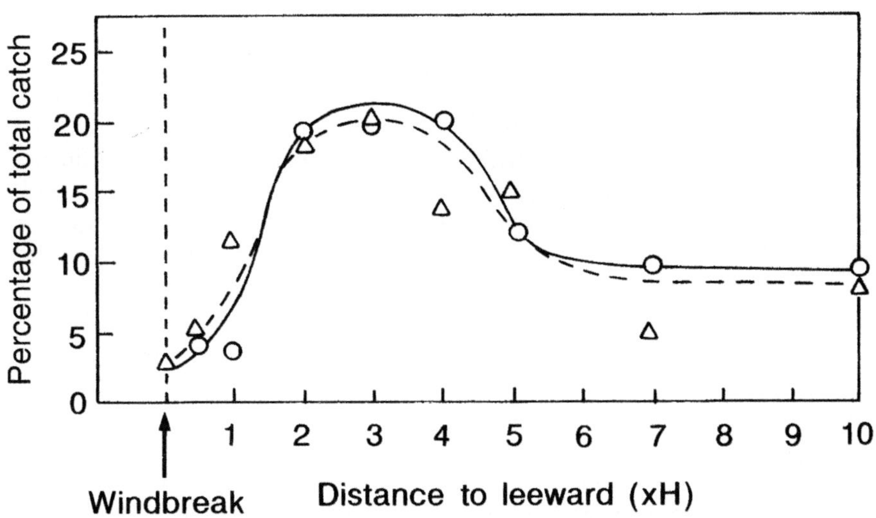

Fig. 5.7. The pattern of airborne *Limothrips cerealium* to leeward of a windbreak of pine trees, 20 m tall, compared with the population of wingless males that developed on the wheat crop 4 weeks later. o —— o airborne females; △ - - - △ males on crop; H, height of windbreak (after Lewis, 1970).

directions (Vité, 1956). Indeed, a single tree protruding above the general level of vegetation may 'sieve' many thrips from the surrounding airborne population. R. Laughlin (S. Australia, 1986, personal communication) estimated that 2000 *Thrips australis* arrived daily on an isolated, 3 m tall, gum tree. Numbers reached approximately 40,000 after 3 weeks. With a mean daily wind run past the tree of 65 km, an aerial density of immigrants as low as 0.0043 m^{-3} would have delivered this number of individuals to the canopy.

Artificial dispersal

The complex natural processes of dispersal by wind described above have been supplemented by commercial activity from the days of the early sailing ships carrying thrips-infested ballast and fodder along trading routes, though many of the species spread in this way were harmless spore-feeders (Mound, 1970, 1983). Among the species spread widely by land and sea routes, the gladiolus thrips, *Thrips simplex*, which probably originated in a Mediterranean climate, had reached Australia, North America, Argentina, Hawaii, Bermuda, Jamaica, South Africa and Europe on bulbs or roots of gladioli and carnations by the early twentieth century; similarly, the western European corn thrips, *Limothrips cerealium*, infests cereals in areas as climatically different from its native home as India and Hawaii (Morison, 1957).

With the development of ever more rapid transport systems and, nowadays, virtually daily long-distance distribution by aircraft of agricultural and horticultural produce and ornamentals, the consequent spread of thrips has expanded greatly the range and pest status of many species (Vierbergen, 1995). Pests of commercial crops frequently transported as fresh plants or cuttings, such as bananas and orchids, are easily carried concealed under bracts and in buds and leaf bases, whereas thrips infesting crops that are normally delivered as dried or processed leaves, such as tobacco and tea, or seeds, are far less likely to be dispersed artifically (Fig. 5.8) (Mound, 1983). For similar reasons, grass, cereal and sugar-cane dwellers hidden in hay, straw or stems are also widespread, as are many polyphagous species, notably *Thrips tabaci* and *Heliothrips haemorrhoidalis*, which can tolerate a broad range of climatic requirements and which, on arrival in temperate countries, are often inadvertently provided with favourable semitropical or tropical environments within glasshouses.

It is extremely difficult to detect adults, larvae or eggs in bulky shipments of plant material and, even if found, specimens are not easily identifiable by non-specialists. Furthermore, their habits and biological characteristics, including embedding the eggs in plant tissue, rapid reproduction often by parthenogenesis, frequent polyphagy, and pupation by some species in soil, all militate against easy detection and facilitate rapid establishment of foci from which further artificial or natural spread may occur. The scale of this dissemination through international trade is well illustrated by the list of 43 terebrantian species and 12 tubuliferan species intercepted at ports of entry into The Netherlands between 1980 and 1993, mainly on ornamental plants imported from some 30 countries (Vierbergen, 1995). Though not all species have the potential to develop into pests, nor are all the countries of origin unequivocally

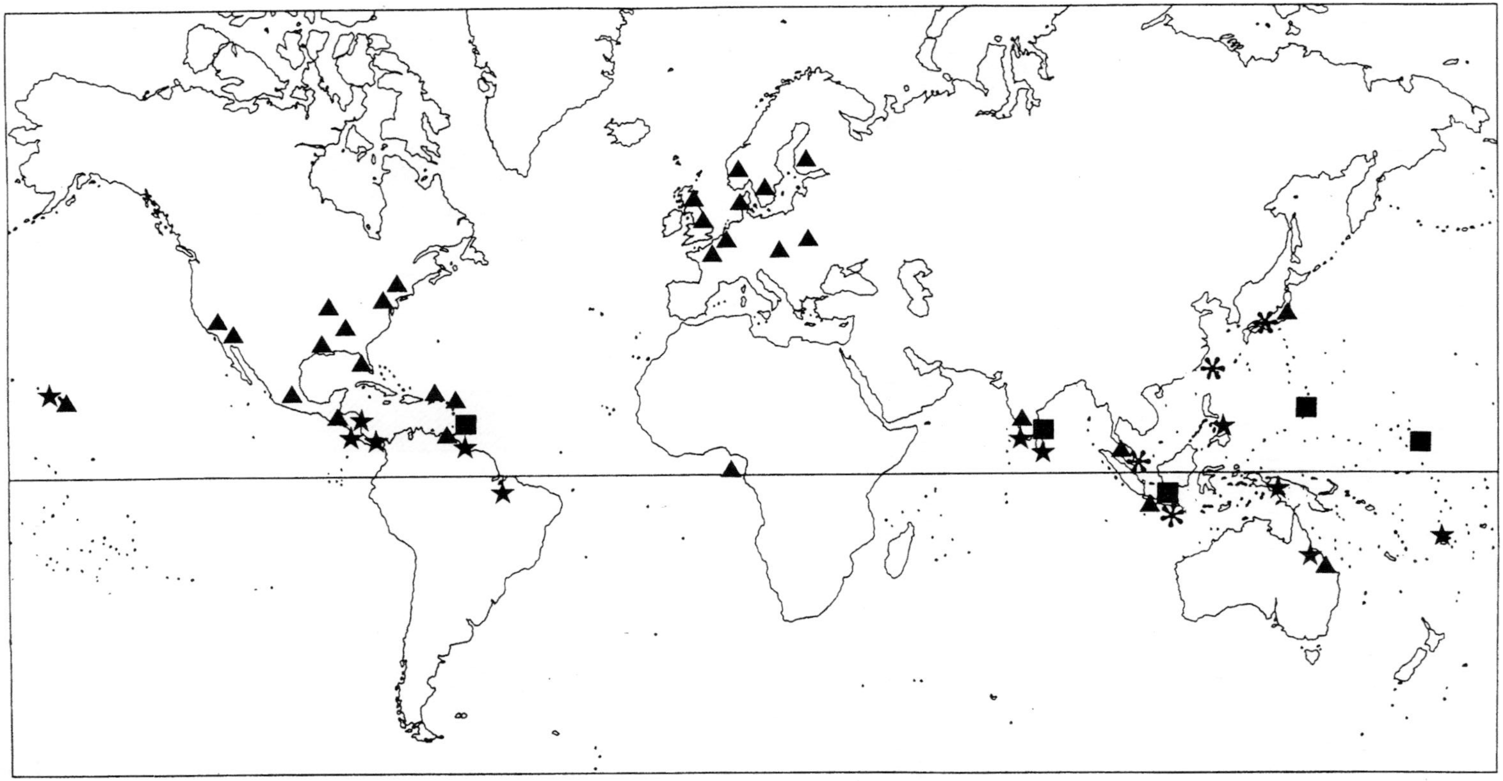

Fig. 5.8. The relatively widespread distribution of *Chaetanaphothrips* species: ■ *leeuweni*, ▲ *orchidii* and ★ *signipennis* on orchids and bananas, compared with the restricted distribution of ✱ *theiperdus* on tea (after Mound, 1983).

confirmed, the abundant opportunities for pest spread by this means are apparent. Quarantine officers in Japan have similarly recorded 18 species of pest thrips on imported produce in recent years (Hayase, 1991). The dynamic nature of this problem is emphasized in Fig. 5.9 which shows the changing proportions of three of the most commonly found species in thrips-infested consignments imported into The Netherlands between 1987 and 1993 (Vierbergen, 1995); over these 6 years, appearances of *Frankliniella occidentalis* have tripled, whereas *Thrips tabaci* and *Frankliniella schultzei* have remained relatively steady. Elsewhere, *Frankliniella occidentalis* has spread extremely rapidly since the mid-1980s and now occurs almost worldwide except in large parts of tropical Africa and South America.

Similar artificial dispersal has probably been largely responsible for the rapid spread of *Thrips palmi*, which has become a serious pest of cucurbits and solanaceous vegetables and has gained a foothold in many tropical regions over the last 15 years. Since its discovery in Japan in 1978 it has invaded many Pacific Islands, including Hawaii, and has been reported from northern Australia, the Caribbean, Guyana, Florida, Mauritius, Reunion, the Canary Islands and Sudan. It may appear soon in temperate glasshouses, having already been intercepted on cut flowers imported into Finland, to become another major cosmopolitan pest (Mound *et al.*, 1992) that causes damage and

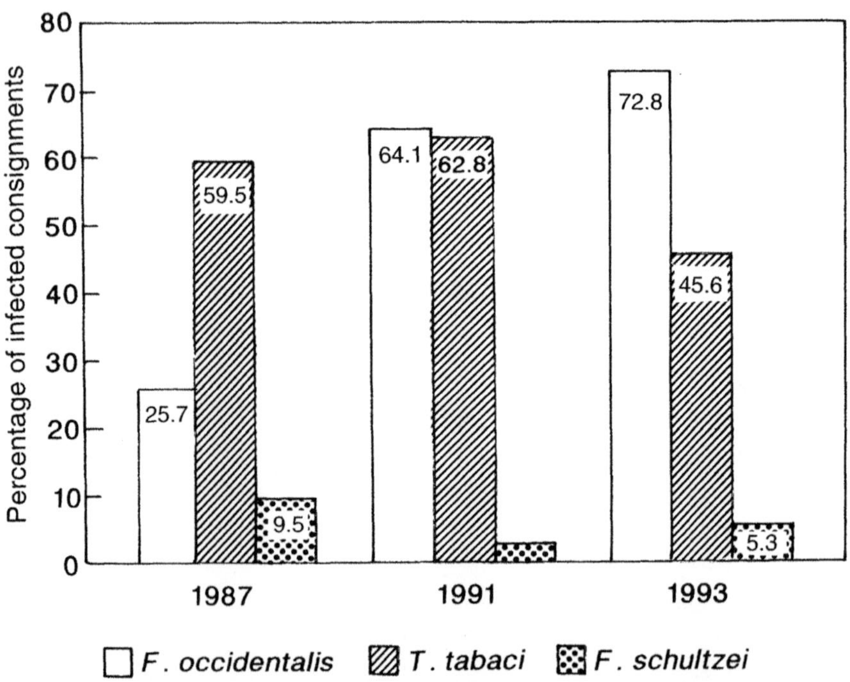

Fig. 5.9. Occurrence of three pest species in thrips-infested consignments of flowers intercepted at port of entry in The Netherlands (after Vierbergen, 1995).

raises production costs for growers comparable with those resulting from outbreaks of *Frankliniella occidentalis*. The spread of these two species underlines both the difficulty of preventing artificial infestations in the course of the modern trade in plants and fresh commodities, and the unusual ability of thrips compared with larger, less secretive, insects to exploit these man-made opportunities.

Whilst the recent spate of new pest problems has arisen largely from the accidental introduction of species to new areas and their adaptation to locally grown crops or cropping systems, there are a few instances of new crops that have been introduced into regions being attacked by endemic species. For example, *Taeniothrips orionis*, whose normal hosts in Alaska are mountain lupin and columbine, damages lettuces, potatoes and cabbages planted in freshly cleared areas (Washburn, 1958). In Hungary, the natural host of *Kakothrips pisivorus* is probably *Lathyrus tuberosus*, a plant with an extended flowering period that affords the larvae food and shelter long enough for most of them to complete development there. It has become a pest on peas and beans (*Fabaceae*) that were introduced from Asia long ago, and have a much shorter flowering period than *Lathyrus*, forcing the larvae to concentrate on the developing pods in damaging numbers (G. Jenser, Budapest, 1995, personal communication).

References

Appanah, S. (1979) The ecology of insect pollination of some tropical rain-forest trees. PhD thesis, University of Malaya, Kuala Lumpur.

Bent, G. (1982) The immediate extraction and display of insect flight trajectories from infra-red remote signals. PhD thesis, Cranfield Institute of Technology, Bedford, UK.

Brødesgaard, H.F. (1990) The effect of anisaldehyde as a scent attractant for *Frankliniella occidentalis* (Thysanoptera: Thripidae) and the response mechanism involved. *Bulletin Organisation Internationale de Lutte Biólogique Section Regional Ouest Palearctic* 8, 36–38.

Cheng, L. and Birch, M.C. (1977) Terrestrial insects at sea. *Journal of the Marine Biological Association, UK* 57, 995–997.

Close, R.A., Moar, N.T., Tomlinson, A.I. and Lowe, A.D. (1978) Aerial dispersal of biological material from Australia to New Zealand. *International Journal of Biometeorology* 22, 1–19.

Cuthbertson, D.R. (1989) *Limothrips cerealium*, an alarming insect. *Entomologist* 108, 246–256.

Czenz, K. (1987) The role of coloured traps in collecting thrips fauna. In: Holman, J., Pelikan, J., Dixon, A.F.G. and Weisman, L. (eds) *Population Structure, Genetics and Taxonomy of Aphids and Thysanoptera.* SPB Academic Publishing, The Hague, pp. 426–435.

Ellington, C.P. (1980) Wing mechanics and take-off preparation of *Thrips* (Thysanoptera). *Journal of Experimental Biology* 85, 129–136.

Felland, C.M., Goulart, B. and Hellerick, B. (1994) Strawberry thrips plague of 1994. *Proceedings of the 70th Cumberland-Shenaendoah Fruit Workers Conference, Winchester, Virginia, 1994*, Rutgers University, Cream Ridge, New Jersey, pp. 1–4.

Foster, S.W. and Jones, P.R. (1915) The life history and habits of the pear thrips in California. *Bulletin of the US Department of Agriculture* 173, 52 pp.

Freitag, R. and Ryder, J.P. (1973) An annotated list of arthropods collected from ring-billed gull nests on Granite Island, Black Bay, Lake Superior, 1972 and 1973. *Proceedings of the Entomological Society of Ontario* 104, 38–46.

Glick, P.A. (1939) The distribution of insects, spiders and mites in the air. *Technical Bulletin US Department of Agriculture* 1222, 28 pp.

Gressitt, J.L., Leech, R.E. and O'Brian, C.W. (1960) Trapping of airborne insects in the Antarctic area. *Pacific Insects* 2, 245–250.

Hardy, A.C. and Milne, P.S. (1938) Studies on the distribution of insects by aerial currents. Experiments in aerial tow-netting from kites. *Journal of Animal Ecology* 7, 199–229.

Hayase, T. (1991) A list of Thysanoptera intercepted in plant quarantine. *Research Bulletin of Plant Protection, Japan* 27, 93–99.

Holzapfel, E.P. and Harrell, J.C. (1968) Transoceanic dispersal studies of insects. *Pacific Insects* 10, 115–153.

Hurst, G.W. (1964) Effects of weather conditions on thrips activity. *Agricultural Meteorology* 1, 130–141.

Irwin, M.E. and Yeargan, K.V. (1980) *Sampling Methods in Soybean Entomology.* Springer-Verlag, New York, pp. 283–303.

Jenser, G. (1973). Observations on the autumn mass flight of *Frankliniella intonsa* Trybom (Thysanoptera, Thripidae). *Acta Phytopathologica Academiae Scientiarum Hungaricae* 8, 227–230.

Johnson, C.G. (1969) *Migration and Dispersal of Insects by Flight.* Methuen, London.

Kirk, W.D.J. (1984) Ecologically selective coloured traps. *Ecological Entomology* 9, 35–41.

Kirk, W.D.J. (1985) Effects of some floral scents on host finding by thrips (Insecta: Thysanoptera). *Journal of Chemical Ecology* 11, 35–43.

Kittel, R. (1958) Untersuchungen über den Einfluss atmosphärishelectrischer Felder auf die Schwarmbildung bei Thysanopteren. *Verhandlungen der Deutschen zoologischen Gesellschaft* 67, 177–181.

Kobatake, H. (1984) Ecology and control of spotted wilt-disease of tomato in Nara Prefecture. (In Japanese.) *Proceedings of the Kansai Plant Protection Society* 26, 23–28.

Körting, A. (1931) Beobachtung über die Fritfliege und einiger Getreidethysanopteren. *Zeitschrift für angewandte Entomologie* 18, 154–160.

Kuethe, A.M. (1975) On the mechanics of flight of small insects. In: Wu, T.Y., Brokoew, C.J. and Brenne, C. (eds) *Swimming and Flying in Nature*, Vol. 2. Plenum, New York, pp. 803–813.

Laughlin, R. (1977) The gum tree thrips *Isoneurothrips australis* Bagnall. Survival at different temperatures and humidities and its relation to capacity for dispersal. *Australian Journal of Ecology* 2, 391–398.

Lewis, T. (1958) The distribution and dispersal of Thysanopteran populations on *Gramineae.* PhD thesis, University of London, UK.

Lewis, T. (1962) The effects of temperature and relative humidity on mortality in *Limothrips cerealium*, Haliday (Thysanoptera) overwintering in bark. *Annals of Applied Biology* 50, 313–326.

Lewis, T. (1963) The effects of weather on emergence and take off of overwintering *Limothrips cerealium*, Haliday (Thysanoptera). *Annals of Applied Biology* 53, 489–502.

Lewis, T. (1964) The weather and mass flights of Thysanoptera. *Annals of Applied Biology* 53, 165–170.

Lewis, T. (1965) The species, aerial density and sexual maturity of Thysanoptera caught in mass flights. *Annals of Applied Biology* 55, 219–225.

Lewis, T. (1970) Patterns of distribution of insects near a windbreak of tall trees. *Annals of Applied Biology* 65, 213–220.

Lewis, T. (1973) *Thrips, their Biology, Ecology and Economic Importance*. Academic Press, London.

Matteson, N.A. and Terry, L.I. (1992) Response to colour by male and female *Frankliniella occidentalis* during swarming and non-swarming behaviour. *Entomologia Experimentalis et Applicata* 63, 187–201.

Maw, M.G. (1962) Some biological effects of atmospheric electricity. *Proceedings of the Entomological Society of Ontario* 92, 33–37.

Maw, M.G. (1963) Behaviour of insects in electrostatic fields. *Proceedings of the Entomological Society of Manitoba* 18, 30–36.

Morishita, M. and Azuma, K. (1989) Seasonal fluctuations of *Thrips palmi* Karny (Thysanoptera: Thripidae) on aubergine in vinyl houses and fields. *Kansai Byochugai Kenkykaido*, No. 31, 1–5.

Morison, G.D. (1957) A review of British glasshouse Thysanoptera. *Transactions of the Royal Entomological Society of London* 109, 467–534.

Morison, G.D. (1973) Observations and records for British Thysanoptera XVI. Thripidae, *Apterothrips secticornis* (Trybom), a species new to Britain. *Entomologists' Monthly Magazine* 109, 208–210.

Moritz, G. (1989) The ontogenesis of Thysanoptera (Insecta) with special reference to the Panchaetothripinae, *Hercinothrips femoralis* (O.M. Reuter, 1891) (Thysanoptera, Thripidae, Panchaetothripinae). *Zoologische Jahrbücher Anatomie* 118, 393–429.

Mound, L.A. (1970) Thysanoptera from the Solomon Islands. *Bulletin of the British Museum (Natural History), Entomology* 24, 83–126.

Mound, L.A. (1983) Natural and disrupted patterns of geographical distribution in Thysanoptera (Insecta). *Journal of Biogeography* 10, 119–133.

Mound, L.A. and Marullo, R. (1996) The thrips of Central and South America: an introduction. *Memoirs on Entomology, International* 6, 1–448.

Mound, L.A. and Walker, A.K. (1982). Terebrantia (Insecta: Thysanoptera). *Fauna of New Zealand* 1, 113 pp.

Mound, L.A., duHeaume, F.J. and Walker, A.K. (1992) Thrips palmi: *a literature survey with annotated bibliography.* International Institute of Biological Control, Ascot, 37 pp.

Murai, T. (1988) Studies on the ecology and control of flower thrips *Frankliniella intonsa* (Trybom). *Bulletin of the Shimane Agricultural Experiment Station* 23, 1–73.

Pedgley, D.E. (1982) *Windborne Pests and Diseases*. Ellis Horwood, Chichester.

Preisner, H. (1964) A monograph of the Thysanoptera of the Egyptian deserts. *Publications de l'Institut du Désert d'Egypte* 13, 549 pp.

Riley, J.R. (1989) Remote sensing in entomology. *Annual Review of Entomology* 34, 247–271.

Riley, J.R. and Reynolds, D. (1993) Radar monitoring of locusts and other migratory insects. In: Cartwright, A. (ed.) *World Agriculture 1993*. Sterling Publications, London, pp. 51–53.

Skinner, M. and Parker, B.L. (1991) Bioecology of pear thrips: distribution in forest soils. In: Parker, B.L., Skinner, M. and Lewis, T. (eds) *Towards Understanding Thysanoptera*. US Department of Agriculture Forest Service, Radnor, PA, pp. 193–211.

Stannard, L.J. (1968) The thrips or Thysanoptera of Illinois. *Bulletin of the Illinois State Natural History Survey* 29, 215–552.

Strauss, S.Y. and Karban, R. (1995) Selection, dispersal and mode of reproduction: attributes contributing to local adaptation by *Apterothrips apteris* to *Erigeron glaucus*. In Parker, B.L., Skinner, M. and Lewis, T. (eds) *Thrips Biology and Management*. Plenum, New York, pp. 537–544.

Taylor, L.R. (1958) Aphid dispersal and diurnal periodicity. *Proceedings of the Linnean Society, London* 169, 67–73.

Terry, L.I. and Gardner, D. (1990) Male mating swarms in *Frankliniella occidentalis*, a terebrantian thrips. *Journal of Insect Behaviour* 3, 133–141.

Teulon, D.A.J. (1988) Pest management of the New Zealand flower thrips (*Thrips obscuratus* (Crawford) Thysanoptera: Thripidae) on stonefruit in Canterbury, New Zealand. PhD thesis, University of Canterbury, New Zealand.

Teulon, D.A.J. and Penman, D.R. (1992) Colour preferences of New Zealand thrips (Terebrantia: Thysanoptera). *New Zealand Entomologist* 15, 8–13.

Vierbergen, G. (1995) International movement, detection and quarantine of Thysanoptera pests. In: Parker, B.L., Skinner, M. and Lewis, T. (eds) *Thrips Biology and Management*. Plenum, New York, pp. 119–132.

Vité, J.P. (1956) Populations studien am Lärchenblasenfuss *Taeniothrips laricivorus* Krat. *Zeitschrift für angewandte Entomologie* 38, 417–488.

Washburn, R.H. (1958) *Taeniothrips orionis* Leh., a thrips destructive to vegetables in Alaska. *Journal of Economic Entomology* 51, 274.

Yoshimoto, C.M., Gressitt, J.L. and Mitchell, C.J. (1962) Trappings of airborne insects in the Pacific-Antarctic area 1. *Pacific Insects* 4, 847–858.

Yudin, L.S., Tabashnik, B.E., Mitchell, W.C. and Cho, J.J. (1991) Effects of mechanical barriers on distribution of thrips (Thysanoptera: Thripidae) in lettuce. *Journal of Economic Entomology* 84, 136–139.

Zenther-Møller, O. (1965) *Taeniothrips laricivorus* (Krat.) (Thripidae, Thysanoptera) in Danish stands of *Larix decidua* (Mill.). *Oikos* 16, 58–69.

1

2

Plate I

1. *Frankliniella occidentalis*, life stages. From left to right: egg at red-eye stage, larva I, larva II, propupa, pupa, adult male, adult female. Thrips from colonies of Diane Ullman. Photos and digital composite by Jack K. Clark, Davis, California.

2. 'Bean pod'and 'cucumber' rearing method for *Frankliniella* spp. *Thrips tabaci* can be reared similarly on french beans, leek or onion shoots, or cabbage. (A.J.M. Loomans.)

3. *Heliothrips haemorrhoidalis* rearing on outer skin of silver-green lemons. (A.J.M. Loomans.)

3

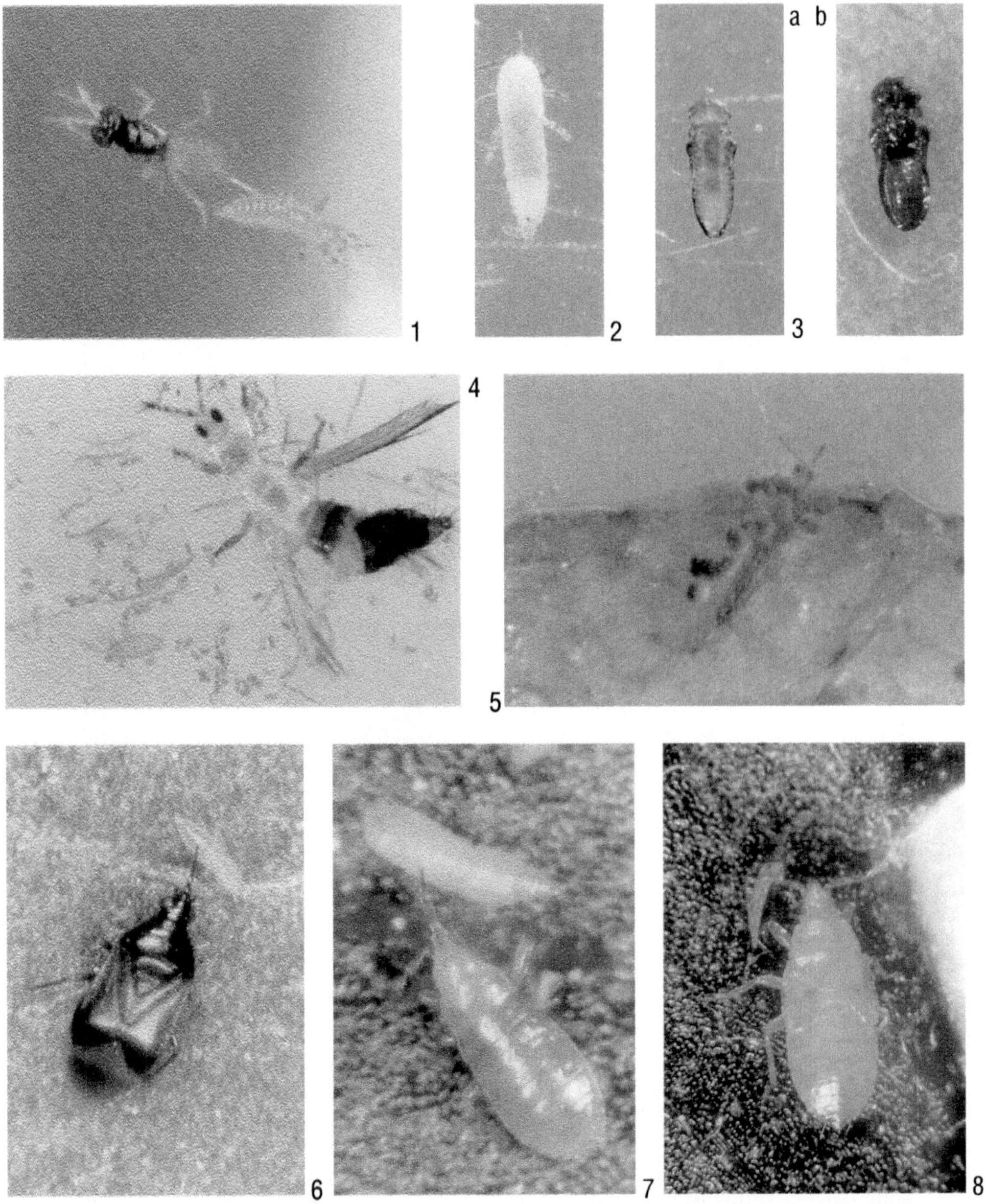

Plate II

1. *Ceranisus menes* stinging *Frankliniella occidentalis* larva. (A.J.M. Loomans.)

2. Larva of *F. occidentalis* parasitized by *Ceranisus menes.* (A.J.M. Loomans.)

3a, b Pupae of *F. occidentalis* parasitized by *C. menes.* (A.J.M. Loomans.)

4. *F. occidentalis*, ruptured abdomen liberating all stages of parasitic nematode *Thripinema nickelwoodii*; eggs, vermiform larvae and infective stage adult females (translucent, eggs visible within). (I.D. Greene.)

5. *F. occidentalis* infected and killed by *Metarhizium anisopliae.* (S. Vestergaard.)

6. *Orius* sp. adult attacking *F. occidentalis* larva. (D. Grasselly.)

7. *Orius* sp. nymph. attacking *F. occidentalis* larva (D. Grasselly.)

8. *Orius laevigatus* nymph attacking *F. occidentalis* adult. (G. Del Bene.)

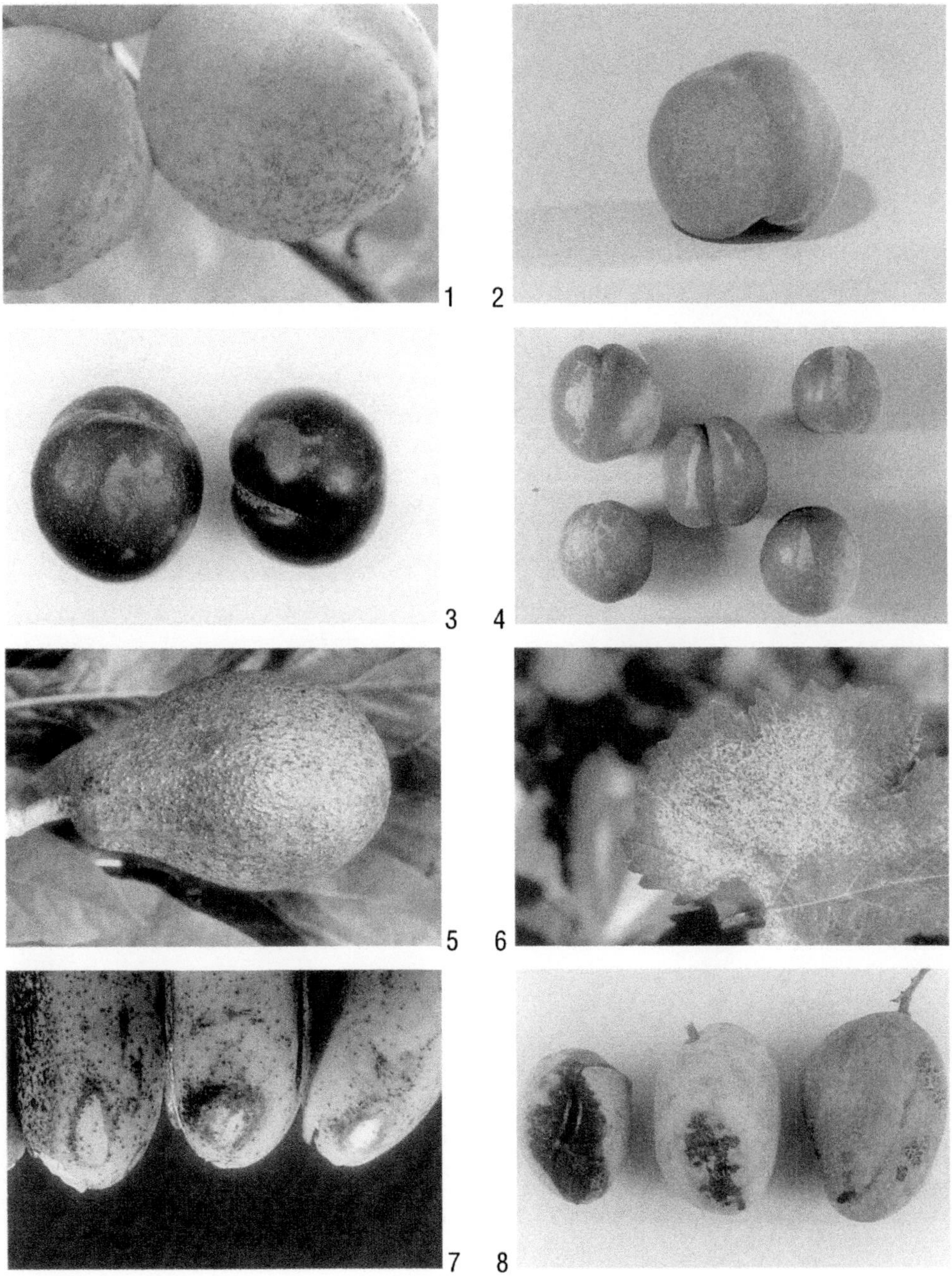

Plate III

1. *Thrips obscuratus*, heavy infestation on peaches. (D.A.J. Teulon).

2. *Thrips obscuratus,* silvering on peach. (D.A.J. Teulon.)

3. *Frankliniella occidentalis*, skin blemish on plums. (M. Wysoki.)

4. *Thrips obscuratus,* silvering on nectarines. (D.A.J. Teulon.)

5. *Heliothrips haemorrhoidalis*, rind blemish on avocado. (M. Wysoki.)

6. *H. haemorrhoidalis*, silvered vine leaf. (P. Del Estal.)

7. *Chaetanaphothrips orchidii*, elongated russetted patches at touch points on bananas. (M. Wysoki.)

8. *F. occidentalis*, scarred and deformed mangoes. (M. Wysoki.)

Plate IV

1. *Frankliniella occidentalis*, degrees of bronzing on strawberry flowers. (D. Grasselly.)

2. *F. occidentalis*, severely bronzed strawberries. (K. Lynch and R. Tremblay.)

3. *F. occidentalis*, moderately bronzed strawberries. (K. Lynch and R. Tremblay.)

4. *F. occidentalis*, close-up of bronzed strawberry surface. (R. Marullo.)

5. *F. occidentalis*, blemished tomatoes. (R. Marullo.)

6. *F. occidentalis*, scarred aubergines. (D. Grasselly.)

7. *F. occidentalis*, blemished table grapes. (R. Marullo.)

8. *F. occidentalis*, damage leading to split fruit in table grapes. (R. Marullo.)

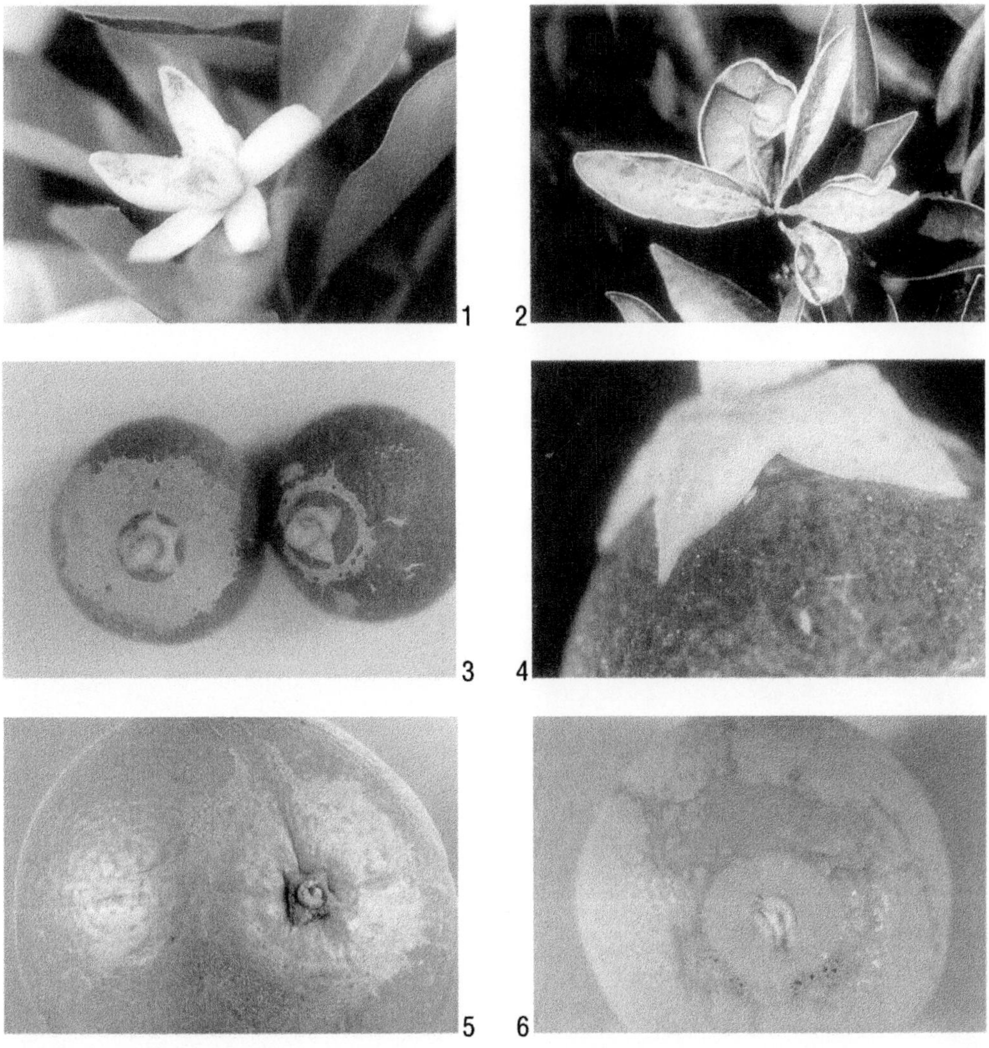

Plate V

1. *Frankliniella bispinosa* in navel orange flower. (C.C. Childers.)

2. *Scirtothrips citri,* injury to orange leaves. (J.G. Morse.)

3. Severe rind blemish on Valencia late fruitlets by *S. inermis.* (A. Lacasa.)

4. *S. citri* larvae beneath sepals on young fruit. (J.G. Morse.)

5. *S. citri,* blemished orange. (J.G. Morse.)

6. *S. inermis,* blemished Thomson's navel orange. (A. Lacasa.)

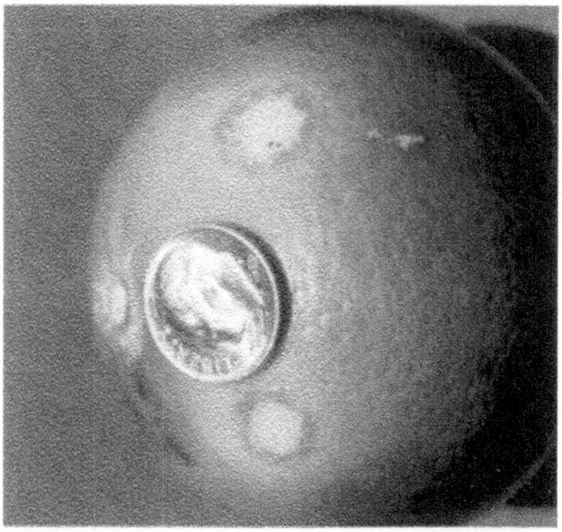

1

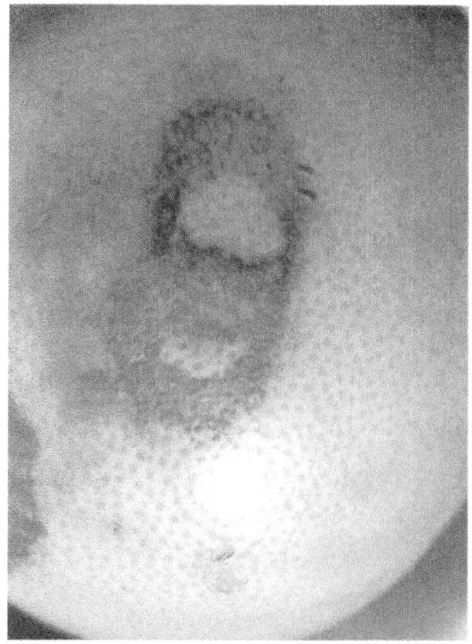

2

3

4

Plate VI

Chaetanaphothrips orchidii, ring spot damage at 'touch points' on red grapefruit cultivars. (C.C. Childers.)

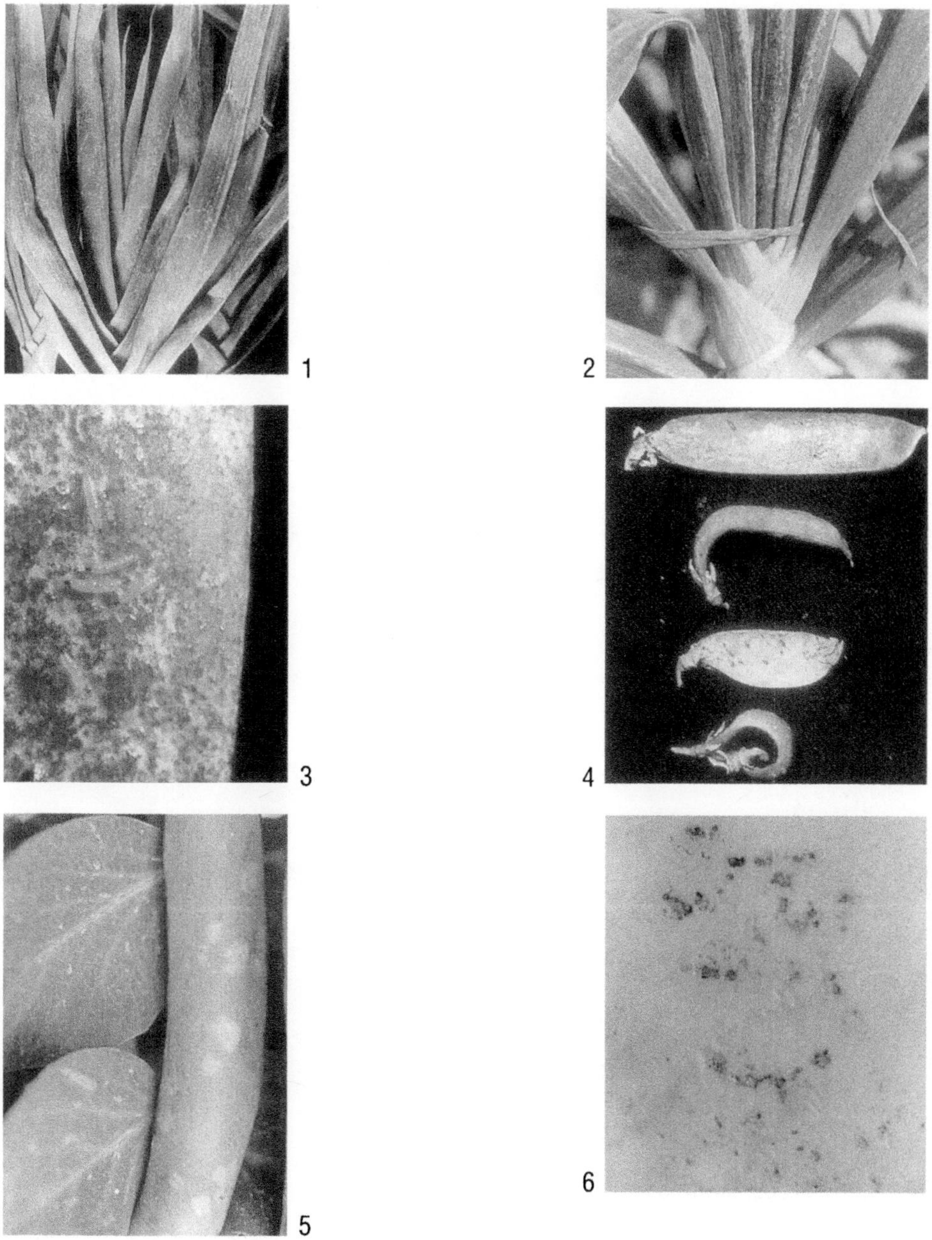

Plate VII

1. *Thrips tabaci*, damage to leek leaves. (H. Legutowska.)
2. *T. tabaci*, damage to onion leaves. (A. Lacasa.)
3. *Kakothrips pisivorus* feeding on pea pod. (T. Lewis.)
4. Distortion due to *K. pisivorus* feeding. (T. Lewis.)
5. *Frankliniella occidentalis,* blotching on pea pods. (A. Lacasa.)
6. *T. tabaci,* feeding blemishes on cabbage head. (H. Legutowska.)

1

2

3

4

5

6

Plate VIII

1. *Thrips angusticeps*, bronzed and distorted sugar beet seedlings. (A. Dewar.)
2. *T. angusticeps*, distortion on young field lupins. (A. Ferguson.)
3. *T. angusticeps*, distorted (L) and healthy (R) shoots of linseed. (A. Ferguson.)
4. *Taeniothrips inconsequens*, partially defoliated canopy of sugar maple. (M. Skinner.)
5. *Taeniothrips inconsequens*, stunted and normal leaves of sugar maple. (M. Skinner.)
6. *Scirtothrips mangiferae*, russetting, necrosis and distortion of mango leaves. (M. Wyoski.)

Plate IX

1. *Corynothrips stenopterus* causing bunchy shoot on cassava, Colombia. (T. Lewis.)
2. *Frankliniella occidentalis,* damaged cucumber foliage. (R. Jacobson.)
3. *F. occidentalis,* distorted young cucumber fruit. (R. Jacobson.)
4. *F. occidentalis,* subsequent growth to produce 'pig-tail' fruit. (R. Jacobson.)
5. *Thrips palmi* scarring and distortion of bitter gourd. (G.V. Pollard.)
6. *T.palmi,* scarred leaves and fruit of sweet pepper. (G.V. Pollard.)
7. *T.palmi,* damaged tip of melongene fruit. (G.V. Pollard.)
8. *T.palmi,* distorted melongene leaves. (G.V. Pollard.)

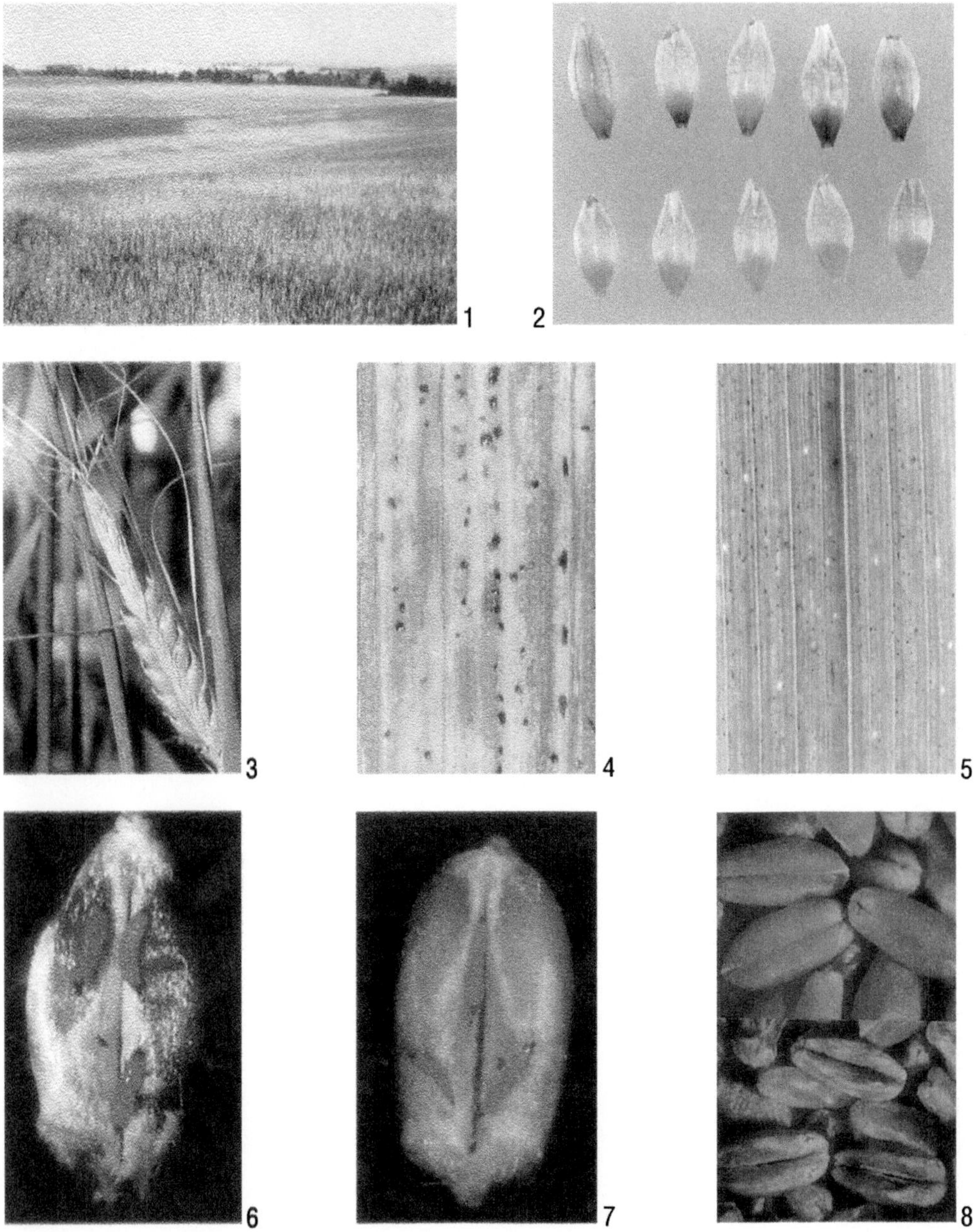

Plate X

1. *Limothrips denticornis* causing severe 'whiteheads' on barley, Idaho. (N. Matteson.)
2. *L. cerealium*, damaged (top row) and healthy barley grains. (T. Lewis.)
3. *L. cerealium*, 'whiteheads' on barley. (T. Lewis.)
4. Larval feeding lesions and faeces on wheat leaf. (T. Lewis.)
5. Eggs (white specks) embedded in wheat leaf. (T. Lewis.)
6. *Haplothrips tritici*, larvae feeding on milky wheat grain. (P. Bielza.)
7. *H. tritici*, larvae on swollen grain. (P. Bielza.)
8. *H. tritici*, healthy (upper) and injured grains. (P. Bielza.)

Plate XI

Feeding damage by *Frankliniella occidentalis* on blooms.

1. *Verbena.* (W.J.D. Kirk.)
2. *Verbena.* (W.J.D. Kirk.)
3. *Verbena.* (R. Jacobson.)
4. *Saintpaulia.* (M. Tatchell.)
5. New Guinea *Impatiens.* (R.D. Oetting.)
6. *Gerbera.* (R.D. Oetting.)
7. Carnations. (A. Lacasa.)

Plate XII

Symptoms of TSWV.

1. Severely infected romain lettuce. (J. Cho.)

2. Severely infected romain lettuce. (J. Cho.)

3. Dahlia. (M. Daughtrey.)

4. Groundnut leaves and shoots. (J.L. Sherwood.)

5. Infected (left-hand side) and healthy (right-hand side) groundnuts. (J.L. Sherwood.)

6. Sweet peppers. (J. Cho.)

7. Tomato fruit. (J. Cho.)

8. INSV in New Guinea *Impatiens* (right-hand side) and control (left-hand side). (M. Daughtrey.)

9. INSV in *Cineraria*. (M. Daughtrey.)

6 Biological Diversity

Laurence A. Mound[1]
Division of Entomology, CSIRO, GPO Box 1700, Canberra, ACT 2601, Australia

Introduction

The purpose of this chapter is to discuss the evolutionary radiation of the Thysanoptera, their systematic and biological diversity including numbers of species and feeding associations, and to consider from this evolutionary perspective what makes a few thrips species so effective as crop pests.

Systematic Diversity

In the course of their evolution thrips are presumed to have shared a common ancestor with the Hemiptera (the bugs), the Psocoptera (the booklice), and the Phthiraptera (the sucking lice) (Kristensen, 1991). The lifestyle of this ancestral group was presumably as detritivores, and the earliest thrips presumably specialized in obtaining liquid food from decaying tissues and fungal hyphae. In discussing the relationships between thrips and crop production it is worth remembering this evolutionary history, because inherited ancestral characteristics both preadapt and limit the pest potential of insects.

Although plant detritus is widely available, it is not uniform in content or quality. The processes of fungal and microbial decay produce rapid changes in microhabitats such that optimum conditions can be relatively brief, sometimes persisting for only a few days judging from changes observed in thrips populations in leaf litter. To succeed under such rapidly changing conditions, several characteristics are likely to be useful to an insect species, such as vagility, a moderately broad food tolerance, a short generation time, a predisposition to parthenogenesis, and possibly a competitive breeding structure that promotes aggregation and thus exploitation of localized optimal conditions. These

[1] Prepared during the tenure of a CSIRO McMaster Fellowship at Canberra, Australia.

biological characteristics are well developed amongst Thysanoptera, and most of them are equally advantageous to these insects in exploiting the relatively transient optimal conditions provided by a growing crop. Thus thrips pests are typified by their ability to invade a wide range of crops and to develop large populations rapidly. The additional ability to act as disease vectors has no evident function within their evolutionary biology, nor does it confer any advantage to the thrips. Indeed, the relatives of the plant viruses carried by thrips are primarily parasites of animal tissues, and tomato spotted wilt virus (TSWV) has probably become a plant pathogen only relatively recently (Goldbach and Peters, 1994).

The primitive detritivore Thysanoptera possibly evolved on to higher plants via senescing leaves, the habitat of many present-day panchaetothripine thrips (e.g. *Heliothrips*). Subsequently, thrips diversified on to flowers and green leaves, but several reversions to fungus feeding have taken place, such that at least half of the present-day Thysanoptera species feed either on fungal hyphae or on fungal spores. There is no evidence that thrips evolved through a phase of feeding on lower plants such as ferns and mosses, although several advanced lineages have radiated on to such plants to a limited extent (Mound and Marullo, 1996).

The thrips species that are most important as crop pests belong to the two genera *Thrips* and *Frankliniella*, and these, together with *Haplothrips* and *Liothrips*, are the largest genera in the Thysanoptera. These four are the only genera of thrips that each include more than 150 species, the approximate number of species currently in each being: *Thrips* – 275, *Liothrips* – 255, *Haplothrips* – 230, and *Frankliniella* – 175. Only one other genus of plant-feeding thrips, *Aeolothrips*, includes more than 100 species, although two genera of fungus-feeding species each include up to 140 species. There are nine genera each with between 50 and 100 species, 95 genera each with 10 to 49 species, 150 genera each with 3 to 9 species, and 450 genera each with only 1 or 2 species. It is interesting that pest species occur randomly in genera of all sizes, from the smallest to the largest, suggesting that the biological characteristics that predispose a species to becoming a pest are widespread in the group.

Variation in biologies between major groups

Currently the 5000 known species of Thysanoptera are classified into eight families (Mound *et al.*, 1980). However, 93 per cent of these species are placed in either the Thripidae or the Phlaeothripidae (Table 6.1), and it is members of these two families that are normally found on crops. The species of Merothripidae are all fungus feeding; they are of interest because they retain the largest number of primitive features of any thrips and presumably represent the closest condition to the ancestral form. The single species in the Uzelothripidae is also fungus feeding. Little is known about the biologies of the four members of the Fauriellidae, but judging from the available collections, four of the five species in the family Adiheterothripidae are associated with the flowers of date palms (*Phoenix dactylifera*) between the Mediterranean and India. The family Heterothripidae includes 70 species from the Americas, where they have been described from the flowers of a wide range of native plants. More commonly

Table 6.1. Thysanoptera family-group names and numbers of species.

Suborders	Families	Subfamilies	Total species
Terebrantia	Uzelothripidae	–	1
	Merothripidae	–	15
	Aeolothripidae	Aeolothripinae	210
		Melanthripinae	50
	Adiheterothripidae	–	5
	Fauriellidae	–	4
	Heterothripidae	–	70
	Thripidae	Thripinae	1400*
		Panchaetothripinae	120*
		Dendrothripinae	70*
		Sericothripinae	120*
Tubulifera	Phlaeothripidae	Phlaeothripinae	2500*
		Idolothripinae	600

* indicates presence of pest species.

encountered on crops, at least in temperate regions, are the 250 species of the worldwide family Aeolothripidae. These are recognized by their forewings which are broad with the apex rounded and usually with black and white transverse bands. The species in this family are not usually considered as pests, but they show a range of feeding habits, from purely phytophagous amongst the less advanced species, such as the 30 species of *Melanthrips* in flowers in Europe, to fully predacious amongst the species of the small genera *Franklinothrips* and *Mymarothrips* that are found on tropical trees. Between these extremes in biologies lie the members of *Aeolothrips*, a genus of about 100 flower-living species in the northern temperate region, many of which appear to be facultative predators, feeding on pollen as well as other arthropods.

Of the two major families in the Thysanoptera, the Phlaeothripidae is the largest with about 3100 described species. About 600 of these, constituting the subfamily Idolothripinae, feed only on fungal spores, but at least half of the species in the second subfamily, Phlaeothripinae, feed only on fungal hyphae. Fungus-feeding thrips live on dead wood or in leaf litter, particularly in tropical countries, and apparently have little or no direct impact on crop production, even as vectors of plant pathogens. However, within the Phlaeothripinae there are two major lineages of plant-feeding species. One lineage includes the many species that feed mainly on the green leaves of tropical trees and shrubs, often inducing galls. The largest genus in this lineage is *Liothrips*, but the species seem to be mainly host specific to particular trees and shrubs and few are known as crop pests. In the tropics, pepper vines and avocado trees are each damaged by particular species of *Liothrips*, as are certain *Liliaceae* and *Orchidaceae* in greenhouses. Also in this lineage is the tropical genus

Gynaikothrips which includes one species important to the horticultural trade because it induces leaf galls on decorative *Ficus* trees (Mound *et al.*, 1996b).

The other major lineage of plant-feeding Phlaeothripinae is the group of genera related to *Haplothrips*, many of which live in flowers, particularly of *Compositae* and *Gramineae*. These thrips are sometimes found associated with crops, often living on weeds. Few are important as pests, although on various cereal crops between Europe and the Philippines *Haplothrips tritici*, *H. aculeatus* or *H. chinensis* are sometimes abundant. Predation, on mite and lepidopteran eggs as well as on the immature stages of scale insects and whitefly, is a lifestyle that has been adopted by several different groups of Phlaeothripinae, and some of these species are locally important as beneficial insects. Finally, at least one group of phlaeothripines has evolved the habit of feeding on mosses (Mound, 1989).

Most pest Thysanoptera are members of the family Thripidae, a group of about 1700 species worldwide (Table 6.1) in which four subfamilies are now recognized. About 120 of these species are placed in the subfamily Panchaetothripinae, and these seem to represent the less advanced members of the family. They feed on the leaves of green plants, and typical members such as *Heliothrips haemorrhoidalis*, the greenhouse thrips, and *Selenothrips rubrocinctus*, the red-banded cocoa thrips, are usually associated with older, even senescing, leaves, although several species in the related genus *Caliothrips* are pests of seedling crops. Members of the small subfamily Dendrothripinae are often minute, feeding on very young leaves (*Dendrothrips*, *Pseudodendrothrips*), particularly in tropical countries. Members of the Sericothripinae live on flowers and on leaves (*Neohydatothrips*); few are considered serious pests, but one, *N. pseudoannulipes*, has recently been recorded as a serious pest of the garden plant *Tagetes*, the African marigold (Mound *et al.*, 1996a).

The majority of Thripidae species are included in the subfamily Thripinae (Table 6.1). A substantial number of these species feed only on grasses, some mainly in the florets (*Chirothrips*, *Limothrips*), and others particularly on leaves (*Aptinothrips*, *Stenchaetothrips*). A few species are predatory on spider mites (*Scolothrips*), and one small group of tropical species seems to be associated with mosses (Mound and Marullo, 1996). A range of genera include species that are found primarily on the leaves of higher plants. Some such as *Mycterothrips* species feed on both young and mature leaves, but others such as *Scirtothrips* species are more particularly associated with very young leaves or even young fruits. Many species feed particularly on pollen and other flower tissues (see Chapter 4), and the flowers attacked range from the male cones of certain Australian cycads (*Cycadothrips*), or the catkins of oak trees (*Oxythrips*), to typically bee-pollinated flowers such as those of many legumes (*Odontothrips*, *Megalurothrips*), and orchid flowers (*Aurantothrips*, *Dichromothrips*).

Many species in the most advanced genera, including *Thrips* and *Frankliniella*, feed on both flower and leaf tissue, and this causes confusion for any simplistic distinction between 'leaf thrips' and 'flower thrips'. In Europe, *Thrips major* is a common species in a wide range of flowers, but if it emerges in spring before suitable flowers are available then it will be found in large numbers within the opening buds of various trees. Similarly, the paradoxical situation arises that *Taeniothrips inconsequens* is a serious pest of sugar-maple young leaves in north-eastern America, but is primarily a 'flower thrips' (Teulon *et al*., 1993). This flexibility of searching behaviour and feeding habits reaches an extreme in a few of the most advanced flower thrips, including the pests *Thrips tabaci, Frankliniella occidentalis* and *F. schultzei* . Each of these is known to feed on both flower and leaf tissues, acting as a vector of the tospovirus TSWV (see Chapter 14), but each of them can also be an important predator on mite eggs and other small arthropods (Wilson *et al*., 1996), suggesting an extreme flexibility of both behaviour and digestive processes. The Australian flower-living phlaeothripid, *Haplothrips victoriensis*, similarly can feed on either mite eggs or pollen grains, and it lays more eggs when fed on pollen (Bailey and Caon, 1986).

Intraspecific variation

Two of the characteristics listed in the section above as probably inherited from the thrips progenitor were short generation time and a competitive breeding system. Rapid development under high temperatures apparently results in a smaller body size. Adult *Thrips tabaci* that develop during summer or at high temperatures are commonly smaller and paler than individuals that develop more slowly during winter under lower temperatures. In other species, under conditions of resource or sexual competition, the body size of one or both sexes varies considerably (Crespi, 1986; Terry, 1995). Selection pressures for large body size and success in combat presumably work antagonistically to selection pressures for small body size and rapid development. The factors affecting body size (inheritance, food, temperature) lead to a wide range of structural variation in some species of thrips (Ananthakrishnan, 1984). This variation is often accompanied by allometric growth patterns in particular structures, and sometimes involves true polymorphism (Mound and Crespi, 1995).

Variation associated with body size seems to be unimodal and is presumably related to the quantity and quality of food taken by an individual larva. In contrast, variation associated with wing length (and with sex) commonly involves distinct morphs. Variation in appearance within such morphs may be remarkable, due to allometry that is itself associated with body size. Allometrical variation and morphs are most common amongst fungus-feeding Phlaeothripidae (Palmer and Mound, 1978), but the males of many thripids, such as *Frankliniella occidentalis*, vary in body size within populations (Terry, 1995). The grass-living species of *Chirothrips* have wingless males, and both sexes often vary considerably in body size from different host plants, resulting in confusion over how many species exist. *Thrips nigropilosus* is unusual (although not unique) in that wing length varies almost continuously, from almost absent to fully developed, without the production of morphs.

Size variation and morph production were probably present in the earliest Thysanoptera, because some of the least advanced species, such as in *Merothrips* and *Cycadothrips*, exhibit allometrical differences in the body armature of males such that the largest males are remarkably different in appearance from the smallest males. In neither of these genera has male behaviour been observed, but male competition has been studied in several larger fungus-feeding thrips (Crespi, 1993). Similarly, competitive behaviour has been studied in gall-forming thrips on *Acacia* trees in the arid areas of Australia, and in some of these species a special soldier morph occurs comprising stoutly built, short-winged individuals of both sexes (Crespi and Mound, 1997).

The advantages of polymorphism and competitive behaviour in gall-forming and fungus-feeding thrips are evident; the resources in such habitats are limited and worth fighting for. However, competition also seems to occur in some flower-living thrips. Males of several of the more common flower-living thrips vary in body size, with larger males having stouter terminal setae. Fighting behaviour between such males has been studied using videocamera techniques (Terry, 1995) in *Frankliniella occidentalis*, but the frequency of, and the advantages incurred by, such behaviour in these small leaf-feeding thripids is far from clear (see Chapter 3). Fighting seems not to occur in this species at high population densities, and in Europe, male–male interactions in flower-living species such as *Frankliniella intonsa* seem to occur only on warm, dry days (W.D.J. Kirk, Keele, 1995, personal communication). Thus any prospect of pest control by manipulating breeding behaviour remains remote.

Variation in biology within species, as opposed to variation in their structure, is more difficult to assess. *Thrips tabaci* is usually rare in the humid tropics and subtropics, but is often abundant at warm, dry sites, particularly when its preferred host plant, onions, is grown as an extensive monoculture as in southern Brazil. In contrast, *Thrips palmi* is most abundant in warm, humid areas, and even in Japan where this species is a major pest it does not extend north of central Honshu (37°N). *Thrips tabaci* thrives in the hot summers of eastern Europe, but the habitat requirements of *T. palmi* suggest that this species is unlikely to become a problem on field crops in Europe. Neither *T. tabaci* nor *T. palmi* seem to vary in their biology throughout their ranges. In contrast, there are indications from around the world that *Frankliniella occidentalis* exists as more than one strain or biotype. A pest strain of this species was introduced into New Zealand in 1993 (Mound and Teulon, 1995), but for the previous 60 years a structurally identical form had been common in that country although restricted largely to lupin flowers. Similarly, in both Colombia and Costa Rica this species is abundant in the shade houses in which cut-flowers are produced, but it is rarely found outside these houses (Mound and Isaza, 1994). Currently, although *Frankliniella occidentalis* is abundant in greenhouses throughout the world, and is also widespread in subtropical areas including the highlands of eastern Africa, it seems to be less of a pest in lowland tropics. In particular it is unknown in Taiwan, despite a flourishing horticultural industry there that supports very large populations of *Frankliniella intonsa*.

Variation in biology with no apparent variation in structure raises the possibility of the existence of sibling species. Zawirska (1978) considered that *Thrips tabaci* exists as two independent subspecies, one parthenogenetic and the other bisexual, and recent work in the Netherlands supports this view. However, in many parts of the world males of this species are rarely found except when populations are very high. Unfortunately, experimental studies on variation within thrips species are remarkably scarce. Bryan and Smith (1956) demonstrated that some of the colour variation in *Frankliniella occidentalis* in California is genetically determined, with a pale form dominant to a dark form. *Thrips imaginis* commonly varies from pale yellow to brown, and Evans (1932) stated that yellow individuals darkened to brown within a few days of emerging from their pupae. In contrast, similar colour variation in *Thrips tabaci* is usually considered to be related to the temperature during larval development, and most taxonomists are familiar with the problem that individuals of several species are likely to be larger and darker in cold areas than in warm areas. This variation is undoubtedly confusing, but biochemical and genetic techniques for investigating intraspecific variation are only now being implemented (Gillings *et al.*, 1996), and the significance of such variation for pest control programmes is likely to be considerable.

Patterns of geographical distribution

Much of the published data on the distributions of thrips species and genera is undoubtedly the result of inadequate sampling (see Chapter 11). Distribution records are heavily dependent on the patterns of behaviour of the few specialists who have studied this group – where they or their colleagues have worked or spent their holidays. Intensive surveys of thrips faunas, involving repetitive collecting throughout different seasons and years, have rarely been attempted, particularly in the tropics where the majority of thrips species live. For example, most of the records of thrips species from Brazil are based on relatively brief studies by just two authors, and the apparently extensive list of thrips known from Panama probably derives from far less than 500 man-hours of sampling activity (see Mound and Marullo, 1996). In particular, it must be emphasized that the thrips fauna of Africa is effectively unknown, despite many species having been described from southern Africa. In contrast, current intensive studies in South-East Asia are radically changing our understanding of thrips distributions. For example, the genus *Merothrips* is regarded as Neotropical with several species widely dispersed, but about five undescribed species of this genus are now available from South-East Asia (S. Okajima, Tokyo, 1995, personal communication). Also, recent work in both South-East Asia (Mound, Okajima and Wang, in preparation) and Central and South America (Mound and Marullo, 1996) has resulted in an increasing number of species being recognized as having extensive and apparently natural distributions.

Cumulative totals of the numbers of thrips species from an area, such as those quoted below, are probably more a measure of collecting effort than of the resident fauna. Comparisons of the true thrips faunas of different parts of the world are thus conjectural, although a few patterns have emerged that are of relevance to thrips as pests. The largest genus, *Thrips*, which includes about 275 species across the northern hemisphere through Asia into Australia, New Zealand and the Pacific, with more species in Africa as yet undescribed, has no species native to America south of Mexico. Indeed, only one member of *Thrips* genus is described from Mexico, although more than 60 are known from North

America (Nakahara, 1994). In contrast, the genus *Frankliniella* has at least 100 species in South America and 50 in North America, but only about 10 species from the rest of the world. Most of the other larger thrips genera, that is genera which include more than 30 species, do not show any particularly restricted geographical distribution. Exceptions are: *Haplothrips* – apparently largely unrepresented in South America; *Aeolothrips* – largely restricted to the northern temperate zone; *Heterothrips* – purely New World; *Melanthrips* and *Odontothrips* – largely European. Another name that was common in older literature on Thysanoptera is *Taeniothrips*, but this now applies only to a group of less than 20 species distributed across the northern hemisphere.

Comparative figures currently available for the species richness of the thrips faunas of most parts of the world, other than Europe, almost certainly do not reflect biological reality. In 1936, R. Takahashi claimed that the thrips fauna of Taiwan, with 99 species, was 'fairly well-known', whereas about 250 species are now known from that island and with intensive collecting that figure continues to rise (Mound, Okajima and Wang, in preparation). Mound and Marullo (1996) indicated that the total numbers of thrips species recorded from Mexico, Costa Rica/Panama, and Brazil are about 300, 300, and 400, respectively; but they pointed out that faunistic comparisons are difficult because studies in Brazil have emphasized leaf-litter Phlaeothripidae, whereas studies in Costa Rica have emphasized flower-living Thripidae. The total number of thrips species recorded from Europe is about 500, with the approximate numbers from Austria, Italy, France and Britain being 200, 220, 260 and 160, respectively (Marullo and zur Strassen, 1995). In contrast, the number of described species recorded from America north of Mexico is about 700 (Nakahara, 1990). But comparisons between the European and North American thrips faunas are unrealistic because the European fauna has been collected and described so much more thoroughly than that of North America. It seems likely that no more than 75 per cent of the North American Thysanoptera fauna has been described. Even in the well-studied genus *Thrips*, 18 species were recently described as new out of a total of 62 species from North America (Nakahara, 1994). In contrast, scarcely 10 species have been added to this genus in Europe during the past 30 years, following Priesner's (1964) recognition of 70 European *Thrips* species (allowing for 1994 taxonomic concepts). Ananthakrishnan and Sen (1980) listed 650 Thysanoptera species from India, and Mound (1996) listed 450 species from Australia, but both of these figures probably represent little more than 60 per cent of the real faunas.

Ecological Diversity

Just as sampling inadequacies limit the generalizations that can be made about the geographical diversity of thrips, so a lack of precision in relating sampling methods to thrips biologies limits conclusions that can be drawn concerning thrips communities or assemblages. The concept of 'community' implies some level of interdependence between its members. Hence to refer to the thrips collected in any recognizable plant community as a 'community' themselves does not have any precise meaning, principally because so many of the species collected will be vagrants that are merely dispersing through the area. To consider the thrips collected from a grass sward, or swept from a cotton crop, or taken in a trap in a maize crop, or even collected by various methods in an oak woodland, as constituting a 'community' is essentially anthropocentric. On the grass sward or maize crop some thrips species will be breeding on these *Gramineae*; some will be living on, and host-specific to, particular weedy non-graminaceous plants that make up the total plant community; some will possibly be feeding on a plant on which they cannot breed; some will probably be searching for a mate; and some will simply be sheltering or resting. The more complex the habitat structure, both physically and botanically, the more niches will be available for thrips to occupy, and the less will each thrips species interact with others. Thus, although the various plants can be considered to constitute a community, in that they provide habitats for each other and compete for resources such as water, nutrients and root room, the thrips on the various species of plants within this community are likely to have little or no interdependence.

Similarly, the concept of 'host plant' for thrips is often so poorly defined as to lack utility. Mound and Marullo (1996) pointed out that no lepidopterist would include the many flowers that a butterfly species visits as 'host plants' of that insect, yet the literature on thrips commonly lists as 'host plants' all the plant species on which adults of a thrips species have been taken. Such plants may play some role in the biology of the thrips, just as a flower does for a butterfly. This is a functional relationship different from the concept of 'vagrants', because nutrients or liquid from a wide range of flowers that are not breeding-hosts may be important to a thrips species. But only in a very general sense can a species specific to grasses be regarded as a member of an 'oak woodland community', and an adult grass thrips such as a *Chirothrips* found on a male catkin of an oak tree does not merit *Quercus* being listed as a host plant of that thrips. These comments become of considerable economic importance when considering the burgeoning literature on thrips assemblages on crops, because the only species requiring control are those that demonstrably reduce yields, not the many other species with more tenuous relationships to a crop. Indeed, as discussed below, the most abundant thrips species on a crop is not always the most damaging. The interests of virologists add a further dimension to this problem, in that although an evolutionary biologist and an economic entomologist might agree that host plants are those plants that are responsible for the maintenance of a thrips population, a virologist must also be interested in any non-hosts to which a feeding thrips adult transmits a tospovirus.

Species richness within crops

Comparisons within and between ecological systems, including crops, are sometimes based on relatively crude sampling techniques that do not take into consideration the precise biology of the species, even to distinguishing residents from vagrants. Gross figures for the numbers of thrips species collected on a crop merely give a momentary 'snapshot' of a set of population changes and dispersal activities that are, at least partially, independent. Again, it is essentially anthropocentric to group all these independent biological activities under one heading, and such grouping is not likely to contribute to our understanding of pest biology and control. Lewis (1973) compiled valuable data on the numbers of thrips species collected on various crops, from a wide variety of sources, and no attempt is made to repeat these data here. But while the index of diversity, α, quoted for such samples, is a statistic that reflects species richness it tells little about the community biology of the thrips.

Grass and cereal crops may support particularly large numbers of thrips species at any one time. *Anaphothrips*, *Aptinothrips*, *Chirothrips*, *Haplothrips*, *Limothrips*, *Melanthrips* and *Stenothrips* can each maintain one or more breeding species on certain *Gramineae* in Europe. However, the distributions of these various thrips species seem to differ from each other independently; *Haplothrips tritici* is dominant on wheat in eastern Europe, but rarer in the west and absent from Britain; *Stenothrips graminum* is widespread in southern Britain but absent from Scotland; *Limothrips cerealium* is common in western Europe but *L. denticornis* is more common further east. Moreover, the interactions between the various thrips species found at any one site on cereal crops and other *Gramineae*, such as the possibility of competitive exclusion on flowering heads, or even predation between species, have scarcely been explored.

Most non-graminaceous crops support considerably fewer breeding thrips species, particularly in temperate regions. Although 18 species have been collected from flax in Britain, only two of these are known to breed on the crop (Lewis, 1973). In the tropics, cassava in South America is unusual in supporting three apparently host-specific thrips species, two on the leaves (*Corynothrips stenopterus* and *Scirtothrips manihoti*) and one mainly on the flowers (*Frankliniella melanommata*) (Mound and Marullo, 1996). Cocoa trees in South America also support at least three thrips species, but these tend to be on different parts of the plants; *Selenothrips rubrocinctus* on mature leaves, *Leucothrips theobromae* on the youngest leaves, and *Frankliniella parvula* in flowers and on young leaves (Mound and Marullo, 1996). Palmer *et al.* (1991) indicated that groundnuts in central India support three species of thrips, *Frankliniella schultzei* in the buds, *Caliothrips indicus* on the leaves, and *Megalurothrips usitatus* in the flowers.

More commonly, cultivated plants have a thrips spectrum that is dominated by a single species. This is often particularly true if insecticides are being used. A relationship between thrips species numbers and spraying becomes particularly obvious in temperate greenhouses, when species such as *Heliothrips haemorrhoidalis* and *Parthenothrips dracaenae* rapidly disappear if insecticides are introduced, only to be replaced by populations of *Frankliniella occidentalis*, a species that can develop resistance to many pesticides (see Chapter 15). Attempts in recent years to reduce the usage of insecticides in greenhouses have led sometimes to a resurgence of *H. haemorrhoidalis*.

Interaction between crops and wild plants

Whereas thrips populations in natural plant stands, or on individual trees, may be largely sedentary (Laughlin, 1970), populations in many crops are largely if not completely derived through immigration. At first sight this seems to be a paradox but, although there is no direct evidence, it seems likely that some proportion of any thrips population has a behavioural tendency to disperse, even when the majority of individuals in that population are non-vagile. Laughlin (1977) indicated that 5 per cent of an experimental population of *Thrips australis* were particularly long-lived individuals, and suggested that these might represent a migratory class within the population. In general, thrips species, particularly pest species, are highly vagile, with some proportion of the large populations produced crawling to the top of their hosts and then flying or being blown away in vast numbers, often dependent on the weather (see Chapter 5). In consequence, if a field is planted with onions, then *Thrips tabaci* will soon arrive. If the same field is then planted with a cereal crop, various grass-living thrips will arrive and the *T. tabaci* will possibly disappear. Thrips populations in such annual crops are evidently dependent on immigration, and low standards of crop hygiene and weed control around such crops commonly lead to severe damage by immigrant thrips and TSWV from surrounding plants and weeds (Yudin *et al.*, 1988).

In non-annual crops, early season populations may, in contrast, be dependent largely on the capacity of resident adults to overwinter within the crop. In southern Africa, overwintering within a citrus orchard largely determines the population size of *Scirtothrips aurantii* in the following season, although surrounding wild hosts supplement these populations later in the season (Gilbert, 1990). Moreover, citrus trees surrounded by windbreaks of *Grevillea* trees have more fruit damaged than trees surrounded by *Pinus* or *Casuarina* trees; citrus thrips do not breed on the latter two plants whereas they do on the former (Grout and Richards, 1990).

The longest-running data set on thrips population fluctuations is that of Davidson and Andrewartha (1948a) who, daily for 14 years, sampled *Thrips imaginis* adults in 20 rose flowers from a hedge in South Australia. This species has been called the plague thrips, because of the large populations that can develop in, and damage, pome-fruit flowers in Australia. However, roses are not a breeding host plant for this thrips, and the flowers sampled by Davidson and Andrewartha were merely a very attractive trap. The authors stated in a second paper (1948b) that ‘there is some evidence that thrips come and go from flowers more or less in waves’, but the cause of these waves was considered to be ‘not readily measurable’. Recently, *Thrips imaginis* was found across the arid centre of Australia wherever one of the many species of *Acacia* or *Maireana* was in flower. The pupae of this thrips have been shown to be unable to survive in dry soil, therefore the adults are presumably highly vagile in order to populate these arid areas. *Thrips imaginis* is therefore probably involved naturally with the sporadic flowering of several semidesert plants across Australia, with populations fluctuating locally depending on the flowers available. Cultivation of certain European plants, including pome-fruits and the pasture plant *Echium plantageneum*, provides additional breeding hosts, and the 'waves' of thrips noted by Davidson and Andrewartha were possibly the result of population fluctuations on surrounding wild and cultivated plants. Their extensive, and much quoted, data set from rose flowers (see above), far from being a study of a 'natural population' as their titles state, contributes little to our understanding of the natural history of *T. imaginis*, because trapping cannot distinguish between population size, trap attractiveness, flight activity, and behaviour including attractability and aggregation (see also Chapter 7).

The natural history of pest thrips is often curiously neglected. Economic entomologists may study population fluctuations on a crop, but neglect to consider the percentage of a population that arrives from elsewhere, or the factors involved in the fluctuations of that part of the total population that is not on the crop. Extreme examples are where the natural host plants of a pest species are not known, such as *Scirtothrips kenyensis*, the tea thrips of Kenya. Indeed, it is only very recently that the natural host plants of such damaging pests as *Scirtothrips citri*, the Californian citrus thrips (*Rhus laurina*, in Morse, 1995), or *Megalurothrips sjostedti*, the African cowpea thrips, have been studied, because within-crop breeding is usually assumed to be more important than immigration.

In the tropics, thrips populations on some annual crops are possibly affected even more by immigration than they are in north temperate regions, due to the north–south movement of the intertropical convergence zone and related wind systems. These wind systems are known to move large populations of insects in parts of Africa (Pedgley, 1982), and Mound and Marullo (1996) suggested that the widespread distributions of many thrips species north or south from Costa Rica, either to Mexico or else to southern Brazil, may have a similar origin. Recent observations in northern Australia indicate that the thrips fauna of that area is similarly derived in part from that of countries to the north (Mound, 1996), and the thrips fauna of Taiwan is also influenced by immigration from Japan and the Philippines (Mound, Okajima and Wang, in preparation).

In contrast, the populations of thrips on crops grown within greenhouses and shade houses are dependent largely on breeding within the crops, and transport between areas is determined largely by human behaviour. For example, it is common practice to produce young chrysanthemum plants by rooting small shoots plucked from the apices of older 'mother plants'. Adult females of *Frankliniella occidentalis* oviposit amongst apical leaves, and it seems likely that many growers inadvertently rear their own pest populations by taking cuttings for the next crop from previously infested 'mother plants'. The flower trade itself is almost certainly responsible for the widespread distribution of this species around the world. The leaf-rolling thrips of decorative *Ficus* trees is similarly largely distributed by the horticultural trade itself (Mound *et al.*, 1996b). The intense thigmotaxis of thrips adults results in their hiding in such small spaces that they can be difficult both to detect visually and to contact with insecticides.

Origins of major pest species

Many thrips species that are considered to be pests are now widely distributed around the world. Although now cosmopolitan, *Thrips tabaci* probably originated in the eastern Mediterranean, from where its preferred host plant, *Allium cepa*, is derived. *Frankliniella occidentalis* was probably distributed fairly widely in the semi-arid parts of south-western North America; one of its nearest relatives is *F. panamensis* from central America, but western flower thrips is now almost cosmopolitan. *Thrips simplex* seems to be part of a complex of African species, and may have come originally from southern Africa, but it is now found wherever *Gladiolus* flowers are grown. *Thrips palmi* was known for more than 50 years (under various names) as widely distributed between Sudan and Taiwan, but was infrequently collected until it suddenly became a pest in Japan in 1978. Since then it has spread throughout the tropics and subtropics as a major pest.

Many grass-living Thripidae, from both temperate and tropical faunas, have been spread widely around the world. *Aptinothrips*, *Chirothrips* and *Limothrips* species from Europe, also the rice thrips, *Stenchaetothrips biformis*, and two sugar cane thrips, *Stenchaetothrips minutus* and *Fulmekiola serrata*, have been carried around the world. Some of these species will have been transported as pupae within the bracts that surround grass seeds, whereas others will have been transported within the straws of hay. Crops whose propagules are large pieces of plant in which a thrips can survive undetected, such as sugar cane, bananas, and orchids, are particularly liable to take their thrips pests with them throughout the world, in contrast to crops like tea. Even the monophagous thrips, *Sciothrips cardamomi*, pest of such a localized crop as cardamoms, has recently found its way from its native India to Costa Rica.

Until about 20 years ago the country of origin of a thrips pest could sometimes be predicted as being that country from which both sexes were known. Presumably males, being short lived, did not survive long sea voyages, and pest species managed to establish and maintain themselves through parthenogenetic reproduction. However, the speed and volume of air traffic in recent years has resulted in the relatively short-lived males of thrips surviving in new countries. The implications of this for the genetic variability of thrips pests, and hence for pest control, are likely to be considerable.

Table 6.2 lists the 35 species of pest thrips that were recorded in the CAB International database as being most commonly cited in research literature during the period 1973 to 1993. This is by no means an exhaustive list of thrips pests or potential pests; it is a measure of research activity rather than pest status. Nakahara (1990) estimated that 45 thrips species were pests in North America, but this included many minor, or occasional, pests. Ananthakrishnan (1984) listed more than 90 thrips species as of economic importance worldwide, but such compilations from published literature inevitably suffer from omissions, misidentifications, and faulty associations with crop damage unless each entry is carefully assessed. In contrast, the data for thrips associated with about 100 different crops listed in Appendix II include only the names of those 65 thrips species that the authors consider are likely to have limited crop production significantly.

Table 6.2. Thysanoptera pest species: distributions, literature citations and crops attacked.

Thrips name	Eur	NAm	SAm	Afr	Ori	Aus	Citation*	Crop
Thripinae								
Anaphothrips obscurus	+	+	+	+	+	+	19	Cereals
Aptinothrips rufus	+	+	+	+	+	+	9	Grasses
Frankliniella fusca	–	+	–	–	–	–	60	Peanuts
Frankliniella intonsa	+	+	–	–	+	–	49	Flowers
Frankliniella occidentalis	+	+	+	+	+	+	327	Polyphage
Frankliniella schultzei	–	+	+	+	+	+	68	Polyphage
Frankliniella tritici	–	+	–	–	–	–	31	Polyphage
Frankliniella williamsi	–	+	+	–	+	+	10	Maize
Kakothrips pisivorus	+	–	–	–	–	–	16	Peas
Limothrips cerealium	+	+	–	+	–	+	26	Cereals
Limothrips denticornis	+	+	–	–	–	+	28	Cereals
Megalurothrips distalis	–	–	–	–	+	+	15	Beans
Megalurothrips sjostedti	–	–	–	+	–	–	57	Cowpeas
Megalurothrips usitatus	–	–	–	–	+	+	24	Beans
Sciothrips cardamomi	–	–	+	–	+	–	16	Cardamom
Scirtothrips aurantii	–	–	–	+	–	–	61	Citrus
Scirtothrips citri	–	+	–	–	–	–	75	Citrus
Scirtothrips dorsalis	–	–	–	+	+	+	137	Polyphage
Stenchaetothrips biformis	+	–	+	–	+	+	83	Rice
Taeniothrips inconsequens	+	+	–	–	–	–	27	Maple
Thrips flavus	+	–	–	–	+	–	35	Polyphage
Thrips florum	–	+	+	–	+	+	12	Polyphage
Thrips hawaiiensis	–	+	+	+	+	+	41	Polyphage
Thrips obscuratus	–	–	–	–	–	+	15	Polyphage
Thrips palmi	–	+	+	+	+	+	136	Polyphage
Thrips simplex	+	+	+	+	+	+	23	Gladiolus
Thrips tabaci	+	+	+	+	+	+	582	Polyphage
Panchaetothripinae								
Caliothrips indicus	–	–	–	–	+	–	40	Polyphage
Heliothrips haemorrhoidalis	+	+	+	+	+	+	57	Polyphage
Selenothrips rubrocinctus	–	+	+	+	+	+	46	Polyphage
Sericothripinae								
Neohydatothrips adolfifriderici	–	–	–	+	–	–	14	Cowpeas
Neohydatothrips variabilis	–	+	–	–	–	–	30	Soybeans
Phlaeothripinae								
Gynaikothrips ficorum	+	+	+	+	+	+	26	Ficus
Haplothrips aculeatus	+	–	–	–	+	–	34	Cereals
Haplothrips tritici	+	–	–	–	–	–	79	Cereals

* Citation totals for the 35 most frequently cited pest thrips in the CAB International database 1973–1993.

Eur, Europe; NAm, North America; SAm, South America; Afr, Africa; Ori, Oriental region; Aus, Australia.

Table 6.2 indicates the importance of the Thripinae as the source of major pests, particularly the genera *Frankliniella* and *Thrips*. However, many other thrips species cause serious crop losses, at least locally, such as *Rhipiphorothrips* and *Retithrips* species on various crops in South-East Asia (Ananthakrishnan, 1984), or *Enneothrips flavens* on groundnuts in Brazil (Monteiro *et al.*, 1996). Similarly, in temperate regions, *Thrips angusticeps* and *T. nigropilosus* are widely distributed periodic pests. Some pests, such as *Thrips imaginis* in Australia, have a period of severity, or at least of great attention, and then fade from the literature. Sometimes the effect of a pest is simply tolerated; cassava plants throughout Latin America have their apical leaves damaged more or less seriously by one or both of the two species *Scirtothrips manihoti* and *Corynothrips stenopterus* in a manner that suggests that yields must be affected. Mulberry plants grown in the Oriental region for the silk industry usually support considerable populations of *Pseudodendrothrips mori*. Sugar cane commonly supports species such as *Stenchaetothrips minutus* and *Fulmekiola serrata*; but on each of these crops thrips do not usually constitute a limiting factor to commercially acceptable yields.

More than 20 species of thrips with some reputation as pests are now largely cosmopolitan or tropicopolitan. However, the reason for such extensive distributions is far from clear, because not all pest thrips have become widely distributed. *Thrips imaginis*, the plague thrips of Australia, is one of the most abundant insects of that continent but, although it has been recorded from New Zealand, it has never become established elsewhere apart from New Caledonia. The two major citrus thrips, *Scirtothrips aurantii* from southern Africa and *S. citri* from California, are particularly notable for not having entered the world trade in thrips species, although two related species, *S. inermis* and *S. longipennis*, have both been reported in low numbers from various parts of the world, particularly in greenhouses, and *S. dorsalis* apparently has a natural range from India to Japan and Australia. The thrips that are pests of cowpeas in Africa and the Oriental region also give no indication of being distributed by human trading. The lack of success of such species in being dispersed by world trade probably owes more to some particular aspect of their host-plant relationship, such as their pupation site and the nature of the plant propagules involved, than to efforts at plant quarantine.

Host ranges and potential as pests

As indicated above, much of the published information on host plant ranges of thrips is unreliable because it fails to distinguish between host plants and ‘finding places’. Even assessing the importance of a thrips species that has been found in large numbers on a crop is not without problems. Large samples of thrips were taken in 1995 from *Solanum melongena* plants in Sao Paulo State, Brazil. Three species were present in the samples, *Thrips palmi*, *Frankliniella schultzei* and *Caliothrips phaseoli*, in the approximate proportions 1:2:3. The initial assumption was that the most abundant species was causing damage to the crop. However, *C. phaseoli* seems to breed only on legume plants, including various weeds on which it develops large populations in the area, but it is exceptionally dispersive in its behaviour. It was concluded that the many brightly coloured *C. phaseoli* adults on the aubergines were probably vagrants, and the observed damage to the crop was due to the relatively fewer and more inconspicuous *T. palmi*, a species that can be difficult to detect and collect from amongst the apical leaves of a plant (Monteiro *et al.*, 1997). Similarly, the reference by Priesner (1964b) quoted by Lewis (1973) of *Thrips australis* as a potential pest in the eastern Mediterranean may have resulted from the conspicuous adults of this species having been collected on various flowers to which they had dispersed, probably after breeding in very large numbers on local *Eucalyptus* tree flowers, their natural host plant.

Despite these comments, there is good evidence that although some species are largely monophagous, others are certainly polyphagous. Moreover, some species are known to exhibit remarkable host shifts, such that they become pests on plants unrelated to their natural hosts. The Oriental species, *Chaetanaphothrips orchidii*, has long been known as a widespread pest of bananas and cultivated orchids, but in recent years it has become a pest of citrus fruits in Florida (Childers and Frantz, 1994). *Neohydatothrips gracilicornis* has been reported as a pest of *Pinus* trees in southern Italy and also in Spain, whereas this species is confined to *Viccia cracca* in northern Europe. *Thrips minutissimus* in Britain is a univoltine species that is restricted to the male flowers of *Quercus robur*, whereas in southern France it not only has more than one generation, it is a pest of peach trees with the larvae occurring on several different tree species. This type of adaptability increases the problem of forecasting the pest potential of a species.

The origin of new pests, and their interactions with pre-existing pests, is scarcely studied. Recent observations in Sao Paulo State, Brazil indicated that *Thrips palmi* was far more common on *Chrysanthemum* plants than the normal pest of this crop, *Frankliniella occidentalis*, and moreover, both *T. palmi* and *F. schultzei* were more common on *Gladiolus* plants than the typical gladiolus thrips, *Thrips simplex*. Competitive exclusion seems unlikely to explain such differences, because populations are rarely anywhere near the carrying capacity of the host plants. Similarly, *F. occidentalis* has yet to establish itself in Taiwan despite the active horticultural trade there, and *F. schultzei* has not yet established itself in Japan, although it is common in other parts of South-East Asia (Okajima *et al.*, 1992). Similarly, *Thrips imaginis*, which is so common in Australia, has not established itself in New Zealand although it has been collected there several times. Given the vigorous trading in living plants around the world, it is probable that such species have been introduced to many areas, but for some reason they have not been able to establish.

The ecology of the establishment of pests is evidently not a simple function of introduction. There must be levels of complexity at which we can as yet only guess, involving interactions between the strains of a species introduced at any one time, the crop strains currently available, and the ecological conditions at the site of introduction. *Thrips palmi* emerged suddenly as a pest in about 1978 in Japan, and swept around the world in tropical latitudes, establishing itself in a wide range of countries. Other members of this genus in South-East Asia are polyphagous and sometimes damage crops; they would appear to have a similar potential to *T. palmi*. However, explanations of future population explosions probably do not lie within the current ecology and behaviour of an organism, and hence the prediction of future pests is likely to remain speculative.

References

Ananthakrishnan, T.N. (1984) *Bioecology of thrips*. Indira Publishing House, Michigan.

Ananthakrishnan, T.N. and Sen, S. (1980) Taxonomy of Indian Thysanoptera. *Zoological Survey of India, Handbook Series* 1, 1–234.

Bailey, P. and Caon, G. (1986) Predation on twospotted mite, *Tetranychus urticae* Koch (Acarina: Tetranychidae) by *Haplothrips victoriensis* Bagnall (Thysanoptera: Phlaeothripidae) and *Stethorus nigripes* Kapur (Coleoptera: Coccinellidae) on seed lucerne crops in South Australia. *Australian Journal of Zoology* 34, 515–525.

Bryan, D.E. and Smith, R.F. (1956) The *Frankliniella occidentalis* (Pergande) complex in California (Thysanoptera: Thripidae). *University of California Publications in Entomology* 10, 359–410.

Childers, C.C. and Frantz, G. (1994) Ring spot damage to Florida citrus fruit caused by thrips feeding injury. *Citrus Industry* May, 38–43.

Crespi, B.J. (1986) Territoriality and fighting in a colonial thrips, *Hoplothrips pedicularius*, and sexual dimorphism in Thysanoptera. *Ecological Entomology* 11, 119–130.

Crespi, B.J. (1993) Sex ratio selection in Thysanoptera. In: Wrensch, D.L. and Ebbert, M. (eds), *Evolution and Diversity of Sex Ratios in Haplodiploid Insects and Mites*. Chapman & Hall, New York, pp. 214–234.

Crespi, B.J. and Mound, L.A. (1997) Ecology and evolution of social behaviour among Australian gall thrips and their allies. In: Choe, J. and Crespi, B.J. (eds) *The Evolution of Social Behaviour in Insects and Arachnids*. Cambridge University Press, 166–180.

Davidson, J. and Andrewartha, H.G. (1948a) Annual trends in a natural population of *Thrips imaginis* (Thysanoptera). *Journal of Animal Ecology* 17, 193–199.

Davidson, J. and Andrewartha, H.G. (1948b) The influence of rainfall, evaporation and atmospheric temperature on fluctuations in the size of a natural population of *Thrips imaginis* (Thysanoptera). *Journal of Animal Ecology* 17, 200–222.

Evans, J.W. (1932) The bionomics and economic importance of *Thrips imaginis* Bagnall. *Council for Scientific and Industrial Research, Commonwealth of Australia* Pamphlet 30, 1–48.

Gilbert, M.J. (1990) Relative population levels of citrus thrips *Scirtothrips aurantii* on commercial citrus and adjacent bush. *South African Journal of Zoology* 25, 72–76.

Gillings, M.R., Rae, D., Herron, G.A. and Beattie, G.A.C. (1996) Tracking thrips populations using DNA-based methods. In: *Proceedings of the 1995 Australia and New Zealand Thrips Workshop*. NSW Agriculture, Gosford, pp. 97–103.

Goldbach, B. and Peters, D. (1994) Possible causes of the emergence of tospovirus diseases. *Seminars in Virology* 5, 113–120.

Grout, A.U. and Richards, G.I. (1990) The influence of windbreak species on citrus thrips (Thysanoptera: Thripidae) populations and their damage to South African citrus orchards. *Journal of the Entomological Society of Southern Africa* 53, 151–157.

Kristensen, N.P. (1991) Phylogeny of extant hexapods. In: CSIRO (ed.) *The Insects of*

Australia, 2nd edn. Melbourne University Press, Carlton, pp. 125–140.
Laughlin, R. (1970) The gum tree thrips, *Isoneurothrips australis*, the population of a single tree. *Entomologia experimentalis et applicata* 13, 247–259.
Laughlin, R. (1977) The gum tree thrips *Isoneurothrips australis* Bagnall. Survival at different temperatures and humidities and its relation to capacity for survival. *Australian Journal of Ecology* 2, 391–398.
Lewis, T. (1973) *Thrips: their Biology, Ecology and Economic Importance*. Academic Press, London.
Marullo, R. and zur Strassen, R. (1995) Thysanoptera. In: Minelli, A., Ruffo S. and La Posta, S. (eds) *Checklist delle Specie della Fauna Italiana*, Fasc. 40. Calderini, Bologna, pp. 1–9.
Monteiro, R.C., Mound, L.A. and Zucchi, R. (1997) Thrips as pests of plant production in Brasil. *Revista Brasiliera de Entomologia* (in press).
Morse, J.G. (1995) Prospects for IPM of citrus thrips in California. In: Parker, B.L., Skinner, M. and Lewis, T. (eds) *Thrips Biology and Management*. Plenum, New York, pp. 371–379.
Mound, L.A. (1989) Systematics of thrips (Insecta: Thysanoptera) associated with mosses. *Zoological Journal of the Linnean Society* 96, 1–17.
Mound, L.A. (1996) Thysanoptera. In: Wells, A. (ed.) *Zoological Catalogue of Australia*, Vol. 26. Psocoptera, Phthiraptera, Thysanoptera. CSIRO, Melbourne, Australia, pp. 249–336, 397–414 (Index).
Mound, L.A. and Crespi, B.J. (1995) Biosystematics of two new species of Australian *Acacia* gall-thrips with soldiers (Insecta; Thysanoptera) *Journal of Natural History* 29, 147–157.
Mound, L.A. and Isaza, J.C. (1994) Trips comunes en la Sabana de Bogota. *Asocolflores* 38, 55–59.
Mound, L.A. and Marullo, R. (1996) The thrips of Central and South America: an introduction. *Memoirs on Entomology, International* 6, 1–488.
Mound, L.A. and Teulon, D.A.J. (1995) Thysanoptera as phytophagous opportunists. In: Parker, B.L., Skinner, M. and Lewis, T. (eds) *Thrips Biology and Management*. Plenum, New York, pp. 3–20.
Mound, L.A., Heming, B.S. and Palmer, J.M. (1980) Phylogenetic relationships between the families of recent Thysanoptera. *Zoological Journal of the Linnean Society of London* 69, 111–141.
Mound, L.A., Goodwin, S. and Steiner, M.Y. (1996a) *Neohydatothrips pseudoannulipes* Johansen (Thysanoptera, Thripidae), a pest thrips on African marigolds new to Australia. *Australian Journal of Entomology* 35, 201–202.
Mound, L.A., Wang, Chin–ling and Okajima, S. (1996b) Observations in Taiwan on the identity of the Cuban laurel thrips (Thysanoptera, Phlaeothripidae). *Journal of the New York Entomological Society* 103, 200–205.
Nakahara, S. (1990) Order Thysanoptera. In: Kosztareb, M. and Schaefer, C.W. (eds) *Systematics of the North American Insects and Arachnids: Status and Needs. Virginia Agricultural Station Information Series* 90–1, pp. 147–149.
Nakahara, S. (1994) The genus *Thrips* Linnaeus (Thysanoptera: Thripidae) of the New World. *United States Department of Agriculture, Technical Bulletin* 1822, 1–183.
Okajima, S., Hirose, Y., Kajita, H., Takagi, M., Napompeth, B. and Buranapanichpan, S. (1992) Thrips on vegetables in South East Asia. *Applied Entomology and Zoology* 27, 300–303.
Palmer, J.M. and Mound, L.A. (1978) Nine genera of fungus-feeding Phlaeothripidae (Thysanoptera) from the Oriental Region. *Bulletin of the British Museum (Natural History), Entomology* 37, 153–215.

Palmer, J.M., Reddy, D.V.R., Wightman, J.A. and Ranga Rao, G.V. (1991) New information about the thrips vectors of tomato spotted wilt virus isolate in groundnut crops in India. *International Arachis Newsletter* 7, 24–25.

Pedgley, D.E. (1982) *Windborne Pests and Diseases. Meteorology of Airborne Organisms*. Ellis Horwood, Chichester.

Priesner, H. (1964a) Ordnung Thysanoptera (Fransenflügler, Thripse) In: Franz, H. (ed.) *Bestimmungsbücher zur Bodenfauna Europas* 2, 1–242.

Priesner, H. (1964b) A monograph of the Thysanoptera of the Egyptian deserts. *Publications de la Institut Desert Egypte* 13, 1–549.

Terry, I. (1995) Competition in western flower thrips males: effects of density on behaviour. In: Parker, B.L., Skinner, M. and Lewis, T. (eds) *Thrips Biology and Management*. Plenum, New York, pp. 433–438.

Teulon, D.A.J., Kolb, T.E., Cameron, E.A., McCormick, L.H. and Hoover, G.A. (1993) Pear thrips, *Taeniothrips inconsequens* (Uzel) (Thysanoptera: Thripidae), on sugar maple, *Acer saccharum* Marsh.: a review. *Zoology (Journal of Pure and Applied Zoology)* 4, 355–380.

Wilson, L.J., Bauer, L.R. and Walter, G.H. (1996) Phytophagous thrips are predators of twospotted spider mites (Acari; Tetranychidae) on cotton in Australia. *Bulletin of Entomological Research* 86, 297–305.

Yudin, L.S., Tabashnik, B.E., Cho, J.J. and Mitchell, W.C. (1988) Colonisation of weeds and lettuce by thrips (Thysanoptera, Thripidae) *Environmental Entomology* 17, 522–526.

Zawirska, I. (1978) Studia nad *Thrips tabaci* Lindeman (Thysanoptera, Thripidae) *Prace naukowe Instytutu Ochrony Roslin* 20, 15–138.

Distribution, Abundance and Population Dynamics

7

William D.J. Kirk
Department of Biological Sciences, Keele University, Keele, Staffs ST5 5BG, UK

Introduction

The size of a population changes as a result of births, deaths, immigration and emigration. If these processes and the factors that influence them are understood, the population dynamics of a pest insect can be predicted to allow the design of effective integrated pest management strategies and provide advance warning of outbreaks.

Thrips are particularly good insects for the study of population dynamics because very large samples can be caught quickly and easily. One of the biggest controversies in population dynamics theory involved data for *Thrips imaginis* in Adelaide, South Australia. The number of thrips in 20 roses was recorded on 2291 days over 14 years from 1932 to 1946, involving counts of some 6 million individuals, which built up a detailed picture of the apparent fluctuations in population (Davidson and Andrewartha, 1948a,b). The research started as a response to massive outbreaks that caused severe damage to fruit, particularly apples, across the southern states of Australia in the 1930s, but some of the conclusions that were drawn from the data by Andrewartha and Birch (1954) sparked debate among ecologists (Nicholson, 1958; Smith, 1961). Current ecology textbooks still discuss these thrips data and the issues raised (Ricklefs, 1990; Begon *et al.*, 1996), so for many ecology students, thrips are synonymous with the fundamentals of population dynamics.

Although it is easy to obtain large samples of thrips, their small size and claustral habits make it difficult to investigate directly the reasons for changes in population size. Thrips that die in the field are rarely found and mass migrations, though sometimes spectacular, can go unnoticed. Very little is known about why some species become pests and others do not, or why a species is common in one year and rare in another.

This chapter considers the main factors that affect thrips numbers, such as the host plant, weather, predation, parasitism, disease and overwintering, and the attempts that have been made to model population dynamics. Other important factors are considered in more detail elsewhere: environmental effects on fecundity and growth of individuals (Chapter 2), host-finding behaviour (Chapter 3), nutrition (Chapter 4), migration (Chapter 5), and the effects of cultural, biological and chemical control (Chapters 15–18).

Plant Effects

Host plants vary in their suitability for thrips, not just between species but between cultivars and with age. As a result, larger infestations occur on some plants than on others, but it is difficult to know from field observations whether this arises from differences in rates of colonization, reproduction, development or mortality. Table 7.1 shows that under the same conditions, rates of population growth of a thrips species can vary enormously between crops. However, the relevance of such laboratory figures to the field is uncertain because the experiments may have omitted some factor that is critical out of doors.

Eggs are vulnerable to desiccation or changes in plant tissue and they cannot defend themselves against predators and parasites, so the oviposition site chosen by a female is important for survival. Some thrips have particular physical requirements for oviposition sites and the availability of suitable surfaces may influence infestation levels between plants. *Taeniothrips inconsequens* will only lay in structures with convex surfaces, such as veins, petioles and stems (Teulon *et al.*, 1994), and *Thrips australis* will only lay in or near cylindrical structures more than about 150 μm wide, such as stamen filaments (Kirk, 1987a). Eggs of *Limothrips denticornis* are only laid in the intervein spaces of leaves, so on spring barley they are restricted to mature leaves in which the spaces are far enough apart for an egg to fit (Bates and Weiss, 1991).

Leaf hairiness is well known as a factor conferring insect resistance to a crop, either because dense hairs prevent access to the leaf surface for feeding and oviposition or because the hairs or trichomes trap or injure the insects. In a study of cultivars of tomato and related species, larvae of *Frankliniella occidentalis* took longer to walk on leaves with dense masses of glandular trichomes. Most of the larvae that did walk onto these leaves either turned back or were greatly hindered or permanently trapped by the trichomes (Kumar *et al.*, 1995). The predatory thrips *Scolothrips sexmaculatus* is also affected and lays about half as many eggs on bean leaves with a high density of hooked trichomes as on sparsely haired cotton leaves (Şengonca and Gerlach, 1984). This species also lays fewer eggs on younger leaves, which have denser hairs. Schoonhoven (1974) surveyed over a thousand clones of cassava and found a strong correlation between thrips resistance and leaf pubescence. However, the presence or absence of hairs on cotton leaves does not seem to affect *Thrips tabaci*, probably because they are widely spaced (Wardle and Simpson, 1927). Hairy leaves can even be beneficial to thrips; for example all stages of the predatory *Haplothrips kurdjumovi* stay within the hairs on the underside of plum leaves, but on peach trees, which have glabrous leaves, the eggs, larvae, prepupae and pupae are usually found in crevices on twigs rather than exposed on leaves (Putman, 1965). Other physical properties of the plant surface, such as epidermal protrusions and surface wax, may provide resistance (Zeier and Wright, 1995). Even if many eggs are laid successfully, hatching mortality can affect population density. Rösingh (1980) found that while *Megalurothrips sjostedti* laid similar numbers of eggs in the calyx of a resistant and non-resistant cultivar of cowpea, fewer larvae hatched and there was higher mortality on the resistant variety. The thicker epidermal cell wall of the calyx of the resistant variety appeared to be reducing egg-hatching and hindering feeding by young larvae.

Plant age and leaf age can greatly affect population growth (see Chapter 4). Nugaliyadde and Heinrichs (1984) studied populations of *Stenchaetothrips biformis* on rice seedlings and found that fecundity almost doubled between the one-leaf stage and the five-leaf stage, but then decreased by the six-leaf stage, although the percentage of eggs hatching was unaffected. The longevity of adult males, but not females, increased with seedling age.

Phytophagous thrips are well known for seeking out narrow spaces in which to live, such as within leaf sheaths or deep within inflorescences. They seem to penetrate to parts of plants that most other insects cannot reach, making them difficult to treat with insecticides and hard to detect in quarantine. This habit provides a microclimate with high relative humidity, making desiccation less likely, and provides protection from larger predators and parasites. It may also allow easier access to food and protection from solar radiation, rain or adverse temperatures. Many cereal- and grass-dwellers have a dorsoventrally flattened body, which allows them to fit between the close layers of leaf sheaths. A correlation has been found between the body length of host-specific flower thrips and the size of their host flower, presumably because smaller thrips can exploit the narrower spaces provided by smaller flowers (Kirk, 1990, 1991). This size relationship suggests that access to restricted spaces has been of evolutionary significance, but it is not known whether the correlation arose through body size adapting to the dimensions of the host flower or by constraints on evolutionary shifts to new hosts. Cereal thrips show a particular attachment to small spaces, with a preference for a crevice width of about 200 to 500 μm (Mallmann, 1959, 1964; Lewis and Navas, 1962). This behaviour is associated with their habit of overwintering as adults deep within small crevices, such as cracks in bark. The behaviour is sometimes described as a thigmotaxis, however the response arises from an increase in the rate of turning (klinokinesis) and slower movement (orthokinesis) with decreasing crevice width (Mallmann, 1959, 1964), so by analogy with the terminology of olfactory responses it is more appropriately described as a thigmokinesis (Kennedy, 1978).

The strength of the preference for small enclosed spaces is further evidence that the behaviour has an adaptive advantage and suggests that thrips populations would do better on plants that are well provided with such spaces. Flower thrips appear to be commoner in more intricate blossoms (Bryan and Smith, 1956), but the direct effect of flower structure on population growth appears never to have been tested. Jones *et al.* (1934) studied the effect of the leaf arrangement of onion varieties on infestation by *Thrips tabaci*, which lives and

Table 7.1. Some estimates of population parameters of pest thrips.

Thrips species Crop species	Temperature (°C)	L:D	R_0	r_m (day^{-1})	T (days)	r_c (day^{-1})	T_c (days)	Reference
Frankliniella fusca								
Peanut	20	14:10	5.07	–	–	0.05	31.2	Lowry *et al.* (1992)
Peanut	25	14:10	4.60	–	–	0.07	21.6	"
Peanut	30	14:10	16.01	–	–	0.16	17.5	"
Peanut	35	14:10	1.67	–	–	0.04	13.2	"
Peanut	27	14:10	–	0.40	–	0.50	11.7	Puche and Funderburk (1992)
Frankliniella occidentalis								
Bean	23	04:20	3.70	0.062	21.2	–	–	Brødsgaard (1994)
Bean	23	08:16	7.60	0.110	18.3	–	–	"
Bean	23	16:08	12.2	0.140	17.9	–	–	"
Chrysanthemum	15	–	42.2	0.056	66.5	–	–	Robb and Parrella (1991)
Chrysanthemum	20	–	86.5	0.094	47.2	–	–	"
Chrysanthemum	25	–	99.5	0.171	26.9	–	–	"
Chrysanthemum	30	–	35.4	0.205	17.4	–	–	"
Chrysanthemum	35	–	2.7	0.056	17.6	–	–	"
Cotton – pollen	27	14:10	30.1	0.157	21.6	–	–	Trichilo and Leigh (1988)
Cotton + pollen	27	14:10	111.8	0.220	23.4	–	–	"
Cucumber	25	16:08	22.1	0.166	–	–	20.1	van Rijn *et al.* (1995)
Peanut	20	14:10	1.08	–	–	0.02	19.7	Lowry *et al.* (1992)
Peanut	25	14:10	2.25	–	–	0.02	15.6	"
Peanut	30	14:10	0.0	–	–	–	–	"
Rose 'Cherry Brandy'	30	16:08	2.54	–	–	0.36	2.67	Gaum *et al.* (1994b)
Rose 'City of Belfast'	30	16:08	0.20	–	–	−0.44	3.60	"

Hercinothrips femoralis								
Bean	22	–	87.3	0.103	–	–	–	Takrony (1973)
Bean	27	16:08	131.0	0.177	–	–	–	"
Thrips palmi								
Cucumber	15	16:08	16.5	–	–	0.035	80.2	Kawai (1985a)
Cucumber	20	16:08	25.9	–	–	0.080	40.7	"
Cucumber	25	16:08	28.0	–	–	0.134	24.8	"
Cucumber	30	16:08	19.1	–	–	0.144	20.5	"
Melon	25	16:08	13.0	–	–	0.111	23.2	Kawai (1986)
Okra	25	16:08	0.5	–	–	−0.027	25.9	"
Pumpkin	25	16:08	7.3	–	–	0.080	24.8	"
Thrips tabaci								
Cucumber	25	16:08	27.5	0.176	–	–	20.4	van Rijn *et al.* (1995)

R_0 is the net reproductive rate; r_m is the intrinsic rate of natural increase; T is the mean generation time; r_c is the capacity for increase; and T_c is the cohort generation time (see Southwood, 1978); L:D is hours of light and dark per day. Note that r_c and T_c are approximations to r_m and T.

breeds between the inner leaves. Susceptible onion varieties had flat leaf surfaces pressed against each other, providing protected areas within which the thrips accumulated. However, 'White Persian', the most resistant variety, had leaves with an almost circular cross-section that diverged at a large angle, so providing far less protection. When the foliage was tied together to create artificial narrow spaces, the population per plant increased rapidly over a few days, indicating the importance of the spaces for population growth.

Correlations have sometimes been noticed between infestation and flower colour. These are most likely to be an effect of attraction of flying adults (Kirk, 1984). For example, *Kakothrips pisivorus* is more numerous in all-white flowers of field bean than in those with black wing-spots, which are not as obvious (Stoddard, 1986), and *Frankliniella occidentalis* differentially infests flowers of various colours (Chyzik *et al.*, 1995), but correlations can also be found for thrips on other parts of the plant. *Thrips simplex* is a pest that breeds and spreads on gladiolus corms, leaves and flowers, but it appears to be more numerous on red-flowered cultivars, whether or not they are in flower (Herr, 1934; Hargreaves and Cooper, 1980). The unusual finding that orange and red sticky traps catch more of this species than other colours (Walker, 1974), suggests that the differences in plant infestation are the result of differences in the attraction of flying adults to the flowers. Comparisons of sister cultivars of florists' chrysanthemum have shown that *Frankliniella occidentalis* is more damaging to leaves of yellow-flowered than white-flowered varieties, even though the comparison was made during the vegetative stage before flowering (Broadbent *et al.*, 1990). The reason is not known, but it is possible that flower colour correlates with a chemical factor in the leaves. De Jager *et al.* (1995) found that the survival of thrips fed on leaf sap or flower sap from a range of chrysanthemum cultivars correlated strongly with their survival on the intact plants, which indicates that resistance is predominantly mediated by chemical factors. Dense aggregations or mating swarms of adult males can sometimes appear temporarily on brightly coloured artificial substrates or on flowers that are not breeding hosts (Kirk, 1985a; Terry and Gardner, 1990). They are usually associated with hot and sometimes humid weather and may only last a few hours. They do not necessarily form any threat to the crops nearby.

Infestation can vary across a field of similar plants. Plant condition may be patchily or directionally affected by soil conditions and microclimate and this can affect population growth, possibly through an effect on thrips nutrition (Chapter 4). Bullock (1970) studied population growth rates of *Thrips nigropilosus* across a field of pyrethrum in East Africa. Several types of population growth curve were recorded patchily across the field, but the differences were thought to be due to soil variation in retention of moisture affecting the viability of pupae in the soil rather than direct plant effects. Wilson *et al.* (1996) found that larvae of *Frankliniella schultzei* and *Thrips tabaci* showed a preference for cotton seedlings infested with mites.

Population growth on a plant can be affected not just by the plant itself but by the surrounding plants. Intercrops (crop mixtures) frequently give higher yields than monocultures through the agronomic benefit of reduced competition, but they are more labour intensive, so commoner in developing countries.

Insect infestations are often lower per plant on intercrops than on monocrops and this 'associational resistance' may arise in many ways (Theunissen, 1994). The most frequent explanations are visual or olfactory disruption of the host-finding behaviour of the pest and increased harbouring of beneficial predators and parasites. Intercropping is perhaps more likely to increase, rather than decrease, pest populations when the pest is polyphagous or the mixed crops are closely related (see Chapter 16).

Short-Term Weather Effects

Rain

Heavy rain can wash thrips off plants and down to the soil surface, causing sudden sharp declines in population density (Bailey, 1933; Harris *et al.*, 1936; Bonnemaison and Bournier, 1964; North and Shelton, 1986a). For example, Buhl (1937) recorded a drop of at least 95 per cent in the numbers of *Kakothrips pisivorus* larvae on field beans during heavy rains. After rain, dead thrips can be found adhering to wet leaves or drowned in wet flowers, whilst survivors persist under leaves or within flowers and buds (Bailey, 1933, 1934a; Buhl, 1937). The habit of feeding on the underside of leaves, found in some species, may lessen mortality from rain (North and Shelton, 1986b), but those that escape being dislodged can still be hindered by becoming stuck to moist surfaces or getting water on their wings, which prevents flight until the fringes have dried and can be arranged for flight (see Chapter 5).

Many thrips seem able to survive a long time on or in water and can even escape from it by overcoming the surface tension of the meniscus. Adult thrips can survive on a water surface for many hours (Wardle, 1927; Hall, 1930) or can move their wings and legs to swim to a solid surface and clamber out (Bailey, 1933), whereas submerged thrips can walk and emerge through the meniscus apparently unharmed (Kirk, 1984). However, thrips in the soil may be trapped under water and be unable to move away. Brødsgaard (1993) submerged individuals of *Frankliniella occidentalis* and found that the time after which 50 per cent died (the LT_{50}) was 45 h for second instar larvae, 40 h for pupae and 21 h for adults. Immature thrips can even moult under water. Larvae of *Taeniothrips inconsequens* can stand as much as 3 days under water, whereas the pupae die within 20 h (Bailey, 1934b). However, Bailey (1933) found that only two adults of *Caliothrips fasciatus* from an original ten survived more than an hour submerged.

Population densities on plants do not recover quickly after rain, so thrips that are washed to the ground do not appear to return to the plants. Some second instar larvae may be able to pupate prematurely and survive, but most thrips appear to be killed. Brødsgaard (1993) found that submergence softened the cuticle of *Frankliniella occidentalis* and made thrips vulnerable to damage while being handled. This softening, in combination with the action of soil particles and becoming embedded in mud as the water drains away, may be more of a problem than drowning for thrips that are washed onto the soil surface. Thrips that are below the surface when rain starts may be sealed in and die in the soil (MacGill, 1929, 1930; Bonnemaison and Bournier, 1964).

Wardle (1927) claimed that the effect of heavy rainfall in reducing thrips infestation was often more apparent than real for *Thrips tabaci* on cotton. The rain was sometimes not so much mechanically removing thrips as stimulating rapid growth of new leaves, which decreased thrips densities per leaf and increased the proportion of undamaged leaves. However, irrigation at the soil surface reduced the population more than rain, probably as a result of soil-caking killing thrips in the soil. Short intervals between waterings and avoidance of furrow irrigation, which does not wet the tops of the ridges, were suggested as ways of preventing crumbly soil and thus increasing mortality.

Drought

Drought is likely to affect thrips through wilting of the host plant, which affects its nutritional value, so can either increase or decrease thrips populations (see Chapter 4). In South Australia, populations of *Thrips imaginis* crash during the summer drought because flowers wither, so there are few places to breed, and soil dries out making pupation difficult (Andrewartha and Birch, 1954). The effect of drought on pupation is likely to be highly dependent on the soil type and depth of pupation (see below).

Dry weather, but not extreme drought, appears to favour increase in some species, such as *Caliothrips fasciatus*, *Diarthrothrips coffeae*, *Selenothrips rubrocinctus*, *Taeniothrips inconsequens* and *Thrips tabaci* (Urich, 1928; Bailey, 1933, 1934a, 1944; Fennah, 1965; Evans, 1967; Wardle, 1927), although it is difficult to know whether the effect is nutritional or just the absence of mortality from rain. Franssen and Huisman (1958) found that soil infestation by the brachypterous generation of *Thrips angusticeps* was 90 per cent less in summers that were rainy and cool rather than dry and hot, and ascribed this to larvae drowning in the soil.

Thrips are highly susceptible to dry microclimates and are liable to die after a few hours of exposure to only moderate relative humidities, although for the survivors there is little or no effect on the rate of development (MacGill, 1931; Lewis, 1962; Cederholm, 1963). For example, the mean time to death for active *Thrips australis* adults at 24°C and 50 per cent relative humidity is 45 h with access to water and 10 h without (Laughlin, 1977). Adults that overwinter in bark are likely to face these conditions, and *Limothrips cerealium* is more tolerant of desiccation. Most thrips seem to do best at about 70 to 90 per cent relative humidity, with a higher relative humidity being needed at higher temperatures, possibly because they are then more active and lose more water (MacGill, 1931; Cederholm, 1963; Laughlin, 1977). Shipp and Gillespie (1993) investigated survival of non-feeding *Frankliniella occidentalis* at a range of relative humidities. Larvae were most susceptible to low relative humidity, followed by adult females and then pupae. The high survival rate of the pupae was related to their low respiration rate. Relative humidity is particularly difficult to measure as it can change rapidly over very small distances; boundary layer effects can maintain high humidity on a plant surface or in small rearing tubes, even when the air is very dry a few centimetres away. Adults and larvae can move to avoid dry conditions (Cederholm, 1963), but eggs cannot. Terebrantian eggs are buried within plant tissue well protected against dry air, whilst tubuliferan eggs laid on the leaf surface are more at risk, although they should benefit from the moist air adjacent to the leaf.

Cold

Table 7.2 shows the survival times of three species of thrips at various low temperatures. *Thrips palmi* is a widespread polyphagous pest in glasshouses and on outdoor crops in warmer climates. It cannot survive subzero temperatures for more than a few days. *Frankliniella occidentalis* is a glasshouse pest in Europe, Canada and the USA. It does not appear to survive the winter outside glasshouses in northern Europe and its poor survival rates at low temperatures probably account for this (Table 7.2). However, the species appears to be able to overwinter in some areas that experience severe frosts, such as the southern interior of British Columbia and the north-eastern USA, where it has survived air temperatures below 0°C on 35 successive days (Felland *et al.*, 1993). Brødsgaard (1993) found that the cold tolerances of an

Table 7.2. Some estimates of the lethal time after which 50 per cent of thrips will have died (LT_{50}) for various acclimation temperatures and exposure temperatures.

Thrips species	Acclimation temperature (°C)	Exposure temperature (°C)	Instar	LT_{50} (h)	Reference
Frankliniella occidentalis	23/15/2	−5	A	57.5	Brødsgaard (1993)
	23/15/2	−5	P	56.3	"
	23/15/2	−5	PP	62.6	"
	23/15/2	−5	L2	59.9	"
	23/15/2	−10	A	32.1	"
	23/15/2	−10	P	15.4	"
	23/15/2	−10	PP	12.1	"
	23/15/2	−10	L2	24.7	"
Limothrips cerealium	Variable	+5	A	3360	Lewis (1962)
	Variable	+1	A	1260	"
	Variable	−1	A	1260	"
	Variable	−5	A	460	"
	10	−10	A	47.4	"
	31	−10	A	46.5	"
Thrips palmi	Variable	0	A	101	Tsumuki *et al.* (1987)
	Variable	0	L2	142	"
	Variable	−5	A	82	"
	Variable	−5	L2	106	"
	Variable	−10	A	14	"
	Variable	−10	L2	26	"

Frankliniella occidentalis was exposed to a succession of three acclimation temperatures.
Instars: A = adult female, P = pupa, PP = prepupa, L2 = second instar larva.

apparently hardy strain from British Columbia and a non-hardy strain from Denmark were the same and suggested that the apparently hardier strain might behave differently and use a different overwintering site, such as deep within the soil rather than in leaf litter or on the soil surface. *Limothrips cerealium* is common in northern Europe and adult females overwinter out of doors, mainly in bark. The species is regularly exposed to subzero temperatures and, predictably, has much longer survival times than *Thrips palmi* and *Frankliniella occidentalis*.

Air temperatures fluctuate in the field, whereas most experiments investigate the effects of constant temperature, and the result is not necessarily the same. Tsumuki *et al.* (1987) found that adult *Thrips palmi* in winter survived longer at a temperature varying between +5°C and −5°C over a 24-h cycle (LT_{95} = 23 days) than at a constant 0°C (LT_{95} = 9 days).

During severe frosts at night, the soil surface can be colder than the air above it as a result of radiant cooling and cold air drainage, but at only a few centimetres deep the temperature can stay above freezing. For example, during a severe frost in northern England, the air was −13°C at a height of 1.5 m, while the soil was −18°C on the surface, +0.1°C at a depth of 0.1 m and +2°C at a depth of 0.5 m (Bale, 1991). The deeper thrips overwinter in the soil, the more they are buffered against extremes of ambient air temperature, with the effect being greatest over the first few centimetres. Snow provides further insulation in cold regions.

The supercooling point at which freezing occurs within an insect is often used as a measure of cold-hardiness. A thermocouple is attached to an individual and the surrounding temperature lowered steadily. The temperature reading decreases gradually, but when the insect freezes, the latent heat of crystallization produces a fractional increase in the body temperature, allowing the supercooling point to be identified (Lee, 1991). Brødsgaard (1993) found that adults and pupae of *Frankliniella occidentalis* reared at 25°C could be supercooled, but died once frozen, so were ‘freeze-intolerant’. Supercooling points were −21.3°C (range: −26.9 to −14.0°C) for adults and −25.1°C (range: −30.8 to −19.5°C) for pupae. The higher supercooling point in adults was thought to result from the presence of ice nucleators in the gut contents, whereas the pupae would not have fed and so would have had empty guts. Murai (1977) investigated *Frankliniella intonsa* in Japan and found supercooling points of −28.8°C (range: −31.0 to −26.4°C) for diapausing adults, −27.5°C (range: −31.2 to −24.8°C) for non-diapausing adults and −24.1°C (range: −27.6 to −17.4°C) for second instar larvae. Figures of about −20°C are typical for many temperate insects (Lee, 1991). However, since thrips appear to have considerable pre-freeze mortality, the supercooling points may not be the main determinants of cold-hardiness.

In some species, such as *Frankliniella occidentalis*, *Thrips australis*, *Thrips imaginis* and *Thrips tabaci*, there is considerable variation in the colour of the adult females. Darker forms are usually slightly larger and tend to increase in numbers during cool seasons or during cooler spells within warm seasons (Bryan and Smith, 1956; Mound, 1967; Kirk, 1984). The dark, light and intermediate colour forms of *Frankliniella occidentalis* are under genetic control (see Chapter 6), and the dark forms appear to be better at surviving cold and wet periods. The variation in the other species might be genetic, but could also depend on the temperature during development. An association between small size and high development temperature is well known in other insects. Whether the underlying factor is heritable or not, variation in the survival, migratoriness and reproductive capacity of populations is likely to occur as the proportions of the different forms change, complicating the modelling of population dynamics.

Heat

In hot dry climates, thrips can escape extremes of heat by retreating to protected sites, such as within inflorescences or below vegetation. Thrips are so small that their high surface area to volume ratio will bring about rapid equilibration with the temperature of their surroundings, but if they are on a surface exposed to the sun this could be dangerously high. However, dryness rather than heat alone is likely to be the main problem for thrips on vegetation.

The development rate of *Frankliniella occidentalis* increases up to about 30°C, but then decreases above this temperature (Robb and Parrella, 1991; Gaum *et al.*, 1994a). Feeding rates may also start to decrease at temperatures above about 30°C (see Chapter 4). MacGill (1937) reared *Thrips tabaci* easily at 27°C and there was no detectable increase in mortality in larvae exposed to 38°C for 2 hours, provided the air was kept moist. *Thrips palmi* cannot develop at temperatures of 35°C and over (Kawai, 1985a) and adults and larvae die after 23 h at 40°C, 30 min at 48°C or 7 min at 55°C (Nonaka *et al.*, 1982). Cederholm (1963) exposed several species of cereal thrips to steadily increasing temperatures and found that they showed uncoordinated behaviour, walked backwards and had symptoms of paralysis in the range 40 to 50°C, finally becoming immobilized at about 50°C. Degree-day models of development might need to incorporate cut-off points for temperatures above about 35°C (Dintenfass *et al.*, 1987; Rhodes *et al.*, 1989; see also Chapter 2).

The possibility of manipulating temperature and relative humidity to control *Frankliniella occidentalis* in glasshouses at the end of the vegetable-growing season has been investigated by Shipp and Gillespie (1993). Larvae were most susceptible to high temperature, followed by adult females and then pupae. Sanitation trials demonstrated that 40°C and a vapour pressure deficit of at least 4.76 kPa (below 36 per cent relative humidity) would control all stages at the end of the season while the glasshouse was being prepared for the next crop.

Short-Term Biotic Effects

The effects of predators, parasites and diseases are not easily observed in thrips. It is therefore unclear whether the shortage of information on them is an artefact of observational difficulties or genuinely reflects that they rarely have much effect on populations. After a long and detailed study of *Thrips imaginis*, Davidson and Andrewartha (1948b) reported that no predators, parasites or diseases of any importance had been observed. The controversial conclusion that the population size of this species was determined by the weather, with no room for other factors (Andrewartha and Birch, 1954), may have given the impression that biotic effects are unimportant for thrips in general. The effect of natural control agents in the field has also been found to be negligible for some other thrips (Fennah, 1963; Bonnemaison and Bournier, 1964; Tamò *et al.*, 1993a), but this is not necessarily so for all species, and in glasshouses biotic agents are becoming progressively more useful and widespread (see Chapter 18).

Predators

The main predators of thrips are mites, heteropteran bugs such as *Orius*, lacewing larvae, ladybird larvae, hoverfly larvae, small spiders and other thrips (see Chapter 8 and Appendix 3b in Lewis, 1973). A few have been used successfully for biological control of thrips (Chapters 16–18). The predators are more likely to feed on the slow-moving immature stages than the more active adults. Although there are many records of predation in the field, it is difficult to know without quantitative studies whether any particular predator is having a significant impact on pest numbers, as no trace is left of the victims. The seasonal cycles of some predators are closely related to those of their prey (Callan, 1943; Isenhour and Marston, 1981), but this in itself does not confirm that the predators are having much impact on thrips densities. Successful biological control can involve low levels of predator and pest at equilibrium.

Thrips are often found with mites attached to them, but the mites are not necessarily predatory. Many mites disperse phoretically by attaching themselves to a flying insect and then detaching themselves later (Binns, 1982). However, Dyadechko (1977) found that the mite *Hauptmannia brevicollis* (Erythraeidae) attaches itself to thrips, mainly to the bases of the legs, and then drops off about 5 or 6 days later, after which the thrips die. In the Ukraine, as many as 30 to 40 per cent of larvae of *Anaphothrips obscurus* from meadow stubble had mites attached and 63 per cent of *Odontothrips phaleratus* on clover. Other species in this mite genus probably also live on Thysanoptera (Haitlinger, 1987).

Some solitary wasps, such as *Spilomena* spp., specialize in foraging on thrips larvae to provide food for their offspring (Danks, 1971). Nests are aggregated and about 50 thrips are needed per cell. Encouragement of nesting, for example by provision of bramble stems, might provide a local decrease in thrips, but is unlikely to be cost-effective on a large scale because of the vast number of wasps that would be needed.

Ants, especially *Iridomyrmex* spp., have sometimes been observed to keep thrips off plants, probably by attacking and eating them (Moulton, 1909a; Froggatt, 1931; Kirk, 1984), but the impact on population dynamics appears never to have been investigated quantitatively.

While most vertebrates, including humans, eat some thrips accidentally from time to time, it is unlikely that foraging for thrips would be energetically cost-effective except for very small species. Buhl (1937) observed various small birds hunting for larvae of *Kakothrips pisivorus* on peas and beans in Germany, and Hamilton (1930) found that thrips constituted 10 per cent of the bulk of the stomach contents of a sample of 400 immature (8 to 12 mm long) toads (*Bufo americanus*) in the USA.

Parasites

The main parasites of thrips are endoparasitoid wasps (see Chapter 9; Ananthakrishnan, 1984; Loomans *et al.*, 1995). Most are in the chalcidoid family Eulophidae, including a range of species from the genera *Ceranisus* (previously known as *Thripoctenus*), *Goetheana*, *Entedonastichus* and *Thripobius*. Adult wasps search plants for a thrips larva and lay an egg inside its body. The wasp larva develops within the thrips, eventually killing it, and finally emerges through the larval, prepupal or pupal body wall. This means that future mortality from parasitoids can be measured by collecting larvae while they are still alive and dissecting them for developing parasitoids.

Sakimura (1937a,b) studied *Ceranisus menes* (= *Thripoctenus brui*), which parasitizes *Thrips tabaci* and several other species of thrips, in Japan. Some species seem to be able to escape by struggling, whereas larvae of *Thrips tabaci* do not. Samples from onion fields revealed up to 88 per cent parasitism, with an average of 34 per cent. There was a positive correlation between thrips density and percentage parasitism, so the wasp was having a density-dependent effect and could be regulating the thrips population (Fig. 9.7). The density of adult wasps was correlated with the density of the prey and the effectiveness of the wasp increased as the season progressed.

High rates of parasitism, which show high eventual mortality rates, have been recorded. In the USA, Russell (1912) reported as much as 70 per cent parasitism of *Caliothrips fasciatus* by *Ceranisus russelli*, although Bailey (1933) estimated only about 5 per cent. In Germany, Buhl (1937) recorded averages of 33 per cent and 23 per cent for parasitism of *Kakothrips pisivorus* larvae by *Ceranisus menes* in successive years, but concluded that parasitism did not play a significant role in limiting population size. Large values for percentage parasitism do not guarantee that parasitism is the key factor responsible for determining population size. To prove this, it would be necessary to analyse life-table data over several years (Varley *et al.*, 1973).

A few parasitoid wasps in the family Trichogrammatidae lay their eggs inside thrips' eggs. The developing wasp kills the host egg and emerges from it. The adult wasps are easily overlooked because they are extremely small, each individual having to grow to maturity with the food provided by only a single thrips egg. Mortalities have been quite high where they have been studied. For example, *Megaphragma mymaripenne* gave 3 to 51 per cent egg parasitism of *Heliothrips haemorrhoidalis* on avocado in California, but did not appear to regulate the thrips population (Hessein and McMurtry, 1988), and two unidentified *Megaphragma* species gave 0.5 to 31 per cent egg parasitism of *Megalurothrips sjostedti* on cowpea in West Africa and as much as 53 per cent on other plants (Tamò *et al.*, 1993a) (see Chapter 9).

Some parasitoids have free-living first-instar larvae (planidia) that locate a host and then attach themselves to the outside of the body. Chalcidoid planidia, some of which have been identified as being in the family Eucharitidae, occur occasionally on pest thrips (Clausen, 1940; Wilson and Cooley, 1972; Beshear, 1974).

Thrips are sometimes parasitized by nematodes, which enter by piercing a hole in the body wall of larvae or pupae (see Chapter 9). Up to 68 per cent parasitism by *Thripinema nicklewoodii* has been recorded in *Aptinothrips rufus* on grasses in England, but the parasitism was thought not to exert a profound influence on the abundance of the thrips (Lysaght, 1936, 1937). Coarseness of growth of the grass appeared to be unfavourable for parasitism. A range of species of thripid, including *Frankliniella occidentalis*, have now been recorded parasitized by this nematode (Greene and Parrella, 1995, see Plate II.4). The nematodes do not seem to affect the appearance or movement of the host, so are easily unnoticed. Their main effect on the population is through parasitic castration of the females, which stops oviposition.

Diseases

Several entomopathogenic fungi, including species of *Aspergillus*, *Beauveria*, *Entomophthora* and *Verticillium* attack thrips (see Chapter 10), although there are few quantitative data on natural infection rates in the field. Fungi can kill, leaving the body mummified or covered in white mould, or may cause parasitic castration without other outward symptoms (Lysaght, 1937). All stages of thrips are attacked by fungi, particularly at high humidity. Dyadechko (1977) found that large numbers of cereal thrips larvae died from *Beauveria bassiana* after stubble ploughing in the Ukraine, and mortality reached 100 per cent after grains on which the fungus had been cultured were distributed.

A microsporidian (protozoan) parasite, *Mrazekia* sp., has been recorded infecting virtually 100 per cent of larvae and adults of *Scirtothrips oligochaetus* and *Thrips flavus* on cotton in India and subsequently the thrips population almost disappeared (Raizada, 1976).

Competition

Interspecific effects of other herbivorous insects appear to be infrequent. However, in field experiments *Aphis gossypii* always eliminated populations of *Thrips palmi* on aubergines in Japan within a few weeks, regardless of the initial density of aphids introduced (Kawai, 1985b).

Intraspecific effects are perhaps more frequent. Although Andrewartha and Birch (1954) claimed that *Thrips imaginis* did 'not appear to be overcrowded except perhaps locally and temporarily' and that 'at no time were the thrips numerous enough to consume more than a fraction of the food', there are growing indications that there is competition between phytophagous thrips for space and food, at least within flowers.

The relatively recent discovery of fighting in flower thrips, fungus thrips and gall thrips suggests that this behaviour is widespread (Crespi 1986, 1992; Terry and Gardner, 1990). It has been observed in flower thrips that have no obvious morphological adaptations to fighting, such as *Frankliniella intonsa* and *Frankliniella occidentalis*, so may be present in many other flower thrips (Kirk, 1996). Structures that are associated with fighting, such as abdominal tubercles, are also found on many thrips that are crop pests, for example

Kakothrips pisivorus and *Limothrips cerealium*, so careful observation of these may reveal fighting behaviour. Competitive interactions are perhaps most likely between males guarding females. A knowledge of such interactions is necessary to understand spacing patterns and effects of density on population growth.

Some behavioural responses to density are known. Adult *Megalurothrips* species chase away thrips that land on the same flower (Appanah and Chan, 1981), and *Microcephalothrips abdominalis* leave flowers sooner when they are more crowded (Gopinathan *et al.*, 1981). The dispersal of adult *Frankliniella occidentalis* in glasshouses has been shown to be density dependent (Puche and Funderburk, 1995). Fecundity can also be affected. As adult density increases, *Frankliniella tenuicornis*, *Stenchaetothrips biformis* and *Thrips fuscipennis* lay fewer eggs, which is possibly an effect of disturbance (Nugaliyadde and Heinrichs, 1984; Malchau, 1991; Kirk, 1994), and similar effects can occur in *Hercinothrips femoralis*, although the responses are mixed (Takrony, 1973). The reductions in fecundity can occur within days at commonly encountered densities. For example, *Stenchaetothrips biformis* lays on average 10.2 eggs per female at one pair per rice seedling, but only 2.6 at five pairs per seedling. Puche and Funderburk (1992) found a much higher mortality of larvae of *Frankliniella fusca* at a density of 117 larvae per bagged peanut leaflet than at 32 larvae per leaflet. Such effects will give large density-dependent reductions in fecundity and survival, but they would be difficult to detect in the field.

It is easy to overlook crowding of thrips because it is most likely to occur in small spaces that are hidden from view; flowers that look empty can reveal dozens of thrips when pulled apart. Often 10 to 30 thrips occur in a flower, and thousands can be found during outbreaks (Zeck and Noble, 1932). Competition for pollen may occur when access to the anthers is spatially and temporally restricted, as when thrips larvae feed mainly at anthers by night. This could occur with as few as 20 thrips in a flower and could produce density-dependent reductions in fecundity and growth (Kirk, 1984). An individual flower thrips can consume about 0.5 per cent of the pollen in a typical flower (Kirk, 1987b), so there may even be a shortage of pollen in infested flowers.

A density effect that is rarely considered is that of frequency of mating. At very low densities, many adult females of arrhenotokous species, such as *Thrips palmi*, may not encounter any males and so produce only male offspring (Kawai, 1987, 1988). The low number of females produced at low density will increase fluctuations in population size and may accelerate population decline. Consideration of the low density effect may allow reduced inputs of insecticides (Kawai and Kitamura, 1987, 1990).

Overwintering and Pupation

Instars and sites

In warm climates, thrips can continue breeding throughout the year without an inactive period. For example, in South Australia, *Thrips imaginis* populations are much smaller in winter, but breeding continues slowly (Andrewartha and Birch, 1954). Populations can also breed continuously in heated glasshouses, completing many generations each year. In cooler climates, thrips overwinter mainly as larvae or adults in soil or as adults on low vegetation, leaf litter or bark. Often, the overwintering stage and site are not known and only a few crop pests appear to have been studied in detail.

Many terebrantian thrips overwinter in soil. For example, *Kakothrips pisivorus* larvae drop from the host plant in summer and burrow into the soil (Williams, 1915; Kutter and Winterhalter, 1933), where they remain as larvae throughout the winter until the following spring, then pass quickly through the prepupal and pupal stages. The adults emerge and force their way to the surface. Each individual spends about 10 months in the soil. Larvae of the second generation of *Thrips angusticeps* also drop from the host plant in autumn and burrow into the soil, but they then pass through the prepupal and pupal stages to spend the winter as adults, forcing their way to the surface in spring (Bonnemaison and Bournier, 1964). They each spend about 9 to 10 months in the soil. However, up to 10 per cent, and sometimes as many as 25 per cent, do not emerge until a year later, having spent about 20 months underground (Franssen and Huisman, 1958). The majority of the few species that have been studied in the soil spend most of the winter as adults, with a few as larvae, but apparently none as prepupae or pupae. As many thrips spend most of their life in soil, this phase is likely to be particularly important for population dynamics.

The depth at which larvae or adults spend the winter varies according to species and soil type (Table 7.3), being typically about 20 to 30 cm, but sometimes as deep as 100 cm. Soils can be classified according to a continuum of increasing particle size, from clay through silt and sand up to gravel. Sandy soils are light (easily tilled) with large particles and a high bulk density, whereas clay soils are heavy with small particles and a low bulk density (Brady, 1974). Thrips larvae are thought to go deeper in lighter soils than in heavy clay, because it is easier to penetrate (Cameron and Treherne, 1918; Bailey, 1944). Larvae of *Kakothrips pisivorus* go deeper in loose than in compacted soil (Kutter and Winterhalter, 1933) and laboratory experiments with cylinders of uniform soil have confirmed that larvae of *Megalurothrips usitatus* go deeper in soils with lower clay content and higher bulk density (Chang, 1989). However, thrips will not burrow into fine sand (Bailey, 1933; Franssen and Huisman, 1958). Foster and Jones (1915) found that larvae of *Taeniothrips inconsequens* penetrated much deeper into a heavy clay soil than into a hard sandy, gravelly soil, possibly because of the greater number of cracks available. Burrowing thrips can also exploit worm holes and old root channels (Bailey, 1944), so the condition and the moisture content of the soil will also affect overwintering depth.

Uniform soils are often used in laboratory experiments, but in the field, soils have a profile with characteristics that change with depth. Thrips appear to stop burrowing either when they can go no further or when suitable physical or microclimatic conditions are found (Bailey, 1933). The soil profile is likely to determine the depth of penetration. For example, larvae of *Taeniothrips*

Table 7.3. Some records of the soil depth at which second instar larvae or adults overwinter in the field.

Thrips species Crop species	Modal depth (cm)	Maximum depth (cm)	Soil type	Region	Reference
Haplothrips tritici	0–10	40–50	–	Kazakhstan	Tansky (1958)
	0	–	–	France	Bournier and Bernaux (1971)
Kakothrips pisivorus	8–30	–	–	England	Williams (1915)
	25–35	35–40	–	Switzerland	Kutter and Winterhalter (1933)
	25–30	45–50	Loam	Germany	Buhl (1937)
Stenothrips graminum	30–50	60–70	Clay	Netherlands	Franssen and Mantel (1965)
	50–70	100–110	River clay	Netherlands	"
Taeniothrips inconsequens	15–23	38–41	Well cultivated	California	Moulton (1909b)
	15–23	38–41	Heavy clay loam	California	Foster and Jones (1915)
	15–18	36–38	Sandy loam	California	"
	13–15	30–33	Gravelly loam	California	"
	15–25	38	Clay	Canada	Cameron and Treherne (1918)
	6–8	26–28	Clay	Vermont	Skinner and Parker (1992)
	4–8	26–28	Fine sandy loam	Vermont	"
	6–8	14–16	Gravelly	Vermont	"
	0–10	40–50	Well drained	Pennsylvania	Brose *et al.* (1993)
	0–10	30–40	Moderately drained	Pennsylvania	"
	0–10	20–30	Poorly drained	Pennsylvania	"
Thrips angusticeps	30–40	70–80	Clay	Netherlands	Franssen and Huisman (1958)
	50–60	70–80	River clay	Netherlands	Franssen and Mantel (1965)

The modal depth is the one at which most thrips were found. The maximum depth was sometimes the greatest depth at which samples were taken, so that thrips may have occurred below this.

inconsequens appear to stop at the transition from the A horizon (mineral soil with organic matter) to the B horizon (mineral soil with accumulation of clays and hydrous oxides), and variation in the thrips distribution between sites can be partly explained by variation in the depth of this transition (Skinner and Parker, 1992). Soils that have been ploughed repeatedly have an undisturbed compacted layer, known as a plough sole, just below the depth reached by the plough share, and thrips stop once they are at or within it (Bailey, 1934b, 1944). Thrips may also remain near the surface among thick grass roots, even in a light soil in which they might have been expected to go deeper (Foster and Jones, 1915; Cameron and Treherne, 1918). Gerdes (1982) found many thrips in the top few centimetres of soil among roots at the bases of common Illinois prairie plants.

Other factors, such as soil moisture, may affect the choice of overwintering site. Skinner and Parker (1992) suggested that larvae of *Taeniothrips inconsequens* might be preferentially selecting well-drained sites, and Brose *et al.* (1993) found that this species goes deeper in better-drained soils. However, *Thrips imaginis* will burrow into the soil whatever the moisture content if given no choice (Andrewartha, 1934).

Some thrips overwinter above the soil and *Limothrips cerealium* has been particularly studied in England (Lewis, 1959, 1962; Lewis and Navas, 1962). In late summer, mated adult females fly from cereal fields to their overwintering sites, which are mainly crevices in rough bark of pine, oak and other trees nearby. They can form aggregations of up to about 60 individuals in a crevice, but many dead thrips that have accumulated from previous years can also be found there. A mature tree can hold as many as 500,000 overwintering thrips in cereal-growing areas of England. As winter progresses, the thrips move from outer crevices to inner ones and then back to outer crevices as spring approaches. If overwintering adults are warmed to 5°C, they become active, but their ovaries do not develop. Other thrips use other sites. For example, *Aptinothrips rufus* moves from standing grass to grass litter and then back to standing grass as the winter progresses; *Caliothrips fasciatus* uses a wide range of plants that remain green; *Chirothrips manicatus* uses dead grass inflorescences and litter; *Frankliniella tenuicornis* uses cereal stubble and shoots of winter rye; *Limothrips denticornis* uses litter and bark; *Thrips laricivorus* moves under the previous year's bud scales of the second uppermost whorl of shoots on spruce trees, but under the current year's bud scales of the terminal needle clusters on Weymouth pine; and *Thrips tabaci* uses winter wheat, lucerne and weeds (Bailey, 1933; Lewis and Navas, 1962; Wetzel, 1963; Maksymov, 1965; Köppä, 1969, 1970; North and Shelton, 1986c). In France, larvae of *Haplothrips tritici* penetrate 25 to 30 cm into the soil to spend the summer, but then move out of the soil to litter and stubble to overwinter (Bournier and Bernaux, 1971).

Although some species have a mixed strategy for overwintering, most species can be categorized as spending winter either above or below ground. Species that are dormant above ground retain more flexibility because they can feed in warm spells and can respond to an early spring by flying promptly to host plants, whereas species in the soil cannot feed and are buffered from changes in air temperature. Flexibility is therefore most likely to be of use to species with evergreen hosts, because these are available at any time. Data for British phytophagous terebrantian thrips support this hypothesis. If species that can be found as adults in December and January according to Mound *et al.* (1976) are assumed to overwinter above ground, and if those that breed on conifers (*Coniferales*) or grasses (*Gramineae*) are considered as having evergreen hosts, 56 per cent of thrips species (10 out of 18) with evergreen hosts overwinter above ground, whereas only 23 per cent of species (16 out of 70) with non-evergreen hosts overwinter above ground. One unusual species that fits this pattern is *Thrips juniperinus*, which overwinters on evergreen juniper, often beneath snow.

Some pest species, such as *Hercinothrips femoralis* and *Frankliniella occidentalis*, can pupate on the plant or at the soil surface. Takrony (1973) found that brighter lighting increases the proportion of *Hercinothrips femoralis* pupating in the soil, suggesting that a negative phototaxis is involved. Although *Scirtothrips citri* was widely assumed to pupate in the soil or in the layer of leaf litter beneath citrus trees, Grout *et al.* (1986) showed that only about 20 to 30 per cent of individuals do this, the rest pupate on the tree, with a higher proportion pupating in larger trees, presumably because these provide more shade and protection.

There are several reports of stages other than adults overwintering above ground, but some of these may be occasional individuals that have not entered diapause and are managing to survive in protected sites or possibly individuals that have been parasitized. Wetzel (1963) recorded larvae of *Aeolothrips albicinctus* and *Aptinothrips rufus* overwintering in eastern Germany. *Scirtothrips citri* overwinters as an egg in California (Horton, 1918; Rhodes *et al.*, 1989), but the eggs may not develop over winter in the cooler parts of California simply because of their high temperature threshold for development (J.G. Morse, University of California, Riverside, 1995, personal communication). Egg or embryonic dormancy does not appear to have been recorded in any other phytophagous thrips.

Dormancy

Dormancy is a state of suppressed development, which can be either an immediate response to adverse environmental conditions (quiescence) or a programmed response that acts through an environmental cue, such as photoperiod, in advance of adverse conditions (diapause) (Danks, 1987). Overwintering thrips are clearly exhibiting dormancy when larvae in the soil do not moult, as in *Kakothrips pisivorus*, or when adults do not become sexually mature or lay eggs, as in *Limothrips cerealium* and *Thrips angusticeps*. Species that overwinter as adults can sometimes become active and feed during mild spells, but if reproductive development remains suppressed, they are still considered as being dormant. A suppression of reproductive development that is programmed in advance, usually by short daylength, is found in many adult female insects and is known as reproductive or ovarian diapause.

This has been investigated for *Frankliniella intonsa* in Japan (Murai, 1977). The adult females overwinter in soil and litter without maturation of the ovaries and this suppression of reproduction can be induced artificially by exposing the larvae to short-day conditions of 10L:14D. Second instar larvae are the most sensitive to photoperiod, although exposure of this instar alone is not sufficient to induce adult diapause. Warming terminates the dormancy, but the response is delayed by several weeks unless the females have previously been exposed to low temperature for at least 50 days.

Reproductive diapause is also found in *Anaphothrips obscurus* in Oregon, USA, and is induced by short days of 10L:14D (Kamm, 1972). In England, adult female *Limothrips cerealium* overwinter with immature ovaries and can become active when the temperature is raised, but their ovaries do not start to develop, indicating a reproductive diapause (Lewis, 1962, 1973). Although dormancy has been investigated for very few species, the results suggest that a short-day-induced reproductive diapause may be common for temperate thrips. However, experiments with short-day photoperiod have not been able to induce reproductive diapause in glasshouse populations of *Frankliniella occidentalis* in Denmark (Brødsgaard, 1994), but this may be a special adaptation to glasshouse conditions.

Geographical variation in diapause response, particularly with latitude, is found in many insects (Danks, 1987). The critical daylength at which 50 per cent of individuals of *Frankliniella intonsa* are induced to enter diapause increases by about 1 h for every 5° of latitude from south to north through Japan (Murai, 1977). Similar variation is likely in other species.

Some species that overwinter as adults, such as *Thrips tabaci* in Texas, have a temperature-induced reproductive quiescence and can lay eggs during warm periods (Sites and Chambers, 1990). Breeding can start as soon as warm weather arrives in spring.

Mortality

Mortality while overwintering in the soil is difficult to measure or attribute to particular causes. However, it can be high and the link between soil type and the probability of infestation for some thrips (see below) suggests that soil mortality can be a key factor in population dynamics. Mortality can also be high in species that pupate in soil during the summer. The mortality of species that overwinter above ground is potentially easier to measure, but rarely appears to have been investigated.

The mortality in turf in North Dakota reaches 60 per cent for *Limothrips denticornis*, 80 per cent for *Frankliniella tritici* and 40 per cent for *Anaphothrips cameroni* (R.L. Post, North Dakota, 1962, personal communication). Bailey (1933) recorded soil mortality in California of 60 per cent for *Caliothrips fasciatus*, with 34 per cent accounted for by death of mature larvae before moulting to the prepupal stage. Tansky (1958) recorded 53 per cent overwintering mortality of *Haplothrips tritici* in fallow fields in Kazakhstan, but only 39 per cent and 12 per cent in fields of wheat and perennial herbs, respectively. Autumn ploughing increased the mortality. Larvae of *Thrips calcaratus* in Wisconsin drop to the soil in spring and develop through to the adult stage in early summer, overwinter and then emerge the following spring (Raffa *et al.*, 1992a,b). As many as 80 per cent of these adults may die in mid- to late summer, well before the onset of cold weather. In laboratory experiments, Andrewartha (1934) recorded mortalities of *Thrips imaginis* in the soil of about 10 per cent under optimum conditions, but up to 99 per cent under adverse conditions, and Chang (1989) recorded mortalities from 7 to 39 per cent for *Megalurothrips usitatus*.

Overwintering mortalities above ground can also be high. Woglum and Lewis (1936) found that 16 per cent of adult *Caliothrips fasciatus* overwintering in the navel end of oranges died by midwinter in California, and 20 to 26 per cent of adult *Thrips laricivorus* died on conifers in Switzerland (Maksymov, 1965). The numbers of dead thrips that accumulate in old bark crevices indicate about 8 per cent average annual mortality of adult *Limothrips cerealium* overwintering in bark in England (Lewis and Navas, 1962).

The main causes of death in soil relate to its water content and the action of rainfall on the surface, and these both interact strongly with soil type. Andrewartha (1934) found that *Thrips imaginis*, in a red loam soil, survived best at a moisture content between about 10 and 34 per cent of dry soil weight (25 and 85 per cent of field capacity) and Chang (1989) found that optimum survival of *Megalurothrips usitatus* in a sandy loam occurred between 5 per cent and 25 per cent of dry soil weight. Thrips probably dehydrate at low moisture content and drown at high moisture content, but this has not been demonstrated. The maintenance of the small cells that thrips form around themselves in the soil may play an important role in survival. Bailey (1944) found that in sandy or gravelly soil, the cells of *Taeniothrips inconsequens* were liable to crumble as the soil dried out and that as these could not be repaired in dry conditions the thrips soon desiccated.

Rainfall can kill larvae making their way into the soil (see above) and heavy rainfall can puddle the surface of a heavy soil and seal it, making it difficult for larvae to enter or for adults to leave (MacGill, 1929, 1930). Wardle (1927) found that for soils with less than 15 per cent clay there was usually no caking after rain, but for 15 to 30 per cent clay, moderate showers would leave the soil compacted for a few days, and for over 30 per cent clay, heavy rain would form a short-lived crust. Irrigation could have the same effect. The high mortality caused by a combination of rain and heavy soil probably explains why infestations of some species are more serious on light soils, for example *Kakothrips pisivorus* on peas and beans in Europe (Williams, 1915; Buhl, 1937), and *Thrips tabaci* on cotton (MacGill, 1930).

However, infestations of other species appear to be more prevalent on heavy soils. Moulton (1909c) found that *Scirtothrips citri* was more abundant on orange trees on clay soils than on sedimentary or loam soils, possibly because of the special soil texture created by frequent irrigation and cultivation. The cells of the thrips would crumble in light soils, but could survive being turned when within clods of clay. *Selenothrips rubrocinctus* on cacao in São Tomé caused most problems on clay soil (Urich, 1928), but since this species pupates mainly on the plant, the reason is more likely to relate to plant stress. Thrips that pupate near the surface in summer are more susceptible to heat and Bailey (1933) found that in California, no pupae of *Caliothrips fasciatus* survived within 8 cm of the soil surface if it was unshaded.

Ploughing can bring larvae, prepupae, pupae and adults to the soil surface, crushing and trapping them or exposing them to drying, cold and disease at a time when they may be too weak to retreat back into the soil. Reductions of up to 81 per cent of wheat thrips have been recorded after ploughing (Dyadechko, 1977).

Thrips that overwinter above ground are susceptible to extremes of heat, cold and low humidity (see above). By remaining above ground, they remain liable to be washed away by rain and Bailey (1933) found heavy winter mortality of *Caliothrips fasciatus*. Bark expands as it cools and takes up moisture and contracts as it warms and dries, so thrips that move deep into bark crevices are liable to be crushed during warm dry weather in spring (Lewis, 1962).

Emergence

Crop damage is often increased when thrips emergence from overwintering coincides with a sensitive plant stage such as bud burst, flowering or young seedlings (Foster and Jones, 1915; Kolb and Teulon, 1991). In such circumstances, it is particularly useful to be able to predict the time of emergence.

In soil-overwintering species, soil temperature is likely to be the best indicator of imminent emergence, as this is the only external cue that insects in the soil can use to detect the arrival of spring. *Thrips angusticeps* in the Netherlands emerge when the soil temperature reaches 6 to 8°C at a depth of 20 cm (Franssen and Huisman, 1958), *Taeniothrips inconsequens* in California emerge rapidly from heavy soil once the temperature has exceeded 11°C at a depth of 25 cm for 2 to 3 consecutive days (Bailey, 1944), and *Stenothrips graminum* in the Netherlands emerge once the soil at 30 cm reaches 13.5°C (Franssen and Mantel, 1965). Warm weather will therefore bring forward the time of emergence, as might deep snow cover, which can insulate the soil (Skinner *et al.*, 1992). Bailey (1934b) found that spring ploughing hastens emergence and a heavy cover crop retards emergence by several days. Both sexes of *Kakothrips pisivorus* emerge from the soil at the same time after overwintering (Buhl, 1937), but female *Caliothrips fasciatus* emerge 10 to 30 h before the males from pupae in the soil in summer (Bailey, 1933).

Raffa *et al.* (1992a) constructed degree-day models, using the Weibull function (Wagner *et al.*, 1984), for both the spring emergence of adult *Thrips calcaratus* and the swelling and opening of the buds of its host, the American basswood tree (*Tilia americana*). There is close phenological synchrony between emergence and bud burst, but with the former slightly ahead of the latter, which explains why thrips concentrate on the earliest opening buds.

Once emergence starts, the rate can fluctuate, increasing in warm dry weather and decreasing when cool and wet (Foster and Jones, 1915; Buhl, 1937; Bailey, 1944). *Taeniothrips inconsequens* takes about 2 to 3 days longer to appear on trees when the soil is covered with weeds and grass than when it is bare, but it is not certain whether this is an effect of the shading and cooling of the soil or of thrips being delayed by feeding on the ground vegetation first. After emergence, there is a maturation feeding period of about a week before *Stenothrips graminum* starts laying (Franssen and Mantel, 1965).

Thrips that overwinter above ground become active as the air temperature increases. Sites and Chambers (1990) calculated that *Thrips tabaci* populations in Texas began to increase after 88 degree-days and *Frankliniella occidentalis* after 114 degree-days (base temperature 6.7°C). However, microclimates are likely to vary at a site and so cause a spread of emergence. For example, *Limothrips cerealium* in England emerge sooner from bark crevices on the warmer south side of trees than from the north side (Lewis, 1963).

Long-Term Effects

Seasonal changes

As temperatures rise in spring, thrips end their dormancy (see above) and start breeding. New plant growth and the opening of flowers provide more and better food and the higher temperatures encourage physiological development, thus increasing the rate of population growth. Degree-day models are widely used to model the development of individuals by integrating the excess temperature above a development threshold over time and this approach can be extended to populations. A small population exposed to the enormous food resources of a crop is liable to increase exponentially at first, but at high densities population growth slows down as factors such as food shortage or natural enemies start to act. The nature of the potential damage to the crop will determine the most appropriate type of modelling for the population dynamics of a particular thrips pest. In some crops, economic damage occurs whenever thrips reach a sufficiently high density (see Chapter 16), in others the timing relative to a sensitive plant stage is critical, while in horticulture or when viruses are involved, very low densities can cause damage. Towards the end of the season, populations decrease or sometimes crash rapidly to a lower level over winter, due mainly to a decrease in quantity and quality of food and lower temperatures.

Population size is often measured by sticky, water or suction traps (see Chapter 11). These can be convenient, but the traps measure flight activity rather than density and the numbers caught are difficult to relate to densities on plants. Much of the variation in the catch might be a function of the weather rather than the population size. However, activity can also affect the densities of thrips on plants, with fewer adults present in plant samples on warm days because more are in flight (Sites *et al.*, 1992). Irwin and Yeargan (1980) found that *Neohydatothrips variabilis* was four times more abundant than *Frankliniella tritici* on soybean when sampling plant parts, but only half as abundant according to trap catches, as a result of flight activity of *Frankliniella tritici* from other host plants. The method of data collection should be kept in mind when interpreting population dynamics data.

Davidson and Andrewartha (1948a) monitored the seasonal changes in populations of *Thrips imaginis* in Adelaide, by counting the number of adults in 20 roses. Day to day fluctuations, which are most likely to be weather-related effects on activity, were reduced by taking a 15-day moving average. The population started from low numbers in late winter (July to August), then grew rapidly in spring and early summer (September to December) to reach a peak that was followed by an abrupt decline in late summer (December to January). The total variability was considered to consist of three components:

natural growth of the population, the influence of weather on activity, and the influence of the weather on the rate of multiplication of the insects. When data for 7 years were analysed, the average trend during the period of increase consistently fitted a logistic function of numbers against time, and usually fitted the first portion of the curve, which approximates to exponential increase.

The peak in the numbers of *Thrips imaginis* was higher in years when it occurred earlier. This was explained in terms of an interaction between the weather and the main wild host-plants, such as Paterson's curse (*Echium plantagineum*). A wet season the previous autumn allows these plants to grow more before winter sets in and as a result flower earlier and more profusely in spring. Thrips populations will increase faster under such conditions, but the earlier start will mean a correspondingly earlier end to flowering. Wet autumns therefore encourage faster population increase, but shift the whole increase curve and its higher peak to earlier in the year. The mechanisms by which flower density affects the rate of increase of the population are not understood.

Dintenfass *et al.* (1987) studied the exponential resurgence of *Frankliniella occidentalis* after onions in Texas were sprayed with insecticide. The increase is similar to that of a population on a new crop except that there are virtually no natural enemies. Instead of using the standard equation for exponential increase, $N_t = N_0 e^{rt}$ (where t is time, N_t is the population size at time t, N_0 is the initial population size, and r is the intrinsic rate of natural increase), time was replaced with the accumulated degree-days or physiological time since spraying (d). A combination of the predicted number from this equation and the accumulated rainfall since spraying, which was detrimental to increase, gave a good fit to the numbers on plants. Over a period of up to 14 days after spraying, there was no indication of a levelling off of numbers, which implied an absence of resource limitation.

Fatzinger and Dixon (1996) have produced degree-day models for the onset of feeding in populations of *Gnophothrips fuscus* and for the development of female strobili (flowers) of slash pine in seed orchards in Florida, USA. Damage occurs when thrips feed during the developing bud stage, so the combination of the two models can be used to predict damage from the relative timing of thrips attack and plant sensitivity.

Modelling populations

The modelling of population dynamics becomes much more difficult when it moves away from classical exponential or logistic increases, because many more factors have to be taken into account. An insufficient knowledge of the biology of thrips means that such models often start with simple assumptions and omit many complicating factors.

Rhodes *et al.* (1989) modelled the development of *Scirtothrips citri* on Californian oranges. This thrips overwinters as an egg and then breeds in cycles that start more or less synchronized across the population. The damage is done by second and third generation larvae during the first 3 weeks after petal-fall and the timing of insecticide treatment is critical. The model incorporated degree-day data with a single sine wave fitted to daily maximum and minimum temperatures. The time spent in each life stage was modelled by a normally distributed random variable. However, there was major variation in seasonal population development when the model was compared with field data and this was attributed to the possible effects of other factors, such as food, predation and cold stress.

Megalurothrips sjostedti is a serious pest of cowpeas in West Africa. Thrips invade the crop and feed on buds and open flowers, causing them to drop off. Tamò *et al.* (1993b) developed a model for the thrips population and a separate model for plant growth and yield, then linked them by relating densities of thrips per flower to flower shedding. A stage-frequency matrix and a time-varying distributed delay model were used for both thrips and plants. For the thrips model, data for stage-specific development, survivorship and fecundity at a range of temperatures were obtained from experiments in environmental cabinets, and rates of predation and parasitism were measured in the field. Immigration rate was incorporated into the model as a function of the number of attracting plant organs, with the parameter for degree of infestation adjusted by eye to match observations. Losses from movement onto green plant parts and emigration were incorporated into larval and adult survivorship according to the proportion of ageing inflorescences. The model gave a close fit to field data for the year and site investigated, and could be used to predict the impact of thrips on yield for different amounts of rainfall and available soil phosphate.

Kawai and Kitamura (1987, 1990) modelled the population of *Thrips palmi* on cucumber, aubergine and sweet pepper in a plastic greenhouse to examine the effectiveness of various spraying strategies. Populations within glasshouses, where factors such as temperature and irrigation can be controlled, lend themselves better to prediction than populations in the field. A Leslie matrix was constructed with parameters for development, survivorship and fecundity obtained from laboratory experiments at 25°C. The thrips on the leaves were considered as a separate population from those on the flowers and fruit, but with both mingling as pupae in the soil. A low density effect by which females mate less often and so are more likely to produce only males was also incorporated and found to have a major effect on the outcome. The model supported the empirical finding that the method of ‘pursuit spraying’, i.e. spraying whenever the population exceeded a threshold and following this with another spray a few days later regardless of the population density, needed fewer sprays in total than the conventional approach.

Once a model for the population dynamics of a thrips pest is available, density predictions can be related to damage thresholds. Thrips densities can be expressed as relative thrips densities (e.g. thrips trap^{-1} day^{-1}), absolute thrips densities (e.g. thrips plant^{-1} or thrips per cm^2 of leaf) or as thrips-days (e.g. thrips-days per cm^2 of leaf). It is therefore important to be able to relate the thrips density measurements used for a population dynamics study to the type of density measurement used for damage thresholds. Welter *et al.* (1990) measured damage to cucumber in thrips-days and found that this correlated well with the peak thrips density. Although the thrips-days approach was preferred, reluctance among growers to adopt the concept meant that a threshold density was more likely to be used in practice (see Chapter 16).

Moderate densities of thrips are harmless if a plant can compensate for damage and might even be beneficial if the thrips provide food for a predator, such as *Orius*, that can reduce populations of other pests later in the season (Irwin and Yeargan, 1980).

Predicting outbreaks

Thrips have an enormous capacity for increase. Vast numbers can arrive at a crop and then breed rapidly, particularly at high temperatures. During outbreaks of *Thrips imaginis*, thousands arrive at individual flowers (Zeck and Noble, 1932) and swarms have even become a nuisance to people on beaches (Bailey, 1936). As many as 200,000 individuals of *Thrips hawaiiensis* have been estimated to be on a single inflorescence of oil palm in Peninsular Malaysia (Syed, 1979); cultured females increased 30-fold in 10 days (Appanah and Chan, 1981).

Andrewartha and Birch (1954) have summarized the classic investigation of the population dynamics of *Thrips imaginis*, which is described in detail by Davidson and Andrewartha (1948a,b). The research was remarkable for producing a simple equation, using weather data, that could be used to predict accurately in advance whether outbreaks would occur. The 'Andrewartha–Birch equation' was very successful and was used until recently as part of an early-warning service to apple growers in the Adelaide Hills to predict whether thrips damage was likely later in the same year (P.E. Madge, Adelaide, 1984, personal communication).

A multiple regression model was produced to predict the relative density at the population peak, based on data collected over 14 years. Since the outbreaks of thrips occurred across very large areas, it had appeared obvious, even before the investigation started, that the weather was determining the size of the outbreaks (Evans, 1932; Davidson and Andrewartha, 1948b). Four measures of rainfall and temperature in the preceding autumn and spring were shown to account for 78 per cent of the variance of the mean density in the samples taken from roses just before the peak (Fig. 7.1). These climatic variables were selected from several that were proposed on the basis of detailed knowledge of the biology of the thrips and its wild host plants:

- x_2 – Total effective degree-days (in units of 100 degree-days on a Fahrenheit scale) from the break of the season to the end of August.
- x_3 – Total rainfall (in inches) for September and October.
- x_4 – Total effective degree-days (in units of 100 degree-days on a Fahrenheit scale) for September and October.
- x_5 – As for x_2 but for the year preceding the one when the samples of thrips were taken.

The multiple regression equation was:

$$y'_2 = -2.390 + 0.125x_2 + 0.202x_3 + 0.187x_4 + 0.058x_5$$

where y'_2 is the logarithm of the geometric mean number of thrips per rose for the 30 days preceding the peak, corrected for the influence of maximum temperature and rainfall on thrips activity on those days. Although the equation generated enormous controversy, it was its interpretation that was

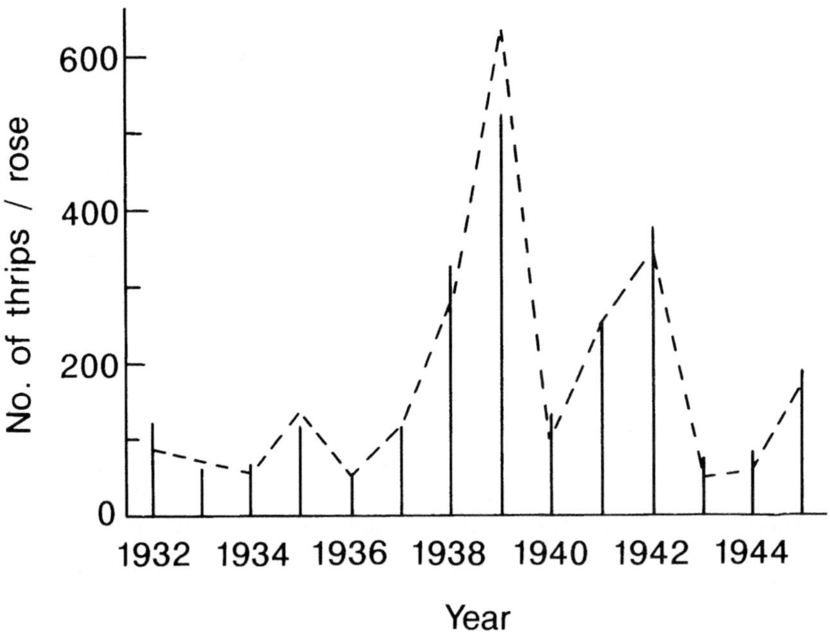

Fig. 7.1. Observed (solid vertical lines) and predicted (dotted line) values of the mean number of *Thrips imaginis* per rose. The observed values are geometric means of the daily counts of thrips per rose and the predicted values are calculated from the equation given in the text (after Davidson and Andrewartha, 1948b).

disputed, not the equation itself. The practical success of the equation showed that weather could be used to predict the size of outbreaks, at least for *Thrips imaginis* in South Australia.

In the 1950s, there was considerable dispute about the relative roles of density-dependent and density-independent factors in population dynamics. Weather was widely considered as a density-independent factor, because the proportion of individuals affected by it is usually independent of the population density. However, theoretical considerations showed that a population could only be regulated by a density-dependent factor that killed a higher proportion of individuals at higher densities. Andrewartha and Birch (1954) concluded from the thrips data that 'not only did we fail to find a "density-dependent factor" but we also showed that there was no room for one'. The subsequent debate was partly confused by differences in the use of the terms 'control', 'regulate' and 'determine'. However, the argument that the ability to account for a high proportion of the variance in population size from density-independent factors left no room for density-dependent factors to be involved was countered by Varley *et al.* (1973). They used a numerical example to show that increasing the impact of a density-dependent factor could increase the proportion of the variance explained by density-independent factors, so the high proportion of the variance explained by density-independent factors could even be seen as evidence for, rather than against, the influence of density-dependent factors.

The Andrewartha–Birch equation predicted numbers accurately without taking account of the numbers in the previous year, so some stabilizing influence must have been present after the widely varying summer peak to return the population to a similar size before the start of each year. Varley *et al.* (1973) have pointed out in support of this that the proportional variation in the summer maxima is much higher than for the winter minima. However, the possible cause of this effect, which may well affect many other species, is unknown.

Methods from a branch of mathematics, known as chaos theory, have been applied to the *Thrips imaginis* data (Schaffer and Kot, 1985, 1986). An analysis of the size of the outbreaks through time showed that the abundance of the thrips exhibited the phenomenon of 'deterministic chaos'. The size of the outbreaks appeared random, but was actually deterministic with considerable sensitivity to the initial conditions. The practical significance of this is that short-term prediction may well be possible and can perhaps be more accurate if methods from chaos theory are applied, but long-term prediction is impossible because the initial conditions can never be known with sufficient precision. Schaffer and Kot (1985) concluded from the mathematical analysis that the population might have been regulated by parasites or predators, but that competition for food was probably not important as a limiting factor.

Pollution, land use and climate change

The effect of pollution on thrips populations has rarely been studied. Vasiliu-Oromulu (1994) has studied the effect of massive industrial pollution in Romania on thrips communities in meadows. Lower densities of thrips were found in the polluted areas and the species diversity decreased. *Frankliniella intonsa*, *Thrips physapus* and *Thrips tabaci* still occurred in the polluted areas, but only *Frankliniella intonsa* survived in the worst area. Overgrazing also affected the thrips community of Romanian meadows and the effect was detectable before species changes in the plant community.

Changes in agricultural practice are bound to affect thrips populations, but there are few data. The increased use of grass for grazing, hay and ensilage removes leaf inflorescences at a critical time for grass thrips and may account for the decline in numbers of *Anaphothrips obscurus* in Britain (Lewis, 1973). Similarly, the trend for growing barley at the expense of wheat and oats from 1950 to 1970 has increased the numbers of *Limothrips denticornis*. Increases in the area of linseed and possibly lupins in Britain as a result of changes in agricultural practices in Europe could increase populations of *Thrips angusticeps* (Ferguson, 1994). Prescribed fires in autumn and spring under sugar maple in Pennsylvania give an earlier and smaller peak emergence of *Taeniothrips inconsequens* (Brose and McCormick, 1992). The move away from burning of cereal stubble in Europe may increase populations of cereal thrips.

There appear to be no recent data on the effects of long-term climate change on thrips populations. However, the years 1900 to 1940 were milder and drier than the three subsequent decades in Britain and this may account for the decrease in abundance and later emergence of aeolothripids in Scotland and the decrease in the numbers of *Limothrips cerealium* in Scotland and northern England over the same period (Lewis, 1973). It is not known whether recent warm years have reversed these trends. *Haplothrips tritici* is a damaging pest of wheat in hot summers in some parts of Europe and its unexpected spread from south-east Hungary to the cooler north-west in 1993 has been associated with general climatic warming (Czencz, 1994).

Distribution on Crops

Population size

The number of thrips present in crops can be astronomical. Even when thrips are not considered as pests and are scarcely noticed, their numbers can be enormous. A few simple calculations illustrate this.

At Keith, South Australia, there are about 1400 ha of lucerne. Each flower contains about four adult *Thrips imaginis* and there are about 400×10^6 flowers ha^{-1}, which gives a thrips density of 1600×10^6 ha^{-1} and a total population of 2×10^{12} individuals, which would have a mass of about 70 t. The thrips are not considered as pests on the crop. Lewis (1973) estimated thrips densities from published data of 740 to 1600×10^6 ha^{-1} for *Thrips tabaci* larvae on onions in the USA, 15 to 17×10^6 ha^{-1} for *Limothrips cerealium* adults on wheat in England and 1 to 10×10^6 ha^{-1} for *Selenothrips rubrocinctus* on cacao in Trinidad.

Thrips reach some of their highest densities in crop monocultures. Densities in mixed vegetation, which might act as sources for crop infestation, are likely to be lower because of the lower density of suitable host plants. Ward (1966) found peak adult densities in chalk grassland of only 66,000 ha^{-1}, 36,000 ha^{-1} and 18,000 ha^{-1} for *Thrips angusticeps*, *Thrips physapus* and *Thrips validus*, respectively. However, for thrips that breed on trees, a single plant can provide an enormous resource. Laughlin (1970) estimated that a small gum tree (*Eucalyptus calycogona*) had about 2000 adult *Thrips australis* arriving from elsewhere per day and 3000 to 8000 adults emerging per day from beneath the tree. The total number of adults on the tree peaked at just over 50,000. In Malaysia, mass blooming of some forest dipterocarp trees gives a rapid population increase on an individual tree. Appanah and Chan (1981) estimated that there were 1.45 million adult thrips on one tree.

Densities of adult *Taeniothrips inconsequens* emerging from the soil have been recorded with emergence traps. Bailey (1944) recorded as many as 14×10^6 ha^{-1} in fruit orchards in California, and Parker and Skinner (1993) recorded densities ranging from 2 to 11×10^6 ha^{-1} in sugar maple forest in Vermont, depending on the type of trap used. Soil extraction of larvae in California has given densities of 14 to 127×10^6 ha^{-1} (Moulton, 1909b) and 13 to 186×10^6 ha^{-1} (Foster and Jones, 1915).

Spatial distribution

Thrips can infest all parts of a crop apart from the roots. However, individual species usually have characteristic sites (Table 7.4), and these are often characteristic not just of species but of genera (see Chapter 4). The sites can differ between adults and larvae, with adults often nearer the top of the plant and larvae lower down. At a finer scale, there can be differences between leaf ages and between the upper and lower sides of leaves. The distribution patterns are mainly related to feeding behaviour (see Chapter 4).

The densities of thrips in flowers depend on the flower stage and sometimes the position of the flower. The number of adults, larvae and sometimes eggs in a flower increases as a bud enlarges and opens, then reaches a peak before decreasing as the flower begins to wilt and finally drops off (Ward, 1966; Laughlin, 1970; Kirk, 1984, 1985a). Significant differences in density can be found between flowers that differ in age by only a few hours. Increases appear to result from an accumulation of thrips at first while arrivals outnumber departures, but the reason for decreases is not known. It may be a combination of an increase in departures in response to wilting and a decrease in arrivals as the appearance of the flower changes. Exposed flowers of *Calystegia sepium* on the surface of vegetation contain more thrips than those within the canopy, presumably because exposure enhances the rate of arrival (Kirk, 1985a).

Thrips larvae often have to leave a flower before they are ready to pupate and so have to locate a new flower to obtain more food. When flowers open sequentially along racemes, a larva can move to the next flower. This appears to happen with second instar larvae of *Kakothrips pisivorus* on field beans, which are more numerous in the last flower on a raceme than in the first (Kirk, 1985b).

Thrips distributions can show diel periodicity (Irwin and Yeargan, 1980; Kirk, 1984; Sites *et al.*, 1992). Decreases in numbers on a crop on warm days are likely to result from flight activity. Changes in position on the crop may relate to responses to sunlight or relative humidity. For example, *Thrips tabaci* roams over petals in full sunlight, while *Thrips imaginis* seeks the shade within flowers (Steele, 1935). Feeding activity can sometimes be greater at night (see

Table 7.4. The distribution of adult thrips by plant part on cotton (Leigh, 1995) and soybean (Irwin *et al.*, 1979).

Crop	On open leaves	On leaf buds, flower buds and flowers
Cotton	*Caliothrips fasciatus*	*Frankliniella fusca*
	Caliothrips impurus	*Frankliniella occidentalis*
	Caliothrips sudanensis	*Frankliniella schultzei*
	Retithrips syriacus	*Frankliniella tritici*
	Thrips palmi	*Neohydatothrips variabilis*
		Thrips tabaci
Soybean	*Neohydatothrips variabilis*	*Frankliniella tritici*

Chapter 4), so distributions may change as thrips move between resting sites and feeding sites. Other activity, such as oviposition, may also alter distributions. Adult female *Kakothrips pisivorus* are found predominantly in flowers at the top of field bean plants by day; however, the adults emerge at night and are then found all over the plant, probably laying eggs (Kirk, 1985b).

The distribution of thrips on plants is of particular relevance for understanding how crop damage occurs and for planning efficient and effective sampling strategies (see Chapter 11). If thrips are highly aggregated, more samples will be needed than if thrips are regularly distributed. The degree of aggregation may also reveal information about thrips behaviour.

The Taylor's power law method, which involves plotting log variance on log mean of sample size, is frequently used to study aggregation (Taylor, 1961; Taylor *et al.*, 1978; Taylor and Woiwod, 1980; see also Chapter 11). A slope greater than unity indicates clumping, whereas a slope less than unity indicates uniformity. A random distribution would have a slope of unity. The method has the advantage that it applies across a wide range of densities. Other commonly used methods of analysis, such as Iwao's mean crowding index or the negative binomial distribution, are described by Southwood (1978).

Plots of log variance (s^2) on log mean (m) for three crop pests are given in Fig. 7.2. The slopes for these species and others (Lewis, 1973; Taylor *et al.*, 1978; Salguero-Navas *et al.*, 1994) are usually just above unity, indicating

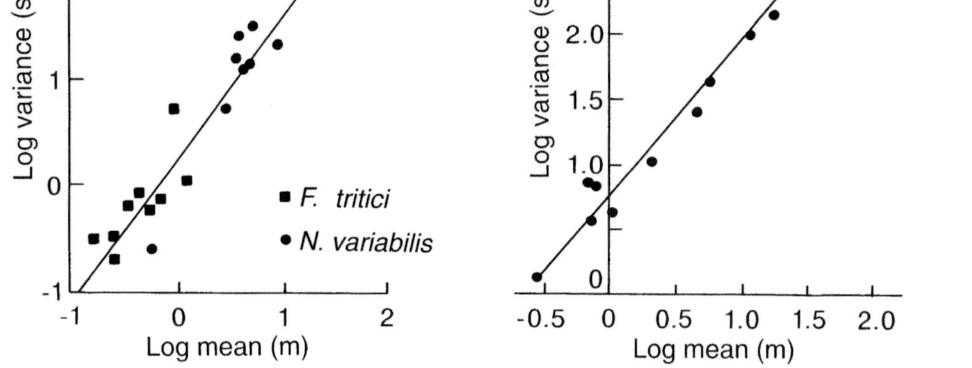

Fig. 7.2. Plots of log variance on log mean for pest thrips on crops. (a) Larvae and adults of *Frankliniella tritici* (■) and *Neohydatothrips variabilis* (●) on leaves and terminals of soybean in the USA (after Irwin and Yeargan, 1980); (b) larvae and adults of *Megalurothrips usitatus* in flowers of adzuki bean in Taiwan (after Chang, 1992).

moderate aggregation. Larvae are often more aggregated than adults. However, the slope is likely to depend not just on the behaviour of the thrips but on the homogeneity of the samples. For example, counts of thrips from a sample of flowers of variable age will have a higher variance and thus a steeper slope than a sample of similarly aged flowers. Heterogeneous samples would therefore be expected to indicate aggregation, even though the individuals might be randomly distributed according to results from homogeneous samples. In practice, it is very difficult to obtain homogeneous samples in order to distinguish behavioural aggregation from environmental heterogeneity.

Although some flower thrips have been described as aggregating or swarming (Kirk, 1985a; Terry and Gardner, 1990), it has not yet been proved that these thrips are responding to each other rather than, for example, to the brightness of the substrate. However, the opposite effect of movement away from high densities has been demonstrated in several species (see above).

References

Ananthakrishnan, T.N. (1984) *Bioecology of Thrips.* Indira Publishing House, Oak Park, Michigan, USA.

Andrewartha, H.G. (1934) Thrips investigation. 5. On the effect of soil moisture on the viability of the pupal stages of *Thrips imaginis* Bagnall. *Journal of the Council for Scientific and Industrial Research, Australia* 7, 239–244.

Andrewartha, H.G. and Birch, L.C. (1954) *The Distribution and Abundance of Animals.* University of Chicago Press, Chicago.

Appanah, S. and Chan, H.T. (1981) Thrips: pollinators of some dipterocarps. *The Malaysian Forester* 44, 234–252.

Bailey, S.F. (1933) The biology of the bean thrips. *Hilgardia* 7, 467–522.

Bailey, S.F. (1934a) A winter study of the onion thrips in California. *Monthly Bulletin of the California Department of Agriculture* 23, 149–152.

Bailey, S.F. (1934b) Factors influencing pear thrips abundance and effectiveness of cultural control. *Journal of Economic Entomology* 27, 879–884.

Bailey, S.F. (1936) Thrips attacking man. *Canadian Entomologist* 68, 95–98.

Bailey, S.F. (1944) The pear thrips in California. *University of California, College of Agriculture, Agricultural Experiment Station Bulletin* 687.

Bale, J.S. (1991) Implications of cold hardiness for pest management. In: Lee, R.E. and Denlinger, D.L. (eds) *Insects at Low Temperature.* Chapman & Hall, New York, pp. 461–498.

Bates, B.A. and Weiss, M.J. (1991) The spatial distribution of *Limothrips denticornis* Haliday (Thysanoptera: Thripidae) eggs on spring barley. *Canadian Entomologist* 123, 205–210.

Begon, M., Harper, J.L. and Townsend, C.R. (1996) *Ecology. Individuals, Populations and Communities*, 3rd edn. Blackwell Science, Oxford.

Beshear, R. (1974) A chalcidoid planidium on thrips larvae in Georgia. *Journal of the Georgia Entomological Society* 9, 265–266.

Binns, E.S. (1982) Phoresy as migration – some functional aspects of phoresy in mites. *Biological Reviews* 57, 571–620.

Bonnemaison, L. and Bournier, A. (1964) Les thrips du lin: *Thrips angusticeps* Uzel et *Thrips linarius* Uzel (Thysanoptères). *Annales des Épiphyties* 15, 97–169.

Bournier, A. and Bernaux, P. (1971) *Haplothrips tritici* Kurdj. et *Limothrips cerealium* Hal. agents de la moucheture des blés durs. *Annales de Zoologie–Écologie Animale* 3, 247–259.

Brady, N.C. (1974) *The Nature and Properties of Soils*, 8th edn. Macmillan, New York.

Broadbent, A.B., Matteoni, J.A. and Allen, W.R. (1990) Feeding preferences of the western flower thrips, *Frankliniella occidentalis* (Pergande) (Thysanoptera: Thripidae), and incidence of tomato spotted wilt virus among cultivars of florist's chrysanthemum. *Canadian Entomologist* 122, 1111–1117.

Brødsgaard, H.F. (1993) Cold hardiness and tolerance to submergence in water in *Frankliniella occidentalis* (Thysanoptera: Thripidae). *Environmental Entomology* 22, 647–653.

Brødsgaard, H.F. (1994) Effect of photoperiod on the bionomics of *Frankliniella occidentalis* (Pergande) (Thysanoptera, Thripidae). *Journal of Applied Entomology* 117, 498–507.

Brose, P.H. and McCormick, L.H. (1992) Effects of prescribed fire on pear thrips in Pennsylvania sugarbushes. *Northern Journal of Applied Forestry* 9, 157–160.

Brose, P.H., McCormick, L.H. and Cameron, E.A. (1993) Distribution of pear thrips (Thysanoptera: Thripidae) in three forest soil drainage classes. *Environmental Entomology* 22, 1118–1123.

Bryan, D.E. and Smith, R.F. (1956) The *Frankliniella occidentalis* (Pergande) complex in California (Thysanoptera: Thripidae). *University of California Publications in Entomology* 10, 359–410.

Buhl, C. (1937) Beiträge zur Kenntnis der Biologie, wirtschaftlichen Bedeutung und Bekämpfung von *Kakothrips robustus* Uz. *Zeitschrift für angewandte Entomologie* 23, 65–113.

Bullock, J.A. (1970) Variation in population growth rates in *Thrips nigropilosus* Thysanoptera. *Tropical Ecology* 11, 107–125.

Callan, E. McC. (1943) Natural enemies of the cacao thrips. *Bulletin of Entomological Research* 34, 313–321.

Cameron, A.E. and Treherne, R.C. (1918) The pear thrips (*Taeniothrips inconsequens* Uzel) and its control in British Columbia. *Bulletin of the Department of Agriculture Canada, Entomology Branch* 15, 1–51.

Cederholm, L. (1963) Ecological studies on Thysanoptera with special reference to species found in mixed grain (oats and barley). *Opuscula Entomologica* Suppl. XXII, 1–215.

Chang, N.T. (1989) Impacts of soil physical factors on the pupation of bean flower thrips, *Megalurothrips usitatus* (Bagnall) (Thysanoptera: Thripidae). *Plant Protection Bulletin* 31, 377–386.

Chang, N.T. (1992) Dispersion patterns of bean flower thrips, *Megalurothrips usitatus* (Bagnall), (Thysanoptera: Thripidae) on flowers of adzuki bean. *Plant Protection Bulletin* 34, 41–53.

Chyzik, R., Klein, M., Ben–Dov, Y. and Cohen, A. (1995) Differential population density of western flower thrips in various colors of gladiolus. In: Parker, B.L., Skinner, M. and Lewis, T. (eds) *Thrips Biology and Management*. Plenum Press, New York, pp. 449–452.

Clausen, C.P. (1940) *Entomophagous Insects*. McGraw-Hill, New York.

Crespi, B.J. (1986) Territoriality and fighting in a colonial thrips, *Hoplothrips pedicularius*, and sexual dimorphism in Thysanoptera. *Ecological Entomology* 11, 119–130.

Crespi, B.J. (1992) Behavioural ecology of Australian gall thrips (Insecta, Thysanoptera). *Journal of Natural History* 26, 769–809.

Czencz, K. (1994) Importance and development of species spectrum of winter wheat damaging thrips in Hungary. *Mededelingen van de Faculteit Landbouwwetenschappen Rijksuniversiteit (Gent)* 59 (2a), 227–233.

Danks, H.V. (1971) Biology of some stem-nesting aculeate Hymenoptera. *Transactions of the Royal Entomological Society of London* 122, 323–399.

Danks, H.V. (1987) *Insect Dormancy: an Ecological Perspective.* Biological Survey of Canada, Ottawa.

Davidson, J. and Andrewartha, H.G. (1948a) Annual trends in a natural population of *Thrips imaginis* (Thysanoptera). *Journal of Animal Ecology* 17, 193–199.

Davidson, J. and Andrewartha, H.G. (1948b) The influence of rainfall, evaporation and atmospheric temperature on fluctuations in the size of a natural population of *Thrips imaginis* (Thysanoptera). *Journal of Animal Ecology* 17, 200–222.

De Jager, C.M., Butôt, R.P.T., Klinkhamer, P.G.L. and van der Meijden, E. (1995) Chemical characteristics of chrysanthemum cause resistance to *Frankliniella occidentalis* (Thysanoptera: Thripidae). *Journal of Economic Entomology* 88, 1746–1753.

Dintenfass, L.P., Bartell, D.P. and Scott, M.A. (1987) Predicting resurgence of western flower thrips (Thysanoptera: Thripidae) on onions after insecticide application in the Texas High Plains. *Journal of Economic Entomology* 80, 502–506.

Dyadechko, N.P. (1977) *Thrips or Fringe-Winged Insects (Thysanoptera) of the European Part of the USSR.* (Translated from the Russian by R.S. Chakravarthy.) Amerind Publishing Company, New Delhi.

Evans, D.E. (1967) Insecticide field trials against the coffee thrips (*Diarthrothrips coffeae* Williams) in Kenya. *Turrialba* 17, 376–380.

Evans, J.W. (1932) The bionomics and economic importance of *Thrips imaginis* Bagnall. *Pamphlet of the Council for Scientific and Industrial Research, Australia* 30.

Fatzinger, C.W. and Dixon, W.N. (1996) Degree-day models for predicting levels of attack by slash pine flower thrips (Thysanoptera: Phlaeothripidae) and the phenology of female strobilus development on slash pine. *Environmental Entomology* 25, 727–735.

Felland, C.M., Hull, L.A., Teulon, D.A.J. and Cameron, E.A. (1993) Overwintering of western flower thrips (Thysanoptera: Thripidae) in Pennsylvania. *The Canadian Entomologist* 125, 971–973.

Fennah, R.G. (1963) Nutritional factors associated with seasonal population increase of cacao thrips, *Selenothrips rubrocinctus* (Giard) (Thysanoptera), on cashew, *Anacardium occidentale*. *Bulletin of Entomological Research* 53, 681–713.

Fennah, R.G. (1965) The influence of environmental stress on the cacao tree in predetermining the feeding sites of cacao thrips, *Selenothrips rubrocinctus* (Giard) on leaves and pods. *Bulletin of Entomological Research* 56, 333–349.

Ferguson, A.W. (1994) Pests and plant injury on lupins in the south of England. *Crop Protection* 13, 201–210.

Foster, S.W. and Jones, P.R. (1915) The life history and habits of the pear thrips in California. *Bulletin of the United States Department of Agriculture* 173.

Franssen, C.J.H. and Huisman, P. (1958) The biology and control of *Thrips angusticeps* Uzel. (In Dutch.) *Verslagen van Landbouwkundige Onderzoekingen* 64, 1–104.

Franssen, C.J.H. and Mantel, W.P. (1965) Thrips in cereal crops (biology, economic importance and control). 1. Biology. (In Dutch.) *Verslagen van Landbouwkundige Onderzoekingen* 662, 1–97.

Froggatt, W.W. (1931) Thrips. The present garden pest. *The Australasian Beekeeper* 33, 130–131.

Gaum, W.G., Giliomee, J.H. and Pringle, K.L. (1994a) Life history and life tables of western flower thrips, *Frankliniella occidentalis* (Thysanoptera, Thripidae), on English cucumbers. *Bulletin of Entomological Research* 84, 219–224.

Gaum, W.G., Giliomee, J.H. and Pringle, K.L. (1994b) Resistance of some rose cultivars to the western flower thrips, *Frankliniella occidentalis* (Thysanoptera, Thripidae). *Bulletin of Entomological Research* 84, 487–492.

Gerdes, C.F. (1982) Distribution of adult overwintering Thysanoptera in an Illinois prairie. *Journal of the Kansas Entomological Society* 55, 418–422.

Gopinathan, K., Varatharajan, R. and Ananthakrishnan, T.N. (1981) Incidence of *Microcephalothrips abdominalis* (Crawford) (Thysanoptera: Thripidae) in relation to the pollination biology of the weed *Ageratum conyzoides* L. (*Compositae*). *Proceedings of the Indian National Science Academy* B47, 505–509.

Greene, I.D. and Parrella, M.P. (1995) Two new natural enemies of western flower thrips in California. In: Parker, B.L., Skinner, M. and Lewis, T. (eds) *Thrips Biology and Management*. Plenum Press, New York, pp. 277–279.

Grout, T.G., Morse, J.G. and Brawner, O.L. (1986) Location of citrus thrips (Thysanoptera, Thripidae) pupation: tree or ground. *Journal of Economic Entomology* 79, 59–61.

Haitlinger, R. (1987) Further information on distribution of species from the genus *Hauptmannia* Oudemans, 1910 (Acari, Prostigmata, Erythraeidae) in Poland. (In Polish.) *Przeglad Zoologiczny* 31, 159–164.

Hall, W.J. (1930) The South African citrus thrips in Southern Rhodesia. *Report of the Mazoe Citrus Experiment Station* 1, 1–55.

Hamilton, W.J. (1930) Notes on the food of the American toad. *Copeia* 18, 45.

Hargreaves, J.R. and Cooper, L.P. (1980) Differential susceptibility of some gladiolus cultivars to gladiolus thrips, *Taeniothrips simplex* (Morison). *Queensland Journal of Agricultural and Animal Sciences* 37, 161–163.

Harris, H.M., Drake, C.J. and Tate, H.D. (1936) Observations on the onion thrips. *Iowa College Journal of Science* 10, 155–172.

Herr, E.A. (1934) The gladiolus thrips, *Taeniothrips gladioli* M. and S. *Bulletin, Ohio Agricultural Experiment Station* 537.

Hessein, N.A. and McMurtry, J.A. (1988) Observations on *Megaphragma mymaripenne* Timberlake (Hymenoptera, Trichogrammatidae), an egg parasite of *Heliothrips haemorrhoidalis* Bouché (Thysanoptera, Thripidae). *Pan-Pacific Entomologist* 64, 250–254.

Horton, J.R. (1918) The citrus thrips. *Bulletin of the United States Department of Agriculture* 616, 1–42.

Irwin, M.E. and Yeargan, K.V. (1980) Sampling phytophagous thrips on soybean. In: Kogan, M. and Herzog, D.C. (eds) *Sampling Methods in Soybean Entomology*. Springer–Verlag, New York, pp. 283–304.

Irwin, M.G., Yeargan, K.V. and Marston, N.L. (1979) Spatial and seasonal patterns of phytophagous thrips in soybean fields with comments on sampling techniques. *Environmental Entomology* 8, 131–140.

Isenhour, D.J. and Marston, N.L. (1981) Seasonal cycles of *Orius insidiosus* (Hemiptera: Anthocoridae) in Missouri soybeans. *Journal of the Kansas Entomological Society* 54, 129–142.

Jones, H.A., Bailey, S.F. and Emsweller, S.L. (1934) Thrips resistance in the onion. *Hilgardia* 8, 215–232.

Kamm, J.A. (1972) Environmental influence on reproduction, diapause and morph determination of *Anaphothrips obscurus* (Thysanoptera: Thripidae). *Environmental Entomology* 1, 16–19.

Kawai, A. (1985a) Studies on population ecology of *Thrips palmi* Karny. VII. Effect of temperature on population growth. (In Japanese.) *Japanese Journal of Applied Entomology and Zoology* 29, 140–143.

Kawai, A. (1985b) Studies on population ecology of *Thrips palmi* Karny. 9. Interspecific competition with *Aphis gossypii* Glover. (In Japanese.) *Proceedings of the Association for Plant Protection Kyushu* 31, 156–159.

Kawai, A. (1986) Studies on population ecology of *Thrips palmi* Karny. X. Differences in population growth on various crops. (In Japanese.) *Japanese Journal of Applied Entomology and Zoology* 30, 7–11.

Kawai, A. (1987) Studies on the population ecology of *Thrips palmi* Karny. XIV. Relationship between the density and rate of copulation. (In Japanese.) *Japanese Journal of Applied Entomology and Zoology* 31, 85–87.

Kawai, A. (1988) Studies on population ecology of *Thrips palmi* Karny. 17. Relationship between the density on egg plant or sweet pepper and rate of copulation. (In Japanese.) *Japanese Journal of Applied Entomology and Zoology* 32, 334–336.

Kawai, A. and Kitamura, C. (1987) Studies on population ecology of *Thrips palmi* Karny. XV. Evaluation of effectiveness of control methods using a simulation model. (In Japanese.) *Applied Entomology and Zoology* 22, 292–302.

Kawai, A. and Kitamura, C. (1990) Studies on population ecology of *Thrips palmi* Karny. 18. Evaluation of effectiveness of control methods of thrips on egg plant and sweet pepper using a simulation model. (In Japanese.) *Applied Entomology and Zoology* 25, 161–175.

Kennedy, J.S. (1978) The concepts of olfactory 'arrestment' and 'attraction'. *Physiological Entomology* 3, 91–98.

Kirk, W.D.J. (1984) Ecological studies on *Thrips imaginis* Bagnall (Thysanoptera) in flowers of *Echium plantagineum* L. in Australia. *Australian Journal of Ecology* 9, 9–18.

Kirk, W.D.J. (1985a) Aggregation and mating of thrips in flowers of *Calystegia sepium*. *Ecological Entomology* 10, 433–440.

Kirk, W.D.J. (1985b) Floral display in *Vicia faba*, and the distribution of a flower thrips, *Kakothrips pisivorus*. *Entomologia Experimentalis et Applicata* 38, 233–238.

Kirk, W.D.J. (1987a) A key to the larvae of some common Australian flower thrips (Insecta: Thysanoptera), with a host-plant survey. *Australian Journal of Zoology* 35, 173–185.

Kirk, W.D.J. (1987b) How much pollen can thrips destroy? *Ecological Entomology* 12, 31–40.

Kirk, W.D.J. (1990) Body size in flower thrips. In: *Proceedings of the Third International Symposium on Thysanoptera, Kazimierz Dolny, Poland, 1990*. Warsaw Agricultural University Press, Warsaw, pp. 65–69.

Kirk, W.D.J. (1991) The size relationship between insects and their hosts. *Ecological Entomology* 16, 351–359.

Kirk, W.D.J. (1994) The effects of density on the oviposition rate of flower thrips. *Courier Forschungsinstitut Senckenberg* 178, 69–73.

Kirk, W.D.J. (1996) *Thrips*. Naturalists' Handbooks 25. The Richmond Publishing Co., Slough.

Kolb, T.E. and Teulon, D.A.J. (1991) Relationship between sugar maple budburst phenology and pear thrips damage. *Canadian Journal of Forest Research* 21, 1043–1048.

Köppä, P. (1969) Studies on the hibernation of certain species of thrips living on cereal plants. *Annales Agriculturae Fenniae* 8, 1–8.

Köppä, P. (1970) Studies on the thrips (Thysanoptera) species most commonly occurring on cereals in Finland. *Annales Agriculturae Fenniae* 9, 191–265.

Kumar, N.K.K., Ullman, D.E. and Cho, J.J. (1995) Resistance among *Lycopersicon* species to *Frankliniella occidentalis* (Thysanoptera: Thripidae). *Journal of Economic Entomology* 88, 1057–1065.

Kutter, H. and Winterhalter, W. (1933) Untersuchungen über die Erbsenschädlinge im St. Gallischen Rheintale. *Landwirtschaftliches Jahrbuch der Schweiz* 47, 273–338.

Laughlin, R. (1970) The gum tree thrips, *Isoneurothrips australis*, the population of a single tree. *Entomologia Experimentalis et Applicata* 13, 247–259.

Laughlin, R. (1977) The gum tree thrips *Isoneurothrips australis* Bagnall: survival at different temperatures and humidities and its relation to capacity for dispersal. *Australian Journal of Ecology* 2, 391–398.

Lee, R.E. (1991) Principles of insect low temperature tolerance. In: Lee, R.E. and Denlinger, D.L. (eds) *Insects at Low Temperature*. Chapman & Hall, New York, pp. 17–46.

Leigh, T.F. (1995) Bionomics of cotton thrips: a review. In: Parker, B.L., Skinner, M. and Lewis, T. (eds) *Thrips Biology and Management*. Plenum Press, New York, pp. 61–70.

Lewis, T. (1959) The annual life cycle of *Limothrips cerealium* Haliday (Thysanoptera) and its distribution in a wheat field. *Entomologia Experimentalis et Applicata* 2, 187–203.

Lewis, T. (1962) The effects of temperature and relative humidity on mortality in *Limothrips cerealium* Haliday (Thysanoptera) overwintering in bark. *Annals of Applied Biology* 50, 313–326.

Lewis, T. (1963) The effect of weather on emergence and take-off of overwintering *Limothrips cerealium* Haliday (Thysanoptera). *Annals of Applied Biology* 51, 489–502.

Lewis, T. (1973) *Thrips. Their Biology, Ecology and Economic Importance*. Academic Press, London.

Lewis, T. and Navas, E. (1962) Thysanopteran populations overwintering in hedge bottoms, grass litter and bark. *Annals of Applied Biology* 50, 299–311.

Loomans, A.J.M., van Lenteren, J.L., Tommasini, M.G., Maini, S. and Riudavets, J. (1995) *Biological Control of Thrips Pests*. Wageningen Agricultural University Papers 95–1, 201 pp.

Lowry, V.K., Smith, J.W. and Mitchell, F.L. (1992) Life-fertility tables for *Frankliniella fusca* (Hinds) and *F. occidentalis* (Pergande) (Thysanoptera: Thripidae) on peanut. *Annals of the Entomological Society of America* 85, 744–754.

Lysaght, A.M. (1936) A note on the adult female of *Anguillulina aptini* (Sharga), a nematode parasitizing *Aptinothrips rufus* Gmelin. *Parasitology* 28, 290–292.

Lysaght, A.M. (1937) An ecological study of a thrips (*Aptinothrips rufus*) and its nematode parasite (*Anguillulina aptini*). *Journal of Animal Ecology* 6, 169–192.

MacGill, E.I. (1929) The biology of Thysanoptera with reference to the cotton plant. 4. The relation between the degree of infestation and surface caking of the soil. *Annals of Applied Biology* 16, 288–293.

MacGill, E.I. (1930) The biology of Thysanoptera with reference to the cotton plant. 5. The relation between the degree of infestation and the type of soil. *Annals of Applied Biology* 17, 150–161.

MacGill, E.I. (1931) The biology of Thysanoptera with reference to the cotton plant. 7. The relation between temperature and humidity and the life cycle. *Annals of Applied Biology* 18, 574–583.

MacGill, E.I. (1937) The biology of Thysanoptera with reference to the cotton plant. 8.

The relation between variations in temperature and the life cycle. *Annals of Applied Biology* 24, 95–109.

Maksymov, J.K. (1965) Die Überwinterung des Lärchenblasenfußes *Taeniothrips laricivorus* Kratochvil und Farsky. *Mitteilungen der schweizerischen Anstalt für das forstliche Versuchswesen* 41, 1–18.

Malchau, W. (1991) Zum Einfluß intraspezifischer Relationen auf die postmetabole Entwicklung von *Frankliniella tenuicornis* (Uzel, 1895) (Thysanoptera, Insecta). *Zoologisches Jahrbuch der Systematik* 118, 265–276.

Mallmann, R.J. de (1959) Observations sur le thigmotactisme de *Limothrips cerealium* (Thysan.). *Bulletin de la Société Entomologique de France* 64, 151–157.

Mallmann, R.J. de (1964) Contribution à l'étude de la thigmotaxie chez les insectes. *Annales de la Société Entomologique de France* 64, 151–157.

Moulton, D. (1909a) Papers on deciduous fruit insects and insecticides: the pear thrips. *United States Department of Agriculture Bureau of Entomology Bulletin* 68, Part I (revised), 1–16.

Moulton, D. (1909b) Papers on deciduous fruit insects and insecticides: the pear thrips and its control. *United States Department of Agriculture Bureau of Entomology Bulletin* 80, Part IV, 1–66.

Moulton, D. (1909c) The orange thrips. *United States Department of Agriculture Bureau of Entomology Technical Series* 12, Part VII, 119–122.

Mound, L.A. (1967) The British species of the genus *Thrips*. Thysanoptera. *Entomologists' Gazette* 18, 13–22.

Mound, L.A., Morison, G.D., Pitkin, B.R. and Palmer, J.M. (1976) *Handbooks for the Identification of British Insects*. Vol. 1, Part 11. *Thysanoptera*. Royal Entomological Society of London.

Murai, T. (1977) Reproductive diapause of flower thrips, *Frankliniella intonsa*. In: Holman, J., Pelikan, J., Dixon, A.F.G. and Weisman, L. (eds) *Population Structure, Genetics and Taxonomy of Aphids and Thysanoptera: Proceedings of International Symposia, Smolenice, Czechoslovakia, 9–14 September 1985*. SPB Academic Publishing, The Hague, pp. 467–479.

Nicholson, A.J. (1958) Dynamics of insect populations. *Annual Review of Entomology* 3, 107–136.

Nonaka, K., Nagai, K. and Yamamoto, E. (1982) Ecology and control of the thrips infesting fruit vegetables. 4. Resistance of *Thrips palmi* to cold and heat. (In Japanese.) *Kyushu Agricultural Research* 44, 119.

North, R.C. and Shelton, A.M. (1986a) Ecology of Thysanoptera within cabbage fields. *Environmental Entomology* 15, 520–526.

North, R.C. and Shelton, A.M. (1986b) Colonization and intraplant distribution of *Thrips tabaci* (Thysanoptera: Thripidae) on cabbage. *Journal of Economic Entomology* 79, 219–223.

North, R.C. and Shelton, A.M. (1986c) Overwintering of the onion thrips, *Thrips tabaci* (Thysanoptera: Thripidae), in New York. *Environmental Entomology* 15, 695–699.

Nugaliyadde, L. and Heinrichs, E.A. (1984) Biology of rice thrips, *Stenchaetothrips biformis* (Bagnall) (Thysanoptera; Thripidae) and a greenhouse rearing technique. *Journal of Economic Entomology* 77, 1171–1175.

Parker, B.L. and Skinner, M. (1993) Field evaluation of traps for monitoring emergence of pear thrips (Thysanoptera: Thripidae). *Journal of Economic Entomology* 86, 46–52.

Puche, H. and Funderburk, J. (1992) Intrinsic rate of increase of *Frankliniella fusca* (Thysanoptera: Thripidae) on peanuts. *Florida Entomologist* 75, 185–189.

Puche, H. and Funderburk, J.E. (1995) Mobility of western flower thrips in tomato. In: Parker, B.L., Skinner, M. and Lewis, T. (eds) *Thrips Biology and Management.* Plenum Press, New York, pp. 563–566.

Putman, W.L. (1965) The predacious thrips *Haplothrips faurei* Hood (Thysanoptera: Phlaeothripidae) in Ontario peach orchards. *Canadian Entomologist* 97, 1208–1221.

Raffa, K.F., Hall, D.J., Kearby, W. and Katovich, S. (1992a) Seasonal life history of introduced basswood thrips (Thysanoptera: Thripidae) in Wisconsin, with observations on associated thrips species. *Environmental Entomology* 21, 771–779.

Raffa, K.F., Hall, D.J., Kearby, W. and Katovich, S. (1992b) *Thrips calcaratus* in Wisconsin: seasonal development and host impact. In: Cameron, E.A., Teulon, D.A.J., McCormick, L.H. and Kolb, T.E. (eds) *The 1991 Conference on Thrips (Thysanoptera): Insect and Disease Considerations in Sugar Maple Management.* General Technical Report NE-161. US Department of Agriculture, Forest Service, Radnor, pp. 29–30.

Raizada, U. (1976) On the occurrence of *Mrazekia* sp. A microsporidian parasite infecting some thrips larvae. *Current Science* 45, 627–628.

Rhodes, A.A., Morse, J.G. and Robertson, C.A. (1989) A simple multigeneration phenology model: application to *Scirtothrips citri* (Thysanoptera: Thripidae) prediction on Californian oranges. *Agriculture Ecosystems and Environment* 25, 299–313.

Ricklefs, R.E. (1990) *Ecology*, 3rd edn. Freeman, New York.

Robb, K.L. and Parrella, M.P. (1991) Western flower thrips, a serious pest of floricultural crops. In: Parker, B.L., Skinner, M. and Lewis, T. (eds) *Towards Understanding Thysanoptera. Proceedings of the International Conference on Thrips, Burlington, Vermont, USA, 1989.* USDA, Radnor, PA, pp. 343–357.

Rösingh, C. (1980) *Untersuchungen über die Resistenz der Kuherbse,* Vigna unguiculata *(L.) Walp., gegen* Megalurothrips sjostedti *(Trybom) (Thysanoptera, Thripidae).* Hochschulsammlung Naturwissenschaft: Biologie, Band 9. Hochschulverlag, Freiburg.

Russell, H.M. (1912) The bean thrips (*Heliothrips fasciatus* Pergande). *Bulletin of the Bureau of Entomology, United States Department of Agriculture* 118, 1–49.

Sakimura, K. (1937a) On the bionomics of *Thripoctenus brui* Vuillet, a parasite of *Thrips tabaci* Lind. in Japan (I). *Kontyû* 11, 370–390.

Sakimura, K. (1937b) On the bionomics of *Thripoctenus brui* Vuillet, a parasite of *Thrips tabaci* Lind. in Japan (II). *Kontyû* 11, 410–424.

Salguero-Navas, V.E., Funderburk, J.E., Mack, T.P., Beshear, R.J. and Olson, S.M. (1994) Aggregation indices and sample size curves for binomial sampling of flower-inhabiting *Frankliniella* species (Thysanoptera: Thripidae) on tomato. *Journal of Economic Entomology* 87, 1622–1626.

Schaffer, W.M. and Kot, M. (1985) Do strange attractors govern ecological systems? *Bioscience* 35, 342–350.

Schaffer, W.M. and Kot, M. (1986) Chaos in ecological systems: the coals that Newcastle forgot. *Trends in Ecology and Evolution* 1, 58–63.

Schoonhoven, A. van (1974) Resistance to thrips damage in cassava. *Journal of Economic Entomology* 67, 728–730.

Şengonca, Ç. and Gerlach, S. (1984) Einfluss der Blattoberfläche auf die Wirksamkeit des räuberischen Thrips, *Scolothrips longicornis* (Thysan.: Thripidae). *Entomophaga* 29, 55–61.

Shipp, J.L. and Gillespie, T.J. (1993) Influence of temperature and water vapor pressure deficit on survival of *Frankliniella occidentalis* (Thysanoptera: Thripidae). *Environmental Entomology* 22, 726–732.

Sites, R.W. and Chambers, W.S. (1990) Initiation of vernal activity of *Frankliniella occidentalis* and *Thrips tabaci* on the Texas South Plains. *Southwestern Entomologist* 15, 339–343.

Sites, R.W., Chambers, W.S. and Nichols, B.J. (1992) Diel periodicity of thrips (Thysanoptera: Thripidae) dispersion and the occurrence of *Frankliniella williamsi* on onions. *Journal of Economic Entomology* 85, 100–105.

Skinner, M. and Parker, B.L. (1992) Vertical distribution of pear thrips (Thysanoptera: Thripidae) in forest soils. *Environmental Entomology* 21, 1258–1266.

Skinner, M., Parker, B.L. and Brownbridge, M. (1992) Regional patterns of pear thrips emergence. In: Cameron, E.A., Teulon, D.A.J., McCormick, L.H. and Kolb, T.E. (eds) *The 1991 Conference on Thrips (Thysanoptera): Insect and Disease Considerations in Sugar Maple Management*. General Technical Report NE-161. US Department of Agriculture, Forest Service, Radnor, p. 41.

Smith, F.E. (1961) Density dependence in the Australian thrips. *Ecology* 42, 403–407.

Southwood, T.R.E. (1978) *Ecological Methods with Particular Reference to the Study of Insect Populations*. Chapman & Hall, London.

Steele, H.V. (1935) Thrips investigation: some common Thysanoptera in Australia. *Pamphlet of the Council for Scientific and Industrial Research Australia* 54, 1–59.

Stoddard, F.L. (1986) The distribution of immature thrips among flowers of faba beans in commercial crops and experimental plots. *Annals of Applied Biology* 109, 61–69.

Syed, R.A. (1979) Studies on oil palm pollination by insects. *Bulletin of Entomological Research* 69, 213–224.

Takrony, M.O. (1973) Bionomics and control of sugar-beet thrips *Hercinothrips femoralis* (Reuter). PhD thesis, University of Reading, UK.

Tamò, M., Baumgärtner, J., Delucchi, V. and Herren, H.R. (1993a) Assessment of three key factors responsible for the pest status of the bean flower thrips *Megalurothrips sjostedti* (Trybom) (Thysanoptera, Thripidae). *Bulletin of Entomological Research* 83, 251–258.

Tamò, M., Baumgärtner, J. and Gutierrez, A.P. (1993b) Analysis of the cowpea agroecosystem in West Africa. II. Modelling the interaction between cowpea and the bean flower thrips *Megalurothrips sjostedti* (Trybom) (Thysanoptera, Thripidae). *Ecological Modelling* 70, 89–113.

Tansky, V.I. (1958) The biology of wheat thrips *Haplothrips tritici* Kurd. (Thysanoptera, Phlaeothripidae) in Northern Kazakhstan and the proposed cultural methods of its control. (In Russian.) *Revue d'Entomologie de l'URSS* 37, 785–797.

Taylor, L.R. (1961) Aggregation, variance and the mean. *Nature* 189, 732–735.

Taylor, L.R. and Woiwod, I.P. (1980) Temporal stability as a density-dependent species characteristic. *Journal of Animal Ecology* 49, 209–224.

Taylor, L.R., Woiwod, I.P. and Perry, J.N. (1978) The density-dependence of spatial behaviour and the rarity of randomness. *Journal of Animal Ecology* 47, 383–406.

Terry, L.I. and Gardner, D. (1990) Male mating swarms in *Frankliniella occidentalis* (Pergande) (Thysanoptera: Thripidae). *Journal of Insect Behavior* 3, 133–141.

Teulon, D.A.J., Groninger, J.W. and Cameron, E.A. (1994) Distribution and host plant associations of *Taeniothrips inconsequens* (Thysanoptera: Thripidae). *Environmental Entomology* 23, 587–611.

Theunissen, J. (1994) Intercropping in field vegetable crops: pest management by agrosystem diversification – an overview. *Pesticide Science* 42, 65–68.

Trichilo, P.J. and Leigh, T.F. (1988) Influence of resource quality on the reproductive fitness of flower thrips (Thysanoptera: Thripidae). *Annals of the Entomological Society of America* 81, 64–70.

Tsumuki, H., Nagai, K. and Kanehisa, K. (1987) Cold hardiness of *Thrips palmi*. 1. Survival period of winter and summer populations at low temperatures. (In Japanese.) *Japanese Journal of Applied Entomology and Zoology* 31, 328–332.

Urich, F.W. (1928) San Thomé cacao industry. *Tropical Agriculture Trinidad* 5, 275–278.

van Rijn, P.C.J., Mollema, C. and Steenhuis-Broers, G.M. (1995) Comparative life history studies of *Frankliniella occidentalis* and *Thrips tabaci* (Thysanoptera: Thripidae) on cucumber. *Bulletin of Entomological Research* 85, 285–297.

Varley, G.C., Gradwell, G.R. and Hassell, M.P. (1973) *Insect Population Ecology. An Analytical Approach*. Blackwell Scientific Publications, Oxford.

Vasiliu-Oromulu, L. (1994) Ecological monitoring in Thysanopterology. *Courier Forschungsinstitut Senckenberg* 178, 101–105.

Wagner, T.L., Wu, H., Sharpe, P.J.H. and Coulson, R.M. (1984) Modeling distributions of insect development time: a literature review and application of the Weibull function. *Annals of the Entomological Society of America* 77, 475–487.

Walker, W.F. (1974) Responses of selected Thysanoptera to colored surfaces. *Environmental Entomology* 3, 295–304.

Ward, L.K. (1966) The biology of Thysanoptera. PhD thesis, University of London.

Wardle, R.A. (1927) The biology of the Thysanoptera with reference to the cotton plant. I. The relation between degree of infestation and water supplied. *Annals of Applied Biology* 14, 482–500.

Wardle, R.A. and Simpson, R. (1927) The biology of the Thysanoptera with reference to the cotton plant. III. The relation between feeding habits and plant lesions. *Annals of Applied Biology* 14, 513–528.

Welter, S.C., Rosenheim, J.A., Johnson, M.W., Mau, R.F.L. and Gusukuma-Minuto, L.R. (1990) Effects of *Thrips palmi* and western flower thrips (Thysanoptera: Thripidae) on the yield, growth, and carbon allocation pattern in cucumbers. *Journal of Economic Entomology* 83, 2092–2101.

Wetzel, T. (1963) Zur Frage der Überwinterung der Gräser-Thysanopteren. *Zeitschrift für angewandte Entomologie* 51, 429–441.

Williams, C.B. (1915) The pea thrips (*Kakothrips robustus*). *Annals of Applied Biology* 1, 222–246.

Wilson, T.H. and Cooley, T.A. (1972) A chalcidoid planidium and an entomophilic nematode associated with the western flower thrips. *Annals of the Entomological Society of America* 65, 414–418.

Wilson, L.J., Bauer, L.R. and Walter, G.H. (1996) 'Phytophagous' thrips are predators of two-spotted spider mites (Acari: Tetranychidae) on cotton in Australia. *Bulletin of Entomological Research* 86, 297–305.

Woglum, R.S. and Lewis, H.C. (1936) Nitrogen trichloride as a fumigant. *Journal of Economic Entomology* 29, 631–632.

Zeck, E.H. and Noble, N.S. (1932) Experiments for the control of *Thrips imaginis* Bagnall. *Agricultural Gazette, New South Wales* 43, 231–236.

Zeier, P. and Wright, M.G. (1995) Thrips resistance in *Gladiolus* spp.: potential for IPM and breeding. In: Parker, B.L., Skinner, M. and Lewis, T. (eds) *Thrips Biology and Management*. Plenum Press, New York, pp. 411–416.

8

Predation by Insects and Mites

Maurice W. Sabelis
and Paul C.J. Van Rijn
University of Amsterdam, Department of Pure and Applied Ecology, Section Population Biology, Kruislaan 320, 1098 SM Amsterdam, The Netherlands

Introduction

Predatory arthropods probably play a prominent role in determining the numbers of plant-feeding thrips on plants under natural conditions. Several reviews have been published listing the arthropods observed to feed and reproduce on a diet of thrips. In chronological order the most notable and comprehensive accounts have been presented by Lewis (1973), Ananthakrishnan (1973, 1979, 1984), Ananthakrishnan and Sureshkumar (1985) and Riudavets (1995) (see also general arthropod enemy inventories published by Thompson and Simmonds, 1965; Herting and Simmonds, 1971; and Fry, 1987). Numerous arthropods, recognized as predators of phytophagous thrips, have proved their capacity to eliminate or suppress thrips populations in greenhouse and field crops of agricultural importance (see Chapters 16 and 18) but a detailed analysis of the relative importance of predators, parasitoids, parasites and pathogens under natural conditions is virtually absent. Such investigations would improve understanding of the mortality factors and selective forces moulding thrips behaviour and life history, and also indicate new opportunities for biological control of thrips. In particular, such studies may help to elucidate the consequences of introducing different types of biological control agents against different pests and diseases in the same crop, many of which harbour food webs of increased complexity.

There are three major reasons why food web complexity on plants goes beyond one-predator-one-herbivore systems. First, the plant exhibits a bewildering variety of traits that promote or reduce the effectiveness of the predator. Plants may provide shelter and alternative food (pollen, extrafloral nectar, exudate) and they signal herbivore attack to the predators of their herbivores (Price *et al.*, 1980; Sabelis and Dicke, 1985; Dicke and Sabelis, 1988, 1989, 1990). In this sense plants use predators as 'bodyguards'. They may also invest in direct defences that not only harm the herbivores, but also the natural enemies of their herbivores. Second, the arthropod predators of thrips are usually generalist feeders. They can feed on many different plant-inhabiting arthropods and even on foods of plant origin. The important consequence of polyphagy is that the impact of predatory arthropods on thrips pests also depends on the abundance of other foods/prey, as well as the details of food/prey preferences. Third, different predators introduced to control the same and/or other pests may not only compete with each other for food, but may also feed on each other (intraguild predation).

This chapter starts with a literature review of arthropods that are predators of thrips. Such an update is much needed because some important groups have never been adequately reviewed (e.g. predatory mites), and because the number of candidate species in several groups is rapidly increasing due to attempts to control thrips that have invaded new continents in the past two decades (e.g. *Frankliniella occidentalis* and *Thrips palmi*). Subsequently, some of the most important groups of predatory arthropods are reviewed with respect to their per capita predation and reproduction rate and with respect to their impact on thrips populations, as derived from experiments on biological control of thrips. Mathematical predator–prey models are presented to analyse how individual-level characteristics influence population phenomena. The chapter ends with a first attempt to review some of the food web complexities arising from polyphagy, plant–predator interactions and intraguild predation.

Predators, Relative Body Size and Prey Vulnerability

One of the most striking features of arthropod predator–prey relations is that predators are similar to or larger than their prey (Warren and Lawton, 1987; Sabelis, 1992; Diehl, 1993 and references therein). In addition, increases in predator body size are correlated with an increase in prey size range. This empirically established pattern arises because maximum prey size increases more steeply with predator body size than minimum prey size. Although such sweeping generalizations across taxa sometimes fail in special cases (e.g. when tested at smaller taxonomic scales or size gradients), they prove to be very useful in classifying arthropod predators of thrips. The arthropod predators recorded in the literature as predators of thrips are listed in Table 8.1; below we discuss how their body size compares with that of the thrips and how this relates to vulnerability to predator attack.

Size, vulnerability and refuges of thrips

Before discussing thrips body size relative to their predators it is important to consider their way of life on a plant. Thrips inhabit various sites on a plant, each affecting the risk of their being eaten. They exhibit one of three major lifestyles (Lewis, 1973). First, there is the highly specialized group whose feeding stimulates the plant to produce galls in which they gain protection from predators (although the degree of protection strongly depends on the structure of the gall). Second, there are the interstitial dwellers that seek protection in narrow spaces on their host plants, such as in bark crevices, in dense

Table 8.1. Predators of thrips arranged according to order (decreasing body size), family, genus and species.

Table 8.1a. Lacewings and flies.

Order	Family	Genus	Species	Author	Thrips prey (*T.=Thrips, F.=Frankliniella*)	References
Neuroptera (lacewings)	Chrysopidae (green lacewings)	*Chrysopa*	*alobana*	Banks	*Selenothrips rubrocinctus*	Callan, 1943
		Chrysopa	*arioles*	Banks	*Selenothrips rubrocinctus*	Callan, 1943
		Chrysopa	*carnea*	Stephens	*Gynaikothrips ficorum, Heliothrips haemorrhoidalis, T. tabaci*	Lewis, 1973; Milbrath *et al.*, 1993
		Chrysopa	*claveri*	Nav.	*Selenothrips rubrocinctus, T. fuscipennis*	Callan, 1943; Carl, 1976
		Chrysopa	*iona*	Banks	*Selenothrips rubrocinctus*	Callan, 1943
		Chrysopa	*montoyana*	Nav.	*Selenothrips rubrocinctus*	Callan, 1943
		Chrysopa	*oculata*	Say	*Liothrips floridensis, Prosopothrips cognatus*	Lewis, 1973
		Chrysopa	*perla*	(L.)	*Odontothrips intermedius, O. phaleratus*	Ananthankrishnan, 1984
		Chrysopa	*plorabuda*	Fitch	*Caliothrips fasciatus, F. tritici*	Lewis, 1973
		Chrysopa	*vulgaris*	Nav.	*Odontothrips intermedius, O. phaleratus, T. tabaci*	Ghabn, 1948; Ananthakrishnan, 1984
		Leucochrysa	*marquesi*	Nav.	*Selenothrips rubrocinctus*	Callan, 1943
		Leucochrysa	*submacula*	Banks	*Selenothrips rubrocinctus*	Callan, 1943
		Leucochrysa	*varia*	Schn.	*Selenothrips rubrocinctus*	Callan, 1943
	Hemerobiidae (brown lacewings)	*Hemerobius*	*californicus*	Banks	*Taeniothrips inconsequens*	Lewis, 1973
		Hemerobius	*pacificus*	Banks	*Taeniothrips inconsequens*	Lewis, 1973
		Hemerobius	*sp.*		*Liothrips setinodis*	Lewis, 1973
Diptera	Cecidomyiidae (gall midges)	*Adelgimyza*	*tripiperda*	Del Guercio	*Liothrips oleae*	Barnes, 1930; Lewis, 1973
		Arthrocnodax	*occidentalis*	Eelt.	*T. palmi*	Chang *et al.*, 1993
		Lestodiplosis	*sp.*		*Gynaikothrips ficorum*	Bennett, 1965
		Thripsobremia	*liothripis*	Barnes	*Liothrips urichi*	Barnes, 1930
		Thripsobremia	*thripivora*	Gagne	*Gynaikothrips ficorum*	Gagné and Bennett, 1993

Table 8.1a. Continued.

Order	Family	Genus	Species	Author	Thrips prey (*T.=Thrips, F.=Frankliniella*)	References
	Asilidae	*Machinus*	*annulipes*	Br.	*Haplothrips* sp.	Kurkina, 1979
	Dolichopodidae	*Condylostylus*	*flavipes*	(Aldrich)	*F. intonsa*	Wheeler, 1977
		Medetera	*ambigua*	Zett.	*Limothrips cerealium, T. tabaci*	Lewis, 1973
		Medetera	*dendrobaena*	Kowarz	*Limothrips cerealium*	Lewis, 1973
		Medetera	*jaculus*	Fall	*Limothrips cerealium*	Lewis, 1973
		Medetera	*truncorum*	Meig.	*Limothrips cerealium, L. denticornis*	Lewis, 1973
	Syrphidae	*Baccha*	*livida*	Schiner	*Gynaikothrips ficorum*	Bennett, 1965
	(hover flies)	*Baccha*	*norina*	Curr.	*Gynaikothrips ficorum*	Bennett, 1965
		Ischiodon	*aegypticus*	Wied	*Caliothrips fasciatus*	Lewis, 1973
		Mesograpta	*marginata*	Say	*T. tabaci*	Lewis, 1973
		Sphaerophoria	*quadrituberculata*	Beazi	*Gigantothrips afer*	Stuckenberg, 1954; Lewis, 1973
		Sphaerophoria	*ruepelli*	(Wiedemann)	*T. tabaci*	Tawfik *et al.*, 1974
		Sphaerophoria	*sulphuripes*	(Thom.)	*Caliothrips fasciatus*	Lewis, 1973
		Syrphus	*corollae*	F.	*Liothrips setinodis, T. tabaci*	Ghabn, 1948; Lewis, 1973
	Chloropidae	*Lioscinella*	*sabroskyi*	Cogan & Smith	*Teuchothrips* sp.	Cogan and Smith, 1982
	Hypotidae	*Platipalpus*	*pallidicornis*	Coll.	–	Kühne and Schrameyer, 1994
		Platipalpus	*picitarsis*		–	Kühne and Schrameyer, 1994
		Stilpon	*nubila*	Coll.	–	Kühne and Schrameyer, 1994

Predator–thrips associations alone have not been taken as evidence, but rather successful predation on thrips and the ability to suppress thrips populations. As the predators of thrips mentioned in Thompson and Simmonds (1965) and Herting and Simmonds (1971) are listed in Lewis (1973), only the latter publication is mentioned as a reference. For brevity, some generic names of thrips are abbreviated e.g. *Caliot* = *Caliothrips*.

Table 8.1b. Wasps, beetles, mantids and crickets.

Order	Family	Genus	Species	Author	Thrips prey *(T.-Thrips, F.=Frankliniella)*	References
Hymenoptera	Sphecidae	*Ammoplanus*	*perrisi*	Giraud	*T. tabaci*	Priesner, 1964; Lewis, 1973
	(digger wasps)	*Microstigmus*	*similis*	Melo	–	De Melo and Evans, 1993
		Microstigmus	*tripoctenus*	Richards	*Leucothrips* sp., *Bradinothrips* sp.	Matthews, 1970
		Microstigmus	*xylicola*	Melo	–	De Melo and Evans, 1993
		Spilomena	*barberi*		*F.* sp., *T.* sp., *Sercothrips* sp.	Ananthakrishnan, 1984
		Spilomena	*elegantula*	Turner	*T. obscuratus*	Vardy, 1987
		Spilomena	*emarginata*	Vardy	–	Vardy, 1987
		Spilomena	*nozela*	Vardy	*Heliothrips haemorrhoidalis*	Vardy, 1987
		Spilomena	*pusilla*	(Say)	*Neohydatothrips variabilis*	Lewis, 1973
		Spilomena	*troglidytes*	Lind.	*F. tenuicornis*	Lewis, 1973
		Spilomena	*vagans*	Bluthgen	*Anaphothrips obscurus*	Lewis, 1973
		Xysma	*sp.*		–	Lewis, 1973
	Vespidae	*Polistes*	*hebraeus*	(Fabricius)	*Rhipiphorothrips cruentatus*	Dhaliwal, 1975
	Formicidae	*Azteca*	*chartifox*		–	Ananthakrishnan, 1984
	(ants)	*Pheidole*	*megacephala*	(Fabricius)	*Liothrips urichi*	Reimer, 1988
		Wasmannia	*auropunctata*	Roger	*Selenothrips rubrocinctus*	Callan, 1943
Coleoptera	Carabidae	*Hexagonia*	*terminalis*	Genom & Har	*Haplothrips sorghicola*	Lewis, 1973
	Coccinellidae	*Adalia*	*bipunctata*	(L.)	*Liothrips setinodis, Thrips laricivorus*	Priesner, 1964; Lewis, 1973
	(ladybirds)	*Adalia*	*conglomerata*	(L.)	*Liothrips setinodis*	Lewis, 1973
		Adonia	*variegata*	Goeze	*Haplothrips tritici*	Lewis, 1973
		Anatis	*ocellata*	(L.)	*Liothrips setinodis, Thrips laricivorus*	Lewis, 1973
		Aphidecta	*obliterata*	(L.)	*Thrips laricivorus*	Lewis, 1973
		Cheilomenes	*vicina*	Muls.	*Caliothrips indicus*	Lewis, 1973

Table 8.1b. Continued.

Order	Family	Genus	Species	Author	Thrips prey *(T.=Thrips, F.=Frankliniella)*	References
		Coccinella	*arcuari*	F.	*Baliothrips biformis*	Mammen and Vasudevan, 1977
		Coccinella	*novemnotata*	Hbst.	*T. tabaci*	Lewis, 1973
		Coccinella	*repanda*	Thunberg	*T. tabaci*	Lewis, 1973
		Coccinella	*septempunctata*	L.	*T. fuscipennis, T. tabaci*	Carl, 1976
		Coccinella	*undecimpunctata*	Hbst.	*T. tabaci*	Ghabn, 1948; Lewis, 1973; Afifi *et al.*, 1976
		Coleomegilla	*maculata*	(DeGeer)	*T. simplex, T. tabaci*	Lewis, 1973
		Cryptomorpha	*desjardinsi*	(Guer.)	*F. occidentalis*	Pena, 1990
		Cycloneda	*sanguinea*	(L.)	*T. simplex*	Lewis, 1973
		Exochomus	*flavipes*	(Thunb.)	*Scirtothrips auranti*	Lewis, 1973
		Exochomus	*quadripustulatus*	(L.)	*Liothrips setinodis, T. laricivorus*	Lewis, 1973
		Hippodamia	*convergens*	Guer.	*Caliothrips fasciatus, Taeniothrips inconsequens, T. tabaci*	Bailey, 1933; Lewis, 1973
		Micraspis	*cardoni*		*Zaniothrips ricini*	Mohandaniel *et al.*, 1983
		Neomysia	*oblongoguttata*	(L.)	*T. laricivorus*	Lewis, 1973
		Propylaea	*quatuordecimpuncta*	(L.)	*Liothrips setinodis*	Lewis, 1973
		Scymnus	*ater*	Kug	*Taeniothrips inconsequens*	Lewis, 1973
		Scymnus	*frontalis*		*Kakothrips robustus, Odontothrips loti, O. phaleratus*	Ananthakrishnan, 1984
		Scymnus	*interruptus*	(Goeze)	*T. tabaci*	Afifi *et al.*, 1976; Habib *et al.*, 1980
		Scymnus	*nubilus*	Muls	*Caliothrips indicus, T. tabaci*	Saxena, 1971, 1977
		Scymnus	*thoracicus*		*Chaetanaphothrips orchidii*	Ananthakrishnan, 1984
		Scymnus	*trepidulus*	Weise	*Scirtothrips aurantii*	Lewis, 1973
	Malachidae	*Laius*	*externotatus*	Pic	*Caliothrips indicus, T. tabaci*	Saxena, 1971, 1977
		Malachius	*viridus*	(F.)	*Haplothrips tritici*	Lewis, 1973; Shurovenkov, 1974

	Staphylinidae (rove beetles)	–			*Acanthothrips nodicornis, Hoplandrothrips pillichianus, Hoplothrips corticis, H. pedicularius*	Ananthakrishnan, 1984
		Gyrophaena	*manca*	Er.	*Taeniothrips inconsequens*	Lewis, 1973
		Paederus	*alfierii*	Koch	*T. tabaci*	Tawfik and Abouzeid, 1977
Dictyoptera	Mantidae	*Haldwania*	*lilliputiana*		*Zaniothrips ricini*	Mohandaniel *et al.*, 1983
Orthoptera	Gryllidae	*Oecanthus*	*longicauda*	Mats	*T. tabaci*	Lewis, 1973
		Oecanthus	*turanicus*	Uv.	*T. tabaci*	Ghabn, 1948; Lewis, 1973

Table 8.1c. Heteropteran bugs.

Order	Family	Genus	Species	Author	Thrips prey (*T.=Thrips, F.=Frankliniella*)	References
Heteroptera	Reduviidae	–			*Varshneyia pasania*	Suzuki *et al.*, 1988
	Miridae (mirid bugs)	*Camptoptera*	*liebknechti*	(Girault)	*F. occidentalis*	Goodwin and Steiner, 1996
		Campylomma	*chinensis*	Schuh	*T. palmi*	Wang, 1995
		Campylomma	*livida*	Reuter	*T. palmi*	Hirose *et al.*, 1993; Chang *et al.*, 1993
		Deraeocorus	*pallens*	Reuter	*T. tabaci*	Zavodchikova, 1974; Abbas *et al.*, 1988
		Deraeocorus	*punctulatus*	Fall	–	Zavodchikova, 1974
		Dicyphus	*tamaninii*	Wagner	*F. occidentalis*	Riudavets *et al.*, 1993; Gabarra *et al.*, 1995; Albajes *et al.*, 1996; Castané *et al.*, 1996
		Dicyphus	*eckerleini*	Wagner	*T. tabaci*	Callan, 1975; Dimitrov, 1975, 1977
		Dicyphus	*rhododendri*	Dolling	*Heterothrips azaleae*	Braman and Beshear, 1994
		Macrolophus	*caliginosus*	Wagner	*F. occidentalis*	Fauvel *et al.*, 1987; Riudavets *et al.*, 1993
		Macrolophus	*rubi*	Woodroffe	*T. tabaci*	Dimitrov, 1975, 1977
		Psalius	sp.		*Megalurothrips distalis*	Ananthakrisnan, 1984
		Rhinocapsus	*vanduzeei*	Uhler	*Heterot. azaleae*	Braman and Beshear, 1994
		Termatophylidea	*pilosa*	Reut & Poppius	*Selenot. rubrocinctus*	Callan, 1943, 1975
		Termatophylidea	*maculata*	Usinger	*Caliot. insularis, Selenot. rubrocinctus*	Callan, 1943; Van Doesburg, 1964; Lewis, 1973; Callan, 1975
		Termatophylidea	*opaca*	Carvalho	*Selenot. rubrocinctus*	Callan, 1975
		Termatophylum	*insigne*	Reuter	*Gynaikot. ficorum*	Lewis, 1973
	Nabidae	*Nabis*	*alternatus*	Parshley	*Aeolot. fasciatus, F. moultoni, F. occidentallis, T. tabaci*	Taylor, 1949; Lewis, 1973; Benedict and Cothran, 1980

(damsel bugs)	*Nabis*	*americoferus*	Caryon	*F. occidentalis*	Benedict and Cothran, 1980; Stolz and McNeal, 1982
	Nabis	*ferus*	(L.)	*T. tabaci*	Dimitrov, 1975
	Nabis	*hinbergii*		*F. occidentalis*	Goodwin and Steiner, 1996
	Nabis	*pseudoferus*	Remane	*T. tabaci*	Dimitrov, 1975
Lygaeidae	*Geocoris*	*atricolor*	Montandon	*F. occidentalis*	Benedict and Cothran, 1980
(lygus bugs)	*Geocoris*	*pallens*	Say	*F. occidentalis*	Benedict and Cothran, 1980; González and Wilson, 1982; Stoltz and Stern, 1978; Yano, 1996
	Geocoris	*ochropterus*	Fabr.	*Ayyaria chaetophora, Caliot. indicus, T. palmi, Scirtot. dorsalis and others*	Chang *et al.*, 1993; Mohandaniel *et al.*, 1983; Sureshkumar and Ananthakrishnan, 1985
	Geocoris	*punctipes*	Say	*Selenot. rubrocinctus*	Callan, 1975
	Ninyas	*torvus*	Distant	*Selenot. rubrocinctus*	Callan, 1943, 1975
Anthocoridae	*Anthocoris*	*austropiceus*	Gross	*Teuchot.* sp.	Cogan and Smith, 1982
(flower bugs)	*Anthocoris*	*nemorum*	(L.)	*F. occidentalis*	Buente *et al.*, 1990; Buxton and Wardlow, 1991; Jacobson, 1991
	Anthocoris	*gallarum-ulmi*	(DeGeer)	*F. occidentalis*	Buente *et al.*, 1990
	Bilia	sp.		*T. palmi*	Hirose *et al.*, 1993
	Carayonocoris	*indicus*	Muraleedharan	*Caliot. indicus, F. schultzei, Haplot. ganglebaueri, Retit. syriacus, Scirtot. dorsalis, Selenot. rubrocinctus, T. tabaci*	Muraleedharan and Ananthakrishnan, 1978; Sureshkumar and Ananthakrishnan, 1984; Ananthakrishnan, 1984
	Cardiastethus	*consors*	White	*Heliot. haemorrhoidalis*	Lewis, 1973
	Cardiastethus	*minutus*		*F. occidentalis*	Goodwin and Steiner, 1996
	Cardiastethus	*poweri*	White	*Heliot. haemorrhoidalis*	Lewis, 1973
	Cardiastethus	*rugicollis*	Champion	*Gynaikot. ficorum*	Bennett, 1965
	Ectemnus	sp.		*Gynaikot. ficorum*	Bennett, 1954
	Ectemnus	*reduvinus*	(H.-S.)	*Liothrips oleae*	Lewis, 1973
	Macrotracheliella	*laevis*	Champion	*Gynaikot. ficorum*	Lewis, 1973

Table 8.1c. Continued.

Order	Family	Genus	Species	Author	Thrips prey *(T.=Thrips, F.=Frankliniella)*	References
	Anthocoridae	*Montandoniola*	*moraguesi*	(Puton)	*Arrhenot. ramakrishnae, F. occidentalis, Gynaikot. ficorum, G. flaviantennatus, Liot. africanus, L. fluggae, L. oleae, L. urichi, T. tabaci*	Tawfik and Nagui, 1965; Muraleedharan and Ananthakrishnan, 1971; Péricart, 1972; Lewis, 1973; Muraleedharan and Ananthakrishnan, 1978; Ananthakrishnan, 1984; Reimer, 1988
		Orius	*albidipennis*	(Reuter)	*F. occidentalis, Gynaikot. ficorum, Megalurot. sjostedti, Retit. syriacus, Selenot. rubrocinctus, T. tabaci*	Lewis, 1973; Saxena, 1977; Ghauri, 1980; Ananthakrishnan and Sureshkumar, 1985; Salim *et al.*, 1987; Pena, 1990; Chyzik *et al.*, 1995a
		Orius	*amnesius*	Ghauri	*Megalurot. sjostedti*	Ghauri, 1980
		Orius	*armatus*	Gross	*F. occidentalis*	Goodwin and Steiner, 1996
		Orius	*indicus*	(L.)	*Megalurot. nigricornis*	Rajasekhara and Chatterji, 1970; Lewis, 1973; Ananthakrishnan and Sureshkumar, 1985
		Orius	*insidiosus*	(Say)	*Anaphot. obscurus, Caliot. phaseoli, Haplot. subtillissimus, F. moultini, F. occidentalis, F. tritici, Hercot. fasciatus, Leptot. mali, Prosopot. cognatus, Neohydatot. variabilis, T. inconsequens, T. simplex, T. tabaci*	Robinson *et al.*, 1972; Ramakers, 1978; Isenhour and Yeargan, 1981b, 1982; Ananthakrishnan and Sureshkumar, 1985; McCaffrey and Horsburgh, 1986a; Van den Meiracker and Ramakers, 1991; Fransen *et al.*, 1993; Coll and Ridgway, 1995; Richards and Schmidt, 1996

Orius	*laevigatus*	(Fieber)	*F. occidentalis, Caliot. fasciatus, T. tabaci*	Pericart, 1972; Tawfik and Ata, 1973b, Afifi *et al.*, 1976; Ananthakrishnan and Sureshkumar, 1985; Tavella, *et al.*, 1991; Vielvieille and Millot 1991; Riudavets *et al.*, 1993; Carnero *et al.*, 1993; Tommasini and Nicoli, 1993; Husseini *et al.*, 1993
Orius	*majusculus*	(Reuter)	*F. occidentalis*	Ramakers, 1990, Trotting Caudal *et al.*, 1991; Fischer *et al.*, 1992; Tommasini and Nicoli, 1993; Riudavets *et al.*, 1995; Jacobson 1995
Orius	*maxidentex*	Ghauri	*Haplot. ganglebaueri, Scirtot. dorsalis, Calio. indicus, F. schultzei, Anaphot. sudanesis, Caliot. graminicola, Microephalot. abdominalis, Retit. syriacus, Stenchaetot. biformis, T. palmi, T. tabaci*	Sureshkumar and Ananthakrishnan, 1984; Ananthakrishnan and Sureshkumar, 1985
Orius	*minutus*	(L.)	*Chirot. manicatus, Drepanot. reuteri, F. intonsa, F. occidentalis, F. schultzei, Limot. schmutzi, L. denticornis, Megalurot. distalis, Haplot. ganglebaueri, H. aculeatus, Parthenot. dracaenae, T. flavus, T. fuscipennis, T. palmi, T. tabaci*	Lewis, 1973; Viswanathan and Ananthakrishnan, 1974; Ramakers, 1978; Ananthakrishnan and Sureshkumar, 1985; Lichtenauer and Sell, 1993
Orius	*niger*	Wolff	*Haplot. aculeatus, H. niger, H. tritici, Thrips fuscipennis, T. simplex, T. tabaci*	Lewis, 1973; Carl, 1976; Ananthakrishnan and Sureshkumar, 1985; Ramakers, 1990; Van de Veire and Degheele, 1992; Tommasini and Nicoli, 1993; Yasunaga and Miyamoto, 1993
Orius	*persequens*	(White)	*T. tabaci*	Ananthakrishnan and Sureshkumar, 1985

Table 8.1c. Continued.

Order	Family	Genus	Species	Author	Thrips prey (*T.=Thrips, F.=Frankliniella*)	References
		Orius	*sauteri*	(Poppius)	*T. palmi, T. setosus, Mycterot. glycines*	Chang *et al.*, 1993; Nakashima *et al.*, 1996; Nagai, 1989, 1990, 1991; Kawai, 1995; Wang, 1995
		Orius	*similis*	Zheng	*T. palmi*	Wei *et al.*, 1984; Kajita, 1986; Yasunaga and Miyamoto, 1993
		Orius	*tantilus*	(Mots.)	*Caliot. indicus, Haplot. ganglebaueri, Microcephalot. abdominalis, Scirtot. dorsalis, Stenchaetot. biformis, T. palmi*	Ananthakrishnan and Sureshkumar, 1985; Mituda and Calilung, 1989; Goodwin and Steiner, 1996
		Orius	*tripoborus*	(Hesse)	*Scirtot. aurantii*	Lewis, 1973
		Orius	*tristicolor*	(White)	*F. occidentalis, F. tritici, F. minuta, F. moultoni, Caliot. fasciatus, Haplot. verbasci, Microcephalot. abdominalis, Odontot. loti, T. abdominalis, T. tabaci, Taeniot. inconsequens, Thrips simplex*	Lewis, 1973; Sales-Aguilar and Ehler, 1977; Hollingsworth and Bishop, 1982; Letourneau and Altieri, 1983; Ananthakrishnan and Sureshkumar, 1985
		Scoloposcelis	*parallelus*	(Motshulski)	*Ecacanthot. tibialis*	Muraleedharan and Ananthakrishnan, 1978
		Tetraphleps	*bicuspis*	H.S.	*Thrips laricivorus*	Nolte, 1951
		Wollastoniella	*parvicuneis*	Yasunaga	*T. palmi*	Yasunaga, 1995
		Wollastoniella	*rotunda*	Yasunga & Miyamoto	*T. palmi*	Yasunaga and Miyamoto, 1993

Table 8.1d. Thrips reported as predators of thrips.

Order	Family	Genus	Species	Author	Thrips prey (*T.=Thrips, F.=Frankliniella*)	References
Thysanoptera (thrips)	Aeolothripidae	*Aeolothrips*	*intermedius*	Bagnall	*Caliot. fasciatus, Haplot. tritici, Heliot. haemorrhoidalis, Kakot. robustus, Sericot. variabilis, Stenot. graminum, Thrips laricivorus, T. simplex, T. tabaci, T. linarius*	Böhm, 1995; Robinson *et al.*, 1972; Lewis, 1973; Ferrari, 1980; El Serwiy *et al.*, 1985; Baker, 1988
		Aeolothrips	*intermedius*	Bagnall	*Heliot. haemorrhoidalis, Odontot. confusus, T. tabaci, and others.*	Bournier *et al.*, 1978, 1979; Lacasa, 1988; Lacasa *et al.*, 1982, 1989
		Aeolothrips	*collaris*	Priesn.	–	Saxena, 1971, 1977
		Aeolothrips	*kuwanai*	Moulton	*T. tabaci*	Lewis, 1973
		Aeolothrips	*melisi*	Priesn.	–	Lacasa, 1988
		Aeolothrips	*tenuicornis*	Bagnall	–	Lacasa, 1988; Lacasa *et al.*, 1989
		Aeolothrips	*vittatus*	Haliday	*T. laricivorus*	Nolte, 1951
		Andrewarthaia	*kellyana*		*T. australis*	Mound, 1972; Goodwin and Steiner, 1996
		Desmothrips	sp.		*F. occidentalis*	Goodwin and Steiner, 1996; Mound, unpublished
		Erythrothrips	*asiaticus*	Ram. & Marg.	*Retit. syriacus, Scirtot. dorsalis, Haplot. ganglebaueri, F. schultzei, Caliot. indicus*	Sureshkumar and Ananthakrishnan, 1987
		Franklinothrips	*caballeroi*	Johansen	–	Johansen, 1981
		Franklinothrips	*megalops*	(Trybom)	*Caliot. indicus, F. schultzei, Heliot. haemorrhoidalis, Retit. syriacus, Scirtot. dorsalis, Zaniot. ricini*	Stannard, 1952; Lewis, 1973; Sureshkumar and Ananthakrishnan, 1987
		Franklinothrips	*tenuicornis*	Hood	*Dinurot. hookeri, Heliot. haemorrhoidalis, Caliot. insularis, Selenot. rubrocinctus*	Callan, 1943
		Franklinothrips	*vespiformis*	(Crawford)	*Caliot. insularis, Dinurot. hookeri, Heliot. haemorrhoidalis, Scirtot. longipennis, Selenot. rubrocinctus, T. palmi*	Callan, 1943; Lewis, 1973; Okajima *et al.*, 1992; Hirose *et al.*, 1993

Table 8.1d. Continued.

Order	Family	Genus	Species	Author	Thrips prey (*T.=Thrips, F.=Frankliniella*)	References
		Mymarothrips	*bicolor*		*Selenothrips rubrocinitus*	Mound, unpublished
	Thripidae	*Scolothrips*	*indicus*	Priesner	–	Ananthakrishnan, 1969
		Scolothrips	*sexmaculatus*	(Pergande)	*T. pini*	Lewis, 1973
	Phlaeothripidae	*Aleurodothrips*	*fasciapennis*	(Franklin)	–	Ananthakrishnan, 1974
		Androthrips	*flavipes*	Schmutz	*Arrhenot. ramakrishnae, Gynaikot. flaviantennatus, Schedot. orientalis*	Varadarasan and Ananthakrishnan, 1981; Ananthakrishnan, 1984; Sureshkumar and Ananthakrishnan, 1987
		Haplothrips	*bedfordi*	Jac.-Guill.	*Scirtot. aurantii*	Lewis, 1973
		Haplothrips	*cahirensis*	(Trybom)	*Gynaikot. ficorum*	Lewis, 1973
		Haplothrips	*victoriensis*		*F. occidentalis*	Goodwin and Steiner, 1996
		Leptothrips	*mali*	(Fitch)	*F. moultoni, Depanot. reuteri*	Bailey, 1940; Lewis, 1973
		Xylaplothrips	*inquilinus*	Priesner	*Gynaikot. uzeli, Liot. brevitubus, L. kuwanai, Mesot. claripennis*	Ananthakrishnan, 1984; Sureshkumar and Ananthakrishnan, 1987

N.B. There are several records of cannibalism in thrips including *Aeolothrips intermedius* (Bournier *et al.*, 1979), *F. occidentalis* (Kirk, 1995), *Scolothrips longicornis* (Gerlach, 1985), *Scolothrips sexmaculatus* (Gilstrap and Oatman, 1976), *Haplothrips niger* (Loan and Holdaway, 1955), *Haplothrips kurdjumovi* (Putman, 1942) and *Haplothrips victoriensis* (Andrewartha, 1936).

Table 8.1e. Mites reported as predators of thrips.

Order	Family	Genus	Species	Author	Thrips prey (*T.=Thrips, F.=Frankliniella*)	References
Heterostigmata	Acaraphenacida	*Adactylidium*	*nicolae*	Krczal	*Gynaikothrips ficorum, Scirtothrips citri*	El-Badry and Tawfik, 1966; Lewis, 1973; Eickwort, 1983
	Pyemotidae	*Pyemotes (=Pediculoides)*	*ventricosus*	Newp. Berl.	*Gynaikothrips ficorum*	Bennet, 1965
Prostigmata	Anystidae	*Anystis*	*agilis*	Banks	*F. occidentalis, Scirtothrips citri*	Mostafa *et al.*, 1975
		Anystis	*astripus*	Karsch	*T. tabaci*	MacGill, 1939; Sorenson *et al.*, 1976
		Anystis	*baccarum*	L.	*F. occidentalis, Kakothrips pisivorus, Scirtothrips aurantii, T. klapaleki*	Franssen, 1960; Morison, 1968; Lewis, 1973
		Actineda	*vitis*	Schrank	*Kakothrips pisivorus*	Franssen, 1960
		Walzia	*australica*	Womersley	*F. occidentalis*	Goodwin and Steiner, 1996
	Trombidiidae	*Allotrombium*	*sp.*		*T. fuscipennis*	Carl, 1976
		Trombidium	*sp.*		*Aptinothrips rufus, Scirtothrips citri*	Sharga, 1933; Lewis, 1973; Eickwort, 1983
	Cheyletidae	*Cheyletus*	*sp.*		*Liothrips oleae*	Tominic, 1950; Lewis, 1973
	Erythraeidae	*Erythrites*	*sp.*		*Scirtothrips citri*	Eickwort, 1983
		Hauptmannia	*brevicollis*	Oud.	*Anaphothrips obscurus, F. intonsa, Haplothrips aculeatus, T. linarius, T. validus, Taeniothrips atratus*	Lewis, 1973; Ananthakrishnan, 1984
Mesostigmata	*Aseidae*	*Lasioseius*	sp.		*F. occidentalis*	Goodwin and Steiner, 1996
	Laelapidae	*Hypoaspis*	*aculeifer*	Canestrini	*F. occidentalis*	Gilkeson *et al.*, 1990; Gillespie and Quiring, 1990; Glockemann, 1992
		Hypoaspis	*miles*	(Berlese)	*F. occidentalis*	Glockemann, 1992; Brødsgaard *et al.*, 1996
		Amblyseius	*addoensis*	VanderMerwe & Ryke	*Heliothrips sylvanus, Scirtothrips aurantii, S. citri, T. tabaci*	Schwartz, 1988; Grout and Richards, 1992b
	Phytoseiidae	*Amblyseius*	*andersoni*	(Chant)	*F. pallida, F. occidentalis, T. tabaci*	Ramakers, 1978; Dicke and Groeneveld, 1986; Rodriques-Reina *et al.*, 1992

Table 8.1e. Continued.

Order	Family	Genus	Species	Author	Thrips prey (*T.=Thrips, F.=Frankliniella*)	References
		Amblyseius	*aripo*	DeLeon	*F. williamsi*	Bakker, 1993
		Amblyseius	*aurescens*	Athias-Henriot	*T. tabaci*	Ramakers, 1978
		Amblyseius	*barkeri*	(Hughes)	*F. intonsa, F. occidentalis, Parthenothrips dracaenae, T. palmi, T. tabaci, T. simplex*	Ramakers, 1978, 1980, 1983, 1988; Kajita, 1986; Hansen, 1988, 1989; Bakker and Sabelis, 1989; Bonde, 1989; Van der Hoeven and Van Rijn, 1990; Rodrigues-Reina *et al.*, 1992; Conijn, 1993; Van Houten *et al.*, 1995a; Johnson, 1996; Piatowski, 1987
		Amblyseius	*californicus (=chilenensis)*	(McGregor)	*F. occidentalis, Retithrips syriacus*	Swirski *et al.*, 1970; Rodrigues-Reina *et al.*, 1992
		Amblyseius	*citri*	VanderMerwe & Ryke	*Scirtothrips aurantii*	Grout and Stephen, 1993; Grout, 1994
		Amblyseius	*cucumeris*	(Oudemans)	*F. tritici, F. occidentalis, T. obscuratus, T. tabaci*	MacGill, 1939; Ramakers, 1978, 1980, 1982, 1988; de Klerk and Ramakers, 1986; Gillespie, 1989; Castagnoli and Simoni, 1990; Steiner, 1990; Shipp and Whitfield, 1991; Van Houten and Van Stratum, 1995; Johnson, 1990
		Amblyseius	*degenerans*	(Berlèse)	*F. occidentalis*	Van Houten and Van Stratum, 1995; Van Houten *et al.*, 1995a
		Amblyseius	*hibisci*	(Chant)	*F. occidentalis, Retithrips syriacus*	Swirski *et al.*, 1970; Van Houten *et al.*, 1995a
		Amblyseius	*lailae*	Schicha	*F. occidentalis*	Goodwin and Steiner, 1996
		Amblyseius	*largoensis*	(Muma)	*F. occidentalis, Heliothrips haemorrhoidalis, Retithrips syriacus*	Kamburov, 1971; Simonishvili, 1976; Goodwin and Steiner, 1996

Amblyseius	*lentiginososus*	Denmark & Schicha	*F. occidentalis*	Goodwin and Steiner, 1996
Amblyseius	*limonicus*	(Garman & McGre	*F. occidentalis, Retithrips syriacus*	Swirski and Dorzia, 1968; Van Houten *et al.*, 1995a; Van Houten, 1996
Amblyseius	*longispinosus*	(Evans)	*T. palmi*	Kajita, 1986; Chang *et al.*, 1993
Amblyseius	*manihoti*	(Moreas)	*F. williamsi*	Bakker and Klein, 1993
Amblyseius	*masiaka*	Biommers & Chaz	*F. occidentalis*	Goodwin and Steiner, 1996
Amblyseius	*montdorensis*	Schicha	*F. occidentalis*	Goodwin and Steiner, 1996
Amblyseius	*multidentatus*	(Swirski & Shechter)	*T. palmi*	Chang *et al.*, 1993
Amblyseius	*okinawanus*	Ehara	*T. palmi*	Kajita, 1986
Amblyseius	*pettatus*	Van der Merwe	*F. occidentalis*	Goodwin and Steiner, 1996
Amblyseius	*scutalis (=rubini =gossipi)*	(Athias-Henriot)	*F. occidentalis, Retithrips syriacus, Scirtothrips citri*	Swirski *et al.*, 1967b; Morse *et al.*, 1986; Bounfour and McMurtry, 1987; Van Houten *et al.*, 1995a
Amblyseius	*sessor*	DeLeon	*Haplothrips subtilissimus, T. tabaci*	Carl, 1976; Sciarappa and Swift, 1977; Beglyarov and Suchalkin, 1983
Amblyseius	*stipulatus*	Athias-Henriot	*Scirtothrips citri*	Rodrigues-Reina *et al.*, 1992; More *et al.*, 1986
Amblyseius	*sullivani*	Schicha	*F. occidentalis*	Goodwin and Steiner, 1996
Amblyseius	*swirskii*	Athias-Henriot	*Retithrips syriacus, T. tabaci*	Swirski *et al.*, 1967b; Hoda *et al.*, 1986; Vartepetov and Vasilev, 1971
Amblyseius	*tularensis (hibisci–1995)*	(Congdon)	*F. occidentalis, Scirtothrips citri*	Tanigoshi *et al.*, 1983, 1984, 1985; Congdon and McMurtry, 1988; More *et al.*, 1986; Grafton-Cardwell and Ouyang, 1995b; Van Houten *et al.*, 1995a
Amblyseius	*waltersi*	Schicha	*F. occidentalis*	Goodwin and Steiner, 1996
Typhlodromus	*athiasae*	Porath & Swirski	*Retithrips syriacus*	Swirski *et al.*, 1967a
Typhlodromus	*caudiglans*	(Schuster)	*Haplothrips kurdjumovi*	Nakahara, 1985
Typhlodromus	*occidentalis*	(Nesbitt)	*Retithrips syriacus*	Swirski and Dorzia, 1969; Rodrigues-Reina *et al.*, 1992
Typhlodromus	*pyri*	Scheuten	*Drepanothrips reuteri*	Engel and Ohnesorge, 1994

inflorescences (grasses, composite flowers), on hairy leaf surfaces, under curled leaf edges and in leaf sheaths. Clearly, when inhabiting galls or interstitial sites, thrips are vulnerable only to predators of similar or smaller size. However, when they move out of their refuge (to forage or disperse), they become vulnerable to a wide range of predators of similar or larger size, just like the third and last group of surface-dwelling thrips species.

Another important point is the change in size and site occupied by the thrips during development. Terebrantian eggs are usually about 0.3 × 0.15 mm (e.g. Aeolothripidae and Thripidae); tubuliferan eggs are somewhat larger. Terebrantia lay their eggs in an incision made in the plant tissue by the ovipositor. This protects them from predation, but the degree of protection depends on how deep the eggs are embedded in the leaf tissue. Tubuliferan species (Phlaeothripidae) do not insert their eggs in the substrate, so as to avoid exposure to predators they deposit them in protected places (in galls, bark crevices, bark beetle galleries and under scales of coccids) or, rarely, protect them by brood care. Despite this, in general the eggs of Tubulifera are thought to be less efficiently protected from predators than those of Terebrantia (Lewis, 1973). On hatching, the soft-bodied and usually slow-moving larvae appear, presenting easy prey for predators unless they inhabit refuges. They cannot jump or run away, but possess some defence mechanisms to deter predators. For example, they can strike a predator with their elongated abdomen and produce rectal droplets which they carry on the raised tip of their abdomen (Lewis, 1973; Bakker and Sabelis, 1986, 1989). These droplets contain a variety of chemicals (Blum, 1991) which deter or irritate their opponent (Howard *et al.*, 1983, 1987; Blum *et al.*, 1992) and/or alarm other (conspecific) thrips nearby (Suzuki *et al.*, 1988; Teerling *et al.*, 1993a,b; Teerling, 1995) (see Table 3.4). Just before entering the second moult, larvae of many species drop to the soil litter beneath the plant or move to a protected place on the plant (crevices, such as bark scales, hollow twigs, bases of leaf stalks, leaf sheaths, leaf spaces where protruding veins branch, or, in some cases (*Aeolothrips*, *Franklinothrips*), a self-made cocoon). At the second moult a propupa emerges which does not feed or excrete, but may exhibit slight mobility upon disturbance. Within the more or less protected site occupied by the propupa, moulting takes place into one or two pupal stages and finally into an adult. Adult size may vary. The largest species, some reaching a length of up to 15 mm, occur in the tropics, but most species (especially in temperate zones) are 1 to 2 mm in length. The adults may escape from predators by jumping or flying, but some litter or bark dwellers retract antennae and legs to feign death (thanatosis).

In summary, the risk of being attacked by predators strongly depends on the developmental stage and on the lifestyle of the thrips species under consideration.

Predatory insects and spiders: miscellaneous taxa

The importance of relative body size is nicely illustrated in two groups of predatory insects, the mantids and the digger wasps (Table 8.1b). Each of these groups covers a wide range of body sizes, much larger than thrips, with the predators of thrips to be found among the smaller species. For example,

Haldwania liliputana (Dictyoptera: Mantidae) is among the smallest species of praying mantid. It is an effective predator of *Zaniothrips ricini* on castor plants in India, consuming almost 100 individuals per day, usually active adult thrips (Mohandaniel *et al.*, 1983). The other example concerns solitary digger wasps in the genus *Spilomena* (Hymenoptera: Sphecidae). Relative to other species of digger wasps they are very small (2 to 4 mm long) and are considered to be genuine thrips hunters (Vardy, 1987; Bohart and Smith, 1994). They seize immature thrips (but also springtails) and carry them to their nests located in abandoned burrows of wood-burrowing beetles or self-excavated holes in pithy twigs. Similarly, other small species of sphecids in the genus *Ammoplanus* and *Microstigmus* forage primarily for thrips (Priesner, 1964; Mathews, 1970; De Melo and Evans, 1993).

Other records of predation on thrips also seem to confirm that the predatory insects are larger than their prey, but among the smaller species in the order and even within the family (Table 8.1). These include larvae of small gall midges (Diptera: Cecidomyiidae) in the genera *Thripsobremia* (Gagné and Bennett, 1993), *Lestodiplosis* (Bennett, 1965) and *Arthrocnodax* (Chang *et al.*, 1993) and mini-ladybeetles (Coleoptera: Coccinellidae), such as *Scymnus* (Afifi *et al.*, 1976; Saxena, 1977; Habib *et al.*, 1980). Other records of thrips predators may include species that belong to the same insect families, but are relatively larger. For example, several larger coccinellids have been recorded, such as *Hippodamia* and *Coccinella* (Table 8.1). Mammen and Vasudevan (1977) observed the ladybeetle *Coccinella arcuata* pushing the curled edges of rice leaves aside to feed on the larvae of *Stenchaetothrips biformis* hiding within. Similar records of feeding on thrips by lacewing and hoverfly larvae have been published (Table 8.1). Stuckenberg (1954) reported small larvae of a hoverfly feeding on thrips (*Sphaerophoria* spp.) and Bennett (1965) found hoverfly larvae (*Baccha* spp.) inside the leaf roll galls induced by *Gynaikothrips ficorum*, showing that leaf rolls do not provide a very effective protection against predators. Also larvae of larger syrphid flies (e.g. *Syrphus corollae*) have been reported as thrips predators (Ghabn, 1948). McMurtry and Badii (1991) found that the numbers of *Heliothrips haemorrhoidalis* were reduced by first-instar lacewing larvae in short (1 to 2 week), small-scale experiments with caged and uncaged fruit clusters of avocado. Possibly, later instars of these lacewings require more food and may prefer to forage on more densely packed and larger prey, such as colonies of aphids. More generally, thrips might be too small for large, adult predators, but this may not be so for the juveniles of these predators. It would be interesting to test whether feeding on thrips occurs more frequently among the younger stages of arthropod predators that as adults are much larger than the thrips. One may also wonder whether the adult females of these larger predators lay their eggs close to thrips-infested plants or prefer to oviposit near colonies of densely packed and/or larger prey.

Other records of predation on thrips (Table 8.1) include crickets, *Oecanthus turanicus* in Egypt (Ghabn, 1948), and adult predatory flies, such as *Stilpon nubila* (Diptera: Hypotidae) (Kühne and Schrameyer, 1994), *Condylostilus flavipes* (Diptera: Dolichopodidae) (Wheeler, 1977) and *Lioscinella sabroskyi* (Cogan and Smith, 1982) (Diptera: Chloropididae). Surprisingly little is known of thrips predation by ants, web-spinning spiders and hunting spiders. As far as the evidence goes, potential predators of thrips are again more likely to be found among the relatively smaller species, such as *Pheidole* ants (Reimer, 1988), *Dictyna* web-spiders (Heidger and Nentwig, 1984) and some small salticid jumping spiders (Lewis, 1973). More research is needed to assess their impact, as they are likely to offer great potential for reducing thrips populations. However, some thrips species manage to ward away predators effectively. For example, some subsocial, mycophagous thrips produce anal droplets containing defensive allomones, such as juglone (Crespi, 1990). This compound aids parental-care behaviour by warding away salticid spiders effectively. Other examples (see Blum, 1991) are anal discharges of *Bagnalliella yuccae* containing γ-decalactone as an effective contact irritant against predatory *Monomorium* ants (Howard *et al.*, 1983), the leaf-roll-gall-inducing thrips *Gynaikothrips ficorum* containing chemicals deterring aggressive myrmicine ants (*Wassmania* spp.) (Howard *et al.*, 1987), and *Haplothrips leucanthemi* containing meillon, an effective repellent against hungry fire ants (*Solenopsis* spp.) (Blum *et al.*, 1992) (see Table 3.4).

Heteropteran predators

There are many generalist predators among the Heteroptera that include thrips among their prey (Table 8.1c). The largest predators in this group belong to the families Pentatomidae and Reduviidae. They exceed thrips in size by an order of magnitude and would probably need many thrips larvae to meet their energy needs. Records of thrips predation by these predators are rare (Callan, 1943). There are more records among predators of intermediate size belonging to the Nabidae in the genus *Nabis* (Taylor, 1949; Dimitrov, 1975; Benedict and Cothran, 1980; Stoltz and McNeal, 1982; Lattin, 1989; Goodwin and Steiner, 1996) and to the Lygaeidae, such as *Geocoris* spp. (Callan, 1975; Stoltz and Stern, 1978; Benedict and Cothran, 1980; González and Wilson, 1982; Mohandaniel *et al.*, 1983; Ananthakrishnan, 1984; Sureshkumar and Ananthakrishnan, 1985; Chang *et al.*, 1993; Goodwin and Steiner, 1996; Yano, 1996). Although the range of prey sizes and species is considerable (e.g. Tamaki and Weeks, 1972; Lattin, 1989; Riudavets, 1995), some field evidence suggests that populations of surface-dwelling thrips in crops like alfalfa, cotton and bean decrease when populations of *Geocoris* spp. and *Nabis* spp. increase (Benedict and Cothran, 1980; Stoltz and McNeal, 1982; Yano, 1996). Sureshkumar and Ananthakrishnan (1985) noted that *Geocoris ochropterus* readily feeds on surface-dwelling thrips on groundnuts. The adults feed most frequently on second, rather than first instar larvae. Perhaps this indicates a general pattern whereby intermediate-sized predators tend to prefer thrips stages with larger body size, but this hypothesis needs scrutiny.

Another group of intermediate-sized heteropteran predators, the Miridae, contain several species that definitely feed on surface-dwelling thrips (Table 8.1). In India, a species of *Psallus* was observed to feed on *Megalurothrips distalis*. The young predators attack mainly larvae by piercing the top of the abdomen, but fourth instars to adults feed both on larval and adult thrips and these attacks are directed to various parts of the victim's body (Rajesekhara and

Chatterji, 1970). In the West Indies and Mexico, a species of *Termatophylidea* attack larvae of *Selenothrips rubrocinctus* (Van Doesburg, 1964; Callan, 1975). In Bulgaria, *Macrolophus* spp. were abundant on tobacco and kept thrips populations at low densities (Dimitrov, 1975, 1977). Some other mirids feed on thrips larvae, but cannot complete their development on a diet of thrips alone. For example, a species of *Deraeocoris* was observed to feed on thrips, but required aphids to reach maturity (Zavodchikova, 1974). During the last decade other mirids from southern Europe were shown to be good predators of surface-dwelling thrips, such as *Dicyphus tamaninii* in Spain (Salamero *et al.*, 1987; Gabarra *et al.*, 1988, 1995; Alomar *et al.*, 1991, 1994; Riudavets *et al.*, 1993; Wang, 1995; Albajes *et al.*, 1996) and *Macrolophus caliginosus* (Fauvel *et al.*, 1987; Riudavets *et al.*, 1995; Castané *et al.*, 1996), and in Asia *Campylomma chinensis* and *C. livida* were also recently recognized as predators of *Thrips palmi* (Chang *et al.*, 1993; Hirose *et al.*, 1993; Wang, 1995). In another recent study two mirid bugs, *Dicyphus rhododendri* and *Rhinocapsus vanduzeei*, were identified as important predators of *Heterothrips azaleae* on Florida azaleas (Braman and Beshear, 1994).

The most compelling evidence for predation on thrips comes from studies on relatively small-sized heteropteran predators, the minute pirate bugs in the family Anthocoridae (Lewis, 1973; Ananthakrishnan and Sureshkumar, 1985; Riudavets, 1995; and other references in Table 8.1). Observations of anthocorids feeding on gall-forming thrips are very rare; Lewis (1973) refers to one case of *Montandoniola moraguesi* feeding on eggs of the gall thrips *Gynaikothrips ficorum* (Tawfik and Nagui, 1965). The majority of studies relate to predation on surface-dwelling thrips. Some of the best known species are *Orius tristicolor* and *Orius insidiosus* in North America (Kelton, 1963), *Orius majusculus*, *Orius minutus* and *Orius niger* in the palaearctic region (Péricart, 1972), *Orius laevigatus* near the Mediterranean coasts and *Orius albidipennis* in Mediterranean countries (Péricart, 1972), *Orius sauteri* in Japan and China (Wang, 1995; Yano, 1996) and *Orius maxidentex*, *Orius tantillus* and *Orius indicus* in India (Ananthakrishnan and Sureshkumar, 1985). Usually, both the nymphs and the adults of these predators feed on first and second instar larvae as well as adult thrips (Bennett, 1965; Isenhour and Yeargan, 1981b; Nagai, 1991; Lichtenauer and Sell, 1993). They pierce thrips either in the head, thorax or abdomen, often holding down the struggling prey by their forelegs. According to Lichtenauer and Sell (1993) *Orius minutus* also kills eggs of *Frankliniella occidentalis* despite the fact that these are embedded in leaves. However, Nagai (1991) found no evidence for predation on eggs of *Thrips palmi* by *Orius sauteri*.

Thripophagous thrips

There are several species of thrips with a predominantly predatory lifestyle (Table 8.1d). Here, emphasis is placed on those predatory thrips that include other species of thrips in their diet. Being of the same size and shape as phytophagous thrips, thripophagous thrips occupy a unique position. For one thing they are expected to reach similar growth capacities (or even higher due to high quality food) and for another they should be able to reach the same sites on a plant. Indeed, their intrinsic growth rates are among the highest for the Thysanoptera (about 0.2 day^{-1} at 25°C; Gilstrap and Oatman, 1976; Coville and Allen, 1976; Selhorst *et al.*, 1991) and they are often found in galls and the typical sites occupied by interstitial dwellers. Lewis (1973, pp. 52–53) provides a description of the thrips fauna inhabiting the galls of *Liothrips kuwanai* on *Piper*, *Gynaikothrips uzeli* on *Ficus* and *Mesothrips claripennis* on *Bladhia*. Apart from several inquilinous thrips, a predatory thrips (*Xylaplothrips inquilinus*) also inhabits the galls. Similarly, the predator *Androthrips flavipes* is found in thrips galls of *Arrhenothrips ramakrishnae* on *Mimusops elengi* (Ananthakrishnan and Varadarasan, 1977). By inhabiting galls, they gain both protection and prey. *Androthrips flavipes* feeds on eggs and larvae of gall thrips (Ananthakrishnan, 1984; Sureshkumar and Ananthakrishnan, 1987). However, when other similar or slightly larger predators manage to enter mature galls, not only gall thrips but also the predatory thrips may be killed. For example, the anthocorid bug *Montandoniola moraguesi* not only feeds on eggs, larvae and adults of gall thrips (Muraleedharan and Ananthakrishnan, 1971), but also on *Androthrips flavipes* in galls of *Gynaikothrips flaviantennatus* on *Casearia tomentosa* (Ananthakrishnan, 1984).

A similar situation applies to thrips as predators of interstitial-dwelling thrips. For example, predatory thrips occur abundantly in the leaf sheaths of graminaceous plants, such as *Aeolothrips intermedius* in winter wheat (Patrzich and Klumpp, 1991). However, *Aeolothrips intermedius* also feeds on surface-dwelling thrips, such as *Thrips tabaci*, *Heliothrips haemorrhoidalis*, *Odontothrips confusus* (Derbeneva, 1967; Bournier *et al.*, 1978, 1979; Lacasa *et al.*, 1982, 1989; Lacasa, 1988) and *Sericothrips variabilis* (Robinson *et al.*, 1972). Adults of predatory thrips, such as *Aduncothrips asiaticus* and *Franklinothrips megalops*, appear to feed primarily on eggs of surface-dwelling thrips on various host plants, whereas the larvae of these predators feed on all mobile stages (Sureshkumar and Ananthakrishnan, 1987).

Phytoseiid mites

Many species of plant-inhabiting, phytoseiid mites feed on thrips (Table 8.1e), yet they are usually smaller than their prey when comparing the adults so can reach sites where thrips may be protected against larger predators. This leads to greater coincidence between predator and prey. To our knowledge there are no published reports on the occurrence of phytoseiid mites in thrips galls, which is surprising as they have been recorded in old and open galls of tiny eriophyid mites. Association with interstitial dwellers is occasionally mentioned. For example, Ramakers (1978) noted that both *Amblyseius cucumeris* and larval *Thrips tabaci* reside in the narrow crevices between calyx and young fruit and in tufts of hair near veins. In addition, Van Houten and Van Stratum (1995) showed that *Amblyseius degenerans* frequents sweet pepper flowers much more than *Amblyseius cucumeris* does and may therefore have a higher coincidence with larval *Frankliniella occidentalis*. All in all, there is still little known of the ability of phytoseiid mites to invade the hiding places of thrips. A disadvantage of the small size of phytoseiid mites is that they have difficulty seizing the relatively larger stages of the thrips. Indeed, adult thrips appear hard for phytoseiids to subdue because they are larger and can jump or fly away (Kajita, 1986; Gillespie, 1989). Thrips larvae are easier prey. They are often successfully attacked when the predatory mite approaches the larvae from the side (MacGill, 1939; Bakker and Sabelis, 1986, 1989; Kajita, 1986), especially when young because then they are smaller than their predator, and less able to counter-attack by jerking or wagging their abdomen. Indeed, phytoseiid mites, such as *Amblyseius cucumeris* and *Amblyseius barkeri* are much more successful at capturing first than second instar larvae (Bakker and Sabelis, 1986, 1989; Kajita, 1986; Van der Hoeven and Van Rijn, 1990). This differential attack success is partially explained by the defensive behaviour of the potential victim; anaesthesia of the thrips larvae improved the capture-success ratio, but it still did not lead to very high capture success and, more importantly, it did not eliminate the differences in capture success between the two larval instars of thrips. Bakker and Sabelis (1989) suggest that first and second instar thrips larvae may differ either in cuticle toughness, or food quality. Alternatively, assuming the predators cannot recognize clearly the anaesthetized state of their prey, the two larval thrips instars may differ in the risks they signal to the phytoseiid mites due to differences in their defensive capacity (when not anaesthetized).

Given that adult phytoseiid mites have difficulty in overcoming the defences of even the smallest larval thrips, juvenile phytoseiids experience these difficulties even more and may be unable to attack thrips larvae (Şengonca and Bendiek, 1988) or less likely to attack them (Shipp and Whitfield, 1991). If so, the juvenile phytoseiids have three options (apart from dispersal): (i) feed on thrips larvae killed by adult phytoseiids, (ii) cannibalize conspecific individuals, or (iii) feed on other prey or alternative foods. Each of these possibilities occurs. Cloutier and Johnson (1993a) found that juveniles stand a better chance of reaching adulthood when close to an adult phytoseiid, and actually observed feeding by the juveniles on thrips larvae partially consumed by the adult phytoseiids. In addition, cannibalism among juvenile phytoseiids is frequently observed when they are starved; the larger stages subdue the smaller ones and the strong subdue weak or lethargic individuals. Finally, alternative prey, such as tetranychid, tarsonemid and eriophyid mites, are consumed and several pollen species can provide food of sufficient quality to complete development and even for oviposition (e.g. Swirski *et al.*, 1967a,b, 1970; Van Rijn and Van Houten, 1991; Duso and Camporese, 1991; Van Houten *et al.*, 1995a) (see Chapter 4).

Attack success on stages other than larvae and adults has been little studied. If pupation occurs in the soil litter beneath the plant and the eggs are deeply inserted into the leaf, then these stages are inaccessible to phytoseiid mites. However, pupation may also take place on the plant and the eggs may vary in the degree to which they protrude from the leaf, depending on the thrips species, the host plant or a combination of these. The extent to which these stages can be fed upon is unknown, but tacitly assumed to be non-existent. This assumption, however, may need scrutiny.

The above features, as illustrated for *Amblyseius cucumeris* and *Amblyseius barkeri* attacking *Thrips tabaci* or *Frankliniella occidentalis*, generally apply to phytoseiid mites in a qualitative sense, but the quantitative details depend on the species or strain of predator and thrips. For example, *Amblyseius cucumeris* has a stronger ability to attack first and second instar larvae of *Thrips tabaci* than has *Amblyseius barkeri* (Bakker and Sabelis, 1986, 1989) and it has a higher rate of oviposition and predation on *F. occidentalis* larvae than a non-diapause strain of *Amblyseius barkeri* (Van Houten *et al.*, 1995a,b). The latter authors found even higher rates of predation and oviposition for *Amblyseius limonicus* compared with *Amblyseius cucumeris* (non-diapause strain), more or less equal rates for *Amblyseius hibisci* and *Amblyseius degenerans*, and much lower rates for *Amblyseius scutalis* and *Amblyseius tularensis*. Although the two latter species perform badly against *F. occidentalis* larvae, they can feed and reproduce readily when offered larvae of the Citrus thrips, *Scirtothrips citri* Moulton (Tanigoshi *et al.*, 1983, 1984, 1985; Bounfour and McMurtry, 1987; Tanigoshi, 1991; Jones and Morse, 1995). Also, predatory mites that proved effective against larvae of several thrips species appear incapable of subduing larvae of other thrips. For example, *Amblyseius barkeri*, *Amblyseius hibisci*, *Amblyseius degenerans* and *Typhlodromus rickeri* do not manage to kill first instar larvae of *Heliothrips haemorrhoidalis* on avocado (McMurtry and Badii, 1991). Apparently, the defence of these larvae is very effective; upon contact, the thrips larvae bend the tip of the abdomen upward toward the predator and often discharge a faecal droplet on the mite, which not only thwarts the attack, but also sticks the mite to the substrate (McMurtry and Badii, 1991).

The overall picture of the interspecific differences is confusing, as there seem to be no obvious morphological adaptations enabling predatory mites to attack thrips larvae and little is known of the differences in chemical composition of the thrips' anal exudates or in defensive behaviour of the thrips larvae. Hence, interspecific differences may be explained by analysing behaviour, tolerance to chemicals from thrips exudate and food quality of thrips, and (given the absence of morphological adaptations) one may expect that selection probably acts on a short time-scale, thereby fine tuning the predator–prey relationship. The results on attack success referred to above stem from predator–prey systems with either a long or virtually no evolutionary history. Just how much genetic variation is present in the populations and how fast selection can lead to improved attack are fundamental ecological questions that may also have great practical importance. A hint on the time scale of selection can be obtained from oviposition experiments showing improved rates of oviposition of thrips-fed predatory mites (*Amblyseius cucumeris*) within three successive generations (Castagnoli and Simoni, 1990). In many cases the predatory mites are reared on entirely different prey or even pollen. Apart from the effects of selection under rearing conditions the food may also create dietary deficiencies that alter attack behaviour toward a novel prey type. For example, *Amblyseius andersoni* reared on two-spotted spider mites does not respond to odours emanating from thrips-infested lima bean plants, irrespective of whether it is satiated or (moderately) starved. However, when reared on a diet of broad bean pollen, they do respond and this response wanes when the carotenoid-deficiency of the pollen is complemented by adding vitamin A or β-carotene to the diet (Dicke *et al.*, 1986; Dicke and Groeneveld, 1986). Similarly, *Amblyseius cucumeris* captures larvae of *Frankliniella occidentalis* much more successfully when reared on broad bean pollen, and this capture success decreases gradually after successive feeds on thrips larvae (Van der Hoeven and Van Rijn, 1990). The rationale of these diet-dependent responses may be that rearing diets create nutritional deficiencies in predatory mites (e.g. β-carotene or vitamin A are indispensable for diapause induction and termination) leading to responses that may help to counterbalance them (feed on other phytophagous arthropods). Clearly caution is necessary in interpreting differential attack success of predatory mites.

Other thrips-eating mites

Several non-phytoseiid mites have been reported to feed on thrips larvae or eggs (Eickwort, 1983; Table 8.1e). Among the largest of these are trombidiid and erythraeid mites, as well as the so-called whirligig mites (Anystidae). *Anystis agilis* is a fast-running, generalist predator that is thought to prefer thrips larvae to spider mites (Mostafa *et al.*, 1975; Sorenson *et al.*, 1976). It develops slowly and has only a few generations per year. It is considered to have an impact on *Scirtothrips* spp. on citrus early in spring when other predators are still sparse.

Other non-phytoseiid predatory mites are similar in size to the phytoseiids, adults matching the size of young second-instar thrips larvae. The laelapid mite, *Hypoaspis aculeifer* (Gillespie and Quiring, 1990) and *Hypoaspis miles* (Glockemann, 1992), have been introduced into greenhouses to control sciarid flies and *Frankliniella occidentalis*. These predators are usually soil-inhabiting, so may kill thrips pupae, as shown for *Hypoaspis miles* by Brødsgaard *et al.* (1996). Predation on thrips eggs and larvae has not been studied. However, the number of predatory mites needed to achieve some impact appears to be exceedingly large (6000 per plant; Gillespie and Quiring, 1990). This suggests that these laelapid soil-inhabiting predators are inefficient at controlling thrips, but may well complement the action of foliage-dwelling predators. Other somewhat smaller, leaf-inhabiting prostigmatic mites, such as *Cheyletus* spp., feed on eggs of *Liothrips oleae* (Tominic, 1950).

Heterostigmatic mites in the families Pyemotidae and Acarophenacidae are the smallest enemies of thrips. With a length of about 0.2 mm their adults are even smaller than thrips eggs. In fact they are characterized more adequately as parasitoids (Eickwort, 1983). They have been found in thrips galls, such as the pyemotid mite, *Pyemotes* sp., in leaf roll galls of *Gynaikothrips ficorum* in Brazil (Bennett, 1965), and adult female mites of *Adactylidium* (Acarophenacidae) have been observed feeding on eggs of *Gynaikothrips* in Egypt (El-Badry and Tawfik, 1966). Unengorged female *Adactylidium* have been collected from adult thrips (Cross, 1965), suggesting a phoretic adaptation.

Predator–Prey Dynamics

Relative body size is probably a key to understanding the dynamics of arthropod predator–prey systems, because usually the maximum rate of predation increases with body size, whereas the intrinsic rate of population increase decreases with body size (Sabelis, 1992). Thus, there are two aspects of increased body size with opposite effects on the ability to control prey populations; on the one hand the functional response to prey density may be increased, but on the other the numerical response to prey density may be retarded relative to the prey's capacity to increase. Whether these opposite effects have a positive or negative influence on the ability to control the prey remains to be elucidated by a more detailed analysis of the specific predator–prey system under study. This is the primary aim of this section.

Intrinsic prey suppression capacity: model and predictions

To estimate the intrinsic potential of predatory arthropods to reduce thrips populations it is necessary to combine predation characteristics and growth capacities of predator and prey into a single measure. This can be derived from a simple one-predator-one-prey model proposed by Sabelis (1992) and Janssen and Sabelis (1992). The model assumes a small spatial scale with a coherent local population of prey (the *prey-pancake* assumption). The prey density is assumed to be sufficient to motivate the colonizing predators and their offspring to stay until virtually all prey are eaten. The predators do not disperse from the local population until after prey extinction (the *arrestment* assumption). The prey distribution within the local population and the foraging behaviour of the predator is such that the rate of predation is continuously at the plateau of the functional response curve (the *pancake-predation* assumption). This requires predators that forage optimally in that they frequent sites harbouring enough thrips to reach maximum predation rates, and spend negligible time travelling between sites. Per capita population growth rates of predator and of prey (in absence of predator) are assumed to be constant (the assumption on *stable age-distribution* and *absence of intraspecific competition*). Clearly, this model estimates the *intrinsic* capacity for prey population suppression, because predator and prey capacities are taken to their extremes. The assumptions lead to linear equations differing from the classic Lotka–Volterra models only in that the predation term now depends only on the number of predators:

$$\frac{dN}{dt} = \alpha N - \beta P$$

$$\frac{dP}{dt} = \gamma P$$

Here, time t = time since start of predator–prey interaction, N_t = number of prey at time t, P_t = number of predators at time t, α = rate of prey population growth, β = maximum predation rate and, γ = rate of predator population growth. Analytical solutions for the number of predators and prey since the start are readily obtained:

$$N_t = N_0 e^{\alpha t} - P_0 \frac{\beta}{\gamma - \alpha} (e^{\gamma t} - e^{\alpha t})$$

$$P_t = P_0 e^{\gamma t}$$

This model exhibits continued exponential growth of the predators until a time τ when all prey are eaten (and the predators move on to another food source):

$$\tau = \frac{1}{\gamma - \alpha} \ln \left(1 + \frac{\gamma - \alpha}{\beta} \frac{N_0}{P_0} \right)$$

Three types of dynamic behaviour of the prey population are possible: (1) continuous decay until extinction, (2) initial increase, followed by decrease until extinction, and (3) continuous increase (but at a pace slower than the intrinsic rate of prey population growth). Clearly, predatory arthropods giving rise to the type-3 behaviour cannot prevent local prey population outbreaks (which in reality would end upon reaching carrying capacity of the local food source). Type-1 or type-2 behaviour requires that the (expression for the) time to extinction has a finite value. This leads to the following condition:

$$\frac{P_0}{N_0} > \frac{\alpha - \gamma}{\beta}$$

Type-1 behaviour (immediate decrease) occurs simply when:

$$\alpha N_0 < \beta P_0 \quad \text{or} \quad \frac{P_0}{N_0} > \frac{\alpha}{\beta}$$

Given the initial numbers of predators and prey, estimates of their intrinsic rate of population increase and their maximum predation rates, it is possible to assess the predators' capacity to suppress the prey population to vanishingly small numbers. The model serves as a template to disguise predators that do no more than 'surf on prey waves', and thereby identify those that may drastically alter local prey population fluctuations. In this sense it is an instrument to further insight into the role of predators in complex food webs and may be of value to find candidates for short-term biological control in simple systems (such as greenhouse crops, their pests and natural enemies). Data on life history from the literature (Table 8.2), on intrinsic growth rates, r_m (and some provisional estimates thereof, based on Table 8.2) and rates of predation on thrips (Table 8.3) were inserted into the preceding model. The outcome shows, indeed, that the intrinsic rate of predator population growth decreases with body size across taxa (Fig. 8.1c), mainly caused by an increased juvenile period (Fig. 8.1a) and despite an increasing ovipositional rate (Fig. 8.1b). Taking group means, body length is 0.4 mm for the phytoseiids, 1.5 mm for the predatory thrips, 2.3 mm for the anthocorids and 4 mm for the mirids, whereas the intrinsic rate of increase is 0.19, 0.14, 0.13 and 0.12 day^{-1}, respectively, and

Table 8.2a. Life history and population growth characteristics of heteropteran bugs known as predators of thrips.

Genus	Species	Adult size (mm)		Food source T: thrips, S: spider mites, A: aphids, L: lepidopteran eggs, P: pollen	Temperature (°C)	Development						Female performance							References
						Eggs	Nymphs	Juv. period	Pre-ovipos.	Egg–egg	Juv. survival	Ovip. rate	Fecundity	Ovip. period	Longevity	Sex ratio	R_0	r_m	
Campylomma	*chinensis*	(3)		T: *Thrips palmi*	25		15.2				0.75				11.5				Wang, 1995
				A: *Aphis gossypii*	25		14.3				0.40				10.6				Wang, 1995
				S: *Tetranychus kanzawai*	25		16.5				0.25				3.4				Wang, 1995
Campylomma	*livida*	(3)		T: *Thrips palmi*	28		10.8								6.5				Chang *et al.*, 1993
Dicyphus	*tamaninii*	4.5	#	T: *F. occidentalis*	25	12.1	18.9	31.0							21.2				Albajes *et al.*, 1996
				Trialeurodes vaporariorum	22		21.8					1.8	88						Albajes *et al.*, 1996
Macrolophus	*caliginosus*	3.3	#	L: *Ephestia kuehniella*	25 [/22]	10.4	15.9 [/22]	(26.3)	3	(29.3)		7.0	122	(17.4)	40.2			(0.117)a	Fauvel, *et al.*, 1987
				Trialeurodes vaporariorum	25 [/22]	10.4	18.9 [/26]	29.3	3	(32.3)		4.0	86	(21.5)					Fauvel, *et al.*, 1987
				A: *Myzus persicae*	25 [/22]		[/27]					2.4	60	(25.0)					Fauvel, *et al.*, 1987
				S: *Tetranychus turkestani*	25 [/22]		[/28]					0.9	20	(22.2)					Fauvel, *et al.*, 1987
Nabis	*alternatus*	(8)		T: *Frankliniella, Thrips, Aeolot.*	23.9	9.0	24–27	(35.5)		(42.6)									Taylor, 1949
			#	A: *Macrosiphum pisi*	23.9	9.0	16–30	(32.0)		(38.4)		(6.4)	127	20.0				(0.093)b	Taylor, 1949
Nabis	*americoferus*	(8)	#	L: *Estigmene acraea*	28			22.1		(26.5)		(9.1)	281	31.0	37.3			(0.131)b	Perkins and Watson, 1972
				L: *Heliothis virescens*	27	7.3	16.8	24.0											Braman *et al.*, 1984
				A: *Acyrthosiphon pisum*	23	10.3	28.6	38.9	7–10	(47.5)		8.2	157	18.6				(0.082)b	Guppy, 1986
Geocoris	*atricolor*	(3)	#	L: *Phthorimaea operculella*	26.7	7.6	25.5	33.1	15.6	48.7		2.7	32	(11.7)				(0.056)b	Dunbar and Bacon, 1972a,b
Geocoris	*pallens*	(3)		*Tetranychus articae*	26.6		26.0												González and Wilson, 1982
			#	L: *Phthorimaea operculella*	26.7	7.1	24.1	31.2	7.6	38.8		4.2	66	(15.6)				(0.082)b	Dunbar and Bacon, 1972b
Geocoris	*punctipes*	(3)		L: *Pseudoplusia includens*	25		30.0												Crocker *et al.*, 1975
			#	L: *Phthorimaea operculella*	26.7	8.7	25.3	34.0	5.5	39.5		7.2	126	(17.5)				(0.093)b	Dunbar and Bacon, 1972a,b

Geocoris	*ochropterus*	(3)		T: *Neohydathrips variabilis*														Sureshkumar and Ananthakrishnan, 1985
			#	Ant: *Oecophylia smaragdina* (pupae)	27	8.3	31.2	39.5	7.6	47.1	0.93	4.6	277	(60.2)	73.8	0.5		(0.072)b	Mukhopadhyay and Sinnigrahi, 1993
Anthocoris	*nemorum*	4.0		A: *Myzus persicae*	25				14.5				96						Lauenstein, 1977
				A: *Drepanosiphum platanoides*	24		28.3												Russel, 1970
			#	A: *Aulocorthum circumflexum*	23	(5.5)	23.7	(29.2)		(35.0)	0.50		73						Anderson, 1962
				S: *Panonychus ulmi*	23		29.7				0.62								Anderson, 1962
Carayonocoris	*indicus*	3.1	#	T: *Scirtothrips dorsalis*		6.0	26.1	32.1	4.8	(36.9)		8.0	38	(4.8)				(0.083)b	Sureshkumar and Ananthakrishnan, 1984
				T: *Haplothrips ganglebaueri*		5.8	25.8	31.6	4–7	(36.6)			32		42.0				Muraleedharan and Ananthakrishnan, 1978
Montandoniola	*moraguesi*	3.1	#	T: *Gynaikothrips ficorum*	28.8	4.2	12.9	17.1	4.5	21.6		3.1	88	28.5	37.6			(0.132)a	Tawfik and Nagui, 1965
				T: *Gynaikothrips flaviantennatus*		4.8	22.7	27.5					54		38.8				Muraleedharan and Ananthakrishnan, 1978
Scoloposcelis	*parallelus*			T: *Ecacanthothrips tibialis* (on bark)		4.5	25.5	30.0	3–6	(34.5)			59		51.2				Muraleedharan and Ananthakrishnan, 1978a
Orius	*albidipennis*	1.9		T: *Thrips tabaci*	25							6.8	217	(31.9)	45.1	0.56			Chyzik *et al.*, 1995a
				S: *Tetranychus urticae*	25							3.9	111	(28.4)	35.1	0.58			Chyzik *et al.*, 1995a
				L: *Ephestia kuehniella*	25							5.8	184	(31.7)	63.0	0.53			Chyzik *et al.*, 1995a
				L: *Ephestia kuehniella*	27	(4.6)	14.5	(19.1)	2.4	(21.5)			53		21.1				Camero *et al.*, 1993
			#	L: *Ephestia kuehniella*	25	(4.6)	11.8	(16.4)		(19.7)	0.97	8.0	211	(26.4)		0.5	98	0.156	Van den Meiracker, unpublished
Orius	*indicus*	(2)	#	T: *Megalurothrips usitatus*	25	5.5	23.5	29.0	5	34.0		(3.6)	50–88	12–23	41.3	0.62	34	(0.090)b	Rajasekhara and Chatterji, 1970

Table 8.2a. Continued.

Genus	Species	Adult size (mm)		Food source T: thrips, S: spider mites, A: aphids, L: lepidopteran eggs, P: pollen	Temperature (°C)	Development						Female performance							References
						Eggs	Nymphs	Juv. period	Pre-ovipos.	Egg–egg	Juv. survival	Ovip.rate	Fecundity	Ovip. period	Longevity	Sex ratio	R_0	r_m	
Orius	*insidiosus*	2.0		T: *F. occidentalis* (L1)	26		13.3											0.130	Cloutier *et al.*, 1995
				T: *F. occidentalis*	26				2			(4.1)	66	(16.0)	17.1				Tommasini and Nicoli, 1993
				L: *Ephestia kuehniella*	26	(5.1)	15.0	(20.1)	2	(22.1)	0.63		144		42.3			(0.143)a	Tommasini and Nicoli, 1993
				T: *F. occidentalis* (20 adults)	24							5.7							Richards and Schmidt, 1996
			#	L: *Ephestia kuehniella*	24	(5.1)	10.0	(15.1)		(18.1)	0.81	5.7	119	(20.9)	28.4			(0.159)b	Richards and Schmidt, 1996
				P: corn (+bean pods)	24	(5.1)	16.1	(21.2)		(25.4)	0.40	1.5	9	(6.1)	13.5			(0.064)b	Richards and Schmidt, 1996
				S: *Panonychus ulmi*	23	5.8	8.1	13.9											McCaffrey and Horsburgh, 1986b
				L: *Ephestia kuehniella*	25	4.6	13.6	18.2		(21.8)	0.77	6.0	213	(35.5)		0.5	86	0.131	Van den Meiracker, unpublished
Orius	*laevigatus*	1.9		T: *F. occidentalis*	26				2			(3.6)	56	(15.5)					Tommasini and Nicoli, 1993
				L: *Ephestia kuehniella*	26	(4.6)	16.0	(20.6)	2	(22.6)	0.22		119					(0.138)a	Tommasini and Nicoli, 1993
			#	L: *Ephestia kuehniella*	25	4.6	13.2	17.8	4	21.8		(7.3)	157	(21.6)	27.0			(0.162)a	Alauzet *et al.*, 1994
				A: *Aphis gossypii*	25.5								113	18–33					Tavella *et al.*, 1991
				P: pine	25	4.6	12.0	16.9		(20.3)		1.8	49	(22.5)	28.1			(0.113)b	Hulshof, 1994
Orius	*limbatus*	(2)	#	L: *Cadra caudella*	25	(4.6)	11.8	(16.4)	3.7	(20.1)		4.9	82	16.4				(0.144)b	Camero *et al.*, 1993
Orius	*majusculus*	2.8		T: *F. occidentalis*	26				2			(5.3)	87	(16.5)	19.7				Tommasini and Nicoli, 1993

				L: *Ephestia kuehniella*	26	(4.0)	15.4	(19.4)	2	(21.4)	0.40		174	(37.6)	47.0	0.55	(0.148)a	Tommasini and Nicoli, 1993, 1995
				T: *F. occidentalis*	25.5		15.2				0.79							Husseini *et al.*, 1993
				L: *Ephestia kuehniella*	26		11.1				0.74		328					Fischer *et al.*, 1992
			#	L: *Ephestia kuehniella*	25	4.0	12.7	16.7	4.4	21.1	0.76	8.7	237	(27.2)			(0.154)b	Alauzet *et al.*, 1990, 1992
				A: *Rhopalosiphum padi*	25	4.0	16.0	20.0										Alauzet *et al.*, 1990
Orius	*maxidentex*	(2)	#	T: *Scirtothrips dorsalis*	25?	4.5	20.2	24.7	4.6	29.3		9.4	43	(4.5)			(0.107)b	Sureshkumar and Ananthakrishnan, 1984
Orius	*minutus*	2.3	#	L: *Sitotroga cerealella*	25	4.1	15.9	20.0		(24.0)	0.92	2.8	31	10.9	20.0		(0.104)b	Niemczyk, 1978
				S: *Tetranychus urticae*	25		14.8				1.00							Niemczyk, 1978
Orius	*niger*	2.0		T: *F. occidentalis*	26				2			(0.8)	17	(21.0)	18.4			Tommasini and Nicoli, 1993
				L: *Ephestia kuehniella*	26	(4.7)	17.8	(22.5)	2	(24.5)	0.06	(2.0)	54	(26.5)	50.0			Tommasini and Nicoli, 1993
			#	L: *Ephestia kuehniella*	25	4.7	15.9	20.6		(24.7)	0.73	6.0	166	(27.7)		0.5 68	0.118	Van den Meiracker, unpublished
Orius	*sauteri*	(2)	#	T: *Thrips palmi* (L2)	25			11.2	2.2	(13.4)	0.84	6.4	42	(6.6)	8.1		(0.201)b	Nakashima *et al.*, 1996
				T: *Thrips palmi*	25	3.9	11.4	15.3		(18.4)								Nagai, 1989
				T: *Thrips palmi*	25		14.9				0.65				36.2			Wang, 1995
				L: *Corcyra cephalonica*	25		17.6				0.26				16.0			Wang, 1995
				S: *Tetranychus kanzawai*	25		23.0				0.32				33.5			Wang, 1995
				L: *Ephestia kuehniella*	25			18.0					<100		30.0			Yanò, 1996
				A: *Myzus persicae*	25	4.3	11.6	15.9		(19.1)								Nakata, 1995
				A: *Aphis gossypii*	25		14.9						29					Funao and Yoshiyasu, 1995
				P: corn	25		17.2						15					Funao and Yoshiyasu, 1995

Table 8.2a. Continued.

Genus	Species	Adult size (mm)	Food source T: thrips, S: spider mites, A: aphids, L: lepidopteran eggs, P: pollen	Temperature (°C)	Development						Female performance					R_0	r_m	References
					Eggs	Nymphs	Juv. period	Pre-ovipos.	Egg–egg	Juv. survival	Ovip.rate	Fecundity	Ovip. period	Longevity	Sex ratio			
Orius	*tristicolor*	2.1 #	T: *F. occidentalis*	26.6	(4.0)	10.4	(14.4)		(17.3)	0.90	4.0	43	(10.9)	11.5			(0.149)b	Salas-Aguilar and Ehler, 1977
			P	26.6		13.0				0.63								Salas-Aguilar and Ehler, 1977
			T + P	26.6	(4.0)	9.3	(13.3)		(16.0)	0.90	3.9	60	(15.4)	15.4			(0.159)b	Salas-Aguilar and Ehler, 1977
			green bean	26.6		15.7				0.43								Salas-Aguilar and Ehler, 1977
			S: *Tetranychus pacificus*	25.5	3–5	14.4	(16.4)	2–3	(20.0)			130						Askari and Stern, 1972

Life history includes juvenile development (days), fraction of eggs successfully developing into adults (juvenile survival), mean oviposition rate (eggs day^{-1}), gross fecundity (eggs), oviposition period (days), mean adult female longevity (days) and proportion of daughters in the offspring when mature, and population growth characteristics concerning the net reproduction rate R_0 and the intrinsic rate of population increase r_m (1 day^{-1}). The table includes life history data obtained with thrips as prey or other relevant food sources (e.g. lepidopteran eggs, spider mites and pollen).

Note: Each food type is represented maximally by two data sets selected for (i) completeness of the life history information, (ii) food type (thrips, spider mites, pollen) and (iii) the highest oviposition rates. All data were measured at 25 to 27°C or the closest temperature available, and refer to the sexually reproducing species/strains. Bracketed values represent our estimates from original data, e.g. the estimates of R_0 and r_m (see Addendum 1 for the two procedures indicated by a and b) and those of the oviposition rates (calculated as the ratio between total fecundity and oviposition period). Data used to construct Figure 8.1 are indicated by a label (#) in the column providing information on predator size.

Table 8.2b. Life history and population growth characteristics of insects (other than bugs) known as predators on thrips.

Genus	Species	Adult size (mm)		Food source T: thrips, S: spider mites, A: aphids, L: lepidopteran eggs, P: pollen	Temperature (°C)	Development							Female performance							References
						Eggs	Larvae	(Pre)-pupae	Juv. period	Pre-ovipos.	Egg–egg	Juv. survival	Ovip.rate	Fecundity	Ovip. period	Longevity	Sex ratio	R_0	r_m	
Aeolothrips	*intermedius*	1.8	#	T: *Thrips tabaci*	26	6.0	8.0	5.0	19.0	1.7	20.7		(2.1)	26	14	19.0	0.58		(0.109)b	Bournier *et al.*, 1978
				T: *Thrips tabaci*	26	(6.0)	10.3		(16.3)											Bournier *et al.*, 1979
				T: *Heliothrips haemorrhoidalis*	26	(6.0)	11.6		(17.6)											Bournier *et al.*, 1979
				T: *Odontothrips confusus*	26	(6.0)	13.6		(19.6)											Bournier *et al.*, 1979
				S: *Tetranychus urticae*	26	(6.0)	20.6		(26.6)											Bournier *et al.*, 1979
				P: luceme	26	(6.0)	14.6	8.0	(28.6)											Bournier *et al.*, 1979
Erythrothrips	*asiaticus*	(1.4)		T: *Caliothrips indicus*		5.1	7.8	7.2	20.1	4.9	25.0			19						Sureshkumar and Ananthakrishnan, 1987
				T: *Frankliniella schultzei*		4.1	5.6	5.5	15.2	3.4	18.6			38						Sureshkumar and Ananthakrishnan, 1987
				T: *Retithrips syriacus*		5.5	6.8	6.9	19.2	4.2	23.4			26						Sureshkumar and Ananthakrishnan, 1987
			#	T: *Scirtothrips dorsalis*		3.5	4.6	5.1	13.2	3.1	16.3			49						Sureshkumar and Ananthakrishnan, 1987
				T: *Haplothrips ganglebaueri*		4.5	6.4	6.1	17.0	3.9	20.9			31						Sureshkumar and Ananthakrishnan, 1987
Franklinothrips	*megalops*	(1.6)		T: *Caliothrips indicus*		5.3	6.6	7.3	19.2	4.6	23.8			24						Sureshkumar and Ananthakrishnan, 1987
				T: *Frankliniella schultzei*		5.5	7.3	7.8	20.6	4.9	25.5			20						Sureshkumar and Ananthakrishnan, 1987
			#	T: *Retithrips syriacus*		3.8	4.8	5.2	13.8	3.6	17.4			39						Sureshkumar and Ananthakrishnan, 1987
				T: *Scirtothrips dorsalis*		4.9	5.8	6.7	13.8	4.5	18.3			27						Sureshkumar and Ananthanrisknan, 1987
				T: *Zaniothrips ricini*		4.2	5.1	5.8	15.1	3.9	19.0			32						Sureshkumar and Ananthakrishnan, 1987

Table 8.2b. Continued.

Genus	Species	Adult size (mm)		Food source T: thrips, S: spider mites, A: aphids, L: lepidopteran eggs, P: pollen	Temperature (°C)	Development: Eggs	Larvae	(Pre)-pupae	Juv. period	Pre-ovipos.	Egg–egg	Juv. survival	Female performance: Ovip.rate	Fecundity	Ovip. period	Longevity	Sex ratio	R_0	r_m	References
Scolothrips	*longicornis*	(0.9)	#	S: *Tetranychus urticae*	25		3.9	2.9						287		62.4			0.199	Gerlach and Şengonca, 1985; Selhorst *et al.*, 1991
Scolothrips	*sexmaculatus*	0.8	#	S: *Tetranychus urticae*	25	(7.1)	4.6	3.5	(15.2)	1.2	16.4		(5.8)	154	26.5		0.67		0.180	Coville and Allen, 1976
				S: *Tetranychus pacificus*	23.9	11.6	7.0	4.5	23.1	< 1	(24.0)	1	6.5	220	33.8	49.0	0.89		0.155	Gilstrap and Oatman, 1976
Leptothrips	*mali*	2.4	#	S: *Panonychus ulmi*	23.9	7.5	11.1	6.5	25.6	6.4	32.0			28	(40.7)	54.2	(0.66)		(0.064)b	Parrella *et al.*, 1982
				P: apple	23.9	(7.5)	23.0		(30.5)											Parrella *et al.*, 1982
Chrysopa	*carnea*	19		A: *Myzus persicae*	21	5.6	14.6	14.9	35.1	(10.5)	(45.6)		(13.6)	477	35				(0.092)b	Sundby, 1966
Hemerobius	*pacificus*	10		A: *Acyrthosiphon pisum*	18.3	6.7	10.2	16.7	33.3	10.1	(43.4)	0.68	10.8	715	(66.2)	71.8			(0.094)a	Neuenschwander, 1975, 1976
				A: *Acyrthosiphon pisum*	24	5.0	7.4	12.3	24.7	(7.5)	(32.2)	0.83	6.6						(0.126)	Neuenschwander, 1975, 1976
Adalia	*bipunctata*	4.5	#	A: *Phorodon humuli*	25–28		8.9		15.8	5–8	(22.0)	0.83	(17.2)	1011	(58.6)	73.3			(0.150)b	Kalushkov, 1994
				A: *Acyrthosiphon pisum*	20					7.8			20.4	1535	(70.0)	79.4				El-Hariri, 1966
Coccinella	*septempunctata*	6	#	A: *Myzus persicae*	21	5.0	15.0	8.5	28.5	(5.7)	(34.2)		(12.5)	814	65				(0.105)b	Sundby, 1966
Coleomegilla	*maculata*	6	#	A: *Myzus persicae*	25	3.0	11.4	3.4	17.9			0.87								Wright and Laing, 1978
				L: *Hyphantria cunea*	26.7	2.8	14.6	3.8	21.2											Warren and Tadic, 1967
				P: wild flower	25		17.9	4.0												Giroux *et al.*, 1994
Hippodamia	*convergens*	5.5		L: *Sitotroga cereallela*	25	3.0	15.2	6.3	24.5											Butler and Dickenson, 1972
			#	A: *Acyrthosiphon pisum*	24	3.4	11.6	4.8	19.8											Obrycki and Tauber, 1982

The table includes life history data obtained with thrips as prey or other relevant food sources (spider mites, aphids, lepidopteran eggs and pollen). See note in legends of Table 8.2a.

Table 8.2c. Life history and population growth characteristics of mites known as predators of thrips.

Genus	Species	Adult size (mm)		Food source T: thrips, S: spider mites, M: acarid mites, P: pollen	Temperature (°C)	Development						Female performance							References
						Eggs+larvae	Nymphs	Juv. period	Pre-ovip	Egg–egg	Juv. survival	Ovip.rate	Fecundity	Ovip. period	Longevity	Sex ratio	R_0	r_m	
Hypoaspis	*aculeifer*	0.80		M: *Tyrophagus putrescentiae*	24.5	7.0	6.1	13.1		15.0	0.55		70.6		55.0	0.52	20.1	0.121	Lobbes and Schotten, 1980
				S: *Tetranychus urticae*	24.5	7.0	14.8	21.8		22.0	0.68		15.6		71.0	0.52	14.9	0.046	Lobbes and Schotten, 1980
			#	M: *Rhizoglyphus echinopus*	25	4.0	7.8	11.6	2.0	13.6		4.6	115.0	26.2	38.6	(0.52)		(0.185)a	Ragusa and Zedan, 1988
Hypoaspis	*miles*	0.80		Sciarid: *Lycoriella solani*	22	4.1	10.5	14.5	5.1	19.6	0.97	(0.9)	(47.9)	53.2	82.0	0.66	27.3	0.078	Enkegaard, *et al.*, 1996
			#	M: *Rhizoglyphus echinopus*	25	(5.2)	9.3	14.5	4.2	18.7		0.8	38.6	49.6	95.4	(0.55)		(0.094)b	Shereef *et al.*, 1980
Amblyseius	*addoensis*			P: iceplant	22							(1.0)							Grout and Richards, 1992a
Amblyseius	*andersoni*	0.38		T: *Frankliniella occidentalis*	20							>1							Rodriques-Reina *et al.*, 1992
			#	S: *Tetranychus pacificus*	23					(9.6)		1.4	57.0	39.5		0.75		(0.184)a	Amano and Chant, 1978
				S: *Eotetranychus carpini*	26.5	2.3	3.0	5.3	1.9	6.8		1.5							Duso and Camporese, 1991
				P: iceplant	26.5	2.2	2.4	4.7	2.0	6.8		1.4							Duso and Camporese, 1991
Amblyseius	*barkeri* (=*mckenziei*)	0.31	#	T: *Thrips tabaci*	25	3.0	3.2	6.2	2.1	8.3		2.3	47.1	20.3	23.4	0.63	27.8	0.220	Bonde, 1989
				T: *Thrips tabaci*	25			5.8			0.84		41.0		24.0				Piatkovski, 1987
				T: *Frankliniella occidentalis*	25							1.5							Van Houten *et al.*, 1995a
				T: *Frankliniella occidentalis*	20							>1.2							Rodrigues-Reina *et al.*, 1992
				S: *Tetranychus cinnabarinus*	25	3.3	3.7	6.9	2.7	9.6	0.85	2.0	43.4	21.8	25.4	0.63		(0.188)a	Lababidi, 1988
				S: *Tetranychus urticae*	25	3.7	4.7	8.5	2.9	11.3		1.9	54.8	30.6	46.1	0.66		(0.183)b	Momen, 1995
				P: broad bean	25					8.3		2.8				0.75		0.240	Van Rijn and Van Houten, 1991
Amblyseius	*californicus* (=*chilenensis*)	0.36		T: *Frankliniella occidentalis*	20							>0.5							Rodriques-Reina *et al.*, 1992
				T: *Retithrips syriacus*	26						0.00	0.0							Swirski *et al.*, 1970
				S: *Tetranychus cinnabarinus*	26						1.00	2.4							Swirski *et al.*, 1970
				P: iceplant	26						1.00	1.1							Swirski *et al., 1970*
Amblyseius	*cucumeris*	0.40		T: *F. occidentalis* (L1 dead)	25	4.3	4.5	8.7				1.5							Gillespie and Ramey, 1988
				T: *F. occidentalis* (L1)	25							2.2							Van Houten *et al.*, 1995a
				T: *F. occidentalis* (L1)	26		6.3											0.160	Cloutier *et al.*, 1995

Table 8.2c. Continued.

Genus	Species	Adult size (mm)		Food source T: thrips, S: spider mites, M: acarid mites, P: pollen	Temperature (°C)	Development: Eggs+larvae	Nymphs	Juv. period	Pre-ovip	Egg–egg	Juv. survival	Female performance: Ovip.rate	Fecundity	Ovip. period	Longevity	Sex ratio	R_0	r_m	References
				T: *F. occidentalis* (L alive/dead)	25		6.3/2.8												Cloutier and Johnson, 1992
				T: *F. occidentalis* (L1+L2, F1–F3)	25			9.2			0.62	1.9							Castagnoli *et al.*, 1990
			#	T: *Thrips tabaci* (L1+L2, F1–F3)	25	3.5	4.7	8.2	2.2	10.3	0.88	2.0	29.2	18.4		0.67	19.5	0.178	Castagnoli and Simoni, 1990
				S: *Tetranychus urticae*	25	3.6	4.4	7.6	2.1	9.7	0.83	(1.6)	39.7	25.6		0.60	21.4	0.174	Castagnoli and Simoni, 1990
				P: oak	25	3.4	3.7	6.8	2.6	9.4	0.96	(1.5)	52.4	34.5		0.61	31.0	0.193	Castagnoli and Simoni, 1990
				S: *Tetranychus cinnabarinus*	25	2.9	3.6	6.5	2.7	9.2	0.88	1.8	37.8	21.2	31.4	0.64		(0.167)a	Lababidi, 1988
				P: broad bean	25					8.3		2.6				0.64		0.220	Van Rijn and Van Houten, 1991
Amblyseius	*degenerans*	0.50		T: *Frankliniella occidentalis*	25							1.4							Van Houten *et al.*, 1995a
			#	S: *Tetranychus pacificus*	25	3.2	2.4	5.7	2.3	8.0		2.2	67.8	30.8	53.3	0.74	50.9	0.248	Takafuji and Chant, 1976
				S: *Tetranychus urticae*	25	2.9	3.8	6.7	2.6	9.3		1.0			45.0	0.55	15.8	0.141	Van Rijn and Tanigoshi, 1997b
				P: broad bean	25	2.9	2.3	5.2	2.1	7.3		2.1			46.0	0.55	29.2	0.207	Van Rijn and Tanigoshi, 1997b
Amblyseius	*hibisci*	0.32		T: *Frankliniella occidentalis*	25							1.7							Van Houten *et al.*, 1995a
				T: *Retithrips syriacus*	26						0.64	0.6							Swirski *et al.*, 1970
				S: *Tetranychus cinnabarinus*	26						0.96	1.1							Swirski *et al.*, 1970
				P: iceplant	26						0.96	1.1							Swirski *et al.*, 1970
			#	S: *Tetranychus pacificus*	25	3.3	2.7	5.9				1.8							Zhimo and McMurtry, 1990
				P: iceplant	25	2.7	2.3	5.0				1.8							Zhimo and McMurtry, 1990
Amblyseius	*largoensis*	0.38		T: *Retithrips syriacus*	26			6.1				2.4							Kamburov, 1971
				S: *Eutetranychus orientalis*	26			4.8				2.0							Kamburov, 1971
				P: iceplant	26			5.1				2.4							Kamburov, 1971
			#	S: *Panonychus citri*	25	2.1	4.4	6.0	2.5	8.5		(1.8)	20.6	11.8	21.0	(0.67)		(0.210)a	Tanaka and Kashio, 1977
				P: *Quercus virgiliana*	25	3.1	3.1	4.3			0.98	(1.9)	34.3		18.3		19.8	0.265	Yue and Tsai, 1996
Amblyseius	*limonicus*	0.37		T: *Retithrips syriacus*	26							1.1							Swirski and Dorzia, 1968
				T: *Frankliniella occidentalis*	25							3.2							Van Houten *et al.*, 1995a

			#	S: *Panonychus citri*	22.2			6.0	2–3	8–9		2.3							McMurtry and Scriven, 1965
				P: iceplant	22.2			6.0	2–3	8–9		2.0							McMurtry and Scriven, 1965
Amblyseius	*longispinosus*	(0.40)	#	S: *Tetranychus cinnabarinus*	25	2.7	1.7	4.4		(5.8)	0.92	2.3	46.8	20.8	33.1	0.70		(0.267)a	Lababidi, 1988
				S: *Tetranychus kanzawai*	24	2.9	2.8	5.7		(6.8)		1.4	22.2	(16.0)		0.60		0.194	Shih and Sheih, 1979
				S: *Tetranychus urticae*	25	3.1	2.6	5.7	2.3	7.9	0.98	2.3	55.6	24.1	42.6	0.64	31.0	0.229	Lo and Ho, 1979
Amblyseius	*okinawanus*			T: *Thrips palmi*															Kajita, 1986
Amblyseius	*scutalis*	0.35		T: *Frankliniella occidentalis*								0.3							Van Houten *et al.*, 1995a
	(*=rubini=gossipi*)			T: *Retithrips syriacus*	26						0.15	0.1							Swirski *et al.*, 1967b
				S: *Tetranychus cinnabarinus*	26						0.85	1.2							Swirski *et al.*, 1967b
				P: almond	26						1.00	2.0							Swirski *et al.*, 1967b
				T: *Scirtothrips citri*	25	3.3	3.2	6.5			0.86	1.4							Bounfour and McMurtry, 1987
				S: *Oligonychus punicae*	25	3.8	2.1	5.9			0.50	1.5							Bounfour and McMurtry, 1987
				P: iceplant	25	3.6	1.8	5.4	2.2	7.6	0.75	2.1	64.8	32.2	37.9	0.56	19.5	0.188	Bounfour and McMurtry, 1987
			#	S: *Eutetranychus orientalis*	25	(4.7)	4.8	(9.5)	2.8	(12.3)		2.1	29.2	14.2	22.9	(0.67)		(0.186)b	Yousef and El-Halawany, 1982
Amblyseius	*sessor*	0.34		T: *Haplothrips subtillissimus*	25							0.2							Sciarappa and Swift, 1977
				T: *Thrips tabaci*	25							0.5							Sciarappa and Swift, 1977
			#	S: *Tetranychus urticae*	25	4.8	2.8	7.6	2.3	9.9		0.8							Sciarappa and Swift, 1977
				P. *Circium vulgare*	25							0.2							Sciarappa and Swift, 1977
Amblyseius	*stipulatus*	0.36		T: *Frankliniella occidentalis*								>0.2							Rodriques-Reina *et al.*, 1992
				S: *Panonychus citri*	25	3.5	3.2	6.9			0.83	1.7							Ferragut *et al.*, 1987
				P: iceplant	25	3.0	2.8	5.5		7.6	0.87	2.0	54.3	30.4		0.63	27.2	0.197	Ferragut *et al.*, 1987
Amblyseius	*swirskii*	(0.40)		T: *Thrips tabaci*	27							1.0							Hoda *et al., 1986*
				S: *Eutetranychus orientalis*	27							1.6							Hoda *et al.*, 1986
				P: date palm	27							1.9							Hoda *et al.*, 1986
				T: *Retithrips syriacus*	26						0.88	1.4							Swirski *et al.*, 1967b
				S: *Tetranychus cinnabarinus*	26						0.95	1.2							Swirski *et al.*, 1967b
				P: almond	26						1.00	1.6							Swirski *et al.*, 1967b
			#	S: *Tetranychus urticae*	27	3.6	2.8	6.4	1.4	7.8		1.7	(55.8)	33.2	48.1	0.66		(0.219)b	Momen and El-Saway, 1993
				P: castor bean	27	3.7	4.0	7.7	2.1	9.9		1.7	(47.7)	28.2	35.5	0.66		(0.196)b	Momen and El-Saway, 1993
Amblyseius	*tularensis*	0.32	#	T: *Scirtothrips citri*	26	3.7	4.6	8.3	2.9	11.2		0.9	17.2	21.7	29.7	0.62		(0.152)a	Tanigoshi *et al.*, 1983
	(*hibisci–1985*)			P: iceplant	26	3.6	3.0	6.6	2.3	8.9		1.3	28.0	22.0	29.9	0.62		(0.210)a	Tanigoshi *et al.*, 1983

Table 8.2c. Continued.

Genus	Species	Adult size (mm)	Food source T: thrips, S: spider mites, M: acarid mites, P: pollen	Temperature (°C)	Development						Female performance							
					Eggs+larvae	Nymphs	Juv. period	Pre-ovip	Egg–egg	Juv. survival	Ovip.rate	Fecundity	Ovip. period	Longevity	Sex ratio	R_0	r_m	References
			T: *Frankliniella occidentalis*	25							0.2							Van Houten *et al.*, 1995a
			P: apple	26.7			6.3			0.87	(2.1)	24.7	11.8		0.76	17.0	0.210	Ouyang *et al.*, 1992
Typhlodrumus	*athiasae*		T: *Retithrips syriacus*	26						0.80	0.1							Swirski *et al.*, 1967a
			S: *Eutetranychus orientalis*	26						1.00	0.7				0.62			Swirski *et al.*, 1967a
			P: iceplant	26						1.00	1.2							Swirski *et al.*, 1967a
Typhlodrumus	*caudiglians*	0.36	S: *Panonychus ulmi*	24.8			6.7	2.8	9.5		1.1							Putnam, 1962
Typhlodrumus	*occidentalis*	0.35	T: *Frankliniella occidentalis*	20							0.0							Rodriques-Reina *et al.*, 1992
			T: *Retithrips syriacus*	26.5						0.87	0.5							Swirski and Dorzia, 1969
			S: *Eutetranychus orientalis*	26.5						1.00	1.7							Swirski and Dorzia, 1969
			P: iceplant	26.5						0.92	1.2							Swirski and Dorzia, 1969
			S: *Tetranychus cinnabarinus*	25	2.8	2.1	4.9	1.8	6.7	0.91	(2.1)	41.2	19.5	24.7	0.60		(0.240)a	Lababidi, 1988
Typhlodromus	*pyri*	0.33	T: *Drepanothrips reuteri*	25							0.6			13.0		2.5	0.050	Engel and Ohnesorge, 1994
			S: *Tetranychus urticae*	25							1.0			27.0		9.4	0.100	Engel and Ohnesorge, 1994
			P: spruce	25							1.3			37.0		16.3	0.145	Engel and Ohnesorge, 1994
			S: *Eotetranychus carpini*	26.5	2.4	4.2	6.6	2.3	8.6		1.2							Duso and Camporese, 1991
			P: iceplant	26.5	2.7	4.4	7.1	2.3	9.4		1.0							Duso and Camporese, 1991

This table includes life history data obtained with thrips as prey or other relevant food sources (juvenile spider mites, not eggs) or pollen as food sources for phytoseiids; and acarid mites, spider mites or sciarid fly larvae as food sources for Laelapidae. See note in Table 8.2a. See Appendix 1, p. 328, for procedure for estimating r_m and meanings of a and b.

Table 8.3. Rate of predation on thrips (prey day^{-1}) by various arthropod predators.

Family	Genus	Species	Prey species	Prey stage	Prey provided (day^{-1})	Substrate	Temp. (°C)	Maximum predation rate per predator stage (thrips day^{-1}) N1	N2	N3	N4	N5*	Adult male	Adult female	Lifetime mean ($female^{-1}$)	References
Mantidae	*Haldwania*	*lilliputana*	*Zaniothrips ricini*	Adult		Castor	28?	—	—	96	—	—				Mohandaniel *et al.*, 1983
Miridae	*Campylomma*	*chinensis*	*Thrips palmi*	Larva			25	3.2	7.4	7.9	11.4	19.5	26.9	29.1	(12.4)	Wang, 1995
	Campylomma	*livida*	*Thrips palmi*	L2			28	8.6						24.1	(10.3)	Chang *et al.*, 1993
	Dicyphus	*tamaninii*	*Frankliniella occidentalis*	Larva			25							12.7	(6.7)	Albajes *et al.*, 1996
Lygaeidae	*Ninyas*	*torvus*	*Selenothrips rubrocinctus*	Larva										several		Callan, 1943
Anthocoridae	*Carayonocoris*	*indicus*	*Scirtothrips dorsalis*	Larva				4.8	6.3	8.1	9.1	10.6	14.3	18.4	(8.8)	Sureshkumar and Ananthakrishnan, 1984
			Haplothrips ganglebaueri	Larva		Filter paper		2.7	2.1	2.6	3.2	5.3				Muraleedharan and Ananthakrishnan, 1978
	Montandoniola	*moraguesi*	*Gynaikothrips ficorum*	Larva		Filter paper	28.8		2.9	2.8	5.6	6.2		5.9	(4.2)	Tawfik and Nagui, 1965
			G. flaviantennatus	Larva/Adult		Casssearia		2.7	3.1	4.4	4.6	5.1				Muraleedharan and Ananthakrishnan, 1978
	Orius	*indicus*	*Megalurothrips usitatus*	L1			25	1.5		4.5						Rajasekhara and Chatterji, 1970
				Adult			25					2.5		3		Rajasekhara and Chatterji, 1970
	Orius	*insidiosus*	*Neohydatothrips variabilis*	L1	20	Soybean	24	4.4		10.3		14	17.2	19	(9.1)	Isenhour and Yeargan, 1981b
				L2	20	Soybean	24	3.4		9.3		12.3	14.3	15.2		Isenhour and Yeargan, 1981b
				Adult	20	Soybean	24	1.9		5.3		10.8	14.5	17.3		Isenhour and Yeargan, 1981b
			Caliothrips phaseoli	Adult	30	Castor							12.2	19.2		Saucedo-Gonzalez and Reyes-Villanueva, 1987
			Frankliniella occidentalis	Adult		Lima bean	25							11.5		Coll and Ridgway, 1995

Table 8.3. Continued.

Family	Genus	Species	Prey species	Prey stage	Prey provided (day^{-1})	Substrate	Temp. (°C)	Maximum predation rate per predator stage (thrips day^{-1})								References
								N1	N2	N3	N4	N5	Adult male	Adult female	Lifetime mean (female^{-1})	
			Frankliniella occidentalis	Adult	80	Bean pod	26							12.1		Tommasini and Nicoli, 1993
	Orius	*laevigatus*	*Frankliniella occidentalis*	Adult	80	Bean pod	26							12.5		Tommasini and Nicoli, 1993
	Orius	*majusculus*	*Frankliniella occidentalis*	Larva		Rose petal	25.5	7.7	6.9	7.3	7.2	8.5				Husseini *et al.*, 1993
			Frankliniella occidentalis	Adult	80	Bean pod	26							11.8		Tommasini and Nicoli, 1993
	Orius	*maxidentex*	*Scirtothrips dorsalis*	Larva				4.0	6.9	9.9	10.0	15.5	19.5	25.3	(12.1)	Sureshkumar and Ananthakrishnan, 1984
	Orius	*minutus*	*Frankliniella occidentalis*	L1	40	Bean	25						11.1	14.5	(6.9)	Lichtenauer and Sell, 1993
				L2	40	Bean	25						10.6	13.1		Lichtenauer and Sell, 1993
				Male	40	Bean	25						6.8	11.6		Lichtenauer and Sell, 1993
				Female	40	Bean	25						6.6	9.8		Lichtenauer and Sell, 1993
	Orius	*niger*	*Thrips tabaci*	Larva										17.5	(8.4)	Carl, 1976
			Frankliniella occidentalis	Adult	80	Bean pod	26							12.5		Tommasini and Nicoli, 1993
	Orius	*sauteri*	*Thrips palmi*	L2	40		25							22	(10.5)	Nagai, 1991
				Adult	40									26		Nagai, 1991
			Thrips palmi	L2			28	—	—	—	—	15	—			Chang *et al.*, 1993
			Thrips palmi	Larva	Ample		25	3.1	2.8	5.7	6.8	6.7	6.4	6.3	(5.0)	Wang, 1995

Table 8.3. Continued.

Family	Genus	Species	Prey species	Prey stage	Prey provided (day⁻¹)	Substrate	Temp. (°C)	Maximum predation rate per predator stage (thrips day⁻¹) L1	L2	Adult female	Lifetime mean (female⁻¹)	References
Chrysopidae	*Chrysopa*	sp.	*Selenothrips rubrocinctus*	Larva				14.2				Callan, 1943
Coccinellidae	*Micraspis*	*cardoni*	*Zaniothrips ricini*	L2/Adult		Castor	28?			20		Mohandaniel *et al.*, 1983
Aeolothripidae	*Aeolothrips*	*intermedius*	*Thrips tabaci* (F1–F4)	L2	Ample	*Datura*	26	— 9.5	—			Bournier *et al.*, 1979
			Heliothrips haemorrhoidalis	L2	Ample	*Datura*	26	— 8.5	—	(0)	(3.3)	Bournier *et al.*, 1979
			Odontothrips confusus	L2	Ample	*Datura*	26	— 5.6	—			Bournier *et al.*, 1979
	Franklinothrips	*megalops*	*Zaniothrips ricini*	Larva		Castor	28?	— 4–5	—			Mohandaniel *et al.*, 1983
			Caliothrips indicus	L2				5.3	9.9		(3.8)	Sureshkumar and Ananthakrishnan, 1987
			Frankliniella schultzei	L2				3.8	6.1			Sureshkumar and Ananthakrishnan, 1987
			Retithrips syriacus	L2				7.1	10.0			Sureshkumar and Ananthakrishnan, 1987
			Scirtothrips dorsalis	L2				9.6	14.8			Sureshkumar and Ananthakrishnan, 1987
			Zaniothrips ricini	L2				8.4	9.9			Sureshkumar and Ananthakrishnan, 1987
	Erythrothrips	*asiaticus*	*Caliothrips indicus*	L2				3.8	7.4		(2.8)	Sureshkumar and Ananthakrishnan, 1987
			Frankliniella schultzei	L2				5.9	9.0			Sureshkumar and Ananthakrishnan, 1987
			Retithrips syriacus	L2				3.5	6.1			Sureshkumar and Ananthakrishnan, 1987
			Scirtothrips dorsalis	L2				15.3	16.1			Sureshkumar and Ananthakrishnan, 1987
			Haplothrips ganglebaueri	L2				3.0	4.9			Sureshkumar and Ananthakrishnan, 1987
Phytoseiidae	*Amblyseius*	*andersoni*	*Frankliniella occidentalis*	L1	5	Judia	20			3.9		Rodriques-Reina *et al.*, 1992
				L2	5	Judia	20			0.8		Rodrigues-Reina *et al.*, 1992
	Amblyseius	*barkeri*	*Frankliniella occidentalis*	L1	12	Cucumber	25			2.6		Van Houten *et al.*, 1995a
			Frankliniella occidentalis	L1	5	Judia	20			1.7		Rodrigues-Reina *et al.*, 1992
				L2	5	Judia	20			0.3		Rodrigues-Reina *et al.*, 1992
			Thrips tabaci	Larva	10	Bean	25			4.3	(1.25)	Bonde, 1989
			Thrips tabaci	L1	20	Bean	25			8.5		Piatkowski, 1987
			Thrips palmi	L1	5	Cucumber	25			2.0		Kajita, 1986

Table 8.3. Continued.

Family	Genus	Species	Prey species	Prey stage	Prey provided (day^{-1})	Substrate	Temp. (°C)	Maximum predation rate per predator stage (thrips day^{-1}) L1	L2*	Adult female	Lifetime mean (female^{-1})	References
	Amblyseius	*californicus*	*Frankliniella occidentalis*	L1	5	Judia	20			1.1		Rodrigues-Reina *et al.*, 1992
				L2	5	Judia	20			0.2		Rodrigues-Reina *et al.*, 1992
	Amblyseius	*cucumeris*	*Frankliniella occidentalis*	L1	12	Cucumber	25			6.0		Van Houten *et al.*, 1995a
			Frankliniella occidentalis	L1	40	Cucumber	26			6.9		Shipp and Whitfield, 1991
			Frankliniella occidentalis	L1	40	Sweet pepper	26			10.0		Shipp and Whitfield, 1991
			Frankliniella occidentalis	L1	10	Artificial	25		1.02	3.2		Castagnoli *et al.*, 1990
			Thrips tabaci	L1	10	Artificial	25		0.80	5.4	(1.56)	Castagnoli *et al.*, 1990
			Frankliniella occidentalis	Larva		Cucumber	25		0.87			Cloutier and Johnson, 1992
	Amblyseius	*degenerans*	*Frankliniella occidentalis*	L1	12	Cucumber	25			4.4	(1.40)	Van Houten *et al.*, 1995a
	Amblyseius	*hibisci*	*Frankliniella occidentalis*	L1	12	Cucumber	25			3.5	(0.99)	Van Houten *et al.*, 1995a
	Amblyseius	*limonicus*	*Frankliniella occidentalis*	L1	12	Cucumber	25			6.9	(1.95)	Van Houten *et al.*, 1995a
	Amblyseius	*okinawanus*	*Thrips palmi*	L1	5	Cucumber	25			1.0		Kajita, 1986
	Amblyseius	*scutalis*	*Frankliniella occidentalis*	L1	12	Cucumber	25			1.3	(0.37)	Van Houten *et al.*, 1995a
	Amblyseius	*swirskii*	*Thrips tabaci*	Larva	10	Castor	27			1.8	(0.50)	Hoda *et al.*, 1986
	Amblyseius	*tularensis*	*Frankliniella occidentalis*	L1	12	Cucumber	25			0.5		Van Houten *et al.*, 1995a

Predation rates are classified according to predator species, predator stage, and prey stage (first and second instar larvae and adults). Lifetime means are calculated by weighting the predation rates of all (female) predator stages by their share in a stable growing population (stable age distribution, r_m as in Table 8.2). For species with no data on predation by juveniles, the ratio between lifetime mean and adult predation is assumed to be similar for all species within the same family (ratios between 0.21 and 0.26). To add the contribution of males in the population, these values are increased by 25 to 100%, depending on sex ratio and relative predation capacity, as discussed in the text. When several data sets are available for a single species, the median value is used to calculate the lifetime means.

*Juvenile prey stages are defined as Nymph stages 1 to 5 (N1–N5) or larval stages 1 and 2 (L1 and L2) depending on species.

Notes: 1. Lifetime means are not calculated for the data of Rodrigues-Reina *et al.* (1992), because prey supply is too low to approach maximum predation rates.

2. The data labelled 'predation efficiency' by Sureshkumar and Ananthakrishnan (1987) are assumed to represent the total prey consumption during one life stage so these are divided by stage duration to obtain predation rate.

the predation rate of adult females (but larvae for predatory thrips) is about 4.5, 8, 17, and 27 young thrips larvae day^{-1}, respectively.

These prey consumption rates of adult female predators cannot be directly used in the model without making further assumptions. First, as the attack rates depend on the prey stage involved (Table 8.3), the mean attack rate should take the prey-stage composition and the prey-stage preference into account. However, it is reasonable to assume that the supply of young thrips larvae is a determinant of the maximum predation rate because (i) the predation rate on young thrips larvae is always higher than on other thrips stages, (ii) the young larvae are the preferred prey stage and (iii) they make up a considerable part of a growing thrips population (about 20 per cent for *Frankliniella occidentalis*). In addition to this assumption on prey-stage selection it is necessary to assume that differential attack on prey stages does not influence the age distribution and hence the growth rate of the prey. Also, it is necessary to correct the predation rate because attack rates depend on the stage of development of the predators (Table 8.3). This correction follows directly from the assumptions on constant predation rates and exponential growth of the predators, as this implies a stable age distribution of the predators. Hence, a stage-weighted lifetime predation rate can be calculated by summing the products of the stage-specific predation rates (= β_i) and the stable proportion of each stage in the predator population (= p_i):

$$\beta = \sum_i p_i \beta_i$$

This weighted predation rate needs further correction for the share of the males in the predator population. If males and females consume prey at equal rates, it suffices to multiply the predation rate with the inverse of the sex ratio (s_i the proportion of females in predator stage i):

$$\beta = \sum_i p_i \beta_i \frac{1}{s_i}$$

If the sex-specific predation rates differ, then the following formula applies:

$$\beta = \sum_i p_i \beta_i \left[1 + \left(\frac{1}{s_i} - 1 \right) q_i \right]$$

where q is the ratio of male-to-female predation rate.

For phytoseiid mites the weighted group mean lifetime predation rate is about 1.5 thrips day^{-1}, assuming a sex ratio of 66%, a stable age distribution of the predators (31% nymphs and 13% adults), a predation rate of nymphs equal to 15% of the adults (Castagnoli *et al.*, 1990; data for *Amblyseius cucumeris* in Table 8.3), with the small males half as voracious as the females (q = 0.5). For anthocorid predators the weighted lifetime predation rate is about 8.7 thrips day^{-1}, assuming a sex ratio of 55%, a stable age distribution of the predators (47% juveniles, 7% adults), a predation rate of the adults equal to 6 thrips larvae day^{-1} and a predation rate of nymphs as reported for

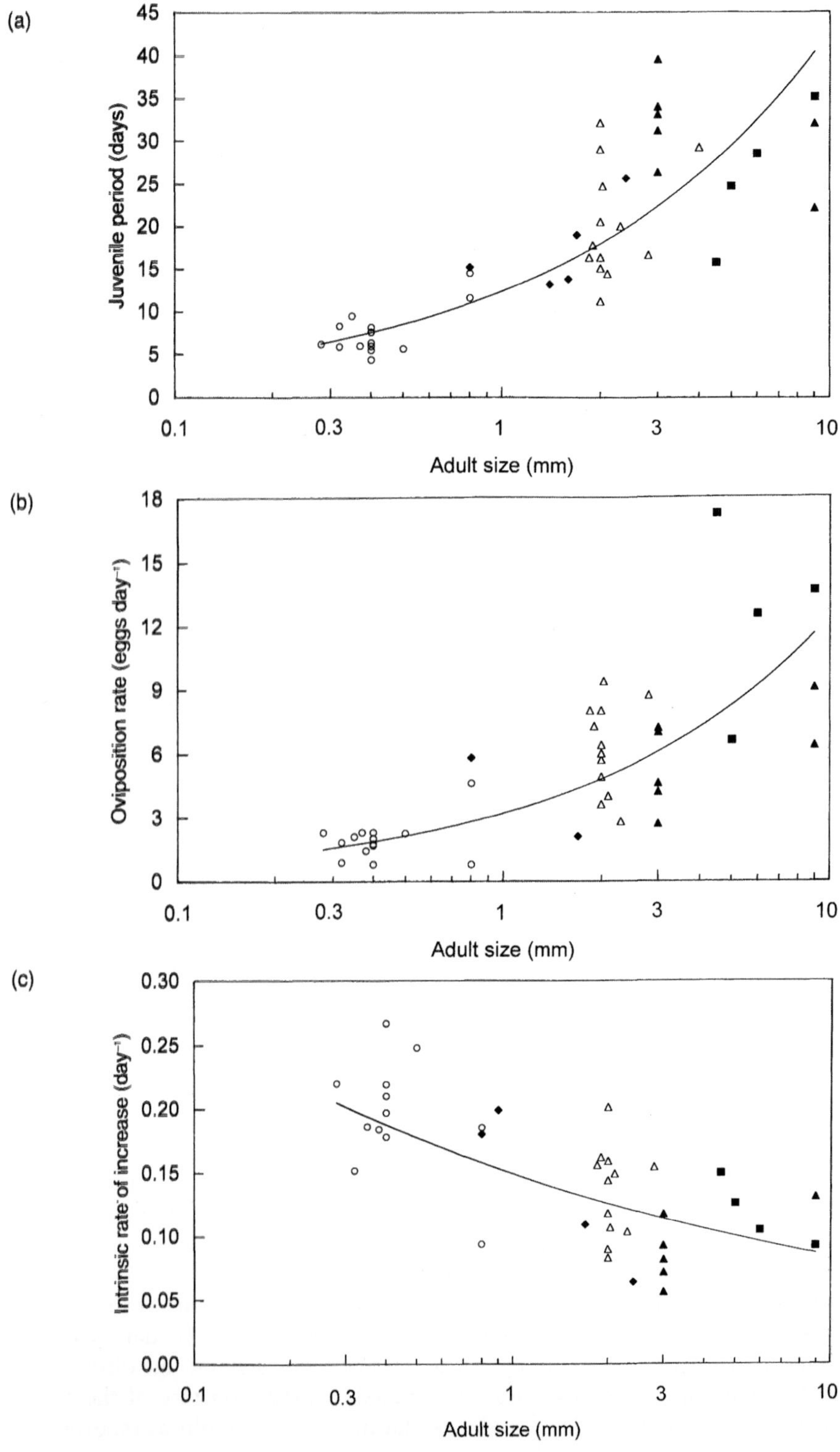
(a)
Juvenile period (days)
45
40
35
30
25
20
15
10
5
0
0.1
0.3
1
3
10
Adult size (mm)
(b)
Oviposition rate (eggs day⁻¹)
18
15
12
9
6
3
0
0.1
0.3
1
3
10
Adult size (mm)
(c)
Intrinsic rate of increase (day⁻¹)
0.30
0.25
0.20
0.15
0.10
0.05
0.00
0.1
0.3
1
3
10
Adult size (mm)

(d)

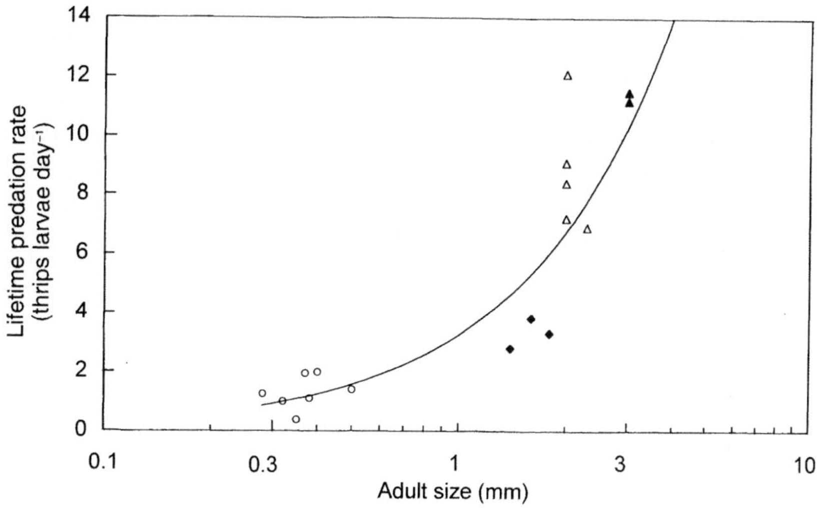

(e)

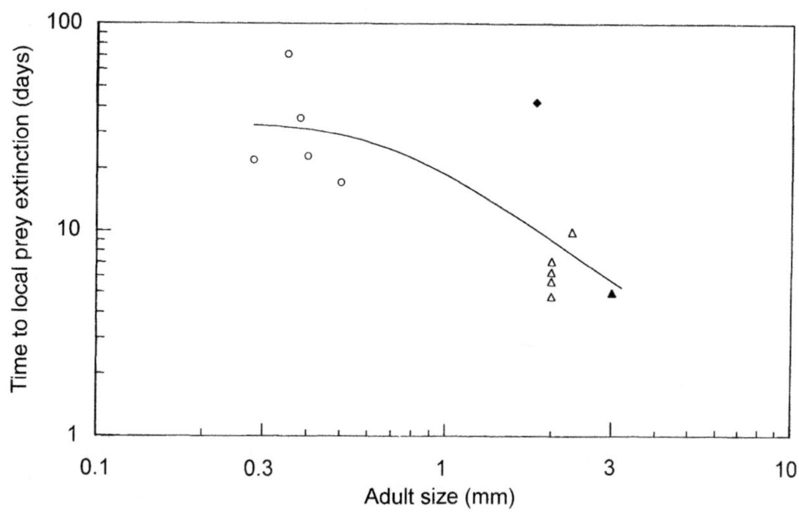

Fig. 8.1. Some life history and population growth characteristics of predators of thrips in relation to their size (provided they are less than 10 mm): (a) developmental period (days), (b) oviposition rate (eggs day^{-1}), (c) intrinsic rate of increase (r_m day^{-1}), (d) predation rate (weighted lifetime mean; thrips larvae day^{-1}). Each point represents one species of predator. The lines in a–d represent best fits of a power function ($y = a\,x^b$). Taking r_m of the thrips equal to 0.17 day^{-1} and the initial predator–prey ratio equal to 1:50, the time to prey extinction was calculated using the 'pancake' model (e), either based on the data for each species (symbols) or on the power function fits in c and d (line).

From Table 8.2 only one data-pair per species was used. Selection was based on completeness of life-history information and on the type of diet (rank order: thrips, spider mites, others). ○, predatory mites (provided that oviposition rate exceeds 1 egg day^{-1}); ◆, predatory thrips; △, anthocorids; ▲, other heteropteran predators; ■, ladybird beetles.

Orius insidiosus (Isenhour and Yeargan, 1981b; Table 8.3) with q equal to unity. For mirid predators the weighted lifetime predation rate is 11.4 thrips day^{-1}, assuming a sex ratio of 50%, a stable age distribution of the predators (46% juveniles; 8% adults), a predation rate of nymphs and adults according to data for *Campylomma chinensis* (Wang, 1995; Table 8.3) with q equal to unity. In a similar way, the lifetime predation rate for each predator species can be calculated (Fig. 8.1d).

Taking 0.17 day^{-1} as α, the intrinsic rate of thrips population growth (Van Rijn *et al.*, 1995), one can calculate the time to thrips extinction for each separate predator species (Fig. 8.1e). For a 1:50 predator–prey ratio the time from predator release to thrips extinction lasts about 26, 24, 6.5 and 5 days for the group mean parameters of phytoseiids, predatory thrips, anthocorids and mirids, respectively. Thrips extinction is predicted to be reached *always* by phytoseiid mites (since $\alpha < \gamma$), whereas it requires a predator–prey ratio above 1:97 for predatory thrips, 1:217 for the anthocorids and 1:228 for the mirids. In addition the thrips population is expected to decline immediately after predator release (type-1 dynamics) when the initial predator–prey ratio exceeds about 1:9 for phytoseiid mites, 1:17 for the predatory thrips, 1:51 for the anthocorids and 1:67 for the mirids. This suggests that type-1 dynamics are more likely to occur for the anthocorid and the mirid predators than for the phytoseiid mites.

Validation of model predictions on short-term and local dynamics

The 'pancake' predation model is at best a good caricature of arthropod predator–prey interactions. It captures the essence whenever local thrips densities are high enough and the spatial scale under consideration is small enough to make strong coupling of the interacting populations a realistic assumption (Sabelis, 1992). Whether it also stands up to scrutiny in more quantitative tests remains to be seen. Nevertheless, its predictions are simple and testable, as will be shown below.

Impact of predator exclusion

One of the obvious predictions of the model is the occurrence of thrips outbreaks when resident predator populations are sufficiently reduced. The literature provides ample evidence for a strong impact of predators. Stoltz and Stern (1978) found increased densities of *Frankliniella occidentalis* on cotton in California and decreased densities of *Geocoris pallens* and *Orius tristicolor* under multiple dimethoate and naled/toxaphene applications. Tanigoshi *et al.* (1984, 1985) used malathion to eliminate early spring populations of *Amblyseius tularensis* in citrus orchards in California. They found that predator exclusion led to much more damage by *Scirtothrips citri* (Tanigoshi, 1991). However, using isoelectric focusing electrophoresis Jones and Morse (1995) rarely found evidence for thrips predation in analysing the diet of field-collected females of *Amblyseius tularensis*. They argue that one should be cautious in inferring predatory activity from laboratory observations on predation and field experiments on predator exclusion alone. Under field conditions, the thrips may partially occupy other areas of the plant than its predator, so they would meet less frequently. Moreover, pesticides used for exclusion may eliminate other predators and perhaps stimulate thrips population growth by hormoligosis. That elimination of predatory mites does not always lead to increased thrips populations was shown by McMurtry *et al.* (1992). They also applied malathion, like Tanigoshi *et al.* (1984, 1985) and found increases of citrus red mites, rather than of *Scirtothrips*. A similar predator exclusion approach, using a miticide, dicofol, suggested that *Amblyseius citri* and *Amblyseius addoensis* have an impact on *Scirtothrips aurantii*, in southern Africa (Grout and Richards, 1992b; Grout, 1994). However, more insight is needed into which predators feed on thrips under field conditions (e.g. by using diet analysis techniques developed by Sopp *et al.* (1992) and Greenstone and Trowell (1994)), and how the relevant natural enemies are affected by dicofol. More stringently controlled experiments were made by Nagai (1990), who used an insecticide (emulsion of fenthion) shown to be harmful to *Orius* sp., but less so to *Thrips palmi* (also suggesting no effects via the host plant, i.e. hormoligosis). Thrips density on potted aubergines in a screen-house with anthocorid predators remained low in the untreated group of plants, but increased sharply on the treated plants. However, Nagai (1990) did not provide precise quantitative information on the change in predator–prey ratio due to the insecticide application.

Minimal conditions for thrips suppression

According to the pancake model thrips populations are predicted to grow indefinitely when the predator–prey ratio is less than $(\alpha - \gamma)/\beta$. Thus, when $\alpha > \gamma$, thrips outbreaks are expected at low predator–prey ratios, as observed for *Orius* sp. by Kawai (1995) only in the experiment with the lowest release rate (= one *Orius* nymph on every other plant). Unfortunately, the precise predator–prey ratio cannot be determined because this requires knowledge on the number of thrips per plant and the author only provided the mean number of thrips per leaf (= 1.3–1.9) and not the number of leaves per plant. Thus, these data are not appropriate for a quantitative test of the model prediction on outbreaks. Another prediction of the pancake model is that thrips populations are decimated as a rule when $\alpha < \gamma$. This condition on the growth rates applies to most of the phytoseiid mites used for control of *Frankliniella occidentalis* and *Thrips tabaci* (Table 8.3). However, there are many cases where phytoseiid mites failed to control these thrips, most of which concern releases of *Amblyseius cucumeris* to control *F. occidentalis* on cucumber (Hansen, 1989; Higgins, 1992; Gerin and Hance, 1993; Jacobson, 1995; Ramakers and Voet, 1995; Van Houten, 1996). These failures appear to be due to much lower predator growth rates on cucumber than expected from life history measurements on a diet of suitable prey, such as mould mites (Castagnoli and Simoni, 1990), or suitable pollen (Van Rijn and Van Houten, 1991). The causes for this retarded growth can be manifold. Apart from the impact of predator diseases (Beerling and Van der Geest, 1991; Steiner and Bjørnson, 1996) there are three major competing explanations: (i) negative microclimatic effects (Shipp *et al.*, 1996; Shipp and Van Houten, 1997), (ii) too low rate of encounter with thrips on cucumber (Higgins, 1992; Van Houten,

1996), and (iii) inability of young predators to develop to a stage in which thrips can be seized with some chance of success. The first explanation is not supported because prevailing temperatures and humidities were not extreme enough (Van Houten and Van Lier, 1995). The second explanation may hold because another species of more active phytoseiid mite, *Amblyseius limonicus* (Van Houten *et al.*, 1995a), showed considerable population growth under exactly the same conditions (Van Houten, 1996). However, at high densities of either first or second instar thrips larvae this phytoseiid species shows higher predation rates than *Amblyseius cucumeris*, which may indicate a better ability to seize thrips larvae upon attack. If this also applies to juvenile mites, or juveniles have more opportunity to attack thrips larvae killed by adult mites, then *Amblyseius limonicus* may suffer less mortality in the juvenile stages. Recently, P.C.J. Van Rijn, Y.M. Van Houten and A.M.M. Van Lier (unpublished results) found that the age class of juveniles of *Amblyseius* sp. on parthenocarpic cucumber was very small, but increased considerably when pollen was provided (Fig. 8.2). Indeed, successful control of *F. occidentalis* with *Amblyseius cucumeris* has usually been obtained in situations where edible pollen was available, e.g. in greenhouse sweet pepper crops that flower continuously after a short growing phase (Ramakers, 1980, 1987, 1988, 1990, 1993; de Klerk and Ramakers, 1986; Tellier and Steiner, 1990; Van de Veire and DeGheele, 1992; Van Houten and Van Stratum, 1995). Whenever *Amblyseius cucumeris* proved successful on parthenocarpic cucumber (in the absence of pollen), releases of predatory mites were either 'inundative and continuous' (Beglyarov and Suchalkin, 1983; Bennison, 1988; Hansen, 1988, 1989; Gillespie, 1989; Bennison *et al.*, 1990; Bennison and Jacobson, 1991), or started before any visible thrips infestation at quite high rates (250 per plant; Brødsgaard and Hansen, 1992). Similarly, releases of *Amblyseius barkeri* to control *Thrips tabaci* on cucumber were successful at high predator–prey ratios (1200 per plant at appearance of first thrips damage symptoms, Ramakers, 1983; fortnightly inundative releases, Ramakers, 1990; four to six introductions of 15 to 20 predators m^{-2}, Lindqvist and Tiittanen, 1989). Inundative releases of *Amblyseius barkeri* were also required to obtain successful thrips control on gladiolus corms (20 to 80 predatory mites per 20 corms lightly infested with *Thrips simplex*; Conijn, 1993), on tobacco seedlings (3 to 4 predatory mites per leaf infested by about 1 *T. tabaci* larva; Trjapitzin, 1995) and on cabbage (about 200 or more predatory mites per plant of which about 10 per cent arrived in cabbage heads harbouring about 40 *T. tabaci* per head; Hoy and Glenister, 1991).

Rate of thrips suppression

In addition to formulating minimal conditions for control, the model provides a precise prediction of the time to prey elimination as well as criteria for the predator–prey ratios required for type-1 dynamics (immediate decrease) and type-2 dynamics (delayed decrease). For phytoseiid mites, type-1 dynamics are expected only when predator–prey ratios are quite high (1:9). Indeed, Bennison and Jacobson (1991) observed type-1 dynamics in a *Frankliniella occidentalis* population at predator–prey ratios that were probably very high because they

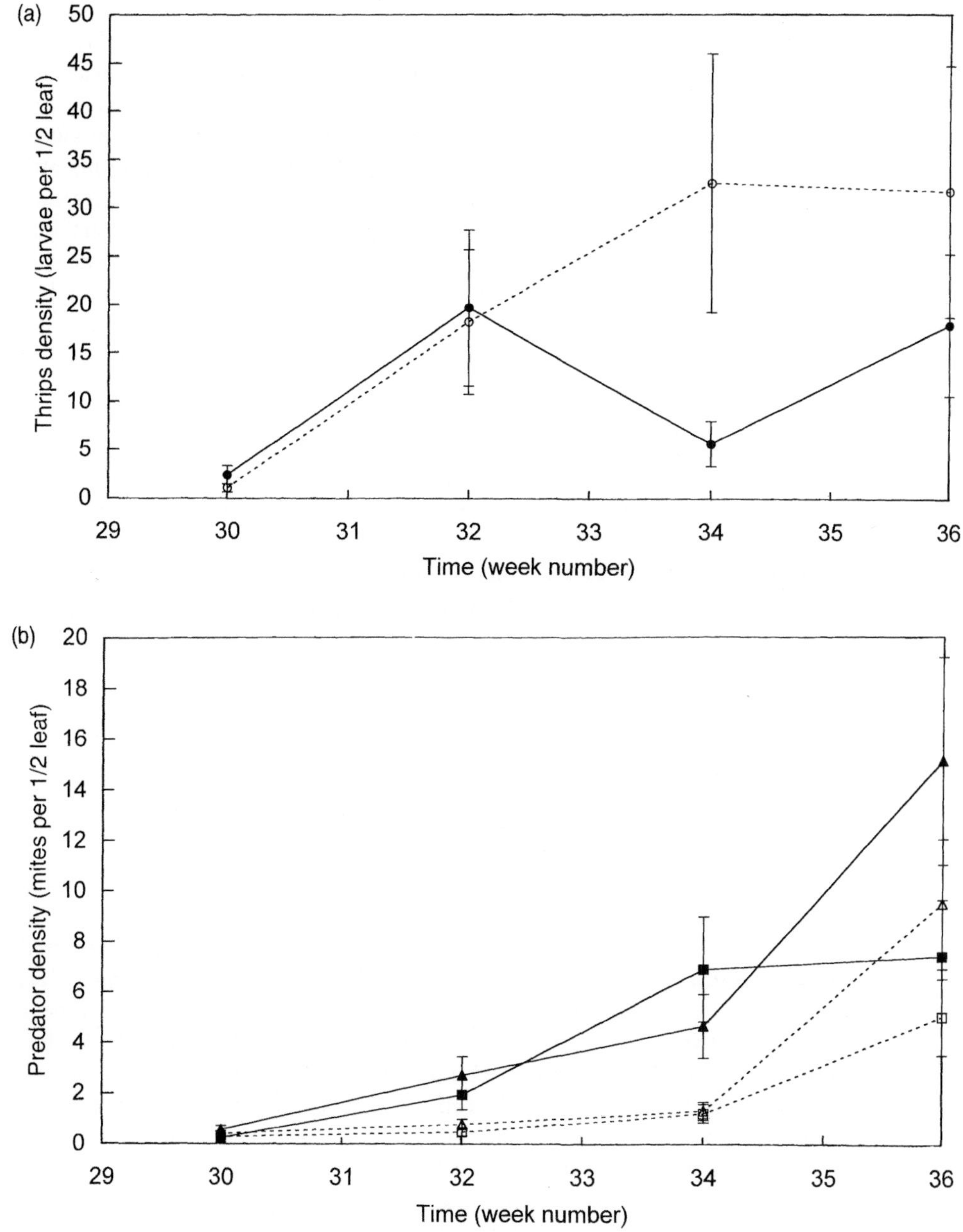

Fig. 8.2. Population dynamics of *Frankliniella occidentalis* larvae (a) and *Amblyseius degenerans* (adults, □ ■; juveniles △ ▲, (b) in absence (□ △, dashed lines) and presence of cattail pollen (■ ▲, solid lines) on cucumber plants. The predatory mites were introduced in week 29 at a rate of eight females per plant. Each point represents a mean of four replicates, obtained by population experiments in separate compartments of a greenhouse (60 plants per compartment) (P.C.J. Van Rijn, Y.M. Van Houten and A.M.M. Van Lier, unpublished data). In the pollen treatment, 50 mg of cattail pollen was provided per plant and per week (totalling about 3 g per greenhouse compartment). Error bars represent 95% confidence intervals.

used a controlled (or rather continuous) release system for *Amblyseius cucumeris* (porous, waxed paper packs with bran and mould mites to rear and release predatory mites continuously). Also as expected for an initial predator–prey ratio close to 1:1, leaf-inhabiting populations of *F. occidentalis* on sweet pepper showed type-1 dynamics after releases of *Amblyseius degenerans* and virtually disappeared in one or a few weeks (Van Houten and Van Stratum, 1995). Releases of *Amblyseius cucumeris* by the same authors led to very low initial recovery of predators on leaves, whereas the density of thrips larvae was close to 1 per leaf. The initial predator–prey ratio is hard to estimate here, but it reached a level between 1:1 and 1:10 within 2 weeks after release and then the thrips population declined as predicted, albeit at a slower rate. In another series of experiments (P.C.J. Van Rijn, Y.M. Van Houten and A.M.M. Van Lier; Fig. 8.2) releases of *Amblyseius degenerans* at a 1:4 predator–prey ratio on cucumber supplied with pollen caused the *F. occidentalis* population to exhibit type-2 dynamics with 2 to 4 weeks to thrips decimation. Both type of dynamics and time to extinction would have been predicted at lower predator–prey ratios (< 1:9). The same applies to experiments of Gillespie (1989) with *Amblyseius cucumeris* on cucumber. Here, predator–prey ratios were between 1:2 and 1:9, but the thrips dynamics were not the expected type 1, but rather like type 2. Lindhagen and Nedstam (1988) also invariably observed type-2 dynamics for initial predator–prey ratios of 20:50 (= 1:2.5), 60:350 (=1:6) and 60:3000 (= 1:50) and time to prey suppression of about 4 weeks for all three cases. Only in the latter case are model predictions satisfactory, suggesting that they hold for very high initial thrips numbers per plant. Thus, for phytoseiid mites the model predictions hold qualitatively, but (perhaps with the exception of high initial thrips densities) not quantitatively, as they tend to overestimate the predator's impact and to underestimate the time to prey extinction.

For predatory thrips attacking pest thrips, there is little information on the consequences of predator–prey ratios for the type of dynamics and the time to prey suppression. For anthocorid predators, type-1 dynamics are predicted above a predator–prey ratio of 1:51 and the time to prey suppression is less than a week. Indeed, releases of anthocorid bugs frequently lead to type-1 dynamics of a thrips population. For example, Nagai *et al.* (1988) found that the population of *Thrips palmi* on young aubergines increased from about 55 to more than 100 individuals per leaf in the absence of predation, but when 6 individuals of *Orius* sp. (5 nymphs and 1 adult) per young aubergine were released the thrips population decreased to about 1 thrips per leaf in a period of 6 to 9 days, which only slightly exceeds the 6.5 days predicted. Without information on the number of leaves per plant it is difficult to assess the initial predator–prey ratio, but given the aubergine was young it seems likely that the ratio was above the critical 1:51. Hence, the experiments of Nagai *et al.* (1988) support the predictions of the 'pancake' model. Similar problems in assessing initial predator–prey ratios arise with most other publications on thrips control with anthocorids. However, given that the conditions for type-1 dynamics are quite broad, it is striking that this type of dynamics and the 1 to 2 week period to suppression were observed in many experiments on thrips control, such as by *Orius insidiosus* on sweet pepper (2 adult predators per plant) (Fig. 8.3a,

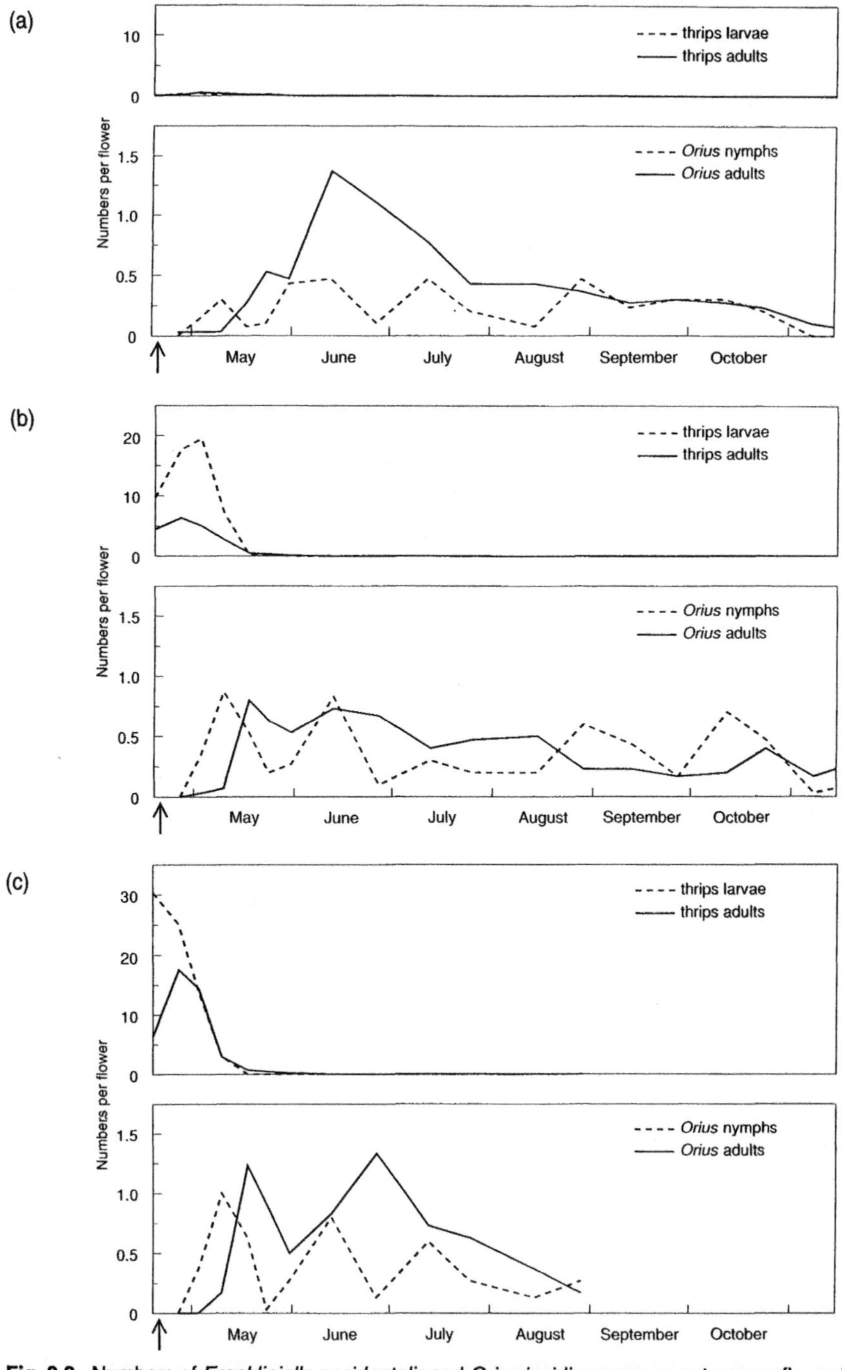

Fig. 8.3. Numbers of *Frankliniella occidentalis* and *Orius insidiosus* per sweet pepper flower in a greenhouse with (a) very low initial thrips density (3 to 15 thrips larvae and adults per flower), (b) intermediate initial thrips density (14 to 24 thrips larvae and adults per flower), and (c) high initial thrips density (36 to 44 thrips larvae plus adults per flower) (Van den Meiracker and Ramakers, 1991). On 20 April, 1990, 500 anthocorid predators (adult females and males with a 1:1 sex ratio) were released.

low initial thrips density; Van den Meiracker and Ramakers, 1991), by three *Orius* spp. (*Orius insidiosus*, *O. albidipennis* and *O. laevigatus*) on melons, strawberries and sweet pepper (Dissevelt *et al.*, 1995), by *Orius laevigatus* on strawberries (Villevieille and Millot, 1991), by *Orius laevigatus* on sweet pepper (under long day conditions in spring and in summer, Chambers *et al.*, 1993; see also Tavella *et al.*, 1991), by *Orius* spp. (predominantly *Orius sauteri*; Y. Hirose, Kyushu University, Japan, personal communication, 1996) on aubergine (1 to 5 predators per plant) (Kawai, 1995) and by *Orius tristicolor* on sweet pepper (data from 1988; Higgins, 1992). All these results of releases of anthocorid predators may be interpreted as support for the pancake model predictions on dynamic behaviour and time to prey suppression, provided that initial predator–prey ratios were indeed above 1:51.

Particularly good information on predator–prey ratios has been provided by the experiments of Trottin-Caudal *et al.* (1991) on biological control of *Frankliniella occidentalis* with *Orius majusculus* on cucumber. When they released 1 predator per 20 thrips, type-1 dynamics were observed with strong thrips suppression within a week and virtual elimination with 4 weeks; but when released in a 1:100 ratio, type-2 dynamics occurred with thrips elimination in 4 to 6 weeks. These observations uphold the model predictions, although the observed time to prey elimination is somewhat longer than the predicted 2 weeks. Additional support comes from observations on type-2 dynamics following predator releases at ratios probably lower than 1:51 (but higher than 1:217): the 1989 data of Higgins (1992) for *Orius tristicolor* in sweet pepper (initial ratio between 1:50 and 1:100), the data of Van den Meiracker and Ramakers (1991) for intermediate and high initial thrips densities (Fig. 8.3b and c) and the data of Jacobson (1991) for *Anthocoris nemorum* on cucumber (initially 192 predators released about 1 month after release of 1056 thrips; given that the actual increase of the thrips population over the prerelease month is about 8.5, the initial predator–prey ratio is close to 1:50). Provided that these predator–prey ratios are correctly gleaned from the information presented, the predictions on type-2 dynamics hold, but the observed time to prey suppression seems to be somewhat longer, being about 1 month rather than a few weeks.

In other cases, model predictions fail. A striking example is the results of Chambers *et al.* (1993); they found a slow, but steady increase of *Frankliniella occidentalis* population on cucumber despite repeated releases of *Orius laevigatus*, frequently leading to predator–prey ratios above 1:10. These authors found a much stronger impact of this anthocorid predator on *F. occidentalis* infesting sweet pepper, suggesting that particular ecological requirements are lacking in parthenocarpic cucumber, but present in sweet pepper. They attribute the difference in predator impact to a higher coincidence of predator and prey in sweet pepper flowers where they may hide and feed on pollen. Another example of poor agreement with model prediction are the experiments of Chyzik and Klein (1995). They found that in the period from May to August 1994, ratios of *Orius laevigatus* to *F. occidentalis* in sunflower heads were above 1:10 and the dynamics, as expected, resembled type 1; yet, the time to thrips suppression was more than 2 months, which is much longer than expected. It would be interesting to study whether thrips larvae can hide in the sunflower heads, thereby partially escaping from predation and lengthening the interaction time. This may also be an explanation for the dynamics of *F. occidentalis* in the flowers (rather than the leaves) of sweet pepper, as reported by Van Houten and Van Stratum (1995). They found that the flower-inhabiting populations of thrips persisted much longer than the leaf-inhabiting population, and that *Amblyseius degenerans* built up much larger populations in flowers and ultimately suppressed their prey far more than *Amblyseius cucumeris*. These experiments are revealing in that they provide evidence for flowers as partial refuges for the thrips and for interspecific differences in flower visitation among phytoseiid species. How this refuge effect arises is still not clear since sites accessible to thrips larvae should also be accessible to predatory mites. However, not only predators feed on pollen, thrips larvae consume it also (see Chapter 4). The effect of pollen feeding by the thrips larvae may be to shorten the developmental phase in which they are vulnerable to attack by predatory mites (Van Rijn, 1997; Van Rijn and Sabelis, 1997). This could lead to increased growth of the thrips population, so lengthening the time to its suppression.

For the mirid predators, type-1 dynamics and less than 1 week to thrips extinction are expected at predator–prey ratios above 1:61. Dimitrov (1975, 1976, 1977) reported immediate suppression of *Thrips tabaci* on tobacco by supplementary releases of *Macrolophus rubi* at a 1:35 predator–prey ratio. Gabarra *et al.* (1995) studied the impact of two release rates of nymphs of *Dicyphus tamaninii* on *F. occidentalis*. In one treatment they released 18 nymphs per plant, but only 6 predators were recovered 6 days later. Total thrips per plant were about 15 larvae and adults. Hence, the initial predator–prey ratio was between 1 and 2.5. The results show that numbers of thrips larvae decrease to a very few within a week as predicted by the pancake model, but the populations of thrips adults decrease much more slowly and only decline to low levels after 8 weeks, suggesting that they are better able to escape from predation. In another treatment Gabarra *et al.* (1995) released only 3 mirid nymphs per plant, but only 1 was recovered per plant, 6 days later. The initial thrips density increased sharply from about 40 to about 100 per plant, perhaps partly due to the introduction of extra thrips 7 days after predator release (= 1 day after first count on plants) and partly due to the predator–prey ratio being lower than the critical ratio, thereby giving rise to type-2 dynamics. As type-1 dynamics would be expected for a ratio of 1:40 and type-2 for a ratio of 1:100, the observations of Gabarra *et al.* (1995) do not contradict the predictions of the pancake model. However, the 8 weeks to suppression of thrips larvae was much longer than the 1 to 2 weeks predicted; the thrips adults were suppressed a few weeks later, again possibly indicating better ability to escape predation. Castané *et al.* (1996) carried out similar experiments with 1:10 and 1:3 ratios of late instar *Dicyphus tamaninii* to *F. occidentalis* at initial densities of 5 thrips per leaf. The establishment of thrips and mirid bugs differed, so the eventual predator–prey ratios were 1:15 and 1:5. Both ratios resulted in immediate decrease of the thrips population during the first 2 weeks, i.e. the type-1 dynamics expected.

However, in the third week the thrips population resurged, but was suppressed a few weeks later. Resurgence was associated with a decrease in the predator population and conversely suppression was preceded by an increase of the predator population. These effects may be due to the delay in predator recruitment and/or dispersal of the adult predator.

All in all, the pancake model provides quite a successful representation of the impact of growth rates, predation rates and initial predator–prey ratios on the local dynamics. The insight gained is mainly of a qualitative nature, leaving quantitative aspects to be improved by more detailed models. It seems safe to conclude that many of the predatory arthropods tested satisfy the arrestment assumption underlying the 'pancake' model. As thrips show aggregated distributions (Petersen, 1990), it pays predators to find and stay in thrips-infested parts of plants (Shipp *et al.*, 1992), thereby creating conditions for maximal predation rates and exponential population growth of the predators. However, the behavioural mechanisms allowing predatory arthropods to detect and stay in these thrips aggregations are largely unexplored. Area-restricted search (phytoseiids: Sabelis and Dicke, 1985; anthocorids: Shields and Watson, 1980), prey kairomones and prey-induced plant synomones (phytoseiids: Dicke and Groeneveld, 1985; Sabelis and Dicke, 1985; anthocorids: Drukker *et al.*, 1995) may well play an important role (Sabelis, 1992). The important point emerging from the validation exercise presented above is that the arrestment behaviour must be very strong, because even small rates of dispersal *during* the interaction in local populations can cause such large (quantitative) deviations from the pancake predictions that they would not easily go unnoticed (Van Baalen and Sabelis, 1995).

Long-term persistence of predator and prey

On a longer time scale the pancake model predictions on local predator–prey dynamics must obviously fail. This is because the assumptions on strong arrestment in local populations and the constant pancake predation rate must unavoidably lead to complete local extinction of the prey. From a theoretical point of view one may object that the model does not allow for persistence of predator and prey. However, while this might be true for a local spatial scale, it is not necessarily true for a larger spatial scale, as shown in predator–prey models with a metapopulation structure (e.g. Sabelis and Diekmann, 1988; Sabelis *et al.*, 1991). There is, however, mounting empirical evidence for population persistence of predators and thrips, even on a local spatial scale. For example, Kawai (1995) found that *Orius* spp. never wiped out local populations of *Thrips palmi*, but instead the thrips densities settled at a more or less fixed level irrespective of the initial predator–prey ratios. Similar evidence emerges from other experiments with anthocorid predators and thrips on strawberries (Villevieille and Millot, 1991), on aubergine (Nagai, 1990) and also from experiments with phytoseiid mites and thrips on parthenocarpic cucumber with addition of pollen (P.C.J. Van Rijn, Y.M. Van Houten and A.A.M. Van Lier, unpublished data) and without pollen (Bennison and Jacobson, 1991), on melon and strawberry (Dissevelt *et al.*, 1995), on sunflower (Chyzik and Klein, 1995) and on sweet pepper (Van Houten and Van Stratum, 1995). To explain local prey persistence four mechanisms may be responsible: (i) reduced predation rates with decreasing prey densities, (ii) escape in space by seeking refuge or dispersal, (iii) escape in time by differential timing of diapause and hibernation and (iv) escape via development by outgrowing the stage vulnerable to predation. Also, there are several cases where predators maintain or even expand their local populations in the virtual absence of thrips as prey. Examples are provided by experiments with anthocorids on sweet pepper (Van den Meiracker and Ramakers, 1991; Dissevelt *et al.*, 1995) and phytoseiid mites on sweet pepper (Ramakers and Voet, 1995; Van Houten and Van Stratum, 1995). The causes of local persistence of predators in the absence of thrips may be due to economizing on energy reserves or feeding on something else, i.e. phytophagy and polyphagy. These explanations are briefly discussed below.

Thrips escape at low thrips densities

When predators decrease prey density in their immediate neighbourhood they reach a point where they cannot maintain maximum predation rate. Then, as assumed in the pancake model, they may move to other more profitable sites nearby (i.e. within the local population), but sooner or later these sites will also become scarce and overall predation rates within a local thrips population will decrease. Concomitantly, development, survival and reproduction of the predator will decrease. Ultimately, the thrips may suffer so little from predation that its population may recover, thereby avoiding extinction. This mechanism leads to predator–prey cycles, as shown by analysis of Lotka–Volterra predator–prey models. Whether the cycles diverge, converge or remain unaltered depends on further details of the system, but in any case the populations will persist beyond a single predator–prey cycle.

For both phytoseiid and anthocorid predators, predation rate on thrips larvae may attain a maximum approached via a decline in predation rate with increasing thrips density. The plateau of the predation rate is due to constraints on the time for searching and handling prey (type-2 Holling curve) or due to satiation (Sabelis, 1990). For arthropod predators, satiation usually causes the maximum rate of predation to be reached at much lower prey densities than predicted from constraints on the time budget (Sabelis, 1992). Hence, it is more relevant to focus on ingestion, digestion and egestion dynamics, and to ignore handling times. A general feature of arthropod predators is that ingestion either stops when the prey is nearly emptied or when the gut is filled to capacity (Sabelis, 1992). This also holds for phytoseiid mites (Sabelis, 1990, 1992) and anthocorid bugs (Rajasekhara and Chatterji, 1970; Lichtenauer and Sell, 1993). For negligible handling times, partial prey ingestion and moderate satiation levels, the relation between the predation rate (F) and prey density (D) can be approximated by the following function (Metz *et al.*, 1988; Sabelis, 1992):

$$F_D = \frac{d}{\left[\dfrac{d}{\ln(m/c)} + \sqrt{\dfrac{bc}{dD}}\right]}$$

Here m represents gut capacity, s is the satiation level or level of gut fullness, c is the level of s above which the rate constant of prey capture ($= g_s$) becomes zero ($g_c = 0$; the so-called capture threshold), d = the rate constant of gut emptying, whereas $b = -0.5\ \pi/g'_c$ and g'_c is the differential of g_s at $s = c$. For high prey densities the predation curve approaches an upper asymptote:

$$F(D \rightarrow \infty) = d/\ln(m/c).$$

Clearly, when $m = c$, there is no maximum because $\ln(1) = 0$; then, F_D reduces to:

$$F_D = \sqrt{\frac{cd}{b} D}$$

Thus, the predation rate increases with the square root of prey density. Although verbal explanations of the precise functional form are inadequate, it is instructive to see that, if the capture success rate becomes zero only at gut capacity ($c = m$), then every morsel of food emptied from the gut will induce a non-zero, positive capture success rate. Thus, increased prey density must necessarily lead to an increased predation rate (until handling time constraints become important, but such high encounter rates are usually not met in reality).

Phytoseiid predators, such as *Amblyseius cucumeris*, appear to refrain from attacking first-instar thrips larvae (about 9 μg) before they reach full satiation (c/m Ã 0.8, with m Ã 4 to 5 μg) (P.C.J. Van Rijn *et al.*, unpublished data) and the rate constant of gut emptying is about 1 per day (at 25°C). This implies that the predation curve has a maximum at about 6 first-instar thrips larvae per day (at 25°C) (Castagnoli *et al.*, 1990; Shipp and Whitfield, 1991; P.C.J. Van Rijn *et al.*, unpublished data). The predation rate is still 66% of the asymptotic value at a density of 1 thrips larva per 16 cm^2 leaf area. Such a density is clearly quite low and, therefore, the assumption of constant, near-maximum predation rate in the pancake model seems to be reasonable for phytoseiid mites.

For anthocorid predators, predation curves also tend to reach a maximum with increasing thrips density when either the predator stage is immature or the thrips has reached a stage with strong defensive or escape capacity. For example, Nagai (1991) found that predation reached a maximum in first to fifth instar nymphs of *Orius sauteri* feeding on second instar larvae of *Thrips palmi*, but not for the adult female predators. Similarly, Isenhour and Yeargan (1981b) found maxima for the nymphal stages of *Orius insidiosus* at high densities of second instar larvae of *Frankliniella occidentalis* and Coll and Ridgway (1995) did not find maxima when the predators were adult females. However, the adult predators reach maxima at high densities of adult thrips on beans (more than 50 per 3.8 cm^2), perhaps because adults are better able to escape predator attack. Adult female anthocorids probably do not have predation rates approaching a plateau at high densities of first- and second-instar thrips larvae. As shown for adult females of *Orius insidiosus*, the success rate of capturing second-instar thrips larvae becomes zero only at full satiation ($c = m$ Ã 62 μg), causing the predation rate to be proportional to the square root of prey density (with a proportionality constant equal to 17.4 at 25°C) (Van den Meiracker and Sabelis, unpublished data). Thus, for example, consider the (low) density of 1 larva per 16 cm^2. Whereas for phytoseiid mites the predation rate at this density equals 4 larvae per day (representing 66% of the maximum), anthocorids consume similar numbers per day, but can consume four to ten times more to reach values reported in the literature at high thrips densities (Rajeskhara and Chatterji, 1970; Isenhour and Yeargan, 1981b; Saucedo-Gonzalez and Reyes-Villanueva, 1987; Mituda and Calilung, 1989; Nagai, 1991; Chang *et al.*, 1993; Lichtenauer and Sell, 1993; Coll and Ridgway, 1995). Thus, anthocorids at low thrips densities consume thrips larvae at rates similar to those of phytoseiid mites, but at higher prey densities the anthocorid predators do not reach a maximum. In fact, one would expect that, if predator and prey are of similar size, the capture threshold (c) is lower than gut capacity, and that c will approach m with increasing predator–prey size or, more generally, as prey are progressively less able to defend themselves (Sabelis, 1992). As a consequence, predation functions change from curves saturating with increasing prey density, to curves of a non-saturating, 'square root' type. Moving from phytoseiid to anthocorid predators this is exactly what is observed. Phytoseiid mites show saturating predation function toward prey of equal or larger size (adult spider mites, thrips larvae), but 'square root' functions towards small and weak prey (young stages of spider mites and rust mites). Similarly, adult females of anthocorid predators show saturating predation functions towards well-defended prey (adult thrips, aphids, caterpillars) (Isenhour *et al.*, 1989; Nagai, 1991), but 'square root' functions towards the small and the weak (adult spider mites, thrips larvae) (McCaffrey and Horsburgh, 1986b; Nagai, 1991; Van den Meiracker and Sabelis, unpublished data). Hence, the assumption of constant predation rate in the pancake model holds for phytoseiid mites, but only partially for anthocorid predators, unless thrips densities within the infested part of a leaf or plant do not vary much under predator-free conditions.

These general observations on predation rates as a function of prey density suggest that at low thrips densities phytoseiid and anthocorid predators will differ little in per capita thrips consumption rates and that the degree of prey suppression will then mainly depend on the abundance of these predators in the trough of the thrips cycle and the tendency to stay in the leaf area infested.

Thrips escape in space

Little information is available on the rate of dispersal of predators during their interaction with local thrips populations. The pancake model assumes no dispersal until all prey have dispersed (the *arrestment* assumption). Given the reasonably good predictions of the time to prey extinction and the type of dynamics, there is reason think that dispersal during the interaction is strongly suppressed. This is because relatively small dispersal rates of predators would lead to large changes in time to extinction and type of dynamics.

Although anthocorids leave at low thrips densities, the threshold is higher than for phytoseiid mites. Anthocorids need more thrips per day for the onset of egg production than phytoseiid mites, so may have a stronger urge to reach new prey patches that meet their energy requirements for reproduction. Moreover, one would expect dispersal to be less risky for the anthocorids, as the adults can fly rather than disperse passively in wind currents, as do the phytoseiid mites. This would lead to a higher chance of recovery for thrips populations suppressed by anthocorids than those suppressed by phytoseiid mites. However plausible this hypothesis may be, there is as yet no evidence to support it. Resurgence of thrips populations seems to occur in populations subject to either of the two types of predator.

Thrips escape in time

An essentially different way for thrips to escape predation is to enter diapause earlier when predation risks are high. The advantage of entering diapause is that it is usually associated with search for a more hidden site for overwintering, often away from the plant (e.g. in the soil, under bark) (see Chapter 7). However, the possibility of predation risk as a diapause trigger or modifier has never been investigated.

Differential timing of diapause in predator and prey can have large consequences for thrips population dynamics, as shown by experiments on biological control of thrips in greenhouses. Various greenhouse experiments in temperate zones aiming at control of (sub)tropical thrips have shown thrips pest resurgence in autumn or winter due to induction of diapause in predators but not in the thrips (anthocorids: Fischer *et al.*, 1992; Chambers *et al.*, 1993; Chyzik and Klein, 1995; phytoseiids: Van Houten *et al.*, 1993, 1995b). Since greenhouses in the temperate zone represent ecological islands where diapause is not needed, phytophagous arthropods without diapause are at an advantage as invaders. If the natural enemy used for biological control is not selected for suppressed diapause, it may become ineffective by the time it moves to hibernation sites. Selection for non-diapause strains of the predator is then a possible solution, as shown for the predatory mites *Amblyseius cucumeris* and *Amblyseius barkeri* by Van Houten *et al.* (1995b), or selection of effective predator species without diapause may help, as shown for anthocorid predators, such as *Orius albidipennis* (Van den Meiracker, 1994) and for phytoseiid mites, such as *Amblyseius limonicus* (New Zealand strain), *Amblyseius degenerans* (Morrocan strain) and *Amblyseius hibisci* (Mexico strain) (Van Houten *et al.*, 1995a). Phytoseiid mites usually overwinter as adults in cracks and crevices on the plant, whereas anthocorids prefer dry, protected places in soil, dead wood or bark. Both types of predator enter a so-called reproductive diapause in the adult phase. The conditions for induction (and termination) of diapause have been studied for anthocorids by Kingsley and Harrington (1982; *Orius insidiosus*), Ruberson *et al.* (1991; *Orius insidiosus*), Van den Meiracker (1994; *Orius insidiosus, O. laevigatus, O. albidipennis* and *O. majusculus*), Tommasini and Nicoli (1995, 1996; *Orius laevigatus*), Wang *et al.* (1994; *Orius sauteri*) and Chyzik *et al.* (1995b; *Orius albidipennis*), and for phytoseiid mites by Morewood and Gilkeson (1991), Van Houten and Van Stratum (1993) and Van Houten *et al.* (1993, 1995b). These predators are sensitive to diapause-inducing conditions (low temperature and short day) in the nymphal stages. High temperatures usually lead to termination of diapause within a few days (Van den Meiracker, 1994; Van Houten *et al.*, 1995b).

Thrips escape in ontogeny

Whereas differential timing of diapause may explain autumn resurgence of the thrips population in several greenhouse experiments, it does not provide an explanation for persistence of predators and thrips at low densities during several months before autumn. Hence other mechanisms may operate. The most likely candidate is the existence of invulnerable stages, as these have a stabilizing effect on predator–prey models of the Lotka–Volterra type (Van den Bosch and Diekmann, 1986; Murdoch *et al.*, 1987; Abrams and Walters, 1996). Phytoseiid mites can only attack young thrips larvae successfully (Bakker and Sabelis, 1986, 1989; Kajita, 1986), whereas anthocorids and mirids also attack thrips eggs and adults, though less readily than the larvae, and thrips pupae may be inaccessable because many thrips species pupate in the soil. Predatory thrips, however, seem to be very good egg predators and when invading thrips galls they attack all stages. Thus, the invulnerable stage of thrips strongly depends on the predator under consideration. It covers most of the lifespan with respect to the phytoseiid mites, it is much shorter with respect to the anthocorids and mirids and it is almost non-existent with respect to predatory thrips-feeding gall thrips. If the invulnerable phase were the only stabilizing mechanism in the predator–prey system, one would expect stable equilibria for interactions with phytoseiid mites and violent predator–prey cycles for interactions with predatory thrips. In addition, one would expect the equilibrium of the thrips population to decrease with the relative length of the invulnerable phase (Van Rijn and Sabelis, 1997). To our knowledge these predictions have not yet been tested experimentally.

Phytophagy of the predator

In particular cases predator populations persist or even increase in virtual absence of thrips and other phytophagous arthropods. For example, Van den Meiracker and Ramakers (1991) found that populations of *Orius insidiosus* persisted on sweet pepper for almost 6 months after they virtually eliminated *Frankliniella occidentalis* from the plants (Fig. 8.3b). In a parallel experiment with extremely low thrips densities from the end of April to mid-November, the anthocorids appeared to increase in abundance in May and June, but decreased later (Fig. 8.3a). Persistence of anthocorids on pollen-bearing plants has also been observed by Tavella *et al.* (1991), Ravensberg *et al.* (1992), Shipp *et al.* (1992), Van de Veire and DeGheele (1992) and Chambers *et al.* (1993). Similar observations on persistence have been published on phytoseiid mites. For example, Van Houten and Van Stratum (1995) found increasing populations of all stages (eggs, juveniles and adult females) of *Amblyseius cucumeris* and *Amblyseius degenerans* on sweet pepper leaves over a period of about 3 months, even though thrips larvae were initially present at low levels and became rare after a few weeks. These authors also observed a much greater increase of *Amblyseius degenerans* in the flowers. These observations may be explained by an ability of anthocorid and phytoseiid predators to feed on food of plant origin. In particular, sweet pepper pollen (Van Houten *et al.*, 1995a) and pollen of various plants (e.g. rosaceous plants, castor bean, cattail, ice plant, broad bean) are known to be food of high quality for phytoseiid mites in that they can increase numerically on a diet of pollen, often just as fast as on their preferred prey (Table 8.2; Swirski and Dorzia, 1968, 1969; Swirski *et al.*, 1967a,b, 1970; Nelson, 1973; Sciarappa and Swift, 1977; Tanigoshi *et al.*, 1981, 1983; Hoda *et al.*, 1986; Van Rijn and Van Houten, 1991; Grout and Richards, 1992a; Ouyang *et al.*, 1992; Momen, 1995; Van Rijn and Tanigoshi, 1997b; P.C.J. Van Rijn, unpublished results). Indeed, addition of pollen promotes the abundance of phytoseiid mites, such as *Amblyseius hibisci* (McMurtry and Scriven, 1966, 1968; Kennett *et al.*, 1979), *Amblyseius cucumeris* and *Amblyseius degenerans* (Ramakers and Voet, 1995, 1996) and removal of pollen-producing flowers in sweet pepper crops decreases the abundance of the latter predators (Ramakers, 1990, 1995). In addition, the experimental results in Fig. 8.2 show that three, weekly supplies of about 3 g of cattail pollen to 60 pollen-free (parthenocarpic) cucumber plants promote the abundance of juvenile and adult phytoseiid mites (*Amblyseius degenerans*) and this contributes significantly to the control of *Frankliniella occidentalis* (P.C.J. Van Rijn, Y.M. Van Houten and A.M.M. Van Lier, unpublished data). Presumably, pollen feeding helps the predatory mites to become adult and thereby reach a stage where they are better able to attack thrips larvae. Possibly, other plant foods play a similar role in promoting survival and reproduction of phytoseiid mites, such as extrafloral nectar (Van Rijn and Tanigoshi, 1997a) and plant exudates (Bakker and Klein, 1992). Feeding on the leaf parenchyma has been suggested for phytoseiid mites, but has never been proven conclusively.

Phytophagy is also common among the Anthocoridae, although it depends on the type of plant, the type of food and the predator species (Naranjo and Gibson, 1995). Feeding on plant juices, observed as they puncture leaves, increases survival and in some cases even development (Askari and Stern, 1972; Fauvel, 1974; Salas-Aguilar and Ehler, 1977; Kiman and Yeargan, 1985; Coll, 1996). Another clear indication of a beneficial impact of plant feeding on the life history of anthocorids is observed on maize. Adults of *Orius insidiosus* feed on maize silks and the nymphs can reach adulthood on this food, but the emerging adults are smaller and less vigorous. Newly hatched nymphs start feeding on maize silks, but as development progresses they prefer thrips larvae as food. Extrafloral nectar of, for example, maize plants is known to attract various predators including anthocorids (Yokoyama, 1978b; Pemberton and Vandenberg, 1993) and the removal of these nectaries leads to a reduction in the abundance of predators (Schuster *et al.*, 1976). Many anthocorids feed on pollen; some apparently almost exclusively (*Orius pallidicornis*), others can complete development and oviposit on a diet of pollen, but also feed on prey (*Orius* spp., Fauvel, 1974; *Orius insidiosus*, Dicke and Jarvis, 1962; Salas-Aguilar and Ehler, 1977; Kiman and Yeargan, 1985; McCaffrey and Horsburgh, 1986a; Richards and Schmidt, 1996; *Orius sauteri*, Zhou and Wang, 1989; Funao and Yoshiyasu, 1995; Yano, 1996; *Orius laevigatus*, Frescata and Mexia, 1995). Dissevelt *et al.* (1995) observed a pronounced difference in population development of various *Orius* spp. in pollen-bearing crops (strawberry, aubergine, sweet pepper and melon), compared with crops without pollen (cucumber). Generally, anthocorids become more abundant in periods of increased pollen availability (Dicke and Jarvis, 1962; Isenhour and Marston, 1981; Mituda and Calilung, 1989; Coll and Bottrell, 1995).

Like the phytoseiids and anthocorids, mirid predators also feed on pollen, but they prefer plant tissues (Fauvel *et al.*, 1987; Salamero *et al.*, 1987; Gabarra *et al.*, 1988; Braman and Beshear, 1994; Naranjo and Gibson, 1995). Plant feeding is also reported for some lygaeid and nabiid predators (Ridgway and Jones, 1968). Much less is known of phytophagy in thripophagous thrips. However, some species of predatory thrips known to feed on mites also develop on a pollen diet. Examples are larvae of *Haplothrips victoriensis* that feed on lucerne pollen (Bailey and Caon, 1986) and *Leptothrips mali* that feeds on apple pollen (Parrella *et al.*, 1982). The latter species also feeds on thrips. The only other reports of pollen feeding by a thripophagous thrips are by Bournier *et al.* (1978, 1979). They found that *Aeolothrips intermedius* required 21 days to mature on a diet of lucerne pollen, as opposed to 10 days when fed with *Thrips tabaci* larvae. They also observed that adults of *Aeolothrips intermedius* feed on flowers (probably also pollen) but not on leaves (see Chapter 4).

The ability to feed on plant foods enhances the survival of predatory arthropods and, especially in the case of pollen, even promotes reproduction. Van Rijn and Sabelis (1997) showed that the equilibrium density of thrips in the vulnerable stage decreases with the addition of pollen, because the prey equilibrium is dominated by parameters determining predator population growth. For much the same reason, this impact of pollen feeding of the predator on the prey equilibrium remains unaltered when thrips feed on pollen and promote their reproduction too. This counterintuitive result arises because the extra population growth of the thrips is simply converted into predator population growth at equilibrium. The principles underlying these results have been pointed out by Holt (1977), who showed that the addition of a new prey species to a one-predator-one-prey system leads to a reduction of the resident prey through their joint impact on the numerical response of the predator. Thus, the two prey reduce each other's equilibrial abundance, even when they do not compete at all. This result is known as the 'apparent competition' hypothesis. If the prey species added to the system is not harmful to the plant, the plant may profit from the impact of the additional prey on the ravenous herbivore via the increase of the predator population. In a sense pollen represents the ideal non-harmful 'prey'. Hence, by modifying the edibility of pollen, plants may influence the impact of the third trophic level to their own benefit, even when the herbivore feeds on pollen too (Van Rijn, 1992; Van Rijn and Sabelis, 1990, 1993, 1997). This mechanism may shed new light on the evolution of pollen edibility.

Polyphagy of the predator

Another factor promoting persistence is the presence of other prey. This may lead to sigmoid functional responses to thrips density. Predation on thrips may become disproportionally lower when thrips densities decrease thereby motivating predators to concentrate foraging on alternative prey. In this way thrips may be relieved of predation pressure and their population may recover until their colonies become profitable again for the predators. This mechanism may stabilize the equilibrium (Murdoch and Oaten, 1975) albeit in a rather small part of the parameter domain (Van Baalen *et al.*, 1997), or it may limit oscillations, thereby creating stable cycles (Van Baalen *et al.*, 1997). This may explain the persistence observed in those experiments where alternative prey are actually present.

Most of the predators reported to attack phytophagous thrips also forage for other prey. The nabids, lygaeids and mirids have been reported to feed on various other phytophagous arthropods, such as caterpillars, seed bugs, aphids, whiteflies, spider mites (Lattin, 1989; Foglar *et al.* 1990; Riudavets, 1995), the anthocorids on small caterpillars and eggs of Lepidoptera, scale insects, leafsuckers, aphids, spider mites and rust mites (Anderson, 1962; Askari and Stern, 1972; Péricart, 1972; McCaffrey and Horsburgh, 1986b; Minks and Harrewijn, 1988; Alauzet *et al.*, 1990; Nagai, 1991; Nakata, 1994; Riudavets, 1995), the predatory thrips on coccids, mealybugs, whiteflies, spider mites and rust mites (Muma, 1955; Selhime *et al.*, 1962; Putman, 1965; Debreneva, 1967; Ananthakrishnan, 1974, 1993; Coville and Allen, 1976; Gilstrap and Oatman, 1976; Bournier *et al.*, 1979; Parrella *et al.*, 1982; Bailey and Caon, 1986; Gerlach and Şengonca, 1985, 1986; Selhorst *et al.*, 1991; Gilstrap, 1995; Zur Strassen, 1995; Lindquist *et al.*, 1996), and predatory mites on eggs of different insects, eggs and crawlers of whiteflies, mealybugs and coccids, spider mites, tarsonemid mites, rust and gall mites (Helle and Sabelis, 1985; Lindquist *et al.*, 1996). How these predators choose between all these prey items is largely unknown, let alone whether the existing prey-choice mechanisms can generate sigmoid functional responses with respect to the density of phytophagous thrips. Therefore, it is as yet impossible to predict the consequences of alternative prey for the persistence of thripophagous predators, as well as for phytophagous thrips.

Complex Interactions

Polyphagy with respect to various phytophagous arthropods is not the only reason why interactions between thripophagous predators and phytophagous thrips should not be studied in isolation. The web of interactions is much more complicated than that. The question arises of how much of what is learned from studying two-species systems in simplified environments (laboratory, greenhouse) still applies under natural conditions, and which of the complexities should be taken into account. The need to understand these aspects does not stem solely from a fundamental interest in mechanisms underlying population dynamics, nor solely from applied interest in managing thrips populations under natural and semi-natural conditions, such as in forests, orchards and field crops. Progress in the application of biological control in greenhouses has led to quite complex systems involving many different pests, each with their own and shared natural enemies (see Chapter 18). Moreover, there is a growing awareness of the role host plants play in influencing the effectiveness of natural enemies of thrips. Some of these potentially important interactions are briefly discussed below.

Plant–predator interactions

Although it is tempting to hypothesize that plants tend to promote the effectiveness of the natural enemies, any attempt to evaluate benefits and costs by a comprehensive ecological analysis is fraught with problems. For example, plants may provide extrafloral nectar and edible pollen, thereby increasing the predator population and possibly decreasing phytophagous arthropods. However, this is not necessarily so, as it depends on the prey preferences for other plant-inhabiting arthropods, such as fungivores or pollenophages. Clearly, plants cannot fully control which arthropod is reaping the benefits of plant-provided foods. A particularly interesting example is that of herbivorous thrips utilizing pollen and promoting their own rate of population increase. Depending on the relative impact of plant food on predator compared with thrips population growth and the rate of predation on thrips, time to prey suppression may become shorter or longer and thrips may or may not escape control by predators (Van Rijn and Sabelis, 1997; P.C.J. Van Rijn, unpublished results).

Plants may also provide shelter. A spectacular example are the 'little mite houses' or acarodomatia that are located where leaf veins branch. Clearly, these structures provide protection to any arthropod small enough to enter and it seems reasonable to suppose that the benefits of protection for phytophagous arthropods are outweighed by the costs in terms of increased predation risk due to increased probability of encounter within the protective plant structures. Yet again, the plant cannot control which arthropods make good use of the domatia and which cannot. A striking example is *Domatiathrips cunninghamii*, a species of thrips so minute (about 1 mm) that it is close to the size of a mite, and hence, can invade acarodomatia on the leaves of an understorey tree in Costa Rica, *Psychotria gracilifora* (Mound, 1993). In fact, specimens were rarely found outside the domatia, and if so they were adults. Less than 10 per cent of the domatia were occupied by usually one thrips in any of the possible life stages (egg to adult). The thrips–domatia association is therefore certainly not a casual one. Mound (1993) rightly challenges hypotheses on mutual benefits for domatia inhabitors and plants, because the thrips clearly has no benefit for the plant. However, one should not forget that most, if not all, mutualistic interactions are open to misuse. Also, he did not report on predators inhabiting the domatia and how they affect the domatia-inhabiting thrips. In fact, the evidence for phytophagous arthropods inhabiting domatia is strikingly scarce, even though gall mites and small (gall) thrips could perhaps make good use of these plant-produced structures. To what extent the small predators of thrips and especially the predatory mites listed in Table 8.1e make use of domatia is almost unknown. Thus, the relevance of domatia for interactions between thrips and predatory mites is unclear.

Whether phytophagous thrips, like spider mites, induce the production and release of volatiles by their host plant, has not yet been demonstrated. Volatiles emanating from the anal discharge of thrips (see Chapter 3) are a serious alternative, although these may contain plant-produced chemicals as well. Responses of predatory mites to odours coming from plants infested by thrips have been noted (Dicke and Groeneveld, 1986). Similarly, some anthocorids respond to odours from plants infested by spider mites (Dwumfour, 1992) and psyllids (Drukker *et al.*, 1995) and in the latter case there is evidence for the involvement of plant-produced monoterpenes and methyl-salicylate eliciting the predator-searching response (Scutareanu *et al.*, 1996). As argued for each route to plant–predator mutualism, plants cannot control which arthropods respond to the release of volatiles upon herbivore attack. Attraction or avoidance of conspecific and heterospecific competitors are among the possibilities (Pallini *et al.*, 1997) and this may affect the benefit–cost balance of the signal-emitting plant.

Although herbivory-induced synomone production seems generally a sensible strategy of the plant to lure predators when they are most needed, there are several reports on predators showing differential arrestment to host plant species and even varieties without an obvious relation to the amount of prey on these plants. Examples are found in Braman and Beshear (1994) for mirid bugs on various species of azalea, in Coll and Ridgway (1995) for anthocorids on maize, lima bean, pepper and tomato crops and in Beekman *et al.* (1991) for the same predators on rose and chrysanthemum in the greenhouse. It is worth considering these examples in more detail.

Beekman *et al.* (1991) found that *Orius insidiosus* spend twice as much time searching on chrysanthemum than on rose and the presence of thrips on either host plant had no effect. However, they reared the predators on other prey (moth eggs) and starved them for 1 day on a rose leaf disc. The predators may therefore have lacked the opportunity to associate plant odours with the presence of prey and they may have associated rose odour with the absence of thrips. These hypotheses on associative learning need further testing as they may well explain the results of Beekman *et al.* (1991) and more generally differential arrestment to host plant species and varieties.

Coll and Ridgway (1995) found that *Orius insidiosus* showed poor functional and numerical response to the density of various pests on tomato, whereas these responses were pronounced on lima bean, pepper and maize. They suggested that the foraging behaviour of the predators was hampered by the glandular hairs on leaves and stems of tomato, and they mentioned high mortality due to collecting the viscous glandular secretions on the predator's legs. However, when given a choice of leaves without prey, *Orius insidiosus* deposited an equal proportion of its eggs on foliage of tomato, pepper and bean, and only very few eggs on maize leaves (Coll, 1996). As these predators were reared on moth eggs, Coll's results may be due to a lack of experience with the host plants. However, he also showed a correspondence between choice and reproductive success on prey-free leaves of these plants, measured as survival of the first instar nymphs, adult longevity and fecundity. Bean leaves were by far the best food source in all respects, tomato leaves offered relatively good prospects for juvenile survival and are marginal food sources for reproduction, and pepper leaves were marginal food sources for survival and fecundity, whereas maize leaves were worst as they do not allow any oviposition. These results seem to contradict the results on predator impact on these crops, since pepper and corn are known to be conducive to biological control and tomato is not. This paradox may well be resolved when realizing that *Orius insidiosus* concentrates foraging on other parts of the plant in preference to the leaves. It frequents the flowers of sweet pepper and it is strongly attracted to the maize silks where it readily oviposits (Reid and Lampman, 1989). Possibly as a consequence of glandular hairs, tomato leaves represent enemy-poor space and are therefore suitable substrates for egg deposition despite the negative impact of tomato hair secretions on foraging behaviour. Females embed their eggs singly in plant tissue with only the opercula protruding. Stems, pods, leaf petioles and veins are preferred oviposition substrates (Rajasekhara and Chatterji, 1970; Isenhour and Yeargan, 1982; Zhou *et al.*, 1991; Chambers and Long, 1993; Van den Meiracker and Sabelis, 1993; Jacobson, 1995; Richards and Schmidt, 1996). On tomato, these plant parts are densely covered with glandular hairs which may help to protect the *Orius* eggs from their natural enemies. Indeed, Ferguson and Schmidt (1996) found quite high rates of egg deposition and egg hatch in petioles of pepper and tomato (1.9–2.7 eggs female^{-1} day^{-1}; 71 to 81 per cent egg hatch), but petioles of cucumber may be even more suitable (3.2–3.4 eggs female^{-1} day^{-1}; 85 per cent egg hatch).

Plants with a dense cover of (glandular) hairs as a direct defence against herbivores unavoidably interfere with the effectiveness of natural enemies of the herbivores. Apart from the examples with respect to anthocorid predators, there are also many examples with respect to phytoseiid mites, such as higher predation rates of *Amblyseius cucumeris* on glabrous sweet pepper leaves than on hairy cucumber leaves (Shipp and Whitfield, 1991). However, the important point emerging is that this mode of defence may have positive and negative effects on the third trophic level. On the one hand it may decrease the availability of herbivorous arthropods to the predators, hinder predator foraging and even cause mortality among the predators, but on the other hand it may provide enemy-poor space and thereby represent a suitable site for egg deposition and juvenile development. In addition, there may be another advantage to the predators, as they can feed on the insects and mites caught in the secretions of the glandular hairs. Such observations on scavenging have been reported by Braman and Beshear (1994) for mirid bugs on Florida azaleas that were repeatedly seen feeding on small insects trapped in sticky glandular secretions on stems and leaves. Direct defences may render other benefits to the predators. For example, retarded growth of the herbivore in resistant species may retard juvenile development of the thrips and thereby increase the length of the developmental stage that is most vulnerable and exposed to predation (Isenhour *et al.*, 1989). All these positive and negative effects of direct defences may cancel out or create a positive or negative balance for the natural enemies, thereby determining selection for attraction or avoidance of well-defended plants.

When predators find more suitable oviposition sites and refuges from their natural enemies on other plants, as well as additional foods for survival (but less preferred than thrips), then the addition of these plants will lead to lower equilibrium herbivore densities and higher predator densities on the target plant species. This is because any factor promoting predator reproduction and survival will lead to higher equilibrium predator densities and lower herbivore densities. Letourneau and Altieri (1983), Letourneau (1990) and Coll and Bottrell (1995) indeed show a trend toward higher densities of *Orius* spp. and lower densities of thrips on target plants in polycultures than in monocultures, but their observations have been made at least partly under non-equilibrium conditions so they do not represent a suitable test of the hypothesis. Intuitively, one would expect positive as well as negative effects of plant diversity on predator and herbivore levels on target plants.

Intraguild and intraspecific predation

Given the intraspecific (e.g. between stage) and interspecific size differences among natural enemies of thrips it is not surprising that within-guild and within-species predation occurs. The larger feed on the smaller and the stronger feed on the weaker. Nabids feed on lygaeids, mirids and anthocorids (Lattin, 1989; Rosenheim *et al.*, 1993, 1995; Riudavets, 1995). Anthocorids in the genus *Orius* feed on predatory thrips (El-Serwiy *et al.*, 1985) and on phytoseiid mites, such as *Phytoseiulus persimilis* (Cloutier and Johnson, 1993b), *Amblyseius cucumeris* (Gillespie and Quiring, 1992), *Amblyseius cucumeris* and, albeit to a lesser extent, *Amblyseius degenerans* (Wittmann and Leather, 1997). Phytoseiid mites feed on young, immobile or otherwise defenceless stages of predatory thrips (Sciarappa and Swift, 1977). This list of examples nicely illustrates the great potential for intraguild predation and the same applies to cannibalism, which occurs in most, if not all, species of thrips predators (Polis, 1981; R.A.F. Van den Meiracker, unpublished results).

However, most observations on intraguild predation stem from laboratory experiments in small arenas (Cloutier and Johnson, 1993b) or from small enclosures in the field (e.g. Rosenheim *et al.*, 1993); under these conditions predators cannot do better than to feed on one another or to share the same prey. Under field conditions intraguild predation may be less intense for at least four reasons. First, increased density of the target prey may lessen predation among predators (Gillespie and Quiring, 1992; Cloutier and Johnson, 1993). Second, the plant may provide refuges, such as domatia, that help predators to avoid hyperpredators. Third, the target prey may create refuges that help predators small enough to enter these refuges to escape from larger predators. An example of such refuges might be those thrips-induced galls that can be invaded by predatory thrips and phytoseiid mites, but not by nabids, mirids and anthocorids. Fourth, predators may avoid each other by feeding on different prey or hiding in the refuges of the alternative prey. For example, *Phytoseiulus persimilis* suffers less predation from *Orius tristicolor* when foraging in the silken web of its preferred prey, the two-spotted spider mite *Tetranychus urticae* (Cloutier and Johnson, 1993). This chaotic web structure has a negative effect on the predatory activity of *Orius* (Cloutier and Johnson, 1993), as it has on various species of phytoseiid mites (e.g. *Amblyseius cucumeris*, *Amblyseius barkeri* and *Amblyseius degenerans*) due to their difficulties in penetrating the web (Sabelis and Bakker, 1992).

Predatory arthropods may be able to perceive direct or indirect (signalled) information on prey food quality, prey defence, refuges from competitors and hyperpredators. By integrating this information they can perhaps determine where to go and on which prey to feed. Such decisions may be taken at some distance from the resource on detection of volatile chemicals signalling the 'state' of the resource. A particularly interesting example is provided by the experiments of Janssen *et al.* (1997) with *Phytoseiulus persimilis*. This predatory mite avoids odours from spider-mite infested leaves occupied by conspecific predators when the alternative is spider-mite infested leaves without these competitors. Janssen *et al.* (1997) suspect that the avoidance is elicited by alarm pheromones produced by the spider mites in response to contact with predators. Such avoidance mechanisms may well exist within predator guilds and perhaps free the thrips predators from the impact of intraguild predation. If avoidance of intraguild predation is the rule rather than the exception, then predators sharing the same prey will have an additive effect on the suppression of the thrips population and the position of the thrips equilibrium will be determined by the relevant traits of each of these predators. If the opportunities for avoidance are lacking, thereby forcing predators to feed on the same prey, then intraguild predation may occur, with important dynamic consequences (Polis *et al.*, 1989; Polis and Holt, 1992). First, intraguild predation may lead to exclusion of the predator with inferior competitive ability or to coexistence of the two predators in the special case where the inferior competitor feeds on the predator superior in competing for thrips. Second, coexistence of the two predators leads to an increase of the thrips equilibrium level, because intraguild predation declines at the expense of the superior competitor (i.e. the predator capable of suppressing the thrips to the lowest level). Third, such systems exhibiting intraguild predation are particularly prone to switch between a steady state with one predator species and one prey species and another steady state with both species of predators and the prey. Such switches between alternative stable states occur as a result of perturbations in numbers alone (e.g. due to non-persistent pesticides with a knock-down effect), and thus not from changes in predator or prey traits.

Clearly, there is every reason to pay more attention to the role of intraguild predation in the dynamics of the predator community affecting thrips.

Population dynamics of two or more predators sharing thrips as prey have been described only in a few cases. Ramakers (1988) observed that simultaneous releases of *Amblyseius cucumeris* and *Amblyseius barkeri* for control of *Thrips tabaci* in sweet pepper crops in greenhouses led to population increase and establishment of the former and very low but persistent populations of the latter. To explain the large difference in abundance he suggested that *A. cucumeris* was the better competitor, but to date there are no published data substantiating whether this is due to exploitative competition or intraguild interference. Brødsgaard and Hansen (1992) also found that *Amblyseius cucumeris* had a stronger numerical response to the density of *Thrips tabaci* than *Amblyseius barkeri*, but in the long run only the latter predator was found in plots of parthenocarpic cucumber plants where both species were released. This may indicate competitive displacement, but there may be alternative explanations, such as differential ability to survive on alternative foods (not pollen, as it is absent in parthenocarpic cucumber) or differential emigration. Ramakers (1993) found little difference in dynamics of *Orius insidiosus* between plots with and without *Amblyseius cucumeris* as a competitor for *Frankliniella occidentalis* as prey. He suggested that the pirate bugs neither suffer nor gain from the presence of this predatory mite. However, the population of *A. cucumeris* persisted in the presence of pirate bugs at low levels and greatly increased in the absence of pirate bugs. Ramakers (1993) suggested that this was mainly due to the increased supply of *F. occidentalis*, but intraguild predation by the pirate bugs cannot be excluded as a cause. Brødsgaard and Enkegaard (1995) performed a series of replicated population experiments in small cabinets with *F. occidentalis* and two-spotted spider mites as herbivores on *Gerbera* plants, and *Orius majusculus* and *Phytoseiulus persimilis* as predators; control of two-spotted spider mites by *P. persimilis* was delayed in the presence of *O. majusculus*. Brødsgaard and Enkegaard (1995) suggest that this is due to the pirate bugs feeding on the predatory mites, in agreement with the laboratory observations of Cloutier and Johnson (1993). Most interestingly, the control of two-spotted spider mites was improved by the addition of *F. occidentalis*, perhaps due to preferential feeding of the pirate bug on thrips and/or to the thrips acting as yet another predator of the two-spotted spider mites (Trichilo and Leigh, 1986).

Indeed, several phytophagous thrips are known to eat spider mites and other phytophagous arthropods. For example, *Frankliniella tritici* feeds on eggs of the alfalfa weevil inside alfalfa stems (Barney *et al.*, 1979) and *Thrips imaginis*, *Thrips tabaci* and *Frankliniella schultzei* feed on eggs of spider mites in cotton fields in Australia (Wilson *et al.*, 1996). *Frankliniella occidentalis* is even considered to be beneficial on cotton (González *et al.*, 1995) because it is a relatively effective natural enemy of tetranychid mites (González and Wilson, 1982; Trichilo and Leigh, 1986; Pickett *et al.*, 1988; Wilson *et al.*, 1991), and because it represents an alternative prey for predators such as *Geocoris pallens*, *G. punctipes* and *Orius tristicolor* (González and Wilson, 1982; González *et al.*, 1995). In agreement with a role of *Frankliniella occidentalis* as a predator of spider mites, Pallini *et al.* (1997) found that two-spotted spider mites avoid the odour emanating from cucumber plants infested by this thrips when the alternative is odour from uninfested cucumber plants. Although these results might just as well indicate a preference for uninfested plants, they are more likely to be explained by avoidance, because two-spotted spider mites have a slight but significant preference for odour from plants damaged to a similar degree by conspecific mites.

Concluding Remarks

The main aim of this chapter on predators of thrips is to detect the most salient traits of predators and prey and to assess whether these traits suffice to understand predator–prey dynamics. We chose to limit our question by focusing on a local spatial scale where interactions between predator and prey populations are strongly coupled. In doing so, we ignored the spatial scale where dynamics at different locations are loosely coupled by long-range dispersal. It is well known that such metapopulation interactions can render the overall predator–prey system persistent, even when local dynamics are unstable (e.g. Sabelis *et al.*, 1991).

Our review shows that information on life history and predation under conditions of ample prey supply give a reasonably clear picture of the capacities of predators to suppress local populations of thrips. To improve the precision of testing these models, more complete information is required on life tables of predators assessed with thrips as prey. Much of the information provided in the literature is only partial (few life history components are measured) and/or obtained on a diet of easy-to-rear prey, other than thrips. This accounts for the disappointing decrease of information going from Fig. 8.1a–d on life history and predation to Fig. 8.1e on model predictions of the time to prey suppression. Moreover, the models can only be subject to critical testing when the initial predator–prey ratios are precisely known. Not only should we know the exact numbers of predators released per plant or per leaf, but also which fraction actually settles on the thrips-infested plants. Moreover, it is necessary to measure the initial thrips density in exactly the same units as the initial density of predators, otherwise predator–prey ratios cannot be assessed.

A very simple predator–prey model extrapolating the dynamic consequences of maximal capacities of predation and population increase proved to be fairly successful in predicting local predator–thrips dynamics. Such an agreement can only be understood if the predators were strongly arrested by the thrips-infested parts of plants and if they foraged so as to keep their predation rate at a maximal level. This suggests that these predators are good at finding thrips and are close to being optimal foragers! Unfortunately, there is virtually no research on the behavioural mechanisms underlying arrestment and foraging. Clearly, such research is essential as it represents an independent test of the above suggestions on optimal foraging and strong arrestment.

Another salient outcome of testing the simple predator–prey model is that, contrary to predictions, the thrips populations are not eradicated but settle at a low equilibrium level after one or more initial cycles. Thus, the local dynamics of predator and prey are not unstable and one may wonder what the stabilizing mechanisms are. Probably, the presence of invulnerable thrips stages and prey refuges plays a major role in stabilizing the interaction, but there may be other mechanisms too, such as the presence of alternative food or prey.

Many of these salient features can be at least partially controlled by the plant. The plant may create refuges (domatia), provide alternative food (pollen, nectar), tolerate less harmful herbivores as alternative prey for predators, and perhaps also release 'alarm calls' to signal the presence of thrips as prey to their predators. In this way plants may facilitate arrestment of predators on thrips-infested plants, they may enable the predators to forage optimally and perhaps provide the conditions for local predator populations to persist. These possible plant–predator interactions have hardly been investigated for predators of thrips and warrant further study. It should be realized also that each of these plant-provided facilities is open to misuse by others, as in any mutualistic interaction.

While food web complexities have been kept to a minimum in the analysis presented, they should not be ignored. There are sound theoretical reasons why systems, characterized by two or more predators feeding on the same prey and on each other, give rise to complex dynamic responses such as a switch to an alternative steady state after perturbation of population size. However, we favour the contention that intraguild predation exists, but that it selects for avoidance responses among predators. Many of the experiments demonstrating intraguild predation did not give an opportunity for such avoidance responses. In other words, there is a need to demonstrate not only how predators balance the benefits of acquiring profitable prey and the risk of being eaten themselves, but also whether intraguild predation occurs often enough to influence simple interactions between one predator and one prey.

Addendum 1: Procedure for Estimating r_m

Assume the non-zero part of the age-dependent reproduction curve can be described by a trapezoid with the position of the corners along the age-axis indicated by:

A = age at first egg deposition,
T_1 = age at start of maximal oviposition,
T_2 = age at end of maximal oviposition, and
W = age of last egg deposition.

Data on these positions were obtained from graphs on age-dependent oviposition (R_{age}) and then used to estimate r_m (labelled by a in Table 8.2). The intrinsic rate of increase was obtained from the Lotka equation which, for this special case of trapezoid reproduction curve, reduces to:

$$R(T)\frac{\left\{\frac{E(A,T_1)}{T_1 - A}+\frac{E(W,T_2)}{W - T_2}\right\}}{r_m^2}+\frac{E(T_1,T_2)}{r_m}=1$$

$$E(X, Y) = \exp(-r_m X) - \exp(-r_m Y)$$

where $R(T) = R(T_1) = R(T_2)$ is the maximum rate of net reproduction.
In case of missing life-history data (see Table 8.2) the following assumptions were made to calculate r_m (labelled by b in Table 8.2):

1. Pre-ovipositional period equals 20% of the juvenile period (egg–adult).
2. Age at first egg deposition (A) equals the mean egg-to-egg developmental time.
3. Sex ratio equals 66% for phytoseiid mites and predatory thrips (overall mean), 55% for anthocorids and 50% for all other species.
4. The position of the four corners of the trapezoid reproduction curve were calculated from the mean oviposition period (P), i.e. for mites and thrips as $T_1 - A = 0.15 \times P$, $T_2 - A = 0.7 \times P$ and $W-A = 1.3 \times P$, and for other insects as $T_1-A = 0.25 \times P$, $T_2-A = 0.75 \times P$ and $W-A = 1.25 \times P$.
5. The height of the trapezoid (maximal oviposition rate) was chosen in such a way that the total fecundity (the surface of the trapezoid) corresponds to the measured value.
6. Juvenile survival was set equal to 80%.

References

Abrams, P.A. and Walters, C.J. (1996) Invulnerable prey and the paradox of enrichment. *Ecology* 77, 1125–1133.

Afifi, A.M., Fraghali, H.T., Rezk, G.N. and Ragab, Z.A. (1976) Seasonal abundance of certain pests and predators in clover fields in Egypt. *Bulletin de Société Entomologique d'Egypte* 60, 273–278.

Alauzet, C., Bouyjou, B., Dargagnon, D. and Hatte, M. (1990) Mise au point d'un élevage de masse d'*Orius majusculus* Reuter (Heteroptera: Anthocoridae). *International Organisation of Biological Control/West Palaearctic Regional Section Bulletin* 13, 118–122.

Alauzet, C., Dargagnon, D. and Hatte, M. (1992) Production d'un Hétéroptère prédateur: *Orius majusculus* (Het.: Anthocoridae). *Entomophaga* 37, 249–252.

Alauzet, C., Dargagnon, D. and Malausa, J.C. (1994) Bionomics of a polyphagous predator *Orius laevigatus* (Het.: Anthocoridae). *Entomophaga* 39, 33–40.

Albajes, R., Alomar, O., Riudavets, J., Castane, C., Arno, J. and Gabarra, R. (1996) The mirid bug *Dicyphus tamaninii*: an effective predator for vegetable crops. *International Organisation of Biological Control/West Palaearctic Regional Section Bulletin* 19, 1–4.

Alomar, O., Castané, C., Gabarra, J., Arino, J. and Albajes, R. (1991) Conservation of native mirid bugs for biological control in protected outdoor tomato crops. *International Organisation of Biological Control/West Palaearctic Regional Section Bulletin* 14, 33–42.

Alomar, O., Goula, M. and Albajes, R. (1994) Mirid bugs for biological control: identification, survey in non-cultivated winter plants, and colonization of tomato fields. *International Organisation of Biological Control/West Palaearctic Regional Section Bulletin* 17, 217–223.

Amano, H. and Chant, D.A. (1978) Some factors affecting reproduction and sex ratios in the two species of predatory mites, *Phytoseiulus persimilis* Athias-Henriot and *Amblyseius andersoni* (Chant) (Acarina: Phytoseiidae). *Canadian Journal of Zoology* 56, 1593–1607.

Ananthakrishnan, T.N. (1969) *Indian Thysanoptera*. Council for Scientific and Industrial Research, New Monograph No 1. New Delhi.

Ananthakrishnan, T.N. (1973) *Thrips Biology and Control*. Macmillan, India.

Ananthakrishnan, T.N. (1974) *Aleurodothrips fasciapennis* (Franklin) predatory on coccids and aleyrodids. *Journal of the Bombay National History Society* 71, 157–160.

Ananthakrishnan, T.N. (1979) Biosystematics of Thysanoptera. *Annual Review of Entomology* 24, 159–183.

Ananthakrishnan, T.N. (1984) *Bioecology of Thrips*. Indira Publishing House, Michigan, USA, pp. 81–100.

Ananthakrishnan, T.N. (1993) Bionomics of thrips. *Annual Review of Entomology* 38, 71–92.

Ananthakrishnan, T.N. and Sureshkumar, N. (1985) Anthocorids (Anthocoridae: Heteroptera) as efficient biocontrol agents of thrips (Thysanoptera: Insecta). *Current Science (Bangalore)* 54, 987–990.

Ananthakrishnan, T.N. and Vandarasan, S. (1977) *Androthrips flavipes* Schmutz (Insecta: Thysanoptera), a predatory inquiline in thrips galls. *Entomon* 2, 105–107.

Anderson, N.H. (1962) Growth and fecundity of *Anthocoris* spp. reared on various prey (Heteroptera, Anthocoridae). *Entomologia Experimentalis et Applicata* 5, 40–52.

Andrewartha, H.V. (1936) Thrips investigation 8. The influence of temperature on the rate of development of immature stages of *Thrips imaginis* Bagnall and *Haplothrips victoriensis* Bagnall. *Journal of the Council for Scientific and Industrial Research, Australia* 9, 57–64.

Askari, A. and Stern, V.M. (1972) Biology and feeding habits of *Orius tristicolor* (Hemiptera: Athocoridae). *Annals of the Entomological Society of America* 65, 96–100.

Bailey, P. and Caon, G. (1986) Predation on twospotted spider mite, *Tetranychus urticae* Koch (Acarina: Tetranychidae) by *Haplothrips victoriensis* Bagnall (Thysanoptera: Phlaeothripidae) and *Stethorus nigriceps* Kapur (Coleoptera: Coccinellidae) on seed lucerne crops in South Australia. *Australian Journal of Zoology* 34, 515–525.

Bailey, S.F. (1933) The biology of the bean thrips. *Hilgardia* 7, 467–522.

Bailey, S.F. (1940) The black hunter *Leptothrips mali* (Fitch). *Journal of Economic Entomology* 33, 539–544.

Baker, J.R. (1988) Biology of the western flower thrips. *Proceedings of the IV Conference on Insect Disease Management on Ornamentals, Kansas City*, pp. 79–83.

Bakker, F.M. and Klein, M.E. (1992) Transtrophic interactions in cassava. *Experimental and Applied Acarology* 14, 293–311.

Bakker, F.M (1993) Selecting phytoseiid predators for biological control, with the emphasis on the significance of tri-trophic interactions. PhD thesis, University of Amsterdam.

Bakker, F.M. and Sabelis, M.W. (1986) Attack success of *Amblyseius mckenziei* and the stage related defensive capacity of thrips larvae. *Mededelingen Faculteit der Landbouwwetenschappen, Universiteit Gent* 51, 1041–1044.

Bakker, F.M. and Sabelis, M.W. (1989) How larvae of *Thrips tabaci* reduce attack success of phytoseiid predators. *Entomologia Experimentalis et Applicata* 50, 47–51.

Barnes, H.F. (1930) A new thrips-eating gall midge, *Thripsobremia liothrips. Bulletin of Entomological Research* 21, 331–332.

Barney, R.J., Roberts, S.J., Pausch, R.D. and Armrust, E.J. (1979) Impact of prey age and temperature on predation by the eastern flower thrips, *Frankliniella tritici*, on eggs of the alfalfa weevil. *Environmental Entomology* 8, 814–815.

Beekman, M., Fransen, J.J., Oetting, R.D. and Sabelis, M.W. (1991) Differential arrestment of the minute pirate bug, *Orius insidiosus* (Say) (Hemiptera: Anthocoridae), on two plant species. *Mededelingen van de Faculteit der Landbouwwetenschappen, Rijksuniversiteit Gent* 56, 273–276.

Beerling, E.A.M. and Van der Geest, L.P.S. (1991) A microsporidiosis (Microspora: Pleistophoridae) in mass-rearings of the predatory mites *Amblyseius cucumeris* and *A. barkeri* (Acarina: Phytoseiidae): analysis of a problem. *International Organisation of Biological Control/West Palaearctic Regional Section Bulletin* 14, 5–8.

Beglyarov, G.A. and Suchalkin, F.A. (1983) A predaceous mite – a potential natural enemy of the tobacco thrips. *Zashcita Rastenii* 9, 24–25.

Benedict, J.H. and Cothran, W.R. (1980) Damsel bugs useful as predators but need help. *California Agriculture* 34, 11–12.

Bennett, F.D. (1965) Observations of the natural enemies of *Gynaikothrips ficorum* Marchal, in Brazil. *Technology Bulletin CIBC* 5, 117–125.

Bennison, J.A. (1988) Integrated control of thrips on cucumber in the UK. *Mededelingen Faculteit Landbouwwetenschappen, Universiteit Gent* 53/3a, 961–966.

Bennison, J.A. and Jacobson, R. (1991) Integrated control of *Frankliniella occidentalis* (Pergande) in UK cucumber crops – evaluation of a controlled release system of introducing *Amblyseius cucumeris. Mededelingen Faculteit Landbouwwetenschappen, Universiteit Gent* 56/2a, 251–258.

Bennison, J.A., Hockland, S. and Jacobson, R. (1990) Recent developments with integrated control of thrips on cucumber in the UK. *International Organisation of Biological Control/West Palaearctic Regional Section Bulletin* 13, 19–26.

Blum, M.S. (1991) Chemical ecology of Thysanoptera. In: Parker, B.L., Skinner, M. and Lewis, T. (eds) *Towards Understanding Thysanoptera*. General Technical Report NE-147, USDA, Forest Service, Radnor, PA, pp. 95–112.

Blum, M.S., Foottitt, R. and Fales, H.M. (1992) Defensive chemistry and the function of the anal exudate of the thrips *Haplothrips leucanthemi. Comparative Biochemistry and Physiology* 102, 209–211.

Bohart, R.M. and Smith, N.J. (1994) Contributions to the knowledge of the genus *Spilomena* Shuckard in America north of Mexico (Hymenoptera, Sphecidae, Pemphredoninae). *Journal of the Kansas Entomological Society* 67, 318–330.

Böhm, O. (1995) Zum Vertilgerkomplex von *Taeniothrips simplex* Mor. *Pflanzenschutzberichte* 22, 4–5.

Bonde, J. (1989) Biological studies including population growth parameters of the predatory mite *Amblyseius barkeri* (Acarina: Phytoseiidae) at 25°C in the laboratory. *Entomophaga* 34, 275–287.

Bounfour, M. and McMurtry, J.A. (1987) Biology and ecology of *Euseius scutalis* (Athias-Henriot) (Acarina: Phytoseiidae). *Hilgardia* 55, 1–23.

Bournier, A., Lacasa, A. and Pivot, Y. (1978) Biologie d'un thrips prédateur *Aeolothrips intermedius* (Thys.: Aeolothripidae). *Entomophaga* 23, 403–410.

Bournier, A., Lacasa, A. and Pivot, Y. (1979) Régime alimentaire d'un thrips prédateur *Aeolothrips intermedius* (Thys.: Aeolothripidae). *Entomophaga* 24, 353–361.

Braman, S.K. and Beshear, R.J. (1994) Seasonality of predaceous plant bugs (Heteroptera: Miridae) and phytophagous thrips (Thysanoptera: Thripidae) as influenced by host plant phenology of native azaleas. *Environmental Entomology* 23, 712–718.

Braman, S.K., Sloderbeck, P.E. and Yeargan, K.V. (1984) Effects of temperature on the development and survival of *Nabis americoferus* and *N. roseipennis* (Hemiptera: Nabidae). *Annals of the Entomological Society of America* 77, 592–596.

Brødsgaard, H.F. and Enkegaard, A. (1995) Interactions among polyphagous anthocorid bugs used for thrips control and other beneficials in multi-species biological pest management systems. *Mededelingen van de Faculteit Landbouwwetenschappen, Rijksuniversiteit Gent* 60/3a, 893–900.

Brødsgaard, H.F. and Hansen, L.S. (1992) The effect of *Amblyseius cucumeris* and *Amblyseius barkeri* as biological control agents of *Thrips tabaci* on glasshouse cucumbers. *Biocontrol Science and Technology* 2, 215–223.

Brødsgaard. H.F., Sardar, M.A. and Enkegaard, A. (1996) Prey preference of *Hypoaspis miles* (Berlese) (Acarina: Hypoaspidae): non-interference with other beneficials in glasshouse crops. *International Organisation of Biological Control/West Palaearctic Regional Section Bulletin* 19, 23–26.

Buente, R., Kuo-Sell, H.L. and Sell, P. (1990) Predation of *Frankliniella occidentalis* (Thysanoptera: Thripidae) by the flower bugs *Anthocoris nemorum* and *Anthocoris gallarum-ulmi* (Heteroptera: Anthocoridae). *Mededelingen Faculteit Landbouwwetenschappen Rijksuniversiteit Gent* 55 (2a), 323–334.

Butler, G.D. and Dickenson, W.A. (1972) Life cycle of convergent lady beetle in relation to temperature. *Journal of Economic Entomology* 65, 1508–1509.

Buxton, J. and Wardlaw, L. (1991) Two years trials work with biological control programmes in year-round chrysanthemums. *OEPP-OILB/SEPP Polonia* 71.

Callan, E. McC. (1943) Natural enemies of the cacao thrips. *Bulletin of Entomological Research* 34, 313–321.

Callan, E. McC. (1975) Miridae of the genus *Termatophylidea* (Hemiptera) as predators of cacao thrips. *Entomophaga* 20, 389–391.

Carl, K.P. (1976) *Report of work carried out at the European Station.* Commonwealth Institute of Biological Control, Delmond, Switzerland.

Carnero, A., Pena, A., Perez-Padrón, F., Garrido, C. and Hernandez García, M. (1993) Bionomics of *Orius albidipennis* and *O. limbatus. International Organisation of Biological Control/West Palaearctic Regional Section Bulletin* 16, 27–30.

Castagnoli, M. and Simoni, S. (1990) Biological observations and life table parameters of *Amblyseius cucumeris* (Oud.) (Acarina: Phytoseiidae) reared on different diets. *Redia* 73, 569–583.

Castagnoli, M., Del Bene, G., Gargani, E. and Simoni, S. (1990) Suitability of *Amblyseius cucumeris* (Oud.) (Acarina: Phytoseiidae) for the biological control of *Thrips tabaci* Lind. and *Frankliniella occidentalis* (Pergande) (Thys.: Thripidae). *Redia* 73, 53–62.

Castañéra, C., Alomar, O. and Ruidavets, J. (1996) Management of western flower thrips on cucumber with *Dicyphus tamaninii* (Heteroptera: Miridae). *Biological Control* 7, 114–120.

Chambers, R.J. and Long, S. (1993) New predators for biological control under glass. *Phytoparasitica* 20 (Suppl.), 57–60.

Chambers, R.J., Long, S. and Helyer, N.L. (1993) Effectiveness of *Orius laevigatus* (Hemiptera: Anthocoridae) for the control of *Frankliniella occidentalis* on cucum-

ber and pepper in the UK. *Biocontrol Science and Technology* 3, 295–307.
Chang, N.T., Hung, C.T., Hua, T. and Ho, C.C. (1993) Notes on predatory natural enemies of *Thrips palmi* Karny (Thysanoptera: Thripidae) on aubergine. *Plant Protection Bulletin (Taiwan)* 35, 239–243.
Chyzik, R. and Klein, M. (1995) Population fluctuations of the western flower thrips *Frankliniella occidentalis* and its predators *Orius* spp., on cultivated sunflower in Israel. *Mededelingen Faculteit Landbouwwetenschappen, Universiteit Gent* 60/3a, 851–856.
Chyzik, R., Klein, M. and Ben-Dov, Y. (1995a) Reproduction and survival of the predatory bug *Orius albidipennis* on various arthropod prey. *Entomologia Experimentalis et Applicata* 75, 27–31.
Chyzik, R., Klein, M. and Ben-Dov, Y. (1995b) Overwintering behaviour of the predatory bug *Orius albidipennis* (Hemiptera: Anthocoridae) in Israel. *Biocontrol Science and Technology* 5, 287–296.
Cloutier, C. and Johnson, S.D. (1993a) Interaction between life stages in a phytoseiid predator: western flower thrips prey killed by adults as food for protonymphs of *Amblyseius cucumeris*. *Experimental and Applied Acarology* 17, 441–449.
Cloutier, C. and Johnson, S.G. (1993b) Predation by *Orius tristicolor* (Hemiptera: Anthocoridae) on *Phytoseiulus persimilis* (Acarina: Phytoseiidae): testing for compatibility between biocontrol agents. *Environmental Entomology* 22, 477–482.
Cloutier, C., Arodokoun, D., Johnson, S.G. and Gélinas, L. (1995) Thermal dependence of *Amblyseius cucumeris* (Acarina: Phytoseiidae) and *Orius insidiosus* (Heteroptera: Anthocoridae) in greenhouses. In: Parker, B.L., Skinner, M. and Lewis, T. (eds) *Thrips Biology and Management*. Plenum Press, New York, pp. 231–235.
Cogan, B.H. and Smith, K.G.V. (1982) A new species of *Lioscinella* (Diptera: Chloropidae) predaceous upon Thysanoptera in Australia. *Memoirs of the Entomological Society of Washington* 10, 42–47.
Coll, M. (1996) Feeding and oviposition on plants by an omnivorous insect predator. *Oecologia* 105, 234–220.
Coll, M. and Bottrell, D.G. (1995) Predator–prey association in mono and dicultures: effect of maize and bean vegetation. *Agriculture, Ecosystems and Environment* 54, 115–125.
Coll, M. and Ridgway, R.L. (1995) Functional and numerical responses of *Orius insidiosus* (Heteroptera: Anthocoridae) to its prey in different vegetable crops. *Annals of the Entomological Society of America* 88, 732–738.
Congdon, B.D. and McMurtry, J.A. (1988) Prey selectivity in *Euseius tularensis* (Acari: Phytoseiidae). *Entomophaga* 33, 281–287.
Conijn, C.G.M. (1993) Biological control of the gladiolus thrips *Taeniothrips simplex* on gladiolus corms in storage rooms with *Amblyseius barkeri* and *Orius insidiosus*. *Mededelingen Faculteit Landbouwwetenschappen, Universiteit Gent* 58/2a, 391–396.
Coville, P.L. and Allen, W.W. (1976) Life table and feeding habits of *Scolothrips sexmaculatus*. *Annals of the Entomological Society of America* 70, 11–16.
Crespi, B.J. (1990) Subsociality and female reproductive success in a mycophagous thrips: an observational and experimental analysis. *Journal of Insect Behaviour* 3, 61–74.
Crocker, R.L., Whitcomb, W.H. and Ray, R.M (1975) Effects of sex, developmental stage, and temperature on predation by *Geocoris punctipes*. *Environmental Entomology* 4, 531–534.
Cross, E.A. (1965) The generic relationships of the family Pyemotidae (Acarina: Trombidiformes). *University of Kansas Scientific Bulletin* 45, 29–275.

De Klerk, M.-L. and Ramakers, P.M.J. (1986) Monitoring population densities of the phytoseiid predator *Amblyseius cucumeris* and its prey after large scale introductions to control *Thrips tabaci* on sweet pepper. *Mededelingen Faculteit Landbouwwetenschappen Rijksuniversiteit Gent* 51/3a, 1045–1048.

De Melo, G.A.R and Evans, H.E. (1993) Two new *Microstigmus* species (Hymenoptera, Sphecidae), with the description of their parasite, *Goniozus microstigmi*, new species (Hymenoptera, Bethylidae). *Proceedings of the Entomological Society of Washington* 95, 258–263.

Derbeneva, N.N. (1967) New data on the biology and structure of preimaginal phases and stages of the predatory thrips *Aeolothrips intermedius* Bagnall (Thysanoptera, Aeolothripidae). *Entomologiczne Obozrenie* 46, 629–644.

Dhaliwal, J.S. (1975) *Polistes hebraeus* (Fabricius) preying upon *Rhipiphorothrips cruentatus* Hood (Thripidae: Thysanoptera). *Current Science (Bangalore)* 44, 368.

Dicke, F.F. and Jarvis, J.L. (1962) The habits and seasonal abundances of *Orius insidiosus* (Say) (Hemiptera–Heteroptera: Anthocoridae) on corn. *Journal of the Kansas Entomological Society* 35, 339–344.

Dicke, M. and Groeneveld, A. (1986) Hierarchical structure in kairomone reference of the predatory mite *Amblyseius potentillae*: dietary component indispensable for diapause induction affects prey location behaviour. *Ecological Entomology* 11, 131–138.

Dicke, M. and Sabelis, M.W. (1988) How plants obtain predatory mites as bodyguards. *Netherlands Journal of Zoology* 38, 148–165.

Dicke, M. and Sabelis, M.W. (1989) Does it pay plants to advertise for bodyguards? Towards a cost–benefit analysis of induced synomone production. In: Lambers, H., Cambridge, M.L., Konings, H. and Pons, T.L. (eds) *Variation in Growth Rate and Productivity of Higher Plants*. SSPB Academic Publishing BV, The Hague, The Netherlands, pp. 341–358.

Dicke, M. and Sabelis, M.W. (1990) Costs and benefits of chemical information conveyance: proximate and ultimate aspects. In: Roitberg, B.D. and Isman, M. (eds) *Evolutionary Perspectives in Insect Chemical Ecology*. Chapman & Hall, New York.

Dicke, M., Sabelis, M.W. and Groeneveld, A. (1986) Vitamin A deficiency modifies response of predatory mite *Amblyseius potentillae* to volatile kairomone of two-spotted spider mite. *Journal of Chemical Ecology* 12, 1389–1396.

Diehl, S. (1993) Relative consumer sizes and strengths of direct and indirect interactions in omnivorous feeding relationships. *Oikos* 68, 151–157.

Dimitrov, A. (1975) Control of tobacco thrips with *Macrolophus costalis*. *Zashcita Rastenii* 23, 34–37.

Dimitrov, A. (1976) Control of tobacco thrips. *Zashcita Rastenii* 24, 34–37.

Dimitrov, A. (1977) A study of the beneficial fauna of Virginia tobacco. *Zashcita Rastenii* 25, 35–58.

Dissevelt, M., Altena, K. and Ravensberg, W.J. (1995) Comparison of different *Orius* species for control of *Frankliniella occidentalis* in glasshouse vegetable crops in the Netherlands. *Mededelingen Faculteit Landbouwwetenschappen, Universiteit Gent* 60/3a, 839–845.

Drukker, B., Scutareanu, P. and Sabelis, M.W. (1995) Do anthocorid predators respond to synomones from *Psylla*–infested pear trees under field conditions? *Entomologia Experimentalis et Applicata* 77, 193–203.

Dunbar, D.M. and Bacon, O.G. (1972a) Feeding, development, and reproduction of *Geocoris punctipes* (Heteroptera: Lygaeidae) on eight diets. *Entomologia Experimentalis et Applicata* 65, 892–895.

Dunbar, D.M. and Bacon, O.G. (1972b) Influence of temperature on development and reproduction of *Geocoris articolor*, *G. pallens* and *G. punctipes* (Heteroptera: Lygaeidae) from California. *Environmental Entomology* 1, 596–599.

Duso, C. and Camporese, P. (1991) Developmental times and oviposition rates of predatory mites *Typhlodromus pyri* and *Amblyseius andersoni* (Acari: Phytoseiidae) reared on different foods. *Experimental and Applied Acarology* 13, 117–128.

Dwumfour, E.F. (1992) Volatile substances evoking orientation in the predatory flower bug *Anthocoris nemorum* (Heteroptera: Anthocoridae). *Bulletin of Entomological Research* 82, 465–469.

Eickwort, G.C. (1983) Potential use of mites as biological control agents of leaf-feeding insects. In: Hoy, M.A., Cunningham, G.L. and Knutson, L. (eds) *Biological Control of Pests by Mites*. Division of Agriculture and Natural Resources, University of California, Special Publication 3304, pp. 41–52.

El-Badry, E.A. and Tawfik, M.S.F. (1966) Life cycle of the mite *Adactylidium* sp. (Acarina: Pyemotidae), a predator of thrips eggs in the United Arab Republic. *Annals of the Entomological Society of America* 59, 458–461.

El-Hariri, G. (1966) Laboratory studies on the reproduction of *Adalia bipunctata* (Coleoptera: Coccinellidae). *Entomologia Experimentalis et Applicata* 9, 200–204.

El-Serwiy, S.A., Razoki, I.A and Ragab, A.S. (1985) Population density of *Thrips tabaci* (Lind.) and the predators *Orius albidipennis* (Reut.) and *Aeolothrips fasciatus* (L.) on onion. *Journal of Agriculture and Water Resources Research* 4, 57–67.

Engel, R. and Ohnesorge, B. (1994) The role of alternative food and microclimate in the system *Typhlodromus pyri* Scheuten (Acari, Phytoseiidae)–*Panonychus ulmi* Koch (Acari, Tetranychidae) on grape vines. I. Laboratory investigations. *Journal of Applied Entomology* 118, 129–150.

Enkegaard, A., Sardar, M.A. and Brodsgaard, H.F. (1996) Life table of *Hypoaspis miles* preying on mushroom sciarid larvae, *Lycoriella solana*, and mould mites, *Tyrophagus putrescentiae*. *International Organisation of Biological Control/West Palaearctic Regional Section Bulletin* 19, 35–38.

Fauvel, G. (1974) Sur l'alimentation pollinique d'un anthocoride prédateur, *Orius (Heterorius) vicinus* Rib (Hémiptère). *Annales Zoologie et Ecologie Animale* 6, 245–258.

Fauvel, G., Malausa, J.C. and Kaspar, E. (1987) Etude en laboratoire des principales caracteristiques biologiques de *Macrolophus caliginosus* (Heteroptera: Miridae). *Entomophaga* 32, 529–543.

Ferguson, G.M. and Schmidt, J.M. (1996) Effect of selected cultivars on *Orius insidiosus*. *International Organisation of Biological Control/West Palaearctic Regional Section Bulletin* 19, 39–42.

Ferrari, R. (1980) *Thrips tabaci* injurious to onion crops in Tuscany. *Informatore Fitopatologico* 10, 27–28.

Fischer, S., Linder, C. and Freuler, J. (1992) Biologie et utilisation de la punaise *Orius majusculus* Reuter (Heteroptera: Anthocoridae) dans la lutte contre les thrips *Frankliniella occidentalis* Perg. et *Thrips tabaci* Lind. en serre. *Revue Suisse de Viticulture, Arboriculture et Horticulture* 24, 119–127.

Foglar, H., Malausa, J.C. and Wajnberg, E. (1990) The functional response and preference of *Macrolophus caliginosus* (Heteroptera: Miridae) for two of its prey: *Myzus persicae* and *Tetranychus urticae*. *Entomophaga* 35, 465–474.

Franssen, C.J.H. (1960) Biology and control of the pea thrips. *Verslagen van Landbouwkundige Onderzoekingen, Rijkslandbouwproefstation* 66, 36pp.

Fransen, J.J., Boogaard, M. and Tolsma, J. (1993) The minute pirate bug, *Orius insidiosus* (Say) (Hemiptera: Anthocoridae), as a predator of western flower thrips, *Frankliniella occidentalis* (Pergande), in chrysanthemum, rose and saintpaulia. *International Organisation of Biological Control/West Palaearctic Regional Section Bulletin* 16, 73–77.

Frescata, C. and Mexia, A. (1995) Biological control of western flower thrips with *Orius laevigatus* (Heteroptera: Anthocoridae) in organic strawberries in Portugal. In: Parker, B.L., Skinner, M. and Lewis, T. (eds) *Thrips Biology and Management*, Plenum Press, New York, pp. 249–250.

Fry, J.M. (1987) *Natural Enemy Databank*. A catalogue of natural enemies of arthropods derived from records on the CIBC Natural Enemy Databank, CAB, IIBC, 185 pp.

Funao, T. and Yoshiyasu, Y. (1995) Development and fecundity of *Orius sauteri* (Poppius) (Hemiptera: Anthocoridae) reared on *Aphis gossypii* Glover and corn pollen. *Japanese Journal of Entomology and Zoology* 39, 84–85.

Gabarra, R., Castané, C., Bordas, E. and Abajes, R. (1988) *Dicyphus tamaninii* as a beneficial insect and pest in tomato crops in Catalonia, Spain. *Entomophaga* 33, 219–228.

Gabarra, R., Castané, C. and Abajes, R. (1995) The mirid bug *Dicyphus tamaninii* as a greenhouse whitefly and western flower thrips predator on cucumber. *Biocontrol Science and Technology* 5, 475–488.

Gagné, R.J. and Bennett, F.D. (1993) Two new species of Lestodiplosini (Diptera: Cecidomyiidae) preying on Homoptera and Thysanoptera in Southern Florida. *The Florida Entomologist* 76, 341–348.

Gerin, C. and Hance, T. (1993) Evolution de l'impact des degats provoqués par *Frankliniella occidentalis* (Pergande) (Thysanoptera: Thripidae) sur la plante hote. Influence de la predation par *Amblyseius cucumeris* (Oudemans) (Acari: Phytoseiidae). *Mededelingen Faculteit Landbouwwetenschappen, Rijksuniversiteit Gent* 58/2a, 409–425.

Gerlach, S. (1985) Biologie, Ökologie und Wirksamkeit der räuberischen thrips, *Scolothrips longicornis* Priesner (Thysanoptera: Thripidae). Dr. agr. Dissertation, Rheinische Friedrich-Wilhelms – Universität, Bonn.

Gerlach, S. and Şengonca, Ç. (1985) Comparative studies on the effectiveness of the predatory mite *Phytoseiulus persimilis* Athias-Henriot and the predatory thrips, *Scolothrips longicornis* Priesner. *Journal of Plant Diseases and Protection* 92, 138–146.

Gerlach, S. and Şengonca, Ç. (1986) Frassaktivität und Wirksamkeit des räuberischen Thrips, *Scolothrips longicornis* Priesner (Thysanoptera: Thripidae). *Journal of Applied Entomology* 101, 444–452.

Ghabn, A.A.E. (1948) Contribution to the knowledge of the biology of *Thrips tabaci* Lind. in Egypt (Thysanoptera). *Bulletin Société Fouad I Entomologie* 32, 123–174.

Ghauri, M.S.K. (1980) Notes on Anthocoridae (Heteroptera) with descriptions of two new species of economic importance from Africa. *Bulletin of Entomological Research* 70, 287–291.

Gilkeson, L.A., Morewood, W.D. and Elliot, D.E. (1990) Current status of biological control in Canadian greenhouses with *Amblyseius cucumeris* and *Orius tristicolor*. *International Organisation of Biological Control/West Palaearctic Regional Section* 5, 71–75.

Gillespie, D.R. (1989) Biological control of thrips (Thysanoptera: Thripidae) on greenhouse cucumber by *Amblyseius cucumeris*. *Entomophaga* 34, 185–192.

Gillespie, D.R. and Quiring, D.J.M. (1992) Competition between *Orius tristicolor* (White) (Hemiptera: Anthocoridae) and *Amblyseius cucumeris* (Oudemans) (Acari:

Phytoseiidae) feeding on *Frankliniella occidentalis* (Pergande) (Thysanoptera: Thripidae). *Canadian Entomologist* 124, 1123–1128.

Gillespie, D.R. and Quiring, D.M.J. (1990) Biological control of fungus gnats, *Bradysia* spp. (Diptera: Sciaridae), and western flower thrips *Frankliniella occidentalis* (Pergande) (Thysanoptera: Thripidae), in greenhouses using a soil-dwelling predatory mite, *Geolaelaps* sp. nr. *Canadian Entomologist* 122, 975–983.

Gillespie, D.R. and Ramey, C.A. (1988) Life history and cold storage of *Amblyseius cucumeris* (Acarina: Phytoseiidae). *Journal of the Entomological Society of British Columbia* 85, 71–76.

Gilstrap, F.E. (1995) Six-spotted thrips: a gift from nature that controls spider mites. In: Parker, B.L., Skinner, M. and Lewis, T. (eds) *Thrips Biology and Management*, Plenum Press, New York, pp. 305–316.

Gilstrap, F.E. and Oatman, E.R. (1976) The bionomics of *Scolothrips sexmaculatus* (Pergande) (Thysanoptera: Thripidae), an insect predator of spider mites. *Hilgardia* 44, 27–59.

Giroux, S., Coderre, D., Vincent, C. and Cote, J.C. (1994) Effects of *Bacillus thuringiensis* var. San Diego on predation effectiveness, development and mortality of *Coleomegilla maculata* Lengi (Col.: Coccinellidae) larvae. *Entomophaga* 39, 61–69.

Glockemann, B. (1992) Biological control of *Frankliniella occidentalis* on ornamental plants using predatory mites. *European Plant Protection Organisation Bulletin* 22, 397–404.

González, D. and Wilson, L.T. (1982) A food-web approach to economic thresholds: a sequence of pest/predaceous arthropods on California cotton. *Entomophaga* 27 (special issue), 31–43.

González, D., Friesen, R., Leigh, T.F., Wilson, T. and Waggoner, M. (1995) Naturally-occurring biological control: western flower thrips impact on spider mites in California cotton. In: Parker, B.L., Skinner, M. and Lewis, T. (eds) *Thrips Biology and Management*. Plenum Press, New York, pp. 317–323.

Goodwin, S. and Steiner, M.Y. (1996) Survey of Australian native natural enemies for control of thrips. *International Organisation of Biological Control/West Palaearctic Regional Section Bulletin* 19, 47–50.

Grafton-Cardwell, E.E. and Ouyang, Y. (1995) Manipulation of the predacious mite *Euseius tularensis* (Acari: Phytoseiidae), with pruning for Citrus thrips control. In: Parker, B.L., Skinner, M. and Lewis, T. (eds) *Thrips Biology and Management*. Plenum Press, New York, pp. 251–254.

Greenstone, M.H. and Trowell, S.C. (1994) Arthropod predation: a simplified immunodot format for predator gut analysis. *Annals of the Entomological Society of America* 87, 214–217.

Grout, T.G. (1994) The distribution and abundance of phytoseiid mites (Acari: Phytoseiidae) on citrus in southern Africa and their possible value as predators of citrus thrips (Thysanoptera: Thripidae). *Experimental and Applied Acarology* 18, 61–71.

Grout, T.G. and Richards, R.I. (1992a) The dietary effect of windbreak pollens on longevity and fecundity of a predacious mite *Euseius addoensis addoensis* (Acari: Phytoseiidae) found in citrus orchards in South Africa. *Bulletin of Entomological Research* 82, 317–320.

Grout, T.G. and Richards, R.I. (1992b) *Euseius addoensis addoensis*, an effective predator of citrus thrips, *Scirtothrips aurantii*, in the eastern Cape Province of South Africa. *Experimental and Applied Acarology* 15, 1–13.

Grout, T.G. and Stephen, P.R. (1993) Predation of alate citrus thrips, *Scirtothrips aurantii* Faure (Thysanoptera: Thripidae), by *Euseius citri* (Van der Merwe and Ryke) (Acari: Phytoseiidae) at low temperatures. *African Entomology* 1, 264–265.

Guppy, J.C. (1986) Bionomics of the damsel bug, *Nabis americoferus* Carayon, a predator of the alfalfa blotch leafminer (Diptera: Agroymyzidae). *Canadian Entomologist* 118, 745–751.

Habib, A., Rezk, G.N., Farghaly, H.T. and Ragab, Z.A. (1980) Seasonal abundance of some predators in cotton fields and its relation to certain pests. *Bulletin du Société Entomologique d'Egypte* 60, 191–196.

Hansen, L.S. (1988) Control of *Thrips tabaci* (Thysanoptera: Thripidae) on glasshouse cucumber using large introductions of predatory mites *Amblyseius barkeri* (Acarina: Phytoseiidae). *Entomophaga* 33, 33–42.

Hansen, L.S. (1989) The effect of initial thrips density (*Thrips tabaci* Lind. [Thysanoptera: Thripidae]) on the control exerted by *Amblyseius barkeri* (Hughes) (Acarina: Phytoseiidae) on glasshouse cucumber. *Journal of Applied Entomology* 107, 130–135.

Heidger, C. and Nentwig, W. (1984) The prey of *Dictynus arundinacea* (Araneae: Dictynidae). *Zoologische Beitragen* 29, 185–192.

Helle, W. and Sabelis, M.W. (1985) *Spider Mites: Their Biology, Natural Enemies and Control.* World Crop Pest Series Vol. 1B. Elsevier, Amsterdam.

Herting, B. and Simmonds, F.J. (1971) *A Catalogue of Parasites and Predators of Terrestrial Arthropods. Section A. Host or Prey/Enemy. 1. Arachnida to Heteroptera.* CAB, Farnham Royal, England.

Higgins, C.J. (1992) Western flower thrips (Thysanoptera: Thripidae) in greenhouses: population dynamics, distribution on plants and associations with predators. *Journal of Economic Entomology* 85, 1891–1903.

Hirose, Y., Kajita, H., Takagi, M., Okajima, S., Napompeth, B. and Buranapanichpan, S. (1993) Natural enemies of *Thrips palmi* and their effectiveness in the native habitat, Thailand. *Biological Control* 3, 1–5.

Hoda, F.M., El-Naggar, M.E., Taha, A.H. and Ibrahim, G.A. (1986) Effect of different types of food on fecundity of predacious mite *Amblyseius swirskii* Athias-Henriot (Acari: Phytoseiidae). *Bulletin du Société Entomologique d'Egypte* 66, 113–116.

Hollingsworth, C.S. and Bishop, G.W. (1982) *Orius tristicolor* (Heteroptera: Anthocoridae) as a predator of *Myzus persicae* (Homoptera: Aphididae) on potatoes. *Environmental Entomology* 11, 1046–1048.

Holt, R.D. (1977) Predation, apparent competition and the structure of predator–prey communities. *Theoretical Population Biology* 12, 197–229.

Howard, D.F., Blum, M.S. and Fales, M.F. (1983) Defense in thrips: forbidding fruitiness of a lactone. *Science* 220, 335–336.

Howard, D.F., Blum, M.S., Jones, T.H., Fales, M.F. and Tomalski, M.D. (1987) Defensive function and chemistry of the anal exudate of the Cuban laurel thrips *Gynaikothrips ficorum* (Marchal). *Phytophaga* 1, 163–170.

Hoy, C.W. and Glenister, C.S. (1991) Releasing *Amblyseius* spp. to control *Thrips tabaci* on cabbage. *Entomophaga* 36, 561–573.

Hulshof, J. (1994) Effect of different pollen species on development, fecundity and longevity on *Amblyseius degenerans*, *Orius laevigatus* and *Frankliniella occidentalis.* MSc thesis, Department of Entomology, Wageningen Agricultural University, The Netherlands.

Husseini, M., Schumann, K. and Sermann, H. (1993) Rearing immature feeding stage of *Orius insidiosus* on the acarid mite *Tyrophagus putrescentiae* as a new alternative prey. *Journal of Applied Entomology* 116, 113–117.

Isenhour, D.J. and Marston, N.L. (1981) Seasonal cycles of *Orius insidiosus* (Hemiptera: Anthocoridae) in Missouri soyabeans. *Journal of the Kansas Entomological Society* 54, 129–142.

Isenhour, D.J. and Yeargan, K.V. (1981a) Effects of temperature on the development of *Orius insidiosus*, with notes on laboratory rearing. *Annals of the Entomological Society of America* 74, 114–116.

Isenhour, D.J. and Yeargan, K.V. (1981b) Predation by *Orius insidiosus* (Hemiptera: Anthocoridae) on the soybean thrips *Sericothrips variabilis* (Thysanoptera: Thripidae): Effect of prey stage and density. *Environmental Entomology* 10, 496–500.

Isenhour, D.J. and Yeargan, K.V. (1981c) Interactive behavior of *Orius insidiosus* (Hemiptera: Anthocoridae) and *Sericothrips variabilis* (Thysanoptera: Thripidae): predator searching strategies and prey escape tactics. *Entomophaga* 26, 213–219.

Isenhour, D.J. and Yeargan, K.V. (1982) Oviposition sites of *Orius insidiosus* (Say) and *Nabis* sp. in soybean (Hemiptera: Anthocoridae and Nabidae). *Journal of the Kansas Entomological Society* 55, 65–72.

Isenhour, D.J., Wiseman, B.R. and Layton, R.C. (1989) Enhanced predation by *Orius insidiosus* (Hemiptera: Anthocoridae) on larvae of *Heliothis zea* and *Spodoptera frugiperda* (Lepidoptera: Noctuidae) caused by prey feeding on resistant corn genotypes. *Environmental Entomology* 18, 418–422.

Jacobson, R.J. (1991) Integrated control of *Frankliniella occidentalis* in UK cucumber crops – use of *Anthocoris nemorum*. *Mededelingen Faculteit Landbouwwetenschappen Rijksuniversiteit Gent* 56/2a, 235–240.

Jacobson, R.J. (1995) Egg laying sites of *Orius majusculus*, a thrips predator, on cucumber. In: Parker, B.L., Skinner, M. and Lewis, T. (eds) *Thrips Biology and Management.* Plenum Press, New York, pp. 241–244.

Janssen, A. and Sabelis, M.W. (1992) Phytoseiid life-histories, local predator-prey dynamics and strategies for control of tetranychid mites. *Experimental and Applied Acarology* 14, 233–250.

Janssen, A., Bruin, J., Jacobs, G., Schraag, R. and Sabelis, M.W. (1997) Predators use volatiles to avoid prey patches with conspecifics. *Journal of Animal Ecology* 66, 223–232.

Johansen, R.M. (1981) *Franklinothrips caballeroi* new-species (Thysanoptera: Franklinothripidae) from eastern Mexico. *Anali Instituto Biologico Universita Nacional Autonomica de Mexico Series Zoologi* 50, 153–158.

Johnson, S.G. (1990) Biology and predacious ability of *Amblyseius barkeri* (Hughes) and *A. cucumeris* (Oudemans) (Acari: Phytoseiidae), predators of the western flower thrips *Frankliniella occidentalis* (Pergande)(Thysanoptera: Thripidae). MSc thesis, McGill University.

Jones, S.A. and Morse, J.G. (1995) Use of isoelectric focusing electrophoresis to evaluate citrus thrips (Thysanoptera: Thripidae) predation by *Euseius tularensis* (Acari: Phytoseiidae). *Environmental Entomology* 24, 1040–1051.

Kajita, H. (1986) Predation by *Amblyseius* spp. (Acarina: Phytoseiidae) and *Orius* spp. (Hemiptera: Anthocoridae) on *Thrips palmi* Karny (Thysanoptera: Thripidae). *Applied Entomology and Zoology* 21, 482–484.

Kalushkov, P.K. (1994) Zur Lebensdauer, Fruchtbarkeit und Entwicklung von *Adalia bipunctata* (L.) (Col., Coccinelidae) bei Aufzucht mit drei verschiedenen Blattlausarten. *Anzeiger für Schädlingskunde, Pflanzenschutz und Umweltschutz* 67, 6–7.

Kamburov, S.S. (1971) Feeding, development, and reproduction of *Amblyseius largoensis* on various food substances. *Journal of Economic Entomology* 64, 643–648.

Kawai, A. (1995) Control of *Thrips palmi* Karny (Thysanoptera, Thripidae) by *Orius* spp. (Heteroptera: Anthocoridae) on greenhouse aubergine. *Applied Entomology and Zoology* 30, 1–7.

Kelton, L.A. (1963) Synopsis of the genus *Orius* Wolff in America north of Mexico (Heteroptera: Anthocoridae). *Canadian Entomologist* 95, 631–673.

Kennett, C.E., Flaherty, D.L. and Hoffmann, R.W. (1979) Effect of wind-borne pollens on the population dynamics of *Amblyseius hibisci* (Acarina: Phytoseiidae). *Entomophaga* 24, 83–98.

Kiman, Z.B. and Yeargan, K.V. (1985) Development and reproduction of the predator *Orius insidiosus* (Hemiptera: Anthocoridae) reared on diets of selected plant material and arthropod prey. *Annals of the Entomological Society of America* 78, 464–467.

Kingsley, P.C. and Harrington, B.J. (1982) Factors influencing termination of reproductive diapause in *Orius insidiosus* (Hemiptera: Anthocoridae). *Environmental Entomology* 11, 461–462.

Kirk, W.D.J. (1995) Feeding behaviour and nutritional requirements. In Parker, B.L., Skinner, M. and Lewis, T. (eds) *Thrips Biology and Management*. Plenum Press, New York, pp. 21–29.

Kühne, K.-St. and Schrameyer, K. (1994) On the occurrence of predatory flies of the family Hypotidae (Dipt.: Empidoidea) in greenhouses and on the predatory efficiency of two species of the genus *Platypalpus* Marquart. *Journal of Applied Entomology* 118, 209–216.

Kurkina, L.A. (1979) Biology of the robber fly *Machimus annulipes* Br. (Diptera: Asilidae). *Entomologicheskoe Obozrenie* 58, 57–63.

Lababidi, M.S. (1988) Biologische Bekämpfungsmoglichkeiten der baumwollspinnmilbe, *Tetranychus cinnabarinus* (Boisduval) (Acari: Tetranychidae). Dissertation Rheinisches Friedrich-Wilhelms-Universität, Bonn, Germany.

Lacasa, A. (1988) Los tisanopteros en horticultura: su implicaion parasitaria. *3e Symposium Nacional de Agroquimicos Sevilla,* 33–53.

Lacasa, A., Bournier, A. and Pivot, Y. (1982) Influencia de la temperatura sobre la biologia de un trips depredador *Aeolothrips intermedius* Bagnall (Thys.: Aeolothripidae). *Annales del Instituto Nacional de Investigaciones Agrarias Ser. Agric.* 20, 87–98.

Lacasa, A., Martinez, M.C and Torres, J. (1989) Los trips asociados a los cultivos protegidos en el sureste espanol. *Cuaderno Fitopatologia,* 81–88.

Lattin, J.D. (1989) Bionomics of the Nabidae. *Annual Review of Entomology* 34, 383–400.

Lauenstein, G. von (1977) Untersuchungen zur Fruchtbarkeit und Eiablage de Räuberischen Blumenwanze *Anthocoris nemorum* L. (Hem.: Heteroptera). *Zeitschrift für Angewandte Entomologie* 83, 355–363.

Letourneau, D.K. (1990) Mechanisms of predator accumulation in a mixed crop system. *Ecological Entomology* 15, 63–69.

Letourneau, D.K. and Altieri, M.A. (1983) Abundance patterns of a predator, *Orius tristicolor* (Hemiptera: Anthocoridae), and its prey, *Frankliniella occidentalis* (Thysanoptera: Thripidae): habitat attraction in polycultures versus monocultures. *Environmental Entomology* 12, 1464–1469.

Lewis, T. (1973) *Thrips: Their Biology, Ecology and Economic Importance*. Academic Press, London.

Lichtenauer, A. and Sell, P. (1993) Erbeutung verschiedener entwicklungsstadien des western flower thrips durch adulte von *Orius minutus* (Heteroptera: Anthocoridae). *Mededelingen Faculteit Landbouwwetenschappen, Universiteit Gent* 58/2a, 397–407.

Lindhagen, M. and Nedstam, B. (1988). Försök med biologisk bekämpning av *Frankliniella occidentalis* (Thysanoptera: Thripidae) med hjälp av rovkvalstret *Amblyseius cucumeris* (Acari: Phytoseiidae). *Växtskyddsnotiser* 52, 13–16.

Lindquist, E.E., Sabelis, M.W. and Bruin, J. (1996). *Eriophyoid Mites. Their Biology, Natural Enemies and Control*, World Crop Pests, Vol. 6. Elsevier Science Publishers, Amsterdam.

Lindqvist, I. and Tiittanen, K. (1989) Biological control of *Thrips tabaci* (Thysanoptera, Thripidae) on greenhouse cucumber. *Acta Entomologici Fennici* 53, 37–42.

Lo, K.C. and Ho, C.C. (1979) Influence of temperature on life history, predation and population parameters of *Amblyseius longispinosus* (Acarina: Phytoseiidae). *Journal of Agricultural Research, China* 28, 237–250.

Loan, C. and Holdaway, F.G. (1955) Biology of the red clover thrips, *Haplothrips niger* (Osborne) (Thysanoptera: Phlaeothripidae). *Canadian Entomologist* 87, 210–219.

Lobbes, P.W. and Schotten, C. (1980) Capacities of increase of the soil mite *Hypoaspis aculeifer* Canestrini (Mesostigmata: Laelapidae). *Zeitschrift für Angewandte Entomologie* 90, 9–22.

MacGill, E.I. (1939) A gamasid mite (*Typhlodromus thripsi* n. sp.), a predator of *Thrips tabaci* Lind. *Annals of Applied Biology* 26, 309–317.

Mammen, K.V. and Vasudevan Naier, K.P. (1977) On the occurence of *Coccinella arcuata* Fabricius (Coleoptera: Coccinellidae) as a predator of rice thrips in Kerala. *Agriculture Research Journal, Kerala* 15, 195–196.

Mathews, R.W. (1970) A new thrips-hunting *Microstigmus* from Costa Rica (Hymenoptera: Sphecidae: Pemphredoninae). *Psyche* 77, 120–126.

McCaffrey, J.P. and Horsburgh, R.L. (1986a) Biology of *Orius insidiosus* (Heteroptera: Anthocoridae): a predator in Virginia apple orchards. *Environmental Entomology* 15, 984–988.

McCaffrey, J.P. and Horsburgh, R.L. (1986b) Functional responses of *Orius insidiosus* (Hemiptera: Anthocoridae) to the European red mite *Panonychus ulmi* (Acari: Tetranychidae), at different constant temperatures. *Environmental Entomology* 15, 532–525.

McMurtry, J.A. and Badii, M.H. (1991) Greenhouse thrips, *Heliothrips haemorrhoidalis*, in California avocado orchards: biological control studies. In: Parker, B.L., Skinner, M. and Lewis, T. (eds) *Towards Understanding Thysanoptera*. General Technical Report NE-147, USDA, Forest Service, Radnor, PA, pp. 393–398.

McMurtry, J.A. and Scriven, G.T. (1965) Life-history studies of *Amblyseius limonicus* with comparative observations on *Amblyseius hibisci* (Acarina: Phytoseiidae). *Annals of the Entomological Society of America* 58, 106–111.

McMurtry, J.A. and Scriven, G.T. (1966) Studies on predator–prey interactions between *Amblyseius hibisci* and *Oligonychus punicae*: (Acarina: Phytoseiidae, Tetranychidae) under greenhouse conditions. *Annals of the Entomological Society of America* 59, 793–800.

McMurtry, J.A., Morse, J.G. and Johnson, H.G. (1992) Studies on the impact of *Euseius* species (Acari: Phytoseiidae) on citrus mites using predator exclusion and predator release experiments. *Experimental and Applied Acarology* 14, 371–382.

Metz, J.A.J., Sabelis, M.W. and Kuchlein, J.H. (1988) Sources of variation in predation rates at high prey densities: an analytic model and a mite example. *Experimental and Applied Acarology* 5, 187–206.

Milbrath, L.R., Tauber, M.J. and Tauber, C.A. (1993) Prey specificity in *Chrysopa*: an interspecific comparison of larval feeding and defensive behavior. *Ecology* 74, 1384–1393.

Minks, A.K. and Harrewijn, P. (1988) *Aphids. Their Biology, Natural Enemies and Control*, World Crop Pest Series, Vol. 2. Elsevier Science Publishers, Amsterdam, The Netherlands.

Mituda, E.C. and Calilung, V.J. (1989) Biology of *Orius tantilus* (Motchulsky) (Hemiptera: Anthocoridae) and its predatory capacity against *Thrips palmi* Karny (Thysanoptera: Thripidae) on watermelon. *The Philippine Agriculturist* 72, 165–184.

Mohandaniel, A., Sureshkumar, N. and Bakthavatsalam, N. (1983) Bioecology of *Zaniothrips ricini* Bhatti (Panchaetothripinae: Terebrantia: Thysanoptera) – A new thrips pest of castor (*Ricinus communis*: Euphorbiaceae). *Proceedings of the Indian Academy of Sciences (Animal Sciences)* 92, 87–94.

Momen, F.M. (1995) Feeding, development and reproduction of *Amblyseius barkeri* (Acarina: Phytoseiidae) on various food substances. *Acarologia* 36, 101–105.

Momen, F.M. and El-Saway, S.A. (1993) Biology and feeding behaviour of the predatory mite, *Amblyseius swirskii* (Acarina: Phytoseiidae). *Acarologia* 34, 199–204.

Morewood, W.D. and Gilkeson, L.A. (1991) Diapause induction in the thrips predator *Amblyseius cucumeris* (Acarina: Phytoseiidae) under greenhouse conditions. *Entomophaga* 36, 253–263.

Morison, G.D. (1968) Observations and records for British Thysanoptera. VII Thripidae. *Thrips klapaleki* Uzel. *Entomologist* 101, 265–268.

Morse, J.G., Elmer, H.S. and Brawner, O.L. (1986) The 1986 control recommendations for California. *Citrograph* 71, 118–120.

Mostafa, A.R., DeBach, P. and Fisher, T.W. (1975) Anystid mite: citrus thrips predator. *California Agriculture* 29, 5.

Mound, L.A. (1972) Further studies on Australian Aeolothripidae (Thysanoptera). *Journal of the Australian Entomological Society* 11, 37–54.

Mound, L.A. (1993) The first thrips species (Insecta) inhabiting leaf domatia: *Domatiathrips cunninghamii* gen. et. sp. nov. (Thysanoptera: Phlaeothripidae). *Journal of the New York Entomological Society* 101, 424–430.

Mukhopadhyay, A. and Sannigrahi, S. (1993) Rearing success of a polyphagous predator *Geocoris ochropterus* (Hem.: Lygaeidae) on preserved ant pupae of *Oecophylla smaragdina*. *Entomophaga* 38, 219–224.

Muma, M.H. (1955) Factors contributing to the natural control of citrus insects and mites in Florida. *Journal of Economic Entomology* 48, 432–438.

Muraleedharan, N. and Ananthakrishnan, T.N. (1971) Bionomics of *Montandoniola moraguesi* (Puton) (Heteroptera: Anthocoridae) a predator of gall thrips. *Bulletin of Entomology* 12, 4–10.

Muraleedharan, N. and Ananthakrishnan, T.N. (1978a) Bioecology of anthocorids predatory on thrips. *Occasional Publications, Zoological Survey of India*, pp. 1–20.

Muraleedharan, N. and Ananthakrishnan, T.N. (1978b) Bioecology of four species of Anthocoridae (Hemiptera: Insecta) predaceous on thrips with key to genera of anthocorids from India. *Records of the Zoological Survey of India. Miscellaneous Publication* no. 11, 1–32.

Murdoch, W.W. and Oaten, A. (1975) Predation and population stability. *Advances in Ecological Research* 9, 1–131.

Murdoch, W.W., Nisbet, R.M., Blythe, S.P., Gurney, W.S.C. and Reeve, J.D. (1987) An invulnerable age class and stability in delay-differential parasitoid–host models. *The American Naturalist* 129, 263–282.

Nagai, K. (1989) Developmental duration of *Orius* sp. (Hemiptera: Anthocoridae) reared on *Thrips palmi* Karny (Thysanoptera: Thripidae). *Japanese Journal of Applied Entomology and Zoology* 33, 260–262.

Nagai, K. (1990) Suppressive effect of *Orius* sp. (Hemiptera: Anthocoridae) on the population density of *Thrips palmi* Karny (Thysanoptera: Thripidae) in aubergine in an open field. *Japanese Journal of Applied Entomology and Zoology* 34, 109–114.

Nagai, K. (1991) Predatory characteristics of *Orius* sp. on *Thrips palmi* Karny, *Tetranychus kanzawai* Kishida, and *Aphis gossypii* Glover. *Japanese Journal of Applied Entomology and Zoology* 35, 269–274.

Nagai, K., Hiramatsu, T. and Henmi, T. (1988) Predatory effect of *Orius* sp. (Hemiptera: Anthocoridae) on the density of *Thrips palmi* Karny (Thysanoptera: Thripidae) on aubergine. *Japanese Journal of Applied Entomology and Zoology* 32, 300–304.

Nakahara, S. (1985) *Haplothrips kurdjumovi* Karny in North America with a new junior synonym (Thysanoptera: Phlaeothripidae). *Proceedings of the Entomological Society of Washington* 87, 894–895.

Nakashima, Y., Hirose, Y. and Kinjo, K. (1996) Rearing *Orius sauteri* (Poppius) on diet of freeze-dried larval powder of melon fly, *Bactrocera cucurbitae* Coquillett. *Japanese Journal of Applied Entomology and Zoology* 40, 80–82.

Nakata, T. (1994) Prey species of *Orius sauteri* (Poppius) (Heteroptera: Anthocoridae) in a potato field in Hokkaido. *Japanese Journal of Applied Entomology and Zoology* 29, 614–616.

Nakata, T. (1995) Effect of rearing temperature on the development of *Orius sauteri* (Poppius) (Heteroptera: Anthocoridae). *Applied Entomology and Zoology* 30, 145–151.

Naranjo, S.E. and Gibson, R.L. (1995) Phytophagy in predaceous Heteroptera: effects on life history and population dynamics. In: Naranjo, S.E. and Gibson, R.L. (eds) *Phytophagy in Predaceous Heteroptera*. Entomological Society of America, Hyattsville, Md.

Nelson, F.R.S. (1973) Reproduction and consumption rate of *Amblyseius hibisci* on a diet of pollen and *Panonychus citri*. *Annals of the Entomological Society of America* 66, 918–919.

Neuenschwander, P. (1975) Influence of temperature on the immature stages of *Hemerobius pacificus*. *Environmental Entomology* 4, 215–220.

Neuenschwander, P. (1976) Biology of the adult *Hemerobius pacificus*. *Environmental Entomology* 5, 96–100.

Niemczyk, E. (1978) *Orius minutus* (L.) (Heteroptera: Anthocoridae): the occurrence in apple orchards, biology and effect of different food on the development. *Polskie Pismo Entomologiczne* 48, 203–209.

Nolte, H.W. (1951) Beitrage zur Morphologie und Biologie des Lärchenblasenfusses (*Taeniothrips laricivorus* Krat.). *Beitrage zur Entomologie* 1, 110–139.

Obrycki, J.J. and Tauber, M.J. (1982) Thermal requirements for development of *Hippodamia convergens* (Coleoptera: Coccinelidae). *Annals of the Entomological Society of America* 75, 678–683.

Okajima, S., Hirose, Y., Kajita, H., Takagi, M., Napompeth, B. and Buranapanichpan, S. (1992) Thrips on vegetables in southeast Asia. *Applied Entomology and Zoology* 27, 300–303.

Ouyang, Y., Grafton-Cardwell, E.E. and Bugg, R.L. (1992) Effects of various pollens on development, survivorship, and reproduction of *Euseius tularensis* (Acari: Phytoseiidae). *Environmental Entomology* 21, 1371–1376.

Pallini, A., Janssen, A. and Sabelis, M.W. (1997) Odour-mediated responses of phytophagous mites to conspecific and heterospecific competitors. *Oecologia* 110, 79–185.

Parrella, M.P., Rowe, D.J. and Horsburgh, R.L. (1982) Biology of *Leptothrips mali*, a common predator in Virginia apple orchards. *Annals of the Entomological Society of America* 75, 130–135.

Patrzich, R. and Klumpp, M. (1991) Comparison of phytophagous and predatory thrips (Thysanoptera) in winter wheat fields in Hesse with different systems of management. *Journal of Plant Diseases and Protection* 98, 464–470.

Pemberton, R.W. and Vandenberg, N.J. (1993) Extrafloral nectar feeding by ladybird beetles (Coleoptera: Coccinellidae). *Proceedings of the Entomological Society of Washington* 95, 139–151.

Pena, M.A. (1990) Tres anos con *Frankliniella occidentalis* en las Islas Canarias. *Phytoma-Espana April*, 33–39.

Péricart, J. (1972) *Hémipteres, Anthocoridae, Cimicidae et Microphsidae de l'ouest-paléartique*. Masson et C Éditeurs, Paris.

Perkins, P.V. and Watson, T.F. (1972) Biology of *Nabis alternatus* (Hemiptera: Anthocoridae). *Annals of the Entomological Society of America* 65, 54–57.

Peterson, B.S. (1990) The effect of host plant on biological control of western flower thrips by the predatory mite, *Amblyseius cucumeris*. MPM thesis, Simon Fraser University, Burnaby, British Columbia.

Piatkowski, J. (1987) Effect of temperature, relative air humidity and kind of food on the development of predatory mite *Amblyseius mckenziei* Sch. et Pr. (Acarina: Phytoseiidae). *International Organisation of Biological Control/West Palaearctic Regional Section Bulletin* 10(3), Suppl.

Pickett, C.H., Wilson, L.T. and González, D. (1988) Population dynamics and within-plant distribution of the western flower thrips: an early season predator of spider mites infesting cotton. *Environmental Entomology* 17, 551–559.

Polis, G.A. (1981) The evolution and dynamics of intraspecific predation. *Annual Review of Ecology and Systematics* 12, 225–251.

Polis, G.A., Myers, C.A. and Holt, R.D. (1989) The ecology and evolution of intraguild predation: potential competitors that eat each other. *Annual Review of Ecology and Systematics* 20, 297–330.

Polis, G.A. and Holt, R.D. (1992) Intraguild predation: the dynamics of complex trophic interactions. *Trends in Ecology and Evolution* 7, 151–154.

Price, P.W., Bouton, C.E., Gross, P., McPheron, B.A., Thompson, J.N. and Weiss, A.E. (1980) Interactions among three trophic levels: influence of plants on interactions between insect herbivores and natural enemies. *Annual Review of Ecology and Systematics* 11, 41–65.

Priesner, H. (1964) *A Monograph of the Thysanoptera of the Egyptian Deserts*. Publications Institute Déserts d'Egypte 13.

Putman, W.L. (1942) Notes on the predaceous thrips *Haplothrips subtilissimus* Hal. and *Aeolothrips melaleucus* Hal. *Canadian Entomologist* 74, 37–43.

Putman, W.L. (1962) Life history and behaviour of the predaceous mite *Typhlodromus* (T.) *caudiglans* Schuster (Acarina: Phytoseiidae) in Ontario with notes on the prey of related species. *Canadian Entomologist* 14, 163–177.

Putman, W.L. (1965) The predacious thrips *Haplothrips faurei* Hood (Thysanoptera: Phlaeothripidae) in Ontario peach orchards. *Canadian Entomologist* 87, 9–33.

Ragusa, S. and Zedan, M.A. (1988) Biology and predation of *Hypoaspis aculeifer* (Canestrini) on *Rhizoglyphus echinopus* (Fom. and Rob.). *Redia* 71, 213–225.

Rajasekhara, K. and Chatterji, S. (1970) Biology of *Orius indicus* (Hemiptera: Anthocoridae), a predator of *Taeniothrips nigricornis* (Thysanoptera). *Annals of the Entomological Society of America* 63, 364–367.

Ramakers, P.M.J. (1978) Possibilities for biological control of *Thrips tabaci* Lind.

(Thysanoptera: Thripidae) in glasshouses. *Mededelingen Faculteit Landbouwwetenschappen Rijksuniversiteit Gent* 43, 463–469.

Ramakers, P.M.J. (1980) Biological control of *Thrips tabaci* (Thysanoptera: Thripidae) with *Amblyseius* spp. (Acari: Phytoseiidae). *International Organisation of Biological Control/West Palaearctic Regional Section Bulletin* 3, 203–307.

Ramakers, P.M.J. (1983) Mass production and introduction of *Amblyseius mckenziei* and *A. cucumeris*. *International Organisation of Biological Control/West Palaearctic Regional Section Bulletin* 6, 203–206.

Ramakers, P.M.J. (1987) Control of spider mites and thrips with phytoseiid predators on sweet pepper. *International Organisation of Biological Control/West Palaearctic Regional Section Bulletin* 10, 158–159.

Ramakers, P.M.J. (1988) Population dynamics of the thrips predators *Amblyseius mckenziei* and *Amblyseius cucumeris* on sweet pepper. *Netherlands Journal of Agricultural Science* 36, 247–252.

Ramakers, P.M.J. (1990) Manipulation of phytoseiid thrips predators in the absence of thrips. *International Organisation of Biological Control/West Palaearctic Regional Section Bulletin* 13/5, 169–172.

Ramakers, P.M.J. (1993) Coexistence of two thrips predators, the anthocorid *Orius insidiosus* and the phytoseiid *Amblyseius cucumeris* on sweet pepper. *International Organisation of Biological Control/West Palaearctic Regional Section Bulletin* 16, 133–136.

Ramakers, P.M.J. (1995) Biological control using oligophagous predators. In: Parker, B.L., Skinner, M. and Lewis, T. (eds) *Thrips Biology and Management.* Plenum Press, New York, pp. 225–229.

Ramakers, P.M.J. and Voet, S.J.P. (1995) Use of castor bean, *Ricinus communis*, for the introduction of the thrips predator *Amblyseius degenerans* on glasshouse-grown sweet peppers. *Mededelingen Faculteit Landbouwwetenschappen, Rijksuniversiteit Gent* 60/3a, 885–891.

Ramakers, P.M.J. and Voet, S.J.P. (1996) Introduction of *Amblyseius degenerans* for thrips control in sweet peppers with potted castor beans as banker plants. *International Organisation of Biological Control/West Palaearctic Regional Section Bulletin* 19, 127–130.

Ramakers, P.M.J., Disseveld, M. and Peeters, K. (1989) Large scale introductions of phytoseiid predators to control thrips on cucumber. *Mededelingen Faculteit Landbouwwetenschappen Rijksuniversiteit Gent* 54, 923–929.

Ravensberg, W.J., Dissevelt, M., Altena, K. and Simonse, M.P. (1992) Developments in the integrated control of *Frankliniella occidentalis* in capsicum and cucumber. *European Plant Protection Organisation Bulletin* 22, 387–396.

Reid, C.D. and Lampman, R.L. (1989) Olfactory responses of *Orius insidiosus* (Hemiptera: Anthocoridae) to volatiles of corn silks. *Journal of Chemical Ecology* 15, 1109–1115.

Reimer, N..J. (1988) Predation on *Liothrips urichi* Karny (Thysanoptera: Phlaeothripidae): a case of biotic interference. *Environmental Entomology* 17, 132–134.

Richards, P.C. and Schmidt, J.M. (1996) The suitability of some natural and artificial substrates as oviposition sites for the insidiosus flower bug, *Orius insidiosus*. *Entomologia Experimentalis et Applicata* 80, 325–333.

Ridgway, R.L. and Jones, S.L. (1968) Plant feeding by *Geocoris pallens* and *Nabis americoferus*. *Annals of the Entomological Society of America* 61, 232–233.

Riudavets, J. (1995) Predators of *Frankliniella occidentalis* and *Thrips tabaci. Wageningen Agricultural University Papers* 95–1, 43–87.

Riudavets, J., Gabarra, R. and Castane, C. (1993) *Frankliniella occidentalis* predation by native natural enemies. *International Organisation of Biological Control/West Palaearctic Regional Section Bulletin* 10, 165–169.

Riudavets, J., Castañé, C. and Gabarra, R. (1995) Native predators of western flower thrips in horticultural crops. In: Parker, B.L., Skinner, M. and Lewis, T. (eds) *Thrips Biology and Management.* Plenum Press, New York, pp. 255–258.

Robinson, A.G., Stannard, L.J. and Armstrong, E.J. (1972) Observations on predators of *Sericothrips variabilis* Beach (Thysanoptera). *Entomological News* 83, 107–111.

Rodriguez-Reina, J.M., Garcia-Mari, F. and Ferragut, F. (1992) Actividad depredadora de varios ácaros fitoseidos sobre distintos estados de desarrollo del trips de las flores *Frankliniella occidentalis* (Pergande). *Boletin Sanitaria Vegetal Plagas* 18, 253–263.

Rosenheim, J.A., Wilhoit, L.R. and Armer, C.A. (1993) Influence of intraguild predation among generalist insect predators on the suppression of an herbivore population. *Oecologia* 96, 439–449.

Rosenheim, J.A., Kaya, H.K., Ehler, L.E., Marois, J.J. and Jaffee, B.A. (1995) Intraguild predation among biological control agents: theory and evidence. *Biological Control* 5, 303–335.

Ruberson, J.R., Bush, L. and Kring, T.J. (1991) Photoperiodic effect on diapause induction and development in the predator *Orius insidiosus* (Heteroptera: Anthocoridae). *Environmental Entomology* 20, 786–789.

Russel, R.J. (1970) The effectiveness of *Anthocoris nemorum* and *A. confusus* (Hemiptera: Anthocoridae) as predators of the sycamore aphid *Drepanosiphum platanoides*. I. The number of aphids consumed during development. *Entomologia Experimentalis et Applicata* 13, 194–207.

Sabelis, M.W. (1990) How to analyse prey preference when prey density varies? A new method to discriminate between effects of gut fullness and prey type composition. *Oecologia* 82, 289–298.

Sabelis, M.W. (1992) Arthropod predators. In: Crawley, M.J. (ed.) *Natural Enemies: The Population Biology of Predators, Parasites and Diseases*. Blackwell, Oxford, pp. 225–264.

Sabelis, M.W. and Bakker, F.M. (1992) How predatory mites cope with the web of their tetranychid prey: a functional view on dorsal chaetotaxy in the Phytoseiidae. *Experimental and Applied Acarology* 16, 203–225.

Sabelis, M.W. and Dicke, M. (1985) Long-range dispersal and searching behaviour. In: Helle, W. and Sabelis, M.W. (eds) *Spider Mites: Their Biology, Natural Enemies and Control*, World Crop Pest Series, Vol. 1B. Elsevier, Amsterdam, pp. 141–160.

Sabelis, M.W. and Diekmann, O. (1988) Overall population stability despite local extinction: the stabilizing influence of prey dispersal from predator-invaded patches. *Theoretical Population Biology* 34, 169–176.

Sabelis, M.W., Diekmann, O. and Jansen, V.A.A. (1991) Metapopulation persistence despite local extinction: predator–prey patch models of the Lotka–Volterra type. *Biological Journal of the Linnean Society* 42, 267–283.

Salamero, A., Gabarra, R. and Albajes, R. (1987) Observations on the predatory and phytophagous habits of *Dicyphus tamaninii* (Thysanoptera: Thripidae). *International Organisation of Biological Control/West Palaearctic Regional Section Bulletin* 10, 165–169.

Salas-Aguilar, J. and Ehler, L.E. (1977) Feeding habits of *Orius tristicolor*. *Annals of the Entomological Society of America* 70, 60–62.

Salim, M., Masud, S.A. and Khan, A.M. (1987) *Orius albidipennis* (Reut.) (Hemiptera:

Anthocoridae) – a predator of cotton pests. *Philippine Entomologist* 7, 37–42.

Saucedo-Gonzalez, J. and Reyes-Villanueva, F. (1987) Respuesta funcional de *Orius insidiosus* (Hemiptera: Anthocoridae) sobre *Caliothrips phaseoli* (Thysanoptera: Thripidae). *Folia Entomologica Mexicana* 71, 27–35.

Saxena, R.C. (1971) Bionomics of *Caliothrips indicus* (Bagnall) infesting onion (Thripidae: Thysanoptera). *Indian Journal of Entomology* 33, 342–345.

Saxena, R.C. (1977) Integrated approach for the control of *Thrips tabaci* Lind. *Indian Journal of Agricutural Science* 45, 434–436.

Schuster, M.F., Lukefahr, M.J. and Maxwell, F.G. (1976) Impact of nectariless cotton on plant bugs and natural enemies. *Journal of Economic Entomology* 69, 400–402.

Schwartz, A. (1988) Population dynamics of *Thrips tabaci* Lindeman (Thysanoptera: Thripidae) on table grapes. *South African Journal of Enology and Viticulture* 9, 19–21.

Sciarappa, W.J. and Swift, F.C. (1977) Biological studies of *Typhlodromips sessor* (Acarina: Phytoseiidae). *Annals of the Entomological Society of America* 70, 285–288.

Scutareanu, P., Drukker, B., Bruin, J., Posthumus, M.A. and Sabelis, M.W. (1996) Leaf volatiles and polyphenols in pear trees infested by *Psylla pyricola*. Evidence of simultaneously induced responses. *Chemoecology* 7, 34–38.

Selhime, A.G., Muma, M.H. and Clancy, D.W. (1962) Biological, chemical and ecological studies on the predatory thrips *Aleurodothrips fasciapennis* in Florida citrus groves. *Annals of the Entomological Society of America* 56, 709–712.

Selhorst, T., Söndgerath, D. and Weigand, S. (1991) A model describing the predator–prey interaction between *Scolothrips longicornis* and *Tetranychus cinnabarinus* based upon the Leslie theory. *Ecological Modelling* 59, 123–138.

Şengonca, Ç. and Bendiek, J. (1988) Suitablity of two predatory mites as a biological control agent of *Frankliniella occidentalis* (Pergande) (Thysanoptera: Thripidae). *Narichtenblatt der Deutschen Pflanzenschutzdienst* 40, 171–175.

Sharga, U.S. (1933) Biology and life history of *Limothrips cerealium* Haliday and *Aptinothrips rufus* Gmelin feeding on Gramineae. *Annals of Applied Biology* 20, 308–326.

Shereef, G.M., Soliman, Z.R. and Afifi, A.M. (1980) Economic importance of the mite *Hypoaspis miles* (Berlese) (Mesostigmata: Laelapidae) and its life history. *Bulletin du Societé Zoologique d'Egypt* 30, 103–108.

Shields, E.J. and Watson, T.F. (1980) Searching behavior of female *Orius tristicolor*. *Annals of the Entomological Society of America* 73, 533–535.

Shih, C.I.T. and Shieh. J.N. (1979) Biology, life table, predation potential and intrinsic rate of increase of *Amblyseius longispinosus* (Evans). *Plant Protection Bulletin Taiwan* 21, 175–183.

Shipp, J.L. and Whitfield, G.H. (1991) Functional response of the predatory mite, *Amblyseius cucumeris* (Acari: Phytoseiidae), on western flower thrips, *Frankliniella occidentalis* (Thysanoptera: Thripidae). *Environmental Entomology* 20, 694–699.

Shipp, J.L. and Van Houten, Y.M. (1997) Influence of temperature and vapor pressure deficit on survival of the predatory mite *Amblyseius cucumeris* (Acari: Phytoseiidae). *Environmental Entomology* 26, 106–113.

Shipp, J.L., Zariffa, N. and Ferguson, G. (1992) Spatial patterns of and sampling methods for *Orius* spp. (Hemiptera: Anthocoridae) on greenhouse sweet pepper. *Canadian Entomologist* 124, 887–894.

Shipp, J.L., Ward, K.I. and Gilespie, T.J. (1996) Influence of temperature and vapor pressure deficit on the rate of predation by the predatory mite, *Amblyseius cucumeris*, on *Frankliniella occidentalis*. *Entomologia Experimentalis et Applicata* 78, 31–38.

Shurovenkov, B.G. (1974) Biology of *Haplothrips tritici* larvae and their predators in the European forest-steppe. *Zoologischeskii Zhurnal* 53, 1096–1098.

Simonishvili, R.A. (1976) Trophic relations of *Amblyseius largoensis* (Muma) and *Typhlodromus pyri* (Phytoseiidae: Gamasoidea). (In Russian.) *Subtropicheskie Kul'tury* 3/4, 149–151.

Sopp, P.I., Sunderland, K.D., Fenlon, J.S. and Wratten, S.D. (1992) An improved quantitative method for estimating invertebrate predation in the field using an enzyme-linked immunosorbent assay (ELISA). *Journal of Animal Ecology* 29, 295–302.

Sorenson, J.T., Kinn, D.N., Doutt, R.L. and Cate, J.R. (1976) Biology of the mite, *Anystis agilis* (Acari: Anystidae): a California vineyard predator. *Annals of the Entomological Society of America* 69, 905–910.

Stannard, L.J. (1952) Peanut-winged thrips. *Annals of the Entomological Society of America* 45, 327–330.

Steiner, M.Y. (1990) Determining population characteristics and sampling procedures for the western flower thrips (Thysanoptera: Thripidae) and the predatory mite *Amblyseius cucumeris* (Acarina: Phytoseiidae) on greenhouse cucumber. *Environmental Entomology* 19, 1605–1613.

Steiner, M.Y. and Bjørnson, S. (1996) Performance of *Phytoseiulus persimilis* and other biological control agents – on what are we basing our standards? – *International Organisation of Biological Control/West Palaearctic Regional Section Bulletin* 19, 163–166.

Stoltz, R.L. and McNeal, C.D.J.R. (1982) Assessment of insect migration from alfalfa hay to bean fields. *Environmental Entomology* 11, 578–580.

Stoltz, R.L. and Stern, V.M. (1978) The longevity and fecundity of *Orius tristicolor* when introduced to increasing numbers of the prey *Frankliniella occidentalis*. *Environmental Entomology* 7, 197–198.

Stuckenberg, B.R. (1954) The immature stages of *Sphaerophoria quadrituberculata* Beazi (Diptera: Syrphidae), a predator of *Cercothrips afer* Priesner. *Journal of the Entomological Society of South Africa* 17, 58–61.

Sundby, R.A. (1966) A comparative study of the efficiency of three predatory insects *Coccinella septempunctata* L. (Coleoptera, Coccinelidae), *Chrysopa carnea* St. (Neuroptera, Chrysopidae) and *Syrphus ribesii* L. (Diptera, Syrphidae) at two different temperatures. *Entomophaga* 11, 395–404.

Sureshkumar, N. and Ananthakrishnan, T.N. (1984) Predator prey interactions with reference to *Orius maxidentex* Ghauri and *Carayonocoris indicus* Muraleedharan. *Proceedings of the Indian National Science Academy Part B* 50, 139–145.

Sureshkumar, N. and Ananthakrishnan, T.N. (1985) *Geocoris ochropterus* Fabr. as a predator of some thrips. *Proceedings of the Indian National Science Academy* 30, 139–145.

Sureshkumar, N. and Ananthakrishnan, T.N. (1987) Biotic interactions in relation to prey–predator relationship with special reference to some thrips species (Thysanoptera : Insecta). *Journal of Entomological Research* 11, 192–202.

Suzuki, T., Haga, K., Kodama, S., Watanabe, K. and Kuwahara, Y. (1988) Secretions of thrips: II. Secretions of three gall–inhabiting thrips (Thysanoptera: Phlaeothripidae). *Applied Entomology and Zoology* 23, 291–297.

Swirski, E. and Dorzia, N. (1968) Laboratory studies on the feeding, development and fecundity of the predacious mite *Amblyseius limonicus* Garman and McGregor

(Acarina: Phytoseiidae) on various kinds of food substances. *Israel Journal of Agricultural Research* 18, 71–75.

Swirski, E. and Dorzia, N. (1969) Laboratory studies on the feeding, development and fecundity of the predacious mite *Typhlodromus occidentalis* Nesbitt (Acarina: Phytoseiidae) on various kinds of food substances. *Israel Journal of Agricultural Research* 19, 143–145.

Swirski, E., Amitai, S. and Dorzia, N. (1967a) Laboratory studies on the feeding, development and oviposition of the predacious mite *Typhlodromus athiasae* Pritchard and Schuster (Acarina: Phytoseiidae) on various kinds of food substances. *Israel Journal Agricultural Research* 17, 213–218.

Swirski, E., Amitai, S. and Dorzia, N. (1967b) Laboratory studies on the feeding, development and reproduction of the predacious mite *Amblyseius rubini* Swirski and Amitai and *A. swirskii* Athias (Acarina: Phytoseiidae) on various kinds of food substances. *Israel Journal of Agricultural Research* 17, 1–119.

Swirski, E., Amitai, S. and Dorzia, N. (1970) Laboratory studies on the feeding habits, post–embryonic survival and oviposition of the predatory mites, *Amblyseius chilenensis* Dosse and *Amblyseius hibisci* Chant (Acarina: Phytoseiidae) on various kinds of food substances. *Entomophaga* 15, 93–106.

Takafuji, A. and Chant, D.A. (1976) Comparative studies of two species of predacious phytoseiid mites (Acarina: Phytoseiidae), with special reference to their responses to the density of their prey. *Researches in Population Ecology* 17, 255–310.

Tamaki, G. and Weeks, R.E. (1972) Biology and ecology of two predators, *Geocoris pallens* Stal. and *G. bullatus* (Say). *USDA, Technical Bulletin* 1446, 1–45.

Tanaka, M. and Kashio, T. (1977) Biological studies on *Amblyseius largoensis* Muma (Acari: Phytoseiidae) as a predator of the citrus red mite, *Panonychus citri* (McGregor) (Acarina: Tetranychidae). *Bulletin of the Fruit Tree Research Station, Japan*, D1, 49–67.

Tanigoshi, L.K. (1991) Biological control of citrus thrips, *Scirtothrips citri*, by predacious phytoseiid mites. In: Parker, B.L., Skinner, M. and Lewis, T. (eds) *Towards Understanding Thysanoptera*. General Technical Report NE-147, USDA, Forest Service, Radnor, PA, pp. 399–418.

Tanigoshi, L.K., Fargerlund, J. and Nishio-Wong, J.Y. (1981) Significance of temperature and food resources to the developmental biology of *Amblyseius hibisci* (Chant) (Acarina, Phytoseiidae). *Zeitschrift für Angewandte Entomologie* 92, 409–419.

Tanigoshi, L.K., Nishio-Wong, J.Y. and Fargerlund, J. (1983) Greenhouse and laboratory rearing studies of *Euseius hibisci* (Chant) (Acarina: Phytoseiidae), a natural enemy of the citrus thrips, *Scirtothrips citri* (Moulton) (Thysanoptera: Thripidae). *Environmental Entomology* 12, 1298–1302.

Tanigoshi, L.K., Nishio-Wong, J.Y. and Fargerlund, J. (1984) *Euseius hibisci*: its control of citrus thrips in southern California citrus orchards. In: Griffiths, D.A. and Bowman, C.E. (eds) *Acarology VI*, Vol. 2. Ellis Horwood, Chichester, pp. 717–724.

Tanigoshi, L.K., Nishio-Wong, J.Y., Fargerlund, J. and Griffiths, H.J. (1985) Biological control of citrus thips, *Scirtothrips citri* (Thysanoptera: Thripidae), in southern California citrus groves. *Environmental Entomology* 14, 733–741.

Tavella, L., Arzone, A.N and Alma, A. (1991) Researches on *Orius laevigatus*, a predator of *Frankliniella occidentalis* in greenhouse. A preliminary note. *International Organisation of Biological Control/West Palaearctic Regional Section Bulletin* 14, 65–72.

Tawfik, M.F.S. and Abouzeid, N.A. (1977) Biology of the rove beetle *Paederus alfierii*

(Coleoptera, Staphylinidae). *Pedobiologia* 17, 51–59.

Tawfik, M.F.S. and Ata, A.M. (1973a) The life history of *Orius albidipennis* (Reut.) (Hemiptera: Anthocoridae). *Bulletin du Societé Entomologique d'Egypte* 57, 117–126.

Tawfik, M.F.S. and Ata, A.M. (1973b) The life history of *Orius laevigatus* (Fieber) (Hemiptera: Anthocoridae). *Bulletin du Societé Entomologique d'Egypte* 57, 145–151.

Tawfik, M.F.S. and Nagui, A. (1965) The biology of *Montandoniella moraguesi* Puton, a predator of *Gynaikothrips ficorum* Marchal in Egypt. *Bulletin du Société Entomologique d'Egypte* 49, 181–200.

Tawfik, M.F.S., Azab, A.K. and Awadallah, K.T. (1974) Studies on the life history and descripton of the immature form of the Egyptian aphidophagous syrphids IV. *Sphaerophoria flavicauda* Zett. (Diptera: Syrphidae). *Bulletin du Societé Entomologique d'Egypte* 58, 103–115.

Taylor, E.J. (1949) A life history study of *Nabis alternatus*. *Journal of Economic Entomology* 42, 991.

Teerling, C.R. (1995) Chemical ecology of western flower thrips. In: Parker, B.L., Skinner, M. and Lewis, T. (eds) *Thrips Biology and Management*. Plenum Press, New York, pp. 436–447.

Teerling, C.R., Gillespie, D.R. and Borden, J.H. (1993a) Utilization of western flower thrips alarm pheromone as a prey-finding kairomone by predators. *Canadian Entomologist* 125, 431–437.

Teerling, C.R., Pierce Jr, H.D., Borden, J.H. and Gillespie, D.R. (1993b) Identification and bioactivity of alarm pheromone in the western flower thrips, *Frankliniella occidentalis*. *Journal of Chemical Ecology* 19, 681–697.

Tellier, A.J. and Steiner, M.Y. (1990) Control of the western flower thrips, *Frankliniella occidentalis* (Pergande) with a native predator *Orius tristicolor* (White) in greenhouse cucumbers and peppers in Alberta, Canada. *International Organisation of Biological Control/West Palaearctic Regional Section Bulletin* 5, 209–211.

Thompson, W.R. and Simmonds, F.J. (1965) *A Catalogue of the Parasites and Predators of Insect Pests. Section 4. Host Predator Catalogue*. CAB, Farnham Royal, UK, 198 pp.

Tominic, A. (1950) Observations on the biology and control of the olive thrips (*Liothrips oleae* Costa). *Review of Applied Entomology* (A) 41, 245–246.

Tommasini, M.G. and Nicoli, G. (1993) Adult activity of four *Orius* species reared on two preys. *International Organisation of Biological Control/West Palaearctic Regional Section Bulletin* 16, 181–184.

Tommasini, M.G. and Nicoli, G. (1995) Evaluation of *Orius* spp. as biological control agents of thrips pests: initial experiments on the existence of diapause in *Orius laevigatus*. *Mededelingen Faculteit Landbouwwetenschappen, Rijksunivrsiteit Gent* 60/3a, 901–908.

Tommasini, M.G. and Nicoli, G. (1996) Evaluation of *Orius* spp. as biological control agents of thrips pests: further experiments on the existence of diapause in *Orius laevigatus*. *International Organisation of Biological Control/West Palaearctic Regional Section Bulletin* 19, 183–186.

Trichilo, P. and Leigh, T.F. (1986) Predation on spider mite eggs by the western flower thrips, *Frankliniella occidentalis*, an opportunist in a cotton agroecosystem. *Environmental Entomology* 15, 821–825.

Trjapitzin, S.V. (1995) Biological control of *Thrips tabaci* on Tobacco seedlings in Ukraine. In: Parker, B.L., Skinner, M. and Lewis, T. (eds) *Thrips Biology and Management*. Plenum Press, New York, pp. 237–240.

Trottin-Caudal, Y., Grassely, D., Trapateau, M., Dobelin, H. and Millot, P. (1991) Lutte biologique contre *Frankliniella occidentalis* avec *Orius majusculus* sur concombre. *International Organisation of Biological Control/West Palaearctic Regional Section Bulletin* 14, 50–56.

Van Baalen, M. and Sabelis, M.W. (1995). The milker–killer dilemma in spatially structured predator–prey interactions. *Oikos* 74, 391–400.

Van Baalen, M., Van Rijn, P.C.J. and Sabelis, M.W. (1997) Alternative food, switching predators, and the persistence of predator–prey systems. *Evolutionary Ecology* (in press).

Van de Veire, M. and DeGheele, D. (1992) Biological control of the western flower thrips, *Frankliniella occidentalis* (Pergande) (Thysanoptera: Thripidae), in glasshouse sweet peppers with *Orius* spp. (Hemiptera: Anthocoridae). A comparative study between *O. niger* and *O. insidiosus* (Say). *Biocontrol Science and Technology* 2, 281–283.

Van den Bosch, F. and Diekmann, O. (1986) Interactions between egg-eating predator and prey: the effect of the functional response and of age structure. *IMA Journal of Mathematics Applied in Medicine and Biology* 3, 53–69.

Van den Meiracker, R.A.F. (1994) Induction and termination of diapause in *Orius* predatory bugs. *Entomologia Experimentalis et Applicata* 73, 127–134.

Van den Meiracker, R.A.F. and Ramakers, P.M.J. (1991) Biological control of the western flower thrips *Frankliniella occidentalis*, in sweet pepper with the anthocorid predator *Orius insidiosus*. *Mededelingen van de Faculteit Landbouwwetenschappen, Rijksuniversiteit Gent* 56/2a, 241–249.

Van den Meiracker, R.A.F. and Sabelis, M.W. (1993) Oviposition sites of *Orius insidiosus* in sweet pepper. *Proceedings IOBC working group, IPM Glasshouses:* 109–112.

Van der Hoeven, W.A.D. and Van Rijn, P.C.J. (1990) Factors affecting the attack success of predatory mites on thrips larvae. *Proceedings of the Dutch Entomological Society, Section Experimental and Applied Entomology* 1, 25–30.

Van Doesburg, P.H. (1964) *Termatophylidea opaca* Carvalho, a predator of thrips (Hem. – Het.). *Entomologische Berichten (Amsterdam)* 24, 248–253.

Van Houten, Y.M. (1996) Biological control of western flower thips on cucumber using the predatory mites *Amblyseius cucumeris* and *A. limonicus*. *International Organisation of Biological Control/West Palaearctic Regional Section Bulletin* 19, 59–62.

Van Houten, Y.M. and Van Lier, A.M.M. (1995) Influence of temperature and humidity on the survival of eggs of the thrips predator *Amblyseius cucumeris*. *Mededelingen Faculteit Landbouwwetenschappen, Universiteit van Gent* 60/3a, 879–884.

Van Houten, Y.M. and Van Stratum, P. (1993) Biological control of western flower thrips in greenhouse sweet peppers using non-diapausing predatory mites. *International Organisation of Biological Control/West Palaearctic Regional Section Bulletin* 16, 77–80.

Van Houten, Y.M. and Van Stratum, P. (1995) Control of western flower thrips on sweet pepper in winter with *Amblyseius cucumeris* (Oudemans) and *A. degenerans* Berlese. In: Parker, B.L., Skinner, M. and Lewis, T. (eds) *Thrips Biology and Management*. Plenum Press, New York, pp. 245–248.

Van Houten, Y.M., Van Rijn, P.C.J., Tanigoshi, L.K. and Van Stratum, P. (1993) Potential of phytoseiid predators to control western flower thrips in greenhouse crops, in particular during the winter period. *International Organisation of Biological Control/West Palaearctic Regional Section Bulletin* 16/8, 98–101.

Van Houten, Y.M., Van Rijn, P.C.J., Tanigoshi, L.K., Van Stratum, P. and Bruin, J. (1995a) Preselection of predatory mites to improve year-round biological control of western flower thrips in greenhouse crops. *Entomologia Experimentalis et Applicata* 74, 225–234.

Van Houten, Y.M., Van Stratum, P., Bruin, J. and Veerman, A. (1995b) Selection for non-diapause in *Amblyseius cucumeris* and *Amblyseius barkeri* and exploration of the effectiveness of selected strains for thrips control. *Entomologia Experimentalis et Applicata* 77, 289–295.

Van Rijn, P.C.J. (1992) Do plants reduce herbivore attack by providing pollen? In: Menken, S.B.J., Visser, J.H. and Harrewijn, P. (eds) *Proceedings 8th International Symposium on Insect–Plant Relationships*, Kluwer Academic Publishers, The Hague, The Netherlands, pp. 351–352.

Van Rijn, P.C.J. and Sabelis, M.W. (1990) Pollen availability and its effect on the maintenance of populations of *Amblyseius cucumeris*, a predator of thrips. *Mededelingen Faculteit Landbouwwetenschappen Rijksuniversiteit Gent* 55/2a, 335–341.

Van Rijn, P.C.J. and Sabelis, M.W. (1993) Does alternative food always enhance biological control? The effect of pollen on the interaction between western flower thrips and its predators. *International Organisation of Biological Control/West Palaearctic Regional Section Bulletin* 16, 123–125.

Van Rijn, P.C.J. and Sabelis, M.W. (1997) Should plants provide food for predators when it also benefits the herbivores? Effects of pollen on a thrips–predatory mite system. In: Mitchell, R., Horn, D.Z., Meedham, G.R. and Welbourn, W.C. (eds) *Acarology IX: volume 2, Proceedings, Ohio Biological Survey, Columbus, Ohio.*

Van Rijn, P.C.J. and Tanigoshi, L.K. (1997a) The contribution of extrafloral nectar to the survival and reproduction of the predatory mite *Iphiseius degenerans* on *Ricinus communis*. *Experimental and Applied Acarology* (in press).

Van Rijn, P.C.J. and Tanigoshi, L.K. (1997b) Life history and pollen diet range of *Iphiseius degenerans* and *Neoseiulus cucumeris* (Acari: Phytoseiidae). *Experimental and Applied Acarology* (in press).

Van Rijn, P.C.J. and Van Houten, Y.M. (1991) Life history of *Amblyseius cucumeris* and *Amblyseius barkeri*, (Acarina: Phytoseiidae) on a diet of pollen. In: Dusbabek, F. and Bukva, V. (eds) *Modern Acarology*, Vol. 2. Academia, Prague and SPB Academic Publishing BV, The Hague, 647–654.

Van Rijn, P.C.J., Mollema, C. and Steenhuis-Broers, M.M. (1995) Comparative life history studies of *Frankliniella occidentalis* and *Thrips tabaci* (Thysanoptera: Thripidae) on cucumber. *Bulletin of Entomological Research* 85, 285–297.

Varadarasan, S. and Ananthakrishnan, T.N. (1981) Population dynamics and prey/predator relationships and gall forming thrips. *Proceedings of the Indian National Science Academy Part B* 47, 321–340.

Vardy, C.R. (1987) The New Zealand fauna of the thrips-hunting solitary wasp genus *Spilomena* Shuckard (Hymenoptera: Sphecidae). *Systematic Entomology* 12, 249–256.

Vartapetov, S.G. and Vasilev, P.A. (1971) *Amblyseius swirskii* Athias–Henriot (Parasitiformes: Phytoseiidae) predator of tobacco thrips (*Thrips tabaci* Lind.). (In Russian.) *Biologickeskii Zhurnal Armyanskoí SSR* 24, 106–107.

Villevieille, M. and Millot, P. (1991) Lutte biologique contre *Frankliniella occidentalis* avec *Orius laevigatus* sur fraisier. *International Organisation of Biological Control/West Palaearctic Regional Section Bulletin* 14, 57–64.

Viswanathan, T.R. and Ananthakrishnan, T.N. (1974) Population fluctuations of 3 species of anthophilous Thysanoptera in relation to the numerical response of their predator *Orius minutus* L. (Anthocoridae: Hemiptera). *Current Science (Bangalore)* 43, 19–20.

Wang, Ch.-L. (1995) Predatory capacity of *Campylomma chinensis* Schuh (Hemiptera: Miridae) and *Orius sauteri* (Poppius) (Hemiptera: Anthocoridae) on *Thrips palmi*. In: Parker, B.L., Skinner, M. and Lewis, T. (eds) *Thrips Biology and Management*. Plenum Press, New York, pp. 259–262.

Wang, F., Zhou, W. and Wang, R. (1994) Studies on the overwintering survival rate of *Orius sauteri* in Beijing. *Journal of Biological Control* 10, 100–102.

Warren, L.O. and Tadic, M. (1967) Biological observations on *Coleomegilla maculata* and its role as a predator of the fall webworm. *Journal of Economic Entomology* 60, 1492–1496.

Warren, P.H. and Lawton, J.H. (1987) Invertebrate predator–prey body size relationships: an explanation for upper triangular food webs and patterns in food web structure? *Oecologia* 74, 231–235.

Wei, C., Peng, Z.J., Yang, G.Q., Huang, B.Z. and Chen, X. (1984) On the flower bug, *Orius similis* Zheng. *Natural Enemies of Insects* 6, 32–40.

Wheeler, A.G. (1977) Studies on the arthropod fauna of alfalfa. VII Predaceous insects. *Canadian Entomologist* 109, 423–427.

Wilson, L.T., Trichilo, P. and González, D. (1991) Natural enemies of spider mites on cotton: density regulation or casual association? *Environmental Entomology* 20, 849–856.

Wilson, L.T., Bauer, L.R. and Walter, G.H. (1996) 'Phytophagous' thrips are facultative predators of two-spotted spider mites (Acari: Tetranychidae) on cotton in Australia. *Bulletin of Entomological Research* 86, 297–305.

Wittmann, E.J. and Leather, S.R. (1997) Compatibility of *Orius laevigatus* Fieber (Hemiptera: Anthocoridae) with *Amblyseius cucumeris* Oudemans (Acari: Phytoseiidae) and *Amblyseius degenerans* Berlese (Acari: Phytoseiidae) for the biological control of *Frankliniella occidentalis* Pergande (Thysanoptera: Thripidae). *Experimental and Applied Acarology* 21, 523–538.

Wright, E.J. and Laing, J.E. (1978) The effects of temperature on development, adult longevity and fecundity of *Coleomegilla maculata lengi* and its parasite *Perilitus coccinellae*. *Proceedings of the Entomological Society of Ontario* 109, 33–47.

Yano, E. (1996) Biology of *Orius sauteri* (Poppius) and its potential as a biocontrol agent for *Thrips palmi*. *International Organisation of Biological Control/West Palaearctic Regional Section Bulletin* 19, 203–206.

Yasunaga, T. (1995) A new species of the genus *Wollastoniella* Reuter (Heteroptera, Anthocoridae), predator of *Thrips palmi* (Thysanoptera) in aubergine gardens of Thailand. *Applied Entomology and Zoology* 30, 203–205.

Yasunaga, T. and Miyamoto, S. (1993) Three anthocorid species (Heteroptera: Anthocoridae), predators of *Thrips palmi* (Thysanoptera) in aubergine gardens of Thailand. *Applied Entomology and Zoology* 28, 227–232.

Yokoyama, V.Y. (1978a) Method for rearing *Geocoris pallens*, a predator in California cotton. *Canadian Entomologist* 112, 1–3.

Yokoyama, V.Y. (1978b) Relation of seasonal changes in extrafloral nectar and foliar protein and arthropod populations in cotton. *Environmental Entomology* 7, 799–802.

Yousef, A.A. and El-Halawany, M.E. (1982) Effect of prey species on the biology of *Amblyseius gossipi* El-Badry (Acari: Phytoseiidae). *Acarologia* 23, 113–117.

Yue, B. and Tsai, J.H. (1996) Development, survivorship and reproduction of *Amblyseius largoensis* (Acari: Phytoseiidae) on selected plant pollens and temperatures. *Environmental Entomology* 25, 488–494.

Zavodchikova, V.V. (1974) The feeding, development and fecundity of *Deraeocoris* (*Camptobrochis*) *punctulatus* Fall. (Heteroptera: Miridae) on different

diets. *Entomological Reviews* 53, 94–97.
Zhimo, Z. and McMurtry, J.A. (1990) Development and reproduction of three *Euseius* (Acari: Phytoseiidae) species in the presence and absence of supplementary foods. *Experimental and Applied Acarology* 8, 233–242.
Zhou, W. and Wang, R. (1989) Rearing of *Orius sauteri* (Hem.: Anthocoridae) with natural and artificial diets. *Chinese Journal of Biological Control* 5, 9–12.
Zhou, W., Wang, R. and Qiu, S. (1991) Use of soybean sprouts as the oviposition material in mass rearing of *Orius sauteri* (Het.: Anthocoridae). *Chinese Journal of Biological Control* 7, 7–9.
Zur Strassen, R. (1995) Binomial data of some predacious thrips. In: Parker, B.L., Skinner, M. and Lewis, T. (eds) *Thrips Biology and Management*. Plenum Press, New York, pp. 325–328.

9

Interactions with Hymenopterous Parasitoids and Parasitic Nematodes

Antoon J.M. Loomans[1], Tamotsu Murai[2] and Ian D. Greene[3]*

[1]*Section Entomology, Plant Protection Service, Geertjesweg 15, PO Box 9102, 6700 HC Wageningen, The Netherlands;* [2]*Research Institute for Bioresources, Okayama University, 2-20-1 Chuo, Kurashiki 710, Japan;* [3]*6735 Mesquite Way, Salinas, California 93907, USA*

Introduction

Recently, there has been renewed interest in the use of hymenopterous parasitoids as potential biological control agents of thrips pests. Classical biological control of thrips, by inoculative releases of parasitoids, has been tried in the past without great success. Efforts to incorporate thrips parasitoids into greenhouse integrated pest management (IPM) programmes by inoculative releases or inundation to control thrips pests are a more recent development. Most attention has been paid to augmentative releases of generalist predators like *Amblyseius* and *Orius* (Riudavets, 1995; Chapter 8), but since control of thrips pests like *Thrips palmi* and *Frankliniella occidentalis* by predators is only partly successful in some crops, the option of using parasitoids is worth consideration (Loomans *et al.*, 1995a).

Hymenopterous parasitoids of thrips all belong to the superfamily Chalcidoidea. Most are solitary endoparasitoids of eggs (Trichogrammatidae, Mymaridae) or larvae (Eulophidae). In the following account, thrips parasitoid species of these two groups will be discussed and characterized by their taxonomy and distribution, their ecology (host range, host plants), behaviour (host selection, searching) and biology (development, life history). This information is used to evaluate the potential role of thrips parasitoids for biocontrol, related to the type of programme and the type of thrips pest involved. Details on how to culture thrips are dealt with in Chapter 12. For a comprehensive review on thrips parasitoids, we refer to Loomans and van Lenteren (1995), on which this chapter is based, with new material added.

A short section at the end of the chapter deals with nematodes parasitic on thrips.

*Contributor on parasitic nematodes.

Larval Parasitoids

Taxonomy

Except for a single species in the genus *Thripastichus* (Tetrastichinae) and *Pediobius* (Entedontinae), all larval parasitoids can be found in four closely related genera: *Ceranisus*, *Goetheana*, *Thripobius* and *Entedonastichus* (Entedontinae) (Boucek, 1976). They vary largely in morphology, behaviour and biology and currently the placement of species is under review (Triapitsyn and Headrick, 1995).

They all are minute (0.5 to 1.1 mm), solitary endoparasitoids of thrips larvae, although sometimes the propupae and/or pupae may be attacked. Currently, 27 species (Table 9.1a,b) are described, parasitizing over 70 species of Thysanoptera (Loomans and van Lenteren, 1995, also for synonyms of genera and species), many of which are known as pests. In some records mentioning *Ceranisus* sp. (e.g. Saxena, 1971; Siddapaji and Reddy, 1974; Hirose, 1989; Greene and Parrella, 1993), specimens have been named, in others (e.g. Chiu, 1984) they await identification. General keys to eulophids, including thrips-parasitizing genera, are provided by, for example, Graham (1959, 1987), Boucek (1988) and Schauff (1991), and to trichogrammatids by Doutt and Viggiani (1968). A description and keys to some species in this group are presented by Graham (1959, 1963), Boucek and Askew (1968), Erdös (1971) and Triapitsyn and Headrick (1995), but no key is available yet to all species of the world.

A few species in the genera *Psilogaster* and *Orasema* (Eucharitidae) are reported as ectoparasites of thrips larvae (Clausen, 1940; Beshear, 1974; Johnson, 1988). Sphecid species (Sphecoidea) in the genera *Spilomena*, *Ammoplanus* and *Xysma* (Muesebeck *et al.*, 1951; Krombein, 1958) and *Microstigmus* (Matthews, 1970) have been reported as preying on thrips larvae (see Table 8.1b) and *Aphanogmus fumipennis* (Thomson) (Ceraphronidae) was reared from *Thrips tabaci* (Dessart and Bournier, 1971), but they will not be further discussed here.

Distribution of species

Species within the genus *Ceranisus* are cosmopolitan (*C. menes*: from tropical, subtropical and temperate regions), palaearctic (*C. lepidotus, pacuvius, planitianus*: European), nearctic (*C. americensis, loomansi, russelli, nubilipennis*), neotropical (*C. nigrifemora*), oriental (*C. bicoloratus, femoratus, javae, maculatus, vinctus*) or Australasian (*C. margiscutum*), the last eight species have been recorded incidentally (references in Loomans and van Lenteren, 1995). *Thripobius* species have been recorded from tropical and subtropical areas of Africa, Asia and Australia (Table 9.1b). Whereas *T. hirticornis* only occurs in Africa (Ghana, Tanzania, Uganda, Zimbabwe, Kenya), *T. semiluteus* has been reported from Africa (Sao Tomé: Boucek, 1976; South Africa: Steyn *et al.*, 1993), Asia (India: Boucek, 1976), Japan (collection of the British Museum, Natural History), Australia (Boucek, 1988) and South America (Brazil: LaSalle and McMurtry, 1989). Between 1986 and 1988 *T. semiluteus* was introduced from Australia (McMurtry *et al.*, 1991) and Brazil (LaSalle and McMurtry,

Table 9.1a. Species of larval endoparasitoids in the genus *Ceranisus* (Hymenoptera: Eulophidae) associated with thrips (Thysanoptera).

Species name	Distribution	Major host genera	Category
Genus *Ceranisus* Walker 1840			
Ceranisus menes (Walker)	Cosmopolitan	*Thrips, Frankliniella, Megalurothrips, Limothrips, Ramaswamiahiella, Microcephalothrips, Craspedothrips, Taeniothrips, Kakothrips*; *Rhipiphorothrips, Zaniothrips, Retithrips*	1 2
Ceranisus lepidotus Graham	Europe	*Limothrips, Frankliniella**	1
Ceranisus vinctus (Gahan)	Philippines	*Megalurothrips*	1
Ceranisus femoratus (Gahan)	Philippines	*Megalurothrips*	1
Ceranisus margiscutum (Girault)	Australia	?	
Ceranisus planitianus Erdös	Central Europe	?	
Ceranisus americensis (Girault)	USA, Canada	*Frankliniella, Thrips*	1
Ceranisus pacuvius Walker	Europe	*Kakothrips, Odontothrips*	1
Ceranisus loomansi Triapitsyn and Headrick	W. USA	*Frankliniella**, *Thrips**	1
Ceranisus nigrifemora DeSantis	Argentina	?	
Ceranisus russelli (Crawford)	USA	*Caliothrips, Heliothrips**, *Hercinothrips**, *Thrips, Frankliniella*	2.1
Ceranisus maculatus (Waterston)	India	*Rhipiphorothrips*	2
Ceranisus javae (Girault)	Indonesia	? *Thrips*	?
Ceranisus bicoloratus (Ishii)	Japan	*Liothrips*	3
Ceranisus nubilipennis (Williams)	N.E. USA	*Cryptothrips, Megalothrips*	3

*Laboratory host; category: 1 = Terebrantia, Thripidae; 2 = Terebrantia, Thripidae, Panchaetothripinae; 3 = Tubulifera.

Table 9.1b. Species of larval endoparasitoids in the genera *Thripobius*, *Goetheana*, *Entedonastichus*, *Pediobius* and *Thripastichus* (Hymenoptera: Eulophidae) associated with thrips (Thysanoptera) (for synonyms see Loomans and van Lenteren, 1995).

Species name	Distribution	Major host genera	Category
Genus *Thripobius* Ferrière 1938			
Thripobius semiluteus Boucek	Subtropical	*Heliothrips*, *Hercinothrips*, *Panchaetothrips*, *Brachyurothrips*	2
Thripobius hirticornis Ferrière	East Africa	*Retithrips*	2
Genus *Goetheana* Girault 1920			
Goetheana shakespearei Girault	Subtropical	*Selenothrips*, *Heliothrips*, *Hercinothrips*, *Caliothrips*	2
		Thrips, *Pseudodendrothrips*, *Echinothrips*	1
Goetheana incerta Annecke	South Africa	*Scirtothrips*	1
Genus *Entedonastichus* Girault 1920			
Entedonastichus albicoxis (Szelényi)	Hungary	?	
Entedonastichus carbonarius (Erdös)	Europe, Australia	?	
Entedonastichus dei (Girault)	Australia	?	
Entedonastichus gaussi (Ferrière)	Central Europe	*Liothrips*, *Phlaeothrips*, *Acanthothrips*, *Cryptothrips*	3
Entedonastichus mirus Girault	Australia	?	
Entedonastichus kaulbarsi (Yoshimoto)	North America	?	
Genus *Thripastichus* Graham 1987			
Thripastichus gentilei (del Guercio)	Subtropical	*Liothrips*, *Gynaikothrips*, *Schedothrips*, *Manothrips*, *Mallothrips*,	3
		Rhipiphorothrips	2
Genus *Pediobius* Walker 1846			
Pediobius thysanopterous (Burks)	Egypt, Israel	*Gynaikothrips*	3

Category: 1 = Terebrantia, Thripidae, Thripinae; 2 = Terebrantia, Thripidae, Panchaetothripinae; 3 = Tubulifera.

1989) into California, and from there into Hawaii (M. Early, personal communication, 1990) and Israel (Wysocki, personal communication, 1991).

Goetheana shakespearei Girault is Australasian, Afrotropical, oriental, neotropical and palaearctic, recorded from Australia (Girault, 1920a; Goodwin and Steiner, 1995), Indonesia (van Heurn, 1923), India (Narayanan *et al.*, 1960) Japan (Takagi, 1988: *Goetheana* sp.), Ghana (Ferrière, 1931), South Africa (Annecke, 1962), Benin (M. Tamò, personal communication), Venezuela (Annecke, 1962) and the Bahamas (Bennett and Baranowski, 1982); *G. shakespearei* has also been found in Spain (Viggiani and Nieves-Aldrey, 1993) and Bulgaria (*Goetheana* sp., V. Pelov, personal collection). *Goetheana incerta* Annecke has been recorded from South Africa (Annecke, 1962) only.

Species belonging to the genus *Entedonastichus* are palaearctic (*E. albicoxus*, *E. gaussi*, *E. carbonarius*), nearctic (*E. kaulbarsi*) or Australasian (*E. dei*, *E. mirus*). Most records are from Central Europe: *E. gaussi* from Germany (Ferrière, 1958) and the Ukraine (Dyadechko, 1964, 1967), *E. carbonarius* from Hungary (Erdös, 1971) and South Slovakia (Boucek and Askew, 1968) and *E. albicoxis* from Hungary (Szelényi, 1982). Specimens belonging to *Entodonastichus* have been found in Sweden (H. Hedqvist, personal collection), Bulgaria (V. Pelov, personal collection) and Russia (S. Triapitsyn, personal communication), but these are not yet identified to the species level. Three species are known from Australia: *E. dei*, *E. mirus* (also in New Zealand) and *E. carbonarius* (Boucek, 1988; Goodwin and Steiner, 1995) and a single brachypterous species (*E. kaulbarsi*) from Canada (collection of the British Museum, Natural History) and Florida (Yoshimoto, 1981).

Thripastichus gentilei is cosmopolitan. Under *Tetrastichus*, it has been recorded under various synonyms (Domenichini, 1966; Graham, 1987; Boucek, 1988) from the Mediterranean (Del Guercio 1911; Bournier, 1967), the Caribbean and Florida (Waterston, 1923; Simmonds, 1933; Dozier, 1937) and India (Narayanan *et al.*, 1960; Ananthakrishnan and Swaminathan, 1977) in the northern hemisphere, and from Brazil in the southern hemisphere (Bennett, 1965). *Pediobius thysanopterous* was reared from material originating from Israel and Egypt (Burks, 1971). Tawfik (1967) recorded a *Pediobius* sp. from Egypt. Other records of a *Pediobius* associated with thrips (Risbec, 1951, 1958) might refer to yet another species (Burks, 1971), but its identity and host relations need to be confirmed.

Host species and host plants

Most thrips parasitoids are host specific to a single subfamily, a few genera or even a few species (Table 9.1a,b). Between species there is considerable variation in host range. That of *Ceranisus menes* is very wide: it has been recorded from over 20 species, Thripinae and Panchaetothripinae (Loomans and van Lenteren, 1995), multivoltine as well as univoltine species. *Ceranisus menes* inhabits a wide range of host plants in different biotopes (Loomans and van Lenteren, 1995; Goodwin and Steiner, 1996; Lacasa *et al.*, 1996), it has been found along roadsides and natural ecosystems (Loomans, 1991), in agricultural ecosystems like gardens (Bühl, 1937; Hirose *et al.*, 1992, 1993), cultivated fields, weeds, fallow land as well as trees and forests. Occasionally it has been found among thrips-infested crops in greenhouses (Loomans, 1991; Rubin and Kuslitzky, 1992). Though present on leaves of onion (van Heurn, 1923; Sakimura, 1937a,b; Carl, 1971; Saxena, 1971), aubergine (Hirose *et al.*, 1992, 1993), castor (Daniel *et al.*, 1983) and waxapple (Chang, 1991), *C. menes* has been collected mostly from flowering plants, representing more than 20 different families. *Ceranisus russelli* also has a relatively wide host range: in addition to its major host *Caliothrips fasciatus* (Russell, 1911, 1912a; Bailey, 1933), infesting bean fields and wild plants surrounding them, it is also recorded as parasitizing *Thrips tabaci*, *Frankliniella tritici* (Russell, 1912b) and *Thrips simplex* (McKenzie, 1935). Other species have a more restricted host range, such as *C. americensis* on *F. occidentalis* (Seamans, 1923) and *T. tabaci* (Loomans and van Lenteren, 1996), *C. lepidotus* on *Limothrips denticornis* and *F. occidentalis* (Lacasa *et al.*, 1996) and *C. vinctus* on *Megalurothrips usitatus* (Fullaway and Dobroscky, 1934). Yet others specialize on a few thrips species with a univoltine life cycle: *C. pacuvius* (and *C. menes*) on *Kakothrips pisivorus* and *Odontothrips* species on *Fabaceae* (Teulon *et al.*, 1996; Thuróczy and Jenser, 1996) and a *Ceranisus* sp. from Turkey to *Taeniothrips inconsequens* (D.A.J. Teulon, collected 1994).

All host records for *Thripobius hirticornis* as well as *T. semiluteus* fall into the subfamily Panchaetothripinae (Thripidae). The former is only known to parasitize *Retithrips syriacus*. For *T. semiluteus*, *Heliothrips haemorrhoidalis* is often reported as a host (Boucek, 1976, 1988; McMurtry *et al.*, 1991; Steyn *et al.*, 1993). Also closely related species like *Panchaetothrips indicus*, *Brachyurothrips anomalus* (Boucek, 1976), *Selenothrips rubrocinctus* (M. Early, personal communication) are mentioned as hosts. In the laboratory it also parasitizes *Hercinothrips femoralis* (M. Early, personal communication), but thripine species like *F. occidentalis* are not accepted as hosts (Loomans, 1991). *Thripobius hirticornis* and *T. semiluteus* have been collected on evergreen shrubs and trees belonging to different families (*T. semiluteus*: cardamom, croton, hibiscus, liquidambar, Valencia oranges, macadamia, avocado; *T. hirticornis*: coffee, rose) (Loomans and van Lenteren, 1995). A *Ceranisus* sp. is known from *Rhipiphorothrips cruentatus* in Taiwan (Chiu, 1984) and also *C. maculatus* is known from this host in India (Rahman and Bhardwaj, 1937).

Host species of *Goetheana shakespearei* all belong to the thripine and panchaetothripine subfamilies (Thripidae) and host plants are related to these records: *Selenothrips rubrocinctus* (on cacao, cashew, mango, tropical almond, *Acalypha wilkesiana*), *Heliothrips haemorrhoidalis* (croton, avocado, orange), *Caliothrips insularis*, *C. indicus* (groundnut, *Achyranthes aspera*), *Hercinothrips femoralis* (amaryllis, passionfruit, bougainvillaea) and *Dinurothrips hookeri* (laboratory). Its first record however was from *T. tabaci* (onion: van Heurn, 1923). It has been reared from this host in the laboratory (Bartlett, 1939; Callan, 1943) but field records are scarce and parasitization levels are low. Adamson (1936) and Bartlett (1939) also reared *G. shakespearei* on two other unidentified species of thrips in Trinidad, Bennett *et al.* (1993) from *Echinothrips caribeanus* on *Clitoria* sp. in Guadeloupe, and Takagi (1988) reported it from *Pseudodendrothrips mori* on mulberry in Japan. *Frankliniella parvula* and *F. intonsa* (as *F. formosa*) have been mentioned as hosts

(Ananthakrishnan, 1984) but Billes (1941) reported that the first was not attacked by *G. shakespearei*, and the latter probably has been erroneously mentioned as such. *Goetheana incerta* is only known to parasitize *Scirtothrips aurantii* (Grout and Stephen, 1995).

From *Entedonastichus*, hosts are only known for *E. gaussi*: all are phlaeothripids living on leaves and bark (*Liothrips setinodis*) (Dyadechko, 1964, 1967), or on or under bark (*Cryptothrips nigripes*, *Acanthothrips nodicornis*, *Phlaeothrips coriaceus* and *Hoplandrothrips pillichianus*) (Dyadechko, 1964) of conifers and broad-leaved trees, on species which often occur in groups in mixed stands. In some years *E. gaussi* is assumed to have great importance in suppressing thrips. In 1957 in the Ukraine, *E. gaussi* parasitized 34 per cent of the total *P. coriaceus* population and 87.6 per cent of the *L. setinodis* population (Dyadechko, 1964, 1967). *Entedonastichus* species have been collected from grassy areas (*E. carbonarius*, *E. albicoxis*, *E. dei*, *E. kaulbarsi*) and forests (*E. gaussi*, *E. mirus*) or from a combination of both. Boucek (1988) mentioned collections of *E. dei* from rose, *Discaria* and *Juncus*. The closely related *Ceranisus nubilipennis* and *C. bicoloratus* have been reared from tubuliferan species as well: the first from *Cryptothrips rectangularis* and *Megalothrips spinosus* from willow galls in a swamp in Massachussets, USA (Williams, 1916); Murai reared the latter from *Liothrips wasabiae* infesting leaves and roots of wasabe (*Cruciferae*) in a forest in Shimane, Japan, August 1990.

Except for a single record on *Rhipiphorothrips cruentatus* (Panchaetothripinae) (Narayanan *et al.*, 1960), thrips host records of *Thripastichus gentilei* fall within the genus *Liothrips* (*floridensis*, *laureli*, *mikaniae*, *oleae*, *urichi*, *varicornis*), *Gynaikothrips ficorum*, *Hoplothrips pedicularius*, and a number of other species from India, all Phlaeothripidae (Table 9.1b). *Liothrips* spp. are leaf feeders, all other hosts inhabit leaf galls of various trees and shrubs (Raman and Ananthakrishnan, 1984). *Thripastichus gentilei* also parasitizes the phlaeothripid predator *Androthrips flavipes* (Ananthakrishnan and Swaminathan, 1977; Varadarasan and Ananthakrishnan, 1981), one of the controlling factors of gall-inhabiting thrips in India, and *Liothrips mikaniae*, a potential biocontrol agent of weeds (Cock, 1982).

Host selection

The actual host range of a parasitoid depends on the probability of a host being found (ecological host range), host defence reactions (behavioural interactions) and host suitability (physiological host range). How thrips parasitoids locate their hosts is not clear. Whereas panchaetothripine hosts generally complete their life cycle in the open on mature leaves and shoots, thripine hosts prefer young plant parts, leaf folds, buds and flowers. *Ceranisus menes* searches in these places (Saxena, 1971) but parasitization is much lower when hosts are concealed than in the open, and searching efficiency of *C. menes* and *C. russelli* declines on hairy leaves (Russell, 1912b; Sakimura, 1937a,b). Semiochemicals seem to be involved in host searching by *C. menes*, wasps searching longer in areas where host larvae have fed (Bazzochi and Santi, 1994; Loomans and Pákozdi, unpublished data). *Thripastichus gentilei* searching for hosts, slowly enters galleries made by scolytids occupied by *Liothrips oleae* larvae or seeks them on branches, leaves and fruits (Del Guercio, 1911). The accessibility of thrips galls plays a role in its parasitization efficiency: in galls of *Schedothrips orientalis*, parasitization reached 20 per cent, but in other species it was restricted to individuals at the base of the galls, or sometimes there was no parasitization at all (Ananthakrishnan and Swaminathan, 1977; Varadarasan and Ananthakrishnan, 1981; Raman and Ananthakrishnan, 1984). *Frankliniella occidentalis* produces an alarm pheromone (see Chapter 3) that acts as a kairomone for polyphagous predators (*Amblyseius*, *Orius*) (Teerling *et al.*, 1993a,b). Whether these substances are used by parasitoid species to locate their hosts is yet unknown.

Host examination and acceptance behaviour is similar for various host–parasitoid combinations (Russell, 1912; Williams, 1916; Bailey, 1933; Sakimura, 1937a; Carl, 1971; Saxena, 1971; Hirose, 1989; Loomans and van Lenteren, 1996): once a host is encountered and examined with the antennae, the ovipositor is thrust forward between the legs and inserted into the host's thorax or abdomen. Most parasitoids stay in this position, but *C. menes* immediately turns 180°, lifting the vigorously moving larvae up in the air (Carl, 1971; Murai, 1988; Loomans, 1991; Galazzi and Bazzocchi, 1993; Fig. 9.1a Plate II.1). With large larvae, wasps often stay standing tail-to-tail to their host (Fig. 9.1b) (Loomans *et al.*, 1992). *Goetheana shakespearei* attacks the host larva by inserting its ovipositor standing laterally to the anterior part of the abdomen (Daniel, 1986; Hessein and McMurtry, 1989), *T. semiluteus* to the posterior part.

Once the ovipositor is inserted, a first stage larva becomes paralysed or immobilized (Daniel, 1986; Loomans, 1991) and a single egg is deposited. Oviposition times vary from 20 to 60 s (range 3–240) for *C. menes* parasitizing *Frankliniella occidentalis* (Loomans *et al.*, 1993), *Thrips palmi* (Hirose, 1989), *T. tabaci* (Sakimura, 1937a; Carl, 1971), *Zaniothrips ricini* and *Retithrips syriacus* (Daniel, 1986). For *C. russelli* attacking *Caliothrips*, oviposition time ranged from 15 to 240 s (Bailey, 1933), but 20 to 50 s according to Russell (1912). *Ceranisus americensis* takes less time to parasitize *F. occidentalis* (15 s) than *T. tabaci* (45 s) larvae of the same age (Loomans and Pákozdi, 1996). For *G. shakespearei* parasitizing *Heliothrips*, oviposition takes 20 to 48 s (Hessein and McMurtry, 1989), but for *T. semiluteus* it is only 5 to 12 s. In *C. nubilipennis* oviposition takes 10 to 15 min, sometimes 30. When host feeding occurs (*C. menes*, *C. americensis*, *C. loomansi*, *C. russelli*, *C. nubilipennis*) no egg is laid, stinging and feeding follow each other repeatedly, and up to 12 min is spent handling the host (Loomans *et al.*, 1992) before it is killed. *Ceranisus menes* is able to discriminate between unparasitized and parasitized hosts (Carl, 1971; Hirose, 1989): after antennal examination and probing, a female recognizes and avoids oviposition in larvae parasitized by herself, but is less likely to avoid those parasitized by conspecifics of the same strain; larvae parasitized by conspecifics of another colour strain are equally accepted as unparasitized one. (A.J.M. Loomans, personal observation).

Within its host range, a preference may occur for certain thrips species (Fig. 9.2i, Loomans and Pákozdi, 1996) or even for different strains of the same host species (Fig. 9.2ii, Loomans, 1997). In some cases parasitoids do not seem to

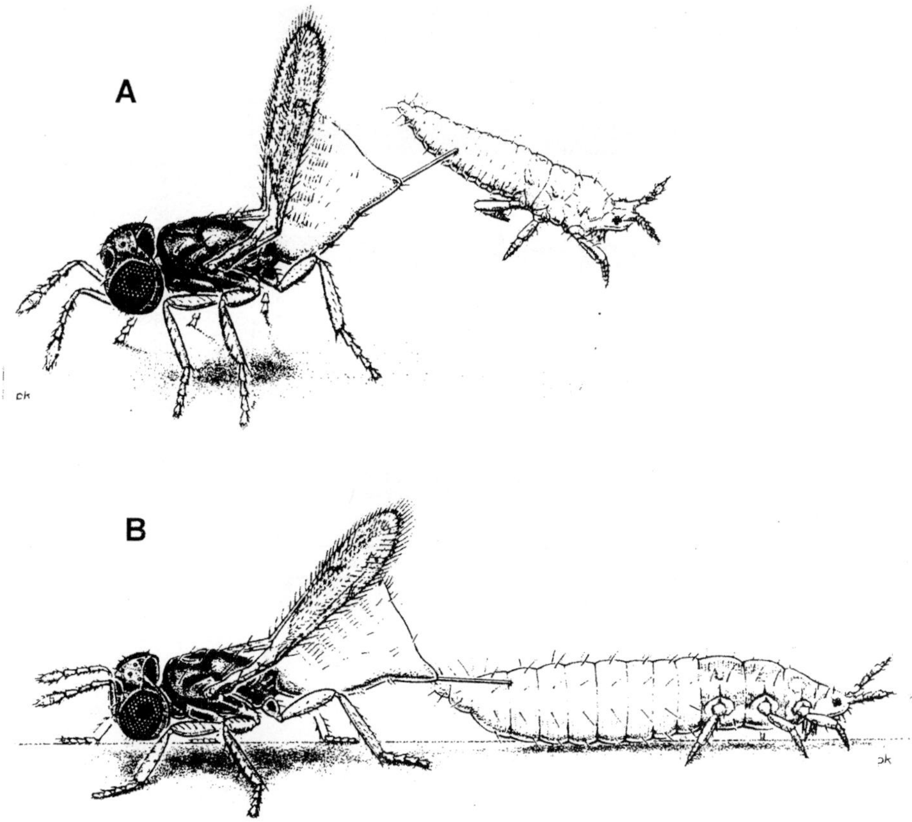

Fig. 9.1. Oviposition postures of *Ceranisus menes* parasitizing a thrips larva. (A) Lifting, (B) tailing or dragging (Loomans and van Lenteren, 1995).

prefer any particular size or age (*C. menes*: *Kakothrips pisivorus*, Bühl, 1937; *Frankliniella intonsa*, Sakimura, 1937a; *Thrips tabaci*, Carl, 1971; *Caliothrips russelli*: *C. fasciatus*, Russell, 1912). Other records state a certain preference even within the same host–parasitoid combination: parasitoids prefer large (*T. tabaci*, Saxena, 1971; *Caliothrips fasciatus*, Bailey, 1933) or medium-sized larvae (*Zaniothrips ricini* and *Retithrips syriacus*, Daniel *et al.* 1983, 1986; Daniel 1986). Detailed behavioural observations show that acceptance of *F. occidentalis* larvae by *C. menes* and *C. americensis* decreases with age, stage and size (Loomans *et al.*, 1992, 1993; Fourez, 1995; Loomans and van Lenteren, 1996) (Fig. 9.3).

Goetheana shakespearei and *T. semiluteus* attack both larval stages of panchaetothripine hosts, but small to medium-sized larvae are required for successful parasitization. The female mostly rejects late second stage larvae, propupae and pupae (Daniel, 1986), in which *T. semiluteus* (McMurtry *et al.*,

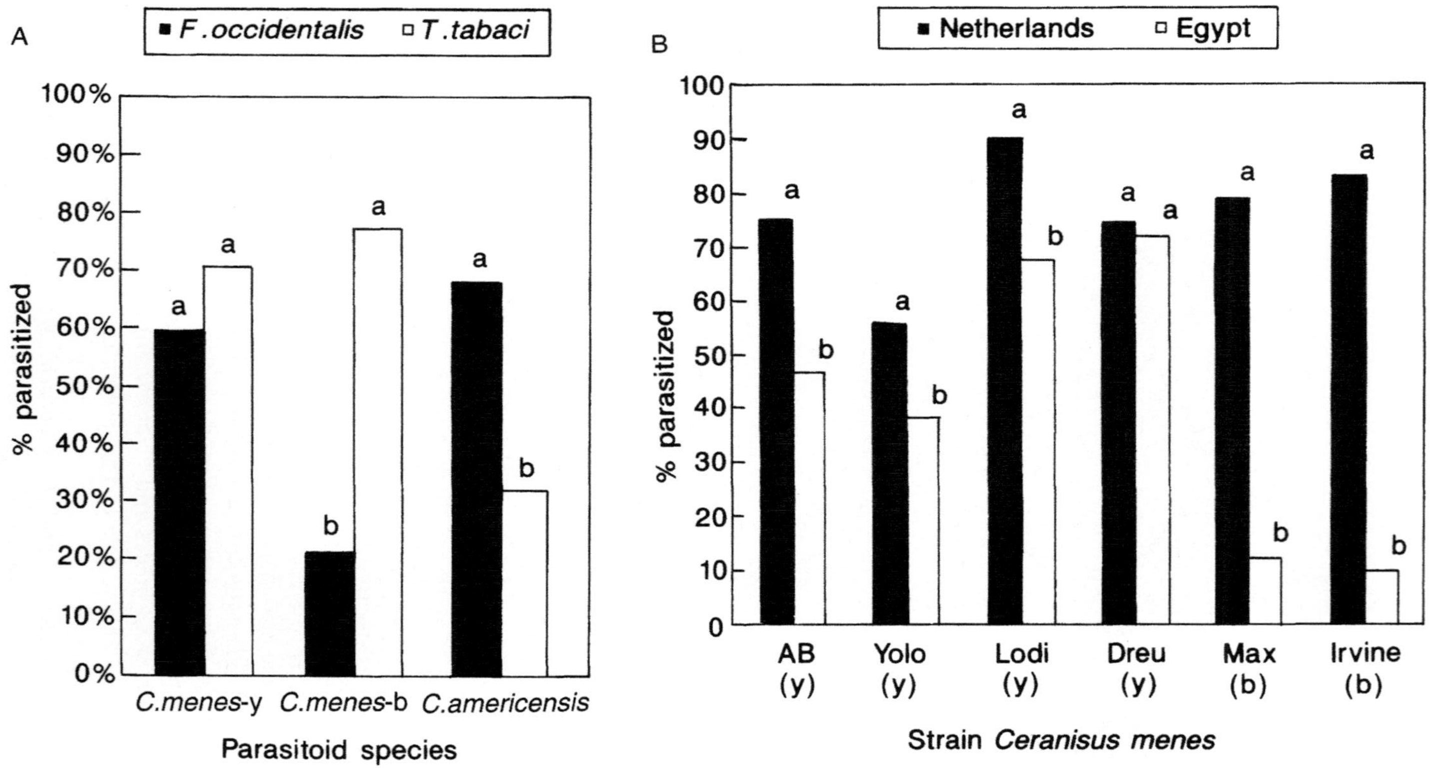

Fig. 9.2. Influence of (A) thrips species (*F. occidentalis*, *T. tabaci*; The Netherlands) and (B) strains (*T. tabaci*: The Netherlands and Egypt) on percentage parasitization by *Ceranisus americensis* and different colour types (yellow, y; brown, b) of *Ceranisus menes* (Loomans and Pákozdi, 1996; Loomans, 1997). n = 10 females, ~100 parasitized larvae per combination. Different letters indicate significances per thrips–parasitoid combination.

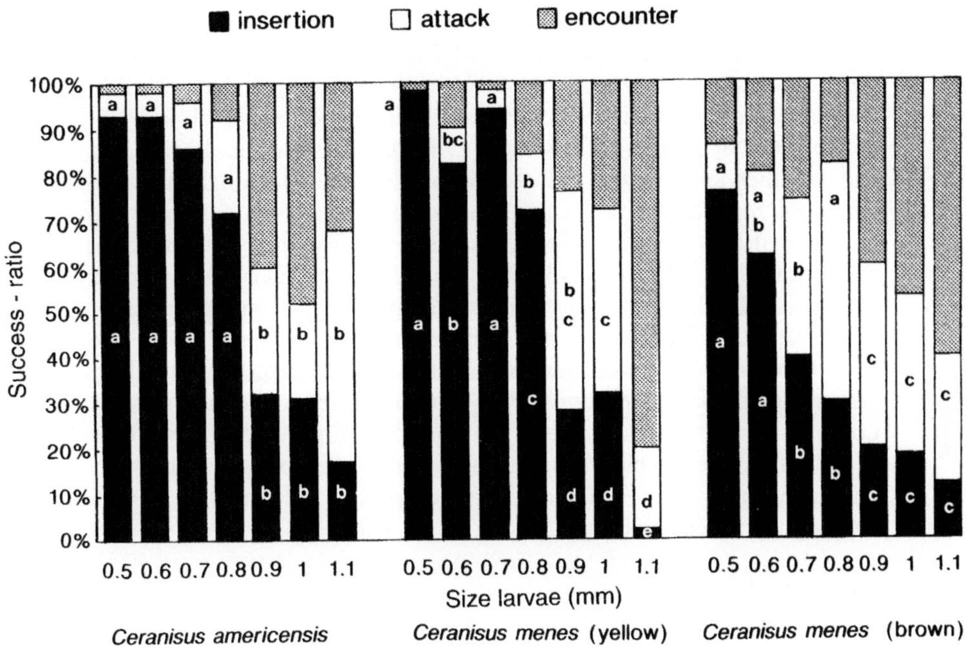

Fig. 9.3. Influence of size of *F. occidentalis* larvae on host acceptance by *Ceranisus americensis* and different colour types of *Ceranisus menes*. Different letters indicate significant differences between sizes of larvae (*C. americensis*: Loomans *et al.*, unpublished data; *C. menes*: Loomans *et al.*, 1993).

1991; Newberger and McMurtry, 1992) and *G. shakespearei* (Dohanian, 1937; Hessein and McMurtry, 1989) fail to complete development. In thripine species, like *Pseudodentrothrips mori*, *G. shakespearei* lays its eggs in first and second instar larvae (Takagi, 1988). Others like *Frankliniella*, which are more active, probably manage to escape from attacks by *G. shakespearei* (Billes, 1941). Female *T. gentilei* lay their eggs preferably inside young larvae and never attack second stage larvae, propupae or pupae (*Gynaikothrips ficorum*: Bournier, 1967). In *Liothrips oleae*, newly hatched and old larvae are not attacked (Del Guercio, 1911). In *Liothrips laureli*, the female lays her eggs inside both larval stages (Mason, 1922).

Behavioural defence reactions (walking away, violent abdominal movements, shaking off the attacker) by old and large thrips larvae can prevent parasitization in many species (Bailey, 1933; Carl, 1971; Saxena, 1971), especially in host species whose larvae are more active and easily disturbed like *T. tabaci*, *F. tritici* (Russell, 1912b), *F. occidentalis* (Loomans *et al.*, 1992), *Megalurothrips usitatus* (Fullaway and Dobroscki, 1934) or *M. sjostedti* (M. Tamò and K. Diop, personal communication). If insertion occurs, the larva drags the wasp along (Williams, 1916), trying to escape (Fig. 9.1b). Sakimura (1937a) noted oviposition failures on *Haplothrips subtilissimus* due to violent defence measures. *Ceranisus menes* attacked *Haplothrips chinensis* (Murai, 1990) but could not develop. Larvae can also defend themselves by excreting anal droplets. Larvae of a number of Panchaetothripinae (*Retithrips*, *Caliothrips*, *Selenothrips*, *Heliothrips*, *Rhipiphorothrips*) constantly keep their abdomen in an uplifted position bearing a droplet of intestinal liquid, enclosed by the whorl of anal bristles (Rivnay, 1935, 1939; Entwistle, 1972; Chiu, 1984) and in other species the abdomen is partly covered with liquid excrement. *Ceranisus russelli*, observed in the open on a croton leaf attempting to oviposit in larvae of *Heliothrips haemorrhoidalis*, was frequently caught by the sticky excrement of its host, and only with great difficulty succeeded in freeing itself (Russell, 1912b). *Ceranisus menes*, attacking early second stage larvae of the same species, ran into the anal excretion droplets, resulting in prolonged periods of preening and unsuccessful attacks. After removing the droplets, *C. menes* stung six out of ten larvae, but oviposition did not occur (A.J.M. Loomans, personal observation). On the other hand, when *T. semiluteus* nears a larva of *H. haemorrhoidalis*, it moves around the tip of the abdomen, lifts its middle leg and puts it on the anal droplet, pushes it down with its middle and hind leg and stings the larva quickly (A.J.M. Loomans, personal observation).

Developmental biology

Thrips parasitoids are all koinobionts, allowing the host to grow beyond the stage attacked. Parasitized larvae move about freely, feed normally and cannot be distinguished externally from unparasitized ones. Parasitism becomes evident at propupation of the host (*Ceranisus*: Fullaway and Dobroscky, 1934; Kutter, 1936; Bühl, 1937; Rahman and Bhardwaj, 1937; Sakimura, 1937a; Daniel, 1986; Murai, 1988; Galazzi and Bazzocchi, 1993; Greene and Parrella, 1993; *Goetheana*: Dohanian, 1937; Takagi, 1988; *Thripobius*: McMurtry *et al.*, 1991): the absence of propupal wing buds, a slightly swollen body, a central spot in the propupa which can vary according to the host–parasitoid combination and varies in colour from creamish white, yellow, orange, crimson to deep red (Fig. 9.4a, Plate II.2,3a). The pupa is almost white (*C. menes*, *C. americensis*, *T. gentilei*) to deep red (*C. maculatus*) when newly formed, becoming dark brown to black in all cases (Plate II.3b).

Pupation takes place outside the larval skin (Fig. 9.4b). The site of pupation is related to that of its host, usually the soil (Thripinae), or it remains on the surface where the parasitized thrips larva has been feeding (Panchaetothipinae: Rahman and Bhardwaj, 1937; Newberger and McMurtry, 1992; Chiu, 1984), sticking to the leaf surface hanging head downwards from the skin of the host (Entwistle, 1972; Daniel, 1986). In *Thripastichus* and *Pediobius*, the black-brown pupa stays inside the host larva integument, sticking on the leaf (*Rhipiphorothrips cruentatus*: Sharma *et al.*, 1965) or hidden under bark or on the underside of dead leaves (*Liothrips oleae*: Melis, 1935). In *Gynaikothrips ficorum* some *T. gentilei* individuals emerge and pupate within the curled leaf (Bennett, 1965). *Thripastichus* larvae exhibit slight up and down movements within the host. Shortly before adult emergence the parasitoid rotates 180° within the host's body, facing its hind end (Del Guercio, 1931; Melis, 1935; Bournier, 1967; Ananthakrishnan and Swaminathan, 1977) (Fig. 9.4c,d). All species mentioned above are solitary, but in *Ceranisus nubilipennis* and *C. bicoloratus* multiple parasitism seems to be the rule: two to three or even more pupae can emerge from the same host, males as well as females (Williams, 1916; Ishii, 1933).

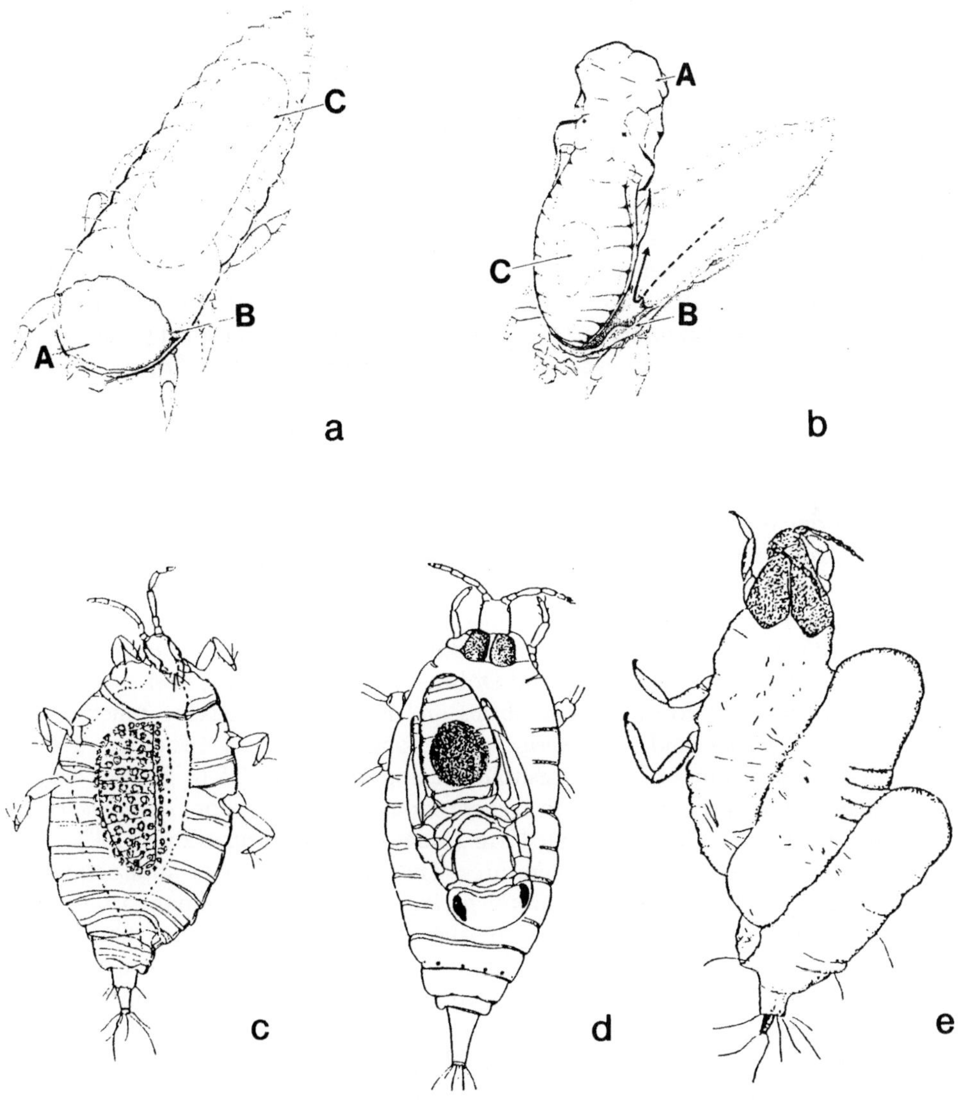

Fig. 9.4. (a) Propupa at the moment of pupation and (b) emerged pupa of *Ceranisus menes*. A = pupa, B = host skin, C = orange/red central spot; arrow indicates emergence from the host skin (Loomans and van Lenteren, 1995); (c) *Liothrips oleae* larva parasitized by *Thripastichus gentilei* and (d) parasitoid pupa inside host body (from Del Guercio, 1931); (e) multiple parasitism in *Ceranisus nubilipennis* (after Williams, 1916).

Life history parameters

Life history parameters (development rate, immature mortality, sex ratio, longevity, oviposition period, fecundity) vary with temperature, the host–parasitoid combination and geographical origin of the parasitoid. Intraspecific differences in qualitative and quantitative aspects of host physiology and nutrition clearly affect development in *C. menes*. In multivoltine hosts, developmental time varies with host age (Sakimura 1937a), sex of the larva on which the egg is laid (Murai, 1988), the host species attacked (Murai and Loomans, 1995) and even between geographical strains of the same host (Loomans, 1997). In general, males emerge shortly before females (*C. menes*: Sakimura, 1937a; Murai, 1990; Castineiras *et al.*, 1996; *C. pacuvius*: Kutter, 1936). Developmental differences occur between strains of *C. menes* according to their phenotypic appearance (abdominal colour types: yellow and brown) and geographical origin (Loomans *et al.*, 1995b; Castineiras *et al.*, 1996). Loomans and Murai (1995) and Galazzi *et al.* (1992) found a large variation in overall developmental time of up to 4 months or more in unisexual yellow strains. Murai (1990) did so in a bisexual strain (Figs 9.5 and 9.6), whereas others did not (Sakimura, 1937a; Castineiras *et al.*, 1996). Daniel (1986) notes an extremely short life cycle of 10.8 days parasitizing *Zaniothrips ricini* and 16.3 days on *Retithrips syriacus*. Pupal duration of the parent had no relation to that of its offspring (Murai, 1988). Parasitizing a univoltine host like *Kakothrips pisivorus*, *C. menes* development is largely governed by host physiology, producing only one generation a year (Bühl, 1937). Size of pupal offspring is related to the host species (Bühl, 1937; Murai, 1988; Loomans and Pákozdi, 1996), host strain (Loomans, 1997) and feeding history of the host larvae.

Developmental times of host and parasitoid can be similar: *C. russelli* and *Caliothrips fasciatus* (Russell, 1912; Bailey, 1933) and *C. vinctus* and *Megalurothrips usitatus* (Fullaway and Dobroscky, 1934). Mostly, host species like those belonging to the genera *Thrips* and *Frankliniella* develop much faster than *C. menes* and *C. americensis* (Loomans and van Lenteren, 1995); development times of the parasitoid are approximately 1.5 to 2 times longer than that of the host, leading to asynchronous generations. For *G. shakespearei* the whole life cycle takes about 17 to 21 days at temperatures above 24°C, but is much longer at lower temperatures. This is much longer or about equal to some of its hosts, such as *Caliothrips indicus* (Ananthakrishnan, 1984) and *Selenothrips rubrocinctus* (Entwistle, 1972), but somewhat faster than *Heliothrips haemorrhoidalis* (Rivnay, 1935; Hessein and McMurtry, 1989). Compared with *H. haemorrhoidalis*, *T. semiluteus* develops much faster (McMurtry *et al.*, 1991): at 23°C the life cycle takes 35.6 (34 to 39) days for the host and 23.6 (22 to 25) days for the parasitoid. Also *T. gentilei* develops much faster than most of its hosts: emergence of *T. gentilei* adults from the host occurs on day 8 to 10 of pupation of *Schedothrips* (Ananthakrishnan and Swaminathan, 1977), and from *Liothrips laureli* after a week (Mason, 1922). In India approximately 12 to 14 days are required to complete the life cycle on *R. cruentatus* (Sharma *et al.*, 1965) and in Europe its life cycle takes about 20 days on *Liothrips oleae* or *Gynaikothrips ficorum*, (Melis, 1935; Bournier, 1967),

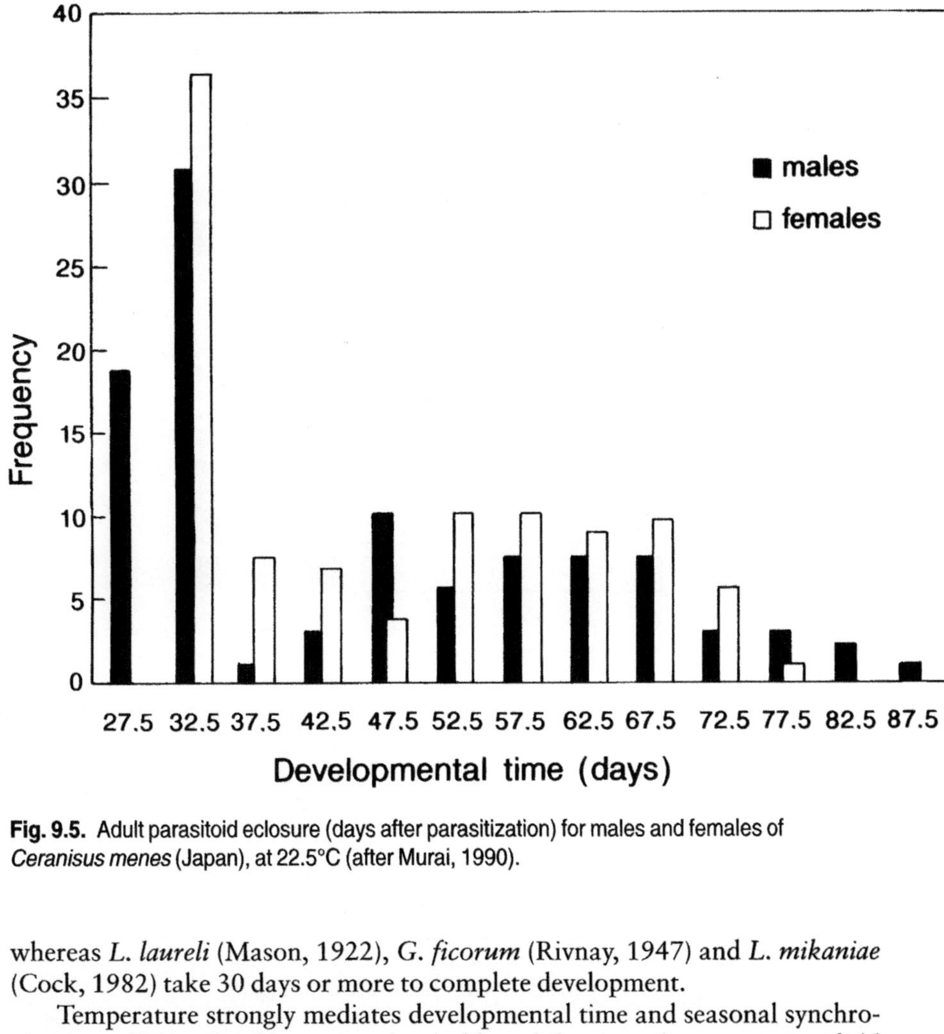

Fig. 9.5. Adult parasitoid eclosure (days after parasitization) for males and females of *Ceranisus menes* (Japan), at 22.5°C (after Murai, 1990).

whereas *L. laureli* (Mason, 1922), *G. ficorum* (Rivnay, 1947) and *L. mikaniae* (Cock, 1982) take 30 days or more to complete development.

Temperature strongly mediates developmental time and seasonal synchronization. Although able to survive cold and hot weather extremes, field temperature requirements for parasitoids are between 20 and 30°C. In univoltine hosts like *Kakothrips pisivorus*, *C. menes* and *C. pacuvius* have one generation a year and overwinter as pupae, whereas the host overwinters as full-grown second stage larvae (Kutter, 1936; Bühl, 1937). In multivoltine host species, parasitoids have more generations a year (four in *C. menes*: Sakimura, 1937b; five to eight in *C. russelli*: Russell 1912b; five in *T. gentilei*: Melis, 1935). At 20°C or lower, life cycles of *C. menes* can be very long (Figs 9.5 and 9.6) (Murai, 1990; Loomans *et al.*, 1995b). At 17.8°C, *G. shakespearei* did not pupate (Hessein and McMurtry, 1989) and at 30°C, high mortality occurred. In *C. menes*, this can vary with the geographical origin of the parasitoid. Castineiras *et al.* (1996) calculated a developmental threshold of 8°C (egg–adult) for Japanese and Thai females of *C. menes* and a thermal constant of 500 degree-days. For Thai males the minimum threshold was 13.7°C and the

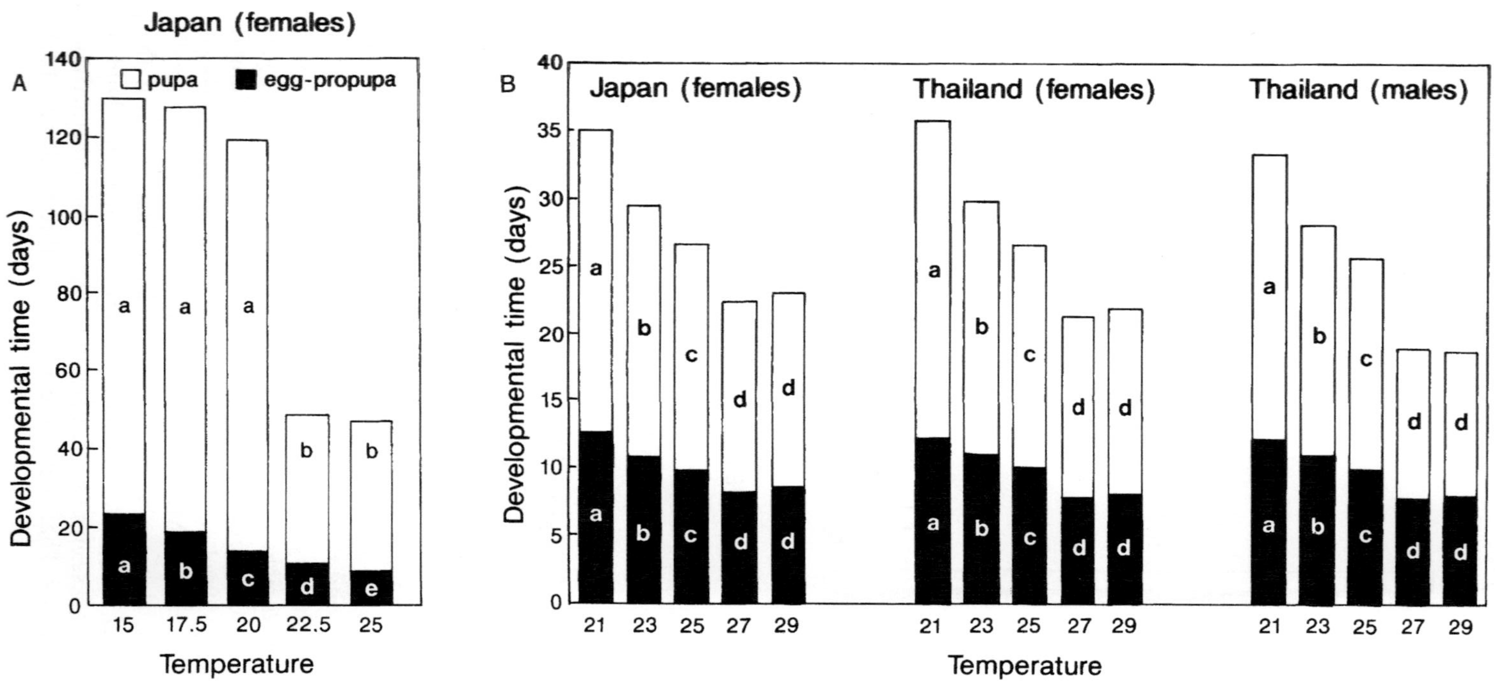

Fig. 9.6. Influence of temperature on (A) the juvenile (egg–propupa/pupa) development of *Ceranisus menes* parasitizing *Frankliniella intonsa* (Murai and Loomans, 1995) and (B) *Thrips palmi* (Castineiras *et al.*, 1996). Different letters indicate significant differences between developmental times.

thermal constant 333 degree-days. With the advance of cold temperatures Sakimura (1937a) noticed an increase in the number of specimens that did not emerge ('hibernated'). At low temperatures *C. menes* pupae probably are not in diapause, but in a quiescent state: pupae readily started hatching when transferred to higher temperatures (Murai, 1988). *Ceranisus maculatus* hibernates as pupae on the host plants and the host (*R. cruentatus*) as pupae in the soil (Rahman and Bhardwaj, 1937). *Thripastichus gentilei* hibernates as a larva or pupa in diapause induced by its host *Liothrips oleae*, which itself passes winter in adult diapause (Del Guercio, 1911; Melis, 1935), but when parasitizing *Gynaikothrips ficorum* in a greenhouse at temperatures of 20 to 25°C, diapause did not occur (Bournier, 1967).

Ceranisus menes (Murai, 1988; Loomans, 1991) and *C. americensis* (Loomans and van Lenteren, 1996) can start ovipositing from the day of emergence onwards, but Sakimura (1937a) recorded a pre-oviposition period of 1 day in summer and 2 days in autumn, Daniel (1986) noted 2 to 3 days for *C. menes*, and Rahman and Bhardwaj (1937) 3 to 5 days for *C. maculatus*. According to Sakimura (1937a), several oocytes mature within 24 h after emergence and full-grown oocytes vary from 29 to 48 per female with an average of 38.6 for *C. menes*. Maturation of eggs continues over the female's lifetime, but in all cases numbers decrease with age. At its peak, *C. menes* can lay 20 to 30 eggs per day (Murai and Loomans, 1995), but this varies with temperature and host species; when parasitizing *Zaniothrips ricini* and *Retithrips syriacus*, Daniel (1986) recorded a maximum of 46 ± 2.6 eggs per day. *Ceranisus americensis* lays 30 to 40 eggs per day at its peak, when supplied with a daily surplus of hosts. Its ovipostion curve shows a peak after a few days and then sharply declines, whereas *C. menes* shows a gradual decline (Loomans and van Lenteren, 1996). Total fecundity (and net reproduction, R_0) is also significantly affected by temperature, and varies between different host–parasitoid combinations: 37 to 61 eggs per female at 20°C and 60 to 170 at 25°C (Loomans and Murai, 1994; Murai and Loomans, 1995) for *C. menes* in *F. occidentalis*/*F. intonsa*, 30 to 105 for *C. maculatus* in *R. cruentatus* (Rahman and Bhardwaj, 1937), 130 to 170 for *C. americensis* in *F. occidentalis* (Loomans and van Lenteren, 1996), and 56 to 91 for *C. russelli* in *C. fasciatus* (Russell, 1912). *Goetheana shakespearei* can produce 24.2 (Bartlett, 1939), 25.5 (Dohanian, 1937) or up to 70 offspring per female (Adamson, 1936) parasitizing *Selenothrips rubrocinctus*. Progeny yield per female was highly variable in *Heliothrips haemorrhoidalis* with 40 to 60 offspring recorded by McMurtry and Johnson (1963); but in the laboratory, reproduction was lower: 25.3 offspring per female for 3 days on avocado leaves, but only 10.1 (7 to 13 at 22.2°C) and 3.43 (1.4 to 5.7 at 29.4°C) on Valencia oranges, suggesting a host plant effect (Hessein and McMurtry, 1989). In India, when parasitizing *Caliothrips indicus* in the laboratory, a single female of *G. shakespearei* laid an average of 78 ± 3.1 eggs per day, with a range of 53 to 105 parasitized offspring per day on *Arachis hypogea* (Daniel, 1986).

Adult longevity is relatively short: females live for 3 to 10 (*C. russelli*) and 10 to 20 (*C. menes*, *C. americensis*) days at 20 to 25°C (Murai, 1990; Loomans and Murai, 1994; Murai and Loomans, 1995). In *G. shakespearei* it was only

3 to 5 days (Daniel, 1986; Hessein and McMurtry, 1989). Total lifespan of *T. gentilei* was 2 to 3 days in the laboratory (Ananthakrishnan and Swaminathan, 1977) and up to 10 days in field cages (Melis, 1935), which is much shorter than most of its hosts. Longevity of females increases when they are fed with sugar or honey solution (Dohanian, 1937; Sakimura, 1937a; Antsiferova and Timraleev, 1974; Hessein and McMurtry, 1989), but it is considerably reduced at high temperatures.

Parasitoid species reproduce parthenogenetically, and are thelytokous (*C. lepidotus*, *C. vinctus*, *C. americensis*, *C. russelli*; *T. semiluteus*), arrhenotokous (*C. maculatus*, *C. pacuvius*, *C. loomansi*, *C. nigrifemora*, *C. planitianus*; *T. hirticornis*; *G. shakespearei*, *G. incerta*) or both (*C. menes*; *T. gentilei*). In other species the type of reproduction is unknown (*C. margiscutum*, *C. femoratus*: both described from a few females). *Ceranisus menes* collected and cultured in Europe, North and South America, and the Near East (Loomans and van Lenteren, 1994) only produce females. In several Asian countries (Ishii, 1933; Saxena, 1971; Tachikawa, 1986) and Australia (collection of the British Museum, Natural History), sexual populations predominate in field-collected material, but non-sexual populations are present as well. In culture, females mostly dominate (sex ratio 3:2, Japan: Sakimura, 1937a; India: Daniel, 1986; Indonesia: van Heurn, 1923), or ratios are equal (1:1, India: Carl, 1971; Thailand: Hirose, 1989; Japan: Murai, 1990), but in occasional laboratory records males predominate (1:2, Thailand: Castineiras *et al.*, 1996). Murai and Loomans (1995) noticed a gradual change in sex ratio in the laboratory to thelytokous parthenogenesis. In *T. gentilei*, males only occur in Europe and India, the sex ratio is nearly 1:1. Sex ratios for *G. shakespearei* may vary from 2:1 (Daniel, 1986) to 6.5:1 (Hessein and McMurtry, 1989), but more females occurred at 30°C (4.8:1) than at 22°C (2.8:1). Temperature did not affect sex ratio (Castineiras *et al.*, 1996) in *C. menes*.

Population growth parameters, the intrinsic rate of increase (r_m) and net reproduction rate (R_0), varied largely with temperature and host species for *C. menes* (Murai, 1990; Loomans and Murai, 1994; Murai and Loomans, 1995) and *C. americensis* (Loomans and van Lenteren, 1996), mainly due to differences in adult longevity, and are usually lower than or equal to that of their hosts (Murai, 1990; Van Rijn *et al.*, 1995).

Egg Parasitoids

Parasitoids attacking thrips eggs, except for *Polynema indica* (Mymaridae), all belong to the genus *Megaphragma* (Trichogrammatidae). Egg parasitoids have been recorded only from terebrantian hosts, that is species belonging to the family Thripidae (Table 9.2), which lay their eggs *inside* the plant tissue. From eggs of terebrantian and tubuliferan species, which lay their eggs *on* the substrate, no parasitoids are known. Keys to some of these are presented by Subba Rao (1969) and Lin (1992), but no key is available to all species.

Taxonomy

The genus *Megaphragma* is characterized by a small flattened body, linear long-fringed wings and an extremely large phragma. Since *Paramegaphragma* was synonymized with *Megaphragma* (Delvare, 1993), 13 species are unified under this genus. Several other specimens have been collected in various parts of the world, but have only been described to the genus level (Table 9.2). *Polynema indica* measures 0.94 mm (Daniel, 1986). *Megaphragma* species belong to the smallest of all insect parasitoids, most of the collected species measure 0.20 to 0.35 mm, *M. caribea* however measures only 0.17 mm (Delvare, 1993), and *M. mymaripenne* 0.17 to 0.20 mm (Dozier, 1932; Doutt and Viggiani, 1968). *Megaphragma priesneri* is the largest at 0.50 mm (Kryger, 1932). The size can partly be explained by the size of the eggs of the thrips host from which they have developed, such as *Leucothrips* (Dozier, 1932) and *Retithrips* (about 0.30 mm: Rivnay, 1939). The circular exit holes, characteristic for the emergence of the parasitoids (avocado: Ebeling, 1959; McMurtry, 1961; aubergine: Hirose, 1989; tea: Takanashi *et al.*, 1996), are minute at 0.08 to 0.09 mm in diameter (Pemberton, 1931) and are found mostly on the underside of leaves. Because of their small size, the collection and research of this group has been greatly hampered.

Distribution

Megaphragma species are tropical and subtropical, distributed roughly between 45°N and 45°S. A *Megaphragma* sp. was first collected in 1920 in Hawaii by Pemberton (1931) and described as *Megaphragma mymaripenne* by Timberlake (1924). Later this species has been recorded from other parts of the nearctic (Dozier, 1932; Hessein and McMurtry, 1988), neotropics (DeSantis, 1965; Delvare, 1993) and the Afrotropical area (A. Polaszek, personal communication). Other species are only known from certain regions (Table 9.2).

Host range and host plants

Megaphragma species have been collected in a wide range of habitats, from cultivated evergreen crops like tea (Kenya, Taiwan, Japan), coffee (Zaire), guava (Guadeloupe, Galapagos Islands), citrus, mulberry, tea (Japan) and avocado (California), from annuals like aubergine (Thailand) and cowpea (Benin), as well as from wild plants belonging to *Leguminosae* (Benin), *Liliacae* (India), *Myrtaceae* and *Caprifoliaceae* (Italy), infested with a wide range of host species (Table 9.2). Others have been collected in yellow pans (Lin, 1992), malaise traps (Noyes and Valentine, 1989), sticky traps (Takagi, 1988), emergence cages (Tamò, 1991) or by sweeping vegetation (Yousuf and Shafee, 1987). *Polynema indica* has only been found in India (Narayanan and Subba Rao, 1961).

Hosts belong to the Panchaetothripinae as well as Thripinae. *Megaphragma mymaripenne* is reported on *Heliothrips haemorrhoidalis* in California, Hawaii, Guadeloupe and Chile, and on *Selenothrips rubrocinctus* in Guadeloupe (Table 9.2), but also on the thripine *Leucothrips* sp. (Dozier, 1932) and *M. sjostedti* (Tamò *et al.*, 1993). Other species are only reported on a single host species: *M. ghesquièrei* on *Panchaetothrips noxius* infesting coffee

Table 9.2. Species of egg parasitoids belonging to the genus *Megaphragma* (Trichogrammatidae): distribution, host species.

Species name	Distribution	Major host genera	Reference
M. mymaripenne Timberlake	Subtropical	*Heliothrips*, *Selenothrips* *Leucothrips*, *Megalurothrips*	Pemberton (1931), Delvare (1993) Dozier (1932), Tamò (1991)
M. ghesquièrei Nowicki	Zaire	*Panchaetothrips*	Ghesquière (1939)
M. priesneri (Kr.) Nowicki	Egypt, Israel	*Retithrips*	Kryger (1932)
M. longicilliatum Subba Rao	India	*Frankliniella*	Subba Rao (1969)
M. aligharensis Yousuf and Shafee	India	Unidentified	Yousuf and Shafee (1987)
M. magniclava Yousuf and Shafee	India	?	Yousuf and Shafee (1987)
M. deflectum Lin	China	?	Lin (1992)
M. decochaetum	China	?	Lin (1992)
M. polychaetum Lin	China	?	Lin (1992)
M. macrostigmum (Lin)	China	?	Lin (1992)
M. stenopterum (Lin)	China	?	Lin (1992)
M. caribea Delvare	Guadeloupe	*Selenothrips*	Delvare (1993)
Megaphragma spp. (unidentified)	Benin	*Megalurothrips*	Tamò (1991)
spp.	Japan	*Pseudodendrothrips*, *Scirtothrips*	Takagi (1988)
sp.	Thailand	*Thrips*	Hirose (1989)
sp.	India	*Heliothrips*, *Thrips*	Narayanan (1971)
sp.	Italy	*Heliothrips*	Viggiani (1992)
spp.	New Guinea	?	British Museum (Natural History) (1986)
spp.	New Zealand	?	Noyes and Valentine (1989)
spp.	Australia	?	Goodwin and Steiner (1995)
sp., sp.	Kenya, Mexico	?	British Museum (Natural History) (1965, 1984)
sp.	Bulgaria	?	Pelov (personal communication)
sp.	South Africa	?	Doutt and Viggiani (1968)

(Ghesquière, 1939) and *M. priesneri* was found within colonies of *Retithrips syriacus* on vine leaves (Kryger, 1932; Rivnay, 1939). In Japan, *Megaphragma* spp. developed from eggs of *Scirtothrips dorsalis* (Takanashi *et al.*, 1996) and *Pseudodendrothrips mori* (Takagi, 1988) and in Thailand, Hirose (1989) and Hirose *et al.* (1993) collected a *Megaphragma* sp. from aubergines infested with *T. palmi*. *Megaphragma longiciliatum* attacked the eggs of *Frankliniella lilivora* on *Polyantes tuberosa* in India (Subba Rao, 1969) and Narayanan (1971) reared a single individual from several eggs of *T. tabaci* exposed to it in the laboratory. *Polynema indica* occurred in relatively large quantities on groundnut (*Arachis hypogea*) as well as an annual weed (*Achyranthes aspera*) present in the crop field, both infested with *Caliothrips indicus* (Daniel, 1986).

Host selection

When searching for hosts *M. mymaripenne* locate the softer parts of an egg blister with their antennae. During oviposition the female puts her weight on the hind legs and the wings, curving her abdomen forward and inserting her ovipositor into the side or sometimes middle of the egg blister. The same egg blister can be stung more than once. After ovipositing in several egg blisters, which can take from 30 to 120 min with intervals for drinking and cleaning, adult females walk or fly to another part of the leaf, where they resume searching. Oviposition takes 0.5 to 7.5 min, but no parasitoids emerge when oviposition time is less than 2.5 min. Eggs that are completely covered with thrips faecal material are also stung, but this takes longer (Hessein and McMurtry, 1988). *Polynema indica* locates the embedded host egg by probing the leaf surface of thrips-infested plants. It moves around the host egg area for 30 to 45 s assessing the suitability of the egg for oviposition, then inserts its ovipositor to lay a single egg into it. Once parasitized, the eggs are not attacked again (Daniel, 1986).

Differences in egg laying strategy by panchaetothripine and thripine host species may affect the searching efficiency of the parasitoid. Semiochemicals are probably involved in host searching by *M. mymaripenne*: excrements (sealing fluid, faeces) of the host (*Heliothrips*) seem to elicit probing and oviposition responses by the female, though egg blisters without faecal material are sometimes also stung (Hessein and McMurtry, 1988).

Life history

Most species are arrhenotokous. In *P. indica* and *M. longiciliatum* only the female is known, in *M. mymaripenne* (Hessein and McMurtry, 1988) and *Megaphragma* spp. (Takanashi *et al.*, 1996) parthenogenetic thelytoky occurred in the laboratory, males were very rare in the field. Developmental time from egg to adult at 22 to 23°C and 10 to 42 per cent relative humidity, ranged from 36 to 46 days, with an average of 41.4 days for *M. mymaripenne* attacking *Heliothrips haemorrhoidalis*. *Heliothrips* has a shorter developmental time: under similar conditions it ranged from 24 to 58 days, average 31 days (Hessein and McMurtry, 1988). Total developmental time of a single adult that emerged from one of the eggs of *T. tabaci* exposed to *M. longiciliatum* was 20 days (Narayanan, 1971) and Takanashi *et al.* (1996) found a similar period for

Scirtothrips doralis. Longevity of adult parasitoids is about 1 to 2 days, their egg laying capacity is unknown. Adult *H. haemorrhoidalis* live 40.6 days on average, ranging from 24 to 36 days, laying 1 to 2 eggs per day, or an average of 38.3 per female (Hessein and McMurtry, 1988) producing 5 to 6 generations in a year in California (Ebeling, 1959).

The timing of the life cycle of *P. indica* is similar to that of its host, *Caliothrips indicus* (Ananthakrishnan, 1984): within 12 to 15 days a single adult emerged from each parasitized host by making a hole in the egg blister; the sex ratio was constant at 3♀:2♂; 18 ± 3.2 adults emerged from the egg mass of a single *C. indicus*, 12 to 17 per cent of which were parasitized on weeds and 9 per cent on the crop itself (Daniel, 1986).

Field Incidences and Natural Control

Rates of parasitism found in the field sometimes exceed 50 per cent (*G. shakespearei*: Cotterell, 1927; Daniel, 1986; Takagi, 1988; *M. mymaripenne*: Hessein and McMurtry, 1988; *C. menes*: Bühl, 1937; Sakimura, 1937b; Hirose, 1989; *C. pacuvius*: Kutter, 1936; unknown: Hukkinen, 1936), but this is in itself not always a good indicator of parasitoid impact (Van Driesche, 1983) and natural control.

In multivoltine hosts, *C. menes* shows a gradual seasonal increase in numbers (Sakimura, 1937b; Hirose *et al.*, 1992, 1993), reaching a maximum following a peak of pest outbreaks. In onion, the rate of parasitism went up to 80 per cent in summer, and almost 100 per cent at some sites in autumn (Sakimura, 1937b). Parasitoid occurrence was well synchronized with the *T. tabaci* population, but only during the latter half of the season were they kept in check. Sakimura's sampling data show a clear positive density-dependent relationship (Fig. 9.7), but the percentage parasitism was not specified per age-classes sampled so it is not clear if this increase in parasitism with host density was due to a numerical response (aggregative or reproductive) or to a functional response (Lewis, 1973). In Taiwan, a *Ceranisus* sp. attacking *R. cruentatus* in waxapple fields (Chiu, 1984) showed a clear negative density-dependent relationship and although parasitism was as high as 52 per cent during summer, it was not able to control thrips population outbreaks.

Parasitism by *C. menes* of *Thrips palmi* larvae infesting leaves of aubergines in home and commercial gardens in Japan and Thailand was very localized and ranged from 0 to 75 per cent (Hirose *et al.*, 1992) and 0 to 60 per cent, respectively (Hirose *et al.*, 1993). A high percentage (40 to 60 per cent) parasitism was frequently observed in unsprayed, small-scale home gardens, while the level was significantly lower in pesticide-treated gardens (Hirose, 1990). On *Ricinus communis* in India, *C. menes* preferred *Zaniothrips ricini* (48 per cent) and *Craspedothrips minor* (50 per cent) over the simultaneously occurring *Retithrips syriacus* (20 per cent). When the first species were absent, it switched over to *R. syriacus* and sustained itself but did not increase (Daniel *et al.*, 1983, 1986; Daniel, 1986). Parasitism levels of *Caliothrips fasciatus* by *C. russelli* ranged as high as 70 per cent in southern California (Russell, 1912),

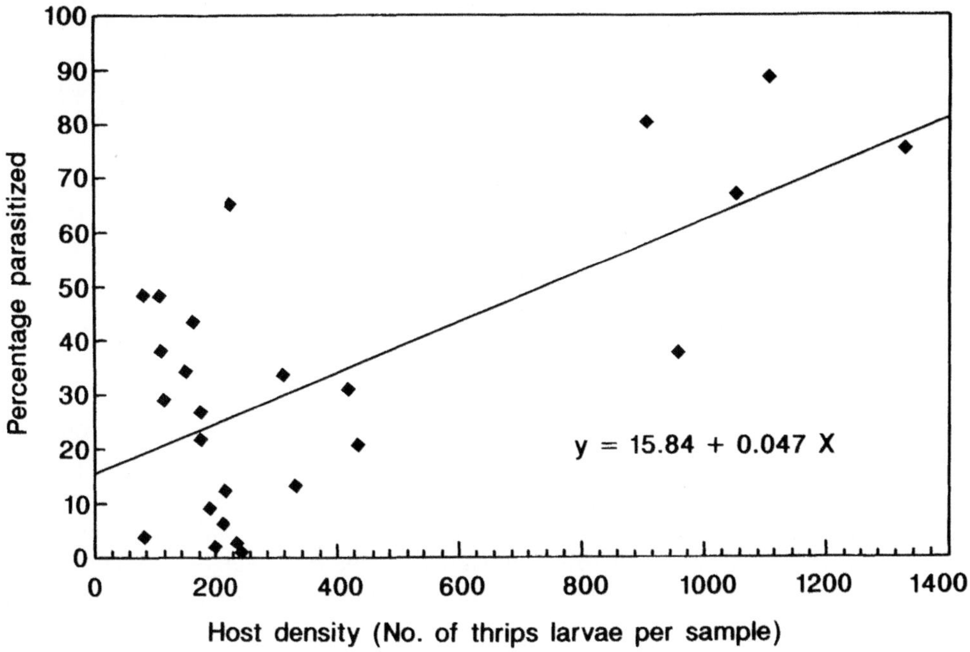

Fig. 9.7. Relationship between the host density of *Thrips tabaci* on onion and percentage parasitism by *Ceranisus menes* (Japan) (data from Sakimura, 1937c).

but only 5 per cent parasitism was found in central California years later (Bailey, 1933, 1937). In field collections of *Thrips tabaci*, parasitization ranged from 15 to 60 per cent, with an average of 33.5 per cent (Russell, 1912). Parasitism rates of *T. tabaci* quoted for *C. menes* from other regions were low (Indonesia: van Heurn, 1923; Franssen and van Heurn, 1932; Japan: Sakimura, 1937b; Europe: Carl, 1971) and varied between 0 to 10 per cent (India: Narayanan, 1971) and 12 to 18 per cent (India: Saxena, 1971, 1981). *Ceranisus menes* is common on flowers of various wild plants, with parasitism levels up to 8 to 25 per cent. In cultivated crops it has been found only occasionally in low numbers (Europe: Carl, 1971; Loomans, 1991; Galazzi *et al.*, 1992) or is absent (California: Heinz *et al.*, 1996).

In univoltine hosts, like *Kakothrips pisivorus* (on pea, *C. pacuvius*: Kutter, 1936; *C. menes*: Bühl, 1937) and *Chirothrips hamatus* (on *Alopecurus*, species unknown: Hukkinen, 1936), parasitoid occurrence is well synchronized with the host population, in others like *Taeniothrips inconsequens* (unknown: Carl *et al.*, 1989) this needs to be ascertained. Rates of parasitism varied with locality (*C. pacuvius*: 0 to 68 per cent and even 92 per cent, Kutter 1936, 1937; unknown: 0 to 58.2 per cent, Hukkinen, 1936). The number of parasitoids found was related to the infestation level of *K. pisivorus* the year before (Kutter, 1937) and fluctuated over the years, but the impact on the thrips population is not clear. In Russia up to 17 per cent of the *Kakothrips* larvae were parasitized by *C. menes* (Antsiferova and Timraleev, 1974), in other countries higher percentages were found (Vuillet, 1914; Bühl, 1937: 35.1 per cent). Although parasitoids were present in pea fields during one month and reached 40 per cent parasitism at their peak, *C. menes* did not contribute significantly to thrips control. Sampling home gardens in northern Germany and Switzerland, Teulon *et al.* (1996) found no signs of parasitism.

Natural incidence of *G. shakespearei* is largely affected by climate, numbers being very low in spring and summer (Bennett, 1970), reaching a peak in the dry season in Ghana (Entwistle, 1972) and Benin (M. Tamò, personal communication) but during wet weather in the Caribbean (Entwistle, 1972). In India, Daniel (1986) recorded parasitism rates up to 92 per cent during winter months by *G. shakespearei* in groundnut fields on the weed *Achyranthes aspera* infested by *Caliothrips indicus*, resulting in a gradual decline of the thrips population in the latter. Compared with the weed, incidence on the crop was low, and mainly governed by host availability. On cacao infested by *Selenothrips rubrocinctus* in Ghana (Cotterell, 1927), up to 70 to 80 per cent were parasitized, leading to its ultimate control. In Benin (M. Tamò, personal communication) *G. shakespearei* controls *S. rubrocinctus* infestations on mango and keeps populations of *Hercinothrips femoralis* in check on bougainvillaea and passionfruit. Attacking *Pseudodendrothrips mori* on mulberry trees in Japan, Takagi (1988) registered parasitization levels up to 59 per cent in one orchard, but only 19.5 per cent in another.

In the field, *Megaphragma* spp. are sometimes abundant. Parasitization reached 51 per cent in avocado orchards in California (McMurtry, 1961) and 20 to 53 per cent on wild *Leguminosae* in south Benin (Tamò *et al.*, 1993). However the parasitization level is highly variable, differing between and within locality (Hessein and McMurtry, 1988; Tamò, 1991), with season (Hessein and McMurtry, 1988; Takagi, 1988; Takanashi *et al.*, 1996), host plant (Tamò, 1991; Tamò *et al.*, 1993), cultural practices (Takanashi *et al.*, 1996) and the use of chemicals (Takagi, 1988). In spite of the sometimes high parasitization levels, Hessein and McMurtry (1988) and Tamò (1991) consider it doubtful that *Megaphragma*, by itself, regulates pests. Its relatively short lifespan of about 2 days and the long developmental time compared with that of its host might limit its ability to suppress increasing pest populations. The incidence of *P. indica* was only seasonal; although the host was present on the weed almost all year, parasitization was only moderate in winter. The presence of *P. indica* coincided with that of *G. shakespearei*, attacking the larvae of *Caliothrips indicus*. Both populations followed a similar trend. However, the impact of *G. shakespearei* was considered much larger (92 per cent) than that of the egg parasitoid (Daniel, 1986).

Thripastichus gentilei was the most important natural control agent of *Liothrips oleae* in the Mediterranean (Del Guercio, 1911, 1931; Melis, 1935). In Italy, it has five generations a year, whereas the host has three. At the end of the season, it parasitized 60 to 70 per cent (Del Guercio, 1931), 75 per cent (Melis, 1935) or 90 per cent of the larvae (Del Guercio, 1911), controlling the thrips population almost completely (Paoli, 1931). Nowadays this species is not considered as a pest and *T. gentilei* is rare. Sharma *et al.* (1965) regarded

T. gentilei as an important natural control agent of *Mallothrips indicus* in the area of New Delhi, parasitizing up to 73.6 per cent. In Trinidad, levels of 40 per cent have been found on *Liothrips mikaniae*, a control agent of the strangler vine, *Mikania micrantha*, but it was rare on *L. urichi* infesting *Clidemia hirta* (Simmonds, 1933). On *Gynaikothrips ficorum*, 75 per cent parasitization was found (Bournier, 1967) in a greenhouse in France. It was introduced from Brazil into Bermuda as a potential biological control agent of *G. ficorum* on *Ficus* spp., but it failed to establish (Bennett, 1965). On *Ficus nitida* in Egypt, *Pediobius* sp. was present most of the year, except when temperatures were above 29°C and relative humidity below 45 per cent, and reached its peak in winter at 14.0°C and 78 per cent relative humidity, 3 months after its host, *G. ficorum* (Tawfik, 1967). In Israel (Rivnay, 1947), an obvious fall in the number of thrips in autumn coincided with an increase in numbers of the parasitoid.

Biological Control

Classical biological control, conservation and augmentation by inoculative releases or inundation are accepted strategies for biological control. Conserving naturally occurring populations of thrips parasitoids by cultural practices has yet to be practised. Reducing sprays (Hirose *et al.* 1992, 1993) or use of multicropping systems (see Daniel, 1986), however, could help to tilt the balance in favour of the natural enemy complex. Classical biological control of thrips pests, by releasing parasitoids in an inoculative way, restoring old and establishing new combinations (Carl, 1982) of pest and parasitoid species has been tried in the past by trial and error (1930s, *T. tabaci* in Hawaii; 1930s, *S. rubrocinctus* and 1960s, *T. tabaci* in the Caribbean; 1970s *T. tabaci* in Europe; 1960s and 1980s, *H. haemorrhoidalis* in California) without great success. Efforts (Sakimura, 1932) to introduce *C. russelli* into Hawaii from California for the control of *T. tabaci* proved unsuccessful (Bailey, 1937) and later attempts with *C. vinctus* also failed (Clausen, 1978). *Ceranisus menes*, introduced from Japan (Sakimura, 1937c), established itself and was subsequently occasionally recovered (Swezey, 1936, 1937, 1950; Yoshimoto, 1965; Hirose, 1990), but no accounts have been published regarding increase and spread, or on its influence in reducing populations of *T. tabaci* (Clausen, 1978). Its effect on *Thrips palmi* and *Frankliniella occidentalis*, which have recently been introduced into Hawaii, is probably low. Attempts to introduce *C. menes* from India into Barbados (Narayanan, 1971; Alam, 1974) for the control of *T. tabaci* failed due to shipping problems, and introductions from Thailand and Japan into Florida to control *T. palmi* (Castineiras *et al.*, 1996) were also unsuccessful.

Several attempts have also been made to control thrips pests with *G. shakespearei* in a classical way. Initial shipments were sent from Ghana to Trinidad (Adamson, 1936) and from there to various islands in the Caribbean for the control of *Selenothrips rubrocinctus*: Puerto Rico (Dohanian, 1937), Jamaica, Bermuda and Granada (Callan, 1943), and into continental USA,

Canada and Hawaii (Callan, 1943; McMurtry and Johnson, 1963; Bennett *et al.*, 1993) for the control of various species of thrips. In a semi-open greenhouse in Trinidad nearly 100 per cent were parasitized, controlling *Selenothrips* populations, but counts in very restricted areas of a few cacao trees gave parasitism rates of only 20 to 30 per cent (Callan, 1943). It successfully established in Trinidad, Jamaica (Cock, 1985) and Puerto Rico (Clausen, 1978), but not elsewhere, either in the field (Caribbean: Bennett, 1970; California: McMurtry *et al.*, 1991) or in a greenhouse (Florida: Bennett *et al.*, 1993). Although not deliberately introduced, it was reported recently from Guadeloupe and Florida (Bennett *et al.*, 1993). Where it has established, the incidence is usually very low (Cock, 1985) or highly seasonal, such as during high infestations of *S. rubrocinctus* on mango, cashew and tropical almond (Bennett, 1970; Bennett and Alam, 1974). No regulatory effects were found on thrips populations (Callan, 1943; Bennett *et al.*, 1993) and economically it is of little importance.

For other thrips pests, classical biological control has never been considered a serious option: for example, *Frankliniella occidentalis*, accidentally introduced into Europe and seasonally important in outdoor crops in the Mediterranean and Central Europe, is a key pest mainly in greenhouses. Attempts to release thrips parasitoids to control thrips pests by seasonal inoculation or inundation in temperate greenhouse ecosystems, a common strategy with predators, are relatively few. In experimental releases of *C. menes* and *C. americensis* to control *F. occidentalis* infestations in sweet pepper, cucumber and roses in Dutch greenhouses, parasitoids spread readily and established themselves throughout the crops but releases did not reduce thrips numbers. Rates of parasitism remained below 10 per cent throughout the season, resulting in severe damage to the crop (Loomans *et al.*, 1995c).

Releases of *T. semiluteus* in avocado orchards in California for the control of *Heliothrips haemorrhoidalis* were more successful and they established in many orchards (McMurtry *et al.*, 1991). Within 3 years of the release of 20,000 *T. semiluteus* at Irvine, California, the parasitoid had established and spread to adjacent fields. In San Diego, widespread establishment and up to 63 per cent parasitization occurred within 2 years, following the release of 11,000 wasps in a 2-ha orchard (McMurtry and Badii, 1991; McMurtry *et al.*, 1991); thrips populations declined when the parasitization rate reached 50 to 60 per cent. However, its ability to spread within and between orchards and its carry-over at low host densities is not enough for sustained control and additional releases are recommended. *Thripobius semiluteus* is now produced commercially (Hunter, 1996) for greenhouse thrips control in avocado and citrus orchards and in ornamentals, dispensed at a rate of 2500 pupae ha^{-1} released in batches of 250 at ten sites (ANBP, 1995).

Prospects for Controlling Pests

Natural control in most cases seems insufficient in itself to prevent damage to a crop, reaching, as it usually does, a maximum late in the growing season following the peak of pest outbreaks. Except for the avocado and *Heliothrips haemorrhoidalis* system, attempts to control thrips pests by parasitoid releases, either in a classical or in a seasonal inoculative way, have failed. From this evaluation of the characteristics of thrips parasitoids as control agents, based on the (still scarce) information available, they do not seem to be promising. However, accumulated information varies greatly in quality and quantity and only for a few species, like *C. menes*, *G. shakespearei*, and *T. semiluteus*, have more elaborate studies been conducted in the laboratory. Greenhouse or field studies are still rare for most species. In contrast to predators, biological characteristics of thrips parasitoids, either qualitative or quantitative, their ecological requirements and their role in thrips control have been poorly investigated. In summary:

1. Many species are poorly described, others have not been identified properly. For most parasitoid species, all that is known are their collection data. Their biology and ecology are poorly understood, thus hampering biological and ecological research. Classification is currently under review and biological data indicate that this is necessary. To institute (inter)national natural-enemy monitoring, more field surveys should be made to identify more efficient species or strains.
2. Levels of parasitism are highly variable, depending on habitat (natural, field, orchard, greenhouse) and season and largely affected by climate and cultural practices. Ecological requirements of thrips parasitoids, the ability to adapt to different host habitats (e.g. greenhouse, cultivated fields, other control measures) and host plants and to switch between hosts are poorly understood. Climatic conditions largely govern thrips–parasitoid activity and population development. Below 20°C, parasitoids remain active, but development either stops or slows down related to their geographic origin (this seems more a state of quiescence than real diapause), whereas thrips hosts still are able to reproduce.
3. All thrips parasitoids are specific to a single subfamily, a few genera or even a few species and, with the exception of *T. gentilei*, negative side-effects like hyperparasitism or attacking beneficials do not occur. The ecological host range, behavioural interactions and the physiological host range have in most cases not been evaluated properly. The existence of intraspecific and geographic variation in both thrips pest and parasitoid (*C. menes*) populations, expressed by variation in colour patterns and sex ratio, differences in host selection behaviour, number and size of the parasitized offspring, developmental time and temperature response, stresses the importance of biotype selection. Matching of various strains of both the parasitoid and host ('pest') can be of importance to possible failure or success in biological control programmes.
4. A high degree of specificity, regarded as very desirable in classical biological control, is considered less important for seasonal inoculative control or inundative releases in greenhouses, because usually one or two thrips pest species occur. Host searching behaviour and the ability to spread and locate hosts efficiently, particularly at low pest densities, is a crucial factor for

successful control, but this has been little studied. *Ceranisus menes* showed a clear positive density-dependent relationship attacking *Thrips tabaci* in onion (Fig. 9.7). However, *Ceranisus* sp. attacking *Rhipiphorothrips cruentatus* on waxapple in Taiwan showed a clear negative density-dependent relationship. The influence of host plant architecture, crop structure and the role of semiochemicals in host location is largely unknown.

5. Seasonal releases for thrips control in greenhouses in temperate climates have not yet been attempted. Except for *T. semiluteus* parasitizing *Heliothrips haemorrhoidalis*, and *T. gentilei* parasitizing *Liothrips* spp., life history parameters suggest that population growth of thrips parasitoids is too low to ensure long-term control of the pest species. Developmental time of parasitoids is equal to or longer than that of their hosts, in particular those belonging to the Thripinae.

6. For seasonal inoculative releases or inundation, requiring many 10,000s of parasitoids ha^{-1} of greenhouses per season or even week, the inability to establish mass rearing of both thrips and parasitoids might be another serious constraint (Chapter 12). Their small size and short lifespan renders egg parasitoids very vulnerable to transport problems and difficult to apply.

Parasitic Nematodes

Records of parasitic nematodes of thrips date back over a century to when Uzel (1895), in Germany, recovered an unnamed specimen from the body cavity of *Thrips physapus*. Although Uzel gave no descriptions or figures of the parasite, he noted some hosts contained over 200 nematodes and that ovaries of infected thrips were both degenerate and non-functional (see Fig. 9.9). Since Uzel's initial record, ten publications have addressed parasitic nematodes of thrips. Central to these reports is the consensus that infected thrips are unable to produce viable eggs and are effectively sterile.

Uzel's discovery was followed by Jones (in Russell, 1912a) who mentioned a nematode 'working in the full-grown larvae of the bean thrips', *Heliothrips fasciatus* in California, but again details were lacking. From Russia, Kolobova (1926) reported an undetermined nematode infecting *Stenothrips graminum* whereby parasitized female hosts were sterilized. Sharga (1932) presented the first detailed life history complete with descriptions and figures of the nematode *Tylenchus aptini* recovered from populations of the wingless *Aptinothrips rufus* near Edinburgh, Scotland. The taxon of this same nematode was revised to *Anguillulina aptini* by Lysaght (1937) who conducted a 3-year ecological study proposing factors that might influence population distribution, abundance and percentage parasitism. Sharga's species was then transferred to the genus *Howardula*. In England, Lewis (1961) reported populations of *Stenothrips graminum* infected by a nematode of similar description. Soon after, in the maritime provinces of eastern Canada, Nickle and Wood (1964) reported *Frankliniella vaccinii* and *Catinathrips vaccinophilus* parasitized by *Howardula aptini*. Noting that ovaries of infected thrips were greatly reduced in size, Nickle and Wood suggested introduction of the nematode into uninfested thrips populations as a potential biological control strategy. In El Paso, Texas, Wilson and Cooley (1972) again recovered *Howardula aptini*, this time from adult female *Frankliniella occidentalis*. A decade later, *Howardula aptini* was reported by Reddy *et al.* (1982) parasitizing *Megalurothrips* sp. in the flowers of southern Indian leguminous field crops. Reddy *et al.* detailed percentage infection and seasonal incidence of the nematode, citing slight morphological differences when compared with descriptions by both Sharga (1932) and Nickle and Wood (1964). Siddiqi (1986) then transferred Sharga's species to a new genus *Thripinema*, placing the Nickle and Wood (1964) and Reddy *et al.* (1982) type specimens into two separate species, *Thripinema nicklewoodii* and *Thripinema renirao*, respectively. *Thripinema nicklewoodii* (formerly *H. aptini*) was again found parasitizing *Frankliniella occidentalis* (Greene and Parrella, 1993), this time in populations throughout California. To date, three nematode species in the genus *Thripinema* have been recovered from eight species among seven genera of thrips.

Biology and life cycle

Thripinema spp. produce one heterosexual generation developing in the haemocoel of the female thrips' abdominal and thoracic cavities. There is no record of a parasitized male thrips. The oviparous adult parasitic female releases eggs ranging in size from 35–60 × 18–28 μm into the host haemocoel. Egg production appears to occur year-round reaching maximum levels during the spring months (Lysaght, 1937; Reddy *et al.*, 1982). The spring incubation period may be between 3 and 4 weeks (Lysaght, 1937). On emergence, young motile vermiform larvae measure 120–140 × 13 μm and feed on the thrips' internal fluids. Mature larvae measure 250–366 × 13–17 μm and are morphologically similar to younger forms.

Sharga (1932) describes a male and female larval form of *Thripinema nicklewoodii*, each with young and mature stages separated by an unspecified number of moults. Males are distinguished by cuticular expansions, or caudal alae, arising anterior to the anus on either side of the nematode's tail end. The exact number of moults varies with reports. For example, Reddy *et al.* (1982) report third and fourth stage larval *Thripinema renirao* leaving the host prior to copulation.

Nematodes leave the host thrips through either the ovipositor (Reddy *et al.*, 1982) or anus (Sharga, 1932; Lysaght, 1936; Greene and Parrella, 1993). First, however, the parasite must move from the abdominal or thoracic haemocoel into the midgut or oviduct. Sharga (1932) twice observed this event: using a slightly protruded stylet, the mature female larva bores through the wall of the host midgut. Once its head enters the lumen of the midgut, the body quickly slides through aided by a wriggling motion. From the midgut, the larva moves to the pyriform rectum where it resides for a period before issuing from the anus. Lysaght (1936) reported that males also left by this route.

Further evidence of anal exit came from Greene and Parrella (1993) who confined *Frankliniella occidentalis* infested with *Thripinema nicklewoodii* to chrysanthemum leaves in modified Munger cells (Munger, 1942; Morse *et al.*, 1986). Observation 24 h later revealed highly active, free-living nematodes moving within freshly deposited faecal droplets. *Thripinema renirao* is unique in that late stage larvae emerge through the ovipositor of *Megalurothrips* sp. (Reddy *et al.*, 1982).

Relatively little is known of the nematodes' free-living stage and while morphological descriptions are included (see Sharga, 1932; Lysaght, 1937; Nickle and Wood, 1964), recovery attempts from plants infested with parasitized thrips have either failed (Lysaght, 1937) or are not mentioned in the literature.

Nickle and Wood (1964) proposed that free-living male and female *Thripinema nicklewoodii* probably mate within leaf galls caused by feeding thrips. Still vermiform in shape, the fertilized female nematode presumably attaches to and penetrates larvae or pupal thrips. Once within the host haemocoel, the infective female nematode swells to a sac-like organism (Fig. 9.8) in which her reproductive system becomes the only visible structure; thus the cycle is complete (see also Plate II.4).

Role of microhabitat

The role of microhabitat on the nematodes' free-living stage is an intriguing aspect of the infection process. A longevity study performed on the free-living stage of the nematode in distilled water showed that males survived for 26 h but females for more than 72 h (Reddy *et al.*, 1982). However, given the absence of water droplets, it is not known how the free-living nematode survived the relatively harsh, dry microclimate outside the host.

Higher relative humidity, as found within a leaf gall, flower perianth, or developing foliage terminal, would greatly favour free-living nematode activity. Since mature nematodes leave the thrips via the anus or ovipositor, these free-living forms would naturally accumulate in and around thrips foraging sites where defecation and oviposition are pronounced.

Early investigators questioned the role of soil habitat in the infection cycle, but the results of several key experiments (Reddy *et al.*, 1982) have helped clarify

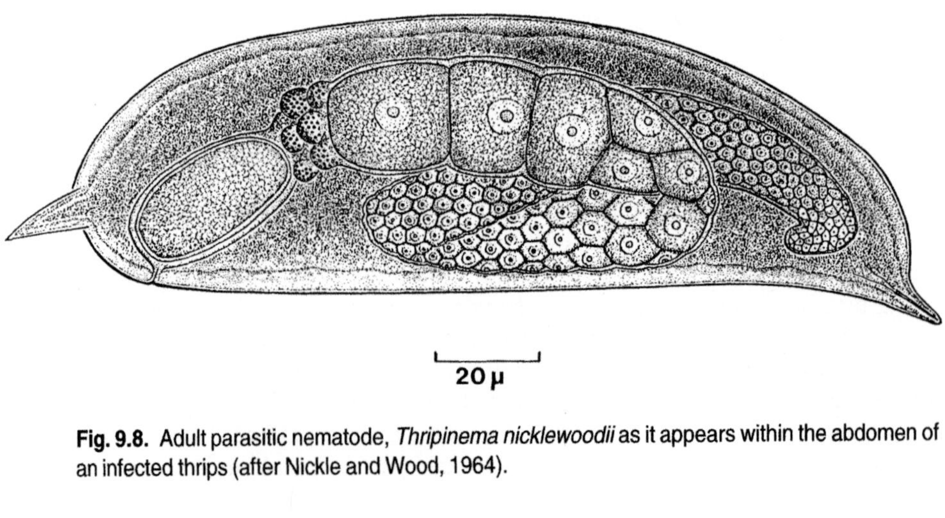

Fig. 9.8. Adult parasitic nematode, *Thripinema nicklewoodii* as it appears within the abdomen of an infected thrips (after Nickle and Wood, 1964).

the infection process. First, male and fourth stage female nematodes were dissected from infected host thrips and released into 'natural conditions'. After 24 h, impregnated female nematodes were observed. The pregnant females were then inoculated into keel petals of washed leguminous flowers along with different larval stages of the host thrips. Within 24 to 36 h dissections of the larval thrips revealed the presence of the impregnated female nematodes.

Dissections of field-collected first and second stage larval *Megalurothrips* sp. revealed parasitization by one to six impregnated female nematodes. Adult thrips within these flowers had been infected by an equal range of fertilized female parasites but were additionally laden with many eggs and vermiform larvae.

Numbers of parasite life stages per host

Gravid female nematodes are highly fecund: in a single host thrips, Lysaght (1936) observed one adult female present with 111 eggs and 18 larvae. When multiple females parasitize a host, all stages of progeny tightly pack the thrips abdomen. Maximum numbers of adult parasitic females recorded from a single host *Thripinema aptini* and *T. renirao* are 6 and 8, respectively; of eggs, 215 and 420; and of juveniles, 255 and 320 (Sharga, 1932; Reddy *et al.*, 1982).

The proportion of male to female juvenile nematodes found within the host is generally low; Sharga (1932) reported typical ratios of 27:228, 4:23, 6:18, 4:38 and 11:73.

Detection and recovery: laboratory methods

It is vital to maintain the integrity of collected thrips prior to dissection. Unsuitable recovery methods often result in a false negative diagnosis. Problems generally occur when thrips are collected and preserved for more than 24 h in various dilutions of water and alcohol. If stored in water, thrips specimens are quickly degraded by fungal or bacterial contaminants, often within 24 h. The viscera of specimens stored in alcohol dilutions congeal and become cloudy making identification of abdominal contents difficult. Given the problems associated with preserving collected material, dissections should be performed on the same day as collection.

While infected specimens can be dissected alive in water (Sharga, 1932), best results are obtained when thrips are anaesthetized (Lysaght, 1937; Reddy *et al.*, 1982) or killed quickly by immersion in alcohol, then transferred immediately to Ringer's solution. Lysaght (1937) used chloroform vapour for anaesthesis but noted that excessive vapour caused thrips blood to coagulate and rapid parasite degeneration. Dissections performed within a fluid medium such as Ringer's solution (Lysaght, 1937; Reddy *et al.*, 1982) provide the most physiologically stable environment for the parasite.

Recovery of the nematode requires the use of a dissecting microscope. The correct relationship between incident-light angle and subject immensely aids parasite viewing. Whether the light source is fibre optic or incandescent, it should illuminate the watchglass from slightly below the horizontal plane. Fine adjustment of the angle of light is important. When properly directed, the parasites are evident in striking contrast to the surrounding medium.

Using two finely pointed probes, the investigator submerges the thrips in the fluid, securing the thorax against the bottom of the watchglass with one probe while lancing the abdominal margin with a second probe. Once the tergal plates rupture, all stages of the nematode are released from the infected thrips almost immediately (Plate II. 4).

Host impact: suppression of oogenesis

Infected thrips show no obvious physical or behavioural signs of the internal parasite, but they are unable to produce viable eggs. Examination of affected internal reproductive organs shows a great reduction in size and change in shape of the oviduct, vagina, and receptaculum seminis (Fig. 9.9). Lysaght's 3-year survey encompassed dissections of 1650 infected thrips: of these, only eight were found to possess normal reproductive organs. Lysaght (1936) suggested that the developing nematodes either deprive the thrips of protein required for normal egg development, or that they secrete a toxin which damages the reproductive organs.

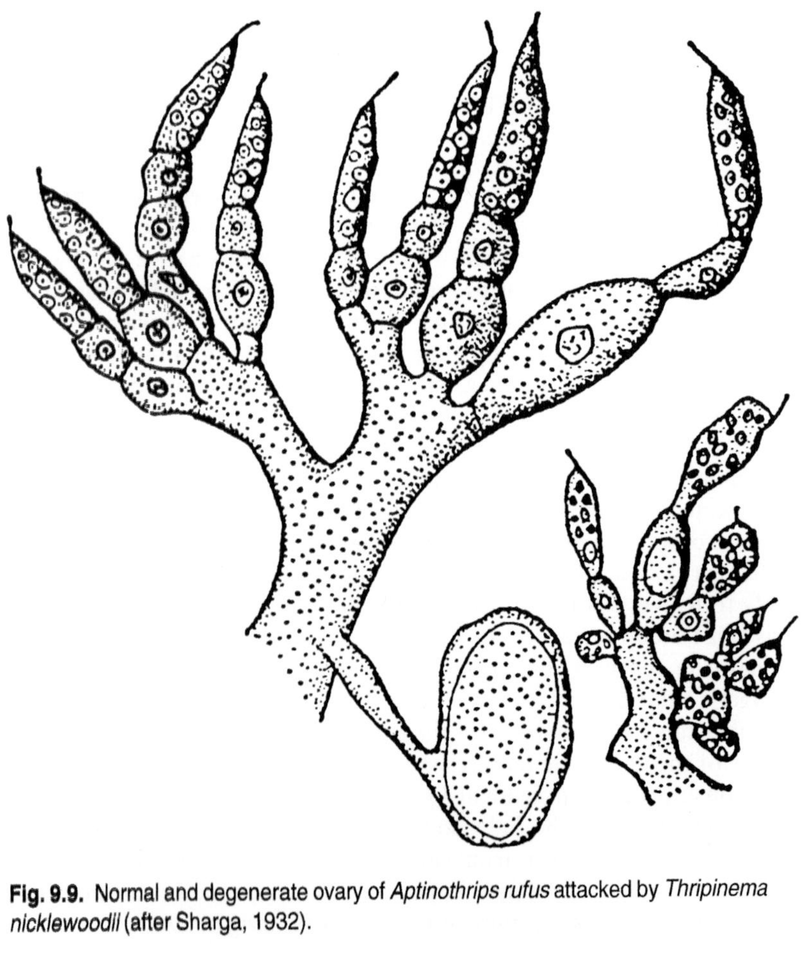

Fig. 9.9. Normal and degenerate ovary of *Aptinothrips rufus* attacked by *Thripinema nicklewoodii* (after Sharga, 1932).

Population dynamics: potential for biological control

Although several studies have identified seasonal incidence and percentage infection, the population dynamics of *Thripinema* spp. appear to follow no distinct pattern. Dissections of weekly thrips samples taken over the course of 1-year (Sharga, 1932; Reddy *et al.*, 1982) and 2-year periods (Lysaght, 1937) show seasonal population shifts that differ between *T. aptini* and *T. renirao*. Despite this variation, percentage parasitism generally appears to reach a maximum during the late spring and summer. Exactly which factors cause these shifts in seasonal incidence remains unknown.

The impact of a natural enemy on the dynamics of its host population is often insignificant when compared with the huge host densities found in the field. Whether or not, for example, a 63 per cent infection of a thrips population would cause a measurable reduction in either crop loss or thrips numbers remains to be determined. With increasing interest in biological control programmes for pest thrips the possibility that *Thripinema* spp. may markedly reduce thrips populations is intriguing (see p. 422).

References

Adamson, A.M. (1936) Progress report on the introduction of a parasite of the cacao thrips from the Gold Coast to Trinidad B.W.I. *Tropical Agriculture* 3, 62–63.

Alam, M.M. (1974) Biological control of insect pests of horticultural crops in Barbados. In: Braithwaite, C.W.D., Phelps, R.H., and Bennett, F.D. (eds) *Crop Protection in the Caribbean, Proceedings of a Symposium on the Protection of Horticultural Crops in the Caribbean, Trinidad, April 8–11*, pp. 253–261.

Ananthakrishnan, T.N. (1984) *Bioecology of Thrips*. Indira Publishing House, Oak Park Michigan, USA.

Ananthakrishnan, T.N. and Swaminathan, S. (1977) Host–parasite and host–predator interactions in the gall thrips *Schedothrips orientalis* Anan. (Insecta: Thysanoptera). *Entomon* 2, 247–252.

Annecke, D.P. (1962) The genus *Goetheana* Girault in South Africa (Eulophidae – Hymenoptera). *South African Journal of Agricultural Science* 5, 273–279.

ANBP (1995) Fact sheet *Thripobius semiluteus*. Association of Natural Bio-control Producers, 1p.

Antsiferova, T.A. and Timraleev, Z.A. (1974) Vrogi gorokhovogo tripsa (Enemies of the pea thrips). *Zashchita Rastenii (Kiev)* 10, 26.

Bailey, S.F. (1933) The biology of the bean thrips. *Hilgardia* 7, 467–522.

Bailey, S.F. (1937) The bean thrips. *Bulletin California Agricultural Experiment Station* 609, 36pp.

Bartlett, K. (1939) The introduction and colonization in Puerto Rico of *Dasyscapus parvipennis* Gahan, a parasite of thrips. *Agricultural Notes, Puerto Rico Experiment Station of the USDA* 87, 1–5.

Bazzocchi, G. and Santi, F. (1994) Comportamento di *Ceranisus menes* (Walker) (Hymenoptera: Eulophidae) parasitoide di *Frankliniella occidentalis* (Pergande) (Thysanoptera: Thripidae). *Atti XVII Congresso Nazionale Italiano di Entomologia, Udine, 13–18 giugnio*, pp. 891–892.

Bennett, F.D. (1965) Observations on the natural enemies of *Gynaikothrips ficorum* Marchal in Brazil. *Technical Bulletin CIBC* 5, 117–125.

Bennett, F.D. (1970) Report on a visit to Jamaica, to collect stocks of the thrips parasite *Goetheana (Dasyscapus) parvipennis* July 1970. *Internal report CIBC*, Ascot, UK, 3 pp.

Bennett, F.D. and Alam, M.M. (1974) Report on a visit to Jamaica, mainly to collect egg parasites of *Exophthalmus* spp. for trials against *Diaprepes abbreviatus* (L.) in Barbados. *Internal Report CIBC–Trinidad*, 5 pp.

Bennett, F.D. and Baranowski, R.M. (1982) First record of the thrips parasite *Goetheana parvipennis* (Gahan) (Eulophidae: Hymenoptera) from the Bahamas. *Florida Entomologist* 65, 185.

Bennet, F.D., Glenn, H. and Baranowski, R.M (1993) *Goetheana shakespearei* (Hymenoptera: Eulophidae) an immigrant parasitoid of thrips in Florida and Guadeloupe? *Florida Entomologist* 76, 395–397.

Beshear, R. (1974) A chalcidoid planidium on thrips larvae in Georgia. *Journal of the Georgia Entomological Society* 9, 265–266.

Billes, D.J. (1941) Pollination of *Theobroma cacao* L. in Trinidad, B.W.I. *Tropical Agriculture* 18, 151–156.

Boucek, Z. (1976) Taxonomic studies on some Eulophidae (Hym.) of economic interest, mainly from Africa. *Entomophaga* 21, 401–414.

Boucek, Z. (1988) *Australasian Chalcidoidea (Hymenoptera): a Biosystematic Revision of Genera of Fourteen Families, with Reclassification of Species.* CAB International, Wallingford, UK.

Boucek, Z. and Askew, R.R. (1968) *Palearctic Eulophidae (excl. Tetrastichinae) (Hym. Chalcidoidea)*, Index of Entomophagous Insects 3. Le François, Paris.

Bournier, A. (1967) Un intéressant parasite de Thysanoptères: *Tetrastichus gentilei* (Hym. Chalcididae). *Annales de la Société Entomologique de France (Nouveau Séries)* 3, 173–179.

Bühl, C. (1937) Beiträge zur Kenntnis der Biologie, wirtschaftliche Bedeutung und Bekämpfung von *Kakothrips robustus* Uz. *Zeitschrift für angewandte Entomologie* 23, 65–113.

Burks, B.D. (1971) A new *Pediobius* parasitic on thrips (Hymenoptera: Eulophidae). *Proceedings of the Entomological Society of Washington* 73, 177–180.

Callan, E. McC. (1943) Natural enemies of the cacao thrips. *Bulletin of Entomological Research* 34, 313–321.

Carl, K.P. (1971) Report of the work carried out at the European station. In: *Investigations of the Natural Enemies of Thrips for the Use as Biological Control Agents Against Glasshouse Thrips in the United Kingdom.* CIBC report, European Station, Delémont-Switzerland, pp. 9–35.

Carl, K.P. (1982) Biological control of native pests by introduced natural enemies. *Biocontrol News and Information* 3, 191–200.

Carl, K.P., Yuill, J., Parsch, M. and Prior, C. (1989) Taeniothrips inconsequens *in Central Europe and its natural enemies.* IIBC Report, European Station, 14 pp.

Castineiras, A., Baranowski, R.M. and Glenn, H. (1996) Temperature response of the two strains of *Ceranisus menes* (Hymenoptera: Eulophidae) reared on *Thrips palmi* (Thysanoptera: Thripidae). *Florida Entomologist* 79, 13–19.

Chang, N.T. (1991) Important thrips species in Taiwan. In: Talekar, N.S. (ed.) *Thrips in Southeast Asia.* AVRDC, Taipei, pp. 40–56.

Chiu, H.T. (1984) The ecology and chemical control of grapevine *thrip* (*Rhipiphorothrips cruentatus*) on wax apple. *Plant Protection Bulletin* 26, 365–378.

Clausen, C.P. (1940) The oviposition habits of the Eucharitidae (Hymenoptera). *Journal of the Washington Academy of Sciences* 30, 504–516.

Clausen, C.P. (1978) Thysanoptera. In: Bartlett, B.R., Clausen, C.P. and DeBach, P. (eds)

Introduced Parasites and Predators of Arthropod Pests and Weeds: a World Review. Agriculture Handbook, USDA, Washington pp. 18–21.

Cock, M.J.W. (1982) The biology and host specificity of *Liothrips mikaniae* (Priesner) (Thysanoptera: Phlaeothripidae), a potential biological control agent of *Mikania micrantha* (*Compositae*). *Bulletin of Entomological Research* 72, 523–533.

Cock, M.J.W. (1985) A review of biological control of pests in the commonwealth Caribbean and Bermuda up to 1982. *CIBC–CAB Technical Communication* 9, 218 pp.

Cotterell, G.S. (1927) A new parasite of *Heliothrips rubrocincta. Bulletin of the Department of Agriculture Gold Coast, Yearbook 1926* 7, 47–48.

Daniel, A.M. (1986) Thrips–parasite interactions in some panchaetothripine Thysanoptera (Insecta: Arthropoda). *Proceedings of the Indian National Science Academy, Part B, Biological Sciences* 52, 437–444.

Daniel, A.M., Suresh Kumar, N. and Bakthavatsalam, N. (1983) Bioecology of *Zaniothrips ricini* Bhatti (Panchaetothripinae: Terebrantia: Thysanoptera) – a new thrips pest of castor (*Ricinus communis*: *Euphorbiaceae*). *Proceedings of the Indian Academy of Sciences (Animal Sciences)* 92, 87–94.

Daniel, A.M., Suresh Kumar, N. and Bakthavatsalam, N. (1986) Thrips–crop–weed interactions of *Thrips palmi* Karny and *Toxothrips ricinus* Bhatti infesting *Sesamum indicum* and *Ricinus communis*. *Proceedings of the Indian National Science Academy, Part B, Biological Sciences* 52, 358–363.

Del Guercio, G. (1911) Il *Tetrastichus gentilei* Del Guerc. nei suoi rapporti col Fleothripide dell'olivo. *Atti della Reale Accademia economico-agraria dei Georgofili di Firenze* 8, 222–227.

Del Guercio, G. (1931) Il pidocchio nero, ver noir o barban dell'olivo (*Phloeothrips oleae* Costa) ed i suoi rapporti con i punteroli. *Redia* 19, 75–195.

Delvare, G. (1993) Sur les *Megaphragma* de Guadeloupe avec la description d'une espèce nouvelle (Hymenoptera, Thrichogrammatidae). *Revue Française d'Entomologie, (Nouveau Séries)* 15, 149–152.

DeSantis, L. (1965) Lucha biológica contra los thrips en América del Sur. *Fitosanitaria* 4, 4–7.

Dessart, P. and Bournier, A. (1971) *Thrips tabaci* Lindman (Thysanoptera), hôte inattendu d'*Aphanogmus fumipennis* (Thomson) (Hym. Ceraphronidae). *Bulletin Annuel de la Société Royale d'Entomologie Belgique* 107, 116–118.

Dohanian, S.M. (1937) Life history of the thrips parasite *Dasyscapus parvipennis* Gahan and the technique for breeding it. *Journal of Economic Entomology* 30, 78–80.

Domenichini, G. (1966) *Palearctic Tetrastichinae (Hym. Eulophidae)*, Index of Entomophagous Insects 1. Le François, Paris.

Doutt, R.L. and Viggiani, G. (1968) The classification of the Trichogrammatidae (Hymenoptera: Chalcidoidea). *Proceedings of the Fourth Series of the California Academy of Sciences* 35, 477–586.

Dozier, H.L. (1932) Descriptions of new trichogrammatid (Hymenoptera) egg parasites from the West Indies. *Proceedings of the Entomological Society of Washington* 34, 29–37.

Dozier, H.L. (1937) Descriptions of miscellaneous chalcidoid parasites from Puerto Rico (Hymenoptera). *Journal of the Agriculture University of Puerto Rico* 21, 121–135.

Dyadechko, N.P. (1964) *Thrips or fringe-winged insects (Thysanoptera) of the European part of the USSR*. Amerind Publishing Company, New Delhi.

Dyadechko, N.P. (1967) Entomophagous insects and the causal agents of fungus diseases in phytophagous species of thrips. *Zashchita Rastenii (Kiev)* 4, 149–151.

Ebeling, W. (1959) *Subtropical Fruit Pests.* University of California Division of Agricultural Sciences, 436 pp.

Entwistle, P.F. (1972) *Pests of Cocoa.* Tropical Science Series, Longmans, London.

Erdös, J. (1971) Chalcidoidea VIII. *Fauna Hungariae* 104, 1–252.

Ferrière, C. (1931) Notes on African Chalcidoidea. *Bulletin of Entomological Research* 22, 127–135.

Ferrière, C. (1958) Un nouveau parasite de thrips en Europe centrale (Hym. Euloph.). *Mitteilungen der Schweizerischen Entomologischen Gesellschaft* 31, 320–324.

Fourez, A. (1995). Influence of age of the hosts *Frankliniella occidentalis* (Thysanoptera: Thripidae) on the parasitism efficiency by *Ceranisus menes* (Hymenoptera: Eulophidae). *Program of the 47th International Symposium on Crop Protection, Gent,* 14.

Franssen, C.J.H. and van Heurn, W.C. (1932) Biologie en bestrijding van de sjalottentrips (*Thrips tabaci* Lindn.). *Korte Mededeelingen van het Instituut voor Plantenziekten* 18, 1–19.

Fullaway, D.T. and Dobroscky, I.D. (1934) A new *Thripoctenus* parasite from the Philippines. *Proceedings of the 5th Pacific Science Congress, Canada 1933* 5, 3439–3444.

Galazzi, D. and Bazzocchi, G. (1993) *Ceranisus menes. Informatore Fitopatologico* 43, 47–50.

Galazzi, D., Maini, S. and Loomans, A.J.M. (1992) *Ceranisus menes* (Walker) (Hymenoptera: Eulophidae): Collection and initial rearing on *Frankliniella occidentalis* (Pergande) (Thysanoptera: Thripidae). *Bollettino dell'Istituto di Entomologia 'Guid Grandi' della Università degli Studi di Bologna* 46, 123–129.

Ghesquière, J. (1939) Contributions à l'étude des Hyménoptères du Congo Belgique. VI. Description d'un Mymaride nouveau et remarques sur le Gn. *Megaphragma* Timb. (Trichogrammidae). *Revue Zoologique et Botanique d'Afrique* 33, 33–41.

Girault, A.A. (1920) New genera and species of Australian Mymaridae (Hymenoptera). *Insecutor Inscitiae Menstruus* 8, 96–100.

Goodwin, S. and Steiner, M.Y. (1996) Biological control of thrips. *Proceedings of the 1995 Australian and New Zealand Thrips Workshop: Methods, Biology, Ecology and Management,* NSW Agriculture, Gosford, Australia, pp. 76–81.

Graham, M.W.R. de V. (1959) Keys to the British genera and species of Elachertinae, Eulophidae, Entedontinae, and Euderinae (Hym. Chalcidoidea). *Transactions of the Society of British Entomology* 13, 169–204.

Graham, M.W.R. de V. (1963) Additions and corrections to the British List of Eulophidae (Hym., Chalcidoidea), with descriptions of some new species. *Transactions of the Society of British Entomology* 15, 167–275.

Graham, M.W.R. de V. (1987) A reclassification of the European Tetrastichinae (Hymenoptera: Eulophidae), with a revision of certain genera. *Bulletin of the British Museum (Natural History), Entomology Series* 55, 1–392.

Greene, I.D. and Parrella, M.P. (1993) An entomophilic nematode, *Thripinema nicklewoodii* and an endoparasitic wasp, *Ceranisus* sp. parasitizing *Frankliniella occidentalis* in California. *International Organisation of Biological Control/West Palaearctic Regional Section Bulletin* 16 (2), 47–50.

Grout, T.G. and Stephen, P.R. (1995) *Goetheana incerta* parasitizing citrus thrips in southern Africa. *Citrus Journal* 5 (4), 30–32.

Heinz, K.M., Heinz, L.M. and Parrella, M.P. (1996) Natural enemies of western flower thrips indigenous to California ornamentals. *International Organisation of Biological Control/West Palaearctic Regional Section Bulletin* 19(1), 51–54.

Hessein, N.A. and McMurtry, J.A. (1988) Observations on *Megaphragma mymaripenne* (Hymenoptera: Trichogrammatidae), an egg parasite of *Heliothrips haemorrhoidalis* Bouché (Thysanoptera: Thripidae). *Pan-Pacific Entomologist* 64, 250–254.

Hessein, N.A. and McMurtry, J.A. (1989) Biological studies of *Goetheana parvipennis* (Gahan) (Hymenoptera: Eulophidae), an imported parasitoid, in relation to the host species *Heliothrips haemorrhoidalis* (Bouché) (Thysanoptera: Thripidae). *Pan-Pacific Entomologist* 65, 25–33.

Hirose, Y. (ed.) (1989) *Exploration for Natural Enemies of* Thrips palmi *in Southeast Asia*. Institute of Biological Control, Faculty of Agriculture, Kyushu University Fukuoka, 58 pp.

Hirose, Y. (1990) Prospective use of natural enemies to control *Thrips palmi* (Thysanop., Thripidae). In: Bay-Petersen, J. (ed.) *The Use of Natural Enemies to Control Agricultural Pests, Proceedings of the International Seminar 'The Use of Parasitoids and Predators to Control Agricultural Pests'*, Tsukuba, Japan, October 2–7, 1989. FFTC book series 40, pp. 135–141.

Hirose, Y., Takagi, M. and Kajita, H. (1992) Discovery of an indigenous parasitoid of *Thrips palmi* Karny (Thysanoptera: Thripidae) in Japan: *Ceranisus menes* (Walker) (Hymenoptera: Eulophidae) on eggplant in home and truck gardens. *Applied Entomology and Zoology* 27, 465–467.

Hirose, Y., Kajita, H., Takagi, M., Okajima, S., Napompeth, B. and Buranapanichpan, S. (1993) Natural enemies of *Thrips palmi* and their effectiveness in the natural habitat, Thailand. *Biological Control* 3, 1–5.

Hukkinen, Y. (1936) Tutkimuksla nurmipuntarpään (*Alopecurus pratensis* L.) siementuholaisista. 1. *Chirothips hamatus* Tryb., putarpääripsiainen. *Valtion Maatalouskoetoiminnan Julkaisuja* 81, 1–132.

Hunter, C.D. (1996) *Suppliers of Beneficial Organisms in North America.* California Environmental Protection Agency, Department of Pesticide Regulation, Environmental Monitoring and Pest Management Branch, Sacramento, California.

Ishii, T. (1933) Notes on two hymenopterous parasites of *Thrips* in Japan. *Kontyû* 7, 13–16.

Johnson, D.W. (1988) Eucharitidae (Hymenoptera: Chalcidoidea): biology and potential for biological control. *Florida Entomologist* 71, 528–537.

Kolobova, A.N. (1926) *Stenothrips graminum* Uzel. (In Russian.) *Review of Applied Entomology* (A) 14, 606–607.

Krombein, K.V. (1958) Miscellaneous records of solitary wasps. III (Hymenoptera, Aculeata). *Proceedings of the Entomological Society of Washington* 71, 21–26.

Kryger, I.P. (1932) One new genus and species, and three new species of Trichogramminae from Egypt with remarks upon *Neocentrobia hirticornis*, *Alaptus minimus*, and *Trichogramma evanescens*. *Bulletin de la Société Royale Entomologique d'Égypte* 16, 38–44.

Kutter, H. (1936) Über einen neuen endoparasiten (*Thripoctenus*, Chalcididae) de Erbsenblasenfusses (*Kakothrips robustus* Uzel), seine Lebensweise und Entwicklung. *Mitteilungen der Schweizerischen Entomologischen Gesellschaft* 16, 640–652.

Kutter, H. (1937) Die Bekämpfung des Konservenerbsenschädlinge im st. gallischen Rheintale, Untersuchungsbericht 1936. *Landwirtschaftliche Jahrbücher Schweiz* 51, 97–119.

Lacasa, A., Contreras, J., Sanchez, J.A., Lorca, M. and Garcia, F. (1996) Ecology and natural enemies of *Frankliniella occidentalis* (Pergande, 1895) in south-east Spain. *Folia entomologica hungarica*, 57 (Suppl.), 67–74.

LaSalle, J. and McMurtry, J.A. (1989) First record of *Thripobius semiluteus* (Hymenoptera: Eulophidae) from the new world. *Proceedings of the Entomological Society of Washington* 91, 634.

Lewis, T. (1961) Records of Thysanoptera at Silwood Park, with notes on their biology. *Proceedings of the Royal Entomological Society of London (A)* 36, 89–95.

Lewis, T. (1973) *Thrips: Their Biology, Ecology and Economic Importance*. Academic Press, London.

Lin, N. (1992) Descriptions of five new species of *Megaphragma* and *Paramegaphragma* gen. nov. (Hymenoptera: Trichogrammatidae) from China. *Entomotaxonomia* 14, 129–138.

Loomans, A.J.M. (1991) Collection and first evaluation of hymenopterous parasites of thrips as biological control agents of *Frankliniella occidentalis. International Organisation of Biological Control/West Palaearctic Regional Section Bulletin* 14, 73–82.

Loomans, A.J.M. (1997) Intraspecific variation in thrips and its relevancy to biocontrol. *Proceedings of the Section Experimental and Applied Entomology of the Netherlands Entomological Society, Amsterdam*, 8.

Loomans, A.J.M. and van Lenteren, J.C. (1994) Evaluation of the effectiveness of parasitic wasps for the biological control of thrips pests in protected crops: state of affairs. *International Organisation of Biological Control/West Palaearctic Regional Section Bulletin* 17, 158–164.

Loomans, A.J.M. and van Lenteren, J.C. (1995) Biological control of thrips pests: a review on thrips parasitoids. *Wageningen Agricultural University Papers* 95–1, 89–201.

Loomans, A.J.M. and van Lenteren, J.C. (1996) Prospects of *Ceranisus americensis* (Girault) as a potential biological control agent of thrips pests in protected crops. *International Organisation of Biological Control/West Palaearctic Regional Section Bulletin* 19, 95–98.

Loomans, A.J.M. and Murai, T. (1995) Comparison of life-histories of different colour types of *Ceranisus menes* (Hymenoptera: Eulophidae), parasitoid of *Frankliniella occidentalis* (Thysanoptera: Thripidae). *Proceedings of the Section Experimental and Applied Entomology of the Netherlands Entomological Society, Amsterdam* 5, 35–40.

Loomans, A.J.M. and Pákozdi, A. (1996) Differential acceptance of *Ceranisus menes* (Walker) and *Ceranisus americensis* (Girault) attacking thrips hosts, *Frankliniella occidentalis* (Pergande) and *Thrips tabaci* (Lind.). *Folia entomologica hungarica* 57 (Suppl.), 83–90.

Loomans, A.J.M., Silva, I. and van Lenteren, J.C. (1992) *Ceranisus menes* (Hymenoptera: Eulophidae), a potential biological control agent of *Frankliniella occidentalis* (Thysanoptera: Thripidae)? *Proceedings of the Section Experimental and Applied Entomology of the Netherlands Entomological Society, Amsterdam* 3, 40–45.

Loomans, A.J.M., Paris, A. and van Lenteren, J.C. (1993) Influence of size of *Frankliniella occidentalis* larvae on host acceptance by *Ceranisus menes. International Organisation of Biological Control/West Palaearctic Regional Section Bulletin* 16, 101–104.

Loomans, A.J.M., van Lenteren, J.C., Maini, S., Riudavets, J. and Tommasini, M.G. (1995a) *Biological Control of Thrips Pests*. Wageningen Agricultural University Papers 95–1, 201p.

Loomans, A.J.M., Murai, T., van Heest, P.J.N.F. and van Lenteren, J.C. (1995b) *Ceranisus menes* (Hymenoptera: Eulophidae) for control of western flower thrips: biology and behavior. In: Parker, B.L., Skinner, M. and Lewis, T. (eds.) 1995. *Thrips Biology and Management*. NATO ASI series. Series A, Life sciences 276, Plenum, New York, pp. 263–268.

Loomans, A.J.M., Tolsma, J., van Heest, P.N.J.F. and Fransen, J.J. (1995c) Releases of parasitoids (*Ceranisus* spp.) as biological control agents of western flower thrips (*Frankliniella occidentalis*) in experimental greenhouses. *Mededelingen van de Faculteit der Landbouwwetenschappen van de Rijksuniversiteit Gent* 60, 869–877.

Lysaght, A.M. (1936) A note on the adult female of *Anguillulina aptini* (Sharga), a nematode parasitizing *Aptinothrips rufus* Gmelin. *Parasitology* 28, 290–292.

Lysaght, A.M. (1937) An ecological study of a thrips (*Aptinothrips rufus*) and its nematode parasite (*Anguillulina aptini*). *Journal of Animal Ecology* 6, 169–192.

Mason, A.C. (1922) *Cryptothrips laureli*, a new thrips from Florida (Thysanop.). *Entomological News* 23, 193–199.

Matthews, R.W. (1970) A new hunting *Microstigmus* from Costa Rica (Hymenoptera: Sphecidae, Pemphredoninae). *Psyche Journal of Entomology (Cambridge Massachussets)* 77, 120–126.

McKenzie, H.L. (1935) Life history and control of the gladiolus thrips in California. *Circular of the California Agricultural Experiment Station* 337, 16 pp.

McMurtry, J.A. (1961) Current research on biological control of avocado insect and mite pests. *California Avocado Society Yearbook* 45, 104–106.

McMurtry, J.A. and Badii, M.H. (1991) Greenhouse thrips, *Heliothrips haemorrhoidalis*, in California avocado orchards: biological control studies. In: Parker, B.L., Skinner, M. and Lewis, T. (eds) *Towards Understanding Thysanoptera, Proceedings International Conference on Thrips, Burlington, Vermont USA, February 21–23, 1989*, General Technical Report NE-147. USDA, Forest Service, North Eastern Forest Experiment Station, pp. 393–398.

McMurtry, J.A. and Johnson, H. (1963) Progress report on the introduction of a thrips parasite from the West Indies. *California Avocado Society Yearbook* 47, 48–51.

McMurtry, J.A., Johnson, H.G. and Newberger, S.J. (1991) Imported parasite of greenhouse thrips established on California avocado. *California Agriculture* 45, 31–32.

Melis, A. (1935) Tisanotteri italiani. Studio anatomo-morphologico e biologico de Liothripidae dell'olivo (*Liothrips oleae* Costa). *Redia* 21, 1–188.

Morse, J.G., Bellows, T.S. and Iwata, Y. (1986) Technique for evaluating residual toxicity of pesticides to motile insects. *Journal of Economic Entomology* 79, 281–283.

Muesebeck, C.F.W., Krombein, K.V. and Townes, H.K. (eds) (1951) Hymenoptera of America north of Mexico: synoptic catalogue. *Agricultural monographs USDA* 2, 1421 pp.

Munger, F. (1942) A method for rearing citrus thrips in the laboratory. *Journal of Economic Entomology* 25, 373–375.

Murai, T. (1988) Studies on the ecology and control of flower thrips, *Frankliniella intonsa* (Trybom). *Bulletin of the Shimane Agricultural Experiment Station* 23, 1–73.

Murai, T. (1990) Rearing method and biology of thrips parasitoid, *Ceranisus menes*. *International Organisation of Biological Control/West Palaearctic Regional Section Bulletin* 13, 142–146.

Murai, T. and Loomans, A.J.M. (1995) Host–parasite interaction between *Frankliniella intonsa*, western flower thrips and *Ceranisus menes* (Hymenoptera: Eulophidae): development and reproduction. In: Parker, B.L., Skinner, M. and Lewis, T. (eds) *Thrips Biology and Management*. Plenum, New York, pp. 269–275.

Narayanan, E. (1971) Report of the work carried out at the Indian station. In: *Investigations of the Natural Enemies of Thrips for Use as Biological Control Agents Against Glasshouse Thrips in the United Kingdom*. CIBC Report, European Station, Delémont-Switzerland, pp. 3–5.

Narayanan, E.S. and Subba Rao, B.R. (1961) Studies on Indian Mymaridae: III

(Hymenoptera: Chalcidoidea). *Beiträge zur Entomologie* 11, 655–671.
Narayanan, E.S., Subba Rao, B.R. and Ramachandra Rao, M. (1960) Some new species of chalcids from India. *Proceedings of the National Institute of Sciences of India* 26, 168–175.
Newberger, S.J. and McMurtry, J.A. (1992) Insectary production of *Thripobius semiluteus* Boucek (Hymenoptera: Eulophidae), an imported parasitoid of the greenhouse thrips, *Heliothrips haemorrhoidalis* (Bouché). Internal Report, University of California at Riverside, 3 pp.
Nickle, W.R. and Wood, G.S. (1964) *Howardula aptini* Sharga (1932) parasitic in the blueberry thrips in New Brunswick. *Canadian Journal of Zoology* 42, 843–846.
Noyes, J.S. and Valentine, E.W. (1989) Chalcidoidea (Insecta: Hymenoptera) – introduction, and review of genera in smaller families. *Fauna New Zealand* 18, 1–91.
Paoli, 1931. *Relazione sull'attività del R. Osservatorio de Fitopathologia per la Liguria in Chiavari nel primo decennio della sua fondazione.* Genoa, Fratelli Treves, 59 pp.
Pemberton, C.E. (1931) An egg parasite of thrips in Hawaii. *Proceedings of the Hawaiian Entomological Society* 7, 481–482.
Raman, A. and Ananthakrishnan, T.N. (1984) Biology of gall thrips (Thysanoptera: Insecta). In: T.N. Ananthakrishnan (ed.) *Biology of Gall-Insects*. Edward Arnold, London, pp. 107–127.
Rahman, K.A. and Bhardwaj, N.K. (1937) The grape-vine thrips (*Rhipiphorothrips cruentatus* Hood) (Thripidae: Terebrantia: Thysanoptera). *Indian Journal of Agricultural Science* 7, 633–651.
Reddy, Y.N., Nickle, W.R. and Rao, P.N. (1982) Studies on *Howardula aptini* (Nematoda – Sphaerulariidae) parasitic in *Megalurothrips* sp. in India. *Indian Journal of Nematology* 12, 1–5.
Risbec, J. (1951) I. Les Chalcidoides d'A.O.F. *Mémoires de l'Institut de France Afrique Noire* 13, 51–67.
Risbec, J. (1958) Contributions à la connaissance des Hyménoptères Chalcidoïdes et Proctrupoïdes de l'Afrique Noire. II. Les *Pleurotropis* d'Afrique et de Madagascar (Chalcidoidae, Eulophidae). *Annales de la Musée Royale Congo Belge, Science et Zoologie* 64, 68–90.
Riudavets, J. (1995) Predators of *Frankliniella occidentalis* (Perg.) and *Thrips tabaci* Lind.: a review. *Wageningen Agricultural University Papers* 95–1, 43–87.
Rivnay, E. (1935) Ecological studies of the greenhouse thrips, *Heliothrips haemorrhoidalis*, in Palestine. *Bulletin Entomological Research* 26, 267–278.
Rivnay, E. (1939) Studies on the biology and ecology of *Retithrips syriacus* Mayet, with special attention to its occurrence in Palestine. *Bulletin de la Société Royale Entomologique d'Égypte* 23, 150–182.
Rivnay, E. (1947) Biology of *Gynaikothrips ficorum* Marsh. in Palestine (Thysanoptera). *Bulletin de la Société Royale Entomologique d'Égypte* 31, 129–140.
Rubin, A. and Kuslitzky, W. (1992) First record of *Ceranisus menes* (Hymenoptera: Eulophidae) in Israel. *Phytoparasitica* 20, 123–124.
Russell, H.M. (1911) A true internal parasite of Thysanoptera. *Proceedings of the Entomological Society of Washington* 13, 235–238.
Russell, H.S. (1912a) The bean thrips (*Heliothrips fasciatus* Pergande). *US Department of Agriculture, Bureau of Entomology Bulletin* No. 118, 1–45.
Russell, H.M. (1912b) An internal parasite of Thysanoptera. *USDA, Bureau of Entomology, Technical Series* 23, 25–52.
Sakimura, K. (1932) Life history of *Thrips tabaci* L. on *Emilia sagittata* and its host plant range in Hawaii. *Journal of Economic Entomology* 25, 884–891.
Sakimura, K. (1937a) On the bionomics of *Thripoctenus brui* Vuillet, a parasite of

Thrips tabaci Lind., in Japan. I. *Kontyû* 11, 370–390.
Sakimura, K. (1937b) On the bionomics of *Thripoctenus brui* Vuillet, a parasite of *Thrips tabaci* Lind., in Japan. II. *Kontyû* 11, 410–424.
Sakimura, K. (1937c) Introduction of *Thripoctenus brui* Vuillet, parasite of *Thrips tabaci* Lind., from Japan to Hawaii. *Journal of Economic Entomology* 30, 799–802.
Saxena, R.C. (1971) Some observations on *Ceranisus* sp. (Hymenoptera, Eulophidae) parasitising *Thrips tabaci* Lind. (Thysanoptera: Thripidae). *Indian Journal of Entomology* 33, 91–92.
Saxena, R.C. (1981) Observations on some predators and parasites of *Thrips tabaci* Lind. *Bulletin of Entomology* 22, 97–100.
Schauff, M. (1991) The holarctic genera of Entedoninae (Hymenoptera: Eulophidae). *Contributions of the American Entomological Institute* 26, 1–109.
Seamans, H.L. (1923) The alfalfa thrips and its effect on alfalfa seed production. *The Canadian Entomologist* 55, 101–105.
Sharga, U.S. (1932) A new nematode *Tylenchus aptini* n. sp. parasite of Thysanoptera (Insecta: *Aptinothrips rufus* Gmelin). *Parasitology* 24, 268–279.
Sharma A.K., Farooqi, S.I. and Subba Rao, B.R. (1965) The biology of *Tetrastichus rhipiphorothripscidis* Narayanan, Rao et Rao (Eulophidae: Hymenoptera), a parasite of *Mallothrips indicus* Ramakrishna. *Mushi* 38, 117–122.
Siddapaji, C. and Reddy, D.N.R.N. (1974) A new record of *Ceranisus* sp. (Eulophidae: Hymenoptera) as a parasite of *Panchaetothrips indicus* Bagnall, (Heliothripinae – Thysanoptera) in cardamom. *Current Research* 3, 125–126.
Siddiqi, M.R. (1986) *Tylenchida: Parasites of Plants and Insects.* Commonwealth Institute of Parasitology, Slough, UK.
Simmonds, H.W. (1933) The biological control of the weed *Clidemia hirta*, D. Don., in Fiji. *Bulletin of Entomological Research* 24, 345–348.
Steyn, W.P., duToit, W.J. and de Beer, M. (1993) *Thripobius semiluteus* (Eulophidae), a potential biocontrol agent of *Heliothrips haemorrhoidalis* and *Selenothrips rubrocinctus* (Thripidae) on avocado. *Proceedings of the 9th Entomology Congress, ESSA, Johannesburg*, 160 pp.
Subba Rao, B.R. (1969) A new species of *Megaphragma* (Hymenoptera: Trichogrammatidae) from India. *Proceedings of the Royal Entomological Society, London (B)* 38, 114–116.
Swezey, O.H. (1936) *Thripoctenus* sp. *Proceedings of the Hawaiian Entomological Society* 9, 361.
Swezey, O.H. (1937) *Thripoctenus brui* Vuillet. *Proceedings of the Hawaiian Entomological Society* 10, 23–24.
Swezey, O.H. (1950) *Thripoctenus brui* Vuillet. *Proceedings of the Hawaiian Entomological Society* 14, 10.
Szelényi, G. (1982) Further data to the Torymid, Pteromalid and Eulophid fauna of the Hortobagý National Park (Hymenoptera: Chalcidoidea). *Acta Zoologica Academiae Scientiarum Hungaricae* 27, 379–388.
Tachikawa, T. (1986) A note on *Ceranisus brui* (Vuillet) in Japan (Hymenoptera: Chalcidoidea, Eulophidae). *Transactions of the Shikoku Entomological Society* 17, 267–269.
Takagi, K. (1988) Natural enemies of thrips. In: Umeya, K., Kudo, I. and Miyazaki, M. (eds) *Pest Thrips in Japan.* Zenkoku Noson Kyoiku Kyokai, Tokyo, Japan, pp. 327–338.
Takanashi, M., Takagi, K. and Hirose, Y. (1996) Egg parasitism by *Megaphragma* sp. of *Scirtothrips dorsalis* and some techniques for determining it. *Shokubutsu boéki* 50, 25–28.

Tamò, M. (1991) The interactions between the cowpea (*Vigna unguiculata* Walp.) and the bean flower thrips (*Megalurothrips sjostedti* Trybom) in Republic of Benin. Dissertation ETH No. 9455, Swiss Federal Institute of Technology, Zurich, 136 pp.

Tamò, M., Baumgärtner, J., Delucchi, V. and Herren, H.R. (1993) Assessment of key factors responsible for the pest status of the bean flower thrips *Megalurothrips sjostedti* (Thysanoptera: Thripidae) in West Africa. *Bulletin of Entomological Research* 83, 251–258.

Tawfik, M.F.S. (1967) Microfauna of the leaf-rolls of *Ficus nitida* Thunb.-Hort. *Bulletin de la Société Royale Entomologique d'Égypte* 51, 483–487.

Teerling, C.R., Pierce Jr, H.D., Borden, J.H. and Gillespie, D.R. (1993a) Identification and bioactivity of alarm pheromone in the western flower thrips, *Frankliniella occidentalis*. *Chemical Ecology* 19, 681–697.

Teerling, C.R., Gillespie, D.R. and Borden, J.H. (1993b) Utilization of western flower thrips alarm pheromone as a prey-finding kairomone by predators. *The Canadian Entomologist* 125, 431–437.

Teulon, D.A.J., Cameron, E.A. and Loomans, A.J.M. (1996) In search of the univoltine Thysanopteran parasitoids, *Ceranisus menes* Walker and *C. pacuvius* Walker (Hym., Eulophidae). *Entomologist's Monthly Magazine* 132, 177–182.

Thuróczy, C. and Jenser, G. (1996) Observations on the occurrence of *Ceranisus pacuvius* (Walker, 1838) (Hymenoptera: Chalcididae) in Hungary. *Folia entomologica hungarica* 57 (Suppl.), 151–153.

Timberlake, P.H. (1924) Descriptions of new Chalcid-flies from Hawaii and Mexico (Hymenoptera). *Proceedings of the Hawaiian Entomological Society* 5, 395–417.

Triapitsyn, S.V. and Headrick, D.H. (1995) A review on the nearctic specis of the thrips-attacking genus *Ceranisus* Walker (Hymenoptera: Eulophidae). *Transactions of the American Entomological Society* 121, 227–248.

Uzel, H. (1895) *Monographie der Ordnung Thysanoptera*. Königgrätz, 472 pp.

Van Driesche, R.G. (1983) The meaning of percentage parasitism in studies of insect parasitoids. *Environmental Entomology* 12, 1611–1622.

Van Heurn, W.C. (1923) Over een waarschijnlijke parasiet (Thripoctenus brui Vuillet) en een vermoedelijke hyperparasiet (Genus? species?) van *Thrips tabaci* Lind. *Tijdschrift voor Entomologie* 66, 119–128.

Van Rijn, P., Mollema, C. and Steenhuis-Broers, G. (1995) Comparative life history studies of *Franklinella occidentalis* and *Thrips tabaci* (Thysanoptera: Thripidae) on cucumber. *Bulletin of Entomological Research* 85, 285–297.

Varadarasan, S. and Ananthakrishnan, T.N. (1981) Population dynamics and prey–predator/parasite relationships of gall-forming thrips. *Proceedings of the Indian National Science Academy B* 47, 321–340.

Viggiani, G. (1992) Notizie preliminari su un *Megaphragma* (Hymenoptera: Trichogrammatidae), ooparassitoide di *Heliothrips haemorrhoidalis* Bouché (Thysanoptera: Thripidae) in Italia. *Bollettino del Laboratorio di Entomologia Agraria 'Filippo Silvestri'* 49, 203–206.

Viggiani, G. and Nieves-Aldrey, J.L. (1993) Prima segnalazione di *Goetheana shakespearei* Girault (Hymenoptera Eulophidae), parassitoide esotico di Thysanoptera, per l'Europa. *Bollettino di Zoologia agaria e di Bachicoltura* (Ser. II) 25, 105–108.

Vuillet, A. (1914) Note sur un Chalcidien parasite de thrips des pois. *Compte rendu Séance de la Société de Biologie de Paris* 75, 552–54.

Waterston, J. (1923) On an internal parasite (Hym. – Chalcidoidea) of a thrips from Trinidad. *Bulletin of Entomological Research* 13, 353–355.

Williams, L.T. (1916) A new species of *Thripoctenus* (Chalcidoidea). *Psyche* 23, 54–61.

Wilson, T.H. and Cooley, T.A. (1972) A chalcidoid planidium and an entomophilic nematode associated with the western flower thrips. *Annals of the Entomological Society of America* 65, 414–418.
Yoshimoto, C.M. (1965) Synopsis of Hawaiian Eulophidae including Aphelininae (Hym.: Chalcidoidea). *Pacific Insects* 7, 665–699.
Yoshimoto, C.M. (1981) *Thripoctenoides kaulbarsi* new record from North America. (Hymenoptera: Eulophidae). *The Canadian Entomologist* 113, 723–726.
Yousuf, M. and Shafee, S.A. (1987) Taxonomy of Indian Trichogrammatidae (Hymenoptera: Chalcidoidea). *Indian Journal of Systematic Entomology* 4, 55–200.

10

Fungal Pathogens of Thrips

Tariq M. Butt[1] and
Michael Brownbridge[2]

[1]*Institute of Arable Crops Research – Rothamsted, Harpenden, Herts AL5 2JQ, UK;* [2]*University of Vermont, Entomology Research Laboratory, PO Box 53400, Burlington, Vermont 05405–3400, USA*

Introduction

Although pathogens have been identified from other insect species there are relatively few data on pathogens of thrips, despite the many pest species and the numerous studies on these insects. Presumably, their small size and cryptic habits have resulted in diseased insects being overlooked. There have been few coordinated searches for pathogens in thrips populations but when these have been undertaken, the most commonly reported pathogens are invariably fungi. This has prompted interest in the possible exploitation for thrips control of entomogenous fungi which, unlike viruses and bacteria, can penetrate host insects directly and do not have to be ingested to cause infection. Already, isolates of *Verticillium lecanii* have been commercialized for control of thrips and other glasshouse pests (e.g. Gillespie, 1986, Ravensberg *et al.*, 1990).

Fungi have several advantages over other biocontrol agents; their activity is unaffected by factors such as day length that can limit the performance of some thrips predators, such as *Orius* spp. (Parrella, 1993), they are active over a range of environmental conditions and can be effective in different climatic zones (Marcandier and Khachatourians, 1987; Moorehouse *et al.*, 1994; Zimmermann, 1994). Development of efficient mass-production procedures, and formulation techniques that promote efficacy and conidial stability in storage and application, make the fungi increasingly attractive management options (Goettel and Roberts, 1992; Bateman *et al.*, 1993; Stathers *et al.*, 1993; Jenkins and Lomer, 1994). This chapter reviews the fungi infecting thrips, their mode of infection and potential development as biocontrol agents.

Fungi Associated with Thrips

Natural incidence

The fungi isolated from or known to infect thrips (Table 10.1) belong either to the class Hyphomycetes or Zygomycetes in the subdivisions Deuteromycotina and Zygomycotina, respectively. The key features of common thrips pathogens are summarized in Table 10.2.

Zygomycetes

Observations of natural epizootics in thrips populations have largely been restricted to Entomophthorales. Bourne and Shaw (1934) and Bourne and Whitcomb (1935) reported that *Thrips tabaci* populations were regulated throughout the onion-growing region of the Connecticut River Valley in Massachusetts, USA, by a fungus, probably *Entomophthora sphaerosperma*. *Neozygites parvispora*, formerly known as *Entomophthora parvispora*, has been found frequently in *Thrips tabaci* and *Thrips fuscipennis* populations on onion crops in central and southern Europe. Attempts to utilize this agent for management under field conditions were not successful, although better control of *Thrips tabaci* was obtained in greenhouses where environmental conditions favoured infection and spread of the disease (Carl, 1975; MacLeod *et al.*, 1976). In the Netherlands, epizootics caused by *Entomophthora thripidum* (Fig. 10.1a and b) have been observed in *Thrips tabaci* infestations on a variety of greenhouse crops, but in field trials the fungus failed to suppress thrips populations below an economically acceptable level (Ramakers, 1976; Samson *et al.*, 1979).

Saito *et al.* (1989) reported a low incidence of *Neozygites parvispora* in field-collected specimens of *Thrips palmi*. This pathogen was also recovered from *Frankliniella occidentalis* infesting glasshouse-grown peppers in Italy (Magano di san Lio *et al.*, 1992). In glasshouse studies, Vacante *et al.* (1994) noted that *N. parvispora* caused up to 60 per cent mortality in motile developmental stages of *Frankliniella occidentalis* and reduced both the insect population density and the proportion of leaves and flowers infested. Development of an epizootic appeared to be less dependent upon high ambient relative humidity than for other isolates of this species.

Hyphomycetes

Verticillium lecanii, *Metarhizium anisopliae* (Fig. 10.1c and d), *Paecilomyces farinosus*, *P. lilacinus* and *Hirsutella* sp. have been isolated from *Taeniothrips inconsequens*, a recurrent pest of sugar maple in the north-eastern USA (Skinner *et al.*, 1991; Brownbridge *et al.*, 1993; Brownbridge, 1995). *Beauveria bassiana* (Fig. 10.1e and f) has been recovered from *Thrips calcaratus*, *Frankliniella occidentalis* and *Haplothrips tritici* (Lyubenov, 1961; Humber, 1992). Hall (1992) and Hall *et al.* (1994) have reported *Hirsutella* sp. nov. causing an epizootic in *Thrips palmi* in Trinidad and the occasional incidence of *Paecilomyces fumosoroseus* in collected specimens. *Hirsutella* sp. was also isolated from *Liothrips mikaniae* (Greenwood and Mills, 1989). Although not generally regarded as a virulent insect pathogen, an *Aspergillus* sp. has also been reported infecting thrips (Dyadechko, 1977).

Table 10.1. Fungi isolated from, or proven to be pathogenic to, thrips.

Pathogen	Host	Reference
Class Zygomycetes		
Neozygites parvispora	*Thrips fuscipennis*	Carl (1975)
	Thrips tabaci	
	Thrips tabaci	MacLeod *et al.* (1976)
	Thrips tabaci	Saito *et al.* (1989)
	Thrips tabaci	Keller and Wuest (1983)
	Thrips spp.	Balazy (1993)
	Limothrips denticornis	Nielsen *et al.* (1996)
	Frankliniella occidentalis	Vacante *et al.* (1994)
		Magano di San Lio *et al.* (1992)
Neozygites cucumeriformis	*Drepanothrips reuteri*	Balazy (1993)
	Thrips tabaci	Bourne and Shaw (1934)
Zoophthora radicans	*Thrips tabaci*	Bourne and Shaw (1934)
	Haplothrips tritici	MacLeod *et al.* (1976)
	Thrips sp.	Thaxter (1888)
	Haplothrips aculeatus	Dyadechko (1977)
	Haplothrips subtilissimus	
	Haplothrips tritici	
	Phlaeothrips coriaceus	
Entomophthora thripidum	*Thrips tabaci*	Samson *et al.* (1979)
	Parthenothrips dracaenae	J. Eilenberg (unpublished)
*Entomophaga grylii**	*Acantothrips nodicornis*	Dyadechko (1977)
	Hoplothrips ulmi	
	Phlaeothrips coriaceus	
	Liothrips hungaricus	
	Taeniothrips atratus	
	Zygothrips minutus	
Coniodiobolus coronatus	*Frankliniella occidentalis*	Humber (1992)
Class Hyphomycetes		
Verticillium lecanii	*Thrips tabaci*	Ekbom (1979)
	Thrips tabaci	Binns *et al.* (1982)
	Thrips tabaci	Gillespie (1986)
	Taeniothrips inconsequens	Skinner *et al.* (1991)
	Frankliniella occidentalis	Ravensberg *et al.* (1990)
	Frankliniella occidentalis	Schreiter *et al.* (1994)
	Frankliniella occidentalis	Brownbridge *et al.* (1994)
	Frankliniella occidentalis	Vestergaard *et al.* (1995)
	Frankliniella occidentalis	Humber (1992)
	Frankliniella tenuicornis	Moritz and Trost (unpublished)
	Thrips tabaci	
	Selenothrips rubrocinctus	Reyne (1921)
	Taeniothrips inconsequens	Skinner *et al.* (1991)
	Taeniothrips inconsequens	Brownbridge *et al.* (1993)
	Limothrips cerealium	Nielsen *et al.* (1996)

continued overleaf

Table 10.1. Continued.

Pathogen	Host	Reference
	Limothrips denticornis	
	Stenothrips graminum	
Beauveria bassiana	*Haplothrips tritici*	Lyubenov (1961)
	Chirothrips manicatus	Dyadechko (1977)
	Limothrips schmutzi	
	Haplothrips tritici	
	Heliothrips haemorrhoidalis	
	Frankliniella occidentalis	Brownbridge *et al.* (1994)
	Thrips tabaci	Fransen (1990)
	Thrips calcaratus	Humber (1992)
	Limothrips cerealium	Nielsen *et al.* (1996)
	Limothrips denticornis	
Aspergillus sp.	*Thrips fuscipennis*	Carl (1975)
	Anaphothrips obscurus	Dyadechko (1977)
	Frankliniella intonsa	
	Limothrips denticornis	
	Odontothrips loti	
	Tenothrips discolor	
	Thrips fuscipennis	
	Thrips physapus	
	Thrips tabaci	
	Thrips validus	
Hirsutella sp.	*Taeniothrips inconsequens*	Humber (1992)
	Thrips palmi	Hall (1992)
	Liothrips mikaniae	Greenwood and Mills (1989)
Cladosporium cladosporioides	*Thrips palmi*	Humber (1992)
Paecilomyces fumosoroseus	*Thrips tabaci*	Fransen (1990)
Paecilomyces lilacinus	*Taeniothrips inconsequens*	Brownbridge *et al.* (1993)
Paecilomyces farinosus	*Taeniothrips inconsequens*	Humber (1992)
	Taeniothrips inconsequens	Brownbridge *et al.* (1993)
Metarhizium anisopliae	*Thrips tabaci*	Fransen (1990)
	Frankliniella occidentalis	Vestergaard *et al.* (1995)
	Frankliniella occidentalis	Brownbridge *et al.* (1994)
Sporothrix sp.	*Thrips tabaci*	Humber (1992)

* *Entomophthora grylli* is a pathogen of orthopteran insects only. It is possible that Dyadechko has misidentified this fungus. The possible cause of infection is *Entomophthora thripidum*.

Table 10.2. Description of some common fungal pathogens of thrips.

Pathogen	Field characters	Conidiophores	Conidia
Zygomycetes			
Entomophthora thripidum	Dead insects attached to plants. Dorsal white collars on abdomen	Simple	Primary conidia 'muscae-type' 10–15 × 8–12 μm, 2 to 4 nuclei, forcibly discharged with halo of protoplasm. Secondary conidia resemble primary conidia (9–12 × 7–9 μm), these are also forcibly discharged
Neozygites parvispora	Dead insects attached to plants, milky when dying, then pale yellow to light brown when fungus produces conidia, black when resting spores present	Clavate, formed directly from rectangular-shaped hyphal bodies. Between 4 and 8 nuclei per cell	Primary ovoid conidia (14.5–16.0 × 12.2–13.9 μm) with broad basal papilla (5 μm wide and 3 μm long). The conidia commonly germinate to form elongate capilliconidia with prominent haptors (17.6–19.7 × 9.1–9.7 μm) on top of fine, capillary conidiophores (26 and 63 μm). Secondary conidia resemble primary conidia, but smaller are also produced. No cystidia or rhizoids produced. Spherical, ornamented, multinucleate resting spores (16.7–18.8 μm diameter) develop from two conjugating hyphal bodies. Resting spores usually dark brown to black, but occasionally hyaline

continued overleaf

Table 10.2. Continued.

Pathogen	Field characters	Conidiophores	Conidia
Hyphomycates			
Metarhizium anisopliae	Pale to dark green conidia often in compact bundles	Abundant, arising from vegetative hyphae, branching irregularly, usually with 2 or 3 branches at each node, 4–14 × 1.5–3.0 μm. Phialides cylindrical or clavate, abruptly tapering towards the apex	Borne in chains at the apices of phialides, narrow cylindrical, tapering and truncate at both ends usually 1.5–3.0 μm wide and 5–8 μm long.
Beauveria bassiana	Fairly compact, white to cream coloured mycelium bearing conidia	Abundant, arising from the vegetative hyphae, 2 μm wide bearing cluster of conidiogenous cells, 3–6 × 3–5 μm, which may give rise to further conidiogenous cells, globose to flask-shaped with a well-developed rachis up to 20 μm long by 1–2 μm wide, geniculate with denticles up to 1 μm long	Blastic, hyaline, smooth, globose to broadly ellipsoidal, sometimes with an apiculate base, 2–3 × 2–2.5 μm
Verticillium lecanii	White to grey mycelium radiating from cadaver	Similar to vegetative hyphae 1–2 μm wide. Phialides borne singly, in pairs or whorls of 3 to 4. These delicate structures may vary in size	Produced singly and aggregated in heads at the tips of phialides, ellipsoidal to cylindrical with rounded ends, varying in size with the strain, usually 2–3 × 1–2 μm

Fig. 10.1. see p.408 for legend

C

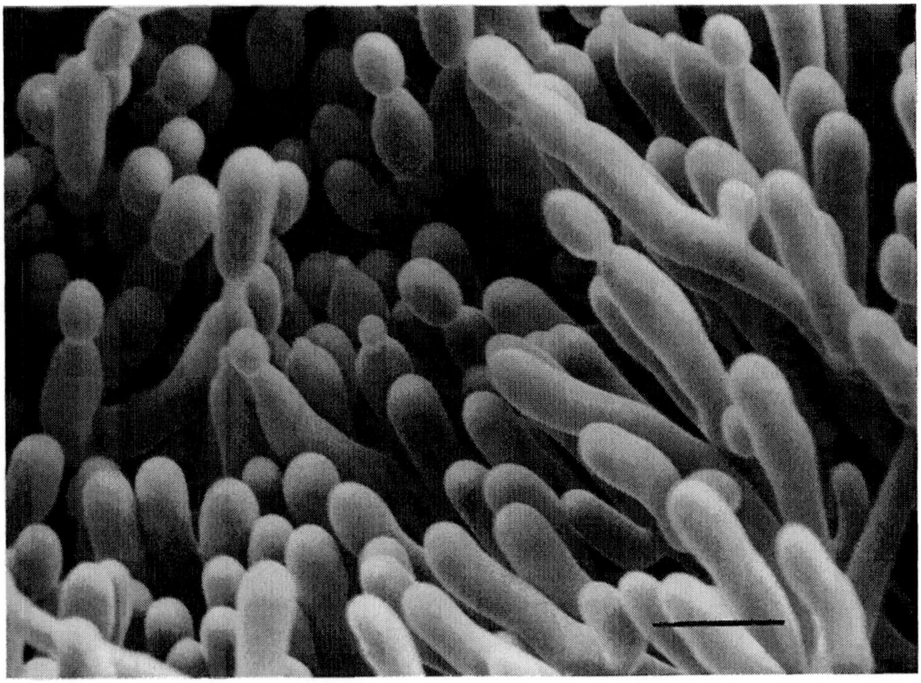

D

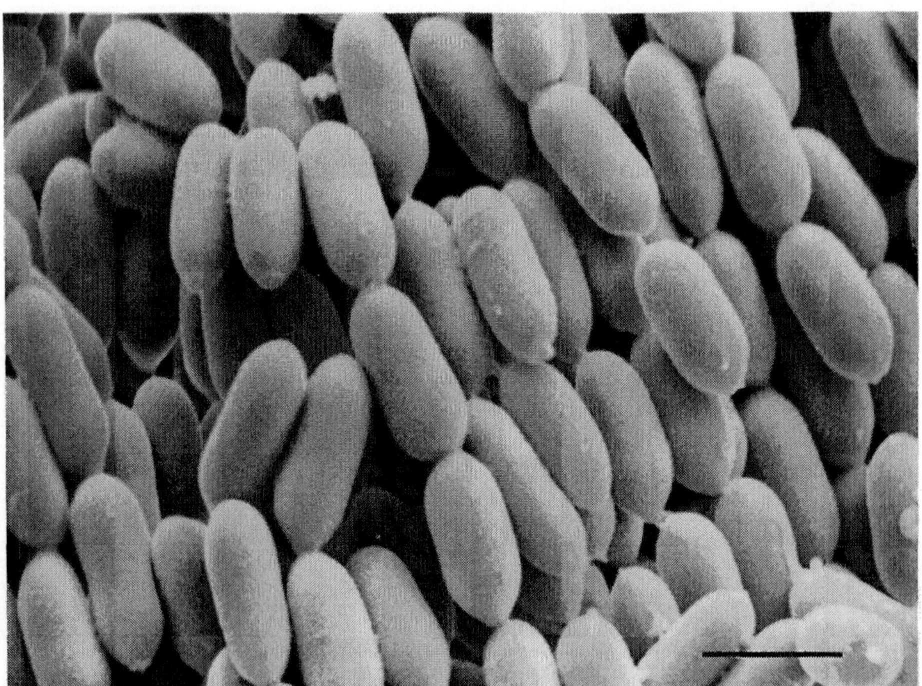

E

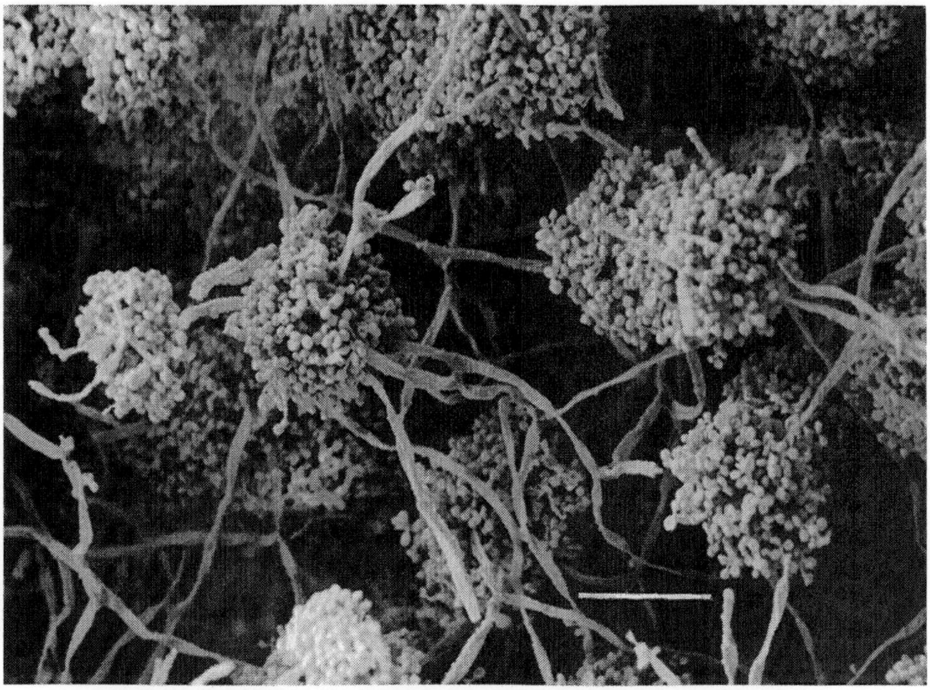

F

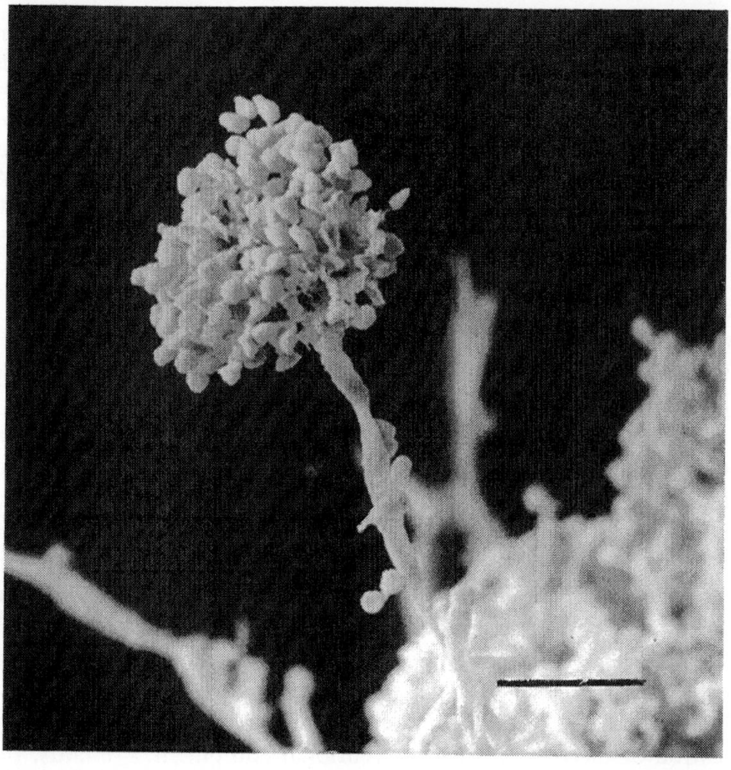

Fig. 10.1. see p.408 for legend

G

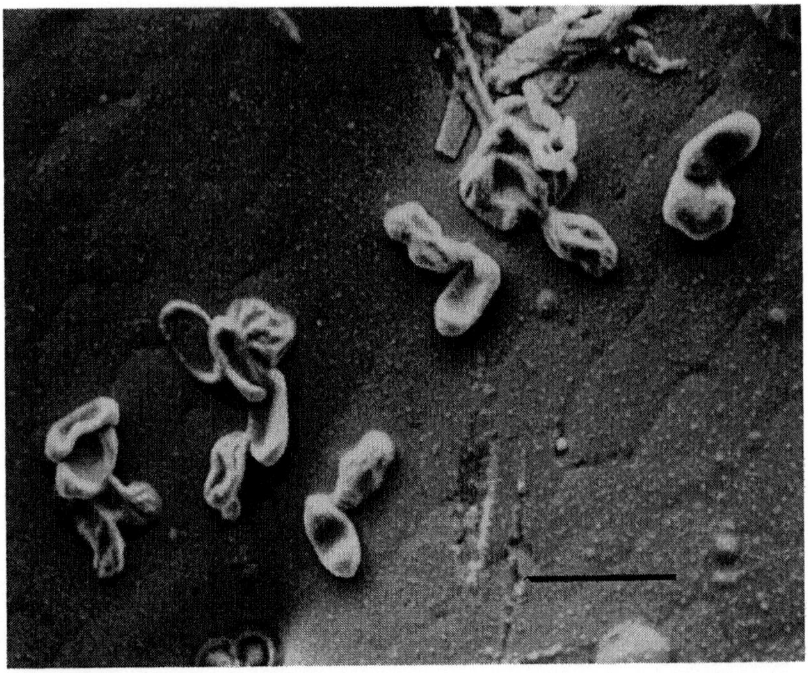

Fig. 10.1. Low temperature scanning electron micrographs of three common pathogens of thrips. (A) *Entomophthora thripidum* emerging from infected thrips. Bar = 113 μm. Photo, R. Samson. (B) Conidia and conidiophores of *Entomophthora thripidum.* Bar = 7.7 μm. Photo, R. Samson. (C) Conidiophores and immature conidia of *Metarhizium anisopliae.* Bar = 5.8 μm. (D) Mature, turgid conidia of *Metarhizium anisopliae.* Bar = 5.9 μm. (E) Surface view of *Beauveria bassiana* showing numerous conidiophores bearing a ball of spherical conidia. Bar = 23 μm. (F) Side view of a conidiophore of *Beauveria bassiana* bearing numerous conidia. Bar = 11.6 μm. (G) *Metarhizium anisopliae* germlings with appressoria. Bar = 11.7 μm.

Fungi assessed for thrips management

Relatively few laboratory and field trials have been undertaken to evaluate fungi specifically for thrips control. While Entomophthorales might be successfully used in a 'classical' biological control approach, where the fungus would gradually spread and be self-sustaining within a thrips population, attempts to utilize these pathogens have met with only limited success. This approach may be suitable for perennial crops, but for seasonal crops a rapid reduction in pest numbers is required to minimize crop damage and losses. Fungi could be utilized as mycoinsecticides, with the aim of maximizing infection and mortality rates from each application of fungal inoculum; several applications may be necessary to provide effective control over an entire growing season (Prior, 1990). For such a management strategy, large amounts of inoculum would have to be produced cheaply, and formulated to facilitate easy application of the fungus with conventional equipment. Presently, this limits the fungi which can be commercialized successfully to the Hyphomycetes, on which most trials have been done.

Often, observations on pathogen efficacy have been made when fungi have been applied to control other pests, such as whiteflies or aphids in glasshouse crops (e.g. van der Schaaf *et al.*, 1991). In laboratory studies, *Thrips tabaci* was susceptible to *V. lecanii*, *B. bassiana*, *M. anisopliae* and *P. fumosoroseus* (Gillespie, 1986; Fransen, 1990). Vestergaard *et al.* (1995) and Brownbridge (1995) showed that *B. bassiana*, *M. anisopliae* and *V. lecanii* were more active against *Frankliniella occidentalis* than *P. fumosoroseus* or *P. farinosus*. *Taeniothrips inconsequens* is particularly sensitive to *V. lecanii*, *B. bassiana* and *M. anisopliae* (Brownbridge, 1995). The work of Hall *et al.* (1994) and Saito (1991) suggests that *Hirsutella* sp., *P. fumosoroseus* and *B. bassiana* may be useful in the management of *Thrips palmi*. Brownbridge *et al.* (1994) demonstrated that several fungal species, including *B. bassiana*, *M. anisopliae* and *V. lecanii*, could control *Frankliniella occidentalis* on chrysanthemum flowers.

In glasshouse trials, *V. lecanii* has been used successfully to control *Thrips tabaci* and *Frankliniella occidentalis* on cucumber and chrysanthemums (Binns *et al.*, 1982; Gillespie, 1986; Helyer *et al.*, 1992). Van der Schaaf *et al.* (1991) indicated that *V. lecanii*, applied primarily to control the glasshouse whitefly, *Trialeurodes vaporariorum*, induced infection rates as high as 60 per cent in *Frankliniella occidentalis* on cucumber. While relative humidity is known to be an important factor affecting *V. lecanii* infection processes (Milner and Lutton, 1986), thrips control was achieved in a cucumber crop even though the ambient relative humidity fell as low as 75 per cent. Higher doses were used to compensate for the negative effects of low humidity on pathogen efficacy, and standard production procedures were maintained. On chrysanthemums, however, use of elevated humidity was advocated to promote efficient infection and control of *Frankliniella occidentalis* (Helyer *et al.*, 1992). Humidity at leaf surfaces appears to vary significantly between different plant types, so some modification of humidity levels may be necessary on certain crops to obtain the desired levels of insect control using this fungus. Where low humidities must be maintained for other reasons, fungal species tolerant of drier conditions, such as strains of *Beauveria* or *Metarhizium*, might be more suitable control agents (Ramoska, 1984; Marcandier and Khachatourians, 1987; Lomer *et al.*, 1994).

Many important thrips pests have a soil-dwelling instar(s) in their life cycle, such as *Frankliniella occidentalis*, *Taeniothrips inconsequens*, *Thrips tabaci* and *Thrips palmi*, and these developmental stages are also susceptible to fungi. Sermann *et al.* (1994) showed that soil applications of *V. lecanii* successfully infected soil-dwelling stages of *Frankliniella occidentalis*, but inoculum levels decreased rapidly in non-sterile soils (Hirte *et al.*, 1994). Brownbridge *et al.* (1994) demonstrated that *M. anisopliae* and *B. bassiana* can suppress *Frankliniella occidentalis* effectively in potting media, and there may be more appropriate strains to use in this type of environment where infective levels of inoculum can be maintained for long periods (Chase *et al.*, 1986; Vänninen and Tyni-Juslin, 1991; Moorehouse *et al.*, 1993). Vestergaard (1995) also obtained good control of *Frankliniella occidentalis* in potting compost and established that *M. anisopliae* survives longer in sterile potting compost than in non-sterile media. Presumably, under field conditions, soil antagonists limit the survival of fungal inocula. Both *B. bassiana* and *V. lecanii* are also under evaluation for control of *Taeniothrips inconsequens* in sugar maple stands (Brownbridge, 1995). In field trials, efficacy against this thrips in forest soil has been observed and enhanced soil populations of *B. bassiana*, applied as a nutrient-based granular formulation, have been maintained for over 18 months (M. Brownbridge, unpublished data).

Host Specificity of Fungal Entomopathogens

The host range of entomophthoralean fungi is usually restricted to a specific insect species, family or order. For example, *Conidiobolus obscurus*, *Neozygites fresenii*, *Entomophthora planchoniana* and *Erynia neoaphidis* are common aphid pathogens which will not infect any other insect species (Butt *et al.*, 1981, 1990). Similarly, the thrips pathogens *Entomophthora thripidum* and *Neozygites parvispora* are restricted to thrips species (Carl, 1975; Samson *et al.*, 1979; Vacante *et al.*, 1994). One exception is *Zoophthora radicans* (= *Entomophthora sphaerosperma*) which is a notable pathogen of lepidopteran (Li *et al.*, 1993) and hemipteran (Butt *et al.*, 1981, 1988) insects but has also been reported infecting thrips (Bourne and Shaw, 1934; Fresenius 1885, cited in Hutchison 1963; Thaxter, 1888). However, unlike hyphomycete fungi, *Zoophthora radicans* isolates are usually more infectious to insect species from which they were originally isolated.

The hyphomycete fungi tend to be less specialized and include common, soil-borne insect pathogens such as *V. lecanii*, *M. anisopliae* and *B. bassiana* which have a wide host range. For example, *M. anisopliae* V90, originally isolated from *Sitona discoideus* (Coleoptera: Curculionidae), will infect aphids, chrysomelid beetles and thrips (Butt *et al.*, 1992; Vestergaard *et al.*, 1995). A strain of *V. lecanii*, originally isolated from glasshouse whitefly and produced commercially by Koppert BV of the Netherlands for control of this pest, is also active against *Frankliniella occidentalis* (van der Schaaf *et al.*, 1991). *Thrips tabaci* is susceptible to a strain of *P. fumosoroseus* under development for control of silverleaf whitefly, *Bemisia argentifolii* (= *B. tabaci*, sweet potato whitefly, strain B) (Fransen, 1990). Thrips-active isolates of *B. bassiana*, *P. farinosus* and *M. anisopliae* have also been shown to be pathogenic to silverleaf whitefly and peach-potato aphid, *Myzus persicae* (Brownbridge, 1995). From a commercial standpoint, fungi with a broader spectrum of activity are more attractive if they facilitate control of more than one pest with a single application. Some fungi, such as *Aspergillus* spp., are not considered to be true pathogens, but are regarded as opportunistic parasites because they normally infect weak or injured insects, and they can infect quite disparate insect species. For example, *A. flavus* has been reported infecting Coleoptera – *Dendroctonus frontalis*, Diptera – *Musca domestica*, *Drosophila melanogaster*, Hemiptera – *Cimex lectularius*, and Lepidoptera – *Trichoplusia ni* (Bell, 1974), in addition to Thysanoptera (Dyadechko, 1977).

Insect infection processes and disease development

As the infection process is genetically regulated, the different host spectra exhibited by the fungi are indicative of the genetic variation that exists within and between species (Bidochka, 1994; Goettel, 1992; Hajek and St Leger, 1994; Moorehouse *et al.*, 1994). Successful exploitation is dependent upon an understanding of their mode of infection, identification of the genes governing this process, and an evaluation of the impact of environmental factors on host–pathogen interactions. This section describes the infection processes of entomogenous fungi with particular reference to *V. lecanii* and *M. anisopliae*, and the implications for thrips control.

Adhesion

The infection processes of thrips mycopathogens follow the same pattern as for other invertebrate mycopathogens: spore attachment to the host surface, germination, differentiation of an appressorium and/or penetration peg and finally penetration of the host cuticle. Conidia of both *M. anisopliae* and *V. lecanii* readily adhere to the surface of the host cuticle *via* non-specific hydrophobic interactions and secretion of a mucilaginous extracellular matrix (Schreiter *et al.*, 1994; Butt *et al.*, 1995).

Since insect mortality is dose related, the number of spores adhering to the cuticle is critical. This influences strain selection and the development of appropriate formulations and application methodology. Poor adhesion is often a feature of hypovirulent isolates (Al-Aidroos and Roberts, 1978). However, inhibitors in the host cuticle may also prevent fungal establishment (Butt *et al.*, 1995). The fact that germination of *M. anisopliae* conidia is faster on dead *Frankliniella occidentalis* than on live insects could be explained by the presence of labile, fungistatic compounds in the epicuticular waxes of live thrips (T.M. Butt, unpublished data). Fungal inoculum is also reduced through preening and ecdysis (T.M. Butt, unpublished data; Vestergaard *et al.*, 1995). The latter may explain why thrips larvae are less susceptible to infection than adults or pupae (Vestergaard *et al.*, 1995).

Germination and penetration

The growth of *Verticillium lecanii* on the surface of the host cuticle differs from that of other entomogenous fungi in that there is extensive growth prior to and concomitant with penetration. In contrast, *Metarhizium anisopliae* and most entomophthoralean fungi penetrate shortly after germination or after limited hyphal growth (Butt, 1987; Butt *et al.*, 1990, 1995; Wraight *et al.*, 1990). In these fungi, differentiation of an appressorium (Fig. 10.1g) is a prerequisite for infection (Butt, 1990), but *V. lecanii* does not appear to produce such specialized structures. The cuticle is usually breached by a narrow hypha or peg produced by the appressorium, but in the case of *V. lecanii* the peg is produced at hyphal tips or laterally from growing hyphae (Schreiter *et al.*, 1994). Differentiation of appressoria and, to some extent, penetration pegs is governed by a combination of physical and chemical factors (St Leger *et al.*, 1989a; Magalhaes *et al.*, 1990; Butt *et al.*, 1995).

All entomopathogenic fungi employ enzymes and mechanical force to penetrate and/or colonize the host cuticle (St Leger *et al*., 1989b; Butt, 1990). Both *V. lecanii* and *M. anisopliae* secrete a wide variety of enzymes which degrade the host cuticle, including proteases, chitinases and lipases. Both fungi secrete an immunologically related subtilisin-like protease, Pr1, which plays an important role in nutrition and pathogenesis (Segers *et al*., 1995, 1996 and references therein).

Enzyme activity of most entomopathogens is usually confined to a region around the penetration hypha or peg, but for *V. lecanii* it may also extend along the length of actively growing hyphae, resulting in numerous lesions in the host cuticle (Schreiter *et al*., 1994). Such damage would not only stress and kill the insect but it could release nutrients which in turn encourage mycelial growth. An excess of nutrients not only represses the production of Pr1 but also promotes surface growth, at least in *M. anisopliae* (St Leger *et al*., 1989a,b). Perhaps the high enzyme activity of *V. lecanii* releases sufficient nutrients from the cuticle to encourage hyphal growth rather than invasion and colonization of the haemocoel. Extensive colonization of the exo- and endocuticles suggests that *V. lecanii* is extremely efficient in extracting nutrients from the cuticle. This is possibly also true for *M. anisopliae* which has been observed growing between the lamellae of the host cuticle (Zacharuk, 1971).

Postpenetration development

Thrips are probably killed by *V. lecanii* as a result of the multiple lesions produced by enzymatic degradation of the cuticle. This is indirectly supported by the fact that little or no fungal material is present in the haemocoel at the time of death and the host tissues are intact. Although *V. lecanii* is known to secrete toxins (Claydon and Grove, 1982), it is not known if toxicosis causes death. In contrast, death caused by *M. anisopliae* isolates is more rapid, presumably due to the action of toxins (Vestergaard *et al*., 1995). Interestingly, there are more reports of suspected toxicosis in insects infected with *M. anisopliae* (Roberts, 1981; Samuels *et al*., 1989; Butt *et al*., 1992). Perhaps the toxins secreted by *M. anisopliae* are produced in higher quantities or are more potent than those of *V. lecanii*.

Verticillium lecanii colonizes the haemocoel as small hyphae or septate filaments containing numerous lipid globules and dense cytoplasm. Formation of hyphal bodies or 'blastospores' *in vivo* is not uncommon, and *Metarhizium anisopliae* (Zacharuk, 1971), *Beauveria bassiana* (Vey and Fargues, 1977), *Nomuraea rileyi* (Boucias and Pendland, 1982), *Tolypocladium cylindrosporum* (Goettel, 1988), *Conidiobolus obscurus* (Butt, 1984) and *Entomophthora coronata* (Prasertphon and Tanada, 1968) all colonize the haemocoel in this form. Hyphal bodies and/or 'blastospores', unlike filamentous hyphae, offer a large surface area to volume ratio for the release of histolytic enzymes and/or for the uptake of nutrients. In contrast, members of the genera *Entomophthora* and *Neozygites* colonize the host haemocoel as protoplasts (e.g. Butt, 1984, Butt *et al*., 1981). The protoplast state offers a large surface area to volume ratio for uptake of nutrients and release of histolytic enzymes and less energy is spent in maintaining a cell wall (Butt *et al*., 1981). Protoplasts, unlike walled cells, do not elicit the host's defences (Butt *et al*., 1996).

Ecology and epizootics

Little is known about the ecology and natural dissemination of entomopathogenic fungi other than the fact that they are widespread (Humber, 1992) and the same species may cause epizootics in different parts of the world. For example, *Neozygites parvispora* has been reported causing epizootics among thrips populations in Europe (Keller, 1991; Vacante *et al.*, 1994) and Australia (Glare and Milner, 1987).

The soil is probably the main reservoir of inoculum, which for most entomopathogens is the conidium. Mycelia of saprophytic forms will periodically produce conidia. Even the thick-walled resting spores of *N. parvispora* will, under favourable environmental conditions, produce conidia.

Conidia are probably transported into the plant canopy by air currents, thrips or other invertebrates. For successful establishment of the pathogen it must be compatible with the host and, because mortality is usually dose related, it must be present in sufficient quantity (e.g. Butt *et al.*, 1992, 1994b; Vestergaard *et al.*, 1995). Provided climatic conditions are favourable and there are enough thrips, then an epizootic may ensue. Climatic conditions favouring epizootics include a relative humidity close to 100 per cent and temperatures between 20 and 28°C. Optimum conditions will vary with the isolate. Some fungi are extremely sensitive to prolonged exposure to high temperatures or to dramatic fluctuations in temperature (Butt *et al.*, 1994a,b, unpublished observations). Viable spore concentrations may diminish on leaves either because rain washes them off the leaf surface or spores are killed by antagonistic organisms and UV radiation (T.M. Butt, unpublished data; Firstencel *et al.*, 1990). Normal leaf expansion during growth also reduces the number of spores per unit area (E. Inyang *et al.*, unpublished data).

If conidia of entomophthoralean fungi land on an unfavourable substrate, including non-hosts, they will produce secondary, tertiary or even quaternary conidia which are forcibly discharged (e.g. Samson *et al.*, 1979; Butt *et al.*, 1990). This is the main means of dispersal of *E. thripidum*. In contrast, the infective unit of *N. parvispora* is the capilliconidium, which is not forcibly discharged. Instead, the fungus depends on insects contacting the sticky capilliconidia which are borne on fine, capillary condiophores.

Development of Fungi for Thrips Management

Reports documenting the evaluation and development of fungi primarily for thrips control are scarce, but a review of the literature on other fungal research suggests opportunities for thrips management.

Evaluation and selection of thrips-active fungi

Selection of strains

It is impossible to predict whether a search among host-related or non-related isolates will yield the most virulent or the best candidate strains. Studies by Vestergaard *et al.* (1995) show isolates of *M. anisopliae* to be far more aggressive against *Frankliniella occidentalis* than those of *V. lecanii*. There is no obvious explanation for this, although Vestergaard *et al.* (1995) speculate that it is due to subtle differences in the mode of infection. For instance, *V. lecanii* often colonizes the host surface before colonizing the haemocoel (Schreiter *et al.*, 1994), whereas *M. anisopliae* appears to penetrate soon after germination (Zacharuk, 1970; Butt *et al.*, 1992, 1995; Schreiter *et al.*, 1994). However, pathogenicity alone should not be the sole criterion governing the selection process. Additional factors which need to be considered are the biology of the target pest, the target habitat, efficacy under field conditions, factors governing ease of mass production and prospects for commercialization.

Laboratory assay

A laboratory assay to demonstrate pathogenicity and determine the relative virulence of the test organisms is generally the first stage in the evaluation and selection process. Ideally, a constant supply of laboratory-reared insects is required, to ensure testing against an even-aged, healthy population. While mass rearing methods have been described for some species (Lewis, 1973; Wang and Chu, 1986; Teulon, 1992; Doane *et al.*, 1995), mass rearing techniques may be problematic for others, particularly for univoltine species such as pear thrips (see Chapter 12). If this is the case, alternative thrips hosts may be considered for an initial series of bioassay tests, followed by confirmative testing against field-collected specimens of the target species. This has proven satisfactory in the selection of fungi for use against pear thrips, where initial assays were done using *Frankliniella occidentalis* (Brownbridge, 1995).

The size and mobility of many thrips makes laboratory assays difficult and labour intensive. Various assay methods have been described (Morse *et al.*, 1986; Hall *et al.*, 1993) and may be adapted according to the thrips host under investigation (see Chapter 11).

Factors affecting efficacy and implications for field use

A variety of factors influence fungal efficacy and must be considered when interpreting laboratory data and extrapolating results to predict activity under field conditions. These factors not only directly affect the activity of the pathogen, (e.g. relative humidity, temperature, dose) but also affect the target pest, which can indirectly affect fungal efficacy. For example, thrips have different activity levels, preferred habitats and behavioral patterns according to the type of host plant they infest, the prevailing ambient conditions and the developmental stage of the host plant and the insect (Higgins, 1992).

Effect of dose, temperature and host age

Vestergaard *et al.* (1995) and Brownbridge *et al.* (1994) showed that mortality of thrips exposed to *B. bassiana*, *M. anisopliae* or *V. lecanii* was dose dependent. High mortality was obtained using fungal concentrations of 10^7 and 10^8 conidia ml^{-1}. In reality, few insects would be expected to receive this dose in glasshouses because of their cryptic habit (i.e. hiding in flowers), but some control was obtained with *M. anisopliae* even at 10^5 conidia ml^{-1} (Vestergaard *et al.*, 1995). Cadavers containing fungus could then act as a source of inoculum for infecting healthy thrips.

Thrips were susceptible to *M. anisopliae* at all temperatures tested, the optimum being 23°C (Vestergaard *et al.*, 1995). A fall in temperature of 3 to 5°C increases the time to death by 1 day, which could be critical in a heavily infested glasshouse, particularly if high humidities could not be sustained long enough for infection to occur (Vestergaard *et al.*, 1995). As pointed out earlier, dramatic fluctuations in temperature (e.g. 10/18 or 10/23°C for 8:16 h periods) will stunt fungal growth.

Larval and pupal stages of *Frankliniella occidentalis* are more resistant than adults to infection by *M. anisopliae*, presumably because fungal inoculum is shed with the exuvium following ecdysis (Vestergaard *et al.*, 1995). Ecdysis is important in the resistance of insects to fungal infection, particularly when the time interval between successive moults is short (Vey and Fargues, 1977). It is not clear why the pupal stage of *Frankliniella occidentalis* was more susceptible to *M. anisopliae* than larvae (41 per cent compared with 27 per cent mortality) but it is unlikely to be due to differences in the chemical nature of the cuticle because germination and appressorium production were similar on adults, larvae and prepupae (Vestergaard *et al.*, 1995). The fact that prepupae shed their cuticle once while larvae shed theirs twice may offer one explanation, but cuticle thickness or metamorphosis may also influence susceptibility. Sporelings of *M. anisopliae* are known to respond to physical and chemical cues (Butt *et al.*, 1992). For example, some will only form appressoria on rigid, hydrophobic substrates (St Leger *et al.*, 1989a). Fransen *et al.* (1987) and Feng *et al.* (1985) also showed that host age was an important factor affecting susceptibility in glasshouse whitefly and European corn borer, respectively. Similar explanations for this variability were suggested, namely intermoult period, differences in cuticle thickness and composition.

Studies by Vestergaard *et al.* (1995) suggest that *M. anisopliae* is a more aggressive thrips pathogen than *V. lecanii*, irrespective of temperature and host age. The most virulent isolates of *M. anisopliae* (275) had LC_{50} values of 3×10^5 conidia ml^{-1} at 5 days and LT_{50} values of 3.0 and 4.5 days at 10^7 and 10^6 conidia ml^{-1}, respectively. Temperature influenced fungal virulence for adult thrips; the LT_{50} at 18 and 20°C was approximately 4 days and at 23 or 26°C it was approximately 3 days. Larvae were less susceptible to infection than adult thrips (27 per cent larval compared with 100 per cent adult mortality), presumably due to the inoculum being shed with the exuvium during ecdysis. Conidia of *M. anisopliae* isolate 275 germinated rapidly on the surfaces of larvae, pupae and adults with most germlings producing appressoria within 24 h postinoculation. Fungal elements were present in significant amounts in the host body 3 days postinoculation.

From Laboratory to Field: Mass Production, Formulation and Application

Once proven effective under laboratory conditions, good performance in the field or glasshouse must be demonstrated, requiring that production, formulation and application protocols be investigated. An understanding of the parameters influencing pathogenicity, and knowledge of the biology of the pest and the pathogen are essential to optimize the timing and targeting of a fungal application. These factors, in addition to pathogenicity, will determine which strains are ultimately utilized, and provide the greatest challenges to the successful development of mycopathogens for use in an effective thrips suppression programme. However, these are also some of the challenges faced in the chemical control of thrips (see Chapter 15), and much of the information relevant to monitoring thrips populations and use of these data to optimize the timing of an insecticide application may be applied to fungal controls.

While the practical performance of some mycopathogens against thrips has been confirmed (Gillespie, 1986; van der Schaaf *et al.*, 1991; Helyer *et al.*, 1992), this has largely been on glasshouse crops, where environmental conditions can be manipulated and application methods ensure thorough coverage of foliage. Activity against thrips in other agricultural situations has been inadequately explored. Certainly, the use of fungal entomopathogens in field crops presents greater problems in terms of combating the effects of ultraviolet radiation and rapid desiccation on viability. In the development of fungal controls for pests in dry climatic regions, such as whitefly on cotton and melon, and grasshopper and locust pests, these difficulties are being overcome through the use of innovative formulations and application strategies, and fungal strains better suited to these harsher conditions.

Production

For field trials and, ultimately, for commercialization of a fungal product, large amounts of viable, non-attenuated and stable inoculum must be produced cheaply. One of the biggest obstacles to the wider use of fungi has been the development of suitable mass production systems. Fungi which grow and conidiate readily on artificial media are, generally, the best candidates for mass production. Some sacrifice in virulence can be tolerated providing the selected strain shows efficient production characteristics. Growth substrates can significantly influence yield, as well as conidial germination and germination rate, pathogenicity and survival of inoculum (Li and Holdom, 1990, 1995; Fargues and Robert, 1991; Hall *et al.*, 1994; Jenkins and Lomer, 1994). Selection of artificial growth substrates is therefore essential to ensure production of an efficacious, reliable preparation. Infective propagules of many fungi can be produced on solid or liquid substrates. Solid culture processes are generally used for the production of stable conidia rather than blastospores, and for pathogens such as *Aschersonia aleyrodis*, infective units are formed only on solid media (Fransen, 1990). Whatever the system and scale of the production operation, it is important that adequate quality controls are incorporated to guarantee that each batch is viable and virulent. This is particularly critical during commercialization, as a product of consistent efficacy is required to promote and maintain user confidence.

For small-scale trials, inoculum can be readily produced on substrates such as rice or wheat bran (Goettel, 1984). Fungi are commercially produced using proprietary solid culture processes, for example *B. bassiana* (Mycotech Corp.;

M.T. Butte) and *V. lecanii* (Koppert BV, Netherlands), and in liquid fermentation, for example *V. lecanii* (Chr. Hansen's BioSystems, Denmark). *Metarhizium flavoviride* has been produced for control of locusts in West Africa on a rice substrate, but a new procedure, in which conidiation occurs on cellulose cloths, appears to be much more efficient (Jenkins and Lomer, 1994). Rice substrates are also used for production of *M. anisopliae* and *P. fumosoroseus* in the West Indies for management of froghoppers, *Bemisia tabaci* and, potentially, *Thrips palmi*, on sugar cane and vegetable crops (Hall *et al.*, 1994). Rice is also the favoured medium for production of *M. anisopliae* and *B. bassiana* in Brazil (Moscardi, 1989); inocula are produced by private industries, state agencies and grower associations, indicating the range of production opportunities for the manufacture of fungal pest management products.

Formulation

Good formulations should maintain conidial viability in storage and during application over a range of conditions and, ideally, promote efficacy and persistence. *Verticillium lecanii* conidia are hydrophilic, and are covered with mucilage which is involved in the attachment of conidia to the insect cuticle (Hall, 1985). The fungus is formulated as a wettable powder, such as Mycotal® (Koppert BV), which has to be soaked before spraying to rehydrate the matrix and other formulation additives. A carbohydrate carrier is contained in the formulation which stimulates the spores to germinate and infect a susceptible host and also enhances the persistence of inoculum on leaves (Gillespie, 1986; Helyer *et al.*, 1992). Persistence of inoculum on a treated surface may be particularly important in thrips control, because they are generally highly mobile and the main route of infection may occur as they pick up conidia when moving over treated leaves or petals rather than through direct contact at the time of spraying.

Conidia produced by *Beauveria*, *Metarhizium* and *Paecilomyces* are hydrophobic but lipophilic, and are readily suspended in oil. The insect is equally hydrophobic so the oil promotes contact between conidium and cuticle, enhancing the infection process (Prior *et al.*, 1988). Several oils (e.g. Shell Sol T®, Ondina®, sunflower) appear to be suitable carriers, offering the potential to apply oil-based formulations in ultra-low-volume applications, or oil–water emulsions for higher volume spray treatments (Stathers *et al.*, 1993; Wright, 1993; Zimmermann, 1994). Oil-based formulations also appear to be effective under dry conditions, provided that direct host contact is obtained (Bateman *et al.*, 1993), but the residual performance of such formulations remains to be determined. Formulation in oil also helps to sustain conidial viability at higher temperatures (above 25°C), and provides some protection against ultraviolet radiation (Moore *et al.*, 1993; Stathers *et al.*, 1993). While chemical sunscreens could be included in a formulation, recent studies suggest that they would provide little additional protection against ultraviolet radiation (Hunt *et al.*, 1994). Some mineral or vegetable oils are phytotoxic, so tests with thrips host plants would have to be conducted in advance to ensure that foliage would not be damaged. This is particularly important for major pests of ornamentals such as *Frankliniella occidentalis*; phytotoxicity could seriously affect the marketability of these aesthetic plants.

Soil application of the fungi could have potential as a cost-effective management technique for many thrips pests, although the impact of other crop management strategies on long-term efficacy must be assessed. This control approach has been contemplated for a number of insects (Keller, 1992; Inglis *et al.*, 1995), and Grivanov (1940) reported that deep ploughing of cereal stubble probably promotes regulation of overwintering thrips by *B. bassiana*.

Formulations have been specifically developed and tested against a variety of soil insects. Drenches containing conidia have been successfully used to control insects such as *Otiorhyncus sulcatus* and *Frankliniella occidentalis* in potting media (Moorehouse *et al.*, 1993; Brownbridge *et al.*, 1994; Sermann *et al.*, 1994). Oil and nutrient-based granular formulations containing *B. bassiana* conidia have exhibited efficacy against pear thrips in maple forest soils, and granular formulations in particular significantly increased inoculum levels in the soil over 18 months (M. Brownbridge *et al.*, unpublished data). Soil inoculations of *B. bassiana*, formulated as a wettable powder and granules, generally had good persistence but relatively poor efficacy against over-wintering Colorado potato beetles, *Leptinotarsa decemlineata* (Storey and Gardner, 1988; Gaugler *et al.*, 1989). Conidia formulated in oil were less liable to leaching from the upper layers of a soil profile than were aqueous suspensions (Storey *et al.*, 1987) and may be a more suitable treatment for pests such as thrips that are largely found in the upper 5 cm of soils. In Tasmania, local grains overgrown with *M. anisopliae* were applied to the soil by direct drilling, at the time of pasture renovation, to control *Adorphorous colouni* (Rath *et al.*, 1992). Using this method, enhanced levels of inoculum were established in an attempt to control this pasture pest without the need to re-inoculate the soil for 5 to 10 years. Mycelial particles and mycelial granules have also been produced and tested for control of soil pests (Andersch *et al.*, 1990; Storey *et al.*, 1990; Kreuger *et al.*, 1992; Pereira and Roberts, 1990, 1991). Infective conidia are produced on these formulations after application. *Metarhizium anisopliae* granules have been incorporated into potting media as a prophylactic treatment to protect glasshouse-grown potted plants against *Otiorhyncus sulcatus* (Zimmermann, 1994). Such treatments may also provide some level of thrips control, but if this were the primary goal, selection and use of thrips-active strains would undoubtedly have a greater impact.

Although the long-term survival and multiplication of fungal inoculum in soil is influenced by physical, chemical and biological factors (e.g. soil moisture, temperature, other soil microbes, fertilizers, pesticides), many insect pathogenic fungi can survive saprophytically in soils, particularly those rich in organic matter (Harney and Widden, 1991; Moorehouse *et al.*, 1993; Li and Holdom, 1994). Establishment and activity of fungi may also be affected by plant root exudates in soils, leaf leachates or other plant derivatives adsorbed onto the surface of insects. Li and Holdom (1995) investigated the effects of various chemicals present in the rhizosphere on conidial germination, growth and sporulation of *M. anisopliae*. They found that various carbohydrates and nitrogen compounds in plant root exudates could cause premature germination in soil, resulting in the rapid loss of inoculum; organic acids might have the opposite effect, contributing to the persistence of conidia in soil but potentially inhibiting germination on an insect host.

Application

Application techniques and treatment regimes need to be thoroughly examined to allow recommendations to be made as to when, where and how fungal treatments can be applied to optimize control efforts. For a foliar treatment, existing spray techniques must be evaluated and guidelines for their effective use documented. The type of sprayer used to apply conidia will affect leaf and insect coverage, which subsequently influences pathogen performance. For example in laboratory studies, two aphid species, *Myzus persicae* and *Aphis gossypii*, were equally susceptible to *V. lecanii*, but when the pathogen was applied in a glasshouse using a conventional high-volume sprayer, *A. gossypii* was controlled less effectively than *M. persicae* (Hall and Papierok, 1982). The effect was attributed to the greater mobility of *M. persicae* over the treated leaf surface leading to increased contact with infective conidia. Satisfactory control of *A. gossypii* was achieved by using a low-volume electrostatic rotary atomizer, which enhanced direct deposition of spores on the aphids (Sopp *et al.*, 1989).

Electrostatic sprayers can provide better penetration and coverage of foliage than conventional hydraulic sprayers. In addition, drift and runoff are reduced and less spray is used, promoting more efficient use of materials and reducing environmental and human hazards. The sprayers produce minute, electronically charged droplets that are particularly attracted to the undersides of plant leaves (Matthews, 1988). Because *Thrips tabaci* are usually more numerous on the underleaf surface of cucumber, use of an electrostatic sprayer could improve spore deposition on the lower leaf surface and enhance the level of control obtained for this pest (Binns *et al.*, 1982).

Despite these benefits, electrostatic sprayers are relatively expensive and so many small-scale growers or farmers are deterred from purchasing them. High volume, hydraulic sprayers are most commonly used for pesticide application. High volume sprays of *V. lecanii* are recommended to control thrips on cucumber and chrysanthemum (Helyer *et al.*, 1992; van der Schaaf *et al.*, 1991). Although larger spray volumes are used, such equipment is comparatively cheap and may prove to be more effective for control of some pests. Ultra-low-volume drift sprays have also been used to apply fungi. *Beauveria bassiana* and *M. flavoviride*, formulated as concentrates in oil and applied using an ultra-low-volume sprayer, are presently under evaluation for control of locusts and grasshoppers (Johnson *et al.*, 1992; Bateman, 1994). It remains to be seen whether this application technique would be of value in thrips control.

Mycopathogens in Integrated Pest Management (IPM)

In a typical crop, a range of pest species, including plant pathogens and insects, are encountered and have to be combated simultaneously. Thus the role of mycopathogens in thrips suppression must be viewed within the context of a total crop management programme. Perhaps the greatest potential at present lies in their use in glasshouse crop protection, and it is from this perspective that their role in IPM will be discussed.

Compatibility with agrochemicals

Agrochemicals will remain necessary in commercial production systems for the foreseeable future. Insecticides may be needed to suppress an expanding thrips population, or control pests not targeted by fungi. Fungicides will be needed to control plant diseases, but the broad spectra of activity of some may also affect entomogenous fungi. Plant growth regulators are also used extensively in ornamental plant production; the compatibility of fungi with these chemicals is unknown. Knowledge of compatibility between beneficial fungi and all these different agrochemicals is crucial, so that growers may select appropriate compounds and schedule fungal or chemical treatments accordingly in order to minimize any deleterious affects on pathogen efficacy.

Germination, fungal growth and sporulation are important in insect infection and disease transmission; the inhibitory effects of agrochemicals on these processes vary depending on the fungal species and strain (Vänninen and Hokkanen, 1988; Anderson *et al.*, 1989; Majchrowicz and Poprawski, 1993; Li and Holdom, 1994). Some insecticides (e.g. carbamates) are fungicidal and analogues are available to control fungal plant pathogens, so some insecticides as well as fungicides may be harmful to beneficial pathogens. Dose is also important; fungal sporulation is less affected at low pesticide concentrations (Anderson *et al.*, 1989; Li and Holdom, 1994). However, pesticides which are inhibitory in the laboratory do not always exhibit the same action in the field. Easwaramoorthy *et al.* (1978) showed that two test insecticides inhibited *Cephalosporium* (= *Verticillium*) *lecanii in vitro*, but the efficacy of the fungus was enhanced when applied in combination with sublethal concentrations of the same insecticides to control coffee green scale. Anderson *et al.* (1989) detected higher insect mortality when *B. bassiana* and sublethal concentrations of insecticides were applied to control Colorado potato beetle, attributing the higher rates to synergism between the two agents. In addition, fungicides and *V. lecanii* have been successfully used together in chrysanthemum and cucumber production (Gardner *et al.*, 1984; van der Schaaf *et al.*, 1991).

As modern agriculture moves to adopt more 'environmentally friendly' practices, there is increasing interest in the use of 'biorational' pesticides, such as insect growth regulators, insecticidal soaps and horticultural oils. These materials have minimal effects on beneficials, low mammalian toxicity and reduced residual activity. Some products are effective against thrips and have excellent potential for inclusion in an IPM strategy (Allen *et al.*, 1993). However, there are few documented studies on interactions between biorationals and entomopathogenic fungi.

As the mode of action of most biorationals relies on disruption of an insect-specific process, they should be relatively innocuous to fungi. While the specific mechanisms of activity differ, all in some way affect the insect cuticle – its formation, deposition, structure or integrity. The cuticle is the primary barrier to infection by entomogenous fungi (Hajek and St Leger, 1994). Insect growth regulators, and botanicals such as neem (azadirachtin), directly interfere with insect development and cuticle formation, or the moulting process (Staal, 1987; Rembold, 1989). If deposition, hardening and tanning of the new cuticle is affected, its ability to serve as an effective barrier to fungal infection is compromised and the chances of a lethal mycosis developing may be enhanced (Zimmermann, 1994). Synergism between *Metarhizium* spp. and teflubenzuron, a benzoylphenyl urea insecticide that interferes with chitin synthesis, has been shown against *Schistocerca gregaria* (Joshi *et al.*, 1992). Alternatively, if the mode of action of the biorational can induce morphogenetic damage, preventing moulting and maturation, then successful fungal invasion is more likely to occur. Insecticidal soaps affect the waxy layers of the insect cuticle. These layers prevent desiccation and are important in the insect defence system; their disruption could predispose insects to fungal infection. Horticultural oils kill insects by asphyxiation; but insecticidal oils have been used in the formulation of several fungal species, enhancing efficacy against the target host insect (Johnson *et al.*, 1992).

Other beneficial interactions may also occur. For example, in spray trials, *P. fumosoroseus* was not effective against glasshouse infestations of the aphids *Aphis gossypii* and *Macrosiphoniella sanborni*, but when applied with azadirachtin (Margosan-O®), fungal efficacy was observed and good control attained (Lindquist, 1993). Such interactions could potentially expand the pest host range of fungal agents, thereby increasing the cost-effectiveness of the control strategy. If greater efficiency could be obtained, fewer applications would be required to achieve the desired degree of pest control.

Use with semiochemicals

When evaluating *V. lecanii* for control of *A. gossypii*, Hockland *et al.* (1986) increased the aphid infection rate by spraying plots with the aphid alarm pheromone (*E*)-(β)-farnesene. The aphids became agitated and increased their movement over the treated leaf surface in response to the pheromone, and in so doing acquired more infective conidia, resulting in higher rates of infection than where no pheromone was applied to the aphid population. Teerling *et al.* (1993) identified an alarm pheromone in droplets of anal fluid produced by *Frankliniella occidentalis*. Both nymphs and adults responded to the pheromone by walking away from the source of application. *Frankliniella occidentalis* are highly thigmotactic and often abundant inside flowers where they are difficult to reach with contact insecticides or with fungi. If the alarm pheromone could stimulate insects to vacate these refugia and increase their exposure to conidia deposited on leaves or petals, better control might be achieved.

Compatibility with other biocontrol agents

Beneficials

IPM also encompasses the use of a variety of other biological control agents such as arthropod predators and parasitoids (see Chapters 8, 9, 16, 17, 18). Several thrips predators are now commercially available, but little work has been done to determine the compatibility of these agents with entomogenous fungi. The pathogenic effects of the fungi are likely to vary according to the fungal species and strain, and the non-target organism.

Some isolates of entomogenous fungi are compatible with some glasshouse beneficials (Magalhaes *et al.*, 1988; Bethke and Parrella, 1989; James and Lighthart, 1994). Laboratory tests can be used as a first step in establishing the susceptibility of non-targets to candidate fungal strains, but there is often poor correlation between these and field test data. In laboratory tests, higher doses of inoculum than those which occur in the field are generally used against stressed hosts, which are held under conditions favouring development of an infection. Under field conditions, the likelihood of high infection rates occurring in non-targets is greatly reduced (Prior, 1990; Goettel, 1994). However, testing is required and suitable precautions taken, if needed, when scheduling releases.

Entomophilic nematodes

Nematodes are increasingly being considered for control of a range of insects, including thrips (Tomalek, 1994) (see Chapter 9). The work of Barbercheck and Kaya (1990, 1991) suggested that dual infections of two nematodes and *B. bassiana* could speed up the lethal infection rate, and cause higher total mortality in the treated pest population than if either agent was used alone. Antagonistic interactions between *B. bassiana* and the nematodes in dually infected hosts, however, could adversely affect pathogen or progeny development and affect the spread of the fungus or the nematode within the pest population. Nevertheless, if inundative releases of both organisms are contemplated in a control strategy, this would not necessarily prevent their combined use.

Host plant resistance

Differences in thrips resistance between chrysanthemum cultivars have been found in plants before and during flowering (de Jager and Butôt, 1993; Dijken *et al.*, 1993). In turn, this could influence thrips susceptibility to a fungal infection. On a non-preferred host plant, insects tend to be more stressed and perhaps more prone to infection. Furthermore, if the developmental rate is affected, then the intermoult period is extended, providing a longer period of susceptibility to infection. Differences in insect susceptibility have been observed depending on the type of host plant on which insects have been reared (Hare and Andreadis, 1983; Ramoska and Todd, 1985). While insect resistance has been bred into many agricultural crops, most plant breeding programmes, particularly in ornamentals, focus on the development of disease resistance. Although this in itself may not directly affect thrips, use of disease-resistant cultivars will decrease the amounts of fungicides needed over a growing season, reducing harmful effects on fungi used against thrips. Resistance to viruses vectored by thrips reduces the need to maintain plants virtually insect free, and allows greater flexibility in the type of control tactics which can be incorporated into a thrips control programme.

Mycopathogens for Thrips Management

Despite the lack of published information, mycopathogens hold great promise for use in thrips management. A range of fungal pathogens exists naturally in populations to which thrips are highly susceptible. The research and commercial aim should be to exploit this potential and develop the fungi into viable and effective products compatible with plant protection/production programmes. Laboratory studies can establish that thrips are susceptible to a particular fungal strain, but to ensure successful use of a fungal product, appropriate mass-production, formulation and application techniques have to be defined, and susceptible life stages and habitats identified. An understanding of the pathogen, the host insect, and interactions with other biotic and abiotic factors is essential to the development of an effective treatment regime.

Fungi will not be a cure-all for thrips problems on all crops and in all agricultural settings and it is unlikely that they will suppress thrips sufficiently alone, or replace chemical insecticides. They represent a valuable management resource to be utilized within the framework of an IPM programme, and could contribute significantly to reductions in pesticide use. Their integration with other management practices will require more compatibility studies, and the development of effective guidelines for their use. Such integration will be more complex than the use of traditional chemical control tactics, and will rely upon efficient monitoring to optimize the levels of control obtained from each application. Recent developments portend well for the future use of fungi in thrips management programmes.

References

Al-Aidroos, K. and Roberts, D.W. (1978) Mutants of *Metarhizium anisopliae* with increased virulence towards mosquito larvae. *Canadian Journal of Genetics and Cytology* 20, 211–219.

Allen, W.R., Tehrani, B. and Luft, R. (1993) Effect of horticultural oil, insecticidal soap, and film-forming products on the western flower thrips and the tomato spotted wilt virus. *Plant Disease* 77, 915–918.

Andersch, W., Hartwig, J., Reinecke, P. and Stenzel, K. (1990) Production of mycelial granules of the entomopathogenic fungus *Metarhizium anisopliae* for biological control of soil pests. In: *Proceedings 5th International Colloquium on Invertebrate Pathology and Microbial Control*, Society for Invertebrate Pathology, Adelaide, Australia, pp. 320–325.

Anderson, T.E., Hajek, A.E., Roberts, D.W, Preisler, H.K. and Robertson, J.L. (1989) Colorado potato beetle (Coleoptera: Chrysomelidae): effects of combinations of *Beauveria bassiana* with insecticides. *Journal of Economic Entomology* 82, 83–89.

Balazy, S. (1993) Entomophthorales. In: *Flora of Poland*. Polska Akademia Nauk, pp. 140–142.

Barbercheck, M.E. and Kaya, H.K. (1990) Interactions between *Beauveria bassiana* and the entomogenous nematodes, *Steinernema feltiae* and *Heterorhabditis heliothidis*. *Journal of Invertebrate Pathology* 55, 225–234.

Barbercheck, M.E. and Kaya, H.K. (1991) Competitive interactions between entomopathogenic nematodes and *Beauveria bassiana* (Deuteromycotina: Hyphomycetes) in soilborne larvae of *Spodoptera exigua* (Lepidoptera: Noctuidae). *Environmental Entomology* 20, 707–712.

Bateman, R. (1994) Physical properties and atomisation of ulv formulations of mycoinsecticides. In: Smits, P.H. (ed.) *Proceedings, 4th European Meeting on Microbial Control of Pests, IOBC Working Group on Insect Pathogens and Insect Parasitic Nematodes, Zurich, Switzerland,* 1993. *International Organisation of Biological Control/West Palaearctic Regional Section Bulletin* 17, pp. 189–192.

Bateman, R.P., Carey, M., Moore, D. and Prior, C. (1993) The enhanced infectivity of *Metarhizium flavoviride* in oil formulations to desert locusts at low humidities. *Annals of Applied Biology* 122, 145–152.

Bell, J.V. (1974) Mycoses. In: Cantwell, G.E. (ed.) *Insect Diseases.* Marcel Dekker Inc., New York, pp. 185–236.

Bethke, J.A. and Parrella, M. (1989) Compatibility of the aphid fungus *Cephalosporium lecanii* with the leaf miner parasite *Diglyphus beginii* (Hymenoptera: Eulophidae). *Pan-Pacific Entomologist* 65, 385–390.

Bidochka, M. (1994) Characterization of entomopathogenic fungi using molecular techniques. In: *Proceedings, VIth International Colloquium on Invertebrate Pathology and Microbial Control*, Society for Invertebrate Pathology, Montpellier, France, pp. 305–310.

Binns, E.S., Hall, R.A. and Pickford, R.J.J. (1982) *Thrips tabaci* Lind. (Thysanoptera – Thripidae) – distribution and behaviour on glasshouse cucumbers in relation to chemical and integrated control. *Entomologist's Monthly Magazine* 118, 55–68.

Boucias, D.G. and Pendland, J.C. (1982) Ultrastructural studies on the fungus, *Nomurea rileyi*, infecting the caterpillar, *Anticarsia gemmatalis. Journal of Invertebrate Pathology* 39, 338–345.

Bourne, A.I. and Shaw, F.R. (1934) Notes on fungus attacking onion thrips. *Journal of Economic Entomology* 27, 860–861.

Bourne, A.I. and Whitcomb, W.D. (1935) Control of onion thrips. *Bulletin of the Massachusetts Experiment Station* 315, 38–52.

Brownbridge, M. (1995) Prospects for mycopathogens in thrips management. In: Parker, B.L., Skinner, M. and Lewis, T. (eds) *Thrips Biology and Management.* Plenum Press, New York, pp. 281–295.

Brownbridge, M., Humber, R.A., Parker, B.L. and Skinner, M. (1993) Fungal entomopathogens recovered from Vermont forest soils. *Mycologia* 85, 358–361.

Brownbridge, M., McLean, D.L., Parker, B.L. and Skinner, M. (1994) Use of fungal pathogens for insect control in greenhouses. In: Robb, K. (ed.) *Proceedings, Tenth Conference on Insect and Disease Management on Ornamentals*, Dallas, Texas. Ball Publishing, Batavia, Illinois, pp. 7–20.

Butt, T.M. (1984) Studies on the invasive and developmental process of five genera of the Entomophthorales with specific reference to *Erynia neoaphidis* on the pea aphid, *Acyrthosiphon pisum* Harris. Ph.D. thesis, University of Bristol, UK.

Butt, T.M. (1987) A fluorescence microscopy method for the rapid localization of fungal spores and penetration sites on insect cuticle. *Journal of Invertebrate Pathology* 50, 72–74.

Butt, T.M. (1990) Fungal infection processes – a mini-review. In: *Proceedings, 5th International Colloquium on Invertebrate Pathology and Microbial Control,* Society for Invertebrate Pathology, Adelaide, Australia, pp. 121–124.

Butt, T.M., Beckett, A. and Wilding, N. (1981) Protoplasts in the *in vivo* life cycle of *Erynia neoaphidis. Journal of General Microbiology* 127, 417–421.

Butt, T.M., Wraight, S.P., Galaini-Wraight, S., Humber, R.A., Roberts, D.W. and Soper, R.S. (1988) Humoral encapsulation of the fungus *Erynia radicans* (Entomophthorales) by the potato leafhopper *Empoasca fabae* (Homoptera: Cicadellidae). *Journal of Invertebrate Pathology* 52, 49–56.

Butt, T.M., Beckett, A. and Wilding, N. (1990) A histological study of the invasive and developmental processes of the aphid pathogen *Erynia neoaphidis* (Zygomycotina: Entomophthorales) in the pea aphid *Acyrthosiphon pisum*. *Canadian Journal of Botany* 68, 2153–2163.

Butt, T.M., Barrisever, M., Drummond, J., Schuler, T.H., Tillemans, F.T. and Wilding, N. (1992) Pathogenicity of the entomogenous, hyphomycete fungus, *Metarhizium anisopliae* against the chrysomelid beetles *Psylliodes chrysocephala* and *Phaedon cochleariae*. *Biocontrol Science and Technology* 2, 325–332

Butt, T.M., Hajek, A.E. and Humber, R.A. (1994a) The effect of temperature on survival of protoplasts of the gypsy moth pathogen *Entomophaga maimaiga*. *Journal of Invertebrate Pathology* 64, 74–75.

Butt, T.M., Ibrahim, L., Ball, B.V. and Clark, S.J. (1994b) Pathogenicity of the entomogenous fungi *Metarhizium anisopliae* and *Beauveria bassiana* against crucifer pests and the honey bee. *Biocontrol Science and Technology* 4, 207–214.

Butt, T.M., Ibrahim, L., Clark, S.J. and Beckett, A. (1995) The germination behaviour of the entomogenous fungus *Metarhizium anisopliae* on the surface of the aphid and flea beetle cuticles. *Mycological Research* 99, 945–950.

Butt, T.M., Hajek, A.E. and Humber, R.A. (1996) Gypsy moth immune defenses in response to hyphal bodies and natural protoplasts of entomophthoralean fungi. *Journal of Invertebrate Pathology* 68, 278–285.

Carl, K.P. (1975) An *Entomophthora* sp. (Entomophthorales: Entomophthoraceae) pathogenic to *Thrips* spp. (Thysanoptera: Thripidae) and its potential as a biological control agent in glasshouses. *Entomophaga* 20, 381–388.

Chase, L.R., Osborne, L.S. and Ferguson, V.M. (1986) Selective isolation of the entomopathogenic fungi *Beauveria bassiana* and *Metarhizium anisopliae* from an artificial potting medium. *Florida Entomologist* 69, 285–292.

Claydon, N. and Grove, J.F. (1982) Insecticidal secondary products from the entomogenous fungus *Verticillium lecanii*. *Journal of Invertebrate Pathology* 40, 413–418.

de Jager, C.M. and Botôt, R.P.T. (1993) Thrips (*Frankliniella occidentalis* Pergande) resistance in chrysanthemum; the importance of pollen as nutrition. In: *Proceedings, IPM in Greenhouse Ornamentals, IOBC/WPRS Working Group 'Integrated Control Glasshouses', Cambridge*, UK, 1992. *International Organisation of Biological Control/West Palaearctic Regional Section Bulletin* 16, pp. 113–115.

Doane, E.N., Parker, B.L. and Pivot, Y. (1995) Method for mass rearing even-aged western flower thrips on beans. In: Parker, B.L., Skinner, M. and Lewis, T. (eds) *Thrips Biology and Management*. Plenum Press, New York, pp. 587–593.

Dyadechko, N.P. (1977) *Thrips or fringe-winged insects (Thysanoptera) of the European part of the USSR*. Amerind Publishing Co., New Dehli.

Easwaramoorthy, S., Regupathy, A., Santharam, C. and Jayaraj, S. (1978) The effect of subnormal concentrations of insecticides in combination with the fungal pathogen, *Cephalosporium lecanii* Zimm. in the control of coffee green scale, *Coccus viridis* Green. *Zeitschrift für angewandte Entomologie* 86, 161–166.

Ekbom, B.S. (1979) Investigations on the potential of a parasitic fungus (*Verticillium lecanii*) for biological control of the greenhouse whitefly (*Trialeurodes vaporariorum*). *Swedish Journal of Agricultural Research* 9, 129–138.

Fargues, J. and Robert, P. (1991) Soil persistence of *Metarhizium anisopliae*: influence of fungal nutrition on both survival and remaining infectivity of conidia. In: Smits,

P.H. (ed.) *Proceedings, 3rd European Meeting on Microbial Control of Pests, IOBC Working Group on Insect Pathogens and Insect Parasitic Nematodes, Wageningen, The Netherlands. International Organisation of Biological Control/ West Palaearctic Regional Section Bulletin* 14, pp. 130–133.

Feng, Z., Carruthers, R.I., Roberts, D.W. and Robson, D.S. (1985) Age-specific dose–mortality effects of *Beauveria bassiana* (Deuteromycotina: Hyphomycetes) on the European corn borer, *Ostrinia nubilalis* (Lepidoptera: Pyralidae). *Journal of Invertebrate Pathology* 46, 259–264.

Firstencel, H., Butt, T.M. and Carruthers, R.I. (1990) A fluorescence microscopy method for determining the viability of entomophthoralean fungal spores. *Journal of Invertebrate Pathology* 55, 258–264.

Fransen, J.J. (1990) Fungi on aphids, thrips and whitefly in the greenhouse environment. In: *Proceedings, 5th International Colloquium on Invertebrate Pathology and Microbial Control.* Society for Invertebrate Pathology, Adelaide, Australia, pp. 376–380.

Fransen, J.J., Winkelman, K. and van Lenteren, J.C. (1987) The differential mortality at various life stages of the greenhouse whitefly, *Trialeurodes vaporariorum* (Homoptera: Aleyrodidae), by infection with the fungus *Aschersonia aleyrodis* (Deuteromycotina: Coelomycetes). *Journal of Invertebrate Pathology* 50, 158–165.

Gardner, W.A., Oetting, R.D and Storey, G.K. (1984) Scheduling of *Verticillium lecanii* and benomyl applications to maintain aphid (Homoptera: Aphididae) control on chrysanthemums in greenhouses. *Journal of Economic Entomology* 77, 514–518.

Gaugler, R., Costa, S.D. and Lashomb, J. (1989) Stability and efficacy of *Beauveria bassiana* soil inoculations. *Environmental Entomology* 18, 412–417.

Gillespie, A.T. (1986) The potential of entomogenous fungi as control agents for onion thrips, *Thrips tabaci. Brighton Crop Protection Conference Monograph No. 34, Biotechnology and Crop Improvement and Protection*, pp. 237–243.

Glare, T.R. and Milner, R.J. (1987) New records of entomophthoran fungi from insects in Australia. *Australian Journal of Botany* 35, 69–77.

Goettel, M.S. (1984) A simple method for mass culturing entomopathogenic hyphomycete fungi. *Journal of Microbiological Methods* 3, 15–20.

Goettel, M. (1988). Pathogenesis of the hyphomycete *Tolypocladium cylindrosporum* in the mosquito *Aedes aegypti. Journal of Invertebrate Pathology* 51, 259–274.

Goettel, M.S. (1992) Fungal agents for biocontrol. In: Lomer, C.J and Prior, C. (eds) *Biological Control of Locusts and Grasshoppers.* CAB International, Wallingford, UK, pp. 122–132.

Goettel, M.S. (1994) Host range and specificity in relation to safety of exotic fungi. In: *Proceedings, VIth International Colloquium on Invertebrate Pathology and Microbial Control*, Society for Invertebrate Pathology, Montpellier, France, pp. 332–333.

Goettel, M.S. and Roberts, D.W. (1992) Mass production, formulation and field application of entomopathogenic fungi. In: Lomer, C.J. and Prior, C. (eds) *Biological Control of Locusts and Grasshoppers.* CAB International, Wallingford, UK, pp. 230–238.

Greenwood, S.R. and Mills, N.J. (1989) The biological control of *Mikania micrantha* for southwest Pacific countries. *Commonwealth Institute of Biological Control* Report, Commonwealth Institute of Biological Control, Silwood Park, Ascot, UK, 28 pp.

Grivanov, K.P. (1940) Deep ploughing instead of burning the stubble for the control of the wheat thrips *Haplothrips tritici. Review of Applied Entomology (A)* 28, 69–70.

Hajek, A.E. and St Leger, R. (1994) Interactions between fungal pathogens and insect hosts. *Annual Review of Entomology* 39, 293–322.

Hall, R.A. (1985) Aphid control by fungi. In: Hussey, N.W. and Scopes, N. (eds) *Biological Pest Control: the Glasshouse Experience*. Cornell University Press, Ithaca, New York, pp. 138–141.

Hall, R.A. (1992) New pathogen on *Thrips palmi* in Trinidad. *Florida Entomologist* 75, 380–383.

Hall, R.A. and Papierok, B. (1982) Fungi as biological control agents of arthropods of agricultural and medical importance. *Parasitology* 84, 205–240.

Hall, R.A., Peterkin, D. and Pollard, G.V. (1993) A new system of caging *Thrips palmi* for laboratory bioassay of pathogens. *Florida Entomologist* 76, 171–175.

Hall, R.A., Peterkin, D. and Ali, B. (1994) Fungal control of whitefly, *Thrips palmi* and sugarcane froghopper in Trinidad and Tobago. In: *Proceedings, VIth International Colloquium on Invertebrate Pathology and Microbial Control.* Society for Invertebrate Pathology, Montpellier, France, pp. 277–282.

Hare, J.D. and Andreadis, T.G. (1983) Variation in the susceptibility of *Leptinotarsa decemlineata* (Coleoptera: Chrysomelidae) when reared on different host plants to the fungal pathogen, *Beauveria bassiana* in the field and the laboratory. *Environmental Entomology* 12, 1892–1897.

Harney, S. and Widden, P. (1991) The ecology of *Paecilomyces farinosus* in two balsam fir forests infested with spruce budworm. *Canadian Journal of Botany* 69, 512–515.

Helyer, N.L., Gill, G., Bywater, A. and Chambers, R. (1992) Elevated humidities for control of chrysanthemum pests with *Verticillium lecanii*. *Pesticide Science* 36, 373–378.

Higgins, C.L. (1992) Western flower thrips (Thysanoptera: Thripidae) in greenhouses: population dynamics, distribution on plants, and associations with predators. *Journal of Economic Entomology* 85, 1891–1903.

Hirte, W., Triltsch, H. and Sermann, H. (1994) Growth and survivability of the entomopathogenic fungus *Verticillium lecanii* in the soil. In: Smits, P.H. (ed.) *Proceedings, 4th European Meeting on Microbial Control of Pests, IOBC Working Group on Insect Pathogens and Insect Parasitic Nematodes*, 1993, *Zurich, Switzerland. International Organisation of Biological Control/West Palaearctic Regional Section Bulletin* 17, pp. 226–229.

Hockland, S.H., Dawson, G.W., Griffiths, D.C., Marples, B., Pickett, J.A. and Woodcock, C.M. (1986) The use of aphid alarm pheromone ((*E*)-(β)-farnesene) to increase effectiveness of the entomophilic fungus *Verticillium lecanii* in controlling aphids on chrysanthemums under glass. In: Samson, R.A, Vlak, J.M. and Peters, D. (eds) *Fundamental and Applied Aspects of Invertebrate Pathology. Foundation of the Fourth International Colloquium on Invertebrate Pathology*, Society for Invertebrate Pathology, Wageningen, Netherlands, pp. 252.

Humber, R.A. (1992) *Collection of Entomopathogenic Fungal Cultures: ARSEF Catalog of Strains, 1992*. US Department of Agriculture, Agricultural Research Service, ARS-110, 117 pp.

Hunt, T.R., Moore, D., Higgins, P.M. and Prior, C. (1994) Effect of sunscreens, irradiance and resting periods on the germination of *Metarhizium flavoviride* conidia. *Entomophaga* 39, 313–322.

Hutchison, J.A. (1963) The genus *Entomophthora* in the western hemisphere. *Transactions of the Kansas Academy of Sciences* 66, 237–254.

Inglis, G.D., Feniuk, R.P., Goettel, M.S. and Johnson, D.L. (1995) Mortality of grasshoppers exposed to *Beauveria bassiana* during oviposition and nymphal emergence. *Journal of Invertebrate Pathology* 65, 139–146.

James, R.R. and Lighthart, B. (1994) Susceptibility of the convergent lady beetle (Coleoptera: Coccinellidae) to four entomogenous fungi. *Environmental Entomology* 23, 190–192.

Jenkins, N.E. and Lomer, C.J. (1994) Development of a new procedure for the mass production of conidia of *Metarhizium flavoviride*. In: Smits, P.H. (ed.) *Proceedings, 4th European Meeting on Microbial Control of Pests, IOBC Working Group on Insect Pathogens and Insect Parasitic Nematodes, Zurich, Switzerland, 1993. International Organisation of Biological Control/West Palaearctic Regional Section Bulletin* 17, pp. 181–184.

Johnson, D.L., Goettel, M.S., Bradley, C., van der Paauw, H., and Maiga, B. (1992) Field trials with the entomopathogenic fungus *Beauveria bassiana* against grasshoppers in Mali, West Africa, July 1990. In: Lomer, C.J. and Prior, C. (eds) *Biological Control of Locusts and Grasshoppers*. CAB International, Wallingford, UK, pp. 296–310.

Joshi, L., Charnley, A.K., Arnold, G., Brain, P. and Bateman, R. (1992) Synergism between entomopathogenic fungi, *Metarhizium* spp. , and the benzoylphenyl urea insecticide, teflubenzuron, against the desert locust, *Schistocerca gregaria*. In: *Proceedings, Brighton Crop Protection Conference, Pests and Diseases*. British Crop Protection Council, Farnham, UK, pp. 369–374.

Keller, S. (1991) Arthropod-pathogenic entomophthorales of Switzerland. II. *Erynia, Eryniopsis, Neozygites, Zoophthora* and *Tarichium*. *Sonderdrucke aus Sywdowia* 43, 39–122.

Keller, S. (1992) The *Beauveria–Melolontha* project: experiences with regard to locust and grasshopper control. In: Lomer, C.J. and Prior, C. (eds) *Biological Control of Locusts and Grasshoppers*. CAB International, Wallingford, UK, pp. 279–286.

Keller, S. and Wuest, J. (1983) Observations sur trois especes de *Neozygites* (Zygomycetes: Entomophthoraceae). *Entomophaga* 28, 123–134.

Kreuger, S.R., Villani, M.G., Martins, A.S. and Roberts, D.W. (1992) Efficacy of soil applications of *Metarhizium anisopliae* (Metsch.) Sorokin conidia, and standard and lyophilized mycelial particles against scarab grubs. *Journal of Invertebrate Pathology* 59, 54–60.

Lewis, T. (1973) *Thrips: Their Biology, Ecology and Economic Importance*. Academic Press, London.

Li, D.P. and Holdom, D.G. (1990) Effects of carbon and nitrogen sources and vitamins on germination, growth and sporulation of *Metarhizium anisopliae* (Metsch.) Sorok. Deuteromycotina: (Hyphomycetes). In: *Proceedings, 5th International Colloquium on Invertebrate Pathology and Microbial Control*, Society for Invertebrate Pathology, Adelaide, Australia, pp. 154.

Li, D.P. and Holdom, D.G. (1994) Effects of pesticides on growth and sporulation of *Metarhizium anisopliae* (Deuteromycotina: Hyphomycetes). *Journal of Invertebrate Pathology* 63, 209–211.

Li, D.P. and Holdom, D.G. (1995) Effects of nutrients on colony formation, growth and sporulation of *Metarhizium anisopliae* (Deuteromycotina: Hyphomycetes). *Journal of Invertebrate Pathology* 65, 253–260.

Li, Z., Butt, T.M., Beckett, A. and Wilding, N. (1993) The structure of dry mycelia of the entomophthoralean fungi *Zoophthora radicans* and *Erynia neoaphidis* following different preparatory treatments. *Mycological Research* 97, 1315–1323.

Lindquist, R.K. (1993) Integrated insect, mite and disease management programs on greenhouse crops: pesticides and application methods. In: Robb, K. and Hall, J. (eds) *Proceedings, 9th Conference on Insect and Disease Management in Orna-*

mentals, Del Mar, California. Society of American Florists, Alexandria, Virginia, USA, pp. 38–44.

Lomer, C.J., Kooyman, C., Prior, C. and Shah, P.A. (1994) Influence of climatic factors on field performance of mycoinsecticides in grasshopper and locust control. In: *Proceedings, VIth International Colloquium on Invertebrate Pathology and Microbial Control*. Society for Invertebrate Pathology, Montpellier, France, pp. 289–291.

Lyubenov, Y. (1961) A contribution to the bionomics of the wheat thrips (*Haplothrips tritici* Kurd) in Bulgaria and possibilities of reducing the injury done by it. *Review of Applied Entomology* (A) 50, 348.

MacLeod, D.M., Tyrrell, D. and Carl, K.P. (1976) *Entomophthora parvispora* sp. nov., a pathogen of *Thrips tabaci*. *Entomophaga* 21, 307–312.

Magalhaes, B.P., Lord, J.C., Wraight, S.P., Daoust, R.A. and Roberts, D.W. (1988) Pathogenicity of *Beauveria bassiana* and *Zoophthora radicans* to the coccinellid predators *Coleomegilla maculata* and *Eriopis connexa*. *Journal of Invertebrate Pathology* 52, 471–473.

Magalhaes, B.P., Butt, T.M., Humber, R.A., Shields, E.J. and Roberts, D.W. (1990) Formation of appressoria *in vitro* by the entomopathogenic fungus *Zoophthora radicans* (Zygomycetes: Entomophthorales). *Journal of Invertebrate Pathology* 55, 284–288.

Magano di San Lio, G., Vacante, V. and Perrotta, G. (1992) *Neozygites parvispora* (MacLeod and Carl) Remaundiére and Keller a pathogen of western flower thrips. *Micologia Italiana* 2, 29–34.

Majchrowicz, I. and Poprawski, T.J. 1993. Effects *in vitro* of nine fungicides on growth of entomopathogenic fungi. *Biocontrol Science and Technology* 3, 321–336.

Marcandier, S. and Khachatourians, G.G. (1987) Susceptibility of the migratory grasshopper, *Melanoplus sanguinipes* (Fab.) (Orthoptera: Acrididae), to *Beauveria bassiana* (Bals.) Vuillemin (Hyphomycete): influence of relative humidity. *Canadian Entomologist* 119, 901–907.

Matthews, G. (1988) The attack of the charged brigade. *New Scientist*, July, 55–57.

Milner, R.J. and Lutton, G.G. (1986) Dependence of *Verticillium lecanii* (Fungi: Hyphomycetes) on high humidities for infection and sporulation using *Myzus persicae* (Homoptera: Aphidae) as host. *Environmental Entomology* 15, 380–382.

Moore, D., Bridge, P.D., Higgins, P.M., Bateman, R.P. and Prior, C. (1993) Ultra violet radiation damage to *Metarhizium flavoviride* conidia and the protection given by vegetable mineral oils and chemical sunscreens. *Annals of Applied Biology* 122, 605–616.

Moorehouse, E.R., Gillespie, A.T. and Charnley, A.K. (1993) Selection of virulent and persistent *Metarhizium anisopliae* isolates to control black vine weevil (*Otiorhynchus sulcatus*) larvae on glasshouse *Begonia*. *Journal of Invertebrate Pathology* 62, 47–52.

Moorehouse, E.R., Gillespie, A.T. and Charnley, A.K. (1994) The influence of temperature on the susceptibility of vine weevil, *Otiorhynchus sulcatus* (Fabricius) (Coleoptera: Curculionidae) larvae to *Metarhizium anisopliae* (Deuteromycotina: Hyphomycetes). *Annals of Applied Biology* 124, 185–193.

Morse, J.G., Bellows, T.S. and Iwata, Y. (1986) Technique for evaluating residual toxicity of pesticides to motile insects. *Journal of Economic Entomology* 79, 281–283.

Moscardi, F. (1989) Production and use of entomopathogens in Brazil. In: Roberts, D.W. and Granados, R.R. (eds) *Biotechnology, Biological Pesticides and Novel Plant-Pest Resistance for Insect Pest Management*. Cornell University Press, Ithaca, New York, pp. 53–60.

Nielsen, C., Madsen, T.U., Eilenberg, J. and Philipsen, H. (1996) Natural occurrence of entomopathogenic fungi on Thrips (Thysanoptera) in a Danish wheat field. *Bulletin of the International Organisation for Biological Control/West Palaearctic Regional Section* 19, 166–171.

Parrella, M.P. (1993) Bugs to the rescue. *Grower Talks*, May, p. 67.

Pereira, R.M. and Roberts, D.W. (1990) Dry mycelium preparations of entomopathogenic fungi, *Metarhizium anisopliae* and *Beauveria bassiana*. *Journal of Invertebrate Pathology* 56, 39–46.

Pereira, R.M. and Roberts, D.W. (1991) Alginate and cornstarch mycelial formulations of entomopathogenic fungi, *Beauveria bassiana* and *Metarhizium anisopliae*. *Journal of Economic Entomology* 84, 1657–1661.

Prasertphon, S. and Tanada, Y. (1968) The formation and circulation, in *Galleria*, of hyphal bodies of entomophthoraceous fungi. *Journal of Invertebrate Pathology* 11, 260–280.

Prior, C. (1990) The biological basis for regulating the release of micro-organisms with particular reference to the use of fungi for pest control. *Aspects of Applied Biology* 24, 231–238.

Prior, C., Jollands, P. and Le Patourel, G. (1988) Infectivity of oil and water formulations of *Beauveria bassiana* (Deuteromycotina: Hyphomycetes) to the cocoa weevil pest *Pantorhytes plutus* (Coleoptera: Curculionidae). *Journal of Invertebrate Pathology* 52, 66–72.

Ramakers, P.M.J. (1976) Preliminary observations on an *Entomophthora* sp. on *Thrips tabaci* Lind. in Dutch glasshouses. *Bulletin of the International Organization for Biological Control, West Palaearctic Regional Section* 4, 180–182.

Ramoska, W.A. (1984) The influence of relative humidity on *Beauveria bassiana* infectivity and replication in the chinch bug *Blissus leucopterus*. *Journal of Invertebrate Pathology* 43, 389–394.

Ramoska, W.A. and Todd, T. (1985) Variation in efficacy and viability of *Beauveria bassiana* in the chinch bug (Hemiptera: Lygaeidae) as a result of feeding activity on selected host plants. *Environmental Entomology* 14, 146–148.

Rath, A.C. (1992) *Metarhizium anisopliae* for control of the Tasmanian pasture scarab *Adoryphorus couloni*. In: Jackson, T.A. and Glare, T.R. (eds) *Use of Pathogens in Scarab Pest Management*. Intercept, Andover, UK, pp. 217–228.

Ravensberg, W.J., Malais, M. and van der Schaaf, D.A. (1990) *Verticillium lecanii* as a microbial insecticide against glasshouse whitefly. In: *Proceedings, Brighton Crop Protection Conference, Pest and Diseases*. British Crop Protection Council, Farnham, UK, pp. 265–269.

Rembold, H. (1989) Azadirachtins: their mode of action. In: Aranson, J.T., Philogene, B.J.R. and Morand, P. (eds) *Insecticides of Plant Origin*. American Chemical Society Symposium Series No. 387, pp. 150–163.

Reyne, A. (1921) De Cacaothrips (*Heliothrips rubrocinctus*, Giard). *Bulletin Department Landbouwkundig Suriname* No. 44.

Roberts, D.W. (1981) Toxins of entomopathogenic fungi. In: Burges, H.D. (ed.) *Microbial Control of Pests and Plant Diseases 1970–1980*. Academic Press, London, pp. 441–464.

Saito, T. (1991) A field trial of an entomopathogenic fungus, *Beauveria bassiana* (Bals.) Vuill., for the control of *Thrips palmi* Karny (Thysanoptera: Thripidae). *Japanese Journal of Applied Entomology and Zoology* 35, 80–81.

Saito, T., Kubota, S. and Shimazu, M. (1989) A first record of the entomopathogenic fungus, *Neozygites parvispora* (MacLeod and Carl) Rem. and Kell., on *Thrips palmi* in Japan. *Applied Entomology and Zoology* 24, 233–235.

Samson, R.A., Ramakers, P.M.J. and Oswald, T. (1979) *Entomophthora thripidum*, a new fungal pathogen of *Thrips tabaci*. *Canadian Journal of Botany* 57, 1317–1323.

Samuels, K.D.Z., Heale, J.B. and Llewellyn, M. (1989) Characteristics relating to the pathogenicity of *Metarhizium anisopliae* toward *Nilaparvata lugens*. *Journal of Invertebrate Pathology* 53, 25–31.

Schreiter, G., Butt, T.M., Beckett, A., Moritz, G. and Vestergaard, S. (1994) Invasion and development of *Verticillium lecanii* in the western flower thrips, *Frankliniella occidentalis*. *Mycological Research* 98, 1025–1034.

Segers, R., Butt, T.M., Kerry, B.R. and Peberdy, J. (1995) The subtilisins of the invertebrate mycopathogens *Verticillium chlamydosporium* and *Metarhizium anisopliae* are serologically and functionally related. *FEMS Microbiology Letters* 126, 227–232.

Segers, R., Butt, T.M., Kerry, B.R., Beckett, A. and Peberdy, J. (1996) The role of the proteinase VCP1 produced by the nematophagous *Verticillium chlamydosporium* in the infection process of nematode eggs. *Mycological Research* 100, 421–428.

Sermann, H., Kästner, U. and Hirte, W. (1994) Effectiveness of a soil application of *Verticillium lecanii* on soilborne stages of *Frankliniella occidentalis*. In: Smits, P.H. (ed.) *Proceedings, 4th European Meeting on Microbial Control of Pests, IOBC Working Group on Insect Pathogens and Insect Parasitic Nematodes, Zurich, Switzerland, 1993. International Organisation of Biological Control/West Palaearctic Regional Section Bulletin* 17, pp. 230–233.

Skinner, M., Parker, B.L. and Bergdahl, D.R. (1991) *Verticillium lecanii*, isolated from larvae of pear thrips, *Taeniothrips inconsequens*, in Vermont. *Journal of Invertebrate Pathology* 58, 157–163.

Sopp, P.I., Gillespie, A.T. and Palmer, A. (1989) Application of *Verticillium lecanii* for the control of *Aphis gossypii* by low volume electrostatic rotary atomiser and a high-volume hydraulic sprayer. *Entomophaga* 34, 417–428.

Staal, G.B. (1987) Juvenoids and anti-juvenile hormone agents as IGRs. In: Delucchi, V. (ed.) *Proceedings, Symposium on Insect Pest Management: Quo Vadis? An International Perspective*. Parasitis, Geneva, Switzerland, pp. 277–292.

Stathers, T.E., Moore, D. and Prior, C. (1993) The effect of different temperatures on the viability of *Metarhizium flavoviride* conidia stored in vegetable and mineral oils. *Journal of Invertebrate Pathology* 62, 111–115.

St Leger, R., Butt, T.M., Goettel, M.S., Staples, R. and Roberts, D.W. (1989a) Production *in vitro* of appressoria by the entomopathogenic fungus *Metarhizium anisopliae*. *Experimental Mycology* 13, 274–288.

St Leger, R., Butt, T.M., Staples, R. and Roberts, D.W. (1989b) Synthesis of proteins including a cuticle-degrading protease during differentiation of the entomopathogenic fungus *Metarhizium anisopliae*. *Experimental Mycology* 13, 253–262.

Storey, G.K. and Gardner, W.A. (1988) Movement of an aqueous spray of *Beauveria bassiana* into the profile of four Georgia soils. *Environmental Entomology* 17, 135–139.

Storey, G.K., Gardner, W.A., Hamm, J.J. and Young, J.R. (1987) Recovery of *Beauveria bassiana* propagules following application of formulated conidia through an overhead irrigation system. *Journal of Entomological Science* 22, 355–357.

Storey, G.K., McCoy, C.W., Stenzel, K. and Andersch, W. (1990) Conidiation kinetics of the mycelial granules of *Metarhizium anisopliae* (B10 1020) and its biological activity against different soil insects. In: *Proceedings, 5th International Colloquium on Invertebrate Pathology and Microbial Control*. Society for Invertebrate Pathology, Adelaide, Australia, pp. 320–325.

Teerling, C.R., Pierce, H.D., Jr, Borden, J.H. and Gillespie, D.R. (1993) Identification and bioactivity of alarm pheromone in the western flower thrips, *Frankliniella occidentalis*. *Journal of Chemical Ecology* 19, 681–697.

Teulon, D.A.J. (1992) Laboratory technique for rearing western flower thrips (Thysanoptera: Thripidae). *Journal of Economic Entomology* 85, 895–899.

Thaxter, R. (1888) The entomophthoreae of the United States. *Memoirs of the Boston Society of Natural History* 4, 133–201.

Tomalek, M. (1994) Genetic improvement of *Steinernema feltiae* for integrated control of the western flower thrips, *Frankliniella occidentalis*. In: Smits, P.H. (ed.) *Proceedings, 4th European Meeting on Microbial Control of Pests, IOBC Working Group on Insect Pathogens and Insect Parasitic Nematodes, Zurich, Switzerland, 1993. International Organisation of Biological Control/West Palaearctic Regional Section Bulletin* 17, pp. 17–20.

Vacante, V., Cacciola, S.O. and Pennisi, A.M. (1994) Epizootiological study of *Neozygites parvispora* (Zygomycota: Entomophthoraceae) in a population of *Frankliniella occidentalis* (Thysanoptera: Thripidae) on pepper in Sicily. *Entomophaga* 39, 123–130.

Van Dijken, F.R., Breugem, G., Dik, M.T.A., Gebala, B., Soria, C. and Jong, J. de. (1993) Effects of flower colour of chrysanthemum on host plant resistance to the western flower thrips (*Frankliniella occidentalis* Pergande) in no-choice situations. *Proceedings of the Section on Experimental and Applied Entomology of the Netherlands Entomological Society* 4, 45–46.

van der Schaaf, D.A. and Ravensberg, W. (1992) Effects of various isolates of *Verticillium lecanii*, against the cotton aphid (*Aphis gossypii*): bio-assays and prospects of use in glasshouse cucumbers. In: *Abstracts, 25th Annual Meeting Society for Invertebrate Pathology*. Society for Invertebrate Pathology, Heidelberg, Germany.

van der Schaaf, D.A., Ravensberg, W.J. and Malais, M. (1991) *Verticillium lecanii* as a microbial insecticide against whitefly. In: Smits, P.H. (ed.) *Proceedings, 3rd European Meeting on Microbial Control of Pests, IOBC Working Group on Insect Pathogens and Insect Parasitic Nematodes, Wageningen, Netherlands. International Organisation of Biological Control/West Palaearctic Regional Section Bulletin* 14, pp. 120–123.

Vänninen, I. and Hokkanen, H. (1988) Effect of pesticides on four species of entomopathogenic fungi *in vitro*. *Annalles Agriculturae Fenniae* 27, 345–353.

Vänninen, I. and Tyni-Juslin, J. (1991) Persistence of *Metarhizium anisopliae* and *Beauveria bassiana* in Finnish agricultural soils: preliminary results. In: Smits, P.H. (ed.) *Proceedings, 3rd European Meeting on Microbial Control of Pests, IOBC Working Group on Insect Pathogens and Insect Parasitic Nematodes, Wageningen, The Netherlands. International Organisation of Biological Control/West Palaearctic Regional Section Bulletin* 14, pp. 67–70.

Vestergaard, S., Butt, T.M., Gillespie, A.T., Schreiter, G. and Eilenberg, J. (1995) Pathogenicity of the hyphomycete fungi *Verticillium lecanii* and *Metarhizium anisopliae* to the western flower thrips, *Frankliniella occidentalis*. *Biocontrol Science and Technology* 5, 185–192.

Vey, A. and Fargues, J. (1977) Histological and ultrastructural studies of *Beauveria bassiana* infection in *Leptinotarsa decemlineata* larvae during ecdysis. *Journal of Invertebrate Pathology* 30, 207–215.

Wang, C.L. and Chu, Y.I. (1986) Rearing method of southern yellow thrips, *Thrips palmi* Karny, in the laboratory. *Plant Protection Bulletin (Taiwan, R.O.C.)* 28, 407–411.

Wraight, S.P., Butt, T.M., Galaini-Wraight, S., Allee, L., Soper, R.S. and Roberts, D.W.

(1990) Germination and infection processes of the entomophthoralean fungus *Erynia radicans* on the potato leafhopper, *Empoasca fabae*. *Journal of Invertebrate Pathology* 56, 157–174

Wright, J.E. (1993) Control of the boll weevil (Coleoptera: Curculionidae) with Naturalis-L: a mycoinsecticide. *Journal of Economic Entomology* 86, 1355–1358.

Zacharuk, R.Y. (1970) Fine structure of the fungus *Metarhizium anisopliae* infecting three species of larval elateridae (Coleoptera). II. Conidial germ tubes and appressoria. *Journal of Invertebrate Pathology* 15, 81–91.

Zacharuk, R.Y. (1971) Fine structure of the fungus *Metarhizium anisopliae* infecting three species of larval elateridae (Coleoptera). IV. Development within the host. *Canadian Journal of Microbiology* 17, 525–529.

Zimmermann, G. (1994) Strategies for the utilization of entomopathogenic fungi. In: *Proceedings, VIth International Colloquium on Invertebrate Pathology and Microbial Control*. Society for Invertebrate Pathology, Montpellier, France, pp. 67–73.

11 Field and Laboratory Techniques

Trevor Lewis
Institute of Arable Crops Research – Rothamsted, Harpenden, Herts AL5 2JQ, UK

Sampling

Thrips pose special sampling problems compared with larger insects inhabiting mainly exposed parts of plants (Southwood, 1979). This section reviews methods that have proved successful for thrips on crops, in the soil and in the air, presenting a selection of approaches as guides to the types of difficulties encountered and possible solutions. It is worth emphasizing that for any sampling attempted the optimal method is that which most cheaply gives the accuracy required, so the sampling of different crops and thrips species usually need different approaches. Decisions about the number and size of sample units can be difficult because the most efficient sampling procedure for a given crop and pest depends on the spatial distribution of the thrips population(s), which may be unclear from preliminary observation. Furthermore, it often changes as the crop develops and insect numbers increase. At many sites, densities differ between the edges and centre of fields (Lewis, 1959a), on different parts of plants (Irwin *et al.*, 1979), from day to day (McPherson *et al.*, 1992) and sometimes even from hour to hour (Tappan, 1986; Sites *et al.*, 1992). Different approaches are required depending on whether the aim is to detect the presence of a pest, estimate absolute or relative numbers of adults or immature instars, determine economic damage or control thresholds, time control measures, assess their effectiveness, or locate concentrations within a crop. The purpose should be clear from the outset to avoid unacceptable time- and labour-intensive field effort, tedious laboratory extraction and counting, and to allow meaningful interpretation of the data collected (Perry, 1994).

Quantitative sampling of field crops and trees

The great diversity in form of crop plants, ranging from low-growing annuals to trees, and the various habits of the many thrips species infesting them, means that sampling is best considered in the context of each crop and its pests rather than as a series of generally applicable methods. Lewis (1973) reviewed these, including sweeping, beating and shaking, suction sampling, knockdown insecticides and the collection of vegetation for subsequent extraction of thrips by washing or dynamic methods. Recently, the pervasive requirements to restrict unnecessary pesticide use, over a period during which thrips problems have greatly increased, have stimulated closer scrutiny of the efficiency and accuracy of these techniques, especially for pest monitoring. This selective review of studies made in particular crops is intended to alert field investigators to the care with which sampling should be undertaken, to potential shortcomings of the various methods, and as a guide to devising procedures suited to specific research, advisory or control requirements.

Simple visual assessment of infestation levels on most growing crops is highly subjective and difficult to interpret in terms of numbers of individuals present unless a sampling plan based on knowledge of the pests' dispersion has been developed. It is sometimes useful on crops where the pests' presence causes easily recognizable symptoms that allow convenient ranking of the severity of infestation. Examples include the percentage of rice plants suffering from leaf-tip curl caused by *Stenchaetothrips biformis* (Chiang, 1977), the discoloration of ornamental blooms caused by petal feeding, or the incidence of virus infection in field tomatoes (Cameron *et al.*, 1992), peanuts (Amin, 1983), pineapples (Linford, 1943) and tobacco (Razvyazkina, 1953). Also for forest trees, visual ranking of infestation may be the only convenient approach. It has been used in New England to categorize damage by *Taeniothrips inconsequens* to sugar maple and to try to relate this to thrips populations in soil. One system used was to rank leaf area reduction as none, light (1 to 30 per cent), moderate (31 to 60 per cent), heavy (61 to 80 per cent) and severe (more than 80 per cent) (Wilmot and Kelley, 1990). A more sophisticated option for forest pest surveying is the use of satellite imaging data to detect severity of attack, an approach also demonstrated on sugar maple in New England (Vogelmann and Rock, 1991) but it is expensive and has not been adopted as a practical approach (see Chapter 17).

A simple method for monitoring sparse populations of *Heliothrips haemorrhoidalis* and *Selenothrips rubrocinctus* in South African avocado orchards is based on the preference these pests show for feeding between touching fruits, which are ten times more susceptible to damage than single fruits (Plate III.5). By restricting tree examination to touching fruit, growers can detect incipient infestations and decide if insecticidal treatment is warranted (Dennill and Erasmus, 1991). Where a thrips population is sparse and spatially clustered, an 'adaptive' sampling procedure (Thomson *et al.*, 1992) may be useful; this combines sampling at random locations with a second stage in which an exhaustive search is made of the units around any unit in which an individual is detected.

In California, scar counts on developing citrus fruit were used to categorize the cumulative damage caused by *Scirtothrips citri*. On each of ten trees on each sampling occasion, all outside fruits from knee to eye level were examined and scarring ranked as slight or severe (i.e. unsuitable for export) to compare with thrips populations assessed by other methods (Grout *et al.*, 1986). Similarly, in

New England, cumulative scarring of nectarines at harvest caused mainly by *Frankliniella occidentalis* was used by Felland *et al.* (1995) to categorize injury, rated as light (less than 1 cm^2), moderate (1 to 4 cm^2) or heavy (more than 4 cm^2).

One of the earlier attempts to compare the precision of different sampling procedures in a field crop was Bullock's (1965) study of *Thrips nigropilosus* on pyrethrum plants which are spaced about 30 cm apart with 1 m between rows. Leaves were collected in four ways: (i) 10 leaves from each of 5 plants chosen at random; (ii) 1 leaf from 28 plants; (iii) 1 from 49, and (iv) 5 from 20 (the three latter being collected in different sequences of traverses across rows), and insects washed from them in 70 per cent alcohol. The sample from the five randomly chosen plants contained fewer thrips than the samples collected by other methods because the thrips were aggregated in the plots, and, by limiting inspection to five plants, some samples included no heavily infested ones. Thus with aggregated distributions, which are far more common in field populations than either uniform or random ones, a sampling design that covers as many patches as possible and includes an effective proportion of the relatively few heavily infested plants must be used. Another factor affecting the accuracy of samples from these populations on pyrethrum, and probably on most other row crops, is that the foliage of adjacent plants may touch, enabling larvae to crawl from plant to plant more easily along rows than between them. Thus regular traverses across several rows, taking sample units from plants not in contact, will help to decrease the errors due to patchy distribution between and within rows.

A different problem was encountered by Irwin *et al.* (1979) for studies on thrips in soybean fields where numbers of the two main pest species, *Neohydatothrips variabilis* and *Frankliniella tritici*, did not vary significantly among plots or over plants but they were unevenly distributed over individual plants. Adults of *N. variabilis* were most common on the uppermost, fully expanded trifoliate or on the one immediately below it, whereas larvae were most abundant on the third to sixth trifoliates below the terminal. All instars of *F. tritici* concentrated in terminal buds and blossoms. Accordingly, for detailed sampling of *N. variabilis* adults, the uppermost fully expanded trifoliate was chosen as the sample unit and collected from three to six plants depending on growth stage; for larvae before blossoming, the fourth from upper fully expanded trifoliate or, if less than four trifoliates were present, the lowest one, was taken. During and after blooming, the fifth from upper fully expanded trifoliate was removed. Collected leaves were washed in detergent and sieved to remove thrips (see p. 445). For rapid sampling as required for pest management decisions, the centre leaflet was carefully turned over and thrips counted with a hand lens. Usually it was only necessary to count larvae to give a reasonable indication of *Neohydatothrips* abundance. By contrast, the recommended sampling unit for *F. tritici* was a terminal bud, which needed dissection and washing to remove thrips. For both pests, border rows were avoided and samples taken between 09.00 and 11.30 h before thrips became too active, from plants in the vegetative stage chosen at random.

By plotting the mean sample size ($\bar{x}$) against the relative variation (se/$\bar{x}$), based on Taylor's (1961) power law, it was possible to calculate the number of samples required to achieve a desired level of accuracy (Fig. 11.1). For example, to achieve an accuracy of 30 per cent of the mean when each sampling unit contained an average of ten thrips, five samples would be required. This graph generated from data gathered for *N. variabilis* with a single leaflet as the sampling unit, and for *F. tritici* with a single terminal bud as the unit, was applicable to both species. The study illustrates the importance of choosing appropriate plant parts as sample units and serves as a guide to the approach required in other crops such as cotton, groundnuts and tobacco which are often infested by several species, each preferring different sites on their hosts (see Appendix II). Sometimes a single thrips species may occupy different niches on different host plants; for example, adults and larvae of *Frankliniella occidentalis* prefer the top third of sweet peppers (Shipp and Zariffa, 1991) but the middle leaves of cucumber (Steiner, 1990), emphasizing the need to devise sampling methods on a crop by crop basis.

Salifu and Hodgson (1987) similarly studied within-plant dispersion in *Megalurothrips sjostedti* infesting cowpeas, using Iwao's (1968) regression procedure and Taylor's (1961) power law analysis. Sample units were five flower buds or, at a later stage of development, five open flowers, taken from

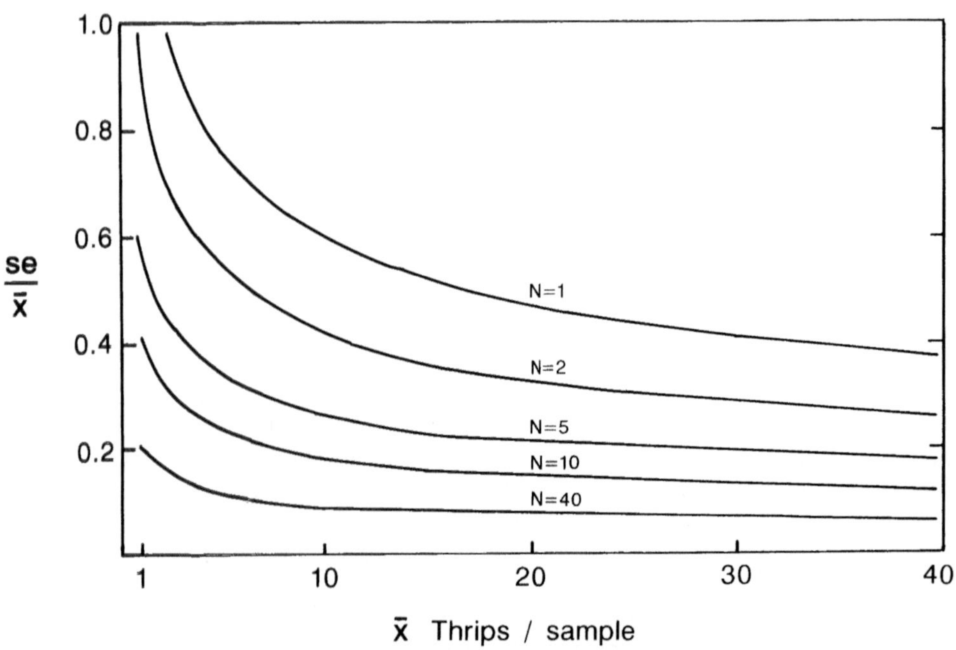

Fig. 11.1. Relationship between number of samples (*N*), thrips per sample and accuracy of the estimate of numbers per sample for thrips on soybean (after Irwin *et al.*, 1979).

Table 11.1. Cost, mean relative variation and relative net precision of different methods of sampling *Megalurothrips sjostedti* on cowpeas (Salifu and Singh, 1987).

Sampling method	Cost	RV	RNP
Absolute standard			
Cut, bag, wash	0.20	8.6	58.3
Relative counts			
Collection in ethanol	0.11	25.4	35.7
Shaking	0.03	9.1	367.5
Sweeping	0.07	28.4	50.2

Cost = mean no. person-hours per sample; RV = mean relative variation = (se/$\bar{x}$)(100); RNP = relative net precision = 100/(RV × cost).

five randomly chosen plants, dropped into vials containing 30 per cent ethanol for laboratory counting. At initial low populations both methods showed that adults were randomly distributed within cowpea plants, but at greater densities later in the season both adults and larvae were aggregated. Using Taylor's power law, sequential count plans were produced to show the number of samples necessary to give a specified level of sampling precision (Taylor, 1984). Similar use of sequential procedures (Binns and Nyrop, 1992) could help lower the cost of sampling in other uniform monocultures, but each crop/pest situation would require detailed quantitative study not usually available to commercial growers.

A revealing comparison of the efficiency of several sampling methods for *M. sjostedti* on cowpeas was made by Salifu and Singh (1987). Three non-trapping methods were compared: (i) collecting directly into ethanol one raceme or flower per plant from ten plants for later dissection and counting; (ii) shaking (tapping) plants five times over gridded wooden boards covered with sticky polythene and counting dislodged thrips *in situ*; and (iii) sweeping along a defined route with a 30 cm diameter oval nylon net, placing the catch in paper-lined polythene containers and killing it with ethyl acetate prior to laboratory sorting and counting. The efficiency of each method was judged on the basis of low coefficient of variation (se/$\bar{x}$)(100), cost (mean number of person-hours to collect and process each sample), and its similarity to absolute counts. This was obtained by cutting and bagging ten whole plants and carefully washing all thrips from racemes and flowers in 30 per cent ethanol and from the remaining parts of the plants using soapy water. Shaking proved the least expensive, the most precise, and was most closely correlated with absolute counts, so it is recommended for the rapid estimation of thrips populations on cowpea (Table 11.1). Nevertheless, it should be noted that collection and washing was the only method that satisfactorily sampled larvae.

Similarly, Hosny (1964) found that shaking cotton plants provided estimates of *Thrips tabaci* populations as accurate as the removal and washing of whole plants, providing 20 to 30 plants were examined; Powell and Landis

(1965) sampled *T. tabaci* on potatoes by tapping single plants five times to dislodge the thrips.

Among the many sampling schemes devised for *T. tabaci* in onion fields (e.g. Suman and Wahi, 1981; Edelsen, 1985), Shelton *et al.* (1987) developed a sequential sampling plan, having first established from absolute counts on three-plant samples that onion thrips were randomly distributed throughout fields during the main period of infestation. Immigrant thrips infest plants at different growth stages (two, four, six, eight or ten leaves), so sampling to indicate whether control measures are required needs to take plant size into consideration. Heavy infestations of about 20 thrips per plant kill seedling onions, whereas similar numbers on mature plants are harmless, or even beneficial, killing the leaves thereby encouraging curing of the onion bulb. In practice, 15 to 20 plants were examined *in situ* weekly in 50 fields, at three to ten sites along a V-shaped transect through each field, to record the number of plants on which the damage threshold of three or more onion thrips was exceeded. Decisions on whether to control the pest were reached on 88 per cent of occasions after inspecting only 15 plants, which was far less labour-intensive, but just as reliable, as the more conventional procedure of taking fixed-size samples of either 50 or 100 plants.

Fournier *et al.* (1995) developed a binomal sequential sampling procedure of similar efficiency based on the presence of five or more *T. tabaci* per onion plant. Acceptance boundaries were calculated for economic thresholds of either 0.9 thrips per leaf (appropriate in seasons with severe drought) or 2.2 thrips per leaf (for dry summers, but no drought). Validation of the plans in fields infested at levels of 0.01 to 32 thrips per leaf showed that correct decisions on whether control would be necessary were reached in 83 per cent of instances by sampling just 12 to 14 plants. This approach was as precise as Iwao's (1968) sequential sampling method that requires the counting of all thrips present in samples taken.

The other significant onion pest, *Frankliniella occidentalis*, has been sampled satisfactorily to establish a no-action threshold by counting thrips on five plants *in situ* (Bender and Morrison, 1989) or by removal of all leaves from single plants, washing and counting the thrips (Dintenfass *et al.*, 1987).

Among other field vegetables, a precision monitoring scheme has been devised for *Frankliniella occidentalis* in tomato flowers based on their presence/absence in a flower, the only limitation being that the proportion of flowers containing thrips be greater than 0.8 (Funderburk, 1993). Most adults of this species and *F. fusca* on tomatoes inhabit flowers on the upper half of the plant, but larvae are more abundant on lower flowers (Salguero Navas *et al.*, 1991), so samples need collecting with this in mind; 20 randomly selected flowers per plant seems a suitable sample size.

The time of day when samples were taken appeared to have no effect on numbers of these species on the tomato flowers, in contrast to *F. fusca* in peanut flowers where numbers rise from 08.00 h to peak between 11.00 and 12.00 h (Tappan, 1986), diurnal activity rhythms clearly influencing the outcome. Species that jump from foliage when disturbed are best sampled at cool times of the day when they are least active.

On field cucumbers infested with *Thrips palmi* and *Frankliniella occidentalis*, Rosenheim *et al.* (1990) sampled the former on foliage by removing the fifth leaf below the tip of ten vines, cutting each in half along the midrib and washing thrips from the ten half leaves; for *F. occidentalis*, samples comprised 12 flowers (10 male and 2 female). The density of *T. palmi* per unit area of plant substrate was greatest on leaves in contrast to *F. occidentalis* which favoured flowers, eight individuals per 200 cm^2 of flower surface being sufficient to increase fruit scarring by 10 per cent. For *F. occidentalis* on squash, visual inspection of single, medium-sized leaves from ten randomly selected plants proved satisfactory (Letourneau and Altieri, 1983).

On small trees and bushes, beating and shaking have been widely used (Kratochvil and Farsky, 1942; MacPhee, 1953; Putman, 1965), but it is difficult to obtain absolute estimates in this way unless 'units' of the tree, such as an 'armful' of branches, are beaten so that the proportion of the whole tree included in the sample can be estimated, and from it the total population computed. To overcome uneven distribution on trees, units representative of different heights and aspects should be taken. To estimate the numbers of *Thrips laricivorus* on young spruce trees, Vité (1956) removed enough twigs to fill a 20-litre beaker from different heights of each tree, then estimated the proportion of each section sampled that was represented by the twigs in the beaker, and computed the numbers on the trees. MacPhee (1953) tried to collect the total population on a small apple tree by suspending a paper cone under it and applying a quick-action toxic dust to foliage and branches with a small power-duster, but this method was tedious and many thrips remained in crevices.

In an attempt to estimate populations of *Taeniothrips inconsequens* on a single, complete sugar maple tree, Teulon *et al.* (1992) tried to count the number of leaves by covering it with a tubular net and collecting them as they were shed in autumn; the intention was to use the data to convert numbers per leaf in spring to numbers per unit area of forest. Another approach to estimating numbers on complete trees, used by Tanigoshi and Moreno (1981) in citrus and by Laughlin (1970) in eucalyptus, was to catch larvae as they fell to pupate in the soil on sticky sheets or in funnels placed beneath trees. Densities of *Taeniothrips inconsequens* larvae falling from sugar maple trees can exceed 30,000 m^{-2} in some years (D. Teulon, Pennsylvania, 1993, personal communication).

The efficiency and accuracy of shaking apple blossom to sample *Frankliniella occidentalis* was studied by Terry and De Grandi-Hoffman (1988). For a 'shake' sample, thrips were dislodged from a selected cluster by flicking it vigorously against the side of an enamel pan for either 3, 6 or 9 s. The accuracy of this procedure for each of three blossom cluster types (pink, open and petalless) was determined by comparison with counts from 'extraction' samples obtained by removing whole clusters from trees and bagging them for later dissection and washing in ethanol. Thrips extraction from combined clusters showed that shaking for 9 and 6 s removed 84 and 74 per cent of thrips, respectively, but a 3 s shake only removed 53 per cent and was more variable. Clusters from the top of tree canopies contained more thrips than lower clusters, and apical clusters more than basal ones during peak bloom. Variance component analysis showed that thrips counts from within-tree clusters were more variable than counts among trees. The degree of spatial aggregation for thrips on each cluster type was established by plotting regressions of mean crowding density (Iwao, 1968) and log variance on log mean (Taylor, 1961) to indicate the optimal number of clusters per tree and the number of trees per plot required to achieve a chosen level of accuracy. This careful approach deserves consideration by researchers concerned to obtain effective sampling with least input. Examples of convenient sample units for a range of crops are listed in Table 11.2.

Quantitative sampling of glasshouse crops

Direct sampling for thrips on high-value ornamentals grown in glasshouses is not practical because by the time symptoms are detected the plants or blooms are probably no longer marketable. Instead, most growers use attractive sticky traps (see p. 453) which can detect sparse populations with less expenditure of effort than is required for routine plant inspection. Glasshouse-grown fruit and vegetables are also commonly monitored for thrips infestation by sticky traps, but leaf and flower sampling is nevertheless useful for the more serious pests.

For *Frankliniella occidentalis* on sweet peppers, Shipp and Zariffa (1991) developed efficient sampling using Taylor's (1961) power law to define the spatial distribution of the pest which was aggregated on and between plants, the majority of thrips occurring on the top third of each plant. Four sampling methods, plant tapping, blossom and leaf removal and blue sticky traps were compared with absolute counts obtained by washing thrips from whole plants in the laboratory. The population trends of adults detected by tapping and blossom removal (1 to 14 blossoms per plant) had the best correlation with whole plant counts (Fig. 11.2). Leaf samples (each comprising three upper, three middle and three lower leaves) and sticky traps, also adequately though less reliably, represented changes in adult population throughout the season. For immature thrips, leaf and plant tapping accurately reflected whole plant counts, but blossom sampling was unreliable and sticky traps caught none. When leaves were used as the sample unit, population trends of adults were better reflected and fewer samples were required if they were taken from the mid-section of the plant rather than from upper or lower sections. As in examples cited above for field crops, a precision sampling programme was devised for adults to indicate the number of samples required for a given level of accuracy (Table 11.3). For larvae, many more samples are needed to assess population density to a given precision, so larval monitoring is probably impractical. Sampling flowers for adults was the most cost-effective method based upon time and number of samples required at the different precision levels.

Steiner (1990) similarly developed a monitoring programme for *F. occidentalis* on cucumber, but based upon presence/absence of thrips on the middle leaves. For a given precision level and mean number of thrips per leaf, the number of leaves to be inspected for presence of thrips was calculated. The general recommendation was to examine 50 leaves per 2000 m^2 of glasshouse

Table 11.2. Examples of convenient sample units and sizes when vegetation is removed for examination.

Crop	Thrips species	Sample unit	Units/sample	References
Apple blossom	*Frankliniella occidentalis*	1 Blossom cluster	8–10	Terry and De Grandi-Hoffman (1988)
Cereals	*Limothrips* spp. *Haplothrips* spp.	1 tiller, ear or panicle	25 or 50	Koppa (1967)
Cocoa	*Selenothrips rubrocinctus*	2 mature leaves of same age and size, or 1 flush (about 6 young leaves)	1	Fennah (1955)
Cotton	*F. occidentalis*	1 whole plant (early season) 1 fruit terminal (late season)	10	Faulkner (1954)
Cowpea	*Megalurothrips sjostedti*	1 raceme or 1 flower	10	Salifu and Singh (1987)
Cucumber	*Thrips palmi*	½ leaf	10	Rosenheim *et al.* (1990)
	F. occidentalis	1 flower	12	Rosenheim *et al.* (1990)
Lettuce	*F. occidentalis*	1 leaf	2 young, 2 medium, 2 old	Faulkner (1954)
Lucerne	*F. occidentalis* *Thrips tabaci*	Terminal 15 cm of shoot	10	Faulkner (1954)
Onions	*T. tabaci* *F. occidentalis*	Plant cut to ground level	10	Faulkner (1954) Hoerner (1947) Shirck (1948) Dintenfass *et al.* (1987)
Pyrethrum	*Thrips nigropilosus*	1 leaf	20–49	Bullock (1965)
Rose blooms	*Frankliniella tritici*	1 flower	5	Ota (1968)
Soybean	*Neohydatothrips variabilis*	1 fully expanded trifoliate	3–6	Irwin *et al.* (1979)
	F. tritici	1 terminal bud	3–6	Irwin *et al.* (1979)
Squash	*F. occidentalis*	1 medium-sized leaf	10	Letourneau and Altieri (1983)

continued overleaf

Table 11.2. Continued.

Crop	Thrips species	Sample unit	Units/sample	References
Sweet pepper	*F. occidentalis*	1 leaf	3 upper, 3 middle, 3 lower	Shipp and Zariffa (1991)
		1 blossom	1–14	
Tobacco	*Frankliniella* spp.	All blooms per plant	10	McPherson *et al.* (1992)
Tomato	*Frankliniella* spp.	1 flower	20	Salguero Navas *et al.* (1991)
Watermelon	*T. palmi*	$\frac{1}{2}$ leaf > 8 cm wide	4	Johnson (1986)
Wheat	*Limothrips cerealium*	1 tiller (before flowering)	10	Lewis (1959a)
		1 ear (after flowering)	10	

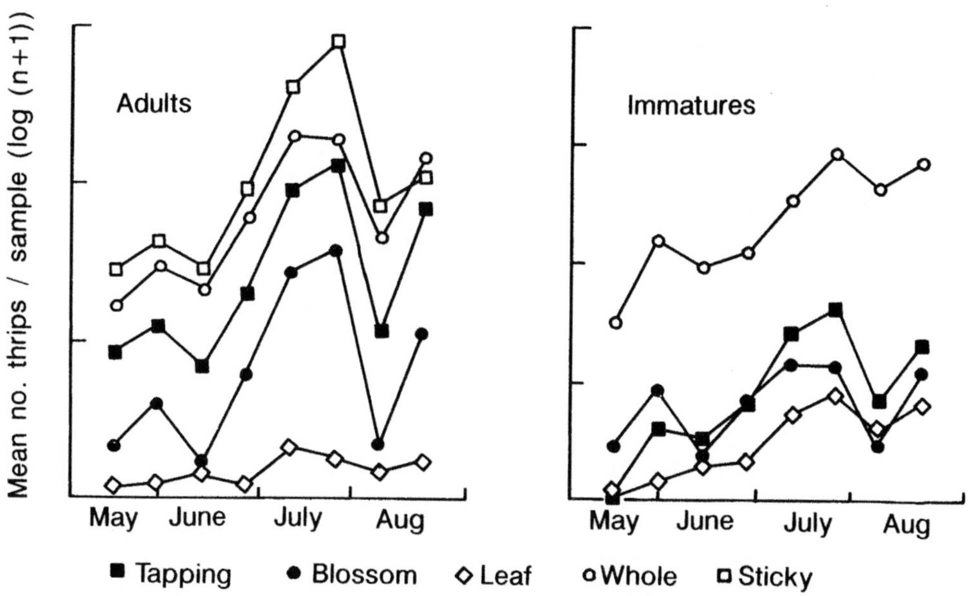

Fig. 11.2. Population trends for adult and immature *Frankliniella occidentalis* on sweet peppers as determined by five sampling methods in a greenhouse (after Shipp and Zariffa, 1991).

at low thrips densities and 25 leaves at high densities. Fifty randomly chosen leaves can be checked in less than an hour.

Extraction of thrips from plant samples

Thrips can be removed from collected vegetation by washing or by one of several dry extraction methods.

Washing

For smooth leaves like coffee, dipping in 70 per cent ethanol is effective (Le Pelley, 1942). For more hairy or wrinkled leaves, blossom or buds, more vigorous shaking in ethanol or detergent solution is necessary after dissection of any compact layers of tissue. For example, Bullock (1963) dislodged thrips from the pinnatisect leaves of pyrethrum by shaking them in jars of 70 per cent ethanol, then separated the coarse vegetation by wet sieving. The thrips and fine debris in the washings were next shaken in a beaker with a little benzene, the beaker placed in a dish and a jet of water introduced below the liquid interface so that thrips which had collected there overflowed into the dish for counting; a metal grid placed in the dish was helpful. Modifications of this procedure have been devised for other crops. For soybean, leaflets collected in the field directly into bottles of detergent solution were subsequently shaken vigorously and the liquid poured through a 100×100 mesh stainless-steel sieve (40 holes cm^{-1}); insects and debris were then washed into vials with 70 per cent ethanol for storage. Alternatively, leaves and stems collected in the field into plastic bags were agitated in 95 per cent ethanol and, after rinsing the bags, the

Table 11.3. Number of samples required for four sampling methods at three precision levels (0.1, 0.15, 0.2) in a 0.2-ha greenhouse to monitor adult *Frankliniella occidentalis* at selected population densities on sweet pepper (Shipp and Zariffa, 1991).

Thrips population density[a]	Leaves[b]			Flowers[c]			Plant tapping			Sticky traps		
	0.1	0.15	0.2	0.1	0.15	0.2	0.1	0.15	0.2	0.1	0.15	0.2
1	840	373	210	203	91	51	56	25	14	51	23	13
5	420	187	105	93	42	23	45	20	12	36	16	9
10	312	139	78	67	30	17	41	19	11	31	14	8
25	210	94	53	43	19	11	36	16	9	25	12	7
50	156	70	39	31	14	8	33	15	9	22	10	6
100	116	52	29	22	10	6	30	14	8	19	9	5
200	86	39	22	16	7	4	27	12	7	16	7	4
300	72	32	18	13	6	4	26	12	7	15	7	4

[a]Expressed on a whole-plant basis.
[b]Based on a mean of three middle-section leaves per plant.
[c]Based on a mean of all flowers per plant (= 1, . . ., 14).

combined washings passed through a 200 × 200 mesh sieve (80 holes cm^{-1}). Insects and debris were washed into a beaker with about 50 ml of water and transferred by vacuum filtration in a Buchner funnel on to a medium grade filter paper with a pencilled grid to assist counting. Buds and flowers needed prior dissection under a microscope to tease out thrips before counting (Irwin *et al.*, 1979). To remove thrips from onion plants, leaves were dissected and washed in ethanol, then the washings passed through a separation funnel to be collected on a gauze cloth from which they were washed into a beaker, a layer of benzene was added and methylene blue dye injected into the lower water layer to improve viewing contrast. Similar washing and sieving procedures are recorded for removing thrips from many crops and flowers including cucumbers (Welter *et al.*, 1990), apple blossom (De Grandi-Hoffman *et al.*, 1988) and roses (Ota, 1968).

For whole plants, leaves and fruit of sweet peppers, Shipp and Zariffa (1991) preferred washing in detergent (0.1% Triton) to ethanol and this too is widely practised and satisfactory.

Dynamic extraction

For separating adults and larvae living in crevices within leaf sheaths or deep inside buds or compact inflorescences, dry, dynamic methods are often effective and less labour intensive than dissection and washing. Simple heated funnels (Fig. 11.3i) placed in an incubator at 43 to 52°C (Hoerner, 1947; Shirck, 1948) or a simple Berlèse funnel with a 40 W bulb (Shelton and North, 1986) work well. Condensation on the inner walls of containers can be prevented by a fine gauze cover over the top of any assembly. Most thrips usually emerge within 8 h, though temperatures need to be adjusted by trial and error for different types of vegetation. Extraction in enclosed funnels using repellents, either turpentine, ethyl acetate, methyl isobutyl ketone or dimethyltoluamide, is much quicker than heating (0.5 to 1 h) and the equipment simpler (Fig. 11.3ii) (Evans, 1933; Lewis, 1960; Patrzich, 1984). Larvae are only partially removed and the immobile pupal stages not at all; they can only be recovered by washing.

Qualitative sampling of vegetation

Qualitative sampling of crops merely to detect the species present requires less statistical precision but nevertheless should be undertaken with care.

Sweep netting

Apart from collecting from individual plants by shaking them over a board or tray, the simplest method is to sweep the crop using a tapering net with a D-shaped opening approximately 30 cm in diameter. The fabric should be smooth, white linen or cotton from which thrips are easily removed. Vegetation should be dry and sweep strokes made with uniform force. The composition of

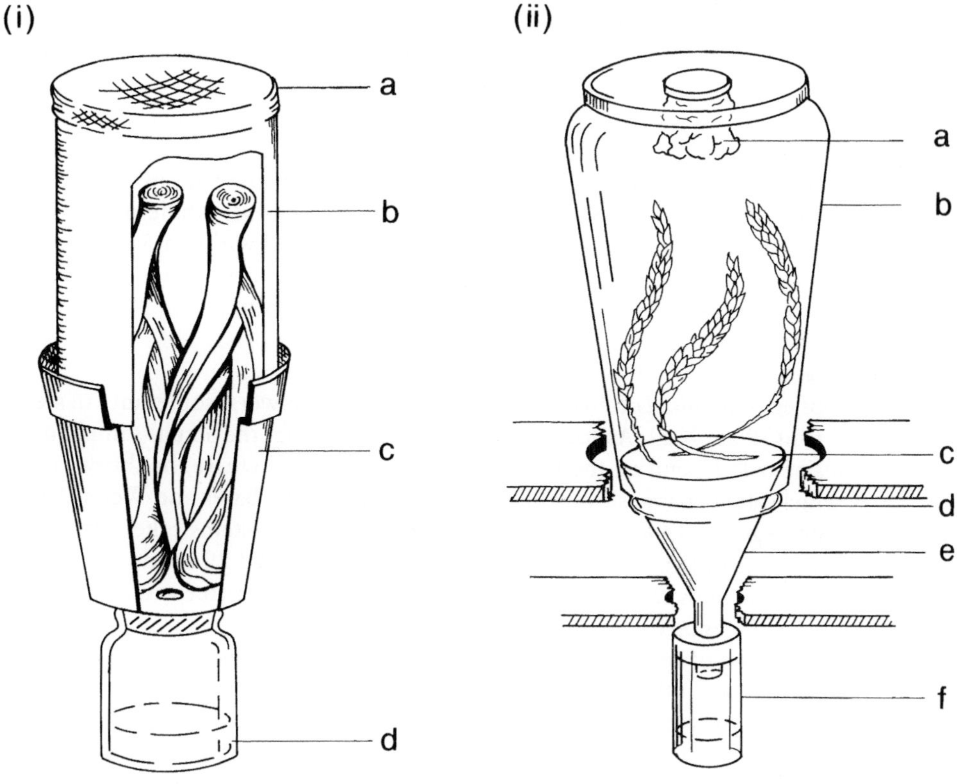

Fig. 11.3. (i) Heat extraction equipment (after Shirck, 1948); (a) muslin cover, (b) cardboard cylinder, (c) plant pot, (d) 5% formalin. (ii) Turpentine vapour extractor (after Lewis, 1960); (a) cotton wool soaked in turpentine, (b) lamp glass, (c) zinc gauze, (d) wax seal, (e) funnel, (f) detergent solution.

catches is influenced by the evenness of the crop canopy, differences in vertical distribution on plants of the thrips species present and the time of day. For most species, sampling is best done when it is warm enough for thrips to be exposed on their hosts. Usually 10 to 50 sweeps suffice for a sample taken while walking along a transect through the crop (Holtmann, 1962; Cederholm, 1963; Koppa, 1967; Salifu and Singh, 1987). Sweeping is not recommended for quantitative studies because of the large variability between samples; Gray and Treloar (1933) calculated that 36 and 897 samples respectively, each of 25 sweeps, were required to restrict the variation of samples of *Anaphothrips obscurus* taken from lucerne to 50 and 10 per cent of the mean, an impractical amount of effort. In a comparison of sampling methods for *Megalurothrips sjostedti* on cowpea, sweeping, taking ten sweeps per sample, proved about twice as expensive and three times more variable than shaking plants just five times over a board (Salifu and Singh, 1987).

D-vac suction samples have some virtue for sampling short vegetation (Johnson *et al.*, 1957) and small citrus trees (Grout *et al.*, 1986). In a comparison of suction sampling and sweeping in cereals and Timothy leys, nets caught about 18 to 25 times more thrips per unit area than suction samplers (Heikinheimo and Raatikainen, 1962) but neither is likely to collect individuals concealed in plant crevices. Perhaps the simplest and quickest method of detection recorded is to brush a moistened palm of the hand across rice leaves, an approach reported by Chiang (1977) to check for *Stenchaetothrips biformis*.

For monitoring faunal composition and phenology of thrips in wheat, oats, lucerne and red clover, Shelton and North (1986) cut all whole plants 1 cm below ground level from four 320 cm^2 patches located within a 5-m radius of a sticky trap that was used to detect complementary aerial movement; vegetation was washed to remove thrips in the laboratory.

Egg sampling

Thrips eggs are most often embedded in leaves and flowers, sometimes in stems and fruit (Plate X.5). As for adult and larval sampling, the actual sample units and size will depend on the form of the host plant and where the thrips lay on it, which often changes over the season as plants grow. Bates and Weiss (1991) counted eggs of *Limothrips denticornis* per ten barley stems taken from two 0.5-m lengths of row, and Leskey *et al.* (1995) examined 20 randomly chosen sugar maple buds taken from the upper third of the outer tree canopy for eggs of *Taeniothrips inconsequens*.

Detection of eggs for counting may require some processing of the collected plant material. In thin leaves and petals they can usually be seen as translucent swellings when viewed against light, either with the naked eye (Lewis, 1973) or under a binocular microscope (Terry, 1991). Thicker tissue needs treating to reveal them. Hall (1930) removed chlorophyll from citrus leaves by immersing them in a series of 50 to 100 per cent ethanol baths then clearing in clove oil. For barley leaves and sheaths (Bates and Weiss, 1991) and bean flowers (Kirk, 1985), 1 to 3 days soaking of tissue in 70 per cent ethanol before microscopic examination sufficed; Laughlin (1970) similarly pickled eucalyptus flowers in alcoholic Bouin's fluid. For detection of eggs of *Taeniothrips inconsequens*, which on sugar maple leaves prefers to lay on convex surfaces of veins, Teulon and Cameron (1991, 1995) and Leskey *et al.* (1995) stained them with acid fuchsin (Backus *et al.*, 1988) (0.2% acid fuchsin in 95% ethanol and glacial acetic acid (1:1, vol:vol), followed by clearing in distilled water, 99% glycerine and 85% lactic acid (1:1:1, vol:vol:vol). On thin-skinned fruit such as tomato (Salguero Navas *et al.*, 1991) and grapes (Jensen, 1973), egg blemishes often appear as indentations surrounded by a faint paler halo, visible to the naked eye.

Soil and litter

Many pest species overwinter in soil and surface litter and during summer others descend as larvae into the soil beneath crops to pupate. The sampling and extraction of these soil dwellers can be extremely tedious and labour intensive, so care should be taken to choose appropriate methods before beginning any extensive survey.

Any living surface vegetation should be cleared before sampling the soil with a tool capable of penetrating as deep as the resting thrips, but not so large as to produce an unwieldy volume of soil. Where populations are sparse, a 10-cm diameter soil corer may be necessary (Bullock, 1964; Edwards and Fletcher, 1970) but where populations are dense, a smaller, more convenient corer should suffice. The depth of samples required depends on the species and season; overwintering larval *Stenothrips graminum* may penetrate to 1 m (Franssen and Mantel, 1965), *Thrips angusticeps* to 70 cm (Franssen and Huisman, 1958) and *Kakothrips pisivorus* to 30 cm (Buhl, 1937), but *Thrips palmi* only to 6 cm to pupate in summer (see Chapter 7). Loose cultivated soil allows deeper penetration than undisturbed ground. Large corers compress the soil least, so thrips can more easily emerge from them if live-extraction methods are subsequently used. For shallow-burrowers (to about 10 cm) in firm ground, such as *Taeniothrips inconsequens* in New England forest soils (Skinner and Parker, 1992), a smaller tulip-bulb planter with sloping sides (approximately 7 cm in diameter, collecting 250 to 300 cm^{-3}) seems ideal, cheap and readily available (Skinner and Parker, 1991). In North Dakota, roadside and shelter-belt sods were sampled, particularly for *Limothrips denticornis, Anaphothrips* sp. and *Frankliniella* sp., by taking soil blocks approximately 400 cm^2 to a depth of 7.5 cm (Post and Thomasson, 1966).

Population densities in soil range from tens to tens of thousands of individuals per cubic metre. Larvae of most species enter the soil beneath the host plant but, under trees especially, spatial distribution may be very patchy, requiring extensive sampling to give a realistic estimate of populations. Skinner and Parker (1991) found the number of *Taeniothrips inconsequens* in soil beneath sugar maples to be very variable from tree to tree and from sample to sample around a single tree, unattributable to any obvious differences in soil or covering vegetation; for standardization, they took samples 2 and 4 m from the south side of tree trunks. At mean densities of five to ten thrips per sample, the population within a stand could be estimated from ten samples with a standard error of $\pm$ 6.5 individuals. Lewis (1973) provided a simple formula for calculation of the number of cores to be bulked to provide a chosen degree of accuracy, especially where populations were sparse. Assuming that each core contains at least one thrips and an error of 10 per cent of the mean is satisfactory, the number of units required should be obtained from preliminary sampling designed to give a rough estimate of the mean and standard deviation(s) of units. These values can then be inserted in the formula:

$$n = \frac{100s^2}{m^2}$$

where n = the number of sample units bulked in each sample, m = the sample average, s = the standard deviation, and an appropriate value for n chosen. For example, when $m = 5$ and $s = 4$

$$n = \frac{1600}{25} = 64$$

The depth to which thrips descend or concentrate in soil can be measured by taking a series of successive cores to different depths. A more accurate approach is to excavate a pit and into one of its vertical faces insert strips of sheet metal to retain the soil as shallow, horizontal slices are removed at 2-cm intervals with a putty knife and placed in bags for later extraction (Fig. 11.4) (Skinner and Parker, 1992). Yet another method, which avoids subsequent separation of the thrips, is to use emergence boxes with transparent plastic bottoms spread on the inside with sticky grease. These are placed upside down over an infested soil surface exposed to different depths; the emerging thrips ascend towards the light (Harrison, 1963). In all studies on vertical distributions the depths of sampling may need modifying during a season as thrips descend on entry and ascend as emergence time approaches.

Thrips can be separated from soil samples by washing and removal of organic matter by flotation using magnesium sulphate (specific gravity = 1.2), a procedure fully described and illustrated by Lewis (1973); thereafter the organic matter is agitated vigorously in a mixture of water and xylene or heptane (11:1), and on settling the thrips picked from the xylene/water interface. Parker *et al.* (1992) devised a modification whereby thrips are separated by washing soil with water through a series of sieves, the lower one with a 250-μm mesh. The residue on this sieve is rinsed into a 1-litre bottle and then transferred to a 2-litre separator funnel. Water is added to a total volume of 550 ml and 50 ml of heptane is added. After swirling and allowing a few minutes for water and detritus to settle, these are drained and discarded and the layer of heptane plus a few millilitres of remaining water is poured onto a disposable prefolded coffee filter over a Buchner funnel attached to a suction pump. The thrips retained on the moist filter are easily counted over a gridded stage under a binocular microscope at × 8 magnification. For *Taeniothrips inconsequens*, recovery rates by this method were 59 per cent (adults) and 84 per cent (larvae) compared with only 14 per cent and 39 per cent, respectively, using magnesium sulphate. This

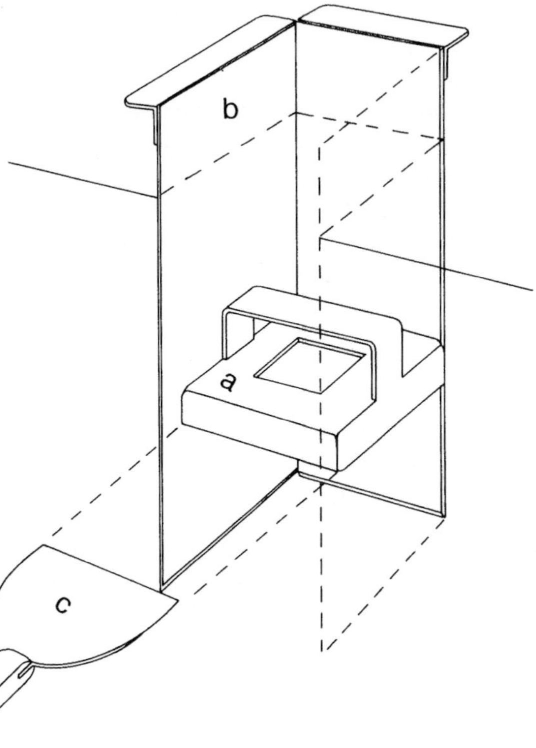

Fig. 11.4. Pit sampling equipment. (a) Sample cutting tool, (b) sheet metal retaining upper soil levels, (c) putty knife used to remove sample (after Skinner and Parker, 1992).

procedure, including counting, took about an hour to process each sample, but it is too complex for use by growers.

An easier method giving even better recovery involved placing individual soil cores in paper food containers (9 cm top diameter, 7.5 cm bottom diameter, 10 cm tall) covered with a clear sticky lid held in place with another translucent plastic lid and kept at 25°C with 12 h of light daily for 35 days. After 25 days 98 per cent of adult thrips emerged (Parker *et al.*, 1995), which is a great improvement on any of the conventional dry funnel methods compared by Edwards and Fletcher (1970).

Another approach to sampling terrestrial populations is to place an emergence trap on the ground surface and await the natural appearance of the thrips beneath. A great variety of traps have been devised to do this, each consisting, essentially, of a wooden, aluminium, plastic or tightly-woven muslin cone or pyramid with dark walls and a glass vial(s) inserted at the top or in the sides, in which the thrips collect as they move towards the light. Using this approach, successful measurement of the onset and duration of emergence has been achieved, for example for *Taeniothrips inconsequens* (Cameron and Treherne, 1918; Smith, 1933; Bailey, 1944; Laudermilch, 1989; Maier, 1991) and for *Thrips calcaratus* (Raffa *et al.*, 1992), although the quantitative reliability of the various devices is rarely recorded. A simpler trap comprising a PVC cylinder placed on the soil and covered with a clear lid coated with sticky grease has proved effective and cheaper than cone traps for monitoring *Taeniothrips inconsequens*. From a range of sizes compared, cylinders with a diameter of 7.6 cm were most satisfactory (Parker and Skinner, 1993). Similar PVC cylinders, approximately 20 cm in diameter and 10 cm tall, were adapted for monitoring *Scirtothrips citri* in California and Arizona. Clear acetate discs with sticky grease on both sides were placed on the upper end of cylinders, thereby catching larvae falling from trees to pupate in the soil and, in different positions, adults emerging from the surface soil and litter. The acetate discs were collected in separate, clear vinyl folders for ease of handling, storage and counting (Tanigoshi and Moreno, 1981). These simple approaches are less cumbersome than the earlier-used cones on emergence boxes (Reed and Rich, 1975), cheaper to construct and quicker to use; all important features because usually many traps are required for ground surveys to overcome the patchy distribution of thrips in soil.

Surface litter and crop stubble, especially near the edges of woodland or in hedgerows, also provide good cover for overwintering adults of cereal and grass feeders (Lewis and Navas, 1962; Wetzel, 1963; Koppa, 1969). Samples of the dead vegetation should be scraped loose to soil level and collected from within a grid of 1000 to 10,000 cm^2, depending on the density of the population. Alternatively, samples of dry litter may be collected by weight; 100 g of dead vegetation is often a convenient amount. Thrips are easily trapped in condensation and killed, so samples should be stored in finely woven fabric bags, or if in plastic bags, kept cool (below 5°C) to inactivate the insects and prevent them walking into moisture on the walls. Extraction of thrips from dry litter is best achieved by a dry method; exposure of samples to turpentine vapour (Lewis and Navas, 1962) or methyl isobutyl ketone (Post, 1947) in a suitable funnel stimulates emergence so that they fall into a collecting bottle containing alcohol placed beneath the funnel (Lewis, 1960). For quantitative studies, the efficiency of extraction for whatever procedure is adopted will need calibration by careful dissection of treated litter to ascertain the percentage of thrips remaining in it.

Airborne populations

Trapping flying thrips above crops is a popular and relatively easy means of monitoring the early presence, population build-up, and seasonal changes in activity of pest species, though catches need to be interpreted with considerable caution. Traps are also used to determine the need for, and timing of, control measures and to assess their effects. In the open field, both sticky and water traps have been widely used; in glasshouses, sticky traps are preferred. The different colour and chemical preferences shown by many species, described in Chapter 3, have been exploited to increase the attractiveness, and thus the sensitivity, of these traps and a profusion of papers on trap design, colour and procedure have been published, most of which do not attempt to relate catches to absolute aerial densities or to infestation levels on surrounding crops (e.g. Lewis, 1959b; Moffit, 1964; Beckham, 1969; Beavers *et al.*, 1971; Walker,

1974; Kirk, 1984; Czenz, 1987; Yudin *et al.*, 1987; Brødsgaard, 1989; Gillespie and Vernon, 1990; Teulon and Penman, 1992).

Sticky traps

Flat sticky surfaces in horizontal and vertical positions are commonly used, but in open fields cylindrical surfaces are more efficient because the airflow around is less turbulent and because they also catch insects blown from all directions. For monitoring, they are best placed within a metre of crop level; they can also be exposed at different heights to estimate relative density profiles, providing catches are corrected for relative wind speeds above the crop (Lewis, 1959b). Failure to do this will produce misleading information and probably indicate spurious specific preferences for height of flight. Because of the edge effects described in Chapter 5, traps should not be placed near the borders of fields or near shelter belts. In glasshouses, where there is less air movement, flat sticky cards hung vertically just above the crop are satisfactory (Shipp, 1995) and easier and cheaper to inspect than cylindrical versions. Sticky tapes are also sometimes used in attempts to control *Franklinella occidentalis* (Shipp, 1995) and *Thrips palmi* (Kawai and Kitamura, 1987).

The size of sticky trap depends on the expected density of the population and the frequency of inspection. Over cereals, where thrips are often abundant, a sticky surface of 100 cm^2 suffices for daily inspection, but over crops where populations are less dense, larger traps are needed. Cylinders with a surface area of about 100 cm^2 have been used over cowpeas (Salifu and Singh, 1987), 250 cm^2 around onion fields (Wolfenbarger and Hibbs, 1958) and 650 cm^2 around cotton fields (Hightower and Martin, 1956).

Outdoors, catches on horizontal traps up to about 500 cm^2 appear to increase on a linear log × log basis, as demonstrated for *Thrips imaginis* (Kirk, 1987), with a marked concentration at the edges of individual traps. In glasshouses, vertical sticky traps between 100 cm^2 and 1000 cm^2, or sticky tapes 5 cm by 60 cm or 120 cm are widely used against *Frankliniella occidentalis* (Higgins, 1992; Shipp, 1995). Cards are preferable for monitoring and can be increased if populations are sparse. There is a similar tendency for catches to be concentrated at the trap edges, so smaller cards are more efficient (Powell and Lindquist, 1992). Sticky tapes employed for mass trapping possibly help to limit 'hot-spots' of *Frankliniella occidentalis*, but simulation models on population control of *Thrips palmi* in cucumber houses suggested that they were only likely to be effective at low pest densities (less than 0.0004 females per leaf; one trap per 2.0 m^2) (see Chapter 18).

Trap shape can affect trap catch. In staked tomato fields, cylindrical traps and cup traps caught more *Frankliniella tritici* per square centimetre than flat traps, but not more *F. fusca* or *F. occidentalis* (Cho *et al.*, 1995). In citrus, more *Scirtothrips citri* were trapped on triangular and elliptical than on rectangular, spherical or square traps (all yellow and similar in surface area), but the differences in numbers ranged from 320 for circular traps to 403 for triangular ones (Moreno *et al.*, 1984). As many *Taeniothrips inconsequens* were trapped on vertical rectangles as on vertical cylinders, but more thrips were caught on traps with upward-facing rather than vertical surfaces (Coli *et al.*, 1992). In contrast, Vernon and Gillespie (1995) found that trap shape (linear, cubic, spherical, rectangular, cylinder, prism) was less important than colour and colour contrast in trap catches of *F. occidentalis*, especially when traps were corrected for area.

The use of transparent or pale brown polyisobutylenes as trap adhesives enables thrips to be recognized and identified most easily. If necessary, they can be removed for identification from the adhesive by dissolving it in a drop of petrol or dichloromethane. Castor oil has been used as an alternative adhesive (Diravian and Uthamasamy, 1992). When sampling total populations of easily identifiable species, glass boiling tubes about 20 cm long and 3 cm in diameter, or plastic conduit tubing, make convenient traps that can be taken to the laboratory for counting and replaced by freshly greased ones. A transparent polythene sheath or envelope covered in grease may be used over a more permanent coloured support. When catches are large, subsampling of the sticky surface is helpful. For example when glasshouse catches of *Frankliniella occidentalis* reach 500 to 1000 individuals per sticky card they become difficult to count and sex, so numbers on half of each card face can be counted and doubled (Higgins, 1992). Heinz *et al.* (1992) found that when small cards (14 by 11.5 cm) were used, it was sufficient to count a 2.3-cm wide strip on one side of each trap and multiply the numbers by five to give a good estimate of total catch.

Water traps

Water traps are useful for catching large numbers of thrips easily and for collecting flying individuals in a condition suitable for dissection to determine whether they contain eggs. Thrips required for this purpose should be removed as soon as possible from the water, and never left in it for more than 24 h. Traps should be about 6 cm deep with a surface area of 250 to 500 cm^2, and preferably round, with the water level about 2 cm below the rim. A few drops of detergent added to the water ensure that thrips sink and do not drift to the edges and escape; a drop of formaldehyde prevents algal and fungal growth. Thrips are easily removed with a pipette. Water traps cannot be used to estimate absolute densities and are unreliable for estimating relative densities at different heights because their efficiency is probably affected by different wind speeds at different heights and by the level of the water in the traps. They are best used at vegetation level, but not where leaves and debris can fall into them because these discolour the water. A coarse wire-netting guard may be necessary to prevent birds bathing in them. As with sticky traps, there appears to be a linear log × log relationship between catch and trap size (Ward, 1966).

Visual and chemical attractants

Polyphagous species that feed on different plant parts including flowers respond to yellows, blues and white with a low ultraviolet content, and these colours can be used to increase both sticky and water trap catches 10- to 100-fold (Lewis, 1961). Thrips with a specific floral host are often attracted to the colour of their host flowers (Kirk, 1984; Czenz, 1987); grass dwellers are less influenced by colour (Lewis, 1959b). Yellow and blue are most used to maximize catches of serious pests and thus enhance trap sensitivity (see Chapter 3).

For *Frankliniella occidentalis*, most growers use yellow because it is the best generally attractive colour for other pests, including whiteflies and aphids, of field, glasshouse, vegetable and ornamental crops and provides a good contrast against which to spot the thrips caught. Some blues catch more individuals than yellows (Brødsgaard, 1989; Mateus and Mexia, 1995; Vernon and Gillespie, 1995), but the proportion of the sexes caught does not consistently differ between these colours exposed in different crops. Competition from surrounding background and differing UV reflectance of trap pigments may account for discrepancies; on lettuce, Yudin *et al.* (1987) found white more attractive than yellow, blue or other colours; in glasshouse-grown roses, carnations and chrysanthemums, yellow and blue were more attractive than white and green (Robb, 1989) and in African violets, blue was the most attractive among white, yellow and a range of pinks and violets.

In field crops, Czencz (1987) showed that the dominant colour of a crop (flowering rape/yellow, peas/white, wheat/green, flax/blue and red clover/red) affected the number of 12 species caught in traps of similar colour exposed against these field backgrounds. Yellow, white and blue were generally preferred, but background colours increased or decreased attractiveness depending on the trap colour and thrips species.

Among other significant pests, blue has been used to catch selectively *Frankliniella intonsa* (Murai, 1988) and *Thrips tabaci* (B. Penzes, Budapest, 1995, personal communication); blue or white for *Thrips palmi* (Kitakata and Yoshida, 1982), white or yellow for *Thrips obscuratus* (Teulon and Penman, 1992) and *Scirtothrips citri* (Beavers *et al.*, 1971; Grout *et al.*, 1986); white for *Megalurothrips sjostedti* (Salifu and Singh, 1987) and *Thrips imaginis* (Kirk, 1987); and green for *F. fusca, F. tritici* and *Neohydatothrips variabilis* (Irwin *et al.*, 1979), and *Limothrips cerealium* (Lewis, 1959b). Traps of any colour will usually catch some individuals; if it is advantageous to increase sensitivity and maximize the catch of target species, a preferred colour should be used.

Another less common approach to increasing trap selectivity is to add chemical attractants to coloured water traps in field crops, and to paint them on sticky traps in glasshouses. The aim is to improve the detection of small initial infestations, and in glasshouses, to use the traps as a contributory control measure. For water traps, attractants can be added to the water or dispersed from a small vial (about 40 mm by 12 mm diameter with a 4 mm cottonwool wick) suspended above each trap with wire. Many organic compounds, particularly aldehydes, are attractive to flower-dwelling species whose catches are often increased 2 to 12-fold (Howlett, 1914; Morgan and Crumb, 1928; Kirk, 1985, 1987) (see Table 3.3). Anisaldehyde, benzaldehyde, cinnamaldehyde and salicylaldehyde are effective for different species including *Thrips tabaci*, *T. imaginis*, *T. fuscipennis*, *Frankliniella intonsa* and *F. occidentalis*. Catches of *T. obscuratus* can be increased more than 100-fold with ethyl nicotinate (Penman *et al.*, 1982).

The prospects of using chemical attractants to sample and assist management of some important pests were considered by Teulon *et al.* (1993). In trials in a grass field in New Zealand, white water traps baited with ethyl nicotinate caught 70 to 150 times more *T. obscuratus* males and females than unbaited traps; anisaldehyde and benzaldehyde were less effective, increasing catches only up to 35-fold. Catches of *T. tabaci* were 5 to 12 times greater in baited traps, but the grass-dwelling *Anaphothrips obscurus* and *Limothrips cerealium* were not attracted. When used for early spring monitoring in a nectarine orchard, white water traps plus ethyl nicotinate caught 27 times as many *T. obscuratus* as plain water. In Dutch trials, yellow water traps plus either of the same three attractants exposed on grass caught 5 to 35 times more *F. occidentalis*, *F. intonsa* and *T. fuscipennis*, but *T. tabaci* appeared not to be attracted. In glasshouses crops of sweet peppers, the effect of anisaldehyde painted on sticky yellow boards was explored as a possible control measure, but catches were only two to six times greater than on untreated boards, so their trapping capacity was probably not improved enough to be useful, except perhaps for limiting 'hot-spots' of *F. occidentalis*.

The apparent greater relative efficiency of traps containing chemical attractants should be interpreted with caution because they may draw thrips away from nearby unbaited traps. The distance over which such a response would occur is unknown, but it will certainly be affected by wind speed and direction.

In laboratory olfactometer trials, the grass thrips *Limothrips cerealium* and *Haplothrips aculeatus* were attracted to capronic and capric acids, acetic ether and geraniol (Holtmann, 1963), but there would be little to be gained by enhancing trap selectivity for these species in the field.

Suction traps

All the trapping methods considered above provide relative estimates of the aerial population at different locations and times, but because of the differential selection of species and variable wind speeds, they often give an erroneous impression of the actual abundance and proportions of each species present in a mixed population flying above a crop. Absolute measurements of population density, numbers taking off from a crop in response to changing weather, or daily flight periodicity can only be obtained reliably with suction traps (Lewis, 1973).

Below 2 m, a small, exposed cone model delivering about 460 $m^3 h^{-1}$ of air (Taylor, 1951) usually catches samples of convenient size (10 to 100 thrips h^{-1}) with individuals clean and easily identifiable. When the orifice of the trap needs to be near the ground, the cone should be lowered into a long slit trench to allow air to escape freely and to prevent recirculation of sampled air. Where populations are more dense than about 1 per m^3, an order of density typical over heavily infested crops, catches in traps delivering this amount of air may be too large. A smaller trap can then be used, or catches may be subsampled for counting and identification by taking aliquots from an agitated suspension of thrips in ethanol.

Wind is the only environmental variable affecting the efficiency of this type of trap, but its effect is small in wind speeds less than 15 km h^{-1}. There is no evidence that these traps attract thrips. The catch can be collected directly into 60 per cent ethanol in a polythene bottle fastened beneath the gauze cone; the bottle should have a small piece of gauze welded into one side to act as an overflow in rain. Alternatively, the catch can be collected in a metal tube and, for studies on periodicity and flight thresholds, segregated automatically into hourly samples by discs that fall into it. The thrips are killed and lightly stuck to the discs by a film of pyrethroid insecticides and light machine oil mixed in 100 ml of petrol ether and applied to the discs before they are loaded into the trap.

Unfortunately, suction traps are expensive and require a power supply so are impractical for widespread routine thrips monitoring except where a large-scale system is installed for other major pests (Matteson *et al.*, 1995), but they can be used, with advantage, to calibrate attractive traps and for studies of the composition, behaviour and dispersal of pest populations (Jenser, 1973, 1981, 1985; Lewis, 1959b, 1964, 1965).

Trap catches and crop infestation

Although there have been many studies on crop and aerial sampling, these include few attempts to ascertain how closely changes on the former are reflected in the latter. Field populations can be huge, ranging from a few million *Selenothrips rubrocinctus* in a hectare of cocoa (Fennah, 1955) to over a billion *Thrips tabaci* in a hectare of flowering onions grown for seed (Carlson, 1964), numbers many orders of magnitude greater than would be caught in traps placed within these crops. The evidence for trap catches being reliable indicators of the size of crop infestations or inflicted damage is mixed.

In Hawaiian lettuce fields, catches of *Frankliniella occidentalis* on seven different coloured sticky traps were highly correlated ($r = 0.83$ to 0.93) with the numbers of adults on lettuce plants (Yudin *et al.*, 1987). By contrast, on Nigerian cowpeas infested with *Megalurothrips sjostedti*, catches in sticky and water traps were poorly correlated with absolute numbers ($r = 0.57$ and 0.47, respectively) and plant shaking proved the most reliable and cheapest sampling method. Experience in Californian citrus groves infested with *Scirtothrips citri* was also conflicting. Over 2 years, Grout *et al.* (1986) used the amount of fruit scarring as an indication of cumulative thrips activity on trees and ranked this with catches on: (i) sticky ground traps, (ii) sticky yellow cards hung on trees (iii) D-vac sampling of the canopy, and (iv) folded pupation papers attached to twigs. In one year, yellow cards were the only method to rank different sites by thrips density in the same order as fruit scarring, but in the other year, pupation papers provided the closest ranking. In cucumbers and sweet peppers grown in glasshouses, Higgins (1992) found that increasing catches of *F. occidentalis* on yellow cards corresponded with increases in adults and immatures on leaves and flowers, but early in the growing season traps caught a preponderance of males whereas later most trapped adults were females, the sex responsible for most of the damage to fruit (Higgins and Myers, 1992). Clearly, the relationship between trap catches, infestation levels and damage is complex and, as with so many aspects of sampling, needs careful study for each combination of pest and crop.

Bioassay

Pesticides

Bioassays are required either to determine the lethal doses of actual or potential insecticides or to measure the amount of resistance in populations previously treated with pesticides. The small size of thrips and the difficulty of handling and confining them in containers presents a challenge, although the principles of bioassay are the same as for larger insects (Busvine, 1957). Most bioassays employ residual-type exposure in which plant material or artificial surfaces are either sprayed or dipped in serial dilutions of insecticide, allowed to dry, then infested with the test insects. For direct topical bioassays, the pesticide is either applied to individual test insects with a microapplicator, or they are totally immersed in it. A less precise method used for pesticides with a predominantly systemic mode of action after absorption by plant roots, particularly granulated formulations, is to evaluate their effect via whole plants in small-plot glasshouse trials. The main features of these various bioassays are illustrated below by studies on *Scirtothrips citri* and *Frankliniella occidentalis*, species against which many insecticides have been tested and in which resistance has developed widely. Procedural details may need to be modified slightly for other species depending on the purpose of the bioassay, the equipment available and the host plant.

In addition to choosing a suitable bioassay method it is equally important that tests be replicated and the data collected be analysed and interpreted properly. The mortality in treated groups of insects should be corrected for mortality in controls (Abbott, 1925) and analysed by probit analysis (Finney, 1971) using appropriate statistical programs, such as the generalized linear model methods of Genstat 5 (Payne *et al.*, 1987) and the Maximum Likelihood Program (MLP) (Ross, 1987). Furthermore, laboratory data should eventually be correlated with expected field efficacy of the tested pesticide because there is no certainty that any laboratory bioassay accurately reflects the field performance of a pesticide or the resistance patterns to it (see Chapter 15, Brattsen *et al.*, 1986; Immaraju *et al.*, 1990; Robb *et al.*, 1995).

Residual bioassays

Commonly, mature whole leaves of the host plant or leaf discs 15 to 20 mm in diameter are either dipped or sprayed in a Potter Precision Spray Tower (Potter, 1952) with serial dilutions of insecticide suspended in water plus wetting/sticking agent (e.g. 0.05 per cent Triton X); control leaves are treated with distilled water and wetter/sticker only (Morse *et al.*, 1986; Immaraju *et al.*, 1990; Robb *et al.*, 1995). After drying, the treated plant tissue is placed in a thrips-proof cell with a moistened basal pad to keep the vegetation fresh (e.g. Munger, 1942; Morse *et al.*, 1986) (Fig. 12.3a and b) (see Chapter 12) or in a small Petri dish with an agar film on the base into which the underside of leaf discs are sealed (Helyer and Brobyn, 1992); this keeps leaf discs turgid and prevents thrips crawling underneath them. Batches of 10 to 30 thrips from the population to be tested are placed in each container using a fine sable brush; active adults can be sucked into a straw and anaesthetized with CO_2 for about

45 s before being tapped into cells. If open dishes or containers are used they must be sealed with tightly stretched clingfilm to prevent escapes. Dead thrips are usually counted 24, 48 and 72 h after incubation at 25 to 27°C.

Brødsgaard (1994) used ventilated glass cells, sandwiched between two glass plates held in place with rubber bands, as bioassay chambers for *F. occidentalis*; moisture was provided internally through a soaked cotton wick (Fig. 12.3c). Before assembly, test pesticides were sprayed on to the upper and lower cover plates in a Potter Tower, and the vertical walls of the cell swabbed with pesticide. When dry, pollen was scattered on the lower plate as food. The spaces between the glass layers were sealed with a drop of deionized water. A less elaborate procedure is to use a Potter Tower to spray known doses of insecticide on to pieces of filter paper and place them inside specimen boxes, or the boxes can be sprayed directly and dried. Then 20 adults are placed in each box with three or four chrysanthemum petals, the boxes sealed and dead insects counted after 24 h (Macdonald, 1995). An even simpler approach is to coat the inside of glass vials with serial doses of pesticide (e.g. 100, 50, 10, 5, 1, 0.5 or 0.1 g per vial) add thrips and, after 6 h, a small piece of leaf or flower, and check mortality after 24 h (A.A. Schreiber, University of Missouri, personal communication).

Topical bioassays

Direct application of pesticide can be achieved by total immersion of individuals or, more precisely, using a microapplicator. For delicate larvae total immersion is easier. Helyer and Brobyn (1992) collected *F. occidentalis* larvae in a 7 ml bottle then added 5 ml of insecticide. After tapping to ensure immersion for 10 s, the contents were poured into a Buchner funnel lined with black filter paper, the pesticide drained and 25 to 30 insects transferred to a repli-plate containing leaf discs sealed to agar and previously sprayed with the same pesticide; the dish was covered with clingfilm. Topical application by microapplicator (Arnold model with a 1 ml syringe and 0.5 ml bore canula) ensures a more precise dose, but thrips first need to be inactivated either by cooling (Robb *et al.*, 1995) or anaesthetizing with CO_2 (Immaraju *et al.*, 1990). Serial dilutions of insecticide made up in distilled acetone are applied as a single 0.4 or 0.5 μl droplet to the thorax of individuals standing on filter paper. Treated thrips are engulfed by this volume of liquid, but excess is absorbed by the filter paper. After treatment, thrips are confined on host tissue in a cell and dead ones counted after 48 h. Droplets of distilled acetone only are used as controls, from which individuals recover rapidly.

A comparison by Immaraju *et al.* (1990) of leaf dip, leaf spray and microapplicator bioassays for *Scirtothrips citri*, using fluvalinate and dimethoate, indicated that the thrips contacted/absorbed more insecticide over a 48 h period following leaf dipping or spraying than during a single dosing with a microapplicator. Whilst the spray method may give more reproducible results because of uniform droplet size and coverage, the leaf dip method is the more versatile because it does not require an expensive standardized spray chamber.

Whole plant bioassays

In attempts to assess the value of granular, systemic formulations, and to overcome some of the difficulties of correlating the results of laboratory bioassays with the impact of the same pesticides or infestations of *F. occidentalis* in glasshouses, Helyer and Brobyn (1992) employed whole plant treatments. Chrysanthemums were grown in peat and grit compost in 1 m^2 plots retained in deep frames lined with polythene. Each plot was infested with 200 adult thrips, mainly females, weekly for 3 weeks starting 2 weeks after planting so that a population of mixed age established before treatment began. A first assessment of population size was made 1 week after the final introduction. Samples of 30 leaves and five growing points were taken on the day of treatment and again 1, 3, 7 and 14 days later. Thrips on leaves were counted directly, then hidden individuals expelled with turpentine for counting (see p. 446). Granular formulations were applied by 'pepper-pot' shaking over plants immediately followed by watering to activate the active ingredient. An equivalent amount of silver sand was shaken on to control plots.

For liquid formulations, similar pretrial infestations were built up on larger 7 m × 1 m plots from each of which larger samples (90 leaves and 20 growing points) were taken to assess the effect of insecticides applied at high volume with a pressurized hydraulic sprayer.

Fungal pathogens

The evaluation of potential fungal pathogens for thrips control likewise requires laboratory bioassays to determine the virulence of organisms, followed by field trials of those chosen as promising (see Chapter 10). Details of procedures for such bioassays have been documented by Gillespie (1986), Morse *et al.* (1986), Hall *et al.* (1993), Schreiter *et al.* (1994), Vestergaard *et al.* (1995) and Gindin *et al.* (1996). Each differs in detail depending on the species of thrips and pathogen involved, but essentially the procedure is to inoculate test insects with serial concentrations of conidial suspensions, place the treated individuals on a food source, incubate them in a moist chamber at a constant temperature and examine them daily for about a week to assess infection and mortality. Initial tests should be made on healthy populations of uniform age, which are best obtained from cultures (see Chapter 12). As an example, the precise procedures followed by Vestergaard *et al.* (1995) for maintaining *Metarhizium anisopliae* and testing it against *Frankliniella occidentalis* are described below.

An isolate of *M. anisopliae* was stored as conidia on silica gel at –40°C (Perkins, 1962). Fresh cultures, initiated using this inoculum, were maintained on Sabouraud dextrose agar in darkness at 23 ± 1°C. Thrips were reared on flowering, potted chrysanthemums at 25 ± 1°C, and 16.8 h photoperiod. Insects at similar stages of development were obtained by incubating adults on fresh, healthy plants, removing the insects after 2 days and allowing the emergent larvae to develop in synchrony. Pupal stages (about 75 per cent prepupae and 25 per cent pupae) were obtained by transferring second instar larvae to a chrysanthemum leaf mounted on 0.9 per cent (w/v) water agar in a Petri dish lid. The lid was inverted over the bottom half of the Petri dish, which was lined with moistened filter paper sandwiched between two layers of black plastic. The larvae migrated from the leaf to the underside of the first and second layers of black plastic to pupate.

A mixture of conidia and hyphae were harvested from 10 to 12-day-old cultures by flooding the plates with sterile 0.05 per cent aqueous (w/v) Triton X-100 and agitating with a glass rod. Conidia were separated by filtering the suspension through tissue paper followed by centrifugation for 5 min at 3000 rpm. The pellet was washed twice with sterile 0.05 per cent Triton X-100 with intervening centrifugation steps. The concentration of the inoculum was then adjusted to 10^7 conidia ml^{-1} with 0.05 per cent Triton X-100. The viability of the conidia was assessed by Hall's (1976) method.

Adults, larvae and pupal stages of *F. occidentalis* were inoculated by immersion for 10 s in 3 ml of conidial suspension. Conidia and thrips were poured on to filter paper in a Buchner funnel, and excess suspension removed by vacuum. Insects were transferred using a paintbrush to 29-mm diameter chambers in plastic trays. A single insect was placed in each chamber, which contained 2 ml of 0.9 per cent (w/v) water agar on which a 23-mm diameter cucumber leaf disc was placed. Trays were then sealed with polyvinyl chloride (PVC) clingfilm which ensured near saturation conditions without condensation, and incubated at 20 ± 1°C with a 16.8 h photoperiod. Controls consisted of thrips treated with 0.05 per cent Triton X-100 only. Mortality was recorded daily. Fungal emergence was determined for each isolate using a stereomicroscope. Each assay consisted of three replicates with 25 ± 5 insects per replicate, and the assay was normally repeated twice for each isolate. The control and standard isolate assays were repeated six times. Trays containing pupal stages were initially kept in the dark for 4 days until newly developed adults had emerged. Mortality studies at different doses of *M. anisopliae* were made using the concentrations 10^5, 10^6, 10^7 and 10^8 conidia ml^{-1}. Mortality was recorded daily for 7 days. The final proportion of dead thrips at each concentration was analysed by probit analysis (using the generalized linear model methods of Genstat 5; Payne *et al.*, 1987). An LC_{50} and standard error were estimated for each isolate using the Genstat 5 procedure 'Fieller' (Lane, 1989).

It should be noted that all instars of the thrips were infected by this pathogen, but pupae and especially larvae were more resistant than adults, presumably because fungal inoculum was shed with the cuticle at moulting.

For greenhouse trials, which represent the next stage in assessing the efficacy of a selected experimental, or commercially available, pathogen, infested plants should be sprayed with the trial product at appropriate concentrations of spores and volumes per plant. The effects on the treated population can be assessed by direct leaf counts or, for thrips species that pupate in the soil, by trapping the descending larvae in dishes containing sticky grease placed beneath treated plants. Fuller details of the mass production, formulation and application of pathogens for field use are given in Chapter 10.

Miscellaneous Techniques

Weighing thrips

It may sometimes be useful to weigh thrips for studies on feeding, growth or bioassay. Zareh and Morse (1989) achieved this by weighing clear plastic straws with corks plugging each end, then collecting into them groups of 100 first or second instar larvae or adult female *Scirtothrips citri* with a vacuum aspirator, corking and weighing on a sensitive balance. Three replicates, each of 100 individuals, were used to calculate the average weight of each life stage as 0.27, 4.23 and 14.53 μg, respectively. T. Lewis (unpublished data) weighed living individual *Limothrips cerealium* and wingless *Aptinothrips rufus* on a sensitive glass-fibre torsion balance by isolating specimens on a small glass platform on which they remained long enough to take a reading. Adult female *L. cerealium* weighed about 40 μg.

Direct observation of feeding behaviour

Many thrips can be induced to feed through artificial membranes (Sakimura and Carter, 1934; Laughlin, 1971) (see also Chapter 12). Chisholm and Doncaster (1982) exploited this behaviour to study how the mouthparts function by devising small chambers, one wall of which comprised a plastic membrane retaining a 10 per cent sucrose solution. Two types of chamber were developed to present either a front or an oblique view of feeding insects (Fig. 11.5). Individual thrips were observed through a microscope and filmed as they fed to reveal the action of the mouthparts and associated muscles, and to measure the rate of ingestion (Chisholm and Lewis, 1984). Larvae and adults probed and fed more readily if starved for up to 4 h beforehand. The light intensity needed careful adjustment to prevent undue agitation of the insects; dark-coloured adults were best viewed by incident light, but for larvae, transmitted light effectively illuminated mouthparts and muscle action through their pale, translucent cuticle. Specifications of the optical equipment are provided by Chisholm and Doncaster (1982), but viewing/feeding chambers of this type could be adapted to use with more modern microscopy and video equipment.

Electrical monitoring of the frequency and nature of feeding has been achieved for *Frankliniella occidentalis* using the electrical penetration graph technique (EPG-DC system) (Harrewijn *et al.*, 1996). Plant penetration was detected by attaching the thorax of a thrips to a 10-μm diameter gold wire with a drop of silver paint so that when its stylets were inserted into tissue an electrical circuit was completed and characteristic waveforms captured via an amplification and recording system. Stylet penetration of separate cells, salivation and ingestion were each detectable on leaves, fruit and pollen grains. The technique offers opportunities to explore many aspects of feeding behaviour relating to sap ingestion and the muscular movements associated with it, virus transmission and plant susceptibility, and resistance to thrips attack.

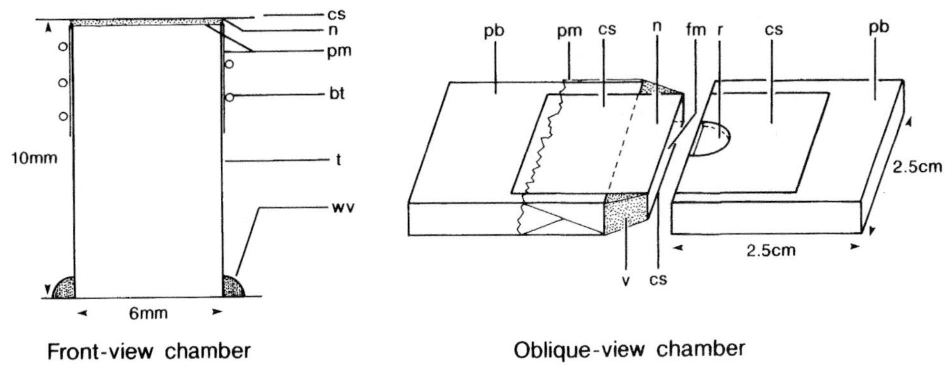

Fig. 11.5. Chambers for viewing feeding thrips. In the oblique-view chamber, the nutrient fluid was injected into the cavity beneath the cover slip with a hypodermic needle inserted through the vaseline seal. The left-hand section containing the nutrient and the right-hand section containing a thrips were then stuck on a microscope slide so that the open side of the thrips chamber abutted against the membrane wall of the nutrient cavity. The whole chamber could be tilted for oblique viewing using a universally adjustable stage. Abbreviations: bt = binding twine; cs = cover slip (5 × 5 mm); fm = front membrane; n = nutrient (10% sugar solution); pb = perspex block; pm = plastic membrane ('cling film'); r = thrips recess (3 × 3 × 1 mm); t = tube; v = vaseline; wv = wax–vaseline mixture.

Electrophysiological recording

Matteson *et al.* (1992) detected the efficiency of thrips vision by means of an electroretinogram (ERG). This was measured in *Frankliniella occidentalis* exposed to flashes of light ranging from 365 to 620 nm. Adults used for testing were cooled for several minutes in a freezer to –18°C to facilitate manipulation with a paint brush. Each thrips was mounted dorsal side up on double-sided sticky tape on a platform (Fig. 11.6) under a stereomicroscope. The insect was further immobilized with two thin strips of the same tape placed across the thorax and the terminal segments of the abdomen. Microelectrodes were pulled from thick-walled borosilicate glass capillaries in a Kopf vertical puller and filled with insect Ringer (Hoyle, 1953). The recording electrode was inserted under the cornea of the left eye and the reference electrode was placed in the thorax. The electrical signal was led through a differential amplifier in the d.c. mode, and observed and recorded on a recording oscilloscope with output onto ultraviolet-sensitive paper.

Thrips were stimulated with ultraviolet (365 nm) and visible (400 to 620) wavelengths flashed from light guides placed within 1 mm of an eye after test insects had been dark adapted. Different visible wavelengths were tested after allowing 1 min dark adaption between each flash. Both sexes showed two peaks of spectral activity, one in the ultraviolet range and one in the visible spectrum with λ_{max} around 540 nm.

Electrophysiological recording from thrips antennae or single antennal cells in response to odours is also probably feasible though not yet reported. The measurement of the bioactivity of pheromones has so far been confined to behavioural assays (Teerling *et al.*, 1993). However, single cell recording from

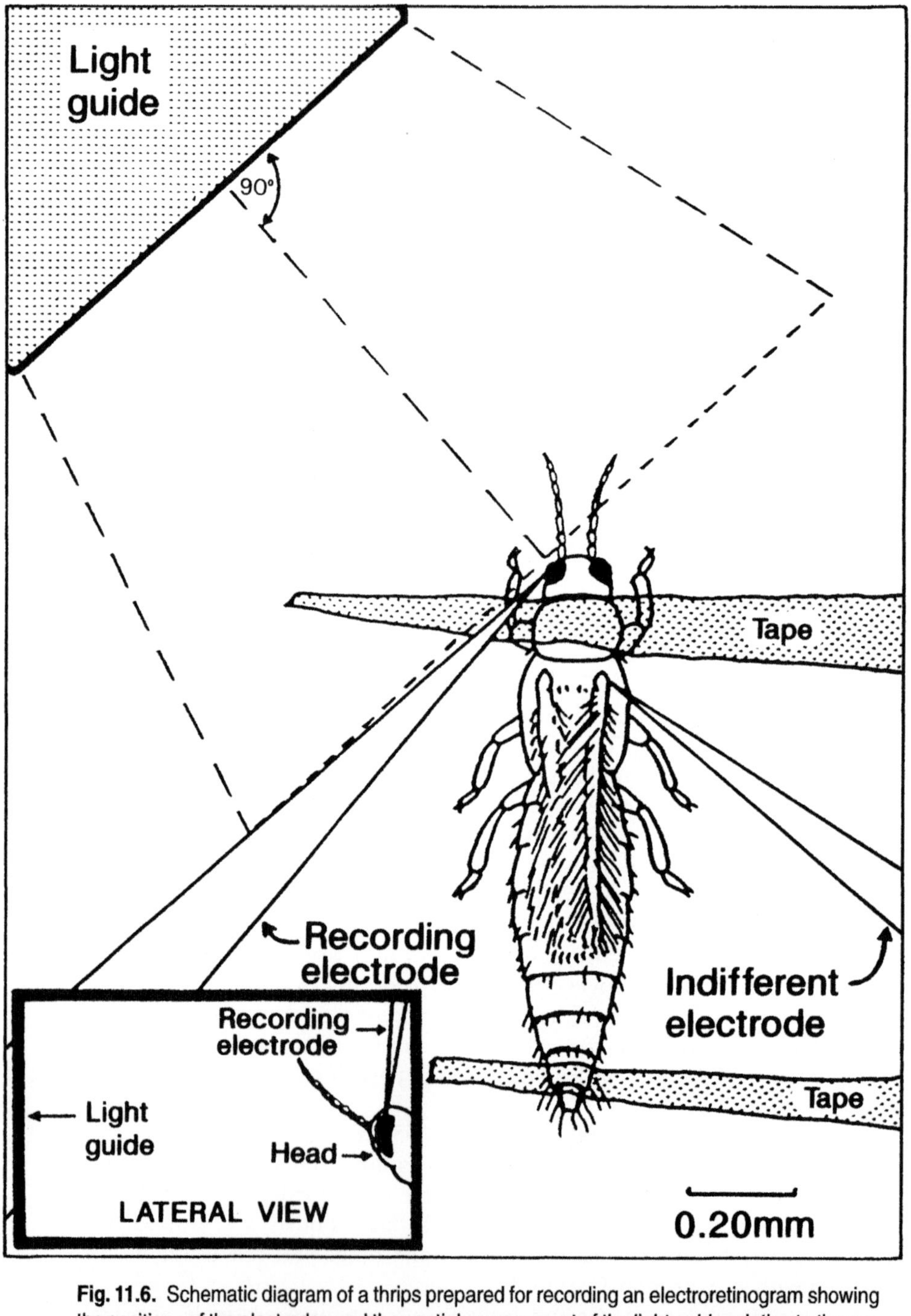

Fig. 11.6. Schematic diagram of a thrips prepared for recording an electroretinogram showing the positions of the electrodes and the spatial arrangement of the light guide relative to the eye. Inset shows a lateral view of the head and indicates the direction of the light guide (after Matteson *et al.*, 1992).

various types of antennal sensillae (see Chapter 2) should be achievable using a slightly modified, gas chromatography–electrophysiology system similar to that used for aphids (Wadhams, 1990) employing very fine tungsten electrodes polished electrolytically to a diameter of a few angstroms (Boeckh, 1962). Hummel and Miller (1984) provide a practical introduction to electrophysiological techniques for pheromone research.

Measuring genetic diversity of populations

Allozyme and DNA-based methods capable of detecting genetic variation within and between insect populations have recently been used on thrips. Lewis *et al.* (1991) tested ten enzymes against *Taeniothrips inconsequens*, *Limothrips cerealium* and *Frankliniella occidentalis*. Individual thrips were prepared for electrophoresis by crushing and macerating in a buffered homogenizing solution and 1 μl aliquots from each crushed insect were applied to cellulose acetate gel sheets. These were exposed to 200 volts for 20 to 30 min in a standard electrophoretic kit to separate the different proteins in the sample, and then stained. Of ten enzymes tested, esterase and peptidase resolved and stained satisfactorily for all three species of thrips and provided a reliable basis for species discrimination.

The genetic diversity within and between populations of the predatory *Aleurodothrips fasciapennis* from locations in China and Australia was studied by Gillings *et al.* (1996) to see whether different populations might be more suited to the climate and ecology of different sites in Australia if this thrips was released to control citrus red scale, *Aonidiella aurantii* (Marshall). Pellets of DNA were prepared from individual thrips and the internal transcribed spacer region of rRNA genes (*ITS1*) was amplified using polymerase chain reaction, and analysed using restriction enzymes, of which *Hin*f revealed polymorphisms within *ITS1*. The same samples were examined using randomly amplified polymorphic DNA (RAPD) and the relationship between the banding patterns analysed. The Chinese samples showed similarities in RAPD patterns which became more obvious the closer collection points were to each other, suggesting that there are distinct populations of these thrips.

ITS restriction analysis was also capable of separating *Frankliniella occidentalis* from *F. schultzei*, a technique that might be more widely developed for discriminating between closely related species. Details of the procedures followed are fully documented in the original paper.

References

Abbott, W.S. (1925) A method of computing the effectiveness of an insecticide. *Journal of Economic Entomology* 18, 265–267.

Amin, P.W. (1983) Studies on arthropod vectors of groundnut viruses, their ecology and control. *Groundnut Entomology Progress Report 2*, International Crops Research Institute for the Semi-Arid Tropics, 167 pp.

Backus, E.A., Hunter, W.A. and Arne, C.M. (1988) Technique for staining leafhopper (Homoptera: Cicadellidae) salivary sheaths and eggs within unsectioned plant

tissue. *Journal of Economic Entomology* 81, 1819–1823.
Bailey, S.B. (1944) The pear thrips in California. *University of California Agriculture Experiment Station Bulletin* 687.
Bates, B.A. and Weiss, M.J. (1991) The spatial distribution of *Limothrips denticornis* Haliday (Thysanoptera: Thripidae) eggs on spring barley. *Canadian Entomologist* 123, 205–210.
Beavers, J.B., Shaw, J.G. and Hampton, R.B. (1971) Color and height preference of the citrus thrips in a navel orange grove. *Journal of Economic Entomology* 64, 1112–1113.
Beckman, C.M. (1969) Color preference and flight habits of thrips associated with cotton. *Journal of Economic Entomology* 62, 591–592.
Bender, D.A. and Morrison, W.P. (1989) Species composition and control of thrips in Texas high plains onions. *Journal of Agricultural Entomology* 6, 257–263.
Binns, M.R. and Nyrop, J.P. (1992) Sampling insect populations for the purpose of IPM decision making. *Annual Review of Entomology* 37, 427–453.
Boeckh, J. (1962) Electrophysiologische untersuchungen an einzelnen Geruchs-Rezeptoren auf den antennen des Totengräbers (*Necrophorus*: Coleoptera). *Zeitschrift für vergleichende Physiologie* 46, 212–248.
Brattsen, L.B., Holyoke, C.W., Leeper, J.R. and Raffa, K.F. (1986) Insecticide resistance: challenge to pest management and basic research. *Science* 231, 1255–1260.
Brødsgaard, H.F. (1989) Coloured sticky traps for *Frankliniella occidentalis* (Pergande) (Thysanoptera: Thripidae) in glasshouses. *Journal of Applied Entomology* 107, 136–140.
Brødsgaard, H.F. (1994) Insecticide resistance in European and African strains of western flower thrips (Thysanoptera: Thripidae) tested in a new residue-on-glass test. *Journal of Economic Entomology* 87, 1141–1146.
Buhl. C. (1937) Beitrage zur Kenntnis der Biologie, wirtschaftliche Bedeutung und Bekämpfung von *Kakothrips robustus* Uz. *Zeitschrift für angewandte Entomologie* 23, 65–113.
Bullock, J.A. (1963) Extraction of Thysanoptera from samples of foliage. *Journal of Economic Entomology* 56, 612.
Bullock, J.A. (1964) A note on the soil fauna of a pyrethrum field. *East Africa Agriculture and Forestry Journal* 30, 8–10.
Bullock, J.A. (1965) The assessment of populations of *Thrips nigropilosus* Uzel on pyrethrum. *Annals of Applied Biology* 55, 1–12.
Busvine, J.R. (1957) *A Critical Review of the Techniques for Testing Insecticides.* Commonwealth Institute of Entomology, London.
Cameron, A.E. and Treherne, R.C. (1918) The pear thrips (*Taeniothrips inconsequens* Uzel) and its control in British Columbia. *Canadian Department of Agriculture* Bulletin 15.
Cameron, P.J., Herman, T.J.B. and Fletcher, J.D. (1992) Incidence of thrips and aphids as potential virus vectors in field tomatoes. In: *Proceedings of the 45th New Zealand Plant Protection Conference, Rotorua*, DSIR Plant Protection, Auckland, pp. 38–42.
Carlson, E.C. (1964) Effect of flower thrips on onion seed plants and a study of their control. *Journal of Economic Entomology* 57, 735–741.
Cederholm, L. (1963) Ecological studies on Thysanoptera. *Opuscula Entomologica* Suppl. XXII, 215 pp.
Chiang, H.C. (1977) Pest management in the People's Republic of China. Monitoring and forecasting insect populations in rice, wheat, cotton and maize. *FAO Plant Protection Bulletin* 25, 1–8.

Chisholm, I.F. and Doncaster, C.C. (1982) Studying and recording the feeding behaviour of thrips. *Entomologia Experimentalis et Applicata* 31, 324–327.

Chisholm, I.F. and Lewis, T. (1984) A new look at thrips (Thysanoptera) mouthparts, their action and effects of feeding on plant tissue. *Bulletin of Entomological Research* 74, 663–675.

Cho, K.J., Eckel, C.S., Walgenbach, J.F. and Kennedy, G.G. (1995) Comparison of colored sticky traps for monitoring thrips populations (Thysanoptera, Thripidae) in staked tomato fields. *Journal of Entomological Science* 30, 176–190.

Coli, W.M., Hollingsworth, C.S. and Maier, C.T. (1992) Traps for monitoring pear thrips (Thysanoptera: Thripidae) in maple stands and apple orchards. *Journal of Economic Entomology* 85, 2258–2262.

Czenz, K. (1987) The role of coloured traps in collecting thrips fauna. In: Holman, J., Pelikan, J., Dixon, A.F.G. and Weisman, L. (eds) *Population Structure, Genetics and Taxonomy of Aphids and Thysanoptera*. SPB Academic Publishing, The Hague, pp. 426–435.

DeGrandi-Hoffman, G., Terry, I. and Huber, R.T. (1988) Incorporating fruit set estimates with thrips management to create a decision support system for apples. *Hortscience* 23, 571–574.

Dennill, G.B. and Erasmus, M.J. (1991) Basis for a practical technique for monitoring thrips in avocado orchards. *Crop Protection* 11, 89–91.

Dintenfass, L.P., Bartell, D.P. and Scott, M.A. (1987) Predicting resurgence of western flower thrips (Thysanoptera: Thripidae) on onions after insecticide application in the Texas high plains. *Journal of Economic Entomology* 80, 502–506.

Diraviam, J. and Uthamasamy, S. (1992) A new sampling technique involving yellow sticky traps for monitoring of thrips infesting different crops. *Journal of Entomological Research* 16, 78–81.

Edelson, J.V. (1985) A sampling method for estimating absolute numbers of thrips on onions. *Southwestern Entomologist* 10, 103–106.

Edwards, C.A. and Fletcher, K.E. (1970) The assessment of terrestrial invertebrate populations. In: *Methods of Study in Soil Ecology, Proceedings of a UNESCO and IBP Symposium, Paris*. UNESCO, Paris, pp. 57–66.

Evans, J.W. (1933) A simple method of collecting thrips and other insects from blossom. *Bulletin of Entomological Research* 24, 349–350.

Faulkner, L.R. (1954) Economic thrips of southern New Mexico. *Bulletin of the New Mexico Agricultural Experiment Station* 387, p. 26.

Felland, C.M., Teulon, D.A.J., Hull, L.A. and Polk, D.F. (1995) Distribution and management of thrips (Thysanoptera: Thripidae) on nectarine in the mid-Atlantic region. *Journal of Economic Entomology* 88, 1005–1011.

Fennah, R.G. (1955) The epidemiology of cacao-thrips on cacao in Trinidad. *Report on Cacao Research, Trinidad* 24, 7–26.

Finney, D.J. (1971) *Probit Analysis*, 3rd edn. Cambridge University Press, London.

Fournier, F., Boivin, G. and Stewart, R.K. (1995) Sequential sampling for *Thrips tabaci* on onions. In: Parker, B.L., Skinner, M. and Lewis, T. (eds) *Thrips Biology and Management*. Plenum Press, New York, pp. 557–562.

Franssen, C.J.H. and Huisman, P. (1958) The biology and control of *Thrips angusticeps* Uzel. *Verslag Landbiouwkundig Onderzoek RijkslandProefstn* 64, 104 pp.

Franssen, C.J.H. and Mantel, W.P. (1965) Thrips in cereal crops (biology, economic importance and control). 1. Biology. *Verslag Landbouwkundig Onderzoek Rijksland Proefstn* 662, 97 pp.

Funderburk, J.E. (1993) Ecology and management of flower thrips in a mixed agronomic and vegetable production system. In: *Memorias XX Congress Sociedad*

Colombiana de Entomologia, Cali, Colombia, pp. 322–329.

Gillespie, A.T. (1986) The potential of entomogenous fungi as control agents for onion thrips, *Thrips tabaci*. *British Crop Protection Council Monograph No. 34, Biotechnology and Crop Improvement and Protection*, 237–243.

Gillespie, D.R. and Vernon, R.S. (1990) Trap catch of western flower thrips (Thysanoptera: Thripidae) as affected by color and height of sticky traps in mature greenhouse cucumber crops. *Journal of Economic Entomology* 83, 971–975.

Gillings, M.R., Rae, D.J., Herron, G.A. and Beattie, G.A.C. (1996) Tracking thrips populations using DNA-based methods. *Proceedings of the 1995 Australia and New Zealand Thrips Workshop*, N.S.W. Agriculture, Gosford, Australia, pp. 97–103.

Gindin, G., Barash, I., Raccah, B., Singer, S., Ben-Ze'ev, I.S. and Klein, M. (1996) The potential of some entomopathogenic fungi as biocontrol agents against the onion thrips, *Thrips tabaci* and the western flower thrips, *Frankliniella occidentalis*. *Folia Entomologica Hungarica* LVII (Supplement) 37–42.

Gray, H. and Treloar, A. (1933) On the enumeration of insect populations by the method of net collection. *Ecology* 14, 356–367.

Grout, T.G., Morse, J.G., O'Connell, N.V., Flaherty, D.L., Goodell, P.B., Freeman, M.W. and Coviello, R.L. (1986) Citrus thrips (Thysanoptera: Thripidae) phenology and sampling in the San Joaquin Valley. *Journal of Economic Entomology* 79, 1516–1523.

Hall, R.A. (1976) A bioassay of the pathogenicity of *Verticillium lecanii* conidiophores on the aphid *Macrosiphoniella sanborni*. *Journal of Invertebrate Pathology* 27, 41–48.

Hall, W.J. (1930) The South African citrus thrips in Southern Rhodesia. *Report of the Mazoe Citrus Experiment Station* 1, 5–55.

Hall, R.A., Peterkin, D. and Pollard, G.V. (1993) A new system of caging *Thrips palmi* for laboratory bioassay of pathogens. *Florida Entomologist* 76, 171–175.

Harrewijn, P., Tjallingii, W.F. and Mollema, C. (1996) Electrical recording of plant penetration by the western flower thrips. *Entomologia Experimentalis et Applicata* 79, 345–353.

Harrison, J.O. (1963) Notes on the biology of the banana flower thrips, *Frankliniella parvula*, in the Dominican Republic. *Annals of the Entomological Society of America* 56, 664–666.

Heikinheimo, O. and Raatikainen, M. (1962) Comparison of suction and netting methods in population investigations concerning the fauna of grass leys and cereal fields, particularly in those concerning the leafhopper, *Calligypona pellucida* (F.). *Valtion Maatalouskoetoiminnan Julkais* No. 191, 31 pp.

Heinz, K.M., Parrella, M.P. and Neunan, J.P. (1992) Time-efficient use of yellow sticky traps in monitoring insect populations. *Journal of Economic Entomology* 85, 2263–2269.

Helyer, N.L. and Brobyn, P.J. (1992) Chemical control of western flower thrips (*Frankliniella occidentalis* Pergande). *Annals of Applied Biology* 121, 219–231.

Higgins, C.J. (1992) Western flower thrips (Thysanoptera: Thripidae) in greenhouses: population dynamics, distribution on plants and association with predators. *Journal of Economic Entomology* 85, 1891–1903.

Higgins, C.J. and Myers, J.H. (1992) Sex ratio patterns and population dynamics of western flower thrips (Thysanoptera: Thripidae). *Environmental Entomology* 21, 322–330.

Hightower, B.G. and Martin, D.F. (1956) Ecological studies on thrips found on cotton in central Texas. *Journal of Economic Entomology* 49, 423–424.

Hoerner, J.L. (1947) A separator for onion thrips. *Journal of Economic Entomology* 40, 755.

Holtmann, H. (1962) Untersuchungen zur Biologie der Getreide-Thysanopteren. Teil I. *Zeitschrift für angewandte Entomologie* 51, 1–41.

Holtmann, H. (1963) Untersuchungen zur Biologie der Getreide-Thysanopteren. Teil II. *Zeitschrift für angewandte Entomologie* 51, 285–299.

Hosny, M.M. (1964) Testing the validity of a simple method for estimating thrips infestation on cotton-seedlings in the field. *Agricultural Research Review, Cairo* 42, 136–140.

Howlett, T.M. (1914) A trap for thrips. *Journal of Economic Biology* 9, 21–23.

Hoyle, G. (1953) Potassium ions and insect nerve muscle. *Journal of Experimental Biology* 30, 121–135.

Hummel, H.E. and Miller, T.A. (eds) (1984) *Techniques in Pheromone Research*. Springer-Verlag, New York.

Immaraju, J.A., Morse, J.G. and Brawner, O.L. (1990) Evaluation of three bioassay techniques for citrus thrips resistance and correlation of the leaf dip method to field mortality. *Journal of Agricultural Entomology* 7, 17–27.

Irwin, M.E., Yeargan, K.V. and Marston, N.L. (1979) Spatial and seasonal patterns of phytophagous thrips in soybean fields with comments on sampling techniques. *Environmental Entomology* 8, 131–140.

Iwao, S. (1968) A new regression method for analysing the aggregation pattern of animal populations. *Researches on Population Ecology* 10, 1–20.

Jensen, F. (1973) Flower thrips damage in the San Joaquin Valley: 1. Timing of halo spotting on table grapes. *California Agriculture* 27, 6–8.

Jenser, G. (1973) Observations on autumn mass flight of *Frankliniella intonsa* Trybom (Thysanoptera, Thripidae). *Acta Phytopathologica Academiae Hungaricae* 8, 227–230.

Jenser, G. (1981) Thysanoptera species collected by a suction trap in Hungary. *Folia Entomologica Hungarica* 34, 91–93.

Jenser, G. (1985) Thysanoptera species collected by suction trap in a maize stand in Hungary. In: Holman, J., Pelikan, J., Dixon, A.F.G. and Weisman, L. (eds) *Population Structure, Genetics and Taxonomy of Aphids and Thysanoptera*. SPB Academic Publishing, The Hague, pp. 436–438.

Johnson, C.G., Southwood, T.R.E. and Entwistle, H.M. (1957) A new method of extracting arthropods and molluscs from grassland and herbage with a suction apparatus. *Bulletin of Entomological Research* 48, 211–218.

Johnson, M.W. (1986) Population trends of a newly introduced species, *Thrips palmi* (Thysanoptera: Thripidae), on commercial watermelon plantings in Hawaii. *Journal of Economic Entomology* 79, 718–720.

Kawai, A. and Kitamura, C. (1987) Studies on population ecology of *Thrips palmi* Karny XV. Evaluation of effectiveness of control methods using a simulation model. *Applied Entomology and Zoology* 22, 292–302.

Kirk, W.D.J. (1984) Ecologically selective colour traps. *Ecological Entomology* 9, 35–41.

Kirk, W.D.J. (1985) Thrips eggs in flowers of Faba beans. *Fabis Newsletter* 11, 23–25.

Kirk, W.D.J. (1987) Effects of trap size and scent on catches of *Thrips imaginis* Bagnall (Thysanoptera: Thripidae). *Journal of the Australian Entomological Society* 26, 299–302.

Kitakata, S.W. and Yoshida, M. (1982) Control of insects in greenhouses by colored ribbon traps. (In Japanese.) *Shokubutu Boueki* 36, 478–481.

Koppa, P. (1967) The composition of the thrips species in cereals in Finland. *Annales Agriculturae Fenniae* 6, 30–45.

Koppa, P. (1969) Studies on the hibernation of certain species of thrips living on cereal plants. *Annales Agriculturae Fenniae* 8, 1–8.

Kratochvil, J. and Farsky, O. (1942) Das Absterben der diesjährigen terminalen Larchentriebe. *Zeitschrift für angewandte Entomologie* 29, 177–218.

Lane, P.W. (1989) Procedure FIELLER. *Genstat Procedure Library Manual.* NAG, Oxford.

Laudermilch, G.E. (1989) Thrips in Pennsylvania. In: Parker, B.L., Skinner, M. and Teillon, H.B. (eds) *Proceedings, Regional Meeting: The 1988 Thrips Infestation of Sugar Maple, Bennington, Vermont. Vermont Agriculture Experiment Station Bulletin* 696.

Laughlin, R. (1970) The gum tree thrips, *Isoneurothrips australis* Bagnall. The population of a single tree. *Entomologia Experimentalis et Applicata* 13, 247–259.

Laughlin, R. (1971) A culture method for *Hercinothrips femoralis* (Reuter) (Thysanoptera). *Journal of the Australian Entomological Society* 10, 301–303.

Le Pelley, R.H. (1942) A method of sampling thrips populations. *Bulletin of Entomological Research* 33, 147–148.

Leskey, T.C., Teulon, D.A.J. and Cameron, E.A. (1995) Position and abundance of pear thrips eggs in sugar maple flowering and vegetative buds. In Parker, B.L., Skinner, M. and Lewis, T. (eds) *Thrips Biology and Management.* Plenum Press, New York, pp. 93–95.

Letourneau, D.K. and Altieri, M.A. (1983) Abundance patterns of a predator, *Orius trisicolor* (Hemiptera: Anthocoridae), and its prey, *Frankliniella occidentalis* (Thysanoptera: Thripidae): habitat attraction in polycultures versus monocultures. *Environmental Entomology* 12, 1464–1469.

Lewis, T. (1959a) The annual cycle of *Limothrips cerealium* Haliday (Thysanoptera) and its distribution in a wheat field. *Entomologia Experimentalis et Applicata* 2, 187–203.

Lewis, T. (1959b) A comparison of water traps, cylindrical sticky traps, and suction traps for sampling thysanopteran populations at different levels. *Entomologia Experimentalis et Applicata* 2, 204–215.

Lewis, T. (1960) A method for collecting Thysanoptera from *Gramineae*. *Entomologist* 93, 27–28.

Lewis, T. (1961) Records of Thysanoptera at Silwood Park, with notes on their biology. *Proceedings of the Royal Entomological Society of London (A)* 36, 89–95.

Lewis, T. (1964) The weather and mass flights of Thysanoptera. *Annals of Applied Biology* 53, 165–170.

Lewis, T. (1965) The species, aerial density and sexual maturity of Thysanoptera caught in mass flights. *Annals of Applied Biology* 55, 219–225.

Lewis, T. (1973) *Thrips, Their Biology, Ecology and Economic Importance.* Academic Press, London.

Lewis, T. and Navas, D.E. (1962) Thysanopteran populations overwintering in hedge bottoms, grass litter and bark. *Annals of Applied Biology* 50, 299–311.

Lewis, T., Loxdale, H.D. and Brookes, C.P. (1991) Prospects for studying persistence and dispersal of pear thrips populations using genetic markers (allozymes). In: Parker, B.L. and Doane, E.N. (eds) *The 1990 Conference on Thysanoptera. Abstracts: South Burlington, Vermont. Vermont Agricultural Experiment Station Bulletin* 698, 1–7.

Linford, M.B. (1943) Influence of plant populations upon incidence of pineapple yellow spot. *Phytopathology* 33, 408–410.

Macdonald, O.C. (1995) Response of western flower thrips to dichlorvos and malathion in the United Kingdom. In: Parker, B.L., Skinner, M. and Lewis, T. (eds)

Thrips Biology and Management. Plenum Press, New York, pp. 347–350.

MacPhee, A.W. (1953) The influence of spray programs on the fauna of apple orchards in Nova Scotia. V. The predacious thrips *Haplothrips faurei* Hood. *Canadian Entomologist* 85, 33–40.

Maier, C.T. (1991) Abundance of pear thrips in Connecticut, with emphasis on spatial distribution in apple orchards. In: Parker, B.L. and Doane, E.N. (eds) *The 1990 Conference on Thysanoptera, Burlington, Vermont. Vermont Agriculture Experiment Station Bulletin* 698.

Mateus, C. and Mexia, A. (1995) Western flower thrips response to color. In: Parker, B.L., Skinner, M. and Lewis, T. (eds) *Thrips Biology and Management.* Plenum Press, New York, pp. 567–570.

Matteson, N., Terry, I., Ascoli-Christensen, A. and Gilbert, C. (1992) Spectral efficiency of the western flower thrips, *Franklinella occidentalis. Journal of Insect Physiology* 38, 453–459.

Matteson, N., Halbert, S.E. and Sandvol, L.E. (1995) Range and in-flight densities of barley thrips in the northwest United Stated. In: Parker, B.L., Skinner, M. and Lewis, T. (eds) *Thrips Biology and Management.* Plenum Press, New York, pp. 583–586.

McPherson, R.M., Beshear, R.J. and Culbreath, A.K. (1992) Seasonal abundance of thrips (Thysanoptera: suborders Terebrantia and Tubulifera) in Georgia flue-cured tobacco and impact of management practices on the incidence of tomato spotted wilt virus. *Journal of Entomological Science*, 27, 257–268.

Moffitt, H.R. (1964) A color preference of the western flower thrips, *Frankliniella occidentalis. Journal of Economic Entomology* 57, 604–605.

Moreno, D.S., Gregory, W.A. and Tanigoshi, L.K. (1984) Flight response of *Aphytis melinus* (Hymenoptera: Aphelinidae) and *Scirtothrips citri* (Thysanoptera: Thripidae) to trap color, size and shape. *Environmental Entomology* 13, 935–940.

Morgan, A.C. and Crumb, S.E. (1928) Notes on the chemotactic responses of certain insects. *Journal of Economic Entomology* 21, 913–920.

Morse, J.G., Bellows, T.S. and Iwata, Y. (1986) Techniques for evaluating residual toxicity of pesticides to motile insects. *Journal of Economic Entomology* 79, 281–283.

Munger, F. (1942) A method for rearing citrus thrips in the laboratory. *Journal of Economic Entomology* 35, 373–375.

Murai, T. (1988) Studies on the ecology and control of flower thrips, *Frankliniella intonsa* (Trybom). *Bulletin of the Shimane Agricultural Experiment Station* 23, 1–73.

Ota, A.K. (1968) Comparison of three methods of extracting flower thrips from rose flowers. *Journal of Economic Entomology* 61, 1754–1755.

Parker, B.L. and Skinner, M. (1993) Field evaluation of traps for monitoring emergence of pear thrips (Thysanoptera: Thripidae). *Journal of Economic Entomology* 86, 46–52.

Parker, B.L., Grehan, J.R. and Skinner, M. (1992) Method for extracting pear thrips (Thysanoptera: Thripidae) from forest soil. *Journal of Economic Entomology* 85, 865–869

Parker, B.L., Skinner, M. and Adamowicz. R.H. (1995) Induced emergence of pear thrips for population monitoring in forest soil. In: Parker, B.L., Skinner, M. and Lewis, T. (eds) *Thrips Biology and Management.* Plenum Press, New York. pp. 579–582.

Patrzich, R. (1984) Methoden zur Erfassung der Thysanopteren-Fauna in Getreidebestanden. *Mitteilungen aus der Biologischen Bundesenstalt fur Land-und-*

Forstwirtschaft, Berlin-Dahlem, No. 223, 346.

Payne, R.W., Lane, P.W., Ainsley, A.E., Bicknell, K.E., Digby, P.G.N., Harding, S.A., Leech, P.K., Simpson, H.R., Todd, A.D., Verrier, P.J. and White, R.P. (1987) *Genstat 5 Reference Manual*. Clarendon Press, Oxford.

Penman, D.R., Osborne, G.O., Worner, S.P., Chapman, R.B. and McLaren, G.F. (1982) A chemical attractant for *Thrips obscuratus* (Thysanoptera: Thripidae) in stone fruit in New Zealand. *Journal of Chemical Ecology* 8, 1299–1303.

Perkins, D.D. (1962) Preservation of *Neurospora* stock cultures with anhydrous silica gel. *Canadian Journal of Microbiology* 8, 591–594.

Perry, J.N. (1994) Sampling and applied statistics for pests and diseases. In Brain, P., Hockland, S.H., Lancashire, P.D. and Sim, C.C. (eds) *Sampling to Make Decisions*; *Aspects of Applied Biology* 37, 1–14.

Post, R.L. (1947) Special apparatus for collecting insects. *Bimonthly Bulletin North Dakota Agricultural Experiment Station* 9, 78–80.

Post, R.L. and Thomasson, G.L. (1966) The relative abundance and overwintering mortality of sod inhabiting thrips. *Department of Entomology Agriculture Experiment Station, North Dakota State University. North Dakota Insects Publication* 6.

Potter, C. (1952) An improved laboratory apparatus for applying direct sprays and surface film with data on the electrostatic charge on atomised spray fluids. *Annals of Applied Biology* 39, 1–28.

Powell, C.C. and Lindquist, R.K. (1992) *Boll pest and disease manual*. Ball Publishing, Geneva, Illinois.

Powell, D.M. and Landis, B.J. (1965) A comparison of two sampling methods for estimating population trends of thrips and mites on potatoes. *Journal of Economic Entomology* 58, 1141–1144.

Putman, W.L. (1965) Paper chromatography to detect predation on mites. *Canadian Entomologist* 97, 435–441.

Raffa, K.F., Hall, D.J., Kearby, W. and Katovich, S. (1992) Seasonal life history of introduced basswood thrips (Thysanoptera: Thripidae) in Wisconsin, with observations on associated thrips species. *Environmental Entomology* 21, 771–779.

Razvyazkina, G.M. (1953) The importance of tobacco thrips in the development of outbreaks of tip chlorosis of makhorka. (In Russian.) *Review of Applied Entomology (A)* 42, 146.

Reed, D.K. and Rich, J.R. (1975) A new survey technique for citrus thrips. *Journal of Economic Entomology*, 68, 739–742.

Robb, K.L. (1989) Analysis of *Frankliniella occidentalis* (Pergande) as a pest in floricultural crops in California greenhouses. PhD thesis, University of California, Riverside.

Robb, K.L., Newman, J., Virzi, J.K. and Parrella, M.P. (1995) Insecticide resistance in western flower thrips. In: Parker, B.L., Skinner, M. and Lewis, T. (eds) *Thrips Biology and Management*. Plenum Press, New York, pp. 341–346.

Rosenheim, J.A.S., Welter, S.C., Johnson, M.W., Gusukuma-Minuto, L.R. and Mau, R.F.L. (1990) Direct feeding damage on cucumber by mixed species infestations of two thrips, *Thrips palmi* and *Frankliniella occidentalis* (Thysanoptera: Thripidae). *Journal of Economic Entomology*, 83, 1519–1525.

Ross, G. (1987) *Maximum Likelihood Program 3.08*. NAG, Oxford.

Sakimura, K. and Carter, W. (1934) The artificial feeding of Thysanoptera. *Annals of the Entomological Society of America* 27, 341–342.

Salguero Navas, V.E., Funderburk, J.E., Beshear, R.J., Olson, S.M. and Mack, T.P. (1991). Seasonal patterns of *Frankliniella* spp. (Thysanoptera: Thripidae) in tomato flowers. *Journal of Economic Entomology* 84, 1818–1822.

Salifu, A.B. and Hodgson, C.J. (1987) Dispersion patterns and sequential sampling plans for *Megalurothrips sjostedti* (Trybom) (Thysanoptera: Thripidae) in cowpeas. *Bulletin of Entomological Research* 77, 441–449.

Salifu, A.B. and Singh, S.R. (1987) Evaluation of sampling methods for *Megalurothrips sjostedti* (Trybom) (Thysanoptera: Thripidae) on cowpea. *Bulletin of Entomological Research* 77, 451–456.

Schreiter, G., Butt, T.M., Beckett, A., Vestergaard, S. and Moritz, G. (1994) Invasion and development of *Verticillium lecanii* in the western flower thrips, *Frankliniella occidentalis*. *Mycological Research* 98, 1025–1034.

Shelton, A.M. and North, R.C. (1986) Species composition and phenology of Thysanoptera within field crops adjacent to cabbage fields. *Environmental Entomology* 15, 513–519.

Shelton, A.M., Nyrop, I.P., North, R.C., Petzoldt, C. and Foster, R. (1987) Development and use of a dynamic sequential sampling program for onion thrips, *Thrips tabaci* (Thysanoptera: Thripidae), on onions. *Journal of Economic Entomology* 80, 1051–1056.

Shipp, J.L. (1995) Monitoring of western flower thrips on glasshouse and vegetable crops. In: Parker, B.L., Skinner, M. and Lewis, T. (eds) *Thrips Biology and Management*. Plenum Press, New York, pp. 547–555.

Shipp, J.L. and Zariffa, N. (1991) Spatial patterns of and sampling methods for western flower thrips (Thysanoptera: Thripidae) on greenhouse sweet pepper. *Canadian Entomologist* 123, 989–1000.

Shirck, F.M. (1948) Collecting and counting onion thrips from samples of vegetation. *Journal of Economic Entomology* 41, 121–123.

Sites, R.W., Chambers, W.S. and Nicholls, B. (1992) Diel periodicity of thrips (Thysanoptera: Thripidae) dispersion and the occurrence of *Frankliniella williamsi* in onions. *Journal of Economic Entomology* 85, 100–105.

Skinner, M. and Parker, B.L. (1991) Bioecology of pear thrips: distribution in forest soils. In: Parker, B.L., Skinner, M. and Lewis, T. (eds) *Towards Understanding Thysanoptera*, USDA Forest Service, Northeastern Forest Experiment Station General Technical Report NE-147, pp. 193–212.

Skinner, M. and Parker, B.L. (1992) Vertical distribution of pear thrips (Thysanoptera: Thripidae) in forest soils. *Environmental Entomology* 21, 1258–1266.

Smith, L.M. (1933) The emergence of pear thrips in Healdsburg area of California in 1932. *University of California Agriculture Experiment Station Bulletin* 562.

Southwood, T.R.E. (1979) *Ecological Methods*. Chapman & Hall, London.

Steiner, M.Y. (1990) Determining population characteristics and sampling procedures for western flower thrips (Thysanoptera: Thripidae) and the predatory mite *Amblyseius cucumeris*. *Environmental Entomology* 19, 1605–1613.

Suman, C.L. and Wahi, S.D. (1981) Sequential sampling plan for the onion thrips, *Thrips tabaci* (Lind.). *Entomon.* 6, 265–269.

Tanigoshi, L.K. and Moreno, D.S. (1981) Traps for monitoring populations of the citrus thrips, *Scirtothrips citri* (Thysanoptera: Thripidae). *Canadian Entomologist* 113, 9–12.

Tappan, W.B. (1986) Relationship of sampling time to tobacco thrips (Thysanoptera: Thripidae) numbers in peanut foliage buds and flowers. *Journal of Economic Entomology* 79, 1359–1363.

Taylor, L.R. (1951) An improved suction trap for insects. *Annals of Applied Biology* 38, 582–591.

Taylor, L.R. (1961) Aggregation, variance and the mean. *Nature* 189, 732–735.

Taylor, L.R. (1984) Assessing and interpreting the spatial distributions of insect populations. *Annual Review of Entomology* 29, 321–357.

Teerling, C.R., Pierce, H.D., Borden, J.H. and Gillespie, D.R. (1993) Identification and bioactivity of alarm pheromone in the western flower thrips, *Frankliniella occidentalis*. *Journal of Chemical Ecology* 19, 681–697.

Terry, L.I. (1991) *Frankliniella occidentalis* (Thysanoptera: Thripidae) oviposition on apple buds: role of bloom state, blossom phenology, and population density. *Environmental Entomology* 20, 1568–1576.

Terry, L.I. and DeGrandi-Hoffman, G. (1988) Monitoring western flower thrips (Thysanoptera: Thripidae) in 'Granny Smith' apple blossom clusters. *Canadian Entomologist* 120, 1003–1016.

Teulon, D.A.J. and Cameron, E.A. (1991) Sampling thrips eggs in maple leaves and flowers. In: Parker, B.L. and Doane, E.N. (eds) *Proceedings of the 1990 Conference on Thysanoptera. Vermont Agricultural Experiment Station Bulletin* 698, p. 51.

Teulon, D.A.J. and Cameron, E.A. (1995) Within-tree distribution of pear thrips (Thysanoptera: Thripidae) on sugar maple. *Environmental Entomology* 24, 233–238.

Teulon, D.A.J. and Penman, D.R. (1992) Colour preferences of New Zealand thrips (Terebrantia: Thysanoptera). *New Zealand Entomologist* 15, 8–13.

Teulon, D.A.J., Cameron, E.A. and Hellested, R.T. (1992) How many leaves are there on a sugar maple tree. In: Cameron, E.A., Teulon, D.A.J., McCormick, L.H. and Kolb, T.E. (1992) *The 1991 Conference of Thrips* (Thysanoptera): *Insect and Disease Considerations in Sugar Maple Management*, General Technical Report NE-161, USDA Forest Service, Radnor, Pennsylvania, p. 55.

Teulon, D.A.J., Penman, D.R. and Ramakers, P.M.J. (1993) Volatile chemicals for thrips (Thysanoptera: Thripidae) host finding and applications for thrips pest management. *Journal of Economic Entomology* 86, 1405–1415.

Thompson, S.K., Ramsey, F.L. and Seber, G.A.F. (1992) An adaptive procedure for sampling animal populations. *Biometrics* 48, 1195–1199.

Vernon, R.S. and Gillespie, D.R. (1995) Influence of trap shape, size and background color on captures of *Frankliniella occidentalis* (Thysanoptera: Thripidae) in a cucumber greenhouse. *Journal of Economic Entomology* 88, 288–293.

Vestergaard, S., Gillespie, A.T., Butt, T.M., Schreiter, G. and Eilenberg, J. (1995) Pathogenicity of the hyphomycete fungi *Verticillium lecanii* and *Metarhizium anisopliae* to the western flower thrips, *Frankliniella occidentalis*. *Biocontrol Science and Technology* 5, 185–192.

Vité, J.P. (1956) Populationsstudien am Lärchenblasenenfuss *Taeniothrips laricivorus* Krat. *Zeitschrift für angewandte Entomologie* 38, 417–488.

Vogelmann, J.E. and Rock, B.N. (1991) Detection of pear thrips damage using satellite imagery data. In: Parker, B.L., Skinner, M. and Lewis, T. (eds) *Towards Understanding Thysanoptera*, General Technical Report NE-147, USDA Forest Service, Northeastern Forest Experiment Station, Radnor, PA, pp. 285–300.

Wadhams, L.J. (1990) The use of coupled gas chromatography: electrophysiological techniques in the identification of insect pheromones. In McCaffery, A.R. and Wilson, I.D. (eds) *Chromatography and Isolation of Insect Hormones and Pheromones*. Plenum Press, New York, pp. 289–298.

Walker, F.W. (1974) Responses of selected Thysanoptera to colored surfaces. *Environmental Entomology* 3, 295–304.

Ward, L.K. (1966) The biology of Thysanoptera. PhD Thesis, University of London.

Welter, S.C., Rosenheim, J.A., Johnson, M.W., Mau, R.F.L. and Gusukuma-Minuto, L.R. (1990) Effects of *Thrips palmi* and western flower thrips (Thysanoptera: Thripidae) on the yield, growth and carbon allocation pattern in cucumbers.

Journal of Economic Entomology 83, 2092–2100.

Wetzel, T. (1963) Zur frages der Überwinterung der Gräser-Thysanoptera. *Zeitschrift für angewandte Entomologie* 51, 429–441.

Wilmot, S.H. and Kelley, R.S. (1990) Stand damage assessment. In: Parker, B.L., Skinner, M., Wilmot, S.H. and Souto, D. (eds) *Pear Thrips Research and Management: Current Methods and Future Plans. Bulletin University of Vermont Agricultural Experiment Station* 697, 117–118.

Wolfenbarger, D. and Hibbs, E.T. (1958) Onion thrips (*Thrips tabaci* Lind.) infesting cabbage. *Journal of Economic Entomology* 51, 394–396.

Yudin, L.S., Mitchell, W.C. and Cho, J.J. (1987) Color preference of thrips (Thysanoptera: Thripidae) with reference to aphids (Homoptera: Aphididae) and leafminers in Hawaiian lettuce farms. *Journal of Economic Entomology* 80, 51–55.

Zareh, N. and Morse, J.G. (1989) Influence of temperature and life stage in monitoring for pesticide resistance in citrus thrips (Thysanoptera: Thripidae) with residual bioassays. *Journal of Economic Entomology* 82, 342–346.

Culturing Thrips and Parasitoids

12

Antoon J.M. Loomans[1]
and Tamotsu Murai[2]
[1] *Section Entomology, Plant Protection Service, Geertjesweg 15, PO Box 9102, 6700 HC Wageningen, The Netherlands;* [2] *Research Institute for Bioresources, Okayama University, 2-20-1 Chuo, Kurashiki 710, Japan*

Introduction

The increase in economic significance of thrips species such as *Thrips palmi* and *Frankliniella occidentalis* as key pests on vegetables and ornamental crops worldwide has also increased interest in culturing thrips for various purposes. The rearing procedures that are followed will depend on the objective and scope of research, whether it be for screening of pesticides (Brødsgaard, 1993), mycopathogens (see Chapter 10) or host plant cultivars for resistance breeding (Mollema *et al.*, 1993; Van Dijken *et al.*, 1994), for producing thrips as beneficials (Maddox and Mayfield, 1972; Cock, 1982) or as hosts for natural enemies (parasitoids: Loomans *et al.*, 1995) or for virus transmission studies (Sakimura, 1961; Lowry *et al.*, 1992; Ullman *et al.*, 1992; Wijkamp and Peters, 1993). For each of these purposes a number of different methods are used for each step in the culturing programme. Mass production of thrips in greenhouses or large cages may be needed to provide a large stock culture as a back-up system. Methods which allow the synchronization of cohorts of even-aged larvae are essential for serial tests and bioassay, for behavioural and life-history studies, to establish the interactions between thrips, host plants and natural enemies, or to perform virus acquisition and transmission studies.

Sakimura (1961), Peterson (1964), Lewis (1973), Bournier (1983), Ananthakrishnan (1984), Chiu (1987) and Murai (1988), reviewed and illustrated different methods for rearing thrips; Bay-Petersen (1990), Anderson and Leppla (1992) and Singh and Moore (1985) provide more general information on rearing other insect groups. In this chapter various techniques for handling, confining and culturing thrips, including newly devised ones, are reviewed and some described and illustrated. Space is too limited to give a full account of all the techniques available, but details can be found in the papers referred to. Almost all methods have been devised for rearing multivoltine pest species which do not diapause in the soil. Univoltine species with only one generation a year are more difficult to rear. A few records mention culturing thrips species during part of their life cycle, ignoring the several months spent in soil or litter as occurs in grass feeders like *Limothrips*, *Anaphothrips*, *Haplothrips* or *Frankliniella* (Köppa, 1970; Bournier and Bernaux, 1971; Lewis, 1973) and tree-infesting species like *Taeniothrips inconsequens*.

All methods require equipment in which specimens can be cultured either in large or small groups or individually, with instars that are mixed or synchronized. Thrips can be either reared on whole plants (within a crop or on individual plants), excised plant parts (leaves, leaf discs, fruits, pods) or on artificial substrates (sugar or honey solution, pollen or dry leaf powder and diets). The choice of rearing method is also related to the feeding habits of the species involved: phytophagous (free-living flower or leaf feeding, or gall inhabiting), mycophagous or entomophagous (parasitoid or predator).

Management of Thrips Rearing

Each rearing programme needs careful management of the work to be carried out (Fig. 12.1). To produce thrips for research in large enough quantities and of constant quality in an efficient and reliable way, a number of key elements need to be fulfilled.

1. The research objectives must be clearly defined.
2. A proper taxonomic identification of the foundation stock of thrips

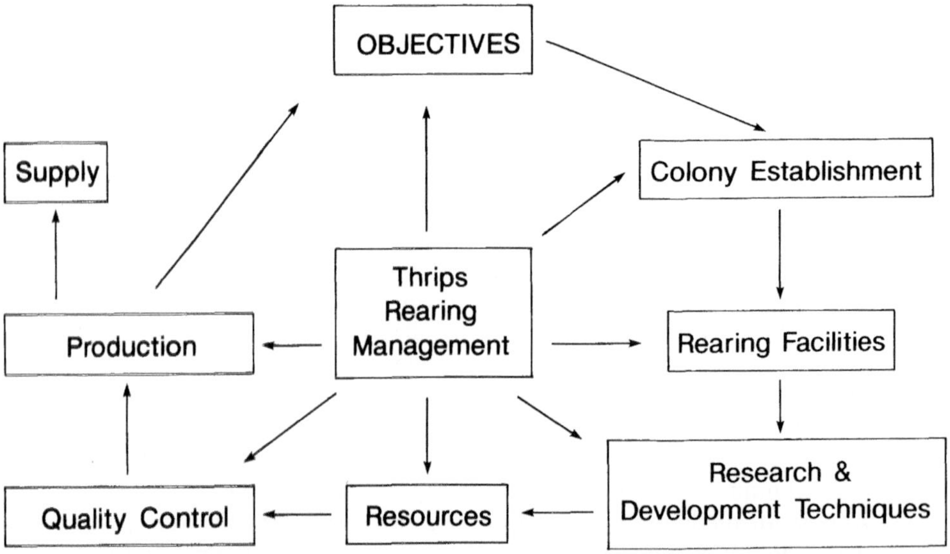

Fig. 12.1. Pathways for a management system for rearing thrips, indicating key elements (after Singh and Moore, 1985).

collected in the field is essential before and during the maintenance of any rearing and testing programme.
3. Rearing facilities and technical equipment should be designed to create a reliable and hygienic environment.
4. Basic biological information (e.g. food, and a system of ageing individuals), obtainable from observations or the literature, needs to be collected for each species to be reared. Before rearing can begin, a culture of the host plant needs to be established and a choice made of the plant parts to be used. Different species vary in their natural host preference and feeding habits. The preferred natural host plant may not necessarily be the most suitable one for culturing or experiments. The criteria needed for good rearing plants include tolerance of thrips feeding, morphological suitability for easy recovery and handling of thrips (Sakimura, 1961), and, for virus studies, good symptom expression (Ullman *et al.*, 1992; Wijkamp and Peters, 1993). Therefore, flowers are often unsuitable. A basic rearing programme usually consists of a sequence of different culture methods (see also Table 12.1) for both thrips and host plant material. This will include breeding host plants, producing stock cultures (of thrips, viruses or prey), synchronizing developmental stages and (individual) testing. For synchronizing different life stages, a rearing protocol must be further subdivided into a series of steps: oviposition and incubation of eggs, larval and nymphal rearing, testing and/or storing thrips, each requiring different culture methods.
5. Evaluation of resources available for maintaining a rearing protocol (space, money, labour, equipment).
6. Regular control of the quality of the cultured specimens and avoidance of contamination by bacteria, fungi, other species or strains of thrips or other insects. Only when the system is reliable can results be reproducible.
7. Estimation of the final output needed for research, colony maintenance and supply to others.

Handling and Confining Thrips

Various techniques for handling thrips have been described by a number of authors (e.g. Sakimura, 1961; Lewis, 1973). Because of their minute size and vulnerability, individuals should be handled as little as possible. Both larvae and adult thrips can be transferred individually from one place to another with a slightly moistened, fine brush (camel or marten hair, size 00 to 0000 trimmed to a few hairs only) either alive, chilled or anaesthetized with CO_2. They can be brushed off plant parts with a larger brush (size 8 to 10). Large numbers of adults or rapid movers can be picked up with an aspirator (Beavers and Ewart, 1971) or a mouth-operated suction device comprising a disposable polypropylene pipette tip covered with a 60 to 120 μm gauze, a mouthpiece and a small electrical pump.

Immature stages and even adults of many panchaetothripine species can be reared uncaged (Callan, 1947; Lewis, 1973) because they are fairly immobile. Many thripine species, like *Frankliniella* and *Thrips*, are extremely active and

Table 12.1. Types of systems used for caging, rearing and testing thrips on various food sources.

Rearing aim	Food source	Devices/method	Sealant
1. Stock culture	Whole plants	Greenhouse, cage	1
	Cuttings	Cage, cylinder	1
	Seedlings	Cylinder, cup	1
	Plant parts	Glass jar	1, 2
2. Synchronization of even-age cohorts	Plant parts	Glass jar	1, 2
	Excised leaves	Cylinder, vial	1, 3
	Pollen + honey	Plastic rings	5
3. Testing individuals	Whole plant	Cylinder, sleeve	1
		Clip-on cage	1, 6
	Seedling	Cylinder, vial	1, 3, 4
	Excised leaves	Munger cell	6
		Plastic rings	5
		Leaf island	1, 4
	Leaf discs	Petri dishes, multidishes	1, 4, 6
		Reaction tubes	
	Pollen + honey	Plastic rings	5

Sealant: 1 = cloth, gauze; 2 = precision paper; 3 = cotton wool; 4 = permeable film (plastic, cellulose); 5 = thin membrane; 6 = glass, acrylate.

need to be confined in an escape-proof cage. Devices for confining thrips are many and ingenious, varying from disposable coffee cups to specialist caging equipment. However, in any rearing device it is important to establish an optimal microclimate, that is, to find a balance between isolation and ventilation, while also maintaining good visibility and accessibility. In particular, it is essential to establish a constant relative humidity to prevent condensation in confined situations. This can be achieved by using saturated solutions for the control of humidity (Winston and Bates, 1960) in enclosed systems, or by culturing thrips in insect growth cabinets, climate rooms or in computer-driven greenhouse systems.

For rearing thrips, various cages, sealing devices and food sources are employed (Table 12.1). Thrips can be reared on host plants grown in metal-, wood- or plastic-framed cages partly covered with fine cloth. Rearing cages must be sealed without gaps larger than 120 µm because many thrips species are small, easily damaged and have a remarkable ability to squeeze through undetectable crevices. Similar cages can be used to sleeve small, growing trees on which thrips are living using synthetic thrips-proof gauze (mesh size 60–120 µm). Small jars can be sealed with gauze, paper (precision wipers: Kimwipes®, Kimberly Clark) or cellulose sheeting (Bailey, 1931; Kinzer *et al.*, 1972) kept in place by a snap-on or screw-on lid. For confining individuals either to restrict movement on plants, or when rearing in glass or plastic tubes, multiwell or

Petri dishes, gauze or cellulose film (Bailey, 1931, 1932) can be used to seal containers; special attention is needed when using cotton wool plugs to ensure that thrips do not get stuck in the fibres. Plastic rings can be sealed by Sealon® film (Fuji Film Co., Japan) or Parafilm M® (American Can Co., USA). Multidishes with 6, 12 or 24 cells (cell diameter greater than 1.5 cm) can be used for individual testing with the ventilation tabs of the lids cut off, the lids secured by large clamps, and either a piece of white tissue paper placed between the dishes and the lids to prevent condensation (Brødsgaard, 1993), or the lids pierced with gauze-covered ventilation holes (Hall *et al.*, 1993). For species whose larvae frequently try to escape, thin (14 μm) transparent permeable plastic film can be used as a cover (Mollema *et al.*, 1990).

Phytophagous Thrips

Rearing on host plants

Rearing on whole plants

Various authors mention mass rearing using host plants in greenhouses (Bailey, 1932; Rivnay, 1935; Sakimura, 1961; Beavers and Ewart, 1971; Tanigoshi and Nishio-Wong, 1981; Sengonca and Gerlach 1983; Nugaliyadde and Heinrichs, 1984). This method is generally used to maintain stock cultures or for mass propagation of all life stages to provide large quantities of thrips at a given time. Usually potted host plants are used, either full-grown or seedlings, such as *Gynaikothrips ficorum* on *Ficus retusa*, *Heliothrips haemorrhoidalis* on *Viburnum tinus*, *Parthenothrips dracaenae* on *Dracaena* or *Cordyline* (Bournier and Bernaux, 1971). Flower thrips, like *Frankliniella occidentalis*, can be reared on caged cut flowers of chrysanthemum, *Thrips tabaci* on caged onion plants.

Tanigoshi and Nishio-Wong (1981) developed a large-scale greenhouse system for rearing *Scirtothrips citri* consisting of four groups of uniformly sized (44 to 55 cm high), 1-year-old Californian sumac shrubs, *Rhus laurina*, at various phases of *S. citri* infestation. Successful germination and propagation of Californian sumac was ensured by autoclaving seed at 0.7 kg cm^{-2} for 10 s. When a group of plants was infested every week, each group potentially supported two generations of citrus thrips over a 4 week period. This schedule provided a continuous cycle of the larval instars and adults. When confined within 3.8 litre containers, *R. laurina* becomes susceptible to water stress and requires daily watering. Both host plant and *S. citri* adapted to greenhouse temperature extremes of 14 to 49°C. Peak population levels for larval and adult life stages per shrub were about 2500 and 515, respectively, during a 4-week infestation cycle.

Nugaliyadde and Heinrichs (1984) mass produced *Stenchaetothrips biformis* using rice seedlings (Fig. 12.2). Thrips were reared on 3- to 4-week-old rice plants in an insectary. Five adult pairs were placed on five plants growing in a pot and the females allowed to oviposit for 3 days. These adults were then removed to reinfest another set of plants. Since females lived about 13 days,

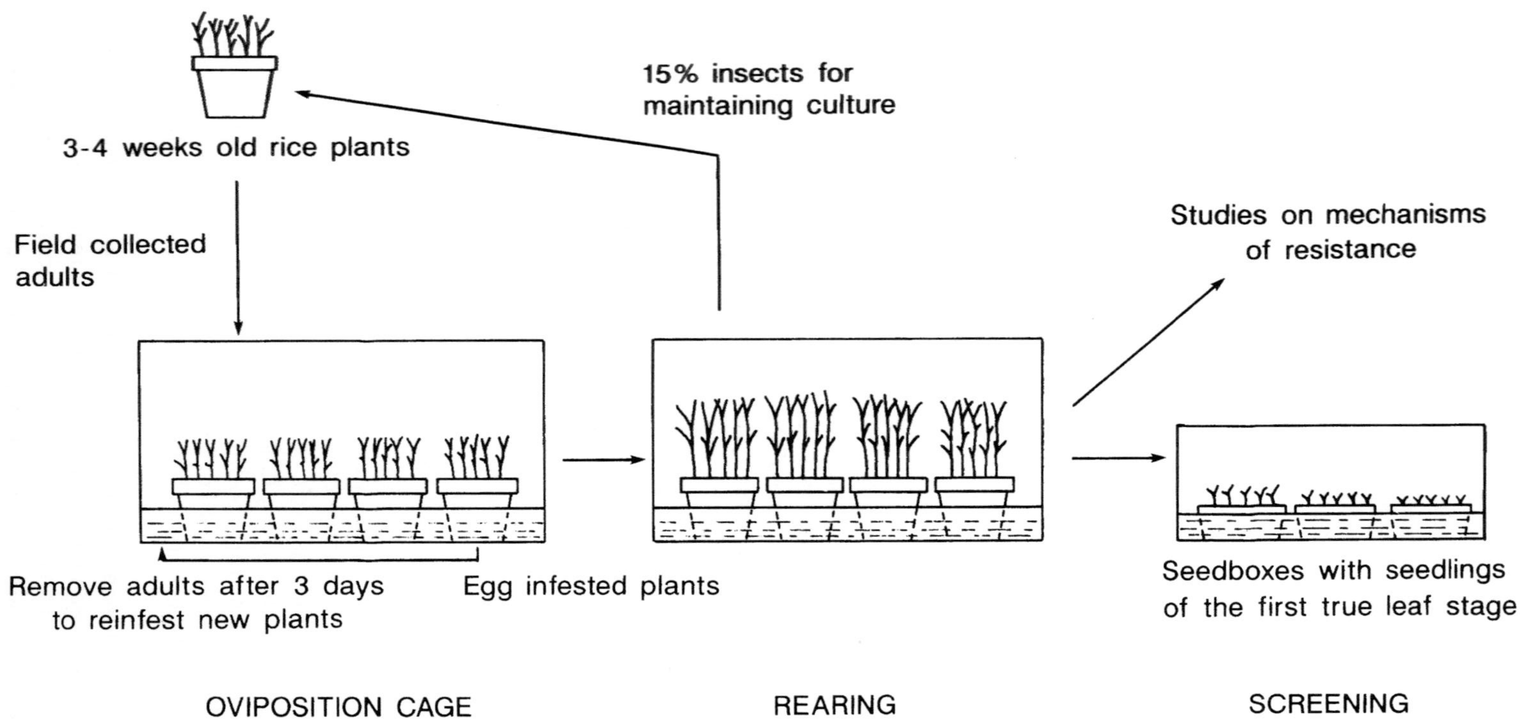

Fig. 12.2. Procedure for rearing and greenhouse screening of *Stenchaetothrips biformis* on rice, for resistance studies (after Nugaliyadde and Heinrichs, 1984).

they were used to oviposit in at least four sets of plants. About 20 adults were reared on each set of potted plants by 17 days after infestation. Since the adults had to be removed very carefully from the plants on which they had laid, about 2 hours were required to infest 25 plants per day, so only 400 to 500 adult thrips were reared daily; 15 per cent of individuals were kept to maintain the colony, the remainder were used for experimental purposes. De Alwis (1983) used a modification of this technique by uprooting heavily infested rice seedlings from the field and planting them in groups of five in shallow plastic trays, surrounded by four groups of 1-week-old uninfested seedlings. The thrips moved from the infested to the uninfested seedlings and reproduced. The colony was maintained by uprooting the newly infested seedlings and planting new uninfested ones.

Van Dijken *et al.* (1994) used a similar technique to maintain a stock culture of *Frankliniella occidentalis* within thrips-proof gauze tents of 15 to 30 m^3, each containing about 200 potted flowering chrysanthemum plants, kept inside an isolated greenhouse at 25 to 30°C with supplementary lighting; 25 per cent of the plants were replaced weekly (Balkema and Van Dijken, personal communication). Mollema *et al.* (1990) and Brødsgaard (1993) used a similar set-up for culturing *F. occidentalis* on cucumber plants resistant to powdery mildew and on french bean plants (*Phaseolus vulgaris*), respectively.

Steiner *et al.* (1996) employed infested dwarf green bean plants grown in 15 cm tall pots as a back-up system for *Frankliniella schultzei*. Four such plants at about the two true-leaf stage were placed on a tray of vermiculite in a sealed cage, screened on the top and two sides with 112 μm mesh. The top of the soil was covered with 5 cm of vermiculite to discourage fungus gnats. The pots were provided with bases to prevent water saturating the vermiculite. Two plants were changed weekly, with the tops cut and left for a few days to allow the thrips to complete development.

For *in situ* studies, for rearing and testing thrips in small groups or individually on the intact plant, both transparent leaf and sleeve cages can be used (Medved and Fleschner, 1970; Tamò *et al.*, 1993; de Jager, 1995). Clip-on sandwich leaf cages, widely used for less mobile insects like whitefly and aphids, are less popular for thrips, because of the difficulty of fixing them to the leaf tightly enough to prevent escape without injuring the leaf. Sakimura (1961) developed a felt cage that can be clipped on a leaf for virus transmission tests by individual *T. tabaci* larvae (Fig. 12.3b). De Jager *et al.* (1995) used a cage consisting of a plastic ring supported by another latex foam ring. The cage of Sharma and Pande (1981) uses foam rubber or cotton wool walls. A glass or clear plastic cover enables the inmates of such cages to be observed without disturbance.

Sticky ring barriers, such as Cellucotton®, Tanglefoot®, Stop® and Stikem®, have been used by some authors (e.g. Sakimura, 1961; Gilstrap and Oaten, 1976) to restrict thrips larvae on a leaf floating in water, but larvae still often manage to escape or run into the barrier and die. Escapes and mortality seem to be far less frequent in arenas on leaves, ringed by a sticky barrier consisting of a mixture of Canada balsam and castor oil (ratio 1.5:1) as used by Tsai *et al.* (1995) for developmental studies on *Thrips palmi*.

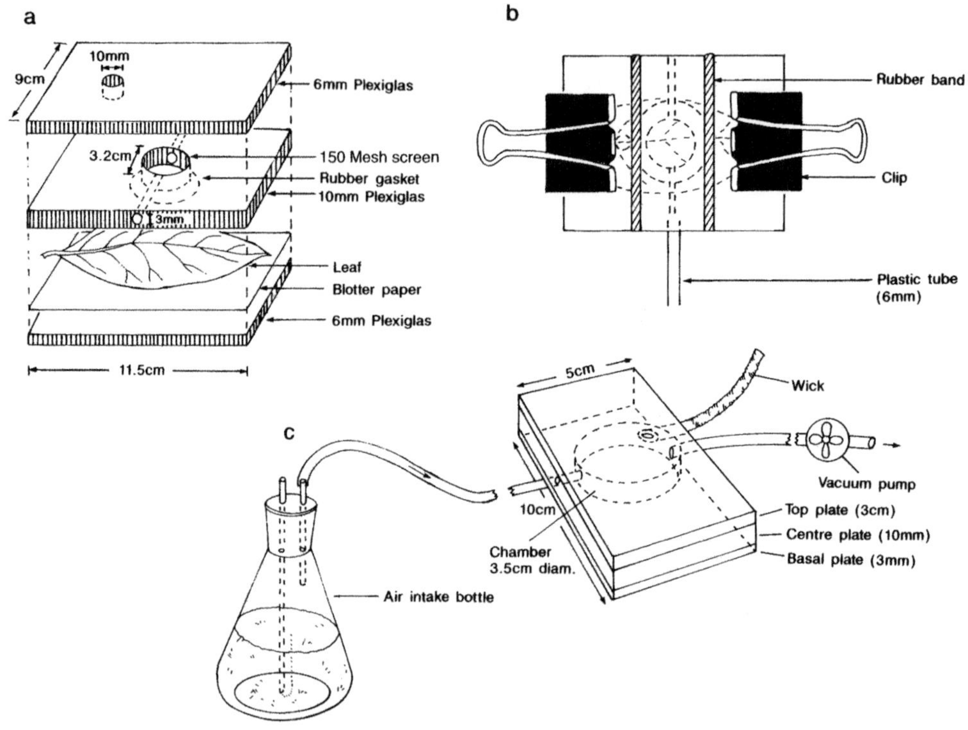

Fig. 12.3. (a) Modified plexiglas Munger cell unit, 'exploded' side view; (b) top view of cell held together by rubber bands and clips with tube to air supply. Used for rearing on excised leaves and especially for bioassay. The cells are placed in racks and each connected via the plastic tube to a blown air supply (after Morse *et al.*, 1986). (c) Ventilated glass cell for monitoring acute insecticide toxicity. The ventilation holes into the chamber are covered with 200-μm mesh nylon screen (after Brødsgaard, 1994) (see Chapter 11).

Rearing on detached plant parts

All these methods use plant parts for rearing thrips, but they differ in their suitability for different purposes. Methods using detached organs (bean pods, cucumber fruits, citrus fruits or parts of the stem) are appropriate for maintaining stock cultures, for mass propagation, and also for synchronization (as a result of a fixed exposure time) of life stages (Table 12.1, rearing aims 1 and 2). The host plant material is easy to handle and tissue containing eggs can be stored for a few days at a low temperature. Methods using excised leaves and leaf discs are time consuming, transfer routines are hazardous, they are more vulnerable to contamination, and the host plant material has a limited lifetime. They are therefore less suitable for whole life-stage studies, but more appropriate for synchronization and serial testing procedures for small groups or individual thrips (Table 12.1, rearing aims 2 and 3).

A well known and popular approach using plant parts for rearing thrips is the bean-pod method. Bailey and Smith (1956) first reared *Frankliniella occidentalis* on bean pods in glass jars. This method has been adopted and slightly modified by several workers, using either bean sections (Sakimura, 1961; Lublinkhof and Foster, 1977; Lowry *et al.*, 1992) or complete beans (Ullman *et al.*, 1992; Loomans *et al.*, 1995; Wijkamp *et al.*, 1995). Details of the procedure followed by the latter are as follows (Plate I.2). Glass jars (0.5 to 1 litre) covered with fine mesh gauze (80 μm) or precision paper for ventilation served as rearing units. Ten to 15 layers of coarse paper were placed on the bottom of each jar to provide a pupation site. Fresh bean pods purchased from a supermarket were placed on the paper. Commercially purchased bee-pollen was supplied as additional food. A few hundred adults were introduced into a jar with several pods on which they laid eggs for 1 to 3 days at 25°C. Then adults were removed (by a brush or shaking) and the bean pods replaced by new ones for further egg laying. The egg-infested pods were transferred to rearing jars, in which eggs hatch within a day or two, to provide larvae suitable for experiments or to develop to adults on added bean pods. At 25°C, the complete process takes about 2 weeks. Pods were replaced twice a week (3 or 4 days apart).

Because of contamination risks, green beans brought into the laboratory were washed in warm soapy water and rinsed with tap water, bleached for 2 to 3 min, rinsed again in water (Lowry *et al.*, 1992) and dipped in fungicide (Steiner *et al.*, 1995). They were then dried on paper towelling and stored in a plastic bag at 4°C until use. Problems with bacteria and fungi were thereby reduced but not eliminated.

The bean-pod method can also be used for rearing other *Frankliniella* species: *F. schultzei* (light and dark forms) (Loomans *et al.*, 1995; Steiner *et al.*, 1995), *F. intonsa* (Wijkamp *et al.*, 1995), *F. fusca* (Lowry *et al.*, 1992) and thrips parasitoids (Loomans *et al.*, 1995). The same procedure is also suitable for rearing *Thrips tabaci* on french bean pods, leek or onion shoots (Loomans, unpublished) or cabbage (*Brassica oleracea*) midribs (Sakimura, 1961), either alone or in combination with different plant parts. For *F. occidentalis*, Mollema *et al.* (1993, 1995) used small cucumber fruits or sliced fruit parts disinfected with 0.1% benomyl, and Doane *et al.* (1995) used bean leaves. The method is useful for mass production and maintaining a stock colony, and also for producing synchronized cohorts of first instars. To ensure this, adults are allowed to lay eggs in new pods or fruits for a short time, such as 1 day; they are then brushed off and pods or fruits incubated or stored at low temperature until required.

Rearing on excised leaves

A second type of method for rearing and testing thrips is a sandwich-cage using detached leaves, originally developed by Munger (1942) for life-history studies of *Scirtothrips citri*. Tashiro (1967) developed this into a self-watering acrylic cage and Morse *et al.* (1986) modified it further into a ventilated cage with a forced air supply. In modified Munger cells (Fig. 12.3 a, b and c), thrips are confined in a circular hole (25 to 35 mm diameter) in the centre of a rectangular piece of glass or plexiglass, approximately 8 × 10 × 1 cm, placed over the detached leaf. A rubber ring of the same diameter is glued to the lower surface of this plate for a tight fit to the leaf. The leaf is placed on top of cloth or blotting paper on the bottom layer which is kept moistened by a cotton wool plug or filter paper inserted in water or a nutrient solution. Large clips or rubber bands are used to hold the cage together and to seal it. Depending on the age and species of the leaf and the number of thrips allowed to feed, leaves can be preserved for 10 to 14 days or longer. This type of cage is widely used for serial tests on pesticide toxicity (Morse *et al.*, 1986; Grasselly *et al.*, 1991; Brødsgaard, 1994) or virus acquisition (Wijkamp and Peters, 1993), for behavioural studies (Loomans *et al.*, 1995) and larval development of various species in small groups of about 25 to 50 individuals.

In yet another approach, glass tubes or acrylic cylinders are used for rearing thrips on detached leaves (Bailey, 1932; Beavers and Ewart, 1971; Köppa, 1970; Lewis, 1973; Parrella and Horsburgh, 1978), twigs (Rivnay,

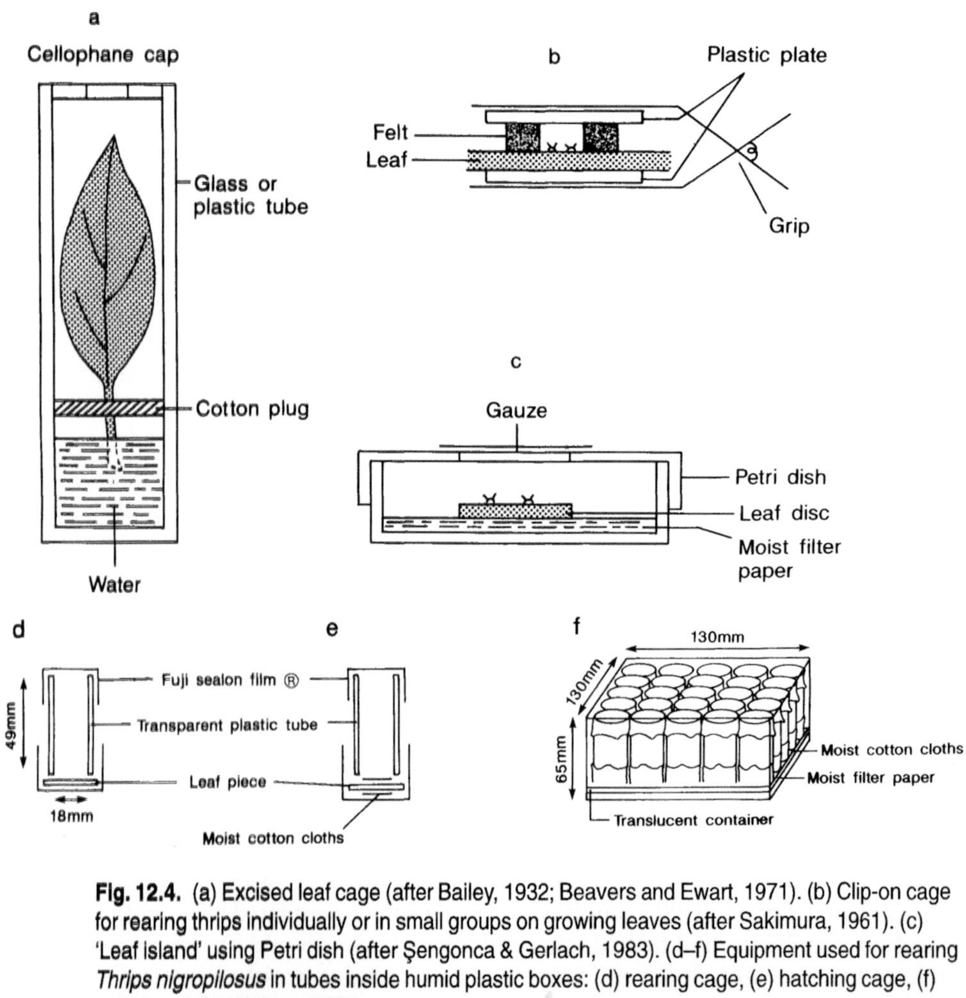

Fig. 12.4. (a) Excised leaf cage (after Bailey, 1932; Beavers and Ewart, 1971). (b) Clip-on cage for rearing thrips individually or in small groups on growing leaves (after Sakimura, 1961). (c) 'Leaf island' using Petri dish (after Şengonca & Gerlach, 1983). (d–f) Equipment used for rearing *Thrips nigropilosus* in tubes inside humid plastic boxes: (d) rearing cage, (e) hatching cage, (f) container (after Nakao, 1993).

1935) or seedlings (Koch, 1979; Steiner *et al.*, 1995) (Fig. 12.4a). Cages usually consist of two parts: a unit containing the plant material with a ventilation hole on top, kept in place by, and isolated from, a watering unit by cotton wool (Bailey, 1932), rubber (Beavers and Ewart, 1971), or adhesive material (e.g. Blu-Tac^{TM}, Steiner *et al.*, 1995). Grass feeders like *Limothrips*, *Anaphothrips*, *Haplothrips* and *Frankliniella* can be reared individually on pieces of leaf or spikelets kept in small glass tubes (Köppa, 1970; Malchau, 1991), which have one end embedded in plaster and the other plugged with cotton wool (Lewis, 1973). When grown from seed or when embedded in plaster of Paris, this prevents condensation on the wall of the vials and thus moulding of plant parts. This method is most applicable for rearing of individuals or small groups of larvae or adults (15 to 20) for serial tests; however, it is time consuming: twigs and leaves need to be changed every 2 to 3 days.

Rearing on leaf discs

A third method is leaf disc flotation, using leaf fragments floating on water or a nutrient solution. This has been used for *Selenothrips rubrocinctus*, *Dinurothrips hookeri*, *Caliothrips insularis* and *Heliothrips haemorrhoidalis* (Callan, 1947) and for *Limothrips*, *Chirothrips* and *Aptinothrips* (Lewis, 1973). Şengonca and Gerlach (1983) developed it into a 'leaf-island' method for life-history studies on *Scolothrips* in the laboratory (Fig.12.4c). Leaf discs 2.5 cm in diameter are laid with the underside uppermost on a layer of moist filter paper in small Petri dishes, 5.5 cm in diameter. Petri dish lids with a 2.5-cm diameter hole in the centre covered with a piece of gauze are placed over the dishes. Every 2 or 3 days the evaporated water is replaced with a syringe. Using this method cotton leaves can be kept turgid for 10 to 15 days, and green bean leaves for 10 to 16 days at 25°C and 50 per cent relative humidity.

The leaf island principle has been adapted in various ways for rearing individuals for life-history studies either in single Petri dishes or multidishes using agar (Brodeur and Cloutier, 1992), florist's oasis discs (Gaum *et al.*, 1994), water (Callan, 1947; Mollema *et al.*, 1990, 1993), plaster of Paris (Lewis, 1973; Hall *et al.*, 1993), moist blotting or filter paper (Callan, 1947; Wang and Chu, 1986) water-soaked cotton (Samsøe-Petersen 1983; Brødsgaard, 1993; Doane *et al.*, 1995) or a sponge pad (Gilstrap and Oaten, 1976), containing a nutrient solution to keep the disc in a good condition. Thrips can be easily handled and observed. In multidishes, bean leaf discs stayed fresh for more than 2 weeks (Brødsgaard, 1993), but often problems are encountered with saprohytic bacteria or fungi growing on the water, agar or leaf discs, so these need to be changed once or twice during larval development (Mollema *et al.*, 1990; Gaum *et al.*, 1994). Mortality can be high and the method is time consuming (Tsai *et al.*, 1995).

Murai (1988) used a similar method for the leaf dwellers *Scirtothrips dorsalis*, *Mycterothrips glycines* and *Thrips setosus*. For oviposition, a leaf disc of 3 cm diameter is put on the side of a ring cage, covered with a double layer of thin plastic membrane on top and a single layer to seal the bottom. Every 2 days, leaf discs are replaced. Oviposited leaf discs are placed in a Petri dish on moist filter paper for incubation and hatched larvae transferred to an excised leaf in a plastic or glass ring cage. Teulon (1992) adopted this method for *F. occidentalis* and Nakao (1993) for rearing individual larvae of *Thrips nigropilosus* (Fig. 12.4 d and e). Rings containing thrips larvae were placed inside humid closed plastic boxes for development (Fig. 12.4f). Wijkamp and Peters (1993) developed a leaf disc assay for virus transmission studies: a leaf disc (15 mm diameter) was enclosed with viruliferous adult thrips in a 2-ml reaction tube; after 24 h the disc was removed and placed in a multiwell system for development of the virus symptoms.

Rearing on artificial food sources

Rearing on solid diets

An artificial diet has one main advantage over natural foods: it avoids the seasonal unavailability of the natural diet and allows continuous laboratory rearing of insects of known 'quality' throughout the year (Singh, 1983). Many species of insects have been reared on various types of artificial diets (Singh, 1977) but, so far, thrips have been reared only incidentally on a complete defined formula. Koch (1978) reared *Hercinothrips femoralis* on a synthetic sugar diet. Kinzer *et al.* (1972) and Lowry *et al.* (1992) reared *Frankliniella fusca* larvae on an artificial diet derived from Vanderzant's modified wheat germ formula (Vanderzant *et al.*, 1962; Table 12.2). Thrips larvae were shaken directly from the plant on to the wheat germ diet in 30 ml medicine cups and

Table 12.2. Composition of artificial diets used for rearing larvae of *Frankliniella fusca* on a solid diet (Kinzer *et al.*, 1972) and *Frankliniella occidentalis* on a liquid diet (de Jager, 1995), modified from multi-insect diets (Singh, 1977, 1983; Vanderzant *et al.*, 1962).

Ingredients and quantity	*F. fusca* solid	*F. occidentalis* liquid	Group
Casein	3.5 g	7.0%	1
Salts (Wesson's)	1.0 g	2.0%	1
Wheat germ (cold-rolled)	3.0 g		1
Cholesterol		0.1%[a]	2
Linoleic acid		0.5%[a]	2
Vitamins			3
B mix (+ inositol) (Vanderzant's mix)	1 ml	4.0%	
Choline chloride	0.1 g	–	
Ascorbic acid	0.4 g	–	
Sucrose	3.5 g	6.0%	3
Distilled water	85 ml[c]	80.4%[b]	
Agar	2.5 g		3

[a]Ingredients of group 2 were dissolved in 50 ml dichloromethane, mixed with group 1 and the dichloromethane was evaporated; [b]ingredients of group 3 were dissolved in water, mixed with groups 1 and 2 and the pH was adjusted to about 6.5 with 4.0 M KOH; [c]for non-aseptic rearing 0.2 g of potassium sorbate and 0.2 g of methyl parahydroxybenzoate were included as antimicrobial agents.

maintained at 26.7°C. Approximately 100 larvae could be reared to adults in each cup if a stable relative humidity was maintained to prevent water condensation. An average of 87 per cent of larvae kept on artificial diet matured to adults compared with 63 per cent survival of larvae confined on peanut leaves, due to slower deterioration of the medium than the plant tissues. Life-cycle duration was approximately the same, but the medium is not suitable for oviposition (Kinzer *et al.*, 1972).

Rearing on artificial membranes

Sakimura and Carter (1934) were the first to use an artificial membrane (fish skin) stretched over a 3 per cent sucrose solution for feeding second instar larvae and adults of *Thrips tabaci*. Later, Day and Irzykiewicz (1954) used a plastic membrane for feeding *Thrips tabaci* and *Thrips imaginis*. Laughlin (1971) developed a culture method using Parafilm M® for feeding both larval and adult *Hercinothrips femoralis*, *T. tabaci*, *T. imaginis* and *Thrips australis*, and was the first to show that eggs could be laid through such a membrane. Many flower dwellers from a wide range of plant species (Kudo, 1971; Teulon and Penman, 1990) feed on pollen (Kirk, 1985) (see Chapter 4). Murai and Ishii (1982) used pollen and honey solution offered through a stretched thin Sealon® film. With this method the complete life cycle of species such as *Frankliniella intonsa*, *Thrips tabaci*, *Thrips hawaiiensis*, *Thrips coloratus*, *Thrips flavus* and *Thrips nigropilosus* could be studied. Their method has been modified for other purposes: Kirk (1985, 1994) for oviposition studies, Teulon and Penman (1986, 1991) to rear *Thrips obscuratus*, Teulon (1992) for individual rearing of *Frankliniella occidentalis* and Galazzi *et al.* (1992) and Loomans *et al.* (1995) to rear *F. occidentalis* and its parasitoid, *Ceranisus menes*. De Jager (1995) modified Singh's solid multi-order diet (Singh, 1983) into a liquid diet to rear and test the effect of different leaf sap fractions on chrysanthemum resistance to *F. occidentalis*. Koyama and Matsui (1993, 1995) modified Ehrhardt's liquid diet developed for aphids (Singh, 1977) to rear adults and larvae of *Thrips palmi*. T. Murai and A.J.M. Loomans (unpublished data) improved the diet to allow mass production of thrips and parasitoids and the description below basically follows their design, if not stated otherwise (Fig. 12.5).

Cages

Two types of cages are required: one for oviposition and one for rearing larvae (Fig. 12.5, I and V). Both consist of a plastic or poly(meth)acrylate cylinder (8 cm diameter, 5 cm tall). In rearing units the bottom end is closed by a double layer of stretched Sealon®. In oviposition units, gauze (60 to 100 μm) glued to the bottom can replace the Sealon. After placing the adults or larvae (200 to 300 per cage) and pollen in the cage, the top end is closed with stretched Sealon film. A few millilitres of water or a nutrient solution is deposited (see below) on top and covered with a Petri dish lid or another layer of Sealon film for oviposition and rearing, respectively. Smaller rings (20 mm diameter, 3 cm tall) can be used for individual rearing. Rearing cages are kept in a translucent closed container and placed above a saturated solution of potassium nitrate, to

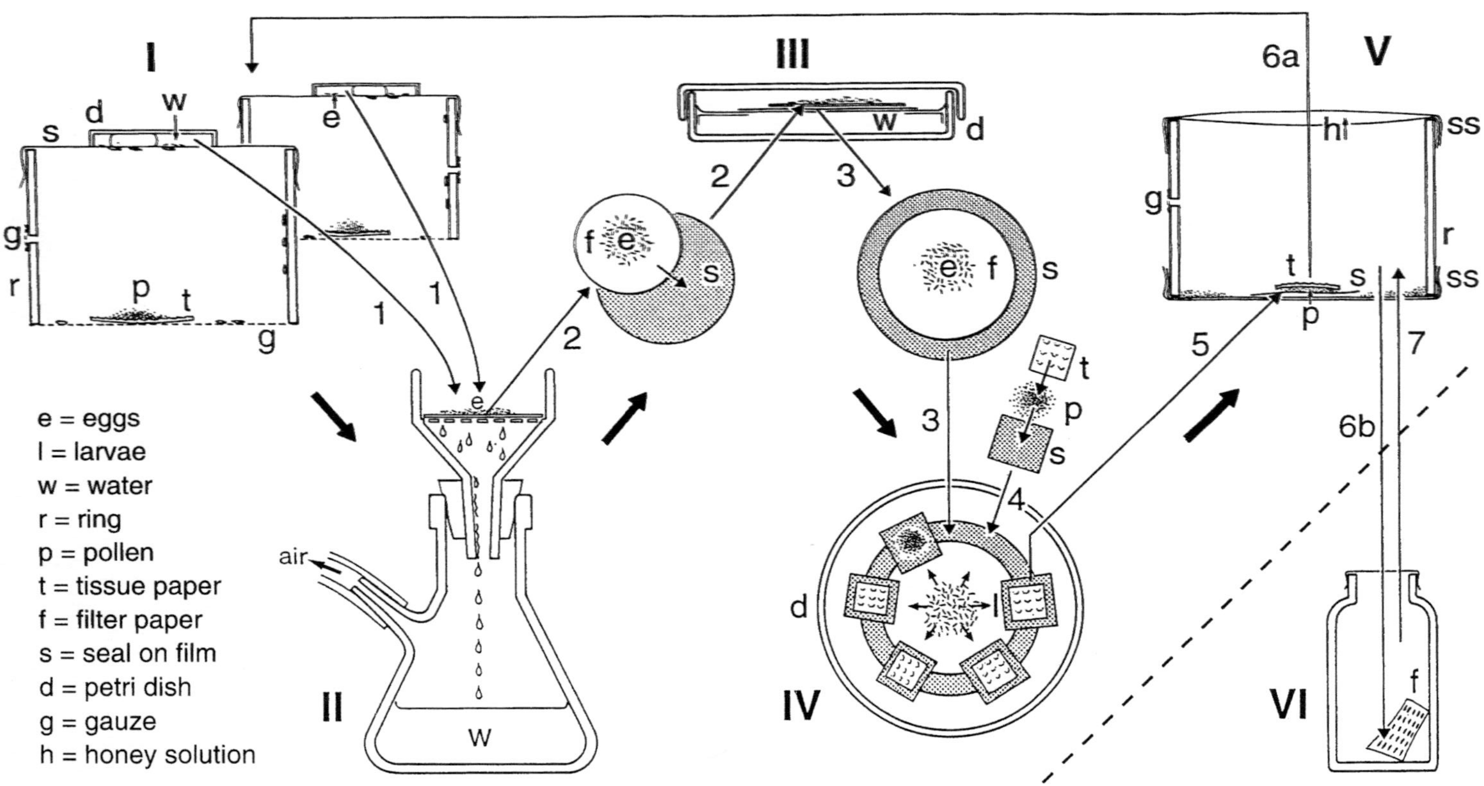

Fig. 12.5. Procedure for mass rearing thrips and parasitoids on an artificial substrate. Units for oviposition I, egg washing II, incubation III, trapping IV, rearing V and storage VI. Eggs laid in water on stretched film (1), are washed in a water filter on filter paper (2) and incubated at high humidity; hatched larvae are trapped (3, 4) with pollen and reared (5) on pollen and 10% honey solution until maturity and transferred to unit I (6a) or parasitoid pupae are stored (6b) until usage (7) (from Murai and Loomans, unpublished).

maintain the relative humidity at 80 per cent (Kirk, 1985), or on moistened filter paper and grease-proof paper placed on the bottom of the box (Teulon *et al.*, 1986, 1991), or moist cotton cloth (Nakao, 1993) placed on top of that to prevent the film from sticking to the box (Fig. 12.4f); relative humidity inside these latter boxes often reaches 100 per cent.

Food supply.

Pollen has a big effect on egg production of flower thrips (Table 12.3) (Andrewartha, 1935; Murai and Ishii, 1982; Kirk, 1985; Annadurai and Morrison, 1987; Trichilo and Leigh, 1988; Teulon and Penman, 1991) (see Chapter 4). Many kinds of pollen, for example tea, pine, tulip, strawberry, pear, maize, gorse, iceland poppy and flax, can be used for rearing thrips (Murai and Ishii, 1982; Kirk 1984, 1985; Teulon and Penman, 1991), but pollen quality and host specificity affect oviposition and development of different species (Murai and Ishii, 1982; Kirk, 1985). Only a few types of pollen can be collected in large enough amounts to sustain stock cultures or to perform mass propagation. Pollen of plants such as pine and maize can be collected very easily in large amounts. For the former, immature cones are collected and spread on paper in the laboratory. After 2 or 3 days, pollen is shed and collected by sieving. Pollen stored at –20°C for more than 2 years can still be used for rearing thrips. Crushed and dried bee-pollen can also be used for rearing larvae or dry leaf powder to rear adults (Koyama and Matsui, 1992a),

Table 12.3. Effect of tea pollen on oviposition of *Frankliniella intonsa*[a] (Murai and Ishii, 1982).

	Number of eggs laid by five females on different diet[b]		
Day	Pollen + honey	Pollen + water	Honey
1	37	32	24
2	23	30	24
3	28	34	7
4	25	43	0
5	40	32	1
6	35	38	0
7	25	38	0
8	45	35	2
9	36	41	0
Total[c]	228	261	10

[a]Five females were tested individually at 20°C.
[b]All females were fed previously on tea pollen and 10% honey solution for 2 days and after the third day they were reared on different diets.
[c]Total number of eggs laid in last 7 days.

but they easily moisten and mould in saturated conditions. In the oviposition unit (Fig. 12.5, I) a little pine pollen (20 to 40 mg) is provided on a piece of tissue paper daily. In the larval rearing unit (Fig. 12.5, V) 100 to 200 mg is supplied only once. Tea pollen is suitable for oviposition and pine pollen for rearing larvae.

Collecting eggs and newly hatched larvae

It is difficult to collect eggs of terebrantian species laid in natural substrates because they are embedded in the host plant tissue. However, many species can be induced to lay eggs in water through a thin membrane, and their eggs can be collected in this way. This method has been used successfully for *Frankliniella intonsa*, *Thrips tabaci*, *Thrips hawaiiensis*, *Thrips coloratus*, *Thrips nigropilosus*, *Thrips palmi*, *Anaphothrips obscurus*, *Microcephalothrips abdominalis* and *Scirtothrips dorsalis* (Murai, 1987), *Frankliniella occidentalis* (Teulon, 1992), *Thrips obscuratus* (Teulon and Penman, 1986), *Frankliniella schultzei* (light and dark forms) (T. Murai and A.J.M. Loomans, unpublished data), *Thrips major* (A.J.M. Loomans, unpublished data), *Thrips fuscipennis*, *Kakothrips pisivorus*, and *Ceratothrips ericae* (Kirk, 1985), *Thrips australis*, and *Hercinothrips femoralis* (Laughlin, 1971) and *Aeolothrips intermedius* (Kirk, 1985). Eggs have also been collected in 3 per cent sucrose (Laughlin, 1971), 10 per cent sucrose (Kirk, 1985) or a nutrient solution (apple/peach juice; Teulon and Penman, 1986). The type of sugar solution used influences adult survival of *Thrips palmi* (Koyama and Matsui, 1992b). *Megalurothrips sjostedti*, *Megalurothrips usitatus* (M. Tamò, personal communication) and some strains of *Thrips tabaci* (A.J.M. Loomans, personal communication) will also lay in water, but only occasionally; specific nutrition requirements and/or egg laying stimuli may play a role in this, but this has not yet been established.

Eggs laid in water can be collected on a filter paper in a water suction bottle (Fig. 12.5, II) every 1 to 3 days. The moist filter paper with eggs is placed on a piece of unstretched Sealon film floating on water in a closed Petri dish (Fig. 12.5, III) until hatching. The emerging larvae walk on the film from where they can be picked-up individually with a fine brush. A quicker way to collect a large number of larvae is to attract them to a small leaf disc (about 1 × 1 cm) or very small amount of pollen (less than 1 mg) placed on a piece of unstretched film in the Petri dish (Fig. 12.5, IV). Larvae gather on this food trap in 1 or 2 hours and can be collected and counted quickly.

Rearing larvae

Newly-hatched larvae are removed into a rearing ring (Fig. 12.5, V), a small ring for individual rearing and a large ring for mass production with suitable pollen and an uneven paper disc (such as kitchen paper, providing cavities) inserted as a pupation site on the bottom. Honey solution, encapsulated between two stretched membranes, is provided, so larvae can feed on pollen directly and honey solution through the thin membrane; alternatively, sugar solutions can be used. Larvae can develop to adults without any additional food and they pupate in the cavities of the paper disc. Many species such as *Frankliniella intonsa*, *F. occidentalis*, *Thrips tabaci*, *Thrips hawaiiensis*, *Thrips*

Table 12.4. Developmental time of some thrips species reared with pollen at 20°C.

Species	Total development (larvae + pupae) (days)	Mortality (%)	Reference
Frankliniella intonsa	12.3 ± 0.6	10.0	Murai and Ishii (1982)
Thrips hawaiiensis	10.5 ± 0.9	35.0	Murai and Ishii (1982)
Thrips obscuratus	10.3 ± 0.3	5.0	Teulon and Penman (1991)
Thrips tabaci	13.0 ± 0.7	5.0	Murai (unpublished)
Thrips palmi	9–11	13.0	Murai (unpublished)

coloratus, *Thrips nigropilosus*, *Anaphothrips obscurus* and *Microcephalothrips abdominalis* can be reared by this method. *Thrips palmi* and *Scirtothrips dorsalis* require additional food of germinated broad bean seeds without the seed coat (T. Murai, unpublished data). Developmental times of some species reared by this method are shown in Table 12.4. The duration of development depends on temperature, but for a given species tends to be fairly uniform. Using this technique, the number of larvae reared depends on the size of the cage used and the number of adult thrips introduced. Up to 400 to 500 larvae of *F. intonsa* and *F. occidentalis*, and 1000 of *T. tabaci*, *T. palmi* and *S. dorsalis*, can be reared in each large ring.

Mycophagous thrips

As in phytophagous thrips, there is a wide range in food and niche resource utilization among mycophagous thrips species (Ananthakrishnan, 1984). Specific microclimatic conditions have to be fulfilled to optimize fungal growth, thus complicating effective rearing. Very few papers describe methods for rearing mycophagous thrips. To study the biology of *Stictothrips fimbriata*, Joshi (1974) cultured its fungal hosts, *Rhizopus stolonifer* and *Pilobolus crystalinus*, on bread and cow dung, respectively. A stock culture of *S. fimbriata* was maintained in 3 lb (1.36 kg) glass jars with a bottom of moist filter paper and fresh cotton leaves for oviposition, containing a culture of the fungus *Rhizopus* at 25°C and 65 per cent relative humidity. For experimental purposes, newly-emerged pairs from the stock culture were released in Petri dishes with a pulverized culture of each fungus and a fresh leaf on a moist filter paper. After every third day fungal cultures as well as the leaves were replaced. The fungal species offered significantly affected fecundity as well as rate of development of *S. fimbriata*.

Ananthakrishnan and Suresh (1983) reared six species, belonging to the genera *Dinothrips*, *Elaphrothrips*, *Loyolaia*, *Priesneriana*, *Hoplothrips* and *Tiarothrips* (Tubulifera), in the laboratory on their natural fungus-infected host material. Specific feeding sites of the host material were cut into pieces, transferred to a 100 ml Erlenmeyer flask containing 25 ml of sterile distilled water and shaken vigorously to transfer the surface microflora into the water; eight or nine repeated host washings were made. Guts of the sporophagous thrips were dissected, cut open and the contents released into sterile distilled water under aseptic conditions. Host plant washings and thrips gut washings were then plated onto potato dextrose agar and oatmeal agar media and maintained at 34°C for 2 to 3 days. Emerging fungal colonies were isolated and transferred to Petri dishes containing fresh media; this procedure was repeated until individual pure cultures were obtained. These fungi were then cultured in test tubes, conical flasks and Petri dishes in potato dextrose agar, oatmeal agar or Czapek's dox agar media, maintained at 25°C. The spores/conidia obtained from the pure cultures were transferred to filter papers inside Petri and plastic vials to feed the thrips (Ananthakrishnan and Suresh, 1983).

Entomophagous thrips

Most oliphagous predators, predatory mites (Acari) and pirate bugs (Anthocoridae), and fungal parasites used for the biological control of thrips pests in greenhouse crops, are not commercially produced on the target pests, but on alternative hosts (for predatory mites see Ramakers and van Lieburg (1982); for pirate bugs see Chapter 8; for entomopathogens see Chapter 10). Predatory thrips are usually cultured on the target pests (e.g. mites, Gilstrap and Oaten, 1976) or on a range of prey (Bournier *et al.*, 1978). For rearing thrips-specific hymenopterous parasites, clearly it is necessary to rear hosts as well (Loomans and van Lenteren, 1995).

Predatory thrips

Approximately 300 species belonging to 24 genera are primarily predatory (zur Strassen, 1995), but the feeding behaviour and nutrition requirements are known for very few. Of the predatory thrips studied, most appear to be predominantly acariphagous, coccidophagous and thripiphagous (Sureshkumar and Ananthakrishnan, 1987). Because of their cannibalistic behaviour, they are usually confined individually or in small groups in laboratory cages, which makes rearing laborious and time consuming. Stock cultures using host-infested plants are rarely used. However, Carl (1971) succeeded in maintaining a colony of *Aeolothrips intermedius* on caged plants for seven generations, providing cages were well aerated.

Parrella and Horsburgh (1978) described a laboratory method for rearing *Leptothrips mali* on mite-infested apple leaves in 120-ml plastic containers for feeding and oviposition, and in plastic vials (41 mm diameter, 13 mm tall) for individual egg incubation and larval rearing. For *Scolothrips longicornis* (Şengonca and Gerlach, 1983) and *Scolothrips sexmaculatus* (Gilstrap and Oaten, 1976; Şengonca and Gerlach, 1983), which are predominantly acariphagous, methods based on the 'leaf-island' principle were used on an experimental scale.

Also Bournier *et al.* (1978) and Lacasa *et al.* (1982) reared *Aeolothrips intermedius* in individual cells (25 mm diameter, 12 mm tall) covered with fine mesh gauze and containing a 21-mm diameter leaf disc of *Datura stramonium* on moist cotton plus larvae of *Thrips tabaci*. In the absence of prey, *A. intermedius* can complete larval growth on a floral diet (T. Murai and A.J.M. Loomans, unpublished data), but pollen feeding is necessary for adults to reproduce (Bournier *et al.*, 1979). Although *Aeolothrips* is known to feed on different host species, *A. intermedius* thrived best when thrips larvae were offered as food sources (Bournier *et al.*, 1979). The feeding behaviour of adult aeolothripids differs from that of the larval stages, as adults only feed on eggs of the prey (Sureshkumar and Ananthakrishnan, 1987) or pollen (Bournier *et al.*, 1978), whereas the larvae feed on the active stages.

Thrips parasitoids

Few thrips parasitoid species have been reared in the laboratory. Methods for rearing them depend largely on the ability to rear their thrips hosts in large cohorts of even-aged larvae. In addition to methods for rearing thrips described above, a parasitoid incubation and storage unit is needed (Fig. 12.5, VI). Good timing and synchronization of the larval host stage is a necessary prerequisite to establish and maintain a successful and continuous parasitoid culture. Here a few methods are described for parasitoids attacking larval hosts belonging to the Panchaetothripinae and Thripinae.

Panchaetothripine hosts

Newberger and McMurtry (1992) described a technique for mass rearing *Thripobius semiluteus*. The procedure is summarized as follows: whole, silver-green lemons (*Citrus limon* cv. Lisbon or Ureka) are first half dipped in wax to ensure that developing larvae are confined on the unwaxed (upper) surface. Then the thrips host, *Heliothrips haemorrhoidalis*, is allowed to oviposit for 2 to 3 days on the lemons in units of about 80 fruits arranged in two layers infested with 1000 adult greenhouse thrips (Plate I.3). Two to four days after thrips larvae hatch, newly emerged parasitoids are introduced. After approximately 3 weeks incubation at 22 to 25°C, the black parasitoid pupae are brushed off the fruit into vials, 250 in each. Pupae are stored for a maximum of 2 weeks, at a temperature of 17 to 18°C, for field colonization, or for establishing new stinging units.

Adamson (1936) and Dohanian (1937) described a technique for rearing *Goetheana shakespearei* in the laboratory. Newly emerged *G. shakespearei* males and females (10 male:15 female) were placed in glass cylinder cages (12.7 cm diameter by 24.5 cm tall) for mating, closed by moist muslin at both ends and provided with food of cut raisins and 10 per cent sugar solution on paper. Cashew (*Anacardium occidentale*) leaves with young *Selenothrips rubrocinctus* larvae collected in the field were introduced into these cylinders, 20 to 200 larvae per parasitoid female. After 24 h exposure, leaves were removed and placed in a humid atmosphere in battery jars of 30.5 × 23 × 18 cm. In this way stocks numbered 10,000 to 12,000 pupae by the fifth to seventh generation. Bartlett (1939) slightly modified this technique by exposing four thrips-infested tropical almond (*Terminalia catappa*) leaves to 5 male and 5 female parasitoids in each cage; four infested leaves were added on the second day and another four on the fourth day. Uninfested leaves were added as food for thrips, all foliage remaining in the cylinder during the entire period. In this way *G. shakespearei* was reared for 33 generations, producing 88,000 pupae. For propagation on *Heliothrips haemorrhoidalis*, McMurtry and Johnson (1963) exposed the thrips host to orange fruits for oviposition over 10 days. A few days after thrips larvae hatched, single fruits were exposed to *G. shakespearei* (10 females:10 males). Parasitoid pupae were collected from the fruits into small vials until adults emerged.

Thripine hosts

Larval parasitoids of multivoltine, thripine hosts (like those belonging to the genera *Frankliniella* and *Thrips*), such as *Ceranisus menes* and *Ceranisus americensis*, can be reared by using either the bean-pod method described above (Loomans *et al.*, 1995) or the 'artificial' method (T. Murai and A.J.M. Loomans, unpublished data). Both *Ceranisus* spp. attack young larvae, so the parasitoids are put into a thrips rearing unit with newly hatched thrips. Parasitized and unparasitized thrips larvae pupate, after about 10 days at 25°C, between layers of paper on the bottom of the unit. When the unparasitized thrips adults have been removed (after about 2 weeks at 25°C), the paper layer is removed (bean-pod method) or parasitoid pupae are collected by an aspirator and put on moist filter paper (artificial method), and placed in a glass jar with 2 per cent agar to maintain a high relative humidity.

Productivity of parasitoids may vary between the thrips host used. Murai and Loomans (1995) found some differences in larval development and fecundity between two strains of *C. menes* on *Frankliniella intonsa* and *Frankliniella occidentalis* as hosts. However, more than 500 parasites in a cage can be reared by the artificial rearing method using *Thrips tabaci* as a host (T. Murai, unpublished data) rising to a maximum of 1000 when using the bean-pod method (Loomans, 1991) with *Frankliniella schultzei* and *C. menes* in a glass jar (2 dm^3).

General problems related to rearing thrips

In managing thrips rearing systems many problems can occur during each step (Fig. 12.1), most are related to the environmental conditions, crowding, thrips quality and contamination.

Environmental conditions

Microclimatic conditions are extremely important for a successful result. Temperature (Andrewartha, 1935; Nakao, 1993), and especially relative humidity can be crucial and requirements vary according to the species and growth stage, for example larvae of *F. occidentalis* need higher relative humidity than those of *T. tabaci*. Adequate light and ventilation are important. Artificial light intensity in rearing rooms is abnormally low in comparison with daylight, so rearing cages should not reduce illumination further than is absolutely necessary. In *Frankliniella intonsa*, reproductive diapause was induced by less than a 13 h photoperiod at 20°C (Murai, 1987). Similar diapause induction occurs in *Anaphothrips obscurus* (Kamm, 1972) and *Thrips nigropilosus* (Nakao, 1994b). The brachypterous form of *Anaphothrips obscurus* was produced under short daylength (Kamm, 1972). Nakao (1993) found that long-day photoperiod induced the development of the macropterous form, short-day photoperiod of the brachypterous form, and high temperatures partly suppress the effect of short-day photoperiods on *Thrips nigropilosus*. In the wing-dimorphic thrips *Anaphothrips obscurus*, Köppa (1970) suggested that wing form was dependent on food types as well as on photoperiod, whereas Nakao (1993, 1994a) found no significant effect of food type on wing form in *T. nigropilosus*. Clearly, photoperiod, temperature and relative humidity should be controlled adequately to avoid such unpredictable responses.

Host quality and crowding

The choice of diet can have a direct effect on reproduction (Trichilo and Leigh, 1988) and survival (Koyama and Matsui, 1992a,b). The quantity and quality of food available affects oviposition rate and larval developmental time (Andrewartha, 1935; Kirk, 1985; Murai and Ishii, 1982; Murai 1987; Annadurai and Morrison, 1987; Trichilo and Leigh 1988; Malchau, 1991; Teulon and Penman, 1991) and longevity (Malchau, 1991), the size of reared adults, and probably sex ratio. In *Thrips fuscipennis*, oviposition rate decreases with density when kept in small cages in the laboratory (Kirk, 1985, 1994) and in *T. nigropilosus*, crowding affects wing form (Nakao, 1994c). At high densities linked with a shortage of food, larvae of species like *Frankliniella occidentalis* that are usually phytophagous may cannibalize each other. In predatory thrips, cannibalism is the factor limiting the establishment of large-scale cultures and the main reason that mass rearing programmes of predatory thrips species have not yet been developed commercially.

Contamination

Clean and tidy working conditions are essential to avoid contamination between stock cultures and infestation by pathogens. Tools used for rearing should be fumigated or heated to prevent accidental infestation by pathogens, and contamination and multispecies or multistrain cultures should be dealt with at a different location and/or time. Host plant material should preferably be cultured on the spot under thrips-free conditions. When bean pods are used these are often contaminated with eggs of other thrips species (or strains of the same species) or predators like *Orius* spp., which also oviposit in the pod tissue. Especially when pods are purchased from local markets or supermarkets, the quality, history and level of contamination is unknown. Pods can be washed in soapy water and disinfected and dried afterwards (Lowrey *et al.*, 1992; Mollema *et al.*, 1993; Steiner *et al.*, 1995), but this only cleans them superficially and fungal and bacterial infections cannot be avoided completely. Heating the beans or exposing them to UV light to avoid contamination is inappropriate but radioactive irradiation (Cobalt-60 γ rays) at 1 kGy for 1 h is helpful (M. Schmidt, Wädenswil, 1996, personal communication).

References

Adamson, A.M. (1936) Progress report on the introduction of a parasite of the cacao thrips from the Gold Coast to Trinidad, B.W.I. *Tropical Agriculture, Trinidad* 13, 62–63.

Ananthakrishnan, T.N. (1984) *Bioecology of Thrips*. Indira Publishing House, Oak Park, MI, USA.
Ananthakrishnan, T.N. and Suresh, G. (1983) Patterns of fungal resource utilization and feeding range in some mycophagous Tubulifera (Insecta: Thysanoptera). *Proceedings of the Indian Academy of Science (Animal Science)* 92, 285–291.
Anderson, T.E. and Leppla, N.C. (eds) (1992) *Advances in Insect Rearing for Research and Pest Management, Studies in Insect Biology*. Westview Press, Oxford and IPH Publ. Co. Pvt.
Andrewartha, H.G. (1935) Thrips investigation 7. On the effect of temperature and food upon egg production and the length of adult life of *Thrips imaginis* Bagnall. *Journal of the Council of Scientific and Industrial Research, Australia* 8, 281–288.
Annadurai, R.S. and Morrison, M.N. (1987) Impact of pollen food on the fecundity and feeding time in two species of thrips (Insecta: Thysanoptera) infesting flowers of Solanaceae. *Current Science* 56, 1032–1033.
Bailey, S.F. (1931) The use of transparent cellulose film in life history studies. *Journal of Economic Entomology* 24, 898–901.
Bailey, S.F. (1932) A method employed in rearing thrips. *Journal of Economic Entomology* 25, 1194–1196.
Bailey, D.E. and Smith, R.F. (1956) The *Frankliniella occidentalis* (Pergande) complex in California (Thysanoptera: Thripidae). *University of California Publications in Entomology* 10, 359–410.
Bartlett, K. (1939) The introduction and colonization in Puerto Rico of *Dasyscapus parvipennis* Gahan, a parasite of thrips. *Agricultural Notes, Puerto Rico Experiment Station USDA*, 87, 1–5.
Bay-Petersen, J. (ed.) (1990) *The Use of Natural Enemies to Control Agricultural Pests*. FFTC Book Series 40, Food and Fertilizer Technology Center for the Asian and Pacific Region, Taipei, Taiwan.
Beavers, J.B. and Ewart, W.H. (1971) Observations on citrus thrips biology and an improved method of rearing them in the laboratory. *Journal of Economic Entomology* 64, 1124–1127.
Bournier, A. (1983) *Les Thrips: Biologie, Importance Agronomique*. INRA, Paris.
Bournier, A. and Bernaux, P. (1971) *Haplothrips tritici* Kurdj. et *Limothrips cerealium* Haliday agents de la moucheture des blés durs. *Annales des Sciences Naturelles Zoologie et Écologie Animale* 3, 247–259.
Bournier, A., Lacasa, A. and Pivot, Y. (1978) Biologie d'un thrips prédateur (*Aeolothrips intermedius* Bag.). *Entomophaga* 23, 403–410.
Bournier, A., Lacasa, A. and Pivot, Y. (1979) Régime alimentaire d'un thrips prédateur *Aeolothrips intermedius* (Thys.: Aeolothripidae). *Entomophaga* 24, 353–361.
Brodeur, J. and Cloutier, C. (1992) A modified leaf disc method for rearing predacious mites (Acarina: Phytoseiidae). *Phytoprotection* 73, 69–72.
Brødsgaard, H.F. (1993) Cold hardiness and tolerance to submergence in water in *Frankliniella occidentalis* (Thysanoptera: Thripidae). *Environmental Entomology* 22, 647–653.
Brødsgaard, H.F. (1994) Insecticide resistance in European and African strains of western flower thrips (Thysanoptera: Thripidae) tested in a new residue-on-glass test. *Journal of Economic Entomology* 87, 1141–1146.
Callan, E. McC. (1947) Technique for rearing thrips in the laboratory. *Nature* 160, 432.
Carl, K. (1971) Report of the work carried out at the European station. In: *Investigations of the Natural Enemies of Thrips for the Use as Biological Control Agents Against Glasshouse Thrips in the United Kingdom*. CAB-CIBC report,

European Station Délémont-Switzerland, pp. 9–35.

Chiu, H.T. (1987) The ecology and rearing methods of thrips. *Chinese Journal of Entomology, Special Publication* 1, 9–26.

Cock M.J.W. (1982) The biology and host specificity of *Liothrips mikaniae* (Priesner) (Thysanoptera: Phlaeothripidae), a potential biological control agent of *Mikania micrantha* (Compositae). *Bulletin of Entomological Research* 72, 523–533.

Day, M.F. and Irzykiewicz, H. (1954) Physiological studies on thrips in relation to transmission of tomato spotted wilt virus. *Australian Journal of Biological Science* 7, 274–281.

de Alwis, H.M. (1983) A simple technique for rearing rice thrips in the glasshouse. *International Rice Research Newsletter* 8, 5.

de Jager, K.M. (1995) Mechanisms of resistance to western flower thrips in chrysanthemum. PhD thesis, Rijksuniversiteit, Leiden.

de Jager, K.M., Butôt, R.P.T. and Guldenmond, A. (1995) Genetic variation in chrysanthemum for resistance to western flower thrips and *Thrips tabaci*. In: Parker, B.L. Skinner, M. and Lewis, T. (eds) *Thrips Biology and Management*. Plenum Press, New York, pp. 403–406.

Doane, E.N., Parker, B.L. and Pivot, Y. (1995) Method for mass rearing even-aged western flower thrips on beans. In: Parker, B.L., Skinner, M. and Lewis, T. (eds) *Thrips Biology and Management*. Plenum Press, New York, pp. 587–593.

Dohanian, S.M. (1937) Life history of the thrips parasite *Dasyscapus parvipennis* Gahan and the technic for breeding it. *Journal of Economic Entomology* 30, 78–80.

Galazzi, D., Maini, S. and Loomans, A.J.M. (1992) *Ceranisus menes* (Walker) (Hymenoptera Eulophidae): collection and initial rearing on *Frankliniella occidentalis* (Pergande) (Thysanoptera Thripidae). *Bolletino Istituto Entomologia* 'Guido Grandi', Università di Bologna 46, 123–129.

Gaum, W.G., Giliomee, J.H. and Pringle, K.L. (1994) Life history and life tables of western flower thrips, *Frankliniella occidentalis* (Thysanoptera: Thripidae), on English cucumbers. *Bulletin of Entomological Research* 84, 219–224.

Gilstrap, F.E. and Oaten, E.R. (1976) The bionomics of *Scolothrips sexmaculatus* (Pergande) (Thysanoptera: Thripidae), an insect predator of spider mites. *Hilgardia* 44, 27–59.

Grasselly, D., Trottin-Caudal, Y. and Trapateau, M. (1991) Lutte chimique contre le thrips *Frankliniella occidentalis*. *Phytoma, La Defénse Végé*tale 433, 54–56.

Hall, R.A., Peterkin, D.D. and Pollard, G.V. (1993) A system of caging *Thrips palmi* for laboratory bioassay of pathogens. *Florida Entomologist* 76, 171–175.

Joshi, G.P. (1974) The biology of *Stictothrips fimbriata* Anan., a mycophagous Thysanoptera on different meals. *Zeitschrift für angewandte Entomologie* 76, 146–149.

Kamm, J.A. (1972) Environmental influence on reproduction, diapause, and morph determination on *Anaphothrips obscurus* (Thysanoptera: Thripidae). *Environmental Entomology* 1, 16–19.

Kinzer, R.E., Young, S. and Walton, R.R. (1972) Rearing and testing tobacco thrips in the laboratory to discover resistance in peanuts. *Journal of Economic Entomology* 65, 782–785.

Kirk, W.D.J. (1984) Pollen feeding of thrips (Insecta: Thysanoptera). *Journal of Zoology, London* 204, 107–117.

Kirk, W.D.J. (1985) Pollen-feeding and the host specificity and fecundity of flower thrips (Thysanoptera). *Ecological Entomology* 10, 281–289.

Kirk, W.D.J. (1994) The effects of density on the oviposition rate of flower thrips. *Curier*

Forschungsinstitut Senckenberg 178, 69–73.
Koch, F. (1978) Eine Methode zur Haltung Pflanzensaft-saugenden-Insekten auf synthetischen Diäten am Beispiel von *Hercinothrips femoralis* Omr. *Wissenschaltliche Hefte der Pädagogischen Hochschule 'W. Ratke' Hochschule Köthen* 2, 125–127.
Koch, F. (1979) Eine neue methode zur zucht von Thysanopteren (Insecta). *Archiv für Phytopathologie und Pflanzenschutz* 15, 217–219.
Köppa, P. (1970) Studies on the thrips (Thysanoptera) species most commonly occurring on cereal in Finland. *Annales Agricultura Fenniae* 9, 191–265.
Koyama, K. and Matsui, M. (1992a) Rearing of adult of *Thrips palmi* on dry leaf powder of cucumber, aubergine and tomato. *Japanese Journal of Applied Entomology and Zoology* 36, 52–54.
Koyama, K. and Matsui, M. (1992b) Survival of the adult of *Thrips palmi* on various sugar solutions. *Proceedings of the Kanto-Tosan Plant Protection Society* 39, 205–208.
Koyama, K. and Matsui, M. (1993) Rearing of adults of *Thrips palmi* Karny, survival period of adults reared on an artificial diet under various temperatures. *Proceedings of the Kanto-Tosan Plant Protection Society* 40, 213–215.
Koyama, K. and Matsui, M. (1995) Nutritional physiology of *Thrips palmi* Karny: survival period of the nymphs reared on various sugar solutions. *Proceedings of the Kanto-Tosan Plant Protection Society* 42, 201–204.
Kudo, I. (1971) Observations on relative abundance, phenology and flower preference of Thysanoptera in Sapporo and the vicinity. *Journal of the Faculty of Science, Hokkaido University, Series VI (Zool.)* 17, 610–627.
Lacasa, A., Bournier, A. and Pivot, Y. (1982) Influencia de la temperatura sobre la biologia de un trips depredador *Aeolothrips intermedius* Bagnall (Thys.: Aeolothripidae). *Anales del Instituto Nacional de Investigaciones Agrarías, Serie Agricola*, 20, 87–98.
Laughlin, R. (1971) A culture method for *Hercinothrips femoralis* (Reuter) (Thysanoptera). *Journal of the Australian Entomological Society* 10, 301–303.
Lewis, T. (1973) *Thrips, Their Biology, Ecology and Economic Importance.* Academic Press, London.
Loomans, A.J.M. (1991) Collection and first evaluation of hymenopterous parasites of thrips as biological control agents of *Frankliniella occidentalis. Bulletin of the International Organisation of Biological Control, West Palaearctic Regional Section* 14, 73–82.
Loomans, A.J.M. and van Lenteren, J.C. (1995) Biological control of thrips pests: a review on thrips parasitoids. *Wageningen Agricultural University Papers* 95–1, 89–201.
Loomans, A.J.M., Murai, T., van Heest, J.P.N.F. and van Lenteren, J.C. (1995) *Ceranisus menes* (Hymenoptera: Eulophidae) for control of western flower thrips: biology and behaviour. In: Parker, B.L., Skinner, M. and Lewis, T. (eds) *Thrips Biology and Management*. Plenum Press, New York, pp. 263–268.
Lowry, V.K., Smith Jr, J.W. and Mitchell, F.L. (1992) Life-fertility-tables for *Frankliniella fusca* (Hinds) and *F. occidentalis* (Pergande) (Thysanoptera: Thripidae) on peanut. *Annals of the Entomological Society of America* 85, 744–754.
Lublinkhof, J. and Foster, D.E. (1997) Development and reproductive capacity of *Frankliniella occidentalis* (Thysanoptera: Thripidae) reared at three temperatures. *Journal of the Kansas Entomological Society* 50, 313–316.
Maddox, D.M. and Mayfield, A. (1972) A method of rearing and studying *Amynothrips andersoni* in the laboratory. *Journal of Economic Entomology* 65, 1521–1523.
Malchau, W. (1991) Zum einfluss intraspezifischer Relation auf die postmetabole

Entwicklung von *Frankliniella tenuicornis* (Uzel, 1895) (Thysanoptera, Insecta). *Zoologische Jahrbücher, Abteilung für Systematik* 118, 265–276.

McMurtry, J.A. and Johnson, H. (1963) Progress report on the introduction of a thrips parasite from the West Indies. *California Avocado Society Yearbook* 47, 48–51.

Medved, R.A. and Fleschner, C.A. (1970) A plastic cage for study of insects on plant stems and leaves. *Journal of Economic Entomology* 64, 342.

Mollema, C., Steenhuis, G. and van Rijn, P. (1990) Development of a method to test resistance to western flower thrips (*Frankliniella occidentalis*) in cucumber. *Bulletin of the International Organisation of Biological Control, West Palaearctic Regional Section* 13, 113–117.

Mollema, C., Steenhuis, M.M., Inggamer, H. and Soria, C. (1993) Evaluating the resistance to *Frankliniella occidentalis* in cucumber: methods, genotypic variation and effects upon thrips biology. *Bulletin of the International Organisation of Biological Control, West Palaearctic Regional Section* 16, 77–82.

Mollema, C., Steenhuis, G. and Inggamer, H. (1995) Genotypic effects of cucumber responses to infestation by western flower thrips. In: Parker, B.L., Skinner, M. and Lewis, T. (eds) *Thrips Biology and Management.* Plenum Press, New York, pp. 397–401.

Morse, J.G., Bellows, T.S. and Iwata, Y. (1986) Technique for evaluating residual toxicity of pesticides to motile insects. *Journal of Economic Entomology* 79, 281–283.

Munger, F. (1942) A method for rearing citrus thrips in the laboratory. *Journal of Economic Entomology* 35, 373–375.

Murai, T. (1987) Studies on the ecology and control of flower thrips, *Frankliniella intonsa* (Trybom). (In Japanese with English summary.) *Bulletin of the Shimane Agriculture Experiment Station* 23, 1–73.

Murai, T. (1988) Rearing method for thrips. In Umeya, K., Kudo, I. and Miyazaki, M. (eds) *Pest Thrips in Japan.* (In Japanese.) Zenkoku Noson Kyoiku Kyokai, Tokyo, Japan, pp. 365–374.

Murai, T. and Ishii, T. (1982) Simple rearing method for flower thrips (Thysanoptera: Thripidae) on pollen. (In Japanese.) *Japanese Journal of Applied Entomology and Zoology* 26, 149–154.

Murai, T. and Loomans, A.J.M. (1995) Host–parasite interaction between *Frankliniella intonsa*, western flower thrips and *Ceranisus menes* (Hymenoptera: Eulophidae): development and reproduction. In Parker, B.L., Skinner, M. and Lewis, T. (eds) *Thrips Biology and Management.* Plenum Press, New York, pp. 269–275.

Nakao, S. (1993) Effects of temperature and photoperiod on wing form determination and reproduction of *Thrips nigropilosus* Uzel (Thysanoptera:Thripidae). *Applied Entomology and Zoology* 28, 463–472.

Nakao, S. (1994a) Effects of food type on wing form determination and development in female *Thrips nigropilosus* Uzel (Thysanoptera: Thripidae). *Applied Entomology and Zoology* 29, 273–278.

Nakao, S. (1994b) Photothermic control of wing form and development in female *Thrips nigropilosus* Uzel (Thysanoptera: Thripidae). *Japanese Journal of Applied Entomology and Zoology* 38, 183–189.

Nakao, S. (1994c) Effect of high-density rearing on the incidence of brachypterae in female *Thrips nigropilosus* Uzel (Thysanoptera: Thripidae). *Applied Entomology and Zoology* 30, 249–251.

Newberger, S.J. and McMurtry, J.A. (1992) Insectary production of *Thripobius semiluteus* Boucek (Hymenoptera: Eulophidae), an imported parasitoid of the greenhouse thrips, *Heliothrips haemorrhoidalis* (Bouché). Internal Paper, University of California, Riverside, 3 pp.

Nugaliyadde, L. and Heinrichs, E.A. (1984) Biology of rice thrips, *Stenchaetothrips biformis* (Bagnall) (Thysanoptera: Thripidae) and a greenhouse rearing technique. *Journal of Economic Entomology* 77, 1171–1175.

Parrella, M.P and Horsburgh, R.L. (1978) Laboratory rearing of *Leptothrips mali* (Fitch) (Thysanoptera: Phlaeothripidae). *Journal of the New York Entomological Society* 86, 312–313.

Peterson, A. (1964) *Entomological Techniques: How to Work with Insects*. Edwards Bros, Ann Arbor, MI, USA.

Ramakers, P.M.J. and van Lieburg, M.J. (1982) Start of commercial production and introduction of *Amblyseius mckenziei* Sch. and Pr. (Acarina: Phytoseiidae) for the control of *Thrips tabaci* Lind. (Thysanoptera: Thripidae) in glasshouses. *Mededelingen Faculteit Landbouwwetenschappen, Rijksuniversiteit, Gent* 47, 541–545.

Rivnay, E. (1935) Ecological studies of the greenhouse thrips *Heliothrips haemorrhoidalis* in Palestine. *Bulletin of Entomological Research* 26, 267–278.

Sakimura, K. (1961) Techniques for handling thrips in transmission experiments with the tomato spotted wilt virus. *Plant Disease Reporter* 45, 766–771.

Sakimura, K. and Carter, W. (1934) Artificial feeding of Thysanoptera. *Annals of the Entomological Society of America* 27, 341–342.

Samsøe-Petersen, L. (1983) Laboratory method for testing side effects of pesticides on juvenile stages of the predatory mite, *Phytoseiulus persimilis* (Acarina, Phytoseiidae) based on detached bean leaves. *Entomophaga* 28, 167–178.

Şengonca, Ç. and Gerlach, S. (1983) A new developed method 'leaf-island' for observations on thrips in the laboratory. *Türkiye Bitki Koruma Dergisi* 7, 17–22.

Sharma, B.L. and Pande, Y.D. (1981) An improved microcage for confining tetranychid mites and small insects. *Science and Culture* 47, 301–303.

Singh, P. (1977) *Artificial Diets for Insects, Mites and Spiders*. Plenum Press, New York.

Singh, P. (1983) A general purpose laboratory diet mixture for rearing insects. *Insect Science and its Application* 4, 357–362.

Singh, P. and Moore, R.F. (eds) (1985) *Handbook of Insect Rearing* (2 vols). Elsevier, New York.

Steiner, M., Goodwin, S. and Wellham, T. (1996) Collecting and rearing thrips and their natural enemies. In: *Proceedings of the Australian and New Zealand Thrips Workshop: Methods, Biology, Ecology and Management*, NSW Agriculture, *Gosford Australia, 25–27 July*, pp. 15–19.

Sureshkumar, N. and Ananthakrishnan, T.N. (1987) Biotic interactions in relation to predator–prey relationship with special reference to some thrips species (Thysanoptera: Insecta). *Journal of Entomological Research* 11, 192–202.

Tamò, M., Baumgärtner, J., Delucchi, V. and Herren, H.R. (1993) Assessment of key factors responsible for the pest status of the bean flower thrips *Megalurothrips sjostedti* (Thysanoptera: Thripidae) in West Africa. *Bulletin of Entomological Research* 83, 251–258.

Tanigoshi, L.K. and Nishio-Wong, J.Y. (1981) Greenhouse rearing of citrus thrips *Scirtothrips citri* for experimental testing. *Journal of Economic Entomology* 74, 213–214.

Tashiro, H. (1967) Self-watering acrylic cages for confining insects and mites on detached leaves. *Journal of Economic Entomology* 60, 354–356.

Teulon, D.A.J. (1992) Laboratory technique for rearing western flower thrips (Thysanoptera: Thripidae). *Journal of Economic Entomology* 85, 895–899.

Teulon, D.A.J. and Penman, D.R. (1986) A rearing method for the New Zealand flower thrips. In: *Proceedings 39th New Zealand Weed and Pest Control Conference*, Control Society, Palmerston North, New Zealand, pp. 244–245.

Teulon, D.A.J. and Penman, D.R. (1990) Host records for the New Zealand flower thrips (*Thrips obscuratus* (Crawford) Thysanoptera: Thripidae). *New Zealand Entomologist* 13, 46–51.

Teulon, D.A.J. and Penman, D.R. (1991) Effects of temperature and diet on oviposition rate and development time of the New Zealand flower thrips, *Thrips obscuratus*. *Entomologia Experimentalis et Applicata* 60, 143–155.

Trichilo, P.J. and Leigh, T.F. (1988) Influence of resource quality on the reproductive fitness of flower thrips (Thysanoptera: Thripidae). *Annals of the Entomological Society of America* 81, 64–70.

Tsai, J.H., Yue, B., Webb, S.E., Funderburk, J.E. and Hsu, H.T. (1995) Effects of host plant and temperature on growth and reproduction of *Thrips palmi* (Thysanoptera: Thripidae). *Environmental Entomology* 24, 1598–1603.

Ullman, D.E., Cho, J.J., Mau, R.F.L., Westcot, D.M. and Custer, D.M. (1992) A midgut barrier to tomato spotted wilt virus acquisition by adult western flower thrips. *Phytopathology* 82, 1333–1342.

Vanderzant, E.S., Richardson, C.D. and Fort, S.W. Jr (1962) Rearing of the bollworm on artificial diet. *Journal of Economic Entomology* 55, 140.

Van Dijken, F.R., Dik, M.T.A., Gabal, B., de Jong, J. and Mollema, C. (1994) Western flower thrips (Thysanoptera: Thripidae) effects on chrysanthemum cultivars: plant growth and leaf scarring in non-flowering plants. *Journal of Economic Entomology* 87, 1312–1317.

Wang, C.L. and Chu, Y.I. (1986) Rearing method of southern yellow thrips, *Thrips palmi* Karny, in laboratory. *Plant Protection Bulletin (Taiwan, ROC)* 28, 407–411.

Wijkamp, I. and Peters, D. (1993) Determination of the median latent period of two tospoviruses in *Frankliniella occidentalis*, using a novel leaf disc assay. *Phytopathology* 83, 986–991.

Wijkamp, I., Almarza, N., Goldbach, R. and Peters, D. (1995) Distinct levels of specificity in thrips transmission of tospoviruses. *Phytopathology* 85, 1069–1074.

Winston, P.W. and Bates, D.H. (1960) Saturated solutions for the control of humidity in biological research. *Ecology* 41, 232–237.

zur Strassen, R. (1995) Binomial data of some predacious thrips. In: Parker, B.L., Skinner, M. and Lewis, T. (eds) *Thrips Biology and Management*. Plenum Press, New York, pp. 325–328.

13

Feeding and Oviposition Injuries to Plants

Carl C. Childers
University of Florida, Citrus Research and Education Center, 700 Experiment Station Road, Lake Alfred, Florida 33850, USA

Introduction

Several hundred thrips species are recognized as serious, sporadic or minor pests of many agricultural, ornamental, plantation and forestry commodities worldwide (Ananthakrishnan, 1971; Lewis, 1973; Bournier, 1983; Childers and Achor, 1995). Some show marked differences in pest status from one area of the world to another; a species may cause damage to a particular cultivated plant in one geographical area and be completely harmless to the same plant elsewhere.

There are many ways of categorizing thrips injury to plants. The approach adopted in this chapter is to consider the types and causes of injury to the main marketable parts of plants, namely leaves, flowers and fruit, and describe symptoms and yield losses recorded on crops representative of the many infested by thrips. Appendix II provides a wider consolidated list of commodities from different geographical regions, a brief summary of the damage inflicted on each, and the main thrips species responsible.

Symptoms of Infestation

Some thrips feed exclusively on vascular plants, fungi, or mosses; others are predacious or omnivorous, and/or feed on pollen (Lewis, 1973; Heming, 1993) (see Chapter 4). Some cause damage to a single plant part (e.g. *Frankliniella bispinosa* on citrus flowers) while others attack flowers, leaf buds and open leaves (e.g. *Taeniothrips inconsequens* on sugar maple), leaves and developing fruit (e.g. *Scirtothrips citri* on citrus, *Heliothrips haemorrhoidalis* on avocado), or leaves, fruit and flowers (e.g. *Frankliniella occidentalis* on various ornamental and vegetable plants). Many species are particularly intractable pests because they are highly polyphagous; *Frankliniella occidentalis* is known to feed on 244 plant species in 62 families worldwide (Anonymous, 1989) and *Thrips obscuratus* has been recorded on 223 plant species from 77 families in New Zealand (Teulon, 1988). Furthermore, in any one locality, abundance from year to year may vary enormously, making prediction of damage and identification of the pest by growers difficult.

Their small size and strongly thigmokinetic behaviour (see Chapter 7) enable many species to occupy narrow crevices within or between plant parts, such as between closed petals or leaflets, in floral or leaf buds, between fronds, sheaths, or adjacent clustered fruit, between a leaf or twig and a fruit surface, or at the bases of young floral ovaries. Damage in restricted places usually remains undetected until some time after injury was inflicted. As affected leaves or fruit mature or flowers develop, damage becomes more apparent, and subsequent rind blemish, fruit or leaf distortion, petal scarring and/or leaf drop may occur (Plates III–XI). Under some conditions, infestation can lead to significant yield loss in fruit crops such as apple, cashew and navel orange (Andrewartha and Kilpatrick, 1951; Igboekwe, 1985; Childers, 1992) to vegetables such as peppers, cucumbers, aubergines or legumes (Sakimura *et al.*, 1986; Rosenheim *et al.*, 1990; Bal, 1991) and to various forage or grain crops (Mikahilova, 1981; Nugaliyadde and Heinrichs, 1984a; Rotrekl, 1985; Bing *et al.*, 1990). Cosmetic damage such as blemished skin, or discoloured petals in ornamental blooms may render produce unsuitable for fresh marketing (Dennill and Erasmus, 1992; Felland *et al.*, 1993). Characteristic types of localized scarring on the surface of fruits, vegetables, stems or leaves (Lewis, 1973; Bournier, 1983; Arpaia and Morse, 1991; Childers and Frantz, 1994) are often accompanied by faecal droplets which turn black as moulds grow on them. Sometimes no injury is observed even though yield declines (Childers, 1992). Most parts of plants except the roots are vulnerable to feeding, but rapidly growing tissues are usually favoured (Lewis, 1973). Specific preferences for different ecological niches may result in several species occurring on different parts of the same plant (Palmer, 1990).

The tissues fed upon, stage of plant development, susceptibility of cultivar or species attacked, depth of feeding and salivary toxicity are all important factors contributing to eventual direct damage (Bournier, 1983). Larvae usually cause more damage than adults because of their greater abundance, the gregariousness of some species and commitment to feeding (Ananthakrishnan, 1971; Wiesenborn and Morse, 1986). They are less mobile than adults and generally confined to protected microhabitats. They concentrate in these protected niches where there is less risk from desiccation and exposure to parasites and predators. Oviposition within plant tissues may also be harmful, for example to the developing fruits of cherries (McMullen, 1972) and tomato (Salguero Navas *et al.*, 1991).

Indirect damage arises in two ways. First, thrips serve as vectors of various pathogenic viruses, fungi and bacteria (Ondrej, 1973; Ananthakrishnan, 1980; Allen and Broadbent, 1986; Ullman *et al.*, 1989). Six species are valid vectors of tomato spotted wilt and related tospoviruses (see Chapter 14). Some thrips act as passive vectors of other plant diseases by carrying bacterial or fungal spores on their bodies from contaminated plant sources to new feeding sites

(Ananthakrishnan, 1980; Ellis *et al.*, 1988) (see Chapter 14). Fruit spoilage diseases in California may be caused by various bacteria, yeasts and fungi spread by thrips (Sakimura, 1947). Second, thrips may predispose plant tissues to subsequent invasion by bacterial or fungal pathogens as a result of feeding injuries inflicted (Lewis, 1973; Ananthakrishnan, 1980; Bournier, 1983). Examples include fusarium ear rot of corn facilitated by *Frankliniella occidentalis* (Farrar and Davis, 1991) and purple blotch on onion by *Thrips tabaci* (McKenzie *et al.*, 1993) (see Chapter 14).

The Biochemical and Physiological Mechanism of Injury

Thrips saliva has reportedly produced toxaemias in many individual plant species. Feeding by adults and larvae of *Pezothrips dianthi* distorts leaves and induces bushy growth on pinks and carnations by shortening the internodes (Bournier, 1983). *Thrips laricivorus* feeding causes malformations in young larch shoots to produce needles resembling early stages of some bud galls (Vité, 1956). Feeding by *Scirtothrips dorsalis* causes substantial damage to chilli peppers; tender shoots, buds and flowers are all attacked, leaves curl and subsequently floral buds and leaves drop (Ananthakrishnan, 1971). *Thrips palmi* feeds on leaves of peppers, cucumbers and aubergines inducing distortion and death of plants in various parts of the world (Sakimura *et al.*, 1986).

Gall induction

Gall production by thrips is another example of the disruptive nature of their saliva to plant tissues, though in a crop pest context, gall production is a relatively minor cause of crop injury. Galls form as a result of saliva either inhibiting tissue differentiation, stimulating cell proliferation, or hypertrophy of the mesophyll cells (Mound, 1994). Most thrips galls occur on leaf blades and axillary and terminal buds. Younger plant tissues are particularly susceptible and, depending on the thrips species, respond by producing a characteristic curling, folding or rolling of the leaf blade (Ananthakrishnan, 1978).

Galls result from aberrant multiplication of growing tissues, particularly of parenchyma, and vascular bundles are destroyed by proliferation of cells in the cambium. Cells of the collenchyma and sclerenchyma develop poorly if at all, and lignification of mesophyll cells may occur along with deposits of tannins and anthocyanin (Bournier, 1983).

Phytohormone induction

Phytohormones, including ethylene, are active at small concentrations within a plant. They are transported from sites of production to sites of action, where they produce appropriate physiological responses (Abeles *et al.*, 1992). Ethylene ($CH_2{=}CH_2$) is a volatile phytohormone that plays important regulatory roles in several physiological processes. Increased ethylene production occurs following stress to plants induced by either abiotic or biological agents. Examples of abiotic stress include chemicals, temperature extremes, drought, hypoxia, γ-irradiation, mechanical wounding, bending and the mechanical impedance of soil to growth. Biological stress can result from disease and insect damage including thrips feeding.

There is also a relationship between leaf, flower and fruit abscission and ethylene production within plants. Increase in the rate of ethylene production after onset of sensitivity to this compound is closely correlated with time of abscission (Jackson and Osborne, 1970). Abscission results from hydrolysis of the middle lamella between two or three layers of cells found at the bases of leaves, petioles, petals, pedicels and fruit. Similar hydrolysis is also involved in cell separation during dehiscence. Not all leaves, flowers or fruits possess an abscission layer or exhibit ethylene-enhanced abscission, but approximately 40 per cent of plant species studied abscind in response to added ethylene. A wide range of sensitivity to ethylene has also been observed in flowers (Abeles *et al.*, 1992).

Significant differences exist between plant species and cultivars in their response to ethylene-releasing compounds. Many chemically diverse compounds, grouped into two classes, induce formation of ethylene by plants (Beaudry and Kays, 1988). Members of one class stimulate ethylene synthesis by affecting the synthesis pathway, mostly 1-aminocyclopropane-1-carboxylic acid (ACC) synthase, the rate-limited step in ethylene synthesis under normal conditions. Auxins and auxin analogues such as indoleacetic acid (IAA), α-naphthaleneacetic acid (NAA), 2,4-dichloro-phenoxyacetic acid (2,4D), and others, act in such a manner. Compounds in the second class act indirectly by wounding or stressing tissue which subsequently accelerates ethylene biosynthesis. For example, ascorbic acid, citric acid, cycloheximide, and iodoacetic acid injure the rind of citrus fruits stimulating the production of ethylene which then weakens the fruit abscission zone (Beaudry and Kays, 1988).

The cotton fleahopper, *Pseudatomoscelis seriatus* (Reuter) (Hemiptera: Miridae) provides an illustration of the possible biochemical consequences to plants of insect feeding. The primary mechanism whereby this insect induces production of ethylene by cotton shoot tips may be by injecting hydrolytic enzymes. Ethylene so generated would be supplemented by wound-induced ethylene resulting from insertion of the insect's mouthparts. Microbes present in the salivary fluid (or on the exterior of the mouthparts) would amplify the primary salivary mechanism. Cotton fleahopper salivary fluid contains pectinases which disrupt the middle lamella of plant cell walls and contribute to the induction of ethylene synthesis in other systems (Baldwin and Pressey, 1987; Campbell and Labavitch, 1987; Martin *et al.*, 1988). Pectinases, or associated enzymes, may hydrolyse oligosaccharide elicitors, thereby mimicking the stress-ethylene-inducing mechanism of many plant pathogens. Salivary homogenates of cotton fleahoppers contain a polygalacturonidase which elicits production of stress ethylene when injected into excised cotton shoot tips (Martin *et al.*, 1988; Abeles *et al.*, 1992).

Over 90 per cent of early-season bud abscission by cotton in Arizona has been related to cotton fleahopper and other insects (Mauney and Henneberry, 1979) and much of this is attributable to production of stress ethylene. The cotton fleahopper plays a dual role in causing ethylene-induced abscission of cotton flower buds. Wounds are first inflicted by the stylets and salivary enzymes while feeding, then tissues are inoculated with microorganisms, or invasion opportunities are provided, through these wounds.

Mechanical damage, plant hormones such as IAA, hydrolytic enzyme elicitors (microbial or plant cell-wall fragments generated by alien enzymes), and microbial products in the insect can elevate ethylene production (Grisham *et al.*, 1987). Grisham *et al.* concluded that an array of inductive signals rather than a single mechanism induces ethylene synthesis by fleahoppers, which may vary with time and place.

Several species of thrips have also been shown to induce ethylene production in plants. In seed legumes, flower abscission is caused by feeding of *Megalurothrips sjostedti* larvae and adults. Additional ethylene was produced by cowpea 4 days after *M. sjostedti* populations averaged three adults and six larvae per peduncle (Wien and Roesingh, 1980). Thrips-infested peduncles produced 640 nmol of ethylene per gram of fresh weight per day compared with 152 nmol in uninfested controls. The most likely source of ethylene was believed to be injured outer cell layers of flower buds and peduncles. Ethylene was also produced by plants in response to salivary constituents injected by the insect. According to Burden *et al.* (1989), the ethylene precursor, ACC, and IAA in cotton fleahoppers (Hemiptera: Miridae) may contribute to the mechanism whereby these insects promote ethylene-mediated abscission of cotton flower buds. A similar mechanism may exist in phytophagous thrips, a view supported by evidence from cowpea. Spray application of 400 ppm of ethephon (2-chloroethylphosphonic acid) on field-grown cowpea plants 1 week before start of flowering caused abscission of young flower buds, but cowpea cultivars resistant to such abscission did not shed their flower buds when attacked by *M. sjostedti* (Wien, 1979).

In another set of experiments (Kendall and Bjostad, 1990), *Thrips tabaci* adults and larvae were allowed to feed on onion plants and a crushed thrips extract was smeared over mechanically wounded areas on another set of similarly-aged onion plants. Both were then compared with onion plants that had been only mechanically injured. Results showed that feeding by *T. tabaci* stimulated production of ethylene from living onion foliage. Thrips feeding damage together with crushed-thrips extract injected into plant tissues stimulated significantly greater ethylene production than either treatment given separately or by intermittent mechanical damage alone (Table 13.1). This suggests that, during feeding, ethylene-inducing cues were introduced that affected the phytohormone physiology of the plant.

Rieske and Raffa (1995) measured ethylene emission from excised and intact American basswood tissue following infestation of seedlings by *Thrips calcaratus* adults and larvae. Mean ethylene production from thrips-damaged tissue was 2.2 times faster than from mechanically damaged tissue, and 3.0 times faster than from control tissue. Ethylene emission from excised, mechanically injured leaves did not differ from controls. All three treatments showed similar ethylene emission 32 h after damage. However, ethylene production from thrips-infested tissue was significantly above that of control (mechanically damaged) tissue at subsequent intervals of 72, 104, 112 and 120 h. Ethylene emission continued to increase in thrips-infested treatments in both

Table 13.1. Ethylene production from infested, uninfested and mechanically wounded intact onion plants (after Kendall and Bjostad, 1990).

	Ethylene production (mol^{-12} g^{-1}) Mean ± se		
Incubation (h)	Infested	Mechanically wounded	Unwounded
28	14.87 ± 3.33a	4.59 ± 0.96b	4.50 ± 0.69b
48	21.27 ± 5.10a	5.23 ± 1.00b	3.64 ± 1.23b

Means followed by the same letter in one row are not significantly different (Duncan's multiple range test, $p > 0.05$).

experiments up to 120 h after initiation of damage. This contrasts with plant responses to mechanical wounding, which resulted in a peak rate of ethylene emission within 6 h.

Feeding Damage to Leaves

In this section, the general effects of leaf feeding, covering malformation of young leaves and growing tips, wilting, silvering, scarring and sometimes bud drop, are reviewed in detail for a selection of major crops and thrips species. The damage described applies generally to crops infested by leaf feeders.

Sugar maple

Adult *Taeniothrips inconsequens* begin feeding during bud burst. Shortly after feeding begins, eggs are laid in succulent tissues, then adults and larvae feed on epidermal and mesophyll leaf cells (Kolb *et al.*, 1991). Feeding sites are identified by concentric areas of green coloration and dark necrotic tissue surrounding a distinct circular puncture hole. Heavily damaged leaves suffered on average 2.7 ± 0.5 puncture sites mm^{-2} compared with lightly damaged leaves with 0.7 ± 0.2 puncture sites mm^{-2} (Shumway *et al.*, 1991).

The introduction of *Taeniothrips inconsequens* adults into caged sugar maple seedlings at all bud-burst stages (Fig. 13.1) decreased total leaf area, particularly at bud-burst stages 3 and 5 (Kolb and Teulon, 1991); the greatest reduction in total leaf area (66 per cent) occurred in seedlings exposed to thrips at stage 3. Foliar damage differed between treatments. The proportion of seedlings with no symptoms was greatest for the control, least for the stage 3 introduction, and intermediate for stage 1 and 5 introductions. Only 14 per cent of control seedlings showed any damage, which consisted of chlorosis only. Nearly all seedlings for stage 3 (98 per cent) and for stage 5 (78 per cent) introductions had chlorotic, tattered or cupped leaves (Fig. 13.2, Plate VIII.4). Most adult pear thrips appear incapable of entering sugar maple buds until bud scales have separated to reveal leaf tissue. Damage was more severe when leaves more than 2.5 cm long (stage 5) were exposed to thrips than when exposure

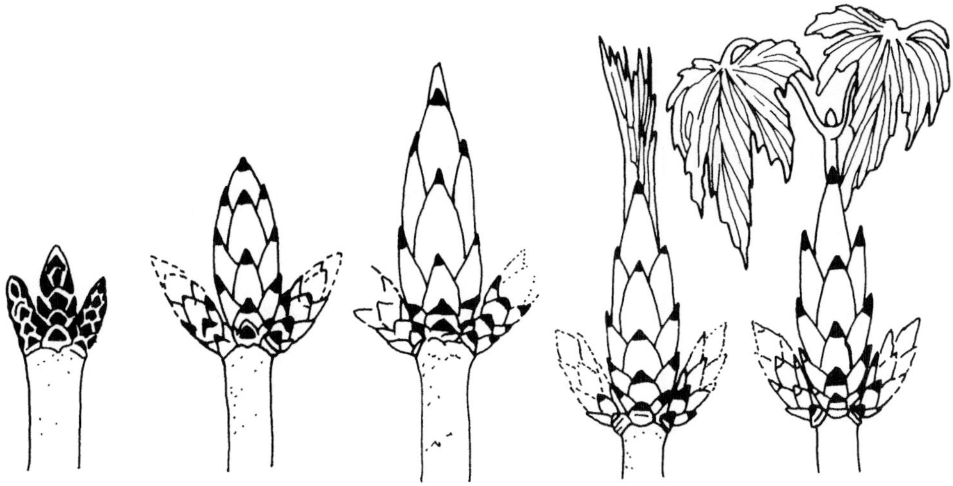

Fig. 13.1. Five stages of sugar maple bud-burst phenology: stage 1, buds swollen from midwinter condition, but bud scales not separated; stage 2, bud scales beginning to separate, green interior of scales visible; stage 3, bud scales loose and leaf margins visible; stage 4, leaves visible, but still folded and less than 2.5 cm long; stage 5, leaves beginning to unfold and more than 2.5 cm long (after Kolb and Teulon, 1991).

occurred at bud-burst (stage 1). Total leaf area and average leaf size of sugar maples were reduced on infested, caged, seedlings following exposure to five adult thrips per bud compared with controls. Damage was greatest on trees whose bud expansion coincided with heavy emergence of adult thrips from soil (Kolb and Teulon, 1991).

Feeding by *Taeniothrips inconsequens* caused rapid water loss over much of the growing season, but differences in chlorophyll concentration or stomatal function between infested and uninfested plants were not detected. High gas exchange through feeding punctures predisposed seedlings to water stress and reduced photosynthesis under conditions of low water availability (Kolb *et al.*, 1991). Pear thrips reduced whole-tree leaf area by approximately 23 per cent, leaf size (both mass and area) by 20 per cent in the upper crown and net CO_2 assimilation rate for fully expanded leaves by approximately 20 per cent, but leaf chlorophyll content and leaf mass per unit area were not affected (Ellsworth *et al.*, 1994). Adult feeding caused interveinal chlorosis, curling of the lamina and ragged or necrotic leaf margins. Oviposition scars were present on damaged trees with 11.3 ± 2.5 oviposition sites per leaf, compared with none on controls (Ellsworth *et al.*, 1994).

Basswood

Feeding by *Thrips calcaratus* adults and larvae caused newly opened leaves of *Tilia americana* to drop, and even a small amount of feeding caused bud shed

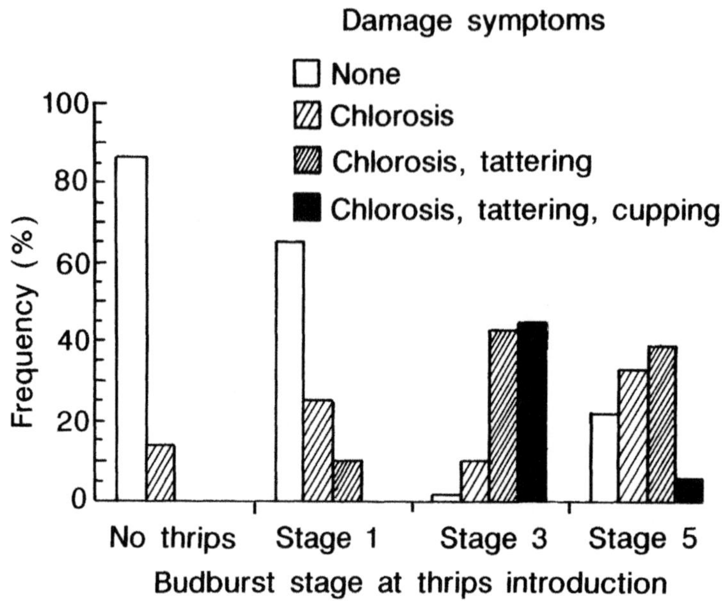

Fig. 13.2. Frequency of damage symptoms on sugar maple seedlings exposed to no pear thrips and to adult pear thrips at different stages of bud burst (after Kolb and Teulon, 1991).

(Raffa and Hall, 1989). Repeated defoliation caused top dieback and sometimes death of the tree.

Grapes

Rhipiphorothrips cruentatus adults and larvae feed mostly on the lower leaf surface of grapevines in India. Leaves turn brown, wither, and drop (Rahman and Bharadwaj, 1937). Feeding first destroys the lower epidermis followed by the spongy and palisade parenchyma. The thick leaves of 'Jaisha' with their hairy lower surfaces confer some resistance to this cultivar.

Drepanothrips reuteri and *Frankliniella occidentalis* cause serious shoot stunting and leaf distortion followed by berry scarring on grapes in northern California. Low populations of one to three thrips per terminal shoot shorten growth significantly. Leaves on unprotected shoots are severely distorted and appear burned. However, shoot growth in infested plots often totally recovers later in the season. Chenin blanc grapes are particularly susceptible to injury earlier in the season (McNally *et al.*, 1985).

Several species infest table grapes in South Africa (Schwartz, 1989). Adults and larvae of *Thrips tabaci* feed on very young berries and new shoots while *Heliothrips sylvanus* adults and larvae prefer the abaxial surface of mature leaves of 'Barlinka' vines and cause silvering or bronzing of foliage (Plates III.6 and IV.7).

Onion

Onions and leeks infested with *Thrips tabaci* develop silvery leaf spots that coalesce into white blotches due to removal of cellular contents and silvery patches develop along the inner angles of onion leaves, which eventually curl (Plate VII.1,2) (Bailey, 1938).

Kendall and Capinera (1987) evaluated the effect of *T. tabaci* on three growth stages of onion (prebulbing, bulbing and sizing). Feeding significantly reduced bulb weight during the bulbing stage. Ten thrips per plant in two field studies in Colorado reduced yield by 2 to 3 per cent and 7 per cent in a greenhouse.

Thrips tabaci is common on dry bulb onions in the north-west United States. Large populations sometime reduce yield, but there is greater concern for thrips increasing the incidence of the fungal pathogen, *Botrytis allii* Munn in stored onions (Mayer *et al.*, 1987).

Onion plants infested with *Thrips tabaci* and inoculated with the fungal pathogen *Alternaria porri* (Ellis) developed more and larger lesions that tended to coalesce as the plant matured (Thind and Jhooty, 1982; McKenzie *et al.*, 1993). Older leaves were more susceptible to this pathogen in the absence of thrips injury, but infection shifted to younger leaves when thrips fed on them. More tissue was killed by *T. tabaci* in the presence of the fungus than in its absence and plant growth was negatively correlated with feeding injury. *Alternaria porri* enters onion leaf tissue through stomata and the epidermis. When both thrips and disease are present, the pathogen penetrates surfaces pierced by thrips and necrotic spots appear on leaves.

McKenzie *et al.* (1993) found that adults and larvae of *T. tabaci* remove the thick wax layer with the entire mouthcone tip before piercing underlying cells with the mandible, and saliva within the area of the mouthcone facilitated digestion of cuticular waxes. Similar injury by *Limothrips cerealium* adults and larvae feeding on wheat leaves showed an imprint of the labral pad around puncture sites (Chisholm and Lewis, 1984) (see Fig. 4.6). These authors thought that the primary function of this pad was to support the maxillary stylets as they protract into plant tissue, but it may also confine and conserve saliva within a localized area to facilitate repeated feeding attempts.

Other vegetables

Severe feeding injury by *Thrips palmi* has been reported from Asia, the Caribbean and Florida on pepper, potato, cucumber, melons, squash and aubergine. Larvae and adults feed primarily on leaves but also on stems, flowers and fruits of many plants. Infestations on cucumber as low as one thrips per leaf resulted in fewer tendrils and leaves, and killed plants (Suzuki and Miyara, 1983). Large populations of *Thrips palmi* caused bronzing of foliage and total destruction of watermelon vine tips (Johnson, 1986).

Thrips tabaci causes a rough, bronzed oedema on leaves inside cabbage heads (Plate VII.6). The original feeding injury is visible only under a microscope, the damage remaining as a brown discoloration after processing for sauerkraut (Hoy and Kretchman, 1991). Susceptibility differs between cultivars (Stoner and Shelton, 1988), those suitable for winter storage being particularly affected (Shelton *et al.*, 1982). Infested cabbage heads may appear clean until the four wrapper leaves are peeled back to reveal the typical bronzing and rough texture of the damage. Thrips penetrate as deep as 11 layers (22 leaves) into the head (North and Shelton, 1986). Insecticides currently used do not control them reliably in the deeper layers (Shelton *et al.*, 1981). Fewer *T. tabaci* were present in August and September in heads of two resistant cultivars, 'Titanic 90' and 'Falcon', compared with two susceptible ones, 'Market Place' and 'Supergreen'. Resistant cultivars supported as many thrips per plant as susceptible ones, but fewer thrips penetrated to the deeper layers (Stoner and Shelton, 1988). The ratio of onion thrips on the head to frame leaves is lower on resistant cultivars than on susceptible ones (Shelton *et al.*, 1988).

Cassava

Thrips are serious pests of cassava and attack plants at all growth stages (Plate IX.1). Heavy infestations of adults and larvae of *Frankliniella* spp. and *Corynothrips stenopterus* feed on lateral buds, killing growing points and inducing excessive sprouting. Severely affected cultivars are stunted with small, deformed leaves (van Schoonhoven, 1974; van Schoonhoven and Pena, 1976). In the rainy season, yield loss for susceptible cultivars was 8 per cent compared with 15 per cent in the dry season. Susceptible clones had glabrous leaves whereas those of resistant clones were pubescent.

Peanut

Frankliniella fusca is a serious pest of seedling peanuts in the United States and feeds on young foliar buds causing small, distorted leaflets and stunted plants (Young *et al.*, 1972).

Soybean

Feeding by *Neohydatothrips variabilis* adults on leaves of soybean destroys epidermal tissues and removes sap from adjoining mesophyll cells (Huckaba and Coble, 1991). Interestingly, feeding and mechanical injury had a minimal effect on penetration of the herbicide, ^{14}C-acifluorfen, whether it was applied to the upper or to both leaf surfaces.

Cotton

The principal thrips attacking cotton in Alabama are *Frankliniella fusca*, *F. tritici* and *Thrips tabaci* (Watson, 1965) along with *F. occidentalis* in Texas (Gaines, 1965). *Thrips tabaci* is an important pest in Asia, Europe and Australia. Small cotton seedlings 3 to 4 cm tall are most susceptible to this thrips in Egypt with up to 19 per cent reduction in seedling size (Hosny and Shoeib, 1973). In Pakistan, they feed on seedling leaves then move upwards to new growth, thereby becoming restricted to the upper half of mature plants by September. Attique and Ahmad (1990) found a positive and linear correlation between damage symptoms and number of thrips present on 10-day-old seedlings.

In North America, thrips attack terminal buds when seedlings are 2 to 4 weeks old. Leaves curl and grow ragged and irregular as they unfold. Injury retards growth of new leaves and stunts plants, though they often recover from moderate attacks (Harp and Turner, 1976). Severe infestations may cause aborted terminals and excessive branching. Stunted plants are more susceptible to seedling disease than normal seedlings (Quisenberry and Rummel, 1979). Severe thrips damage to cotton in Arkansas reduced plant stand by 19 per cent; leaves were destroyed, deformed or ragged with their total area reduced by up to 88 per cent (Tugwell and Cochran, 1989).

Similarly, in Louisiana, *Frankliniella fusca*, *F. tritici*, *Neohydatothrips variabilis* and *Thrips tabaci* are important early season cotton pests alongside serious seedling fungal diseases caused by *Rhizoctonia solani* Kuhn, *Thielaviopsis basicola* (Berk. and Br.), *Pythium ultimum* Trow, and *Fusarium* spp. Colyer *et al.* (1991) measured cotton stand reduction, number of thrips per ten plants, plant height and the root-hypocotyl disease index (RHDI) over 3 years. Different densities of thrips and disease were obtained on plots receiving applications of insecticides only, insecticides plus fungicide, fungicide only or no treatment. Thrips infestation increased severity of seedling diseases based on stand reduction and RHDI. The smooth-type Egyptian cottons are more susceptible to thrips attack than pubescent Indian or American varieties (Ananthakrishnan, 1971; Quisenberry and Rummel, 1979).

Cereals and grains

Typically, thrips feeding turns cereal leaf sheaths and blades silvery to white, then brown and brittle. As leaves dry, feeding is transferred to the ears which often wither to produce characteristic 'whiteheads' or 'silvertop' (Plate X). Symptoms on leaves and ears are more pronounced in dry weather.

Limothrips cerealium is a widespread European and North American pest appearing on grasses and young cereals in spring. It feeds on leaf bases and oviposits in leaf sheaths, moving to more tender leaves as old leaves harden (Lewis, 1959). When this species feeds, it removes leaf surface waxes causing eventual desiccation of epidermal and mesophyll cells adjacent to puncture sites (Chisholm and Lewis, 1984). Wound periderm is formed by the wheat plant after removal of cellular contents. Feeding ultimately results in shrunken grains, white ears, yield loss and reduced viability of seed. *Limothrips denticornis* has similar effects on the yield and quality of barley (Post, 1955).

Stenchaetothrips biformis is a major pest of rice in Asia, especially of young plants. Larvae and adults infest seedlings within a few weeks of transplanting or sowing. Feeding causes leaf curling along the margins, scorching, wilting, leaf-tip drying and eventually stunting or death of seedlings (Nugaliyadde and Heinrichs, 1984a,b; Chatterjee, 1987).

High levels of resistance to this thrips exist in certain cultivars and wild species and include non-preference and antibiosis, the latter manifested in low survival, fecundity and reduced longevity of adults. Reddy *et al.* (1987) reported *S. biformis* causing severe damage to rice in India during prolonged dry periods, late rice being more prone to infestation than early crops.

Stenchetothrips biformis has become increasingly important with the spread of high-yielding cultivars, and frequent thrips outbreaks and crop losses now occur in Bangladesh, India and Sri Lanka. High-yielding susceptible cultivars support large populations of this thrips ranging from 68 to 75 per seedling at 20 days after planting, whereas on seedlings of resistant cultivars, Ptb21 and IR62, only about 7 and 10 thrips, respectively were recorded (Velusamy, 1990). Oviposition was greater on all high-yielding cultivars except IR62. Populations increased ninefold on susceptible compared with resistant cultivars and adult longevity was shorter on Ptb21 and IR62 than on other resistant cultivars tested. Damaged leaves curl inwards from both margin and tip and ultimately dry, larvae remaining within the rolled leaves. *Sorghothrips jonnaphilus* has similar effects on rice, wheat and sorghum in Asia (Ananthkrishnan, 1971).

Ornamentals

Thrips simplex adults and larvae feed on gladiolus foliage and stalks causing silvering or blasting of leaves. Green chlorophyll from emptied cells is clearly visible through the body wall of larvae (Herr, 1934). Removal of cellular contents by various species of thrips and consequent stippling or chlorosis of foliar ornamentals causes serious economic losses to nurseries (Oetting *et al.*, 1993).

Feeding Damage to Flowers and Buds

In this section, the general effects of flower and bud feeding, covering abscission and malformation of buds or flowers, are reviewed for some major crops and thrips species. Fruit can become distorted or develop rind blemishes if thrips feed during floral bud formation or bloom. The damage described applies generally to crops infested by flower or bud feeders.

Grasses

Thrips impair quality and yield of grasses grown for seed by feeding particularly on inflorescences. Type and extent of damage depends on the species of thrips and on characteristics of the grass, but most develop shrivelled glumes and seed, within whitened seed heads.

Anaphothrips obscurus is considered the most destructive species in Oregon grasslands (Kamm, 1972). It produces a distinct type of silvertop in bluegrass and bentgrass by feeding on developing inflorescences while still in the boot stage (Kamm, 1971). In silvertop, the inflorescence turns white due to culm injury and seeds are sterile. Loss of seed among 'Emerald' bentgrass heads with silvertop was 16 per cent of total yield and 64 per cent of heads were damaged to varying degrees by thrips.

Several species of *Chirothrips* are also serious grass seed pests in North America and Europe (see Appendix II).

Ornamentals

Anthurium andraeanum Lind. and Andre is a commercially grown herbaceous perennial with colourful flowers and *Chaetanaphothrips orchidii* infests unopened buds soon after their emergence from leaf axils (Hata and Hara, 1992). The thrips usually feed within the narrow, concealed spaces of rolled young leaves or floral buds causing white streaks and distortion on both upper and lower surfaces of the spathe (Sakimura, 1975; Hara *et al.*, 1988; Hata and Hara, 1992).

Larvae and adults of *Thrips simplex* cause silvery white spots on gladiolus flowers which later turn brown (Denmark and Poe, 1972). *Frankliniella tritici* damages flowers and flower buds of outdoor roses in spring and early summer. Petals turn brown and affected buds fail to open (Henneberry *et al.*, 1961). Feeding injury by *Frankliniella occidentalis* on *Gloxinia* flowers likewise causes brown streaks on petals. Buds fail to develop, turn brown, and abort (Cummiskey, 1988). *Frankliniella occidentalis* causes streaking, browning and distortion of leaves and/or petals of various flowering ornamentals including roses, gerberas, chrysanthemums, carnations, geraniums, pansies and marigolds (Oetting *et al.*, 1993) (Plate XI), reducing aesthetic quality and marketability of many commercial flowers (Parrella and Jones, 1987).

Gladiolus stalks become brittle and prone to breakage following feeding by *Thrips simplex*, which also infests stored corms and causes seepage from injured cells; epidermal cells harden or suberize (Herr, 1934). Severe feeding injury to corms affects developing roots resulting in small or deformed flowers. Feeding around the eyes or developing bud shoots of corms may prevent growth totally. Corms and bulbs of other ornamental and greenhouse plants are attacked by several other thrips species (Wang, 1987; Mantel and van de Vrie, 1988).

Pine

According to Fatzinger and Dixon (1991), *Gnophothrips fuscus* adults may destroy up to a quarter of flowers (cones) present in slash pine seed orchards in the southern United States. Infestations are more prevalent on young female strobili (flowers) of open grown trees than on trees in forest stands. This thrips feeds externally on succulent flowers only between bud stage and pollination. Feeding sites can be identified by small beads of oleoresin oozing from the seed cones. Severe feeding injury destroys scales and bracts, flowers are distorted or killed and fall from the tree with consequent reduced seed yields.

Pyrethrum

Thrips nigropilosus attacks flowers and developing floral buds of *Chrysanthemum cinerariaefolium* causing shrinkage, distortion and/or death (Bullock, 1963). Heavy infestations provided access for the pathogenic fungi, *Alternaria gossypina* and *Cladosporium* sp., through damaged flowers. Decreases in seed yields can be serious, especially in dry seasons.

Onion

Thrips tabaci adults and larvae feed on onion flower pedicels and developing flower buds on first opening of bracts that envelop the immature umbel (Elmore, 1949). Injury varies from slightly damaged pedicels to complete destruction. The initial small, silvery, shrunken areas enlarge and eventually the entire pedicel collapses. The number of buds in each umbel may be greatly reduced causing lower seed yield. All floral parts of onion flowers are attacked by *Frankliniella occidentalis* as they open. Both filaments and anthers can be severely injured, reducing available pollen (Elmore, 1949). Feeding by ten *F. occidentalis* per onion floret was sufficient to injure flower filaments and anthers, causing premature loss of pollen; however, low populations of *F. occidentalis* actually improved pollination and seed set compared with plants caged to exclude completely thrips and other insects (Carlson, 1964).

Avocado

In Florida, adults and larvae of both *Frankliniella bispinosa* and *F. kelliae* feed in floral tissues of avocado and produce lesions on the surface of the ovary that extend three to four cells below the epidermis, with subsequent cell proliferation (Fisher and Davenport, 1989). The wound response differs from typical wound periderm formation because of proliferation of cells far beneath the surface and subsequent growth of the ovary wall. Feeding injury on avocado by these two species is characterized by numerous large bumps and ridges on the surface of developing fruits. Damage is most common on the West Indian cultivar 'Simmonds' (Plate III.5).

Apple

Adults and larvae of *Thrips imaginis* sometimes reduce apple yields through injury to floral parts in Australia (Lloyd, 1973). This species infests many native and introduced plants and is found in apple blossoms every season, but only sporadically in damaging numbers (see Chapter 7). When conditions are favourable, large populations of adults fly into orchards during early flowering. The South Australian apple industry experienced a 56 per cent reduction in apple production during the 1950–51 season due to heavy infestations of this thrips on blossoms (Kilpatrick, 1955).

The cell contents of stamens and pistils are removed, which results in the death of attached pollen and interference with successful pollination by honey bees. Attacked buds in the 'pink stage' turn brown and shrivel without opening (Andrewartha and Kilpatrick, 1951). Larvae of *Thrips imaginis* also feed on stamens and styles, which wither, and petals are 'scorched'. Damage is worse in hot, dry weather when the floral organs are stressed. In cooler conditions, up to 20 thrips per flower are not injurious. The cultivar 'Granny Smith' appeared to be more susceptible to thrips feeding injury than 'Jonathan', whose flowers have shorter, more robust styles than the long delicate styles of 'Granny Smith' flowers (Lloyd, 1973).

Adults of *Thrips flavus* oviposit in floral buds of the cultivar 'Red Delicious' in India. Larvae feed on petals, pistils and stamens, so fruit either fail to set or fall prematurely (Sharma and Bhalla, 1963; Singh, 1989).

Between 1985 and 1988, significant injury to apple blossoms by *Taeniothrips inconsequens* occurred in Massachusetts and southern Vermont (Coli *et al.*, 1992). Injury to both petals and pistils was observed with reduced flowering on heavily infested trees (W.M. Coli, University of Massachusetts, Amherst, 1994, personal communication).

Citrus

In Florida, adults and larvae of *Frankliniella bispinosa* feed on anthers, petals, pistils and calyces of citrus flowers (Plate V.1). Initial injury on navel orange results from punching of a hole 2.5 to 5.0 μm in diameter through the cuticle and epidermis with the mandibular stylet, followed by penetration of underlying cells by the paired maxillary stylets. Differences in colour and vigour of citrus flowers between infested and uninfested trees are evident following substantial increase in thrips numbers. Petals on infested trees untreated with insecticides were withered, yellow and sickly in appearance compared with healthy, white and vigorous flowers in treated plots (Childers, 1992).

Following *F. bispinosa* feeding, extensive depressed or sunken areas developed on the surface of the ovary and style (Fig. 13.3). Types of injury included cellular evacuation, necrosis, plasmolysis, and cellular collapse on the calyx, floral disk, ovary, petals and pistil (Childers and Achor, 1991). In light micrographs, areas of injury were dark staining, dense, and frequently occurred after 6 days of exposure to thrips, appearing to trace the feeding paths of the

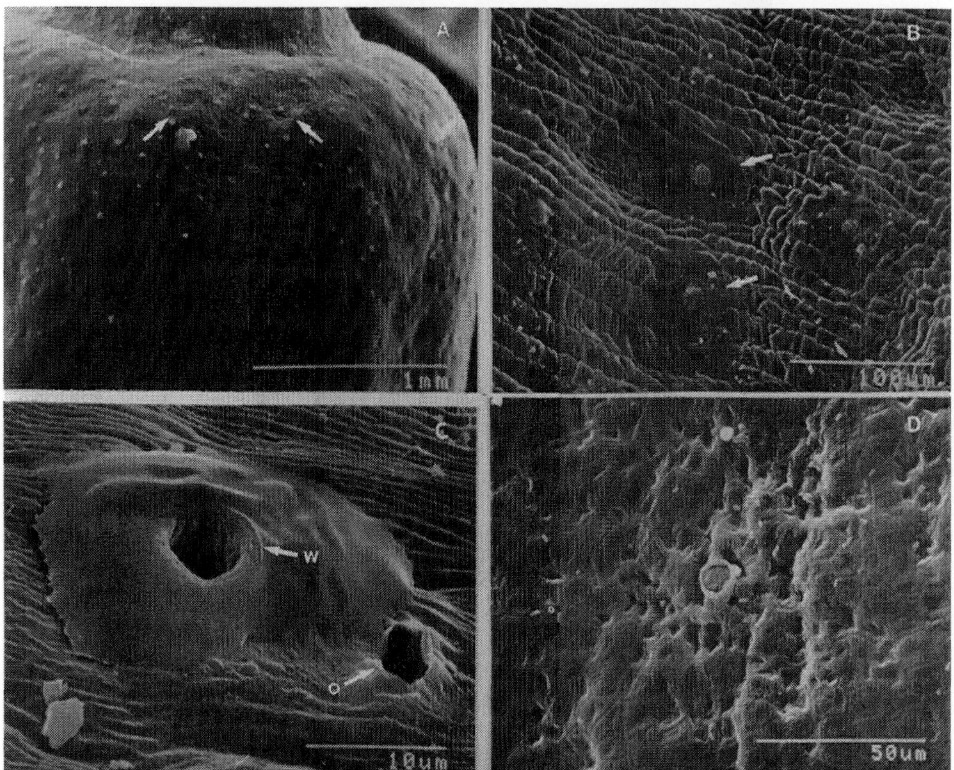

Fig. 13.3. (A) SEM preparations of surface injury from *Frankliniella bispinosa* feeding on navel orange showing irregular depressions on ovary. (B) Sunken areas on navel orange ovary with individual, dried saliva plugs. (C) Feeding punctures with (w) and without (o) a saliva plug. (D) Extensive surface punctures on ovary of navel orange caused by *F. bispinosa*.

stylets. Salivary secretions of feeding thrips probably caused this injury since they were absent in corresponding control tissues (Fig. 13.4). Depth of injury ranged from the epidermis (12 to 20 μm) to necrosis and evacuation extending 20 to 84 μm into the tissues. Mean depth of injury was 63 μm (37 to 96 μm) (Fig. 13.5) (Childers and Achor, 1991).

Injured citrus flowers and buds probably produce more ethylene and so abort, which would mask any sign of previous physical injury. As described above, the stress-induced ethylene production following insect or disease

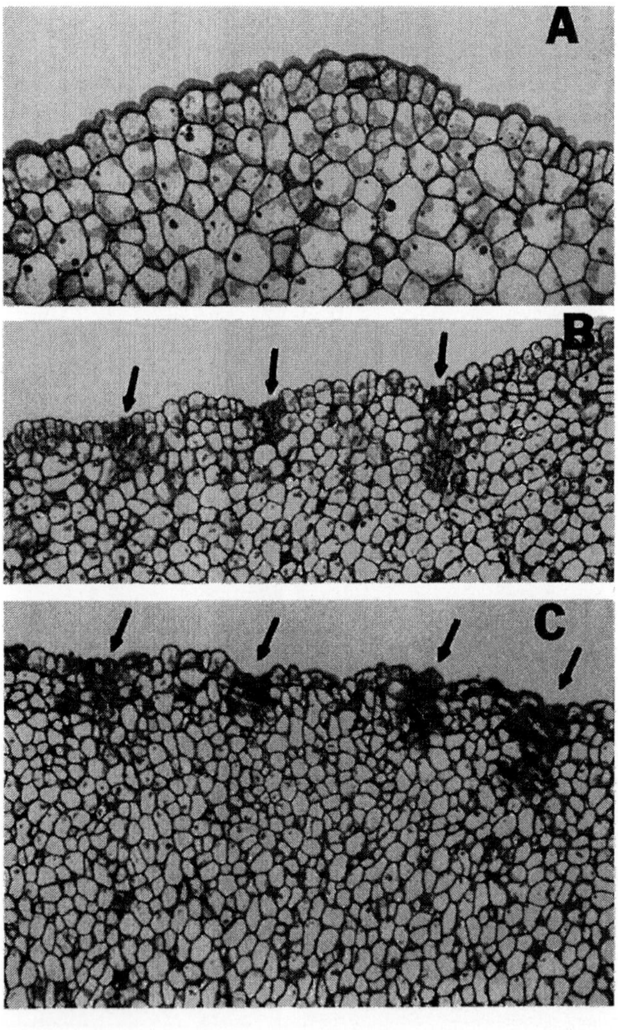

Fig. 13.4. Light micrographs of sections of navel orange tissues. (A) Nectary control without thrips for 96 h. (B) Nectary tissues after 96 h of exposure to *Frankliniella bispinosa* feeding. Arrows show collapsed and necrotic areas in epidermis. (C) Floral disk after 96 h of exposure to *F. bispinosa* feeding. Arrows show feeding sites with injured and collapsed tissues.

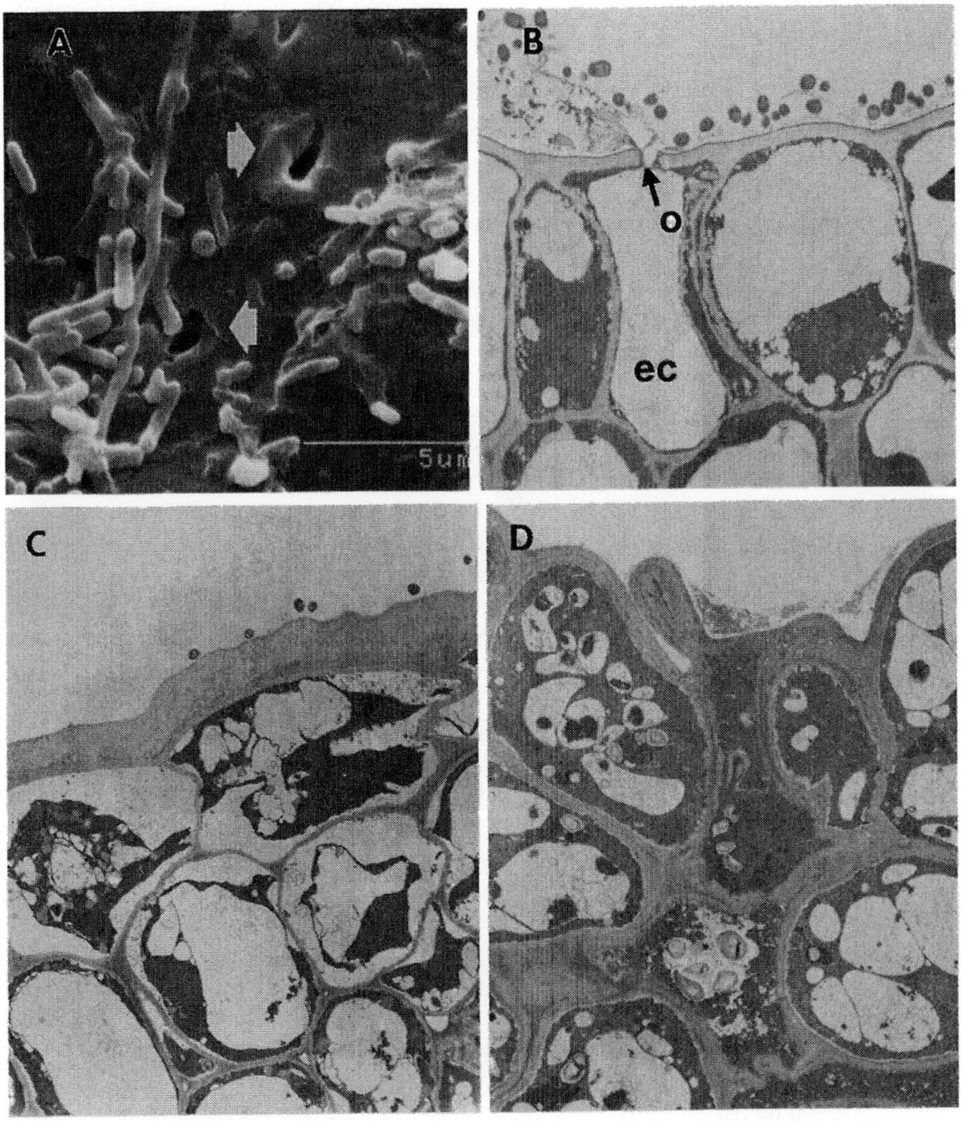

Fig. 13.5. (A) SEM preparation of surface injury on navel orange pistil from *Frankliniella bispinosa* feeding punctures (arrows) and surface contamination of fungi and bacteria. (B) TEM of puncture site in floral disk of navel orange; o, stylet puncture; ec, evacuated cell. (C) TEM of pistil tissues showing plasmolysis and cytoplasmic darkening following feeding by *F. bispinosa*. (D) TEM of navel orange nectary tissues showing cellular collapse and necrosis following feeding by *F. bispinosa*.

infestation accelerates senescence and the shedding of plant organs (Abeles *et al.*, 1992).

Legumes

In Africa, females of *Megalurothrips sjostedti* cause distortion, malformation and discoloration of floral parts, especially the anthers and filaments of cowpea, *Vigna unguiculata* (Singh and Taylor, 1978). This reduces pollination and seed production, while heavily infested plants produce no flowers (Taylor, 1965). Injury to flowers begins before they open when larvae and adults attack the terminal buds and bracts. Infested flowers and buds appear dried, brown and may eventually abort leaving dark red scars (Alghali, 1992). Cowpea yields have been increased by 30 to 90 per cent by controlling this thrips with insecticides (Bal, 1991).

In India, *Megalurothrips distalis* feeds on the pedicles and stigma of the flowers of mung bean, *Vigna radiata* (Chhabra and Kooner, 1985). Feeding begins at bud initiation and damage peaks at full bloom (Chhabra *et al.*, 1986). Flowers may be shed prematurely in severe infestations and all yield lost.

In Europe, adult *Kakothrips pisivorus* feeds on pollen grains of faba beans and peas and oviposits on flowers; larvae feed on flowers and young pods (Stoddard, 1986). Movement by adult thrips among flowers can spread the pathogenic fungus, *Botrytis fabae* Sardiña, which was often found in flowers damaged by *K. pisivorus* and heaviest infections occurred on plants with most thrips (Ondrej, 1973).

Odontothrips confusus, *Thrips flavus* and *Frankliniella intonsa* all cause lucerne flowers to shed, with subsequent yield reduction (Rotrekl, 1985). Insecticide-treated lucerne yielded four times as much seed as untreated plots.

Feeding Damage to Fruit

In this section, the general effects of feeding on fruits covering various types of fruit malformations, stunting and rind blemishes (russetting, silvering, fruit cracking or splitting, and/or abscission) are reviewed for a selection of major crops and thrips species. The damage described applies generally to crops infested by fruit feeders.

Avocado

Heliothrips haemorrhoidalis and *Selenothrips rubrocinctus* are important pests of avocado fruit in South Africa (Dennill and Erasmus, 1992). They feed on the pericarp and cause bronzing of the surface and, in severe infestations, skin cracking. Mouldy thrips excreta are visible on discoloured parts of the pericarp. Such fruits are unattractive and unsuitable for export and up to 80 per cent crop loss has been recorded.

Most avocados hang singly but some are clustered, encouraging damage to the sides of more mature fruit (Dennill and Erasmus, 1992). Infested, clustered fruits were consistently more damaged than single fruits (Table 13.2). In Israel,

Table 13.2. Occurrence of *Heliothrips haemorrhoidalis* and *Selenothrips rubrocinctus* on single and touching avocado fruits, and the number of thrips per pair of touching fruits (from Dennill and Erasmus, 1992).

		Percentage of thrips		
Date (1990)	Sample	Single fruits	Touching fruits	n^a
9 June	1	11.11	88.89	18
9 June	2	2.94	97.06	34
4 July	3	7.69	92.31	48
Mean		7.25	92.75	33

[a]The number of fruit (single or pairs) on which thrips occurred.

H. haemorrhoidalis attacks avocado fruits with the same effect (Plate III.5) (Swirski *et al.*, 1988).

Bananas

Females of *Frankliniella parvula* oviposit and larvae feed on the stem, bracts, pistils, petals, and banana fingers in the Dominican Republic (Harrison, 1963).

A superficial skin injury of bananas called 'corky spot' occurs in Australia (Trochoulias *et al.*, 1984), caused by *Thrips hawaiiensis*. Adults are attracted to emerging inflorescences, then they oviposit and feed on developing fruit while bunches are still wrapped closely in the bracts. Damage is usually confined to lower parts of the fingers, and to the rachis and attached cushion. In hot, dry weather, injury extends to the outer surface of the fruit. Injured tissues expand as fruit develops, resulting in a grey-brown corky covering when it matures. Feeding by adults and larvae of *Chaetanaphothrips signipennis* produces an unsightly skin blemish referred to as 'banana rust' in Australia, where thrips colonize and feed in the protected areas between adjacent fruits (Franzmann, 1974; Pinese, 1987). *Chaetanaphothrips orchidii* produces the same injury in the French Antilles (Delattre and Torregrossa, 1978) and on fruit grown around houses in Florida (Plate III.7).

Hercinothrips bicinctus is an important pest of banana in South Africa. Infestation causes coarse stippling, irregular silvery areas and black faecal droplets on the fruit that lower its value (Keetch, 1972).

In Yemen, adults and larvae of both *Scirtothrips aurantii* and *Thrips pusillus* cause fruit spotting. Small circular spots appear on the surface of fruit, which gradually enlarge and blacken into oily, water-soaked lesions, sometimes followed by skin splitting which allows entry of pathogens (El-Bashir and Al-Zabidi, 1985).

Citrus

Feeding injury by *Scirtothrips citri* adults and larvae in California and Arizona and by *S. aurantii* in South Africa results in development of a fairly uniform ring encircling the stem end of developing fruitlets. Adults and larvae feed under the protective covering of the sepals, and as the fruit matures, its diameter increases and the injured areas become increasingly exposed (Plate V). Second instars of *S. citri* feed more than adult females or first instars, in part because of their larger size and more sedentary habit (Wiesenborn and Morse, 1985, 1986). Young twigs, leaves and leaf buds are also attacked by these thrips with damage especially prevalent in young trees (Gallehugh, 1982; Gilbert, 1990). Navel oranges are a preferred host in California and cosmetic injury results in rejection of the fruit for the fresh market. Scarred oranges also lose weight more rapidly, have higher percentages of soluble solids, lower titratable acidity, and higher sugar:acid ratios than undamaged fruit (Arpaia and Morse, 1991).

In Florida, adults and larvae of *Chaetanaphothrips orchidii*, *Danothrips trifasciatus* and *Heliothrips haemorrhoidalis* damage developing citrus fruits, especially red grapefruit.

Mature fruit of 'Valencia' oranges are attacked by *H. haemorrhoidalis* in Australia. According to Beattie and Jiang (1990), susceptibility of fruit is related to the length of time it remains on trees. Similarly, mature grapefruit held past maturity have been damaged by *H. haemorrhoidalis* in Florida. Feeding of adults and larvae causes silvering of the rind. Affected areas are circular or irregular in shape and their cellular contents have been removed. Infestation of immature fruit increases in the presence of infested mature fruit on the same trees (Childers and Frantz, 1994).

Red grapefruit cultivars tend to have more clustered fruit inside the canopy than other citrus varieties in Florida. *Chaetanaphothrips orchidii* is the prevalent species on Florida citrus with *Danothrips trifasciatus* less abundant; *Heliothrips haemorrhoidalis* inflicts similar damage. A characteristic, brown ring of rough or smooth russetting occurs where clustered fruit touch or between a leaf surface or twig and a fruit (Plate VI). Recently, in several Florida grapefruit groves, nearly all clustered fruit has suffered ring spot damage and has been rejected for the fresh market (Childers and Frantz, 1994).

Peaches and nectarines

Thrips obscuratus is an important pest of stone fruit in New Zealand feeding both on developing fruit causing silvering and scarring, and in flowers (Plate III.1,4) (Teulon and Penman, 1991). Eggs are laid in floral tissues, leaves and ripe fruit and the principal economic problem has been contamination of fruit for export by eggs, larvae and adults (Teulon and Penman, 1988). In addition, the fungal pathogen *Monilinia fructicola* is transferred to flowers and fruit by this thrips (Ellis *et al.*, 1988).

Yonce *et al.* (1990) reviewed species of thrips injuring peaches and nectarines including *Frankliniella occidentalis* in California, *F. occidentalis* and *F. tritici* in Georgia and *F. intonsa* and *Thrips meridionalis* in Greece. Typical injury to peaches consists of a smooth, superficial, pale surface blemish referred to as 'silvering'. Nectarines develop more serious, roughened, brown blemishes referred to as 'russeting'. Populations of *F. occidentalis* and *F. tritici* peak during final swelling of early ripening peach cultivars in Georgia and this appears to be when silvering injury occurs.

Nectarines were grown in Pennsylvania without observable thrips damage until 1991 (C. Felland, Pennsylvania, 1993, personal communication). Since then, silvering injury on 'Red Gold' and 'Fantasia' nectarines and 'Redhaven' peaches has always been associated with the presence of *F. occidentalis* (Felland *et al.*, 1993, 1995) and was so extensive in some orchards in 1991 that the crop was not harvested. *Frankliniella occidentalis* comprised 92 per cent of thrips present on nectarines in south central Pennsylvania during 1992 where 16 per cent of fruit was silvered. The whitish skin patches caused by removal of cell contents resembled spray residue that could not be washed off (Rice, 1992).

By contrast, in north central Pennsylvania where increased fruit scarring was also observed, *Taeniothrips inconsequens* was the only injurious thrips collected from nectarine flowers. Silvering of nectarines was correlated with number of thrips per fruit and damage occurred primarily in the lower third of the tree canopy.

Similar injuries to peaches and nectarines are produced by *Thrips meridionalis* and *Thrips major* in Italy (Cravedi *et al.*, 1983) and by *F. occidentalis* in Israel (M. Wysoki, The Hebrew University, Jerusalem, personal communication). In Greece, *F. intonsa* and *Thrips meridionalis* cause malformation, scarring, pitting and gum exudation of nectarines (Kourmadas *et al.*, 1982).

Strawberry

A complex of species including *Thrips atratus*, *T. major* and *T. tabaci* is found in flowers of late season strawberries in England and has been implicated as the cause of fruit distortion (Buxton and Easterbrook, 1988; Easterbrook, 1991). Elsewhere in Europe, *Frankliniella intonsa* and *F. occidentalis* feed on immature fruit when the petals drop, leading to small deformed berries. Damage to mature fruit resembles 'catface' injury caused by certain hemipteran insects. The taste of thrips-damaged fruit is also inferior (Bournier, 1983). Fruit bronzing is caused by feeding of *F. tritici*, spider mites and cyclamen mites in New Brunswick Province in Canada (Lynch and Tremblay, 1995). Adults and larvae of *F. tritici* begin feeding on seeds and on the inner surfaces of the hull soon after buds open, eventually producing necrotic flecking or bronzing, especially at the base of fruit normally covered by the hull. Sometimes the entire unripened berries appear bronzed, but this becomes less evident as colour develops (Plate IV.1–4).

Cucumber

In Hawaii, *Frankliniella occidentalis* adults and larvae aggregate in male and female cucumber flowers where ovary feeding causes silvery, web- or streak-like scarring on immature fruits. *Thrips palmi* stunts plants, causes premature drop of developing fruits and buds and deforms fruit in Puerto Rico (Pantoja *et al.*, 1988). Two classes of damage are caused by thrips feeding on cucumber: general scarring of fruit and changes in yield, growth, or fruit shape. *Frankliniella occidentalis* was primarily responsible for scarring the fruit and *T. palmi* for damaged foliage. Mixed infestations of thrips (about 6 per cent *F. occidentalis* and 94 per cent *T. palmi*) significantly reduced total yield, mean fruit size and total fruit (Welter *et al.*, 1990) (Plate IX 2–4).

Peas

Silver mottling or blotching on the epidermis of pea pods is caused by feeding of *Thrips tabaci* in New York (Shelton and North, 1987), a common type of injury often accompanied by severe distortion caused widely by *Kakothrips pisivorus* in Europe (Lewis, 1973). Similar damage was recently reported on snow peas in Guatemala (H.A. Denmark, Florida Department of Agriculture and Consumer Services, Gainesville, 1993, personal communication) (Plate VII.3–5).

Grapes

In California, *Frankliniella occidentalis* larvae feed under the persisting caps (calyptras) of berries, eventually causing sunburst or starfish-shaped scars. Without caps or other flower parts, little or no scarring occurs. Gibberellin-thinning sprays applied during bloom accentuated scarring more than did shatter-period, berry-enlarging sprays (Jensen and Luvisi, 1973). Yokoyama (1977) found that *F. occidentalis* adults and larvae scarred the vegetative parts of the cluster by feeding on young tissues, though fortunately, scarring from feeding and oviposition on table grapes ('Thompson seedless' and 'Calmeria') in California is cosmetic and has no effect on measurable quality of fruit (Yokoyama, 1979). *Frankliniella occidentalis* also causes scarring of table grapes in north-western Mexico (Guerra-Sobrevilla, 1989); scarred berries sometimes split during ripening, then rot, especially in damp weather (Plate IV.7–8).

Oviposition Injuries to Plants

In this section, the general effects of oviposition in leaves, flowers and fruits are reviewed. Oviposition injury by thrips can result in economic loss in some plants while producing no measurable damage in others.

Apples

Oviposition by *Frankliniella occidentalis* in 'Granny Smith' apple blossoms causes a central russet area surrounded by a white halo on the fruit (Terry and DeGrandi-Hoffman, 1988; Terry, 1991). This condition, called 'pansy spot', was first reported by Childs (1927) in the Pacific Northwest on 'Northern Spy', 'McIntosh' and 'Transparent' apples. Swift and Madsen (1956) reported similar damage on 'Golden Delicious' and 'Red Delicious' in California, most injury occurring between late pink stage and bloom. Madsen and Jack (1966) also found this skin discoloration sometimes severe on 'McIntosh' and 'Spartan'.

Boyce (1955) reported that oviposition by *Frankliniella tritici* reduced fruit set in apple, sweet and sour cherries, plum and peach due to premature flower drop.

Avocados

In Florida, females of *Frankliniella bispinosa* and *F. kelliae* oviposit in floral pedicels, receptacle bases and outer surfaces of sepals and pistils. Injury is localized compared with the more widespread damage following thrips feeding (Fisher and Davenport, 1989).

Citrus

Frankliniella bispinosa oviposits in the pistil, calyx, petals and filaments of navel orange flowers, and eggs, embryos and empty egg cavities can be found in all these tissues (Childers and Achor, 1991). Injury is limited to formation of a cavity in the tissue and an entry point, both cut by the ovipositor (Fig. 13.6). Surrounding plant cells do not react to this type of intrusion as they do when thrips feed, but instead collapse.

Bananas

Females of *Frankliniella parvula* prefer to oviposit in the epidermis of young banana fruits and less frequently in flowers (DeLeon and Ibarra, 1974). Fruits with more than 100 0.5-mm diameter oviposition spots are rejected for export.

Cherries

Taeniothrips orionis females oviposit in ovaries of cherry flowers and in immature fruits between May and early July in British Columbia (McMullen, 1972), causing fruit dimpling during the last 2 weeks of June. Dimples turn deep red but are less noticeable when the cherries ripen.

Grapes

Jensen (1973) reported oviposition by *Frankliniella occidentalis* females in grape berries creating small dark scars surrounded by halos of whitish tissue. Most injury occurs over a 2-week period between 95 per cent bloom, when the largest berries are up to 3 mm in diameter, and the shatter stage when they have grown to 3 to 5 mm.

Legumes

Salifu *et al.* (1988a) evaluated differences in oviposition by *Megalurothrips sjostedti* females in racemes of cowpea. The thrips-resistant cowpea genotypes 'TVx 3236' and 'TVu 1509' had significantly less larval emergence compared with others tested. Non-preference for oviposition by *M. sjostedti* was shown clearly in choice experiments with the resistant cultivar 'TVx 3236' compared with susceptible 'VITA 7'. Mean numbers of eggs obtained from the resistant cultivar were 12.9 ± 2.7 (choice) and 16.2 ± 1.4 (no choice) compared with 23.8 ± 2.6 (choice) and 22.9 ± 1.9 (no choice), respectively, confirming the existence of non-preference or antixenosis for oviposition. Differences in suitability appeared to be related to the internal organization of calyx tissue; resistant cultivars had smaller air spaces below the parenchyma than susceptible ones. In addition, antibiosis was demonstrated by extended larval development (19.9 ± 0.3 days) and decreased adult longevity (19.1 ± 0.8 days) for the resistant 'TVx 3236' compared with the susceptible 'VITA 7' (18.5 ± 0.4 days and 23.0 ± 0.7 days, respectively) (Salifu *et al.*, 1988b).

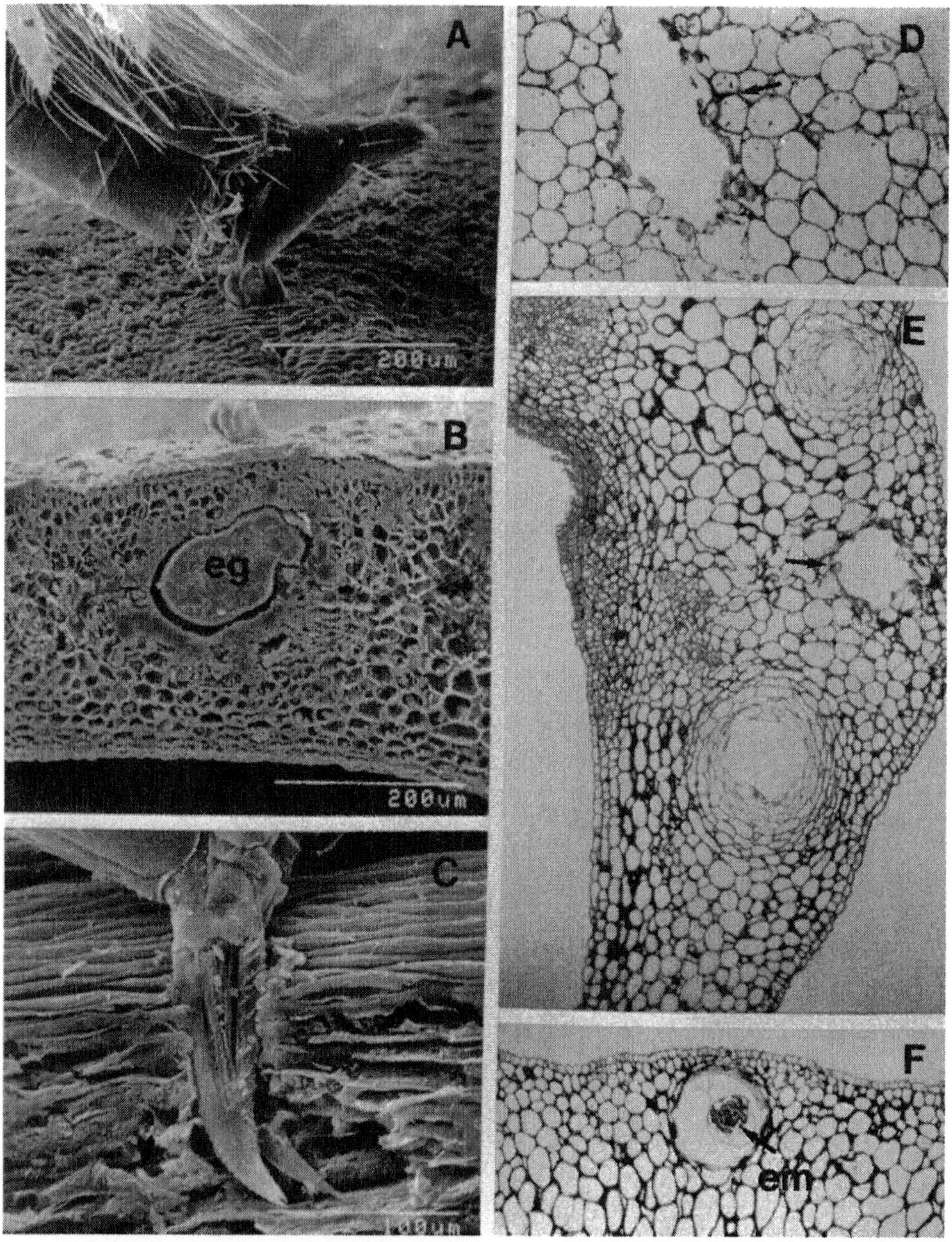

Fig. 13.6. (A) *Frankliniella bispinosa* ovipositing in navel orange pistil. (B) Thrips egg (eg) in navel orange filament tissue. (C) *F. bispinosa* ovipositor and exposed cavity for egg in navel orange petal tissue. *F. bispinosa* egg cavities (arrows) in navel orange (D) pistil and (E–F) calyce tissues and developing embryo (em) in (F). Note collapse of localized cells and a lack of necrotic areas around egg sites.

Tomatoes

Frankliniella occidentalis females oviposit in small tomato fruits and a pale halo sometimes surrounds the puncture site resulting in downgrading or culling of fruit. Other *Frankliniella* species in north Florida did not have this effect (Salguero Navas *et al.*, 1991).

Peas

Frankliniella occidentalis females ovipositing in pea pods (*Pisum sativum* L.) during January in Texas produced slightly raised, light-coloured circular areas on the surfaces of pods (Riherd, 1942).

References

Abeles, F.B., Morgan, P.W. and Saltveit, M.E. Jr (1992) *Ethylene in Plant Biology*, 2nd edn. Academic Press, San Diego, CA.

Alghali, A.M. (1992) On-farm evaluation of control strategies for insect pests in cowpea with emphasis on flower thrips, *Megalurothrips sjostedti* (Trybom) (Thysanoptera: Thripidae). *Tropical Pest Management* 38, 420–424.

Allen, W.R. and Broadbent, A.B. (1986) Transmission of tomato spotted wilt virus in Ontario greenhouses by *Frankliniella occidentalis*. *Canadian Journal of Plant Pathology* 8, 33–38.

Ananthakrishnan, T.N. (1971) Thrips (Thysanoptera) in agriculture, horticulture and forestry – diagnosis, bionomics and control. *Journal of Science and Industry Research* 30, 113–146.

Ananthakrishnan, T.N. (1978) Thrips galls and gall thrips. *Zoological Survey of India, Technical Monograph* No. 1., Calcutta.

Ananthakrishnan, T.N. (1980) Thrips. In: Harris, K.F. and Maramorosch, K. (eds) *Vectors of Plant Pathogens*. Academic Press, New York.

Ananthakrishnan, T.N. (1984) *Bioecology of Thrips*. Indira Publishing House, Oak Park, USA.

Andrewartha, H.G. and Kilpatrick, D.T. (1951) The apple thrips. *Journal of the Department of Agriculture South Australia* 54, 586–592.

Anonymous (1989) *Frankliniella occidentalis* (Perg.) Thysanoptera: Thripidae. EPPO Data Sheets on Quarantine Organisms. *European Plant Protection Organisation Bulletin* 19, 725–731.

Arpaia, M.L. and Morse, J.G. (1991) Citrus thrips *Scirtothrips citri* (Moulton) (Thysanoptera, Thripidae) scarring and navel orange fruit quality in California. *Journal of Applied Entomology* 111, 28–32.

Attique, M.R. and Ahmad, Z. (1990) Investigation of *Thrips tabaci* Lind. as a cotton pest and the development of strategies for its control in Punjab. *Crop Protection* 9, 469–473.

Bailey, S.F. (1938) Thrips of economic importance in California. *California Agricultural Experiment Station Circular* 346.

Bal, A.B. (1991) Action threshold for flower thrips on cowpea (*Vigna unguiculata* (L.) Welp.) in Senegal. *Tropical Pest Management* 37, 363–367.

Baldwin, E.A. and Pressey, R. (1987) Induction of ethylene production in tomato by cell-wall lysing enzymes separated from pectolyase. *Plant Physiology* 83, S-95.

Beattie, G.A.C. and Jiang, L. (1990) Greenhouse thrips and its parasitoids in coastal New South Wales. *General Applied Entomology* 22, 21–24.

Beaudry, R.M. and Kays, S.J. (1988) Applications of ethylene-releasing compounds in agriculture. In: Neuman, P.M. (ed.) *Plant Growth and Leaf Applied Chemicals*. CRC Press, Boca Raton, Florida, pp. 127–155.
Bing, J.W., Dicke, F.F. and Guthrie, W.D. (1990) Genetics of resistance in maize to a complex of three species of *Thrips* (Thysanoptera: Thripidae). *Journal of Economic Entomology* 83, 621–624.
Bournier, A. (1983) *Les Thrips. Biologie Importance Agronomique*. Institute National Recherche Agronomique, Paris, France.
Boyce, H.R. (1955) Notes on injury to tree fruits by *Frankliniella tritici* (Fitch) (Thysanoptera: Thripidae). *Canadian Entomologist* 87, 238–239.
Bullock, J.A. (1963) Thysanoptera associated with pyrethrum, and the control of *Thrips tabaci* Lind. *Tropical Agriculture* 40, 329–335.
Burden, B.J., Morgan, P.W. and Sterling, W.L. (1989) Indole-acetic acid and the ethylene precursor, ACC, in the cotton fleahopper (Hemiptera: Miridae) and their role in cotton square abscission. *Annals of the Entomological Society of America* 82, 476–480.
Buxton, J. and Easterbrook, M.A. (1988) Thrips as probable cause of severe fruit distortion in late-season strawberries. *Plant Pathology* 37, 278–280.
Campbell, A. and Labavitch, J.M. (1987) Induction of ethylene biosynthesis by oligosaccharides. *Plant Physiology* 83, S-140.
Carlson, E.C. (1964) Effect of flower thrips on onion seed plants and a study of their control. *Journal of Economic Entomology* 57, 735–741.
Chatterjee, P.B. (1987) Rice thrips infestations in west Bengal. *International Rice Research Newsletter* 12, 41–42.
Chhabra, K.S. and Kooner, B.S. (1985) Problem of flower shedding caused by thrips, *Megalurothrips distalis* (Karny), on summer mungbean, *Vigna radiata* (L.) Wilczek, and its control. *Tropical Pest Management* 31, 186–188.
Chhabra, K.S., Kooner, B.S. and Sharma, A.K. (1986) Check flower drop in summer mungbean by controlling thrips. *Indian Farming*, December, pp. 25–26.
Childers, C.C. (1992) Suppression of *Frankliniella bispinosa* (Thysanoptera: Thripidae) and the fungal pathogen *Colletotrichum gloeosporioides*, with pesticides during the bloom cycle and improved fruit set on 'navel' orange in Florida. *Journal of Economic Entomology* 85, 1330–1339.
Childers, C.C. and Achor, D.S. (1991) Feeding and oviposition injury to flowers and developing floral buds of 'navel' orange by *Frankliniella bispinosa* (Thysanoptera: Thripidae) in Florida. *Annals of the Entomological Society of America* 84, 272–282.
Childers, C.C. and Achor, D.S. (1995) Thrips feeding and oviposition injuries to economic plants, subsequent damage and host responses to infestation. In: Parker, B., Skinner, M. and Lewis, T. (eds) *Thrips Biology and Management*. Plenum Press, New York, pp. 31–52.
Childers, C.C. and Frantz, G. (1994) Ring spot damage to Florida citrus fruit caused by thrips feeding injury. *Citrus Industry Magazine* 75, 38–40, 42–43.
Childs, L. (1927) Two species of thrips injurious to apples in the Pacific Northwest. *Journal of Economic Entomology* 20, 805–808.
Chisholm, I.F. and Lewis, T. (1984) A new look at thrips (Thysanoptera) mouthparts, their action and effects of feeding on plant tissue. *Bulletin of Entomological Research* 74, 663–675.
Coli, W.M., Hollingsworth, C.S. and Maier, C.T. (1992) Traps for monitoring pear thrips (Thysanoptera: Thripidae) in maple stands and apple orchards. *Journal of Economic Entomology* 85, 2258–2262.

Colyer, P.D., Micinski, S. and Vernon, P.R. (1991) Effect of thrips infestation on the development of cotton seedling diseases. *Plant Disease* 75, 380–382.

Cravedi, P., Molinari, F. and Gianluigi, S. (1983) Control of *Taneiothrips meridionalis* Priesn. (Thysanoptera: Terebrantia: Thripidae) damaging nectarines. (In Italian.) *Estratto da Redia* 66, 37–46.

Cummiskey, P. (1988) Controlling thrips on gloxinias. *Greenhouse Grower*, December, 75–77.

Delattre, P. and Torregrossa, J.P. (1978) Abondance saisonnière, distribution et deplacement des populations du thrips de la rouille de la banane *Chaetanaphothrips orchidii* (Moulton) (Thysanoptera: Thripidae) aux Antilles Francaises. (In French.) *Annales Zoologie* 10, 149–169.

DeLeon, M.J.R. and Ibarra, D.G.E. (1974) Biology, population fluctuation and control of the banana thrips, *Frankliniella parvula* (Hood) in Chantalpa, Tabasco. (In Spanish.) *Folia Entomologia Mexicana* 74, 21–26.

Denmark, H.A. and Poe, S.L. (1972) The gladiolus thrips, *Taeniothrips simplex* (Morison), in Florida (Thysanoptera: Thripidae). *Florida Department of Agriculture and Consumer Services, Division of Plant Industry Entomology Circular* No. 119.

Dennill, G.B. and Erasmus, M.J. (1992) The insect pests of avocado fruits – increasing pest complex and changing pest status. *Journal of the Entomological Society of South Africa* 55, 51–57.

Easterbrook, M.A. (1991) Species of thrips associated with flowers of late-flowering strawberries. *The Entomologist* 110, 5–10.

El-Bashir, S. and Al-Zabidi, A.H. (1985) Fruit spotting of banana in the Yemen Arab Republic. *FAO Plant Protection Bulletin* 33, 113–118.

Ellis, E.C., Penman, D.R. and Gaunt, R.E. (1988) Thrips as potential vectors of brown rot in stonefruit in New Zealand. *Proceedings of the 41st New Zealand Weed and Pest Control Conference*, New Zealand Weed and Pest Control Society, Aukland, New Zealand, pp. 286–287.

Ellsworth, D.S., Tyree, M.T., Parker, B.L. and Skinner, M. (1994) Photosynthesis and water-use efficiency of sugar maple (*Acer saccharum*) in relation to pear thrips defoliation. *Tree Physiology* 14, 619–632.

Elmore, J.C. (1949) Thrips injury to onions grown for seed. *Journal of Economic Entomology* 42, 756–760.

Farrar, J.J. and Davis, R.M. (1991) Relationships among ear morphology, western flower thrips, and fusarium ear rot of corn. *Phytopathology* 81, 661–666.

Fatzinger, C.W. and Dixon, W.N. (1991) Development of sampling methods for the slash pine flower thrips, *Gnophothrips fuscus* (Morgan), (Thysanoptera: Phlaeothripidae). In: Parker, B.L., Skinner, M. and Lewis, T. (eds) *Towards Understanding Thysanoptera, Burlington, VT, 21–23 February 1989*, USDA Forest Service General Technical Report NE-147, Radnor, Pennsylvania.

Felland, C.M., Teulon, D.A., Hull, L.A., Kleiner, W.C. and Polk, D.K. (1993) Western flower thrips on peach and nectarine. *Pennsylvania Fruit News Proceedings* 73, 50–55.

Felland, C.M., Teulon, D.A., Hull, L.A. and Polk, D.K. (1995) Distribution and management of thrips (Thysanoptera: Thripidae) on nectarine in the mid-Atlantic region. *Journal of Economic Entomology* 88, 1004–1011.

Fisher, J.B. and Davenport, T.L. (1989) Structure and development of surface deformations on avocado fruits. *HortScience* 24, 841–844.

Franzmann, B.A. (1974) Banana rust thrips in north Queensland. *Queensland Agriculture Journal* 100, 595–596.

Gaines, J.C. (1965) Cotton insects. *Texas Agricultural Extension Service Bulletin* No. 933.

Gallehugh, J.F. Jr (1982) Citrus thrips. *The Citrus Industry Magazine* 63, 10–11.

Gilbert, M.J. (1990) Relative population levels of citrus thrips *Scirtothrips aurantii* on commercial citrus and adjacent bush. *South African Journal of Zoology* 25, 72–76.

Grisham, M.P., Sterling, W.L., Powell, R.D. and Morgan, P.W. (1987) Characterization of the induction of stress ethylene in cotton caused by the cotton fleahopper (Hemiptera, Miridae) and its microorganisms. *Annals of the Entomological Society of America* 80, 411–416.

Guerra-Sobrevilla, L. (1989) Effectiveness of aldicarb in the control of the western flower thrips, *Frankliniella occidentalis* (Pergande), in table grapes in northwestern Mexico. *Crop Protection* 8, 277–279.

Hara, A.H., Nishijima, W.T. and Sato, D.M. (1988) Impact on anthurium production of controlling an orchid thrips (Thysanoptera: Thripidae), an anthurium whitefly (Homoptera: Aleyrodidae), and a burrowing nematode (Tylenchida: Tylenchidae) with certain insecticide–nematicides. *Journal of Economic Entomology* 81, 582–585.

Harp, S.J. and Turner, V.V. (1976) Effects of thrips on cotton development in the Texas blacklands. *Southwestern Entomologist* 1, 40–45.

Harrison, J.O. (1963) Notes on the biology of the banana flower thrips, *Frankliniella parvula,* in the Dominican Republic (Thysanoptera: Thripidae). *Annals of the Entomological Society of America* 56, 664–666.

Hata, T.Y. and Hara, A.H. (1992) Anthurium thrips, *Chaetanaphothrips orchidii* (Moulton): biology and insecticidal control in Hawaiian anthuriums. *Tropical Pest Management* 38, 230–233.

Henneberry, T.J., Taylor, E.A. and Smith, F.F. (1961) Foliage and soil treatments for the control of flower thrips in outdoor roses. *Journal of Economic Entomology* 54, 233–235.

Herr, E.A. (1934) The gladiolus thrips, *Taeniothrips gladioli* M. and S. *Ohio Agricultural Experiment Station Bulletin* No. 537, Wooster.

Hosny, M.M. and Shoeib, M.M. (1973) Population fluctuations of *T. tabaci* Lind. on cotton seedlings in Egypt. *Bulletin de la Société Entomologique d'Égypte* 56, 247–252.

Hoy, C.W. and Kretchman, D.W. (1991) Thrips (Thysanoptera: Thripide) injury to cabbage cultivars in Ohio. *Journal of Economic Entomology* 84, 971–977.

Huckaba, R.M. and Coble, H.D. (1991) Effect of soybean thrips (Thysanoptera: Thripidae) feeding injury on penetration of acifluorfen in soybean. *Journal of Economic Entomology* 84, 300–305.

Igboekwe, A.D. (1985) Injury to cashew plants, *Anacardium occidentale* L. by the red-banded thrips, *Selenothrips rubrocinctus* Girard (Thysanoptera: Thripidae). *Agriculture, Ecosystems and Environment* 13, 25–32.

Jackson, M.B. and Osborne, D.J. (1970) Ethylene, the natural regulator of leaf abscission. *Nature* 225, 1019–1022.

Jensen, F. (1973) Timing of halo spotting by flower thrips on table grapes. *California Agriculture*, October, 6–8.

Jensen, F. and Luvisi, D. (1973) Flower thrips nymphs involved in scarring of Thompson seedless grapes. *California Agriculture*, October, 8–9.

Johnson, M.W. (1986) Population trends of a newly introduced species, *Thrips palmi* (Thysanoptera: Thripidae), on commercial watermelon plantings in Hawaii. *Journal of Economic Entomology* 79, 718–720.

Kamm, J.A. (1971) Silvertop of bluegrass and bentgrass produced by *Anaphothrips*

obscurus. Journal of Economic Entomology 64, 1385–1387.
Kamm, J.A. (1972) Thrips that affect production of grass seed in Oregon. *Journal of Economic Entomology* 65, 1050–1055.
Keetch, D.P. (1972) Thrips and red spider mite a nuisance in Natal. *Farming in South Africa*, August, 19–21.
Kendall, D.M. and Bjostad, L.B. (1990) Herbivory by *Thrips tabaci* induces greater ethylene production in intact onions than mechanical damage alone. *Journal of Chemical Ecology* 16, 981–991.
Kendall, D.M. and Capinera, J.L. (1987) Susceptibility of onion growth stages to onion thrips (Thysanoptera: Thripidae) damage and mechanical defoliation. *Environmental Entomologist* 16, 859–863.
Kilpatrick, D.T. (1955) There is an answer to biennial bearing in apples. *Journal of the Department of Agriculture of South Australia* 58, 471–474.
Kolb, T.E. and Teulon, D.A.J. (1991) Relationship between sugar maple budburst phenology and pear thrips damage. *Canada Journal of Forestry Research* 21, 1043–1048.
Kolb, T.E., McCormick, L.H. and Shumway, D.L. (1991) Physiological responses of pear thrips-damaged sugar maples to light and water stress. *Tree Physiology* 9, 401–413.
Kourmadas, A.L., Zestas, T. and Argyriou, L.C. (1982) Timing of spraying for control of thrips in nectarine trees. *Annals of the Institute of Phytopathology, Benaki (N.S.)* 13, 120–129.
Lewis, T. (1959) The annual cycle of *Limothrips cerealium* Haliday (Thysanoptera) and its distribution in a wheat field. *Entomologia Experimentalis et Applicata* 2, 187–203.
Lewis, T. (ed.) (1973) *Thrips, Their Biology, Ecology and Economic Importance.* Academic Press, London.
Lloyd, N.C. (1973) Plague thrips a pest of apples. *Agriculture Gazette of New South Wales* 84, 249–251.
Lynch, K. and Tremblay, R. (1995) *Fruit Bronzing of Strawberry.* New Brunswick Department of Agriculture, Plant Industry Branch Fact Sheet.
Madsen, H.F. and Jack, I.D. (1966) The relation of thrips to pansy spot on apples. *Canadian Entomologist* 98, 903–908.
Mantel, W.P. and van de Vrie, M. (1988) A contribution to the knowledge of Thysanoptera in ornamental and bulbous crops in the Netherlands. *Acta Phytopathologica et Entomologica Hungarica* 23, 301–311.
Martin, W.R. Jr, Morgan, P.W., Sterling, W.L. and Kenerley, C.M. (1988) Cotton fleahopper and associated microorganisms as components in the production of stress ethylene by cotton. *Plant Physiology* 87, 280–285.
Mauney, J.R. and Henneberry, T.J. (1979) Identification of damage symptoms and patterns of feeding of plantbugs in cotton. *Journal of Economic Entomology* 72, 496–501.
Mayer, D.F., Lunden, J.D. and Rathbone, L. (1987) Evaluation of insecticides for *Thrips tabaci* (Thysanoptera: Thripidae) and effects of thrips on bulb onions. *Journal of Economic Entomology* 80, 930–932.
McKenzie, C.L., Cartwright, B., Miller, M.E. and Edelson, J.V. (1993) Injury to onions by *Thrips tabaci* (Thysanoptera: Thripidae) and its role in the development of purple blotch. *Environmental Entomologist* 22, 1266–1277.
McMullen, R.D. (1972) *Taeniothrips orionis* oviposition and feeding injury on cherries. *Journal of the Entomological Society of British Columbia* 69, 3–5.
McNally, P.S., Fogg, C., Flynn, J. and Horenstein, J. (1985) Effects of thrips

(Thysanoptera: Thripidae) on shoot growth and berry maturity of 'Chenin Blanc' grapes. *Journal of Economic Entomology* 78, 69–72.

Mikahilova, N.A. (1981) Influence of structural peculiarities of different species of wheat on the attack by *Haplothrips tritici* and *Trigonotylus coelestialium*. *Acta Phytopathologica Academiae Scientiarum Hungaricae* 16, 199–202.

Mound, L.A. (1994) Gall-inducing Thysanoptera: a search for patterns. In: Williams, M.A.J. (ed.) *Plant Galls: Organisms, Interactions, Populations*. *Systematics Association Special Volume* 49. Clarendon Press, Oxford, pp. 131–149.

North, R.C. and Shelton, A.M. (1986) Colonization and intraplant distribution of *Thrips tabaci* (Thysanoptera: Thripidae) on cabbage. *Journal of Economic Entomology* 79, 219–223.

Nugaliyadde, L. and Heinrichs, E.A. (1984a) Biology of rice thrips, *Stenchaetothrips biformis* (Bagnall) (Thysanoptera: Thripidae) and a greenhouse rearing technique. *Journal of Economic Entomology* 77, 1171–1175.

Nugaliyadde, L. and Heinrichs, E.A. (1984b) Resistance of *Oryza* spp. to thrips, *Stenchaetothrips biformis* (Bagnall) (Thysanoptera: Thripidae). *Crop Protection* 3, 305–313.

Oetting, R.D., Beshear, R.J., Liu, T., Braman, S.K. and Baker, J.R. (1993) Biology and identification of thrips on greenhouse ornamentals. *Georgia Agricultural Experiment Station Research Bulletin* 414, 20 pp.

Ondrej, M. (1973) Paraziticka houba *Botrytis fabae* Sard. ve vztahu K savemu hmyzu. (In Czechoslvakian.) *Biologica (Bratislava)* 28, 57–63.

Palmer, J.M. (1990) Identification of the common thrips of tropical Africa (Thysanoptera: Insecta). *Tropical Pest Management* 36, 27–49.

Pantoja, A., Segarro, A., Ruiz, H. and Medina-Gaud, S. (1988) *Thrips palmi* (Thysanoptera: Thripidae): a new insect pest for Puerto Rico. *Journal of Agriculture, University of Puerto Rico* 72, 327–329.

Parrella, M.P. and Jones, V.P. (1987) Development of integrated pest management strategies in floricultural crops. *Bulletin of the Entomological Society of America* 33, 28–34.

Pinese, B. (1987) Soil and bunch applications of insecticides for control of banana rust thrips. *Queensland Journal of Agriculture and Animal Sciences* 44, 107–111.

Post, R.L. (1955) The barley thrips in North Dakota. *North Dakota Seed Journal* 24, 2–3.

Quisenberry, J.E. and Rummel, D.R. (1979) Natural resistance to thrips injury in cotton as measured by differential leaf area reduction. *Crop Science* 19, 879–881.

Raffa, K.F. and Hall, D.J. (1989) *Thrips calcaratus* Uzel (Thysanoptera: Thripidae), a new pest of basswood trees in the Great Lakes region. *Canadian Journal of Forestry Research* 19, 1661–1662.

Rahman, K.A. and Bharadwaj, N.K. (1937) The grapevine thrips, *Rhipiphorothrips cruentatus* Hood (Thripidae: Terebrantia: Thysanoptera). *Indian Journal of Agricultural Sciences* 7, 633–651.

Reddy, K.L., Srinivas, C., Nageshwar Rao, T.G. and Srinivas Rao, P. (1987) Rice thrips, a new rice pest in northern Telangana, India. *International Rice Research Newsletter* 12, 28–29.

Rice, J. (1992) Our recent experiences with handling and marketing nectarines. *Pennsylvania Fruit News Proceedings* 72, 48–49.

Riherd, P.T. (1942) An unusual injury to peas by thrips. *Journal of Economic Entomology* 35, 453.

Rieske, L.K. and Raffa, K.F. (1995) Ethylene emission by a deciduous tree, *Tilia americana*, in response to feeding by introduced basswood thrips, *Thrips calcar-*

atus. *Journal of Chemical Ecology* 21, 187–197.
Rosenheim, J.A., Welter, S.C., Johnson, M.W., Mau, R.F.L. and Gusukuma-Minuto, L.R. (1990) Direct feeding damage on cucumber by mixed-species infestations of *Thrips palmi* and *Frankliniella occidentalis* (Thysanoptera: Thripidae). *Journal of Economic Entomology* 83, 1519–1525.
Rotrekl, J. (1985) The harmfulness of thrips (Thysanoptera: Terebrantia) to seed lucerne. (In Czechoslovakian with abstract in English.) *Sbornik Uvtiz. Ochrana Rostlin* 21, 255–260.
Sakimura, K. (1947) Thrips in relation to gall-forming and plant disease transmission: a review. *Proceedings of the Hawaiian Entomological Society* 13, 59–96.
Sakimura, K. (1975) *Danothrips trifasciatus*, a new species, and collection notes on the Hawaiian species of *Danothrips* and *Chaetanaphothrips* (Thysanoptera: Thripidae). *Proceedings of the Hawaiian Entomological Society* 22, 125–132.
Sakimura, K., Nakahara, S. and Denmark, H.A. (1986) A thrips, *Thrips palmi* Karny. *Florida Department of Agriculture and Consumer Services, Division of Plant Industry Entomology Circular* No. 280.
Salguero Navas, V.E., Funderburk, J.E., Olson, S.M. and Beshear, R.J. (1991) Damage to tomato fruit by the western flower thrips (Thysanoptera: Thripidae). *Journal of Entomological Science* 26, 436–442.
Salifu, A.B., Hodgson, C.J. and Singh, S.R. (1988a) Mechanism of resistance in cowpea (*Vigna unguiculata* (L.) Walp.) genotype, TVx 3236, to the bean flower thrips, *Megalurothrips sjostedti* (Trybom) (Thysanoptera: Thripidae). 1. Ovipositional nonpreference. *Tropical Pest Management* 34, 180–184.
Salifu, A.B., Singh, S.R. and Hodgson, C.J. (1988b) Mechanism of resistance in cowpea (*Vigna unguiculata* (L.) Walp.) genotype, TVx 3236, to the bean flower thrips, *Megalurothrips sjostedti* (Trybom) (Thysanoptera: Thripidae). 2. Nonpreference and antibiosis. *Tropical Pest Management* 34, 185–188.
Schwartz, A. (1989) The incidence of *Heliothrips sylvanus* Faure (Thysanoptera: Thripidae) on table grapes. *South African Journal of Enology and Viticulture* 10, 59–61.
Sharma, P.L. and Bhalla, O.P. (1963) Occurrence of the thrips damaging apple blossoms in Himachal Pradesh. *Indian Journal of Entomology* 25, 85–86.
Shelton, A.M. and North, R.C. (1987) Injury and control of onion thrips (Thysanoptera: Thripidae) on edible podded peas. *Journal of Economic Entomology* 80, 1325–1330.
Shelton, A.M., Wilsey, W.T. and Luetchford, C.M. (1981) Control of thrips on cabbage, 1980. *Insecticide and Acaricide Tests* 6, 60.
Shelton, A.M., Stoner, J.R., Wilsey, W.T., Stoyla, B.O. and Andaloro, J.T. (1982) Onion thrips (Thysanoptera: Thripidae) damage and contamination in sauerkraut. *Journal of Economic Entomology* 75, 492–494.
Shelton, A.M., Hoy, C.W., North, R.C., Dickson, M.H. and Barnard, J. (1988) Analysis of resistance in cabbage varieties to damage by Lepidoptera and Thysanoptera. *Journal of Economic Entomology* 81, 634–640.
Shumway, D.L., Teulon, D.A.J. and Kolb, T.E. (1991) Identification of thrips (Thysanoptera: Thripidae) feeding sites on sugar maple leaves. *Journal of Economic Entomology* 84, 1771–1773.
Singh, M. (1989) Damage and economic threshold level of *Thrips flavus* Schrank to apple blossoms in Himachel Pradesh. *Indian Journal of Entomology* 25, 85–86.
Singh, S.R. and Taylor, T.A. (1978) Pests of grain legumes and their control in Nigeria. In: Singh, S.R., van Emden, H.F. and Taylor, T.A. (eds) *Pests of Grain Legumes: Ecology and Control*. Academic Press, London, pp. 99–111.

Stoddard, F.L. (1986) The distribution of immature thrips among flowers of faba bean in commercial crops and experimental plots. *Annals of Applied Botany* 109, 61–69.

Stoner, K.A. and Shelton, A.M. (1988) Effect of winter storage on thrips damage to cabbage. *New York's Food and Life Sciences Bulletin* No. 121. New York State Agricultural Experiment Station, Geneva.

Suzuki, H. and Miyara, A. (1983) Integrated control of *Thrips palmi* using silver-colored materials. 1. Loss assessment on cucumber. (In Japanese.) *Proceedings of the Association of Plant Protection Kyushu* 29, 77–80.

Swift, J.E. and Madsen, H.F. (1956) Thrips damage to apple. *Journal of Economic Entomology* 49, 398–399.

Swirski, E., Wysoki, M. and Izhar, Y. (1988) Integrated pest management in the avocado orchards of Israel. *Applied Agricultural Research* 3, 1–7.

Taylor, T.A. (1965) Observations on the bionomics of *Laspeyresia ptychora* Meyr. (Lepidoptera, Eucosmidae) infesting cowpea in Nigeria. *Bulletin of Entomological Research* 55, 761–774.

Terry, L.I. (1991) *Frankliniella occidentalis* (Thysanoptera: Thripidae) oviposition in apple buds: role in bloom state, blossom phenology, and population density. *Environmental Entomologist* 20, 1568–1576.

Terry, L.I. and DeGrandi-Hoffman, G. (1988) Monitoring western flower thrips (Thysanoptera: Thripidae) in 'Granny Smith' apple blossom clusters. *Canadian Entomologist* 120, 1003–1016.

Teulon, D.A.J. (1988) Pest management of the New Zealand flower thrips, *Thrips obscuratus* (Crawford) (Thysanoptera: Thripidae) on stonefruit in Canterbury, New Zealand. PhD thesis, Lincoln College, University of Canterbury, New Zealand.

Teulon, D.A.J. and Penman, D.R. (1988) Thrips control on stonefruit at harvest with fluvalinate. *Proceedings of the 40th New Zealand Weed and Pest Control Conference on Fruit Crops*, New Zealand Weed and Pest Control Society, Palmerston North, New Zealand, pp. 262–266.

Teulon, D.A.J. and Penman, D.R. (1991) Effects of temperature and diet on oviposition rate and development time of the New Zealand flower thrips, *Thrips obscuratus*. *Entomologia Experimentalis et Applicata* 60, 143–155.

Thind, T.S. and Jhooty, J.S. (1982) Association of thrips with purple blotch infection on onion plants caused by *Alternaria porri*. *Indian Phytopathology* 35, 696–698.

Trouchoulias, T., Wright, W.E. and Ross Loebel, M. (1984) Use of insecticides to reduce corky scab injury in bananas. *Tropical Agriculture (Trinidad)* 61, 193–195.

Tugwell, P. and Cochran, M. (1989) Cotton's response to thrips damage. In: Oosterhuis, D.M. (ed.) *Proceedings of the 1988 Cotton Research Meeting*, Arkansas Experiment Station Special Report 132, University of Arkansas, Fayetteville, USA, pp. 57–59.

Ullman, D.E., Westcot, D.M., Hunter, W.B. and Mau, R.F.L. (1989) Internal anatomy and morphology of *Frankliniella occidentalis* (Pergande) (Thysanoptera: Thripidae) with special reference to interactions between thrips and tomato spotted wilt virus. *International Journal of Insect Morphology and Embryology* 18, 289–310.

van Schoonhoven, A. (1974) Resistance to thrips damage in cassava. *Journal of Economic Entomology* 67, 728–730.

van Schoonhoven, A. and Pena, J.E. (1976) Estimation of yield losses in cassava following attack from thrips. *Journal of Economic Entomology* 69, 514–516.

Velusamy, R. (1990) Relative susceptibility of high-yielding rice varieties to the thrips, *Stenchaetothrips biformis* (Bagnall) (Thysanoptera: Thripidae). *Crop Protection* 9, 193–196.

Vité, J.P. (1956) Populationsstüdien am Lärchenblasefüss *Taeniothrips laricivorus* Krat. *Zeitschrift für angewandte Entomologie* 38, 417–488.

Wang, C.L. (1987) The infestation of thrips on floriculture and their control. (In Chinese with abstract in English.) *Chinese Journal of Entomology, Special Publication* 1, 37–43.

Watson, T.F. (1965) Influence of thrips on cotton yields in Alabama. *Journal of Economic Entomology* 58, 1118–1122.

Welter, S.C., Rosenheim, J.A., Johnson, M.W., Mau, R.F.L. and Gusukuma-Minuto, L.R. (1990) Effects of *Thrips palmi* and western flower thrips (Thysanoptera: Thripidae) on the yield, growth, and carbon allocation pattern in cucumbers. *Journal of Economic Entomology* 83, 2092–2101.

Wien, H.C. (1979) The use of ethephon (2-chloroethylphosphonic acid) to screen for abscission resistance among cowpea (*Vigna unguiculata* L. Welp.) cultivars. *HortScience* 14, 422.

Wien, H.C. and Roesingh, C. (1980) Ethylene evolution by thrips-infested cowpea provides a basis for thrips resistance screening with ethephon sprays. *Nature* 283, 192–194.

Wiesenborn, W.D. and Morse, J.G. (1985) Feeding rate of Thysanoptera estimating ^{14}C inulin. *Journal of Economic Entomology* 78, 1515–1518.

Wiesenborn, W.D. and Morse, J.G. (1986) Feeding rate of *Scirtothrips citri* (Moulton) (Thysanoptera: Thripidae) as influenced by life stage and temperature. *Environmental Entomologist* 15, 763–766.

Yokoyama, V.Y. (1977) *Frankliniella occidentalis* and scars on table grapes. *Environmental Entomologist* 6, 25–30.

Yokoyama, V.Y. (1979) Effects of thrips scars on table grape quality. *Journal of the American Society for Horticultural Sciences* 104, 243–245.

Yonce, C.E., Payne, J.A., Beshear, R.J. and Horton, D.L. (1990) Thrips (Thysanoptera: Thripidae) associated with unsprayed and sprayed peaches in Georgia. *Journal of Economic Entomology* 83, 511–518.

Young, S., Kinzer, R.E., Walton, R.R. and Matlock, R.S. (1972) Field screening for tobacco thrips resistance in peanuts. *Journal of Economic Entomology* 65, 828–832.

Thrips as Vectors of Plant Pathogens

14

Diane E. Ullman[1], John L. Sherwood[2] and Thomas L. German[3]

[1]*University of California, Department of Entomology, Davis, California 95616, USA;* [2]*University of Georgia, Department of Plant Pathology, Athens, Georgia 30602–7274, USA;* [3]*University of Wisconsin – Madison, Department of Plant Pathology, Russell Laboratory, 1630 Linden Drive, Madison, Wisconsin 53706, USA*

Introduction

Severe epidemics in food, fibre and ornamental crops are caused by thrips-transmitted pathogens in tropical, subtropical and temperate regions throughout the world. These problems are especially difficult to control in crops grown in mild climates and greenhouses that favour nearly continuous cropping and year-round presence of thrips vector populations and plant pathogens. The diseases caused by thrips transmission of pathogens are thought to cost growers worldwide billions of dollars in control costs and lost productivity. Spread of thrips-transmitted plant viruses remains one of the most intractable unsolved problems in management of cropping systems. Indeed, the lack of suitable control measures leaves a major gap in virtually all integrated pest management (IPM) and low input sustainable agricultural (LISA) systems. Control of thrips-transmitted plant pathogens can rarely be achieved with pesticides for several reasons. First, relatively small numbers of vector thrips can result in high rates of pathogen spread. Second, many thrips species are intensely resistant to pesticides, so their populations are not well controlled. Third, inoculation of the plant pathogens transmitted by thrips occurs so quickly that the insects are not killed by the pesticide until after they have transmitted the pathogen. Finally, many epidemics are caused by dispersing thrips that are transient in the affected crop. Pesticides do not control these transient populations unless applied at an unacceptable frequency. Resistant crop plants are effective management tools when they are available but, unfortunately, resistance genes for thrips-transmitted pathogens are not always present in existing germplasm (Cho, 1996). Also, when resistance genes are located they do not always protect plants from the diversity of isolates present concomitantly in crop production areas.

Epidemics of thrips-transmitted plant pathogens require the interaction of three basic components: the thrips vector, the pathogen host plant and the pathogen. Intricate and complex biological relationships occur between thrips and some of the pathogens they transmit, while in other cases they merely facilitate mechanical transmission. Although thrips are individually minute creatures, their tendency to occur in large populations, across broad host and climatic ranges makes them especially significant as direct crop pests and as pathogen vectors. The past decade has been witness to a great surge in our understanding of thrips biology and the role of these fascinating insects as vectors of plant pathogens. This chapter reviews thrips transmission of viruses, fungi and bacteria and provides an understanding of the specific biological relationships occurring between thrips and the pathogens they transmit.

Thrips Transmission of Plant Viruses

Among the 5000 or so known species of thrips (zur Strassen, 1960), only ten species, all of which are members of the Thripidae, have been confirmed as vectors of plant viruses (Tables 14.1 and 14.2). The known vector species represent four genera, three of which constitute the largest genera of the Thripidae. Most of these species do not form closely related groups phylogenetically, however they do share important ecological characteristics such as extreme polyphagy and the ability to reproduce on a broad range of host plants (Mound, 1996). In addition, the feeding behaviour of thrips (see Chapter 4) contributes very significantly to the propensity of these insects to serve as vectors.

Thrips are known to transmit viruses belonging to at least four virus groups including bunyaviruses (family Bunyaviridae, genus *Tospovirus*), ilarviruses, sobemoviruses and carmoviruses. They have been implicated as vectors of maize chlorotic mottle virus (MCMV) (an ungrouped plant virus), although this relationship has not been confirmed. Among these plant viruses, the tospoviruses have the most complex biological relationship with their thrips vectors and are transmitted primarily from leaf to leaf by thrips. In contrast, the ilarviruses, sobemoviruses and carmoviruses are pollen-borne and although thrips are generally essential in facilitating transmission, these viruses have a less complex relationship with their vectors and leaf-to-leaf transmission seldom occurs.

Thrips Transmission of Pollen-Borne Viruses

Until the last decade, thrips were not known to transmit any viruses other than tomato spotted wilt (TSWV) (see next section). The first reproducible reports of thrips transmission of a virus other than TSWV came in 1982 and 1987, when transmission of tobacco streak virus (TSV), a pollen-borne ilarvirus, by *Thrips tabaci* (Kaiser *et al.*, 1982; Sdoodee and Teakle, 1987) and *Frankliniella occidentalis* (Kaiser *et al.*, 1982) was documented. Reports of thrips transmission of additional ilarviruses, sobemoviruses and carmoviruses quicky followed (Greber *et al.*, 1991a,b, 1992; Hardy and Teakle, 1992; Sdoodee and Teakle,

1993; Krczal *et al.*, 1995). Currently, at least five thrips species are associated with transmission of five pollen-borne viruses in three separate virus groups. The relationships between these viruses and particular thrips species are summarized in Table 14.1.

The mechanism by which thrips transmit pollen-borne viruses is not understood. It has been generally thought that pollen-borne viruses are seldom, if ever, transmitted leaf to leaf, although sowbane mosaic sobemovirus (SoMV) was transmitted leaf to leaf by *Thrips tabaci* in the absence of infected pollen (Hardy and Teakle, 1992). In general, infected pollen is probably moved between plants by wind and pollinators (e.g. bees, flies) and thrips are thought to facilitate transmission through mechanical wounding of the plant during feeding. The thrips species associated with transmission of pollen-borne viruses are all known to include pollen in their diets and may have pollen on their body surface as they move between flowers and plants. In addition, debris from disintegrating pollen containing virus particles may adhere to the body surface of thrips (Mink, 1992), which also feed by piercing and sucking the contents from foliage and floral plant parts (Chisholm and Lewis, 1984; Hunter and Ullman, 1989; Huckaba and Coble, 1991; Kumar *et al.*, 1995b). During the feeding process thrips puncture the plant surface with their stylets. These feeding punctures are on some occasions filled with saliva when the feeding bout ends and in others are left apparently open (see Chapter 4). Experiments with prune dwarf virus (PDV) and prunus necrotic ringspot virus (PNRSV) have shown that virus transmission occurs readily when infected pollen and

Table 14.1. Thrips vectors of pollen-borne ilarviruses, sobemoviruses and carmoviruses.

Virus	Thrips vector	Reference
Prunus necrotic ringspot ilarvirus	*Frankliniella occidentalis*	Greber *et al.*, (1992)
	Thrips tabaci	Greber *et al.*, (1991a)
	Thrips imaginis[a]	Greber *et al.*, (1991a)
	Microcephalothrips abdominalis[a]	
	Thrips australis[a]	
Prune dwarf ilarvirus	*Frankliniella occidentalis*	Greber *et al.*, (1992)
Tobacco streak ilarvirus	*Thrips tabaci*	Sdoodee and Teakle (1987, 1993)
	Microcephalothrips abdominalis	Greber *et al.*, (1991b)
Sowbane mosaic sobemovirus[b]	*Thrips tabaci*	Hardy and Teakle (1992)
Pelargonium flower break carmovirus	*Frankliniella occidentalis*	Krczal *et al.*, (1995)

[a]These three species transmitted prunus necrotic ringspot from plum pollen when placed together with *Thrips tabaci* on cucumber seedlings. Their possible role as vectors of prunus necrotic ringspot has not been further examined.

[b]This virus was also transmitted leaf to leaf by *T. tabaci* in the absence of infected pollen.

thrips are present on the leaf surface (Greber *et al.*, 1991a,b; 1992; Mink, 1992). These authors also demonstrated that transmission occurs if the leaf surface is slightly wounded by rubbing with abrasive materials followed by infected pollen. Thrips are known to ingest pollen and to suck the liquid contents from pollen grains (Lewis, 1973); however, thrips that have been fed infected pollen or infected plant foliage have not been shown to transmit any of the pollen-borne viruses (Mink, 1992) with the exception mentioned above for *Thrips tabaci* and SoMV (Hardy and Teakle, 1992).

Most of the details surrounding thrips transmission of pollen-borne viruses in nature are entirely unknown. Because of the minute size and grooming habits of thrips, they probably do not play a major role in transporting infected pollen between plants. Bees, wind and rain may have a more important role in movement of infected pollen, although this has not been demonstrated experimentally. Thrips and infected pollen may also be transported by bees (G. Mink, Washington, 1995, personal communication), although there are not yet direct observations to support this hypothesis. In many of the systems where pollen-borne viruses are important, i.e. stone fruits such as peach, prune and cherry, the plants flower before foliage is present, yet flower-to-flower transmission has not been demonstrated. Experimental systems have involved moving infected pollen to leaves of alternate hosts (i.e. cucumber) and have not documented the circumstances leading to natural spread. Thrips may facilitate transmission of pollen-borne viruses when infected pollen enters wounds in the plant surface made by thrips stylets during feeding. Certainly, developing an understanding of how thrips contribute to transmission of ilarviruses, sobemoviruses and carmoviruses provides a rich and exciting area of study. For the moment, evidence suggests that the relationship between thrips and pollen-borne viruses is essentially mechanical and that other insects (i.e. bees) or conditions may also contribute to virus spread in nature. Control strategies suggested include removing alternate virus hosts near crops and close-mowing of weeds to limit flowering and production of infected pollen (Greber *et al.*, 1991b).

Thrips Transmission of Tospoviruses

The tospoviruses have received international attention because they cause severe diseases of food, fibre and ornamental crops throughout the world (see Appendix II) and epidemics are extremely difficult to control. The symptoms of virus infection vary with plant species and tospovirus. Examples of symptoms caused by TSWV and INSV in a variety of vegetable crops, peanuts and ornamentals are shown in Plate XII. Several thrips species serve as vectors of tospoviruses (Pittman, 1927; Sakimura, 1940; Sakimura, 1962, 1963, 1969; Amin *et al.*, 1981; Iwaki *et al.*, 1984; German *et al.*, 1992; Yeh *et al.*, 1992; Wijkamp *et al.*, 1995a) (Table 14.2). This section explores the interactions between thrips and tospoviruses, changing vector relationships and possibilities for current and future management strategies. Attention is given to the history of thrips/tospovirus research, as well as a detailed summary of recent advances in the understanding of interactions between these viruses and their thrips vectors. Finally, potential directions for future research are discussed. For detailed discussions of thrips–tospovirus relationships, tospovirus diagnosis, molecular biology and phylogeny several reviews and articles may be consulted (German *et al.*, 1992; Ullman *et al.*, 1992a; de Avila *et al.*, 1993; Heinze *et al.*, 1995; Yeh and Chang, 1995).

The disease 'spotted wilt' was first observed in 1916 in Australia

Table 14.2. Thrips vectors of tomato spotted wilt tospovirus (TSWV) (serogroup I, type member of the genus *Tospovirus*), tomato chlorotic spot tospovirus (TCSV) and groundnut ringspot tospovirus (GRSV) (serogroup II), impatiens necrotic spot tospovirus (INSV) (serogroup III), and groundnut bud necrosis tospovirus (GBNV) and watermelon silver mottle tospovirus (WSMV) (serogroup IV)[a].

Species	Tospovirus	Reference
Frankliniella occidentalis	TSWV	Gardner *et al.*, (1935); Allen and Broadbent (1986); Cho *et al.*, (1986)
	INSV	DeAngelis *et al.*, (1993); Wijkamp and Peters (1993); Wijkamp *et al.*, (1995a,b)
	GRSV	Wijkamp *et al.*, (1995a,b)
	TCSV	Wijkamp *et al.*, 1995a,b)[b,c]
Frankliniella schultzei[b]	TSWV	Samuel *et al.*, (1930); Sakimura (1969); Cho *et al.*, (1988)
	TCSV, GRSV	Wijkamp *et al.*, (1995a,b)
Frankliniella fusca	TSWV	Sakimura (1963)
Frankliniella intonsa[c]	TSWV, TCSV, GRSV	Wijkamp *et al.*, (1995a,c)[c]
Thrips tabaci[d]	TSWV	Pittman (1927); Lindford (1932), Sakimura (1940, 1963)
Thrips setosus	TSWV	Kobatake *et al.*, (1984)
Thrips palmi	WSMV	Iwaki *et al.*, (1984); Kameya-Iwaki *et al.*, (1988)
	GBNV	Lakshmi *et al.*, (1995)

[a]These viruses share 99% N protein amino acid sequence identity suggesting they may be one species, however, they have different host ranges which may warrant their designation as different species.

[b]Dark colour forms transmitted TSWV, TCSV and GRSV, while pale forms transmitted only TSWV.

[c]Transmission was measured by assessing local lesion responses on *Petunia hybrida* 'Blue Magic' and not by transmission to a systemic host.

[d]Many populations of *T. tabaci* do not transmit contemporary tospoviruses (i.e., the Hawaiian TSWV isolate HAW: R.F.L. Mau, unpublished data; the Brazilian TSWV isolate BR-01, TCSV or GRSV: Wijkamp *et al.*, 1995c; however, an arrhenotokous population from Japan was shown to transmit the Brazilian TSWV isolate BR-01 at a low rate (Wijkamp *et al.*, 1995c). This species was formerly an important vector of TSWV worldwide (Sakimura, 1962); however, further experimentation to clarify its current vector status is warranted.

(Brittlebank, 1919) and associated with transmission by thrips several years later (Pittman, 1927). In the following years the disease was shown to have a viral aetiology (Samuel *et al.*, 1930). TSWV and TSWV-like viruses have been the cause of severe pandemics that have generated international concern. At the time of Sakimura's landmark review of the status of thrips-transmitted viruses (Sakimura, 1962), four thrips species, *Thrips tabaci, Frankliniella occidentalis, Frankliniella schultzei and Frankliniella fusca*, had been associated with transmission of TSWV. TSWV was considered the sole member of the tomato spotted wilt virus group of plant viruses and was thought to be the only virus transmitted by thrips. A distinctive characteristic of the thrips–TSWV interaction that had been recognized was that infective adults could only be produced when virus acquisition occurred during the larval stage of the insect (Linford, 1932) (Fig. 14.1). The primary vector worldwide was *Thrips tabaci* and this thrips species was thought to transmit every known strain of the virus. The three species of *Frankliniella* reported as vectors were associated with transmission of particular virus strains,

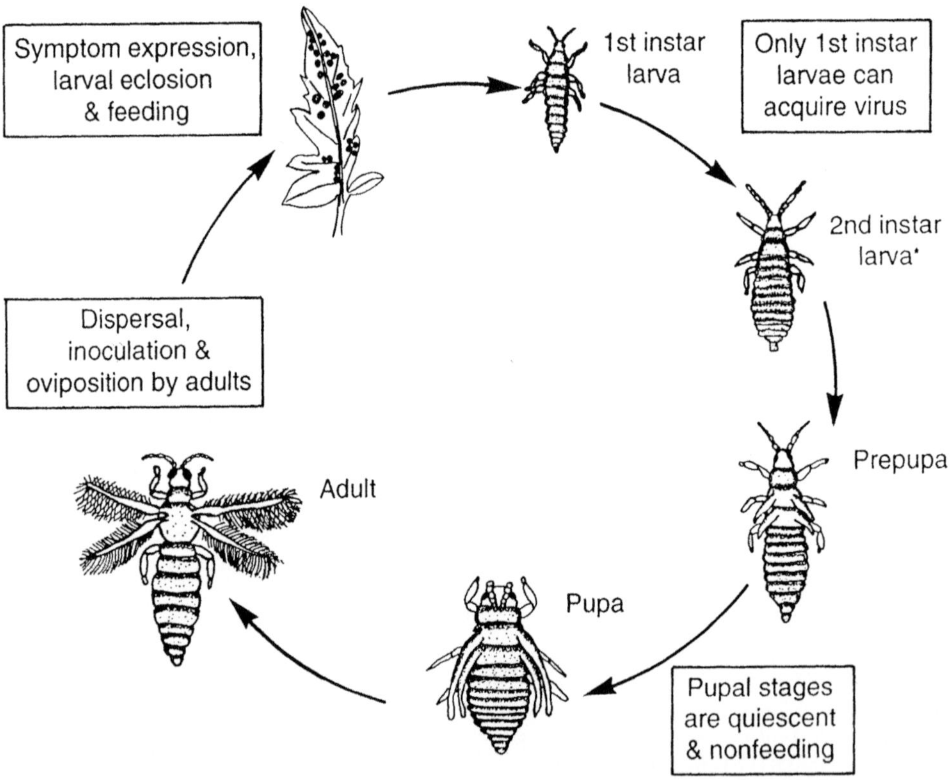

Fig. 14.1. Line drawing depicting the relationship between thrips and tospoviruses. Transmission by adults only occurs if larvae acquire the virus. *Second instar larvar occasionally transmit tospoviruses, but are wingless and seldom disperse widely. The adult is the primary disperal stage of the insect and thought to be of greatest epidemiological significance.

although there was little complete evidence on possible vector specificities.

In the 20 years following Sakimura's review of thrips-transmitted viruses, TSWV pandemics continued and *Thrips setosus* and *Scirtothrips dorsalis* (see later comments on accepted vector species; Mound, 1996) were reported as new vector species (Amin *et al.*, 1981; Kobatke *et al.*, 1984). A burst of interest in TSWV and thrips as vectors was generated during the 1980s because of the frequency and severity of epidemics occurring worldwide. The interest in this topic became so keen that two international conferences on TSWV and thrips–TSWV relationships were held in 1988 and 1990. The proceedings published from the latter conference provide a useful summary of the general body of knowledge at the end of the last decade (Hsu and Lawson, 1990). TSWV was still considered the sole member of the tomato spotted wilt virus group of plant viruses; however, several distinct new isolates had been recognized (Chanekar *et al.*, 1979; Kameya-Iwaki *et al.*, 1988; Law and Moyer, 1989, 1990; Reddy *et al.*, 1990) and the similarities between TSWV and animal-infecting viruses in the family Bunyaviridae noted (Milne and Francki, 1984; de Haan *et al.*, 1989). *Frankliniella occidentalis* (Plate I.1) had emerged as the primary vector of TSWV isolates in most parts of the world and the vector status of *Thrips tabaci* had become questionable (Mau *et al.*, 1990; German *et al.*, 1992). A good polyclonal antibody and enzyme-linked immunosorbent assay (ELISA) for TSWV had been produced (Gonsalves and Trujillo, 1986) paving the way for many research groups to produce a plethora of serological tools. Considerable effort was focused on understanding TSWV genome organization and strategies, as well as obtaining complete nucleotide sequences for the viral proteins of isolates thought to be in different serogroups. Antibodies and riboprobes to TSWV-like viruses from many geographic locations were being generated by laboratories worldwide. This progress paved the way to a revised view of TSWV classification, phylogeny and vector relationships.

Tospovirus phylogeny

Based on detailed knowledge of virus structure, genomic organization and strategy, TSWV has been classified as the type member of the genus *Tospovirus* in the family Bunyaviridae (Francki *et al.*, 1991). Following this classification and introduction of the species concept in virus taxonomy, it was logical to ask whether adequate variation occurred between the many isolates of TSWV to distinguish some as different species. Traditionally, the classification of viruses in the family Bunyaviridae has been based mainly on serology, with viruses being divided into serogroups and serotypes. Comparisons of this nature, using polyclonal and monoclonal antibodies against the TSWV N protein, demonstrated distinct serological relationships between virus isolates from different geographical areas and crops and suggested the existence of at least four serogroups (Kameya-Iwaki *et al.*, 1988; Law and Moyer, 1989, 1990; de Avila *et al.*, 1990; Reddy *et al.*, 1990, 1992; Wang and Gonsalves, 1990) (Table 14.3). As the complete nucleotide sequences of the N protein genes of different isolates were obtained, it was proposed that the importance of this protein in viral function supports use of N protein nucleotide and amino acid sequence comparisons for the purpose of defining species within the genus *Tospovirus*

Table 14.3. Current classification of viruses in the genus *Tospovirus* (family Bunyaviridae). Tomato spotted wilt tospovirus is the type member of the genus.

Tospovirus sp.	Serogroup	Reference
Tomato spotted wilt tospovirus (TSWV)	I	de Avila *et al.*, (1993)
Tomato chlorotic spot tospovirus (TCSV) Groundnut ringspot tospovirus (GRSV)	II	de Avila *et al.*, (1993)
Impatiens necrotic spot tospovirus (INSV)	III	Law and Moyer (1989, 1990); Law *et al.*, (1991, 1992); de Avila *et al.*, (1992)
Groundnut bud necrosis tospovirus (GBNV)[a]	IV	Chanekar *et al.*, (1979); Reddy *et al.*, (1992); Adam *et al.*, (1993); Heinze *et al.*, 1995)
Watermelon silver mottle tospovirus (WSMV)		Yeh *et al.*, (1992); Adam *et al.*, (1993); Yeh and Chang (1995)

[a]GBNV and WSMV share 99% N protein amino acid sequence identity suggesting they may be one species, however, they have different host ranges which may warrant their designation as different species. Their final classification requires further study.

(de Avila *et al.*, 1993). Indeed, comparisons of nucleotide and amino acid sequence identities support the existence of at least five tospovirus species (Table 14.3). The following phylogenetic scheme has been proposed (de Avila *et al.*, 1993; Heinze *et al.*, 1995; Yeh and Chang, 1995). Those tospoviruses that were formerly placed in serogroup I (TSWV, the type member of the genus *Tospovirus*) share 99% identity with one another and should be called TSWV. The two virus isolates in serogroup II share only 81% N protein sequence identity with each other, 74.4–78.2% identity with viruses in serogroup I and 54–56% identity with viruses in serogroup III. Therefore, these isolates are thought to be separate species and have been called tomato chlorotic spot tospovirus (TCSV) (formerly serotype I: BR-03) and groundnut ringspot tospovirus (GRSV) (formerly serotype II: SA-05). Viruses in serogroup III have also diverged significantly from TSWV (serogroup I) sharing only 55.4% sequence identity. These viruses have also been given species status and have been named impatiens necrotic spot tospovirus (INSV) (Law and Moyer, 1990; Law *et al.*, 1991, 1992; de Avila *et al.*, 1992). At least two tospovirus isolates have been identified and placed in a fourth serogroup (Chanekar *et al.*, 1979; Iwaki *et al.*, 1984; Kameya-Iwaki *et al.*, 1988; Reddy *et al.*, 1992; Yeh *et al.*, 1992; Adam *et al.*, 1993; Heinze *et al.*, 1995; Yeh and Chang, 1995). These isolates include a Taiwanese tospovirus isolate from tomato (TSWV-To), a Taiwanese tospovirus isolate from watermelon (TSWV-W), a Japanese tospovirus isolate from watermelon and an Indian tospovirus isolate from groundnut (bud necrosis virus). The designation groundnut bud necrosis tospovirus (GBNV) was proposed for these isolates because of the strong serological reactions observed with different antisera among the N proteins (Adam *et al.*, 1993). Although TSWV-To and TSWV-W share 99% identity with each other, the designation GBNV for both of them is probably not appropriate because the isolates from watermelon in Japan and Taiwan did not originate from groundnut and they have the unique ability to infect cucurbits. Currently, GBNV (Heinze *et al.*, 1995) is proposed as the species name for the former TSWV-To and BNV, and watermelon silver mottle tospovirus (WSMV) (Yeh and Chang, 1995) has been proposed as the name for the former TSWV-W. Preliminary reports of at least an additional eight tospovirus species have been made by various researchers and was published as the proceedings of the International Symposium on Tospoviruses and Thrips of Floral and Vegetable Crops held in Taiwan (Kuo, 1996). Among these viruses, some have been assigned isolate names and preliminary information is available regarding their host ranges, relationships to other tospoviruses and thrips vectors. Some examples are BR-09, identified in Brazil from onion (de Resende, 1996) (serologically similar to a tospovirus isolate identified from onion in California: J. Moyer, Raleigh, North Carolina, personal communication; vector relationships unknown) and several new viruses from watermelon (watermelon mottle spot virus) and peanuts in India (peanut bud necrosis virus, peanut yellow fan virus) thought to be transmitted by *Thrips palmi*. As the characterization of tospoviruses continues to develop, numerous proposals regarding virus species names can be expected. Species designations for tospoviruses where only serogroup and serotype designations formerly occurred may clarify other biological relationships, such as thrips–tospovirus interactions. Serious reconsideration is given to these relationships in the following section and a summary of thrips–tospovirus associations is provided (Table 14.2). Tospovirus classification has been thoroughly discussed in the literature, which can be consulted for details (German *et al.*, 1992; de Avila *et al*, 1993; Heinze *et al.*, 1995; Yeh and Chang, 1995).

Thrips/tospovirus relationships

Among the plant viruses transmitted by thrips, the tospoviruses have the most complex biological relationship with their vectors and are transmitted primarily from leaf to leaf. The relationship between thrips–tospovirus interactions and tospovirus transmission are summarized in Fig. 14.1. The recent advances in our understanding of tospovirus phylogeny and molecular biology make it critical to reconsider the association of each tospovirus with its thrips vectors. Known associations between thrips vector species and currently recognized tospovirus species are summarized in Table 14.2. Strong evidence of virus transmission supports the validity of six thrips species as vectors of contemporary tospoviruses. Although *Thrips tabaci* was a key vector of TSWV worldwide 30 years ago, this species does not transmit contemporary TSWV isolates (i.e. the Brazilian TSWV isolate BR-01 or the Hawaiian TSWV isolate HAW), TCSV or GRSV (Mau *et al.*, 1990; Wijkamp *et al.*, 1995a). There is no question regarding the validity of earlier reports regarding the vector status of this thrips species (Pittman, 1927; Linford, 1932; Sakimura, 1940, 1963), thus this represents a dramatic change in vector status. Considering the previous global importance of *Thrips tabaci* as a vector, it seems appropriate that possible associations between other tospoviruses (i.e. onion-infecting isolates) and *Thrips tabaci* should be considered before this thrips species is fully relegated to non-vector status.

Some previously reported vector–tospovirus associations (i.e. *Scirtothrips dorsalis*, *Frankliniella schultzei* and *Frankliniella tenuicornis*) require experimental validation. *Scirtothrips dorsalis* and *Frankliniella schultzei* were initially reported as vectors of peanut bud necrosis virus (currently classified as GBNV) (Amin *et al.*, 1981). Nine years later, it was discovered that this report was probably based on an incorrect species identification (Mound, 1996), in which *Thrips palmi* was identified as the pale form of *Frankliniella schultzei*. Consistent with this finding, subsequent research demonstrated that *Thrips palmi* was the primary vector of GBNV in India, *Frankliniella schultzei* transmitted GBNV at a very low rate (2%) and *S. dorsalis* was a non-vector (Lakshmi *et al.*, 1995). Clearly, *Thrips palmi* is a valid and important vector of GBNV, while *Scirtothrips dorsalis* should not be considered a tospovirus vector until further experimental confirmation can be provided. Similarly, the validity of *Frankliniella schultzei* as a vector of GBNV requires confirmation because of the extremely low rate of transmission reported (Lakshmi *et al.*, 1995). Inclusion of *Frankliniella tenuicornis*, a grass-living species, as a vector of TSWV (Zitter *et al.*, 1989; de Haan *et al.*, 1991; Kormelink, 1994; Kormelink *et al.*, 1994; Hunter *et al.*, 1995) is presumed to be an error as there is no published experimental evidence that this species transmits any tospovirus.

Emergence of new thrips vector species

To circumvent the tedium associated with thrips transmission of tospoviruses to systemic hosts, a leaf disc assay using *Petunia X hybrida* 'Blue Magic' has been developed (Wijkamp and Peters, 1993; Wijkamp *et al.*, 1995b). This method is based on a local lesion response in *Petunia X hybrida* 'Blue Magic' to tospoviruses and results obtained with this method and thrips transmissions to systemic hosts are not thought to differ significantly (Wijkamp *et al.*, 1995a,b). Unfortunately, *Petunia X hybrida* 'Blue Magic' is no longer commercially available, but can be replaced with 'Burgundy Madness' or 'Celebrity Blue' (D.E. Ullman, California, 1996, unpublished data). Using the leaf disc method, several new thrips vector–tospovirus relationships and specificities between virus species and thrips species have been demonstrated (Wijkamp *et al.*, 1995a,b). These findings are very significant and indicate a need to re-evaluate thrips–tospovirus relationships on a broad scale. *Frankliniella occidentalis* was shown to transmit TSWV, TCSV, GRSV and INSV; *Frankliniella intonsa* was shown to be a new vector of TSWV, TCSV and GRSV, a particularly interesting finding because this species was previously shown to be a non-vector of TSWV (Bonnemaison, 1937). Significant differences in the vector propensity of dark and pale forms of *Frankliniella schultzei* were demonstrated, in which dark forms efficiently transmitted TSWV, TCSV and GRSV, while the pale form transmitted only TSWV and TCSV and with poor efficiency. Previous research suggests that the pale form of *Frankliniella schultzei* is not a vector of TSWV (Sakimura, 1969), so the latter association may require further validation. There are at least two complicating factors confusing the vector status of *Frankliniella schultzei*. First, dark and pale forms of *Frankliniella schultzei* may actually be different species, and dark forms are pale when they emerge from their pupal cases and may be readily confused with pale forms (L.A. Mound, London, 1995, personal communication; see also Chapter 6). This is an excellent example of why it is critical to take voucher specimens when conducting transmission experiments, particularly when new vector–virus associations are investigated. The leaf disc method of assessing thrips as vectors is clearly convenient; however, it would seem well advised to use transmissions to systemic hosts to validate new vector–virus relationships (see Table 14.2 for associations established solely on leaf disc assays).

Identification of the Thripidae can be quite difficult and the number of systematists able to verify identifications currently is limited. A key to the major vector species (Mound, 1996) will be helpful to those studying the epidemiology of tospoviruses and vector–virus relationships. In light of the difficulties surrounding identification of thrips, the importance of collecting and depositing voucher specimens cannot be overemphasized. It is essential that the voucher specimens be collected during and not after completion of relevant experiments. Vouchers should be properly prepared for identification and storage, verified and deposited in an accessible collection. When publishing experimental results, the location of voucher specimens should be stated so other researchers can access them if questions arise. Strict adherence to this procedure will alleviate confusion due to misidentifications and provide a record that can be referred to as vector–virus associations change.

Thrips vector specificities and tospoviruses as emerging pathogens

Viruses in the family Bunyaviridae are considered among the most dangerous of emerging pathogens and are responsible for haemorrhagic fevers in many parts of the world (Le Guenno, 1995). In the realm of plant-infecting viruses, the tospoviruses are seemingly among the most aggressive emerging viruses, with the identification of at least five new species during the past 5 years and the proposal of at least eight additional potentially distinct species. It is not known whether novel tospoviruses are emerging or whether advances in virological identification combined with changing environmental conditions have brought to light viruses that could have existed for millions of years. Certainly, the dramatic changes that have occurred in agricultural practices and rapid distribution of seed, propagules and products worldwide has provided unprecedented selection pressures and avenues of spread for both viruses and vectors. Close scrutiny of thrips–tospovirus relationships does provide some evidence that evolution of tospoviruses and/or selection pressures on both the viruses and vectors contributes to changing vector specificities. The most striking example supporting this notion is the altered status of *Thrips tabaci* as a vector of TSWV. Thirty years ago, this species transmitted all known isolates of TSWV worldwide (Sakimura, 1962), yet today is apparently a non-vector of modern TSWV isolates and newly emerged tospovirus species (TCSV, GRSV) (Mau *et al.*, 1990; Wijkamp *et al.*, 1995a). There are several hypotheses that might explain this phenomenon. For example, TSWV isolates around the world may have evolved such that they no longer infect *Thrips tabaci*, isolates associated with the highly polyphagous and widely distributed *Frankliniella occidentalis* may have displaced those transmitted by *Thrips tabaci*, *Thrips tabaci* may have evolved such that it can no longer acquire TSWV, or both insect and virus may have evolved such that the complex events leading to acquisition and transmission of tospoviruses can no longer occur. Another example relevant to emergence of new tospoviruses and new thrips–virus associations is the apparently very specific relationship between *Thrips palmi* and GBNV (Lakshmi *et al.*, 1995) and WSMV (Yeh *et al.*, 1992). These viruses were first detected in the original distribution area of *Thrips palmi* (from Mauritius to Japan and New Caledonia) (Chanekar *et al.*, 1979; Amin *et al.*, 1981; Iwaki *et al.*, 1984; Kameya-Iwaki *et al.*, 1988; Yeh *et al.*, 1992) in some locations up to 50 years after the first detection of *Thrips palmi* (Palmer, 1992). *Thrips palmi* does not transmit any of the TSWV isolates that have been tested (the Hawaiian TSWV isolate HAW: Mau *et al.*, 1990; the Brazilian TSWV isolate BR-01: Wijkamp *et al.*, 1995b); but it commonly feeds upon and completes its life cycle in association with TSWV plant hosts and other tospovirus vectors. Theories of evolution suggest that adaptation of one species to another is likely to occur when species interact over a relatively long time. The mechanisms underlying the emergence of GBNV, WSMV and *Thrips palmi* as their vector are unknown; however, one feasible hypothesis could be that *Thrips palmi* coevolved with TSWV such that new host relationships arose, i.e. once viral changes enabled infection of *Thrips palmi*, dispersal of the insect to other hosts would bring large numbers of virions into contact with new potential host plants (i.e. cucurbits). Certainly, the evolution of tospoviruses and specific associations between tospoviruses and their thrips vectors, in particular the molecular determinants of thrips acquisition, warrant intensive future investigation.

Cellular interactions between thrips vectors and tospoviruses

Availability of antibodies to all the TSWV proteins and riboprobes for detection of TSWV RNA greatly facilitated localization and understanding of the interactions occurring within thrips cells. Most of the studies done on a cellular level to date have been conducted with TSWV and *Frankliniella occidentalis* because of the global importance of this tospovirus–vector combination. Recently, investigations of cellular interactions between *Thrips setosus* and TSWV have shown that the thrips midgut is the first site of TSWV infection and that infection of the salivary glands coincides with pupation (Tsuda and Ohnishi, 1996). The latter study is the first done beyond the *Frankliniella occidentalis*–TSWV relationship. It will be important to compare findings from future studies of various vector–tospovirus relationships. Investigations of *Frankliniella occidentalis* that acquired TSWV during the larval stages, reveal that TSWV enters the thrips midgut epithelial cells and the midgut is the first site of viral infection (Ullman *et al.*, 1993) (Fig. 14.2). Following infection of the midgut, TSWV circulates to other thrips organs and is localized in the midgut epithelial cells, muscle cells surrounding the alimentary canal and salivary glands (Ullman *et al.*, 1992a,c, 1993; 1995b; Wijkamp *et al.*, 1993). Virus circulation to the salivary gland and subsequent egestion is essential for inoculation to occur and requires virus passage across at least four membrane barriers (Fig. 14.2). The processes

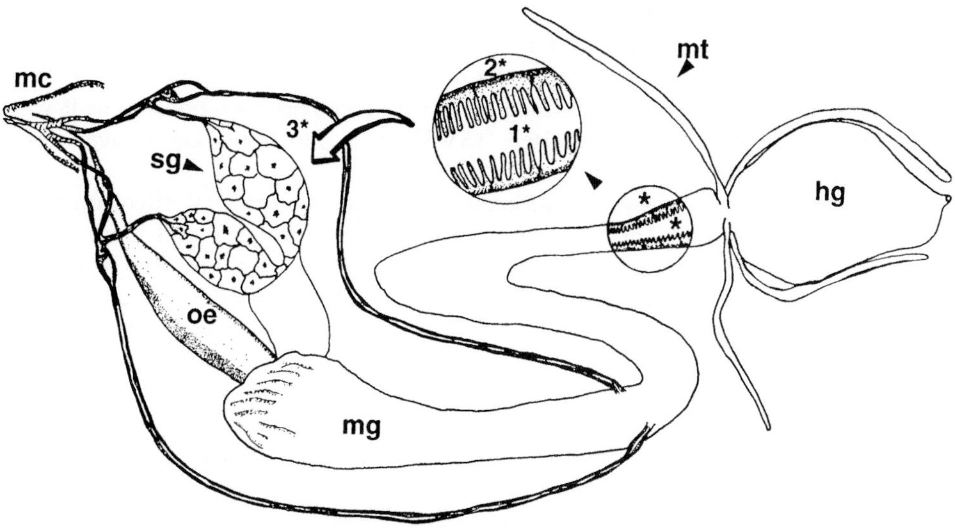

Fig. 14.2. Line drawing of dissected thrips organs depicting the primary cellular barriers (1*–3*) encountered by tospoviruses prior to reaching the only epidemiologically significant organ, the salivary gland. hg = hindgut, mc = mouthcone, mg = midgut, mt = Malpighian Tubules, oe = oesophagus, sg = salivary gland, 1* = apical membrane of the midgut, 2* = basement membrane of the midgut, 3* = basement membrane of the salivary gland.

governing virus circulation are little understood (see later section on determinants of tospovirus acquisition) and thrips vector competence is probably also affected by virus multiplication (Ullman *et al.*, 1992c, 1993; Wijkamp *et al.*, 1993). Virus multiplication and circulation to the salivary gland underly the median latent period, which may be defined as the time from acquisition to successful inoculation. The median latent period in *Frankliniella occidentalis* is temperature dependent (Wijkamp and Peters, 1993; Wijkamp *et al.*, 1995b), thus under higher temperature regimes, second instar larvae may inoculate plants, while at lower temperature regimes only adults transmit the virus (Fig. 14.1). Under most field conditions, adults thrips are probably most important to virus spread because they are winged and disperse widely. In certain situations, such as production of greenhouse ornamentals, where cuttings are transferred from stock areas to propagation sites, second instar larvae may be very significant in virus spread.

The finding that TSWV multiplies in *Frankliniella occidentalis* was very significant in terms of virus classification, epidemiology and vector–virus relationships. With regard to virus classification, all the bunyaviruses replicate in their arthropod vectors, thus the finding that TSWV replicated in at least one thrips vector provided additional evidence supporting placement of the tospoviruses in the family Bunyaviridae. In terms of epidemiology, tospovirus replication in thrips suggests that thrips can inoculate plants for the duration of their lives as well as serving as alternate hosts for the viruses. Thus, virus isolates may be spread between locations by infected thrips hidden in otherwise healthy plant shipments. This is a very serious concern where transplants are shipped between states and countries because thrips are more difficult to detect from a quarantine point of view than infected plants. The finding that TSWV replicated in thrips also redefined the thrips–tospovirus relationship as it was thought that thrips transmitted TSWV in a persistent, circulative, non-propagative fashion (Sakimura, 1962).

Other bunyaviruses are thought to have no detrimental effect on their arthropod vectors. The effect of tospoviruses on fecundity and longevity of their thrips hosts is controversial and in need of further study. Two investigations, one with TSWV (Robb, 1989) and one with INSV (DeAngelis *et al.*, 1993), have demonstrated lowered survival and reproductive potential and slower development rates for thrips exposed to tospovirus-infected plants. These investigations evaluated thrips exposed to virus throughout their lifespan (Robb, 1989) and for the first 24 h of their lives (DeAngelis *et al.*, 1993). A more recent investigation (Peters, 1996) sought to determine whether the negative effects that had been documented (Robb, 1989; DeAngelis *et al.*, 1993) were due to virus infection or possible reduced nutritional quality of infected plant hosts. In these studies, first instar thrips were exposed to TSWV infected *Datura* for 6 h and then transferred to healthy plant material every day until pupation. Upon adult emergence, insects were transferred daily as individuals to discs of *Petunia* (as described earlier) which allowed evaluation of virus transmission status, as well as their fecundity and longevity. In contrast to the findings of Robb (1989) and DeAngelis *et al.* (1993), these experiments showed no significant difference in the fecundity or longevity of virus-infected and non-infected thrips. *Petunia* leaves are good indicators for TSWV infection, but are not generally good hosts for *Frankliniella occidentalis*. It would be interesting to determine if the same result would occur on a host that supported both virus and thrips. As evaluation of thrips–tospovirus interactions advances, it will be important to understand better this aspect of their relationship, particularly if tospoviruses have a negative impact on their vectors and the mechanism by which this occurs can be determined and used in thrips control.

In actuality, little is known about replicative processes in thrips cells or how tospovirus infection affects thrips cell function. Several types of inclusions are produced in infected cells and viral proteins are compartmentalized in these structures (Ullman *et al.*, 1995a,b). In this respect, TSWV infections in thrips cells do not differ significantly from tospovirus infections of plant cells (de Haan *et al.*, 1990; Kormelink *et al.*, 1991; Urban *et al.*, 1991; Kitajima *et al.*, 1992; de Avila *et al.*, 1993). In contrast with plant cell infections, TSWV proteins apparently enter autophagic pathways in thrips cells, suggesting cellular responses to infection (Ullman *et al.*, 1995a,b). It has been proposed that the non-structural protein on the mRNA (NSm) is a viral movement protein assisting TSWV in crossing the plasmodesmata between plant cells (Kormelink *et al.*, 1994), that NSm is an adaptation to plant infection, and that this protein is not important relative to infection of animal cells. NSm has been detected in infected thrips cells, although its role in movement between thrips cells has not been demonstrated (Ullman *et al.*, 1995a). By analogy to other bunyaviruses, movement between animal cells is generally thought to involve accumulation of mature virions within Golgi cisternae or smooth membrane vesicles, with final release occurring by exocytosis (Anderson and Smith, 1987). In thrips cells, mature virions have only been observed in the salivary glands, although the cytoplasmic inclusions indicative of infection are evident in midgut and associated muscle cells (Ullman *et al.*, 1995a,b). These data strongly suggest an alternative method of virus movement in thrips cells, such as translocation of viral ribonucleoprotein (vRNP) complexes across membranes (Ullman *et al.*, 1995a,b). The potential role of vRNPs in virus movement certainly seems worthy of further exploration.

***Determinants of TSWV acquisition by* Frankliniella occidentalis**

It has long been recognized that adult thrips can only inoculate TSWV if acquisition occurred during the larval stages (Linford, 1932). Recently, a biotype of *Frankliniella occidentalis* from The Netherlands was shown to acquire TSWV only during the first larval instar (Wettering, 1996; Wettering *et al.*, 1996; Fig. 14.1). Further investigations comparing *Frankliniella occidentalis* biotypes from several European countries, the United States and Israel demonstrated that the latter relationship is apparently limited to the biotype from The Netherlands and that all the other biotypes examined were able to acquire TSWV during the first and second instars (D. Peters, Wageningen, The Netherlands, 1996, unpublished data). The mechanisms underlying this phenomenon are not known; however, it has been demonstrated that TSWV enters the thrips haemocoel at the midgut during larval acquisition (Ullman *et al.*, 1995a) (Fig. 14.2) and that a midgut barrier prevents acquisition by adult thrips (Ullman *et al.*, 1992b). During additional analyses of the midgut barrier in adults, it has been demonstrated that contrary to initial findings (Ullman *et al.*, 1992b), virions do not enter the midgut epithelium of adult thrips (D.E. Ullman, California, 1995, unpublished data), suggesting that the barrier is at the midgut apical plasmalemma. By analogy to other bunyaviruses (Beaty *et al.*, 1982; Ludwig *et al.*, 1989), tospovirus entry to the thrips midgut is probably mediated by an interaction between receptor(s) at the thrips midgut apical plasmalemma and viral attachment proteins (probably one or both TSWV glycoproteins). There are at least two processes by which receptor-mediated cell entry may occur and these are illustrated in Fig. 14.3. The first and most common process is receptor-mediated endocytosis (Fig 14.3, pathway 1) during which viral attachment protein(s) interact with receptor(s) at the cell membrane causing the cell to engulf the virus in an endosome. The acidic nature of the endosome induces fusion between the virus and the endosome membrane (uncoating) followed by deposition of the virus core (viral genes and various enzymes, e.g. RNA-dependent RNA polymerase). This process of entry to animal cells has been well demonstrated for membrane-bound and capsid-bound viruses (Ludwig *et al.*, 1989, 1991; Gildow, 1993). The second process for virus entry to cells (Fig. 14.3, pathway 2) has been demonstrated for enveloped viruses such as the human immunodeficiency virus (HIV) (Haseltine and Wong-Staal, 1988), and is again mediated by an interaction between viral attachment protein(s) and cellular receptor(s). Instead of inducing formation of an endosome, this interaction results in virus fusion with the cell membrane and immediate deposition of the viral core. Some viruses, such as HIV, may use both entry pathways. Observations from electron microscopy suggest that TSWV entry to thrips midgut cells is by fusion at the cell membrane. Immunolabelling of thrips midgut cells also indicates that TSWV membrane glycoproteins remain at the membrane following fusion, while the N protein, presumably with the viral RNA and polymerase, enters the cell (Ullman *et al.*, 1995a) An interaction between thrips midgut proteins and TSWV glycoproteins has also been demonstrated (Bandla *et al.*, 1996). The possibility that the thrips midgut protein under investigation is the cell receptor mediating virus entry to the midgut is compelling and is the topic of ongoing studies. The actual interaction between the glycoproteins and thrips midgut receptors is probably complicated and could be influenced by the pH, enzymes present in the thrips gut or even other viral proteins (i.e. NSs).

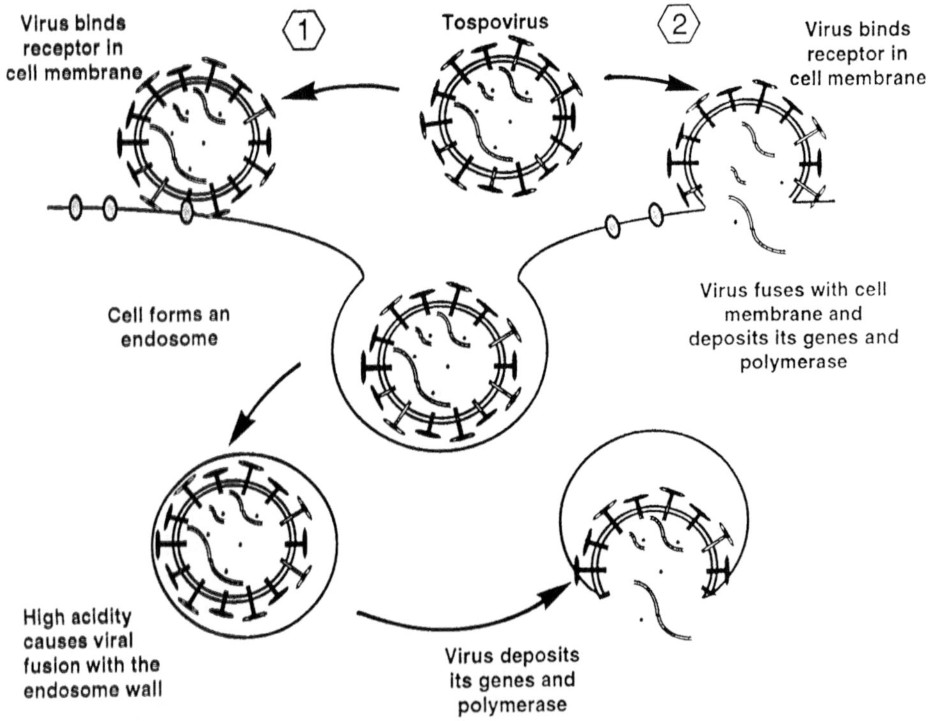

Fig. 14.3. Line drawing depicting two possible models for tospovirus entry to thrips cells. Both models begin with viral proteins binding specific receptor proteins in the cell membrane. Model 1 is a classic model for virus entry involving endosome formation followed by virus uncoating. Model 2 involves virus fusion and uncoating at the cell membrane surface. The latter model most closely resembles observations of interactions between thrips midgut cells and tomato spotted wilt tospovirus.

Current strategies for management of tospovirus epidemics

None of the currently available methods for managing tospovirus epidemics is completely satisfactory, but there are a number of strategies that, when integrated into a pest management programme, can reduce the pressure of tospovirus epidemics. Epidemics of tospoviruses can only occur when conditions promote the intimate relationship between the viruses, thrips vectors and plant hosts. Integrated pest management programmes for tospoviruses must aim to interrupt this relationship and, therefore, revolve around thrips management through exclusion, cultural and chemical controls and methods for preventing virus infection (i.e. use of resistant cultivars, cultural methods for reducing virus inoculum). This section focuses on some of the methods that may be used currently and provides a management framework into which methods developed in the future may be inserted. Even as bioengineered strategies become readily available, it is unlikely that any of them will serve as the 'magic bullet' and eliminate the need for the more integrated approaches discussed in the following.

Management through prevention is critical to succeeding in controlling thrips vector species and tospovirus spread, because insecticides and natural enemies are seldom efficacious in reducing thrips populations and plants cannot be cured once they are infected. Gaining control of thrips populations with pesticides is extremely difficult because eggs within leaf tissue, larvae deep within developing buds and flowers and pupae in the soil/leaf litter are protected from most sprays. In addition, development of pesticide resistance in many thrips populations makes it increasingly difficult to obtain control through spraying of conventional insecticides. Some of the newer pesticides such as entomopathogenic fungi (see Chapter 10) and plant natural products (M.P. Parrella, California, 1996, unpublished data) are promising, particularly when used in conjunction with control methods aimed at prevention. Biological control using natural enemies has not been broadly successful, although some thrips control has been obtained using the predaceous mite, *Amblyseius cucumeris* on cucumbers in greenhouses and the generalist predator, *Orius* spp., on a number of crops (see Chapter 18). Exclusion of thrips through screening is of increasing importance for success in production of greenhouse crops; however, this option is costly, requiring retrofitting of older greenhouses and additional expense when new greenhouses are built. Preventing thrips immigration in field crops is exceedingly difficult, but some benefit can certainly be obtained by considering time of planting, crop choice and placement (see Chapter 16). Although difficult to practise, a combination of cultural and sanitation methods aimed at reducing the incidence of virus inoculum can significantly reduce the potential for tospovirus epidemics (Cho *et al.*, 1989). Several articles have reviewed thrips management and approaches to limiting tospovirus spread and can be consulted for details (Cho *et al.*, 1989; Parrella, 1995a,b,c).

Availability of commercial cultivars with resistance to thrips and tospovirus infection would provide an excellent avenue for managing this difficult problem; however, research to develop crops with durable resistance to tospoviruses using traditional methods has been limited. Large-scale screening of genetic sources of resistance to tospoviruses has been frustrated by difficulties associated with effective inoculation of the virus, few sources of known genetic resistance and the wild nature of resistant germplasm (requiring lengthy backcrossing programmes and specialized techniques to overcome intra- and interspecific incompatibility barriers). Perhaps the most successful efforts in the search for resistance have been with tomato and pepper for which several sources of resistance have been identified (Cho *et al.*, 1989; Boiteux *et al.*, 1993; Kumar *et al.*, 1993, 1995a,b; Boiteux and de Avila, 1994; Cho, 1996). Identification of a single, dominant gene for TSWV resistance in tomato (SW-5) resulted in production of several resistant hybrids that could be used successfully on the farm, even under high levels of disease pressure (Cho, 1996). Unfortunately, recent evidence revealed the presence of a new TSWV pathotype capable of breaking the SW-5 gene (Cho, 1996). This finding and the diversity currently recognized among tospoviruses, emphasize the importance of incorporating a range of tospovirus isolates and species into programmes screening for resistance and the need for pyramiding resistance genes. Advances in molecular technology and development of molecular markers linked to known resistance genes should vastly improve the opportunities for pyramiding resistance genes to different tospoviruses and accelerating breeding programmes. Progress in this arena will be discussed in the following section on future possibilities for management of tospoviruses.

Future possibilities for management of tospovirus epidemics

Current knowledge of the tospovirus genome provides an unprecedented opportunity to investigate fully the interactions between tospoviruses and thrips vectors on a molecular level. It is critical that future research be aimed at understanding the molecular determinants of virus acquisition, infectious processes and movement in thrips vectors. Much basic information can be learned through comparative analysis of tospovirus infections in thrips and plant cells. There are two broad areas in which this type of comparison will certainly be fruitful. One is to explore the fundamental differences in the mechanisms governing virus movement in plant and thrips cells. The second is to investigate fundamental differences in how tospoviruses replicate and mature in thrips and plant cells. Perhaps the most promising area of future research is elucidation of the molecular determinants of virus acquisition by thrips and isolation of the receptors mediating virus entry to thrips cells. This line of research is particularly attractive because it can be applied readily to controlling virus spread. For example, plants transformed with receptor-blocking agents could be used to prevent thrips from acquiring the virus, thus limiting the development of infective insects in a crop and consequent secondary spread of the virus, within the crop and to other areas to which the adult thrips may ultimately disperse.

Future research should continue to focus on applications of basic findings to strategies that will aid in management of tospovirus epidemics. Given that direct control of thrips and development of tospovirus-resistant host plants is difficult, tactics that contain tospovirus epidemics by limiting or preventing development of infective thrips populations should receive research priority.

Several types of strategies have potential in this regard. For example, NSs detection in thrips by ELISA is a good indicator of infective thrips populations (Bandla *et al.*, 1994). This technology has potential to be extended as a means of quantifying infective thrips populations for forecasting tospovirus epidemics, as well as directing other IPM strategies by identifying sources of infective thrips. With an understanding of how viral components enter thrips cells, replicate and move between cells, it should be possible to develop antiviral strategies that block these essential processes, thus preventing development of infective thrips. For example, factors preventing virus translocation could be introduced to plants such that tospoviruses ingested by thrips feeding on these plants cannot disseminate to epidemiologically important target organs such as the salivary glands. Another possible approach is to use pathogen-derived protection strategies to alter thrips such that they no longer acquire and transmit TSWV. While the latter technology is not yet established, it is feasible and is being considered in other virus–vector systems. For example, the use of antisense technology to alter genetically the vector competence of mosquito populations is currently under consideration for control of La Crosse virus, a mosquito-transmitted *Bunyavirus* that causes a serious disease of humans and is in the same virus family as TSWV (Aldhous, 1993b). Genetically altering vector populations to interfere with vector competence is also being considered as a means of preventing mosquitoes from acquiring the protozoan aetiological agent of malaria (Aldhous, 1993c) and in preventing tsetse flies from transmitting trypanosomes causing human disease (Aldhous, 1993a). Certainly, advances in introducing genes to insects via their endosymbionts suggest 'pseudo-transgenic' insects will soon be possible (Aldhous, 1993a; Beard, 1993). It seems only appropriate that future research should provide an arsenal of antiviral genes that can be used when transgenic technology for altering insect populations becomes available.

Given the advances in molecular technology, there is considerable hope for bioengineered host plant resistance as a durable and efficacious control measure. Molecular markers tightly linked with TSWV-resistance have been identified in tomato lines (Brommonschenkel *et al.*, 1994) and should assist the search for resistance in tomato and allow characterization and cloning of resistance genes. Ultimately, these efforts will provide an arsenal of resistance genes with efficacy against various tospoviruses and provide plant breeders with the tools they need to pyramid tospovirus resistance genes and accelerate development of commercially acceptable plant crop cultivars (Cho, 1996).

Thrips as Vectors of Fungi and Bacteria

Thrips of many species are known to transmit fungi and bacteria, presumably by mechanical contact and movement of the pathogen on their body surface. Although several diseases of economic importance are transmitted in this way by thrips, their role as vectors of fungi and bacteria has not been extensively studied nor has much new information become available since the last major text on thrips (Lewis, 1973) was published. In addition, thrips are not considered particularly important as vectors of these pathogens relative to dissemination occurring by wind and/or rain (Lewis, 1973; McManus and Jones, 1994). Many thrips species feed on fungal spores and hyphae (Mound and Teulon, 1995) and are known to transfer pathogenic fungi to healthy plants by carrying conidia and spores on their bodies (Granovsky and Levine, 1932; Bailey, 1935; Yarwood, 1943; Ondrej, 1973; Adis *et al.*, 1985; Ellis *et al.*, 1988; Farrar and Davis, 1991). *Thrips tabaci* has been associated with powdery mildew infections on grape, rose, strawberry, canteloupe, clover and *Oenothera* (Yarwood, 1943), and the greenhouse thrips *Hercinothrips femoralis* can transmit uredinia of the cereal rust *Puccinia graminis* (Granovsky and Levine, 1932). *Thrips obscuratus*, the New Zealand flower thrips, has been shown to carry spores of *Monilinia fructicola*, the causal organism of brown rot of stone fruit, on their bodies under laboratory and field conditions (Ellis *et al.*, 1988), although direct evidence of transmission has not been provided. In the central Amazon, *Liothrips adisi* was shown to carry the conidia of *Calonectria rigidiuscula* (Berk. and Br.) Sacc. (the condial state is known as *Fusarium decemcellulare* Brick) and was demonstrated to transmit this organism between guarana plants resulting in the disease called 'supersprouting' (Adis *et al.*, 1985). *Frankliniella occidentalis* has also been associated with mechanical transmission of the fungi causing *Fusarium* ear rot of corn (Farrar and Davis, 1991). Although nearly all reports of thrips association with transmission of fungal pathogens suggest that the mechanism of transmission is purely mechanical, the recent work of Marullo (1995) demonstrates that numerous thrips species, including *Chirothrips manicatus*, *Limothrips denticornis* and *Stenothrips graminum*, common in leaf litter of forests in a mountain community of Italy, ingest conidia of pathogenic fungi. She proposes that these conidia may remain viable as they pass through the insect digestive system and that transmission via feeding or faeces should be considered further.

Perhaps the most significant bacterium transmitted by thrips is *Erwinia amylovora*, the fire-blight bacterium which has been isolated from the bodies of *Frankliniella occidentalis* and *Taeniothrips inconsequens*; thrips are thought to be among the complex of pollinators important in spreading this disease (Bailey, 1935). Thrips have also been associated with bacterial wilt of corn (*Pantoea stewartii*) (Poos and Elliot, 1936; Elliot and Poos, 1940) and a bacterial disease of beans (*Pseudomonas medicaginis* var. *phaseolicola*) (Buchanan, 1932). Again, the mechanism of transmission is presumed to be mechanical.

References

Adam, G., Yeh, S.-D., Reddy, D.V.R. and Green, S.R. (1993) Serological comparison of tospovirus isolates from Taiwan and India with impatiens necrotic spot virus and different tomato spotted wilt virus isolates. *Archives of Virology* 130, 237–230.

Adis, J., Fonseca, A.J. and zur Strassen, L.A. (1985) A thrips as a vector of the 'supersprouting' fungus disease in Guarana plants of the central Amazon. *Fitopatologia Brasileira* 10, 677–679.

Aldhous, P.A. (1993a) Bacteria may provide access to the tsetse fly. *Science* 261, 548.

Aldhous, P.A. (1993b) Insect-borne viruses: help from plants. *Science* 261, 547.

Aldhous, P.A. (1993c) Malaria: focus on mosquito genes. *Science* 261, 546–548.

Allen, W.R. and Broadbent, A.B. (1986) Transmission of tomato spotted wilt virus in Ontario greenhouses by the western flower thrips, *Frankliniella occidentalis* (Pergande). *Canadian Journal of Plant Pathology* 8, 33–38.

Amin, P.W., Reddy, D.V.R. and Ghanekar, A.M. (1981) Transmission of tomato spotted wilt virus, causal agent of bud necrosis of peanut, by *Scirtothrips dorsalis* and *Frankliniella schultzei. Plant Disease* 65, 663–665.

Anderson, G.W. and Smith, J.F. (1987) Immunoelectron microscopy of Rift Valley fever viral morphogenesis in primary rat hepatocytes. *Virology* 161, 91–100.

Bailey, S.F. (1935) Thrips as vectors of plant disease. *Journal of Economic Entomology* 28, 856–863.

Bandla, M.D., Westcot, D.M., Ullman, D.E., German, T.L. and Sherwood, J.L. (1994). Use of monoclonal antibody to the nonstructural protein encoded by the small RNA of the tomato spotted wilt tospovirus to identify viruliferous thrips. *Phytopathology* 84, 1427–1431.

Bandla, M., Westcot, D.M., Ullman, D.E. and Sherwood, J.L. (1996) A thrips midgut protein that binds to tomato spotted wilt virus glycoprotein. In: Kuo, G. (ed.) *Tospoviruses and Thrips of Floral and Vegetable Crops: An International Symposium, Taichung, Taiwan. Acta Horticulturae, Technical Communications of International Society for Horticultural Science* 431, 341–349.

Beard, C.B. (1993) Modifications of arthropod vector competence via symbiotic bacteria. *Parasitology Today* 9, 179.

Beaty, B.J., Miller, B.R., Shope, R.E., Rozhon, E.J. and Bishop, D.H.L. (1982) Molecular basis of bunyavirus per os infection of mosquitoes: role of the middle-sized RNA segment. *Proceedings of the National Academy of Sciences (USA)* 79, 1295–1297.

Boiteux, L.S. and de Avila, A.C. (1994) Inheritance of a resistance specific to tomato spotted wilt tospovirus in *Capsicum chinense* (EPI 159236) *Euphytica* 75, 139–142.

Boiteux, L.S., Nagata, T., Dutra, W.P. and Fonseca, M.E.N. (1993) Sources of resistance to tomato spotted wilt virus (TSWV) in cultivated and wild species of *Capsicum. Euphytica* 67, 89–94.

Bonnemaison, L. (1937) Role des insectes piquers dans la transmission des maladies a virus des vegetaux. *Annales des Epiphyties et de Phytogenetique* 3, 282.

Brittlebank,C.C. (1919) Tomato diseases. *Journal of Agriculture, Victoria* 17, 231–235.

Brommenschenkel, S.H., Cho, J.J., Custer, D. and Tanksley, S.D. (1994) Molecular tagging of tomato spotted wilt tospovirus (TSWV) resistance derived from *Lycopersicon peruvianum. Journal of Cell Biology* 18A, 116.

Buchahan, D. (1932) A bacterial disease of beans transmitted by *Heliothrips femoralis* Reut. *Journal of Economic Entomology* 24, 49–53.

Chanekar, A.M., Reddy, D.V.R., Iizuka, N., Amin, P.W. and Gibbous, R.W. (1979) Bud necrosis of groundnut (*Arachis hypogaea1*) in India caused by tomato spotted wilt virus. *Annals of Applied Biology* 93, 173–179.

Chisholm, I.F. and Lewis, T. (1984) A new look at thrips (Thysanoptera) mouthparts, their action and effects of feeding on plant tissue. *Bulletin of Entomological Research* 74, 663–675.

Cho, J.J. (1996) Conventional breeding: host plant resistance and the use of molecular maker assisted selection in the development of TSWV-resistant commercial plants. In: Kuo, G. (ed.) *Tospoviruses and Thrips of Floral and Vegetable Crops: An International Symposium, Taichung, Taiwan. Acta Horticulturae, Technical*

Communications of International Society for Horticultural Science 431, 367–378.
Cho, J.J., Mau, R.F.L. and Mitchell, J. (1986) Reservoir weed hosts of tomato spotted wilt virus. *Plant Disease* 70, 1014–1017.
Cho, J.J., Mau, R.F.L., Gonsalves, D. *et al.*, (1988) Detection of tomato spotted wilt virus in individual thrips by enzyme linked immunosorbent assay. *Phytopathology* 78, 1348–1352.
Cho, J.J., Mau, R.F.L., German, R.L., Hartmann, R.W. and Yudin, L.S. (1989) A multidisciplinary approach for tomato spotted wilt virus (TSWV) management in Hawaii. *Plant Disease* 73, 375–383.
de Avila, A.C., Huguenot, C., de Resende, R.O., Kitajima, E.W., Goldbach, R.W. and Peters, D. (1990) Serological differentiation of 20 isolates of tomato spotted wilt virus. *Journal of General Virology* 71, 2801–2807.
de Avila, A.C., de Haan, P., Kitajima, E.W., Kormelink, R., de Resende, R.O., Goldbach, R.W. and Peters, D. (1992) Characterization of a distinct isolate of tomato spotted wilt virus (TSWV) from *Impatiens* sp. in the Netherlands. *Journal of Phytopathology* 134, 133–151.
de Avila, A.C., de Haan, P., Kormelink, R., de Resende, R.O., Goldbach, R.W. and Peters, D. (1993) Classification of tospoviruses based on phylogeny of nucleoprotein gene sequences. *Journal of General Virology* 74, 153–159.
de Haan, P., Wagemakers, L., Goldbach, R. and Peters, D. (1989) Tomato spotted wilt virus, a new member of the Bunyaviridae? In: Kolakofsky, B.W. and Mahy, H. (eds) *Pathogenicity of Negative Strand Viruses*. Elsevier, Amsterdam, pp. 287–290.
de Haan, P., Wagemakers, L., Peters, D. and Goldbach, R. (1990) The S RNA segment of tomato spotted wilt virus has an ambisense character. *Journal of General Virology* 71, 1001–1007.
de Haan, P., Kormelink, R., de Resende, R.O., van Poelwijk, F., Peters, D. and Goldbach, R. (1991) Tomato spotted wilt virus L RNA encodes a putative RNA polymerase. *Journal of General Virology* 71, 2207–2216.
de Resende, R.O. (1996) Diversity of tospoviruses in Brazil. In: Kuo, G. (ed.) *Tospoviruses and Thrips of Floral and Vegetable Crops: An International Symposium, Taichung, Taiwan. Acta Horticulturae, Technical Communications of International Society for Horticultural Science* 431, 78–89.
De Angelis, J.D., Sether, D.M. and Rossignol, P.A. (1993) Survival, development, and reproduction in western flower thrips (Thysanoptera: Thripidae) exposed to impatiens necrotic spot virus. *Environmental Entomology* 22, 1308–1312.
Elliot, C. and Poos, F.W. (1940) Seasonal development, insect vectors, and host range of bacterial wilt of sweet corn. *Journal of Agricultural Research* 60, 645–686.
Ellis, E.C., Penman, D.R. and Gaunt, R.E. (1988) Thrips as potential vectors of brown rot of stonefruit in New Zealand. In: *Proceedings of the 41st New Zealand Weed and Pest Control Conference, New Zealand*, Wellington, New Zealand, pp. 286–287.
Farrar, J.J. and Davis, R.M. (1991) Relationships among ear morphology, western flower thrips and *Fusarium* ear rot of corn. *Phytopathology* 81, 661–666.
Francki, R.I.B., Fauquet, C.M., Knudson, D.L. and Brown, F. (1991) *Classification and Nomenclature of Viruses. Representing ICTV Archives of Virology*, 5th edn. Springer-Verlag. New York.
Gardner, M.W., Tompkins, C.M. and Whipple, O.C. (1935) Spotted wilt of truck crops and ornamental plants. *Phytopathology*, 25, 17.
German, T.L., Ullman, D.E. and Moyer, J.W. (1992) Tospoviruses: diagnosis, molecular biology, phylogeny, and vector relationships. *Annual Review of Phytopathology* 30, 315–348.

Gildow, F. (1993) Evidence for receptor-mediated endocytosis regulating luteovirus acquisition by aphids. *Phytopathology* 83, 270–277.

Gonsalves, D. and Trujillo, E.E. (1986) Tomato spotted wilt virus in papaya and detection of the virus with ELISA. *Plant Disease* 70, 501–506.

Granovsky, A.A. and Levine, M.N. (1932) The dissemination of cereal rust spores in the greenhouse by terrestrial invertebrates. *Phytopathology* 22, 9–10.

Greber, R.S., Klose, M.J., Milne, J.R. and Teakle, D.S. (1991a) Transmission of prunus necrotic ringspot virus using plum pollen and thrips. *Annals of Applied Biology* 118, 589–593.

Greber, R.S., Klose, M.J., Teakle, D.S. and Milne, J.R. (1991b) High incidence of tobacco streak virus in tobacco and its transmission by *Microcephalothrips abdominalis* and pollen from *Ageratum houstonianum*. *Plant Disease* 75, 450–452.

Greber, R.S., Teakle, D.S. and Mink, G.I. (1992) Thrips-facilitated transmission of prune dwarf and prunus necrotic ringspot viruses from cherry pollen to cucumber. *Plant Disease* 76, 1039–1041.

Hardy, V.G. and Teakle, D.S. (1992) Transmission of sowbane mosaic virus by *Thrips tabaci* in the presence and absence of virus-carrying pollen. *Annals of Applied Biology* 121, 315–320.

Haseltine, W.A. and Wong-Staal, F. (1988) The molecular biology of the AIDS virus. *Scientific American*, October, 52–62.

Heinze, C., Maiss, E., Adam, G. and Casper, R. (1995) The complete nucleotide sequence of the S RNA of a new *Tospovirus* species, representing serogroup IV. *Phytopathology* 85, 683–690.

Hsu, H. and Lawson, R.H. (1990) *Virus–Thrips–Plant Interaction of Tomato Spotted Wilt Virus, Proceedings of a USDA Workshop, Beltsville, Maryland*. Agricultural Research Service, 170 pp.

Huckaba, R.M. and Coble, H.D. (1991) Effect of soybean thrips (Thysanoptera: Thripidae) feeding injury on penetration of acifluorfen in soybean. *Journal of Economic Entomology* 84, 300–305.

Hunter, W.B. and Ullman, D.E. (1989) Analysis of mouthpart movements during feeding of *Frankliniella occidentalis* (Pergande) and *F. schultzei* Trybom (Thysanoptera: Thripidae). *International Journal of Insect Morphology and Embryology* 18, 161–171.

Hunter, W.B., Hsu, H.T. and Lawson, R.H. (1995) A novel method for tospovirus acquisition by thrips. *Phytopathology* 85, 480–483.

Iwaki, M., Honda, Y., Hanada, K., Tochihara, H., Yonaha, T., Hokama, K. and Yokoyama, T. (1984) Silver mottle disease of watermelon caused by tomato spotted wilt virus. *Plant Disease* 68, 1006–1008.

Kaiser, W.J., Wyatt, S.D. and Pesho, G.R. (1982) Natural hosts and vectors of tobacco streak virus in eastern Washington. *Phytopathology* 72, 1508–1512.

Kameya-Iwaki, M., Hanada, K., Honda, Y. and Tochira, H. (1988) A watermelon strain of tomato spotted wilt virus (TSWV-W) and some properties of its nucleic acid. *Fifth International Congress of Plant Pathology*, Kyoto, Japan, Abstract.

Kitajima, E.W., de Avila, A.C., de Resende, R.O., Goldbach, R. and Peters, D. (1992) Comparative cytological and immunogold labelling studies on different isolates of tomato spotted wilt virus. *Journal of Submicroscopic Cytology and Pathology* 24, 1–14.

Kobatake, H., Osaki, T. and Inouye, T. (1984) Reservoirs of tomato spotted wilt virus in Nara Prefecturev. *Annals of the Phytopathological Society of Japan* 50, 541–544.

Kormelink, R.J.M. (1994) Structure and expression of the tomato spotted wilt virus genome, a plant-infecting Bunyavirus. PhD thesis, Agricultural University, Wageningen, 134 pp.

Kormelink, R. Kitajima, E.W., de Haan, P., Zuidema, D., Peters, D. and Goldbach, R. (1991) The nonstructural protein (NSs) encoded by the ambisense S RNA segment of tomato spotted wilt virus is associated with fibrous structures in infected plant cells. *Virology* 181, 459–468.

Kormelink, R., Storms, M., Van Lent, J., Peters, D. and Goldbach, R. (1994) Expression and subcellular location of NSm protein of tomato spotted wilt virus (TSWV), a putative viral movement protein. *Virology* 200, 56–65.

Krczal, G., Albouy, J., Damy, I., Kusiak, C., Deogratias, J.M., Moreau, J.P., Berkelmann, B. and Wohanka, W. (1995) Transmission of pelargonium flower break virus (PFBV) in irrigation systems and by thrips. *Plant Disease* 79, 163–166.

Kumar, N.K.K., Ullman, D.E. and Cho, J.J. (1993) Evaluation of *Lycopersicon* germ plasm for tomato spotted wilt tospovirus resistance by mechanical and thrips transmission. *Plant Disease* 77, 938–941.

Kumar, N.K.K., Ullman, D.E. and Cho, J.J. (1995a) *Frankliniella occidentalis* (Thysanoptera: Thripidae) landing and resistance to tomato spotted wilt tospovirus among *Lycopersicon* accessions with additional comments on *Thrips tabaci* (Thysanoptera: Thripidae). *Environmental Entomology* 24, 513–520.

Kumar, N.K.K., Ullman, D.E. and Cho, J.J. (1995b) Resistance among *Lycopersicon* species to *Frankliniella occidentalis* (Thysanoptera: Thripidae). *Journal of Economic Entomology* 88, 1057–1065.

Kuo, G. (ed.) (1996) *Proceedings of the International Symposium on Tospoviruses and Thrips of Floral and Vegetable Crops, Taichung, Taiwan. Acta Horticulturae, Technical Communications of the International Society for Horticultural Science* 431, 531 pp.

Lakshmi, K.V., Wightman, J.A., Reddy, D.V.R., Ranga Rao, G.V., Buiel, A.A.M. and Reddy, D.D.R. (1995) Transmission of peanut bud necrosis virus by *Thrips palmi* in India. In: Parker, B.L., Skinner, M. and Lewis, T. (eds) *Thrips Biology and Management.* Plenum Press, New York, pp. 179–184.

Law, M.D. and Moyer, J.W. (1989) Physiochemical analysis of a serologically distinct tomato spotted wilt virus strain. *Phytopathology* 79, 933–938.

Law, M.D. and Moyer, J.W. (1990) A tomato spotted wilt-like virus with a serologically distinct N protein. *Journal of General Virology* 72, 2597–2601.

Law, M.D., Speck, J. and Moyer, J.W. (1991) Nucleotide sequence of the 3' non-coding region and N gene of the S RNA of a serologically distinct tospovirus. *Journal of General Virology* 72, 2597–2601.

Law, M.D., Speck, J. and Moyer, J.W. (1992) The M RNA of impatients necrotic spot tospovirus (Bunyaviridae) has an ambisense genomic organization. *Virology* 188, 732–741.

Le Guenno, B. (1995) Emerging viruses. *Scientific American* 273, 56–64.

Lewis, T. (1973) *Thrips, Their Biology, Ecology and Economic Importance.* Academic Press, London.

Linford, M.B. (1932) Transmission of the pineapple yellow-spot virus by *Thrips tabaci. Phytopathology* 22, 301–324.

Ludwig, G.V., Christensen, B.M., Yuill, T.M. and Schultz, K.T. (1989) Enzyme processing of La Crosse virus glycoprotein G1: a bunyvirus-vector infection model. *Virology* 171, 108–113.

Ludwig, G.V., Israel, B.A., Christensen, B.M., Yull, T.M. and Schultz, K.T. (1991) Role of La Crosse virus glycoproteins in attachment of virus to host cells. *Virology* 181, 564–571.

Marullo, R. (1995) Possible dissemination of pest fungi by thrips. In: Parker, B.L., Skinner, M. and Lewis, T. (eds) *Thrips Biology and Management*. Plenum Press, New York, pp. 201–202.

Mau, R.F.L., Bautista, R., Cho, J.J., Ullman, D.E., Gusukuma Minuto, L. and Custer, D.M. (1990) Factors affecting the epidemiology of TSWV in field crops: comparative virus acquisition efficiency of vectors and suitability of alternate hosts to *Frankliniella occidentalis* (Pergande). In: *Virus–Thrips–Plant Interaction of Tomato Spotted Wilt Virus, Proceedings of a USDA Workshop, Beltsville, Maryland*. Agricultural Research Service, pp. 21–27.

McManus, P.S. and Jones, A.L. (1994) Role of wind-driven rain, aerosols, and contaminated budwood in incidence and spatial pattern of fire blight in an apple nursery. *Plant Disease* 78, 1059–1066.

Milne, R.G. and Francki, R.I.B. (1984) Should tomato spotted wild virus be considered as a possible member of the family Bunyaviridae? *Intervirology*, 22, 72–76.

Mink, G.L. (1992) Ilarvirus vectors. In: Harris, K.F. (ed.) *Advances in Disease Vector Research* 9. Springer-Verlag, New York, pp. 261–281.

Mound, L. (1996) The thysanopteran vector species of tospoviruses. In: Kuo, G. (ed.) *Tospoviruses and Thrips of Floral and Vegetable Crops: An International Symposium, Taichung, Taiwan, Acta Horticulturae, Technical Communications of International Society for Horticultural Science* 431, 298–309.

Mound, L.A. and Teulon, D.A.J. (1995) Thysanoptera as phytophagous opportunists. In: Parker, B.L., Skinner, M. and Lewis, T. (eds) *Thrips Biology and Management*, Plenum Press, New York, pp. 3–20.

Ondrej, M. (1973) Parazitick a houba *Botrytis fabae* Sard. ve vatahu k savemu hmyzu. (The parasitic fungus *Botrytis fabae* Sard. in connection with sucking insects.) *Biologia (Bratislava)* 28, 57–63.

Palmer, J.M. (1992) Thrips (Thysanoptera) from Pakistan to the Pacific: a review. *Bulletin of the British Museum of Natural History (Entomology)* 61, 1–76.

Parrella, M.P. (1995a) IPM – approaches and prospects. In: Parker, B.L., Skinner, M. and Lewis, T. (eds) *Thrips Biology and Management*. Plenum Press, New York, pp. 357–364.

Parrella, M.P. (1995b) Thrips management guide, Part II: proper identification. *Grower Talks* 59, 82–94.

Parrella, M.P. (1995c) Thrips management guide, Part 1: prevention and control. *Grower Talks* 58, 30–38.

Peters, D. (1996) Thrips as vectors of tospoviruses. In: Kuo, G. (ed.) *Tospoviruses and Thrips of Floral and Vegetable Crops: An International Symposium, Taichung, Taiwan, Acta Horticulturae, Technical Communications of International Society for Horticultural Science* 431, 29–43.

Pittman, H.A. (1927) Spotted wilt of tomatoes. Preliminary note concerning the transmission of the 'spotted wilt' of tomatoes by an insect vector (*Thrips tabaci* Lind.). *Australian Journal of the Council for Scientific and Industrial Research* 1, 74–77.

Poos, F.W. and Elliot, C. (1936) Certain insect vectors of *Aplanobacter stewarti*. *Journal of Agricultural Research* 52, 585–608.

Reddy, D.V.R., Sudarshana, M.R., Ratna, A.S., Reddy, A.S., Amin, P.W., Kumar, I.K. and Murthy, A.K. (1990) The occurrence of yellow spot virus, a member of tomato spotted wilt virus group, on peanut (*Arachia hypogaea* L.) in India. In: *Virus–Thrips–Plant Interactions of Tomato Spotted Wilt Virus, Proceedings of a USDA Workshop, Beltsville, Maryland*. Agricultural Research Service, pp. 77–88.

Reddy, D.V.R., Ratna, A.S., Sudarshana, M.R., Poul, F. and Kiran Kumar, I. (1992)

Serological relationships and purification of bud necrosis virus, a tospovirus occurring in peanut (*Arachis hypogaea* L.) in India. *Annals of Applied Biology* 120, 279–286.

Robb, K.L. (1989) Analysis of *Frankliniella occidentalis* (Pergande) as a pest of floriculture crops in California greenhouses. PhD thesis, University of California, Riverside, 135 pp.

Sakimura, K. (1940) Evidence for the identity of the yellow-spot virus with the spotted-wilt virus: experiments with the vector, *Thrips tabaci. Phytopathology* 30, 281–299.

Sakimura, K. (1962) The present status of thrips-borne viruses. In: Maramorosch, K. (ed.) *Biological Transmission of Disease Agents*. Academic Press, New York, pp. 33–40.

Sakimura, K. (1963) *Frankliniella fusca*, an additional vector for the tomato spotted wilt virus, with notes on *Thrips tabaci*, another vector. *Phytopathology* 53, 412–415.

Sakimura, K. (1969) A comment on the color forms of *Frankliniella schultzei* (Thysanoptera: Thripidae) in relation to transmission of the tomato spotted wilt virus. *Pacific Insects* 11, 761–762.

Samuel, G., Bald, J.G. and Pittman, H.A. (1930) Investigations on 'spotted wilt' of tomatoes in Australia. *Commonwealth Council for Scientific and Industrial Research Bulletin* No. 44, 8–11.

Sdoodee, R. and Teakle, D.S. (1987) Transmission of tobacco streak virus by *Thrips tabaci*: a new method of plant virus transmission. *Plant Pathology* 36, 377–380.

Sdoodee, R. and Teakle, D.S. (1993) Studies on the mechanism of transmission of pollen associated tobacco streak ilarvirus by *Thrips tabaci. Plant Pathology* 42, 88–92.

Tsuda, S. and Ohnishi, J. (1996) The thysanopteran vector species of tospoviruses. In: Kuo, G. (ed.) *Tospoviruses and Thrips of Floral and Vegetable Crops: An International Symposium, Taichung, Taiwan. Acta Horticulturae, Technical Communications of International Society for Horticultural Science* 431, 325–332.

Ullman, D.E., Cho, J.J., Mau, R.F.L., Hunter, W.B., Westcot, D.M. and Custer, D.M. (1992a) Thrips-tomato spotted wilt virus interactions: morphological, behavioral and cellular components influencing thrips transmission. In: Harris, K.F. (ed.) *Advances in Disease Vector Research* 9. Springer-Verlag, New York, pp. 195–240.

Ullman, D.E., Cho, J.J., Mau, R.F.L., Westcot, D.M. and Custer, D.M. (1992b) Midgut epithelial cells act as a barrier to tomato spotted wilt virus acquisition by adult western flower thrips. *Phytopathology* 82, 1333–1342.

Ullman, D.E., German, T.L., Sherwood, J.L., Westcot, D.M. and Cantone, F.A. (1993) *Tospovirus* replication in insect vector cells: immunocytochemical evidence that the nonstructural protein encoded by the S RNA of tomato spotted wilt tospovirus is present in thrips vector cells. *Phytopathology* 83, 456–463.

Ullman, D.E., German, T.L., Sherwood, J.S. and Westcot, D.M. (1995a) Thrips transmission of tospoviruses: future possibilities for management. In: Parker, B.L., Skinner, M. and Lewis, T. (eds) *Thrips Biology and Management*. Plenum Press, New York, pp. 135–152.

Ullman, D.E., Westcot, D.M., Chenault, K.D., Sherwood, J.L., German, T.L., Bandla, M.D., Cantone, F.A. and Duer, H.L. (1995b) Compartmentalization, intracellular transport and autophagy of tomato spotted wilt tospovirus proteins in infected thrips cells. *Phytopathology* 85, 644–654.

Ullman, D.E., Westcot, D.M., Cantone, F.A., Sherwood, J.L. and German, T.L. (1992c) Immunocytochemical evidence for tomato spotted wilt virus (TSWV) replication in cells of the western flower thrips, *Frankliniella occidentalis* (Pergande). *Phytopathology* 82, 1087 (Abstract).

Urban, L.A., Huang, P. and Moyer, J.W. (1991) Cytoplasmic inclusion in cells infected with isolates of L and I serogroups of tomato spotted wilt virus. *Phytopathology* 81, 525–529.

Wang, M. and Gonsalves, D. (1990) ELISA detection of various tomato spotted wilt virus isolates using specific antisera to structural proteins of the virus. *Plant Disease* 74, 154–158.

Wettering, F. (1996) Acquisition of tomato spotted wilt virus occurs only during the first larval instar. In: Kuo, G. (ed.) *Tospoviruses and Thrips of Floral and Vegetable Crops: An International Symposium, Taichung, Taiwan. Acta Horticulturae, Technical Communications of International Society for Horticultural Science* 431, 350–366.

Wettering, F., Golbach, R. and Peters, D. (1996) Tomato spotted wilt tospovirus ingestion by first instar larvae of *Frankliniella occidentalis* is a prerequisite for transmission. *Phytopathology* 86, 900–905.

Wijkamp, I. and Peters, D. (1993) Determination of the median latent period of two tospoviruses in *Frankliniella occidentalis*, using a novel leaf disk assay. *Phytopathology* 83, 986–991.

Wijkamp, I., Almarza, N., Goldbach, R. and Peters, D. (1995a) Distinct levels of specificity in thrips transmission of tospoviruses. *Phytopathology* 85, 1069–1074.

Wijkamp, I., Almarza, N. and Peters, D. (1995b) Median latent period and transmission of tospoviruses vectored by thrips. In: Parker, B.L., Skinner, M. and Lewis, T. (eds) *Thrips Biology and Management*. Plenum Press, New York, pp. 153–156.

Wijkamp, I.N., Almarza, R., Goldbach *et al.* (1995c) Distinct levels of specificity in thrips transmission of topoviruses. *Phytopathology* 85, 1069–1074.

Wijkamp, I., van Lent, J. Kormelink, R., Goldbach, R. and Peters, D. (1993) Multiplication of tomato spotted wilt virus in its insect vector, *Frankliniella occidentalis*. *Journal of General Virology* 74, 341–349.

Yarwood, C.E. (1943) Association of thrips with powdery mildews. *Mycologia* 35, 189–191.

Yeh, S.D. and Chang, T.F. (1995) Nucleotide sequence of the N gene of watermelon silver mottle virus, a proposed new member of the genus *Tospovirus*. *Phytopathology* 85, 58–64.

Yeh, S., Lin, Y., Cheng, Y., Jih, C., Chen, M. and Chen, C. (1992) Identification of tomato spotted wilt virus infecting watermelon in Taiwan. *Plant Disease* 76, 835–840.

Zitter, T.A., Daughtrey, M.L. and Sanderson, J.P. (1989) Tomato spotted wilt virus. *Cornell Cooperative Extension Fact Sheet* 735, 6 pp.

Zur Strassen, R. (1960) Catalogue of the known species of South African Thysanoptera. *Journal of the Entomological Society of Southern Africa* 23, 321–367.

Chemical Control 15

Trevor Lewis
Institute of Arable Crops Research – Rothamsted, Harpenden, Herts AL5 2JQ, UK

Introduction

As thrips infesting field, orchard and plantation crops are so responsive to cultural methods of control, the need to resort to insecticides to limit direct damage often implies that some aspect of crop or farm management has been neglected, except perhaps in dry weather when plants may succumb to infestations that would normally be harmless. However, insecticides are sometimes required to prevent significant virus spread by populations too sparse to cause severe direct feeding damage and, in the artificial environment of glasshouses, populations often increase too rapidly for cultural control to prevent disfiguring of ornamentals. Also, in some fruit and vegetable crops, cosmetic blemishes inflicted at relatively low population densities are commercially unacceptable and pesticides may be necessary to limit them. They thus have an important role but, wherever possible, should be used as components in integrated pest management (IPM) programmes rather than as a single control element, an approach that is becoming progressively less reliable in the short term and unsustainable over longer periods. This chapter, therefore, considers the history, practices and consequences of chemical control, but should be read as background to Chapters 16, 17 and 18 on integrated pest management in field, tree and glasshouse crops.

Numerous insecticides kill individual thrips collected from previously untreated populations when administered topically or by surface-exposure bioassays, but infestations on crops are more difficult to control than might be expected for insects susceptible to so many compounds. Problems arise as a result of the great numbers that infest individual plants, the ability of field populations to increase rapidly through breeding and airborne immigration, and the secretive habits of some species. Small numbers of individuals living hidden in the crevices of flowers, bracts or leaf sheaths may require a different method of control from large populations feeding on the exposed surfaces of leaves, and species that spend part of their life cycle in the soil may be vulnerable at times different from species that remain above ground. Such difficulties can be partly overcome by using appropriate insecticides, formulations and methods of application timed to attack the pests at the most vulnerable stage of their life cycle, when they are most accessible on host plants, and before crops have been seriously damaged. Because thrips feed on the contents of plant cells, simple stomach poisons on leaf surfaces are rarely satisfactory. Persistent, contact insecticides are usually unsuitable for controlling young and adults protected in leaf sheaths or flowers. They may be applied to control exposed leaf or fruit feeders but are safer and more useful for destroying stages in the soil. Many non-persistent systemic or translaminar insecticides are effective against thrips living in inaccessible places on plants and against sluggish species living on the underside of leaves where sprays may not reach and these are now the most widely used. Unfortunately, the availability of appropriate compounds is constantly changing as products are superseded and particularly as resistance develops.

Historical Background

A brief review of the chemical measures used against thrips since the beginning of the century provides a useful background to more recent insecticide control and its effects. At the outset it should be emphasized that no thrips-specific compound has been marketed because the scale of development and registration costs to target such a relatively minor use would be uneconomic for the agrochemical industry. The pesticides available for thrips control have therefore all been produced and tested primarily against pests in other major taxa, so many have had limited efficacy against thrips, which exhibit atypical behaviour and feeding habits compared with the majority of pest insects.

Before the advent of advanced synthetic insecticides in the mid-1940s, a wide range of inorganic and organic compounds, and plant derivatives, were tried in field and orchard crops, with varying degrees of success. Hinds (1902) and particularly Bailey (1944) list most of the compounds, formulations (mainly sprays) and rates used. They included kerosene on grassy borders, whale oil soap, crude carbolic and creosote, calcium cyanamide, sulphur dust and lime sulphur, quicklime, tartar emetic, nicotine sulphate, sabadilla, derris, ryania, pyrethrum, thiocyanates and dinitro-compounds on fruit trees, with nicotine fumigation popular in glasshouses.

Between 1945 and 1960, persistent chlorinated hydrocarbons, mainly BHC, DDT, aldrin, dieldrin and toxaphene, were applied generously as sprays and dusts to kill thrips living in exposed positions. Predictably, they were most successful against leaf feeders on plants with a fairly open form of growth including coffee (Anonymous, 1963), cotton (Lincoln and Leigh, 1957), tea (Rao, 1970), young onions (Richardson, 1957), tomatoes (MacGillivray *et al.*, 1950) and flax (Franssen and Huisman, 1958). Species feeding in crevices were less affected; Riherd (1954) obtained good control of *Chirothrips* spp. feeding in grass heads in a nursery, but heavy doses of insecticide were necessary.

Some chlorinated hydrocarbons proved repellent. For example, against *Heliothrips haemorrhoidalis*, the toxicity of chlordane, and at strong doses DDT and heptachlor, was inversely related to their repellency (Rogoff, 1952). Where other pests occurred on a crop the persistence of chlorinated hydrocarbons proved a disadvantage because natural enemies were killed (MacPhee, 1953; Tapley, 1960; Evans, 1967).

When formulated as dusts or granules some chlorinated hydrocarbons were more effective used as soil insecticides against immature and adult thrips living in the ground than against thrips on the plant. Granulated aldrin controlled infestations of *Frankliniella fusca* and *Frankliniella tritici* on peanuts (Dogger, 1956) and BHC dust incorporated into the upper 10 cm of soil in seed beds before sowing, followed by monthly dustings with DDT, controlled *Thrips tabaci* and the spread of tomato spotted wilt in tobacco crops (Ivancheva-Gabrovska, 1959).

Semi-persistent systemic insecticides, mostly organophosphorus derivatives or carbamates, gradually replaced the chlorinated hydrocarbons and were used for thrips control on most growing crops. Of the early organophosphates, parathion applied as sprays or dusts probably gave the best results over a wide spectrum of species. Routine sprays of parathion were used for many years to control pests in citrus, including thrips (Hodgson, 1970) and were commonly used against species attacking cotton, groundnuts and onions. Damage to grass seed crops by *Chirothrips* spp. was notably prevented by parathion dust (Roney, 1949). Other systemic insecticides sometimes applied as dusts included ethion and carbophenothion used against *Frankliniella vaccinii* on blueberries because they did less harm to bees than dieldrin (Boulanger and Abdalla, 1966).

An important advance in the use of systemics against thrips was their formulation as seed dressings and granules for application to the soil at sowing or planting time to kill thrips attacking the developing seedlings. Demeton and phorate seed dressings protected young cotton plants for 4 to 6 weeks after planting. Applied in this way these compounds were slightly phytotoxic and retarded germination, seedling emergence and plant growth, but eventual yield was unaffected (Hanna, 1958; Mistric and Spyhalski, 1959). Onion seeds treated with phorate before sowing produced seedlings immune to attack by *Thrips tabaci* for about a month (Shazli and Zazou, 1959). Granules of phorate or dimethoate sprinkled on mature onion plants intended for seed production lodged in the angle of the leaves and protected the plants from *Frankliniella occidentalis* for 4 to 21 days; longer protection, lasting for a month, was obtained when phorate granules were applied in a shallow trench to one side of the plant row (Hale and Shorey, 1965). Granules of aldicarb applied in the furrow at planting gave excellent control of *F. fusca* on cotton without the harmful effects on young plants caused by phorate (Pfrimmer, 1966). Latterly, treatment of rice seed with carbosulfan or the broadcasting of carbofuran granules into standing water have both proved effective against *Stenchaetothrips biformis* attacking rice (Sathiyanandam *et al.*, 1987).

Pesticide Usage and Efficacy

Almost all modern synthetic insecticides have been evaluated against one or more thrips pests, although commercial field use is governed in different countries by local regulatory requirements with varying degrees of permissiveness and enforcement. The result is a bewildering list of claim and counter-claim for the efficacy of compounds and formulations against many species on different crops in diverse climates using a range of application methods. Generalizations are thus difficult and of dubious value; nevertheless this section attempts to elicit some order from key studies concentrating on currently economically important species and crops, and pesticides available. There is no universally ideal thripsicide and over all looms the apparently inexorable development of resistance (see below) with the consequent need to devise and employ resistance-retarding strategies. It is essential to distinguish between toxicity and mortality of insecticides tested in laboratory bioassays on exposed insects, and control in the field where so many other factors intervene. These include plant and crop structure, pest behaviour, feeding and pupation sites, application methods and pesticide persistence, previous pesticide treatments and weather, and possibly whether control is directed against direct feeding damage by heavy infestations or virus spread by lighter ones. Furthermore, the effect of different compounds on larvae, pupae and adults almost always differs; there is no reliable ovicide. Regrettably, many published field trials disregard or do not report such important details. The difficulties can be illustrated by examination of the response to different insecticides of *Scirtothrips citri* and *Frankliniella occidentalis*, two species which have received intense treatment and study.

Scirtothrips citri has been an important pest of Californian citrus groves since about 1900. The larvae and adults feed around the sepals of young fruit and the initial lesions develop into characteristic ring-shaped rind scars as the fruit matures. Dimethoate, applied in foliar sprays was the predominant pesticide used against the pest from 1969 to the mid-1980s, but more than 25 other synthetic compounds have been used since 1946 including other organophosphates, carbamates and pyrethroids. Some 40 per cent of pesticides currently used on citrus are directed against thrips (Morse, 1995). With this history of treatments and exposure, Morse and Brawner (1986) produced toxicity data for three colonies of *Scirtothrips citri* collected between 1981 and 1984 from navel orange groves in the San Joaquin Valley. Adult females were exposed to a leaf dip bioassay (see Chapter 11) of nine pesticides: dimethoate and acephate (organophosphates); formentanate, carbosulfan and bendiocarb (carbamates); DDT (chlorinated hydrocarbon); fluvalinate and flucythrinate (pyrethroids); and abermectin (macrocyclic lactone acaricide) and the LC_{90} for each assessed. Toxicity levels differed between compounds and colonies depending on the degree of tolerance present in each arising from selection pressure exerted by previous pesticide treatments. By comparing the laboratory LC_{90}s with the respective field rates in use, tentative comparisons of the field efficacy of the nine insecticides were obtained. All thrips were susceptible to acephate, formentanate and carbosulfan, two of the three colonies were also susceptible to dimethoate and bendiocarb, but all showed considerable, though variable, tolerance to DDT, fluvalinate, flucythrinate and abamectin. Extrapolation of laboratory bioassays to expected field efficacy must be cautious because the larvae are smaller than adult females so may have a lower tolerance to pesticides, and some compounds may therefore provide more sustained residual activity to larvae than is indicated by bioassays on adults. The most useful commercial pesticides for *S. citri* control are those with a residual efficacy against both immigrant adults and first instar larvae that hatch from the eggs adults lay within leaves, rather than those compounds that are most effective over a brief 2-day bioassay. Furthermore, some pesticides may be repellents, feeding deterrents, or affect fecundity; Morse and Zareh (1991) found that certain thrips females laid fewer or more eggs depending on the residual concentrations of dicofol, malathion, fluvalinate or formentanate.

Since the early 1990s, *Frankliniella occidentalis* has been subject to similar intensive study on the use and efficacy of insecticides to control its rapid spread, especially in glasshouses. Its secretive habits and preference for pupation in soil pose control problems. Helyer and Brobyn (1992) evaluated the effects of direct pesticide contact for 51 compounds on larvae in laboratory bioassays, and further assessed the most promising products, and granular formulations and systemic compounds, against larvae, pupae and adults in glasshouse trials. For laboratory screening, pesticides were applied to chrysanthemum leaf discs at the rates recommended for thrips on glasshouse ornamentals; glasshouse trials were done in 1 m^2 plots (see Chapter 11).

The initial screen showed that 14 compounds killed more than 75 per cent of larvae after 3 days, with the organophosphates, chlorpyrifos (98.1 per cent) and quinalphos (99.8 per cent) the most effective. When these 14 compounds were then tested against larvae, pupae and adults, chlorfenvinphos, chlorpyrifos, chlorpyrifosmethyl and malathion were effective against all three stages but pupae were more susceptible than larvae or adults. In the follow-up glasshouse trials of 24 pesticides applied as foliar sprays, most had a rapid effect within a day, but of those approved for use on protected crops in the UK only malathion provided continuous control for more than 2 weeks; carbofuran, an unapproved systemic, was similarly persistent. In practice, the compounds permitted by state or national regulatory schemes for glasshouse use differ depending on whether or not the crop is edible and on the susceptibility of the infesting species (see below).

The trials outlined above were all made on one strain of *Frankliniella occidentalis* thought to represent the 'average' pesticide tolerance of this species, but unfortunately this varies widely between localities. For example, in central, eastern and southern England, 30-fold differences in susceptibility to malathion were detected between different populations of this thrips, and 25-fold differences to dichlorvos (Macdonald, 1995). Also, as with citrus thrips, extrapolation of topical application or leaf dip bioassays to field situations must be speculative. A single dose placed on an individual thrips is not comparable with an insect encountering residues on plants over a period, because it ignores the possible effects of repellency, attractancy or ingestion. Such factors may contribute to the unexpected performance of some compounds in the field; for example, deltamethrin is claimed to be an antifeedant, which may not be useful in practice if disturbed insects merely move to nearby crops.

An even wider range of compounds than those tested on *Scirtothrips citri* and *Frankliniella occidentalis* is being used in many parts of South-East Asia against species attacking many vegetables, ornamentals and cereals (Talekar, 1991). Over 50 different pesticides and formulations, including all major chemical groups, are applied widely and frequently, and often as 'cocktails', to try to enhance control, a practice that is unlikely to prove sustainable for long (see below).

Against four species infesting seedling cotton, *Frankliniella fusca*, *F. tritici*, *Neohydatothrips variablis* and *Thrips tabaci*, Burris *et al.* (1989) compared foliar applications of 15 compounds presented in 22 different formulations and doses, as well as aldicarb applied as in-furrow granules (see below). Apart from phosphamidon, all the other compounds significantly reduced populations; in general, the pyrethroids gave faster knockdown but populations recovered more rapidly from them than from organophosphates.

Various less conventional pesticides have been evaluated from time to time. Insect growth regulators have been tried with variable success against *Thrips palmi*. Pyripoxyfen, a juvenile hormone mimic, had no effect on adults when applied to leaves, but increased pupal mortality (Nagai, 1990). In laboratory tests another growth regulator, flufenoxuron, inhibited moulting of first instar larvae and the metamorphosis of second instars to prepupae, but did not affect survival and fecundity of females; in the field it was less effective than the organophosphate sulprofos (Nagai *et al.*, 1988). By contrast, in other laboratory tests, flufenoxuron, chlorfluazuron, diflubenzuron and teflubenzuron all prevented pupation, but the two former in particular were more effective on greenhouse-grown musk melons than the conventional organophosphate, methidathion, or the carbamate fenobucarb (Kubota, 1989). Abamectin, a slow-acting nerve poison, was initially reasonably effective against *Scirtothrips citri* (Morse and Brawner, 1986), but in later trials some colonies were highly tolerant. *Bacillus thuringiensis* formulated in mineral oil proved effective against *Frankliniella occidentalis* (Helyer and Brobyn, 1992), but probably as a result of suffocation from the oil rather than as a direct effect of the bacterial toxin. Mineral oils mixed with insecticides have been reported as having both positive and neutral effects; Moustafa and El-Attal (1985) claimed that they enhanced the efficacy of insecticidal sprays against thrips on cotton, whereas Heugens *et al.* (1989) found no evidence for such improvement against *Frankliniella occidentalis* on chrysanthemums. The botanical insecticides nicotine, rotenone, ryania and sabadilla have sometimes proved effective but have now been largely superseded. Neem oil is often used, especially in India where it is claimed to be as effective as some organophosphates for controlling *Stenchaetothrips biformis* on rice (Pillai and Ponniah, 1988); in Israel several neem formulations prevented the development of first instar to second instar larvae in both *Thrips tabaci* and *Frankliniella occidentalis* (Ascher *et al.*, 1992, 1993).

In summary, it is impossible to recommend with certainty any compound with guaranteed long-term efficacy for a given crop and pest. Most of the pesticides available and specifically recommended for thrips control by manufacturers are listed in Table 15.1, but the list is not exhaustive, and there are other compounds of proven local value widely used by growers (see Tables 15.4 and 15.5). All major chemical types of insecticide are represented, though cholinesterase inhibitors predominate.

Resistance to Pesticides

Insecticide resistance in thrips, as in many other insects, is a complex and intractable problem. In a practical context it can be defined as the stage beyond which control with a given pesticide is no longer economic (Sawicki, 1981). At the level of the individual insect it requires understanding of the genetic, biochemical and toxicological properties of resistance mechanisms, which have barely been studied for any thrips species except *S. citri* and latterly *F. occidentalis*.

As early as 1941, sprays of tartar emetic plus sucrose were becoming ineffective against citrus thrips (Persing *et al.*, 1942) and against *Thrips simplex*. Signs of resistance to chlorinated hydrocarbons appeared in the mid-1950s; of 18 insecticides tested against *T. tabaci* in Texas in 1955, dieldrin, heptachlor, toxaphene, aldrin and endrin gave no control when used alone (Richardson and Wene, 1956).

As an example of the typical chronological evolution of chemical control against a thrips on a major crop, Table 15.2 lists the series of compounds used against *Scirtothrips citri* in California over 70 years and, of great significance, the year in which failure was first reported (Morse and Brawner, 1986). The rapidity with which this pest developed resistance to different compounds after their introduction into commercial use was remarkable.

The overall outcome of this important series of Californian investigations on *Scirtothrips citri* (e.g. Morse and Brawner, 1986; Immaraju *et al.*, 1989; Haney *et al.*, 1992; Morse, 1995) demonstrates the gradual failure of chemical control as resistance develops, albeit at different rates, to virtually all insecticides introduced. For example, it took only 4 years before it became apparent to DDT and dieldrin, but about 18 years before failure was detected in the organophosphate dimethoate, which must therefore rank as a highly successful compound against this pest on this crop. Latterly, resistance to other organophosphates, carbamates and pyrethroids has been detected after considerably shorter periods of exposure. By 1992, the only broad-spectrum insecticides recommended for citrus thrips were dimethoate and formentanate, despite there being in some localities 90-fold resistance (at LC_{50}) to the former and 30-fold to the latter (Immaraju *et al.*, 1989; Haney *et al.*, 1992).

There is also strong cross-resistance (Immaraju and Morse, 1990). In glasshouse studies a strain of *Scirtothrips citri* resistant to dimethoate was selected with the pyrethroid, fluvalinate. After ten selections over 10 months the strain was 482-fold resistant to fluvalinate (at LC_{50}), strongly resistant to all other pyrethroids tested (LC_{50} ratios ranging from 10-fold for cypermethrin

Table 15.1. Principal insecticides specifically recommended by manufacturers against thrips, application methods and main crop usage (subject to National Approval Schemes, permitted maximum residue levels, and crop and environmental safety restrictions).

Compound	Application method(s)*	Principal crop usage
Acephate – systemic, contact and stomach organophosphate	1	General, especially fruit (including citrus), cotton, peanuts, soybeans
Acrinathrin – contact and stomach pyrethroid	1	Fruit trees, vegetables, vines
Alanycarb – contact and stomach oxime carbamate	1,2,3	Vines
Aldoxycarb – systemic oxime carbamate	1,3,4	Cotton, peanuts, tobacco, vegetables
Bendiocarb – contact stomach and systemic carbamate	1	General
Bensultap – contact and stomach benzenethiosulfonate	1	Tea
Butocarboxim – systemic oxime carbamate	1	Fruit trees (including citrus), cotton, ornamentals, tobacco, vegetables
Carbofuran – systemic and stomach carbamate	1,2	General
Chlorfluazuron – benzoylurea growth regulator	1	Fruit, ornamentals, potatoes, tea
Deltamethrin – contact and stomach pyrethroid	1	General, cereals, citrus, cotton
Dimethoate – systemic and contact organophosphate	1	General
Disulfoton – systemic organophosphate	2,3	Fruit, ornamentals, tobacco, vegetables
Etofenprox – contact and stomach pyrethroid	1	Cereals, fruit trees (including citrus), rice, soybeans, tea, vegetables
Fenitrothion – contact organophosphate	1	Peas, salad crops, sweet corn
Fenobucarb – contact carbamate	1	Cereals, fruit trees, lucerne, ornamentals, sugar cane, rice, vegetables
Fipronil – contact and stomach phenyl pyrazole	1,2,3	Cotton, rice, vegetables
Formetanate – contact and stomach carbamate	1	General, cotton, fruit trees (including citrus), hops, ornamentals, rice, sugar cane, tobacco, vegetables
Formothion – systemic and contact organophosphate	1	General
Heptachlor – contact and stomach organochlorine with fumigant action	1,2,3	General
Heptenophos – contact and organophosphate	1	General

Table 15.1. Continued.

Compound	Application method(s)*	Principal crop usage
Imidacloprid – systemic, contact and stomach action nitromethylene	1,2,3	Cereals, cotton, fruit trees (including citrus), potatoes, rice, sugar beet, vegetables
Isofenphos – systemic, contact and stomach organophosphate	2	Citrus, onions
Lamda-cyhalothrin – contact and stomach pyrethroid	1	General
Malathion – contact organophosphate	1	Ornamentals, vegetables
Mecarbam – systemic, contact and stomach carbamate	1	Cotton
Mephosfolan – systemic organophosphate	1	Cotton
Nicotine – contact alkaloid with fumigant action	1,5	Protected ornamentals, salads and vegetables
Omethoate – systemic, contact and organophosphate	1	General
Phenthoate – contact and stomach organophosphate	1	Citrus, coffee, cotton, fruit, rice, tea, tobacco, vegetables
Phosalone – contact and stomach organophosphate	1	Cardamom, pome and stone fruit, tea
Phosmet – contact organophosphate	1	Olives
Phosphamidon – systemic organophosphate	1	Cotton
Pirimiphos-methyl – systemic organophosphate with fumigant action	1	General, bulbs, citrus, glasshouse salads, vegetables
Pyridaphenthion – contact and stomach organophosphate	1	Fruit, ornamentals, rice, vegetables
Quinalphos – contact and stomach organophosphate	1	General
Sulfotep – contact organophosphate with fumigant action	5	Glasshouse ornamentals, vegetables
Sulprofos – contact and stomach organophosphate	1	Cotton, soybeans, tobacco, tomatoes, vegetables
Triazophos – contact and stomach organophosphate	1	Bananas, cereals, coffee, fruit trees, soybeans, sugar cane
Tau-fluvalinate – contact and stomach pyrethroid	1	Cotton, fruit trees (including citrus), ornamentals, vegetables, vines
Verticillium lecanii – fungus	1	Glasshouse salad crops, ornamentals

*1 = foliar spray; 2 = soil treatment; 3 = seed treatment; 4 = transplant water treatment; 5 = fumigant (compiled from Tomlin, 1994; Whitehead, 1995 – and commercial sources).
NB Other broad-spectrum insecticides, not specifically recommended for thrips control, may nevertheless be effective, e.g. abamectin, aldicarb and methiocarb.

Table 15.2. Pesticides used for control of *Scirtothrips citri* in California from 1912 to 1982 (from Morse and Brawner, 1986).

Pesticide	Year first used commercially	Year field failure first reported
Lime sulphur, lime sulphur + nicotine sulphate	*c.* 1912	–
Dusting sulphur	1926	–
Tartar emetic + sugar	1939	1941
Nicotine sulphate + sugar	1944	–
DDT	1946	1949
Sabadilla + sugar	1948	–
Parathion	1949	–
Dieldrin	1953	1954
Malathion	1954	1961
Dioxathion, dioxathion + sugar	1961	–
Dimethoate	1962	1980
Phosphamidon	1963	–
Ryania + sugar	1964	–
Azinphosmethyl	1969	–
Formetanate	1972	–
Biothion	1973	–
Methomyl	1973	–
Acephate	1982	–

to 128-fold for bifenthrin) and to DDT (54-fold). Among organophosphates and carbamates cross-resistance was moderate to parathion, low to bendiocarb and marginal or absent to a range of other compounds. Such tests demonstrate how regular heavy field use of pesticides might rapidly lead to control failures. Already, the practical consequences are that in some areas residual activity has been reduced from about six to three weeks (Haney *et al.*, 1992), which is serious because fruit needs protection for 4 to 6 weeks, and consequent repeated applications tend to exacerbate the problem. The recently introduced fipronil (see Table 15.1) appears to show no cross-resistance with either organophosphates, carbamates or pyrethroids.

In the field, untreated resistant populations often gradually revert due to fitness disadvantages associated with resistance and immigration of susceptible genotypes from untreated areas, but this may take a year or more. The rate of reversion in more-isolated greenhouse populations is likely to be slower, a pointer to resistance problems to which populations of other species infesting greenhouse crops would be subject.

Not surprisingly, resistance in *Frankliniella occidentalis* has also been the focus of much study following its establishment and spread as a glasshouse pest of ornamentals, fruit and vegetables since the mid-1980s. First reports of resistance were to toxaphene on cotton (Race, 1961), but since then there have been many instances of the diminishing efficacy of insecticides of all major chemical groups, encouraged by the frequent treatments to protected crops (Robb, 1989; Robb *et al.*, 1995). Immaraju *et al.* (1992) evaluated the extent of resistance to four classes of insecticides: organophosphates, carbamates, pyrethroids and the macrocyclic lactone, abermectin, and the effects of pyrethroids synergized with piperonyl butoxide. Using a laboratory leaf-dip assay (see Chapter 11, Topical Bioassays) high levels of resistance were found to all four classes of compound. Compared with a susceptible strain not known to have been exposed to insecticides for more than 20 years, populations from various commercial greenhouses were 138-fold and more than 8000-fold resistant to the pyrethroids bifenthrin and permethrin, respectively, at the LC_{90} level. There was modest to high resistance to the carbamoyloxime methomyl, lower resistance to chlorpyrifos, but high levels to abermectin even though it had not then been registered for western flower thrips control in California. Greenhouse corroboration of these laboratory tests to give a broad impression of field performance did not always closely reflect the laboratory findings. Unexpectedly, permethrin was the most effective at the rates used; its performance was not improved by addition of piperonyl butoxide though the synergist may enhance efficacy of lower doses. Abermectin performed well; chlorpyrifos did not. All this emphasizes the complexity and unpredictability of the development of resistance and the dangers of extrapolating laboratory tests to field situations, where the many factors mention above affect the outcome.

Another series of bioassays compared a susceptible, wild, Californian population of *Frankliniella occidentalis* with others from glasshouses in Europe where the pest cannot survive outdoors (Brødsgaard, 1994). Populations in these commercial glasshouses are thus likely to be isolated and subject to insecticides on a regular basis but without the opportunity for dilution with wild, susceptible individuals from outside. Resistance was detected in these glasshouse populations to organophosphates (acephate), chlorinated hydrocarbons and carbamates, with cross-resistance to methiocarb from unknown previous exposure. Once established, resistance proved extremely persistent, and was still apparent in one culture for 100 generations' after exposure to any pesticides. Similarly, Robb (1989) identified a strain which retained resistance to dimethoate for 7 years without further exposure, indicating that resistance in field populations is not necessarily eliminated by the cessation of pesticide treatments.

Clearly, these detailed evaluations of pesticides against *Scirtothrips citri* in the USA and *Frankliniella occidentalis* in the USA and Europe and the limited registrations for their use have not prevented the onset of resistance to insecticides in field populations of these two species. The profligate use of compounds on various vegetable crops in South-East Asia is even more self-defeating. In attempts to control *Thrips tabaci*, *Thrips palmi* and *Thrips parvispinus* on chilli, onion and potato in Indonesia, some farmers apply up to 16 sprays of quinalphos, endosulfan, mercaptodimethur, acephate, formentanate and cypermethrin at 3 to 4 day intervals throughout the growing season; on chilli, insecticidal treatments can account for more than half the total variable production costs (Sastrosiswojo, 1991). In the Philippines, farmers depend on insecticides to control *T. palmi* and the bean flower thrips

Megalurothirps usitatus on most vegetables, using two to four chemicals in succession applied at 3 to 14 day intervals, chosen from the wide range of pyrethroids, organophosphates and carbamates obtainable. Despite this, yield losses are large and sometimes total (Bernardo, 1991). The compounds used include profenofos, cypermethrin and methamidophos, the latter requiring a pre-harvest interval of 7 days, but such precautions are frequently ignored if thrips populations are perceived to be building up too rapidly. Similar excessive reliance on pesticides is rife in Thailand, where 'cocktails' of two to five chemicals are mixed (Bansiddhi and Poonchaisri, 1991), in Taiwan (Chang, 1991) and Japan (Hirose, 1991). Multi-pesticide treatments specifically aimed against *Thrips palmi* in other regions seem to be following the same pattern and already populations of this species on diazinon- and profenofos-treated aubergines have been shown to exceed numbers on untreated plots (Etienne *et al.*, 1990).

Such heavy pesticide usage encourages the rapid development of resistance and the destruction of natural enemies. Indeed, it is probably the fundamental cause of the rapid spread of *Thrips palmi* in parts of South-East Asia and elsewhere since about 1980. After being under intense selection pressure from a range of pesticides of all major types applied against thrips and other pests, large resistant populations have greatly increased the chances of its spread. It remains to be seen whether the most recent useful compounds, imidocloprid and fipronil, with different modes of action to other recent synthetic insecticides and currently effective against *T. palmi*, *T. tabaci* and *Frankliniella schultzei* on cotton (Elbert *et al.*, 1990; Prudent, 1990; N. Hamon, S. Carolina, personal communication) will retain their efficacy for more than the typical 4 years. Fortunately, in the USA there are few reports of resistance occurring to any compounds in the 12 or so species that infest cotton, presumably because invasions arrive from host sources that are seldom treated with insecticides (Leigh, 1995). In greenhouses, imidocloprid applied at high doses via the soil to ornamentals effectively controls leaf feeders, especially *T. palmi*, so it is extremely useful for clearing plants in quarantine, but it is less effective against flower and pollen feeders including *F. occidentalis*. Being systemic, its impact declines as the size of plants increases (G. Vierbergen, Wageningen, 1995, personal communication). Some growers find an effective way to clear greenhouses of *F. occidentalis* is to heat the house for 4 to 5 days at 30°C and then wash down the structure with disinfectant.

Mechanisms of Resistance

The genetic, biochemical and toxicological properties of resistance mechanisms in individual insects have only been studied closely in *Scirtothrips citri* and in *Frankliniella occidentalis*. Work on *Scirtothrips citri* followed the approach of using synergists, compounds that increase the toxicity of insecticides by inhibiting potential detoxifying mechanisms, to determine the relative contribution made by various enzyme systems in one strain resistant to the pyrethroid fluvalinate and to DDT, and in another resistant to the organophosphate dimethoate (Immaraju *et al.*, 1990a). Mixed function oxidases proved to be largely responsible for the resistance to fluvalinate, with a significant, though smaller contribution by esterases; the involvement of glutathione S-transferases was uncertain. The thrips strain selected by exposure to fluvalinate was cross-resistant to DDT (48-fold at LC_{90}), but this resistance was not synergized with piperonyl butoxide with added dichlorophenyl ethanol, suggesting that a target site insensitivity mechanism (knockdown resistance or kdr) was also involved. Further studies using enzyme assays (Ferrari *et al.*, 1993) also showed there to be higher levels of esterase activity and a higher degree of acetylcholinesterase insensitivity in some field populations resistant to organophosphates and carbamates than in either resistant laboratory strains or in a susceptible strain from Mexico. Although the differences may not be responsible for the observed levels of field resistance, they indicate that enzyme mechanisms which have been associated with resistance to these two classes of insecticide in other orders of insects are present in citrus thrips.

The eventual identification of resistance mechanisms in field populations should help in the choice of appropriate combinations of pesticides and synergists to retard the development of resistance to pyrethroids in certain thrips, and sesame seed oil seems a cheap alternative to piperonyl butoxide. However, the possible contribution of kdr to resistance is discouraging because, unlike metabolic resistance, it is still impossible to overcome using synergists (Sawicki *et al.*, 1984).

Permethrin treatments of *Frankliniella occidentalis* are also significantly synergized with piperonyl butoxide, indicating that mixed function oxidases are a component of the resistance to pyrethroids in this species as well (Immaraju *et al.*, 1992). The biochemical basis of resistance to organophosphates in western flower thrips has been examined in most detail for diazinon (Liu *et al.*, 1994; Zhao *et al.*, 1994). Penetration, metabolism and excretion of the pesticide proved more rapid in a resistant than in a susceptible strain, and the combination of rapid oxidative metabolism plus insensitive acetylcholinesterase conferred the resistance. The resistant strain tested had less carboxylesterase activity than the susceptible strain, but comparable acetylcholinesterase and butyrylcholinesterase activity; no difference in glutathione S-transferase activity between strains was detected. The role of butyrylcholinesterase was not established, but it may function as a scavenger enzyme to protect acetylcholinesterase from anticholinesterase insecticides. Incidentally, striking similarities were noted between the cholinesterases in western flower thrips and those occurring in aphids, suggesting closer phylogenetic affinities between thrips and aphids than are indicated by traditional systematics.

Robb *et al.* (1995) suggested that highly polyphagous species such as *Frankliniella occidentalis*, *Thrips tabaci* and *Thrips palmi* are presumably able to detoxify many types of plant toxins, a facility which might predispose them to survive exposure to other xenobiotic compounds such as insecticides. However, *Scirtothrips citri* with a much more limited host range seems just as capable of developing detoxifying mechanisms, suggesting that this is a broad family-wide ability rather than a relatively recently-acquired evolutionary response at species level. Species with a short life cycle and high fecundity conferring the ability to colonize growing crops rapidly may, nevertheless have the greatest potential to develop resistance. Its widespread presence in *T. tabaci*, *T. palmi* and *F. occidentalis*, all 'opportunistic' species (Mound and Teulon, 1995) that have been subject to intense selection pressure, supports this.

In *Thrips palmi*, the structure of the egg chorion has been suggested as a possible resistance mechanism against DDT by Rajula and Gowri (1988). They identified three distinct layers in the egg shell in susceptible strains, the middle layer is claimed to be rich in sulphydryl groups and the inner layer in protein-containing aromatics, whereas in resistant strains the middle layer incorporates disulphide groups and the inner layer disulphides and aromatics. Whether or not this confers some resistance, it is unlikely to be the sole or major mechanism involved because eggs are embedded in plant tissue so are largely protected from contact insecticides such as DDT.

Retarding development of resistance

The foregoing sections illustrate that sole reliance on chemicals for thrips control will almost inevitably lead to resistance, as has happened with hundreds of other crop pests (Georghiou, 1986). The most obvious way of delaying this is to use insecticides only when required, relying whenever possible on other forms of control within the framework of IPM to prevent crop damage. Unfortunately, such methods, especially routine monitoring to ascertain timing of treatments for maximum effect, are often more expensive and difficult to achieve than control with insecticides alone, but they do offer the most sustainable approach.

As well as using less insecticide, it is undesirable to introduce new chemicals where prolonged use of older materials has consistently given satisfactory control with no detectable resistance in the pest (Sawicki, 1981), as was the case with dimethoate used against citrus thrips from 1969 to 1981 (Morse and Brawner, 1986). It is also sensible to avoid accidental or deliberate treatment of refugia, non-host plants or non-vulnerable stages of crops, so that susceptible strains of thrips remain available to dilute the strains on crops exposed to strong selection pressure.

In practice, many resistance-retarding regimes are implemented to remedy a serious resistance problem rather than to delay or prevent its initial occurrence. This is unfortunate because after resistance becomes established, the benefits to be gained from such strategies are limited (Immaraju *et al.*, 1990b); it would be better to introduce them to prevent the problem arising at all. Procedures available to extend the useful impact of insecticides include the use of synergists (see above) and particularly the rotation or mixture of unrelated compounds. The relative merits of these two approaches against many non-thrips pests is controversial (Leeper *et al.*, 1986), but their respective merits for citrus thrips control were assessed directly in field trials in 1987 (Immaraju *et al.*, 1990b). Five spray regimes, each comprising five applications spaced at 2-week intervals, were compared. The best regime for retarding resistance over a single season was three fluvalinate (pyrethroid) treatments interspaced with two formentanate (carbamate) treatments. The next most successful was three formentanate treatments interspaced with two fluvalinate. Repeated sole applications of either insecticide produced resistance and mild cross-resistance to the other compound. Rotation was superior to the use of mixtures of these two products applied at half the recommended field rate for each insecticide. Whilst it is unwise to generalize on the relative merits of different treatment regimes on the basis of a single species, it seems likely from experiences with several other groups of pests that most long-term thrips problems would be best treated by rotating insecticides of different chemical types rather than by repeating and/or increasing doses of a single compound, a practice pursued, regrettably, on many glasshouse and high-value field crops. In Uzbekistan, attempts to combat serious resistance in *Thrips tabaci* on cotton grown in a cotton–lucerne rotation have been made by alternating organophosphate and pyrethroid insecticides (Deryabin, 1991). To control western flower thrips in glasshouses, Robb and Parrella (1995) recommend rotation of classes of compounds every 4–6 weeks, thereby exposing two or three generations only to the same type of insecticide.

As no insecticides marketed are thrips specific, their availability for thrips control will continue to be determined by the requirement to control pests in other taxa. Nevertheless, because of the high costs of pesticide production, research and registration, it is in the interests of the agricultural, horticultural and agrochemical industries to preserve the useful life of existing compounds by limiting their use against all pests. This raises problems for insecticide manufacturers who need to recoup investment in production costs by pressing for sales of new compounds, even though this tactic may jeopardize IPM systems which rely on the *status quo* of well-established control methods for their success (Sawicki, 1981).

Application Methods

Most compounds used to control thrips on outdoor food crops are applied by standard ground or air equipment as foliar sprays, which are often directed against other pests at the same time. Practical advice and guidance on safe and effective application methods is available in several handbooks (e.g. Anonymous 1989, 1991, 1992). In reality, most farmers use available machinery appropriate for the crop to obtain the quickest treatment, whatever pest species is present. For ground applications to field crops the coverage and penetration of flowers, buds and dense foliage necessary to reach thrips is best achieved by using low volumes and small droplet sizes. There is some evidence that electrostatic sprayers may be more effective than conventional machines, presumably because of the better coverage by the droplets of lower as well as upper surfaces of leaves. Cypermethrin, used to control *Enneothrips flavens* on peanuts in Brazil, was as effective when applied at 1.7 litre ha^{-1} with an electrohydrodynamic sprayer as when 280 litre ha^{-1} was applied by conventional manual machines (Galli and Arruda, 1989). The 'electrodyn' sprayer (Coffee, 1979) has added advantages for the small farmer in dry regions because it does not need water or preparation of the spray, so treatments could be applied rapidly after detection of an infestation. As a complete contrast, in fruit orchards where high volume spraying is still practical, wind-blower velocity should be regulated to ensure outside coverage of each tree (Haney *et al.*, 1992). In the 1980s Californian citrus growers were even advised how to modify weed sprayers as an alternative to conventional air blast sprayers for applying pesticides at about 3000 litre ha^{-1} against citrus thrips. Spraying by all methods should be confined to dry weather to ensure correct dilutions of pesticide.

Careful timing to coincide with pest immigration or emergence from eggs or soil is critical. If compounds with low residual activity are used, two applications may be needed to clear some crops because eggs laid in leaf tissue and pupae in the soil will be unaffected by the first spray. The spraying interval depends on the rate of development of the species and would need to be shorter the warmer the weather. Timing is also important for the control of virus infection and unless sprays are applied very soon after the first thrips are detected they are unlikely to be useful because infective adults require only 5 to 15 min feeding to transmit virus to a healthy plant (Sakimura, 1962) (see Chapter 14). Unless arriving vectors are killed before they feed and sparse populations are prevented from multiplying, initial infection will spread rapidly. There is some evidence that pesticides and repellents may increase shallow probing and dispersal by adults, thereby aggravating epidemics (German *et al.*, 1992).

The dusting of crops, common in the 1950s and 1960s (see above) has now been superseded, but on crops vulnerable as seedlings such as cotton, tobacco, peanuts and ornamentals, and on tree nursery stock, seed or soil treatments can be effective. As an illustration, seed treatments, in-furrow sprays and granules, and foliar sprays were compared against thrips on cotton, mainly *F. fusca* and *Neohydatothrips variabilis*, whose feeding curtails apical dominance causing excessive branching in seedlings (Burris *et al.*, 1989). Acephate, aldoxycarb and disulfoton performed well as seed treatments (2 ml of water plus insecticide solution per 450 g of seed); overall, aldicarb was the most consistent material as an in-furrow spray or granule, uptake of which is usually improved if the crop is irrigated (Matthews, 1994). Seed treatments normally provided adequate control for 2 to 3 weeks but sometimes needed supplementing with foliar sprays if weather retarded growth. In preliminary trials, fipronil applied in a polymer coating to leek seeds has given good protection against *Thrips tabaci*. In-furrow applications of insecticides as aqueous sprays or granules were normally effective for 4 to 6 weeks, but foliar sprays provided little residual activity and plants were often reinfested within a week. Other compounds often used elsewhere for seed treatments against cotton thrips include isofenphos, carbosulfan, furathiocarb, thiocarb and imidocloprid, with oxamyl, carbofuran, carbosulfan and diazinon formulated as granules (Bournier, 1994). Much of the heavy use of insecticides on cotton thrips, especially in the USA is successfully aimed at the prevention of stunting and distortion of the young seedlings, but though terminal buds may be severely damaged or destroyed in the first few weeks after emergence, control of thrips at this stage seldom increases eventual yields (Leigh, 1995). Similarly, Tappan and Gorbet (1981) reduced foliage injury by *F. fusca* on peanuts for up to 80 days after planting with broadcast, band and granular applications of systemic compounds, but yield increases were barely justified by the cost of treating for thrips alone.

Aerial spraying of large tracts of forests against thrips pests is impractical and environmentally undesirable. This was tried using carbaryl in kerosene at 4.6 litre ha^{-1} against pear thrips on sugar maple in Vermont to coincide with peak emergence of adults from soil in spring, but was not consistently effective (Teillon and Parker, 1991; Wilmot *et al.*, 1991). Subsequently, the use of carbaryl on sugar bushes has been banned because sugar maple is classified as a food crop and there is danger of pesticide drifting to nearby homes and properties. Cythion and acephate sprays are registered (1991) for control of the slash pine flower thrips (*Gnophothrips fuscus*) in slash pine seed orchards (Fatzinger and Dixon, 1991).

Systemic insecticides have been applied successfully to individual larch trees through the trunks to control *Thrips pini* attacking the young shoots. Heavy infestations on 14-year-old trees were controlled by applying demeton-S-methyl on a band of cellulose wadding tied round the scraped bark of the trunks, and covered with a plastic sheet to prevent evaporation. A 50% emulsion concentrate was used, diluted to contain 5% toxicant, and 250 ml of the diluent applied to each tree in summer. In early autumn the crowns and leading shoots of treated trees were healthy and showed little damage, whereas leading shoots on untreated trees were scarred and stunted (Martignoni and Zemp, 1956). The treatment was less effective the greater the girth of the tree, but was unaffected by tree height. Insecticide can also be injected into trees by hammering either a hollow cartridge or a hollow knife with openings in the cutting edge, into the trunk to a depth of 2 to 3 cm, and pouring it into the cavity of the tool (Vité, 1955, 1956). A 10% solution of demeton methyl sprayed directly on to the trunks of young trees before the summer generation of larch thrips hatched also killed the young nymphs when they emerged, but it was ineffective on old trees because the chemical took too long to reach the vulnerable shoots. On bananas in Queensland, *Chaetanaphothrips signipennis* and the incidence of 'rust' it causes on fruit can be reduced by applying systemic sprays (chlorpyrifos, pirimiphos-ethyl, prothiophos) to bases of plants, suckers and surrounding soil, or granules of ethoprophos to soil surrounding plant butts (Pinese, 1987).

In glasshouses, application methods are more specialized than in field or tree crops, and pesticide usage generally has to comply with stricter regulatory requirements. For example, on edible protected crops in the UK the only acceptable pesticide against *Frankliniella occidentalis* is dichlorvos, usually applied at high volume by pressurized hydraulic sprayers or as a fog with nevolin as a carrier. Formulations of dichlorvos which contain dibutylphthalate are unsuitable because they are phytotoxic. In The Netherlands, slightly less restrictive regulations permit sprays or fogs of dichlorvos, sprays of methiocarb (with massaral) or imidocloprid, abermectin (in propagation houses) and oxamyl through drip irrigation systems. On ornamentals in the UK all these compounds and malathion, or deltamethrin plus heptenophos, are used against *F. occidentalis* and applied as sprays, fogs, or as granular formulations broadcast or incorporated in compost. Other greenhouse species including *Thrips tabaci*, *Heliothrips haemorrhoidalis*, *Parthenothrips dracaenae* and *Hercinothrips femoralis* are much easier to control by a wider range of compounds including most pyrethroids, or nicotine, applied as sprays and fogs, or low volume mists of heptenophos. Aerosols of dichlorvos or naled vaporized from heating pipes have proved effective against thrips on ornamentals if used often enough.

'Thripstick', a formulation of deltamethrin in a sticky polybutene base can be used on floors or soil surfaces in glasshouses to intercept larvae as they fall off the plant to pupate. It is not intended to entangle the insects permanently, though it often does when freshly applied, but to ensure that a lethal dose is absorbed. Also, polyisobutylene plus emulsifier sprayed on to onions forms a non-phytotoxic film on the leaf surface which deters feeding by *Thrips tabaci* without causing them to stick (den Ouden *et al.*, 1987). Table 15.3 lists significant temperate glasshouse crops and the thrips species most commonly infesting them. Table 15.4 identifies the compounds best suited for control in given crops.

The potential to dislodge larvae of *Frankliniella occidentalis* from plants by exposing them to the species' alarm pheromone (decyl acetate and dodecyl acetate in a molar ratio of approximately 1.5:1) was investigated by Teerling *et al.* (1993). Unfortunately the response is weak, only 5 to 6 per cent of individuals falling from treated plants, so it is unlikely to improve the efficacy of 'Thripstick'. However, by formulating the pheromone with an insecticide and/or a fungal pathogen of thrips, an approach that has been tried successfully with aphids (Griffiths and Pickett, 1987), even the limited extra movement

Table 15.3. Temperate glasshouse crops and main associated thrips pests (after Scopes and Stables, 1989).

Crop	Thrips
Arum lily	*Frankliniella occidentalis* *Heliothrips haemorrhoidalis* *Thrips tabaci*
Asparagus fern	*T. tabaci*
Begonia	*F. occidentalis* *H. haemorrhoidalis* *H. femoralis* *T. tabaci*
Carnations	*T. tabaci* *Thrips atratus*
Chrysanthemum	*F. occidentalis*
Cucumber	*T. tabaci* *F. occidentalis*
Melon	*T. fuscipennis*
Cyclamen	*F. occidentalis* *H. haemorrhoidalis* *T. fuscipennis* *T. tabaci*
Orchids	*Anaphothrips orchidaceous* *F. occidentalis* *H. haemorrhoidalis*
Sweet peppers	*F. occidentalis* *T. tabaci*
Tomatoes	*T. tabaci*

Table 15.4. Compounds of proven effectiveness for thrips control in glasshouses (after Scopes and Stables, 1989).

Compound	Crop: Ornamentals	Crop: Cucumbers/ melons	Crop: Tomato
High volume sprays			
Carbaryl	✓	✓	✓
Cypermethrin	✓	✓	✓
Deltamethrin	✓	✓	✓
Diazinon	×	✓	✓
Gamma-HCH	✓	×	×
Heptenophos	✓	✓	✓
Malathion	✓	×	✓
Nicotine	✓	×	✓
Permethrin	✓	✓	✓
Atomizations			
Gamma-HCH	✓	×	✓
Malathion	✓	×	✓
Smokes (s) and fogs (f)			
Gamma-HCH (s)	✓	×	✓
Nicotine (s)*	✓	✓	✓
Permethrin (f)	✓	✓	✓
Permethrin (s)	✓	✓	✓
Pirimiphos-methyl (s)	✓	✓	✓
Pirimiphos-methyl (f)	✓	✓	✓
Tecnazene/gamma-HCH (s)	✓	×	✓
Compost treatments			
Aldicarb granules	✓	×	×

*Shreds

stimulated by the pheromone acting as a bioirritant may enhance pick-up of toxic agent or pathogen and thereby improve control. This approach would be worth investigating for other species in which the alarm pheromone may stimulate greater movement.

Postharvest fumigation has long been used to disinfest shipments of fresh produce for export. Traditionally methyl bromide has been favoured but this is dangerous to operators and phytoxic to some flowers. As an alternative, to clear carnations of *Frankliniella occidentalis*, Tjosvold and Ali (1995) tested naled but achieved only 80 to 90 per cent kill because some individuals remained inaccessible to the fumigant even after a 1-h exposure at a dose of 4.49 g a.i. 100 m^{-3}. Higher doses or longer exposure were no improvement; naled discoloured the petals of one out of ten cultivars (cv. 'Chianti').

Another post-production treatment now under investigation to disinfest flowers, and possibly fresh edible produce, is the use of controlled atmospheres with elevated CO_2 and reduced O_2 followed by anoxic conditions attained by flushing with pure N_2. As growing plants and cut flowers tolerate levels of CO_2 which are lethal to insects, it may be possible to develop this technique and eventually avoid the present need to produce crops virtually free of pests (Parrella, 1995).

Economics of Pesticide Usage

Despite extensive literature on the damage inflicted by pest species (see Chapters 1, 13, and 14) there are very few published assessments of the costs and financial returns of applying pesticides. The effect of infestations on produce range from the occasional total loss of high-value glasshouse ornamental crops on which there is no tolerance of even minor cosmetic injury, to undetectable effects on yield in field crops grown for processing, even when the non-harvestable foliage may have been heavily infested during early growth.

In The Netherlands, aerial spraying of flax infested with *Thrips angusticeps* allegedly increased profits by US\$146 ha^{-1} (Franssen and Kerssen, 1962) and Post and McBride (1966) claimed that barley growers in North Dakota could increase profits by US\$5 ha^{-1} by spraying against *Limothrips denticornis*. Savings on control costs in field crops can often be made by applying insecticides with other routine treatments such as herbicides. For example, when demeton was added to applications of 2,4-D herbicide on cereals in Bulgaria, adult *Haplothrips tritici* were controlled, but it is doubtful whether the costs of spraying against the thrips alone would have been worthwhile (Lyubenov, 1961). Smith and Sams (1977) thought that control directed specifically at thrips attacking peanuts in the USA was rarely profitable, and although Tappan and Gorbet (1981) obtained greater net returns in each of 3 years by applying systemics to control *Frankliniella fusca* on this crop, the gains were not reliably correlated with pesticide use; they recommended that any decision to use pesticides in peanut production should not depend entirely on the prospect of thrips control. Similarly, Leigh (1995) thought there was little return from spraying cotton seedlings in California, but in Louisiana, control of infestation during the 4 weeks after seedling emergence eventually produced yield increases of 10 per cent (Burris *et al.*, 1989).

In Californian citrus crops, the total cost of pesticides for thrips control alone reached almost US\$12 million in 1990, equivalent to US\$116 ha^{-1} excluding application costs (Morse, 1995). Such expenditure is considered justifiable by growers to prevent scarring, particularly of small fruit, which may advance maturation and cause water loss. Unblemished fruit attracts a US\$2 bonus per carton, so prevention of this damage is a major factor affecting profitability. Nevertheless, the maintenance of such high, largely cosmetic standards results in extremely low economic thresholds in the month before petal fall, and as described earlier, the intensive spray programmes that ensue are gradually becoming counterproductive (Morse and Brawner, 1986).

Late preharvest applications to fresh edible crops must be balanced against the dangers of exceeding maximum residue levels (MRLs) and the resulting unmarketability of produce. Table 15.5 shows the widely variable application rates and minimum harvest interval for eight compounds commonly used against pea thrips, *Kakothrips pisivorus*, in the UK emphasizing the importance of complying with label instructions. Acceptable residue levels are usually strict; for example in New Zealand the MRL for fluvalinate used to control *Thrips obscuratus* on peaches is nil at harvest; in France there is a 7-day withholding period for peaches contaminated with 0.1 ppm of this compound (Teulon, 1988), regulations which could each lead to rejection of consignments.

Pesticide contamination is less of a problem in non-edible crops, such as pyrethrum, in which the economic injury level caused by *Thrips nigropilosus* appears to be very low and justifies insecticidal control virtually as soon as one individual adult or larva is detected on the leaves. Infestations reduce the number of flowers per plant rather than affecting the pyrethrum content of seeds, but overall yield loss of up to 43 per cent can occur. Average monetary losses for three cultivars were calculated to be US\$284 ha^{-1} in a crop potentially capable of producing returns exceeding US\$1000 ha^{-1}. Control of this pest with dimethoate costs only US\$34 ha^{-1} so it would clearly pay to spray. However, on small-scale farms in Kenya where this is an important crop, farmers would need training to recognize thrips damage and to differentiate it from drought effects apparent in the dry season, and also to have available a safe, inexpensive insecticide that could be applied without a backpack sprayer (Smith and Hanson, 1991). Thrips control in specialized grass seed production may also be profitable. As long ago as 1948, yields of Bermuda grass seed in Arizona were increased 12-fold, from about 40 to 500 kg ha^{-1}, by controlling *Chirothrips* spp. with parathion dust. The extra financial returns were about 70 times the cost of insecticide (Roney, 1949).

Another drawback to some pesticides alluded to above is their phytotoxicity to many glasshouse ornamentals. For example, very few cultivars of chrysanthemums can tolerate dimethoate or formothion sprays; malathion

Table 15.5. Sprays for *Kakothrips pisivorus* used in the UK (after Scopes and Stables, 1989).

Insecticide	Application rate ($g\ ha^{-1}$)	Minimum harvest interval (days)
Azinphos-methyl	308	14
Carbaryl	1700	2
Cypermethrin	75	0
Demeton-S-methyl	240	21
Dimethoate	340	2
Fenitrothion	700	14
Fenvalerate	25	0
Nicotine	1000	2

sprays or dicofol smokes should not be used on open blooms. Many other compounds are harmful to just one or a few cultivars (Scopes and Stables, 1989).

The overall impression of chemical control of thrips in the majority of field and glasshouse crops grown for food or fibre, and in plantation crops and forestry, is that as a sole approach to control it is becoming either less effective or less acceptable environmentally or to the public. The future for these valuable materials lies in their timely but restricted use as part of IPM systems described in Chapters 16, 17 and 18.

References

Anonymous (1963) *An Atlas of Coffee Pests and Diseases.* Coffee Board, Nairobi, Kenya, 146 pp.

Anonymous (1989) *Hand-operated Sprayers Handbook.* British Crop Protection Council, Farnham, UK, 44 pp.

Anonymous (1991) *Boom Sprayers Handbook.* British Crop Protection Council, Farnham, UK, 60 pp.

Anonymous (1992). *Fruit Sprayers Handbook.* British Crop Protection Council, Farnham, UK, 51 pp.

Ascher, K.R.S., Klein, M. and Meisner, J. (1992) Azatin, a neem formulation, acts on nymphs of the western flower thrips. *Phytoparasitica* 20, 305–306.

Ascher, K.R.S., Klein, M., Meisner, J., Ben-Moshe, E. and Caspy, I. (1993). Neemazal, an insecticidal formulation from neem seed kernel enriched with azadirachtin, controls nymphs of the onion thrips, *Thrips tabaci*, on systemic application. (In Hebrew with English summary.) *Hassadeh*, 73, 1262–1265.

Bailey, S.F. (1944). The pear thrips in California. *California Agriculture Experiment Station Bulletin* 687, 1–55.

Bansiddhi, K. and Poonchaisri, S. (1991) Thrips of vegetables and other commercially important crops in Thailand. In: Talekar, N.S. (ed.) *Thrips in Southeast Asia.* Asian Vegetable Research and Development Center, AVRDC, Publication No. 91-342, pp. 34–39.

Bernardo, E.N. (1991) Thrips on vegetable crops in the Philippines. In: Talekar, N.S. (ed.) *Thrips in Southeast Asia.* Asian Vegetable Research and Development Center, AVRDC, Publication No. 91–342, pp. 5–11.

Boulanger, L.W. and Abdalla, D.A. (1966) Control of blueberry thrips in Maine. *Journal of Economic Entomology* 59, 1007–1008.

Bournier, J.P. (1994) Thysanoptera. In: Matthews, G.A. and Tunstall, J.P. (eds) *Insect Pests of Cotton.* CAB International Wallingford, UK, pp. 381–391.

Brødsgaard, H.F. (1994) Insecticide resistance in European and African strains of western flower thrips (Thysanoptera: Thripidae) tested in a new residue-on-glass test. *Journal of Economic Entomology* 87, 1141–1146.

Burris, E., Ratchford, K.J., Pavloff, A.M., Boquet, D.J., Williams, B.R. and Rogers, R.L. (1989) Thrips on seedling cotton: related problems and control. *Louisana Agricultural Experiment Station Bulletin* 811, 19 pp.

Chang, N.T. (1991) Important thrips species in Taiwan. In: Talekar, N.S. (ed.) *Thrips in Southeast Asia.* Asian Vegetable Research and Development Center, AVRDC, Publication No. 91–342 pp. 40–56.

Coffee, R.A. (1979) Electrodynamic energy – a new approach to pesticide application.

In: *Proceedings of the 1979 Brighton Crop Protection Conference, Pests and Diseases*. British Crop Protection Council, Farnham, UK, pp. 883–889.

den Ouden, H., Theunissen, J. and Shelton, A.M. (1987) Prevention of plant injury by gall midge (*Contarinia nasturtii* Kieffer) and onion thrips (*Thrips tabaci* Lindemann) using emulsions of polyisobutylene. *Journal of Applied Entomology* 104, 313–318.

Deryabin, V.I. (1991) Rotation of insecticides to overcome resistance. *Zashcihita Rastenii Moskva* 3, 45–46.

Dogger, J.R. (1956) Thrips control on peanuts with granulated insecticides. *Journal of Economic Entomology* 49, 632–635.

Elbert, A., Overbeck, H., Iwaya, K. and Tsuboi, S. (1990) Imidocloprid, a novel systemic nitromethylene analogue insecticide for crop protection. In: *Proceedings of the 1990 Brighton Crop Protection Conference, Pests and Diseases*. British Crop Protection Council, Farnham, UK, pp. 21–28.

Elmer, H.S., Brawner, O.L. and Morse, J.G. (1983) Modifying weed sprayers for citrus thrips control. *California Agriculture* March–April, 15–16.

Etienne, J., Guyot, J. and Van Waetermeulen, X. (1990) Effect of insecticides, predation and precipitation on populations of *Thrips palmi* on aubergine (eggplant) in Guadeloupe. *Florida Entomologist* 73, 339.

Evans, D.E. (1967) Insecticide field trials against the coffee thrips (*Diarthrothrips coffeae* Williams) in Kenya. *Turrialba* 17, 376–380.

Fatzinger, C.W. and Dixon, W.N. (1991) Development of sampling methods for the slash pine flower thrips *Gnophothrips fuscus* (Morgan), (Thysanoptera: Phlaeothripidae). In: Parker, B.L., Skinner, M. and Lewis, T. (eds) *Towards Understanding Thysanoptera*, General Technical Report NE-147, USDA Forest Service, Northeastern Forest Experiment Station, Radnor, PA, pp. 149–161.

Ferrari, J.A., Morse, J.G., Georghiou, G.P. and Sun, Y. (1993) Elevated esterase activity and acetylcholinesterase insensitivity in citrus thrips (Thysanoptera: Thripidae) populations from the San Joaquin Valley of California. *Journal of Economic Entomology* 86, 1645–1652.

Franssen, C.J.H. and Huisman, P. (1958) The biology and control of *Thrips angusticeps* Uzel. *Verslag Landbouwkundig Onderzoek Rijkslandb Proefstn* 64, 104 pp.

Franssen, C.J.H. and Kerssen, M.C. (1962) Aerial control of thrips on flax in the Netherlands. *Agricultural Aviation* 4, 50–54.

Galli, J.C. and Arruda, A.C. (1989) Aplicacao de cypemethrin 30ED em controle experimental de *Enneothrips flavens* Moulton, 1941 (Thysanoptera, Thripidae) em ultra baixo volume em cultivo de amendoim. *Revista de Agricultura Piracicaba* 64, 21–34.

Georghiou, G.P. (1986) The magnitude of the resistance problem. In: *Pesticide Resistance: Strategies and Tactics for Management*. National Academy of Sciences, Washington, DC, pp. 14–43.

German, T.L., Ullman, D.E. and Moyer, J.W. (1992) Tospoviruses: diagnosis, molecular biology, phylogeny and vector relationships. *Annual Review of Phytopathology* 30, 315–348.

Griffiths, D.C. and Pickett, J.A. (1987) Novel chemicals and their formulation for aphid control. In: Ping, I.L. and Leonhardt, B. (eds) *Proceedings of the 14th International Symposium on Controlled Release of Bioactive Materials*. Controlled Release Society Inc., Toronto, pp. 243–244.

Hale, R.L. and Shorey, H.H. (1965) Systemic insecticides for the control of western flower thrips on onion bulbs. *Journal of Economic Entomology* 58, 793–794.

Haney, P.B., Morse, J.G., Luck, R.F., Griffiths, H., Grafton-Cardwell, E.E. and

O'Connell, N.V. (1992) *Reducing Insecticide Use and Energy Costs in Citrus Pest Management.* University of California Statewide IPM Project Publication 15, Davis, CA, 62 pp.
Hanna, R.L. (1958) Insecticidal seed treatments for cotton. *Journal of Economic Entomology* 51, 160–163.
Helyer, N.L. and Brobyn, P.J. (1992) Chemical control of western flower thrips (*Frankliniella occidentalis* Pergande). *Annals of Applied Biology* 121, 219–231.
Heugens, A., Buysse, G. and Vermaerke, D. (1989) Control of *Frankliniella occidentalis* on *Chrysanthemum indicum* with pesticides. (i). *Mededelingen van de Faculteit Landbouwwetenschappen* 54, 975–981.
Hinds, W.E. (1902) Contribution to a monograph of the insects of the order Thysanoptera inhabiting North America. *Proceedings of the US National Museum* 26, 79–247.
Hirose, Y. (1991) Pest status and biological control of *Thrips palmi* in Southeast Asia. In: Talekar, N.S. (ed.) *Thrips in Southeast Asia.* Asian Vegetable Research and Development Centre, AVRDC, Publication No. 91–342, pp. 57–60.
Hodgson, C.J. (1970) Pests of citrus and their control. *Pest Articles and News Summaries (PANS)* 16, 647–666.
Kubota, S. (1989) Evaluation of the effect of some chitin synthesis inhibitors against *Thrips palmi* Karny (Thysanoptera: Thripidae) infesting musk melons. *Applied Entomology and Zoology* 24, 349–357.
Immaraju, J.A. and Morse, J.G. (1990) Selection for pyrethroid resistance, reversion and cross-resistance with citrus thrips (Thysanoptera: Thripidae). *Journal of Economic Entomology* 83, 698–704.
Immaraju, J.A., Morse, J.G. and Kersten, D.J. (1989) Citrus thrips (Thysanoptera: Thripidae) pesticide resistance in the Coachella and San Joaquin Valleys of California. *Journal of Economic Entomology* 82, 374–380.
Immaraju, J.A., Morse, J.G. and Gaston, L.K. (1990a) Mechanisms of organophosphate, pyrethroid, and DDT resistance on citrus thrips (Thysanoptera: Thripidae). *Journal of Economic Entomology* 83, 1723–1732.
Immaraju, J.A., Morse, J.G. and Hobza, R.F. (1990b) Field evaluation of insecticide rotation and mixtures as strategies for citrus thrips (Thysanoptera: Thripidae) resistance management in California. *Journal of Economic Entomology* 82, 306–314.
Immaraju, J.A., Paine, T.D., Bethte, J.A., Robb, K.L. and Newman, J.P. (1992) Western flower thrips (Thysanoptera: Thripidae) resistance to insecticides in coastal California greenhouses. *Journal of Economic Entomology* 85, 9–14.
Ivancheva-Gabrovska, T. (1959) Tomato spotted wilt (*Lycopersicum* virus 3 Smith) on tobacco in Bulgaria. (In Bulgarian.) *Review of Applied Entomology (A)* 49, 339–340.
Leeper, J.R., Roush, R. T. and Reynolds, H.T. (1986) Preventing or managing resistance in arthropods. In: *Pesticide Resistance: Strategies and Tactics for Management.* Natural Academy of Sciences, Washington, DC, pp. 335–346.
Leigh, T.F. (1995) Bionomics of cotton pests: a review. In: Parker, B.C., Skinner, M. and Lewis, T. (eds) *Thrips Biology and Management.* Plenum Press, New York, pp. 61–70.
Lincoln, C. and Leigh, T.F. (1957) Timing insecticide applications for cotton insect control. *Bulletin Arkansas Agricultural Experiment Station* 588, 47 pp.
Liu, W., Zhao, G. and Knowles, C.O. (1994) Substrate specificity and inhibitor sensitivity of cholinesterases in homogenates of western flower thrips. *Pesticide Biochemistry and Physiology* 49, 121–131.

Lyubenov, Y. (1961) A contribution to the bionomics of the wheat thrips *(Haplothrips tritici* Kurd.) in Bulgaria and possibilities of reducing the injury done by it. *Review of Applied Entomology (A)* 50, 348.

Macdonald, O.C. (1995) Response of western flower thrips to dichlorvos and malathion in the United Kingdom. In: Parker, B.L., Skinner, M. and Lewis, T. (eds) *Thrips Biology and Management*. Plenum Press, New York, pp. 347–350.

MacGillivray, J.H., Michelbacher, A.E. and Scott, C.E. (1950) Tomato production in California. *Circular of the California Agricultural Experiment Station Extension Service* 167.

MacPhee, A.W. (1953) The influence of spray programs on the fauna of apple orchards in Nova Scotia V. The predacious thrips *Haplothrips faurei* Hood. *Canadian Entomologist* 85, 33–40.

Martignoni, M.E. and Zemp, H. (1956) Versuch zur Bekämpfung des Larchenblasenfusses *Taeniothrips laricivorus* Kratochvil and Farsky (Thripidae, Thysanoptera) mit einem systemischen Insectizid. *Mittelungen der Schweizerischen Anstalt für das forstliche Versuchswesen* 32, 1–22.

Matthews, G.A. (1994) Chemical control. In: Matthews, G.A. and Tunstall, J.P. (eds) *Insect Pests of Cotton*. CAB International, Wallingford, UK, pp. 535–557.

Mistric, W.J. and Spyhalski, E.J. (1959) Response of cotton and cotton pests to thimet seed treatment. *Journal of Economic Entomology* 52, 807–811.

Morse, J.G. (1995) IPM of citrus thrips in California. In: Parker, B.L., Skinner, M. and Lewis, T. (eds) *Thrips Biology and Management*. Plenum Press, New York pp. 371–379.

Morse, J.G. and Brawner, O.L. (1986) Toxicity of pesticides to *Scirtothrips citri* (Thysanoptera: Thripidae) and implications to resistance management. *Journal of Economic Entomology* 79, 565–570.

Morse, J.G. and Zareh, N. (1991) Pesticide induced hormoligosis of citrus thrips (Thysanoptera: Thripidae) fecundity. *Journal of Economic Entomology* 84, 1169–1174.

Mound, L.A. and Teulon, D.A.J. (1995) Thysanoptera as phytophagous opportunists. In: Parker, B.L., Skinner, M. and Lewis, T. (eds) *Thrips Biology and Management*. Plenum Press, New York, pp. 3–19.

Moustafa, O.K. and El-Attal, Z.M. (1985) Enhancement of the efficiency of some insecticides against thrips and cotton leaf worm by mineral oils. *Journal of Agricultural Science* 105, 63–66.

Nagai, K. (1990) Effects of a juvenile hormone mimic material, 4 phenoxyphenyl (*RS*)-2-(2-pyridyloxy) propyl ether, on *Thrips palmi* Karny (Thysanoptera: Thripidae) and its predator *Orius* sp. (Hemiptera: Anthocoridae). *Applied Entomology and Zoology* 25, 199–204.

Nagai, K., Hiramatsu, T. and Henmi, T. (1988) Effect of flufenoxuron (benzoylphenylurea) on *Thrips palmi* Karny (Thysanoptera: Thripidae) (In Japanese, with English summary.) *Japanese Journal of Applied Entomology and Zoology* 32, 297–299.

Parrella, M. (1995) IPM – Approaches and prospects. In: Parker, B.L., Skinner, M. and Lewis, T. (eds) *Thrips Biology and Management*. Plenum, New York, pp. 357–363.

Persing, C.O., Barnhart, C.S., Worthy, W.L. and Boyce, A.M. (1942) The resistance of citrus thrips to tartar emetic treatment in San Fernando Valley. *California Citrus Grower* 28, 24–25.

Pfrimmer, T.R. (1966) Systemic insecticides for cotton insect control in 1965. *Journal of Economic Entomology* 59, 1113–1118.

Pillai, M.A.K. and Ponniah, S. (1988) Neem for control of rice thrips. *International Rice Research Newsletter* 13, 33–34.

Pinese, B. (1987) Soil and bunch applications of insecticides for control of the banana rust thrips. *Queensland Journal of Agricultural and Animal Sciences* 44, 107–111.

Post, R.L. and McBride, D.K. (1966) Barley thrips: biology and control. *Circular of the North Dakota State University Extension Service* A292.

Prudent, P. (1990) Efficacite de divers insecticides appliques en traitement des semences, contre les thrips du cotonnier au Paraquay. *Rapport annuel IRCT* (unpublished).

Race, S.R. (1961) Early season thrips control on cotton in New Mexico. *Journal of Economic Entomology* 36, 493–500.

Rajulu, G.S. and Gowri, N. (1988) A comparative study of the chorion of the DDT-resistant and susceptible strains of *Thrips palmi* Karny. *Acta Phytopathologica et Entomologica Hungarica* 23, 331–341.

Rao, G.N. (1970) Tea pests in southern India and their control. *Pest Articles and News Summary* (*PANS*) 16, 667–672.

Richardson, B.H. (1957) Control of onion thrips in the winter garden area of Texas. *Journal of Economic Entomology* 46, 92–95.

Richardson, B.H. and Wene, G.P. (1956) Control of onion thrips and its tolerance to certain chlorinated hydrocarbons. *Journal of Economic Entomology* 49, 333–335.

Riherd, P.T. (1954) Thrips as limiting factors in grass seed production. *Journal of Economic Entomology* 47, 709–719.

Robb, K.L. (1989) Analysis of *Frankliniella occidentalis* (Pergande) as a pest of floricultural crops in California greenhouses. PhD thesis, University of California, Riverside, 135 pp.

Robb, K.L. and Parrella, M.P. (1995) IPM of western flower thrips. In: Parker, B.L., Skinner, M. and Lewis, T. (eds) *Thrips Biology and Management.* Plenum Press, New York, pp. 365–370.

Robb, K.L., Newman, J., Virzi, J.K. and Parrella, M.P. (1995) Insecticide resistance in western flower thrips. In: Parker, B.L., Skinner, M., and Lewis, T. (eds) *Thrips Biology and Management.* Plenum Press, New York, pp. 341–346.

Rogoff, W.M. (1952) The repellency of chlordane, DDT and other residual insecticides to greenhouse thrips. *Journal of Economic Entomology* 45, 1065–1071.

Roney, J.N. (1949) Bermuda seed grass insects in Arizona. *Journal of Economic Entomology* 42, 555.

Sakimura, K. (1962) The present status of thrips-borne viruses. In: Maramorosch, K. (ed.) *Biological Transmission of Disease Agents.* Academic Press, New York, pp. 33–40.

Sastrosiswojo, S. (1991) Thrips on vegetables in Indonesia. In: Talekar, N.S. (ed.) *Thrips in Southeast Asia.* Asian Vegetable Research and Development Center, AVRDC, Publication No. 91-342, pp. 12–17.

Sathiyanandam, V.K.R., Venugopal, M.S., Logiswaran, G., Babu, P.C.B. and Gopalan, M. (1987) Efficacy of seed treatment against thrips. *Rice Research Newsletter – CRRI* 8, 2–3.

Sawicki, R.M. (1981) Problems in countering resistance. *Philosophical Transactions of the Royal Society of London, B* 195, 143–151.

Sawicki, R.M., Farnham, A.W., O'Dell, K. and Denholm, I. (1984) Factors affecting resistance to insecticides in houseflies (*Musca domestica* L.). 3. Relationships between the level of resistance to pyrethroids, control failure and the frequency of the gene *kdr*. *Bulletin of Entomological Research* 74, 581–589.

Scopes, N. and Stables, L. (1989) *Pest and Disease Control Handbook.* British Crop Protection Council, Thornton Heath, UK, 732 pp.

Shazli, A. and Zazou, M.H. (1959) Control of onion thrips by surface and systemic insecticides. *Bulletin of the Entomological Society of Egypt* 43, 185–191.

Smith, L.M. and Hanson, P.M. (1991) Yield reduction in pyrethrum caused by *Thrips nigropilosus* Uzel. (Thysanoptera, Thripidae) *Tropical Agriculture (Trinidad)* 68, 235–238.

Smith, J.W. and Sams, R.L. (1977) Economics of thrips control on peanuts in Texas. *Southwestern Entomologist* 2, 149–154.

Talekar, N.S. (ed.) (1991) *Thrips in Southeast Asia.* Asian Vegetable Research and Development Center, AVRDC Publication No 91–342, 74 pp.

Tapley, R.G. (1960) Coffee leaf miner epidemics in relation to the use of persistent insecticides. *Annual Report of the Coffee Research Station, Lyamungu, Tanganyika*, pp. 43–55.

Tappan, W.B. and Gorbet, D.W. (1981) Economics of tobacco thrips, *Frankliniella fusca* control with systemic pesticides, aldicarb, carbofuran, disulfoton, on Florunner peanuts in Florida. *Journal of Economic Entomology* 74, 283–286.

Teerling, C.R., Pierce, H.D., Borden, J.H. and Gillespie, D.R. (1993) Identification and bioactivity of alarm pheromone in the western flower thrips, *Frankliniella occidentalis*. *Journal of Chemical Ecology* 19, 681–697.

Teillon, H.B. and Parker, B.L. (1991) Aerial spray trials for pear thrips management, fall 1988. In: Parker, B.L., Skinner, M. and Lewis, T. (eds) *Towards Understanding Thysanoptera*, General Technical Report NE-147. USDA Forest Service, Northeastern Forest Experiment Station, Radnor, PA, pp. 231–240.

Teulon, D.A.J. (1988) Pest management of the New Zealand flower thrips, *Thrips obscuratus* (Crawford) (Thysanoptera: Thripidae) on stonefruit in Canterbury, New Zealand. PhD thesis, Lincoln College, University of Canterbury, New Zealand.

Tomlin, C. (ed.) (1994) *The Pesticide Manual*, 10th edn. British Crop Protection Council, Farnham, UK and Royal Society of Chemistry, Cambridge, 1341 pp.

Tjosvold, S.A. and Ali, A.D. (1995) Effect of postharvest Naled and Sulfotep fumigation on western flower thrips infesting carnation. In: Parker, B.L., Skinner, M. and Lewis, T. (eds) *Thrips Biology and Management*. Plenum Press, New York, pp. 351–353.

Vitê, J.P. (1955) Attempts at simultaneous control of thrips and larch case bearers: its fundamental principles and prospects. *Hofchen-Breife, Bayer Pflanzenschutz Nachrichten* 1, 34–52.

Vitê, J.P. (1956) Populationsstudien am Lärchenblassenfuss *Taeniothrips laricivorus* Krat. *Zeitschrift für angewandte Entomologie* 38, 417–488.

Whitehead, R. (ed.) (1995) *The UK Pesticide Guide*, 8th edn. CAB International and British Crop Protection Council, Farnham, UK.

Wilmot, S.H., Teillon, H.B. and Parker, B.L. (1991) The use of carbaryl for management of pear thrips in sugarbushes. In: Parker, B.L. and Doane, E.N. (eds) The 1990 Conference of Thysanoptera, *University of Vermont, Burlington. Vermont Agricultural Experiment Station Bulletin* 698, pp. 47.

Zhao, G., Liu, W. and Knowls, C.O. (1994) Mechanisms associated with diazinon resistance in western flower thrips. *Pesticide Biochemistry and Physiology* 49, 13–23.

16

Integrated Pest Management (IPM) in Field Crops

Michael P. Parrella[1] and Trevor Lewis[2]

[1]*Department of Entomology, University of California, Davis, California 95616-8584, USA;* [2]*Institute of Arable Crops Research – Rothamsted, Harpenden, Herts AL5 2JQ, UK*

Introduction

Thrips infestations on field crops have become more important over the past 40 years. Often considered only as incidental or secondary pests, today they are responsible for causing considerable damage to many field crops including cotton, cabbage, oilseed rape, peas, beans, lucerne, peanuts, onions, cucumbers, melons and chrysanthemum (grown for pyrethrum production). The recent worldwide redistribution of *Frankliniella occidentalis* and the tospoviruses it vectors is one reason for the enhanced pest status of thrips in field crops, but this is not a complete explanation because a far wider range of thrips species attacking many different agricultural crops are now considered major pests.

Damage inflicted varies with the stage of crop development. Seedlings are usually the most sensitive though they often survive and outgrow attack, but feeding injury can be serious at any time depending on the density of thrips and whether or not virus transmission is involved. In reviewing the considerable literature on thrips as pests of field crops, one major consistency appears: it is critical to understand the regional population dynamics of the pest thrips in the affected and surrounding crops and weeds. Migration of pests into an agricultural crop can disrupt the most refined management strategies. Dealing with this phenomenon is particularly difficult with thrips because of their great tendency to migrate on prevailing winds and storm fronts, their small size, the difficult nature of quantifying airborne populations, and the large numbers that may move into a field over a short time (see Chapter 5). Estimates of damage by thrips to field crops abound (see Chapter 1, Table 1.2); for example, in Georgia several species were listed as serious pests on tomatoes, snap beans, peppers, watermelons and southern peas with an estimated yearly loss of more than $US20 million (McPherson and Douce, 1992).

Pest management programmes are built on a number of principles including: (i) the biology of the thrips on the crop and the damage caused, including economic thresholds; (ii) the spatial distribution within and between plants which determines the development of efficient sampling schemes (see Chapter 11); (iii) the influence of host plant resistance on thrips biology; (iv) the importance of natural enemies in regulating thrips populations; (v) the utility of cultural and physical control options; and (vi) the utility of chemical control strategies. The following sections provide more information in each of these categories building on the biology, population dynamics and control options described in previous chapters. The intent is not to provide an exhaustive review of all literature, but rather to focus on those citations that best illustrate the most important aspects. Attention is also drawn to some of the more traditional approaches to thrips control, the value of which has been appreciated anew as problems have arisen with excessive pesticide usage.

Biology and Ecology in Field Crops

The basic biology of only a few thrips pests infesting field crops has been studied in depth. These include *Thrips tabaci* on onions (Waterhouse and Norris, 1989), *Thrips angusticeps* on flax, cereals and peas (Franssen and Huisman, 1958), *Limothrips cerealium* on wheat (Lewis, 1973), *Frankliniella intonsa* on tomatoes (Murai, 1988), *Frankliniella* spp. on peanuts (Lowry *et al.*, 1992) and an extended study of *Thrips palmi* on cucurbits and field-grown vegetables (e.g. Kawai, 1986, 1990). Such studies emphasize the importance of weather, soil type and susceptibility of host plants on the development of large thrips populations. For example, Bernardo (1991) found that *Thrips palmi* had the highest fecundity and longest adult survival on watermelon. He speculated that this could be responsible for the large populations and the severe damage to this crop. Kawai (1986) found differences between the fecundity of females feeding on cucumbers, melons, aubergines and pumpkins. Although the generation times did not differ between crops, there were significant differences in net reproductive rate on each host and hence on the intrinsic rate of natural increase in the respective populations; consequently cucumbers and melons are more vulnerable than aubergines or pumpkins.

A number of studies have examined the biology/ecology of thrips in a variety of field cropping systems focusing on the damage they cause, the mechanism of that damage, and the interaction with crop development, abiotic factors, other crops and weedy habitats. For example, early season cotton was found to be very sensitive to injury by *Frankliniella* spp. (Terry and Barstow, 1988) and weather strongly influenced the susceptibility of cotton plants to injury. Cool, wet weather that slowed down the development of cotton seedlings made them more sensitive to thrips injury (Rummel and Hatfield, 1988). Using several sampling techniques specifically designed to monitor early season populations of *F. fusca* on seedling cotton, Burris *et al.* (1990) found that adult and immature thrips peaked 20 and 28 days after cotton was planted in Louisiana. However, despite severe damage to early growth, yield reduction in cotton may be slight or undetectable (Sadras and Wilson, 1997). The short-term effects of weather on the development of crop infestations, especially rain, drought, cold and heat are described for many other species in Chapter 7.

Many harmful thrips can survive on weeds growing in and around fields in the absence of a suitable crop, and poor crop hygiene and weed control around such crops may lead to heavy infestations, severe direct damage and virus infection. It is important to distinguish between true host plants on which thrips breed and the many plants on which migratory individuals temporarily alight (see Chapter 6). Nevertheless, many of the more serious pests are polyphagous, so weed control helps to lessen infestations. Wild grasses in areas that produce grass seed provide a permanent and uncontrollable source of *Chirothrips* spp., so host grasses and volunteer plants on which thrips can hibernate within fields should be destroyed.

Understanding the ability of thrips to develop in adjacent crops or fields and then to migrate into susceptible crops is of paramount importance in devising IPM programmes. For species with a wide host range the probability of large infestations developing on a young crop can sometimes be predicted by examining populations on older, different crops nearby. For example, when populations of *Frankliniella occidentalis* were sparse in onion, lucerne and lettuce fields bordering cotton fields that were soon to be planted, the emerging cotton seedlings were likewise lightly infested when they eventually appeared (Race, 1965).

One of the most extensively studied thrips–crop systems has been *Thrips tabaci* infesting cabbage in New York State. North and Shelton (1986a) found that although there were many thrips in cabbage fields during certain periods, most were not *T. tabaci* and need not concern cabbage growers. Clearly, correct identification of thrips is vital to an effective management system. In cabbage, North and Shelton (1986a) found that thrips species composition and phenology varied dramatically through the season and was related to phenology and management of adjacent field and forage crops. By sampling thrips populations in the foliage and utilizing sticky traps to monitor aerial populations, they had some success in predicting the population build-up in adjacent crops and ultimately the timing and size of migration into cabbage.

The importance of overwintering populations in non-cropping areas adjacent to susceptible crops has proved to be important for a wide range of species (North and Shelton, 1986b; Chambers and Sites, 1989; Nagai and Tsumuki, 1990; Chamberlin *et al.*, 1992). Migration of thrips from these sites as well as from adjacent fields or crops where large thrips populations have developed is frequently cited as a major cause of thrips problems in sensitive crops (North and Shelton, 1986b; Weiss and Beshear, 1987; Tamò *et al.*, 1993; Bautista *et al.*, 1996). For a few species the overwintering situation is complicated by the ability of a proportion of the population to remain in diapause in the soil for more than 1 year. For example, in parts of northern Europe, 5 to 10 per cent of larvae of *Thrips angusticeps* may persist in soil beneath a previously infested crop (e.g. flax, peas, cereals) for two winters, to emerge as brachypterous adults in spring (Franssen and Huisman, 1958). This makes the choice of likely pest-free sites for susceptible crops extremely difficult. In 1996 extensive areas planted with sugar beet in eastern England had to be resown in spring due to the unexpected emergence of this thrips in cool, dry weather that was unsuitable for the growth of seedlings, which succumbed to attack. The species disperses widely in summer when a macropterous generation is produced, but then vigorous crop growth masks any damage they may inflict. The offspring of this migratory generation overwinter so it is difficult to predict fields harbouring potentially harmful populations before sowing.

Life-fertility tables generated for *Frankliniella occidentalis* and *F. fusca* on peanuts (Lowry *et al.*, 1992) demonstrated that the latter was the more serious pest of this crop and this coincides with field studies in the south-eastern USA on peanuts which showed that *F. fusca* dominated and was the principal vector of viruses (Todd *et al.*, 1996). These thrips also overwinter in a number of sites adjacent to fields used for peanut production in the summer (Chamberlin *et al.*, 1992); the destruction of volunteer crops and annual weeds on which these thrips overwinter was suggested as a strategy to manage virus in peanut fields. In general, isolation of crops susceptible to virus infection from infected crops or alternate weed hosts lessens the chance of early or heavy infestation. Safe distances are difficult to predict because they depend on the direction of wind at the time vector species are migrating, but growers should plan for host-free zones of at least several hundred metres wide.

Fundamental to all biological/ecological studies relating to population growth and size are appropriate methods of measuring abundance and spatial distribution on crops and on individual parts of plants, and the density of airborne immigrants. The principles and procedures for quantitative sampling of thrips in soil, on plants and in the air are described in detail in Chapter 11 and should constitute an essential basis of any IPM programme.

Economic Injury Thresholds

Given the widespread and frequent presence of thrips on crops, growers must decide when infestations are heavy enough to merit control. Such decisions depend not simply on the number of insects present but on the value of the crop, the cost of control measures required and the degree of tolerable damage. Thus the economic threshold, that is, the population density at which control measures should be initiated to prevent an increasing pest population from reaching an economic injury level, varies considerably for each combination of crop growth stage, plant part attacked, pest, season and climate. There is unlikely to be a single economic threshold for a given pest on a given crop, but many, depending on the market and climatic conditions (Pedigo *et al.*, 1986). Most growers err on the side of caution choosing to control an infestation rather than risk serious losses or total rejection of a crop, but some economic thresholds derived empirically or by modelling are available for a few thrips species and crops.

Decisions on whether or not to apply control measures are easiest to make when zero or low tolerance levels of thrips or their damage are demanded in a crop when sold. The strict absence of any trace of thrips (or any other insect) required by some countries, merchants and importers clearly leaves no alternative but to apply chemical control. As a policy this has drawbacks because it encourages the heavy use of insecticides with consequent resistance, environmental and political problems (Vierbergen, 1995). For thrips, so easily overlooked in fresh plant material, it is probably unrealistic. Slightly less stringent criteria are laid down in the production of some fruits for export. For example in New Zealand, nectarines and peaches produced for export to Australia are unacceptable if more than 1 per cent of fruit harbours a living thrips; the figure is raised to 2 per cent for export worldwide (Anonymous, 1988). At these levels control of *Thrips obscuratus* in orchards is imperative.

For many potent pests on highly susceptible crops, economic losses can occur at very low densities and control decisions are difficult (Table 16.1). For example, no more than one adult *Thrips palmi* per ten leaves of aubergine or ten flowers of sweet pepper is tolerable in Japan (Kawai, 1986). By contrast, in India the presence of 20 *Thrips flavus* per apple flower is necessary to cause crop loss (Singh, 1989); in Ghana, comparable figures for *Megalurothrips sjostedti* on cowpeas are five individuals per raceme on 'resistant' cultivars and three on susceptible ones (Salifu and Singh, 1987). Cereals can carry heavier infestations before control is worthwhile. Each larva of *Haplothrips tritici* on a wheat grain has been estimated to reduce individual grain weight by 0.5 mg (Kamenchenko, 1988), 1 mg (Pavlov, 1937) and 5.3 mg (Bielza and Torres-Vila, 1995). At the lower estimate about 80 larvae per ear would need to be present

Table 16.1. Examples of economic injury thresholds for different field crops and thrips species.

Crop	Species	No. of thrips	Reference
Aubergine	*Thrips palmi*	0.08/leaf	Kawai (1990)
Cotton	*Thrips tabaci*	8–10/leaf	Bughio *et al.* (1986)
Cucumber	*Thrips palmi*	4.4/leaf	Kawai (1990)
Cowpea			
(resistant cv)	*Megalurothrips sjostedti*	5/raceme	Salifu and Singh (1987)
(susceptible cv)		3/raceme	
Onion	*Thrips tabaci*	0.9–2.2/leaf	Edelson *et al.* (1989)
Pigeon pea	*Megalurothrips sjostedti*	15/flower	Peter and Govindarajulu (1990)
Pyrethrum	*Thrips nigropilosus*	416 thrips-days/25 leaves	Smith and Hanson (1991)
Rye	*Limothrips cerealium*	> 20/tiller	Larsson (1988)
Strawberry	*Frankiniella occidentalis*	5–15/flower	D. Grasselly, Bellegarde France (personal communication, 1996)
Sweet pepper	*Thrips palmi*	> 0.11/flower	Kawai (1986)
Vines	*Drepanothrips reuteri*	50–250 thrips-days/flower	Boller and Candolfi (1990)

to cause 5 per cent weight loss, whereas at the higher estimate a mere 8 larvae would have the same effect. Furthermore, the actual number of larvae feeding appears to affect not only the weight but the baking quality of grain for which a premium is often paid (Bielza and Torres-Vila, 1995). On strawberry flowers, 5 to 8 *Frankliniella occidentalis* cause bronzing of the fruit and over 15 result in a crop so damaged that it has no commercial value (see Plate IV.1–4) (D. Grasselly, Bellegarde, France, 1996, personal communication). Clearly, generalizations on economic thresholds may be misleading. To be realistic they need to be assessed for each individual instance.

Onions infested by *Thrips tabaci* represent the crop–thrips combination for which economic thresholds have most often been assessed, for example in Michigan (Quartey, 1982), Texas (Harding, 1961; Edelson *et al.*, 1989), Colorado (Kendall and Capinera, 1987), New York (Shelton *et al.*, 1987) and Quebec (Fournier *et al.*, 1995). Problems with determining realistic economic threshold for thrips are exemplified for this crop where direct feeding on the bulb is not a concern. Rather, the injury is caused indirectly by thrips feeding on leaves, thereby reducing the photosynthetic ability of the plant and possibly interfering with transportation of foliar nutrients to the bulb. In addition, there is evidence of greater ethylene production in onions damaged by thrips (Kendall and Bjostad, 1990) (see Chapter 13). In general, onion plants seem relatively insensitive to thrips feeding early and late in the season, but water stress greatly increases the amount of injury. Edelson *et al.* (1989) showed insecticidal control to be worthwhile in Texas, with returns exceeding costs. Shelton *et al.* (1987) established an action threshold at 3 thrips per leaf in New York. For this crop and pest in Quebec, Fournier *et al.* (1995) combined data on yield, duration and level of infestation, costs of pesticides and their application, storage costs and losses, and fluctuating prices of onions to calculate economic thresholds corresponding to 2.2 thrips per leaf in a summer with a slight water deficit, and 0.9 per leaf in a season with early and severe drought.

As indicated in the previous example, the outcome of an infestation depends not only on the number of thrips present but for how long they persist, so economic thresholds are probably most usefully expressed as cumulative 'thrips-days' over which a particular plant part has been exposed to attack; for example, a mean level of 416 thrips-days per 25 leaves is quoted as necessary before *Thrips nigripilosus* causes economic losses to pyrethrum in Kenya (Smith and Hanson, 1991). Unfortunately, the extensive data needed to calculate reliable economic thresholds (e.g. Fournier *et al.*, 1995) are rarely available. In addition, for such data to be valid across seasons, regions and crops, extensive inquiry involving many years of work is necessary.

Management Strategies

Host plant resistance

In general, host plant resistance to insect pests is likely to play an ever increasing role in pest management programmes (Smith, 1990), but the development and incorporation of resistant cultivars into pest management programmes to control or supplement control of thrips is still in its infancy. There is considerable ongoing research in this topic as a result of the rise in pest status of many thrips species (see Chapter 3). In addition, development of plants resistant to tospoviruses vectored by some species of thrips indirectly reduces the direct damage thrips cause to a wide range of agricultural crops (Culbreath *et al.*, 1996; Ullman, 1996). Resistant cultivars of aubergine have been investigated in Japan, the Philippines and Taiwan (Yasuda and Momonoki, 1988; Bernardo, 1991) and ovipositional non-preference is suggested as the mechanism of thrips resistance in some cultivars of cowpea (Salifu *et al.*, 1988). The levels of resistance to flower thrips and other pests in cowpea justified, in economic terms, the use of these cultivars and the manipulation of planting date as primary pest management tools. This was because of the high cost of pesticides and the comparatively poor control obtained after their application. Resistance in several cultivars of cowpea less favoured for oviposition by *Megalurothrips sjostedti* has been found (Salifu *et al.*, 1988); in this case the physical basis for the non-preference seems to be associated with reduced airspace in the stems. A cultivar of maize highly resistant to *Frankliniella fusca*, *F. tenuicornis* and *Anaphothrips obscurus* has been identified and may be useful as a donor parent in breeding (Guthrie *et al.*, 1991).

Plant and cultivar resistance to *Thrips tabaci* has received much attention. Morphological characters, especially round or flat-sided leaves and open plant architecture, have been associated with low thrips densities on onion cultivars (Coudriet *et al.*, 1979) (see also Chapter 7). Lack of protection from pesticides, natural enemies and adverse abiotic conditions have been suggested as mechanisms of this resistance (Fournier *et al.*, 1995). White cultivars of onion seem less susceptible to attack than red ones (Verma, 1966; Lall and Singh, 1968). Glossy foliage offers a degree of resistance probably due to differences in the chemistry of the leaf waxes (Molenaar, 1984). Cotton cultivars with hairy lower leaf surfaces have long been thought to suffer less from thrips than more glabrous cultivars (Wardle and Simpson, 1927; Zareh, 1985), but the relationship is a complex one. For thrips infesting cotton buds, including *Thrips tabaci* and several *Frankliniella* spp., differences in severity of bud damage appear to be expressions of antixenosis and antibiosis because they have been noted within groups of hairy genotypes and within groups of glabrous breeding lines (Leigh, 1995). More features influencing the suitability of plants as hosts for thrips are described in Chapters 4 and 7.

Varietal resistance was found in processing cabbage to *Thrips tabaci* (Shelton *et al.*, 1983), although no cultivars were completely immune to injury. In later studies, this resistance was confirmed as innate and uninfluenced by external factors such as planting date (Stoner and Shelton, 1988). Therefore, plant resistance may eventually contribute to pest management for onion thrips on this crop (Shelton *et al.*, 1983). The use of peanut cultivars with high levels of resistance to tospovirus was found to be the most successful of many cultural techniques explored to reduce injury by *Frankliniella fusca* (Brown *et al.*, 1996). In winter wheat, the depth of grain grooves was found to influence populations of *Haplothrips tritici* (Czencz, 1992) and screening rice germplasm for resistance to *Stenchaetothrips biformis* utilizing traditional breeding techniques is underway (Velusamy, 1990; Velusamy *et al.*, 1992).

Chemical control

Management of thrips with pesticides is problematic. As with any pest, repeated applications have caused high levels of resistance and cross-resistance (Robb *et al.*, 1995). Chapter 15 reviews the many difficulties and attempted solutions and should be studied as background to the role of pesticides in IPM programmes for thrips. Suffice here to emphasize that in the field many factors affect the performance of pesticides, including plant and crop structure, pest behaviour, feeding and pupation sites, application methods and pesticide persistence, previous treatments and weather. In addition, pesticide application can exacerbate thrips problems by destroying natural enemies (Leggett, 1992). The constant threat of migration, the ability to transmit tospoviruses, the propensity of thrips to prefer tight, protected areas on the plant, and the failure of many systemic pesticides to translocate to the flowers where many thrips feed, all combine to make control with chemicals exceedingly difficult. Where chemical control is attempted, frequent applications are required. A typical practice for control of both thrips and tospovirus is the application of a systemic insecticide such as aldicarb at planting and then weekly applications of acephate or carbofuran through the season (McPherson *et al.*, 1992). In some cropping systems, even this is not sufficient to improve yields over untreated controls (Todd *et al.*, 1996). A number of new insecticides are under development and these may offer different opportunities for control in IPM programmes. For example, research has begun on antifeedants for use against thrips (Yasumi *et al.*, 1991; Hirano *et al.*, 1993) and the potential for new growth regulators is reviewed by Mound *et al.* (1990) (see Chapter 15, Pesticide Usage and Efficacy).

Physical/mechanical controls

The utility of mechanical barriers and UV-reflective mulches has been evaluated to see whether they decrease thrips and virus in selected field crops. Yudin *et al.* (1991) found that 1.5-m tall mechanical barriers made of aluminium polyvinyl net placed around lettuce fields reduced interfield movement of *Frankliniella occidentalis* by almost 10 per cent. However, these barriers had no effect on within-field movement of thrips and were considered to be of little utility in reducing virus in lettuce fields. There is also the potential danger of barriers around fields actually causing airborne migrants to accumulate on their lee side in response to the shelter and the modified air currents created (Lewis, 1970, 1973) (see Chapter 5, Settling and infestation patterns in crops).

Aluminium-surfaced mulch dramatically decreases thrips and virus in tomato, pepper and tobacco (Scott *et al.*, 1989; Greenhough and Black, 1990; Kring and Schuster, 1992). Compared with the non-mulched treatments, aluminium mulch reduced the number of trapped thrips in tomato by 68 per cent and the incidence of tomato spotted wilt virus (TSWV) by 64 per cent; in bell pepper the figures were 60 per cent and 78 per cent, respectively (Greenhough and Black, 1990). The effectiveness of the mulch generally disappeared later in the season with increased shading by lower leaves and as traps were raised to accommodate increased plant height. In studies where thrips were monitored using sticky traps placed above the crops, mulches brought about a marked reduction in thrips numbers. However, when Kring and Schuster (1992) compared thrips in flowers from mulched and non-mulched plots, no differences were found. The experimental use of aluminized polyethylene as a direct covering on young leeks early in the season deterred thrips and would probably reduce the need for insecticidal treatment (Benoit and Ceustermans, 1990).

Cultural/sustainable management practices

The basis of sound IPM should be good culture to promote rapid plant growth, and clean management; together these can often lessen the risk of thrips attack and sometimes completely avoid the need for pesticides. Useful, long-established methods include rotations, cultivations, irrigation, time of planting and harvesting, plant spacing, and more recently judicious intercropping.

Crop rotation is helpful against species that survive in soil (see Chapter 7) or on crop residues. In north-west Europe, *Thrips angusticeps* attacks several crops including flax, brassicas, lupins, sugar beet and peas (Plate VIII.1–3), and a brachypterous generation hibernates in the soil. When susceptible crops are grown in consecutive years, serious damage often results; this can be prevented completely by following a three-course rotation, such as red clover or potatoes, oats, flax. Peas should not be grown after flax or brassicas, especially mustard (Gough, 1955; Franssen and Huisman, 1958). In warmer areas, *T. tabaci* may survive in plant remains left in fields, and new crops should not be planted on or near the sites of old ones. In Turkey, Ileri (1947) stressed the importance of destroying plant residue in tobacco fields and of ploughing fields immediately after one crop was harvested to prevent infestation of the next. In eastern Europe and Asia, populations of *Haplothrips tritici* on cereals are smaller on crops grown after perennial crops than after other cereals (Tansky, 1958).

Soil-dwelling stages are vulnerable to mechanical cultivations. Ploughing cereal stubble to a depth of about 25 cm destroys some species of cereal thrips, either because the adults that emerge from the buried immature stages cannot reach the surface, or because the soil at this depth is too moist for survival (Lyubenov, 1961). Harrowing the stubble before ploughing ensures that the straw is buried and no thrips are left on the surface. Populations of *Haplothrips tritici* can be decreased by 30 to 60 per cent by this treatment (Tansky, 1958). The hitherto common practice in Europe and central Asia of burning stubble before ploughing helps control this thrips by killing larvae overwintering in dead stalks on the surface of the ground (Shurovenkov, 1961). However, this is probably less effective than supposed because most larvae hibernate about 10 cm below the surface and at this depth the soil never reaches lethal temperatures. The recent banning of straw burning in some western European countries for environmental reasons may exacerbate cereal thrips infestations. Stubble burning is probably a more effective means of control for grass thrips, because many of the harmful species of *Chirothrips* overwinter in dead stubble and culms lying on the ground (Lewis and Navas, 1962). Care should be taken when harvesting grass seed crops to leave as few heads as possible in the field, and to burn the chaff and straw after threshing.

Because plants under water stress are so susceptible to direct injury, adequate irrigation in dry areas and in hot seasons alone decreases losses. Bullock (1963) mentions a striking example in a field of chrysanthemums grown for pyrethrum production. In this field there was a heavy and fairly uniform infestation of *Thrips nigropilosus*, and only one edge of the field received overhead irrigation. The watered plants were unharmed whereas the unwatered ones were seriously damaged. Similarly, mulching and overhead irrigation of coffee in Kenya has lessened trouble from thrips (Evans, 1967). In dry parts of the Lockyer Valley, Queensland, *Thrips tabaci* was a perennial problem on onions. By irrigating to ensure rapid growth, profitable crops can now be grown despite the presence of the thrips. In fact, irrigation has so decreased the harmful effects of this pest that the benefits which might have been derived from the use of modern insecticides under older dry-farming conditions have been largely forestalled (Passlow, 1957).

As discussed earlier for cotton (Micinski *et al.*, 1990) and cabbage (Stoner and Shelton, 1988), planting date can have a dramatic impact on thrips densities and the amount of damage thrips cause. Well-established crops can withstand attack better than newly planted ones, so early planting is usually desirable. It is especially beneficial on light, dry soils where it is common for plants to suffer from water shortage as the growing season progresses. In the central rainlands of the Sudan where upland strains of cotton may be grown with rainfall as the only source of water, July sowing used to be essential to prevent thrips damage, prior to the use of insecticides. Plants sown much later had to be grown under regular irrigation to enable them to withstand heavy thrips infestations (Pearson, 1958). Most horticultural varieties of peas flower before agricultural types, so where *Kakothrips pisivorus* is abundant it is advisable to grow only the former, and to sow them early to avoid attack (Franssen, 1960).

In north Khazakhstan the slower-developing ears of spring-sown cereals are much more likely to be infested by *Haplothrips tritici* than ears of autumn-sown wheat (Tansky, 1958). However, the benefits of early sowing depend on the climate, soil and perhaps variety of wheat. In contrast, this same thrips, in Bulgaria, lays more eggs on early varieties of winter wheat than on spring wheat (Lyubenov, 1961). In Belarus, a combination of cultural practices including the use of mineral fertilizers, optimal planting times and ploughing after harvest is reported to prevent damage from thrips attacking cereals, especially rye (Samersov and Yachenya, 1980).

Careful timing of harvesting can also diminish loss of yield. Yields of some grass seed crops have the potential to be increased by grazing the grass early in the season to delay flowering so that plants have no attractive inflorescences available when *Chirothrips* migrate. Early harvesting of cereals may increase mortality in the second generation larvae of *Limothrips* spp., but unless practised over a large area it would not noticeably lessen populations of adults the following spring. Damage to lucerne seed crops by *Odontothrips confusus* is partly prevented by taking the seed from the first cut, before flowers are sterilized by heavy infestations of thrips later in the summer (Bournier and Kochbav, 1965).

Other factors such as row pattern and seeding rate have been examined for their impact on reducing thrips and virus in peanuts (Brown *et al.*, 1996); twin row planting resulted in significantly less virus than conventional single rows, but this effect was not consistent from year to year.

By manipulating cover crops and nitrogen levels in cabbage, Roberts and Cartwright (1991) were able to reduce thrips populations. In addition, when a threshold-based pesticide application programme was used, it produced high quality heads with fewer pesticide applications compared with timed prophylactic treatments.

A comparison of conventional and organic farming systems with respect to the impact on thrips populations has provided mixed results. Helenius (1990) found population densities of *Frankliniella tenuicornis* were almost double in ears of barley grown organically compared with a conventional system. When thrips control on potato was investigated using onions in an intercropping strategy, the populations of aphids and leafhoppers declined, but there was an increase in the numbers of *Thrips* spp., perhaps because onions are such a good host. There is considerable anecdotal evidence that *Frankliniella occidentalis* does not become a serious pest on organically grown field vegetables and fruit in Europe.

Intercropping appears to be most helpful in warm dry situations, but successful examples have also been reported in more temperate climates. In Egypt, intercropping onion and garlic with tomato reduced infestations of *Thrips tabaci* by almost 80 per cent, but the yield from both crops declined (Afifi and Haydar, 1990). In England, infestation of *Thrips tabaci* on onions was halved when intercropped with carrots (Uvah and Coaker, 1984). The effect was greatest with closely alternating single rows of each and was attributed to the masking of the onions by the carrot foliage. Infestations of *Thrips tabaci* on cabbage were reduced tenfold by growing clovers, either *Trifolium repens* or *T. subterraneum*, between rows (Theunissen *et al.*, 1995). In Kenya, populations of *Megalurothrips sjostedti* and *Hydatothrips adolfifriderici* on cowpea buds were almost halved by intercropping the cowpea with sorghum and maize (Dissemond and Hindorf, 1990). As in the example described for *Thrips tabaci* (Uvah and Coaker, 1984), these effects are probably caused through shading of the lower crop by the taller intercrop, which influences the abundance and activity of the thrips (Kyamanywa and Ampofo, 1988); however, the practice is likely to incur a yield penalty and will not always reduce the infestation on the vulnerable crop. Populations of *Thrips palmi* and *T. parvispinus* have been shown to increase on potato when intercropped with shallot and garlic (Potts and Gunatti, 1991), as is *Caliothrips indicus* on groundnut intercropped with pigeon pea and mung bean (Singh *et al.*, 1990).

A mixed cropping habitat has been shown to encourage a major thrips predator, *Orius tristicolor* (White), compared with monocultures (Letourneau and Altieri, 1983). The underlying causes of this predator accumulation were shown to be independent of prey density and plant diversity and dependent on plant architecture and density (Letourneau, 1990). The ratio of predacious to phytophagous thrips was compared in fields of winter wheat grown in a conventional 'high input' manner and in a sustainable 'low input' production system (Klump, 1991). The ratios were 1:20 in sustainable fields and 1:422 in conventional production. The presence of numerous weeds in the alternative production fields may have been the reason for the large numbers of predatory thrips such as *Aeolothrips intermedius*.

Postharvest control

The presence of thrips in harvested commodities presents two problems: first contamination of the commodity that leads to rejection for consumer-health reasons (Shelton *et al.*, 1982), and second, products shipped to countries with quarantine regulations that specify fumigation when live insects are detected (Hinsch *et al.*, 1991). For these reasons, considerable attention has focused on postharvest disinfestation methods for thrips. A thorough review of procedures for thrips (although focused on ornamental crops) can be found in Hara (1994) (see also Chapter 15, Application methods). Rather than relying solely on disinfestation, Carpenter and Epenhuijsen (1993) advocate the development of a risk reduction strategy where growers identify important control points during the growth of the crop. If pests are controlled in the field or prevented from becoming established in the crop, there is less likelihood of quarantine problems.

Biological control

Although much has been written about natural enemies and biological control of thrips (see chapters 8, 9 and 10) and there are successful programmes in greenhouses (see Chapter 18), little has been integrated into actual pest management programmes for field crops. In some situations, key natural enemies are thought to be lacking (Tamò *et al.*, 1993) or the native complex of natural enemies is deemed insufficient to do the job (Gonzalez *et al.*, 1992). Manipulation of natural enemies using inoculative/inundative procedures has been tried in only a few situations. Both Gonzalez *et al.* (1992) and Hoy and Glenister (1991) released predatory mites in the genus *Amblysieus* with little success. The use of natural enemies for thrips control is made even more remote when there is a threat of virus transmissions to crops. Clearly, intense insecticide applications for control of thrips are incompatible with biological control, as demonstrated by Leggett (1992). The potential compatibility of thrips natural enemies with insecticides is being investigated (Nagai, 1990) and may occur with some of the newer biorational pesticides (Ascher *et al.*, 1992). However, it is clear in pest management programmes in field crops today that natural enemies play a very small role. The problems are epitomized by the effort expended over 60 years to establish successful biological control agents for *Thrips tabaci* in many areas, so far without marked success in field crops. About 90 natural enemies of this species are listed from various parts of the world, including predators, parasites and pathogens (Waterhouse and Norris, 1989). Between 1936 and 1962 there were nine attempts to introduce hymenopterous parasites in places as diverse as California, Japan, the Philippines, China and Trinidad. Although in four instances the parasites established, they never provided total control; they still persist as residual populations, probably exerting some control over the pest, but rarely enough to prevent crop damage. Predators, especially *Orius* spp. and *Chrysoperla carnea*, likewise seem ineffective alone. This is probably due to the ability of thrips to migrate to a crop in such numbers that they can cause damage before beneficial organisms exert significant control.

Fortunately, there are now areas of research which might draw together some of the natural control methods available to produce a more promising package. More prominence is being given to improving the virulence and ecological fitness of fungal inocula through manipulation of cultural, formulation and storage conditions (see Chapter 10) to be used alongside resistant cultivars, and possibly mass-reared selected predators and parasites. There may also be opportunities for synergy between thrips fungal pathogens, natural enemies, and insect growth-regulating chemicals.

Conclusions

Pest management programmes for thrips attacking a wide range of field crops are still evolving. In some crops, such as peanuts and cabbage, they have been under development for many years and have reached a point where solid recommendations can be made, based on good research and backed by demonstration plots with cooperating growers. Even with these refined management programmes, thrips control is not easy and requires considerable effort on the part of the grower. The methods mentioned in this chapter, such as using resistant cultivars, changing the planting time, altering the row spacing, sampling the plants and fields and making threshold calculations, and correctly identifying the thrips, all require a major commitment. The temptation to utilize insecticides with the promised 'quick fix' is considerable despite all the disadvantages associated with chemical dependency, such as resistance, destruction of natural enemies, environmental contamination, compromising worker health and safety, and residues on the crop. However, growers cannot be unduly criticized if they turn to a pesticide management strategy in the face of such a formidable enemy.

It is unfortunate that natural enemies have been shown to play a rather insignificant role in regulating thrips populations in field crops. Much more research needs to be done on this topic. The use of intercropping and other sustainable approaches to crop production and pest management which foster the development of natural enemy populations needs more attention. More research is needed to lead to recommendations that growers can follow to enhance the natural enemy fauna in and around their crops.

Thrips seem to be increasing as pests faster than researchers can respond, although there is probably more research in progress today aimed at developing good pest management strategies than has occurred during the last 100 years. New information is rapidly becoming available and this should result in comprehensive pest management programmes for many more thrips species on a diversity of crops.

References

Afifi, F.M.L. and Haydar, M.F. (1990) Effect of different intercropping systems on onion and garlic infestation with the cotton and onion thrips, *Thrips tabaci* Lind. *Bulletin of the Faculty of Agriculture, University of Cairo* 41, 639–648.

Anonymous (1988) *Acceptable Quality Levels. Summerfruit.* Ministry of Agriculture and Fisheries, New Zealand.

Ascher, K.R.S., Klein, M. and Meisner, J. (1992) Azatin, a neem formulation, acts on nymphs of the western flower thrips. *Phytoparasitica* 20, 305–306.

Bautista, R.C., Mau, R.F.L., Cho, J.J. and Custer, D.M. (1996) Thrips, tospovirus, and host plant associations in Hawaiian farm ecosystem: prospects for reducing disease losses. In: Kuo, G. (ed.) *Proceedings of the International Symposium on Tospoviruses and Thrips of Floral and Vegetable Crops, Taichung, Taiwan. Acta Horticulture, Technical Communications of the International Society for Horticulural Science*, 431, 477–490.

Benoit, F. and Ceustermans, N. (1990) Some findings from Belgian research on leeks. *Revue de l'Agriculture* 43, 33–41.

Bernardo, E.N. (1991) Thrips on vegetable crops in the Philippines. In: Talekar, N.S. (ed.) *Thrips in Southeast Asia: Proceedings of a Regional Consultation Workshop.* Asian Vegetable Research and Development Center, Publication No. 91-342, pp. 5–11.

Bielza, P. and Torres-Vila, L.M. (1995) Efecto del nivel poblacional de *Haplothrips tritici* Kurd. (Thysanoptera: Phlaeothripidae) en la produccion de trigo en Castilla–La Mancha. *XIV Journadas de la Asociacion espanola de Entomologia*, Cuenca, Spain.

Boller, E. and Candolfi, M.P. (1990) Thrips in vineyards of eastern Switzerland: 3. The effect of thrips on yield formation of the grapevine. *Schweizerische Zeitschrift für Obst und Weinbau* 126, 253–258.

Bournier, A. and Kochbav, A. (1965) *Odontothrips confusus* Pr. nuisible à la luzerne. *Annals des Epiphyties* 16, 53–69.

Brown, S.L., Todd, J.W. and Culbreath, A.K. (1996) Effects of selected cultural practices on the incidence of tomato spotted wilt in peanuts. In: Kuo, G. (ed.) *Proceedings of the International Symposium on Tospoviruses and Thrips of Floral and Vegetable Crops, Taichung, Taiwan. Acta Horticultural, Technical Communications of the International Society for Horticultural Science*, pp. 491–498.

Bughio, A.R., Hussain, T. and Qureshi, Z.A. (1986) Incidence and population density of sucking complex on cotton. *Proceedings of Pakistan Congress of Zoology* 6, 93–98.

Bullock, J.A. (1963) Thysanoptera associated with pyrethrum and the control of *Thrips tabaci* Lind. *Tropical Agriculture (Trinidad)* 40, 329–335.

Burris, E., Pavloff, A.M., Leonard, B.R., Graves, J.B. and Church, H. (1990) Evaluation of two procedures for monitoring populations of early season insect pests (Thysanoptera; Thripidae and Homoptera; Aphididae) in cotton under selected management strategies. *Journal of Economic Entomology* 83, 1064–1068.

Carpenter, A. and Epenhuijsen, K.V. (1993) A systems approach to quarantine entomology: using asparagus infestation as a model. In: *Proceedings of the 46th New Zealand Plant Protection Conference*, New Zealand Plant Protection Society, Rotorua, pp. 53–56.

Chamberlin, J.R., Todd, J.W., Beshear, R.J., Culbreath, A.K. and Demski, J.W. (1992) Overwintering hosts and wingforms of thrips, *Frankliniella* spp., in Georgia (Thysanoptera; Thripidae): implications for management of spotted wilt disease.

Environmental Entomology 21, 121–128.
Chambers, W.S. and Sites, R.W. (1989) Overwintering thrips fauna in croplands of the Texas south plains. *Southwestern Entomologist* 14, 325–328.
Coudriet, D.L., Kisaba, A.N., McCreight, J.D. and Bohn, G.W. (1979) Varietal resistance in onion to thrips. *Journal of Economic Entomology* 72, 614–615.
Culbreath, A.K., Todd, J.W., Branch, W.D., Gorbet, D.W., Holbrook, C.C., Shokes, F.M. and Demski, J.W. (1996) Variation in susceptibility to TSWV among advanced breeding lines of peanut (*Arachis hypogaea*). In: Kuo, G. (ed.) *Proceedings of the International Symposium on Tospoviruses and Thrips of Floral and Vegetable Crops, Taichung, Taiwan. Acta Horticulturae, Technical Communications of the International Society for Horticultural Science* 431, 402–410.
Czencz, K. (1992) A comparative investigation of the infestation of winter wheat types by *Haplothrips tritici* Kurd. (Thysan., Phlaeothripidae). *Journal of Applied Entomology* 113, 209–213.
Dissemond, A. and Hindorf, H. (1990) Influence of sorghum/maize/cowpea intercropping on the pest, disease and weed situation at Mbita/Kenya. In: *Proceedings of the Integrated Pest Management in Tropical and Subtropical Cropping Systems* No. 3. Deutsche Landwirtschafts-Gesellschaft, Frankfurt am Main, pp. 843–851.
Edelson, J.V., Cartwright, B. and Royer, T.A. (1989) Economics of controlling onion thrips (Thysanoptera: Thripidae) on onions with insecticides in south Texas. *Journal of Economic Entomology* 82, 561–564.
Evans, D.E. (1967) Insecticide field trials against the coffee thrips (*Diarthrothrips coffeae* Williams) in Kenya. *Turrialba* 17, 376–380.
Fournier, F., Boivin, G. and Stewart, R.K. (1995) Effect of *Thrips tabaci* (Thysanoptera: Thripidae) on yellow onion yields and economic thresholds for its management. *Journal of Economic Entomology* 88, 1401–1407.
Franssen, C.J.H. (1960) Biology and control of pea thrips. *Verslag Landbouwkundig Onderzoek Rijksland Proefstn* 66, 36 pp.
Franssen, C.J.H. and Huisman, P. (1958) The biology and control of *Thrips angusticeps* Uzel. *Verslag Landbouwkundig Onderzoek RijkslandProefstn* 64, 104 pp.
Gonzalez, J.E., Garcia-Maria, F., Benages, E. and Orenga, S. (1992) Biological control of the western flower thrips *Frankliniella occidentalis* in strawberries. *Bulletino Sanitario Vegetale Plagas* 18, 265–288.
Gough, H.C. (1955) *Thrips angusticeps* Uzel attacking peas. *Plant Pathology* 4, 53.
Greenhough, D.R. and Black, L.L. (1990) Aluminum-surfaced mulch; an approach to the control of tomato spotted wilt virus in solanaceous crops. *Plant Disease* 74, 805–808.
Guthrie, W.D., Russell, W.H., Bing, J.W. and Dicke, F.F. (1991) Registration of B96 germplasm line of maize. *Crop Science* 31, 231–240.
Hara, A.H. (1994) Ornamentals and flowers. In: Paull, R.E. and Armstrong, J.W. (eds) *Insect Pests and Fresh Horticultural Products: Treatments and Responses*. CAB International, Wallingford, UK, pp. 329–347.
Harding, J.A. (1961) Studies on the control of thrips attacking onion. *Journal of Economic Entomology* 54, 1254–1255.
Helenius, F. (1990) Conventional and organic cropping systems at Suitia VI: Insect populations in barley. *Journal of Agricultural Science in Finland* 62, 349–355.
Hinsch, R., Vail, P.V., Tebbets, J.S. and Hoffman, D.F. (1991) Live insects and other arthropods on California iceberg head and shredded lettuce. *Southwestern Entomologist* 16, 261–266.
Hirano, C., Itoh, E., Yasumi, K. and Horike, M. (1993) Short-distance walking responses of *Thrips palmi* Karny (Thysanoptera: Thripidae) to tomato leaves.

Applied Entomology and Zoology 28, 233–234.

Hoy, C.W. and Glenister, C.S. (1991) Releasing *Amblyseius* spp. (Acarina: Phytoseiidae) to control *Thrips tabaci* (Thysanoptera: Thripidae) on cabbage. *Entomophaga* 36, 561–573.

Ileri, M. (1947) The bionomics, mode of infection and control of the tobacco thrips. (In Turkish.) *Ziraat Dergisí* 8, 19–57. (*Review of Applied Entomology* (*A*) 38, 142–143.)

Kamenchenko, S.E. (1988) Assessment of the injuriousness of phytophages of spring wheat and prospects for using it in an intensive technology under irrigation conditions. *Sibirskii Vestnik sel' Skokhoziaistuennoi Nauki* 1, 33–37.

Kawai, A. (1986) Studies on population ecology of *Thrips palmi* Karny. X. Differences in population growth on various crops. *Japanese Journal of Applied Entomology and Zoology* 30, 7–11.

Kawai, A. (1990) Life cycle and population dynamics of *Thrips palmi*. *Japan Agriculture Research Quarterly* 23, 282–288.

Kendall, D.M. and Bjostad, L.B. (1990) Herbivory by *Thrips tabaci* induces greater ethylene production in intact onions than mechanical damage alone. *Journal of Chemical Ecology* 16, 981–991.

Kendall, D.M. and Capinera, J.L. (1987) Susceptibility of onion growth and stages to onion thrips (Thysanoptera: Thripidae). *Journal of Economic Entomology* 16, 859–863.

Klump, R.P.M. (1991) Comparison of phytophagous and predatory thrips (Thysanoptera) in winter wheat fields in Hesse with different management systems. *Journal of Plant Diseases and Protection* 98, 464–470.

Kring, J.B. and Schuster, D.J. (1992) Management of insects on pepper and tomato with UV-reflective mulches. *Florida Entomologist* 75, 119–129.

Kyamanywa, S. and Ampofo, J.K.O. (1988) The effect of cowpea/maize mixed cropping on the incident light at the cowpea canopy and flower thrips (Thysanoptera: Thripidae) population density. *Crop Protection* 7, 186–189.

Lall, B.S. and Singh, L.M. (1968) Biology and control of the onion thrips in India. *Journal of Economic Entomology* 61, 676–679.

Larsson, H. (1988) Economic damage caused by cereal thrips in winter rye in Sweden. *Acta Phytopathologica et Entomologica Hungarica* 23, 291–293.

Leggett, J.E. (1992) The influence of ULV malathion, applied for boll weevil control, on other pest and beneficial species in Arizona cotton fields 1989–1990. *Southwestern Entomologist* 17, 49–61.

Leigh, T.F. (1995) Bionomics of cotton thrips: a review. In: Parker, B.L., Skinner, M. and Lewis, T. (eds) *Thrips Biology and Management*. Plenum Press, New York, pp. 61–70.

Letourneau, D.K. and Altieri, M.A. (1983) Abundance patterns of a predator, *Orius tristicolor* (Hemiptera: Anthocoridae) and its prey, *Frankliniella occidentalis* (Thysanoptera: Thripidae): habitat attraction in polycultures versus monocultures. *Environmental Entomology* 12, 1464–1469.

Letourneau, D.K. (1990) Mechanisms of predator accumulation in a mixed crop system. *Ecological Entomology* 15, 63–69.

Lewis, T. (1970) Patterns of distribution of insects near a windbreak of tall trees. *Annals of Applied Biology* 65, 213–220.

Lewis, T. (1973) *Thrips, Their Biology, Ecology and Economic Importance*. Academic Press, London.

Lewis, T. and Navas, D.E. (1962) Thysanopteran populations overwintering in hedge bottoms, grass litter and bark. *Annals of Applied Biology* 50, 299–311.

Lowry, V.K., Smith, J.W. and Mitchell, F.L. (1992) Life-fertility tables for *Frankliniella fusca* (Hinds) and *F. occidentalis* (Pergande) (Thysanoptera: Thripidae) on peanut. *Annals of the Entomological Society of America* 85, 744–754.

Lyubenov, Y. (1961) A contribution to the bionomics of the wheat thrips (*Haplothrips tritici* Kurd.) in Bulgaria and possibilities of reducing the injury done by it. (In Bulgarian.) *Review of Applied Entomology (A)* 50, 348.

McPherson, R.M., Beshear, R.J. and Culbreath, A.K. (1992) Seasonal abundance of thrips (Thysanoptera: Suborders Terebrantia and Tubilifera) in Georgia flue-cured tobacco and impact on management practices on the incidence of tomato spotted wilt virus. *Journal of Economic Science* 27, 257–268.

McPherson, R.M. and Douce, G.K. (1992) *Summary of Losses from Insect Damage and Costs of Control in Georgia, 1992*. The Georgia Agricultural Experiment Station, College of Agriculture and Environmental Sciences, University of Georgia.

Micinski, S., Colyer, P.D., Nguyen, K.T. and Konce, K. (1990) Effects of planting date and early-season pest control on yield in cotton. *Journal of Production Agriculture* 3, 597–602.

Molenaar, N.D. (1984) Genetics, thrips (*Thrips tabaci* L.) resistance and epicuticular wax characteristics of nonglossy and glossy onions (*Allium cepa* L.) (Abstract). *Dissertation Abstracts, B (Sciences and Engineering)* 45, 1075B.

Mound, L.A., du Heaume, G.J. and Walker, A.K. (1990) *Thrips palmi*. A literature survey with an annotated bibliography. International Institute of Biological Control, Silwood Park, UK.

Murai, T. (1988) Studies on the ecology and control of flower thrips, *Frankliniella intonsa* (Trybom). *Bulletin of the Shimane Agricultural Experimental Station* No. 23, 1–73.

Nagai, K. (1990) Effects of insecticides on *Orius* sp. the natural enemy of *Thrips palmi* Karny. *Japanese Journal of Applied Entomology and Zoology* 34, 105–108.

Nagai, K. and Tsumuki, H. (1990) Search for winter host plants of *T. palmi* in winter. *Japanese Journal of Applied Entomology and Zoology* 34, 321–324.

North, R.C. and Shelton, A.M. (1986a) Ecology of Thysanoptera within cabbage fields. *Environmental Entomology* 15, 520–526.

North, R.C. and Shelton, A.M. (1986b) Overwintering of the onion thrips, *Thrips tabaci* (Thysanoptera: Thripidae), in New York. *Journal of Economic Entomology* 15, 695–699.

Passlow, T. (1957) Control of *Thrips tabaci* Lind. in onion crops in the Lockyer Valley. *Queensland Journal of Agricultural Science* 14, 53–72.

Pavlov, I.F. (1937) Noxiousness of *Haplothrips tritici* Kurd. *Review of Applied Entomology (A)* 25, 143–144.

Pearson, E.O. (1958) *The Insect Pests of Cotton in Tropical Africa*. CAB, London.

Pedigo, L.P., Hutchins, S.H. and Highey, L.G. (1986) Economic injury levels in theory and practice. *Annual Review of Entomology* 31, 342–368.

Peter, C. and Govindarajulu, V. (1990) Management of blossom thrips, *Megalurothrips usitatus* on pigeonpea. *Tropical Pest Management* 36, 312–313.

Potts, M.J. and Gunadi, N. (1991) The influence of intercropping with *Allium* on some insect populations in potato (*Solanum tuberosum*). *Annals of Applied Biology* 119, 207–213.

Quartey, S.W. (1982) Population dynamics of the onion thrips, *Thrips tabaci* (Thysanoptera: Thripidae), on onions. Michigan State University, East Lansing.

Race, S.R. (1965) Predicting thrips populations on seedling cotton. *Journal of Economic Entomology* 58, 1013–1014.

Robb, K.L., Newman, J., Virzi, J.K. and Parrella, M.P. (1995) Insecticide resistance in the western flower thrips. In: Parker, B.L., Skinner, M. and Lewis, T. (eds) *Thrips Biology and Management*. Plenum Press, New York, pp. 341–346.

Roberts, B.W. and Cartwright, B. (1991) Alternative soil and pest management practices for sustainable production of fresh-market cabbage. *Journal of Sustainable Agriculture* 1, 21–35.

Rummel, D.G.B. and Hatfield, J. (1988) Interaction of weather and thrips injury during the early season cotton growing season. In: *Beltwide Cotton Production Research Conference, Memphis, Tennessee*. National Cotton Council and the Cotton Foundation.

Sadras, V.O. and Wilson, L.J. (1997) Recovery of cotton crops after early season damage by thrips (Thysanoptera). *Bulletin of Entomological Research* (in press).

Salifu, A.B. and Singh, S.R. (1987) Evaluation of sampling methods for *Megalurothrips sjostedti* (Trybom) (Thysanoptera: Thripidae) on cowpea. *Bulletin of Entomological Research* 77, 451–456.

Salifu, A. B., Hodgson, C.J. and Singh, S.R. (1988) Mechanism of resistance in cowpea (*Vigna unguiculata* (L.) Walp) genotype, TVx3236, to the bean flower thrips, *Megalurothrips sjostedti* (Trybom) (Thysanoptera: Thripidae). 1. Ovipositional non-preference. *Tropical Pest Management* 34, 180–184.

Samersov, V.F. and Yachenya, S.V. (1980) The effect of some agritechnical practices on the abundance and harmfulness of cereal thrips. In: Lopatin, I.K. (ed.) *Vliyanie Khozyaistvennoi Deyatel Nosti Cheloveka na Bespozvonochnykh*. Minsk. (In Russian.)

Scott, S.J., McLeod, P.J., Montgomery, F.W. and Hander, C.A. (1989) Influence of reflective mulch on incidence of thrips (Thysanoptera: Thripidae: Phlaeothripidae) in staked tomatoes. *Journal of Entomological Science* 4, 422–427.

Shelton, A.M., Stammer, J.R., Wilsey, W.T., Stoyla, B.O. and Andalora, J.T. (1982) Onion thrips (Thysanoptera: Thripidae) damage and contamination in sauerkraut. *Journal of Economic Entomology* 75, 492–494.

Shelton, A.M., Becker R.F. and Andalora, J.T. (1983) Varietal resistance to onion thrips (Thysanoptera: Thripidae) in processing cabbage. *Journal of Economic Entomology* 76, 85–86.

Shelton, A.M., Nyrop, J.P., North, R.C., Petzoldt, C. and Foster, R. (1987) Development and use of a dynamic sequential sampling program for onion thrips, *Thrips tabaci* (Thysanoptera: Thripidae) on onion. *Journal of Economic Entomology* 80, 1051–1056.

Shurovenkou, B.G. (1961) Biological peculiarities of the larvae of *Haplothrips tritici* Kurd. under the conditions of the Siberian Trans-Ural. *Zoologicheskii Zhurnal* 40, 1568–1571.

Singh, M. (1989) Damage and economic threshold level of *Thrips flavus* Schrank to apple blossom in Himachal Pradesh. *Indian Journal of Entomology* 51, 288–293.

Singh, T.V.K., Singh, K.M. and Singh, R.N. (1990) The groundnut pest complex: IV, Regression studies to determine the association between jassid and thrips and weather parameters. *Indian Journal of Entomology* 52, 693–701.

Smith, L.M. and Hanson, P.M. (1991) Yield reduction in pyrethrum caused by *Thrips nigropilosus* Uzel. (Thysanoptera: Thrypidae). *Tropical Agriculture (Trinidad)* 68, 235–238.

Smith, M.C. (1990) Adaptation of biochemical and genetic techniques to the study of plant resistance to insects. *American Entomologist* 36, 141–146.

Stoner, K.A. and Shelton, A.M. (1988) Effect of planting date and timing of growth stages on damage to cabbage by onion thrips (Thysanoptera: Thripidae). *Journal of Economic Entomology* 81, 1186–1189.

Tamò, M., Baumgartner, J., Delucchi, V. and Herren, H.R. (1993) Assessment of key factors responsible for the pest status of the bean flower thrips *Megalurothrips sjostedti* (Thysanoptera: Thripidae) in West Africa. *Bulletin of Entomological Research* 83, 251–258.

Tansky, V.I. (1958) The biology of wheat thrips *Haplothrips tritici* Kurd. (Thysanoptera, Phlaeothripidae) in Northern Kazakhstan and the proposed cultural methods of its control. (In Russian.) *Revue entomologie URSS* 37, 785–797.

Terry, L.I. and Barstow, B.B. (1988) Susceptibility of early season cotton floral bud types to thrips (Thysanoptera: Thripidae) damage. *Journal of Economic Entomology* 81, 1785–1791.

Theunissen, J., Booji, C.J.H. and Lotz, L.A.P. (1995) Effects of intercropping white cabbage with clovers on pest infestation and yield. *Entomologia Experimentalis et Applicata* 74, 7–16.

Todd, J.W., Culbreath, A.K. and Brown, S.L. (1996) Dynamics of vector populations and TXWV disease progress relative to insecticide use in peanut. In: Kuo, G. (ed.) *Proceedings of the Symposium on Tospoviruses and Thrips of Floral and Vegetable Crops, Taichung, Taiwan. Acta Horticulturae, Technical Communications of the International Society of Horticultural Science*, pp. 483–490.

Ullman, D.E. (1996) Thrips and tospoviruses; advances and future directions. In: Kuo, G. (ed.) *Proceedings of the Symposium on Tospoviruses and Thrips of Floral and Vegetable Crops, Taichung, Taiwan. Acta Horticulturae, Technical Communications of the International Society of Horticultural Science*, pp. 310–324.

Uvah, I.I.I. and Coaker, T.H. (1984) Effect of mixed cropping on some insect pests of carrots and onions. *Entomologia Experimentalis et Applicata* 36, 159–167.

Velusamy, R.T. (1990) Using rice nurseries to collect thrips for use in screening rice germplasm. *International Rice Research Newsletter* 15, 16.

Velusamy, R., Mohankumar, S., Manoharan, S. (1992) Resistance to rice thrips in breeding lines derived from *Oryza officinalis*. *International Rice Research Newsletter*, 17, 10.

Verma, S.K. (1966) Studies on the host preference of the onion thrips, *Thrips tabaci* Lindeman, to the varieties of onion. *Indian Journal of Entomology* 28, 396–398.

Vierbergen, G. (1995) International movement, detection and quarantine of Thysanopteran pests. In: Parker, B.L., Skinner, M. and Lewis, T. (eds) *Thrips Biology and Management*. Plenum Press, New York, pp. 119–132.

Wardle, R.A. and Simpson, S.R. (1927) The biology of Thysanoptera with reference to the cotton plant. 3. The relation between feeding habits and plant lesions. *Annals of Applied Biology* 14, 513–528.

Waterhouse, D.F. and Norris, K.R. (1989) Biological control: Pacific prospects – Supplement 1. *Australian Centre for Agricultural Research (ACIAR) Monograph* No. 12, 125 pp.

Weiss, M.J. and Beshear, R.J. (1987) Seasonal distribution of the flower thrips (*Frankliniella tritici* (Fitch)) and onion thrips (*Thrips tabaci* Lindeman) (Thysanoptera: Thripidae) on safflower in Montana. *Canadian Entomologist* 119, 399–401.

Yasuda, T. and Momonoki, T. (1988) Varietal resistance of eggplant (aubergine) introduced from southeast Asia to *Leucinodes orbonalis* Guenée and *Thrips palmi* Karny. *Proceedings of the Association for Plant Protection of Kyushu* 34, 139–140.

Yasumi, K., Shinohara, T., Horike, M. and Hirano, C. (1991) Effect of tomato leaf constituents on survival of *Thrips palmi* Karny (Thysanoptera: Thripidae).

Japanese Journal of Applied Entomology and Zoology 35, 311–316.
Yudin, L.S., Tabashnik, B.E., Mitchell, W.C. and Cho, J.J. (1991) Effects of mechanical barriers on distribution of thrips (Thysanoptera: Thripidae) in lettuce. *Journal of Economic Entomology* 84, 136–139.
Zareh, N. (1985) Evaluation of six cotton cultivars for their resistance to thrips and leafhoppers. *Iran Agricultural Research* 4, 89–97. (*Review of Applied Entomology* (*A*) 75, 1944.)

17

Integrated Pest Management (IPM) in Tree Crops

Bruce L. Parker and
Margaret Skinner
University of Vermont, Entomology Research Laboratory, PO Box 53400, Burlington, Vermont 05405-3400, USA

Introduction

Due to population expansion and greater land utilization, the importance of trees to human welfare, national economies and the environment as a whole steadily grows. Thrips have also increased in importance and as pests of trees have become of concern to many dealing with the long-term health of forest and urban trees, and the quality and quantity of tree fruit production. Even in easily accessible crops, thrips are a difficult group of insects to manage because of their small size, thigmotactic behaviour and reproductive capacity. In tree crops they present a still greater management challenge. Population monitoring is complicated by tree size and height, effective coverage of insecticide sprays is difficult to achieve and many of the cultural controls used by growers of annual field- or greenhouse-grown crops (e.g. rotation, fallowing, flooding, exclusion screening) are not feasible for perennial tree crops. Thus the options for thrips management available to growers of tree crops are fairly limited compared with other crops. In all agricultural settings, IPM, an integrated approach for managing insects and diseases utilizing cultural, biological and chemical techniques, is gradually replacing more simplistic, agrochemical-based methods. However, comprehensive, formalized IPM programmes for thrips in tree crops are rare. More commonly, individual management components are each addressed separately, and seldom viewed holistically or combined to provide growers with a complete management plan. For example, chemical insecticides have been tested extensively for thrips management in tropical tree fruits (Niann, 1991), but effective sampling methods or economic thresholds that inform growers when to make pesticide applications are often lacking. The role of natural enemies may be assessed, ignoring the potential impact of insecticides or other management practices.

Despite the general lack of IPM programmes for thrips on tree crops, there is considerable information available on particular management options which have implications for multiple pest situations. Appendix III lists the tree crops for which an associated thrips pest has been noted, indicating specific IPM tactics available for each species, though the existence of an appropriate management practice does not imply that it is commonly used by growers as part of an IPM plan. Though several thrips species may be classified as pests of tree crops, relatively few pose a serious enough economic threat to warrant management action.

To demonstrate the multiple components of IPM for thrips pests on trees we have selected two crops that suffer from thrips attack, citrus and sugar maple, for which IPM approaches have been devised or are under development. These tree crops represent two ends of this agricultural spectrum – fruit orchards and forests. The current status of IPM for these two tree crops is considered in depth, and many of the techniques developed for them can be generalized to other thrips pests of tree crops.

Fruit Tree Crop: Citrus

Pest status

A complex of different thrips species attack citrus in different regions: *Scirtothrips citri*, *S. aurantii*, *S. dorsalis*, *Frankliniella bispinosa*, *Heliothrips haemorrhoidalis*, *Chaetanaphothrips signipennis*, *C. orchidii* and *Thrips hawaiiensis* (see Appendix III). Whereas *F. bispinosa* feeds exclusively on flowers, other species feed on and cause damage to the rind of immature fruit (see Chapter 13 and Plate V). Different IPM approaches must be developed that specifically address each of these situations. Various components of IPM have been investigated for *S. citri*, *S. aurantii*, *H. haemorrhoidalis* and *F. bispinosa*.

Scirtothrips citri has been a serious economic problem on all varieties of citrus in California for almost 100 years. Feeding by this pest does not affect internal fruit quality, but causes external scarring and silvering which results in fruit of a lower market grade and value (Elmer and Brawner, 1988). Most scarring damage is caused by second instar larvae (Wiesenborn and Morse, 1986) and the susceptible fruit stage spans a 3 to 8 week period after fruit begins to form (Rhodes *et al.*, 1986). Little new economically important scarring occurs beyond 3 weeks after petal fall (Lewis, 1935). Most fruit damage occurs shortly after petal fall and decreases at a constant rate thereafter. Ambient temperatures may have a significant effect on the amount of fruit scarring that occurs. Temperatures of 29 to 30°C are optimum for thrips feeding and at higher temperatures first and second instars feed more slowly (Wiesenborn and Morse, 1986). Damage cannot be anticipated in advance without specific information on the population dynamics of thrips. Rhodes *et al.* (1986) examined the effect of temperature on thrips injuriousness, using degree-days above a base of 14.6°C for predicting crop response to thrips feeding. Thrips larvae are involved almost exclusively in feeding and thus degree-days for larval activity as related to fruit damage were considered of prime importance. Rhodes *et al.* (1986) concluded that the amount of thrips scarring damage can be anticipated from data on accumulated larval days (Fig. 17.1).

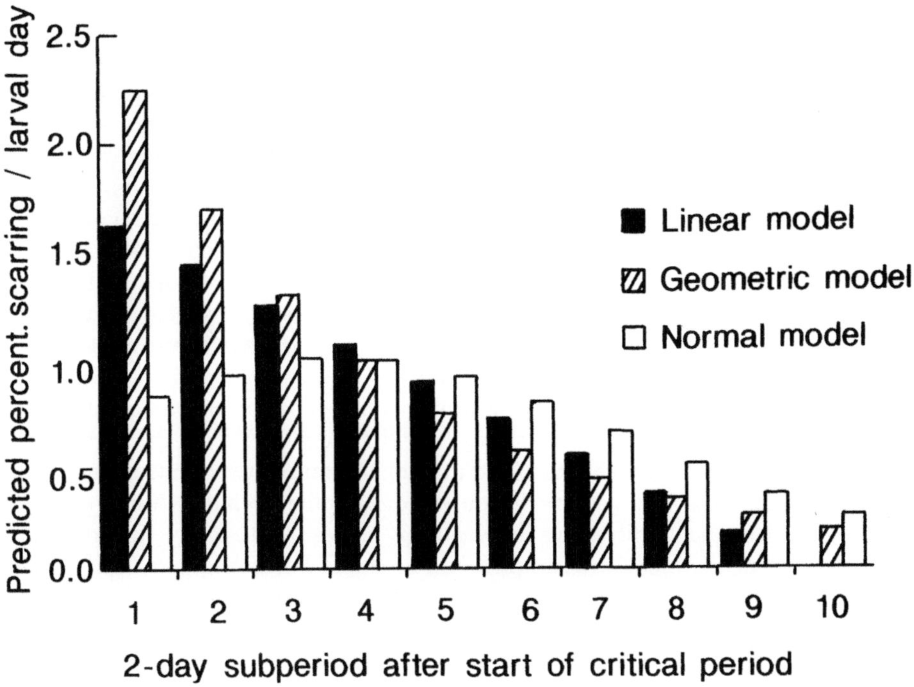

Fig. 17.1. Predicted percentage of fruit scarred per additional larval day in each of the ten subperiods of 2 days in the critical period reported by Lewis (1935) compared for damage functions incorporating the linear, geometric and normal models of changing fruit susceptibility (after Rhodes *et al.*, 1986).

In addition to causing fruit damage, *Scirtothrips citri* also feeds on the foliage, which on young trees may delay the onset of initial fruit production. A closely related species, *S. aurantii*, causes damage similar to *S. citri* in South Africa (Grout and Stephen, 1993) and on Reunion Island (Quilici *et al.*, 1988).

The greenhouse thrips, *Heliothrips haemorrhoidalis*, is an occasional pest of citrus, preferring coastal climates where temperatures are mild with moderate humidity. It is especially damaging to coastal Valencia oranges following a mild winter. Significant *H. haemorrhoidalis* mortality occurs when temperatures fall below 0°C (Anonymous, 1994).

Before 1992, *Colletotrichum gloeosporioides*, a fungal plant pathogen, was considered the cause of post-bloom fruit drop of Florida citrus (Fagan, 1984). Research by Childers (1992) revealed that potentially there were two organisms involved in post-bloom fruit drop, the fungus and *Frankliniella bispinosa*. Childers *et al.* (1990) found that 92 per cent of all thrips collected in closed buds and open flowers of citrus in Florida were *F. bispinosa*; the next most common species was *F. kelliae* (7 per cent). Following *F. bispinosa* feeding, flowers often abort and yield declines (Childers and Achor, 1995).

Survey and damage assessment

Scirtothrips citri should be monitored on a regular basis for 6 weeks after petal fall to predict damage accurately and determine the need for, and appropriate timing of, pesticide applications. Traps for sampling thrips populations on leaves (McGregor, 1926) and monitoring thrips emergence and dispersal in citrus groves (Reed and Rich, 1975) have been developed. Tanigoshi and Moreno (1981) modified the Reed and Rich trap, constructing theirs from a 20-cm diameter polyvinyl chloride (PVC) pipe, and fitting it with a clear acetate cover coated on both sides with a sticky substance. When placed on the ground beneath citrus trees, it captured adult thrips emerging from the soil and larvae falling from the tree canopy (see Chapter 11). These traps provide useful information on when and how many adults emerge and when larvae are falling out of trees to seek pupation sites in the soil.

For *Scirtothrips citri* and their natural enemies, monitoring fruit and leaves is recommended. To sample for thrips, 25 fruits are picked from the corner of each block (a block may be 4 to 20 ha in size) and one or two fruits from trees located about four rows from the outer edge of the orchard (Anonymous, 1994). For beneficial mites, five inner canopy leaves from each of five trees in each corner of the block are sampled and counted. For oranges in California this sampling should be carried out for 6 to 8 weeks when the fruit is susceptible (March to May) and for lemons from June to October (Anonymous, 1994).

Several methods for monitoring the aerial activity of adult thrips have been tested (see Chapter 11). In an experiment to test the best colour and position of sticky traps for monitoring *Scirtothrips citri* in flight, most adults were captured on white traps positioned 2.7 m above ground on the south quadrant of the tree (Beavers *et al.*, 1971). Yellow fluorescent sticky traps are recommended for monitoring *S. aurantii* in citrus orchards on Reunion Island (Quilici *et al.*, 1988). *Thrips hawaiiensis*, a problem in Taiwan on lemon and lime, shows a preference for blue traps and the highest catches were recorded from early morning to 10.00 h. These results coincided with field activity of this species (Chiu and Wu, 1993). Seasonality of *S. dorsalis* was monitored from 1984 to 1988 in citrus in Japan using sticky suction traps; there were seven or eight generations per year and their developmental rate was linearly related to temperature (Tatara, 1994). Because *Heliothrips haemorrhoidalis* populations tend to be patchy within an orchard, responding to the more favourable microclimates that may exist, fruit must be inspected in those areas where damage occurred the previous year (Anonymous, 1994). This species can be easily found in California in mid-May by looking at adjacent touching fruits.

Monitoring of citrus growth and development is also important for management and a model of the phenology of navel orange and *Scirtothrips citri* has been developed by Tanigoshi and Nishio-Wong (1982) (Fig. 17.2). The critical period to implement management for *S. citri* spans the time from petal fall, when fruit is first formed, until it hardens off at walnut size. No *S. citri* scarring damage is initiated on fruit with a diameter greater than 4 cm (Tanigoshi and Moffitt, 1984), and fruit of 25 mm in diameter is no longer susceptible to attack (Grafton-Cardwell *et al.*, 1995).

In Florida, Childers (1993) tested three trap designs for monitoring

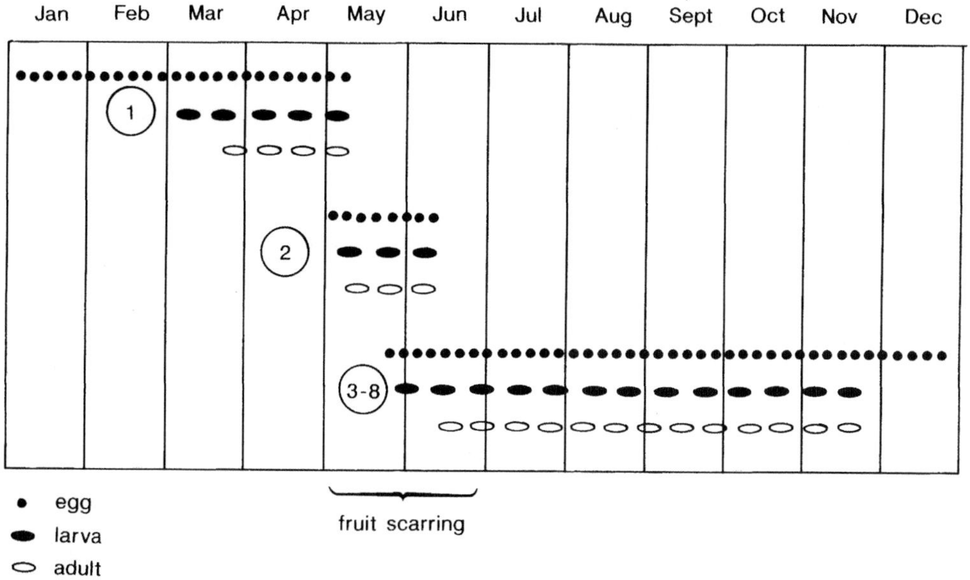

Fig. 17.2. Seasonal development of *Scirtothrips citri* on navel orange in the San Joaquin Valley. *S. citri* produce up to eight generations per year, but the most important generation is the second (2) because it scars the small fruit (after Anonymous, 1991a).

Frankliniella bispinosa beneath citrus trees with results comparable to those of Tanigoshi and Moreno (1981). Significantly more thrips were caught when traps were placed 15 cm from the tree boles or inside the drip lines than when they were located between the rows of trees.

Biological and cultural control

Seven species of the predatory mite *Amblyseius* have been collected on citrus in Zimbabwe, Swaziland and South Africa. *Amblyseius citri*, although not as effective as *A. addoensis*, which occurs further south, is capable of controlling *Scirtothrips aurantii* (Grout, 1994). Grout and Richards (1992) in trials in several orange groves noted that *A. addoensis* significantly reduced populations of *S. aurantii* and subsequent scarring damage of fruit. When populations of predatory mites were greater than 1 per inside leaf at petal fall, thrips scarring resulted in only about 1 per cent rejection of fruit at the packing house.

The predatory role of the mite, *Amblyseius hibisci*, in Californian citrus was evaluated by Tanigoshi *et al.* (1985). This predator was successfully colonized and introduced into orchards for control of *Scirtothrips citri* in 1978 (Tanigoshi and Griffiths, 1982). From 1978 to 1981 in an experimental citrus grove, an 85 per cent reduction in damage by *S. citri* was attributed to the influence of *A. hibisci*. Applications of dimethoate, methomyl and parathion were toxic to these mites, whereas chlorobenzilate and carbaryl at about 4 weeks post-application were not (Tanigoshi *et al.*, 1985). The hymenopterous parasite *Thripobius semiluteus*, originally from Brazil and Australia, is the only natural enemy recorded attacking *Heliothrips haemorrhoidalis* in coastal citrus in southern California (Anonymous, 1994).

In South Africa, Grout and Stephen (1993) suggested, on the basis of laboratory observations, that during periods of cold weather the predatory mite, *Amblyseius citri*, may be effective against *Scirtothrips aurantii*. This mite becomes active following cold periods faster than its thrips prey, giving it a predatory advantage.

Several cultural control strategies have been tested to manage *Scirtothrips citri*. Early infestations of *S. aurantii* do not come from non-cultivated lands adjacent to orange orchards but originate within the orchard (Gilbert, 1990). Later in the season thrips may move into orchards from surrounding vegetation to increase resident populations.

Although not normally considered, windbreaks near citrus groves appear to have an influence on populations of *Scirtothrips aurantii*. Fruit scarring by this thrips was greater on oranges growing adjacent to *Grevillea robusta* than on those in the centre of an orchard, but this relationship did not hold when windbreaks were created with *Pinus radiata* or *Casuarina cunninghamiana* (Grout and Richards, 1990). The same authors also reported that populations of the thrips predator, *Amblyseius addoensis*, were larger on oranges next to windbreaks of *P. radiata* and *C. cunninghamiana* than in the centre trees.

Most cultural practices, such as fertilization or irrigation, in orange groves do not appear to have a significant influence on the amount of damage by *Scirtothrips citri* to fruit (Hare *et al.*, 1990). Decisions on these practices should be made on the basis of tree health, yield and quality of fruit. However, populations of *Amblyseius tularensis*, a predatory mite, increased following tree pruning and responded to the new flush of foliage (Grafton-Cardwell and Ouyang, 1995a). Given a cost of US$432 to treat a hectare with insectary-reared mites, releases of this predator at rates of 500 per tree to every other tree in an orchard were deemed not cost-effective. This species shifts its feeding preferences, from pollen to *S. citri* larvae, when trees are beginning to bloom (Tanigoshi, 1991). In some Californian orchards with augmented releases of *A. tularensis*, larger mite populations were significantly correlated with improved predation of *S. citri* and less scarring of fruit (Grafton-Cardwell and Ouyang, 1995b). These authors concluded that in citrus managed with selective pesticides that favour the natural enemy complex, *A. tularensis* may play an important role in reducing damage by *S. citri*.

Damage by *Heliothrips haemorrhoidalis* is cumulative over the growing season and therefore one way of limiting this injury is to harvest early (Anonymous, 1994). This practice also reduces injury in subsequent seasons because the thrips tend to congregate on the fruit and are removed from fields at harvest.

Economic and action thresholds

In Japan, *Scirtothrips dorsalis* causes silver-scar damage to the stylar end of the fruit. This damage first appears in early June and reaches a maximum in early August. When 8 per cent of sampled fruit are infested with thrips larvae, the

economic threshold for damage to occur to the fruit pedicel has been reached (Tatara and Furuhashi, 1992).

The University of California IPM programme for citrus suggests a number of different treatment thresholds for *Scirtothrips citri*. For example, for growers mainly managing with broad-spectrum pesticides, the threshold in the absence, or for low populations, of predatory mites is when 5 per cent of navel fruit are infested with thrips larvae. For growers using selective pesticides, the threshold for treatment of navel oranges is when 10 per cent of fruit are infested. With the latter approach there is a threshold difference depending on the variety grown, for example with variety Valencia the treatment threshold is when 20 per cent of the fruit are infested with thrips larvae (Anonymous, 1991a).

In South Africa the threshold population level of *Scirtothrips aurantii* for 1 per cent damage on citrus fruit is based on counts from fluorescent yellow sticky traps. In orchards in the Eastern Transvaal as few as three traps for more than 5000 trees was practical and thresholds were 9 thrips per three traps during the central critical period (1 October to 21 November) and 20 thrips per three traps during the late risk period (22 November to 31 December) (Samways, 1986).

There is no established economic threshold for *Heliothrips haemorrhoidalis* on oranges but, in areas of the fruit where thrips and damage are seen, populations will increase and therefore damage will increase over the season unless management is initiated (Anonymous, 1994).

Chemical control

IPM of *Scirtothrips citri* is an evolving process necessitated by rapid development of insecticide resistance, fewer insecticides available, more stringent government regulations and greater public concern for the safe use of insecticides with less impact on the environment (Morse, 1995) (see Chapter 15). The control of thrips on citrus predictably will change from traditional heavy reliance on insecticides to the usage of selective materials in harmony with other biotic factors. Morse *et al.* (1988) anticipated this transition and suggested pesticides such as abamectin, sabadilla and ryania be seriously considered for use with the predator *A. tularensis* for management of *Scirtothrips citri* in California.

Because other citrus pests, such as *Aonidiella aurantii* and *Panonychus citri*, are being controlled by biological agents, it is critical before using new pesticides against *Scirtothrips aurantii* that they be screened for effects on non-target organisms (Bedford *et al.*, 1987).

Scirtothrips citri develops resistance readily and repetitive use of insecticides should be avoided. To illustrate the severity and speed at which resistance can develop, after ten laboratory selections and 10 months pressure with fluvalinate, *S. citri* was 482 to 1380-fold more resistant to this material (LC_{50} and LC_{90}, respectively). They also exhibited high cross-resistance to other pyrethroids, DDT, and about a 17-fold increase in resistance to parathion (Immaraju and Morse, 1990) (see Chapter 15).

To delay the onset of resistance and prolong the effective life of insecticides, resistance management, which works best in populations that are still susceptible, should be practised (Georghiou, 1983). Two common strategies employed are the use of insecticide rotations and mixtures. In a field trial comparing these strategies with the insecticides fluvalinate and formetanate, Immaraju *et al.* (1990a) recommended resistance management of *Scirtothrips citri* with rotations. Additionally they stressed using biological controls and educating consumers to accept fruit with a few blemishes. For effective field management of resistance when using new insecticides, dosage rates must be considered. High dosages, although producing heavy mortality of thrips, may accelerate resistance development (Immaraju *et al.*, 1990b).

In an attempt to develop alternative strategies with pesticides, Milne and de Villiers (1975) tested a number of systemic insecticides applied in irrigation water to the soil. Applications of dimethoate were effective and populations of *Scirtothrips citri* were controlled for 10 to 12 weeks after application. Citrus red mite populations were also suppressed and no apparent reduction in the degree of parasitism by *Aphytis* sp. was observed. Soil applications of various organophosphate and carbamate insecticides for management of *S. citri* were compared in California orange groves. Tanigoshi *et al.* (1982) determined that emulsifiable concentrates and flowable formulations were more efficacious than granules when used at bloom. Timing of applications was critical and the authors stressed that continual surveying of citrus phenology and thrips development was essential. One subtle factor to consider is that pre-bloom sprays may stimulate surviving thrips to produce more offspring. This new generation conceivably would be feeding actively during petal fall and create more damage than the original population (Morse and Zareh, 1991). In southern Californian coastal citrus, pyrethrum sprays in early spring are recommended for control of *Heliothrips haemorrhoidalis* (Anonymous, 1994). These sprays have a minimal impact on the natural enemy complex.

The value of soil applications of insecticides was questioned by Grout *et al.* (1986a), who concluded that most *Scirtothrips citri* found in citrus trees in California pupated within the tree rather than in the soil.

In California, peak population levels of first-generation second instar *Scirtothrips citri* were used as a 'biofix', and 125 degree-days later (using a lower threshold of 14.6°C) the optimal time for spraying was reached (Grout *et al.*, 1986b). If insecticides are not used selectively, foliar sprays for control of *S. citri* usually have a negative effect on the natural enemy complex that controls citrus red mite. Prophylactic insecticide treatments for management of *S. citri* should be discouraged because this strategy may increase *S. citri* fecundity (Morse and Zareh, 1991).

In Thailand permethrin, cyhalothrin, carbosulfan and monocrotophos are recommended for controlling thrips damaging flowers and scarring fruit of pommellos (Suwanbutr *et al.*, 1992). The optimal time for spraying and spray impact on non-target organisms are unknown.

IPM of citrus is complicated. For example, Californian growers who sprayed early in the season with broad-spectrum insecticides had less fruit scarring, but secondary pest populations increased and resistance problems surfaced. If they used selective materials coupled with natural enemies, secondary pest resurgence was minimized yet scarring by *Scirtothrips citri* was not always controlled (Grafton-Cardwell *et al.*, 1995). Pesticide applications made with ground sprayers are better than applications made by airplane. Volumes of 200 gallons per acre (757 litres per 0.404 ha) are more efficacious than lower or higher volumes, except when sugar or molasses are used as baits with sabadilla and ryania (Anonymous, 1994).

In Florida, *Frankliniella bispinosa* was controlled effectively for 3 to 7 days during citrus flowering and fruit set with applications of chlorpyrifos, dimethoate or formetanate hydrochloride. The greatest increases of fruit set were obtained when insecticide sprays were made during early flowering accompanied by dry weather (Childers, 1992).

Forest Tree Crop: Sugar Maple

Pest status and biology

Taeniothrips inconsequens, pear thrips, was first recognized as a pest of sugar maple, *Acer saccharum* Marsh., in 1979 in Pennsylvania and gained national recognition in 1988 when over 600,000 ha of maple trees in the eastern United States were severely defoliated (Parker, 1991). It continues to cause damage intermittently, though the particular factors favouring population increase and damage are not fully understood. Sugar maple is one of the most abundant tree species in the northern hardwood forest in North America with a natural range from Tennessee to Nova Scotia (Collingwood and Brush, 1979). In Vermont, which is about 75 per cent wooded (1.8 million ha), sugar maple makes up 23.1 per cent of all forest trees (Frieswyk and Malley, 1985). This tree species contributes significantly to the economy of the north-eastern United States, over US$24 million annually in Vermont alone not including tourism (Parker, 1991). The maple syrup industry is an important source of revenue for farmers in the north-east. Syrup is produced from sap collected from sugar maple trees in forests managed to increase the sap yield and sugar concentration. These 'sugarbushes', a term for natural forest stands predominating in sugar maple where sap is collected, range in size from 2 to 100 ha. To maintain production, dead or dying trees and some of the non-maple species are removed to obtain sufficient numbers of well-spaced, high-quality maples. To collect sap for boiling into syrup, several holes are drilled into the trunk in late winter, into which a spout (tap) is placed to direct its flow. Sap begins to flow as soon as temperatures rise in the spring (February to March) and is usually collected for a 1 to 2-month period. Maple trees are about 40 years old before they are tapped and can be productive for over 100 years, thereby representing a long-term investment to syrup producers (sugarmakers). The widespread damage observed in 1988 precipitated a regional research effort to develop IPM strategies for *T. inconsequens* in sugarbushes.

Taeniothrips inconsequens in the north-eastern forest environment presents a unique IPM challenge. A favoured perennial host plant, sugar maple, is virtually ubiquitous throughout the pest range, providing an inexhaustible food source. Sugarbushes often cover a large area and are inaccessible to standard ground-spray equipment. Mature maple trees attain a height of over 30 m, making the buds in which *T. inconsequens* feed difficult to reach for treatment. Unlike pests which predominate in specific habitats within an ecosystem, *T. inconsequens* is found almost everywhere in the north-east, making spot treatment of infested areas ineffectual. It lives below ground for 10 months of the year, protected within an earthen cell, making survey and detection difficult and limiting control options. It is above ground in the early spring, when temperatures are cool, and before most natural enemies have emerged. Maple trees, though contributing significantly to north-eastern agricultural revenues, do not generate sufficient income to enable sugarmakers to spend much on protection. It is impossible to assign a monetary value to the impact of thrips damage on sugar maple. Though no mature trees are known to have been killed directly as a result of thrips damage, it stresses the trees, which can reduce photosynthate production, sap yield or sugar content, or renders them susceptible to secondary, lethal disease organisms (Houston *et al.*, 1990; Kolb *et al.*, 1992; Ellsworth *et al.*, 1994). The level of starch in the roots provides an indication of the overall health of the tree and an estimate of the food reserves available to the tree over the winter (Wargo, 1988). Based on root starch testing in maple trees, those that were severely defoliated by thrips and then refoliated had a higher starch content than those that suffered light or moderate defoliation but did not refoliate (Burns, 1991). Because the damage caused by *T. inconsequens* occurs early in the season, trees have an opportunity to recover the lost food reserves if no additional climatic stress or pest attack occurs later in the summer.

Attention has been focused primarily on the impact of damage by *Taeniothrips inconsequens* to the health and sap production of mature maple trees, but its damage to flowers and the subsequent seed crop has also been noted. In Pennsylvania, where sugar maple provides essential species diversity in hardwood forests predominating in oak, forest managers have expressed concern about the decline in maple regeneration which may be a result of thrips feeding in the flowers or on seedlings (Laudermilch, 1988; McCormick, 1990). In addition, an association between *T. inconsequens* and the foliar fungus *Discula campestris* has been detected (Nash *et al.*, 1991; Stanosz, 1993). This plant pathogen enters leaf tissue through wounds, causing leaves to become necrotic, and in severe cases to fall off (Nash *et al.*, 1994). Because a high incidence of this disease has been found in foliage damaged by *T. inconsequens*, it is believed that wounding from thrips feeding provides suitable sites of entry for the fungus. Maple seedlings which often have only two or four leaves are particularly susceptible to this disease (Stanosz, 1993).

Taeniothrips inconsequens was introduced into California from Europe in 1902, on soil accompanying rootstocks (Bailey, 1944). Though known to occur in New York as early as 1911 where it caused damage on pears (Parrott, 1913), confirmation of the presence of *T. inconsequens* in Vermont was not made until 1985 (Teillon *et al.*, 1985). Whether it occurred there earlier is unknown. Because this is a relatively new pest for sugar maple, no efforts at management had been attempted before 1985.

To design effective IPM plans an understanding of the biology of the target pest is essential. The life cycle of *Taeniothrips inconsequens* in Californian and

British Columbian fruit orchards is well documented (Foster and Jones, 1915; Cameron and Treherne, 1918; Bailey, 1944), but was essentially unknown for north-eastern maple forest environments prior to 1988. Recent research has shown that though the life cycle basically mirrors that of the Californian population, the timing of key life events varies slightly (Skinner *et al.*, 1991a; Teulon and Cameron, 1991). Adult *T. inconsequens* emerge from the soil in early spring, responding to warming temperatures (Fig. 17.3). They fly to host trees, seeking partially open buds. Once in the buds they feed on tender leaf and flower tissue. Because several leaves or flowers are contained within each bud, only a few thrips are needed to cause significant damage. Larvae hatch from eggs laid in the leaf veins and feed for a few weeks in the upper tree canopy before dropping to the forest floor in early summer. They feed for a short time on understorey vegetation, sometimes causing heavy damage to the first-year maple seedlings (Gardescu, 1991), and then enter the soil, where they construct an earthen cell in which to overwinter. In well-drained forest soils, over 65 per cent of the *T. inconsequens* are found in the top 10 cm (Skinner and Parker, 1992).

Basswood thrips, *Thrips calcaratus*, is the only other pest thrips species known to occur in north-eastern maple forests. It is primarily a pest of American basswood, *Tilia americana*, in the Great Lakes states (Raffa and Hall, 1989), and its role in defoliation of sugar maple is unknown (Raffa *et al.*, 1992). In soil samples taken in Vermont maple stands, *Thrips calcaratus* generally comprised less than 10 per cent of the emerging thrips population, suggesting that it is not a major threat to maple trees (Parker and Skinner, 1993). No specific IPM strategies have been developed for this thrips pest in maple or basswood forests (K. Raffa, University of Wisconsin, 1996, personal communication).

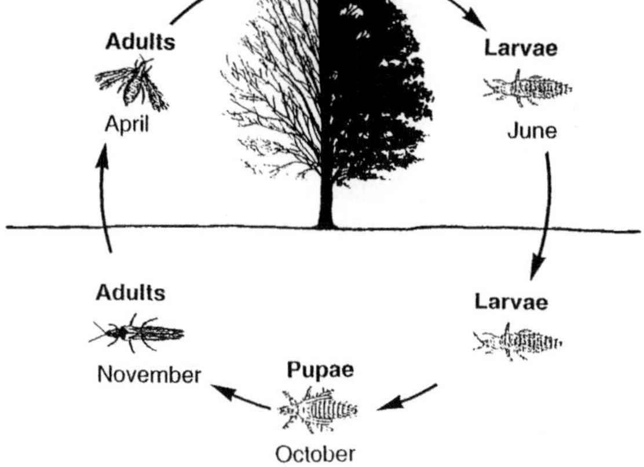

Fig. 17.3. Life cycle of *Taeniothrips inconsequens* in sugar maple forests in the north-eastern United States (after Skinner and Parker, 1995).

Pest survey, damage assessment and action threshold

To be appropriate for forest IPM, a survey method must provide accurate information on pest levels and their relationship to future damage in time for sugarmakers or pest managers to take appropriate action to protect trees from damage. The method must be simple, inexpensive and quick.

Several methods of evaluating adult population levels have been used. For several years yellow sticky cards were placed in forests before adult emergence (Hollingsworth *et al.*, 1995). This proved effective for detecting the presence of *Taeniothrips inconsequens* in an area, but the number of thrips per trap showed no correlation with damage (Coli *et al.*, 1992). It is still used to assess the possible relationships between adult thrips activity and abundance, maple bud development and subsequent damage, but for sugarmakers it is of little practical value (Wilmot and Scherbatskoy, 1994). Using suction traps and towers equipped with clear sticky cylinders, spring movement of large numbers of *T. inconsequens* into and out of sugarbushes was detected (B.L. Parker, Vermont, 1990, unpublished data). It is unknown whether they are merely carried by strong prevailing winds or actively migrating in search of suitable hosts. The reliability of any method for monitoring thrips populations within individual stands is confounded by the fact that widespread relocation may occur. This has obvious implications for management as well and suggests that a regional rather than site-specific approach would be the most effective.

A method for sampling maple buds (100 per sugarbush) and dissecting them to determine the number of thrips within has been devised (Burns, 1990a). When repeated every year, this method gives a picture of annual thrips population fluctuations. In Vermont where sampling has been carried out since 1986, an average of less than two thrips per bud was usually found, compared with nine per bud in 1988, when damage was severe (Teillon *et al.*, 1994). Without a pole pruner or tree climber, sampling is restricted to understorey vegetation, and the relationship between populations at this level and in the tree canopy is unknown. A correlation between the number of thrips in the buds and the amount of damage to the understorey was observed, but the reliability of this association has never been established (Burns, 1990a).

Emergence traps of various sizes, styles and colours have been tested to quantify thrips populations as they emerge in spring and monitor the timing of emergence in maple forests (see Chapter 11). The most basic style was one made from a brown, opaque, 15-litre plastic tub, placed upside down on the soil, with an inverted funnel and glass jar attached to the top (Laudermilch, 1989). Though it was easy and inexpensive to construct this trap, its accuracy was never determined. Two types of traps were tested under forest conditions – a metal cone-shaped trap (Maier, 1991) and cylindrical traps constructed from different colours (black and white) and diameters of PVC pipe (7.6, 10.2, 15.2 and 20.3 cm in diameter) which were covered with a clear sticky plastic lid (Parker and Skinner, 1993). Several variables were considered, including trap cost, capture efficiency and sampling time, to determine the most suitable trap for *Taeniothrips inconsequens* IPM. The white, 7.6-cm diameter trap was found to be the least expensive to construct and took the least amount of time to inspect in the field and laboratory. Fewer thrips per square centimetre were captured in the 7.6-cm trap than in the 15.2- and 20.3-cm traps, though the magnitude of difference was no more than 0.13 thrips cm^{-2}.

Using the 7.6-cm emergence trap, a correlation was obtained between the number of thrips that emerged from soil in the spring and resultant thrips damage, from which a thrips threshold was established (Skinner and Parker, 1996). Over the 3 years of this study (1989 to 1991), the predicted number of thrips per soil sample for moderate damage to occur ranged from 8 to 18. Why such large annual fluctuations in this threshold occurred is unknown, but may relate to climatic conditions that affected thrips activity or the rate of bud development. Despite the broad threshold range, sugarmakers can use this information as a rough guide for predicting damage.

Though each of these sampling methods has unique advantages in terms of IPM, all share one major disadvantage: the data generated are not available in time to help the producer make management decisions prior to damage. To overcome this problem, a method was developed to extract thrips from forest soil collected in the autumn. Because most of the thrips occur in the top 10 cm of the soil, a tulip bulb planter (7.6-cm diameter, 10-cm length) was selected as an inexpensive but appropriate sample tool. Studies of the spatial distribution of *Taeniothrips inconsequens* in a sugarbush revealed that in forests with an average of 5 to 10 thrips per sample, thrips population levels could be estimated from ten samples, allowing for an error rate of ± 6.5 thrips (Skinner and Parker, 1991). A standardized protocol based on these results was established and has been used for several years in regional surveys of *T. inconsequens* (Skinner and Parker, 1990).

Two methods of extracting *Taeniothrips inconsequens* from forest soil have been used. A heptane flotation process was developed that removed live and dead thrips (Parker *et al.*, 1992) (see Chapter 11). Using this method, thrips population levels can be determined prior to their emergence in the spring, from which damage predictions can be made. A quicker and easier method is to take samples with a bulb planter, as described above, in late November after *T. inconsequens* have reached adulthood, place each in a paper container covered with a clear plastic lid coated with a sticky material, and hold at room temperature for 35 days to allow thrips to emerge naturally (Parker *et al.*, 1995; Skinner and Parker, 1995). Using this method, information about thrips population levels and potential damage is available to sugarmakers by January, prior to the start of tree tapping. Because it takes a minimum of time and equipment, it is recommended to sugarmakers as part of a total approach for long-term forest health management. A project to transfer this technology to sugarmakers was initiated in 1995 (M. Skinner and B.L. Parker, Vermont, 1996, unpublished data).

For the past 8 years in Vermont, soil populations of *Taeniothrips inconsequens* have been monitored in over 100 sugarbushes around the state in a joint project by the University of Vermont and the State of Vermont Department of Forest and Parks. This service has supplied sugarmakers statewide with essential information about trends in thrips populations, prediction of damage on a regional scale and management recommendations before the sugaring season begins.

A standardized method for assessing thrips damage to foliage from the ground has been developed for understorey and mature maple trees (Wilmot and Kelley, 1990). When used in conjunction with soil sampling, sugarmakers can evaluate the relationship between the number of thrips obtained in soil samples with the resulting damage in spring (Skinner and Parker, 1995). Because thrips damage may have an effect on the long-term health of their trees, sugarmakers are encouraged to inspect them for damage annually and consider this information when making silvicultural decisions.

Though damage assessment by sugarmakers is essential for sound management decisions, a survey of damage on a regional scale also provides valuable information on the distribution and extent of the problem. A protocol and damage rating system for aerial sketchmapping of *Taeniothrips inconsequens* damage were developed for this purpose and have been used by forestry departments in Vermont and Pennsylvania (Burns, 1990b). The patchy nature of the damage, which reflects the distribution of sugar maple in a forested area, makes aerial mapping of thrips damage very difficult. It must be done in early June before damaged trees refoliate. The use of satellite imagery has also been evaluated to detect damage by *T. inconsequens* and, in general, results from this method agreed with those obtained from sketchmapping (Vogelmann and Rock, 1991). Because analysis of this type of data is very expensive, and cloud-free satellite images are sometimes unobtainable, this survey approach was judged unacceptable.

As demonstrated by the wide range in the thrips damage threshold obtained from the 3-year Vermont study, prediction of damage based on thrips population levels has proved difficult. Climate, the timing of thrips emergence and bud development may be important factors that influence damage. In Californian fruit orchards, Bailey (1944) noted that damage was more severe in years when budbreak was delayed or prolonged, which gave thrips more time to feed within the buds. Kolb and Teulon (1991) found that damage to greenhouse-grown maple seedlings was greater if thrips were introduced when the buds were slightly open than if they were completely closed or open (see Chapter 13 and Fig. 13.2). In a study of forest trees, no relationship between damage and the timing or duration of emergence was detected (Skinner and Parker, 1996). However, a picture of the complexity of the relationship between thrips, their damage and bud phenology was revealed. For example, damage was 1.5 times greater in study sites in one region than it had been in previous years, though 33 per cent fewer thrips emerged. In contrast, in another region where thrips populations based on emergence traps were relatively high, damage was very light. It is likely that several natural factors affect the extent of damage that occurs, and most of them are beyond human management. Only with continued sampling and damage assessment over many years will a more complete understanding of these factors be obtained.

Cultural control

In contrast to an orchard crop where the primary goal is protection of the flowers or fruit, sugarmakers manage their maple forests to ensure the long-term health of their trees, thereby maximizing sap production. Vigorous, fast growing trees with large crowns usually produce more sap than slower growing trees. For example, a tree that increases its trunk diameter by 2.5 cm every 2 to 5 years yields up to 30 per cent more sap than one that takes 7 to 10 years to attain this diameter growth (Houston *et al.*, 1990). Trees with large healthy crowns have the capacity to produce and accumulate greater food reserves during the summer, which translates to higher sap yields the following spring. These factors must be considered when developing IPM for *Taeniothrips inconsequens*. A study in Pennsylvania suggested that thrips damage may result in lower sap volume and sugar concentration the following year (Kolb *et al.*, 1992).

As a general rule, sugarmakers must make every effort to reduce potential stresses to maple trees that have suffered from thrips damage or are expected to in the coming year. Drilling a hole in a tree to collect its sap is a recognized stress and, according to long-standing recommendations, no more than two tap holes should be drilled into mature maple trees (over 45 cm in diameter at a height of 1.5 m). More than that may weaken a tree, making it susceptible to secondary decay organisms. When damage is expected based on thrips numbers above the damage threshold, sugarmakers are encouraged to decrease the number of taps per tree, and refrain from tapping stressed or unhealthy trees. Maple stands are regularly thinned to reduce competition and stimulate growth among favoured trees. Because this initially places an unusual stress on the released trees, thinning should be delayed in years when thrips damage has been severe (Burns and Myott, 1989).

In fruit orchards several cultural methods have been used to kill the soil-inhabiting, immature *Taeniothrips inconsequens* or delay adult emergence, including flooding or cultivating the soil beneath the trees, or planting heavy covercrops (Bailey, 1944). The value of these methods in orchards was questionable, and would not be feasible in a forest. Prescribed burning of the understorey vegetation in maple stands in the early spring was tested as a method to kill thrips before they emerged (Brose and McCormick, 1992) but, because of potential damage to the shallow root system of maple and the possible indirect environmental impact of this technique, it is not generally recommended.

Biological control

Taeniothrips inconsequens is rarely, if ever, a pest problem in Europe, suggesting that one or more biological or environmental factors keep populations below damaging levels. Though a few arthropod predators were identified in studies in Californian fruit orchards, natural enemies were thought to play an insignificant role in suppression of *T. inconsequens* (Essig, 1920; Bailey, 1944). In a recent survey of the natural enemies of *T. inconsequens* in central Europe, the rate of parasitism among field-collected larvae was less than 4 per cent (Carl *et al.*, 1989). Predation of *T. inconsequens* larvae by arthropods has also been observed infrequently, but not quantified, in eastern maple forests (M. Skinner and B.L. Parker, Vermont, 1989, unpublished data; Teulon *et al.*, 1993). Therefore, the impact of predators is presumed to be minimal in maple forest environments. Because

T. inconsequens is univoltine and lives above ground for only 2 months of the year, there is little opportunity for a natural enemy population to increase in response to this pest. Thus, it is doubtful whether augmentative releases of a predator would ever be effective.

In 1988, the entomopathogenic fungus *Verticillium lecanii* was first discovered infecting *Taeniothrips inconsequens* in Vermont (Skinner *et al.*, 1991b), and later found in many other north-eastern forests. Natural infection rates as high as 57 per cent were observed in some Vermont sugarbushes in 1989, and infected thrips larvae were found in 76 per cent of the 111 sites sampled. Several other fungi were later isolated from forest soil or insects, which have been found to be pathogenic to *T. inconsequens* (Carl *et al.*, 1989; Brownbridge *et al.*, 1993). Studies of the natural infection cycle proved that most infection of *T. inconsequens* occurs after larvae fall from the leaves (Brownbridge *et al.*, 1992). They probably come in contact with the infective spores as they crawl across the forest floor, or as they descend into the soil to overwinter. Therefore, for a fungal application to be effective, it would need to be made so that high concentrations of infective material were present before larval descent.

Research into the development of entomopathogenic fungi for management of *Taeniothrips inconsequens* is currently underway. Indigenous strains of *Beauveria bassiana* and *Verticillium lecanii*, known to be adapted to the cold northern climates, have been selected that grow and reproduce at temperatures which commonly occur in late spring. These strains have been mass-produced and formulated into emulsifiable concentrates (EC), wettable powders (WP) and nutrient-based granules (NBG) for testing under field conditions. Their efficacy against *T. inconsequens* in the forest is still being evaluated, including the appropriate dosage and timing of application. For field testing, fungal materials are applied to plots with a CO_2 sprayer, a method unsuitable for treating large acreages of maple forests. Recognizing the difficulties of gaining access into maple forests, which are often located on steep terrain lacking good roads, researchers believe aerial applications would be necessary. Studies to date have demonstrated that infective material from the NBG formulations persists at very high levels for over 16 weeks after application (Parker *et al.*, 1996). Given this fact, there would be a reasonable period for making applications before larval drop. In other studies, elevated fungal concentrations were detected 1 year after application (B.L. Parker, Vermont, 1994, unpublished data).

In contrast to orchard crops, where pest managers seek to protect the current flowers or fruit, the primary goal of *Taeniothrips inconsequens* IPM in sugarbushes is to minimize stress from foliage damage caused by thrips. Assuming a suitable fungal product was available, a scenario has been proposed for utilizing it effectively as part of an IPM plan. When thrips populations, based on soil sampling, showed a regional trend over 1 or 2 years of increasing to a damaging level, an aerial application of a granular fungal formulation would be made in selected stands in the spring, targeting the larvae as they returned to the soil. Because, with a granular formulation, elevated spore concentrations would persist for more than 1 year, the treatment effect would be maximized. Alternatively, sugarmakers with small sugarbushes could make applications of a granular formulation by hand. Given the abundance and widespread occurrence of *T. inconsequens* in northeastern maple forests, complete suppression is inconceivable. However, with the carefully timed, intermittent use of fungi, local populations could be maintained at low levels.

Chemical control

In the early and mid-1900s when *Taeniothrips inconsequens* was a major orchard pest in California, chemical insecticides were the predominant means of control. A broad array of chemical insecticides were used, many of which, such as DDT and lead arsenate, are no longer registered in the United States (Bailey, 1944). The decline in the pest status of *T. inconsequens* in California has been attributed to the use of synthetic organic insecticides to control the codling moth (Anonymous, 1991b).

Use of chemical insecticides for crop protection against damage from *Taeniothrips inconsequens* is far more complex in a forest than in an orchard habitat. In Vermont, sugar maple trees occur widely throughout the urban, rural and forested regions, as does *T. inconsequens*. Among the 100 sugarbushes sampled from 1988 to 1995 in Vermont, *T. inconsequens* was found in 85 to 100 per cent of the sites (M. Skinner and B.L. Parker, Vermont, 1995, unpublished data). Given the abundance of a preferred host, prevalence of the pest and possible negative environmental impacts to non-target organisms, the potential for control of *T. inconsequens* in forests with chemical insecticides is questionable. Despite these drawbacks, an aerial spray trial was conducted in 1989 to test the efficacy of carbaryl against *T. inconsequens* in maple forests. Though mortality of *T. inconsequens* was observed, significant differences in their population levels or the rate of damage in treated and untreated sites were undetectable (B.L. Parker and M. Skinner, Vermont, 1990, unpublished data). Insecticides have been used with some success to protect the foliage of individual backyard or urban trees, using carbaryl applied with hydraulic sprayers from ladder trucks (F. Marden, Vermont, 1996, personal communication). In this case, the adult must be killed before it enters the bud, or as the new leaves begin to expand. Because the foliage is still tender at this time, problems with phytotoxicity resulting from these applications have been encountered.

Concluding Remarks

The issues discussed in relation to development of management strategies for thrips on citrus and sugar maple trees demonstrate how very difficult and complex IPM implementation can be. Furthermore, thrips do not exist alone; there are many other insect and disease pests which must also be considered when preparing a comprehensive IPM plan for any crop. There are many tree crops for which thrips pose a serious threat. For most of these crops, one or more IPM components have been investigated. Despite the logic of IPM, the individual components of pest management are rarely consolidated into a total plan that is suitable for the grower to use. This is a challenge that, because of the complexity of today's cropping systems, requires a team approach and awaits the scientists, applied entomologists and pest managers of the future. IPM must be a fluid and evolving process, responding to ever-changing environmental, political and economic situations as well as consumer demands. As new pest management strategies are developed and old ones are perfected, they must be incorporated into existing IPM programmes and the technology transferred to the grower. Thrips demonstrate an uncanny capacity to overcome most traditional management tactics and, for long-term success, a multifaceted IPM approach to thrips management holds the greatest promise.

References

Anonymous (1991a) *Integrated Pest Management for Citrus*, University of California Publication 3303. IPM Education and Publications, University of California, Davis, 144 pp.

Anonymous (1991b) *Integrated Pest Management for Apples and Pears*, University of California Publication 3340. IPM Education and Publications, University of California, Davis, 214 pp.

Anonymous (1994) *Citrus Pest Management Guidelines*. University of California Publication 28, 108 pp.

Bailey, S.F. (1944) *The Pear Thrips in California*. University of California Bulletin 687. Berkeley, CA, 55 pp.

Beavers, J.B., Shaw, J.G. and Hampton, R.B. (1971) Color and height preference of the citrus thrips in a navel orange grove. *Journal of Economic Entomology* 64, 1112–1113.

Bedford, E.C.G., Vercueil, S. and Deacon, V.E. (1987) The effect of isofenphos, profenofos and triazophos on the biological control of non-target citrus pests in field trials, when used to control citrus thrips, *Scirtothrips aurantii* Faure. *Citrus and Subtropical Fruit Journal* 639, 6–11.

Brose, P. and McCormick, L.H. (1992) Effects of prescribed fire on pear thrips in Pennsylvania sugarbushes. *Northern Journal of Applied Forestry* 9, 157–160.

Brownbridge, M., Parker, B.L. and Skinner, M. (1992) Infection of pear thrips with *Verticillium lecanii* – timing of the natural infection process and its relation to temperature. In: Cameron, E.A., McCormick, L.H., Teulon, D.A.J. and Kolb, T.E. (eds) *The 1991 Conference on Thrips (Thysanoptera): Insect and Disease Considerations in Sugar Maple Management, University Park, PA, 21–22 November 1991*. US Department of Agriculture, Forest Service, Northeastern Forest Experiment Station General Technical Report NE-161, Radnor PA, p. 33.

Brownbridge, M., Humber, R.A., Parker, B.L. and Skinner, M. (1993) Fungal entomopathogens recovered from Vermont forest soils. *Mycologia* 85, 358–361.

Burns, B. (1990a) A quick measure of thrips populations using buds of sugar maple regeneration. In: Parker, B.L., Skinner, M., Wilmot, S.H. and Souto, D. (eds) *Pear Thrips Research and Management: Current Methods and Future Plans*, Vermont Agricultural Experiment Station Bulletin 697. University of Vermont, Burlington, pp. 111–112.

Burns, B. (1990b) Aerial sketchmapping: a record of where damage has occurred. In: Parker, B.L., Skinner, M., Wilmot, S.H. and Souto, D. (eds) *Pear Thrips Research*

and Management: Current Methods and Future Plans, Vermont Agricultural Experiment Station Bulletin 697. University of Vermont, Burlington, pp. 113–114.

Burns, B.S. (1991) Root starch in defoliated sugar maples following thrips damage. In: Parker, B.L., Skinner, M. and Lewis, T. (eds) *Towards Understanding Thysanoptera, Proceedings of the International Conference on Thrips, Burlington, VT, 21–23 February 1989*. General Technical Report NE-147, US Department of Agriculture, Forest Service, Radnor, PA, pp. 257–265.

Burns, B. and Myott, L. (1989) *The Sugar Maple and Pear Thrips: A Landowner and Sugarmaker Guide for Use in 1989–90*. Vermont Department of Forests, Parks and Recreation and University of Vermont Extension Service No. FS-153, 4 pp.

Cameron, E.A. and Treherne, R.C. (1918) *The Pear Thrips (*Taeniothrips inconsequens *Uzel) and its Control in British Columbia*. Canadian Department of Agriculture, Entomology Branch Bulletin 15.

Carl, K., Yuill, J., Parsch, M. and Prior, C. (1989) Taeniothrips inconsequens *in Central Europe and its Natural Enemies*. CAB International Institute of Biological Control, European Station Report, Delémont, Switzerland, 17 pp.

Childers, C.C. (1992) Suppression of *Frankliniella bispinosa* (Thysanoptera: Thripidae) and the fungal pathogen *Colletotrichum gloeosporioides*, with pesticides during the bloom cycle and improved fruit set on 'navel' orange in Florida. *Journal of Economic Entomology* 85, 1330–1339.

Childers, C.C. (1993) Trap designs for monitoring emergence of *Frankliniella bispinosa* (Morgan) (Thysanoptera: Thripidae) from soil in citrus groves. *Canadian Entomologist* 125, 449–456.

Childers, C.C. and Achor, D.S. (1995) Thrips feeding and oviposition injuries to economic plants, subsequent damage and host responses to infestation. In: Parker, B.L., Skinner, M. and Lewis, T. (eds) *Thrips Biology and Management*. Plenum Press, New York, pp. 31–52.

Childers, C.C., Beshear, R.J. Bushwein, J.R. and Denmark, H.A. (1990) Thrips (Thysanoptera) species, their occurrence and seasonal abundance on developing buds and flowers of Florida citrus. *Journal of Entomological Science* 25, 601–614.

Chiu, H.T. and Wu, M.Y. (1993) Attractiveness of color traps to *Thrips hawaiiensis* (Morgan) (Thysanoptera: Thripidae) in the field. *Chinese Journal of Entomology* 13, 229–234.

Coli, W.M., Hollingsworth, C.S. and Maier, C.T. (1992) Visual trap for monitoring pear thrips, *Taeniothrips inconsequens* (Uzel) (Thysanoptera: Thripidae) in maple stands and apple orchards. *Journal of Economic Entomology* 85, 2258–2262.

Collingwood, G.H. and Brush, W.D. (1979) *Knowing Your Trees*. American Forestry Association, Washington, D.C.

Ellsworth, D.S., Tyree, M.T., Parker, B.L. and Skinner, M. (1994) Photosynthesis and water-use efficiency of sugar maple (*Acer saccharum*) in relation to pear thrips defoliation. *Tree Physiology* 14, 619–632.

Elmer, H.S. and Brawner, O.L. (1988) The effect of citrus thrips and citrus red mite on navel orange fruit production in California's San Joaquin Valley. In: Goren, R. and Mendel, K. (eds) *Proceedings of the Sixth International Citrus Congress, Tel Aviv, Israel, 6–11 March*. Balaban Publishers, Philadelphia, pp. 1089–1095.

Essig, E.O. (1920) *The Pear Thrips*. University of California Agricultural Experiment Station Circular 223.

Fagan, H.J. (1984) Postbloom fruit drop of citrus in Belize. I. Disease epidemiology. *Turrialba* 34, 173–177.

Foster, S.W. and Jones, P.R. (1915) *The Life History and Habits of the Pear Thrips in California*. US Department of Agriculture Bulletin No. 173.

Frieswyk, T.S. and Malley, A.M. (1985) *Forest Statistics for Vermont – 1973 and 1983*. US Department of Agriculture, Forest Service Resource Bulletin Northeast 87, Broomall, PA, 102 pp.

Gardescu, S. (1991) Pear thrips and sugar maple seedling establishment in central New York. In: Parker, B.L. and Doane, E.N. (eds) *The 1990 Conference on Thysanoptera, Abstracts*, Vermont Agricultural Experiment Station Bulletin 698. University of Vermont, Burlington, p. 39.

Georghiou, G.P. (1983) Management of resistance in arthropods. In: Georghiou, G.P. and Saito, T. (eds) *Management of Resistance in Arthropods*. Plenum Press, New York, pp. 769–792.

Gilbert, M.J. (1990) Relative population levels of citrus thrips, *Scirtothrips aurantii* on commercial citrus and adjacent bush. *South African Journal of Zoology* 25, 72–76.

Grafton-Cardwell, B., Eller, A. and O'Connell, N. (1995) Integrated citrus thrips control reduces secondary pests. *California Agriculturalist* 49, 23–28.

Grafton-Cardwell, E.E. and Ouyang, Y. (1995a) Manipulation of the predacious mite, *Euseius tularensis* Congdon (Acarina: Phytoseiidae) with pruning for citrus thrips control. In: Parker, B.L., Skinner, M. and Lewis, T. (eds) *Thrips Biology and Management*. Plenum Press, New York, pp. 251–255.

Grafton-Cardwell, E.E. and Ouyang, Y. (1995b) Augmentation of *Euseius tularensis* (Acari: Phytoseiidae) in citrus. *Environmental Entomology* 24, 738–747.

Grout, T.G. (1994) The distribution and abundance of phytoseiid mites (Acari: Phytoseiidae) on citrus in southern Africa and their possible value as predators of citrus thrips (Thysanoptera: Thripidae). *Experimental and Applied Acarology* 18, 61–71.

Grout, T.G. and Richards, G.I. (1990) The influence of windbreak species on citrus thrips (Thysanoptera: Thripidae) populations and their damage to South Africa citrus orchards. *Journal of the Entomological Society of Southern Africa* 53, 151–157.

Grout, T.G. and Richards, G.I. (1992) *Euseius addoensis addoensis*, an effective predator of citrus thrips, *Scirtothrips aurantii*, in the eastern Cape Province of South Africa. *Experimental and Applied Acarology* 15, 1–13.

Grout, T.G. and Stephen, P.R. (1993) Predation of alate thrips, *Scirtothrips aurantii* Faure (Thysanoptera: Thripidae), by *Euseius citri* (Van der Merwe and Ryke) (Acari: Phytoseiidae) at low temperatures. *African Entomologist* 1, 264–265.

Grout, T.G., Morse, J.G. and Brawner, O.L. (1986a) Location of citrus thrips (Thysanoptera: Thripidae) pupation: tree or ground. *Journal of Economic Entomology* 79, 59–61.

Grout, T.G., Morse, J.G., O'Connell, N.V., Flaherty, D.L., Goodell, P.B., Freeman, M.W. and Coviello, R.L. (1986b) Citrus thrips (Thysanoptera: Thripidae) phenology and sampling in the San Joaquin Valley. *Journal of Economic Entomology* 79, 1516–1523.

Hare, J.D., Morse, J.G., Menge, J.L., Pehrson, J.E., Coggins Jr, C.W., Embleton, T.W., Jarell, W.M. and Meyer, J.L. (1990) Controlling citrus mite and thrips. *Citrus*, April, 11–14.

Hollingsworth, C.S., Knodel, J.J., Coli, W.M. and Weaver, J.S. (1995) Patterns of pear thrips activity in the northeastern United States, 1990–1992. In: Parker, B.L., Skinner, M. and Lewis, T. (eds) *Thrips Biology and Management*. Plenum Press, New York, pp. 77–80.

Houston, D.R., Allen, D.C. and Lachance, D. (1990) *Sugarbush Management: A Guide to Maintaining Tree Health*. US Department of Agriculture, Forest Service

Northeast Forest Experiment Station General Technical Report NE-129, Radnor, PA, 55 pp.

Immaraju, J.A. and Morse, J.G. (1990) Selection for pyrethroid resistance, reversion, and cross-resistance with citrus thrips (Thysanoptera: Thripidae). *Journal of Economic Entomology* 83, 698–704.

Immaraju, J.A., Morse, J.G. and Hobza, R.F. (1990a) Field evaluation of insecticide rotation and mixtures as strategies for citrus thrips (Thysanoptera: Thripidae) resistance management in California. *Journal of Economic Entomology* 83, 306–314.

Immaraju, J.A., Morse, J.G. and Brawner, O.L. (1990b) Evaluation of three bioassay techniques for citrus thrips resistance and correlation of the leaf dip method to field mortality. *Journal of Agricultural Entomology* 7, 17–27.

Kolb, T.E. and Teulon, D.A.J. (1991) Relationship between sugar maple budburst phenology and pear thrips damage. *Canadian Journal of Forest Research* 21, 1043–1048.

Kolb, T.E., McCormick, L.H., Simons, E.E. and Jeffery, D.J. (1992) Impacts of pear thrips damage on root carbohydrate, sap, and crown characteristics of sugar maples in a Pennsylvania sugarbush. *Forest Science* 38, 381–392.

Laudermilch, G. (1988) Thrips in Pennsylvania. In: Parker, B.L., Skinner, M. and Teillon, H.B. (eds) *The 1988 Thrips Infestation of Sugar Maple, Proceedings of a regional meeting, 23 June 1988, Bennington, VT.* Vermont Agricultural Experiment Station Bulletin 696, University of Vermont, Burlington, pp. 36–62.

Laudermilch, G.E. (1989) *Pear Thrips,* Taeniothrips inconsequens *(Uzel), Adult Emergence Trapping Study.* Final report. Pennsylvania Division of Forest and Pest Management, Wellsboro, PA.

Lewis, H.C. (1935) Factors influencing citrus thrips damage. *Journal of Economic Entomology* 28, 1011–1015.

Maier, C.T. (1991) Abundance of pear thrips in Connecticut, with emphasis on spatial distribution in apple orchards. In: Parker, B.L. and Doane, E.N. (eds) *Proceedings of the 1990 Conference on Thysanoptera: Abstracts, S. Burlington, VT, 23–24 October 1990.* Vermont Agricultural Experiment Station Bulletin 698, University of Vermont, Burlington, 68 pp.

McCormick, L.H. (1990) Research overview: pear thrips impact to sugar maples. In: Parker, B.L., Skinner, M., Wilmot, S.H. and Souto, D. (eds) *Pear Thrips Research and Management: Current Methods and Future Plans, Proceedings of a regional meeting, 15–16 November.* Vermont Agricultural Experiment Station Bulletin 697, University of Vermont, Burlington, pp. 28–31.

McGregor, E.A. (1926) A device for determining the relative degree of insect occurrence. *Pan-Pacific Entomologist* 3, 29–33.

Milne, D.L. and de Villiers, E.A. (1975) Citrus thrips control using systemic insecticides applied to the soil. *Citrus and Sub-Tropical Fruit Journal*, March, 16–18.

Morse, J.G. (1995) Prospects for IPM of citrus thrips in California. In: Parker, B.L., Skinner, M. and Lewis, T. (eds) *Thrips Biology and Management.* Plenum Press, New York, pp. 371–380.

Morse, J.G. and Zareh, N. (1991) Pesticide-induced hormoligosis of citrus thrips (Thysanoptera: Thripidae) fecundity. *Journal of Economic Entomology* 84, 1169–1174.

Morse, J.G., Immaraju, J.A. and Brawner, O.L. (1988) Citrus thrips: looking to the future. *Citrograph* 73, 112–115.

Nash, B., Stanosz, G., Taylor, G. and Davis, D. (1991) Association of a fungus with overstory sugar maple leaves injured by pear thrips in Pennsylvania. (Abstract.) *Phytopathology* 81, 123.

Nash, B.L., Stanosz, G.R., Taylor, G. and Davis, D.D. (1994) *Discula campestris* infection of sugar maple leaves associated with pear thrips injury. *Plant Disease* 78, 285–289.

Niann, T.C. (1991) Important thrips species in Taiwan. In: Talekar, N.S. (ed.) *Thrips in Southeast Asia, Proceedings of a Regional Consultation Workshop, Bangkok, Thailand, 13 March*, Asian Vegetable Research and Development Center (AVRDC) Publication No. 91–342, Taipei, Taiwan, pp. 40–56.

Parker, B.L. (1991) The pear thrips problem. In: Parker, B.L., Skinner, M. and Lewis, T. (eds) *Towards Understanding Thysanoptera, Proceedings of the International Conference on Thrips, Burlington, Vermont, 21–23 February 1989*. General Technical Report NE-147, US Department of Agriculture, Forest Service, Radnor, PA, pp. 179–188.

Parker, B.L. and Skinner, M. (1993) Field evaluation of traps for monitoring emergence of pear thrips (Thysanoptera: Thripidae). *Journal of Economic Entomology* 86, 46–52.

Parker, B.L., Grehan, J.R. and Skinner, M. (1992) Method for extracting pear thrips (Thysanoptera: Thripidae) from forest soil. *Journal of Economic Entomology* 85, 865–869.

Parker, B.L., Skinner, M. and Adamowicz, R.H. (1995) Induced emergence of pear thrips for population monitoring in forest soil. In: Parker, B.L., Skinner, M. and Lewis, T. (eds) *Thrips Biology and Management*. Plenum Press, New York, pp. 579–582.

Parker, B.L., Skinner, M., Brownbridge, M., Adamowicz, A. and Fu-cheng, Su (1996) Entomopathogenic fungi for thrips management: recent advances. *Acta Horticulturae* 431, 521–534.

Parrott, P.J. (1913) New destructive insects in New York. *Journal of Economic Entomology* 6, 61–66.

Quilici, S., Geslin, P. and Trahais, B. (1988) Population fluctuations of *Scirtothrips aurantii* Faure in citrus orchards in Reunion Island. In: Goren, R. and Mendel, K. (eds) *Proceedings of the Sixth International Citrus Congress, Tel Aviv, Israel, 6–11 March*. Balaban Publishers, Philadelphia, pp. 1291–1297.

Raffa, K.F. and Hall, D.J. (1989) *Thrips calcaratus* Uzel (Thysanoptera: Thripidae), a new pest of basswood trees in the Great Lakes region. *Canadian Journal of Forest Research* 19, 1661–1662.

Raffa, K.F., Hall, D.J., Kearby, W. and Katovich, S. (1992) Seasonal life history of introduced basswood thrips (Thysanoptera: Thripidae) in Wisconsin, with observations on associated thrips species. *Environmental Entomology* 21, 771–779.

Reed, D.K. and Rich, J.R. (1975) A new survey technique for citrus thrips. *Journal of Economic Entomology* 68, 739–742.

Rhodes, A.A., Baritelle, J.L. and Morse, J.G. (1986) Method for predicting crop response to pest attack over time: application to citrus thrips (Thysanoptera: Thripidae) scarring on navel oranges. *Entomological Society of America Bulletin* Fall, 153–156.

Samways, M.J. (1986) Spatial distribution of *Scirtothrips citri* Faure (Thysanoptera: Thripidae) and threshold level for one per cent damage on citrus fruit based on trapping with fluorescent yellow sticky traps. *Bulletin of Entomological Research* 76, 649–659.

Skinner, M. and Parker, B.L. (1990) Vermont soil survey for pear thrips: protocol for regional soil sampling program. In: Parker, B.L., Skinner, M., Wilmot, S.H. and

Souto, D. (eds) *Proceedings of the Regional Meeting: Pear Thrips Research and Management: Current Methods and Future Plans.* Vermont Agricultural Experiment Station Bulletin 679, University of Vermont, Burlington, pp. 94–100.

Skinner, M. and Parker, B.L. (1991) Bioecology of pear thrips: distribution in forest soils. In: Parker, B.L., Skinner, M. and Lewis, T. (eds) *Towards Understanding Thysanoptera, Proceedings of the International Conference on Thrips, Burlington, Vermont, 21–23 February 1989.* General Technical Report NE-147, US Department of Agriculture, Forest Service, Radnor, PA, pp. 193–212.

Skinner, M. and Parker, B.L. (1992) Vertical distribution of pear thrips (Thysanoptera: Thripidae) in forest soils. *Environmental Entomology* 21, 1258–1266.

Skinner, M. and Parker, B.L. (1995) *Sugarmaker's Guide to Pear Thrips Monitoring.* Vermont Agricultural Experiment Station Research Report 72 and Vermont Monitoring Cooperative Research Report 9. University of Vermont, Burlington, 13 pp.

Skinner, M. and Parker, B.L. (1996) Emergence of pear thrips (Thysanoptera: Thripidae) and its relation to foliar damage. *Environmental Entomology* 25, 350–358.

Skinner, M., Parker, B.L. and Wilmot, S.H. (1991a) The life cycle of pear thrips, *Taeniothrips inconsequens* (Uzel) in Vermont. In: Parker, B.L., Skinner, M. and Lewis, T. (eds) *Towards Understanding Thysanoptera*, Proceedings of the International Conference on Thrips, Burlington, Vermont, 21–23 February 1989. US Department of Agriculture, Forest Service, Northeast Forest Experiment Station General Technical Report NE-147, Radnor, PA, pp. 435–441.

Skinner, M., Parker, B.L. and Bergdahl, D.R. (1991b) *Verticillium lecanii*, isolated from larvae of pear thrips, *Taeniothrips inconsequens*, in Vermont. *Journal of Invertebrate Pathology* 58, 157–163.

Stanosz, G.R. (1993) Symptoms, association, and pathogenicity of *Discula campestris*, a cause of sugar maple seedling anthracnose. *Plant Disease* 77, 1022–1026.

Suwanbutr, C., Tongklad, W., Uhnchit, P., Thayamanon, P., Witthayarug, W. and Khewpoompung, P. (1992) A field trial on the efficacy of some insecticides for controlling thrips attacking pommello. *Acta Horticulturae* 2, 876–879.

Tanigoshi, L.K. (1991) Biological control of citrus thrips, *Scirtothrips citri,* by predacious phytoseiid mites. In: Parker, B.L., Skinner, M. and Lewis, T. (eds) *Towards Understanding Thysanoptera.* US Department of Agriculture, Forest Service, Northeast Forest Experiment Station, General Technical Report NE-147. Radnor, PA, pp. 399–418.

Tanigoshi, L.K. and Griffiths, H.J. (1982) A new look at biological control of citrus thrips. *Citrograph* 67, 157–158.

Tanigoshi, L.K. and Moffitt, L.J. (1984) Citrus thrips, *Scirtothrips citri* (Thysanoptera: Thripidae) and 'Washington' navel orange injury. *Environmental Entomology* 13, 1312–1318.

Tanigoshi, L.K. and Moreno, D.S. (1981) Traps for monitoring populations of the citrus thrips, *Scirtothrips citri* (Thysanoptera: Thripidae). *Canadian Entomologist* 113, 9–12.

Tanigoshi, L.K. and Nishio-Wong, J.Y. (1982) *Citrus Thrips: Biology, Ecology, and Control.* US Department of Agriculture Technical Bulletin No. 1668, 17 pp.

Tanigoshi, L.K., Moreno, D.S., Nishio-Wong, J.Y. and Fargerlund, J. (1982) Soil applications of insecticides for control of *Scirtothrips citri* (Moulton) (Thysanoptera: Thripidae) on citrus. *Zeitschrift für Angewondte Entomologie* 93, 17–24.

Tanigoshi, L.K., Fargerlund, J., Nishio-Wong, J.Y. and Griffiths, H.J. (1985) Biological control of citrus thrips, *Scirtothrips citri* (Thysanoptera: Thripidae), in Southern California citrus groves. *Environmental Entomology* 14, 733–741.

Tatara, A. (1994) Effect of temperature and host plant on the development, fertility and longevity of *Scirtothrips dorsalis* Hood (Thysanoptera: Thripidae). *Applied Entomology and Zoology* 29, 31–37.

Tatara, A. and Furuhashi, K. (1992) Analytical study on damage to satsuma mandarin fruit by *Scirtothrips dorsalis* Hood (Thysanoptera: Thripidae), with particular reference to pest density. *Japanese Journal of Agricultural Entomology* 10, 21–29.

Teillon, H.B., Burns, B.S. and Kelley, R.S. (1985) *Forest Insect and Disease Conditions in Vermont, Calendar Year 1985*. Agency of Environmental Conservation, Department of Forests, Parks and Recreation, Division of Forestry, Waterbury, VT, 38 pp.

Teillon, H.B., Burns, B.S. and Kelley, R.S. (1994) *Forest Insect and Disease Conditions in Vermont, Calendar Year 1994*. Agency of Environmental Conservation, Department of Forests, Parks and Recreation, Division of Forestry, Waterbury, VT, 132 pp.

Teulon, D.A.J. and Cameron, E.A. (1991) The life history of the pear thrips (*Taeniothrips inconsequens* (Uzel)) in a sugar maple stand in Pennsylvania. In: Parker, B.L. and Doane, E.N. (eds) *The 1990 Conference on Thysanoptera, Abstracts*. Vermont Agricultural Experiment Station Bulletin 698, University of Vermont, Burlington, p. 34.

Teulon, D.A.J., Kolb, T.E., Cameron, E.A., McCormick, L.H. and Hoover, G.A. (1993) Pear thrips, *Taeniothrips inconsequens* (Uzel) (Thysanoptera: Thripidae), on sugar maple, *Acer saccharum* Marsh.: a review. *Advances in Thysanopterology, Journal of Pure and Applied Zoology* 4, 355–380.

Vogelmann, J.E. and Rock, B.N. (1991) Detection of pear thrips damage using satellite imagery data. In: Parker, B.L., Skinner, M., and Lewis, T. (eds) *Towards Understanding Thysanoptera*, Proceedings of the International Conference on Thrips, Burlington, Vermont, 21–23 February 1989, US Department of Agriculture, Forest Service, Northeastern Forest Experiment Station General Technical Report NE-147, Radnor, PA, pp. 285–300.

Wargo, P.M. (1988) *Judging Vigor of Deciduous Hardwoods*. Agriculture Information Bulletin No. 418. US Department of Agriculture, Forest Service, Combined Forest Pest Research and Development Program, 16 pp.

Wiesenborn, W.D. and Morse, J.G. (1986) Feeding rate of *Scirtothrips citri* (Moulton) (Thysanoptera: Thripidae) as influenced by life stage and temperature. *Environmental Entomolology* 15, 763–766.

Wilmot, S.H., and Kelley, R.S. (1990) Stand damage assessment. In: Parker, B.L., Skinner, M., Wilmot, S.H. and Souto, D. (eds) *Proceedings of the Regional Meeting, Pear Thrips Research and Management: Current Methods and Future Plans*. Vermont Agricultural Experiment Station Bulletin 679, University of Vermont, Burlington, pp. 117–118.

Wilmot, S.H. and Scherbatskoy, T.D. (eds) (1994) *Vermont Monitoring Cooperative: 1994 Work Plan*. Vermont Monitoring Cooperative Work Plan No. 4. Vermont Department of Forests, Parks and Recreation, Waterbury, 42 pp.

18

Integrated Pest Management (IPM) in Glasshouses

Robert J. Jacobson
Horticulture Research International, Stockbridge House, Cawood, Selby, North Yorkshire YO8 OTZ, UK

Introduction

Modern commercial greenhouse and indoor amenity plantings are subject to frequent invasion by indigenous thrips in most countries, and in temperate climates they are also highly susceptible to the introduction of alien pest species on plant material imported from warmer climes (Morison, 1957; Vierbergen, 1995). Traditionally, minor, sporadic infestations of most of these pests, including *Thrips tabaci*, *Chaetanaphothrips orchidii*, *Dichromothrips* spp., *Heliothrips haemorrhoidalis*, *Hercinothrips bicinctus*, and *Parthenothrips dracaenae* have been controlled by localized spray treatments or fumigation (see Chapter 15). Unfortunately, over the last decade, there has been a widespread occurrence in greenhouses and interior landscapes of other species, especially *Frankliniella occidentalis*, *Hercinothrips femoralis*, *Thrips fuscipennis* and *Thrips palmi*, each capable of causing severe damage to food and ornamental crops and of disrupting the sophisticated, intensive production systems now practised. The situation has been exacerbated by the development of resistance to insecticides by *F. occidentalis* (see Chapter 15) and the stringent quality requirements which allow minimum/zero pesticide residues and little or no damage, or even the presence of insects, on marketable produce. To combat these relatively recent problems much effort has been devoted to developing integrated pest management (IPM) to control pests, including thrips in greenhouses and large-scale indoor plantings.

What is IPM?

The term IPM is commonly used as a synonym for biological control but this understates a complex approach to pest control. While biological control techniques play a crucial part in IPM, the strategies also draw on physical and cultural control measures, combined with minimal and selective use of target-specific chemicals.

It is impossible to consider IPM of thrips in isolation from other pests and most protected crops are host to many species of insects and mites. For example, the pests infesting some important protected crops in the UK are listed in Table 18.1. The importance of each is indicated depending on severity of damage and difficulty of control within an IPM programme. Sometimes a pest may have a different status in an IPM programme than in a chemical-based alternative, depending on the availability of specific and compatible control agents. Clearly, an IPM programme for any crop must include compatible control measures for all pests that attack that crop. The failure of any one component in the programme, which necessitates the resort to remedial treatment with a non-specific insecticide, could ultimately result in the complete breakdown of the whole system.

With so many control measures operating simultaneously in IPM, the systems are inevitably more difficult to manage than chemical-based strategies. The introduction of a new component in an IPM programme usually affects other components, thereby leading to some disruption in the overall programme. To make sound decisions, managers must have a good understanding of all the pests, their control measures and the complex interactions between them. However, this background knowledge is only useful if applied in a topical context, so it is vital to have information on current pest levels based on detailed crop monitoring.

The need for separate control agents against every pest species and the requirement for additional management input means IPM systems are often more expensive than chemical alternatives. The financial implications of IPM are thus important when evaluating existing programmes or planning new ones.

Despite problems associated with the complex nature of IPM in protected crops, there are significant benefits to be gained from limiting pesticide use. The production of plants with minimal use of chemicals helps to avoid phytotoxicity, provides a better working environment for greenhouse staff and reduces the risk of chemical residues in food. It is better for crops, workers and consumers alike, and will, progressively, form the basis of pest control in the future. There is still much to learn about the interactions between cropping environments, plants, pests and beneficial organisms but, as new information becomes available, IPM systems should become more reliable, easier to manage and attractive to more growers.

Components of IPM

Biological control

Biological control may be defined as the use of a living organism (the beneficial) for the regulation of populations of another (the pest). There are now beneficial insects, mites, nematodes, fungal pathogens, bacteria and viruses available commercially for a wide range of pests. The principle agents of biological

Table 18.1. The pests of five important protected crops in the UK.

		Sweet peppers	Cucumbers	Tomatoes	Bedding plants	Chrysanthemums
Thrips	*Frankliniella occidentalis*	✓	✓	+	✓	✓
	Thrips tabaci	✓	✓	0	✓	✓
	Thrips fuscipennis	0	+	0	0	+
Whiteflies	*Trialeurodes vaporariorum*	0	✓	✓	✓	+
	Bemisia tabaci	*	*	*	*	*
Aphids	*Myzus persicae*	✓	0	+	✓	✓
	Aphis gossypii	0	✓	0	+	✓
	Other species	+	+	+	+	✓
Mites	*Tetranychus urticae*	✓	✓	✓	✓	✓
	Tetranychus cinnabarinus	+	0	✓	+	✓
	Polyphagotarsonemus latus	+	+	0	+	0
	Other species	0	0	+	0	0
Leaf miners	*Liriomyza bryoniae*	+	+	✓	+	✓
	Other species	+	+	+	+	✓
Caterpillars	Various species	+	+	+	+	+
Other	Leaf hoppers	+	+	+	+	+
	Sciarids/shore flies	+	+	+	+	+
	Capsid bugs	+/✓	+/✓	+	+	+/✓

✓ = Pest of major importance, + = pest of minor importance, 0 = pest of negligible importance, * = potential threat.

control used against thrips in commercial greenhouses are predatory mites and bugs (see Chapter 8) and to a much lesser degree pathogenic fungi (see Chapter 10). It may eventually be possible to use hymenopterous parasites and parasitic nematodes (see Chapter 9), but they are difficult to exploit at present. The detailed biology and production of these organisms are considered in Chapters 8, 9, 10 and 12. This chapter concentrates on their use in IPM programmes within the practical constraints of greenhouse management.

Physical control

Physical control measures utilized against thrips in greenhouses include screens to prevent their entry, barriers to intercept them as they move locally and traps to attract them from plants. These apparently simple techniques are based on knowledge of the organism's biology and behaviour which has identified a vulnerable point in its life cycle.

Screens

The sudden invasion of a greenhouse by thrips is probably the most difficult situation to manage within an IPM programme. These pests can rapidly overwhelm a crop, perhaps before the grower has a chance to react, and severe damage may occur before biological control agents make any impact. This situation arises in intensive growing areas when pests migrate from neighbouring greenhouses and, in warmer climates, when the pest invades from vegetation in the surrounding countryside.

Growers have two solutions to this problem. If the invasion is of only short duration, chemical pesticides can be used to reduce the number of adults surviving on plants before releasing biological control agents. This is only realistic if an effective chemical with short persistence is available and can be used in a way which will not disrupt any other component in the overall IPM system. However, this approach is unacceptable if the invasion is continuous, as occurs in Mediterranean areas and the south-west USA, where a better option is exclusion by screening doors and ventilators.

The use of screens against larger insects, such as whiteflies, is well advanced (Berlinger *et al.*, 1991), but it is more difficult to exclude thrips because they can penetrate such small apertures. Screens which exclude them purely by mesh size also impede air flow, the primary purpose of ventilators. However, the colour of a screen has been shown to enhance its effect (Berlinger *et al.*, 1993). A loose shading net, through which whiteflies and thrips passed freely in the laboratory, reduced penetration of thrips if it were 'aluminium coloured'. These results were reinforced in field trials in which an 'aluminium-coloured' screen performed much better than an otherwise identical white one (see Chapter 3). Screens and covers will probably make a substantial contribution to future IPM programmes, particularly in warmer climates.

Traps

Sticky traps are widely used in greenhouses, particularly in temperate regions, both as a means of reducing pest numbers and of monitoring the development of insect populations (see Chapter 11). They are particularly useful for detecting thrips at very low population densities, thereby aiding the timing of other control measures.

Traditionally, yellow traps have been used because they are attractive to a wide range of herbivorous insects. However, different species prefer different colours (see Chapters 3 and 11) and this preference may vary according to the crop type and the position of the trap. In one outdoor experiment *Thrips tabaci* preferred white to yellow and blue (Kirk, 1984), although in another these colours were equally attractive (Czencz, 1987). In protected cucumbers, Brødsgaard (1993a) demonstrated that shades of yellow and blue were more attractive to *Thrips tabaci* than white, and that blue was the most attractive to *Frankliniella occidentalis*. The latter suggests that blue traps will suffice to monitor these two important glasshouse thrips in cucumber crops. Blue traps have also been shown to be effective for monitoring *F. occidentalis* in ornamental crops (Brødsgaard, 1993b).

Vernon and Gillespie (1995) examined the influence of trap shape and background colour on the capture of *F. occidentalis* by comparing different combinations of yellow and violet traps, of flat, cubical, spherical, rectangular prism and cylindrical shape, against yellow and violet backgrounds. There appeared to be no advantage in changing from flat traps to more expensive three-dimensional shapes. However, their results indicated that trapping might be more efficient with a distinct colour contrast between trap and background, for instance, yellow traps against violet background.

Brødsgaard (1993b) also noted that the numbers of *F. occidentalis* recorded were influenced by trap position. In cucumbers, most thrips were caught on traps placed just above the crop canopy, although many were also recorded on traps just above ground level. Very few were caught in the middle strata. Possibly, the insects caught at the top of the crop are dispersing or seeking a mate, while those at the bottom have just emerged from pupae on the floor.

Several researchers have attempted to correlate trap catches with the size of the population in the crop (Shipp and Zariffa, 1991; Brødsgaard 1993a) (see Chapter 11). A reliable correlation would be a useful aid to decision-making, but so far results have been inconsistent, probably because the catches depend on so many factors including plant species, stage of crop growth, position of trap, orientation of trap in relation to light, environmental conditions and population density of insects. The number caught may be dependent more on the pest's behaviour and activity patterns than any other factor. For example, a disproportionate number of males have often been observed on yellow traps at very low population densities early in the season and this has been attributed to their increased activity in seeking females. The presence of predators may also influence the catch by increasing pest activity.

The attractiveness of traps may be improved by incorporating scents, such as anisaldehyde (Brødsgaard, 1990) (see Chapter 11). Yellow water traps baited with *p*-anisaldehyde trapped over ten times more *F. occidentalis* than non-scented traps in tomato and sweet pepper crops (Teulon *et al.*, 1993b). Another study in sweet peppers found up to five to seven times more female *F. occidentalis* on yellow sticky traps baited with anisaldehyde (Teulon *et al.*,

1993a). The behavioural response is not understood but the technique has potential for the development of mass-trapping as a control method (see Chapter 11, Visual and Chemical attractants). Currently, traps serve five roles within IPM programmes: (i) to detect thrips at an early stage as an aid to timing control measures; (ii) to compare activity before and after treatments; (iii) to trap some thrips as they enter glasshouse ventilators, thereby reducing invasion pressure; (iv) to mop up surviving thrips in empty glasshouses – a practice that can be particularly effective in glasshouses heated for short periods between crops; and (v) to reduce numbers locally by deploying many traps or large sticky strips, although care must be taken not to catch beneficials.

Barriers

Fully grown nymphs of some thrips species drop from leaves to pupate on the ground. Where plants are grown in soil the mortality of thrips may be high, particularly if the soil is wet, due to drowning or the presence of fungal pathogens such as *Entomophthora* sp. (Samson *et al.*, 1979) (see Chapter 10). It is now common in temperate regions to grow protected salad crops in artificial media (e.g. foam, peat, rockwool, perlite, sawdust), the soil being covered with polythene. The floor is drier and thrips dropping to pupate have a better chance of survival, resulting in faster population increase on the plants above.

Pickford (1984) reported a novel floor treatment based on a sticky polybutene-based product incorporating the active ingredient deltamethrin, now available commercially as 'Thripstick' (see Chapter 15). When applied to the plastic floor covering, it intercepts thrips and prevents completion of their life cycle. It has been particularly effective against *Thrips tabaci*, especially when applied before planting as this helps to prevent their establishment. Control persists for over 10 weeks, but the effectiveness gradually declines due to contamination with crop debris. Repeat applications during cropping must be made with care as any drift onto the plants can harm beneficial species. Polybutene-based floor treatments are less effective against the other two most important thrips in temperate greenhouses, *Thrips fuscipennis* and *Frankliniella occidentalis* (Jacobson, 1995) probably because a significant proportion of their populations pupate on the plant.

Cultural control

Cultural pest control may be defined as the adjustment of growing practices and environmental conditions to the disadvantage of the pest. Techniques range from basic nursery hygiene to sophisticated manipulation of temperature and humidity. When modifying growing systems it is important to ensure that the changes are not detrimental to the plants or provide other pests or diseases with an advantage.

Nursery hygiene

Most thrips species causing problems in greenhouses have a broad host range which includes weeds as well as the cropped plants. Weeds serve as reservoirs for pests and allow continuity of infestation by providing a 'green bridge' between crops. The effects are most significant in short-duration crops where pest populations do not have sufficient time to increase to damaging levels, unless there are already substantial numbers in the greenhouse. Removal of weeds in and around greenhouses can greatly reduce the likelihood of pest invasion.

It is good practice to remove or destroy crop trimmings and plant residues after cropping is completed, because their presence in waste heaps on-site provides another important source of reinfestation.

If infested plants are brought onto a nursery during cropping they will cause a sudden increase in pest numbers which can disrupt an established equilibrium between pests and beneficials, perhaps resulting in the total collapse of an IPM programme. This can be avoided by holding newly purchased plants in a quarantine area where they can be inspected and, if necessary, treated with insecticide before transfer to production houses.

Resistant cultivars

Host plant resistance to pests and diseases is common in the wild but some traits have been lost as cultivars have been bred for maximum yield or aesthetic appearance. Breeding programmes have now reintroduced some genetic resistance, particularly to diseases, in modern cultivars. The results of a screening programme for resistance to *Frankliniella occidentalis* in cucumbers are particularly exciting (Mollema *et al.*, 1993a,b). The reproductive fitness of *F. occidentalis* on resistant accessions appeared to decline enough to prevent an exponential growth of the pest population. Reduced oviposition on the resistant plants was considered to be due to malnutrition caused by altered feeding behaviour, possibly a result of a repellent or nutritional deficiency in the plants.

The benefits gained from careful choice of plant cultivar in an IPM programme need not be limited to their direct resistance to the pest. For instance, cucumber cultivars which produce vigorous, lush growth, aid the establishment of the predatory mite, *Amblyseius cucumeris*, by providing a more humid environment. Other cucumber cultivars, with tolerance to mildew, minimize the use of fungicides, some of which are harmful to beneficials. The choice of plant cultivar is likely to become a much more important part of IPM programmes as any trait which limits pest development or enhances the effect of another component in the package will be advantageous.

Manipulation of the crop environment

Manipulation of the cropping environment helps to optimize the effects of beneficial organisms. Quite small modifications to temperature, daylength, light intensity, humidity and crop maintenance provide significant benefits, but such fine tuning requires a thorough understanding of the behaviour and biology of all the organisms involved. It is important that any changes do not adversely affect plant growth, other beneficial organisms or the control of plant pathogens.

There are several documented examples of manipulation of the cropping environment influencing the management of thrips in greenhouses, although none of the techniques has yet been developed into robust control methods.

Verticillium lecanii has been used to control thrips but it requires high relative humidity to be effective: 85 to 90 per cent for a prolonged period to allow germination of spores (Hall and Burges, 1979) (see Chapter 10). Good control of *F. occidentalis* has been achieved in trials in chrysanthemums by elevating relative humidity either on four consecutive nights each week or by cycling two nights of high humidity with two nights of ambient conditions. This is achieved by applying water automatically through high pressure misting equipment (Helyer *et al.*, 1992), but there remains concern over the increased risk of plant diseases, in particular white rust and botrytis.

The predatory bugs *Orius majusculus* and *O. laevigatus* (Plate II.6–8) appear to enter reproductive diapause in daylengths less than 16 h, which limits their use in UK crops during the early season. This can be overcome with supplementary lighting which enables the predators to establish on cucumber and sweet pepper crops at daylengths below the estimated natural threshold (Chambers *et al.*, 1993; Jacobson, 1993a). The performance of *O. majusculus* in cucumbers has never matched expectations, even at adequate daylength. This is because most eggs are laid in the youngest 500 mm of growth of the main laterals and side shoots (Jacobson, 1993b) and many of these are removed as part of the routine crop maintenance. A change in growing practice would improve the establishment of this predator.

Chemical control

There are two methods of integrating chemicals into IPM programmes; either as highly selective products which are specific to the target pest or less specific products applied in a manner, or at a time, which has minimal effect on any beneficial organisms in the crop. There are currently no truly selective insecticides available for thrips, so growers are limited to careful use of more broad-spectrum products (see Chapter 15).

Insecticides with very short persistence, such as dichlorvos, can be integrated most successfully. Dichlorvos is applied commonly in cucumbers and sweet peppers, soon after planting and before biological control agents are released, to kill adult *Frankliniella occidentalis* which have survived between crops on the structure of the greenhouse. This reduces the initial pressure of pest invasion and allows subsequent introductions of biological control agents a better chance of success.

Short-persistence insecticides may also be helpful after biological control agents are established if their use is limited to localized areas of crop. Spot treatments in small areas of excessive pest activity can avoid a major problem by preventing the insects from spreading. Biological control agents can then either be re-released or allowed to move back onto the plants naturally from the surrounding areas. Dichlorvos may be used in this way against *F. occidentalis*, as may heptenophos against *Thrips tabaci* and *T. fuscipennis*.

Systemic insecticides applied to the growing medium through the irrigation/feeding system will probably have a more important role in IPM in future because they are effective against plant-feeding insects with minimal harm to beneficials.

Crop Monitoring and Making Decisions in IPM

To make sensible decisions about the cost-effective use of expensive control measures within IPM systems in protected crops, growers must have accurate and up to date knowledge of the pest abundance and the establishment of beneficials throughout their crops. This knowledge is gained through regular crop monitoring or, as it is known in parts of North America, crop scouting.

There are some short cuts; for example, the use of sticky traps for the early detection of thrips and certain other flying pests, but generally these techniques are no substitute for regular examination of plants. However, crop walking and plant examinations are very time consuming, even tedious if approached as a single task. The most successful crop monitoring systems enlist the help of nursery staff as they perform their day-to-day crop work. They see and handle the plants regularly, often daily, and are best placed to detect the arrival and subsequent development of pests. To make the most of this important resource, nursery management must have a strong commitment to training and allow staff time to include this extra duty in their daily routines.

In its simplest form, a staff-based crop monitoring system requires workers to report their findings verbally to management, who then make decisions and take any necessary action. However, managers are not always accessible and unless the verbal report is made soon after the observation, the recollection can become vague, distorted or even forgotten.

More advanced crop monitoring systems involve the installation of charts in each section of the nursery, upon which staff promptly mark their findings. The information is then openly displayed for the decision-makers, even if they visit after working hours.

Two of the most sophisticated crop monitoring systems in operation have been developed independently, in the UK and Canada by teams from quite different backgrounds. The first was developed by a horticultural business in the north of England for their own use but should eventually become more widely available (T. Salmon and S. Wade, Hedon Salads, Newport, 1991, personal communication). Nursery staff on this site are highly trained in the recognition of pests and beneficial species and carry hand held data loggers to record the activity of both. These data are downloaded regularly onto a central nursery computer which is programmed with software developed in-house. The nursery managers have ready access to topical information and sequential historical data showing how pest populations developed and were brought under control. The latter is a valuable aid to decision-making in subsequent years. This system has worked well for several years on two large mixed cucumber and pepper nurseries and is continually being refined and updated.

The second is an 'expert' system developed in Ontario (Shipp *et al.*, 1993). It is targeted less at the collection of data from the crop and more towards the interpretation of the gathered information. It is designed to be used in two modes; the first to diagnose problems and to recommend preventative and remedial action, while the second is a reminder of daily management tasks. The programme reviews recent environmental conditions and alerts the user to conditions conducive to pest and/or disease outbreaks.

Unfortunately, most growers do not have the benefit of computerized expert systems to aid their decisions. After assessing the size of a pest population they must make a personal judgement on the best course of action, taking into account any side-effects on other components in the IPM system. They have few hard and fast rules to follow and must learn when it is appropriate to use more biological material and when to resort to remedial sprays. The decisions can be complex and mistakes are costly for high-value crops, so growers often reduce the risks by employing specialist consultants.

In most cases the crop workers who record pest incidence are also in the best position to take the required action, such as distributing biological material or applying spot treatments of pesticides. By following this through to monitor the efficacy of their treatments they further develop their own expertise, thus benefiting the business and adding to their own job satisfaction.

Whichever method is adopted it is vital that the pests are detected as early as possible and action taken promptly. Any delay will allow the pests to multiply, requiring more biological material, and this will ultimately become uneconomic. Though time consuming, an efficient pest monitoring system should pay for itself in terms of reduced crop damage, and reduced expenditure on materials and on labour for their application.

Advantages and Disadvantages of IPM Systems

Advantages

The advantages associated with IPM systems in commercial crops all relate to the reduction in use of broad-spectrum pesticides. Consumers worldwide are becoming increasingly sensitive to the potential risks of pesticide residues in food and many people actively seek produce which has received no, or minimal, applications of chemicals. In the UK, several leading food retailers have introduced new marketing standards to convince customers that the goods they sell have been produced in an environmentally acceptable way and these standards include minimal use of pesticides (Walton, 1992). The retailers liaise closely with their suppliers and some have introduced codes of practice which set demanding but achievable standards of food production aimed at structured, sustainable and scientific systems to minimize any impact on the environment. This often raises production costs with no guaranteed price premium but the growers must conform if they wish to gain access to these important market outlets.

Pesticides can cause phytotoxic effects to crop plants resulting in reduced yield or downgraded quality. Young ornamental plants are particularly vulnerable and chemical treatments to prevent the establishment of *Frankliniella occidentalis* often cause significant production losses. The prospect of reducing such losses is a strong incentive for growers to move away from intensive chemical control.

The use of bumble-bees as pollinators in tomato crops has been rapidly adopted in north-west Europe. The savings in labour compared with hand pollination and the increased yield from improved fruit set have persuaded the

entire tomato industry to use this technique. To gain these benefits, growers must maintain a low pesticide input regime which excludes the use of persistent broad-spectrum insecticides. There is growing interest in biological pollination systems in other parts of the world and in other crops, such as sweet peppers, strawberries and beans. It will be equally important for these growers to adapt their pest control strategies.

Spraying protected crops in full protective clothing, particularly in hot climates, is an unpleasant task. For many growers the prospect of eliminating this aspect of their work and generally reducing toxic residues in their working environment is highly desirable and a major driving force in their move towards IPM.

The complete removal of non-specific pesticides from growing systems allows naturally occurring parasitoids and predators to enter greenhouses and to operate on the plants, thereby contributing to overall pest control.

All the advantages of IPM detailed so far have considered the reduction of pesticide use by choice. However, the number of useful pesticides available to horticulture is declining, forcing growers to seek alternative methods of pest control. This reduction is partly due to the development of pest resistance but also to the withdrawal of products for commercial reasons. The prospects of obtaining new products specifically for this industry are remote because the markets are considered too small to justify the manufacturer's development costs. Furthermore, it is relatively expensive to register pesticides in many countries and this prohibits the release of chemicals which are already used on the major international crops. Control of *Frankliniella occidentalis* in protected edible crops in the UK illustrates the overall problem. There are currently no pesticides available in this country which are both effective against this pest and safe to use in the crops; this situation is driving cucumber and pepper growers towards biological techniques.

Disadvantages

Most biological control programmes require intensive crop monitoring throughout the season to ensure that pests are detected early and action is timed correctly. There are many variables, the decisions are complex and risks high because of the value of the crop. Pest problems develop very rapidly in the greenhouse environment and if a small population is overlooked during monitoring or if action is not taken promptly against a sudden invasion, the quantity of biological control material required subsequently can be prohibitively expensive. Often cucumber and pepper growers using IPM systems in temperate regions spend over 12 h week^{-1} ha^{-1} collecting information on pest activity which, in the UK, equates to £2200 (US$3300) per season. The collected information must then be interpreted by more experienced staff, often with the help of specialist consultants. The whole procedure can become expensive, which deters many people. In contrast, the treatments in prophylactic chemical control programmes are applied according to predetermined schedules which provide constant protection from pests. These programmes require minimal monitoring and few decisions during the growing season, a very attractive alternative for busy managers.

An IPM programme requires a compatible control agent for every pest which attacks the crop. This must include provision against 'secondary pests', such as capsid bugs, which are not usually a problem in a chemical control regime because they are inadvertently killed by products used against the major pests, but the removal of routine treatments allows them to survive. Should application of a non-specific pesticide become necessary it could destroy introduced beneficials and cause the entire programme to fail.

The need for separate control measures against every pest increases the cost of IPM programmes compared with those based on broad-spectrum pesticides.

Supplies of biological control agents which have been in production for many years for pests other than thrips (e.g. *Phytoseiulus persimilis*, *Encarsia formosa*) are generally reliable and the material of good quality. The biology of these beneficials and interaction with their prey is well understood and their rates of introduction are well defined for most situations. However, newer products for use against pests such as *F. occidentalis* are less well researched. For instance, *Orius* spp. products were released early in north-west Europe because of pressure from growers who were desperate for additional control measures. The development work was still being done in parallel with commercial use, so failures in crops were common.

An efficient biological system takes time to reach an equilibrium between the beneficials and the prey/hosts, so some pest activity persists on the plants after treatment. Some markets demand that produce is completely free of invertebrates, whether they are pests or beneficials, and this deters growers. This requirement is understandable for leafy vegetables where the presence of a beneficial insect would be as unacceptable to a consumer as a pest, but it is less acceptable in ornamental markets.

Examples of Successful IPM Programmes

The most widely adopted IPM programmes, which include control of thrips, are those practised in commercial sweet pepper and cucumber crops in greenhouses throughout north-west Europe and Canada. Growers in these temperate regions are able to manage pest populations successfully because they do not suffer the huge invasions from surrounding vegetation experienced in hotter climates. Furthermore, these crops can withstand a degree of pest activity on foliage before fruit is affected. This means there is time for biological control agents to establish and the continued pest activity in the beneficial–pest equlibrium can be tolerated. Effective IPM programmes have also been devised and used commercially in aubergines and chilli peppers, but have not been as widely adopted.

Although IPM programmes are being developed for ornamentals, progress has not matched that in edible crops and the uptake by commercial growers has been slow because of the extremely low tolerance of pests and their damage within the ornamental markets. For example, in The Netherlands, a quality standard for export material known as 'zero-tolerance' has almost precluded growers from using any measures other than frequent, routine applications of broad-spectrum pesticides. However, there have been notable successes in experimental situations, particularly in cut flowers and bedding plants, and in many cases these have been scaled up for use on production nurseries. Emphasis has been placed on preventing pest establishment with repeated prophylactic introductions of relatively large numbers of parasites and predators, starting before the pests are even detected. In pot plants, the greatest success with IPM has generally been achieved on species which are not a good host for thrips, for example, poinsettia. This illustrates the fact that thrips are perhaps the most difficult pests to control by IPM.

Interior landscapes offer an opportunity for IPM quite different from plant production in commercial horticulture. Once restricted to conservatories in parks and gardens, it is now common for large public buildings to have extensive plantings extending over several floors, even including mature trees. It is very difficult to apply pesticides in such situations due to public access and often interior landscape contractors have little choice but to develop biological systems. Fortunately, the stable environment favours the establishment of a good, long-term beneficial–prey equilibrium.

IPM of thrips can rarely be considered in isolation from action required against other pests attacking the same crop. Unless compatible control agents are available for all pests, the measures developed for thrips are merely academic to commercial growers. For this reason, the examples given below are of entire programmes only. In all these examples, regular and intensive monitoring is required to provide information which forms the basis for all decision-making within the programme. The monitoring includes assessments of pest populations, establishment of beneficial species and the complex interactions between them, achieved both by plant examination and the use of traps.

IPM in UK sweet peppers

IPM is used in all sweet pepper crops grown in the UK. The components of programmes are summarized in Table 18.2. They are so successful that growers rarely apply any pesticides other than localized treatments of highly selective products.

Thrips are controlled by two species of *Amblyseius* predators which are usually released in February when flowers first appear. *Amblyseius cucumeris* are introduced in culture packs at rates of one pack to every five to seven plants (Fig. 18.1). The mites establish rapidly and can usually be found on every plant within 3 weeks. Fewer *A. degenerans* need to be released, just one per two plants, but they also establish quickly. It is not uncommon to find more than ten in each flower and on each of the upper leaves by mid-season. These single introductions are often sufficient to suppress the development of thrips and further introductions should be unnecessary unless the predators are harmed by chemicals.

Occasionally thrips become a localized problem and these hot spots of activity are controlled with *Orius* spp. (Fig. 18.2). Both *O. majusculus* and *O. laevigatus* have given very good results, particularly in summer.

The relative ease with which the thrips predators (mites and bugs) become

Table 18.2. Components of IPM programmes in UK peppers.

Pest status	Pest	Control measure
Major pests	*Frankliniella occidentalis* *Thrips tabaci*	*Amblyseius cucumeris* in culture packs and *A. degenerans* sprinkled on the plants in a loose material when flowers first appear. *Orius majusculus* and *O. laevigatus* at first sight of thrips, usually concentrated in hot spots of pest activity
	Myzus persicae	The parasitic wasp, *Aphidius colemanii*
	Spider mites	The predatory mite, *Phytoseiulus persimilis* and spot sprays of the selective acaricide, fenbutatin oxide
	Caterpillars	The bacterial insecticide based on spores and toxins of *Bacillus thuringiensis*
Minor pests	Other aphids	The selective aphicide, pirimicarb
	Capsid bugs	Spot treatments of heptenophos
	Leafhoppers	The egg parasite, *Anagrus atomus* and the insect growth regulator, buprofezin
	Tarsonemid mites	*A. cucumeris*

established has been attributed to the abundance of pollen which provides a supplementary food source at low prey densities (van Rijn and Sabelis, 1993). Similar IPM programmes are used in other north-western European countries.

IPM in UK cucumbers

IPM is used by many cucumber growers in the UK; components are summarized in Table 18.3. The programmes are usually successful but require more intensive management and more biological material than those in sweet peppers. Despite this extra input, growers are often forced to accept some loss of yield. *Frankliniella occidentalis* presents the greatest challenge and failure to control it is the most common reason for overall failure. Crops which are replanted mid-season are most vulnerable to serious damage. Thrips surviving in the greenhouse structure attack the crop immediately after planting, often concentrating on young, main stem fruits which become distorted and unmarketable (Plate IX.2–4).

Since the early 1980s the standard treatment for *Thrips tabaci* has been 'Thripstick' – applied to the plastic covering the floor beneath the crop to intercept the thrips as they drop to pupate. Thripstick is less effective against *F. occidentalis* because a significant proportion of the population avoid the treatment by pupating on the foliage. Even those which do drop may survive

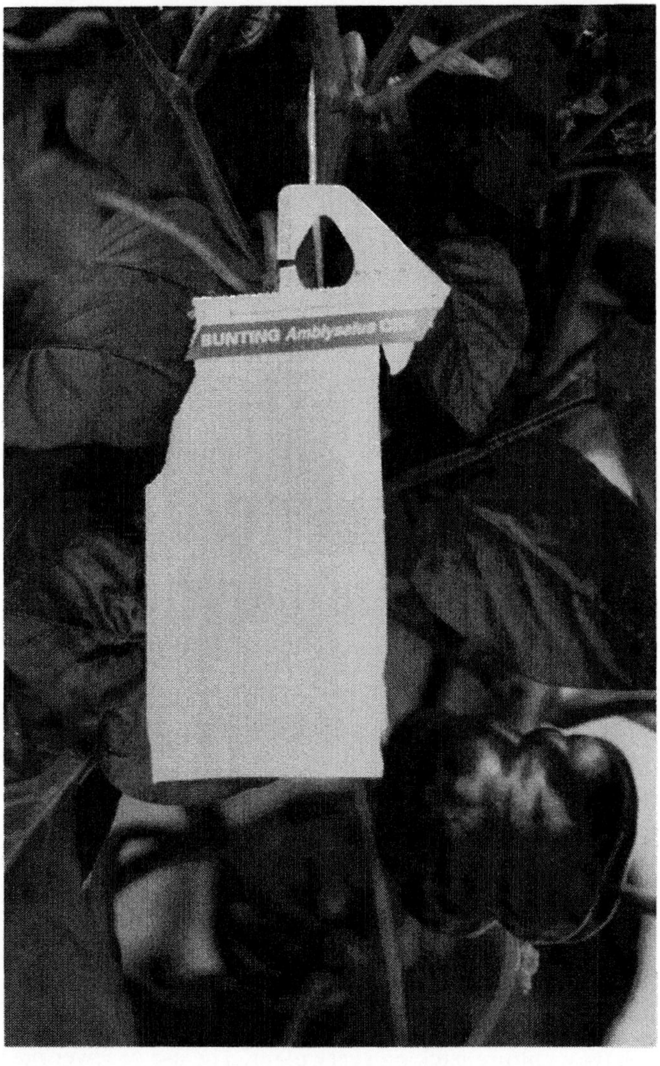

Fig. 18.1. *Amblyseius* sp. culture pack suspended on peppers.

the exposure to deltamethrin (Helyer and Brobyn, 1992).

Amblyseius cucumeris has formed the basis of the control of *F. occidentalis* for several years. The predators are introduced in culture packs which usually remain active for 6 to 8 weeks and then require replacement. The best results are obtained when packs are placed in the crop immediately after planting (Jacobson, 1995). Any delay allows thrips the opportunity to establish and jeopardizes control. It is usually 3 to 4 weeks before all adjacent plants touch, allowing the predators to move freely among the crop, and until then it is

Fig. 18.2. Distributing *Orius* sp. in a buckwheat/vermiculite carrier.

necessary to have a culture pack on every plant. The use of fewer packs allows thrips establishment on the unprotected plants and also jeopardizes control.

It is important that preventative treatments are used to best effect because there are no reliable remedial treatments available in the UK. *Orius* spp. and *Verticillium lecanii* have both been employed successfully to combat established thrips populations but with inconsistent results. *Orius* spp. frequently fail because they breed slowly and lay eggs in parts of the plant which are trimmed during routine crop work (Jacobson, 1993b). *Verticillium lecanii* often fails because the relative humidity is too low to allow spores to germinate.

Similar IPM programmes are used elsewhere although the components differ slightly in each country. In The Netherlands, growers use abamectin

Table 18.3. Components of IPM programmes in UK cucumbers.

Pest status	Pest	Control measure
Major pests	*Frankliniella occidentalis*	*Amblyseius cucumeris* in culture packs. The best results obtained when placed one per plant immediately after planting *Orius* spp. are used occasionally as a remedial treatment, with introductions concentrated in hot spots of pest activity
	Thrips tabaci	Polybutene-based floor treatments
	Aphis gossypii	*Aphidius colmani*, nicotine
	Tetranychus urticae	*Phytoseiulus persimilis*, fenbutatin oxide
	Trialeurodes vaporariorum	The parasitic wasp, *Encarsia formosa* and buprofezin
Minor pests	*Thrips fuscipennis*	Spot treatments of heptenophos
	Other aphids	Pirimicarb
	Tarsonemid mites	*A. cucumeris*
	Leaf miners	The parasitic wasps *Dacnusa sibirica* and *Diglyphus isaea*
	Capsid bugs	Heptenophos

against spider mites and thrips in propagation, dichlorvos as an aerosol immediately after planting, and an additional selective acaricide, hexythiazox. Canadian growers place more dependence on *Orius* spp., in particular, *O. insidiosus*, and the predatory mite *Hypoaspis miles* against thrips pupae beneath the crop.

IPM in UK spring and summer bedding plants

There is usually a diverse range of ornamental plants grown on a bedding plant nursery, with many different species and cultivars within single greenhouses. The individual crops are sometimes on the nursery for less than 6 weeks. During that time they are moved around the site as they progress from propagation to growing-on areas, and are then re-spaced or replanted in trays, pots or decorative containers. There may eventually be several species of plants within a single container despite their differences in susceptibility to both pests and chemical sprays.

Each plant species has its own spectrum of pests. *Frankliniella occidentalis*, *Trialeurodes vaporariorum* and *Myzus persicae* are the most common, but *Thrips tabaci* and *Aulocorthum solani* are also frequently encountered. Strict marketing standards demand that plants be virtually free from these pests at the time of sale. This unique combination of factors causes difficulties in pest control whatever strategy is chosen.

Chemical control consists typically of routine weekly space treatments, alternating between malathion and heptenophos/deltamethrin mixtures, plus occasional treatments with pyrethrin/resmethrin mixtures against *Trialeurodes vaporariorum*. If pest numbers build up during warmer weather this may be increased to two or three treatments per week. Such an intensive spray programme commonly causes phytotoxicity, particularly to the younger stages of the more sensitive plants.

An IPM package designed to reduce phytotoxicity in a bedding plant nursery was tested in northern England over two consecutive seasons (Jacobson, 1993c). The total package included strategic planning, pre-crop preparation and a pest monitoring system, as well as the selection of appropriate formulations and application rates of beneficials and selective chemicals. The biological control agents comprised *Amblyseius cucumeris*, *Encarsia formosa* and *Aphidius* spp. with additional, spot treatments of the selective aphicide pirimicarb, to combat sudden, large invasions of aphids. The beneficials were applied prophylactically in generous numbers to prevent pest establishment, thereby satisfying the market requirements.

One of two formulations of *Amblyseius cucumeris* was chosen, depending on the growing system employed. Where plants susceptible to thrips were grown in 3-m wide beds of trays, modules or pots, the *A. cucumeris* were distributed weekly, with the stored-product mites upon which they had been reared, in a specially formulated mixture of bran and vermiculite. A dusting machine driven by bellows was used to facilitate efficient, safe and even distribution of the material; dusting machines with mechanical action cause a high mortality of both species of mites. Where susceptible plants were grown in hanging baskets, tubs or other containers, protection was provided with breeding colonies of *A. cucumeris* in a small version of the culture packs used in edible crops. It was rarely necessary to use more than one pack per container.

The overall results with IPM compared favourably with the previously used chemical control strategy. No plants were lost due to direct pest damage or to phytotoxicity. The only invertebrate presence noted on the plants at the time of sale was parasitized aphids, so no plants were rejected by the nurseries' own quality control checks, or by customers.

Progress with IPM in cut flowers

IPM in cut flower crops is still largely restricted to experimental crops although there have been attempts in several countries to establish programmes on commercial nurseries. Most of the work has been done in chrysanthemum and gerbera.

Chrysanthemum

Most pests have been controlled satisfactorily by IPM programmes but thrips, and in particular *Frankliniella occidentalis*, have caused difficulties (Buxton and Finlay, 1993; Fransen *et al.*, 1993; Ravensberg and Altena, 1993). While various combinations of *Amblyseius* spp., *Orius* spp. and *Verticillium lecanii* have suppressed thrips development, crops have not been marketable due to continued pest activity and damaged flower heads. Control of *F. occidentalis* has been improved with *V. lecanii* by elevating relative humidity to aid spore germination (Helyer *et al.*, 1992), although this may deter plant growth and encourage disease.

Further development in chrysanthemums is dependent on a much improved understanding of the interactions between host plant, pest and beneficial organisms. Partial resistance to thrips in the host plants will probably become much more important and could form the basis of the whole IPM programme by slowing down pest development and allowing the various beneficials better chance of success. Progress is being made through studies of resistance mechanisms to *F. occidentalis* and measurements of genetic variation in chrysanthemum cultivars (de Jager *et al.*, 1995; van Dijken *et al.*, 1995). Meanwhile, pesticide use in chrysanthemum crops is declining where intensive crop monitoring ensures that chemicals are only applied when really needed (Guldemond and den Belder, 1993; Murphy and Broadbent, 1993).

Gerbera

Only the flowers are harvested in gerbera crops so, like peppers, tomatoes and cucumbers, some leaf damage can be tolerated. The crop has therefore attracted more attention for biological control than most other ornamentals. Most pests infest the leaves, only damaging flowers at very large population densities, and can be suppressed with minimal use of pesticides. Thrips are the main problem because even a few can spoil flowers (Plate XI.6). Integrated control of thrips has been based on the regular release of relatively large numbers of *Amblyseius* spp. and *Orius* spp. (Ravensberg and Altena, 1993; Sorenson and Nedstram, 1993). Both predators perform well, probably using pollen as a supplementary food, and control has been acceptable at certain times of the year. However, the programmes are not yet sufficiently robust to be adopted widely on commercial nurseries.

IPM in interior landscapes

Interior landscapes vary greatly in terms of both plant species and environmental conditions. At one extreme are large public conservatories which house diverse mature exotic plants and sometimes incorporate butterflies, birds and small mammals. These plantings are usually long term with an unkempt, wild appearance. The environment is designed to provide optimum conditions for plant growth, which encourages pest development. Due to the size of the plants it is difficult to monitor pest populations properly and it is destructive to remove them should they become badly infested. Fortunately, a relatively high degree of pest damage can be tolerated allowing scope for biological systems that establish an equilibrium between pests and natural enemies.

At the other end of the scale are small displays in shops and offices, usually of hardy plants which can maintain attractive foliage in an artificial environment with air conditioning, low humidity and supplementary lighting for up to 24 h daily. They tend to be small neat groupings which receive infrequent attention yet must remain aesthetically pleasing, so individual plants are replaced regularly and do not provide a very stable environment for classical biological control.

Between these extremes are many variations and it is necessary to tailor each IPM programme to the needs of the individual situation. Despite these difficulties, surprisingly good results are obtainable once operators learn to be patient and experiment to find the most appropriate biological control agents and methods of introduction.

Table 18.4 lists the commonly encountered pests in interior landscapes and the control agents which may be used against them in IPM programmes (Steiner and Elliot, 1987; Sopp, 1993). A notable difference between this pest complex and those previously discussed is the presence of mealy bugs and scale insects, both of which have an extensive host range and thrive on the more succulent growth in longer-term plantings. Diapsids (hard or armoured scales) are particularly difficult to control without using broad-spectrum insecticides. Petroleum oils kill them but care must be taken to avoid harming plants and beneficials.

Several species of thrips may establish in interior landscapes. *Frankliniella occidentalis* and *Thrips tabaci* are the most common in all situations, while

Table 18.4. Commonly encountered pests in interior landscapes and components of IPM programmes.

Pest status	Pest	Control measure
Major pests	*Frankliniella occidentalis*	*Amblyseius* spp.
	Thrips tabaci	*Orius* spp.
	Thrips palmi	
	Hercinothrips femoralis	
	Coccids (soft scales)	*Metaphycus helvolus*
		*Chilocorus nigritus**
		*Lindorus lophanthae**
	Diapsids (hard scales)	*Aphytis melinus**
		Petroleum oils
	Mealybugs	*Cryptolaemus montrouzieri*
		Leptomastix dactylopii
		Petroleum oils
	Aphids	*Aphidius colemani*
		Aphidoletes aphidimyza
		Pirimicarb
	Tetranychus urticae	*Phytoseiulus persimilis*
		Fenbutatin oxide
	Trialeurodes vaporariorum	*Encarsia formosa*
		Buprofezin
Minor pests	Caterpillars	*Bacillus thuringiensis*
	Vine weevil	Entomopathogenic nematodes
	Tarsonemid mites	*Amblyseius cucumeris*
		Petroleum oils
	Leaf miners	*Dacnusa sibirica*
		Diglyphus isaea

*Check host/prey specificity before use.

Hercinothrips femoralis and *Parthenothrips dracaenae* are troublesome in long-term plantings. *Thrips palmi* remains a serious potential threat in the UK.

The predatory mites, *Amblyseius* spp., are effective against all the most commonly found thrips species. However, control may take time and a residual population of the pests may have to be tolerated. The method of introduction varies depending on the individual situation. Culture packs have given good results on large mature plants while regular scatterings of loose, bran-based material have proved successful on low foliage. Culture packs are impractical in some well-established conservatories where infestations of cockroaches may destroy the contents overnight. An annual release of *Orius* spp. offers a second line of defence against thrips in conservatories, but they should not be used where there are breeding colonies of butterflies as they also feed on caterpillars. It is also inadvisable to release these predators in office complexes because they sometimes 'bite' people.

Economics of IPM

One of the most important considerations determining the choice of control strategy is cost. All the costs involved in chemical and IPM programmes need calculating and then balancing against any financial benefits. This is difficult because the component costs are extremely variable and open to wide interpretation. Before studying any programme in detail it is worth considering the factors involved in preparing a cost–benefit analysis and highlighting the most important variables.

Difficulties in preparing costs

The components in a cost–benefit analysis are summarized in Table 18.5. The main variables are: (i) capital items such as spraying or fogging equipment which may be written off over periods of up to 10 years – if a short period is used, costs are loaded heavily against chemical strategies which use this type of equipment; (ii) the cost of consumables varies depending upon the supplier and the purchasing power of the grower, for example, during 1993 the price of *A. cucumeris* culture packs varied from £0.04 to £0.08 each (US$0.06 to $0.12), a difference of over £600 (US$900) for a single treatment of one pack per cucumber plant ha^{-1}; (iii) labour costs are often ignored, particularly on small family-run nurseries where the owner does not price his own time; (iv) the value of produce is dependent upon 'supply and demand' and is rarely known until the day of sale – fluctuations in market price can mask large benefits of yield and quality; and (v) to optimize the effects of biological control agents it may be necessary to modify the cropping environment (e.g. by providing a temperature boost), and if this is done the extra cost must be built into the calculations.

By selecting different extremes of these variables, costs can be manipulated to support arguments for any particular control strategy and must be viewed with caution.

Table 18.5. Components to include in a cost–benefit analysis.

Expenditure	Income
Capital items (e.g. application equipment)	Effect on yield
Consumable items (e.g. chemicals, biocontrol agents, heating fuel)	Effect on quality
Labour	Value of produce

Comparative costs in spring and summer bedding

Spring and summer bedding plants are one of the few crops for which there are effective biological and chemical control strategies, so this crop provides a good example of comparative costs. The components of typical programmes are detailed in Table 18.6. In these calculations the capital costs of application equipment were written off over 5 years.

At £3800 (US$5700) per hectare per season IPM is about 50 per cent more expensive than the chemical alternative. The most significant difference is the cost of labour, which reflects the complexity of the pest monitoring operations in the IPM programme, but consumable items are also more expensive.

Production losses in the chemical programme may arise due to phytotoxicity. These are difficult to quantify but sometimes tilt the balance of costs back in favour of the IPM system. Despite this, many growers still consider IPM to be too expensive.

Table 18.6. Components of costs of typical IPM and chemical control programmes in spring and summer bedding plant crops (£k ha^{-1} $season^{-1}$).

	Control strategy	
Expense category	IPM	Chemical
Capital		0.1
Consumables		
Biologicals against thrips	0.93	
Biologicals against other pests	0.20	
Chemicals used in crop	0.06	0.8
Chemicals used as clean-up	0.10	
Labour		
Monitoring	1.80	0.47
Application	0.70	1.10
Total	3.79	2.47

Costs in protected peppers and cucumbers

Full chemical control strategies are not available for either cucumber or pepper crops so growers must use IPM. The pests and control agents, which are largely biological, are summarized in Tables 18.2 and 18.3. The actual costs from cucumber and pepper crops grown under very similar management regimes in 1992 and 1993 are detailed in Table 18.7. The most significant difference between the crops was the cost of controlling *Frankliniella occidentalis*. In peppers, *Amblyseius* spp. and *Orius majusculus* established quickly at low introduction rates and were effective, but in cucumbers they were slower and control was less reliable. Therefore, more biological material was used in cucumbers and the cost of both consumables and labour for application was greater.

In peppers, the costs represented about 3 per cent of the total value of the crop, which is acceptable because the programme was successful. However, in cucumbers, the costs were 6 per cent of the crop value which, considering there was still a decrease in yield or quality due to thrips, was not satisfactory.

Uptake of IPM in Glasshouses Around the World

In 1994, key research and extension workers from around the world were asked to comment on the use of IPM programmes within the commercial greenhouse industry in their region. This survey showed that IPM is often used in protected edible crops throughout north-west Europe and Canada but is much less common elsewhere.

The crops in which IPM techniques were most widely adopted were protected tomatoes, sweet peppers and cucumbers; all long-season monocrops grown in stable environments. In temperate regions, they do not suffer continual pressure from invading pests, nor do they yet suffer serious problems

Table 18.7. Comparative costs of IPM in cucumber and pepper crops in the UK (£k ha^{-1} $season^{-1}$).

	Crop	
Expense category	Cucumber	Pepper
Consumables		
Biologicals against thrips	4.73	1.40
Biologicals against other pests	2.47	2.20
Floor treatment	0.80	0.47
Chemicals used in crop	1.67	0.13
Chemicals used as clean-up	1.13	0.20
Labour		
Monitoring	2.00	0.60
Application	2.20	2.20
Total	15.00	7.20

caused by insect-transmitted diseases. The situation was quite different in leafy vegetable and ornamental crops in these temperate lands. Although many growers experimented with biological techniques, few had adopted full programmes and IPM was still the exception because of the complexities and cost of IPM on mixed ornamental nurseries and the need to adhere to strict marketing standards. However, many growers of these crops had reduced their pesticide usage in recent years by adopting supervised pest control.

Elsewhere in the world, chemicals still formed the basis of pest control in most of the commercial greenhouse crops. The main factors limiting the uptake of IPM in hotter climates in the northern hemisphere were the difficulties in combating sudden and huge pest invasions, the speed of development in the crops and the fear of virus infections spread by insects, particularly *Frankliniella occidentalis* and *Bemisia tabaci*.

In Australia and Japan, the unavailability of biological control agents limited the uptake of IPM.

The Future of IPM in Protected Crops

IPM will form the basis of pest control in glasshouse crops in the future, although progress will depend on a better understanding of the systems currently in use as well as on the development of new techniques and strategies. The research and development required for this ranges from applied biological studies in crops to basic molecular research. The former may be funded by the commercial horticultural and biocontrol supply industries, but the more basic and strategic work will need to be funded by governments.

In addressing the more strategic issues it is important that research workers retain sight of the wider concept of IPM. Although studies on individual subjects will inevitably be done in isolation, researchers must remain mindful of the need for complete IPM programmes covering all pests of each crop and in particular the compatibility of control agents.

Short-term research and development needs

There are many examples of potential biological control agents which have failed to provide acceptable control in commercial crops because there is inadequate information on how to use them effectively. The immediate priority of research and development should be to generate data packages to support all biological products. These should include information on duration of life cycles at relevant temperatures, rates of population growth and host consumption (see Chapter 8), as well as factors which could cause a temporary breakdown in control, such as winter diapause or aestivation. This information should then be incorporated into broader studies to elucidate the complex interactions between plants, pests and beneficial organisms, leading to improved control by manipulation of the growing environment.

Another high priority should be the search for new biological control agents, particularly against the minor or secondary pests which appear when routine use of pesticides stops. These pests are often ignored in the early development of IPM, only to find that the continued need to apply broad-spectrum insecticides against them prevents the implementation of the wider programme.

The release rates of biological control agents are sometimes limited by their cost and much better results could be obtained if more were used. The modelling approaches described in Chapter 8 could lead to better prediction of pest–predator ratios necessary to give adequate control. Larger applications in the early stages of ornamental crops could eliminate pests and enable growers to meet the strict marketing standards. Lower prices need to be achieved by the development of more efficient production systems. Cheaper biological control agents would also encourage the development of prophylactic strategies which would simplify overall systems and require less management. Such a move towards less intensive crop monitoring can only occur if growers have faith in the quality of the biological material. It is thus important that the International Organisation for Biological Control (IOBC) guidelines for quality control (van Lenteren, 1993) are extended to all biological control agents.

The integration of selective insecticides and acaricides in IPM programmes also eases the pressure on managers by providing a second line of defence and reducing the risk of complete failure. However, such products are currently only available against a small proportion of greenhouse pests and not all are universally obtainable. It would greatly assist the development of IPM programmes if more selective products could be used. Currently, pesticide manufacturers cannot always justify the cost of generating data required for registration of new products for horticultural crops, or the actual registration fees, in all countries.

Longer-term research needs

The selection of plant cultivars with an inherent tolerance to pests will also contribute to IPM programmes by retarding pest development and allowing beneficials to achieve control before significant damage occurs. Three attributes of plants that may contribute to partial resistance are: (i) antixenosis – deterrent compounds reducing host plant selection; (ii) antibiosis – failure of the feeding pest to thrive due to poor nutrition or toxic compounds (Mollema and van der Schoot, 1990; Soria and Mollema, 1995); and (iii) physical properties – reduced pest mobility caused by sticky trichomes or excessive hairiness of leaves (van Haren *et al.*, 1987). Research to incorporate these features into commercial cultivars is long term; pest-tolerant plant species or cultivars must be found, the morphological and biochemical characteristics responsible for partial resistance identified, before these characteristics can be bred or engineered into commercial cultivars.

Baculoviruses offer another option to replace selective chemicals within IPM programmes. Being natural pathogens of invertebrates they exhibit strong host specificity, but none is yet available against the commonly encountered glasshouse pests. The baculovirus genome could perhaps be modified, using genetic engineering to incorporate foreign genes encoding other insect-specific toxins, hormones or enzymes (Possee *et al.*, 1993).

The type of chemicals used in IPM is likely to change gradually with invertebrate poisons being phased out to be replaced by synthetic analogues of pheromones, allomones and kairomones (see Chapter 3). The use of such semiochemicals will require thorough knowledge of the tritrophic interactions between plants, herbivores and beneficial organisms. Perhaps the ultimate goal in pest control should be the use of genetic engineering to incorporate antifeedants or invertebrate alarm pheromones into plant cultivars.

References

Berlinger, M.J., Mordechi, S. and Leeper, A. (1991) Application of screens to prevent whitefly penetration into greenhouses in the Mediterranean Basin. *International Organisation of Biological Control/West Palaearctic Regional Section Bulletin* 14, 105–110.

Berlinger, M.J., Lebuish-Mordechi, S., Fridja, D. and Mor, N. (1993) The effects of types of greenhouse screens on the presence of western flower thrips: a preliminary study. *International Organisation of Biological Control/West Palaearctic Regional Section Bulletin* 16(2), 13–16.

Brødsgaard, H.F. (1990) The effect of anisaldehyde as a scent attractant for *Frankliniella occidentalis* (Thysanoptera: Thripidae) and the response mechanism involved. *International Organisation of Biological Control/West Palaearctic Regional Section Bulletin* 13(5), 36–38.

Brødsgaard, H.F. (1993a) Coloured sticky traps for thrips monitoring on glasshouse cucumbers. *International Organisation of Biological Control/West Palaearctic Regional Section Bulletin* 16(2), 19–22.

Brødsgaard, H.F. (1993b) Monitoring thrips in glasshouse pot plant crops by means of blue sticky traps. *International Organisation of Biological Control/West Palaearctic Regional Section Bulletin* 16(8), 29–32.

Buxton J.H. and Finlay, R. (1993) Integrated pest management in AYR chrysanthemums. *International Organisation of Biological Control/West Palaearctic Regional Section Bulletin* 16(8), 33–41.

Chambers, R.J., Long, S. and Helyer, N. (1993) Effectiveness of O. *laevigatus* for the control of *F. occidentalis* on cucumber and pepper in the UK. *Biocontrol Science and Technology* 3, 295–307.

Czencz, K. (1987) The role of coloured traps in collecting thrips fauna. In: Holman J., Pelikan, J., Dixon, A.F.G. and Weismann, L. (eds) *Population Structure, Genetics and Taxonomy of Aphids and Thysanoptera.* SPB Academic Publishing, The Hague, pp. 426–435.

de Jager, K.M., Butot, R.P.T. and Guldemond, A. (1995) Genetic variation in chrysanthemum for resistance to western flower thrips and *Thrips tabaci.* In: Parker, B.L., Skinner, M. and Lewis, T. (eds) *Thrips Biology and Management.* Plenum Press, New York, pp. 403–406.

Fransen, J.J., Boodgaard, M. and Tolsma J. (1993) The minute pirate bug, *Orius insidiosus* (Say) (Hemiptera; Anthocoridae), as a predator of western flower thrips, *Frankliniella occidentalis* (Pergande), in chrysanthemum, rose and saintpaulia. *International Organisation of Biological Control/West Palaearctic Regional Section Bulletin* 16(8), 51–54.

Guldemond, J.A. and den Belder, E. (1993) Supervised control in chrysanthemums: one years experience. *International Organisation of Biological Control/West Palaearctic Regional Section Bulletin* 16(8), 51–54.

Hall, R.A. and Burges, H.D. (1979) Control of aphids in glasshouses with the fungus

V. lecanii. Annals of Applied Biology 93, 235–246.

Helyer, N. and Brobyn, P.J. (1992) Chemical control of western flower thrips (*Frankliniella occidentalis* Pergande). *Annals of Applied Biology* 121, 219–231.

Helyer, N., Gill, G., Bywater, A. and Chambers, R.J. (1992) Elevated humidities for control of chrysanthemum pests with *V. lecanii. Pesticide Science* 36, 373–378.

Jacobson, R.J. (1993a) Control of *Frankliniella occidentalis* with *O. majusculus*: experiences during the first full season of commercial use in the UK. *International Organisation of Biological Control/West Palaearctic Regional Section Bulletin* 16(2), 81–84.

Jacobson, R.J. (1993b) Egg laying sites of *Orius majusculus*, a thrips predator, on cucumber. In: Parker, B.L., Skinner, M. and Lewis, T. (eds) *Thrips Biology and Management*. Plenum Press, New York, pp. 241–244.

Jacobson, R.J. (1993c) IPM in spring and summer bedding plants: a successful package for commercial crops. *International Organisation of Biological Control/West Palaearctic Regional Section Bulletin* 16(8), 105–112.

Jacobson, R.J. (1995) IPM in cucumbers – prevention of establishment of *Frankliniella occidentalis*. Mededelingen van de Faculteit van de Landbouwwentenschappen, University of Gent, 60/3a, 857–863.

Kirk, W.D.J. (1984) Ecologically selective coloured traps. *Ecological Entomology* 9, 35–41.

Mollema, C. and van der Schoot, M. (1990) Investigation of the resistance in tomato to the leafminer *Liriomyza trifolii. International Organisation of Biological Control/West Palaearctic Regional Section Bulletin* 13, 117–121.

Mollema, C., Steenhuis, M.M., Inggamer, H. and Soria, C. (1993a) Evaluating the resistance to *F. occidentalis* in cucumber: genotypic variation and effects upon thrips biology. *International Organisation of Biological Control/West Palaearctic Regional Section Bulletin* 16(5), 77–82.

Mollema, C., Steenhuis, M.M. and Inggamer, H. (1993b) Genotypic effects of cucumber responses to infestation by WFT. In: Parker, B.L., Skinner, M. and Lewis, T. (eds) *Thrips Biology and Management*. Plenum Press, New York, pp. 397–401.

Morison, G.D. (1957) A review of British glasshouse Thysanoptera. *Transactions of the Royal Entomological Society of London* 109, 467–534.

Murphy, G.D. and Broadbent, A.B. (1993) Development and implementation of IPM in greenhouse floriculture in Ontario, Canada. *International Organisation of Biological Control/West Palaearctic Regional Section Bulletin* 16(2), 113–116.

Pickford, R.J.J. (1984) Evaluation of soil treatment for control of *Thrips tabaci* on cucumbers. Tests of Agrochemicals and Cultivars. *Annals of Applied Biology* 5, 18–19.

Possee, R.D., Calyley, P.J., Cory, J.S. and Bishop, D.H.L. (1993) Genetically engineered viral insecticides with improved phenotypes. *Pesticide Science* 39, 109–115.

Ravensberg, W.J. and Altena, K. (1993) IPM in glasshouse ornamentals: The first steps in practice. *International Organisation of Biological Control/West Palaearctic Regional Section Bulletin* 16(8), 119–122.

Samson, R.A., Ramakers, P.M.J. and Oswald, T. (1979) *Entomophthora thripidum*, a new fungal pathogen of *Thrips tabaci. Canadian Journal of Botany* 57, 1317–1323.

Shipp, J.L. and Zariffa, N. (1991) Spatial patterns of and sampling methods for western flower thrips (Thysanoptera: Thripidae) on greenhouse sweet pepper. *Canadian Entomologist* 123, 989–1000.

Shipp, J.L., Clark, N.D., Jarvis, W.R. and Papadopoulos, A.P. (1993) Expert system for integrated crop management of greenhouse cucumber. *International Organisation of Biological Control/West Palaearctic Regional Section Bulletin* 16(2), 149–152.

Sopp, P. (1993) IPM in interior landscapes. *International Organisation of Biological Control/West Palaearctic Regional Section Bulletin* 16(8), 126–128.

Sorenson, A. and Nedstram, B. (1993) Effect of *Amblyseius cucumeris* and *Orius insidiosus* on *Frankliniella occidentalis* in ornamentals. *International Organisation of Biological Control/West Palaearctic Regional Section Bulletin* 16(8), 129–132.

Soria, C. and Mollema, C. (1995) Life-history components of western flower thrips on susceptible and resistant cucumber genotypes. *Entomologia Experimentalis et Applicata* 74, 177–184.

Steiner, M.Y. and Elliot, D.P. (1987) *Biological Pest Management for Interior Plantscapes*, 2nd edn. Alberta Environment Centre, AECV87-E1. Vegreville, AB.

Teulon, D.A.J., Hollister, B. and Cameron, E.A. (1993a) Behavioural responses of western flower thrips to anisaldehyde, and implications for trapping in greenhouses. *International Organisation of Biological Control/West Palaearctic Regional Section Bulletin* 16(8), 177–180.

Teulon, D.A.J., Penman, D.R. and Ramakers, P.M.J. (1993b) Volatile chemicals for thrips (Thysanoptera: Thripidae) host-finding and applications for thrips pest management. *Journal of Economic Entomology* 86, 1405–1415.

van Dijken, F.R., Mollema, C. and de Jong, J. (1995) Host plant resistance to western flower thrips in chrysanthemum. In: Parker, B.L., Skinner, M. and Lewis, T. (eds) *Thrips Biology and Management.* Plenum Press, New York, pp. 407–410.

van Haren, R.J.F., Steenhuis, M.M., Sabelis, M.W. and de Ponti, O.M.B. (1987) Tomato stem trichomes and dispersal success of *Phytoseilus persimilis* relative to its prey *Tetranychus urticae. Experimental and Applied Acarology* 3, 115–121.

van Lenteren, J.C. (1993) Quality control for natural enemies used in greenhouses. *International Organisation of Biological Control/West Palaearctic Regional Section Bulletin* 16(2), 89–92.

van Rijn, P.C.J. and Sabelis, M.W. (1993) Does alternative food always enhance biological control? The effect of pollen on the interaction between WFT and its predators. *International Organisation of Biological Control/West Palaearctic Regional Section Bulletin* 16(8), 123–125.

Vernon, R.S. and Gillespie, D.R. (1995) Influence of trap shape, size, and background color on captures of *Frankliniella occidentalis* (Thysanoptera: Thripidae) in a cucumber greenhouse. *Journal of Economic Entomology* 88, 288–293.

Vierbergen, G. (1995) International movement, detection and quarantine of Thysanoptera pests. In: Parker, B.L., Skinner, M. and Lewis, T. (eds) *Thrips Biology and Management.* Plenum Press, New York, pp. 119–132.

Walton, S. (1992) Safer foods on the shelf. *Grower*, July, 11–12.

Appendix I

Thrips Species Cited, with Authorities and Common Names*

Species	Common name
Acanthothrips nodicornis (Reuter)	
Aduncothrips asiaticus (Ramakrishna and Margabandhu)	
Aeolothrips albicinctus Haliday	
Aeolothrips astutus Priesner	
Aeolothrips collaris Priesner	
Aeolothrips intermedius Bagnall	Banded thrips
Aeolothrips kawanaii Moulton	
Aeolothrips melisi Priesner	
Aeolothrips tenuicornis Bagnall	
Aeolothrips vittatus Haliday	
Aleurodothrips fasciapennis (Franklin)	
Anactinothrips gustaviae Mound and Palmer	
Anaphothrips cameroni (Bagnall)	
Anaphothrips obscurus (Müller)	American grass thrips
Anaphothrips sudanensis Trybom	
Anaphothrips woodi Pitkin	
Andrewarthaia kellyana (Bagnall)	
Androthrips flavipes Schmutz	
Anisopilothrips venustulus Priesner	
Apterothrips apteris (Daniel)	
Apterothrips secticornis (Trybom)	
Aptinothrips rufus (Haliday)	Grass thrips
Aptinothrips stylifer Trybom	Grass thrips

continued overleaf

*NB. Not all pest species have a common name, neither does a name associated with a particular host necessarily imply that the species is monophagous. These current names are used throughout the text replacing earlier synonyms.

Species	Common name
Arorathrips mexicanus (Crawford)	
Arrhenothrips ramakrishnae Hood	
Aurantothrips orchidaceous (Bagnall)	
Australothrips bicolor Bagnall	
Ayyaria chaetophora Karny	
Bactrothrips brevitubus Takahashi	
Bagnalliella yuccae (Hinds)	Yucca thrips
Baliothrips biformis (Bagnall)	
Baliothrips dispar (Haliday)	
Brachyurothrips anomalus Bagnall	
Caliothrips fasciatus (Pergande)	Bean thrips
Caliothrips gramicola (Bagnall and Cameron)	
Caliothrips impurus (Priesner)	
Caliothrips indicus (Bagnall)	Sesbania or groundnut thrips
Caliothrips insularis (Hood)	
Caliothrips phaseoli (Hood)	
Caliothrips sudanensis (Bagnall and Cameron)	Cottonleaf thrips
Catinathrips vaccinophilus (Hood)	
Ceratothrips ericae (Haliday)	Heather thrips
Ceratothrips frici (Uzel)	
Chaetanaphothrips leeuweni (Karny)	
Chaetanaphothrips orchidii (Moulton)	
Chaetanaphothrips signipennis (Bagnall)	Banana thrips
Chaetanaphothrips theiperdus (Karny)	
Chirothrips crassus Hinds	
Chirothrips hamatus Trybom	Meadow foxtail thrips
Chirothrips manicatus Haliday	
Chirothrips pallidicornis Priesner	
Corynothrips stenopterus Williams	Cassava thrips
Craspedothrips minor (Bagnall)	
Cryptothrips nigripes (Reuter)	
Cryptothrips rectangularis Hood	
Danothrips trifasciatus Sakimura	
Dendrothrips ornatus (Jablonowski)	Privet thrips
Diarthrothrips coffeae Williams	Coffee thrips
Dichromothrips corbeti (Priesner)	Orchid thrips
Dichromothrips phalaenopsidis Sakimura	Orchid thrips
Dinurothrips hookeri Hood	
Domatiathrips cunninghamii Mound	
Drepanothripoides innoxius (Karny)	
Drepanothrips reuteri Uzel	Grape or vine thrips
Ecacanthothrips tibialis (Ashmead)	
Echinothrips americanus Morgan	
Echinothrips caribeanus Hood	
Elaphrothrips tuberculatus (Hood)	
Enneothrips flavens Moulton	
Erythrothrips asiaticus Ramakrishna and Marghabandhu	–

Species	Common name
Eurylaprothrips crassus Ramakrishna and Marghabandhu	
Florithrips traegardhi (Trybom)	
Frankliniella bispinosa (Morgan)	
Frankliniella bruneri Watson	
Frankliniella cestrum Moulton	
Frankliniella fusca (Hinds)	Tobacco thrips
Frankliniella insularis (Franklin)	
Frankliniella intonsa (Trybom)	Flower thrips
Frankliniella kelliae Sakimura	
Frankliniella lilivora Kurosawa	
Frankliniella melanommata Williams	
Frankliniella minuta (Moulton)	
Frankliniella moultoni Hood	
Frankliniella occidentalis (Pergande)	Western flower thrips
Frankliniella pallida (Uzel)	
Frankliniella panamensis Hood	
Frankliniella parvula Hood	Banana flower thrips
Frankliniella schultzei (Trybom)	Common blossom or cotton bud thrips
Frankliniella tenuicornis (Uzel)	
Frankliniella tritici (Fitch)	Eastern flower thrips
Frankliniella vaccinii Morgan	Blueberry thrips
Frankliniella williamsi Hood	
Franklinothrips caballeroi Johansen	
Franklinothrips megalops (Trybom)	
Franklinothrips tenuicornis Hood	
Franklinothrips vespiformis (Crawford)	
Fulmekiola serrata (Kobus)	Cane thrips
Gigantothrips afer (Priesner)	
Gnophothrips fuscus (Morgan)	
Gynaikothrips ficorum (Marchal)	Cuban laurel thrips
Gynaikothrips flaviantennatus Moulton	
Gynaikothrips uzeli (Zimmermann)	
Haplothrips aculeatus (Fabricius)	
Haplothrips bedfordi Jacot-Guillarmond	
Haplothrips cahirensis (Trybom)	
Haplothrips ceylonicus Schmutz	
Haplothrips chinensis Priesner	
Haplothrips cottei (Vuillet)	
Haplothrips ganglebaueri Schmutz	
Haplothrips globiceps Bagnall	
Haplothrips kurdjumovi Karny	
Haplothrips leucanthemi (Schrank)	
Haplothrips niger (Osborn)	
Haplothrips senecionis Bagnall	
Haplothrips sorghicola Bagnall	

continued overleaf

Species	Common name
Haplothrips statices (Haliday)	
Haplothrips subtilissimus (Haliday)	
Haplothrips tardus Priesner	
Haplothrips tenuipennis Bagnall	
Haplothrips tritici (Kurdjumov)	Wheat thrips
Haplothrips verbasci (Osborn)	Mullein thrips
Haplothrips victoriensis Bagnall	
Heliothrips haemorrhoidalis (Bouché)	Greenhouse thrips
Heliothrips sylvanus Faure	
Hercinothrips bicinctus (Bagnall)	Banana silvering thrips
Hercinothrips femoralis (Reuter)	Banded greenhouse or sugar beet thrips
Heterothrips arisaemae Hood	
Heterothrips azaleae Hood	Azalea thrips
Holothrips hagai Okajima	
Holothrips japonicus Okajima	
Hoplandrothrips bredoi Priesner	
Hoplandrothrips coffeae Bagnall	
Hoplandrothrips marshalli Karny	
Hoplandrothrips pillichianus Priesner	
Hoplothrips corticis (De Geer)	
Hoplothrips japonicus (Karny)	
Hoplothrips karnyi (Hood)	
Hoplothrips pedicularius (Haliday)	
Hoplothrips ulmi (Fabricius)	
Hydatothrips adolfifriederici Karny	
Kakothrips pisivorus (Westwood)	Pea thrips
Larothrips dentipes Pitkin	
Leeuwenia pasanii (Mukaigawa)	
Lefroyothrips lefroyi (Bagnall)	
Leptothrips mali (Fitch)	Black hunter
Leucothrips theobromae Priesner	
Limothrips cerealium Haliday	Cereal, corn or grain thrips
Limothrips denticornis Haliday	Barley thrips
Limothrips schmutzi Priesner	
Liothrips adisi Zur Strassen	
Liothrips africanus Vuillet	
Liothrips brevitubus Kono	
Liothrips floridensis (Watson)	
Liothrips fluggae Bournier	
Liothrips hungaricus (Priesner)	
Liothrips karnyi (Bagnall)	Pepper leaf gall thrips
Liothrips kuwanai (Moulton)	
Liothrips laureli (Mason)	
Liothrips mikaniae (Priesner)	
Liothrips oleae (Costa)	Olive thrips
Liothrips piperinus Priesner	
Liothrips setinodis (Reuter)	

Species	Common name
Liothrips urichi Karny	
Liothrips vaneeckei Priesner	Lily bulb thrips
Liothrips varicornis Hood	
Liothrips wasabiae Haga and Okajima	
Mallothrips indicus Ramakrishna	
Megalothrips spinosus Hood	
Megalurothrips distalis (Karny)	
Megalurothrips nigricornis (Schmutz)	
Megalurothrips sjostedti (Trybom)	Bean flower thrips
Megalurothrips usitatus (Bagnall)	
Merothrips brunneus Ward	
Mesothrips claripennis Moulton	
Microcephalothrips abdominalis (Crawford)	Composite thrips
Mycterothrips consociatus (Targioni-Tozzetti)	
Mycterothrips glycines (Okamoto)	
Mymarothrips bicolor zur Strassen	
Neohydatothrips gracilicornis (Williams)	
Neohydatothrips pseudoannulipes Johansen	
Neohydatothrips variabilis (Beach)	
Odontothrips confusus Priesner	
Odontothrips intermedius (Uzel)	
Odontothrips loti (Haliday)	Birdsfoot trefoil thrips
Odontothrips phaleratus (Haliday)	
Organothrips bianchi Hood	Taro thrips
Panchaetothrips indicus Bagnall	
Panchaetothrips noxius Priesner	
Parthenothrips dracaenae (Hegeer)	Palm thrips
Pezothrips dianthi (Priesner)	
Phlaeothrips coriaceus Haliday	
Ponticulothrips diospyrosi Haga and Okajima	
Prosopothrips cognatus Hood	
Pseudodendrothrips mori (Niwa)	
Pseudodendrothrips ornatissimus Schmutz	
Retithrips syriacus (Mayet)	Castor thrips
Rhipiphorothrips cruentatus Hood	Grapevine thrips
Rubiothrips vitis (Priesner)	
Schedothrips orientalis (Ananthakrishnan)	
Sciothrips cardamomi (Ramakrishna)	Cardamon thrips
Scirtothrips aurantii Faure	South African citrus thrips
Scirtothrips bispinosus (Bagnall)	Coffee thrips
Scirtothrips citri (Moulton)	Citrus thrips
Scirtothrips dorsalis Hood	Assam or chilli thrips
Scirtothrips inermis Priesner	
Scirtothrips kenyensis Mound	Tea thrips
Scirtothrips longipennis (Bagnall)	Begonia thrips

continued overleaf

Species	Common name
Scirtothrips mangiferae Priesner	
Scirtothrips manihoti Bondar	
Scirtothrips oligochaetus (Karny)	
Scolothrips acariphagus Yakhontov	
Scolothrips indicus Priesner	
Scolothrips longicornis Priesner	
Scolothrips sexmaculatus (Pergande)	Six-spotted thrips
Selenothrips rubrocinctus (Giard)	Cacao or red-banded thrips
Sorghothrips jonnaphilus (Ramakrishna)	
Sporothrips amplus Hood	
Stenchaetothrips biformis (Bagnall)	Oriental rice thrips
Stenchaetothrips minutus (Deventer)	
Stenothrips graminum (Uzel)	Oats thrips
Stictothrips fimbriata (Ananthakrishnan)	
Taeniothrips inconsequens (Uzel)	Pear thrips
Taeniothrips major Bagnall	
Taeniothrips orionis Treherne	
Tenothrips discolor (Karny)	
Tenothrips frici (Uzel)	
Thlibothrips isunoki Okajima	
Tiarothrips subramanii (Ramakrishna)	
Thrips alliorum (Priesner)	
Thrips angusticeps Uzel	
Thrips atratus Haliday	
Thrips australis (Bagnall)	Gum tree thrips
Thrips calcaratus Uzel	Basswood thrips
Thrips carthami Shumsher	
Thrips coloratus Schmutz	
Thrips flavus Schrank	Honeysuckle thrips
Thrips florum Schmutz	
Thrips fuscipennis Haliday	
Thrips hawaiiensis (Morgan)	
Thrips imaginis Bagnall	Apple blossom or plague thrips
Thrips juniperinus Linnaeus	
Thrips laricivorus (Kratochvil and Farsky)	Larch thrips
Thrips linarius Uzel	Flax thrips
Thrips madroni Moulton	
Thrips major Uzel	Rose thrips
Thrips meridionalis (Priesner)	
Thrips minutissimus Linnaeus	
Thrips nigropilosus Uzel	Chrysanthemum thrips
Thrips obscuratus Crawford	New Zealand flower thrips
Thrips palmi Karny	
Thrips parvispinus (Karny)	
Thrips physapus Linnaeus	Dandelion thrips
Thrips pillichi Priesner	
Thrips pini (Uzel)	
Thrips pusillus Bagnall	

Species	Common name
Thrips setosus Moulton	
Thrips simplex (Morison)	Gladiolus thrips
Thrips tabaci Lindeman	Onion or potato thrips
Thrips validus Uzel	
Thrips vulgatissimus Haliday	
Trichinothrips breviceps (Bagnall)	
Varshneyia pasania (Mukaigawa)	
Xylaplothrips inquilinus Priesner	
Zaniothrips ricini Bhatti	
Zygothrips minutus Uzel	

Appendix II

Major Crops Infested by Thrips with Main Symptoms and Predominant Injurious Species*

This appendix provides a guide to the main food, commodity and glasshouse (GH)/protected crops and ornamentals susceptible to thrips injury in different parts of the world, with brief descriptions of injury symptoms and names of species likely to cause them. It is confined to species recorded as producing repeated damage and does not include sporadic injury from highly polyphagous species which often feed unpredictably on crops to cause minor damage, or the many harmless vagrants which temporarily alight on crops and as a result are often misleadingly recorded as 'pests'. Serious pests are aserisked. Locations are confined to broad geographic regions, except where important species appear in relatively localized areas, or appear to have been introduced recently and have become well established.

Generally, thrips feeding damage to plants is detectable initially as a silvery sheen due to air entering damaged cells, which later turn yellow, greenish-brown or even red. Often discoloured areas coalesce and whole leaves may dry, shrivel and fall prematurely. Black dots of excrement which may become covered with brown fungal growths often further disfigure infested parts. The silvering and yellowing of leaves is often termed 'white blight', 'white spot' or 'yellow spot' by growers. Buds and young leaves of some crops turn reddish where thrips have fed and as the foliage expands it appears scarred and distorted. A few phlaeothripids induce marginal leaf-curling galls. On flowers, the petals, stamens or pistil may be attacked separately and develop silver, white, then brown patches or streaks; often the whole flower withers. Infested inflorescences of cereals wither and turn silver-white, producing 'whiteheads' or 'silvertop'. Stiffer organs like pine needles, stems and fruits grow distorted, scarred or with russeted skin, and young shoots may die.

The information presented in this appendix is solely qualitative and the

*Compiled by T. Lewis, assisted by L.A. Mound, S. Nakahara (Systematic Entomology Laboratory, USDA/ARS, Beltsville) and C.C. Childers.

different infestation levels required to produce the symptoms described may need to be measured for each thrips species, crop and location using the principles outlined in Chapters 11, 16, 17 and 18 to decide whether control is worthwhile. For each crop listed the most characteristic symptoms are given, including the effects of thrips-transmitted viruses, with key references to further details where available.

Food and commodity crops

Crop	Location	Symptoms	Thrips	References
Alfalfa – see forage legumes				
Apple (*Malus pumila*: *Rosaceae*)	N. America	Flowers and stems discoloured; fruit pitted by oviposition at pink bud and on young fruit	*Frankliniella occidentalis*	Swift and Madsen, 1956; Singh, 1989; Terry, 1991
	California	"	*Thrips madroni**	
	E. Australia	Flowers with white and silvery patches. Poor fruit set and distorted, scarred fruit	*Thrips imaginis*	
	S. Asia	"	*Thrips flavus* *T. carthami*	
Apricots – see stonefruit				
Asparagus (*Asparagus officinalis*: *Liliaceae*)	Worldwide	Shoot tips silvered and shrunken	*Thrips tabaci*	Banham 1968; Townsend *et al.*, 1984
	New Zealand	"	*Thrips obscuratus*	
	N. America	Contaminated tips reduce marketability	*Thrips tabaci* *Frankliniella tritici* *Frankliniella occidentalis*	
	Taiwan	"	*Frankliniella intonsa*	
Aubergine (*Solanum melongena*: *Solanaceae*)	S., S.E., E. Asia, Caribbean, Florida, Hawaii, S. America	Leaves silvered and bronzed; fruit skin scarred and shed prematurely	*Thrips palmi**	Kawai, 1986a, 1990
	S., S.E., E. Asia, European glasshouses	"	*Thrips tabaci* *Frankliniella occidentalis*	

Continued

Crop	Location	Symptoms	Thrips	References
	N., S. Africa, Argentina	"	*Frankliniella schultzei*	
	S. Asia	"	*Caliothrips indicus*	
Avocado (*Persea americana*: *Lauraceae*)	S. Africa, Caribbean, Florida, Oceania	Interveinal leaf yellowing and browning; premature leaf-shed; skin bronzed and cracked, especially on touching fruits	*Selenothrips rubrocinctus**	Bailey, 1938; Dennill and Erasmus, 1992
	Israel, W., N. America S. America	"	*Heliothrips haemorrhoidalis*	
	Middle East	Lower leaf surfaces discoloured	*Retithrips syriacus*	
	N. America	Flowers discoloured; developing fruit with bumps and ridges	*Frankliniella bispinosa* *F. kelliae*	
Banana (*Musa paradisiaca*: *Musaceae*)	C., S. America, Caribbean	Smokey-red discoloration between fruits; skin cracks	*Chaetanaphothrips signipennis* *Chaetanaphothrips orchidii** *Hercinothrips femoralis*	Anonymous, 1984; El-Bashir and Al-Zabidi, 1985; Feakin, 1971; Treverrow, 1987
		Fruit lightly spotted; unsightly egg punctures	*Frankliniella parvula*	
	Australia	Fruit spotted on tip and outer curve	*Thrips hawaiiensis*	
		Fruit spotted, turning to red blotches	*Hercinothrips femoralis* *Chaetanaphothrips signipennis*	

		Fruit silvered and brown coarse stippling leading to 'corky scab'	*Hercinothrips bicinctus**	
	E., S. Africa, Canary Is., Mediterranean, S. Asia, S. America	"	*H. bicinctus* *H. femoralis*	
	S. Asia	Smokey-red discoloration between fruits, skin cracks	*Chaetanaphothrips signipennis**	
	Yemen	Fruit spotted, enlarged black lesions	*Scirtothrips aurantii* *Thrips pusillus*	
Barley (*Hordeum vulgare*: *Poaceae*)	W. Europe, N. America	Florets silvery-white and shrivelled; young leaves silvery streaked; ears with white spikelets and shrivelled grains	*Limothrips denticornis** *L. cerealium* *Frankliniella tenuicornis*	Bailey, 1948; Post, 1957; Koppa, 1970; Bates and Weiss, 1991; Bates *et al.*, 1991
	E. Europe, W. Asia	"	*Haplothrips aculeatus* *H. tritici*	
	S. Australia, New Zealand	"	*Limothrips cerealium*	
Basswood (**Lime**) (*Tilia americana*: *Tiliaceae*)	N. America	Buds and newly opened leaves drop; repeated defoliation causes dieback	*Thrips calcaratus*	Raffa and Hall, 1989
Beans (Various spp.: *Fabaceae*) Kidney bean (*Phaseolus vulgaris*)	S. America	Flowers discoloured and distorted; leaves initially with white flecks, later dry and brittle	*Caliothrips phaseoli*	Karel and Mghogho, 1985; Stoddart, 1986; Hachiya, 1990
	Africa	"	*Megalurothrips sjostedti*	
	E. Asia, Caribbean, Florida, Hawaii	"	*Thrips palmi*	

Continued

Crop	Location	Symptoms	Thrips	References
	N. America	Leaves bronzed and distorted; seedlings stunted	*Frankliniella occidentalis* *Thrips tabaci*	
Mung bean (*Vigna radiata*)	S. Asia	Shoots and flowers distorted; plants often killed	*Megalurothrips distalis*	
	Oceania	Flowers and leaves discoloured	*Thrips nigropilosus*	
Adzuki bean (*Vigna angulari*)	E. Asia	Flowers discoloured; pod formation restricted and distorted	*Megalurothrips usitatus* *Frankliniella intonsa*	
Broad bean (*Vicia faba*)	Europe	Flowers discoloured and pods distorted	*Kakothrips pisivorus*	
Hyacinth bean (*Lablab purpureus*)	S.E. Asia	Flowers discoloured and distorted	*Thrips hawaiiensis*	
Cabbage (*Brassica oleracea*: *Brassicaceae*)	N., S. America, E. Europe, Asia	Leaves with rough bronzed blisters penetrating many layers into head	*Thrips tabaci*	Wolfenbarger and Hibbs, 1958; Stoner and Shelton, 1988a,b; Hoy and Kretchman, 1991
	Europe	Leaves slightly bronzed and distorted	*Thrips angusticeps*	
Cardamom (*Elettaria cardamomum*: *Zingiberaceae*)	S. Asia, C. America	Flowers, leaves, fruit and shoot tips distorted and discoloured	*Sciothrips cardamomi*	Ananthakrishnan, 1984; Gonzalez and Rodriguez, 1990
Cassava (*Manihot esculenta*: *Euphorbiaceae*)	C., S. America, Caribbean	Leaves and stem yellowed; growing point killed; excessive sprouting; internodes shortened	*Corynothrips stenopterus*	van Schoonhoven and Pend, 1976; Lozano and Nolt, 1989; Mound and Marullo, 1996

	"	Young leaves curled and distorted	*Scirtothrips manihoti*	
	"	Flowers distorted	*Frankliniella melanommata*	
	E., W. Africa, S. Asia, Middle East	Leaves discoloured and shrivelled	*Retithrips syriacus*	
Cashew (*Anacardium occidentale*: *Anacardiaceae*)	S. America, Caribbean, W. Africa, S. Asia	Interveinal leaf yellowing and browning. Seedlings distorted	*Selenothrips rubrocinctus*	Fennah, 1963
	S. Asia	Leaves discoloured	*Rhipiphorothrips cruentatus*	
Castor bean (*Ricinus communis*: *Euphorbiaceae*)	S. Asia	Leaf margins with downward curving edges heavily speckled	*Craspedothrips minor* *Scirtothrips dorsalis* *Zaniothrips ricini* *Retithrips syriacus* *Rhipiphorothrips cruentatus*	Ananthakrishnan, 1984; Gilbert, 1986
	S. Africa	"	*Scirtothrips dorsalis*	
Ceara rubber (*Manihot glaziovii*: *Euphorbiaceae*)	S. America	Leaves discoloured and shrivelled	*Retithrips syriacus*	Bastos *et al.*, 1981
Celery (*Apium graveolens*: *Apiaceae*)	Worldwide	TSWV causing yellow outer stalks with necrotic pockets inside petioles	*Thrips tabaci*	
Clover – see forage legumes				
Citrus (*Citrus* spp.: *Rutaceae*)	Africa	Flowers browned and shrivelled. Fruit with silvery brown ring rind scars around calyx, sometimes deformed. Bud rosettes in nursery stock	*Scirtothrips aurantii**	Hall, 1930; Bailey, 1938; Mound and Palmer, 1981; Childers and Achor, 1991
	W., N. America	"	*Scirtothrips citri**	
	S., E., S.E. Asia, N. Australia	"	*Scirtothrips dorsalis**	

Continued

Crop	Location	Symptoms	Thrips	References
	Florida	Flowers discoloured and young fruit distorted by oviposition. Brown ring spot on grapefruit	*Frankliniella bispinosa** *Chaetanaphothrips orchidii* *Heliothrips haemorrhoidalis* *Danothrips trifasciatus*	
	S., S.E. Asia	Flowers discolored and young fruit distorted	*Thrips hawaiiensis*	
Cocoa (*Theobroma cacao*: *Sterculiaceae*)	C., S. America, Caribbean, S., S.E. Asia	Mature leaves with widespread interveinal yellow and brown patches; pods bronzed	*Selenothrips rubrocinctus* *Heliothrips haemorrhoidalis*	Fennah, 1955; Entwistle, 1972
	C. Africa	"	*Heliothrips haemorrhoidalis* *Selenothrips rubrocinctus*	
	Caribbean	Young leaves yellowed on lower surface	*Leucothrips theobromae*	
	S. Asia	Young leaves with pink-fringed discoloration on upper surface. Mature leaves brown with necrotic patches	*Scirtothrips dorsalis*	
	S. America	Young leaves and flowers discolored	*Frankliniella parvula*	
Coconut (*Cocos nucifera*: *Arecaceae*)	C., S. America, Caribbean	Flowers distorted	*Frankliniella bruneri* *Frankliniella kelliae* *Frankliniella insularis*	Sakimura, 1986
		Fruit scarred	*Anisopilothrips venustulus* *Heliothrips haemorrhoidalis* *Hercinothrips femoralis*	

	Asia	Flowers distorted	*Haplothrips ceylonicus*	
Coffee (*Coffea* spp.: *Rubiaceae*)	E., C. Africa	Leaves bronzed and shed; twigs and berries scarred	*Diarthrothrips coffeae**	Le Pelley, 1968
			Panchaetothrips noxius	
			Retithrips syriacus	
		Leaves slightly rolled	*Hoplandrothrips coffeae*	
			H. marshalli	
			H. bredoi	
	S., S.E. Asia	Young leaves distorted	*Scirtothrips bispinosus*	
			Heliothrips haemorrhoidalis	
	S. America	"	*H. haemorrhoidalis*	
Custard apple (*Annona squamosa*: *Annonaceae*)	Asia	Young leaves distorted	*Retithrips syriacus*	
			Rhipiphorothrips cruentatus	
Cotton (*Gossypium* spp.: *Malvaceae*)	N. America	Buds distorted; seedling terminals killed; floral and leaf buds shed; petals scarred	*Frankliniella tritici**	Bournièr, 1994; Leigh, 1995
			*F. fusca**	
			*F. occidentalis**	
			Neohydatothrips variabilis	
	Worldwide	"	*Thrips tabaci*	
	S. America, Africa, Australia	"	*Frankliniella schultzei**	
	S. Asia	"	*Scirtothrips dorsalis*	
	N., C. America	Leaves curled, ragged, silvered, bronzed and shed	*Caliothrips fasciatus**	
	Africa-	"	*Caliothrips impurus*	
			Caliothrips sudanensis	
			Retithrips syriacus	
	S. America	"	*Caliothrips phaseoli*	
	S.E. Asia, Caribbean	"	*Thrips palmi*	
Cowpea (*Vigna unguiculata*: *Fabaceae*)	Africa	Flowers distorted, discoloured and shed	*Megalurothrips sjostedti**	Palmer, 1987; Sweeden and McLeod, 1993; Tamò *et al.*, 1993
			Hydatothrips adolfifriderici	

Continued

Crop	Location	Symptoms	Thrips	References
	N. America	Leaves crinkled with ragged edges, especially on young plants	*Frankliniella fusca* *F. tritici* *F. occidentalis* *F. bispinosa* *Neohydatothrips variabilis*	
Cucumber (*Cucumis sativus*: *Cucurbitaceae*)	N. America, W. Europe (GH)	Flower feeding causes fruit to be silvered with webbed or streaked scars	*Frankliniella occidentalis*	Kawai, 1983; Rosenheim *et al.*, 1990; Welter *et al.*, 1990
	Worldwide	Leaves silvered and bronzed; flowers silvered; fruit scarred on some cvs	*Thrips tabaci*	
	E., S., S.E. Asia, Oceania, Caribbean, Florida, Hawaii	"	*Thrips palmi*	
	Europe	"	*Thrips fuscipennis*	
Egg plant – see aubergine				
Flax (*Linum usitatissimum*: *Linaceae*)	Europe	Flowers and leaves withered; growing tips of young plants killed producing branched stems; stems silvered	*Thrips angusticeps** *T. linarius*	Bonnemaison and Bouriér, 1964
Forage grasses (*Alopecurus*, *Agrostis*, *Dactylis*, *Lolium*, *Phleum*: *Poaceae*)	Europe	Inflorescences white and silvered with shrivelled glumes and shrunken seeds	*Chirothrips manicatus** *C. hamatus** *Haplothrips aculeatus* *Aptinothrips rufus* *Anaphothrips obscurus*	Hukkinen, 1936; Wetzel, 1964; Kamm, 1971

	N. America	"	*Arorathrips mexicanus* *Anaphothrips obscurus*	
	New Zealand	"	*Chirothrips manicatus*	
Forage legumes (*Medicago sativa, Melilotus* spp., *Trifolium* spp.: *Fabaceae*)	Europe	Red clover and lucerne flowers scarred with rusty-brown patches on petals; stamens stunted, often sterile; leaves mildly silvered	*Haplothrips niger*	Loan and Holdaway, 1955; Bournier and Kochbav, 1963; Noll and Rohr, 1966; Perju and Muset, 1982
	N. America	"	*Haplothrips leucanthemi*	
	Europe	" (mainly on lucerne)	*Odontothrips confusus** *Thrips flavus*	
	Europe	White clover flowers discoloured and distorted, some sterile	*Thrips tabaci* *Thrips fuscipennis* *Frankliniella intonsa*	
	Europe	" (mainly on sweet clover)	*Odontothrips loti*	
	New Zealand	Growing tips of lucerne curled after mowing	*Apterothrips apteris*	
	N. America	Mechanical transmission, through pollen, of sweet clover necrotic mosaic dianthovirus	*Frankliniella occidentalis*	
Garlic (*Allium sativum*: *Alliaceae*)	Worldwide	Leaves silvered and distorted	*Thrips tabaci*	
Grapes (*Vitis vinifera*: *Vitidacae*)	Europe, N. America	Leaves distorted with 'burned' margins; stems and berry pedicels scarred; shoots stunted; berries sometimes with halo scars caused by oviposition	*Drepanothrips reuteri** *Frankliniella occidentalis** *Thrips tabaci*	Yokoyama, 1977; Bournièr, 1983; McNally *et al.*, 1985; Redmund and Boller, 1989

Continued

Crop	Location	Symptoms	Thrips	References
	S. America	"	*Retithrips syriacus*	
	S.E. Europe	"	*Retithrips syriacus*	
			Haplothrips globiceps	
	S. Africa	"	*Thrips tabaci*	
			Heliothrips sylvanus	
	E. Asia	"	*Scirtothrips dorsalis*	
	S. Asia	Lower surface of leaves discoloured and distorted	*Rhipiphorothrips cruentatus*	
Groundnut (*Arachis hypogaea*: *Fabaceae*)	S. Asia	Terminal leaf buds pale, white and scarred; leaves with puckered margins and stippled silver or yellow patches	*Frankliniella schultzei*† *Caliothrips indicus* *C. impurus* *Thrips palmi*†	Poos, 1941; Feakin, 1973; Ghanekar *et al.*, 1979; Amin and Palmer, 1985; Lynch and Mack, 1995
	S.E. Asia	Flowers discoloured and scarred	*Megalurothrips usitatus*	
	N. America	Flowers discoloured and scarred; terminal leaf buds blackened and distorted after unfolding	*Frankliniella tritici* *F. fusca*† *F. occidentalis*†	
	S. America	"	*Enneothrips flavens*	
	W. Africa	"	*Frankliniella schultzei*†	

	E. Africa	Flowers discoloured and scarred	*Megalurothrips sjodesti*	
	Worldwide	"	*Thrips tabaci*†	
Larch (*Larix decidua*: *Pinaceae*)	C. Europe	Needles stunted and hook-shaped; fissures on axis of young shoots in upper crown; tree axis sometimes destroyed causing shrubby form	*Thrips laricivorus*	Vité, 1956; Zenther-Müller, 1965
Leeks (*Allium porrum*: *Alliaceae*)	Worldwide	Leaves silvered and streaked on upper part of shaft	*Thrips tabaci*	Theunissen and Legutowska, 1991
Lettuce (*Lactuca sativa*: *Asteraceae*)	N. America, Europe (GH) Tasmania	TSWV seen as yellowed, retarded leaves with brown blemishes on heart, parchment-like spots with brown margins	*Frankliniella occidentalis* *Thrips tabaci*	
	S. Africa, S. America	"	*Frankliniella schultzei*	
Lucerne – see forage legumes				
Maize (*Zea mays*: *Poaceae*)	N. America	Seedling leaves silvered, whitened and discoloured	*Anaphothrips obscurus* *Frankliniella fusca* *F. tenuicornis*	Bing *et al.*, 1990
	S. America	"	*F. williamsi*	
	S. Asia	"	*Anaphothrips sudanensis* *Florithrips traegardhi*	
	S.E. Asia, Australia	"	*Frankliniella williamsi*	
	Africa	"	*Florithrips traegardhi*	
Mango (*Manifera indica*: *Anacardiaceae*)	N., C. America, Caribbean, S., S.E. Asia	Interveinal areas of leaves discoloured and puckered	*Selenothrips rubrocinctus*	Russell, 1912; Ben-David *et al.*, 1985

Continued

Crop	Location	Symptoms	Thrips	References
	S. Asia	"	*Rhipiphorothrips cruentatus* *Scirtothrips dorsalis*	
	S. Asia	Flowers discoloured and distorted	*Haplothrips tenuipennis* *Thrips palmi*	
	Middle East	"	*Frankliniella occidentalis* *Scirtothrips mangiferae*	
Melon (*Cucumis melo*: *Cucurbitaceae*)	S., S.E. Asia, Caribbean	Leaves silvered and bronzed; terminal shoots stunted; fruits scarred and deformed	*Thrips palmi**	
	Worldwide	"	*Thrips tabaci*	
	N. America, Europe (GH)	"	*Frankliniella occidentalis*	
	N.W., N. America	"	*Thrips fuscipennis*	
Millet (*Panicum miliaceum*: *Poaceae*)	E. Europe	Inflorescences whitened with shrunken grains	*Haplothrips aculeatus* *Frankliniella tenuicornis* *F. intonsa*	
Nectarines – see stone fruit				
Oats (*Avena sativa*: *Poaceae*)	W. Europe, N. America	Leaf sheaths and blades silvered; oviposition scars on blades; ears with whitened spikelets, either empty or with shrivelled grains	*Limothrips cerealium* *L. denticornis* *Frankliniella tenuicornis* *Anaphothrips obscurus*	Hewitt, 1914; Kolobova, 1926; Bailey, 1948
	W. Europe	"	*Stenothrips graminum**	
	E. Europe	"	*Haplothrips aculeatus*	
Okra (*Abelmoschus esculentus*: *Malvaceae*)	E., S.E. Asia	Fruit scarred	*Thrips palmi* (low survival on this host)	Kawai, 1986b

Olives (*Olea europaea*: *Oleaceae*)	S. Europe, N. Africa	Leaves deformed often with lateral clefts and constrictions; flower drop; fruit distorted with necrotic patches	*Liothrips oleae* *Frankliniella occidentalis*	Bournièr, 1983
Onion (*Allium cepa*: *Alliaceae*)	Worldwide	Leaves with longitudinal silvery mottles or blotches	*Thrips tabaci*	Bailey, 1938
	E. Asia	"	*Thrips alliorum*	
	N. America	Leaf blades silvered; florets killed on seed heads	*Frankliniella occidentalis*	
Pawpaw (*Carica papaya*: *Caricaceae*)	S.E. Asia	Leaves lightly mottled and streaked. Mechanical transmission of *Cladosporium oxysporum* causing 'bunchy-top', i.e. malformed leaves and shot holes on new flushes	*Thrips parvispinus*	Lim, 1989
Pea (*Pisum sativum*: *Fabaceae*)	Europe	Leaves silvered and yellowed; pods sometimes severely silvered and distorted	*Kakothrips pisivorus*	Williams, 1915; Franssen, 1960; Bournier, 1983
	E. Asia	Flowers discoloured	*Frankliniella intonsa*	
	S. Asia	"	*Thrips flavus*	
	Worldwide	Leaves mildly silvered. TSWV causes purplish necrotic spots on leaves and stems, circular spots and wavy lines on pods	*T. tabaci*	
	Europe	Seedling leaves silvered and distorted	*T. angusticeps*	

Peaches – see stonefruit
Peanut – see groundnut

Continued

Crop	Location	Symptoms	Thrips	References
Pear (*Pyrus communis*: *Rosaceae*)	E., N. America, W. Europe	Young leaves curled and ragged; blossoms stunted and blackened by feeding and oviposition; fruit with short, scarred stems, distorted with rough skin	*Taeniothrips inconsequens*	Bailey, 1938
	W., N. America	Leaves silvered and shed; lower half of mature fruit speckled	*Caliothrips fasciatus*	
Pepper, black (*Piper nigrum*: *Piperaceae*)	S. Asia	Leaves with marginal folded galls	*Liothrips karnyi*	
Pepper, sweet and chilli (*Capsicum annuum* and spp.: *Solanaceae*)	S., E. and S.E. Asia	Bud necrosis virus causes poorly expanded blackened leaves with chlorotic ring spots and distorted, shortened internodes; flowers discoloured; fruits distorted	*Thrips palmi* *Thrips parvispinus*	Talekar, 1991; Vos *et al.*, 1991; Higgins, 1992; Nuessly and Nagata, 1995
	Europe (GH)	"	*Frankliniella occidentalis** *Thrips tabaci*	
	N. America	Flowers discoloured; fruit calyx scarred, fruit surface bronzed and silvered	*Frankliniella occidentalis** *F. bispinosa* *Thrips tabaci*	
	N. America, Europe Cosmopolitan (GH)	TSWV – leaf vein thickening, bronze rings; fruit pale with irregular or ringed mottles	*T. tabaci* *Frankliniella occidentalis*	
	N. America	"	*F. fusca*	

	S. Africa, S. Asia	"	*Frankliniella schultzei*	
Pigeon Pea (*Cajanus cajan*: *Fabaceae*)	S. Asia	Flowers discoloured	*Megalurothrips distalis*	
Pineapple (*Ananas comosus*: *Bromeliaceae*)	C. America, Hawaii	TSWV causing yellow spots on leaves coalescing into streaks and bands with paler halos. Fruit distorted, rotted or absent	*Frankliniella occidentalis** *Thrips tabaci**	Linford, 1932
	S. Africa	"	*Frankliniella schultzei** *Thrips tabaci**	
	Australia	"	*T. tabaci**	
Potato (*Solanum tuberosum*: *Solanaceae*)	E., S.E. Asia	TSWV causing necrotic spots on upper leaves, streaks on stems which collapse; plants stunted; yield small	*T. palmi* *T. tabaci*	Oria, 1985
		Minor leaf silvering	*Megalurothrips usitatus*	
	S. Asia, N. Africa	"	*Thrips flavus*	
		Also subject to TSWV	*T. tabaci*	
Pyrethrum (*Tanacetum cinerariifolium*: *Steraceae*)	E. Africa	Leaves silvered, necrotic dry patches	*Thrips nigropilosus**	Bullock, 1963; Smith and Hanson, 1991
		Flowers distorted and browned	*T. tabaci*	
		Flowers slightly discoloured	*Microcephalothrips abdominalis*	
Rape (*Brassica campestris*: *Brassicaceae*)	N. America	Flowers discoloured	*Frankliniella tritici*	Burgess and Weegar, 1988
	Canada	Seedlings slightly silvered Flowers occasionally discoloured	*Thrips tabaci* *T. vulgatissimus*	

Continued

Crop	Location	Symptoms	Thrips	References
Rice (*Oryza sativa*: *Poaceae*)	S., S.E., E. Asia, Guyana, Venezuela	Young leaf tips rolled inwards, silvered turning yellow with brown streaks; rarely in panicles	*Stenchaetothrips biformis*	Anonymous 1976; Vidyasagar and Kulshreshtha, 1983; Nugaliyadde and Heinrichs, 1984
	S. Asia	Rachilla sessile, sometimes distorted, dry and whitish	*Haplothrips ganglbaueri*	
		Leaves bleached	*Florithrips traegardhi*	
Rubber (*Hevea brasiliensis*: *Euphorbiaceae*)	S.E. Asia	Lower surface of tender leaves becomes curled, then leaves shed; feeding concealed within terminal bud of young plants causing corkiness on stems and upper surface of leaf stalks; old leaves with minor brown patches on underside	*Scirtothrips dorsalis** *Craspedothrips minor* *Heliothrips haemorrhoidalis*	Rao, 1965; Sujan *et al.*, 1985
Rye (*Secale cereale*: *Poaceae*)	N. Europe	Leaf sheaths and blades silvered; oviposition scars on blades; ears with whitened spikelets, either empty or with shrivelled grains	*Haplothrips aculeatus** *H. tritici* *Limothrips cerealium* *L. denticornis* *Frankliniella tenuicornis* *Anaphothrips obscurus*	Koppa, 1970; Larsson, 1988

	N. America	"	*Limothrips cerealium* *L. denticornis* *Frankliniella tenuicornis* *Anaphothrips obscurus*	
	W. Europe	Seedling leaves slightly silvered	*Thrips angusticeps*	
Safflower (*Carthamus tinctorius*: *Asteraceae*)	N. America	Flower buds bronzed and 'blasted'; dark margins on bracts of immature seed heads	*Frankliniella tritici* *Frankliniella occidentalis* *Thrips tabaci*	Carlson, 1964; Weiss and Beshear, 1987
Slash pine (*Pinus elliottii*: *Pinaceae*)	E., N. America	Flowers on upper crown exude small resinous beads; scales and bracts dry and shed or grow distorted	*Gnophothrips fuscus*	Ebel, 1961; Fatzinger and Dixon, 1991
Sorghum (*Sorghum bicolor*: *Poaceae*)	S. Asia, Africa	Leaf blades and sheaths bleached	*Florithrips traegardhi*	Ananthakrishnan, 1971
	S. Asia	Leaves slightly silvered and streaked	*Sorghothrips jonnaphilus*	
Soybean (*Glycine max*: *Fabaceae*)	N. America	Young leaves discoloured and distorted	*Neohydatothrips variabilis*	Irwin and Yeargan, 1980; Singh and Singh, 1991
		Flowers and buds distorted	*Frankliniella tritici** *F. fusca**	
		Tobacco ring spot virus causing bud 'blight', curling and necrosis of shoot terminals	*Thrips tabaci*	
	C., S. America	Young leaves discoloured and distorted	*Caliothrips phaseoli* *Frankliniella schultzei* *Thrips tabaci*	

Continued

Crop	Location	Symptoms	Thrips	References
	S.E. Asia	"	*Scirtothrips dorsalis** *Megalurothrips usitatus** *Frankliniella intonsa* *Ayyaria chaetophora*	
	S. Asia	"	*Caliothrips indicus* *Ayyaria chaetophora*	
	N. Africa	Flowers and buds distorted	*Frankliniella schultzei*	
	C. Africa	"	*Hydatothrips adolfifriderici*	
Stone fruit (*Prunus armeniaca*, *P. persica* v. *nucipersica*, *P. persica*: *Rosaceae*)	W., N. America	Feeding and oviposition on flowers of apricots, peaches and nectarines causes depressed scars, pits and distortion of fruit; fruit shed early	*Frankliniella occidentalis**	Bailey, 1938; Bournier, 1983; Teulon, 1988; Teulon and Penman, 1994
		Leaves distorted and fruit on nectarines and peaches russetted	*Taeniothrips inconsequens*	
	E., N. America	Skin of nectarines and peaches with silvered surface blemishes	*Taeniothrips inconsequens* *Frankliniella tritici* *F. occidentalis*	
	New Zealand	Feeding on nectarine and peach flowers and fruitlets causes russetted and distorted fruit. Silvering and egg contamination of mature fruit, mainly peaches	*Thrips obscuratus**	

	Australia	Flowers of nectarines discoloured, fruit shed	*Thrips imaginis**	
	S. Europe	Nectarine and peach flowers discoloured; fruit distorted and shed. Fruit skin contaminated by eggs	*Thrips major* *T. flavus* *T. meridionalis** *T. minutissimus* *Taeniothrips inconsequens** *Frankliniella occidentalis*	
	W., S. America	Leaves and shoots scarred	*Frankliniella cestrum* *Thrips tabaci*	
(*Prunus cerasus*, *P. avium*)	W. Canada	Leaf chlorosis; fruit russetted and split	*Taeniothrips orionis*	
Strawberries (*Fragaria* spp.: *Rosaceae*)	Europe	Petals shrivelled and eventually blackened, seriously misshapen berries	*Frankliniella occidentalis* *F. intonsa* *Thrips atratus* *T. major*	Bailey, 1938; Buxton and Easterbrook, 1988
	N. America	"	*Frankliniella tritici* *F. dorsalis*	
	Australia	"	*Scirtothrips dorsalis*	
Sugar apple – see custard apple				
Sugar beet (*Beta vulgaris*: *Chenopodiaceae*)	N. Europe	Young leaves silvered, browned and reddened, often curl and fail to expand	*Thrips tabaci* *T. angusticeps*	Anonymous, 1982; Bailey, 1938
	N. America	"	*Thrips tabaci* *Caliothrips fasciatus*	

Continued

Crop	Location	Symptoms	Thrips	References
Sugar maple (*Acer saccharum*: *Aceraceae*)	E., N. America	Opening buds deformed; leaves pale yellow mottled and stunted, sometimes severely tattered. Flower damage less obvious, some seed reduction	*Taeniothrips inconsequens**	Parker, 1991; Teulon *et al.*, 1992
Sugar cane (*Saccharum* cvs: *Poaceae*)	E., S. Asia	Tips of young leaves, especially within spindle, yellowed and fail to unroll, often remain joined at tips. Older leaves silvered	*Fulmekiola serrata** *Haplothrips aculeatus* *Anaphothrips sudanensis*	Des Vignes, 1987
	Caribbean, Venezuela	"	*Fulmekiola serrata*	
	S. Asia	"	*Florithrips traegardhi*	
	Africa, Mauritius	"	*Fulmekiola serrata* *Florithrips traegardhi* *Anaphothrips sudanensis*	
	S. America	"	*Stenchaetothrips minutus*	
Sunflower (*Helianthus annuus*: *Asteraceae*)	S. America	Leaves initially whitened, then dry and brittle	*Caliothrips phaseoli* *Frankliniella schulzei*	
Sweet potato (*Ipomoea batatas*: *Convolvulaceae*)	N. Australia	Leaves distorted and discoloured	*Drepanthripoides innoxius*	
Taro (*Colocasia esculenta*: *Araceae*)	C. Pacific, S.E. Asia	Stem bases, stems and young leaves discoloured and distorted	*Organothrips bianchi*	Sakimura and Krauss, 1945

Tea (*Camellia sinensis*: *Theaceae*)	E. Africa	Young leaves and shoots stunted, brittle, shed; rolled leaf margins brown, surface streaked with linear thickenings; stem internodes scarred	*Scirtothrips kenyensis** *Heliothrips haemorrhoidalis*	Dev, 1964; Mound, 1968; Okada and Kudo, 1982; Sudoi and Ruto, 1992
	E., S. Asia	"	*Scirtothrips dorsalis**	
	S. Asia	"	*Scirtothrips bispinosus*	
	S.E. Asia	"	*Chaetanaphothrips theiperdus*	
	E. Asia	Flowers slightly discoloured	*Lefroyothrips lefroyi* *Thrips hawaiiensis*	
Tobacco (*Nicotiana tabacum*: *Solanaceae*)	Worldwide	Leaves discoloured, but TSWV more damaging causing concentric necrotic rings on young leaves, bent stem apex, stunted and uneven growth	*Thrips tabaci*	Zawirska, 1978; Akehurst, 1981; McPherson *et al.*, 1992
	S. Africa, Australia, S. America	"	*Frankliniella schultzei*	
	N. America	"	*Frankliniella occidentalis* *F. fusca*	
	N. America	Flower discoloration	*Frankliniella bispinosa* *F. tritici*	
	Australia	Mechanical transmission of tobacco streak virus via pollen	*Thrips tabaci*	Sdoodee and Teakle, 1987
Tomato (*Lycopersicon esculentum*: *Solanaceae*)	Worldwide	TSWV causing stunting and rosetting in early infected plants, leaves with severe mosaic, necrosis and distortion. Fruits with concentric pale red, yellow and white rings	*Thrips tabaci**	Kobatake, 1984; Murai, 1988; Salguero-Navas *et al.*, 1991

Continued

Crop	Location	Symptoms	Thrips	References
	Africa, Australia, S. America, S., S.E. Asia	"	*Frankliniella schultzei**	
	N. America, Europe (GH)	"	*Frankliniella occidentalis**	
	N. America	"	*Frankliniella fusca*	
	S.E. Asia, Australia, Caribbean, S. America	(Vector species, but cannot complete life cycle on this host)	*Thrips palmi*	
	N. America, Japan (GH)	Fruit with white oviposition swelling surrounded by halo	*Frankliniella occidentalis* *F. intonsa*	
Water melon (*Citrullus lanatus*: *Cucurbitacae*)	E., S., S.E. Asia, Caribbean, Australia	Leaves bronzed; vine tips killed. TSWV causes silver mottling on stunted leaves, fruits with an irregular surface and chlorotic mottling	*Thrips palmi**	Iwaki *et al*., 1984; Johnson, 1986
	S. America	"	*Frankliniella schultzei*	
Wheat (*Triticum sativum*: *Poaceae*)	W. Europe, N. America	Leaf sheaths and blades silvered; oviposition scars on blades; ears with whitened spikelets, either empty or with shrivelled grains – 'whiteheads', 'silvertop'	*Limothrips cerealium** *L. denticornis* *Frankliniella tenuicornis* *Anaphothrips obscurus*	Bailey, 1948; Lewis, 1959; Koppa, 1970; Czenz, 1992
	E. Europe, W. Asia	"	*Haplothrips tritici** *H. aculeatus**	
	S. Asia	"	*Haplothrips ganglebaueri*	
	S. Australia, New Zealand	"	*Limothrips cerealium*	

Ornamentals

Aluminium plant (*Pilea cadierei*: *Urticaceae*)	Worldwide (GH)	Leaves severely silvered, sometimes plant killed	*Hercinothrips femoralis*	Oetting and Beshear, 1980
Calla lily (*Zantedeschia aethiopica*: *Araceae*)	Worldwide (GH)	Leaves silvery and white spotted; flowers silvered, turning brown	*Frankliniella occidentalis* *Hercinothrips femoralis* *Heliothrips haemorrhoidalis* *Thrips tabaci*	
Carnations (*Dianthus caryophyllus*: *Caryophyllaceae*)	Worldwide (GH)	Leaves and buds scarred and distorted leading to stunted plants. Flowers with silvery-white patches and streaks, turning brown	*Thrips tabaci** *T. simplex** *Frankliniella occidentalis**	Pelikan, 1951; Bournièr, 1983
	W., S., S.E. Europe	"	*Pezothrips dianthii** *Thrips atratus** *T. flavus* *T. meridionalis* *T. angusticeps* *Haplothrips cottei*	
Christmas cactus (*Schlumbergera bridgesii*: *Cactaceae*)	Worldwide (GH)	Leaves silvered and plant sometimes killed	*Hercinothrips femoralis*	Oetting and Beshear, 1980
Chrysanthemum (*Chrysanthemum* spp. and cvs: *Asteraceae*)	Worldwide (GH)	Leaves silvered, scarred and distorted; plants stunted; flowers lightly to heavily discoloured, turning brown. TSWV – see Footnote 2	*Frankliniella occidentalis** *Thrips tabaci* *Heliothrips haemorrhoidalis* *Hercinothrips femoralis*	Frey, 1993; Miyashita and Soichi, 1993
	S.E. Asia, Caribbean, Florida, W. Europe (GH)	"	*Thrips palmi*	

Continued

Crop	Location	Symptoms	Thrips	References
Eucalyptus (*Eucalyptus* spp.: *Myrtaceae*)	Australia	Leaves discoloured and distorted	*Australothrips bicolor*	
	S. America	"	*Retithrips syriacus*	
			Selenothrips rubrocinctus	
Figs (*Ficus retusa*, *F. microcarpa*, *F. benjamina*: *Moraceae*)	N., S. America; M. East, S., S.E. Asia, S. Europe	Young leaves with marginal roll galls; growth arrested	*Gynaikothrips ficorum*	Avidov and Harpaz, 1969; Bournièr, 1983; Mound *et al.*, 1996
	Africa, S.E. Asia	"	*Gynaikothrips ficorum*	
	Africa, Europe, N., C. America	Leaves slightly streaked	*Hercinothrips femoralis*	
Ficus elastica	Worldwide (GH)	Leaves silvered, browned and scarred	*Parthenothrips dracaenae*	
			Hercinothrips femoralis	
Ficus benjamina	Europe (GH)	"	*Thrips palmi*	
Gladiolus (*Gladiolus* spp., also *Freesia*, *Iris* spp.: *Iridaceae*)	Worldwide	Leaves and stems with pale yellow or silver streaks turning brown. Flowers flecked white, may brown and die. Corms in store with dry rough red-brown patches	*Thrips simplex*	Bournièr, 1983; Conijn and Groen, 1990a,b; Chyzik *et al.*, 1995
	Worldwide (GH)	Flowers with yellow or silver streaks and white flecks; leaves silvered in absence of flowers	*Frankliniella occidentalis*	
	E., S.E. Asia, Oceania	Flowers lightly flecked	*Thrips hawaiiensis*	
Gloxinia (*Gloxinia* spp.: *Gesneriaceae*)	Worldwide (GH)	Petals with brown streaks	*Frankliniella occidentalis*	Cummiskey, 1988

Lilies (*Lilium* spp.: Liliaceae)	Worldwide	Rust-coloured shrunken areas at base of outer bulb scales; aerial parts unaffected	*Liothrips vaneeckei* *Thrips simplex*	Bailey, 1939; Mantel and van der Vrie, 1988
	Worldwide (GH)	Flowers with yellow or silvery streaks and white flecks; leaves silvered in absence of flowers	*Frankliniella occidentalis*	
Orchids (numerous spp.: *Orchidaceae*)	Worldwide (GH)	Green tuberous parts of plants marked with brown patches; flowers and leaves silvered and browned	*Aurantothrips orchidaceous* *Chaetanophothrips orchidii* *Heliothrips haemorrhoidalis* *Hercinothrips femoralis* *Frankliniella occidentalis* *Dichromothrips corbetti* *D. phalaenopsidis* *Thrips tabaci*	Mound, 1976; Mantel and van der Vrie, 1988
	S. America	"	*Aurantothrips orchidaceous*	
	Europe (GH)	"	*Thrips palmi*	
Privet (*Ligustrum* spp.: *Oleaceae*)	N. America, Europe	Leaves severely silvered, speckled, puckered and shed	*Dendrothrips ornatus*	Schread, 1969
Roses (*Rosa* spp.: *Rosaceae*)	Europe	Leaves silvered; petals with raised white spots then brown patches	*Thrips fuscipennis*	Henneberry *et al.*, 1961; Bournièr, 1983
	E., N. America	Flower discoloured	*Frankliniella tritici*	
	S. America	"	*Retithrips syriacus*	

*Predominant injurious species

†Transmit bud necrosis virus (BNV) causing chlorotic ring spots and severe stunting. See note 2 and Chapter 14

[1]Almost all greenhouse plants are susceptible to direct feeding damage from one or several of the polyphagous species listed under 'ornamentals' above. Palms are susceptible to *Parthenothrips dracaenae* and ferns to *Leucothrips nigripennis*.

[2]Tomato spotted wilt virus in ornamentals (Gofflot and Verhoyen, 1990). Many ornamentals are susceptible to TSWV, symptoms varying between hosts but usually appearing as purple-brown surface blotches or streaks on stems; leaves develop concentric rings or irregular blotches outlined with halos, and blossoms are discoloured with circular patterns in the pigment. Susceptible garden and especially greenhouse plants include amaryllis, aster, begonia, calendula, calla, chrysanthemum, dahlia, delphinium, fuschia, gaillardia, geranium, gloxinia, hydrangea, nasturtium, petunia, phlox, primrose, salvia, stephanotis, stock, verbena and zinnia. Transmission in greenhouses is mainly by *Frankliniella occidentalis* and outdoors by *F. shultzei*, *Thrips palmi* and *T. setosus* in S. and S.E. Asia. *Thrips tabaci* does not transmit contemporary TSWV isolates, except reportedly in Australia, but it remains on the list of accepted vectors due to its former importance as a vector worldwide (see Chapter 14).

References

Akehurst, B.C. (1981) *Tobacco*. Longmans, London, 764 pp.

Amin, P.W. and Palmer, J.M. (1985) Identification of groundnut Thysanoptera. *Tropical Pest Management* 31, 286–291.

Ananthakrishnan, T.N. (1971) Thrips (Thysanoptera) in agriculture, horticulture and forestry diagnosis, bionomics and control. *Journal of Science and Industry Research* 30, 113–146.

Ananthakrishnan, T.N. (1984) *Bioecology of Thrips*. Indira, Oak Park, Michigan, 233 pp.

Anonymous (1976) Pest control in rice. *PANS Manual* No. 3. Centre for Overseas Pest Research, London, 295 pp.

Anonymous (1982) *Pests, Diseases and Disorders of the Sugar Beet*. Deleplanque and Cie, Maisons Laffite, 166 pp.

Anonymous (1984) Insecticide control of corky scab. *Banana Bulletin* 48, 12–13.

Avidov, Z. and Harpaz, I. (1969) *Plant Pests of Israel*. University Press, Jerusalem.

Bailey, S.F. (1938) Thrips of economic importance in California. *Circular of the California Agriculture Experiment Station* 346, 77 pp.

Bailey, S.F. (1939) The lily thrips. *Bulletin of the California Department of Agriculture* 28, 479–483.

Bailey, S.F. (1948) Grain and grass-infesting thrips. *Journal of Economic Entomology* 41, 701–706.

Banham, F.L. (1968) Thrips infesting the tips of asparagus spears. *Journal of the Entomological Society of British Colombia* 65, 16–18.

Bastos, J.A.M., Alves, V.P.O., Magalhaes-Bastos, J.A.and Oliveira-Alves, V.R. (1981) Survey of the population of the red 'manicoba' thrips, *Retithrips syriacus* (Mayet, 1890) on manicoba. *Fitossanidade* 5, 1–6.

Bates, B.A. and Weiss, M.J. (1991) Seasonal abundance of *Limothrips denticornis* (Thysanoptera: Thripidae) on spring barley. *Environmental Entomology* 20, 577–581.

Bates, B.A., Weiss, M.J. and McBride, D.K. (1991) Biology and management of barley thrips. *North Dakota State University Extension Bulletin* E-1007.

Ben-David, T., Venezian, A. and Ben-Dov, Y. (1985) Observations on the phenology, biology and control of the mango thrips, *Scirtothrips mangiferae* Priesner, in the Arava of Israel. (in Hebrew.) *Hassadeh* 65, 1826–1830.

Bing, J.W., Dicke, F.F. and Guthrie, W.D. (1990) Genetics of resistance in maize to a complex of three species of thrips (Thysanoptera: Thripidae). *Journal of Economic Entomology* 83, 621–624.

Bonnemaison, L. and Bournier, A. (1964) Les thrips du lin *Thrips angusticeps* Uzel et *Thrips linarius* (Thysanoptères). *Annales des Epiphyties* 15, 97–169.

Bournier, A. (1983) *Les Thrips, Biologie, Importance Agronomique*. Institut National de la Recherche Agronomique, Paris.

Bournier, A. and Kochbav, A. (1963) Action stérilisante d'un Thysanoptère sur les fleurs de luzerne. *Bulletin de la Socièté Entomologique de France* 68, 28–30.

Bournier, J.P. (1994) Thysanoptera. In: Matthews, G.A. and Tunstall, J.P. (eds) *Insect Pests of Cotton*. CAB International, Wallingford, pp. 381–391.

Bullock, J. (1963) Thysanoptera associated with pyrethrum, and the control of *Thrips tabaci* Lind. *Tropical Agriculture* 40, 329–335.

Burgess, L. and Weegar, H.H. (1988) Thrips (Thysanoptera) in canola crops in Saskatchewan. *Canadian Entomologist* 120, 815–819.

Buxton, J.H. and Easterbrook, M.A. (1988) Thrips as a probable cause of severe fruit

distortion in late season strawberries. *Plant Pathology* 37, 278–280.
Carlson, E.C. (1964) Damage to safflower plants by thrips and lygus bugs and a study of their control. *Journal of Economic Entomology* 57, 140–145.
Childers, C.C. and Achor, D.S. (1991) Feeding and oviposition injury to flowers and developing floral buds of 'navel' orange by *Frankliniella bispinosa* (Thysanoptera: Thripidae) in Florida. *Annals of the Entomological Society of America* 84, 272–282.
Chyzik, R., Klein, M., Ben-Dov, Y. and Cohen, A. (1995) Differential population density of western flower thrips in various flower colours of gladiolus. In: Parker, B.L., Skinner, M. and Lewis, T. (eds) *Towards Understanding Thrips Management.* Plenum Press, New York, pp. 449–453.
Conijn, C.G.M. and Groen, N.P.A. (1990a) Physical, biological and chemical control of *Taeniothrips simplex* in gladiolus. *Acta Horticulturae* 266, 539–546.
Conijn, C.G.M. and Groen, N.P.A. (1990b) Low temperature treatment to control *Taeniothrips simplex* Morison on gladiolus corms in storage rooms. *Mededelingen-von-de Faculteit Landbouwwetenschappen, Rijksuniversiteit-Gent* 54, 931–935.
Cummiskey, P. (1988) Controlling thrips in gloxinias. *Greenhouse Grower*, December, 75–77.
Czencz, K. (1992) A comparative investigation of some winter wheat types by *Haplothrips tritici* Kurd. *Journal of Applied Entomology* 113, 209–213.
Dennill, G.B. and Erasmus, M.J. (1992) Basis for a practical technique for monitoring thrips in avocado orchards. *Crop Protection* 11, 89–91.
Des Vignes, W. (1987) The assessment of economic damage caused by *Fulmekiola serrata* (Kobus) (Thysanoptera: Thripidae) on sugar cane in Trinidad. *Journal of the Agricultural Society of Trinidad and Tobago* 87, 67–81.
Dev, H.N. (1964) Preliminary studies on the biology of the Assam thrips, *Scirtothrips dorsalis* Hood, on tea. *Indian Journal of Entomology* 26, 184–194.
Ebel, B.H. (1961) Thrips injure slash pine female flowers. *Journal of Forestry* 59, 374–375.
El-Bashir, S. and Al-Zabidi, A.H. (1985) Fruit spotting of banana in the Yemen Arab Republic. *FAO Plant Protection Bulletin* 33, 113–118.
Entwistle, P.F. (1972) *Pests of Cocoa.* Longman, London.
Fatzinger, C.W. and Dixon, W.N. (1991) Development of sampling methods for the slash pine flower thrips *Gnophothrips fuscus* (Morgan), (Thysanoptera: Phlaeothripidae). In: Parker, B.L., Skinner, M. and Lewis, T. (eds) *Towards Understanding Thysanoptera.* USDA Forest Service, Northeastern Forest Experimental Station General Technical Report NE-147, 149–161.
Feakin, S.D. (1971). Pest control in bananas. *PANS Manual* No.1. Centre for Overseas Pest Research, London.
Feakin, S.D. (1973) Pest control in groundnuts. *PANS Manual* No. 2. Centre for Overseas Pest Research, London.
Fennah, R.G. (1955) The epidemiology of cacao-thrips on cacao in Trinidad. *Report of Cacao Research* 24, 7–26.
Fennah, R.G. (1963) Nutritional factors associated with seasonal population increases of cacao thrips, *Selenothrips rubrocinctus* (Giard) (Thysanoptera) on cashew, *Anacardium occidentale. Bulletin of Entomological Research* 53, 681–713.
Franssen, C.J.H. (1960) Biology and control of pea thrips. *Verslag Landbouwkundig Onderzoek Rijkslandproefstn* 66, 36 pp.
Frey, J.E. (1993) Damage thresholds for western flower thrips *Frankliniella occidentalis* (Perg.) (Thysanoptera: Thripidae) on ornamentals. *Bulletin OILB/SROB* 16, 78–81.

Ghanekar, A.M., Reddy, D.V.R., Iizuka, N., Amin, P.W. and Gibbons, R.W. (1979) Bud necrosis of groundnut (*Arachis hypogaea*) in India caused by tomato spotted wilt virus. *Annals of Applied Biology* 93, 173–179.

Gilbert, M.J. (1986) First African record of *Scirtothrips dorsalis* Hood (Thysanoptera: Thripidae) a potential pest of citrus and other crops in southern Africa. *Journal of the Entomological Society of Southern Africa* 49, 159–161.

Gofflot, A. and Verhoyen, M. (1990) Speedy development of tomato spotted wilt virus infection transmitted by *Frankliniella occidentalis* in ornamental greenhouse cultivars in Belgium. *Parasitica* 46, 85–88.

Gonzalez, H.L. and Rodriguez, G.C. (1990) *Sciothrips cardamomi* (Ramk.) (Thysanoptera: Thripidae) nueva plaga del cardamomo (*Elettaria cardamomum* Maton) en Costa Rica. *Maneja Integrado de Playas* 16, 38–40.

Hachiya, K. (1990) Effect of injury by the flower thrips *Frankliniella intonsa* Trybom, on pod formation and yield of adzuki bean *Phaseolus angularis* L. *Japanese Journal of Applied Entomology and Zoology* 34, 326–329.

Hall, W.J. (1930) The South African citrus thrips in Southern Rhodesia. *Report Mazoe Citrus Experiment Station* 1, 5–55.

Henneberry, T.J., Taylor, E.A. and Smith, F.F. (1961) Foliage and soil treatments for control of flower thips in outdoor roses. *Journal of Economic Entomology* 54, 233–235.

Hewitt, C.G. (1914) Sterility in oats caused by thrips. *Journal of Economic Entomology* 7, 211.

Higgins, C.J. (1992) Western flower thrips (Thysanoptera: Thripidae) in greenhouses: population dynamics, distribution on plants, and associations with predators. *Journal of Economic Entomology* 85, 1891–1903.

Hoy, C.W. and Kretchman, D.W. (1991) Thrips (Thysanoptera: Thripidae) injury to cabbage cultivars in Ohio. *Journal of Economic Entomology* 84, 971–977.

Hukkinen, Y. (1936) Tutkimuksia Nurmipuntarpään (*Alopecurus pratensis* L.) Siementuholaisista. 1. *Chirothrips hamatus* Tryb. *Valtion Maatalouskoetoiminnan Julkaisuja* 81, 131 pp.

Irwin, M.E. and Yeargan, K.V. (1980) Sampling phytophagous thrips on soybean. In: Kogan, M. and Herzog, D.C. (eds) *Sampling Methods in Soybean Entomology.* Springer Verlag, New York, pp. 239–259.

Iwaki, M., Honda, Y., Hanada, K., Tochihara, H. and Yanaha, T. (1984) Silver mottle disease of watermelon caused by tomato spotted wilt virus. *Plant Discoverer* 68, 1006–1008.

Johnson, M.W. (1986) Population trends of newly introduced species, *Thrips palmi*, (Thysanoptera: Thripidae) on commercial watermelon plantings in Hawaii. *Journal of Economic Entomology* 79, 718–720.

Kamm, J.A. (1971) Silvertop of bluegrass and bentgrass produced by *Anaphothrips obscurus*. *Journal of Economic Entomology* 64, 1385–1387.

Karel, A.K. and Mghogho, R.M.K. (1985) Effects of insecticide and plant populations on the insect pests and yield of common bean (*Phaseolus vulgaris* L.) *Journal of Economic Entomology* 78, 917–921.

Kawai, A. (1983) Studies on population ecology of *Thrips palmi* Karny. 1. Population growth and distribution pattern on cucumber in the greenhouse. (In Japanese.) *Japanese Journal of Applied Entomology and Zoology* 17, 261–264.

Kawai, A. (1986a) Studies on population ecology of *Thrips palmi*. XII. Analysis of damage to egg plant and sweet pepper. *Japanese Journal of Applied Entomology and Zoology* 30, 178–187.

Kawai, A. (1986b) Studies on the population ecology of *Thrips palmi* Karny X.

Differences in population growth on various crops. (In Japanese.) *Japanese Journal of Applied Entomology and Zoology* 30, 7–11.

Kawai, A. (1990) Life cycle and population dynamics of *Thrips palmi* Karny. *Japanese Agricultural Research Quarterly* 23, 282–288.

Kolobova, A.N. (1926) *Stenothrips graminum* Uzel. (In Russian.) *Review of Applied Entomology* (A) 14, 606–607.

Kobatake, H. (1984) Ecology and control of spotted wilt disease of tomato in Nasa Prefecture. (In Japanese.) *Proceedings of the Kansai Plant Protection Society* 26, 23–28.

Koppa, P. (1970) Studies on the thrips (Thysanoptera) species most commonly occurring on cereals in Finland. *Annales Agriculturae Fenniae* 9, 191–265.

Larsson, H. (1988) Economic damage caused by cereal thrips in winter rye in Sweden. *Acta Phytopathologica et Entomologica Hungarica* 23, 291–293.

Leigh, T.F. (1995) Bionomics of cotton pests: a review. In: Parker, B.L., Skinner, M. and Lewis, T. (eds) *Towards Understanding Thrips Management.* Plenum Press, New York, pp. 61–70.

Le Pelley, R.H. (1968) *Pests of Coffee.* Longmans, London.

Lewis, T. (1959) The annual cycle of *Limothrips cerealium* Haliday (Thysanoptera) and its distribution in a wheat field. *Entomologica Experimentalis et Applicata* 2, 187–203.

Lim, W.H. (1989) Bunchy and malformed top of papaya cv Eksotika caused by *Thrips parvispinus* and *Cladosporium oxysporium. Malaysian Agricultural Research Development Institute Research Journal* 17, 200–207.

Linford, M.B. (1932) Transmission of the pineapple yellow-spot virus by *Thrips tabaci. Phytopathology* 22, 301–324.

Loan, C. and Holdaway, F.G. (1955) Biology of the red clover thrips, *Haplothrips niger* (Osborn). *Canadian Entomologist* 87, 210–219.

Lozana, J.C. and Nolt, B.L. (1989) Pests and pathogens of cassava. In: Kahn, R.P. (ed.) *Plant Protection and Quarantine, Vol. II. Selected Pests and Pathogens of Quarantine Significance.* CRC Press Inc., Boca Raton, Florida, pp. 169–182.

Lynch, R.E. and Mack, T.P. (1995) Biological and biotechnical advances for insect management in peanut. In: Stalker, H.T. and Pattee, H.E. (eds) *Advances in Peanut Science.* American Peanut Research and Education Society Inc., Yoakum, Texas, pp. 95–159.

Mantel, W.P. and van der Vrie, M. (1988) A contribution to the knowledge of Thysanoptera in ornamental and bulbous crops in The Netherlands. *Acta Phytopathologica et Entomologica Hungarica* 23, 301–311.

McNally, P.S., Fogg, C., Flynn, J. and Horenstein, J. (1985) Effects of thrips (Thysanoptera: Thripidae) on short growth and berry maturity of 'Chenin Blanc' grapes. *Journal of Economic Entomology* 78, 69–72.

McPherson, R.M., Beshear, R.J. and Culbreath, A.K. (1992) Seasonal abundance of thrips (Thysanoptera: suborders Terebrantia and Tubulifera) in Georgia flue-cured tobacco and impact of management practices on the incidence of tomato spotted wilt virus. *Journal of Entomological Science* 27, 257–268.

Miyashita, T. and Soichi, N. (1993) Studies on the varietal difference in injury caused by *Thrips palmi* in chrysanthemum. 3. Relationship between infested part, density and injury. (In Japanese.) *Japanese Journal of Applied Entomology and Zoology* 37, 227–233.

Mound, L.A. (1968) A new species of *Scirtothrips* from Kenya attacking tea with synonymic notes on two related pest species. *Bulletin of Entomological Research* 57, 533–538.

Mound, L.A. (1976) Thysanoptera of the genus *Dichromothrips* on Old World Orchidaceae. *Biological Journal of the Linnean Society* 8, 245–265.

Mound, L.A. and Marullo, R. (1996) The thrips of Central and South America: an introduction. *Memoirs on Entomology, International* 6, 1–448.

Mound, L.A. and Palmer, J.M. (1981) Identification, distribution and host plants of the pest species of *Scirtothrips* (Thrysanoptera: Thripidae). *Bulletin of Entomological Research* 71, 467–479.

Mound, L.A., Wang, C.-L. and Okajima, S. (1996) Observations in Taiwan on the identity of the Cuban laurel thrips (Thysanoptera, Phlaeothripidae). *Journal of the New York Entomological Society* 103, 200–205.

Murai, T. (1988) Studies on the ecology and control of flower thrips, *Frankliniella intonsa* (Trybom). *Bulletin of the Shimane Agricultural Experiment Station* 23, 1–73.

Noll, J. and Rohr, C. (1966) Thysanopterenschäden an Luzerneblüten. *Nachrichtenblatt für den Deutschen Pflanzenschutzdienst* 20, 47–52.

Nuessly, G.S. and Nagata, R.T. (1995) Pepper varietal response to thrips feeding. In: Parker, B.L., Skinner, M. and Lewis, T. (eds) *Towards Understanding Thrips Management*. Plenum Press, New York, pp. 115–118.

Nugaliyadde, L. and Heinrichs, E.A. (1984) Biology of rice thrips, *Stenchaetothrips biformis* (Bagnall), (Thysanoptera: Thripidae) and a greenhouse rearing technique. *Journal of Economic Entomology* 77, 1171–1175.

Oetting, R.D. and Beshear, R.J. (1980) Host selection and control of the banded greenhouse thrips on ornamentals. *Journal of the Georgia Entomology Society* 15, 475–479.

Okada, T. and Kudo, I. (1982) Relative abundance and phenology of Thysanoptera on a tea field. (In Japanese.) *Japanese Journal of Applied Entomology and Zoology* 26, 96–102.

Orita, S. (1985) Damage caused by *Thrips palmi* to the autumn potato crop. (In Japanese.) *Proceedings of the Association for Plant Protection of Kyushu* 31, 135-155.

Palmer, J.M. (1987) *Megalurothrips* in the flowers of tropical legumes in a morphometric study. In: Holman, J., Pelikan, J., Dixon, A.G.F. and Weismann, L. (eds) *Population Structure, Genetics, and Taxonomy of Aphids and Thysanoptera*, SPB Academic Publishing, The Hague, pp. 480–495.

Parker, B.L. (1991) The pear thrips problem. In: Parker, B.L., Skinner, M. and Lewis, T. (eds) *Towards Understanding Thysanoptera*. USDA Forest Service, Northeastern Forest Experimental Station General Technical Report, NE-147, 179–188.

Pelikan, J. (1951) On carnation thrips *Taeniothrips dianthi* Pr. (In Czech.) *Entomologické listy* 14, 5-38.

Perju, T. and Muset, D. (1982) Red clover thrips (*Haplothrips niger* Osb. Phlaeothripidae – Thysanoptera): bioecology and control. *Probleme de Protectia Plantelor* 10, 113–122.

Poos, F.W. (1941) On the causes of peanut 'pouts'. *Journal of Economic Entomology* 34, 727–728.

Post, R. (1957) The barley thrips in North Dakota. *The North Dakota Seed Journal* 26, 4.

Raffa, K.F. and Hall, D.J. (1989) *Thrips calcaratus* Uzel (Thysanoptera: Thripidae), a new pest of basswood trees in the Great Lakes region. *Canadian Journal of Forestry Research* 19, 1661–1662.

Rao, B.S. (1965) *Pests of Hevea Plantations in Malaya*. Rubber Research Institute, Kuala Lumpur, 98 pp.

Redmund, U. and Boller, E. (1989) Thripse im Ostchweizer Rebbau: 1. Problemstelung,

Artenspektrum und Lebensweise. *Schweizerische Zeitschrift für Obst und Weinbau* 125, 173–188.
Rosenheim, J.A., Welter, S.G., Johnson, M.W., Mau, R.F.L. and Gusukuma-Minuto, L.R. (1990) Direct feeding damage in cucumber by mixed-species infestations of *Thrips palmi* and *Frankliniella occidentalis* (Thysanoptera: Thripidae). *Journal of Economic Entomology* 83, 1519–1525.
Russell, H.M. (1912) The red-banded thrips. *Bulletin of the US Department of Agriculture Bureau of Entomology* 99, 17–29.
Sakimura, K. (1986) Thrips in and around the coconut plantations in Jamaica, with a few taxonomical notes. *Florida Entomologist* 69, 348–363.
Sakimura, K. and Krauss, N.L.H. (1945) Collections of thrips from Kauai and Hawaii. *Proceedings of the Hawaii Entomological Society* 22, 319–331.
Salguero-Navas, V.E., Funderburk, J.E., Olson, S.M. and Beshear, R.J. (1991) Damage to tomato fruit by the western flower thrips (Thysanoptera: Thripidae). *Journal of Entomological Science* 26, 436–442.
Schread, J.C. (1969) Privet thrips. *Circular Connecticut Agricultural Experiment Station* No. 230, 4 pp.
Sdoodee, R. and Teakle, D.S. (1987) Transmission of tobacco streak virus by *Thrips tabaci:* a new method of plant virus transmission. *Plant Pathology* 36, 377–380.
Singh, M. (1989) Damage and economic threshold level of *Thrips flavus* Schrank to apple blossom in Himachal Pradesh. *Indian Journal of Entomology* 51, 288–293.
Singh, O.P. and Singh, K.J. (1991) Seasonal incidence, distribution and chemical control of thrips, *Caliothrips indicus* (Bagnall), on soybean. *Indian Journal of Plant Protection* 19, 197–200.
Smith, L.M. and Hanson, P.M. (1991) Yield reduction in pyrethrum caused by *Thrips nigropilosus* Uzel (Thysanoptera: Thripidae). *Tropical Agriculture* 68, 235–238.
Stoddard, F.L. (1986) The distribution of immature thrips among flowers of faba beans in commercial crops and experimental plots. *Annals of Applied Biology* 109, 61–69.
Stoner, K.A. and Shelton, A.M. (1988a) Effect of planting date and timing of growth stages on damage to cabbage by onion thrips (Thysanoptera: Thripidae). *Journal of Economic Entomology* 81, 1186–1189.
Stoner, K.A. and Shelton, A.M. (1988b) Influence of variety on abundance and within-plant distribution of onion thrips (Thysanoptera: Thripidae) on cabbage. *Journal of Economic Entomology* 81, 1190–1195.
Sudoi, V. and Ruto, J.K. (1992) An economic assessment of permethrin spraying programmes on control of thrips *Scirtothrips kenyensis* Mound (Thysanoptera: Thripidae) and effect on other arthropods in Kenya. *Tea* 13, 111–119.
Sujan, A., Lim, T.M. and Leong, C.P. (1985) Thermal fogging of pesticides against flower and leaf pests on rubber. *Planters Bulletin of the Rubber Research Institute of Malaysia* 182, 13–15.
Sweeden, M.B. and McLeod, P.J. (1993) Seasonal occurrence of thrips (Thysanoptera) on cowpeas in western Arkansas and northeast Oklahoma. *Journal of Entomological Science* 28, 427–432.
Swift, J.E. and Madsen, H.F. (1956) Thrips damage to apple. *Journal of Economic Entomology* 49, 398–399.
Talekar, N.S. (1991) Thrips on pepper: AVRDC's research strategy. In: Talekar, N.S. (ed.) *Thrips in Southeast Asia: Proceedings of a Regional Consultation Workshop, Bangkok, Thailand, 13 March*. Asian Vegetable Research and Development Centre, AVRDC Publication No. 91-342, 74 pp.
Tamò, M., Baumgartner, J., Delucchi, V. and Herren, H.R. (1993) Assessment of key factors responsible for the pest status of the bean flower thrips *Megalurothrips*

sjostedtii (Thysanoptera: Thripidae) in West Africa. *Bulletin of Entomological Research* 83, 251–258.

Terry, L.I. (1991) *Frankliniella occidentalis* (Thysanoptera: Thripidae) oviposition in apple buds: role of bloom state, blossom phenology and population density. *Environmental Entomology* 20, 1568–1576.

Teulon, D.A.J. (1988) Pest management of the New Zealand flower thrips, *Thrips obscuratus* (Crawford) (Thysanoptera: Thripidae) on stonefruit in Canterbury, New Zealand. PhD thesis, Lincoln College, University of Canterbury, New Zealand.

Teulon, D.A.J. and Penman, D.R. (1994) Phenology of the New Zealand flower thrips *Thrips obscuratus* (Crawford) (Thysanoptera: Thripidae), on nectarine and peach flowers. *New Zealand Entomologist* 17, 70–77.

Teulon, D.A.J., Kolb, T.E., Cameron, E.A., McCormick, L.H. and Hoover, G.A. (1992) Pear thrips, *Taeniothrips inconsequens* (Uzel) (Thyanoptera: Thripidae) on sugar maple, *Acer saccharum* Marsh: a review. In: Bhatti, J.S. (ed.) *Advances in Thysanopterology.* Scienta Publishing, New Delhi, pp. 355–379.

Theunissen, J. and Legutowska, H. (1991) *Thrips tabaci* Lindeman (Thysanoptera: Thripidae) in leek: within plant distribution. *Journal of Applied Entomology* 112, 309–316.

Townsend, R., Watson, R.N., Hartley, M.J., Popay, A.J. and Popay, A.I. (1984) Stand management of asparagus for control. In: Hartley, M.J., Popay, A.J. and Popay, A.I. (eds) *Proceedings of the Thirty-seventh New Zealand Weed and Pest Control Conference, Christchurch*, pp. 151–155.

Treverrow, N.L. (1987) Banana pest thrips and banana silvering thrips. *Agfacts* H6. AE.2, 2 pp.

van Schoonhoven, A. and Pena, J.E. (1976) Estimation of yield losses in cassava following attack from thrips. *Journal of Economic Entomology* 69, 514–516.

Vidyasagar, P.S.P.V. and Kulshreshtha, J.P. (1983) Observations on the nature of damage to rice earheads by thrips *Haplothrips gangebaueri* Schmutz. *Current Science* 52, 173-174.

Vité, J.P. (1956) Populationsstudien am Lärchenblasenfuss *Taeniothrips laricivorus* Krat. *Zeitschrift für angewandte Entomologie* 38, 417–488.

Vos, J.G.M., Sastrosiswojo, S., Uhan, T.S. and Setiawate, W. (1991) Thrips on hot peppers in Java, Indonesia. In: Talekar, N.S. (ed.) *Thrips in Southeast Asia: Proceedings of a Regional Consultation Workshop, Bangkok, Thailand*, 13 March. Asian Vegetable Research and Development Centre, AVRDC Publication No. 91–342, pp. 18–28.

Weiss, M.J. and Beshear, R.J. (1987) Seasonal distribution of the flower thrips (*Frankliniella tritici* Fitch) and onion thrips (*Thrips tabaci* Lindeman) (Thysanoptera: Thripidae) on safflower in eastern Montana. *Canadian Entomologist* 119, 399–401.

Welter, S.C., Rosenheim, J.A., Johnson, M.W., Mau, R.F.L. and Gusukuma-Minuto, L.R. (1990) The effects of *Thrips palmi* and western flower thrips (Thysanoptera: Thripidae) on the yield, growth and carbon allocation pattern in cucumbers. *Journal of Economic Entomology* 83, 2092–2101.

Wetzel, T. (1964) Untersuchungen zum Auftreten, zur Schudwirkung und zur Bekämpfung von Thysanopteren in Grassamenbeständen. *Beiträge zur Entomologie* 14, 427–500.

Williams, C.B. (1915) The pea thrips (*Kakothrips robustus*). *Annals of Applied Biology* 1, 222–246.

Wolfenbarger, D. and Hibbs, E.T. (1958) Onion thrips (*Thrips tabaci*. Lind.) infesting cabbage. *Journal of Economic Entomology* 51, 394–396.

Yokoyama, V.A. (1977) *Frankliniella occidentalis* and scars on table grapes. *Environmental Entomology* 6, 25–30.
Zawirska, I. (1978) Studia nad *Thrips tabaci* Lindeman (Thysanoptera: Thripidae). *Prace Naukowe Instytutu Ochrony Roslin* 20, 15–138.
Zenther-Müller, O. (1965) *Taeniothrips laricivorus* (Krat) in Danish stands of *Larix decidua* (Mill.) 16, 58–69.

Appendix III

Tree Crops, Associated Thrips and Components of Control*

Crop	Associated thrips species	IPM components	References
Forest trees			
Slash pine	*Gnophothrips fuscus*	SD, BE, CI, I	19, 21, 22, 23, 24, 25, 26, 27
Japanese black pine	*Gnophothrips fuscus*	BE	20
Basswood	*Thrips calcaratus*	BE	83, 84, 85
Larch	*Thrips laricivorus*	BE, SD	28, 29
Sugar maple	*Taeniothrips inconsequens*	See Chapter 17	
Avocado	*Heliothrips haemorrhoidalis*	PR, PA, BE, SD, I, CI, CC	53, 54, 55, 57, 58, 59, 60, 61, 62
	Retithrips syriacus	BE, BR	56
	Selenothrips rubrocinctus	SD, BE, PR, I, CI	57, 58, 60, 61
Apple	*Frankliniella occidentalis*	BE, SD, CI	8, 10, 11, 12, 14, 16, 17
	Haplothrips ceylonicus	CI	5, 6
	Taeniothrips inconsequens	SD, CI, BE	4, 7, 15, 17
	Thrips carthami	CI	4, 7
	Thrips cinnabarinus	PR	9
	Thrips flavus	CI, PR, ET	1, 2, 4, 5, 6, 7, 13
Banana	*Frankliniella parvula*	BE	50
	Thrips hawaiiensis	CI, BE, CC, SD	38, 39, 40, 43, 45, 51

* Compiled by B.L. Parker and M. Skinner.

Continued

Crop	Associated thrips species	IPM components	References
	Scirtothrips aurantii	BR, CI, BE, ET	41, 52
	Thrips pusillus	SD	52
	Hercinothrips femoralis	BE, I	42
Cacao	*Selenothrips rubrocinctus*	BE	46
Cashew	*Selenothrips rubrocinctus*	BE	47
Macadamia	*Scirtothrips aurantii*	I	18
Mulberry	*Pseudodendrothrips ornatissimus*	CI	44
Papaya	*Thrips parvispinus*	BE, CI	48, 49
Citrus	*Scirtothrips citri*	See Chapter 17	
	Scirtothrips aurantii		
	Scirtothrips dorsalis		
	Frankliniella bispinosa		
	Heliothrips haemorrhoidalis		
	Chaetanophothrips signipennis		
	Chaetanophothrips orchidii		
	Thrips hawaiiensis		
Stone fruit	*Frankliniella cestrum*	BE, CI, SD	72, 81, 82
	Frankliniella occidentalis	BE, CI, SD, CC	63, 64, 65, 66, 67, 69, 70, 71, 76, 78, 80
	Frankliniella tritici	BE, SD, CI	63, 69, 70
	Neohydatothrips variabilis	CI, BE	69, 70
	Taeniothrips inconsequens	BE, SD, I	63, 68
	Thrips major	BE, CI	79
	Thrips meridionalis	BE, CC, ET, CI	66, 79
	Thrips obscuratus	BE, CA, CI, CC	64, 73, 74, 75, 77
	Thrips tabaci	BE, CC, ET, CI, SD	66, 72, 80, 81, 82
Mango	*Frankliniella occidentalis*	BE, CI, SD	9, 30, 36
	Haplothrips tenuipennis	BE	35
	Megalurothrips distalis	BE	35
	Rhipiphorothrips cruentatus	BE, CI	33, 34, 45
	Scirtothrips dorsalis	BE, CI, PR	33, 37, 44, 45
	Thrips hawaiiensis	BE, CI	33, 35, 45

Crop	Associated thrips species	IPM components	References
	Thrips palmi	BE, SD	31, 32
Rambutan	*Thrips hawaiiensis*	CI	44
	Scirtothrips dorsalis	CI	44
Waxapple	*Rhipiphorthrips cruentatus*	CI	45

Codes for IPM components: SD = survey and detection, CI = chemical insecticides, PR = predators, PA = parasites, CC = cultural controls, BE = biology and ecology, CA = chemical attractant, BR = biorational pesticides, I = identification, ET = economic thresholds. Compiled by B.L. Parker and M. Skinner (see Chapter 17).

References

1. Gupta, P.R. and Bhalla, O.P. (1993) Conservation of biotic agents in apple orchards of Himachal Pradesh. *Journal of Insect Science* 6, 204–206.
2. Singh, M. (1990) Damage and economic threshold level of *Thrips flavus* Schrank to apple blossom in Himachal Pradesh. *Indian Journal of Entomology* 51, 288–293.
3. Fatzinger, C.W. and Dixon, W.N. (1996) *User's guide for SeedCalc: A Decision-support System for Integrated Pest Management in Slash Pine Seed Orchards.* General Technical Report SE-92. US Department of Agriculture, Forest Service, Southern Research Station, Asheville, NC, 64 pp.
4. Singh, M. (1989) Chemical control of blossom thrips in apple. *Indian Journal of Entomology* 49, 580–582.
5. Singh, M. (1988) Field evaluation of selected insecticides against apple blossom thrips. *Indian Journal of Entomology* 48, 391–394.
6. Singh, M. (1986) Field evaluation of selected insecticides against apple blossom thrips. In: *Proceedings of the National Symposium on Temperate Fruits, 15–18 March 1984.* Himachal Pradesh Agriculture University, Solan, India, pp. 363–366.
7. Mishra, R.C., Verma, A.K. and Gupta, P.R. (1987) Insecticides for the control of apple blossom thrips. *Indian Journal of Horticulture* 44, 115–120.
8. Terry, L.I. (1991) *Frankliniella occidentalis* (Thysanoptera: Thripidae) oviposition in apple buds: role of bloom state, blossom phenology, and population density. *Environmental Entomology* 20, 1568–1576.
9. Ben–Dov, Y., Wysoki, M., Tomer, E., Ofek, G. and Avin, Z. (1992) Observations on the phenology and control of the western flower thrips *Frankliniella occidentalis* Pergande of mango inflorescence in Israel. *Alon Hanotea* 46, 497–503.
10. Terry, L.I. and Kelly, C.K. (1993) Patterns of change in secondary and tertiary sex ratios of the Terebrantian thrips, *Frankliniella occidentalis. Entomologia Experimentalis et Applicata* 66, 213–225.
11. Terry, I. (1990) Bionomics of *Frankliniella occidentalis* in apple orchards in the southwestern USA. In: *Proceedings of the Third International Symposium on Thysanoptera, Kazimiera Dolny, Poland 11–16 June*, Warsaw Agricultural University Press, Warsaw, Poland, pp. 91–95.
12. Bradley, S.J. and Mayer, D.F. (1994) Evaluation of monitoring methods for western

flower thrips, *Frankliniella occidentalis* (Thysanoptera; Thripidae), during the blossom period of 'Granny Smith' apples. *Journal of the Entomological Society of British Columbia* 91, 63–68.

13. Gupta, P.R. and Bhalla, O.P. (1993) Conservation of biotic agents in apple orchards of Himachal Pradesh. *Journal of Insect Science* 6, 204–206.
14. DeGrandi-Hoffman, G., Terry, I. and Huber, R.T. (1988) Incorporating fruit set estimates with thrips management to create a decision support system for apples. *Horticultural Science* 23, 571–574.
15. Paradis, R.O. (1969) Le thrips du poirier, *Taeniothrips inconsequens* (Uzel) (Thysanoptera: Thripidae), dans une pommeraie du sud-ouest Quebec. *Annals of the Entomological Society of Quebec* 14, 83–84.
16. Terry. I.L. and DeGrandi-Hoffman, G. (1988) Monitoring western flower thrips (Thysanoptera: Thripidae) in 'Granny Smith' apple blossom clusters. *Canadian Entomologist* 120, 1003–1016.
17. Anonymous (1991) Integrated pest management for apples and pears. *University of California Publication* 3340. Oakland, CA, 214 pp.
18. Wentzel, P.C. (1971) Thrips discovered on macadamia trees. *Farming in South Africa* 47, 57.
19. Ranasinghe, M.A.S.K. (1981) Bionomics of slash pine flower thrips, *Gnophothrips fuscus* (Morgan), in pine seed orchards in Florida. PhD thesis, University of Florida, Gainesville, FL, 120 pp.
20. Cartier, J.P. (1964) Description and life history in Rhode Island of the thrips, *Gnophothrips piniphilus* Cwfd. MSc thesis, University of Rhode Island, Kingston. 70 pp.
21. Merkel, E.P., Fatzinger, C.W. and Dixon, W.D. (1994) Keys for distinguishing thrips (Thysanoptera) commonly found on slash pine in Florida. *Journal of Entomological Science* 29, 93–99.
22. Ranasinghe, M.A.S.K. and Wilkinson, R.C. (1988) Seasonal occurrence of *Gnophothrips fuscus* (Thysanoptera: Phlaeothripidae) on slash pine in Florida. *The Florida Entomologist* 71, 384–387.
23. Ebel, B.H. (1963) Insects affecting seed production of slash and longleaf pines – their identification and biological annotation. *US Department of Agriculture, Forest Service Paper SE-6*. Southeast Forest Experiment Station, Asheville, NC, 24 pp.
24. Ebel, B.H. (1965) Control of thrips on slash pine female strobili. *Journal of Forestry* 63, 287–288.
25. DeBarr, G.L. (1969) The damage potential of a flower thrips in slash pine seed orchards. *Journal of Forestry* 67, 326–327.
26. DeBarr, G.L., Merkel, E.P., O'Gwynn, C.H. and Zoerb Jr, M.H. (1972) Differences in insect infestations in slash pine seed orchards due to phorate treatment and clonal variation. *Forest Science* 18, 56–64.
27. DeBarr, G.L. and Williams, J.A. (1971) Nonlethal thrips damage to slash pine flowers reduces seed yields. *US Department of Agriculture, Forest Service Research Note SE–160*. Southeast Forest Experiment Station, Asheville, NC, 4 pp.
28. Zethner–Møller, O. (1965) *Taeniothrips laricivorus* (Krat.) (Thripidae: Thysanoptera) in Danish stands of *Larix decidua* (Mill.). *Oikos* 16, 58–69.
29. Vité, J.P. (1956) Populationsstudien am Larechenblasenfuß *Taeniothrips laricivorus* Krat. *Zeitschrift für Angewandte Entomologie* 38, 420–448.
30. Ben-Dov, Y., Wysoki, M., Tomer, E., Ofek, G. and Avin, Z. (1992) Observations on the phenology and control of the western flower thrips, *Frankliniella occidentalis* Pergande of mango inflorescences in Israel. *Alon Hanotea* 46, 497–503.

31. Verghese, A., Tandon, P.L. and Rao, G.S.P. (1988) Ecological studies relevant to the management of *Thrips palmi* Karny on mango in India. *Tropical Pest Management* 34, 55–58.
32. Tandon, P.L. and Verghese, A. (1987) New insect pests of certain fruit crops. *Indian Journal of Horticultural Research* 44, 121–122.
33. Lee, H.S. and Wen, H.C. (1982) Season occurrence of and injury caused by thrips and their control on mangoes. *Plant Protection Bulletin, Taiwan* 24, 179–187.
34. Khuhro, R.D., Soomro, A.H., Ikhlas, A., Dharejo, A.G., Ansari, H.U. and Baloch, H.B. (1991) Bioefficacy of some contact and systemic insecticides in the control of mango thrips, *Rhipiphorothrips cruentatus*. In: *Proceedings of the Pakistan Congress of Zoology* 7, 95–98.
35. Ramasubbarao, V. and Thammiraju, N.B. (1994) New record of blossom thrips *Megalurothrips distalis* on mango (*Mangifera indica*). *Indian Journal of Agriculture Science* 64, 417–418.
36. Murguido, C., Jiminez, S. Ruiz, I. and Delgado, M. (1987) Captur de insectos con trampas de colores y atrayentes alimenticios. *Ciencia y Tecnica en la Agricultura, Proteccion de Plantas* 10, 47–59.
37. Kumar, N.S. and Ananthakrishnan, T.N. (1984) Predator–thrips interactions with reference to *Orius maxidentrex* Ghauri and *Carayonocoris indicus* Muraleedharan (Anthocoridae: Heteroptera). In: *Proceedings of the Indian National Science Acadamy B*. 50, 139–145.
38. Anonymous (1984) Insecticide control of corky scab. *Banana Bulletin* 48, 12–13.
39. Trochoulias, T., Wright, W.E. and Loebel, M.R. (1984) Use of insecticides to reduce corky scab injury in bananas. *Tropical Agriculture* 61, 193–195.
40. Su, C.Y., Hwang, M.T. and Wang, H.C. (1982) Tests of attractants for the banana thrips, *Thrips hawaiiensis* Morgan, in the laboratory. *National Taiwan University (NTU) Phytopathologist and Entomology* 9, 92–97.
41. Nasseh, O.M. and Mughni, A.A.A. (1990) Efficacy of chemical and natural insecticides for suppression of *Scirtothrips aurantii* (Faure) (Thripidae: Thysanoptera) causing banana fruit spotting in the Yemen Arab Republic. In: *Proceedings: Integrated Pest Management in Tropical and Subtropical Cropping Systems. 3. Bad Durkheim, Germany, 8–15 February 1989*, Deutsche Landwirtschafts-Gesellschaft, Frankfurt am Main, Germany, pp. 749–756.
42. Houston, K.J., Mound, L.A. and Palmer, J.M. (1991) Two pest thrips (Thysanoptera) new to Australia, with notes on the distribution and structural variation of other species. *Journal of the Australian Entomological Society* 30, 231–232.
43. Tsai, Y.P., Hwang, M.T. and Chen, H.P. (1992) Occurrence and damage of *Thrips hawaiiensis* in banana orchards. *Chinese Journal of Entomology* 12, 231–237.
44. Bansiddhi, K. and Poonchaisri, S. (1991) Thrips of vegetables and other commercially important crops in Thailand. In: Talekar, N.S. (ed.) *Proceedings of a Regional Consultation Workshop: Thrips in Southeast Asia, Bangkok, Thailand, 13 March*. AVDRC Publication No. 91–342, pp. 34–39.
45. Niann, T.C. (1991) Important thrips species in Taiwan. In: Talekar, N.S. (ed.) *Proceedings of a Regional Consultation Workshop: Thrips in Southeast Asia, Bangkok, Thailand, 13 March*. AVDRC Publication No. 91–342, pp. 40–56.
46. Fennah, R.G. (1965) The influence of environmental stress on the cacoa tree in predetermining the feeding sites of cacao thrips, *Selenothrips rubrocinctus* (Giard), on leaves and pods. *Bulletin of Entomological Research* 56, 333–349.
47. Fennah, R.G. (1963) Nutritional factors associated with seasonal population increase of cacao thrips, *Selenothrips rubrocinctus* (Giard), on cashew, *Anacardium occidentale. Bulletin of Entomological Research* 53, 681–713.

48. Lim, W.H. (1989) Bunchy and malformed tops of papaya cv. Ekostika caused by *Thrips parvispinus* and *Cladosporium oxysporum*. *Malaysian Agriculture Research and Development Institute Research Journal* 17, 200–207.
49. Vijayasegaran, S. (1986) *Thrips parvispinus* Karny (Thripidae: Thysanoptera) – a potential new pest of papaya. In: *Proceedings of the 2nd International Conference on Plant Protection in the Tropics, Genting Highlands, Malaysia, 17–20 March*. Malaysian Agriculture Research and Development Institute, Serdang, Malaysia pp. 140–142.
50. Harrison, J.O. (1963) Notes on the biology of the banana flower thrips, *Frankliniella parvula*, in the Dominican Republic (Thysanoptera: Thripidae). *Annals of the Entomological Society of America* 56, 664–666.
51. Mohanasundaram, M. and Sivakumar, C.V. (1973) Varietal susceptibility of banana to the banana fruit thrips (*Thrips florum* Schm.). *Madras Agricultural Journal* 60, 624–625.
52. El-Bashir, S. and Al-Zabidi, A.H. (1985) Fruit spotting of banana in the Yemen Arab Republic. *FAO Plant Protection Bulletin* 33, 113–118.
53. Swirski, E., Wysoki, M. and Izhar, Y. (1996) Avocado pests in Israel. In: *Proceedings of the World Avocado Congress III, Tel Aviv, Israel, 22–27 October 1995*. *Phytoparasitica* 24, 70.
54. Wysoki, M., Kuslitzky, W., Swirski, E., Hadar, D. and Ben-Yehuda, S. (1995) Introduction and establishment of the parasitoid of the greenhouse thrips in Israel. In: *Proceedings of the World Avocado Congress III, Tel Aviv, Israel. 22–27 October 1995*. *Phytoparasitica* 24, 72.
55. Izhar, Y., Swirski, E., Ben–Yehuda, S., Wysoki, M. and Hadar, D. (1994) Susceptibility of the avocado cultivar 'Ardit' to the greenhouse thrips, *Heliothrips haemorrhoidalis*. *Scientific Activities 1988–1993. Institute of Plant Protection Special Publication* 255, Bet Dagan, Israel.
56. Izhar, Y., Ben-Yehuda, S., Swirski, E., Wysoki, M. and Dagan, M. (1994) Occurrence of the black vine thrips, *Retithrips syriacus* Mayet, on avocado in Israel and trials to control the pest with insecticides of plant origin. *Alon Hanotea* 46, 523–529.
57. Dennill, G.B. and Erasmus, M.J. (1992) Basis for a practical technique for monitoring thrips in avocado orchards. *Crop Protection* 11, 89–91.
58. Dennill, G.B. and Erasmus, M.J. (1991) A packhouse survey of insect damage to avocados in the Nelspruit/Hazyview area during 1990. *Yearbook – South African Avocado Growers' Association* 14, 79–82.
59. McMurtry, J.A.S., Johnson, H.G. and Newberger, S.J. (1991) Imported parasite of greenhouse thrips established in California avocado. *California Agriculturalist* 45, 31–32.
60. Dennill, G.B. (1992) *Orius thripoborus* (Anthocoridae), a potential biocontrol agent of *Heliothrips haemorrhoidalis* and *Selenothrips rubrocinctus* (Thripidae) on avocado fruits in eastern Transvaal. *Journal of the Entomological Society of Southern Africa* 55, 255–258.
61. Steyn, W.P. (1991) Blaaspootjies op avokado's. *Inlightingsbulletin – Navorsingsinstitut vir Sitrus en Subtropiese Vrugte* 224, 10–13.
62. Phillips, P.A., Bekey, R.S. and Bailey, J.B. (1993) An effective strategy for management of greenhouse thrips in 'Hass' avocado orchards. *California Avocado Society Yearbook* 76, 99–107.
63. Felland, C.M., Teulon, D.A.J., Hull, L.A. and Polk, D.F. (1995) Distribution and management of thrips (Thysanoptera: Thripidae) on nectarine in the mid–atlantic region. *Journal of Economic Entomology* 88, 1004–1011.

64. Teulon, D.A.J., Penman, D.R. and Ramakers, P.M.J. (1993) Volatile chemicals for thrips (Thysanoptera: Thripidae) host-finding and applications for thrips pest management. *Journal of Economic Entomology* 86, 1405–1415.
65. Lacasa, A., Torres, M. and Martinez, C. (1992) Ensayo de control quimico de *Frankliniella occidentalis* en nectarinas tempranas. *Phytoma España* 37, 35–41.
66. Lacasa, A., Martinez, C. and Aliaga, J. (1990) Los trips, una importante plaga de la nectarina. *Cuadernos de Fitopatologia* 23, 73–78.
67. Lacasa, A., Martinez, M.C., Torres, J. and Aliaga, J. (1991) Datos preliminares sobre *Frankliniella occidentalis* como plaga del la nectarina y el melocotonero en el Sureste Espanol. *Fruticultura Profesional* 36, 39–46.
68. Bournier, A. and Blache, M. (1956) Thysanopteres nuisibles aux pechers. *Revue de Zoologie Agricole et Appliquée* No. 1–3, 1–8.
69. Yonce, C.E., Payne, J.A., Beshear, R.J. and Horton, D.L. (1990) Thrips (Thysanoptera: Thripidae) associated with unsprayed and sprayed peaches in Georgia. *Journal of Economic Entomology* 83, 511–518.
70. Yonce, C.E., Beshear, R.J., Payne, J.A. and Horton, D.L. (1990) Population distribution of flower thrips and the western flower thrips (Thysanoptera: Thripidae) in nectarines and their relative association with injury to fruit in the southeastern United States. *Journal of Entomological Science* 25, 427–438.
71. Juarez, J.C. and Byerly–Murphy, K.F. (1988) Comportamiento y evaluacion de danos del trips *Frankliniella* sp. (Thysanoptera: Thripidae) sobre la calidad y cantidad de lod frutos de 'nectarino' (*Prunus* sp.) en Nuevo Casas Grandes, Chihuahua, Mexico. *Folia Entomologica Mexicana* 76, 25–36.
72. Ripa, S.R. (1988) Damage to plum and apricot shoots associated with thrips. *Agricultura Tecnica* 48, 46–47.
73. Penman, D.R., Osborne, G.O., Worner, S.P., Chapman, R.B. and McLaren, G.F. (1982) Ethyl nicotinate: a chemical attractant for *Thrips obscuratus* (Thysanoptera: Thripidae) in stonefruit in New Zealand. *Journal of Chemical Ecology* 8, 1299–1303.
74. Teulon, D.A.J. and Penman, D.R. (1987) Thrips control on stonefruit at flowering in Canterbury. In: *Proceedings of the New Zealand Weed and Pest Control Conference*. New Zealand Weed and Pest Control Society, Palmerston, NZ, pp. 153–160.
75. Teulon, D.A.J. and Penman, D.R. (1990) Host records for the New Zealand flower thrips (*Thrips obscuratus* (Crawford) Thysanoptera: Thripidae. *New Zealand Entomologist* 13, 46–51.
76. Felland, C.M., Hull, L.A., Teulon, D.A.J. and Cameron, E.A. (1993) Overwintering of western flower thrips (Thysanoptera: Thripidae) in Pennsylvania. *Canadian Entomologist* 125, 971–973.
77. Lay-Yee, M. and Rose, K.L. (1993) The effect of heat treatments for insect control on the quality of Redgold and Fantasia nectarines. *Orchardist of New Zealand* 66, 20.
78. Lacasa, A., Torres, J. and Martinez, M.C. (1993) *Frankliniella occidentalis* (Perg.) en los cultivos de nectarina de Murcia. Evolucion de las poblaciones y comportamiento de variedades. *Boletin de Sanidad Vegetal, Plagas* 19, 335–344.
79. Cinti, S., de Cristofaro, A. and Vigilante, G. (1993) La difesa del pesco: lotta ai principali insetti ed acari. *Terra e Sole* 48, 215–219.
80. Gonzalez, E., Alvarado, M., Berlanga, E., Serrano, A. and de la Rosa, A. (1994) Danos producidos por trips en nectarinas en el Valle del Guadalquivir. *Boletin de Sanidad Vegetal, Plagas* 20, 229–241.
81. Ripa, S. and Rodriguez, A.F. (1993) Relacion entre la presencia de trips durante la floracion de nectarinos e incidencia de 'russet' en los frutos en la cosecha.

Agricultura Tecnica (Santiago) 53, 23–28.
82. Prado, C.E. (1991) Estudio de la poblacion de trips (Thysanoptera: Thripidae) en huertos de carozo. *Agricultura Tecnica (Santiago)* 51, 131–137.
83. Raffa, K.F. and Hall, D.J. (1989) *Thrips calcaratus* Uzel (Thysanoptera: Thripidae), a new pest of basswood trees in the Great Lakes region. *Canadian Journal of Forest Research* 19, 1661–1662.
84. Raffa, K.F., Hall, D.J., Kearby, W. and Katovich, S. (1992) Seasonal life history of introduced basswood thrips (Thysanoptera: Thripidae) in Wisconsin, with observations on associated thrips species. *Environmental Entomology* 21, 771–779.
85. Rieske, L.K. and Raffa, K.F. (1995) *Thrips calcaratus* – induced defoliation and subsequent foliar suitability. In: Parker, B.L., Skinner, M. and Lewis, T. (eds) *Thrips Biology and Management*. Plenum Press, New York, pp. 97–100.

Index

abdomen 31, 32
Acanthothrips nodicornis 361
acarodomatia 321
Acarophenacidae 273, 283
Adalia bipunctata 263, 292
Adiheterothripidae 198, 199
Adunkothrips asiaticus 280
Aeolothripidae 19, 26, 143, 199, 276
Aeolothrips
- *albicinctus* 235
- *astutus* 50
- *collaris* 271
- *intermedius* 17, 20, 22, 25, 32, 42, 67, 68, 98, 141, 153, 271, 280, 291, 299, 319, 494
- *kawanaii* 271
- *melisi* 271
- *tenuicornis* 271
- *vittatus* 271

Aleurodothrips fasciapennis 141, 156, 157, 272, 465
aldehydes
- *see* *p*-anisaldehyde

alfalfa
- *see* forage legumes

allomones
- *see* pheromones

Alternaria porri 513
aluminium plant 699
Amblyseius
- *addoensis* 273, 293, 619, 620
- *andersoni* 273, 282, 293, 299
- *aripo* 274
- *aurescens* 274
- *barkeri* 274, 281, 293, 299, 307
- *californicus* 274, 293, 300
- *citri* 274, 619
- *cucumeris* 90, 91, 274, 280, 293, 300, 306, 308, 314, 645, 651, 652
- *degenerans* 274, 280, 294, 300, 308, 651
- *hibisci* 274, 282, 294, 300, 619
- *lailae* 274
- *largoensis* 274, 294
- *lentiginososus* 275
- *limonicus* 275, 282, 294, 300
- *longispinosus* 275, 295
- *manihoti* 275
- *masiaka* 275
- *montdorensis* 275
- *multidentatus* 275
- *okinawanus* 275, 295, 300
- *scutalis* 275, 282, 295, 300
- *sessor* 275, 295
- *stipulatus* 275, 295
- *swirskii* 275, 295, 300
- *tularensis* 275, 282, 295, 300, 620
- *waltersi* 275

American grass thrips
- *see Anaphothrips obscurus*

Anactinothrips gustaviae 93
anal droplets 90, 139, 276, 278, 282, 322

Anaphothrips
cameroni 236
obscurus 4, 181, 185, 210, 228, 244, 516, 684, 685, 687, 688, 692, 693, 698
sudanensis 687, 696
woodi 183
anatomy 33
abdominal glands 36, 86
alimentary tract 41, 44–46, 125, 554
circulatory system 41
compound eyes 19, 35
head 19, 22, 23, 35
legs 26, 30
male genitalia 36, 40
malphighian tubules 46, 91
musculature 35, 39, 40, 42, 179, 188
nervous system 33, 38
ovaries 46, 179, 382, 386
reproductive systems 46–49
respiratory system 41
salivary glands 46, 125, 127, 134, 551
testes 47
thorax 37
Andrewarthaia kellyana 271
Androthrips flavipes 272, 280, 361
Anisopilothrips venustulus 682
p-anisaldehyde 77, 78, 80, 187, 455, 643
antennae
see morphology
antibiosis 66, 80, 84, 85, 601, 663
antifeedant 82
antixenosis 66, 80, 85, 601, 663
Anthocoridae 267, 297, 304, 315, 317, 318
see also Anthocoris, Orius
Anthocoris austropiceus 267
nemorum 267, 287
gallarum-ulmi 267
anthophilous thrips 67, 71, 73, 75, 94, 134, 198
ants 228, 263
see also Formicidae
Anystidae 273, 283
apple 441, 443, 711
feeding injury 518, 677
oviposition injury 102, 526, 677
Apple blossom thrips
see Thrips imaginis
apricots
see stonefruit
Apterothrips apteris 49, 99, 100, 183, 685
secticornis 185
Aptinothrips rufus 210, 234, 235, 386, 684
stylifer 185
aromatic amino acids 146, 147
Arorathrips mexicanus 685
Arrhenothrips ramakrishnae 280
arrhenotoky 49, 50, 372, 375
arum lily 584
Asilidae 262
asparagus 584, 677
Aspergillus sp. 402
Assam thrips
see Scirtothrips dorsalis
attractants 77–80, 454–456, 643
aubergine 160, 212, 308, 599, 601, 677, Plate IV (6)
Aurantothrips orchidaceous 701
Australothrips bicolor 700
avocado 381, 436, 678, 711
feeding injury 518, 522, 523, Plate III (5)
oviposition injury 526
Ayyaria chaetophora 267, 694
Azalea thrips
see Heterothrips azaleae
azaleas 74

Bacillus thuringiensis 572
Bactrothrips brevitubus 88, 89
baculoviruses 663
Bagnalliella yuccae 21, 88, 90, 278
baking quality 600
Baliothrips biformis 264
Baliothrips dispar 185
banana 191, 209, 212, 583, 711
feeding injury 523, 678, Plate III (7)
oviposition injury 527, 678
rust 523, 583
Banana flower thrips
see Frankliniella parvula
Banana silvering thrips
see Hercinothrips bicinctus
Banana thrips
see Chaetanaphothrips signipennis

Banded greenhouse thrips
see Hercinothrips femoralis
Banded thrips
see Aeolothrips intermedius
barley 6, 443, 586
feeding injury 515, 679, Plate X (1,2)
Barley thrips
see Limothrips denticornis
basswood 511, 625, 711
feeding injury 509, 511, 679
Basswood thrips
see Thrips calcaratus
Bean flower thrips
see Megalurothrips sjostedti
Bean thrips
see Caliothrips fasciatus
Beans 679, 680
adzuki 103, 680
broad 102, 680
faba 75, 160, 220, 522
hyacinth 160, 680
kidney 679
lima 159
Beauveria bassiana 230, 400, 402, 404, 407, 422, 630
begonia 584
Begonia thrips
see Scirtothrips longipennis
benzaldehyde 77, 78, 80, 455
Bilia sp
bioassay 458–461, 477
fungal pathogens 460, 461
pesticides 458, 570, 571
residual 458
topical 459
whole plant 460
see also insecticidal treatments
biological control 228, 379, 380, 660
classical 355, 379, 408
compatibility of agents 155, 650
fungal agents 399, 630
see also entomopathogenic fungi
of *Frankliniella occidentalis* 310
in greenhouses 157, 316, 380, 651–657
of *Heliothrips haemorrhoidalis* 379, 380
by nematodes 387, 422
prospects for 380–382, 660
of *Thrips palmi* 308, 379
of *Thrips tabaci* 307, 311, 377, 379
of *Selenothrips rubrocinctus* 379
see also parasitoids, predators
birds 185, 228
Birdsfoot trefoil thrips
see Odontothrips loti
bitter gourd Plate IX (5)
Black hunter
see Leptothrips mali
Blueberry thrips
see Frankliniella vaccinii
Brachyurothrips anomalus 360
bulbs
see gladiolus, lillies

cabbage 597
feeding injury 513, 514, 680, Plate VII (6)
resistance in 84, 85
Cacao thrips
see Selenothrips rubrocinctus
Caliothrips fasciatus 224, 234, 236, 360, 376, 683, 690, 695
impurus 5, 683, 686
indicus 206, 210, 360, 678, 686, 694
insularis 360
phaseoli 183, 211, 679, 683, 693, 696
russelli 376
sudanensis 5, 683
calla lily 699
Camptoptera liebknechti 266
Campylomma chiensis 266, 297
livida 266, 286, 297
Cane thrips
see Fulmekiola serrata
cannibalism 142, 281
Carabidae 263
Carayoncornis indicus 267, 287, 297
cardamom 680
Cardamom thrips
see Sciothrips cardomomi
Cardiostethus
consors 267
poweri 267
rugicollis 267
carnations 74, 584, 699, Plate XI (7)
carrot 7, 605
cashew 4, 154, 681, 711

cassava 6, 7, 206
  feeding injury 514, 680, Plate IX (1)
  resistance in 218
Cassava thrips
  see Corynothrips stenopterus
castor bean 277, 681
Castor thrips
  see Retithrips syriacus
Catinathrips vaccinophilus 328
Ceara rubber 681
Cecidomyiidae 261, 277
celery 681
Ceranisus 356, 357
  americensis 357, 362, 364, 365, 496
  bicoloratus 357
  femoratus 357
  javae 357
  lepidotus 357
  loomansi 357, 362
  maculatus 357
  margiscutum 357
  menes 229, 357, 363–365, 367, 369, 370, 376, 377, 496, Plate II (1–3)
  nigrifemora 357
  nubilipennis 357, 362, 367
  pacuvius 357
  planitianus 357
  russelli 357, 362
  vinctus 357
Ceratothrips ericae 81, 158, 162
  frici 67, 69
Cereal thrips
  see Haplothrips aculeatus, H. tritici, Limothrips cereallium
Cereals
  see barley, oats, rice, wheat
Chaetanaphothrips leeuweni 191
  orchidii 191, 212, 516, 523, 678, 682, 701, 712, Plates III (7), VI
  signipennis 191, 523, 583, 678, 679, 712
  theiperdus 191, 697
check lists 2
chemical control 567–588
  historical background 568, 569
  see also insecticides, insecticidal treatments
cherries
  feeding injury 695
  oviposition injury 527
  see stonefruit
Cheyletidae 273
Chillie thrips
  see Scirtothrips dorsalis
Chirothrips
  crassus 7, 101
  hamatus 377, 684
  manicatus 68, 181, 184, 186, 234, 684, 685
  pallidicornis 7
Chloropidae 262
christmas cactus 699
chrysanthemum 76, 85, 212, 220, 409, 584
  feeding injury 222, 699
  IPM 656, 657
  resistance in 83, 133
Chrysanthemum thrips
  see Thrips nigropilosus
Chrysopa
  alobana 261
  arioles 261
  carnea 261, 292
  claveri 261
  iona 261
  montoyana 261
  oculata 261
  perla 261
  plorabunda 261
  vulgaris 261
Chrysopidae 261, 299
Cineraria Plate XII (9)
cinnamaldehyde 77, 455
citral 77
citrus 4–7, 102, 207, 211, 212, 457, 570
  feeding injury 436, 519–521, 523, 524, 586, 681, 682, Plates V (1–6), VI
  IPM in 616–623
  oviposition injury 527, 682
Citrus thrips
  see Scirtothrips citri
Cladosporium cladosporiodes 402
clover 7
Coccinella septempunctata 264, 292
Coccinellidae 263, 277
cocksfoot 7
cocoa 4, 160, 206, 245, 443, 682, 711
coconut 682
coffee 132, 682
Coffee thrips
  see Diarthrothrips coffeae,

Scirtothrips bispinosus
cold hardiness 225–227
see also overwintering
Coleomegilla maculata 264, 292
colour
attractive flowers 222
electroretinograms 73
hues, response and host selection 67–76, 222
traps 67, 187, 454–456
U.V. reflectance 67, 71, 72, 187
colour forms/strains 201, 226, 362, 543, 548
Common blossom thrips
see Frankliniella schultzei
common names 667–673
Composite thrips
see Microcephalothrips abdominalis
compound eyes
see anatomy, morphology
Coniobiolus coronatus 401
Control measures
evolution of 4
see also biological control, entomopathogenic fungi, insecticidal treatments, IPM, parasitoids, predators
corn thrips
see Limothrips cerealium
Corynothrips stenopterus 6, 7, 206, 514, 680, Plate IX (1)
cotton 5, 155, 220, 246, 438, 443, 569
feeding injury 508, 514, 515, 596, 599, 683
resistance in 83, 84, 133, 218
seedlings 222, 572, 586, 596
Cotton bud thrips
see Frankliniella schultzei
Cotton thrips
see Caliothrips sudanensis
countries, regions & states
Africa 2
Australia 2, 18, 192, 203, 204, 208, 211, 212, 514, 523, 661, 677, 678, 681, 683, 691, 697, 698, 700
Queensland 94, 583, 604
South Australia 207, 217, 242, 518
Tasmania 687
Austria 204
Bangladesh 516
Bengal 7
Brazil 203, 204, 208, 211
Britain 2, 204, 206, 212, 597, 647, 655
Bulgaria 279, 586
Canada 6, 7, 233, 525, 647, 654, 661, 691,
British Columbia 527, 625
Quebec 600
Canary Islands 192, 679
Caribbean 4, 192, 279, 379, 513, 677–684, 687, 688, 696, 698, 699
Central America 2, 183, 678, 680, 682, 683, 687, 691, 693, 700
China 157, 606
Colombia 6, 7, 202
Costa Rica 183, 184, 202, 204, 208
Cuba 7
Denmark 8
Egypt 84, 364, 514
England
see Britain, U.K.
Europe 2, 6, 200, 204, 206, 603, 661, 677, 679, 680, 684–689, 692, 695, 698, 699–701
Fiji 157
Finland 8
France 204, 212, 233, 587
French Antilles 523
Germany 186, 233
Ghana 379
Greece 524, 525
Guadeloupe 380
Guyana 192, 692
Hungary 186, 193, 245
India 2, 198, 204, 230, 379, 512, 516, 518
Indonesia 577
Israel 522, 678
Italy 204, 212, 525
Jamaica 2
Japan 2, 192, 208, 212, 236, 370, 379, 550, 578, 606, 661, 698
Kenya 7, 208, 600, 604
Khazakhstan 233, 236, 604
Malaysia 97, 182
Mauritius 192, 550, 696
Mediterranean 198, 679
Mexico 203, 204, 208, 279, 526

Countries, regions & states *continued*
Netherlands 7, 190, 192, 202, 233, 364, 586, 654
New Caledonia 211, 550
New Zealand 2, 7, 148, 183, 202, 203, 211, 524, 587, 599, 677, 679, 688, 694, 698
North America 2, 4, 6, 204, 677–701
see also U.S.
Pakistan 514
Panama 2, 203, 204
Philippines 84, 200, 208, 578, 606
Puerto Rico 184, 379
Reunion Is. 192, 617
Rhodesia
see Zimbabwe
Romania 244
Sao Tomé 237
Scandinavia 4
South Africa 4, 5, 7, 207, 211, 512, 522, 523, 617, 678, 679, 681, 683, 686, 687, 691, 697
South America 2, 203, 677–683, 686, 687, 693, 696–701
South-east Asia 4, 203, 513, 572, 577, 677–694, 696–701
Spain 7, 212
Sri-Lanka 516
Sudan 192, 604
Sweden 7
Switzerland 233, 237
Taiwan 6, 202, 208, 212, 578, 677
Thailand 370, 379
Trinidad 184, 379, 606
Turkey 603
U.K. 7, 8
see also Britain
Ukraine 228, 230
U.S. 84
Alabama 514
Alaska 193
Arizona 186, 508, 523, 587
Arkansas 515
California 5, 6, 7, 185, 203, 211, 233, 235, 379, 512, 523, 524, 526, 570, 573, 586, 606, 616, 677
Colorado 513
Florida 5, 7, 192, 212, 380, 513, 518, 519, 523, 677–679, 681, 684, 699
Georgia 524, 595
Hawaii 379, 525, 677, 679, 684
Illinois 184
Iowa 7
Louisiana 515, 586, 596
New York 597, 600
North Dakota 236, 586
Oregon 236, 516
Pennsylvania 6, 8, 233, 525
Texas 514, 600
Vermont 6, 518, 625, 627
Wisconsin 236
U.S.S.R. 3
Venezuela 692, 696
West Indies
see Caribbean
Yemen 523, 679
Zimbabwe 6
cowpea 7, 102, 160, 211, 439, 443, 457
feeding injury 509, 522, 599, 683, 684
oviposition injury 527
resistance in 219, 527, 601
Craspedothrips minor 376, 681, 692
crop loss 7, 8, 586, 587, 595
costs 7, 8, 586, 587, 595
percentage losses 6, 7
see also feeding
crop–weed interactions 207, 597, 598
Cryptothrips
nigripes 17, 361
rectangularis 361
Cuban laurel thrips
see Gynaikothrips ficorum
cucumber 7, 85, 220, 221, 307, 409, 442, 443, 525, 584, 599, 652–655, 660, 661, 684, Plate IX (2–4)
cultural control
see IPM
culturing
see rearing
custard apple 683
cyclamen 584

Dahlia Plate XII (3)
Dandelion thrips
see Thrips physapus
Danothrips trifasciatus 524, 682

date palms 198
γ-decalactone 88, 278
degree days 227, 238–240, 371, 616, 622
Dendrobium cvs. 74, 76
Dendrothrips ornatus 701
Deraeocoris
pallens 266
punctulatus 266
development 3, 15, 51–58, 201, 227, 493, Plate I (1)
see also life history
diapause 100, 316, 597
see also reproductive diapause
Diarthrothrips coffeae 224, 683
Dichromothrips
corbeti 701
phalaenopsidis 701
Dicyphus
eckerleini 266
rhododendri 266, 279
tamaninii 266, 286, 297, 311
diel periodicity 181
Dinurothrips hookeri 360
dispersal 66, 181–-187, 231
artificial 190–193, 208, 209
flight 182–187, 208
international trade 190, 191, 209
phoresy 185, 283
of predators 316
see also flight
diversity
biological 198–201
ecological 204–206
genetic 465
index of 206
systematic 197–204
DNA 5, 465
Dolichopodidae 262
Domatiathrips cunninghamii 321
Drepanothripoides innoxius 696
Drepanothrips reuteri 512, 685

Eastern flower thrips
see Frankliniella tritici
Ecacanthothrips tibialis 270
Echinothrips caribeanus 360
economic thresholds 241, 440, 586, 598–600, 620, 621, 626
Ectemnus reduvinus 267
egg plant
see aubergine
eggs 2, 33, 51, 81, 145
cannibalism 142
cavities 101
collecting 492
distribution on plant 2, 102, 103, 218, 276
embryogenesis 48, 50, 51
endoparasitoids 229, 355, 372–376
guarding 93
hatching 51
injury 103, 506, 526–529
sampling 448, 449
structure 2, 47, 51, 100, 580
see also oviposition/ovipositor
Elaphrothrips tuberculatus 88, 89, 93
electrical penetration graph 85, 462
electrophysiological recording 463–465
electroretinograms 73, 463, 464
electrostatic sprayers 419, 581
embryogenesis 40, 50, 51
endoparasitoids 229, 230, 355
egg 229
larval 229, 356
species lists 357, 358
Enneothrips flavens 211, 581, 686
Entedonastichus 359, 361
albicoxis 358
carbonarius 358
dei 358
gaussi 358
kaulbarsi 358
mirus 358
Entomophaga grylii 401
Entomophthora thripidum 400, 401, 403, 405
entomopathogenic fungi 230, Plate II (5)
application 409, 419
compatibility with agrochemicals 421
descriptions 403, 404
dose 414
efficacy 414
epizootics 412, 413
formulation 417, 418
host specificity 410
infection processes 410–413
laboratory assay 414
in management 401, 408–410, 413–415, 423, 630

entomopathogenic fungi *continued*
penetration and development 411, 412
production 416
species list 401, 402
see also biological control
Erythraeidae 273
Erythrothrips asiaticus 271, 291, 299
ethylene 81, 127, 507–510, 520, 600
ethyl nicotinate 78, 80, 187, 455
eucalyptus 441
eugenol 78
Eucharitidae 356
Eulophidae 355
Eurylaprothrips crassus 88, 89
excretory system 46

Fauriellidae 198, 199
feeding
antifeedants 82
bacterial pathogens 506
behaviour 128–142, 462, 463
cannabalism 142, 281
cues 81, 85, 128
diet 142–150
direct damage 5, 132, 505–526
see also crop loss, injury to plants
distribution patterns 146, 153–155
egestion time 139, 152
electrical monitoring 85, 127, 132, 462
evolutionary patterns 148–150, 198
fruit spoilage 507
fungal pathogens 506, 557, 558
on fungus 142, 143, 198, 199
on grasses 67, 68, 200, 205, 516, 587
ingestion rates 140, 150–152, 153
injury to plants 132–134, 139, 152, 505–526, 675–701
on leaves 128–134, 146, 198–200, 510–516
on liquids 121, 142
through membranes 81, 133, 462, 489
on mosses 198, 200
nutrition 142–150
on pollen 81, 121, 134–139, 145, 152, 158, 200, 319
polyphagy 71, 73, 137, 149, 505
on prey 139–142, 279, 280
resistant cultivars 82, 83, 132, 516
saliva 46, 125–128, 132, 134, 507, 520
on sugar 147, 151
sources of food 143
virus transmission 82, 477, 540–555
see also crop loss, galls, mouthparts, predatory thrips
figs 208, 700
fighting 30, 92, 93, 95–97, 202, 230
flax 160, 206, 586, 684
Flax thrips
see Thrips linarius
flight
airborne populations 182, 183–186
capability 161, 175
exodus 182–185
mass flights 185, 186
mating swarms 186
mechanism 175–181
migratory/migration 182–187
muscles 26, 39, 40, 179
radar 183, 185
settling/immigration 187–190, 207, 208
shelter effects 188, 189, 207, 620
speed 177, 187
take-off 177, 180, 181, 182
teneral period 81, 179
thunder 35, 95, 186
triggering fire alarms 186, 187
weather, affect on 179, 181, 182, 186
wind 182–185, 187, 208
see also wings
Florithrips traegardhi 687, 692, 693, 696
flower thrips 67, 74, 134, 145, 158, 200, 201, 219, 230
see also anthophilous thrips
forage grasses 244, 587, 597
feeding injury 516, 684
forage legumes 7, 160, 245, 443, 604, 677, 685
Formicidae 263
Frankliniella
bispinosa 5, 69, 102, 122, 518, 519–521, 623, 678, 682, 684, 690, 697, 712, Plate V (1)
bruneri 682
cestrum 695, 712

fusca 5, 21, 68, 72, 83, 84, 189, 210, 220, 231, 440, 455, 514, 543, 596, 598, 683, 684, 686, 687, 690, 693, 697, 698
insularis 682
intonsa 69, 179, 181, 202, 210, 236, 370, 493, 543, 677, 680, 688, 689, 694, 695, 698
kelliae 518, 682
melanommata 206, 681
occidentalis 3, 7, 8, 210, 225
aggregations 86, 94, 95, 186
attractive colours 67, 69, 71, 72, 75, 76
biological control 310, 652–657, 711, 712
chemical control 571, 572, 585, 644, 646, 649
copulation 98
development 3, 227, Plate I(1)
eggs 50, 102
feeding 122, 126, 127, 132, 134, 139, 140, 147, 152, 155, 156
feeding injury 222, 512, 514, 518, 524–526, 677, 680, 683–691, 693–695, 697–701, Plates III (3, 8), IV (1–7), VII (5), IX (2–4), XI (1–7)
fighting 95–97
fungal control 409, 653
glasshouses 8, 202, 206, 208, 227, 400, 409, 454, 652–657
mating 87, 95, 98
morphology 18, 22, 23, 25, 28, 29, 34, 36, 54–56
natural enemies 281, 283, 304, 306, 308, 360, 362, 364, 365, 400, 409 Plate II (1–8)
occurrence 192, 202, 208
origins 209
pheromones 89, 91
population parameters 220
resistance in 576, 577
resistance to 83, 84, 218, 645
sampling 438, 440, 441, 442, 445, 453–455, 457
ex ratios 95
strains 226
virus transmission 541, 543, 550–553, 598, 701
UV response to 71, 72–74
panamensis 209
parvula 206, 523, 678, 682, 711
schultzei 87, 94, 98, 156, 192, 210, 211, 543, 678, 683, 686, 687, 691, 693, 696–698
tenuicornis 21, 67, 68, 234, 679, 687, 688, 692, 693, 698
tritici 5, 69, 71, 72, 84, 183, 184, 188, 189, 210, 437, 455, 677, 683, 684, 686, 693, 694, 679, 701
vaccinii 382
williamsi 210, 687
Franklinothrips
caballeroi 271
megalops 271, 280, 291, 299
tenuicornis 156, 271
vespiformis 271
Fulmekiola serrata 6, 209, 211, 696
fungi
see entomopathogenic fungi, plant pathogens

galls 202, 507
garlic 605, 685
genetic diversity 465
Geocoris
atricolor 267, 286
bullatus 267
ochropterus 267, 286
pallens 267, 286
punctipes 267, 286
geographical distribution 203, 204
geraniol 77, 78
gerbera 657, Plate XI (6)
gladiolus 75, 212, 222
feeding injury 516, 517, 700
resistance in 83
Gladiolus thrips
see Thrips simplex
gloxinia 517, 700
Gnophothrips fuscus 240, 517, 693, 711
Goetheana
incerata 358
shakespearei 358, 359, 495
Grain thrips
see Limothrips cerealium
Grape thrips
see Drepanothrips reuteri

grapes
  feeding injury 512, 526, 599, 685, Plates III (6), IV (7, 8)
  oviposition injury 527, 685
Grapevine thrips
  see Rhipiphorothrips cruentatus
Grass thrips
  see Aptinothrips rufus
grasses
  see forage grasses
greenhouses/glasshouses 206, 208, 227, 316, 571, 577, 583–585, 639–664
  see also IPM
Greenhouse thrips
  see Heliothrips haemorrhoidalis
grooming 159
groundnut 161, 206, 220, 436, 438, 586, 598
  feeding injury 514, 686, Plate XII (4, 5)
  resistance in 83
Groundnut thrips
  see Caliothrips indicus
Gryllidae 265
Gum tree thrips
  see Thrips australis
gut
  symbionts 147
  tospovirus entry 554
  see anatomy, alimentary tract
Gynaikothrips
  ficorum 18, 50, 88, 89, 91, 210, 277, 361, 700
  flaviantennatus 280
  uzeli 88, 89, 280

Haplothrips
  aculeatus 7, 210, 679, 684, 688, 692, 696, 698
  bedfordi 272
  cahirensis 272
  ceylonicus 683, 711
  chiensis 366
  cottei 699
  ganglebaueri 7, 692, 698
  globiceps 686
  kurdjumovi 139, 156, 218
  leucanthemi 88, 89, 90, 278, 685
  niger 7, 685
  senecionis 148
  subtilissimus 366
  tenuipennis 75, 688, 712
  tritici 7, 103, 200, 206, 210, 233, 234, 236, 586, 599, 679, 692, 698, Plate X (6–8)
  verbasci 21, 130, 131
  victoriensis 141, 156, 319
head
  see anatomy, morphology
Heather thrips
  see Ceratothrips ericae
Heliothrips
  haemorrhoidalis 5, 7, 90, 101, 132, 148, 190, 200, 206, 210, 282, 360, 381, 522, 682, 683, 692, 697, 699, 701, 711, Plates I (3), III (5, 6)
  sylvanus 512, 686
Hemerobiidae 261
Hemerobius
  californicus 261
  pacificus 261, 292
Hercinothrips
  bicinctus 523, 679
  femoralis 5, 17, 18, 21, 38, 42, 48, 50, 53, 146, 176, 221, 678, 679, 682, 699, 700, 701, 712
Heterothripidae 19, 199
Heterothrips
  arisaemae 159
  azaleae 74, 279
Hippodamia convergens 264, 292
Hirsutella sp 402
Holothrips hagai 88, 89
Honeysuckle thrips
  see Thrips flavus
Hoplandrothrips
  bredoi 683
  coffeae 683
  marshalli 683
  pillichianus 265, 361
Hoplothrips
  corticus 265
  japonicus 88, 89
  karnyi 92
  pedicularius 92, 265, 361
host plant
  anthophilous thrips 67, 71, 73, 75, 94, 134, 198
    see also flower thrips

attractive flowers 74–76, 222
colour 67–76, 222
concept of 205
defensive compounds 81
detection 65
landing on 76, 187, 188
morphology and acceptability 81, 218
odours 76–80, 81, 322
post-alighting effects 80
range 198, 211–213
resistance in 66, 80, 81, 83, 132, 218, 422, 663
shape 76
size 76
suitability 218–223
see also crop loss, feeding, resistant cultivars
hover fly
see Syrphidae
Hyphomycetes 400–402
Hypoaspis
aculeifer 273, 293
miles 273, 293, 655
Hypotidae 262

identification 1
index
of diversity 206
of nomadism 66
injury to plants 132–134, 139, 152, 505–526
see also crop loss, feeding, viruses
infestation patterns 188, 189, 207, 620
insecticidal treatments
aerosols 584
antifeedant 82
application methods 581–586, 587
atomisation 585
bioassay 458–461, 570, 571, 577
costs 586–588
dusts 582
efficacy 570–573
electrostatic sprayers 581
fumigants 575, 585
granules 569, 572, 582, 584
in greenhouse/glasshouse 571, 577, 583, 646
growth regulators 572
historical background 568, 569
lists 574, 575
phytotoxicity 587, 640, 648
resistance 573, 576–581, 621
resistance mechanisms 578–581
retarding resistance 580, 581, 621, 622
seed dressings 569, 575
smokes 585
soil insecticides 569, 575
sprays 575, 582, 585
synergists 577, 579, 580
Thripstick 584, 644, 652
insecticides
abamectin 572, 583, 654
acephate 570, 574, 576, 582, 583
acrinathrin 574
alanycarb 574
aldicarb 585
aldoxycarb 574, 582
aldrin 568, 569
azadirachtin 421
azinphosmethyl 376
Bacillus thuringiensis 572
bendiocarb 571, 574
bensultap 574
BHC 568
bifenthrin 576
biothion 576
butocarboxim 574
calcium cyanide 568
carbaryl 583, 585
carbofuran 569, 574
carbosulfan 569, 570, 622
chlorfuazuron 574
chlorpyrifos 571, 623
CO_2 586
creosote 568
cyhalothrin 622
cypermethrin 573, 578, 585
cythion 622
DDT 568, 569, 576, 580
deltamethrin 574, 583, 585
demeton 569, 583
derris 568
diazinon 585
dichlorvos 571, 583, 646, 654
dicofol 305, 571
dieldrin 568, 576
dimethoate 304, 459, 571, 574, 576, 579, 622, 623
dioxathion 576

insecticides *continued*
disulfoton 582
entofenprox 574
fenitrothion 574
fipronil 574, 582
fluvalinate 571, 578, 621
formentanate 570, 571, 574, 576, 579, 580, 623
HCH 585
heptachlor 574
hetpenophos 574, 583, 585, 646
imidocloprid 575, 583
isofenphos 575
kerosene 568
Lamda-cyhalothrin 575
lime sulphur 568, 576
malathion 304, 571, 575, 576, 583, 585
mephosfolan 575
mercarbam 575
methadimophos 578
methiocarb 583
methomyl 576
monocrotophos 622
mycoinsecticides 408
naled 304, 584, 585
neem oil 572
nicotine 575, 585
nicotine sulphate 568, 576
omethoate 575
parathion 576
permethrin 579, 585, 622
phenthoate 575
phorate 569
phosalone 575
phosmet 575
phosphamidon 575, 576
pirimiphos-methyl 575, 585
profenofos 578
pyridaphenthion 575
quinalphos 571, 575
ryania 568, 576
sabadilla 568, 576
sulfotep 575
sulphur 568, 576
sulprofos 575
tartar emetic 568
tau-fluvalinate 459, 575
tecnazene 585
Thripstick 584, 644, 652
toxaphene 304, 568, 576
triazophos 575
integrated pest management (IPM) 4, 639, 640, 648–651, 661
barriers 602, 644
in bedding plants 655, 656
biological control 379, 421, 606–607, 619, 620, 629–631, 640, 642
chemical control 567– 588, 602, 621–623, 631, 646
in citrus 616–623
crop monitoring 618, 626, 647, 648
in cucumbers 645, 652–655, 660
cultural control 603–605, 619, 620, 628, 629, 644
economic thresholds 241, 440, 586, 598–600, 620, 621, 626
economics 649, 659–661
field crops 595–607
in glasshouses 639–664
host plant resistance 66, 80, 81, 83–85, 132, 133, 218, 219, 222, 422, 516, 527, 600, 601
intercropping 222, 605
interior landscapes 657, 658
irrigation 604
management strategies 600–607
mulches 72, 602
mycopathogens 419–422, 630
see also entomopathogenic fungi
organic farming 605
in peppers 651–652, 660
physical control 602, 642–644
principles 595, 596
research and development 662–664
rotation 603
screens 642
tospoviruses 602, 605
in tree crops 615–632, 711–713
in sugar maple 623–631
weed control 597, 598
intercropping/interplanting 222, 605
intraspecific variation 201–203

Japanese black pine 711
juglone 89, 278

Kakothrips pisivorus 7, 66, 81, 148, 154, 159, 183, 193, 210, 232, 233,

360, 587
flower colour preferences 75
oviposition sites 102
plant injury 522, 680, 689, Plate VII (3, 4)
keys 2, 372, 549

lacewings
see Chrysopidae
Laelapidae 273, 283
larch 4, 507, 583, 687, 711
Larch thrips
see Thrips laricivorus
Larothrips dentipes 161
larva
morphology 52–56
overwintering 232–234, 449, 625
in soil 237, 449, 625
see life history
leeks 102, 582, 687, Plate VII (1)
Lefroyothrips lefroyi 697
legs
see anatomy, morphology
Leptothrips mali 141, 155, 156, 272, 292, 319, 494
lettuce 83, 85, 443, 457, 602, 687, Plate XII (1, 2)
Leucochrysa
marquesi 261
submacula 261
varia 261
Leucothrips theobromae 206, 682
life histories 2, 619, 625, Plate I (1)
duration 3
generations 2
of nematodes 383–384
see also parasitoids, predators
lillies 701
Lilybulb thrips
see Liothrips vaneeckei
Limothrips
cerealium 68, 97, 122, 128, 129, 131, 133, 190, 206, 210, 225, 234, 245, 455
feeding injury 131, 513, 515, 679, 688, 692, 693, 698, Plate X (2–5)
flight 179, 181, 183, 184, 185, 187, 189
denticornis 68, 97, 103, 181, 183, 184, 186, 206, 210, 515, 586, 679, 692, 693, 698, Plate X (1)
linseed Plate VIII (3)
Liothrips
africanus 268
brevitubus 272
floridensis 361
fluggae 268
karnyi 690
kuwanai 88, 89, 280
laureli 361, 365
mikaniae 361, 379
oleae 283, 361, 365, 378, 689
piperinus 88, 89
setinodis 264
urichi 268, 361
vaneeckei 90, 701
varicornis 361
wasabiae 361
LT_{50} 225
lucerne
see forage legumes
lupins Plate VIII (2)
Lygaeidae 267, 297

macadamia 712
Macrolophus
caliginosus 266, 286
rubi 266, 311
Macrotracheliella laevis 267
mahogany 160
maize 79, 323, 601, 687
Malachiidae 264
Mallothrips indicus 379
mandible
see mouthparts
mango 160, 687, 712, Plates III (8), VIII (5)
Mantidae 265, 277, 297
mating 91–100, 231
aggregations 86, 93, 95, 186
copulation 87, 94, 98–100
fighting 92, 93, 95–97, 202
pheromones 86–89
in terebrantians 92
in tubuliferans 92
Meadow foxtail thrips
see Chirothrips hamatus
Megalothrips spinosus 361
Megalurothrips
distalis 210, 522, 680, 712

Megalurothrips continued
- *sjostedti* 7, 80, 102, 210, 219, 241, 439, 457, 522, 679, 683
- *usitatus* 69, 206, 210, 237, 578, 680, 686, 691

Megaphragma
- *aligharensis* 374
- *caribea* 374
- *decochaetum* 374
- *deflectum* 374
- *ghesquièrei* 374
- *longicilliatum* 374
- *macrostigmum* 374
- *magniclava* 374
- *mymaripenne* 299, 374, 375
- *polychaetum* 374
- *priesneri* 374
- *stenopterum* 374

meillon 89, 278
melon 221, 584, 688
melongene Plate IX (7, 8)
Merothripidae 19, 87, 198, 199
Merothrips brunneus 25, 36, 87
Mesothrips claripennis 272, 280
Metarhizium anisopliae 400, 402, 404–406, 408, 411, 460, Plate II (5)
methyl-salicylate 322
Micraspis cardoni 264, 299
Microcephalothrips abdominalis 231, 691
microsporidian parasite 230
migration
- *see* flight

millet 688
Miridae 266, 278, 297, 304, 317
mites 228, 280–284
- *see also Amblyseius, Anystidae, Cheyletidae, Erythraeidae, Hypotidae, Laelapidae, Phytoseiidae, Trombidiidae*

models
- chaos theory 244
- degree-day 227, 238, 239, 240, 616, 622
- Lotka-Volterra 284, 313, 317
- mathematical 240, 244, 247, 260, 284, 285, 304–320
- pancake 284, 304, 305, 312
- predicting outbreaks 242–244, 617
- Taylor's power law 247, 438, 439, 442

Monilinia fructicola 524, 558
monoterpenes 322
Montandoniola moraguesi 268, 279, 287, 297
morphology 15
- abdomen 31, 32, 87
- antennae 17, 19
- compound eyes 19, 35
- head 17, 19, 22, 23
- larva 52-57
- legs 26, 30, 92
- mandible
  - *see* mouthparts
- maxillae
  - *see* mouthparts
- ocelli 19, 24, 35
- ovipositor 2, 33, 34, 40
- propupa 58
- pupa 58
- sense organs 19, 21, 35, 121, 125
- thorax 26
- wings 26, 29, 30, 175–181

mouthparts 24, 119–128
- action/function 24, 26, 121, 123–133, 135
- structure 23, 24, 26, 52, 119–125
- *see also* feeding

mulberry 361, 712
Mullein thrips
- *see Haplothrips verbasci*

Mycterothrips
- *consociatus* 17, 19
- *glycines* 487

Mymaridae 355
myrcene 78, 89

Nabidae 266, 278
Nabis
- *alternatus* 266, 286
- *americoferus* 267, 286
- *ferus* 267
- *pseudoferus* 267

natural enemies
- *see* biological control, mites, nematodes, parasitoids, predators

nectarines 599
- feeding injury 437, 524, 525, Plate III (4)
- *see also* stonefruit

nematodes 230, 382–387

biological control 387, 422
host impact 386
laboratory methods 385, 386
life cycle 383–385
population dynamics 387
Neohydatothrips
adolfifriderici 210, 683, 694
gracilicornis 212
pseudoannulipes 200
variabilis 68, 72, 188, 189, 210, 437, 455, 514, 683, 684, 693, 712
Neozygites
cucumeriformis 401
parvispora 400, 401, 403
New Guinea Impatiens Plates XI (5), XII (8)
New Zealand flower thrips
see Thrips obscuratus
Ninyas torvus 267, 297
nomenclature 666–673

oats 688
Oats thrips
see Stenothrips graminum
ocelli 19, 24, 35
Odontothrips
confusus 604, 685
loti 264, 685
phaleratus 228
oil palm 160
okra 220, 688
olfactometer trials 80
Olive thrips
see Liothrips oleae
olives 689
onion 6, 7, 160, 245, 377, 440, 443, 569
feeding injury 127, 509, 513, 517, 600, 689, Plate VII (2)
resistance in 84, 222, 601
Onion thrips
see Thrips tabaci
ontogeny 317
Orchid thrips
see Chaetanaphothrips orchidii, Dichromothrips spp
orchids 191, 209, 212, 584, 701
see also Dendrobium
Organothrips bianchi 696
Oriental rice thrips
see Stenchaetothrips biformis
Orius
albidipennis 268, 279, 287
amnesius 268
armatus 268
indicus 268, 279, 287, 297
insidiosus 86, 268, 279, 288, 297, 310, 315, 655
laevigatus 269, 279, 288, 298, 311, 646, 651, Plate II (6–8)
limbatus 288
majusculus 269, 288, 298, 646, 651
maxidentex 269, 279, 289, 298
minutus 269, 279, 289, 298
niger 269, 289, 298
persequens 269
sauteri 270, 279, 289, 298, 315
similis 270
tantilus 270, 279
tripoborus 270
triscolor 270, 279, 289
ornamentals 192
feeding injury 516, 699–701, Plates XI (1–7), XII (3, 8, 9)
IPM 655–657
susceptible to viruses 701
see carnations, chrysanthemum, gerbera, gladiolus, Saintpaulia
ovarian diapause
see reproductive diapause
ovaries
see anatomy
overwintering 207, 226, 231–235, 597
in bark 226, 238
dormancy 235–236
emergence from 30, 238, 239
instars 231
in litter 234
mortality 236–238
ploughing 237, 238, 603
reproductive diapause 236
in soil 226, 232–234, 237, 449, 597, 625
see also cold hardiness
oviposition 51, 100–103
in artificial surfaces 101, 492
injury 506, 526–529
rate 146
resistant *cultivars* 527
by parasitoids 323, 363, 375

oviposition *continued*
sites 100
ovipositor
see morphology

Paecilomyces sp 400, 402
Palm thrips
see Parthenothrips dracaenae
Panchaetothrips
indicus 360
noxius 683
Panicum maximum 7
parasitoids
biological control 355, 379, 380
development 366–369, 375, Plate II (1–3)
distribution 356–359, 373
endoparasitoids
see eggs
field incidence 376–379
host selection 361–366, 375
host species 359–361, 373
life history 368–372, 375, 376
longevity 372
multiple parasitism 376
natural control 376–379
oviposition 362, 363, 375
population growth 372
rates of parasitism 376–379
rearing 490, 495, 496
reproduction 371
parthenogenesis 49, 372
Parthenothrips dracaenae 18, 23, 28, 54, 206, 700
papaya 160, 689, 712
pawpaw
see papaya
Pea thrips
see Kakothrips pisivorus
peaches 212, 587, 599
feeding injury 524, Plate III (1, 2)
see also stonefruit
peanut
see groundnut
pear 690
Pear thrips
see Taeniothrips inconsequens
peas 7, 102
feeding injury 526, 689, Plate VII (3–5)
oviposition injury 529
Pedobius thysanopterous 358, 359
Pepper leaf gall thrips
see Liothrips karnyi
peppers
black 690
sweet 7, 310, 323, 442, 444, 445, 584, 599, 651, 652, 660, 661, 690, Plates IX (6), XII (6)
perillene 91
pest genera 198
origins 209
species, distribution list 210
status 5, 616, 623
Pezothrips dianthi 87, 507, 699
pheromones 86–91, 663
alarm 90, 91, 362, 421, 584
chemical constituents 88, 89
defensive allomones 90, 91, 276, 278
kairomones 362
secretory glands 49, 87, 91
semiochemicals 421, 663
sex 86, 89
synonome 322
Phlaeothripidae 2, 30, 86, 142, 143, 198, 199, 276
Phlaeothrips coriaceus 361
phytohormones 507–510
see also ethylene
Phytoseiidae 273, 280–283, 304, 317
pigeon pea 161, 599, 691
pineapple 436, 691
Plague thrips
see Thrips imaginis
plant pathogens 539–558
Alternaria porri 513
bacterial 558
Calonectrice rigidiuscula 558
Colletotrichum gloeosporioides 617
Discula campestris 624
Erwinia amylovora 558
fire blight 558
fungal 557–558
Fusarium ear rot 507, 558
Monilia fructicola 524, 558
Pantoea stewarti 558
powdery mildew 558
Pseudomonas medicaginis 558
Puccinia graminis 558
thrips vectors 506, 507, 541, 557, 558

pollen 80, 135, 137, 138
feeding by predators 282, 283, 301–312
by thrips 81, 134–139, 145, 152
nutrition 145
rearing thrips 491, 492
see also anthophilous thrips, flower thrips
pollen-borne viruses 540–542
pollinators 150, 157–162
polymorphism 201, 202
Polynema indica 372, 375
pommellos 622
Ponticulothrips diospyrosi 88, 89
population dynamics/fluctuations 217–248
agricultural practice, affects on 244
climatic change 244–245
cold 225–227
competition 230–231
drought 224–225
heat 227
life-fertility tables 598
modelling
see models
overwintering mortality 236–238
parameters 211, 220, 221, 241, 242–244
parasitism 229, 230
pollution, affects of 244–245
predation 228, 259–329
rain 223–224, 237
seasonal changes 239–240
see also parasitoids, predators
population size 184, 185, 190, 234, 242, 245, 457
potato 440, 691
Potato thrips
see Thrips tabaci
predators 228, 259-329, Plate II (6–8)
biological control 307, 308, 310, 316
birds 228
body size 260, 276–278
cannibalism 281
culture packs 651, 653, 654, 659
defence mechanisms 276
development parameters 286–296
dispersal 316
dynamics, predator-prey 283–320
galls 260
intraguild predation 324–327
intraspecific predation 324–327
life histories 286–300
mites 228, 283
see also species/family lists
models 284, 285, 304–320
persistence 312–320
phytophagy 313, 317, 320
plant–predator interactions 321–324
pollen 282, 283, 306, 318
polyphagy 260, 313, 320
prey consumption rate 140, 141, 297–300, 301, 313
prey suppression 284, 285, 301–312
species family lists 261–275
toads 229
thanatosis 276
see also Amblyseius, *Anthocoridae*, *Orius*
predatory thrips 139–142, 143, 144, 155–157, 199, 271, 272, 279, 280, 304, 319
cannibalism 142
rearing 494, 495
privet 701
Privet thrips
see Dendrothrips ornatus
propupa
see morphology
Pseudodendrothrips
mori 211, 361, 378
ornatissimus 712
Pyemotidae 273, 283
pyrethrum 161, 222, 437, 443, 587
feeding injury 517, 599, 600, 691
see also insecticides
pumpkin 221
pupa
see morphology

quarantine 192, 211, 219, 552, 606

radish 161
rambutan 713
rape 160, 691
rearing 477–497, Plate I (2, 3)
artificial food 488, 489
artificial membranes 489

rearing *continued*
bean pod method 484, 485, Plate I (2)
cages 479–481, 482, 484, 486, 489
contamination 497
diapause 496
diets 488, 489
egg collection 492
environmental conditions 496, 497
handling 479–481
host quality 497
larvae 492, 493
leaf discs 487
on lemons 495, Plate I (3)
management 478, 479
Munger cells 484, 485
mycophagous thrips 493, 494
parasitoids 490, 495, 496
phytophagous thrips 481–493
pollen 491, 492
predatory thrips 494, 495
Red-banded thrips
see Selenothrips rubrocinctus
Reduviidae 266, 278
reflective mulches 72, 602
reproduction 2, 49, 104
see also arrhenotoky, parthenogenesis, thelytoky
reproductive diapause 236, 496
reproductive system
see anatomy
resistant cultivars
aubergine 601
cabbage 85, 601
chrysanthemum 83
cotton 83, 84, 133, 218, 601
cowpea 219, 527, 601
cucumber 85, 645
gladiolus 83
maize 601
onion 84, 222, 601
peanut 83, 601
rice 84, 148, 516, 601
roses 84
to tospoviruses 556, 601
Retithrips syriacus 84, 98, 101, 360, 368, 678, 681, 683, 686, 700, 701, 711
Rhinocapsus vanduzeei 266, 279
Rhipiphorothrips cruentatus 98, 101, 512, 681, 683, 686, 688, 712, 713
rice 7, 219, 436, 482, 569
feeding injury 515, 516, 692
resistance in 84, 148, 516
Rose thrips
see Thrips major
roses 74, 207, 220, 443, 701
rubber 161, 692
rye 6, 7, 599, 692

safflower 693
Saintpaulia 74, Plate XI (4)
salicylaldehyde 77, 455
saliva 26, 125–128, 132, 148
galls 134, 507
plugs 127, 134, 519
toxaemias 507
salivary glands
see anatomy
sampling 435–458
airborne populations 452–457
beating 441
D-vac 448
efficiency 439
eggs 448, 449
emergence boxes 451
extraction 446, 447, 450, 627
field crops 435–444
glasshouse crops 442–445
litter 449–452
precision 204, 437, 432, 445, 627
qualitative 447, 448
quantitative 435–445
satellite imaging 436, 628
sequential 440
shaking 441
soil 449–452, 627
sweeping 447, 448
trees 436, 441, 442, 618, 626
see also traps
Schedothrips orientalis 362
Sciothrips cardamomi 210, 680
Scirtothrips
aurantii 5, 7, 122, 207, 210, 211, 523, 617, 620, 621, 679, 681, 712
bispinosus 683, 697
citri 5, 6, 68, 131, 139, 159, 210, 211, 235, 240, 481
control 459, 570, 571, 573, 580, 616–623, 712

natural enemies 282, 305
plant injury 523, 616, 681, Plate V (2, 4, 5)
sampling 436, 452, 453, 457, 618–619
seasonal development 619
traps 452, 453, 618
dorsalis 68, 72, 210, 487, 681, 682, 683, 686, 688, 692, 694, 695, 697, 712, 713
inermis 211, Plate V (3, 6)
kenyensis 208, 697
longipennis 211, 271
mangiferae Plate VIII (6)
manihoti 206, 680
oligochaetus 230
Scolopscelis parallelus 270, 287
Scolothrips acariphagus 156
indicus 271
longicornis 141, 146, 153, 157, 292, 494
sexmaculatus 99, 140, 141, 153, 155, 156, 218, 271, 292, 494
Selenothrips rubrocinctus 152, 154, 200, 206, 210, 224, 237, 245, 360, 378, 457, 522, 678, 681, 682, 687, 700, 712
semiochemicals
see pheromones
sense organs
see morphology
Sesbania thrips
see Caliothrips indicus
sex ratios 49, 92, 93, 94
sexual dimorphism 15, 92
Six-spotted thrips
see Scolothrips sexmaculatus
slash pine 240, 711
feeding injury 517, 693
smoke detectors 186
Sorghothrips jonnaphilus 516, 693
sorghum
feeding injury 516, 693
South African citrus thrips
see Scirtothrips aurantii
soybean 188, 246, 437, 438, 443
feeding injury 514, 693
spatial distribution 246–248
species richness 205, 206
Sphecidae 263, 277, 356
spiders 276
spinach 148
Sporothrips amplus 92
Sporothrix sp 402
spruce 441
squash 443
Staphylinidae 265
Stenchaetothrips
biformis 84, 148, 209, 210, 219, 231, 277, 436, 481, 482, 569, 692
minutus 209, 211, 696
Stenothrips graminum 68, 179, 206, 233, 382, 688
Stictothrips fibriatus 493
stonefruit 712
feeding injury 694, 695, Plate III (3)
see also nectarines, peaches
strawberries 184
feeding injury 525, 599, 600, 695, Plate IV (1–4)
stylets
see mouthparts
sugar apple
see custard apple
sugar beet 8, 160, 597, 695, Plate VIII (1)
Sugar beet thrips
see Hercinothrips femoralis
sugar cane 6, 209, 696
sugar maple 4, 6, 102, 189, 245, 436, 441, 623–631
bud-burst phenology 511
feeding injury 510–512, 623, 696, Plate VIII (4, 5)
sampling 626–628
sunflower 311, 696
sweet potato 696
symbionts 147
Syrphidae 262, 277

Taeniothrips inconsequens 4, 36, 68, 148, 159, 185, 189, 200, 210, 244, 245, 400, 409, 690, 711, 712
biology 49, 102, 232, 233, 234, 623–625
feeding injury 510–512, 624, 628, 694, 695, Plate VIII (4, 5)
IPM 623–631
life cycle 625
pest status 6, 623–625

Taeniothrips inconsequens continued
sampling 441, 449, 450, 626-628
Taeniothrips orionis 193, 527, 695
taro 696
Taro thrips
see Organothrips bianchi
tea 7, 191, 697
Tea thrips
see Scirtothrips kenyensis
testes
see anatomy
Tetraphleps bicuspis 270
thanatosis 276
thelytoky 49, 50, 372, 375
thigmokinesis/thigmotaxis 219, 506
Thlibothrips isunoki 88, 89
thorax
see anatomy, morphology
Thripastichus gentilei 358, 359, 367, 378
Thripidae 2, 8, 86, 94, 128, 143, 198, 199, 276
Thripinema
nicklewoodi 230, 383, 384, 386, Plate II (4)
renirao 384
Thripobius 356, 358
hirticornis 358
semiluteus 358, 380, 495, 619
Thrips
alliorum 689
angusticeps 8, 69, 211, 224, 232, 233, 238, 244, 586, 597, 603, 680, 684, 689, 693, 695, 699, Plate VIII (1–3)
atratus 68, 69, 93, 525, 699
australis 179, 207, 224
calcaratus 100, 236, 509, 511, 625, 679, 711
carthami 711
flavus 210, 230, 518, 677, 685, 689, 691, 695, 699, 711
florum 210
fuscipennis 93, 137, 644, 646, 684, 685, 701
hawaiiensis 69, 75, 210, 242, 493, 678, 680, 682, 697, 700, 711–713
imaginis 69, 75, 144, 153, 156, 207, 211, 217, 224, 236, 239, 242–244, 245, 455, 518, 677, 695
juniperinus 235
laricivorus 4, 132, 237, 441, 507, 687, 711
linarius 70, 684
madroni 677
major 67, 70, 93, 200, 525, 695, 712
meridionalis 525, 695, 699, 712
minutissimus 212, 695
nigropilosus 211, 222, 437, 517, 587, 600, 680, 691
obscuratus 67, 70, 80, 145, 210, 455, 456, 493, 587, 677, 694, 712, Plate III (1, 2, 4)
palmi 7, 202, 210, 212, 362, 370, 376, 400, 441, 453, 572, 577, 596, 698, 701, 713
attractive colours 71, 74
cold survival 225, 226
development 227, 493
dispersal 192, 212, 550, 578
feeding deterrent 82, 132
mating 99
plant injury 513, 525, 677, 679, 683, 684, 686, 688, 690, 691, 699, 700, Plate IX (5–8)
population parameters 211, 221, 241, 308, 596
sex ratio 94
virus transmission 543, 550, 701
UV response to 71, 72
paravispinus 72, 577, 689, 712
physapus 20, 70, 176, 178, 180, 382
pillichi 67, 70
pini 583
pusillus 523, 679, 711
setosus 487, 543
simplex 69, 72, 83, 190, 209, 210, 222, 307, 360, 516, 517, 699, 700. 701
tabaci 3, 5, 6, 7, 54, 90, 102, 139, 146, 156, 190, 192, 210, 440, 455
control 569, 577, 644, 646, 652, 712
development 201, 227, 493
in glasshouses 400, 409, 644, 652
natural enemies 281, 307, 400, 409

origins 202, 209
parasitism 360, 362, 364, 376, 377
plant injury 84, 134, 509, 513, 514, 517, 525, 600, 677, 680, 681, 683–691, 693, 695, 697, 699, 701, Plate VII (1, 2, 6)
population parameters 221, 245, 457
traps 69
UV response to 72
variation 201, 202, 203, 226
virus transmission 541, 543, 549–550, 701
validus 42, 87
vulgatissimus 70, 691
thunderstorms 35, 95, 186
Tiarothrips subramanii 98
toads 229
tobacco 7, 311, 436, 438, 444, 697
Tobacco thrips
see Frankliniella fusca
tomato 436, 440, 441, 444, 584, 649
feeding injury 697, 698, Plate IV (5)
oviposition injury 529
resistance in 82, 132, 218, 323
TSWV 540, 602, 697, Plate XII (7)
Tospoviruses Plate XII (1–8)
classification 543, 546
leaf disc assay 548
management of epidemics 555–557
phylogeny 545–547
transmission 205, 542–557
see also viruses
traps 182, 452–457
catches v. infestation 457
chemical attractants 77–79, 80, 454–456, 643
colour 67, 73, 187, 454–456, 618, 643
emergence 451, 626
in glasshouses 642–644
odours
see chemical attractants
shape 453
sticky 445, 453, 454
suction 73, 159, 456, 457
U.V. reflectance 67–74, 187
visual attractants
see colour
water 159, 454
Trichinothrips breviceps 141
Trichogrammatidae 355, 372
Trombidiidae 273
Typhlodromus
athiasae 275, 296
caudiglans 275, 296
occidentalis 275, 296
pyri 275, 296

Ultra violet (U.V.)
mulches 71, 72
reflectance 67–74, 187
Uzelothripidae 198, 199

Varshneyia pasania 88, 89, 90, 91
Verbena Plate XI (1–3)
Verticillium lecanii 399, 401, 404, 411, 575, 646
Vespidae 263
Vine thrips
see Drepanothrips reuteri
viruses 8, 540–557, Plate XII (1–8)
acquisition 477, 553
Bunyaviridae 540, 549
carmoviruses 540
cellular interactions 550–553
control 555, 556, 602
crop losses 8, 85
distribution on crop 188, 207
ELISA 545, 557
groundnut bud necrosis virus 546
groundnut ringspot tospovirus (GRSV) 546
ilarviruses 540
impatiens necrotic spot tospovirus (INSV) 546, Plate XII (8)
pelargonium flower break carmovirus 541
pollen-borne 540–542
prune dwarf virus (PDV) 541
prunus necrotic ringspot virus (PNRSV) 541
resistance to 556, 601
sobemoviruses 540
sowbane mosaic sobemovirus (SoMV) 541
susceptible ornamentals 701
tobacco streak ilarvirus (TSV) 540

viruses *continued*
tomato chlorotic spot tospovirus (TCSV) 546
tomato spotted wilt tospovirus (TSWV) 8, 82, 85, 198, 543, 544, 546, 547, 549, Plate XII (1–8)
transmission 82, 477, 540–555
vectors 4, 541, 543, 548-553
water melon silver mottle tospovirus (WSMV) 546
see also Tospoviruses
visual cues 67–76
voucher specimens 549

water melon 444, 513, 698
waxapple 382, 713
weather
population growth 223–227, 239
rain 223–224, 237
wind/windspeed 182–185, 187, 208
see also flight, population dynamics, thunderstorms
weighing thrips 462
Western flower thrips
see Frankliniella occidentalis
Wheat 6, 185, 245, 444
feeding injury 515, 599, 698, Plate X (3–8)
Wheat thrips
see Haplothrips tritici
windbreaks 188, 189, 207, 620
wings 26, 29, 30, 175–181, 188
brachypterous forms 66, 175
morphs 175, 201
muscles
see anatomy, musculature, thorax
preparation for flight 176–178
structure
see morphology
Wollastoniella parvicuneis 270
rotunda 270

Xylaplothrips inquilinus 272, 280

yam 161
yield loss
see crop loss
Yucca thrips
see Bagnalliella yuccae

Zaniothrips ricini 277, 362, 368, 376, 681
Zoophthora radicans 401
Zygomycetes 400, 401